EX LIBRIS
SOUTH ORANGE
PUBLIC LIBRARY

About Island Press

Island Press is the only nonprofit organization in the United States whose principal purpose is the publication of books on environmental issues and natural resource management. We provide solutions-oriented information to professionals, public officials, business and community leaders, and concerned citizens who are shaping responses to environmental problems.

In 2005, Island Press celebrates its twenty-first anniversary as the leading provider of timely and practical books that take a multidisciplinary approach to critical environmental concerns. Our growing list of titles reflects our commitment to bringing the best of an expanding body of literature to the environmental community throughout North America and the world.

Support for Island Press is provided by the Agua Fund, The Geraldine R. Dodge Foundation, Doris Duke Charitable Foundation, Ford Foundation, The George Gund Foundation, The William and Flora Hewlett Foundation, Kendeda Sustainability Fund of the Tides Foundation, The Henry Luce Foundation, The John D. and Catherine T. MacArthur Foundation, The Andrew W. Mellon Foundation, The Curtis and Edith Munson Foundation, The New-Land Foundation, The New York Community Trust, Oak Foundation, The Overbrook Foundation, The David and Lucile Packard Foundation, The Winslow Foundation, and other generous donors.

The opinions expressed in this book are those of the authors and do not necessarily reflect the views of these foundations.

Ecosystems and Human Well-being:
Scenarios, Volume 2

Ecosystems and Human Well-being: Scenarios, Volume 2

Edited by:

Steve R. Carpenter
University of Wisconsin-Madison
USA

Prabhu L. Pingali
Food and Agriculture Organization of the UN
Italy

Elena M. Bennett
University of Wisconsin-Madison
USA

Monika B. Zurek
Food and Agriculture Organization of the UN
Italy

Findings of the Scenarios Working Group
of the Millennium Ecosystem Assessment

Washington • Covelo • London

The Millennium Ecosystem Assessment Series

Ecosystems and Human Well-being: A Framework for Assessment
Ecosystems and Human Well-being: Current State and Trends, Volume 1
Ecosystems and Human Well-being: Scenarios, Volume 2
Ecosystems and Human Well-being: Policy Responses, Volume 3
Ecosystems and Human Well-being: Multiscale Assessments, Volume 4
Our Human Planet: Summary for Decision-makers

Synthesis Reports (available at MAweb.org)

Ecosystems and Human Well-being: Synthesis
Ecosystems and Human Well-being: Biodiversity Synthesis
Ecosystems and Human Well-being: Desertification Synthesis
Ecosystems and Human Well-being: Human Health Synthesis
Ecosystems and Human Well-being: Wetlands and Water Synthesis
Ecosystems and Human Well-being: Opportunities and Challenges for Business and Industry

Library of Congress Cataloging-in-Publication data.

Ecosystems and human well-being : scenarios : findings of the Scenarios Working Group, Millennium Ecosystem Assessment / edited by Steve R. Carpenter . . . [et al.].
p. cm.—(The Millennium Ecosystem Assessment series ; v. 2)
Includes bibliographical references and index.
ISBN 1-55963-390-5 (cloth : alk. paper)—ISBN 1-55963-391-3 (pbk. : alk. paper)
1. Human ecology. 2. Ecosystem management. 3. Environmental policy. 4. Biological diversity. I. Carpenter, Stephen R. II. Millennium Ecosystem Assessment (Program). Scenarios Working Group. III. Series.
GF50. E268 2005
333.95—dc22

2005017195

British Cataloguing-in-Publication data available.

Printed on recycled, acid-free paper ♻

Book design by Maggie Powell
Typesetting by Coghill Composition, Inc.

Manufactured in the United States of America
10 9 8 7 6 5 4 3 2 1

The Scenarios Working Group dedicates this volume to the memory of our valued colleague, Dr. Tsuneyuki Morita. We deeply regret his loss.

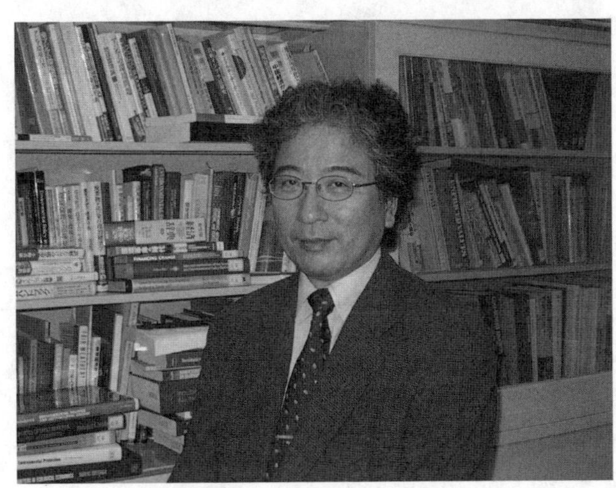

Millennium Ecosystem Assessment: Objectives, Focus, and Approach

The Millennium Ecosystem Assessment was carried out between 2001 and 2005 to assess the consequences of ecosystem change for human well-being and to establish the scientific basis for actions needed to enhance the conservation and sustainable use of ecosystems and their contributions to human well-being. The MA responds to government requests for information received through four international conventions—the Convention on Biological Diversity, the United Nations Convention to Combat Desertification, the Ramsar Convention on Wetlands, and the Convention on Migratory Species—and is designed to also meet needs of other stakeholders, including the business community, the health sector, nongovernmental organizations, and indigenous peoples. The sub-global assessments also aimed to meet the needs of users in the regions where they were undertaken.

The assessment focuses on the linkages between ecosystems and human well-being and, in particular, on "ecosystem services." An ecosystem is a dynamic complex of plant, animal, and microorganism communities and the nonliving environment interacting as a functional unit. The MA deals with the full range of ecosystems—from those relatively undisturbed, such as natural forests, to landscapes with mixed patterns of human use and to ecosystems intensively managed and modified by humans, such as agricultural land and urban areas. Ecosystem services are the benefits people obtain from ecosystems. These include *provisioning services* such as food, water, timber, and fiber; *regulating services* that affect climate, floods, disease, wastes, and water quality; *cultural services* that provide recreational, aesthetic, and spiritual benefits; and *supporting services* such as soil formation, photosynthesis, and nutrient cycling. The human species, while buffered against environmental changes by culture and technology, is fundamentally dependent on the flow of ecosystem services.

The MA examines how changes in ecosystem services influence human well-being. Human well-being is assumed to have multiple constituents, including the *basic material for a good life*, such as secure and adequate livelihoods, enough food at all times, shelter, clothing, and access to goods; *health,* including feeling well and having a healthy physical environment, such as clean air and access to clean water; *good social relations,* including social cohesion, mutual respect, and the ability to help others and provide for children; *security,* including secure access to natural and other resources, personal safety, and security from natural and human-made disasters; and *freedom of choice and action,* including the opportunity to achieve what an individual values doing and being. Freedom of choice and action is influenced by other constituents of well-being (as well as by other factors, notably education) and is also a precondition for achieving other components of well-being, particularly with respect to equity and fairness.

The conceptual framework for the MA posits that people are integral parts of ecosystems and that a dynamic interaction exists between them and other parts of ecosystems, with the changing human condition driving, both directly and indirectly, changes in ecosystems and thereby causing changes in human well-being. At the same time, social, economic, and cultural factors unrelated to ecosystems alter the human condition, and many natural forces influence ecosystems. Although the MA emphasizes the linkages between ecosystems and human well-being, it recognizes that the actions people take that influence ecosystems result not just from concern about human well-being but also from considerations of the intrinsic value of species and ecosystems. Intrinsic value is the value of something in and for itself, irrespective of its utility for someone else.

The Millennium Ecosystem Assessment synthesizes information from the scientific literature and relevant peer-reviewed datasets and models. It incorporates knowledge held by the private sector, practitioners, local communities, and indigenous peoples. The MA did not aim to generate new primary knowledge but instead sought to add value to existing information by collating, evaluating, summarizing, interpreting, and communicating it in a useful form. Assessments like this one apply the judgment of experts to existing knowledge to provide scientifically credible answers to policy-relevant questions. The focus on policy-relevant questions and the explicit use of expert judgment distinguish this type of assessment from a scientific review.

Five overarching questions, along with more detailed lists of user needs developed through discussions with stakeholders or provided by governments through international conventions, guided the issues that were assessed:

- What are the current condition and trends of ecosystems, ecosystem services, and human well-being?
- What are plausible future changes in ecosystems and their ecosystem services and the consequent changes in human well-being?
- What can be done to enhance well-being and conserve ecosystems? What are the strengths and weaknesses of response options that can be considered to realize or avoid specific futures?
- What are the key uncertainties that hinder effective decision-making concerning ecosystems?
- What tools and methodologies developed and used in the MA can strengthen capacity to assess ecosystems, the services they provide, their impacts on human well-being, and the strengths and weaknesses of response options?

The MA was conducted as a multiscale assessment, with interlinked assessments undertaken at local, watershed, national, regional, and global scales. A global ecosystem assessment cannot easily meet all the needs of decision-makers at national and sub-national scales because the management of any

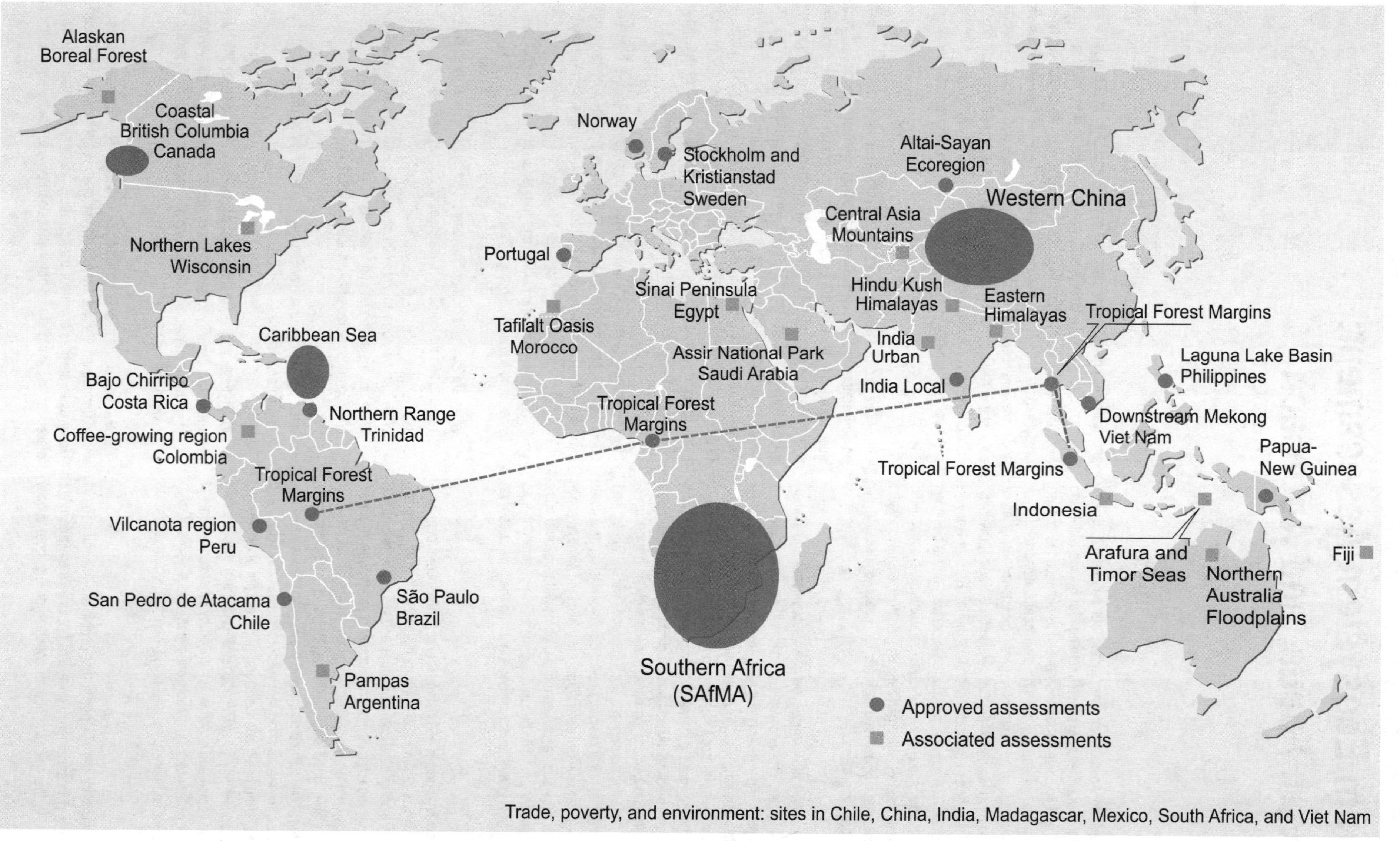

Eighteen assessments were approved as components of the MA. Any institution or country was able to undertake an assessment as part of the MA if it agreed to use the MA conceptual framework, to centrally involve the intended users as stakeholders and partners, and to meet a set of procedural requirements related to peer review, metadata, transparency, and intellectual property rights. The MA assessments were largely self-funded, although planning grants and some core grants were provided to support some assessments. The MA also drew on information from 16 other sub-global assessments affiliated with the MA that met a subset of these criteria or were at earlier stages in development.

SUB-GLOBAL ASSESSMENT	ECOSYSTEM TYPES										ECOSYSTEM SERVICES								
	COASTAL	CULTIVATED	DRYLAND	FOREST	INLAND WATER	ISLAND	MARINE	MOUNTAIN	POLAR	URBAN	FOOD	WATER	FUEL and ENERGY	BIODIVERSITY-RELATED	CARBON SEQUESTRATION	FIBER and TIMBER	RUNOFF REGULATION	CULTURAL, SPIRITUAL, AMENITY	OTHERS
Altai-Sayan Ecoregion			●	●	●			●			●		●	●		●		●	
San Pedro de Atacama, Chile			●		●						●	●		●			●	●	●
Caribbean Sea	●					●	●				●	●		●				●	
Coastal British Columbia, Canada	●			●	●			●			●			●		●	●	●	
Bajo Chirripo, Costa Rica		●		●	●						●	●		●		●		●	●
Tropical Forest Margins		●		●							●	●		●	●	●	●		●
India Local Villages		●		●	●						●	●	●	●		●	●	●	●
Glomma Basin, Norway		●		●	●			●			●		●			●		●	●
Papua New Guinea	●	●				●	●				●	●	●	●		●	●	●	●
Vilcanota, Peru		●	●					●			●	●		●			●	●	●
Laguna Lake Basin, Philippines		●		●	●						●	●		●	●			●	●
Portugal	●	●	●	●	●	●	●	●		●	●	●		●	●	●	●	●	●
São Paulo Green Belt, Brazil	●	●		●	●					●	●	●		●	●	●	●	●	●
Southern Africa	●	●	●	●	●					●	●	●	●	●		●		●	●
Stockholm and Kristianstad, Sweden		●			●					●	●	●		●	●	●	●	●	●
Northern Range, Trinidad	●			●	●			●			●	●		●		●	●	●	●
Downstream Mekong Wetlands, Viet Nam	●	●			●						●	●	●	●	●	●	●	●	●
Western China		●	●	●	●			●			●	●		●	●		●		●
Alaskan Boreal Forest				●	●						●					●		●	●
Arafura and Timor Seas	●					●	●				●			●	●				●
Argentine Pampas		●									●	●						●	●
Central Asia Mountains								●			●	●		●					●
Colombia coffee-growing regions		●						●			●	●		●				●	
Eastern Himalayas				●				●			●	●	●	●				●	
Sinai Peninsula, Egypt			●					●						●			●	●	●
Fiji	●					●					●	●	●						●
Hindu Kush-Himalayas					●			●				●		●			●	●	●
Indonesia	●					●	●				●			●					●
India Urban Resource										●	●	●	●	●	●			●	●
Tafilalt Oasis, Morocco		●	●								●	●						●	●
Northern Australia Floodplains					●						●	●		●			●	●	●
Assir National Park, Saudi Arabia		●		●				●			●						●	●	●
Northern Highlands Lake District, Wisconsin				●	●							●				●	●	●	●

particular ecosystem must be tailored to the particular characteristics of that ecosystem and to the demands placed on it. However, an assessment focused only on a particular ecosystem or particular nation is insufficient because some processes are global and because local goods, services, matter, and energy are often transferred across regions. Each of the component assessments was guided by the MA conceptual framework and benefited from the presence of assessments undertaken at larger and smaller scales. The sub-global assessments were not intended to serve as representative samples of all ecosystems; rather, they were to meet the needs of decision-makers at the scales at which they were undertaken. The sub-global assessments involved in the MA process are shown in the Figure and the ecosystems and ecosystem services examined in these assessments are shown in the Table.

The work of the MA was conducted through four working groups, each of which prepared a report of its findings. At the global scale, the Condition and Trends Working Group assessed the state of knowledge on ecosystems, drivers of ecosystem change, ecosystem services, and associated human well-being around the year 2000. The assessment aimed to be comprehensive with regard to ecosystem services, but its coverage is not exhaustive. The Scenarios Working Group considered the possible evolution of ecosystem services during the twenty-first century by developing four global scenarios exploring plausible future changes in drivers, ecosystems, ecosystem services, and human well-being. The Responses Working Group examined the strengths and weaknesses of various response options that have been used to manage ecosystem services and identified promising opportunities for improving human well-being while conserving ecosystems. The report of the Sub-global Assessments Working Group contains lessons learned from the MA sub-global assessments. The first product of the MA—*Ecosystems and Human Well-being: A Framework for Assessment,* published in 2003—outlined the focus, conceptual basis, and methods used in the MA. The executive summary of this publication appears as Chapter 1 of this volume.

Approximately 1,360 experts from 95 countries were involved as authors of the assessment reports, as participants in the sub-global assessments, or as members of the Board of Review Editors. The latter group, which involved 80 experts, oversaw the scientific review of the MA reports by governments and experts and ensured that all review comments were appropriately addressed by the authors. All MA findings underwent two rounds of expert and governmental review. Review comments were received from approximately 850 individuals (of which roughly 250 were submitted by authors of other chapters in the MA), although in a number of cases (particularly in the case of governments and MA-affiliated scientific organizations), people submitted collated comments that had been prepared by a number of reviewers in their governments or institutions.

The MA was guided by a Board that included representatives of five international conventions, five U.N. agencies, international scientific organizations, governments, and leaders from the private sector, nongovernmental organizations, and indigenous groups. A 15-member Assessment Panel of leading social and natural scientists oversaw the technical work of the assessment, supported by a secretariat with offices in Europe, North America, South America, Asia, and Africa and coordinated by the United Nations Environment Programme.

The MA is intended to be used:

- to identify priorities for action;
- as a benchmark for future assessments;
- as a framework and source of tools for assessment, planning, and management;
- to gain foresight concerning the consequences of decisions affecting ecosystems;
- to identify response options to achieve human development and sustainability goals;
- to help build individual and institutional capacity to undertake integrated ecosystem assessments and act on the findings; and
- to guide future research.

Because of the broad scope of the MA and the complexity of the interactions between social and natural systems, it proved to be difficult to provide definitive information for some of the issues addressed in the MA. Relatively few ecosystem services have been the focus of research and monitoring and, as a consequence, research findings and data are often inadequate for a detailed global assessment. Moreover, the data and information that are available are generally related to either the characteristics of the ecological system or the characteristics of the social system, not to the all-important interactions between these systems. Finally, the scientific and assessment tools and models available to undertake a cross-scale integrated assessment and to project future changes in ecosystem services are only now being developed. Despite these challenges, the MA was able to provide considerable information relevant to most of the focal questions. And by identifying gaps in data and information that prevent policy-relevant questions from being answered, the assessment can help to guide research and monitoring that may allow those questions to be answered in future assessments.

Contents

Foreword

The Millennium Ecosystem Assessment was called for by United Nations Secretary-General Kofi Annan in 2000 in his report to the UN General Assembly, *We the Peoples: The Role of the United Nations in the 21st Century.* Governments subsequently supported the establishment of the assessment through decisions taken by three international conventions, and the MA was initiated in 2001. The MA was conducted under the auspices of the United Nations, with the secretariat coordinated by the United Nations Environment Programme, and it was governed by a multistakeholder board that included representatives of international institutions, governments, business, NGOs, and indigenous peoples. The objective of the MA was to assess the consequences of ecosystem change for human well-being and to establish the scientific basis for actions needed to enhance the conservation and sustainable use of ecosystems and their contributions to human well-being.

This volume has been produced by the MA Scenarios Working Group and examines possible changes in ecosystem services during the twenty-first century by developing four global scenarios exploring plausible future changes in drivers, ecosystems, ecosystem services, and human well-being. The material in this report has undergone two extensive rounds of peer review by experts and governments, overseen by an independent Board of Review Editors.

This is one of four volumes (*Current State and Trends, Scenarios, Policy Responses,* and *Multiscale Assessments*) that present the technical findings of the Assessment. Six synthesis reports have also been published: one for a general audience and others focused on issues of biodiversity, wetlands and water, desertification, health, and business and ecosystems. These synthesis reports were prepared for decision-makers in these different sectors, and they synthesize and integrate findings from across all of the working groups for ease of use by those audiences.

This report and the other three technical volumes provide a unique foundation of knowledge concerning human dependence on ecosystems as we enter the twenty-first century. Never before has such a holistic assessment been conducted that addresses multiple environmental changes, multiple drivers, and multiple linkages to human well-being. Collectively, these reports reveal both the extraordinary success that humanity has achieved in shaping ecosystems to meet the need of growing populations and economies and the growing costs associated with many of these changes. They show us that these costs could grow substantially in the future, but also that there are actions within reach that could dramatically enhance both human well-being and the conservation of ecosystems.

A more exhaustive set of acknowledgements appears later in this volume but we want to express our gratitude to the members of the MA Board, Board Alternates, Exploratory Steering Committee, Assessment Panel, Coordinating Lead Authors, Lead Authors, Contributing Authors, Board of Review Editors, and Expert Reviewers for their extraordinary contributions to this process. (The list of reviewers is available at www.MAweb.org.) We also would like to thank the MA Secretariat and in particular the staff of the Scenarios Working Group Technical Support Unit for their dedication in coordinating the production of this volume, as well as the University of Wisconsin-Madison, the Food and Agriculture Organization of the United Nations, and the International Maize and Wheat Improvement Center, which housed this TSU.

We would particularly like to thank the Co-chairs of the Scenarios Working Group, Dr. Stephen Carpenter and Dr. Prabhu Pingali, and the TSU Coordinators, Dr. Elena Bennett and Dr. Monika Zurek, for their skillful leadership of this working group and their contributions to the overall assessment.

Dr. Robert T. Watson
MA Board Co-chair
Chief Scientist, The World Bank

Dr. A.H. Zakri
MA Board Co-chair
Director, Institute for Advanced Studies,
United Nations University

Preface

Scenarios is one of four central volumes of the Millennium Ecosystem Assessment, a four-year international program designed to meet the needs of decision-makers for scientific information on the links between ecosystem change and human well-being. Leading scientists from around the world have been involved with the development of the scenarios and the writing of this book.

Scenarios are plausible, challenging, and relevant sets of stories about how the future might unfold. They are generally developed to help decision-makers understand the wide range of potential futures, confront critical uncertainties, and understand how decisions made now may play out in the future. They are intended to widen perspectives and illuminate key issues that might otherwise be missed or dismissed. The goal of developing scenarios is often to support more informed and rational decision-making that takes both the known and the unknown into account.

We developed four scenarios that focus on ecosystem change and the impacts on human well-being. Each scenario demonstrates development pathways commonly discussed today by decision-makers around the world. They address assumptions that people hold about how the world works and the best paths to a sustainable future. By comparing different scenarios, readers can understand the potential impact of today's decisions on tomorrow's ecosystems and human well-being. The probability of any one of our scenarios being the real future is low: the real future is likely to be some mix of the scenarios that we present. The future could be far worse or far better than any of the individual scenarios, depending on the choices made by decision-makers as well as on unforeseeable events.

The scenarios could be presented in many different ways. We have chosen to present them in three sections. Part I presents the background material for the scenarios. **Chapter 1** summarizes the MA conceptual framework. It describes the assumptions that underlie the MA and explains the basic framework for analysis and decision-making. It was developed through interactions of the experts involved in the MA as well as stakeholders who will use the findings of the MA. **Chapter 2** explores the history of global environmental scenario building for sustainable development. While scenarios first emerged as a war planning technique in the 1950s, the first ones that explicitly included environmental issues were not developed until the 1970s.

Although scenarios have been developed to improve understanding of the environment, **Chapter 3** explains that even these focus primarily on socioeconomic changes and have rarely taken ecological dynamics into account. The authors show that incorporating ecosystem dynamics could radically alter the outcome of some scenarios, and they make the case that including ecosystem knowledge into scenarios about ecosystem change and human well-being is critical.

Quantitative projections using models are an important element of the MA scenarios. Models are used to add quantitative dimensions to scenarios, compare outcomes, evaluate the consistency of scenarios with known conditions and trends, and assess plausibility in relation to generally accepted mechanisms of ecosystem change. Models exist to quantify many, but not all, aspects of the MA scenarios. Even in cases where models exist, however, there may be critical uncertainties or other weaknesses. **Chapter 4** explores the strengths and weaknesses of the models that are available to quantify the MA scenarios in nine areas: forecasting land cover change, impacts of land cover changes on local climates, changes in food demand and supply, changes in biodiversity and extinction rates, impacts of changes in nitrogen and phosphorus cycles, fisheries and harvest, alterations of coastal ecosystems, and impacts on human health. The ninth area considered is integrated assessment models that seek to piece together many different trends by predicting the consequences of changes in critical drivers.

The next four chapters form Part II, the presentation of the scenarios themselves. There are an infinite number of interesting scenarios about ecosystem change and human well-being, but we chose to present four specific ones. **Chapter 5** explains the rationale for choosing these four particular areas and how decision-maker concerns and ecosystem management dilemmas led us to that focus. We also present brief versions of each of the scenarios and some ideas about the potential benefits and risks of each scenario. In **Chapter 6** we present the methods by which the scenarios were developed, including both qualitative and quantitative aspects of scenario development. The qualitative part of the chapter describes how we considered user needs and questions when outlining four storylines, and how the scenarios grew and were modified from this beginning. The quantitative part of the chapter describes the various models that were used to quantify the scenarios as well as the process by which these models were soft-linked. Finally, we describe how we addressed uncertainty in both the qualitative and quantitative parts of the scenarios and the sensitivity analysis for the quantitative aspect of the scenarios.

Chapter 7 presents some of the key input information needed to determine the outcome of the scenarios—the material about the key drivers of ecosystem change. The chapter examines two of the main elements of the MA conceptual framework, indirect and direct drivers. The goal of the chapter is to provide an overview at the global level of key drivers of ecosystem change and the ability to deliver services that improve human well-being. The scenario outlines presented in Chapter 5 can be used to infer changes in the drivers presented in Chapter 7. In turn, the changes in these drivers will go on to determine the outcomes for ecosystem change, which are presented later. The final chapter in this section, **Chapter 8,** is the full presentation of the scenario storylines. Chapter 8 also details the differences and similarities among the four scenarios, as well as providing an in-depth examination of the potential risks and benefits of each of our four scenarios.

The last six chapters, Part III, delve into the implications of the scenarios for ecosystem change and changes in human well-being as well as for managing socioecological systems. In **Chapter 9,** we present estimates of changing ecosystem services in the form of both qualitative and quantitative information. The qualitative information is based on our interpretation of the storylines in Chapters 5 and 8, while the quantitative information is based on the related modeling analysis. Quantification provides insight into demand for food, water, and other ecosystem services and the potential effects on future capacity of ecosystems to provide these services.

Chapter 10 looks specifically at changes in biodiversity across the scenarios. Despite management efforts to stem losses, biodiversity has continued to decline in many parts of the world. This chapter examines what the scenarios tell us about how biodiversity is likely to change in the future and what actions we can take to help maintain biodiversity. Because biodiversity is necessary for the provision of many other ecosystem services, changes in biodiversity in the future may have important implications for the provision of key ecosystem services. Because ecosystems underpin human well-being through supporting, provisioning, regulating, and cultural services, changes in ecosystem services also affect human well-being. Well-being also depends on the supply and quality of human services, technology, and institutions. We examine changes in human well-being across the scenarios in **Chapter 11,** which also looks at the resilience and vulnerability of human well-being to adverse surprises across the scenarios.

Once we understand the similarities and differences in the provision of ecosystem services and human well-being across the scenarios, we can begin to think about ecosystem management. The final three chapters address ecosystem management options and their consequences. We examine the implications of the scenarios for trade-offs between ecosystem services in **Chapter 12.** Trade-offs are reductions in one ecosystem service that accompany increased use of another service or increased intensity of some non-ecosystem-based human activity. The scenarios indicate that major policy decisions in the next 50–100 years will have to address trade-offs among ecosystem services. Many trade-offs, such as the one between agricultural production and water quality, are consistent across all scenarios. We provide a synthesis of the lessons of the MA scenario development in **Chapter 13.** This chapter is directed primarily at the global assessment community. Finally, **Chapter 14** synthesizes the results of the MA scenarios for policy-makers, focusing on the Convention on Biological Diversity, the RAMSAR convention on wetlands, the Convention to Combat Desertification, national governments, communities and nongovernmental organizations, and the private sector.

Elena Bennett and Steve Carpenter
University of Wisconsin-Madison
United States

Prabhu Pingali and Monika Zurek
Food and Agriculture Organization of the United Nations
Rome, Italy

Acknowledgments

First and foremost, we would like to thank the MA Scenarios Working Group for their hard work, and for all the stimulating and fun discussions we had over the course of the project. It was truly a pleasure to work with a group of people who were so eager and excited about the project.

Writing this report would not have been possible without the many comments and useful insights of the members of the MA Assessment Panel and we would like to thank all of them. We are also very grateful to Dr. Walter Reid, the MA Director, for the numerous helpful discussions and his continuous support of the group. Many thanks also go to the reviewers of this report, who ensured that we answered the right questions in a scientifically sound way.

The advice and assistance of Veronique Plocq-Fichelet at SCOPE were invaluable to us throughout this project. We would also like to thank the Figure designers—Pille Bunnell, Philippe Rekacewicz, and Emmanuelle Bournay—who were essential for making different Chapters in this volume more attractive and compelling.

Special thanks are due to the MA Secretariat staff who worked tirelessly on this project:

Administration
Nicole Khi—Program Coordinator
Chan Wai Leng—Program Coordinator
Belinda Lim—Administrative Officer
Tasha Merican—Program Coordinator

Sub-global
Marcus Lee—Technical Support Unit (TSU) Coordinator and MA Deputy Director
Ciara Raudsepp-Hearne—TSU Coordinator

Condition and Trends
Neville J. Ash—TSU Coordinator
Dalène du Plessis—Program Assistant
Mampiti Matete—TSU Coordinator

Scenarios
Elena M. Bennett—TSU Coordinator
Veronique Plocq-Fichelet—Program Administrator
Monika B. Zurek—TSU Coordinator

Responses
Pushpam Kumar—TSU Coordinator
Meenakshi Rathore—Program Coordinator
Henk Simons—TSU Coordinator

Engagement and Outreach
Christine Jalleh—Communications Officer
Nicolas Lucas—Engagement and Outreach Director
Valerie Thompson—Associate

Other Staff
John Ehrmann—Lead Facilitator
Keisha-Maria Garcia—Research Assistant
Lori Han—Publications Manager
Sara Suriani—Conference Manager
Jillian Thonell—Data Coordinator

Interns
Emily Cooper, Elizabeth Wilson, Lina Cimarrusti

We would like to acknowledge the contributions of all the authors of this book and the support provided by their institutions that enabled their participation. We would like to thank the host organizations of the MA Technical Support Units—WorldFish Center (Malaysia); UNEP-World Conservation Monitoring Centre (United Kingdom); Institute of Economic Growth (India); National Institute of Public Health and the Environment (Netherlands); University of Pretoria (South Africa), Food and Agriculture Organization of the United Nations (Italy), World Resources Institute, Meridian Institute, and Center for Limnology of the University of Wisconsin-Madison (all in the United States); Scientific Committee on Problems of the Environment (France); and International Maize and Wheat Improvement Center (Mexico)—for the support they provided to the process. The Scenarios Working Group was established as a joint project of the MA and the Scientific Committee on Problems of the Environment, and we thank SCOPE for the scientific input and oversight that it provided.

We thank several individuals who played particularly critical roles: Linda Starke and Noreen McAuliffe for editing the report; Hyacinth Billings and Caroline Taylor for providing invaluable advice on the publication process; Maggie Powell for preparing the page design and all the Figures; and Elizabeth Wilson and Julie Feiner for helping to proof the Figures and Tables. And we thank the other MA volunteers, the administrative staff of the host organizations, and colleagues in other organizations who were instrumental in facilitating the process: Mariana Sanchez Abregu, Isabelle Alegre, Adlai Amor, Emmanuelle Bournay, Herbert Caudill, Habiba Gitay, Helen Gray, Sherry Heileman, Norbert Henninger, Toshi Honda, Francisco Ingouville, Humphrey Kagunda, Brygida Kubiak, Nicolas

Lapham, Liz Leavitt, Christian Marx, Stephanie Moore, John Mukoza, Arivudai Nambi, Laurie Neville, Carolina Katz Reid, Liana Reilly, Philippe Rekacewicz, Carol Rosen, Anne Schram, Jeanne Sedgwick, Tang Siang Nee, Darrell Taylor, Tutti Tischler, Dan Tunstall, Woody Turner, Mark Valentine, Elsie Velez Whited, and Mark Zimsky.

We thank the members of the MA Board and its chairs, Robert Watson and A.H. Zakri, the members of the MA Assessment Panel and its chairs, Angela Cropper and Harold Mooney, and the members of the MA Review Board and its chairs, José Sarukhán and Anne Whyte, for their guidance and support for this working group. We also thank the current and previous Board Alternates: Ivar Baste, Jeroen Bordewijk, David Cooper, Carlos Corvalan, Nick Davidson, Lyle Glowka, Guo Risheng, Ju Hongbo, Ju Jin, Kagumaho (Bob) Kakuyo, Melinda Kimble, Kanta Kumari, Stephen Lonergan, Charles Ian McNeill, Joseph Kalemani Mulongoy, Ndegwa Ndiang'ui, and Mohamed Maged Younes. We thank the past members of the MA Board whose contributions were instrumental in shaping the MA focus and process, including Philbert Brown, Gisbert Glaser, He Changchui, Richard Helmer, Yolanda Kakabadse, Yoriko Kawaguchi, Ann Kern, Roberto Lenton, Corinne Lepage, Hubert Markl, Arnulf Müller-Helbrecht, Seema Paul, Susan Pineda Mercado, Jan Plesnik, Peter Raven, Cristián Samper, Ola Smith, Dennis Tirpak, Alvaro Umaña, and Meryl Williams. We wish to also thank the members of the Exploratory Steering Committee that designed the MA project in 1999–2000. This group included a number of the current and past Board members, as well as Edward Ayensu, Daniel Claasen, Mark Collins, Andrew Dearing, Louise Fresco, Madhav Gadgil, Habiba Gitay, Zuzana Guziova, Calestous Juma, John Krebs, Jane Lubchenco, Jeffrey McNeely, Ndegwa Ndiang'ui, Janos Pasztor, Prabhu L. Pingali, Per Pinstrup-Andersen, and José Sarukhán. We thank Ian Noble and Mingsarn Kaosa-ard for their contributions as members of the Assessment Panel during 2002.

We would particularly like to acknowledge the input of the hundreds of individuals, institutions, and governments (see list at www.MAweb.org) who reviewed drafts of the MA technical and synthesis reports. We also thank the thousands of researchers whose work is synthesized in this report. And we would like to acknowledge the support and guidance provided by the secretariats and the scientific and technical bodies of the Convention on Biological Diversity, the Ramsar Convention on Wetlands, the Convention to Combat Desertification, and the Convention on Migratory Species, which have helped to define the focus of the MA and of this report.

We also want to acknowledge the support of a large number of nongovernmental organizations and networks around the world that have assisted in outreach efforts: Alexandria University, Argentine Business Council for Sustainable Development, Arab Media Forum for Environment and Development, Asociación Ixacavaa (Costa Rica), Brazilian Business Council on Sustainable Development, Charles University (Czech Republic), Chinese Academy of Sciences, European Environmental Agency, European Union of Science Journalists' Associations, EIS-Africa (Burkina Faso), Forest Institute of the State of São Paulo, Foro Ecológico (Peru), Fridtjof Nansen Institute (Norway), Fundación Natura (Ecuador), Global Development Learning Network, Indonesian Biodiversity Foundation, Institute for Biodiversity Conservation and Research–Academy of Sciences of Bolivia, International Alliance of Indigenous Peoples of the Tropical Forests, IUCN office in Uzbekistan, IUCN Regional Offices for West Africa and South America, Northern Temperate Lakes Long Term Ecological Research Site (USA), Permanent Inter-States Committee for Drought Control in the Sahel, Peruvian Society of Environmental Law, Probioandes (Peru), Professional Council of Environmental Analysts of Argentina, Regional Center AGRHYMET (Niger), Regional Environmental Centre for Central Asia, Resources and Research for Sustainable Development (Chile), Royal Society (United Kingdom), Stockholm University, Suez Canal University, Terra Nuova (Nicaragua), The Nature Conservancy (United States), United Nations University, University of Chile, University of the Philippines, Winslow Foundation (USA), World Assembly of Youth, World Business Council for Sustainable Development, WWF-Brazil, WWF-Italy, and WWF-US.

We are extremely grateful to the donors that provided major financial support for the MA and the MA Sub-global Assessments: Global Environment Facility; United Nations Foundation; David and Lucile Packard Foundation; World Bank; Consultative Group on International Agricultural Research; United Nations Environment Programme; Government of China; Ministry of Foreign Affairs of the Government of Norway; Kingdom of Saudi Arabia; and the Swedish International Biodiversity Programme. We also thank other organizations that provided financial support: Asia Pacific Network for Global Change Research; Association of Caribbean States; British High Commission, Trinidad & Tobago; Caixa Geral de Depósitos, Portugal; Canadian International Development Agency; Christensen Fund; Cropper Foundation, Environmental Management Authority of Trinidad and Tobago; Ford Foundation; Government of India; International Council for Science; International Development Research Centre; Island Resources Foundation; Japan Ministry of Environment; Laguna Lake Development Authority; Philippine Department of Environment and Natural Resources; Rockefeller Foundation; U.N. Educational, Scientific and Cultural Organization; UNEP Division of Early Warning and Assessment; United Kingdom Department for Environment, Food and Rural Affairs; United States National Aeronautic and Space Administration; and Universidade de Coimbra, Portugal. Generous in-kind support has been provided by many other institutions (a full list is available at www.MAweb.org). The work to establish and design the MA was supported by grants from The Avina Group, The David and Lucile Packard Foundation, Global Environment Facility, Directorate for Nature Management of Norway, Swedish International Development Cooperation Authority, Summit Foundation, UNDP, UNEP, United Nations Foundation, United States Agency for International Development, Wallace Global Fund, and World Bank.

Reader's Guide

The four technical reports present the findings of each of the MA Working Groups: Condition and Trends, Scenarios, Responses, and Sub-global Assessments. A separate volume, *Our Human Planet*, presents the summaries of all four reports in order to offer a concise account of the technical reports for decision-makers. In addition, six synthesis reports were prepared for ease of use by specific audiences: Synthesis (general audience), CBD (biodiversity), UNCCD (desertification), Ramsar Convention (wetlands), business and industry, and the health sector. Each MA sub-global assessment will also produce additional reports to meet the needs of its own audiences.

All printed materials of the assessment, along with core data and a list of reviewers, are available at www.MAweb.org. In this volume, Appendix A contains color maps and figures. Appendix B lists all the authors who contributed to this volume. Appendix C lists the acronyms and abbreviations used in this report and Appendix D is a glossary of terminology used in the technical reports. Throughout this report, dollar signs indicate U.S. dollars and ton means tonne (metric ton). Bracketed references within the Summary are to chapters within this volume.

In this report, the following words have been used where appropriate to indicate judgmental estimates of certainty, based on the collective judgment of the authors, using the observational evidence, modeling results, and theory that they have examined: very certain (98% or greater probability), high certainty (85–98% probability), medium certainty (65%–58% probability), low certainty (52–65% probability), and very uncertain (50–52% probability). In other instances, a qualitative scale to gauge the level of scientific understanding is used: well established, established but incomplete, competing explanations, and speculative. Each time these terms are used they appear in italics.

Ecosystems and Human Well-being:
Scenarios, Volume 2

Summary: Comparing Alternate Futures of Ecosystem Services and Human Well-being

Core Writing Team: Elena Bennett, Steve Carpenter, Prabhu Pingali, Monika Zurek
Extended Writing Team: Scenarios Working Group

Envisioning the Future for Ecosystems and People

The capacity of Earth's ecosystems to provide life-support services is changing rapidly, at a time when human pressures on ecosystems are also increasing.

These changes in ecosystems have enormous implications for life on Earth. Yet they can seem bewildering because of their complexity, speed, surprises, and demands on human ingenuity.

Scenarios organize information about plausible causes of and responses to long-term change. The central idea is to categorize outcomes into a few plausible futures, making the complex more comprehensible. Contrasts among scenarios illuminate key linkages and probable outcomes of various approaches or decisions.

Ecosystems are always changing, but the rate and magnitude of change are not constant over time. Most of the time, change is gradual, incremental, and perhaps reversible. However, some changes in ecosystems and their services are large in magnitude and can be difficult, expensive, or impossible to reverse (*high certainty*). Examples of ecosystems subject to large, important changes are pelagic fisheries (economic collapse), freshwater lakes and reservoirs (toxic blooms, fish kills), pastoral lands (conversion to woodland with overgrazing and fire suppression), and dryland agriculture (desertification). The thresholds and triggering events for these large changes are often difficult to predict. [3, 5]

Slow losses of resilience set the stage for large changes that occur after the ecosystem crosses a threshold or is subjected to a random event such as a climate fluctuation (*established but incomplete*). For example, incremental buildup of phosphorus in soils gradually increases the vulnerability of lakes and reservoirs to runoff events that trigger oxygen depletion, toxic algae blooms, and fish kills. Cumulative effects of overfishing and nutrient runoff make coral reefs susceptible to severe deterioration triggered by storms, invasive species, or disease. Slow decrease in grass cover crosses a threshold so that grasslands can no longer carry a fire, allowing woody vegetation to dominate and severely decreasing forage for livestock. [3, 5] These long-lasting and costly changes from seemingly random events pose a daunting challenge for decision-makers concerned with ecosystems as well as for people whose livelihoods depend on ecosystems.

Recent trends in human use of ecosystem services reveal rapid changes and great uncertainty about future changes. (See MA *Current State and Trends* volume.) While many ecosystem services are renewable, current rates of use are often greater than the renewal rates, leading to degradation and declines in the future capacity of ecosystems to provide services. Dryland agricultural areas around the world are threatened by desertification. Freshwater supplies have been stressed by increasing withdrawals of groundwater and surface water, as well as by pollution. Marine fish harvest has declined since the late 1980s, and one quarter of marine fish stocks are overexploited or depleted. Despite growing global timber production, the condition of forests is diminishing. The observed rates of species extinction in modern times are as much as 1,000 times higher than the average observed for comparable taxonomic groups from the fossil record. These and many other losses have occurred in the course of using ecosystem services. The capacity of Earth's ecosystems to provide life-support services is changing rapidly, at a time when human pressures on ecosystems are also increasing. **The *Scenarios* volume explores the implications of different approaches for sustaining ecosystem services in the face of growing demand.** [8, 9, 11, 14]

In order to plan for a changing and uncertain future, we must have tools for organizing extensive information about socioecological systems. Scenarios are such a tool. **Scenarios are plausible, provocative, and relevant stories about how the future might unfold. They can be told in both words and numbers. Scenarios are not forecasts, projections, predictions, or recommendations, though model projections may be used to quantify some aspects of the scenarios.** The process of building scenarios is intended to widen perspectives and illuminate key issues that might otherwise be either missed or dismissed. By offering insight into uncertainties and the consequences of current and possible future actions, scenarios support more informed and rational decision-making in situations of uncertainty. Scenarios are a powerful way of exploring possible consequences of different policies. They force us to state our assumptions clearly, enabling the consequences of those assumptions to be analyzed. Scenarios, and the products of scenarios, are not predictions. Rather, they explore consequences of different policy choices based on current knowledge of underlying socioecological processes. [2, 3, 5]

This summary explores the scenarios, how we developed them, and what we have learned in the process. The first section describes the methods and the assumptions behind the scenarios. This is followed by four sections that explore the results for ecosystem services, trade-offs among ecosystem services, biodiversity, and human well-being. We conclude with a section describing research needs for improving future development of scenarios for ecosystem services and human well-being.

Developing the Millennium Ecosystem Assessment Scenarios

The MA scenarios assess the consequences of contrasting development paths for ecosystem services.

Because stresses on ecosystems are increasing, it is likely that large, costly, and even irreversible changes will become more common in the future. This will lead to reduced services provided by ecosystems or increased costs of maintaining services. Management that deliberately maintains resilience of ecosystems can reduce the risk of large, costly, or irreversible change.

Proactive or anticipatory management of ecosystems is particularly important under rapidly changing or novel conditions.

The MA developed a set of global scenarios to address the effects of different development paths on ecosystem services and human well-being. The scenarios extend into the future from the situation described in the MA *Current State and Trends* volume. Three of the four pathways involve major positive actions taken to move toward sustainable development. The alternate pathways of the four contrasting scenarios illustrate many of the tools described in the MA *Policy Responses* volume. Although the scenarios focus on the global scale, many implications for regional and local ecosystems were examined. These provide a bridge to the MA *Multiscale Assessments* volume. **The contrasts among the global scenarios are designed to illuminate key risks and benefits of each pathway and to examine the interaction among drivers of ecosystem change, ecosystem services, and human well-being.**

The MA scenarios explore the potential consequences of alternate pathways to development, and they inform decision-makers about the consequences for ecosystem services. **The scenarios were designed to explore contrasting transitions of society as well as contrasting approaches to policies about ecosystem services.** (See Figure S1). We explore two kinds of transitions—one in which the world becomes increasingly globalized and another in which it becomes increasingly regionalized. Furthermore, we address two different approaches for governance and policies related to ecosystems and their services. In one case, management of ecosystems is reactive, and most problems are addressed only after they become obvious. In the other case, management of ecosystems is proactive, and policies deliberately seek to maintain ecosystem services for the long term.

Framed in terms of these contrasts, the four scenarios developed by the MA were named Global Orchestration (socially conscious globalization, with an emphasis on equity, economic growth, and public goods and with a reactive approach to ecosystems), Order from Strength (regionalized, with an emphasis on security and economic growth and with a reactive approach to ecosystems), Adapting Mosaic (regionalized, with an emphasis on proactive management of ecosystems, local adaptation, and flexible governance), and TechnoGarden (globalized, with an emphasis on using technology to achieve environmental outcomes and with a proactive approach to ecosystems). **The focus on ecosystem services and effects of ecosystems on human well-being distinguish the MA scenarios from previous global scenario exercises.** [2, 3, 5, 8]

The future will represent a mix of approaches and consequences described in the scenarios, as well as events and innovations that have not yet been imagined. No scenario will match the future as it actually occurs. No scenario represents business as usual, although all begin from current conditions and trends. None of the MA scenarios represents a "best" or a "worst" path. Instead, they illustrate choices and trade-offs. There could be combinations of policies that produce significantly better, or worse, outcomes than any of the scenarios. Each of the scenarios begins in 2000 and ends in 2050. Each emphasizes different pathways of development. [2] (See Box S1.)

Interviews with stakeholders and a literature review of major ecological dilemmas were used to identify focal questions, key uncertainties, and cross-cutting assumptions behind the scenarios. (See Figure S2). These focal questions, uncertainties, and assumptions, which are explored in more detail in the next paragraphs, were used to develop the four plausible, alternative futures. Scenarios were then constructed by working through the MA conceptual framework (indirect drivers, direct drivers,

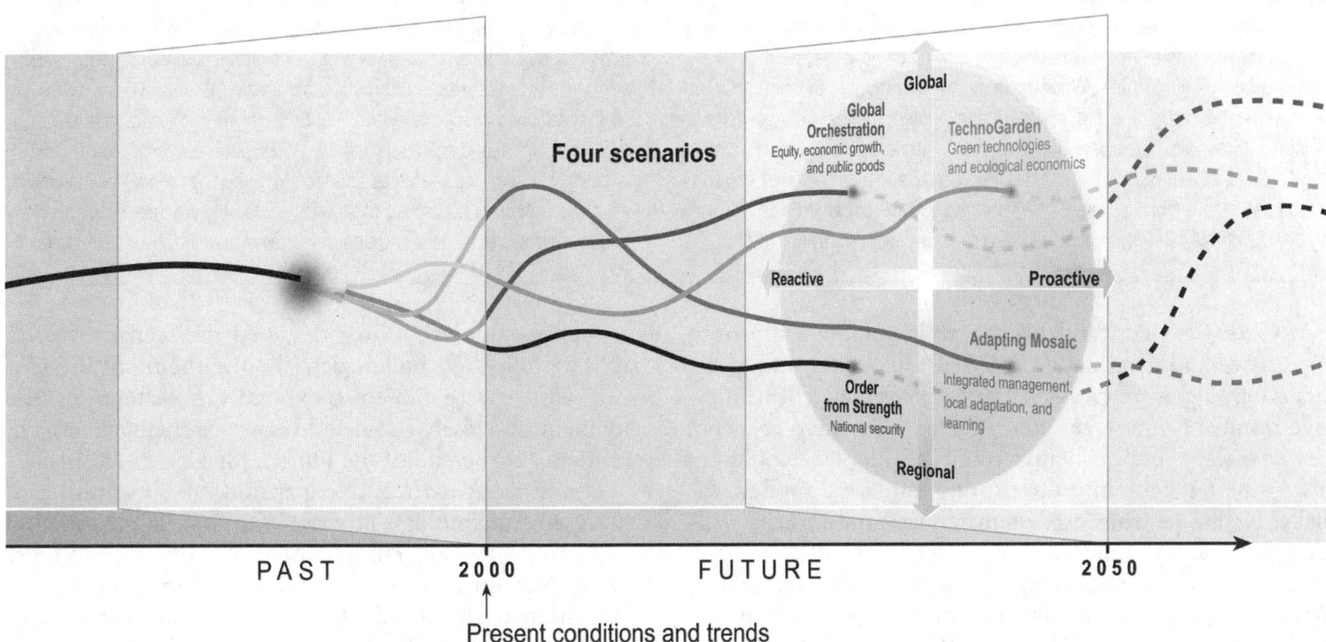

Figure S1. Millennium Ecosystem Assessment Scenarios: Plausible Future Development Pathways until 2050. The scenario differences are based on the approaches pursued toward governance and economic development (regionalized versus globalized) and ecosystem service management (reactive versus proactive).

BOX S1

Global Scenarios of the Millennium Ecosystem Assessment

The Global Orchestration scenario depicts a globally connected society in which policy reforms that focus on global trade and economic liberalization are used to reshape economies and governance, emphasizing the creation of markets that allow equitable participation and provide equitable access to goods and services. These policies, in combination with large investments in global public health and the improvement of education worldwide, generally succeed in promoting economic expansion and lift many people out of poverty into an expanding global middle class. Supranational institutions in this globalized scenario are well placed to deal with global environmental problems such as climate change and fisheries. However, the reactive approach to ecosystem management favored in this scenario makes people vulnerable to surprises arising from delayed action. While the focus is on improving human well-being of all people, environmental problems that threaten human well-being are only considered after they become apparent.

Growing economies, expansion of education, and growth of the middle class leads to demand for cleaner cities, less pollution, and a more beautiful environment. Rising income levels bring about changes in global consumption patterns, boosting demand for ecosystem services, including agricultural products such as meat, fish, and vegetables. Growing demand for these services leads to declines in other services, as forests are converted into cropped areas and pasture, and the services formerly provided by forests decline. The problems related to increasing food production, such as loss of wildlands, are remote to most people because they live in urban areas. These problems therefore receive only limited attention.

Global economic expansion expropriates or degrades many of the ecosystem services poor people once depended on for their survival. While economic growth more than compensates for these losses in some regions by increasing our ability to find substitutes for particular ecosystem services, in many other places it does not. An increasing number of people are affected by the loss of basic ecosystem services essential for human life. While risks seem manageable in some places, in other places there are sudden, unexpected losses as ecosystems cross thresholds and degrade irreversibly. Loss of potable water supplies, crop failures, floods, species invasions, and outbreaks of environmental pathogens increase in frequency. The expansion of abrupt, unpredictable changes in ecosystems, many with harmful effects on increasingly large numbers of people, is the key challenge facing managers of ecosystem services.

The Order from Strength scenario represents a regionalized and fragmented world concerned with security and protection, emphasizing primarily regional markets, and paying little attention to common goods. Nations see looking after their own interests as the best defense against economic insecurity, and the movement of goods, people, and information is strongly regulated and policed. The role of government expands as oil companies, water systems, and other strategic businesses are either nationalized or subjected to more state oversight. Trade is restricted, large amounts of money are invested in security systems, and technological change slows due to restrictions on the flow of goods and information. Regionalization exacerbates global inequality.

Agreements on global climate change, international fisheries, and the trade in endangered species are only weakly and haphazardly implemented, resulting in degradation of the global commons. Local problems often go unresolved, but major problems are sometimes handled by rapid disaster relief to at least temporarily resolve the immediate crisis. Many powerful countries cope with local problems by shifting burdens to other, less powerful countries, increasing the gap between rich and poor. In particular, natural resource–intensive industries are moved from wealthier nations to poorer and less powerful ones. Inequality increases considerably within countries as well.

Ecosystem services become more vulnerable, fragile, and variable in Order from Strength. For example, parks and reserves exist within fixed boundaries, but climate change crosses them, leading to the unintended extirpation of many species. Conditions for crops are often suboptimal, and the ability of societies to import alternative foods is diminished by trade barriers. As a result, there are frequent shortages of food and water, particularly in poor regions. Low levels of trade tend to restrict the number of invasions by exotic species; however, ecosystems are less resilient and invaders are therefore more often successful when they arrive.

In the Adapting Mosaic scenario, hundreds of regional ecosystems are the focus of political and economic activity. This scenario sees the rise of local ecosystem management strategies and the strengthening of local institutions. Investments in human and social capital are geared toward improving knowledge about ecosystem functioning and management, which results in a better

ecosystem services, and human well-being), using both qualitative and quantitative analyses. Qualitative and quantitative results were cross-checked at every stage. Quantitative results of one stage often affected qualitative results of the next stage, but qualitative results of one stage could not always be fed back into the existing numerical models. Finally, feedbacks from ecosystem services and human well-being played an important role in development of indirect and direct driver trajectories for the qualitative assessment. Such feedbacks are difficult to incorporate in the quantitative models, however. [6]

Interviews identified many benefits, risks, opportunities, and threats from contrasting paths of globalization and governance for ecosystem management. While some advantages and disadvantages are clear, many have not been thoroughly explored, so we designed the scenarios to do that. The following bullets describe the theme of the scenarios, which were chosen to explore various tensions (the storyline most closely associated with each theme appears in parentheses at the end of the bullet). [8, 11, 12, 13, 14]

- Economic growth and expansion of education and access to technology increases the capacity to respond effectively when environmental problems emerge. However, if the focus on reducing poverty and increasing human and social capital overwhelms attention to the environment, and if proactive environmental policies are not pursued, there is increased risk of regional or even global interruptions in the provision of ecosystem services. Severe and irreversible declines in ecosystem services and human well-being may occur if we do not

understanding of resilience, fragility, and local flexibility of ecosystems. There is optimism that we can learn, but humility about preparing for surprises and about our ability to know everything about managing ecosystems.

There is also great variation among nations and regions in styles of governance, including management of ecosystem services. Many regions explore actively adaptive management, investigating alternatives through experimentation. Others use bureaucratically rigid methods to optimize ecosystem performance. Great diversity exists in the outcome of these approaches: some areas thrive, while others develop severe inequality or experience ecological degradation. Initially, trade barriers for goods and products are increased, but barriers for information nearly disappear (for those who are motivated to use them) due to improving communication technologies and rapidly decreasing costs of access to information.

Eventually, the focus on local governance leads to some failures in managing the global commons. Problems like climate change, marine fisheries, and pollution grow worse, and global environmental problems intensify. Communities slowly realize that they cannot manage their local areas because global and regional problems are infringing, and they begin to develop networks among communities, regions, and even nations to better manage the global commons. Solutions that were effective locally are adopted among networks. These networks of regional successes are especially common in situations where there are mutually beneficial opportunities for coordination, such as along river valleys. Sharing good solutions and discarding poor ones eventually improves approaches to a variety of social and environmental problems, ranging from urban poverty to agricultural water pollution. As more knowledge is collected from successes and failures, provision of many services improves.

The TechnoGarden scenario depicts a globally connected world relying strongly on technology and highly managed, often engineered ecosystems to deliver ecosystem services. Overall efficiency of ecosystem service provision improves but is shadowed by the risks inherent in large-scale human-made solutions and rigid control of ecosystems.

Technology and market-oriented institutional reform are used to achieve solutions to environmental problems. These solutions are designed to benefit both the economy and the environment. These changes co-develop with the expansion of property rights to ecosystem services, requiring people to pay for pollution they create and paying people for providing key ecosystem services through actions such as preservation of key watersheds. Interest in maintaining, and even increasing, the economic value of these property rights, combined with an interest in learning and information, leads to an increase in the use of ecological engineering approaches for managing ecosystem services.

Investment in green technology is accompanied by a significant focus on economic development and education, improving people's lives and helping them understand how ecosystems make their livelihoods possible. A variety of problems in global agriculture are addressed by focusing on the multifunctional aspects of agriculture and a global reduction of agricultural subsidies and trade barriers. Recognition of the role of agricultural diversification encourages farms to produce a variety of ecological services rather than simply maximizing food production. The combination of these movements stimulates the growth of new markets for ecosystem services, such as trade in carbon storage, and the development of technology for increasingly sophisticated ecosystem management. Gradually, environmental entrepreneurship expands as new property rights and technologies co-evolve to stimulate the growth of companies and cooperatives providing reliable ecosystem services to cities, towns, and individual property owners.

Innovative capacity expands quickly in lower-income nations. The reliable provision of ecosystem services as a component of economic growth, together with enhanced uptake of technology due to rising income levels, lifts many of the world's poor into a global middle class. While the provision of basic ecosystem services improves the well-being of the world's poor, the reliability of the services, especially in urban areas, is increasingly critical and increasingly difficult to ensure. Not every problem has succumbed to technological innovation. Reliance on technological solutions sometimes creates new problems and vulnerabilities. In some cases, we seem to be barely ahead of the next threat to ecosystem services. In such cases, new problems often seem to emerge from the last solution, and the costs of managing the environment are continually rising. Environmental breakdowns that affect large numbers of people become more common. Sometimes new problems seem to emerge faster than solutions. The challenge for the future will be to learn how to organize socioecological systems so that ecosystem services are maintained without taxing society's ability to implement solutions to novel, emergent problems.

address natural capital at the same time that we address social capital. (Global Orchestration)

- A focus on strong national security, which restricts the flow of goods, information, and people, coupled with a reactive approach to ecosystem management, can create great stress on ecosystems, particularly in poorer countries. While there may be some opportunities for conservation of biodiversity in wealthy or highly prized areas, in general a focus on security in wealthy nations leads to a loss of biodiversity in developing ones, as they often lack the resources to create measures for biodiversity protection. Without active, proactive management of ecosystems in a world like this, pressure on the environment increases; there is greater risk of large disturbances of ecosystem services and vulnerability to interruptions in provision of ecosystem services. Severe and irreversible declines in ecosystem services and human well-being may occur if we do not address ecosystem management where we live, in addition to focusing on reserves. (Order from Strength)
- When regional ecosystem management is proactive and oriented around adapting to change, ecosystem services become more resilient and society becomes less vulnerable to disturbances of ecosystem services. However, a regional focus can diminish attention to the global commons and exacerbates global environmental problems, such as climate change and declining oceanic fisheries. An adaptive approach may also have high initial costs and an initially slower rate of environmental improvement. If the focus on natural capital overwhelms attention to

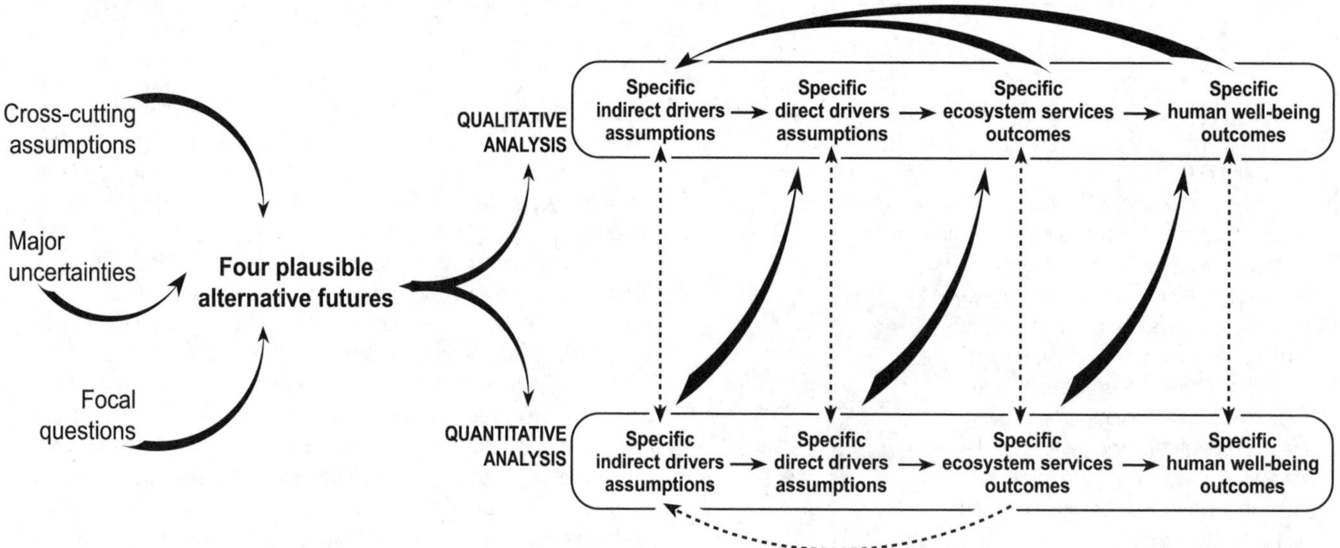

Figure S2. Flow Chart of MA Scenario Development. The focal questions, major uncertainties, and cross-cutting assumptions were used to develop basic ideas about four plausible alternative futures. These futures were elaborated using qualitative and quantitative methods. At each step, quantitative and qualitative results were cross-checked (the dotted lines between boxes). Quantitative results of each step were used to help determine qualitative results of the next step (diagonal arrows). Finally, feedbacks from qualitative ecosystem services and human well-being outcomes were used to re-evaluate assumptions about indirect drivers. This feedback procedure was also done in a qualitative way for some quantitative ecosystem services outcomes.

immediate human well-being, poverty alleviation may be somewhat slower. (Adapting Mosaic)

- Technological innovations and ecosystem engineering, coupled with economic incentive measures to facilitate their uptake, can lead to highly efficient delivery of provisioning ecosystem services. However, technologies can create new environmental problems, and in some cases the resulting disruptions of ecosystem services affect large numbers of people. In addition, efficient provision of ecosystem services may lead to greater demand for ecosystem services rather than less pressure on ecosystems to provide the same amount of service. (TechnoGarden)

The scenarios were also designed to explore key ecosystem management dilemmas. One such dilemma is that ecosystem management that neglects slow changes in resilience or vulnerability of ecosystems increases the susceptibility of ecosystems to large, rapid changes (*established but incomplete*). For example, government subsidies to agriculture have allowed farmers to continue harmful practices that eventually lead to larger losses of ecosystem services. When fish stocks decline, subsidies that sustain fishing effort prevent recovery of the stocks. Dependency on biocides can increase the vulnerability of agroecosystems to evolution of biocide-resistant pests. Because stresses on ecosystems are increasing, it is likely that large, costly, and even irreversible changes will become more common in the future. On the other hand, management that deliberately maintains resilience of ecosystems can reduce the risk of large, costly, or irreversible change (*established but incomplete*). The scenarios were constructed to explore this dynamic. [5, 8, 9, 10]

Managing for surprise is another dilemma explored by the scenarios. **The MA scenarios differ in the frequency and magnitude of surprising changes in ecosystem services due to the management undertaken in each scenario, not due to any underlying ecological differences across the scenarios.** Each scenario implies different distributions of extreme events. (See Figure S3.) Examples of extreme events that affect ecosystem services are famines, technological failure of systems for quality control of food or water, massive floods, or serious and long-lasting heat waves or storms. The impact of an extreme event is driven by both the chance of an event happening and the vulnerability of people to the event. Extreme events

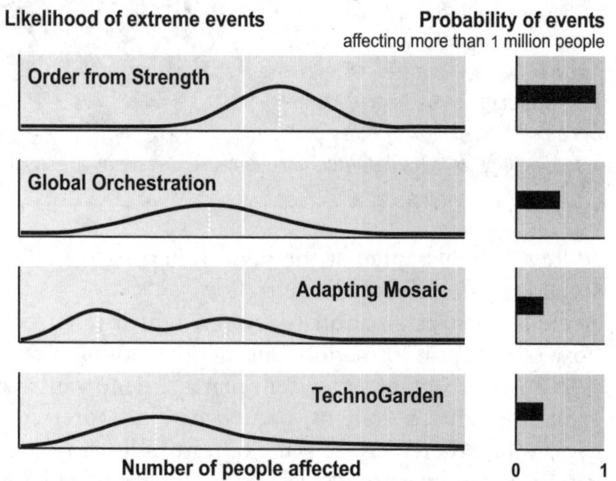

Figure S3. Probabilities of Extreme Events That Involve Ecosystem Services in MA Scenarios. Left column: Magnitude of extreme event (measured as the number of people affected) on the x-axis versus likelihood of events of a given magnitude, on the y-axis. Right column: Length of the bar indicates the annual probability of events that affect more than 1 million people.

affecting at least 1 million people are most common in Order from Strength and least common in Adapting Mosaic and TechnoGarden. [5, 8]

Proactive or anticipatory management of ecosystems is particularly important under rapidly changing or novel conditions. (See Table S1.) Ecological surprises are inevitable. Currently well understood phenomena that were surprises of the past century include the ability of pests to evolve resistance to biocides, the contribution to desertification of certain types of land use, biomagnification of toxins, and the increase in vulnerability of ecosystems to eutrophication and invasion due to removal of keystone predators. While we do not know which surprises will arise in the next 50 years, we can be certain that some will occur. Restoration of ecosystems or ecosystem services following degradation is usually time-consuming and expensive, if possible at all, so anticipatory management to build resilient, self-maintaining ecosystems is likely to be extremely cost-effective. This is particularly true when conditions are changing rapidly, when conditions are variable, when control of ecosystems is limited, or when uncertainty is high. [3]

The MA scenarios examine the need to develop and expand mechanisms of ecosystem management that avoid large ecosystem changes (by reducing stress on ecosystems), allow for the possibility of large ecosystem changes (by choosing reversible actions, experimenting cautiously, and monitoring appropriate ecological indicators), and increase the capacity of societies to adapt to large ecosystem changes (diversifying the portfolio of ecosystem services and developing flexible governance systems that adapt effectively to ecosystem change). [3, 5]

Table S1. Costs and Benefits of Proactive Management as Contrasted with Reactive Ecosystem Management

	Proactive Ecosystem Management	Reactive Ecosystem Management
Payoffs	benefit from lower risk of unexpected losses of ecosystem services, achieved through investment in more-efficient use of resources (water, energy, fertilizer, and so on), more innovation of green technology, the capacity to absorb unexpected fluctuations in ecosystem services, adaptable management systems, and ecosystems that are resilient and self-maintaining do well under changing or novel conditions build natural, social, and human capital	avoid paying for monitoring efforts do well under smoothly or incrementally changing conditions build manufactured, social, and human capital
Costs	technological solutions can create new problems costs of unsuccessful experiments costs of monitoring some short-term benefits are traded for long-term benefits	expensive unexpected events persistent ignorance (repeating the same mistakes) lost option values inertia of less flexible and adaptable management of infrastructure and ecosystems loss of natural capital

Quantitative and qualitative results for drivers, ecosystem services, and human well-being are presented in Tables S2 and S3. Indirect drivers are generally the result of group consensus and represent our assumptions about the factors that underlie each of the scenarios. Direct drivers are most often model outcomes based on the indirect drivers. For example, model outcomes show carbon emissions to be quite high in the scenarios with high economic growth, especially if proactive climate policies are not adopted. (See Figure S4.) Ecosystem service outcomes are a mixture of model outcomes and qualitative estimates, both based on the direct drivers. Most human well-being outcomes, determined largely by the ecosystem services outcomes while taking into account other social conditions, such as wealth and education, are qualitative estimates.

For some drivers, ecosystem services, and human well-being indicators, quantitative projections were calculated using established, peer-reviewed global models. Quantifiable items include drivers such as economic growth and land use change and ecosystem services such as water withdrawals, food production, and carbon emissions. Other drivers (such as rates of technologic change), ecosystem services (particularly supporting and cultural services such as soil formation and recreational opportunities), and human well-being indicators (such as human health and social relations) for which there are no appropriate global models were estimated qualitatively. Qualitative estimates were the consensus professional judgment of experts in relevant fields.

We explored the status of quantitative modeling in at least nine areas relevant to the MA: land cover change, impacts of land cover changes on local climates, changes in food demand and supply, changes in biodiversity and extinction rates, impacts of changes in nitrogen/phosphorus cycles, fisheries and harvest, alterations of coastal ecosystems, and impacts on human health as well as the use of integrated assessment models that seek to piece together many different trends by predicting the consequences of changes in critical drivers. **All these models have weaknesses, but the alternative is no quantification whatsoever. Therefore, we used appropriate models with caution and explicitly stated our uncertainties.** Key uncertainties include limitations on the spatial or temporal resolution of input data, bias or random error in input data, poor or unknown correspondence between modeled mechanisms and natural processes (model uncertainty), lack of information about model parameters, limited experience with linking the different models, and the impossibility of predicting human events and individual choices (which may be altered by the forecasts themselves). [4]

In general, models address incremental changes but fail to address thresholds, risk of extreme events,

Table S2. Main Assumptions about Indirect and Direct Driving Forces across the Scenarios [8, 9]

	Global Orchestration	Order from Strength: Industrial Nations[a]	Order from Strength: Developing Nations[a]	Adapting Mosaic	TechnoGarden
Indirect Driving Forces					
Demographics	high migration; low fertility and mortality levels; 2050 population: 8.1 billion	relatively high fertility and mortality levels (especially in developing countries); low migration, 2050 population: 9.6 billion		high fertility level; high mortality levels until 2010 then to medium by 2050; low migration, 2050 population: 9.5 billion	medium fertility levels, medium mortality; medium migration, 2050 population: 8.8 billion
Average income growth	high	medium	low	similar to Order from Strength but with increasing growth rates toward 2050	lower than Global Orchestration, but catching up toward 2050
GDP growth rates/capita per year until 2050 (global)	1995–2020: 2.4% per year 2020–50: 3.0% per year	1995–2020: 1.4% per year 2020–50: 1.0% per year		1995–2020: 1.5% per year 2020–50: 1.9% per year	1995–2020: 1.9% per year 2020–50: 2.5% per year
Income distribution	becomes more equal	similar to today		similar to today, then becomes more equal	becomes more equal
Investments into new produced assets	high	medium	low	begins like Order from Strength, then increases	high
Investments into human capital	high	medium	low	begins like Order from Strength, then increases in tempo	medium
Overall trend in technology advances	high	low		medium-low	medium in general; high for environmental technology
International cooperation	strong	weak—international competition		weak—focus on local environment	strong
Attitude toward environmental policies	reactive	reactive		proactive—learning	proactive
Energy demand and lifestyle	energy-intensive	regionalized assumptions		regionalized assumptions	high level of energy-efficiency
Energy supply	market liberalization; selects least-cost options; intensified use of technology	focus on domestic energy resources		some preference for clean energy resources	preference for renewable energy resources and rapid technology change
Climate policy	no	no		no	yes, aims at stabilization of CO_2-equivalent concentration at 550 ppmv
Approach to achieving sustainability	economic growth leads to sustainable development	national-level policies; conservation; reserves, parks		local-regional co-management; common-property institutions	green-technology; eco-efficiency; tradable ecological property rights
Direct Driving Forces					
Land use change	global forest loss until 2025 slightly below historic rate, stabilizes after 2025; ~10% increase in arable land	global forest loss faster than historic rate until 2025, near current rate after 2025; ~20% increase in arable land compared with 2000		global forest loss until 2025 slightly below historic rate, stabilizes after 2025; ~10% increase in arable land	net increase in forest cover globally until 2025, slow loss after 2025; ~9% increase in arable land
Greenhouse gas emissions by 2050	CO_2: 20.1 GtC-eq CH_4: 3.7 GtC-eq N_2O: 1.1 GtC-eq other GHGs: 0.7 GtC-eq	CO_2: 15.4 GtC-eq CH_4: 3.3 GtC-eq N_2O: 1.1 GtC-eq other GHGs: 0.5 GtC-eq		CO_2: 13.3 GtC-eq CH_4: 3.2 GtC-eq N_2O: 0.9 GtC-eq other GHGs: 0.6 GtC-eq	CO_2: 4.7 GtC-eq CH_4: 1.6 GtC-eq N_2O: 0.6 GtC-eq other GHGs: 0.2 GtC-eq

Air pollution emissions	SO_2 emissions stabilize, NO_x emissions increase from 2000 to 2050	both SO_2 and NO_x emissions increase globally	SO_2 emissions decline; NO_x emissions increase slowly	strong reductions in SO_2 and NO_x emissions
Climate change	2.0°C in 2050 and 3.5°C in 2100 above pre-industrial	1.7°C in 2050 and 3.3°C in 2100 above pre-industrial	1.9°C in 2050 and 2.8°C in 2100 above pre-industrial	1.5°C in 2050 and 1.9°C in 2100 above pre-industrial
Nutrient loading	increase in N transport in rivers	increase in N transport in rivers	increase in N transport in rivers	decrease in N transport in rivers

[a] "Industrial " and "developing " refer to the countries at the beginning of the scenario; some countries may change categories by 2050.

Table S3. Outcomes for Ecosystem Services and Human Well-being in 2050 Compared with 2000 across the Scenarios [8, 9]

	Global Orchestration		Order from Strength		Adapting Mosaic		TechnoGarden	
	Industrial[a]	Developing[a]	Industrial[a]	Developing[a]	Industrial[a]	Developing[a]	Industrial[a]	Developing[a]
ECOSYSTEM SERVICES								
Provisioning Services								
Sufficient access to food	↑	↑	↔	↓	↔	↓	↑	↑
Fuel	↑	↑	↑	↑	↑	↑	↑	↑
Genetic resources	↔	↔	↓	↓	↑	↑	↔	↑
Biochemicals/Pharmaceuticals discoveries	↓	↑	↓	↓	↔	↔	↑	↑
Ornamental resources	↔	↔	↔	↓	↑	↑	↔	↔
Freshwater	↑	↑	↔	↓	↑	↓	↑	↔
Regulating Services								
Air quality regulation	↔	↔	↔	↓	↔	↔	↑	↑
Climate regulation	↔	↔	↓	↓	↔	↔	↑	↑
Water regulation	↔	↓	↓	↓	↑	↑	↔	↑
Erosion control	↔	↓	↓	↓	↑	↑	↔	↑
Water purification	↔	↓	↓	↓	↑	↑	↔	↑
Disease control: Human	↔	↑	↔	↓	↔	↑	↑	↑
Disease control: Pests	↔	↓	↓	↓	↑	↑	↔	↔
Pollination	↓	↓	↓	↓	↔	↔	↓	↓
Storm protection	↔	↓	↔	↓	↑	↑	↑	↔
Cultural Services								
Spiritual/religious values	↔	↔	↔	↓	↑	↑	↓	↓
Aesthetic values	↔	↔	↔	↓	↑	↑	↔	↔
Recreation and ecotourism	↓	↑	↓	↑	↓	↓	↑	↑
Cultural diversity	↓	↓	↓	↓	↑	↑	↓	↓
Knowledge systems (diversity and memory)	↔	↓	↓	↓	↑	↑	↔	↔
HUMAN WELL-BEING								
Material well-being	↑	↑	↑	↓	↔	↑	↑	↑
Health	↑	↑	↑	↓	↑	↑	↑	↑
Security	↑	↑	↓	↓	↑	↑	↑	↑
Social Relations	↔	↑	↓	↑	↑	↑	↓	↓
Freedom and Choice	↔	↑	↓	↓	↑	↑	↑	↑

[a] "Industrial " and "developing " refer to the countries at the beginning of the scenario; some countries may change categories by 2050.

Key: ↑ = increase in ecosystems' ability to provide the service, ↔ = ability of ecosystem to provide the service remains the same as in 2000, ↓ = decrease in ecosystems' ability to provide the service

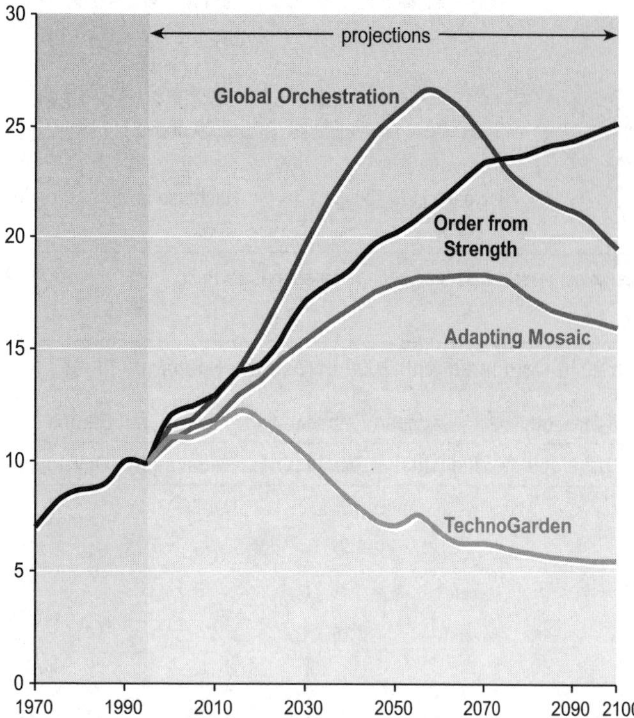

Figure S4. Total Greenhouse Gas Emissions in CO_2 Equivalents per Year versus Time in the MA Scenarios (equivalent emissions based on 100-year GWPs) [9]

or impacts of large, extremely costly, or irreversible changes in ecosystem services. We addressed these phenomena qualitatively by considering the risks and impacts of large but unpredictable ecosystem changes in each scenario. Some ecosystem services and aspects of human well-being could not be quantified and could be assessed only qualitatively. [4]

The Future of Ecosystem Services

The capacity of ecosystems to provide services in the future is jeopardized by rates of use that exceed rates of renewal and by degradation of regulating ecosystem services.

Although the current flow of many ecosystem services to people has increased, the status of many ecosystems, including stocks of provisioning ecosystem services, has shifted to degraded conditions (*well established*). These include losses in marine fish stocks and dryland agriculture; emergence of diseases that threaten plants, animals, and humans; deterioration of water quality in fresh waters and coastal oceans; and regional climate changes and increased climate variability. Such shifts are likely to increase in the future (*established but incomplete*). The impact of unexpected ecosystem changes depends on the intensity of stress on ecosystems as well as societal expectations about reliability of ecosystem services and the capacity of societies to cope with changes in the provision of ecosystem services. [8, 9, 13]

For some components of the future state of human-ecosystem interactions, all four scenarios make similar projections:

- Demand for provisioning services, such as food, fiber, and water, increases due to growth in population and economies (*high certainty*).
- Food security remains out of reach for many people, and child malnutrition will be difficult to eradicate even by 2050 (*low to medium certainty*), despite increasing food supply under all four scenarios (*medium to high certainty*) and more diversified diets in poor countries (*low to medium certainty*). (See Figure S5.)
- Vast changes with great geographic variability occur in freshwater resources and their provisioning of ecosystem services in all scenarios. (See Figure S6.) Climate change will lead to increased precipitation over more than half of Earth's surface and this will make more water available to society and ecosystems (*medium certainty*). However, increased precipitation is also likely to increase the frequency of flooding in many areas (*high certainty*). Increases in precipitation will not be universal, and climate change will also cause a substantial decrease in precipitation in some areas, with an accompanying decrease in water availability (*medium certainty*). These areas could include highly populated arid regions such as the Middle East and Southern Europe (*low to medium certainty*). While water withdrawals decrease in most industrial countries, water withdrawals and wastewater discharges are expected to increase enormously in Africa and some other developing regions, and this will intensify their water stress and overshadow the possible benefits of increased water availability (*medium certainty*).
- The services provided by freshwater resources (such as aquatic habitat, fish production, and water supply for households, industry, and agriculture) deteriorate severely in developing countries under the scenarios that are reactive to environmental problems. Less severe but still important declines are expected in the scenarios that are more proactive about environmental problems (*medium certainty*).
- Growing demand for fish and fish products leads to an increasing risk of a major and long-lasting decline of

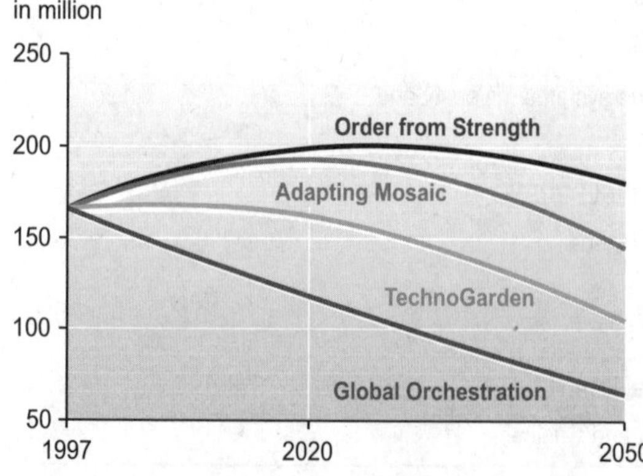

Figure S5. Number of Malnourished Children in Developing Countries over Time in MA Scenarios [9]

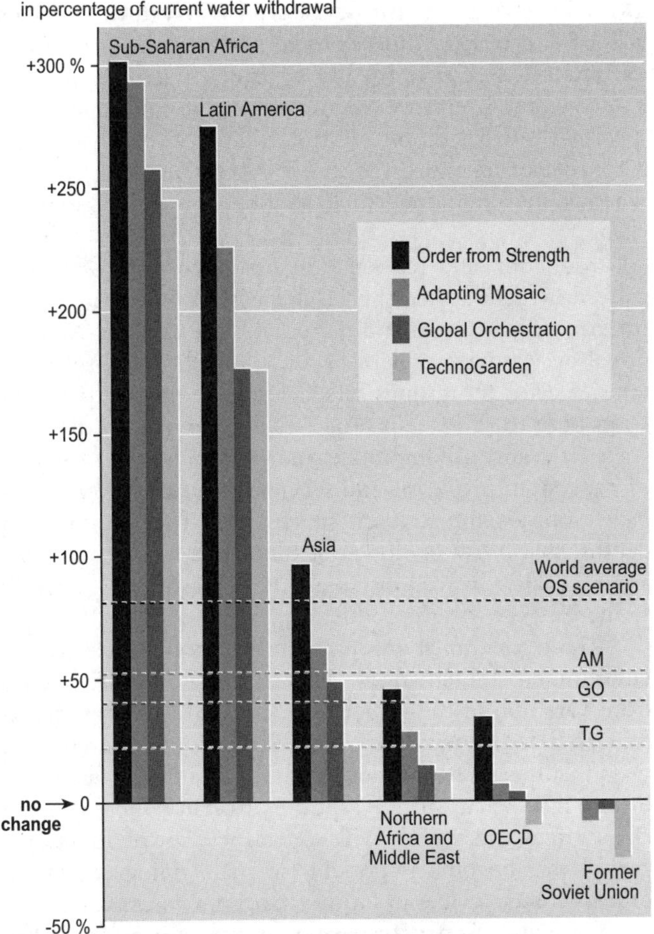

Figure S6. Change in Water Withdrawals from 2000 to 2050 in MA Scenarios, Globally and for Six Groups of Nations [9]

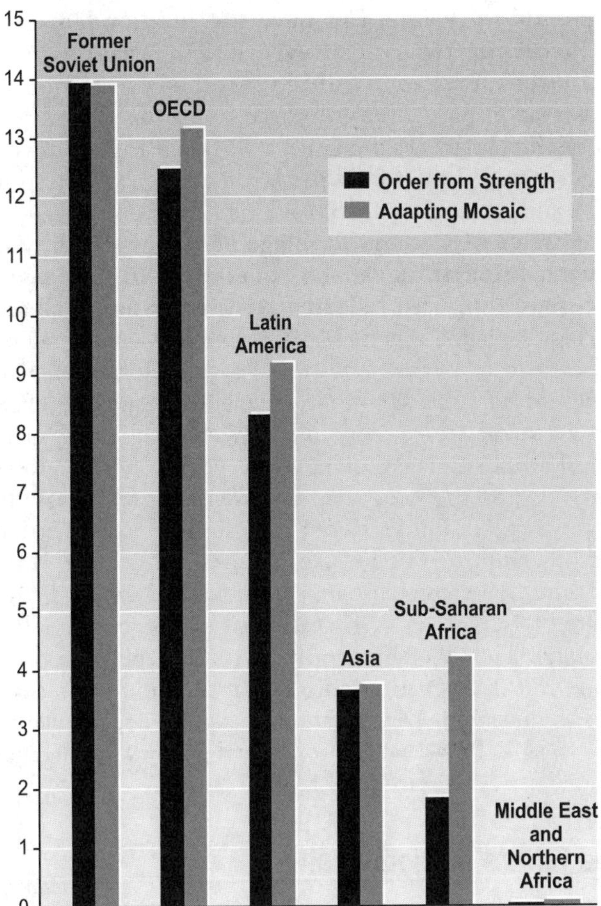

Figure S7. Forest Area in 2050 in Adapting Mosaic and Order from Strength Scenarios in Six Groups of Nations. Forest area is the net result of losses of pre-existing forest and establishment of new forest on land that was formerly used for something else [9]

regional marine fisheries (*medium to high certainty*). Aquaculture cannot relieve this pressure so long as it continues to rely heavily on marine fish as a food source.

Land use change is expected to continue to be a major driver of changes in the provision of ecosystem services up to 2050 (*medium to high certainty*) [9]. The scenarios indicate (*low to medium certainty*) that 10–20% of current grassland and forestland will be lost between now and 2050. This change occurs primarily in low-income and arid regions. (See Figure S7.) The provisioning services associated with affected biomes (such as genetic resources, wood production, and habitat for terrestrial biota) will also be reduced. The degree to which natural land is lost differs among the scenarios. Order from Strength has the greatest land use changes, with large increases in both crop and grazing areas. The two proactive scenarios, TechnoGarden and Adapting Mosaic, are the most land-conserving ones because of increasingly efficient agricultural production, lower meat consumption, and lower population increases. Existing wetlands and the services they provide (such as water purification) are faced with increasing risk in some areas due to reduced runoff or intensified land use in all scenarios.

Threats to drylands are multiscale—ranging from global climate change to local pastoral practices. In addition, dryland ecosystem services are particularly vulnerable to substantial and persistent reductions in ecosystem services driven by climate change, water stress, and intensive use. For example, sub-Saharan Africa is projected to expand water withdrawals rapidly to meet needs for development. Under some scenarios, this causes a rapid increase in untreated return flows to freshwater systems, which could endanger public health and aquatic ecosystems (*medium certainty*). Expansion and intensification of agriculture in this area may lead to loss of natural ecosystems and higher levels of surface and groundwater contamination. Loss of ecosystem services related to these changes could undermine the future provision of ecosystem services in this region, eventually leading to increased poverty. **Global institutions to address dryland problems (such as desertification) need to consider responses at multiple scales, such as mitigation of climate change, technological development, and trade and resource transfers that foster local adaptation.** [14]

In our scenarios, continued population growth, improving economic conditions, and climate change over the next decades exert additional pressure on land resources and pose additional risk of desertification in dryland regions. **Subsi-**

dizing food production and water development in vulnerable drylands can have the unintended effect of increasing the risk of even larger breakdowns of ecosystem services in future years. Local adaptation and conservation practices can mitigate some losses of dryland ecosystem services, although it will be difficult to reverse trends toward loss of food production capacity, water supplies, and biodiversity in drylands. [14]

Threats of wetland drainage and conversion, with adverse impacts on capacity of ecosystems to provide adequate supplies of clean water, increased in all scenarios. Reductions in trade that accompany greater regionalization can increase pressure on agricultural land and water withdrawals. To some extent, these adverse effects can be mitigated by economic growth, technology, or regional adaptive management. However, economic growth without proactive ecosystem management can increase the risk of large disturbances of water supplies, water quality, and other aquatic resources such as fish and wildlife. [14]

Terrestrial ecosystems are currently a net sink of CO_2 at a rate of 1.2 (+/− 0.9) gigatons of carbon per year (*high certainty*). They thereby contribute to the regulation of climate. But the scenarios indicate that the future of this service is uncertain. Deforestation is expected to reduce the carbon sink. Proactive environmental policies can maintain a larger terrestrial carbon sink. [9]

The Future of Biodiversity

Present goals for reduced rates of biodiversity loss will be difficult to achieve because of changes in land use that have already occurred and ongoing stresses from climate change and nutrient enrichment.

Ecosystem management practices that maintain response diversity, functional groups, and trophic levels while mitigating chronic stress are more likely to increase the supply of ecosystem services and decrease the risk of large losses of ecosystem services than practices that ignore these factors.

The scenarios indicate that present goals for reduced rates of biodiversity loss, such as the 2010 targets of the Convention of Biological Diversity, will be difficult to achieve because of changes in land use that have already occurred, ongoing stresses from climate change, and nutrient enrichment. In all scenarios, projections indicate significant negative impacts on biodiversity and its related ecosystem services. However, these scenarios were not designed to optimize the path for preserving biodiversity. Negative impacts on biodiversity can be reduced by proactive steps to, for example, decrease the rate of land conversion, integrate conservation practices with landscape planning, restore ecosystems, and mitigate emissions of nutrients and greenhouse gasses. It is important to note that decreasing rates of land conversion may impair our ability to meet increased demands for food or other ecosystem services. [10, 14]

Significant decline of ecosystem services can occur from species loss even if species do not become globally extinct. Some terrestrial ecosystem services will be lost (*very certain*) as local native populations are extirpated (become locally extinct). Examples include loss of cultural services when a culturally important forest species is extirpated, loss of supporting services when pollinator species are extirpated, and loss of provisioning services when an important medicinal plant becomes locally extinct. [10]

Production and resilience of ecosystems are often enhanced by genetic and species diversity as well as by spatial patterns of landscapes and temporal cycles (such as successional cycles) with which species evolved. Within ecosystems, species and groups of species perform functions that contribute to ecosystem processes and services in different ways. Diversity among functional groups increases the flux of ecosystem processes and services (*established but incomplete*). For example, plant species that root at different depths, that grow or flower at different times of the year, and that differ in seed dispersal and dormancy act together to increase ecosystem productivity.

Within functional groups, species respond differently to environmental fluctuations. This response diversity derives from variation in the response of species to environmental drivers, heterogeneity in species distributions, differences in ways that species use seasonal cycles or disturbance patterns, or other mechanisms. Response diversity increases the chance that ecosystems will contain species or functional groups that become important for maintaining ecosystem processes and services in future changed environments (*medium certainty*). **Ecosystem management practices that maintain response diversity, functional groups, and trophic levels while mitigating chronic stress will increase the supply and resilience of ecosystem services and decrease the risk of large losses of ecosystem services** (*established but incomplete*). [5]

Habitat loss in terrestrial environments is projected to lead to decline in local diversity of native species in all four scenarios by 2050 (*high certainty*). (See Figure S8.) Loss of habitat results in the immediate extirpation of local populations and the loss of the services that these populations provided. [10]

Decreases in river flows from water withdrawals and climate change (decreases occur in 30% of all major river basins) are projected to result in loss of species under all scenarios (*low certainty*). Rivers that are forecast to lose fish species are concentrated in poor tropical and sub-tropical countries, where the needs for human adaptation are most likely to exceed governmental and societal capacity to cope. The current average GDP in countries with diminishing river flows is about 20% lower than in countries whose rivers are not drying. [10]

Habitat loss will eventually lead to global extinctions as species approach equilibrium with the remnant habitat. Although there is *high certainty* that this will happen eventually, the time to equilibrium is *very uncertain*, especially given continued habitat loss through time. Between 10% and 15% of vascular plant species present in 1970 were lost across the four scenarios when species numbers reached equilibrium with reduced habitat (*low certainty*). This may be an under-

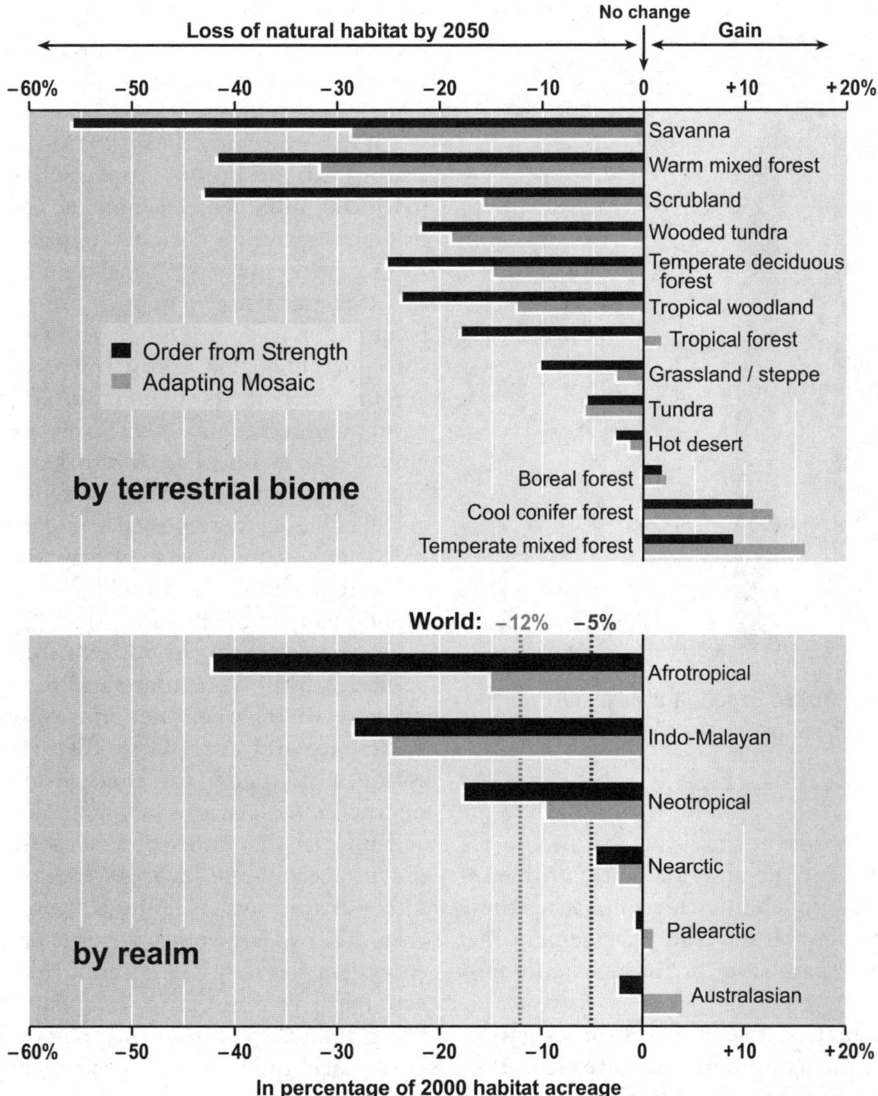

Figure S8. Loss or Gain of Natural Habitat from 1970 to 2050 in Adapting Mosaic and Order from Strength Scenarios. Habitat changes are indicated by biome and by biogeographic realm. [9, 10]

estimate because it addresses only those changes due to habitat loss and does not consider the effects of other stressors such as climate change or nutrient deposition. Time lags between habitat reduction and extinction provide a precious opportunity for humans to rescue those species that otherwise may be on a trajectory toward extinction. [10]

Trade-offs among Ecosystem Services

Increasing the flow of provisioning services often leads to reductions in supporting, regulating, and cultural ecosystem services. This may reduce the future capacity of ecosystems to provide services.

Building understanding about how ecosystems provide services will increase society's capacity to avert large disturbances of those services or to adapt to them rapidly when they do occur.

Trade-offs exist in all of the MA scenarios between food and water and between food and biodiversity. Each scenario takes a slightly different approach to addressing these trade-offs. By comparing these approaches and their outcomes, we can learn about managing trade-offs. [12]

- **In all four MA scenarios, application of fertilizers, including manure, in excess of crop needs caused large nutrient flows into fresh waters and estuaries** (*high certainty*). (See Figure S9.) This overenrichment of water causes serious declines in ecosystem services (food, recreation, fresh water, and biodiversity) provided by aquatic ecosystems. There are possibilities for mitigating these trade-offs through technological enhancements such as agricultural efficiency (in the use of land, water, and fertilizers) and through productivity-enhancing, resource-conserving technologies, which combine natural capital conservation with yield improvement techniques.
- **In all four MA scenarios, conversion of land to agricultural uses for food production reduced biodiversity.** Clearing diverse land cover for crop production reduces biodiversity by eliminating local populations.

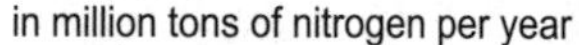

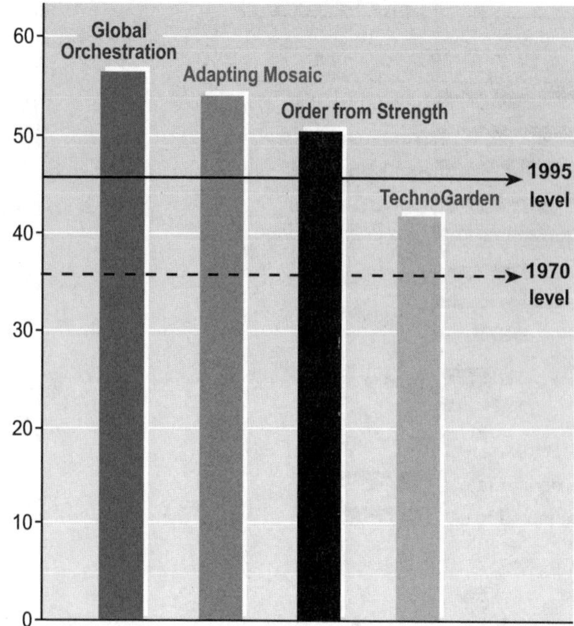

Figure S9. Global River Nitrogen Export in 2030 in MA Scenarios. Reference lines show global river nitrogen export in 1970 and 1995.

Removing water from lakes and rivers for use can reduce aquatic biodiversity because less aquatic habitat is available. There are possibilities for mitigating these trade-offs through agricultural land management that explicitly maintains biodiversity or through more efficient use of water.

- **In all four MA scenarios, use of water for irrigation of crops reduced the availability of water for other uses, such as household or industrial use or the maintenance of other ecosystem services.** Although water is a renewable resource, the amount available in any one place at any one time is finite. Thus, excessive use of water for irrigation can restrict the amount of water for other important uses.

All scenarios show the general tendency of management to focus intensely on increasing the availability of provisioning services, which often leads to reductions in the provision of supporting, regulating, and cultural ecosystem services. (See Figure S10.) Efforts to increase the short-term provision of services typically reduce the capacity of ecosystems to provide the full array of services in the future. This vulnerability can be difficult to detect because ecosystems often exhibit threshold behavior that can mask declines in regulating and supporting services until a collapse occurs. Such trade-offs have far-reaching consequences for maintaining ecosystem functioning in the long term. For example, decisions about fertilizer use in the 1960s are still affecting water quality in the twenty-first century.

Scenarios in which long-term consequences of trade-offs are not taken into consideration exhibit the largest risk of declines in supporting and regulating services (such as climate change and biodiversity loss). Scenarios with a proactive approach to ecosystem management via flexible ecosystem governance mechanisms and learning or technological innovations are more likely to sustain ecosystem services in the future. [12]

At every scale, there are opportunities for combining advantageous approaches to achieve synergistic benefits. For example, actions to preserve marine fish species have been shown to make coral reefs more resistant to the pressures associated with declines in other species or excess nutrients. Actions to preserve local fisheries have been shown to have positive benefits on human well-being through enhancing social interactions and networking among fishers in the region. Advantages can be found by combining techniques from each of the scenarios. For example, combining the advantages of green technology (TechnoGarden) with fairer markets (Global Orchestration) and flexible ecosystem management that encourages local creativity (Adapting Mosaic) may lead to improvements in ecosystem services and human well-being beyond those found in any individual scenario. [12]

In the scenarios in which monitoring was a focus, societies built an understanding of large changes in ecosystem services that increased their capacity to anticipate and avert large disturbances of ecosystem services or to adapt to them more rapidly if they did occur. In the scenarios in which monitoring was not done and policies that anticipate the possibility of large breakdowns in ecosystem services were not implemented (Global Orchestration and Order from Strength), societies faced increased risk of large impacts from unexpected disruptions of ecosystem services. The greatest risks of large, unfavorable ecological changes arise in dryland agriculture, marine fisheries, quality of fresh and coastal marine waters, disease emergence, and regional climate change. [8, 12, 14]

The Future of Human Well-being

Attempts to improve human well-being that do not actively take ecosystems into account can cause unintended but rapid, severe, and persistent degradation of ecosystem services.

Most of the 2015 targets established for the Millennium Development Goals were not achieved in the MA scenarios. The scenarios also indicate that some strategies for achieving goals such as poverty reduction and hunger reduction quickly could increase pressures on ecosystems, thereby compromising the ability to continue progress toward these goals in the future and undermining progress toward the MDG of environmental sustainability. Although the MA scenarios were not designed to chart an optimal path to meeting the MDGs, they provide useful information about plausible paths. Attempts to meet the MDGs by 2015, which will largely involve increased use of provisioning ecosystem services, may lead to ecosystem degradation and reductions in regulating and supporting services that undermine future ecosystem capacity to supply provisioning services. This

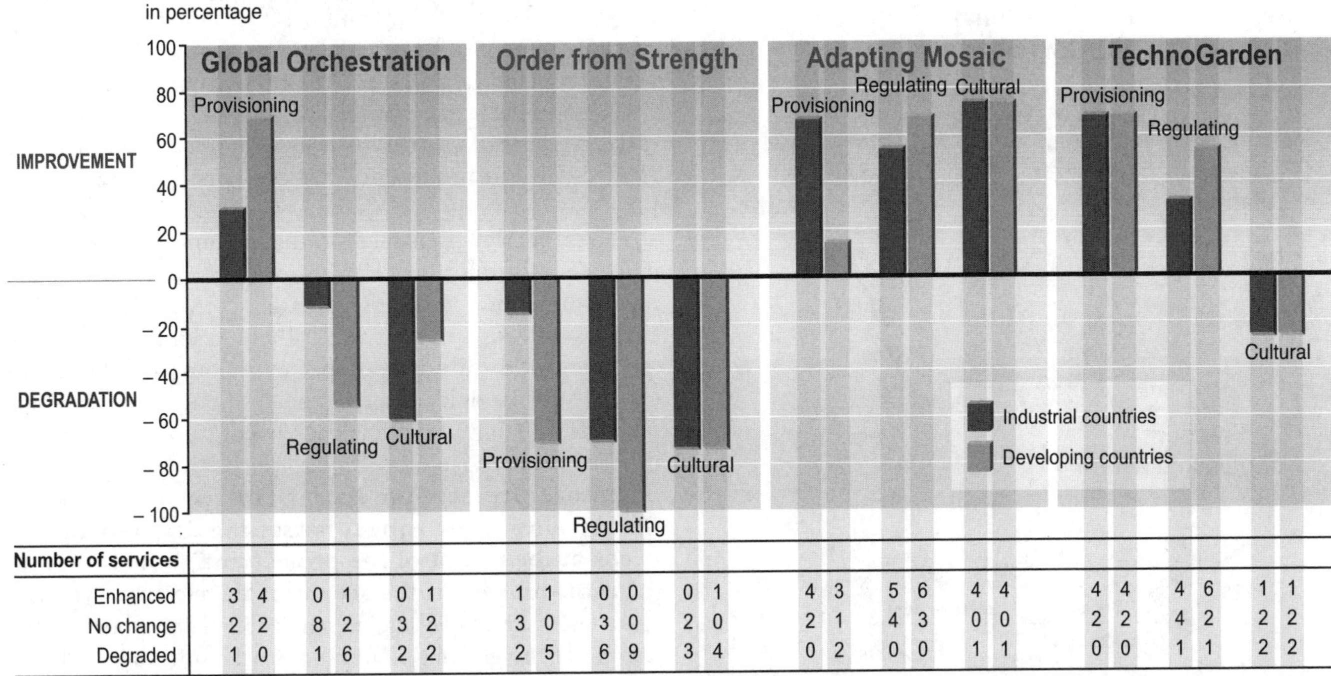

Number of services												
Enhanced	3 4	0 1	0 1	1 1	0 0	0 1	4 3	5 6	4 4	4 4	4 6	1 1
No change	2 2	8 2	3 2	3 0	3 0	2 0	2 1	4 3	0 0	2 2	4 2	2 2
Degraded	1 0	1 6	2 2	2 5	6 9	3 4	0 2	0 0	1 1	0 0	1 1	2 2

Figure S10. Net Changes in Availability of Provisioning, Regulating, and Cultural Ecosystem Services by 2050 in MA Scenarios for Industrial and Developing Countries. The y-axis is the net percentage of ecosystem services enhanced or degraded. For example, 100% degradation of the six provisioning ecosystem services would mean that all of these were degraded in 2050 relative to 2000, while 50% enhancement could mean that three were enhanced and the other three were unchanged, or that four were enhanced, one was degraded, and the other two were unchanged. The data used to calculate the y-axis are presented beneath the figure.

degradation may increase the risk of regime shifts and other surprises that seriously undermine human well-being. [14]

Ecosystem services are essential for human well-being. However, the relationship between human well-being and ecosystem services is discontinuous. Above some threshold, a marginal increase in ecosystem services contributes only slightly to human well-being, but below that threshold, a small decrease in ecosystem services can substantially reduce it. [11]

Across the dimensions of human well-being, each scenario yields a different package of gains, losses, and vulnerabilities for different regions and populations. (See Figure S11.) In our scenarios, actions that focused on improving the lives of the poor by reducing barriers to international flows of goods, services, and capital tended to lead to the most improvement for those who are currently the most disadvantaged. Health and social relations improve, but human vulnerability to ecological surprises is high.

Globally integrated approaches that focused on technology and property rights for ecosystem services generally improved human well-being in terms of health, security, social relations, and material needs. When those same technologies were used globally, however, local culture was lost or undervalued. High levels of trade lead to more rapid spread of emergent diseases, somewhat reducing the gains in health in all areas. Locally focused, learning-based approaches led to the largest improvements in social relations, but with variability by region. Order from Strength, which focuses on reactive policies in a regionalized world, has the least favorable outcomes for human well-being, as the global distribution of ecosystem services and human resources that underpin human well-being are increasingly skewed. [11]

Toward Future Assessments of Ecosystem Services

The future capacity of ecosystems to provide services is often determined by feedbacks at multiple scales. Future projects on ecosystem service scenarios should explicitly nest or link assessments at several scales from the beginning.

Active adaptive ecosystem management (experimentation with monitoring and analysis to learn more-sustainable management methods) could greatly improve outcomes for ecosystem services and human well-being.

In considering multiple aspects of ecosystem services and feedbacks with human well-being, this assessment is the first of its kind. Lessons learned from the MA suggest many opportunities to improve the development of ecosystem service scenarios in the future.

The future capacity of ecosystems to provide services is often determined by feedbacks at multiple scales. Future projects on ecosystem service scenarios should explicitly nest or link assessments at several scales from the beginning. This innovation would pro-

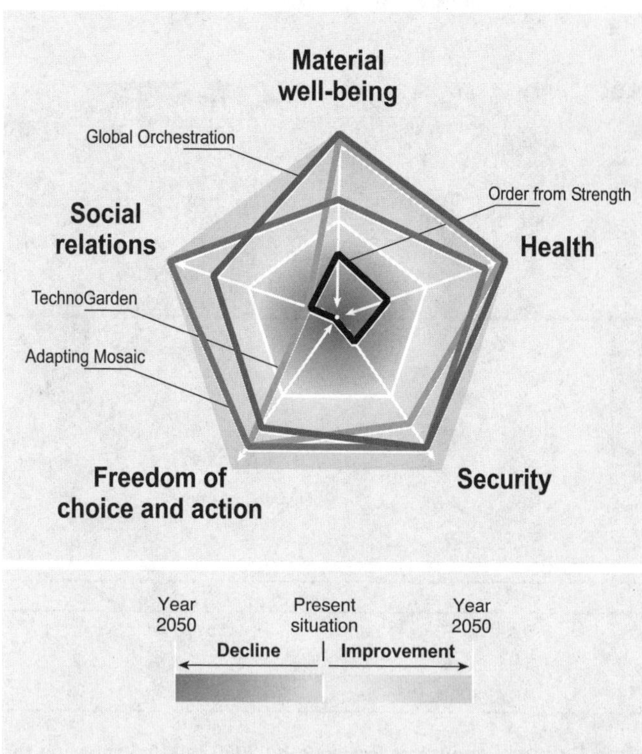

Figure S11. Changes in Components of Human Well-being in 2050 in MA Scenarios. The pentagon in the middle represents the situation in 2000. Moving outward from that indicates an improvement in that component of human well-being in that scenario by the year 2050. Moving inward from the pentagon indicates a decline in that aspect of human well-being since 2000.

vide decision-makers with information that links local, national, regional, and global futures of ecosystem services directly. In addition, future projects should allow more time for iterations between qualitative and quantitative assessments of the storylines. This additional work would improve the harmonization of qualitative and quantitative assessments and allow for a more diverse set of simulations to address risks and regime shifts. [4, 6, 13]

Active adaptive ecosystem management (experimentation with monitoring and analysis to learn more-sustainable management methods) could greatly improve outcomes for ecosystem services and human well-being. Existing assessment models for most ecosystem services do not account for effects of active adaptive management at local to regional scales. Thus most of our projections of ecosystem services represent outcomes in the *absence* of local-to-regional adaptive change. Actively adaptive management could significantly improve the outcomes relative to the projections presented here. (See the MA *Policy Responses* volume.) [4, 5, 13]

There are important gaps between the processes depicted in the MA conceptual framework and the existing capacity of ecosystem modeling. Major elements of the conceptual framework that are not well addressed by models include the effects of changes in ecosystems on flows of ecosystem services and the effects of changes in ecosystem services on changes in human well-being. In addition, existing models focus mainly on a subset of provisioning and regulating ecosystem services, largely neglecting cultural and supporting ecosystem services. Cultural ecosystem services, together with the other ecosystem services, play a critical role in adaptive responses and changes in human attitudes and behaviors toward nature. [4, 13]

The underlying chapters in this volume list many specific needs for improved models. Models are needed to address thresholds and the risk of large, costly, or irreversible changes in ecosystem services. There is emerging understanding that the diversity of species response and the heterogeneity of landscapes affect the resilience of ecosystem services. This important feedback needs to be incorporated in ecosystem service models. [4, 5, 9, 10, 13]

Future projects on ecosystem service scenarios should allow more time for assessing decision-maker needs at the outset of the project and should include decision-makers in the scenarios development team. Differences among disciplines in core beliefs about functioning of the global system are a crucial uncertainty that is addressed in the scenarios. Better interdisciplinary communication would make it easier to understand and assimilate these differences in future scenario exercises. Finally, communication of scenarios requires development of synthetic graphics, nontechnical narratives, and nontechnical illustrations. Future projects on ecosystem service scenarios should allocate more time for creation of these important communication and outreach products. [13]

Synthesis

Future conditions of ecosystem services could be worse or better than in the present, depending on policy choices.

None of the MA scenarios represents an optimal outcome. A selected mix of policies from several scenarios may yield better outcomes than any single scenario.

The Millennium Ecosystem Assessment scenarios show that the condition of ecosystem services in the future could be significantly worse or better than in the present. Scenarios that improve the condition of ecosystem services and human well-being involve substantial changes in policy. Examples include:

- major investments in public goods and poverty reduction, together with elimination of harmful trade barriers and subsidies (Global Orchestration);
- widespread use of actively adaptive ecosystem management and investment in education (Adapting Mosaic); and
- significant investments in technologies to use ecosystem services more efficiently, along with widespread inclusion of ecosystem services in markets (TechnoGarden).

Although examples of all these policies are known from the world of today, such policies are not widespread at the present time.

The MA scenarios were not designed to determine optimal policies for any specific locale, nation, international bloc, or Earth as a whole. Different combinations of policies may produce significantly better results than any of the scenarios presented here. Successful hybrid policies may capitalize on the advantages of several scenarios while avoiding the risks. For example, combining the local-learning approach of Adapting Mosaic with the global coordination and technological advances of TechnoGarden may capitalize on the benefits of both scenarios while avoiding the loss of cultural services found in TechnoGarden and the global commons problems found in Adapting Mosaic.

PART I

State of Knowledge Concerning Ecosystem Forecasts and Scenarios

Chapter 1

MA Conceptual Framework

BOXES

FIGURES

This chapter provides the summary of Millennium Ecosystem Assessment, *Ecosystems and Human Well-being: A Framework for Assessment* (Island Press, 2003), pp. 1–25, which was prepared by an extended conceptual framework writing team of 51 authors and 10 contributing authors.

Main Messages

Human well-being and progress toward sustainable development are vitally dependent upon improving the management of Earth's ecosystems to ensure their conservation and sustainable use. But while demands for ecosystem services such as food and clean water are growing, human actions are at the same time diminishing the capability of many ecosystems to meet these demands.

Sound policy and management interventions can often reverse ecosystem degradation and enhance the contributions of ecosystems to human well-being, but knowing when and how to intervene requires substantial understanding of both the ecological and the social systems involved. Better information cannot guarantee improved decisions, but it is a prerequisite for sound decision-making.

The Millennium Ecosystem Assessment was established to help provide the knowledge base for improved decisions and to build capacity for analyzing and supplying this information.

This chapter presents the conceptual and methodological approach that the MA used to assess options that can enhance the contribution of ecosystems to human well-being. This same approach should provide a suitable basis for governments, the private sector, and civil society to factor considerations of ecosystems and ecosystem services into their own planning and actions.

1.1 Introduction

Humanity has always depended on the services provided by the biosphere and its ecosystems. Further, the biosphere is itself the product of life on Earth. The composition of the atmosphere and soil, the cycling of elements through air and waterways, and many other ecological assets are all the result of living processes—and all are maintained and replenished by living ecosystems. The human species, while buffered against environmental immediacies by culture and technology, is ultimately fully dependent on the flow of ecosystem services.

In his April 2000 Millennium Report to the United Nations General Assembly, in recognition of the growing burden that degraded ecosystems are placing on human well-being and economic development and the opportunity that better managed ecosystems provide for meeting the goals of poverty eradication and sustainable development, United Nations Secretary-General Kofi Annan stated that:

> *It is impossible to devise effective environmental policy unless it is based on sound scientific information. While major advances in data collection have been made in many areas, large gaps in our knowledge remain. In particular, there has never been a comprehensive global assessment of the world's major ecosystems. The planned Millennium Ecosystem Assessment, a major international collaborative effort to map the health of our planet, is a response to this need.*

The Millennium Ecosystem Assessment was established with the involvement of governments, the private sector, nongovernmental organizations, and scientists to provide an integrated assessment of the consequences of ecosystem change for human well-being and to analyze options available to enhance the conservation of ecosystems and their contributions to meeting human needs. The Convention on Biological Diversity, the Convention to Combat Desertification, the Convention on Migratory Species, and the Ramsar Convention on Wetlands plan to use the findings of the MA, which will also help meet the needs of others in government, the private sector, and civil society. The MA should help to achieve the United Nations Millennium Development Goals and to carry out the Plan of Implementation of the 2002 World Summit on Sustainable Development. It has mobilized hundreds of scientists from countries around the world to provide information and clarify science concerning issues of greatest relevance to decision-makers. The MA has identified areas of broad scientific agreement and also pointed to areas of continuing scientific debate.

The assessment framework developed for the MA offers decision-makers a mechanism to:

- ***Identify options that can better achieve core human development and sustainability goals. All countries and communities are grappling with the challenge of meeting growing demands for food, clean water, health, and employment.*** And decision-makers in the private and public sectors must also balance economic growth and social development with the need for environmental conservation. All of these concerns are linked directly or indirectly to the world's ecosystems. The MA process, at all scales, was designed to bring the best science to bear on the needs of decision-makers concerning these links between ecosystems, human development, and sustainability.
- ***Better understand the trade-offs involved—across sectors and stakeholders—in decisions concerning the environment.*** Ecosystem-related problems have historically been approached issue by issue, but rarely by pursuing multisectoral objectives. This approach has not withstood the test of time. Progress toward one objective such as increasing food production has often been at the cost of progress toward other objectives such as conserving biological diversity or improving water quality. The MA framework complements sectoral assessments with information on the full impact of potential policy choices across sectors and stakeholders.
- ***Align response options with the level of governance where they can be most effective.*** Effective management of ecosystems will require actions at all scales, from the local to the global. Human actions now directly or inadvertently affect virtually all of the world's ecosystems; actions required for the management of ecosystems refer to the steps that humans can take to modify their direct or indirect influences on ecosystems. The management and policy options available and the concerns of stakeholders differ greatly across these scales. The priority areas for biodiversity conservation in a country as defined based on "global" value, for example, would be very different from those as defined based on the value to local communities. The multiscale assessment framework developed for the MA provides a new approach for analyzing

policy options at all scales—from local communities to international conventions.

1.2 What Is the Problem?

Ecosystem services are the benefits people obtain from ecosystems, which the MA describes as provisioning, regulating, supporting, and cultural services. (See Box 1.1.) Ecosystem services include products such as food, fuel, and fiber; regulating services such as climate regulation and disease control; and nonmaterial benefits such as spiritual or aesthetic benefits. Changes in these services affect human well-being in many ways. (See Figure 1.1.)

The demand for ecosystem services is now so great that trade-offs among services have become the rule. A country can increase food supply by converting a forest to agriculture, for example, but in so doing it decreases the supply of services that may be of equal or greater importance, such as clean water, timber, ecotourism destinations, or flood regulation and drought control. There are many indications that human demands on ecosystems will grow still greater in the coming decades. Current estimates of 3 billion more people and a quadrupling of the world economy by 2050 imply a formidable increase in demand for and consumption of biological and physical resources, as well as escalating impacts on ecosystems and the services they provide.

The problem posed by the growing demand for ecosystem services is compounded by increasingly serious degradation in the capability of ecosystems to provide these services. World fisheries are now declining due to overfishing, for instance, and a significant amount of agricultural land has been degraded in the past half-century by erosion, salinization, compaction, nutrient depletion, pollution, and urbanization. Other human-induced impacts on ecosystems include alteration of the nitrogen, phosphorous, sulfur, and carbon cycles, causing acid rain, algal blooms, and fish kills in rivers and coastal waters, along with contributions to climate change. In many parts of the world, this degradation of ecosystem services is exacerbated by the associated loss of the knowledge and understanding held by local communities—knowledge that sometimes could help to ensure the sustainable use of the ecosystem.

This combination of ever-growing demands being placed on increasingly degraded ecosystems seriously diminishes the prospects for sustainable development. Human well-being is affected not just by gaps between ecosystem service supply and demand but also by the increased vulnerability of individuals, communities, and nations. Productive ecosystems, with their array of services, provide people and communities with resources and options they can use as insurance in the face of natural catastrophes or social upheaval. While well-managed ecosystems reduce risks and vulnerability, poorly managed systems can exacerbate them by increasing risks of flood, drought, crop failure, or disease.

Ecosystem degradation tends to harm rural populations more directly than urban populations and has its most direct and severe impact on poor people. The wealthy control access to a greater share of ecosystem services, consume those services at a higher per capita rate, and are buffered from changes in their availability (often at a substantial cost) through their ability to purchase scarce ecosystem services or substitutes. For example, even though a number of marine fisheries have been depleted in the past century, the supply of fish to wealthy consumers has not been disrupted since fishing fleets have been able to shift to previously underexploited stocks. In contrast, poor people often lack access to alternate services and are highly vulnerable to ecosystem changes that result in famine, drought, or floods. They frequently live in locations particularly sensitive to environmental threats, and they lack financial and institutional buffers against these dangers. Degradation of coastal fishery resources, for instance, results in a decline in protein consumed by the local community since fishers may not have access to alternate sources of fish and community members may not have enough income to purchase fish. Degradation affects their very survival.

Changes in ecosystems affect not just humans but countless other species as well. The management objectives that people set for ecosystems and the actions that they take are influenced not just by the consequences of ecosystem changes for humans but also by the importance people place on considerations of the intrinsic value of species and ecosystems. Intrinsic value is the value of something in and for itself, irrespective of its utility for someone else. For example, villages in India protect "spirit sanctuaries" in relatively natural states, even though a strict cost-benefit calculation might favor their conversion to agriculture. Similarly, many countries have passed laws protecting endangered species based on the view that these species have a right to exist, even if their protection results in net economic costs. Sound ecosystem management thus involves steps to address the utilitarian links of people to ecosystems as well as processes that allow considerations of the intrinsic value of ecosystems to be factored into decision-making.

BOX 1.1
Key Definitions

Ecosystem. An ecosystem is a dynamic complex of plant, animal, and microorganism communities and the nonliving environment interacting as a functional unit. Humans are an integral part of ecosystems. Ecosystems vary enormously in size; a temporary pond in a tree hollow and an ocean basin can both be ecosystems.

Ecosystem services. Ecosystem services are the benefits people obtain from ecosystems. These include provisioning services such as food and water; regulating services such as regulation of floods, drought, land degradation, and disease; supporting services such as soil formation and nutrient cycling; and cultural services such as recreational, spiritual, religious and other nonmaterial benefits.

Well-being. Human well-being has multiple constituents, including basic material for a good life, freedom of choice and action, health, good social relations, and security. Well-being is at the opposite end of a continuum from poverty, which has been defined as a "pronounced deprivation in well-being." The constituents of well-being, as experienced and perceived by people, are situation-dependent, reflecting local geography, culture, and ecological circumstances.

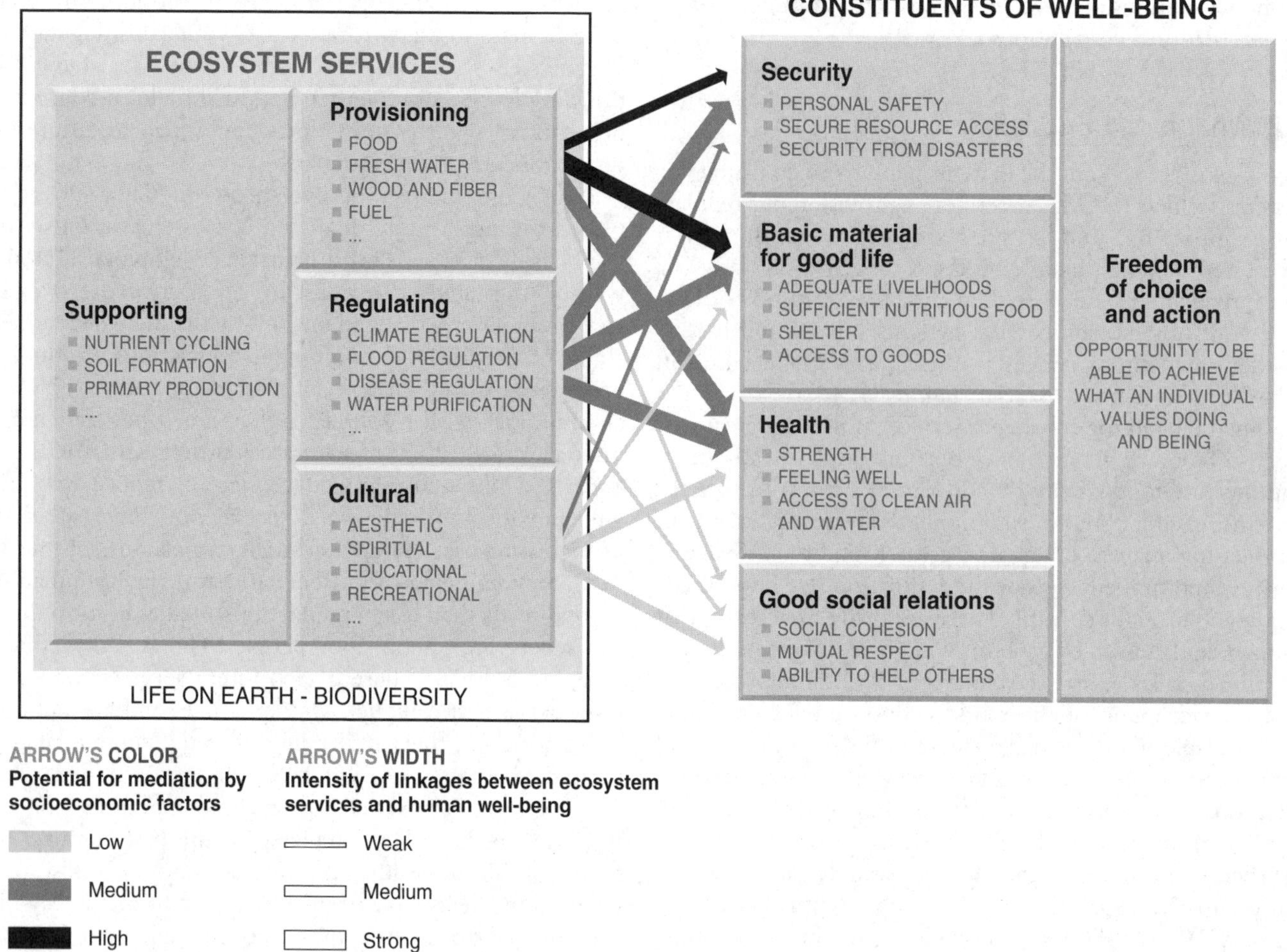

Figure 1.1. Linkages between Ecosystem Services and Human Well-being. This Figure depicts the strength of linkages between categories of ecosystem services and components of human well-being that are commonly encountered and includes indications of the extent to which it is possible for socioeconomic factors to mediate the linkage. (For example, if it is possible to purchase a substitute for a degraded ecosystem service, then there is a high potential for mediation.) The strength of the linkages and the potential for mediation differ in different ecosystems and regions. In addition to the influence of ecosystem services on human well-being depicted here, other factors—including other environmental factors as well as economic, social, technological, and cultural factors—influence human well-being, and ecosystems are in turn affected by changes in human well-being.

The degradation of ecosystem services has many causes, including excessive demand for ecosystem services stemming from economic growth, demographic changes, and individual choices. Market mechanisms do not always ensure the conservation of ecosystem services either because markets do not exist for services such as cultural or regulatory services or, where they do exist, because policies and institutions do not enable people living within the ecosystem to benefit from services it may provide to others who are far away. For example, institutions are now only beginning to be developed to enable those benefiting from carbon sequestration to provide local managers with an economic incentive to leave a forest uncut, while strong economic incentives often exist for managers to harvest the forest. Also, even if a market exists for an ecosystem service, the results obtained through the market may be socially or ecologically undesirable. Properly managed, the creation of ecotourism opportunities in a country can create strong economic incentives for the maintenance of the cultural services provided by ecosystems, but poorly managed ecotourism activities can degrade the very resource on which they depend. Finally, markets are often unable to address important intra- and intergenerational equity issues associated with managing ecosystems for this and future generations, given that some changes in ecosystem services are irreversible.

The world has witnessed in recent decades not just dramatic changes to ecosystems but equally profound changes to social systems that shape both the pressures on ecosystems and the opportunities to respond. The relative influence of individual nation-states has diminished with the growth of power and influence of a far more complex array of institutions, including regional governments, multinational companies, the United Nations, and civil society organizations. Stakeholders have become more involved in decision-making. Given the multiple actors whose decisions now strongly influence ecosystems, the challenge of providing information to decision-makers has grown. At the same time, the new institutional landscape may provide an unprecedented opportunity for information concerning ecosystems to make a major difference. Improvements in ecosystem management to enhance human well-being will require new institutional and policy arrangements and changes in rights and access to resources that may be more possible today under these conditions of rapid social change than they have ever been before.

Like the benefits of increased education or improved governance, the protection, restoration, and enhancement of ecosystem services tends to have multiple and synergistic benefits. Already, many governments are beginning to recognize the need for more effective management of these basic life-support systems. Examples of significant progress toward sustainable management of biological resources can also be found in civil society, in indigenous and local communities, and in the private sector.

1.3 Conceptual Framework

The conceptual framework for the MA places human well-being as the central focus for assessment, while recognizing that biodiversity and ecosystems also have intrinsic value and that people take decisions concerning ecosystems based on considerations of well-being as well as intrinsic value. (See Box 1.2.) The MA conceptual framework assumes that a dynamic interaction exists between people and other parts of ecosystems, with the changing human condition serving to both directly and indirectly drive change in ecosystems and with changes in ecosystems causing changes in human well-being. At the same time, many other factors independent of the environment change the human condition, and many natural forces are influencing ecosystems.

The MA focuses particular attention on the linkages between ecosystem services and human well-being. The assessment deals with the full range of ecosystems—from those relatively undisturbed, such as natural forests, to landscapes with mixed patterns of human use and ecosystems intensively managed and modified by humans, such as agricultural land and urban areas.

A full assessment of the interactions between people and ecosystems requires a multiscale approach because it better reflects the multiscale nature of decision-making, allows the examination of driving forces that may be exogenous to particular regions, and provides a means of examining the differential impact of ecosystem changes and policy responses on different regions and groups within regions.

This section explains in greater detail the characteristics of each of the components of the MA conceptual framework, moving clockwise from the lower left corner of the Figure in Box 1.2.

1.3.1 Ecosystems and Their Services

An ecosystem is a dynamic complex of plant, animal, and microorganism communities and the nonliving environment interacting as a functional unit. Humans are an integral part of ecosystems. Ecosystems provide a variety of benefits to people, including provisioning, regulating, cultural, and supporting services. Provisioning services are the products people obtain from ecosystems, such as food, fuel, fiber, fresh water, and genetic resources. Regulating services are the benefits people obtain from the regulation of ecosystem processes, including air quality maintenance, climate regulation, erosion control, regulation of human diseases, and water purification. Cultural services are the nonmaterial benefits people obtain from ecosystems through spiritual enrichment, cognitive development, reflection, recreation, and aesthetic experiences. Supporting services are those that are necessary for the production of all other ecosystem services, such as primary production, production of oxygen, and soil formation.

Biodiversity and ecosystems are closely related concepts. Biodiversity is the variability among living organisms from all sources, including terrestrial, marine, and other aquatic ecosystems and the ecological complexes of which they are part. It includes diversity within and between species and diversity of ecosystems. Diversity is a structural feature of ecosystems, and the variability among ecosystems is an element of biodiversity. Products of biodiversity include many of the services produced by ecosystems (such as food and genetic resources), and changes in biodiversity can influence all the other services they provide. In addition to the important role of biodiversity in providing ecosystem services, the diversity of living species has intrinsic value independent of any human concern.

The concept of an ecosystem provides a valuable framework for analyzing and acting on the linkages between people and the environment. For that reason, the "ecosystem approach" has been endorsed by the Convention on Biological Diversity, and the MA conceptual framework is entirely consistent with this approach. The CBD states that the ecosystem approach is a strategy for the integrated management of land, water, and living resources that promotes conservation and sustainable use in an equitable way. This approach recognizes that humans, with their cultural diversity, are an integral component of many ecosystems.

In order to implement the ecosystem approach, decision-makers need to understand the multiple effects on an ecosystem of any management or policy change. By way of analogy, decision-makers would not make a decision about financial policy in a country without examining the condition of the economic system, since information on the economy of a single sector such as manufacturing would be insufficient. The same need to examine the consequences of changes for multiple sectors applies to ecosystems. For instance, subsidies for fertilizer use may increase food production, but sound decisions also require information on whether the potential reduction in the harvests of downstream fisheries as a result of water quality degradation from the fertilizer runoff might outweigh those benefits.

For the purpose of analysis and assessment, a pragmatic view of ecosystem boundaries must be adopted, depending on the questions being asked. A well-defined ecosystem has strong interactions among its components and weak interactions across its boundaries. A useful choice of ecosystem boundary is one where a number of discontinuities coincide, such as in the distribution of organisms, soil types,

BOX 1.2

Millennium Ecosystem Assessment Conceptual Framework

Changes in factors that indirectly affect ecosystems, such as population, technology, and lifestyle (upper right corner of figure), can lead to changes in factors directly affecting ecosystems, such as the catch of fisheries or the application of fertilizers to increase food production (lower right corner). The resulting changes in the ecosystem (lower left corner) cause the ecosystem services to change and thereby affect human well-being. These interactions can take place at more than one scale and can cross scales. For example, a global market may lead to regional loss of forest cover, which increases flood magnitude along a local stretch of a river. Similarly, the interactions can take place across different time scales. Actions can be taken either to respond to negative changes or to enhance positive changes at almost all points in this framework (black cross bars).

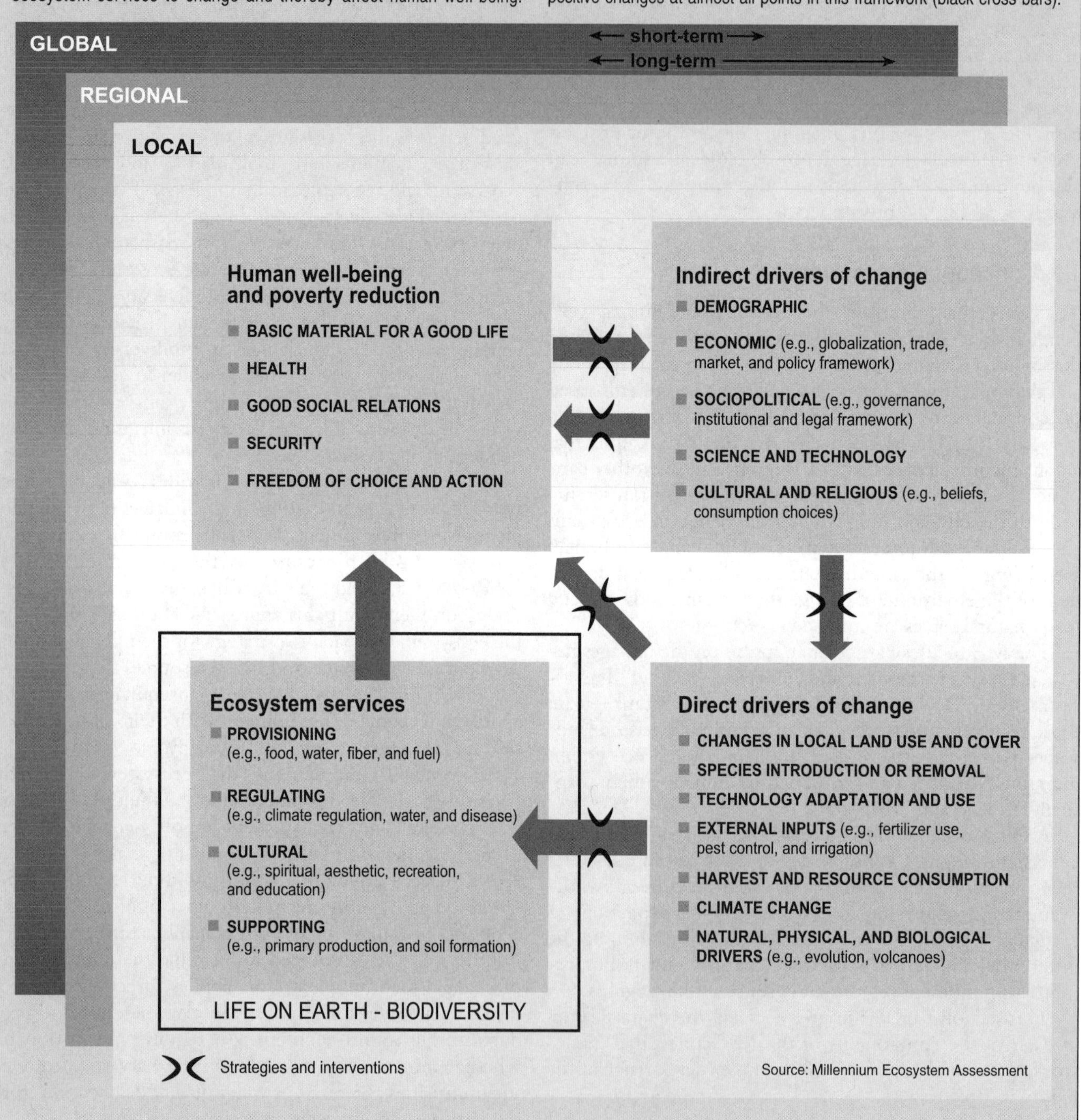

drainage basins, and depth in a waterbody. At a larger scale, regional and even globally distributed ecosystems can be evaluated based on a commonality of basic structural units. The global assessment being undertaken by the MA reports on marine, coastal, inland water, forest, dryland, island, mountain, polar, cultivated, and urban regions. These regions are not ecosystems themselves, but each contains a number of ecosystems. (See Box 1.3.)

People seek multiple services from ecosystems and thus perceive the condition of given ecosystems in relation to their ability to provide the services desired. Various methods can be used to assess the ability of ecosystems to deliver

BOX 1.3

Reporting Categories Used in the Millennium Ecosystem Assessment

The MA used 10 categories of systems to report its global findings. (See Table.) These categories are not ecosystems themselves; each contains a number of ecosystems. The MA reporting categories are not mutually exclusive: their areas can and do overlap. Ecosystems within each category share a suite of biological, climatic, and social factors that tend to differ across categories. Because these reporting categories overlap, any place on Earth may fall into more than one category. Thus, for example, a wetland ecosystem in a coastal region may be examined both in the MA analysis of "coastal systems" as well as in its analysis of "inland water systems."

Millennium Ecosystem Assessment Reporting Categories

Category	Central Concept	Boundary Limits for Mapping
Marine	Ocean, with fishing typically a major driver of change	Marine areas where the sea is deeper than 50 meters.
Coastal	Interface between ocean and land, extending seawards to about the middle of the continental shelf and inland to include all areas strongly influenced by the proximity to the ocean	Area between 50 meters below mean sea level and 50 meters above the high tide level or extending landward to a distance 100 kilometers from shore. Includes coral reefs, intertidal zones, estuaries, coastal aquaculture, and seagrass communities.
Inland water	Permanent water bodies inland from the coastal zone, and areas whose ecology and use are dominated by the permanent, seasonal, or intermittent occurrence of flooded conditions	Rivers, lakes, floodplains, reservoirs, and wetlands; includes inland saline systems. Note that the Ramsar Convention considers "wetlands" to include both inland water and coastal categories.
Forest	Lands dominated by trees; often used for timber, fuelwood, and non-timber forest products	A canopy cover of at least 40% by woody plants taller than 5 meters. The existence of many other definitions is acknowledged, and other limits (such as crown cover greater than 10%, as used by the Food and Agriculture Organization of the United Nations) are also reported. Includes temporarily cut-over forests and plantations; excludes orchards and agroforests where the main products are food crops.
Dryland	Lands where plant production is limited by water availability; the dominant uses are large mammal herbivory, including livestock grazing, and cultivation	Drylands as defined by the Convention to Combat Desertification, namely lands where annual precipitation is less than two thirds of potential evaporation, from dry subhumid areas (ratio ranges 0.50–0.65), through semiarid, arid, and hyper-arid (ratio <0.05), but excluding polar areas; drylands include cultivated lands, scrublands, shrublands, grasslands, semi-deserts, and true deserts.
Island	Lands isolated by surrounding water, with a high proportion of coast to hinterland	Islands of at least 1.5 hectares included in the ESRI ArcWorld Country Boundary dataset.
Mountain	Steep and high lands	As defined by Mountain Watch using criteria based on elevation alone, and at lower elevation, on a combination of elevation, slope, and local elevation range. Specifically, elevation $>2{,}500$ meters, elevation 1,500–2,500 meters and slope >2 degrees, elevation 1,000–1,500 meters and slope >5 degrees or local elevation range (7 kilometers radius) >300 meters, elevation 300–1,000 meters and local elevation range (7 kilometers radius) >300 meters, isolated inner basins and plateaus less than 25 square kilometers extent that are surrounded by mountains.
Polar	High-latitude systems frozen for most of the year	Includes ice caps, areas underlain by permafrost, tundra, polar deserts, and polar coastal areas. Excludes high-altitude cold systems in low latitudes.
Cultivated	Lands dominated by domesticated plant species, used for and substantially changed by crop, agroforestry, or aquaculture production	Areas in which at least 30% of the landscape comes under cultivation in any particular year. Includes orchards, agroforestry, and integrated agriculture-aquaculture systems.
Urban	Built environments with a high human density	Known human settlements with a population of 5,000 or more, with boundaries delineated by observing persistent night-time lights or by inferring areal extent in the cases where such observations are absent.

particular services. With those answers in hand, stakeholders have the information they need to decide on a mix of services best meeting their needs. The MA considers criteria and methods to provide an integrated view of the condition of ecosystems. The condition of each category of ecosystem services is evaluated in somewhat different ways, although in general a full assessment of any service requires considerations of stocks, flows, and resilience of the service.

1.3.2 Human Well-being and Poverty Reduction

Human well-being has multiple constituents, including the basic material for a good life, freedom of choice and action, health, good social relations, and security. Poverty is also multidimensional and has been defined as the pronounced deprivation of well-being. How well-being, ill-being, or poverty are experienced and expressed depends on context and situation, reflecting local physical, social, and personal factors such as geography, environment, age, gender, and culture. In all contexts, however, ecosystems are essential for human well-being through their provisioning, regulating, cultural, and supporting services.

Human intervention in ecosystems can amplify the benefits to human society. However, evidence in recent decades of escalating human impacts on ecological systems worldwide raises concerns about the spatial and temporal consequences of ecosystem changes detrimental to human well-being. Ecosystem changes affect human well-being in the following ways:

- ***Security*** is affected both by changes in provisioning services, which affect supplies of food and other goods and the likelihood of conflict over declining resources, and by changes in regulating services, which could influence the frequency and magnitude of floods, droughts, landslides, or other catastrophes. It can also be affected by changes in cultural services as, for example, when the loss of important ceremonial or spiritual attributes of ecosystems contributes to the weakening of social relations in a community. These changes in turn affect material well-being, health, freedom and choice, security, and good social relations.
- ***Access to basic material for a good life*** is strongly linked to both provisioning services such as food and fiber production and regulating services, including water purification.
- ***Health*** is strongly linked to both provisioning services such as food production and regulating services, including those that influence the distribution of disease-transmitting insects and of irritants and pathogens in water and air. Health can also be linked to cultural services through recreational and spiritual benefits.
- ***Social relations*** are affected by changes to cultural services, which affect the quality of human experience.
- ***Freedom of choice and action*** is largely predicated on the existence of the other components of well-being and are thus influenced by changes in provisioning, regulating, or cultural services from ecosystems.

Human well-being can be enhanced through sustainable human interactions with ecosystems supported by necessary instruments, institutions, organizations, and technology. Creation of these through participation and transparency may contribute to freedoms and choice as well as to increased economic, social, and ecological security. By ecological security, we mean the minimum level of ecological stock needed to ensure a sustainable flow of ecosystem services.

Yet the benefits conferred by institutions and technology are neither automatic nor equally shared. In particular, such opportunities are more readily grasped by richer than poorer countries and people; some institutions and technologies mask or exacerbate environmental problems; responsible governance, while essential, is not easily achieved; participation in decision-making, an essential element of responsible governance, is expensive in time and resources to maintain. Unequal access to ecosystem services has often elevated the well-being of small segments of the population at the expense of others.

Sometimes the consequences of the depletion and degradation of ecosystem services can be mitigated by the substitution of knowledge and of manufactured or human capital. For example, the addition of fertilizer in agricultural systems has been able to offset declining soil fertility in many regions of the world where people have sufficient economic resources to purchase these inputs, and water treatment facilities can sometimes substitute for the role of watersheds and wetlands in water purification. But ecosystems are complex and dynamic systems and there are limits to substitution possibilities, especially with regulating, cultural, and supporting services. No substitution is possible for the extinction of culturally important species such as tigers or whales, for instance, and substitutions may be economically impractical for the loss of services such as erosion control or climate regulation. Moreover, the scope for substitutions varies by social, economic, and cultural conditions. For some people, especially the poorest, substitutes and choices are very limited. For those who are better off, substitution may be possible through trade, investment, and technology.

Because of the inertia in both ecological and human systems, the consequences of ecosystem changes made today may not be felt for decades. Thus, sustaining ecosystem services, and thereby human well-being, requires a full understanding and wise management of the relationships between human activities, ecosystem change, and well-being over the short, medium, and long term. Excessive current use of ecosystem services compromises their future availability. This can be prevented by ensuring that the use is sustainable.

Achieving sustainable use requires effective and efficient institutions that can provide the mechanisms through which concepts of freedom, justice, fairness, basic capabilities, and equity govern the access to and use of ecosystem services. Such institutions may also need to mediate conflicts between individual and social interests that arise.

The best way to manage ecosystems to enhance human well-being will differ if the focus is on meeting needs of the poor and weak or the rich and powerful. For both groups, ensuring the long-term supply of ecosystem services is essential. But for the poor, an equally critical need is to provide more equitable and secure access to ecosystem services.

1.3.3 Drivers of Change

Understanding the factors that cause changes in ecosystems and ecosystem services is essential to designing interventions that capture positive impacts and minimize negative ones. In the MA, a "driver" is any factor that changes an aspect of an ecosystem. A direct driver unequivocally influences ecosystem processes and can therefore be identified and measured to differing degrees of accuracy. An indirect driver operates more diffusely, often by altering one or more direct drivers, and its influence is established by understanding its effect on a direct driver. Both indirect and direct drivers often operate synergistically. Changes in land cover, for example, can increase the likelihood of introduction of alien invasive species. Similarly, technological advances can increase rates of economic growth.

The MA explicitly recognizes the role of decision-makers who affect ecosystems, ecosystem services, and human well-being. Decisions are made at three organizational levels, although the distinction between those levels is often diffuse and difficult to define:

- by individuals and small groups at the local level (such as a field or forest stand) who directly alter some part of the ecosystem;
- by public and private decision-makers at the municipal, provincial, and national levels; and
- by public and private decision-makers at the international level, such as through international conventions and multilateral agreements.

The decision-making process is complex and multidimensional. We refer to a driver that can be influenced by a decision-maker as an endogenous driver and one over which the decision-maker does not have control as an exogenous driver. The amount of fertilizer applied on a farm is an endogenous driver from the standpoint of the farmer, for example, while the price of the fertilizer is an exogenous driver, since the farmer's decisions have little direct influence on price. The specific temporal, spatial, and organizational scale dependencies of endogenous and exogenous drivers and the specific linkages and interactions among drivers are assessed in the MA.

Whether a driver is exogenous or endogenous to a decision-maker is dependent upon the spatial and temporal scale. For example, a local decision-maker can directly influence the choice of technology, changes in land use, and external inputs (such as fertilizers or irrigation), but has little control over prices and markets, property rights, technology development, or the local climate. In contrast, a national or regional decision-maker has more control over many factors, such as macroeconomic policy, technology development, property rights, trade barriers, prices, and markets. But on the short time scale, that individual has little control over the climate or global population. On the longer time scale, drivers that are exogenous to a decision-maker in the short run, such as population, become endogenous since the decision-maker can influence them through, for instance, education, the advancement of women, and migration policies.

The indirect drivers of change are primarily:

- demographic (such as population size, age and gender structure, and spatial distribution);
- economic (such as national and per capita income, macroeconomic policies, international trade, and capital flows);
- sociopolitical (such as democratization, the roles of women, of civil society, and of the private sector, and international dispute mechanisms);
- scientific and technological (such as rates of investments in research and development and the rates of adoption of new technologies, including biotechnologies and information technologies); and
- cultural and religious (such as choices individuals make about what and how much to consume and what they value).

The interaction of several of these drivers, in turn, affects levels of resource consumption and differences in consumption both within and between countries. Clearly these drivers are changing—population and the world economy are growing, for instance, there are major advances in information technology and biotechnology, and the world is becoming more interconnected. Changes in these drivers are projected to increase the demand for and consumption of food, fiber, clean water, and energy, which will in turn affect the direct drivers. The direct drivers are primarily physical, chemical, and biological—such as land cover change, climate change, air and water pollution, irrigation, use of fertilizers, harvesting, and the introduction of alien invasive species. Change is apparent here too: the climate is changing, species ranges are shifting, alien species are spreading, and land degradation continues.

An important point is that any decision can have consequences external to the decision framework. These consequences are called externalities because they are not part of the decision-making calculus. Externalities can have positive or negative effects. For example, a decision to subsidize fertilizers to increase crop production might result in substantial degradation of water quality from the added nutrients and degradation of downstream fisheries. But it is also possible to have positive externalities. A beekeeper might be motivated by the profits to be made from selling honey, for instance, but neighboring orchards could produce more apples because of enhanced pollination arising from the presence of the bees.

Multiple interacting drivers cause changes in ecosystem services. There are functional interdependencies between and among the indirect and direct drivers of change, and, in turn, changes in ecological services lead to feedbacks on the drivers of changes in ecological services. Synergetic driver combinations are common. The many processes of globalization lead to new forms of interactions between drivers of changes in ecosystem services.

1.3.4 Cross-scale Interactions and Assessment

An effective assessment of ecosystems and human well-being cannot be conducted at a single temporal or spatial scale. Thus the MA conceptual framework includes both of these dimensions. Ecosystem changes that may have little impact on human well-being over days or weeks (soil erosion, for instance) may have pronounced impacts over years or decades (declining agricultural productivity). Similarly, changes at a local scale may have little impact on some services at that scale (as in the local impact of forest loss on water availability) but major impacts at large scales (forest loss in a river basin changing the timing and magnitude of downstream flooding).

Ecosystem processes and services are typically most strongly expressed, are most easily observed, or have their dominant controls or consequences at particular spatial and temporal scales. They often exhibit a characteristic scale—the typical extent or duration over which processes have their impact. Spatial and temporal scales are often closely related. For instance, food production is a localized service of an ecosystem and changes on a weekly basis, water regulation is regional and changes on a monthly or seasonal basis, and climate regulation may take place at a global scale over decades.

Assessments need to be conducted at spatial and temporal scales appropriate to the process or phenomenon being examined. Those done over large areas generally use data at coarse resolutions, which may not detect fine-resolution processes. Even if data are collected at a fine level of detail, the process of averaging in order to present findings at the larger scale causes local patterns or anomalies to disappear. This is particularly problematic for processes exhibiting thresholds and nonlinearities. For example, even though a number of fish stocks exploited in a particular area might have collapsed due to overfishing, average catches across all stocks (including healthier stocks) would not reveal the extent of the problem. Assessors, if they are aware of such thresholds and have access to high-resolution data, can incorporate such information even in a large-scale assessment. Yet an assessment done at smaller spatial scales can help identify important dynamics of the system that might otherwise be overlooked. Likewise, phenomena and processes that occur at much larger scales, although expressed locally, may go unnoticed in purely local-scale assessments. Increased carbon dioxide concentrations or decreased stratospheric ozone concentrations have local effects, for instance, but it would be difficult to trace the causality of the effects without an examination of the overall global process.

Time scale is also very important in conducting assessments. Humans tend not to think beyond one or two generations. If an assessment covers a shorter time period than the characteristic temporal scale, it may not adequately capture variability associated with long-term cycles, such as glaciation. Slow changes are often harder to measure, as is the case with the impact of climate change on the geographic distribution of species or populations. Moreover, both ecological and human systems have substantial inertia, and the impact of changes occurring today may not be seen for years or decades. For example, some fisheries' catches may increase for several years even after they have reached unsustainable levels because of the large number of juvenile fish produced before that level was reached.

Social, political, and economic processes also have characteristic scales, which may vary widely in duration and extent. Those of ecological and sociopolitical processes often do not match. Many environmental problems originate from this mismatch between the scale at which the ecological process occurs, the scale at which decisions are made, and the scale of institutions for decision-making. A purely local-scale assessment, for instance, may discover that the most effective societal response requires action that can occur only at a national scale (such as the removal of a subsidy or the establishment of a regulation). Moreover, it may lack the relevance and credibility necessary to stimulate and inform national or regional changes. On the other hand, a purely global assessment may lack both the relevance and the credibility necessary to lead to changes in ecosystem management at the local scale where action is needed. Outcomes at a given scale are often heavily influenced by interactions of ecological, socioeconomic, and political factors emanating from other scales. Thus focusing solely on a single scale is likely to miss interactions with other scales that are critically important in understanding ecosystem determinants and their implications for human well-being.

The choice of the spatial or temporal scale for an assessment is politically laden, since it may intentionally or unintentionally privilege certain groups. The selection of assessment scale with its associated level of detail implicitly favors particular systems of knowledge, types of information, and modes of expression over others. For example, non-codified information or knowledge systems of minority populations are often missed when assessments are undertaken at larger spatial scales or higher levels of aggregation. Reflecting on the political consequences of scale and boundary choices is an important prerequisite to exploring what multi- and cross-scale analysis in the MA might contribute to decision-making and public policy processes at various scales.

1.4 Values Associated with Ecosystems

Current decision-making processes often ignore or underestimate the value of ecosystem services. Decision-making concerning ecosystems and their services can be particularly challenging because different disciplines, philosophical views, and schools of thought assess the value of ecosystems differently. One paradigm of value, known as the utilitarian (anthropocentric) concept, is based on the principle of humans' preference satisfaction (welfare). In this case, ecosystems and the services they provide have value to human societies because people derive utility from their use, either directly or indirectly (use values). Within this utilitarian concept of value, people also give value to ecosystem services that they are not currently using (non-use values). Non-use values, usually known as existence values, involve the case where humans ascribe value to knowing that a resource exists even if they never use that resource directly. These often involve the deeply held historical, national, ethical, religious, and spiritual values people ascribe to ecosystems—the values that the MA recognizes as cultural services of ecosystems.

A different, non-utilitarian value paradigm holds that something can have intrinsic value—that is, it can be of value in and for itself—irrespective of its utility for someone else. From the perspective of many ethical, religious, and cultural points of view, ecosystems may have intrinsic value, independent of their contribution to human well-being.

The utilitarian and non-utilitarian value paradigms overlap and interact in many ways, but they use different metrics, with no common denominator, and cannot usually be aggregated, although both paradigms of value are used in decision-making processes.

Under the utilitarian approach, a wide range of methodologies has been developed to attempt to quantify the benefits of different ecosystem services. These methods are particularly well developed for provisioning services, but recent work has also improved the ability to value regulating and other services. The choice of valuation technique in any given instance is dictated by the characteristics of the case and by data availability. (See Box 1.4.)

Non-utilitarian value proceeds from a variety of ethical, cultural, religious, and philosophical bases. These differ in the specific entities that are deemed to have intrinsic value and in the interpretation of what having intrinsic value means. Intrinsic value may complement or counterbalance considerations of utilitarian value. For example, if the aggregate utility of the services provided by an ecosystem (as measured by its utilitarian value) outweighs the value of converting it to another use, its intrinsic value may then be complementary and provide an additional impetus for conserving the ecosystem. If, however, economic valuation indicates that the value of converting the ecosystem outweighs the aggregate value of its services, its ascribed intrinsic value may be deemed great enough to warrant a social decision to conserve it anyway. Such decisions are essentially political, not economic. In contemporary democracies these decisions are made by parliaments or legislatures or by regulatory agencies mandated to do so by law. The sanctions for violating laws recognizing an entity's intrinsic value may be regarded as a measure of the degree of intrinsic value ascribed to them. The decisions taken by businesses, local communities, and individuals also can involve considerations of both utilitarian and non-utilitarian values.

The mere act of quantifying the value of ecosystem services cannot by itself change the incentives affecting their use or misuse. Several changes in current practice may be required to take better account of these values. The MA assesses the use of information on ecosystem service values in decision-making. The goal is to improve decision-making processes and tools and to provide feedback regarding the kinds of information that can have the most influence.

BOX 1.4

Valuation of Ecosystem Services

Valuation can be used in many ways: to assess the total contribution that ecosystems make to human well-being, to understand the incentives that individual decision-makers face in managing ecosystems in different ways, and to evaluate the consequences of alternative courses of action. The MA uses valuation primarily in the latter sense: as a tool that enhances the ability of decision-makers to evaluate trade-offs between alternative ecosystem management regimes and courses of social actions that alter the use of ecosystems and the multiple services they provide. This usually requires assessing the change in the mix (the value) of services provided by an ecosystem resulting from a given change in its management.

Most of the work involved in estimating the change in the value of the flow of benefits provided by an ecosystem involves estimating the change in the physical flow of benefits (quantifying biophysical relations) and tracing through and quantifying a chain of causality between changes in ecosystem condition and human welfare. A common problem in valuation is that information is only available on some of the links in the chain and often in incompatible units. The MA can make a major contribution by making various disciplines better aware of what is needed to ensure that their work can be combined with that of others to allow a full assessment of the consequences of altering ecosystem state and function.

The ecosystem values in this sense are only one of the bases on which decisions on ecosystem management are and should be made. Many other factors, including notions of intrinsic value and other objectives that society might have (such as equity among different groups or generations), will also feed into the decision framework. Even when decisions are made on other bases, however, estimates of changes in utilitarian value provide invaluable information.

1.5 Assessment Tools

The information base exists in any country to undertake an assessment within the framework of the MA. That said, although new data sets (for example, from remote sensing) providing globally consistent information make a global assessment like the MA more rigorous, there are still many challenges that must be dealt with in using these data at global or local scales. Among these challenges are biases in the geographic and temporal coverage of the data and in the types of data collected. Data availability for industrial countries is greater than that for developing ones, and data for certain resources such as crop production are more readily available than data for fisheries, fuelwood, or biodiversity. The MA makes extensive use of both biophysical and socioeconomic indicators, which combine data into policy-relevant measures that provide the basis for assessment and decision-making.

Models can be used to illuminate interactions among systems and drivers, as well as to make up for data deficiencies—for instance, by providing estimates where observations are lacking. The MA makes use of environmental system models that can be used, for example, to measure the consequences of land cover change for river flow or the consequences of climate change for the distribution of species. It also uses human system models that can examine, for instance, the impact of changes in ecosystems on production, consumption, and investment decisions by households or that allow the economy-wide impacts of a change in production in a particular sector like agriculture to be evaluated. Finally, integrated models, combining both the environmental and human systems linkages, can increasingly be used at both global and sub-global scales.

The MA incorporates both formal scientific information and traditional or local knowledge. Traditional societies have nurtured and refined systems of knowledge of direct value to those societies but also of considerable value to assessments undertaken at regional and global scales. This information often is unknown to science and can be an expression of other relationships between society and nature in general and of sustainable ways of managing natural resources in particular. To be credible and useful to decision-makers, all sources of information, whether scientific, traditional, or practitioner knowledge, must be critically assessed and validated as part of the assessment process through procedures relevant to the form of knowledge.

Since policies for dealing with the deterioration of ecosystem services are concerned with the future consequences of current actions, the development of scenarios of medium- to long-term changes in ecosystems, services, and drivers can be particularly helpful for decision-makers. Scenarios are typically developed through the joint involvement of decision-makers and scientific experts, and they represent a promising mechanism for linking scientific information to decision-making processes. They do not attempt to predict the future but instead are designed to indicate what science can and cannot say about the future consequences of alternative plausible choices that might be taken in the coming years.

The MA uses scenarios to summarize and communicate the diverse trajectories that the world's ecosystems may take in future decades. Scenarios are plausible alternative futures, each an example of what might happen under particular assumptions. They can be used as a systematic method for thinking creatively about complex, uncertain futures. In this way, they help us understand the upcoming choices that need to be made and highlight developments in the present. The MA developed scenarios that connect possible changes in drivers (which may be unpredictable or uncontrollable) with human demands for ecosystem services. The scenarios link these demands, in turn, to the futures of the services themselves and the aspects of human welfare that depend on them. The scenario building exercise breaks new ground in several areas:

- development of scenarios for global futures linked explicitly to ecosystem services and the human consequences of ecosystem change,
- consideration of trade-offs among individual ecosystem services within the "bundle" of benefits that any particular ecosystem potentially provides to society,
- assessment of modeling capabilities for linking socioeconomic drivers and ecosystem services, and
- consideration of ambiguous futures as well as quantifiable uncertainties.

The credibility of assessments is closely linked to how they address what is not known in addition to what is known. The consistent treatment of uncertainty is therefore essential for the clarity and utility of assessment reports. As part of any assessment process, it is crucial to estimate the uncertainty of findings even if a detailed quantitative appraisal of uncertainty is unavailable.

1.6 Strategies and Interventions

The MA assesses the use and effectiveness of a wide range of options for responding to the need to sustainably use, conserve, and restore ecosystems and the services they provide. These options include incorporating the value of ecosystems in decisions, channeling diffuse ecosystem benefits to decision-makers with focused local interests, creating markets and property rights, educating and dispersing knowledge, and investing to improve ecosystems and the services they provide. As seen in Box 1.2 on the MA conceptual framework, different types of response options can affect the relationships of indirect to direct drivers, the influence of direct drivers on ecosystems, the human demand for ecosystem services, or the impact of changes in human well-being on indirect drivers. An effective strategy for managing ecosystems will involve a mix of interventions at all points in this conceptual framework.

Mechanisms for accomplishing these interventions include laws, regulations, and enforcement schemes; partnerships and collaborations; the sharing of information and knowledge; and public and private action. The choice of options to be considered will be greatly influenced by both the temporal and the physical scale influenced by decisions, the uncertainty of outcomes, cultural context, and the implications for equity and trade-offs. Institutions at different levels have different response options available to them, and special care is required to ensure policy coherence.

Decision-making processes are value-based and combine political and technical elements to varying degrees. Where technical input can play a role, a range of tools is available to help decision-makers choose among strategies and interventions, including cost-benefit analysis, game theory, and policy exercises. The selection of analytical tools should be determined by the context of the decision, key characteristics of the decision problem, and the criteria considered to be important by the decision-makers. Information from these analytical frameworks is always combined with the intuition, experience, and interests of the decision-maker in shaping the final decisions.

Risk assessment, including ecological risk assessment, is an established discipline and has a significant potential for informing the decision process. Finding thresholds and identifying the potential for irreversible change are important for the decision-making process. Similarly, environmental impact assessments designed to evaluate the impact of particular projects and strategic environmental assessments designed to evaluate the impact of policies both represent important mechanisms for incorporating the findings of an ecosystem assessment into decision-making processes.

Changes also may be required in decision-making processes themselves. Experience to date suggests that a number of mechanisms can improve the process of making decisions about ecosystem services. Broadly accepted norms for decision-making process include the following characteristics. Did the process:

- bring the best available information to bear?
- function transparently, use locally grounded knowledge, and involve all those with an interest in a decision?

- pay special attention to equity and to the most vulnerable populations?
- use decision analytical frameworks that take account of the strengths and limits of individual, group, and organizational information processing and action?
- consider whether an intervention or its outcome is irreversible and incorporate procedures to evaluate the outcomes of actions and learn from them?
- ensure that those making the decisions are accountable?
- strive for efficiency in choosing among interventions?
- take account of thresholds, irreversibility, and cumulative, cross-scale, and marginal effects and of local, regional, and global costs, risk, and benefits?

The policy or management changes made to address problems and opportunities related to ecosystems and their services, whether at local scales or national or international scales, need to be adaptive and flexible in order to benefit from past experience, to hedge against risk, and to consider uncertainty. The understanding of ecosystem dynamics will always be limited, socioeconomic systems will continue to change, and outside determinants can never be fully anticipated. Decision-makers should consider whether a course of action is reversible and should incorporate, whenever possible, procedures to evaluate the outcomes of actions and learn from them. Debate about exactly how to do this continues in discussions of adaptive management, social learning, safe minimum standards, and the precautionary principle. But the core message of all approaches is the same: acknowledge the limits of human understanding, give special consideration to irreversible changes, and evaluate the impacts of decisions as they unfold.

Chapter 2

Global Scenarios in Historical Perspective

Coordinating Lead Author: Paul Raskin
Lead Authors: Frank Monks, Teresa Ribeiro, Detlef van Vuuren, Monika Zurek
Review Editors: Antonio Alonso Concheiro, Christopher Field

BOXES

FIGURES

TABLES

Main Messages

Scenarios are plausible, challenging, and relevant stories about how the future might unfold, which can be told in both words and numbers. Scenarios are not forecasts, projections, predictions, or recommendations. They are about envisioning future pathways and accounting for critical uncertainties.

The process of building scenarios is about asking questions as well as providing answers and guidance for action. It is intended to widen perspectives and illuminate key issues that might otherwise be missed or dismissed. By offering insight into uncertainties and the consequences of current and possible future actions, scenarios support more informed and rational decision-making.

Scenarios address real-world questions of systems dynamics, policy choices, technological evolution, and consumption and production patterns. They reflect the modern worldview that the future is not preordained but rather is subject to human actions and choices. Yet the age-old drive to ponder the possibilities for our collective future, and to draw relevant lessons for how to live today, remains.

The international commitment to sustainable development that has emerged in recent decades gives the study of the future a new urgency and direction. Global scenario analysis has evolved rapidly over the past 10 years in response to this challenge. Although maintaining the long-term resilience of the world's ecosystems is fundamental to sustainable development, ecosystem dynamics themselves have not yet been central to this research. Adding such a focus is a primary aim of the Millennium Ecosystem Assessment scenarios. This chapter sets the historical context and point of departure for the ecosystem outlook.

This chapter reviews the historical context of scenarios, beginning with brief sketches of early scenario activity, from its post–World War II origins up to about a decade ago. While heterogeneous, these studies all take a global perspective in attempting to link social, economic, and environmental issues. The literature can be split into two largely nonoverlapping streams—quantitative simulation and qualitative narrative. Scenario analysis as a professional undertaking first surfaced with strategic planning and war games during the early years of the cold war. But the direct antecedents of contemporary scenarios lie with the future studies of the 1970s. At this time, scenario analysis was also first used as a corporate strategic management technique. After a considerable lull during the early 1980s, a second round of integrated global analysis began in the late 1980s and 1990s, prompted by concerns with climate change and sustainable development. The first decades of scenario assessment paved the way by showing the power—and limits—of both deterministic modeling and descriptive future analyses. A central challenge of contemporary global scenario exercises is to unify these two aspects by blending the replicability and clarity of quantification with the richness of narrative.

Today, scenarios are being applied in an expanding array of business, community, policy, and research contexts with highly varied aims—better management, consciousness raising, conflict resolution, policy advice, and research. Scenarios can be forward-looking, exploring how futures might unfold from current conditions and uncertainties, or backward-looking, beginning with a normative vision of the future and asking whether there is a plausible path to it. Scenario building can include the active engagement of targeted audiences through participatory processes and game playing, deliberation among expert scenario panels, and quantitative simulations by modeling groups.

Scenarios relevant for the MA process are those that have a public policy and scientific orientation. The ideal attributes for such scenarios are: integration across social, economic, and environmental dimensions; regional disaggregation of global patterns; multiple futures that reflect the deep uncertainties of long-range outcomes; and quantification of key variables linked to ecosystem conditions.

The following global scenario building exercises after 1995 were considered for building the MA scenarios: Global Scenarios Group, Global Environment Outlook, Special Report on Emissions Scenarios, World Business Council for Sustainable Development, World Water Vision, and the Organisation for Economic Co-operation and Development. A scan of these studies suggests great variation in the way each exercise was structured. Yet beneath the diversity, the scenarios are rooted in a common set of archetypal visions of the future—worlds that evolve gradually, shaped by dominant driving forces; worlds that are influenced by a strong policy push for sustainability goals; worlds that succumb to fragmentation, environmental collapse, and institutional failure; and worlds where new human values and forms of development emerge.

In the coming years, the enrichment of global scenarios, often through participatory processes, will define an important agenda for policy analysis, scientific research, and education. Improving environmental scenario building will require the enhancement of the role of ecosystems in both scenario narrative and quantification. Narratives will need to more richly reflect ecosystem descriptors, impacts, and feedbacks. Models will need to simulate ecosystem services and ecosystem dynamics, including feedbacks from ecosystem processes to ecosystem services and human well-being.

2.1 Sustainability and the Future

Since ancient times, speculation about human destiny has infused culture through myth and religious cosmology. Prophetic tales told of what is to come, of apocalypse and resurrection, often conveying moral messages for the here and now. These stories gave voice to the powerful impulse to give meaning to the human condition and to act worthily. In a sense, they were the first scenarios.

Of course, where ancient mythology sought to divine the workings of spirits, gods, and cosmic forces, contemporary global scenarios are a good deal more prosaic, harnessing the imagination to the secular insights of the social and natural sciences. They address real-world questions of systems dynamics, policy choices, technological evolution, and consumption and production patterns. They reflect the modern worldview that the future is not preordained but rather is subject to human actions and choices. Yet the age-old drive to ponder the possibilities for our collective future, and to draw relevant lessons for how to live today, remains.

Basically, scenarios are plausible, challenging, and relevant stories about how the future might unfold, which can be told in both words and numbers. Scenarios are not forecasts, projections, or predictions. They are about envisioning future pathways and accounting for critical uncertainties. The process of building scenarios is about asking questions as well as suggesting answers and guidance for action. It is intended to widen perspectives and illuminate key issues that might otherwise be missed or dismissed.

The international commitment to sustainable development that has emerged in recent decades gives the study of the future a new urgency and direction. The essence of sustainability is to harmonize economic development with social goals and environmental preservation. At its core is the moral imperative that current generations should pass along an undiminished world to their descendants. To a large degree, sustainability is a challenge to think about the long-range future and, in so doing, to rethink the present. Sustainable development brings the question of the future to the strategic forefront of scientific research, policy deliberation, forward-thinking organizations, and the concerns of citizens.

The challenge of sustainability poses fundamental questions: How might global development evolve over the coming decades? Are we currently on a sustainable path? What surprises could deflect the global system in novel directions? How do environmental, social, and economic processes interact, dampening or amplifying change? How do global and sub-global processes interact? What actions, policies, and value changes can best ensure a sustainable future?

By offering insight into uncertainties and the consequences of current actions, scenarios support more informed and rational decision-making. Global scenario analysis has evolved rapidly over the past 10 years in response to this challenge. Although maintaining the long-term resilience of the world's ecosystems is fundamental to sustainable development, ecosystems have not yet been central to this research. Adding such a focus is a primary aim of the Millennium Ecosystem Assessment scenarios. This chapter sets the historical context and point of departure for the ecosystem outlook. It sketches early scenario efforts in Section 2.2, introduces key concepts in Section 2.3, reviews recent global scenario studies in Section 2.4, and draws lessons for ecosystem scenarios in Section 2.5. The rationale and methods used in the MA scenarios are addressed in subsequent chapters of this volume.

2.2 Scenarios Then

To provide historical context, we briefly sketch scenario activity prior to 1995. (See Box 2.1.) While heterogeneous, these studies all take a global perspective in attempting to link social, economic, and environmental issues. The literature can be split into two largely nonoverlapping streams—quantitative modeling and qualitative narrative. This dualism mirrors the twin challenges of providing systematic and replicable quantitative representation, on the one hand, and contrasting social visions and nonquantifiable descriptors, on the other hand.

Scenario analysis as a professional undertaking first surfaced with strategic planning and war games during the early years of the cold war, as popularized by Herman Kahn and his colleagues (Kahn and Wiener 1967). But the direct antecedents of contemporary scenarios lie in the future studies of the 1970s. These responded to emerging concerns about the long-term sufficiency of natural resources to support expanding global populations and economies. This first wave of global scenarios included ambitious mathematical simulation models (Forrester 1971; Meadows et al. 1972; Mesarovic and Pestel 1974) as well as speculative narrative (Kahn et al. 1976). At this time, scenario analysis was first used at Royal Dutch/Shell as a strategic management technique (Wack 1985), an effort that spawned a small industry of consultants working with major corporations to broaden perspectives on how to position the firm in a changing world (Schwartz 1991). At the same time, the French "prospective" school of strategic scenario analysis evolved a sharp critique of conventional techniques and a rich conceptual framework for exploratory future assessments (Godet 1987).

After a considerable lull during the early 1980s, a second round of integrated global analysis began in the late 1980s and 1990s, prompted by concerns with climate change and sustainable development. These included narrative scans of alternative futures (Burrows et al. 1991; Milbrath 1989), an optimistic analysis by the Central Planning Bureau of the Netherlands (1992), a pessimistic vision by Kaplan (1994), and a consideration of surprising futures (Svedin and Aniansson 1987; Toth et al. 1989). The long-term nature of the climate change issue spawned countless world energy scenarios, the most important of which were those of the Intergovernmental Panel on Climate Change (Leggett et al. 1992), which generally explored technological change and economic policy within a conventional "business-as-usual" framework.

The first decades of scenario assessment paved the way by showing the power—and limits—of both deterministic modeling and descriptive future analyses. A central challenge of contemporary global scenario exercises is to unify these two aspects by blending the replicability and clarity of quantification with the richness of narrative.

2.3 Scenarios Now

Scenarios are being applied in an expanding array of business, community, policy, and research contexts with highly varied aims—better management, consciousness raising, conflict resolution, policy advice, and research. "Business strategy scenarios" explore uncertainty in a world that the business does not control in order to test the robustness of decision-making and to identify opportunities and challenges. "New conversation scenarios" explore new and unknown topics and can be used as an educational tool for wide audiences. "Groups-in-conflict scenarios" use scenario techniques to understand differences and jointly explore consequences of actions. "Public interest scenarios" aim to shape the future by articulating a common agenda and highlighting potential actions and their consequences. "Scientific scenarios" examine the possible long-range behavior of biophysical systems as perturbed by human influence.

There are many methods for scenario building. They can be forward-looking, exploring how futures might unfold from current conditions and uncertainties, or backward-looking, beginning with a normative vision of the future and asking whether there is a plausible path to it. The approach to generating scenarios can include the active engagement of targeted audiences through participatory processes and game playing, deliberation among expert sce-

BOX 2.1

Selected Global Scenarios to 1995

Modeling-based

Meadows et al. (1972). *The Limits to Growth* report and the ensuing controversy surrounding its results was a seminal moment in global modeling. A systems dynamics model was used to assess the limits of the world system and the constraints these limits place on human numbers and activity. The model was global, with five sectors: population, capital, agriculture, nonrenewable resources, and pollution. Results were presented for 14 scenarios with varying assumptions on technical progress, social policy, and value changes from 1900 through 2100. The report emphasized that present trends will lead to major crises; however, concerted effort could alter these trends.

Mesarovic and Pestel (1974). *Mankind at the Turning Point* was a follow-up project to *Limits to Growth,* aiming to provide a more regionally disaggregated analysis. The Mesarovic/Pestel Model, or World Integrated Model, organized the world into 10 regions. Instead of a unified systems dynamics approach, WIM used five different linked sub-models for economy, population, food, energy, and environment. The model featured an interactive mode that allowed choices to be entered during a model run. The report underscores an impending global crisis and the need for significant societal changes.

Leontief (1976). *The Future of the World Economy* relied on an input-output model to analyze the impact of prospective economic policies for the United Nations. It tracked economic flows among 15 world regions. The report focused on a relatively optimistic scenario in which the income gap between industrial and developing countries decreases by one half by the year 2000. The report concluded that the limits to sustained economic growth and accelerated development are political, social, and institutional, not physical. However, this growth would require very large capital investments in developing regions and significant political, social, and institutional changes.

Herrera et al. (1976). *Catastrophe or New Society? A Latin American Model* emphasized sociopolitical rather than physical issues. It relied on the so-called Bariloche Model, an optimization model with four global regions and five sectors (agriculture, nutrition, housing, capital goods, and other) that ran for a 100-year period from 1960. The simulations addressed such questions as "what future global order would be best for humankind?" It asked how a human-oriented global society could grow, meet basic human needs, and manage resources wisely. This exercise was unique for its time in its explicit normative purpose of defining a future that the authors considered desirable and examining pathways for getting there.

Barney (1980). The *Global 2000* scenario assumed that existing trends would continue into the future. The analysis relied on a set of linked models to project global changes in population, natural resources, and the environment. The model covered 11 sectors, including population, economy, climate, technology, food, fisheries, forestry, water, and energy. The report argued that a projected global population of 10 billion by 2000 and 30 billion by 2100 would be dangerously close to Earth's carrying capacity and would lead to persistent global poverty.

Central Planning Bureau (1992). In *Scanning the Future,* the Dutch Central Planning Bureau considered four scenarios through the year 2015—global crisis, balanced growth, European renaissance, and global shift—using a macroeconomic model. The global crisis scenario assumed an overall economic decline. The European renaissance and the global shift scenarios projected economic stagnation in some regions and rapid economic expansion in others. All scenarios were optimistic compared with *The Limits to Growth.* In fact, all the scenarios projected significant convergence between rich and poor regions by the beginning of the twenty-first century.

Narrative-based

Kahn et al. (1976). *The Next 200 Years* presented an optimistic scenario in response to the pessimistic *Limits to Growth,* which was receiving considerable publicity at the time. The basic message of the scenario was that the world could vastly expand its population and economic scale and remain far from the natural limits for most resources. The authors also discussed the possibility of considerable human activities occurring in outer space.

Robertson (1983). *The Sane Alternative* presented five futures: business as usual; disaster; authoritarian control; hyper-expansionist; and sane, humane, and ecological. The latter scenario incorporated significant changes in the direction of human activities and policies, emphasizing the need for development in psychological and social spheres rather than economic and technical growth. The author argued that the future is likely to be a mix of all five scenarios.

Burrows et al. (1991). *Into the 21st Century* examined three scenarios: a pessimistic scenario, a piecemeal scenario, and an optimistic scenario. The pessimistic scenario explored what the future might be like if present trends continue unchecked. The piecemeal scenario assumed a determined but fragmented attempt to find solutions to environmental and social problems. The optimistic scenario featured dramatic changes in attitudes and values toward altruism, cooperation, and ecology. The authors argued that the holistic approach, together with an appropriate planet management system, is the only way to solve growing problems, and that we must act urgently lest the world reach a point of no return soon.

Kaplan (1994). *The Coming Anarchy* presented a dark vision of a dramatic increase in demographic, environmental, and social stresses worldwide. The author argued that the current critical situation in West Africa is a premonition of the future. He underscored the critical links between environmental and social stresses, foreseeing surging populations, spreading disease, resource degradation, water depletion, air pollution, and rising sea levels. These stresses will cause mass migrations and, in turn, incite group conflicts. This was later expanded into a book (Kaplan 2000).

Svedin and Aniansson (1987). *Surprising Futures* presented the results of a workshop of social and natural scientists that explored a range of "surprise-rich" futures through the year 2075. Four alternative scenarios were developed: the big load, the big shift, history lost, and hope regained. These explored, respectively, dominant trends, a shift toward new centers of power such as China and India, a future of crises, and environmentally and socially balanced world development. The workshop participants generated numerous variations on these scenarios along with hypothetical "future histories." Quantitative sketches were presented in Toth et al. (1989).

nario panels, and quantitative simulations by modeling groups.

For this report, we focus on the subset of environmental global scenario projects that have a public policy and scientific orientation, since these are of greatest interest to the MA. The ideal attributes for such scenarios are *integration* across social, economic, and environmental dimensions; *regional disaggregation* of global patterns; *multiple futures* that reflect the deep uncertainties of long-range outcomes; and *quantification* of key variables linked to ecosystem conditions. We briefly expand on each of these.

2.3.1 Integration

Integration is needed because multiple anthropogenic stressors have an impact on the environment or lead to changes in the provision of ecological services. These stressors (or direct drivers) include pollution, climate change, hydrological change, resource extraction, and land degradation and conversion. In turn, these direct drivers result from long causal chains of indirect socioeconomic drivers, such as demographic, economic, and technological developments. Finally, changing patterns of human values, culture, interest, and power set the conditioning framework (or ultimate drivers) for unfolding socioecological systems. To capture this nexus of interactions, a systemic framework is required that includes key economic, social, and environmental subsystems and links. (See Figure 2.1.)

2.3.2 Regional Disaggregation

Such systems can be meaningfully defined at different scales—global, regional, national, and local. A planetary panorama reveals global economic, cultural, and environmental phenomena, and becomes more critical as global connectivity increases. A regional perspective brings the problems of acid rain, water allocation, trade, and migration into focus. A national viewpoint sheds light on policy, environmental, and security issues. A local standpoint is needed for detailed assessment of land change patterns, biodiversity, and ground-level pollution.

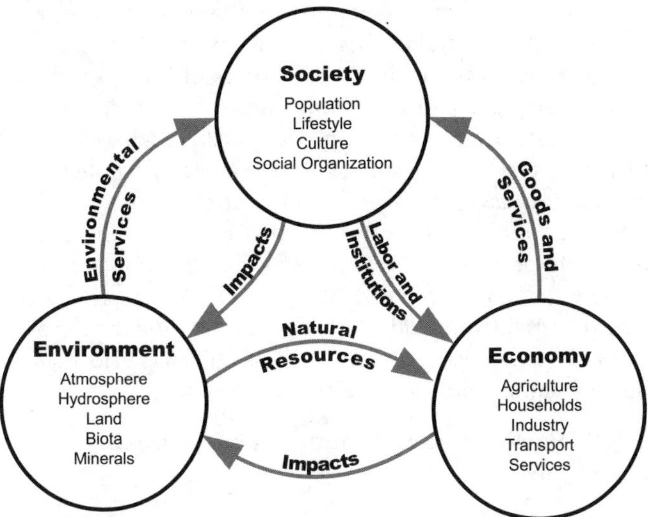

Figure 2.1. The Socioecological System and Its Components

These different spatial scales provide mutually enriching windows of perception into a unitary world system. Globalization links these different scales through processes that increase economic, cultural, social, and geopolitical interdependence. In particular, the factors that directly affect ecosystems are conditioned by far-flung global influences —patterns of production and consumption, the character of economic globalization, cultural influence, migration, and global environmental change. Ecosystem futures are an aspect of the wider question of global futures.

One day, perhaps, scenario-building techniques will evolve to allow analysts to seamlessly zoom across levels, representing each spatial unit as an interacting component of an integrated global system. But in these early years of this analytic discipline, the state-of-the-art is far more modest —the disaggregation of the global system into a single layer comprised of major multinational regions.

2.3.3 Multiple Futures

Since the issues that need to be considered in the context of sustainable development or the long-term provisioning of ecological services embody multigenerational concerns, and since certain ecological consequences only become visible over long time periods, the scenario outlook must span at least several decades. The MA horizon is 50 years. Over such an extended time frame, current trends can evolve in unexpected ways, all the while subject to new phenomena, events, and human influence. At critical thresholds, the planetary socioecological system can branch into unique pathways. Thus, global outlooks that do not consider a broad range of plausible long-range visions are incomplete.

Three distinct sources of indeterminacy are *ignorance, surprise,* and *volition* (Raskin et al. 2002). *Ignorance* refers to limits of scientific knowledge on current conditions and dynamics. Even if the global system were deterministic in principle, this classic form of uncertainty would lead to a statistical dispersion over possible future states. *Surprise* is the uncertainty due to the inherent indeterminism of complex systems, which can exhibit emergent phenomena and structural shifts.

Volition refers to the unique uncertainty that is introduced when human actors are internal to the system under study—the future is subject to human choices that have not yet been made. Moreover, the very process of ruminating on the future can influence these choices. Through this reflexivity, this double role of humans as observers and actors, scenario studies become internal to the story they tell to the degree they alter awareness, behavior, and the future.

Alternative global futures can result from the accumulation of gradual incremental changes. Or perhaps a threshold of instability will be crossed in which the global trajectory bifurcates into very different possible outcomes. Massive unexpected events could change the course of development —a world war, a pandemic, a large-scale act of terrorism, a systemic economic breakdown, abrupt climate change, a technological wildcard, and so on. The exploration of multiple futures is fundamental to the scenario enterprise.

2.3.4 Quantification and Narrative

In view of this complexity and uncertainty, scenario analysis requires approaches that transcend the limits of conventional deterministic models of change. Predictive modeling is appropriate for simulating well-understood systems over sufficiently short times (Peterson et al. 2003). But as complexity increases and the time horizon lengthens, the power of prediction diminishes. Quantitative forecasting is legitimate only to the degree the system state can be well specified, the dynamics governing change are known and persistent, and mathematical algorithms can be devised to validly represent these relationships.

These conditions are violated when it comes to assessing the long-range future of socioecological systems—state descriptions are uncertain, causal interactions are poorly understood and may change by unknown ways in the future, and nonquantifiable factors are significant. Probabilistic forecasts of a given future state, or a spectrum of possible states, are simply not feasible when structurally unique futures can emerge from current conditions and trends, and novel behavior can be expected. To take but one example, the combined effects of abrupt climate change, geopolitical conflict, and global economic instability could drive the planetary system into a new state that exhibits historically unprecedented institutional and biophysical processes.

The development of methods to blend quantitative and qualitative insight effectively is at the frontier of scenario research today. (See Figure 2.2.) The scenario narrative gives voice to important qualitative factors shaping development such as values, behaviors, and institutions, providing a broader perspective than is possible from mathematical modeling alone. Narrative offers texture, richness, and insight, while quantitative analysis offers structure, discipline, and rigor. The most relevant recent efforts are those that have sought to balance these.

2.4 Major Studies

The catalogue of recent global scenarios encompasses hundreds of greenhouse gas emissions projections (IPCC 2001), regional and national scenarios (Raskin 2000), and sectoral outlooks on energy, food, economy, demography, and technology (Glenn and Gordon 2002). But only a few studies have the comprehensive scope and analytic detail to satisfy the criteria of integration, regionalization, multiple futures, and quantification. We introduce six major efforts below—a subset of global exercises that have a scientific and public policy orientation—and summarize their salient features in Table 2.1.

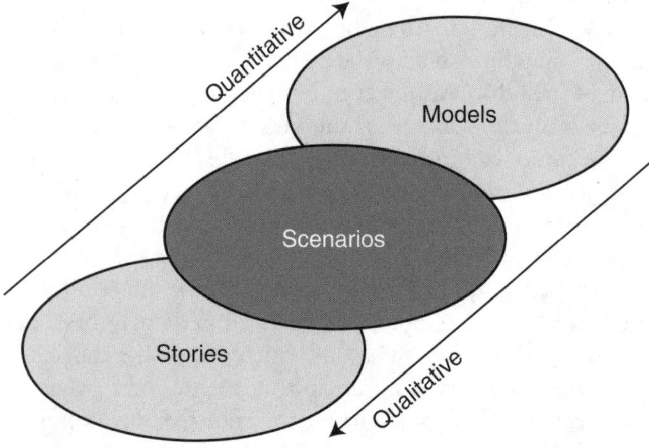

Figure 2.2. Scenarios, Models, and Stories (Nakićenović et al. 2000)

2.4.1 Global Scenario Group

Convened in 1995 by the Stockholm Environment Institute, the Global Scenario Group is an independent, international, interdisciplinary body that has been developing integrated global and regional scenarios (Raskin et al. 1998, 2002; Gallopín et al. 1997). The GSG scenario narratives are quantified with the use of the PoleStar System, a transparent tool for synthesizing global data sets, organizing sectoral linkages, and introducing assumptions (Raskin et al. 1999). This work has been used by a number of international assessments, including several discussed below.

2.4.2 Global Environment Outlook

The United Nations Environment Programme's third Global Environment Outlook placed greater emphasis than previous editions on integrated global and regional scenarios (UNEP 2002). The scenarios were developed through a lengthy collaborative process that began with four of the GSG scenarios, which were then refined through a series of regional and global meetings (Raskin and Kemp-Benedict 2002), with input from the IPCC's Special Report on Emissions Scenarios. The emphasis of the process was on refining the narratives and giving them regional texture. A consortium of modeling teams elaborated on different aspects of the scenarios (Potting and Bakkes 2004).

2.4.3 Special Report on Emissions Scenarios

The IPCC's Special Report on Emissions Scenarios (Nakićenović et al. 2000) was a significant advance over prior IPCC scenarios. Its purpose was to develop a wide range of emissions scenarios as input to ongoing climate change research. A set of alternative social visions was linked to assumptions on the main driving forces of human-induced climate change, and the implications for future energy-related and land-use emissions were analyzed. Based on its mandate, the SRES scenarios did not include policies for greenhouse gas mitigation in the scenarios and thus only simulated emissions in the absence of such policies. Six modeling groups analyzed each family, thus giving a wide range of outcomes for each storyline. An open process solicited participation and feedback. Several modeling groups went on to publish variants of the SRES scenario that do include climate policy (Morita et al. 2001).

2.4.4 World Business Council for Sustainable Development

The World Business Council for Sustainable Development constructed a set of three scenarios to engage the business community in the debate on sustainable development

Table 2.1. Description of Selected Global Scenario Studies since 1995 and Their Structure (Raskin, in press)

Study	Horizon	Regions	Focus	Scenario Structure
GSG	2050	11	Environment; poverty reduction; human values	1. *Conventional Worlds:* gradual convergence in incomes and culture toward dominant market model a) *Market Forces:* market-driven globalization, trade liberalization, institutional modernization b) *Policy Reform:* strong policy focus on meeting social and environmental sustainability goals 2. *Barbarization:* social and environmental problems overwhelm market and policy response a) *Breakdown:* unbridled conflict, institutional disintegration, and economic collapse b) *Fortress World:* authoritarian rule with elites in "fortresses," poverty and repression outside 3. *Great Transitions:* fundamental changes in values, lifestyles, and institutions a) *Eco-Communalism:* local focus and a bio-regional perspective b) *New Sustainability Paradigm:* new form of globalization that changes the character of industrial society
GEO-3	2032	6	Environment	*Markets First; Policy First; Security First; Sustainability First* (correspond, respectively, to 1a, 1b, 2b, and 3b above)
SRES	2100	4	Climate change	*A1:* rapid market-driven growth with convergence in incomes and culture *A2:* self-reliance and preservation of local identities, fragmented development *B1:* similar to *A1*, but emphasizes global solutions to sustainability *B2*: local solutions to economic, social, and environmental sustainability
WBCSD	2050	n.a.	Business and sustainability	*FROG!:* market-driven growth, economic globalization *GEOpolity:* top-down approach to sustainability *Jazz:* bottom-up approach to sustainability, ad hoc alliances, innovation
WWV	2025	18	Freshwater crisis	*Business-as-usual:* current water policies continue, high inequity *Technology, Economics, and the Private Sector:* market-based mechanisms, better technology *Values and Lifestyles:* less water-intensive activities, ecological preservation
OECD	2020	10	Environment in OECD countries	*Reference* with policy variants (e.g., subsidy removal, eco-taxes)

Key: GSG Global Scenario Group, GEO-3 Global Environment Outlook, SRES Special Report on Emissions Scenarios, WBCSD World Business Council on Sustainable Development, WWV World Water Vision, OECD Organisation for Economic Co-operation and Development

(WBCSD 1997). The focus is on the scenario narratives, which span a broad spectrum of possible futures. For each narrative, the authors present a set of challenges to business and lessons to be drawn. The scenarios were developed in an open process involving representatives from 35 organizations.

2.4.5 World Water Vision

The World Water Vision was conducted by the World Water Council to increase awareness of a rising global water crisis (Cosgrove and Rijsberman 2000). The WWV presents three global water scenarios that focus on issues of water supply and demand, conflict over water resources, and water requirements for nature (Gallopín and Rijsberman 1999). While only a subset of water-related issues and variables were quantified, the scenario narratives extend beyond issues specific to water, including lifestyle choice, technology, demographics, and economics. Some of these additional themes were explored quantitatively in background studies.

2.4.6 Organisation for Economic Co-operation and Development

The *Environmental Outlook* of the Organisation for Economic Co-operation and Development developed a baseline scenario based on development projections to 2020, complemented by several policy variants (OECD 2001). The *Outlook* examined drivers of environmental change, specific sectors that put the greatest pressure on the environment, and resulting environmental impacts. The focus of the *Outlook* is the critical environmental concerns facing OECD countries, but the study is global in scope. Global economic patterns were modeled using the OECD's JOBS model. These drivers were then used as inputs to the PoleStar System to assess potential environmental impacts in the scenarios.

2.4.7 Study Outputs

Two of the studies—GSG and SRES—stand out as presenting both a broad range of scenario visions and a wide set of quantitative indicators. Table 2.2 summarizes the kinds of outputs provided by these studies, including drivers of environmental change, resource requirements, and environmental stressors. The Table also references the IMAGE 2.2 model, which has been used to update and expand the SRES scenarios (RIVM 2001; Alcamo et al. 1998), providing input to GEO-3.

A scan of these studies suggests great variation in the way each exercise was structured. Yet beneath the diversity, the

Table 2.2. Variables Included in Scenario Simulations by GSG, IPCC-SRES, and IMAGE

Variable	GSG	IPCC-SRES	IMAGE
Demographics			
Population	X	X	X
Distribution	urban/rural		age and sex
Poverty	X		
Economics			
GDP	X	X	X
Sectors	X		X
Income distribution/poverty	X		
Agriculture and forestry			
Diets	X		X
Yields	X		X
Livestock practices	X		X
Inputs	X		X
Timber production	X		X
Fish production	X		
Cropland degradation	X		X
Water			
Withdrawals	X		
Resources	X		
Stress	X		
Energy			
Requirements by fuel	X	X	X
Production	X	X	X
Land use			
Built environment	X		gridded
Cropland	X	X	gridded
Grazing/grassland	X	X	gridded
Forest	X	X	gridded
Plantation	X	X	gridded
Pollution/waste			
Air	GHGs; SO_x	GHGs; SO_x	GHGs; SO_x; NO_x; VOC
Water	X		
Toxics	X		
Solid waste	X		

Key: GSG Global Scenario Group, SRES Special Report on Emissions Scenarios

scenarios are rooted in a common set of archetypal visions of the future—worlds that evolve gradually, shaped by dominant driving forces; worlds that are influenced by a strong policy push for sustainability goals; worlds that succumb to fragmentation, environmental collapse, and institutional failure; and worlds where new human values and forms of development emerge. The scenarios from the various studies are mapped into a common framework in Table 2.3, using the GSG scenario structure as a template. Thumbnail sketches of narratives used in the GSG, SRES, and WBCSD studies are presented in Box 2.2.

2.5 The Past as Prelude

Over the past decade, global scenarios of increasing sophistication have influenced the policy discussion of sustainable development, sharpened perspectives, and broadened awareness. But they have yet to focus directly and systematically on the role of ecosystem conditions and management as a critical component of the global future, nor have they fully taken ecosystem dynamics into account. The scenario experience to date provides a useful point of departure for injecting this dimension. To do so, storylines must be enriched with an ecosystem perspective, and quantifications expanded to include measures of ecosystem condition.

The recent global scenario literature covers an immense array of detailed findings, conclusions, and lessons, far more than can be summarized here. But some broad lessons can be drawn that cut across these diverse studies. For example, collectively they suggest that a global future that excessively relies on a "market forces" vision of economic globalization and on the consumer society as the model for successful development would be a perilous basis for global development. The risk is that social polarization, persistent poverty, and environmental degradation would undercut sustainability by eroding social cohesion, ecosystem resilience, and the global economy. Then, rather than mitigating current tendencies toward global polarization and conflict, a full descent into a fragmented "fortress world" or other unpleasant possibilities becomes a real danger.

The studies tend to find great scope for "policy reform" scenarios for reducing such risk. The ambitious social and environmental goals articulated in such high-level international appeals as *Agenda 21* (UN 1993) and the Millennium Development Goals (UN 2000) are seen as feasible, at least in principle. A host of specific actions could accelerate the deployment of nature-sparing technology, alleviate poverty, and reduce social tension. But the reform strategy is problematic in practice. It requires mobilizing a comprehensive

Table 2.3. Comparing Selected Global Scenarios after 1995

GSG	SRES	WBCSD	GEO-3	WWV	OECD
Conventional worlds					
Market forces	A1	FROG!	Markets first	B-a-u	Reference
Policy reform	B1	GEOpolity	Policy first	Technology and economics	Policy variants
Barbarization					
Breakdown	A2				
Fortress world			Security first		
Great transitions					
Eco-communalism	B2				
New sustainability paradigm		Jazz	Sustainability first	Lifestyles and values	

Key: GSG Global Scenario Group, GEO-3 Global Environment Outlook, SRES Special Report on Emissions Scenarios, WBCSD World Business Council on Sustainable Development, WWV World Water Vision, OECD Organisation for Economic Co-operation and Development

BOX 2.2

Narrative Sketches

Global Scenario Group. GSG scenarios are organized into three classes—*Conventional Worlds, Barbarization,* and *Great Transitions*—with two variants for each class. *Conventional Worlds* envisions the spread of dominant values and development patterns with the gradual convergence of developing regions toward rich-region patterns. In the *Market Forces* variant, powerful global actors advance the priority of growth, liberalization, and privatization. In the *Policy Reform* variant, concern over environmental deterioration, social conflict, and economic instability leads to comprehensive government action for sustainable development. The *Barbarization* class of scenarios envisions the eventual deterioration of civilization, as crises overwhelm the coping capacity of both markets and policy reform. The *Breakdown* variant spirals toward unbridled conflict and institutional collapse. The *Fortress World* variant features an authoritarian response to this threat, with elites in protected enclaves and an impoverished majority outside. The *Great Transitions* class depicts fundamental changes in the global development model. Human values emphasize ecology, dematerialized lifestyles, and strong social solidarity. Regions pursue diverse strategies building on unique cultural, ecological, and institutional attributes. The *Eco-Communalism* variant is a highly localist vision that emphasizes regional self-reliance. The *New Sustainability Paradigm,* highlighted by the GSG (Raskin et al. 2002), would build a more humane global civilization rather than retreat into localism.

Special Report on Emissions Scenarios. In the SRES notation, "A" and "B" signify little and high commitment to sustainable development, respectively, and "1" and "2" signify regional integration or fragmentation, respectively. Thus, A1 is an integrated unsustainable world of rapid economic growth, stabilizing populations, rapid technological change, and convergence among regions. The scenario has three variants that assume different energy mixes. A2 is a fragmented unsustainable world in which regions and nations stress self-reliance and preservation of local identities with relatively high population growth, slow income convergence, and heterogeneous local development patterns. B1 is a regionally convergent world with global population peaking in mid-century and declining thereafter, as in A1. B1 is also an integrated sustainable world that features a rapid shift to a service economy and clean technologies, and the pursuit of global solutions to economic, social, and environmental problems (excluding climate change). B2 is a fragmented sustainable world in which regions and nations pursue plural models of development with diverse local initiatives that balance economic, social, and environmental goals.

World Business Council for Sustainable Development. *FROG!* begins as a business-as-usual scenario, where economic growth is the major concern and action on sustainable development is weak. However, the reliance on technology does not deliver environmental or social health, and eventually increased inequity and unrest threaten basic survival. In *GEOpolity,* a new global consensus welcomes technocratic solutions, sanctions, and more direct control of the market to ensure environmental preservation and social cohesion. New global governance institutions are created, and governments become rejuvenated as focal points in shifting the structure of the economy toward sustainable development. *Jazz* is an innovative world of ad hoc alliances of diverse stakeholders, experimentation, adaptation, and a dynamic global market. Government activity is largely at the local level, as ad hoc global governance institutions address particular problems, and environmental and social goals are achieved largely out of corporate self-interest and partnerships among nongovernmental organizations, governments, consumers, and businesses.

array of incremental adjustments to gradually counteract underlying trends that are highly unsustainable. The plausibility of such scenarios rests on the difficult task of providing an account of how the necessary political will would emerge for such a massive effort.

In view of these difficulties, many of the studies go on to explore scenarios that include more fundamental transformations of the underlying values and institutions of development. Several find that an eventual transition toward a form of global development based on "sustainability first" may be both the necessary and desirable condition for ensuring an ecologically sound and humane global future. This alternative paradigm draws attention to institutional and value innovations. Multiscale, adaptive, and democratic governance processes include the constructive engagement of civil society in balancing economic, social, and environmental concerns. Human values come to emphasize qualitatively rich lifestyles, encompassing material sufficiency, a strong sense of human solidarity from local to global levels, and an ecological sensibility.

In the coming years, the enrichment of global scenarios, often through participatory processes, will define an important agenda for policy analysis (EEA 2001), scientific research, and education. This will require the enhancement of the role of ecosystems in both scenario narrative and quantification. Narratives will need to more richly reflect ecosystem descriptors, impacts, and feedbacks. Models will need to simulate ecosystem services within global assessment frameworks. Previous global scenario studies are the prelude and platform for beginning to address these challenges.

References

Alcamo, J., R. Leemans, and E. Kreileman (eds.), 1998: *Global Change Scenarios of the Twenty-First Century: Results from the IMAGE 2.1 Model.* Elsevier, Kidlington, U.K.

Barney, G. (ed.), 1980: *The Global 2000 Report to the President of the U.S.: Entering the Twenty-First Century,* vols. IIII. Pergamon Press, New York.

Burrows, B., A. Mayne, and P. Newbury, 1991: *Into the 21st Century: A Handbook for a Sustainable Future.* Adamantine, Twickenham, UK.

Central Planning Bureau, 1992: *Scanning the Future: A Long-term Scenario Study of the World Economy 1990–2015.* SDU Publishers, The Hague.

Cosgrove, W. and F. Rijsberman, 2000: *World Water Vision: Making Water Everybody's Business.* Earthscan, London.

EEA (European Environment Agency), 2001: *Scenarios as Tools for International Environmental Assessments.* Environmental Issue Report No 24. Office for Official Publications of the European Communities, Luxembourg. Available at http://www.eea.eu.int.

Forrester, J., 1971: *World Dynamics.* Wright-Allen Press, Cambridge, MA.

Gallopín, G. and F. Rijsberman, 1999: *Three Global Water Scenarios.* World Water Council, Paris. Available at http://www.worldwatervision.org.

Gallopín, G., A. Hammond, P. Raskin, and R. Swart, 1997: *Branch Points: Global Scenarios and Human Choice.* Stockholm Environment Institute, Stockholm. Available at http://www.gsg.org.

Glenn, J.C. and T.J. Gordon, 2002: *State of the Future at the Millennium.* American Council for the United Nations University, Washington, DC.

Godet, M., 1987: *Scenarios and Strategic Management.* Elsevier, The Netherlands.

Herrera, A., H. Scolnic, G. Chichilnisky, G. Gallopín, J. Hardoy, D. Mosovich, E. Oteiza, G. de Romero Brest, C. Suarez, and L. Talavera, 1976: *Catastrophe or New Society? A Latin American World Model.* IDRC, Ottawa.

IPCC (Intergovernmental Panel on Climate Change), 2001: *Climate Change 2001: Mitigation. Third Assessment Report of the Intergovernmental Panel on Climate Change.* University of Cambridge Press, Cambridge, UK.

Kahn, H. and A. Weiner, 1967: *The Year 2000: A Framework for Speculation on the Next Thirty-three Years.* Macmillan, New York.

Kahn, H., W. Brown, and L. Martel, 1976: *The Next 200 Years: A Scenario for America and the World.* Morrow, New York.

Kaplan, R., 1994: The Coming Anarchy. *The Atlantic Monthly,* **273(2),** 44–76.

Kaplan, R., 2000: *The Coming Anarchy. Shattering the Dreams of the Post Cold War.* Random House, New York.

Leggett, J., W.J. Pepper, and R.J. Swart, 1992: Emissions Scenarios for IPCC: An Update. In: *Climate Change 1992. The Supplementary Report to the IPCC Scientific Assessment,* J.T. Houghton, B.A. Callander, and S.K. Varney (eds.), Cambridge University Press, Cambridge.

Leontief, W., 1976: *The Future of the World Economy: A Study on the Impact of Prospective Economic Issues and Policies on the International Development Strategy.* United Nations, New York.

Meadows, D., D. L. Meadows, J. Randers, and W. Behrens, 1972: *The Limits to Growth.* Universe Books, New York.

Mesarovic, M. and E. Pestel, 1974: *Mankind at the Turning Point.* Dutton, New York.

Milbrath, L., 1989: *Envisioning a Sustainable Society: Learning Our Way Out.* SUNY Press, Albany, New York.

Morita, T. and Robinson J., 2001: Greenhouse gas emission mitigation scenarios and implications. In *Climate Change 2001: Mitigation.* B. Metz, O. Davidson, R. Swart, and J. Pan (eds.), Cambridge University Press. Cambridge.

Nakićenović, N., J. Alcamo, G. Davis, B. de Vries, J. Fenhann, S. Gaffin, K. Gregory, A. Grübler, T.Y. Jung, T. Kram, E. Lebre La Rovere, L. Michaelis, S. Mori, T. Morita, W. Pepper, H. Pitcher, L. Price, K. Riahi, A. Roehrl, H.-H. Rogner, A. Sankovski, M. Schlesinger, P. Shukla, S. Smith, R. Swart, S. van Rooijen, N. Victor, and Z. Dadi, 2000: *Special Report on Emissions Scenarios: A Special Report of Working Group III of the Intergovernmental Panel on Climate Change.* Cambridge University Press, Cambridge.

OECD (Organisation for Economic Co-operation and Development), 2001: *OECD Environmental Outlook.* Paris: OECD.

Peterson, G., G. Cumming, and S. R. Carpenter, 2003: Scenario planning: a tool for conservation in an uncertain world. *Conservation Biology* **17**:358–366.

Potting, J. and J. Bakkes (eds.), 2004: *The GEO-3 Scenarios 2002–2032. Quantification and Analysis of Environmental Impacts.* UNEP/RIVM: Nairobi and Bilthoven, The Netherlands.

Raskin, P., in press: Global Scenarios and the Millennium Ecosystem Assessment: An Historic Overview. *Ecosystems.*

Raskin, P., 2000: Regional Scenarios for Environmental Sustainability: A Review of the Literature. UNEP, Nairobi.

Raskin, P. and E. Kemp-Benedict, 2002: GEO Scenario Framework Background Paper for UNEP's Third Global Environmental Outlook Report. UNEP, Nairobi.

Raskin, P., T. Banuri, G. Gallopín, P. Gutman, A. Hammond, R. Kates and R. Swart, 2002: *Great Transition: The Promise and Lure of the Times Ahead.* SEI-B/Tellus Instittues, Boston. Available at http://www.tellus.org/seib/publications/Great_Transitions.pdf.

Raskin P., C. Heaps, J. Sieber, and E. Kemp-Benedict, 1999: *PoleStar System Manual for Version 2000.* SEI-B/Tellus Institute. Available at http://www.tellus.org/seib/publications/ps2000.pdf.

Raskin, P., G. Gallopín, P. Gutman, A. Hammond, and R. Swart, 1998: *Bending the Curve: Toward Global Sustainability.* Stockholm Environment Institute, Stockholm. Available at http://www.tellus.org/seib/publications/bendingthecurve.pdf.

RIVM (Rijksinstituut voor Volksgezondheid en Milieu), 2001: The IMAGE 2.2 implementation of the SRES scenarios: A comprehensive analysis of emissions, climate change and impacts in the 21st century. CD-ROM. RIVM, Bilthoven, The Netherlands.

Robertson, J., 1983: *The Sane Alternative—A Choice of Futures.* River Basin, St. Paul, Minnesota.

Schwartz, P., 1991: *The Art of the Long View: Planning for the Future in an Uncertain World.* Doubleday, New York.

Svedin, U. and B. Aniansson (eds.), 1987: *Surprising Futures: Notes From an International Workshop on Long-term World Development.* Swedish Council for Planning and Coordination of Research, Stockholm.

Toth, F., E. Hizsnyik, and W. Clark (eds.),1989: *Scenarios of Socioeconomic Development for Studies of Global Environmental Change: A Critical Review.* International Institute for Applied Systems Analysis, Laxenburg, Austria.

UN (Untied Nations), 2000: *We the Peoples—The Role of the United Nations in the 21st Century.* United Nations, New York.

UN, 1993: *Report of the United Nations Conference on Environment and Development.* United Nations, New York.

UNEP (United Nations Environment Programme), 2002: *Global Environmental Outlook 2002.* Earthscan, London.

Wack, P., 1985: Scenarios: Shooting the Rapids. *Harvard Business Review,* **63,** 135–150.

WBCSD (World Business Council for Sustainable Development), 1997: *Exploring Sustainable Development. Summary Brochure.* WBCSD, Geneva. Available at http://www.wbcsd.org/newscenter/reports/1997/exploringscenarios.pdf.

Chapter 3

Ecology in Global Scenarios

Coordinating Lead Authors: Graeme Cumming, Garry Peterson
Review Editors: Antonio Alonso, Christopher Field, Robin Reid

---

BOXES

FIGURES

TABLES

Main Messages

Ecosystems are essential to the survival of human societies and economies. Ecosystems provide a range of economic and cultural services to humans. These include such basic necessities as clean air, clean water, and the production of food. Ecosystems also enhance human well-being through a diverse range of services that include climate and disease regulation, flood and erosion control, pollination, recreational areas, and enhancement of spiritual and aesthetic experiences.

The inclusion of ecology in past global scenario exercises has been limited. Previous global scenario exercises (see Chapter 2) have largely focused on social and economic drivers and consequently have presented an incomplete picture of the world.

Ecological change affects scenario outcomes. Ecosystems have a significant influence on societies and economies, and people modify ecosystems. One of the goals of the Millennium Ecosystem Assessment is to develop the first set of global scenarios to explore the importance of ecosystems and ecological change for human well-being while maintaining an awareness of the importance of social and economic change.

There are substantial risks that ecological degradation will diminish the future well-being of humanity. Much of our current socioeconomic progress is not sustainable because it reduces the capacity of the biosphere to provide the ecological services that we depend on. Irreversible ecological changes, such as extinctions and species invasions, are of particular concern. It is likely that changes in production systems, ecological management, and social organization will be necessary if we are to sustain human well-being.

Regime shifts in ecosystems cause rapid, substantial changes in ecosystem services and human well-being. Ecosystem services that have been impaired by regime shifts include fisheries and food production in drylands and the quality of fresh waters. Other types of ecological regime shifts with important effects on people include regional climate changes and the emergence of disease. Increasing pressure on these ecosystems will increase the frequency of regime shifts that affect ecosystem services and human well-being.

Ecological feedbacks may accentuate human modifications of ecosystems. Changes in ecological functioning produced by unintended ecological feedbacks from human actions appear likely to amplify climate change, decrease agricultural productivity, reduce human health, and increase the vulnerability of ecosystems to invasive species.

Although ecological theory is well developed, an improved understanding of the relationships between ecosystems and human well-being would facilitate sustainability. There are numerous ecological theories, described in this chapter, that help us understand ecological processes and their relevance for thinking about ecosystem services in global scenarios. Recent developments in complex systems theory offer further insights into the relationships between ecosystems, economies, and societies. Research on resilience, adaptive management, political ecology, and ecological economics offers guidance on linkages between ecosystems, societies, and economies. Although we believe that the inclusion of ecology in global scenarios is a big step forward, further research is needed to better understand the connections among the production of multiple ecosystem services, the local and global impact of ecological processes, and the determinants of ecological resilience.

3.1 Introduction

It is easy for us to take for granted the complex environment that has given rise to our species. Although life on Earth has persisted far longer than we have, and will probably outlast us, recent years have brought an awareness that ecosystems may be more fragile than we had thought. Some of the changes that humans have caused in ecosystems are now affecting people directly. Continuing human impacts on ecosystems cast doubt on the capacity of ecosystems to continue to provide the goods and services that we depend on. We need to pay attention to changes in ecosystems, even if only because our social and economic systems are embedded within them.

The direct importance of ecosystem services to humans is explained in Chapter 1 and is summarized in the MA conceptual framework. Ecosystem services emerge from the interactions of diverse ecological structures and processes. They are not independent of one another; what may be most important for people is the continued existence, or resilience, of an entire bundle of interdependent services. It is possible to affect a range of ecosystem services when attempting to manage or change only a single service.

Many ecosystem services interact with one another through trade-offs, in which increasing the provision of one service causes declines in provision of another service. Decisions concerning the economic benefits of ecosystem modification often require us to address trade-offs between different types of ecosystem service. For example, Fearnside (2000) describes how climate regulation (carbon storage, evapotranspiration) may conflict with food production (such as clearing of woodlands to create pastures); similarly, the use of river systems as conduits for the removal of wastes can have severe impacts on water quality and human health (e.g., Donnison and Ross 1999).

Changes in ecosystems may have both direct and indirect effects on human health and well-being. These changes are often more complicated than direct provision of food and fiber, recreational areas, or clean water. For example, a decrease in flow variability caused by an impoundment on the Vaal River in South Africa contributed to an outbreak of the blackfly *Simulim chutteri,* the vector of river blindness (Carr 1983; Chutter 1968), and destruction of wetlands has resulted in higher levels of heavy metals in table fishes (Brant et al. 2002; King et al. 2002). In Ecuador, destruction of mangroves for the aquaculture of shrimps for the export market has contributed to declining food security through the loss of coastal fisheries (Parks and Bonifaz 1994).

Ecosystem services are intricately related to poverty (Martinez-Alier 2002). People with few financial resources are more likely to rely on the direct provisioning services of ecosystems, such as bushmeat and unpurified water. They may also be less able to manage resources effectively if they have been resettled in unknown areas (Angelsen and Kaimowitz 1999; Deininger and Binswanger 1999), are denied full tenure (Lawrence 2003; Parks and Bonifaz 1994; Robinson and Bennett 2002), or lack the political power to prevent imports of externally generated pollutants (Martinez-Alier 2002). Effective ecosystem management will require policies that take poverty into account. Similarly, effective poverty alleviation requires realistic policies that take into account the capabilities of different ecosystems to provide bundles of ecosystem services.

In this chapter we describe the future of ecosystem services, the motivation for developing scenarios that consider ecological services, some of the ecological theories that may be useful for integrating ecosystem services into scenarios, the integration between ecology and related disciplines, and the relevance of ecological theories and scenarios for the development of management and policy approaches. Our aim is to provide a cohesive summary of relevant ecological thinking (and its relationship to other disciplines) for readers who are interested in understanding the motivation for the MA scenarios and the current limitations and future needs for the development of ecological scenarios.

3.2 The Future of Ecosystem Services

There is increasing evidence that the activities of humans can alter a range of ecosystem services at global and regional scales. Well-documented impacts of human activities on ecosystem services at a variety of scales include changes in Earth's climate (Watson and Team 2001), the number and distribution of species (Chapin et al. 2000; Higgins et al. 2003; Sala et al. 2000), the quality and quantity of fresh water (Meyer et al. 1999; Brinson and Malvarez 2002), and air quality and pollution levels (Sinha et al. 2003). Human activities also affect ecosystems in ways that have diverse effects on bundles of ecosystem services, for instance through changes in the ability of organisms to disperse (Hill and Curran 2003) and by disrupting food webs through species translocations (Simon and Townsend 2003; Zavaleta et al. 2001).

Sustainable development has become a mantra for many development organizations, although (or perhaps because) the concept of sustainability has proved difficult to pin down and apply (Goldman 1995). Given projected increases in human population and the slow rate of change of many human behaviors, it seems increasingly likely that human impacts on ecosystem services will affect the quality of life of the majority of the human population within the next 50 years. Our current lack of knowledge concerning the resilience of ecosystem services makes it difficult to assess the degree to which we should be concerned about this. If ecosystems are relatively robust, it is possible that current trends may not greatly alter the provision of the more vital ecosystem services. By contrast, if ecosystems are relatively brittle and if the relationship between ecological impacts and ecosystem services is nonlinear, we run the risk that cumulative human impacts will some day push ecosystems over one or more thresholds, resulting in the collapse of a bundle of ecosystem services (Peterson et al. 2003a).

The true state of affairs probably lies somewhere between these two extremes and will differ for different ecosystem services. Current understanding suggests that there are high levels of uncertainty concerning the relative magnitude of human impacts on ecosystems, that rates of habitat destruction and species extinctions are higher than they have ever been in the history of humanity (McNeill 2000), and that ecosystem services may be intricately linked to one another in surprising or unforeseen ways. For example, in Australia, deforestation has led to the unexpected rise of a saline water table, severely affecting food production (Keating et al. 2002).

One of the most worrying aspects of the loss and modification of natural habitats is that we risk damaging our own life-support systems irreversibly. This is particularly true in situations where cross-scale interactions (and other kinds of nonlinearity) are possible. Cross-scale interactions occur from broad scales to fine scales, and vice versa. For example, a broad-scale process such as the formation of clouds may be tightly linked to a fine-scale process such as evapotranspiration (Heck et al. 2001; Wang and Eltahir 2000). Rainfall affects the moisture that is available to plants, driving evapotranspiration. At the same time, increases in evapotranspiration make the air more humid, affecting circulation patterns and potentially making rainfall more likely. Although we typically assume that the broad-scale process drives (or constrains) the small-scale process, this is not necessarily the case in every instance or at all times. Small-scale disturbances can affect broad-scale processes either by individual action (for example, a single highway blocks an important migration corridor for the Florida black bear) or, more commonly, by the combined effects of small-scale contagion (for example, a single lightning strike starts a fire that burns a vast area of forest).

Cross-scale interactions occur between fine- and broad-scale processes, as in the rainfall-evapotranspiration example. Where the effect influences the cause, these interactions are termed cross-scale feedbacks. Cross-scale feedbacks often start with large-scale stressors (such as droughts, glaciers, or floods) that cause local ecosystem change. Local change leads in turn to a contagious spread of ecological responses that collectively cause an upscaling of the problem. Positive feedback loops, in which fine- and broad-scale processes amplify one another, can lead to escalating changes. For example, Foley et al. (2003) and Higgins et al. (2002) explore the ways in which land use and land cover change may affect the global climate. (See Box 3.1.)

A second example of a cross-scale feedback involves schistosomiasis, a debilitating parasitic disease in the Lake Malawi area (Stauffer et al. 1997). Until the early 1990s, schistosomiasis was thought to be absent from Lake Malawi. By 1994, however, nearly 80% of all schoolchildren evaluated had schistosomiasis. The change in human schistosomiasis levels was caused by an increase in the abundance of snails, the intermediate hosts of the *Schistosoma* parasite, in the nearshore regions of the lake. Snail populations increased following a decline in the fish that preyed on them, which in turn occurred as a result of introductions of nonnative fish and intensified fishing. Ironically, intensive fishing was facilitated by a program that was intended to protect local people from malaria-carrying mosquitoes, when mosquito nets were converted to fishing nets by enterprising fishers.

3.3 Why We Need to Develop Ecological Scenarios

3.3.1 Ecological Critique of Existing Scenarios and Statement of What Value We Add

Chapter 2 of this volume presents the motivations for developing scenarios and the main tenets of scenario building.

BOX 3.1

Green Surprises: Climate, Ecology, and Carbon

The dynamics of terrestrial ecosystems can have both physical and chemical influences on climate. Albedo, which is the proportion of incident radiation that is reflected by Earth's surface, is modified by changes in land cover. For example, ice caps and bare sand (high albedo) tend to reflect far more radiation than a multilayered tree canopy and understory (low albedo). Changes in albedo affect energy exchange between atmosphere and land, and so they can modify surface temperatures. Changes in surface temperatures affect heat transfer via evapotranspiration and air movement. Surface structure also affects air currents, altering the movement and mixing of the atmosphere near Earth's surface. Rougher surfaces produce more mixing, and so cool Earth's surface more effectively. Changes in vegetation, such as deforestation, affect both albedo and surface structure and thus can influence climate.

Chemical transformations in terrestrial ecosystems influence the climate by changing the composition of the atmosphere. The amount of carbon absorbed by the biosphere is the difference between the amount of carbon plants absorb through photosynthesis and the amount released to the atmosphere by plant and microbial respiration. Disturbances such as fire, wind, and insect outbreaks, in conjunction with human modification of land cover, alter carbon absorption and respiration and frequently release additional carbon. The terrestrial biosphere appears to have acted as a net carbon sink for the last few decades, absorbing nearly 20% of anthropogenic emissions (Prentice et al. 2001). Whether terrestrial ecosystems continue to provide this service will depend on land use and land cover change, as well as on changes in climate and atmospheric CO_2 concentrations.

The cumulative impacts of local changes in land cover can combine to produce regional or global changes (Brovkin et al. 2004). Agriculture currently occupies about 35% of Earth's land surface, and deforestation continues to reduce the area of the world's remaining large areas of forest. The impacts of these patterns on regional climate (and hence on the dynamics of terrestrial ecosystems) are not well understood. However, the presence of climate-vegetation feedbacks creates the potential that changes in land cover may trigger a cascade of biophysical feedbacks to climate. For instance, the northward expansion of forest could decrease the albedo of northern areas, enhancing global warming (Levis et al. 2000).

Modeling studies have shown that anthropogenically forced climate change could cause the biosphere to switch from being a net sink to a net source of CO_2, further accelerating the process. This positive feedback would occur as a consequence of changes in rainfall that reduce forest productivity and increase soil respiration, causing some regions to switch from being sinks of CO_2 to sources (Cox et al. 2000).

The relative importance of the interaction between biogeochemical and biophysical processes appears to vary by region. Changes in terrestrial ecosystems may either dampen or amplify the effects of anthropogenic climate change (Foley et al. 2003). An integrated model that includes both biogeochemical and biophysical feedbacks shows that deforestation in the tropics tends to result in warming due to biophysical feedbacks, while boreal deforestation tends to result in cooling due to biogeochemical feedbacks (Claussen et al. 2001). These loops suggest that climate-ecosystem feedback processes may act to resist change driven by external forcing, but when change occurs it can be abrupt and surprising, as positive feedback processes move regional climate and vegetation away from its historical state. There is some evidence that Amazonian deforestation may be approaching such a threshold (Rial et al. 2004).

As it makes clear, although there are a number of detailed, carefully constructed global scenarios in existence, their focus is largely on social, economic, and immediate environmental issues. Scenarios are usually designed to differ in a way that is important to the issue being addressed (van der Heijden 1996; van Notten et al. 2003). Where the central issue relates to the environment, however, previous scenarios have tended to downplay the importance of ecosystem dynamics. Environmental changes, as distinct from ecosystem dynamics, are incorporated in many existing global scenarios. They are explicitly included in the biodiversity scenarios of Sala et al. (2000) and the Intergovernmental Panel on Climate Change's global climate change scenarios (Watson and Team 2001). They are implicitly incorporated as drivers of societal change in most of the Global Scenarios Group scenarios (Raskin et al. 2002).

While the IPCC's global emissions take into account global feedbacks between climate, land use, and emissions, the many complex feedbacks that characterize real ecosystems (Higgins et al. 2002) are not explored or tested in detail in existing global scenarios. Such feedbacks can result in nonlinear system behaviors that differ profoundly from those of models that do not include feedbacks. Local ecosystem feedbacks may be important for global processes in several ways. (See Box 3.2.) In general, although previous scenario exercises have been environmentally aware, they have largely ignored the role of ecological feedbacks. A more detailed discussion of the role of ecology in previous global scenario exercises is presented in Cumming et al. (in press).

Ecological feedbacks matter for global scenarios because the continued provision of ecological services is central to human well-being. Ecological science has demonstrated that certain kinds of anthropogenic ecological change can radically transform the ability of ecosystems to provide ecosystem services (Turner et al. 1993). The unintended consequences of attempts to increase food production include alterations to rainfall patterns, increases in soil erosion and populations of agricultural pests, introductions of new diseases, and reduced water quality and quantity. These changes represent feedbacks from one part or scale of the ecosystem to another; perturbations in one part of the system are translated by ecological dynamics into environmental changes. Ecological feedbacks have the potential to become important drivers of human action, although they may be difficult to include quantitatively in scenario exercises because their likelihood and their strength are uncertain (Bennett et al. 2003).

From the perspective of scenario development, one type of ecosystem change that is of greatest concern involves regime shifts. These occur when an entire system flips into an alternative stability domain or stable state (Scheffer et al. 2001). Such changes occur rapidly and are often a strongly nonlinear response to gradual changes in the variables that

BOX 3.2

Local and Global Ecosystem Feedbacks

One of the central messages of this chapter is that ecosystems are active entities that can cause extensive changes in socioeconomic systems. Although the relevance of ecosystems for human societies at a local scale is well documented, the relevance of ecology at a global scale (and hence for global scenarios) is less obvious. We suggest that ecosystem feedbacks matter most for global scenarios when they are cumulative, nonlinear, or interactive.

Cumulative feedbacks. Small-scale changes become broad-scale changes when they are sufficiently widespread. For most socioeconomic drivers, such as stock markets and human population growth rates, local fluctuations have little significance until a global trend emerges. The same is true of ecological drivers like deforestation or infectious diseases. Local ecosystem changes that are usually considered to have local impacts will have global impacts if a global trend develops. Global scenarios will need to consider the overall scale of ecological feedbacks, rather than continuing to categorize all ecosystem feedbacks as local.

Nonlinear feedbacks. Gradual changes in ecosystems may either elicit gradual and corresponding changes in global systems (linear responses) or appear to have no effect until a threshold is crossed and a strong, relatively sudden global response occurs (nonlinear responses). System responses that are strongest at particular scales are typically nonlinear. For example, the total area that a fire can burn from a single ignition is a nonlinear function of the number of flammable habitat patches in the landscape. The potential exists for small-scale ecosystem feedbacks to have disproportionately large effects, particularly as the effective scale of the problem changes from one hierarchical level to another. One of the challenges for global scenarios is to try to determine where important thresholds are located and at what point the net impact of local and regional feedbacks becomes global in nature.

Interactive feedbacks. Ecosystem feedbacks have the potential to interact with one another and to compound socioeconomic problems at global scales. Ecosystem feedbacks are unlikely to occur singly or in a way that is independent of context. For example, extensive clearing of woodlands for sheep pastures in Australia has resulted in a rise of the saline water table and a widespread soil salinity problem. Technological replacement of the ecological service of groundwater regulation (through pumping and water recycling) has been expensive and has reduced the profitability of farming in the area, with ramifications for social and economic systems that were already in a state of flux.

Although it is currently difficult to make rigorous predictions in each of these areas for most ecosystem feedbacks, we can at least conclude that local ecosystem feedbacks with effects that are cumulative, interactive, or nonlinear will lead to greater uncertainties in global models. One of the challenges for the future quantitative development of global scenarios will be to establish which ecosystem feedbacks are significant enough to warrant inclusion in global and regional models and which can safely be ignored.

have defined a system's stability domain. The stability domain can be conceptualized as a cup in which the ecosystem moves like a rolling ball. If the shape of the cup is altered, or the ball is knocked hard enough by some external agent, the system can escape its current stability domain and enter a new state. For example, gradual increases in phosphorus levels in a shallow lake can result in a regime shift that propels the lake from a clear water system to a turbid water system in a relatively short period of time (Carpenter et al. 1999b). Regime shifts have been documented in lakes, woodlands, deserts, coral reefs, and oceans (Scheffer et al. 2001). They are of high importance for scenarios because they usually have large impacts on ecosystem services and human well-being; because they often occur rapidly and with little warning, making them hard to predict and manage; and because they may be irreversible or extremely expensive to reverse, raising the possibility of long-term ecosystem degradation and the effective loss of ecosystem services.

The kinds of ecological feedbacks that either maintain system stability or result in regime shifts are incompletely understood. Three key areas in which further understanding would be valuable are the connectivity of ecosystem services, the role of cross-scale connections, and the question of what determines ecological resilience. It is unclear to what extent ecosystem services can be examined in isolation or should be considered as a coproduced ensemble. For example, differentiation in species (biodiversity) is the basis for variety in ecosystem services (Kinzig et al. 2001), because no single organism provides all ecosystem services; groundwater depletion may affect certain components of the biota but leave others untouched; and many organisms may be able to cope relatively well with a change in the variability of the global climate. Some case studies of trade-offs between different ecosystem services are described in Chapter 12.

Cross-scale interactions complicate the analysis of ecological feedbacks. Systems may be highly resilient to human impact at one scale and very brittle at others. It is difficult to establish definitively the probability and plausibility of different scenarios without a more comprehensive understanding of the cross-scale properties of resilience, a topic that is currently a frontier in ecological research. The ways in which ecological processes interact to determine ecological resilience are poorly understood, and we have little ability to identify, detect, or monitor changes in the resilience of ecosystems—especially in the face of novel types of disturbance (Carpenter et al. 2001).

Difficulties also emerge in assessing the contribution of ecological feedbacks at a global scale. Global systems are hugely complex and may be highly resistant to change. It took a large amount of research to demonstrate that anthropogenic activities have influenced carbon dioxide levels in the atmosphere, and even more to demonstrate that these changes have altered global temperatures (Weart 2003). The significance of the contribution to global processes made by any system component, including ecosystems, is difficult to establish because of the multivariate complexity of the global system. Furthermore, many ecological processes are essentially homeostatic within a wide range of conditions, serving to regulate global systems and maintain their stability, and hence there may be few global signals that can be strongly linked to ecosystem change unless a major regime shift occurs. The lack of extensive evidence for global ecological feedbacks should not, therefore, be interpreted as implying that no such feedbacks exist.

One of the main themes of several previous global scenarios is that people will respond to environmental change rather than ignoring it. Societal responses to change, and the potential for changes in human values, are of central importance in the future. Existing values may themselves be either highly resilient or refreshingly adaptive. For example, the development of China's current forestry policy was initiated in part by severe dust storms in the seat of government, Beijing (Zhang et al. 2000; Zhuang et al. 2001).

If governments continue to act only when such clear and unmistakable signals of ecological degradation become apparent, then changes to the status quo may only come through an ecological perturbation of a magnitude greater than anything that we have yet experienced—which implies that the change in values that would be necessary for the formation of a true ethic of sustainability might only occur in response to the local destruction of a significant component of the environment. This Catch-22 situation is not explicit in the scenario literature and may make some existing global scenarios excessively optimistic. On the other hand, remedial actions have been taken rapidly in the past without a fundamental change in values. For example, the specter of a hole in the ozone layer led to the speedy adoption of the Montreal Protocol, which restricted the use of ozone-destroying chemicals (Beron et al. 2003; Mullin 2002; Powell 2002), although this response occurred in conjunction with the understanding that the industry creating the problem would in turn be the one to most profit from its solution.

Adoption of the Montreal Protocol required a rapid and coordinated institutional response. It was fortunate that the capacity for such a response existed. The capacity of institutions to respond to ecological change across a variety of scales is central to the creation of global scenarios. A mismatch in the scales of social, economic, and ecological processes may create inconsistencies in scenarios. For example, a true reformation of global markets (to incorporate evaluations of ecosystem services) would require a substantial rearrangement of the institutions responsible for managing economies and ecosystems. This would have to occur at many different scales, from global to local. Scenarios that envisage the emergence of ecological sustainability under the invisible guidance of the market would require that local institutions were able to manage ecosystem processes at the appropriate scales in order to avoid the kinds of scale mismatch problems described here (Spaargaren and Mol 1992).

To some extent, the idea that development will result in ecological concerns being addressed assumes that there are relatively direct and manageable links between ecological cause and effect that allow the causes of negative ecological outcomes to be identified and addressed. Ecological science suggests that while this is true in some cases, frequently ecological causation is complex, with impacts being produced at locations distant in time and space from their causes. If such complex causation is common, the costs of reactive approaches to ecological issues are likely to be high.

The capacity of institutions to respond to change often depends on the resources that are available to them and hence may be linked to the wealth of the society in which they occur. Economists have proposed that an inverted U-shaped relationship exists between the per capita income of a country and the environmental impact of its economic activities. This relationship, which has aroused controversy, is termed the Environmental Kuznets Curve after the relationship that the economist Kuznets found between income inequality and per capita income (Canas et al. 2003).

The model underlying the Kuznets Curve views environmental services as a luxury. It assumes that when people are poor the environment provides many services, and that development represents a trade-off between ecosystem services and economic growth; as societies become richer and can afford to sustain a wider range of environmental services, they are assumed to invest more money in environmental protection. Empirical research has supported the environmental Kuznets model for a number of regional atmospheric pollutants (Selden and Song 1994), but not for many other types of environmental degradation (Agras and Chapman 1999; Arrow et al. 1996; Dietz and Adger 2003). Gergel et al. (2004) show that an Environmental Kuznets Curve may be more likely for phenomena that are ecologically reversible or of concern for human health. Our current understanding of the Environmental Kuznets Curve may simply reflect the absence of studies designed to distinguish between alternative hypotheses; for instance, Bruvoll and Medin (2003) identify a range of other relevant covariates that do not necessarily parallel economic growth.

3.3.2 Value of Ecological Scenarios

Ecological scenarios will take into account existing ecological knowledge while recognizing the uncertainties that are present in any complex scientific analysis. One of the greatest contributions to be made by ecological scenario exercises will be to thoroughly work through the likely impacts of current resource exploitation and habitat conversion on the long-term sustainability of future human societies. Various estimates of the current human impact on Earth suggest that it is impossible to greatly expand human consumption of ecological production (Haberl et al. 2002; Rojstaczer et al. 2001; Vitousek et al. 1986). Wackernagel et al. (2002) have estimated that humanity is already exceeding the carrying capacity of the biosphere.

Contrary to the assumptions made by many economic models, continued increases in the production of ecosystem services over the long term will simply not be possible. Global ecological scenarios will highlight the regions in which declines in ecosystem goods and services are most likely to have significant impacts on human health and well-being. In particular, they will clarify the trade-offs that must be made between ecological, economic, and social capital, while also identifying the key ecological processes and ecological thresholds that can be used to guide policy responses. We also anticipate that ecological scenarios will aid in the development of appropriate measures (indicators) of change, relevant monitoring programs, and realistic policy and management goals by working through the range of multivariate interactions that may occur between ecosystems and people.

3.4 Relevant Ecological Theories and Ideas for Global Scenarios

Ecosystems have been defined by Tansley (1935) as "the fundamental concept appropriate to the biome considered together with all the effective inorganic factors of its environment." (See also the MA Glossary in Appendix D.) Ecology is defined as the study of the distribution and abundance of organisms (Andrewartha and Birch 1954) or, more broadly, as "the scientific study of the processes influencing the distribution and abundance of organisms, the interactions among organisms, and the interactions between organisms and the transformation and flux of energy and matter" (IES 2004).

Ecology has produced numerous ideas that are relevant to the development of global scenarios. The ecological theories and ideas that are of the greatest importance for global scenarios are those that relate to global processes and broad-scale spatial and temporal patterns. They can be categorized in four main groups: fundamental frameworks that underpin ecological thinking; community ecology theories, especially those that relate specifically to biodiversity, abundance, and other aspects of community composition that are of particular importance in the provision of ecosystem services; landscape and ecosystem ecology theories, especially those that deal with broad-scale spatial patterns and the movements of organisms or substances; and a more general set of ideas relating to prediction, forecasting, and uncertainty. This section summarizes the theories and ideas from ecology that we consider most important for global scenario exercises. In each instance we explain a little about the theory and its relevance to the development of scenarios. This list is necessarily incomplete; in particular, we have focused on ecological theories that are not described in detail elsewhere in this volume. We conclude this section with a short discussion of some of the topics that we still need to learn more about.

3.4.1 Fundamental Frameworks

3.4.1.1 Evolution

The theory of evolution is the central organizing idea in biology (Mayr 1991). We understand and interpret the diversity of organisms in the world according to the principles of descent, variation, and selection. Evolution gives us numerous insights into the nature of change in the natural world. Macroevolution (the origin and extinction of species) and microevolution (the adaptation of species to their environment) are both important to ecological scenarios.

Although the rate of speciation is slow by comparison to the time frames for which we design global scenarios, the rate of extinction is not. Rates of extinction currently exceed rates of speciation by around four orders of magnitude (Lawton and May 1995). This asymmetry in evolutionary processes explains why ecologists have become increasingly concerned about the recent accelerations in extinction rates; once lost, species that perform particular functions cannot be replaced at time scales that have any meaning for human society. Despite recent advances in genetics and biotechnology, we do not consider it plausible that these technologies will be able to restore extinct species effectively within the time period of this assessment. The many failures of attempts at species reintroductions from small numbers of captive individuals serve to underline this point.

Microevolutionary theory describes how natural selection drives changes in species attributes. Through natural selection, changes in the environment can produce changes within populations of species over a relatively short time period. Consequently, microevolutionary theory is particularly important for understanding possible changes in the behavior of short-lived species in response to anthropogenic modification of their environment, such as in predicting how ecological change will alter the epidemiology of disease (Anderson and May 1991; Daily and Ehrlich 1996) and how changes in ecosystems will affect the evolution of agricultural pests and their predators (Conway 1997). For example, many species of pests have rapidly evolved resistance to pesticides and appear to be evolving resistance to transgenic crops that incorporate the organic insecticide *Bacillus thuringiensis* (Wolfenbarger and Phifer 2000). Numerous examples of microevolution also exist for plants.

Few scenario exercises have considered either macro- or microevolution directly. Speciation usually occurs slowly enough that it is not perceived as relevant over the time scales of most assessments. However, since evolution is the sole mechanism for the replacement of biodiversity, reductions in genetic diversity (including species extinctions and loss or reduction of distinct populations) are of extremely high concern. Microevolution is also an important issue for scenarios in which organisms with short life spans and relatively simple genomes may play an important role. In particular, the emergence of new infectious diseases and more virulent or drug-resistant pathogen strains has the potential to influence global scenarios, not only because of the possibility for the occasional massive epidemic but also because pathogen microevolution places a continuing burden on the economies of developing nations. (See Box 3.3.)

3.4.1.2 Hierarchy Theory

The issue of scale lies at the center of ecology (Levin 1992). As an ecosystem is examined at larger or smaller scales, the apparent magnitudes and rates of ecological processes change. The relationships between pattern (variation, heterogeneity) and process may also change as a function of scale. Hierarchy theory offers a way of organizing and visualizing the world as a series of scale-dependent units (Allen and Starr 1982).

The units that make up hierarchies are typically ordered from big to small or from fast to slow. In most hierarchies, the general principle applies that "upper levels constrain, lower levels explain." In other words, the mechanisms that

BOX 3.3
Ecology of Emerging Infectious Disease

The ecology of infectious disease has shaped human history. Diseases can have large effects on human populations, and humans have often facilitated the emergence of new diseases (McNeill 1976). Human diseases are concentrated in the tropics; about 75% come from other animals (Taylor et al. 2001). Disease has both direct and indirect social and economic consequences. For example, malaria kills and incapacitates millions of individuals every year and greatly reduces the economic growth of countries where it is endemic (Sachs and Malaney 2002). Understanding epidemiology depends not only on medicine and molecular biology, but also on disease ecology: the ways in which transformation of ecosystems alters the distribution and abundance of pathogens. Human interactions with ecosystems have changed over time through four main eras of disease (McMichael 2004):

- Agriculture brought people in close contact with domestic animals, such as cows and pigs, and parasitic species that occupied agricultural settlements, such as lice and rats. This contact provided the opportunity for the ancestors of many of the pathogens that cause disease (such as influenza, tuberculosis, leprosy, cholera, and malaria) to adapt from their animal hosts to infect humans.
- Conflict and trade among civilizations in Eurasia connected populations, allowing the spread of epidemic disease, and began a process that led to the co-evolution of people and their pathogens. Many epidemic diseases became endemic diseases, and urban populations developed disease resistance.
- European colonization connected diverse populations more tightly, spreading infectious disease to people with little previous exposure and causing horrific epidemics of measles, smallpox, and influenza on small oceanic islands, Australia, and most famously in the Americas. These epidemics affected entire civilizations and facilitated the European colonization of the temperate Americas, Australia, and New Zealand (Crosby 1986).
- Over the twentieth century, the expansion and increasing mobility of the human population produced a globalized community of pathogens. The ecology of infectious disease is currently being shaped by four main drivers: land use and land cover change, urbanization, human migration and trade, and diet.

People alter their disease environment in many ways, of which road construction, water control systems, and the conversion of forest to agriculture are of particular importance (MA *Current State and Trends,* Chapter 14). These ecological changes affect the abundance and distribution of both pathogens and their hosts, changing the timing and location of encounters between people and pathogens and altering disease dynamics.

Many emerging infectious diseases have spread from their animal hosts to people as people have cleared disease-rich tropical forests. Clearing disrupts existing host-parasite interactions and encourages the selection of strains suited to new, human-dominated environments by increasing the exposure of people and their domestic animals to diseases (Daszak et al. 2001). For instance, deforestation has coincided with increases in malaria in Africa, Asia, and Latin America. This increase is due in part to the creation of new areas of mosquito habitat in cleared land (Patz et al. 2004), as has occurred during the expansion of irrigation in India (MA *Current State and Trends,* Chapter 14). Leaky irrigation systems increase standing water, fields are often leveled to improve production, and irrigation has raised the water table (Tyagi 2004). Roads further impede the flow of water, creating pools of standing water that can increase populations of disease-transmitting mosquitoes and snails; people are also more likely to come into contact with water near roads and may encounter or introduce pathogens as they enter new areas (Patz et al. 2000).

Urbanization is an important component of recent patterns of land use/land cover change. The world's urban population has been steadily increasing since reaching 1.7 billion in 1980 and is expected to reach 5 billion by 2030. At this time, 30% of humanity is projected to be living in cities of more than 5 million people. (See Chapter 7.) Drainage and water supplies are critical factors that determine the extent to which many diseases are either contained or propagated in urban communities. A combination of poverty and rapid, unplanned growth of urban populations can produce high-density areas that lack infrastructure for the safe storage and distribution of water and the drainage of wastewater. Failure to collect garbage increases the number of small pools of water that provide habitat for mosquitoes and can, for example, lead to epidemics of dengue fever (Patz et al. 2004).

Tourists and business travelers can carry infectious diseases from one region of the world to another, as has been the case with AIDS and SARS. The introduction of new diseases and new disease-transmitting organisms into a region is a form of "pathogen pollution" that places an increased pressure on public health efforts (Daszak et al. 2001). Furthermore, some researchers have suggested that today's rapid changes in the distribution of pathogens could favor the evolution of virulent diseases (Ewald 1994).

Human dietary demands and production practices can also influence disease emergence. Bushmeat hunting—the commercial hunting of wild animals—has lead to outbreaks of Ebola and monkeypox and has been linked to the emergence of HIV 1 and 2 (MA *Current State and Trends,* Chapter 14). Human-animal contact in production systems has been implicated in the emergence of Avian flu and SARS. Feeding herbivores to other herbivores that humans then eat has further contributed to the emergence of diseases of both livestock and humans. In recent times, one of the most notable of these has been the prion-caused bovine spongiform encephalitis, which manifests itself as Creutzfeldt-Jakob disease in humans and is thought to have been caused by feeding cattle with protein obtained from sheep with scrapie. All these emerging diseases have been facilitated by an increasing societal demand for meat.

explain a particular event usually originate at smaller scales (faster rates, smaller areas) while the potential of a particular unit is constrained by the levels above it in the hierarchy. For example, outbreaks of spruce budworm (a herbivorous caterpillar) in the northeastern United States and Canada can result in the defoliation and subsequent death of spruce trees over large areas. The lower-level mechanisms that explain budworm population dynamics are the reproductive rates of budworms and predation by birds (Holling 1988). Once the spruce trees grow sufficient foliage to provide protection from predation for the budworms, the budworm population can increase rapidly. The amount of food available for budworms to consume and the degree to which they are protected from predation by slow-growing foliage act as upper-level constraints on the ultimate size of the budworm population (Holling 1986).

For global scenarios, awareness of the hierarchical arrangement of the world is essential. In many cases the regional properties of a particular location will constrain possible events at that location. Crop production in higher latitudes is constrained by the number of growing days in a season; growth rates of small towns are constrained by the national economy; and innovative management of natural resources may be constrained by tenure and property systems that operate at a higher hierarchical level than the individual. Many of the impacts in which we are interested involve top-down effects, such as where changes in a nation's economy can influence small-scale mining activities in remote regions (Heemskerk 2001). There are also bottom-up effects where the cumulative impacts of small-scale changes result in changes at larger scales. Examples include the effects of individual car engines on the gas composition of the atmosphere, the fragmentation of landscapes by individual clear-cuts, and, in political systems, individual discontent rising in turbulent revolutions such as those in China and East Germany (Kuran 1989). It is essential that global scenarios are not naive about the possibilities for cross-scale effects, meaning both that such effects are invoked only where they are plausible and that their potential as agents of sweeping change is not ignored.

As an ecosystem changes, its dynamics vary in rate. Periods of slow accumulation of natural capital, such as biomass or soil, are interrupted by its abrupt release and reorganization (Holling 1986). Ecological disturbance releases natural capital that was tightly bound in accumulations of biomass and nutrients. Rare events, such as hurricanes or the arrival of invading species, can unpredictably shape structure at critical times or at locations of increased vulnerability. As resources enter and leave the system, and as system components enter new relationships with one another, ecological innovation can occur.

This dynamic tension between growth and destruction, between stabilization and disruption, appears to represent a key aspect of ecological dynamics. Stabilizing forces (those that push a system toward an equilibrium) maintain productivity and biogeochemical cycles. Destabilizing forces (those that push systems away from equilibrium conditions) serve to maintain diversity and create opportunity by removing portions of a population, reducing competition, making habitats available for colonization, and creating new niches (Gunderson and Holling 2002). For example, organisms may take advantage of unusual climatic events, fluctuating habitat conditions, or predator-free environments to achieve rapid increases in numbers (e.g., Bakun and Broad 2003). Similarly, forested areas that are cleared by fires or landslides offer opportunities for early successional species.

From the perspective of the MA, the key aspect of this conceptualization of ecological dynamics is that the connections between an ecosystem and the context in which it is embedded will change over time. Although ecosystems are typically constrained by top-down processes, there will be some periods during which they are vulnerable to disruption from bottom-up change (Peterson 2000b). A small-scale disturbance can trigger a larger-scale collapse if the larger system is vulnerable to disturbance. The introduction of shrimp into lakes in the Columbia River Basin provides an example of a small event triggering large-scale reorganization. The shrimp have caused the reorganization of the lake and surrounding ecosystems, as salmon populations and the species feeding upon them have declined and been replaced by bottom-feeding fish (Spencer et al. 1991).

As an ecosystem reorganizes following a disturbance, the remaining ecosystem legacies and surrounding large-scale systems provide the components and constraints out of which a system reorganizes. For example, the 1934 destruction of a dam on the Salmon River allowed salmon from neighboring watersheds to colonize the restored river and establish new populations (Wilkinson 1992). Without the maintenance of source populations in neighboring watersheds, recolonization would have been extremely unlikely.

We are not aware of any previous scenario exercises that have explicitly considered hierarchy theory. However, choosing spatial and temporal scales for analysis is a continual issue in any modeling exercise. Since processes at different scales can interact with one another in complex and unexpected ways, awareness of the hierarchical arrangement of ecosystems is essential for scenario exercises. Hierarchy theory will also provide the conceptual basis for models that predict the cumulative effects of local ecosystem feedbacks.

3.4.2 Theories from Community Ecology

3.4.2.1 Island Biogeography

Since the early work of Darwin and Wallace, island communities have been used as model systems in ecology (Quammen 1996). The theory of island biogeography has been the inspiration for many of the quantitative approaches currently used in population and landscape ecology. MacArthur and Wilson (1967) noted that small oceanic islands tended to have communities that were composed of a subset of the species that were present on nearby mainland areas. They argued that community composition would be limited by the dispersal ability of its constituent species; poor dispersers would not be able to travel from mainland to island. As the distance of islands from the mainland increased, colonization by new species would become increasingly less likely. Similarly, species living on larger islands would be able to maintain larger populations and would be less likely to become extinct.

MacArthur and Wilson (1967) proposed that the community of species living on an island would be determined by the balance that was reached between the processes of colonization and extinction. They argued that island size was the principal determinant of the overall species extinction rate on the island and that the distance of the island from the mainland was the prime factor driving colonization. According to their framework, variations in these two factors would explain differences in community composition among islands. While a number of the extensions of this theory (such as the importance of the arrival sequence of new species on an uncolonized island) have been contested, its basic predictions have been strongly supported.

Island biogeography was one of the first formal, quantitative recognitions of the role of space and dispersal in determining community composition. Many theories that are currently used to predict the long-term persistence of communities rely on the same basic mechanisms of reproduction and distance-dependent dispersal. Islands of species habitat are not identical to oceanic islands because the degree of isolation is less, the area surrounding a patch of habitat on the mainland is likely to be habitable by many terrestrial species, and changes in terrestrial vegetation will not present the same type of barrier as the ocean provides to the dispersal of terrestrial species. Despite these differences, however, the basic tenets of island biogeography have been used to predict species richness and changes in biodiversity on continents as well as on oceanic islands (e.g., Davis et al. 2002; Fragoso et al. 2003; Lomolino and Weiser 2001; Sanchez and Parmenter 2002).

Global scenarios inevitably involve changes in the location and spatial pattern of human settlement and either the destruction or restoration of natural areas. Some areas are naturally patchy while others are naturally continuous but may be fragmented by humans. Island biogeography tells us how different populations of organisms will respond to these different conditions as a function of their dispersal ability and their proximity to potential sources of colonization. There are numerous models that allow quantitative estimation of the likelihood of population persistence in patchy landscapes (Bascompte 2003; Husband and Barrett 1996; Wennergren et al. 1995).

In scenarios, recognition of the impacts of land cover change on the distribution and abundance of species is integral to making connections between economic and social changes and likely changes in the provision of ecosystem services. This point is further elaborated in Chapter 10, where the species-area relationship and its relevance for the estimation of biodiversity are described in detail. Island biogeography makes it clear that scenarios must consider not only the amount of habitat change, but also its spatial pattern, since equivalent amounts of habitat reduction that occur in different spatial configurations can have very different implications for the provision of ecosystem services.

As far as we are aware, island biogeography has not been incorporated in previous scenario exercises. The relevance of island biogeography and related ideas (such as metapopulation theory and the design of corridors and reserve networks) for studies of ecosystem function is becoming increasingly apparent as humans fragment systems that were formerly continuous (Sanderson et al. 2002). Predictions about the sustainability of biodiversity and the continued provision of ecosystem services in fragmented landscapes will have to rely on island biogeography theory. Island biogeography and its offshoots will also provide the bridge for linking broad-scale satellite remote sensing assessments of land cover change directly to populations, communities, and ecosystems. Although the methods are at an early stage, Chapter 10 raises the possibility that future scenario exercises will be able to link quantitative simulations of land cover change to changes in biodiversity and the provision of ecosystem services.

3.4.2.2 Disturbance, Succession, and Patch Dynamics

One of the debates that has surrounded studies of small islands is whether they are more vulnerable to disturbances, in which case they would have an elevated extinction rate and should contain fewer species than expected from their location (Herwitz et al. 1996; Jones et al. 2001; Komdeur 1996; Whittaker 1995). Island biogeography recognizes disturbance as a major influence on ecosystems. The importance of disturbance has also been apparent in studies of vegetation succession as an answer to the question of why old-growth forests tend to be highly diverse instead of dominated by a single, highly competitive species.

The continual disturbance of areas within the boundaries of a particular ecosystem or community creates a mosaic of vegetation patches, each at different stages of succession. Successional processes create predictable temporal changes in communities, where hardy earlier colonizers are gradually replaced through time by slower-growing competitors (Vanandel et al. 1993). The spatial and temporal diversity that is produced by disturbance and succession allows a range of species to survive within the system (Levin 1992), even though individual patches may tend toward homogeneity. The development of patchiness (heterogeneity, variation) within an ecosystem, and the ways in which patches change through time and interact with one another, is termed patch dynamics.

Disturbance, succession, and patch dynamics are integral components of ecosystems. Human managers are often uncomfortable with processes that are not strongly regulated or controllable. Consequently, many management strategies have resulted in reductions in the number, intensity, and duration of natural disturbances such as floods, fires, and pest outbreaks. The net consequence of such decreases in natural disturbances is frequently to create a system that becomes increasingly vulnerable to other kinds of disturbance (Holling and Meffe 1996). For example, fire suppression in the United States in the middle of the last century allowed fuel loads to increase beyond their normal densities, resulting in huge and potentially catastrophic fires.

For global scenarios, it is important to recognize not only that disturbance regimes are integral parts of ecosystems, but also that systems tend to cope well with some kinds of disturbance but not others. Feedbacks may occur between the properties of landscapes and the kinds of disturbance that they experience. Disturbance regimes and their interactions with ecosystems can be major sources of surprises and shocks in scenario storylines.

Broad-scale ecological disturbances have been considered as drivers of change in previous scenario exercises. However, the focus of these analyses has typically been on anthropogenic drivers of change, such as CO_2 emissions, and abiotic responses, such as changes in the frequencies of extreme rainfall events or hurricanes. Ecosystems are often perceived as dependent on socioeconomic forces rather than as independent systems that can cause change in their own right. This perspective ignores the degree to which ecosystem characteristics influence their susceptibility to disturbance. For example, changes in albedo in Alaskan and Australian habitats can influence the frequency of lightning strikes that forests and grasslands experience and hence the number of fires that occur (Bonan et al. 1995; Higgins et al. 2002; Kasischke et al. 1995; Lafleur and Rouse 1995;

Rabin et al. 1990). The vital role of natural disturbance regimes and patch dynamics in maintaining biodiversity and the continued provision of ecosystem services, as well as the two-way interactions between ecosystems and disturbance regimes, have not been considered in depth in any global scenario exercise.

3.4.2.3 Food Webs, Bioaccumulation, and Trophic Cascades

Each species within an ecosystem eats and is eaten by a limited set of the other species within an ecosystem. This network of feeding relationships constitutes a food web. The relative position of a species within a food web is termed its trophic level. Photosynthetic organisms, which receive their energy directly from the sun, are at the lowest trophic level. Trophic levels increase as organisms become more removed from primary production. Organisms at different trophic levels play different roles in ecosystems. Species at lower trophic levels tend to be abundant producers; those at higher trophic levels tend to be rarer and to act more as regulators of other populations.

The trophic level of a species predicts, to some extent, the response of the entire system to changes in the population of that species. For example, overfishing of Caribbean coral reefs has lowered populations of many herbivorous fish species. When sea urchin populations were suddenly affected by pathogens and hurricanes around the same time as several coral bleaching events, many reefs became dominated by algae (McClanahan et al. 2002). Changes in the composition and abundance of species at the top of the food web can have consequences that resonate through the food web in surprising ways (Pinnegar et al. 2000; Schmitz 2003; Snyder and Wise 2001). Species at high trophic levels are often large, long-lived predators with slow population growth rates. A decline in the populations of these species can initiate a trophic cascade, in which the abundance of species at lower levels of the food web increases as they are released from predation and species in the next lower trophic level in the food web are suppressed (Carpenter and Kitchell 1993b).

The trophic level of a species in the food web can also be used as a guide to the vulnerability of that species to contaminants in the food web. For example, mercury is concentrated in living tissue as it moves up the food web. Small fish in a lake may be unaffected by their low concentrations of mercury, but birds that eat piscivorous fish will accumulate a much higher level of mercury (Brant et al. 2002). This process is called bioaccumulation and is often associated with contaminants that are fat-soluble. Species at the top of a food web are more vulnerable to bioaccumulation than those at lower trophic levels. Humans, for example, are potentially vulnerable to the consumption of biomagnified contaminants that have accumulated in farmed salmon (Fairgrieve and Rust 2003).

An understanding of food web interactions, the feeding relationships between organisms, is important for global scenarios because disruption of individual food web components may have surprising effects on other organisms. For example, the removal of birds from a system can lead to increases in the abundance of the insect species on which they feed, resulting in pest outbreaks and reduced productivity of agriculture (Battisti et al. 2000; Crawford and Jennings 1989; Mols and Visser 2002). Similarly, the removal of large predators has resulted in increases in herbivore densities in many areas, reducing densities of plants (Terborgh et al. 2001).

Many trophic interactions have immediate importance for humans. An interesting example comes from the c. 50-year periodic masting (flowering and fruiting) of *Melocanna bambusoides* bamboo plants in India. The ready availability of nutritious bamboo seeds after masting events can lead to rapid increases in rodent populations. High rodent abundance creates a subsequent problem for farmers, whose grain crops are vulnerable to rats once the brief pulse of bamboo production is over. Plagues of rats associated with bamboo masting have been blamed for famines in the northeastern state of Mizoram in 1861, 1911, and 1959 (John and Nadgauda 2002).

Trophic cascades and food web dynamics have entered into previous global scenarios where depletion of food stocks has been considered important, particularly in scenarios that have considered fisheries; but in general, the potential for nonlinear food web change (and its impacts) has been ignored.

3.4.3 Systems Approaches: Landscape Ecology and Ecosystem Ecology

Landscape and ecosystem ecology are focused on the study of broad-scale processes and patterns in ecology. Ecosystem ecology has traditionally focused on the movements of matter and energy through ecosystems. It has already made many important contributions to global scenarios, including models that describe fluxes of carbon, nitrogen, and phosphorus from soils through plants, animals, and decomposers.

Landscape ecology has been less on the agenda of scenario planners, although it also has potentially valuable contributions to make. One of its central tenets is the idea that the locations at which ecosystem processes occur and the spatial relationships between locations are important. Although habitat amount is of primary importance in determining the size and ultimately the persistence of populations, habitat arrangement becomes increasingly more important as habitat is lost (Flather and Bevers 2002). Percolation theory (Stauffer 1985) and neutral landscape models (Gardner et al. 1987) predict that the ease of movement of animals through a given habitat type should follow a logistic function, with a rapid decline in connectivity once 30–50% of habitat is lost (Plotnick and Gardner 1993). Recent studies have suggested that habitat arrangement may also affect equilibrial population densities (Cumming 2002; Flather and Bevers 2002) and predator-prey dynamics (Cuddington and Yodzis 2002). Ecosystem services are provided by populations of organisms. Consequently, global scenarios that seek to link ecosystems and human well-being will have to take into account the potential for local extinctions and population changes as a consequence of habitat arrangement, not just as a function of habitat amount.

The impacts of anthropogenic activities can be pervasive at broad scales. For example, Forman (1999) has estimated that up to a fifth of the United States is affected by roads. The relevance of this kind of habitat fragmentation will differ for different species, depending on their dispersal capabilities and habitat requirements (Poiani et al. 2000). Animals perceive and move through landscapes at distinct scales that relate to their body size (Holling 1992; Roland and Taylor 1997); habitat fragmentation is likely to have different effects on animals at different trophic levels. Larger terrestrial species will have larger home ranges and require more habitat; the relatively coarse grain at which they perceive the landscape suggests that they will be among the first species to be affected by habitat fragmentation. However, larger species may also be less vulnerable to predation and more capable of traveling through areas of suboptimal habitat (as witnessed by the persistence of the Florida black bear, for example).

Anthropogenic changes in landscapes have altered the ways in which plants and animals disperse. Human modification of the landscape has separated areas that were formerly continuous. For example, roads and cities create barriers to dispersal for a variety of organisms, forestry clearcuts and agricultural land conversion may disrupt landscapes that were formerly continuous, and impoundments reduce the connectivity of streams and lakes. Humans have also created novel connections between ecosystems that were formerly separate. Roads and trade (both terrestrial and marine) have resulted in the translocation of many species into new habitats, with huge consequences for people and the world's ecosystems (Crosby 1986).

Understanding the flows of energy, material, and organisms across landscapes integrates ecosystem and landscape ecology. In many instances, the continued provision of ecosystem services in a given area is dependent on exchanges of organisms, substances, or materials with other areas. This effect is termed a spatial subsidy. For example, many cities are built on the banks of large rivers. The continued provision of water by the river depends on the spatial subsidy provided by the upper watershed. Changes in the upper watershed, such as deforestation or increased numbers of livestock, can result in changes in the quality and quantity of water provided downstream, as well as affecting siltation, nutrient influxes to floodplains, and eutrophication of lakes. In small oceanic islands, soil fertility may be maintained by dust blown in from mainland areas (Chadwick et al. 1999), and recolonization by tree species after a hot fire may depend on dispersal from nearby forests.

There are a number of approaches to thinking about these kinds of phenomena, including ideas about boundaries and flows of substances and organisms through landscapes (Cadenasso et al. 2003); the spread of invasions (Muller-Landau et al. 2003), colonization, metapopulation, and island biogeography; and biogeochemical cycles that describe the movements of essential substances (such as water, carbon, calcium, nitrogen, and phosphorus) through ecosystems (Krug and Winstanley 2002; Newman 1995; Schimel et al. 1991; Singh and Tripathi 2000). Subsidies may also be temporal, such as through seeds that are stored in the soil.

Global scenarios will inevitably depict a variety of spatial patterns of anthropogenic activity and different degrees of infrastructure development, human settlement, and urbanization. They will also vary in levels of resource exploitation and the ability of communities or governmental organizations to cope with the management issues that are raised by changes in ecosystem connectivity. As connectivity between different areas changes, ecological processes will be influenced by increases or decreases in the variety and amount of spatial subsidies that they receive.

The responses of people to such changes may in turn create either positive or negative feedbacks between management actions and ecosystem services. For example, declines in water quality and the increased likelihood of flooding in rivers such as the Yangtse have largely been blamed on environmental changes in the upper catchment. The Chinese response to this problem has taken several forms. The ban on logging on the Tibetan plateau, which will serve to stabilize soils and improve water quality, may result in further positive feedbacks toward ecological enhancement (Zhang et al. 2000). By contrast, the construction of large impoundments such as the Three Gorges Dam is likely to create further ecological and social problems (resettling 2 million people, creating an impassable barrier for fish and mussel species, altering the natural variability of the downstream flow regime, affecting coastal fisheries and food security) while potentially solving the problem of flooding. This type of destabilizing ecological feedback, in which the anthropogenic modification of one set of ecological subsidies alters another set , can have important implications for ecological scenarios. Ecological trade-offs are described in more detail in Chapter 13 of the MA *Multiscale Assessments* volume.

Although global climate models and emissions scenarios have taken account of spatial patterns and flows of substances, in general the roles of ecosystem subsidies and changes in the configuration of habitats have been ignored in global scenario exercises.

3.4.4 Prediction, Forecasting, and Uncertainty

Human action now dominates the dynamics of many ecosystems. People generally make decisions based on their current knowledge and their expectations about the future. The heterogeneity, nonlinear dynamics, and cross-scale feedbacks that occur within ecosystems make ecosystem behavior difficult to predict (Holling 1978; Levin 1999). Although management decisions are often constrained by the amount of information that is available about the system, monitoring is frequently perceived as an irrelevant or excessively costly activity. In reality, people seldom have enough information to make reliable forecasts of ecosystem behavior (Sarewitz et al. 2000).

Even in situations where large amounts of data exist and there are relatively reliable and accepted ecosystem models, unexpected environmental variation can falsify predictions. Exogenous variables, such as changes in climate or disturbance regimes, can have enormous impacts on ecosystems and are often difficult to predict with great accuracy. For example, the El Niño-Southern Oscillation is a global weather pattern driven by the interaction between the ocean and the atmosphere in the central and eastern Pacific. ENSO alternates on a two- to seven-year period and exerts a strong influence on the productivity of fisheries in the eastern Pacific (Bakun and Broad 2003). Although our understanding of ENSO events is improving, it is still difficult to make precise and accurate predictions about its onset and impacts.

A considerable amount of variation in different variables can also be generated by processes that are endogenous to ecosystems. For instance, relatively high levels of variability in the relationship between phosphorus and chlorophyll production in freshwater lakes can be generated by changes in food web structure (Carpenter 2002), and ecosystem-climate coupling can produce complex behaviors in weather systems (Higgins et al. 2002). Predictions about social systems may also be falsified by both exogenous and endogenous drivers. For example, fluctuations in the global market (an exogenous driver) can have unexpected effects on local communities, and the formation of new political organizations (an endogenous driver) can result in broader societal change, such as when the organization of rubber tappers in the Amazon stimulated new approaches to forest management.

The uncertainty associated with ecological statements about the future is seldom evaluated in a rigorous manner. In particular, the problem of model uncertainty is often ignored in ecology, even though statistical methods are available to address the issue (Clark et al. 2001). Rigorous evaluation of the uncertainty associated with an ecological prediction usually indicates that a forecast is quite uncertain, meaning that it assigns roughly equal probability to a wide range of different outcomes. The weaknesses of ecological predictions are typically exacerbated by their reliance on drivers that are difficult to predict, such as human behavior. The reflexivity of human behavior further constrains the reliability of ecological predictions (Funtowicz and Ravetz 1993); if predictions are made and taken seriously, people will change their actions in response to the predictions, making accurate forecasts difficult (Carpenter et al. 1999a). For example, a coordinated global response to climate change could make current predictions based on high-emissions scenarios incorrect.

Despite the difficulties of producing reliable forecasts, people need to make decisions about the future. Carpenter (2002) suggests that three ways in which science can contribute to decision-making include obtaining a better understanding of ecological thresholds and dynamics, assessing uncertainty more rigorously, and using scenarios as tools for thinking through the possible consequence of decisions and the ways in which unexpected events may influence their outcomes. The narrative form of scenarios makes them more accessible than many other kinds of scientific information. Their accessibility provides a forum for dialogue between scientists, the public, and decision-makers, which can be useful for addressing complex issues of high public concern (Funtowicz and Ravetz 1993; Kinzig et al. 2003).

Questions of prediction, forecasting, and uncertainty have been major concerns in several past scenario exercises, most notably the IPCC scenarios (Nakićenović and Swart 2000). These questions relate more to the applications of scenario planning than to their internal consistency. One of the main benefits of attaching estimates of uncertainty to the events that are envisaged in scenarios is that uncertainty estimates give scenario users an indication of the degree of scientific confidence in individual forecasts. The risk of presenting uncertainty estimates is that they may become an excuse for failing to act. In general, since the risks of mismanaging ecosystems are so large, the precautionary principle should be applied (Harremoes et al. 2001); managers should try, where possible, to keep systems well clear of key thresholds that might lead to ecosystem degradation.

3.4.5 The Application of Ecological Theories in Scenarios

The ecological theories just described are relevant to global scenarios. We envision that they will be applied in different ways and at different scales of analysis in different contexts. Evolution and hierarchy theory provide a basic context for thinking about ecosystems and their interactions with social systems. Evolution (including the study of the fossil record) offers a long-term perspective on environmental change and the ways in which species responded to it in the past and provides a frame of reference for thinking about how species may respond in the future. Microevolution is a likely source of ecological feedbacks, particularly those relating to pests and pathogens. Hierarchy theory is relevant in any context in which some kind of change in space or time is posited. Hierarchies offer a structured approach to problems of scale and for thinking about the interactions of processes and patterns that occur at the same scales or different ones. Scenarios will need to justify the lower-level mechanisms that create system changes and to take into account the upper-level constraints on what is possible. Coping with the concept of scale and the dynamics that are generated by cross-scale interactions, particularly the possibility for broad-scale regime shifts, remains a major challenge for scenario development.

Theories from community ecology, ecosystem ecology, and landscape ecology are especially relevant for scenarios that incorporate anthropogenic impacts on the environment. They will be applicable in situations where humans extract resources, alter the flows and movements of energy and materials, or change land use or land cover. Hunting and fishing, logging, fruit and nut extraction, and other activities that have focused impacts on particular components of the ecosystem will set in train a series of knock-on effects that may be transmitted through the food web. Alterations in temperature and rainfall regimes will have profound effects on nutrient cycles, the domain of ecosystem ecology. Changes in land use and land cover will affect the broader-scale context in which communities of organisms live and may disrupt processes such as migration and gene flow. Although many of these feedbacks are typically characterized as local, they may have larger impacts under certain conditions.

3.4.6 What Don't We Know?

Consideration of uncertainty makes it clear that in many cases we know less than we think we do. There are also areas of ecology about which we are spectacularly ignorant or whose true importance we have only recently started to recognize. One of the most critical areas for global scenarios concerns the connections between ecosystem patterns and ecosystem processes. Current global models for ecological variables produce estimates of changes in patterns based on a mixture of correlative and mechanistic understanding. We urgently need better models that link likely changes in landscape patterns to likely changes in essential ecosystem processes, including nutrient cycles, primary production, and community dynamics (such as predator-prey cycles, trophic cascades, and pest outbreaks).

The loss of species and the functions that they perform is closely related to changes in habitat. There has been considerable debate over the question of whether higher species diversity results in greater community stability and/or resilience (Ives and Hughes 2002; Ives et al. 2000; Pimm 1984). Relevant questions include whether more diverse communities are better able to survive extreme disturbances, whether more diverse communities are more vulnerable to invasion by introduced species, and whether ecosystem function is more likely to be maintained in a diverse community, assuming that diversity includes "redundant" species that perform similar functions to one another but have different environmental tolerances (Huston 1997; Loreau et al. 2001; Naeem 2002; Tilman et al. 1996; Walker et al. 1999). Furthermore, there is little understanding of how changes in the interactions between species at different scales influence ecosystem function (Peterson et al. 1998).

Our current understanding of ecology is also weak in the area of long-term and large-area ecological dynamics (Carpenter 2002). Studies of ecological processes at very large spatiotemporal scales are rare, partly because the necessary data are so hard to obtain. Our understanding of the relative importance of different variables may change when analyses are undertaken at broader scales. For example, local studies of the Caribbean Sea often ignore the impact of the Amazon and Orinoco outflows on water quality and circulation patterns (Hellweger and Gordon 2002). At broad scales, the magnitude and even the direction (positive versus negative correlation) of relationships that have been established at finer scales may change (Allen and Starr 1982). Broad-scale processes are of high importance for global scenarios because they often provide the slowly changing variables that can force ecosystems from one state to another (Bennett et al. 2003).

3.5 Placing Ecology in a Socioeconomic Context

Ecological knowledge arises and is applied in a socioeconomic context. In each of the MA scenarios, the causes and consequences of ecological change depend on the nature of the interactions between ecosystems and socioeconomic systems. The scenarios explore not only the importance of ecological dynamics for human societies, but also the consequences of alternative approaches to ecosystem management. Approaches to ecological management can be organized using the concepts of uncertainty and controllability.

3.5.1 Ecological Uncertainty and Control

In human societies there are different ways of knowing, ranging from the formal structures of science to less formal knowledge systems such as customs and traditions. Regardless of the variety of knowledge in question, however, knowledge is used whenever decisions are made. People who make decisions about natural resource management generally take into account both what they know and what they are capable of achieving. It is difficult to track ecological knowledge through global scenarios, but it is clear that ecological knowledge is more likely to increase in scenarios in which people work with ecosystems and have structured ways of learning from their experiences. (See MA *Multiscale Assessments*, Chapter 5.)

High levels of uncertainty correspond to a lack of ecological knowledge and hence an inability to predict future aspects of system behavior. The degree to which aspects of system behavior are predictable affects both the likelihood that a given management action will achieve its desired aim and the ease of obtaining social approval for the action to be taken. Where uncertainty is high, costly interventions are less likely to be approved and a command-and-control management approach is unlikely to be successful. Depending on the context, high levels of uncertainty may have different effects on management. Uncertainty can lead to inaction because it can be hard to determine the best course of action when uncertainty is high. Uncertainty can also provide opportunities that inspire action by fostering the belief that the future is malleable and that desired futures are attainable (Ney and Thompson 2000). Last, uncertainty can encourage humility and tolerance, because managers and stakeholders are ignorant of what the future will bring and may find that the plans and beliefs of others are more effective or correct than their own.

The controllability of ecological processes by management actions depends on aspects of both the ecosystem and the social system. Available technologies influence the controllability of ecosystems; for example, it is currently easier to add nutrients to a system than to remove them, to control access to an island than to an offshore fishery, and to monitor and regulate a stream rather than groundwater. A second component of controllability is the willingness and ability of a group of people to coordinate their ecological management actions. Changes in ecological controllability can occur due to social change, such as increased agreement on what constitutes fair or good management, or changes in technologies relating to ecosystem services (e.g., Kiker et al. 2001). Throughout history, groups of people have organized ways of managing water, game, and fisheries (Berkes 1999), with varying degrees of success. Governments have frequently expropriated resources from local people and then been unable to manage the resources effectively, as a consequence of passive or active resistance to their policies. The increased level of interest in community-based conservation in recent decades is largely due to the failure of coercive and nonparticipatory environmental management practices (Agrawal and Gibson 1999).

The appropriateness of a given approach to ecosystem management depends largely on the degree of uncertainty about a system's behavior and the degree to which the system can be controlled. (See Figure 3.1.) Optimizing approaches make sense when a system is controllable and known. Resilience-building approaches to management are more appropriate when a system is difficult to control but understanding of its dynamics is high. When understanding is lacking, learning-based approaches are appropriate. If control is possible, adaptive management can be useful; if control is difficult, however, more exploratory and dialogue-centered techniques are likely to be needed and the focus shifts from ecosystem management to societal adjustment. Many of our most pressing environmental problems, such as concern over the ecological impacts of transgenic organisms or the local impacts of climate change, are situations in which control is difficult and uncertainty is high. These problems appear best suited to open ecological management practices that engage an extended community in defining and analyzing the socioecological context (Funtowicz and Ravetz 1993).

3.5.2 Command and Control

Managers have historically tended to view ecosystems as places in which isolated, individual provisioning ecosystem services exist and can be enhanced. This optimization approach has largely been implemented via the goal of "maximum sustained yield." The MSY approach combines quantification and technical understanding with command-and-control management to attempt to produce the maximum achievable continuous supply of an ecosystem service.

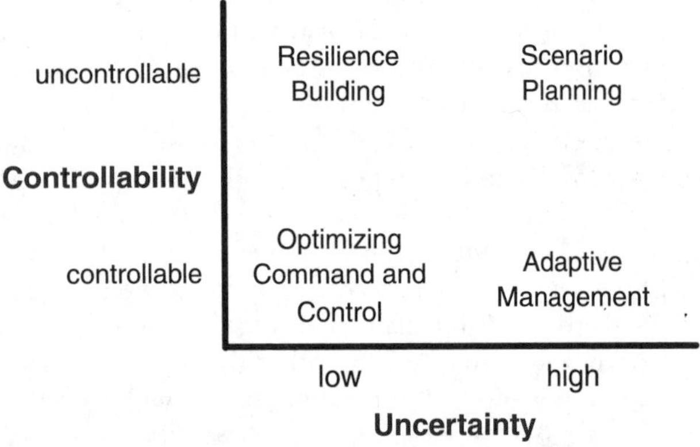

Figure 3.1. Uncertainty and Controllability in Ecological Management. Ecological management situations can be represented in a two dimensional space defined by the uncertainty that surrounds our knowledge of the system and the degree to which the system is controllable by management. (Adapted from Peterson et al. 2003b)

It has been the guiding philosophy of agriculture, forestry, hunting, and fishing.

Fisheries management provides many rich examples of MSY applications. The concept of MSY in fisheries was developed in the early twentieth century but was formalized, extended, and extensively applied following World War II (Clark 1985; Ricker 1975; Schaefer 1954). MSY approaches were largely based on fitting a population growth curve using estimates of current numbers of fish and their reproductive rates and then setting a level of exploitation that maximized the biomass or the monetary value (or some other criterion determined by the manager) of the catch. Difficulties in measuring fish populations, identifying stocks, enforcing regulations, and coping with environmental variation all present challenges to the MSY approach, as does managing political intervention in the process of setting sustainable catches.

Despite good progress in fisheries stock assessment in addressing many of these challenges (Hillborn 1992), it is difficult to find a case where MSY fisheries management has unequivocally succeeded. Indeed, the concept of MSY appears to be an idea that is more resilient than the fisheries it has been used to manage. For example, Larkin (1977), a prominent fisheries scientist, argued nearly 30 years ago that MSY should be abandoned because it risks the catastrophic decline of populations, it fails to recognize the role of trophic interactions, and it is not necessarily desirable in economic terms. Despite these warnings and the poor track record of MSY, it has continued to dominate fisheries management.

More generally, the command-and-control approach adopted by MSY views ecological management as a straightforward process of problem definition, solution development, and solution implementation (Holling and Meffe 1996). Solutions are expected to be direct, appropriate, feasible, and effective over relevant scales. Command and control is expected to solve the problem either through control of the processes that lead to the problem (such as hygiene to prevent disease) or through mitigation of the problem after it occurs (such as pathogens killed by antibiotics). A command-and-control approach assumes that the problem is well bounded, clearly defined, relatively simple, and follows a linear or nearly linear relationship between cause and effect. Most of the problems with command and control arise when it is applied to complex, nonlinear systems that show low levels of predictability. Unfortunately, many ecosystems (and most ecological problems) fit this description. Societal recognition of the weaknesses of command and control approaches to natural resource management, and the degree to which the search for alternatives is successful, is a key aspect of the MA scenarios.

3.5.3 Managing for Resilience

Managing for resilience is intended to increase the ability of a system to cope with stress or surprise. It is an approach that has been advocated in situations where control is difficult but where there is understanding about how the system works. This approach has arisen in response to failures of command-and-control management. It is based on the argument that rather than maximizing production of individual ecosystem services, the central goal of ecological management should be to maintain a range of supporting and regulating ecosystem services to ensure the reliable supply of provisioning services. Resilience theory offers a framework for understanding the supporting and regulating systems that maintain ecosystem organization (Holling 1973; Peterson et al. 1998).

The aim of management for resilience is to maintain ecosystems that can persist despite environmental changes, management mistakes, and unexpected events (Gunderson and Holling 2002). Managers can do so by enhancing the ecosystem services that regulate and maintain the ecosystem. For example, lakes in the U.S. Midwest can be managed for resilience by the manipulation of lake food webs. Many of the agricultural areas in the Midwest have experienced large increases in soil phosphorus, and lakes in the region are vulnerable to eutrophication from high-nutrient runoff. Controlling runoff is very difficult. An alternative approach to coping with the increased stress on lake ecosystems is to increase lake resilience to phosphorus loading. This can be done by ensuring that lakes have a robust food web that includes substantial populations of piscivorous (fish-eating) fish (Carpenter and Kitchell 1993). An increase in these leads, via a trophic cascade, to increases in populations of the large herbivorous zooplankton that prey on lake algae. Increases in zooplankton populations decrease the likelihood that increased phosphorus loading will tip lakes into an alternate state where undesirable algal blooms occur.

The ability of a service to persist depends heavily on its response diversity—the variation of responses to environmental change among species that contribute to the same ecosystem service (Elmqvist et al. 2003). Increasing response diversity, such as by allowing the recovery of a diverse set of fish species with different responses to environmental change, can further increase resilience. While ecosystem management is increasingly aimed at managing for resilience, the capacity for managers to do so has been limited by a lack of models and tools for understanding resilience in ecosystems (Carpenter 2002). Socioecological researchers are actively working to fill this gap (Berkes et al. 2003).

3.5.4 Adaptive Management

A second alternative to command and control is adaptive management. This is a systematic process for continually improving management policies and practices by learning from the outcomes of operational programs. It is particularly appropriate when there is uncertainty about how an ecosystem functions and managers have some ability to manipulate the environment.

Adaptive management regards policies as alternative hypotheses and management actions as experiments (Holling 1978; Lee 1993; Walters 1986; Walters and Hilborn 1978; Walters and Holling 1990). This approach is very different from the typical "informed trial-and-error" approach, which uses the best available knowledge to generate a risk-averse, "best guess" management strategy that is only changed as new information becomes available. Practicing adaptive management involves identifying uncertainties and then establishing ways to reduce them. It is a tool not only to change the system, but also to learn about the system. The key scientific and social aspects of adaptive management include the following: a link to appropriate temporal and spatial scales; a focus on statistical power and controls; use of computer models to synthesize and build an embodied ecological consensus; use of embodied ecological consensus to evaluate strategic alternatives; and communication of alternatives to the political arena for the negotiation and selection of a management action (Holling 1978; Lee 1993; Walters 1986). In its strongest form—"active" adaptive management—interventions are designed to experimentally evaluate alternative hypotheses about the system being managed (e.g., Prato 2003).

Adaptive management is particularly useful in situations where management intervention is possible and there is a focus on the development of scientific knowledge for ecological intervention. These processes are appropriate in social contexts where technical understanding is used as the basis for ecosystem manipulation, but they are less likely to be successful in situations where ecological dynamics are not considered in decision-making or where ecosystem manipulation is unfeasible.

3.5.5 Social Learning

The degree to which learning, adaptation, and innovation can occur in socioecological systems shapes the ability of that system to cope and respond to the emergence of poorly defined and understood ecological problems. Resilience theory identifies three types of social learning (Gunderson and Holling 2002): incremental, lurching, and transformational.

Incremental learning occurs during phases of gradual system change. In this instance, the process of learning involves the collection of data or information to refine existing models. It is based on the assumption that models of how the world works are structurally correct, but imprecise. Incremental learning is similar to the process of single-loop learning described by Argyris and Schoen (1978). In many cases, organizations view this type of change and learning as problem solution (Westley 1995).

Lurching learning is episodic, discontinuous, and surprising. It often occurs when a system changes, making inadequacies in a previously acceptable model more apparent. For example, inadequacies in food production and distribution systems often emerge during drought years. Lurching learning is frequently stimulated by an environmental crisis that makes policy failure undeniable (Gunderson et al. 1995). In this case, where the underlying model is questioned and rejected, the learning process is described as double-loop (Argyris and Schoen 1978). It is also characterized as problem reformulation. In organizations, lurching learning is frequently facilitated by outside groups or charismatic leaders.

Transformational learning is the most profound type of learning and is often a consequence of dramatic system changes. It is characterized by the emergence of novel or unexpected outcomes from complex, nonlinear, and/or cross-scale interactions. Transformational learning involves the identification of the variables that define the domain, in multivariate space, of the system of interest (Ludwig 2001). Defining or bounding variables are typically broad-scale and slow to change. For example, phosphorus levels in sediments are a bounding variable for lake eutrophication (Carpenter et al. 1999b). Transformational learning occurs via the assimilation of knowledge about slowly changing variables into the views of managers and policy-makers, including recognition of the possibility that slow variables may create surprises (such as the nonlinear shift from clear to turbid lake water). Examples of transformational learning in the ecological sciences include the discovery of and response to the Antarctic ozone hole and the discovery of the bioaccumulation of DDT and the resulting control of the use of DDT and other persistent organic pollutants. Transformational learning has also been described as evolutionary learning (Parsons and Clark 1995), where not only new models but also new paradigmatic structures are developed (Kuhn 1962). Transformational learning differs from double-loop learning in that it involves substantial alterations to a dominant worldview.

Social learning processes allow groups of people to develop new adaptive responses to various types of surprising situations. Consequently, the possibilities for social learning present in each scenario will determine the capacity of people to respond to ecological surprises.

3.6 Ecosystem Management and Economics

Ecological and environmental economics have sought to understand how economies shape people's interactions with ecosystems, and how economic incentives can be used to improve ecological management. The complexities of both human behaviors and ecosystems make the application of economic theory to economic management difficult. For example, indicators commonly monitored by governments are unlikely to accurately reflect ecosystem resilience (Deutsch et al. 2003). Three of the more active areas of research at the nexus of ecology and economics are the use of economics to improve the efficiency of ecological management, the assessment of the value of ecosystem services to improve decision-making, and conflicts over property rights to nature.

3.6.1 Economics and Ecology

Understanding human behavior is important for natural resource management because ecology alone is insufficient to explain the dynamics of human-dominated ecosystems. Humans, individually or in groups, can anticipate and prepare for the future to a much greater degree than ecological systems can (Brock and Hommes 1997; Westley et al. 2002). Human views of the future are based on mental models of varying complexity and completeness. People have developed elaborate ways of exchanging, influencing, and updating their mental models of both the past and the future. Individual and societal perspectives on the future can create complicated dynamics that are influenced by access to information, ability to organize, and power.

By contrast, although some components of ecosystems are capable of "anticipating" future changes—for instance, many bats undergo reproductive delays that allow their offspring to be born at a time of year when food is abundant (Bernard and Cumming 1997)—the behavior of ecological systems is based primarily on the past. Ecological dynamics are the products of the mutual reinforcement of many interacting structures and processes. The behavior of ecosystems emerges from the successes of past evolutionary experimentation at the species level. The fundamental differences between human and ecological behavior mean that understanding the role of people in ecological systems requires not only understanding how people have acted in the past, but also what they think about the future.

Many economic ideas have been applied to understanding the dynamics of socioecological systems. One key distinction that economists have drawn from coupled economic-ecological models is the need to consider that economic activity anticipates the future. Economic criticisms of the early global environmental modeling work *The Limits to Growth* (Meadows et al. 1972) argued that the conclusions were flawed because people's views of the future were not incorporated into the models. Specifically, economists argued that people will shift their spending and investments as their perceptions of their current and future situation change. Many global models of ecosystem and economic change have not included these dynamics.

The economic concept of rational expectations proposes that people's actions are based on what they think will change in the future and how other people will respond to those changes. If people's behaviors are based on their expectation of what will happen, and this expectation is based on a prediction of the behaviors of other people, then when the world is well understood, such expectations will cause individual behaviors to converge rapidly. However, when the world is poorly understood many possible behaviors become equally likely. Consequently, when the world is unknown and difficult to understand, the consequences of individual rational behavior can make the future more difficult to predict (Brock and Hommes 1997). Game theory is one area of economics that allows for an analysis of this type of socioecological dynamic, but game theory for ecological management is still at an early stage (Brock and de Zeeuw 2002; Roth 2002; Supalla et al. 2002).

Another key insight from economics is the value of markets for distributing knowledge, observations, and decision-making for ecological management (Scott 1998). The successful development of markets for ecosystem services is an exciting advance in economics that shares important similarities (and some differences) with the economics of public and club goods, such as policing and intellectual property rights. Recent work has examined the design of markets for pollution emissions and genetically modified crops (Batie and Ervin 2001).

3.6.2 Valuation of Ecosystem Services

A second area in which the integration of ecology and economics may make a large contribution to scenarios is through the valuation of ecosystem services. Many decision-making processes involve some sort of cost-benefit analysis that attempts to convert all costs and benefits into a single currency that can easily be compared. Doing this with ecosystem services is difficult, however, as most ecosystem services are not traded in markets and consequently do not have prices. To improve people's ability to evaluate ecological management decisions involving a mix of market and nonmarket values, it is sometimes useful to illustrate the nonmarket economic value of services provided by ecosystems. Valuation is not necessary for many types of decisions, as people do not place economic values on many things they prize, such as freedom or democracy, and do not necessarily conserve things that they value. The valuation of ecosystem services is difficult, and only appropriate in specific situations, but it can illustrate the value of investing in natural capital. (For more details on ecological valuation, see MA 2003, Chapter 6.)

Some types of evaluation use market prices to estimate the value of ecosystem services. Hedonic prices, travel costs, and replacement costs all use techniques that estimate the marginal value people attach to a service (Heal 2000; Wilson and Carpenter 1999). That is, these approaches estimate how much a small change in the supply of a specific ecosystem service would be worth. They can be difficult to apply in cases where market data are lacking. In such situations, contingent valuation—the statistical analysis of questionnaires that ask people how much they would pay or spend for a specific ecosystem service—is used to estimate the marginal value of services.

Ecological valuation usually differentiates between use and non-use values. Use values derive from the use of a service, such as clean water. Non-use valuation of ecosystem services arises from diverse cultural, religious, ethical, and philosophical sources. Some of these values are strongly held and have endured for centuries. Some have decreased over time, while other new values have emerged. Worldwide concern for animal rights is an example of a relatively new movement that has had major impacts on how many societies view their relationship with animals. Intrinsic values can complement or counterbalance utilitarian values. For example, the Endangered Species Act in the United States is an expression of the view that human action should not directly cause extinction. This value is distinct from the economic value of the species that it protects. Similarly, many people donate money for tiger conservation because they value the existence of tigers in the wild, without expecting that they will derive an economic benefit from the presence of tigers.

Valuation does not solve the problem of who should have rights to use ecosystem services. Nor does it define good management or answer the question of how to construct institutions or markets that provide economic incentives to manage ecosystem services well (Martinez-Alier 2002). These issues, along with the technical and definitional problems surrounding ecosystem services, can lead to large differences between the values placed on an ecosystem service by different parties (Wilson and Carpenter 1999). Despite these difficulties, even the gradual and partial assignment of new property rights to ecosystem services (such as carbon credits and emissions trading) is likely to have substantial impacts on future scenarios.

One concern is that the partial incorporation of property rights for ecosystem services may have perverse or unexpected impacts on ecological services that remain open access. For example, assigning property rights to forests based solely on their role as producers of timber has been partially responsible for the undervaluation of the many other ecosystem services that forests provide (Scott 1998). However, valuation can provide useful information for dialogues about complex ecological management issues, which may help people develop better assessments of the trade-offs and synergisms among different sets of ecosystem services.

The complex interrelationships of ecosystem components complicate the creation and allocation of property rights that provide social and economic benefits. Ecosystems produce many different services, often at different scales, and the maintenance of ecosystem function may also depend on spatial or temporal subsidies that occur between systems. For example, a forested watershed can simultaneously provide clean water to downstream ecosystems, a habitat for migrating songbirds, and timber for a property owner. Conflicts over ecosystem change and use frequently relate to issues of who should own or control different ecosystem services (Martinez-Alier 2002). These questions are largely political; little economic theory has been developed to cope with them, although there has been substantial research on understanding common pool resources (Committee on the Human Dimensions of Global Change 2002; Levin 1992; Ostrom 2003).

3.6.3 Ecosystem Management and Political Ecology

Political ecology—the study of the relationship between nature and society—arose out of a theoretical need to integrate local situations into a political economy that often transcended the local (Blaikie and Brookfield 1987; Peet and Watts 1993; Schmink and Wood 1992; Watts 1983; Wolf 1972). Its basic theoretical framework encompasses "the constantly shifting dialectic between society and land-based resources, and also within classes and groups within society itself" (Blaikie and Brookfield 1987).

A focus on the structure of human systems has dominated much recent writing about political ecology (Martinez-Alier 2002; Pred and Watts 1992; Rocheleau et al. 1996). These approaches could be described as the political economy of natural resources, rather than political ecology, because they consider ecosystems primarily as passive objects that are transformed by human actors. An ecological political ecology should incorporate the active role of ecosystems as agents of political change, and an understanding of their diversity and dynamics (Peterson 2000a; Robbins 2004). The ecological services and resources that are available at a given time and place determine the alternatives that are available to people. This set of alternatives shapes the politics, economics, and management of ecosystems. However, constraints imposed by ecosystems are fluid, because ecosystems are dynamic and variable.

Ecological approaches to management will be strengthened by an understanding of political dynamics as they relate to human actions. Natural scientists frequently disregard the politics of human societies (Martinez-Alier 2002). This attitude can lead to scientific recommendations that ignore important determinants of human behavior, such as the political forces that influence what and how people learn, the political dimensions that determine which events are considered crises, and what kinds of things are considered to be property. Such blind spots may cause scientists to provide advice or formulate policy that is either spectacularly inadequate or may be open to disastrous misuse (Gunderson et al. 1995; Ludwig et al. 1993). The social consequences of such failures can be severe.

3.7 Application of Theory to Scenario Storylines

The ecological concepts described in this chapter are relevant to the MA scenario storylines in many different ways. A number of valuable insights relating to the role of ecology and ecosystem services in scenario exercises have emerged from the MA process (summarized in Table 3.1). This list is not exhaustive; it is intended as a summary for decision-makers who are wondering why they should be concerned about ecosystems.

Differences in the relationships between people and ecosystems are the main driver of differences among the MA scenarios. Key aspects of the relationship between people and ecosystems include the ways in which people learn about ecosystems, the approaches people take toward ecological management, and the extent to which ecosystem services are incorporated in economies and economics. The nature of ecosystem management will inevitably change as societies accumulate knowledge. Approaches to ecological management depend on people's abilities to control ecosystems as well as their certainty about ecosystem dynamics and their confidence or risk adversity in applying this knowledge. The degree to which future decision-making considers ecological trade-offs will be an important determinant of ecosystem and societal change. The scenarios explore these differences by considering alternative futures under different degrees of societal learning.

3.8 Synthesis

The importance of ecosystems as a sustaining, interactive partner to human social and economic systems emerges strongly from this volume. In Chapter 1 the necessity of ecosystem services for human well-being is described. Society has not always given enough thought to its future need for ecosystem services. In recognition of this failing, the MA scenarios have built on past scenario exercises (see Chapter 2), acknowledging both their strengths and their weaknesses.

In this chapter, we first explain why the future of ecosystem services should be of particular concern as the human population increases and resource scarcity becomes increasingly more likely. Since ecosystem services play an essential role in our societies, greater consideration of ecosystems is needed in policy and management decisions. The rigorous inclusion of ecology in global scenarios is an important step toward bringing ecosystems back onto the stage of global decision-making. Previous scenario exercises have not given ecosystems adequate consideration or recognized the potential for the disruption of social and economic processes that can occur when the flow of ecosystem services is reduced or removed. The discipline of ecology has made considerable progress over the last 50–100 years in developing and testing quantitative approaches and conceptual frameworks that can be useful in assessing and understanding the impacts of anthropogenic modification of our environment, although ecological theory needs further development in many areas to address newly emerging global issues.

Knowledge of ecology is not sufficient on its own to produce effective and sustainable management of natural resources. The future of ecosystems is also dependent on our achieving social, political, and economic awareness of their importance, and on placing ecology in a socioeconomic context, so that decision-makers who are not ecologists can apply ecological theory effectively. The need for interdisciplinary approaches to management and policy decisions that affect multiple spheres is in many ways self-evident. However, achieving the balanced view that we consider necessary for long-term sustainability will require that societies develop the capacity to learn and to adopt flexible management approaches that can be modified as environmental conditions change. Fostering a flexible learning approach is one of the greatest challenges facing managers and policy-makers. Ultimately, although the social, ecological, and economic issues described in this chapter could play out in many different ways in the future, a number of key principles emerge that will be relevant in all instances.

The MA has used many of the same quantitative models (see Chapters 4–7) that have been applied in past scenario exercises, although the MA storylines attempt to introduce a greater awareness of ecological relevance into the process. Unfortunately, the majority of existing quantitative approaches for making socioeconomic projections at broad scales do not explicitly incorporate ecosystem feedbacks. Many of the principles that are described in this chapter are thus applied qualitatively rather than quantitatively in the scenario storylines. (See Chapters 8 and 9.) Ecologists have not always made the relevance of their research clear to practitioners in other disciplines and have frequently been naive about the causes of anthropogenic impacts.

Making detailed projections of the consequences of human impacts on biodiversity is difficult in its own right, and we are far from being able to make similar projections about the impacts of biodiversity loss on ecosystem services. A general principle that emerges from Chapter 10 is that ecosystem services depend on the abundance of individuals in populations of species rather than on simple species pres-

Table 3.1. Relevance of Ecological Principle or Insight to the Development of Global Scenarios

Ecological Principle or Insight	Relevance for Global Scenarios	Illustrative Example
Current rates of change (habitat destruction, extinctions) are extremely high by comparison to historical rates	ecosystems are more likely to be near to boundary conditions than they were historically baseline data collected in the last 50 years do not necessarily reflect unperturbed state	Chapin et al. 2000
Ecosystem services are interdependent	need to consider "bundles" of services and their relevance to society interactions among ecological processes can lead to surprises attempts at making trade-offs between ecosystem services may not be successful	interactions between climate and forests; conflicts over climate regulation, timber production, and harvesting of non-timber forest products
Levels of ecological uncertainty may be higher than traditional models have suggested; thresholds are difficult to quantify precisely	risk associated with different magnitudes of human impacts is uncertain current global models are naive about human impacts on ecosystems	collapse of major marine fisheries (cod, Atlantic salmon, sea turtle) (Jackson et al. 2001; Pauly et al. 1998)
Relationship between biodiversity and ecosystem function is unclear	uncertain whether projected losses of biodiversity will have high or low impact on provision of ecosystem services	trophic cascades in lakes have demonstrated high interconnectedness of aquatic food webs
Many ecosystems exhibit nonlinear dynamics	ecological shocks and surprises are likely to emerge from unexpected threshold effects	shallow lakes can switch rapidly from clear to turbid with a slight, linear increase in P load (Carpenter et al. 1999b)
Cross-scale dynamics, particularly those driven by the interactions of variables with different scale-dependent rates and magnitudes, can produce feedbacks and cascades	ecological impacts and drivers must be considered at a variety of scales context of ecological impacts is key, especially when considering likelihoods of positive vs. negative feedbacks constraints and mechanisms needed to explain storylines will come from different scales human learning about ecosystems is made harder by relevance of large-scale processes and slow variables	multiple small-scale N inputs from farms on the Mississippi are creating dead zone in Gulf of Mexico decline in molluscivorous fishes leading to increases in snails that act as secondary hosts to *Schistosoma* spp. in Lake Malawi; resulting increase in schistocomiasis in human population (Stauffer et al. 1997)
Spatial and temporal variations are essential components of ecosystems	changes in mean trends may be less important than changes in timing and magnitude of variations	major impacts of climate change on biota will come from extremes, rather than from changes in means
Evidence for "ecological Kuznets" is lacking	economic theory cannot be applied indiscriminately to relationship between society and ecological services in scenarios	Bruvoll and Medin 2003
Command-and-control management approaches often decrease system resilience	inherent or unexpected vulnerabilities are more likely to influence storylines in systems where command and control is or has been practiced	fires in California; development of resistant strains of antibiotics (Holling and Meffe 1996)
Successful application of ecology to management/policy depends on political context	scenarios must ensure that political context is appropriate for the ecosystem management actions that are envisaged	Walters (1997) presents a number of examples of situations in which adaptive management has succeeded or failed

ence or absence. Species loss is worrying, but declines in ecosystem services will become evident well in advance of species extinctions. Hence, the management of vital populations of organisms (and the abiotic environment they depend on) may be a more appropriate focus than entire species for decision-makers who are concerned about the contribution of ecosystems to human well-being.

As many of the chapters in this volume make clear, human well-being is intricately connected to the components and functions of ecosystems. Changes in ecosystems are likely to have a number of important impacts on human societies. (See Chapter 11.) Decision-makers must often balance short-term economic or societal gains against long-term ecosystem costs. (See Chapter 12.) By attempting to include ecology in the process of scenario development, we have learned many lessons about the relationship between ecosystem services and human well-being. (See Chapter 13.) These are translated into a set of possible responses and recommendations for policy-makers and managers. (See Chapter 14.)

Although ecology and related disciplines have much to offer in this context, there are a number of areas in which further exploration of ecosystem dynamics would be useful. For example, the ecological scope of the scenarios could have been greatly strengthened if we had a stronger quantitative understanding of such things as diversity-function relationships, the endogenous dynamics of ecosystems and the circumstances under which they cause unexpected perturbations, the role of cross-scale variation in sustainability, and the locations of thresholds in the provision of ecosystem services. The scenarios would also have benefited from more extensive quantification and analysis of the links between resource value, resource use, and resource management.

In conclusion, we have argued that the consideration of ecosystems in scenario exercises and in policy and management decisions is vital to the long-term sustainability of human society. Despite the progress that the MA has made in tackling the complexity of the global socioecological system, it is clear that this volume represents a beginning rather than an end in the ongoing process of learning to manage ecosystems to increase human well-being sustainably.

References

Agras, J. and D. Chapman, 1999: A dynamic approach to the Environmental Kuznets Curve hypothesis. *Ecological Economics,* **28(2),** 267–277.

Agrawal, A. and C.C. Gibson, 1999: Enchantment and disenchantment: The role of community in natural resource conservation. *World Development,* **27(4),** 629–649.

Allen, T.F.H. and T.B. Starr, 1982: *Hierarchy: Perspectives for Ecological Complexity.* The University of Chicago Press, Chicago, 310 pp.

Anderson, R.M. and R.M. May, 1991: *Infectious Diseases of Humans: Dynamics and Control.* Oxford University Press, Oxford.

Andrewartha, H.G. and L.C. Birch, 1954: *The Distribution and Abundance of Animals.* University of Chicago Press, Chicago.

Angelsen, A. and D. Kaimowitz, 1999: Rethinking the causes of deforestation: lessons from economic models. *World Bank Research Observer,* **14(1),** 73–98.

Argyris, C. and D.A. Schoen, 1978: *Organizational Learning.* Addison-Wesley, Reading, MA.

Arrow, K., B. Bolin, R. Costanza, P. Dasgupta, C. Folke, C.S. Holling, B.O. Jansson, S. Levin, K.G. Maler, C. Perrings, and D. Pimentel, 1996: Economic growth, carrying capacity, and the environment. *Ecological Applications,* **6(1),** 13–15.

Bakun, A. and K. Broad, 2003: Environmental "loopholes" and fish population dynamics: comparative pattern recognition with focus on El Niño effects in the Pacific. *Fisheries Oceanography,* **12(4/5),** 458–473.

Bascompte, J., 2003: Extinction thresholds: insights from simple models. *Annales Zoologici Fennici,* **40(2),** 99–114.

Batie, S.S. and D.E. Ervin, 2001: Transgenic crops and the environment: missing markets and public roles. *Environment and Development Economics,* **6,** 435–457.

Battisti, A., M. Bernardi, and C. Ghiraldo, 2000: Predation by the hoopoe (*Upupa epops*) on pupae of *Thaumetopoea pityocampa* and the likely influence on other natural enemies. *Biocontrol,* **45(3),** 311–323.

Bennett, E.M., S.R. Carpenter, G.D. Peterson, G.S. Cumming, M. Zurek, and P. Pingali, 2003: Why global scenarios need ecology. *Frontiers in Ecology and the Environment,* **1,** 322–329.

Berkes, F. (ed.), 1999: *Sacred Ecology: Traditional Ecological Knowledge and Resource Management.* Taylor and Francis, Philadelphia, PA.

Berkes, F., J. Colding, and C. Folke, editors. 2003. *Navigating Social-ecological Systems: Building Resilience for Complexity and Change.* Cambridge University Press, Cambridge, UK.

Bernard, R.T.F. and G.S. Cumming, 1997: African bats: evolution of reproductive patterns and delays. *Quarterly Review of Biology,* **72(3),** 253–274.

Beron, K.J., J.C. Murdoch, and W.P.M. Vijverberg, 2003: Why cooperate? Public goods, economic power, and the Montreal Protocol. *Review of Economics and Statistics,* **85(2),** 286–297.

Blaikie, P. and H. Brookfield, 1987: *Land Degradation and Society.* Methuen, London.

Bonan, G.B., F.S. Chapin, and S.L. Thompson, 1995: Boreal Forest and Tundra Ecosystems as Components of the Climate System. *Climatic Change,* **29,** 145-167.

Brant, H.A., C.H. Jagoe, J.W. Snodgrass, A.L. Bryan, and J.C. Gariboldi, 2002: Potential risk to wood storks (*Mycteria americana*) from mercury in Carolina Bay fish. *Environmental Pollution,* **120(2),** 405–413.

Brinson, M.M. and A.I. Malvarez, 2002: Temperate freshwater wetlands: types, status, and threats. *Environmental Conservation,* **29(2),** 115–133.

Brock, W.A., 2000: Whither Nonlinear? *Journal of Economic Dynamics and Control,* **24,** 663–678.

Brock, W.A. and A. de Zeeuw, 2002: The repeated lake game. *Economics Letters,* **76(1),** 109–114.

Brock, W.A. and C.H. Hommes, 1997: A rational route to randomness. *Econmetrica,* **65(5),** 1059–1095.

Brovkin, V., S. Sitch, W. von Bloh, M. Claussen, E. Bauer, and W. Cramer, 2004: Role of land cover changes for atmospheric CO_2 increase and climate change during the last 150 years. *Global Change Biology,* **10(8),** 1253–1266.

Bruvoll, A. and H. Medin, 2003: Factors behind the environmental Kuznets curve—A decomposition of the changes in air pollution. *Environmental & Resource Economics,* **24(1),** 27–48.

Cadenasso, M.L., S.T.A. Pickett, K.C. Weathers, and C.G. Jones, 2003: A framework for a theory of ecological boundaries. *BioScience,* **53(8),** 750–758.

Canas, A., P. Ferrao, and P. Conceicao, 2003: A new environmental Kuznets curve? Relationship between direct material input and income per capita: evidence from industrialized countries. *Ecological Economics,* **46(2),** 217–229.

Carpenter, S.R., 2002: Ecological futures: building an ecology of the long now. *Ecology,* **83(8),** 2069–2083.

Carpenter, S. R., and J. F. Kitchell 1993: *The Trophic Cascade in Lakes.* Cambridge University Press, Cambridge.

Carpenter, S., W. Brock, and P. Hanson. 1999a: Ecological and social dynamics in simple models of ecosystem management. *Conservation Ecology,* **3:**4.

Carpenter, S.R., D. Ludwig, and W.A. Brock, 1999b: Management of eutrophication for lakes subject to potentially irreversible change. *Ecological Applications,* **9(3),** 751–771.

Carpenter, S.R., M. Walker, J.M. Anderies, and N. Abel, 2001: From metaphor to measurement: resilience of what to what? *Ecosystems,* **4,** 765–781.

Carr, M., 1983: The Influence of Water-Level Fluctuation on the Drift of Simulium-Chutteri Lewis, 1965 (Diptera, Nematocera) in the Orange River, South Africa. *Onderstepoort Journal of Veterinary Research,* **50(3),** 173–177.

Chadwick, O.A., L.A. Derry, P.M. Vitousek, B.J. Huebert, and L.O. Hedin, 1999: Changing sources of nutrients during four million years of ecosystem development. *Nature,* **397(6719),** 491–497.

Chapin, F.S., E.S. Zavaleta, V.T. Eviner, R.L. Naylor, P.M. Vitousek, H.L. Reynolds, D.U. Hooper, S. Lavorel, O.E. Sala, S.E. Hobbie, M.C. Mack, and S. Diaz, 2000: Consequences of changing biodiversity. *Nature,* **405(6783),** 234–242.

Chutter, F.M., 1968: On Ecology of Fauna of Stones in Current in a South African River Supporting a Very Large Simulium (*Diptera*) Population. *Journal of Applied Ecology,* **5(3),** 531.

Clark, C.W., 1985: *Bioeconomic Modelling and Fisheries Management.* John Wiley & Sons, New York, 291 pp.

Clark, J.S., S.R. Carpenter, M. Barber, S. Collins, A. Dobson, J.A. Foley, D.M. Lodge, M. Pascual, R. Pielke, W. Pizer, C. Pringle, W.V. Reid, K.A. Rose, O. Sala, W.H. Schlesinger, D.H. Wall, and D. Wear, 2001: Ecological forecasts: An emerging imperative. *Science,* **293(5530),** 657–660.

Claussen, M., V. Brovkin, and A. Ganopolski, 2001: Biogeophysical versus biogeochemical feedbacks of large-scale land cover change. *Geophysical research letters,* **28(6),** 1011–1014.

Committee on the Human Dimensions of Global Change, 2002: *Drama of the Commons.* National Academy Press, Washington, DC.

Conway, G., 1997: *The Doubly Green Revolution: Food for All in the Twenty-first Century.* Comstock Pub. Assoc, Ithaca, NY.

Cox, P.M., R.A. Betts, C.D. Jones, S.A. Spall, and I.J. Totterdell, 2000: Acceleration of global warming due to carbon-cycle feedbacks in a coupled climate model. *Nature,* **408(6809),** 184–187.

Crawford, H.S. and D.T. Jennings, 1989: Predation by Birds on Spruce Budworm Choristoneura-Fumiferana—Functional, Numerical, and Total Responses. *Ecology,* **70(1),** 152–163.

Crosby, A.W., 1986: *Ecological Imperialism: The Biological Expansion of Europe, 900–1900.* Cambridge University Press, Cambridge, 368 pp.

Cuddington, K. and P. Yodzis, 2002: Predator-prey dynamics and movement in fractal environments. *American Naturalist,* **160,** 119–134.

Cumming, G.S., 2002: Comparing climate and vegetation as limiting factors for species ranges of African ticks. *Ecology,* **83(1),** 255–268.

Cumming, G.S., J. Alcamo, O. Sala, R. Swart, E.M. Bennett, and M. Zurek, in press: Are existing global scenarios consistent with ecological feedbacks? *Ecosystems.*

Daily, G. and P.R. Ehrlich, 1996: Impacts of development and global change on the epidemiological environment. *Environment and development economics,* **1(3),** 311–346.

Daszak, P., A.A. Cunningham, and A.D. Hyatt, 2001: Anthropogenic environmental change and the emergence of infectious diseases in wildlife. *Acta Tropica,* **78(2),** 103–116.

Davis, A.L.V., C.H. Scholtz, and T.K. Philips, 2002: Historical biogeography of *scarabaeine* dung beetles. *Journal of Biogeography,* **29(9),** 1217–1256.

Deininger, K. and H. Binswanger, 1999: The evolution of the World Bank's land policy: Principles, experience, and future challenges. *World Bank Research Observer,* **14(2),** 247–276.

Deutsch, L., C. Folke, and K. Skanberg, 2003: The critical natural capital of ecosystem performance as insurance for human well-being. *Ecological Economics,* **44(2–3),** 205–217.

Dietz, S. and W.N. Adger, 2003: Economic growth, biodiversity loss and conservation effort. *Journal of Environmental Management,* **68(1),** 23–35.

Donnison, A.M. and C.M. Ross, 1999: Animal and human faecal pollution in New Zealand rivers. *New Zealand Journal of Marine and Freshwater Research,* **33(1),** 119–128.

Elmqvist, T., C. Folke, M. Nystrom, G. Peterson, J. Bengtsson, B. Walker, and J. Norberg. 2003: Response diversity, ecosystem change, and resilience. *Frontiers in Ecology and the Environment,* **1**:488–494.

Ewald, P.W., 1994: *Evolution of Infectious Disease.* Oxford University Press, Oxford.

Fairgrieve, W.T. and M.B. Rust, 2003: Interactions of Atlantic salmon in the Pacific northwest V. Human health and safety. *Fisheries Research,* **62(3),** 329–338.

Fearnside, P.M., 2000: Global warming and tropical land-use change: Greenhouse gas emissions from biomass burning, decomposition and soils in forest conversion, shifting cultivation and secondary vegetation. *Climatic Change,* **46(1–2),** 115–158.

Flather, C.H. and M. Bevers, 2002: Patchy reaction-diffusion and population abundance: the relative importance of habitat amount and arrangement. *American Naturalist,* **159,** 40–56.

Foley, J.A., M.H. Costa, C. Delire, N. Ramankutty, and P. Snyder, 2003: Green surprise? How terrestrial ecosystems could affect earth's climate. *Frontiers in Ecology and the Environment,* **1(1),** 38–44.

Forman, R.T.T., 1999: Estimate of the area affected ecologically by the road system in the United States. *Conservation Biology,* **14,** 31–35.

Fragoso, J.M.V., K.M. Silvius, and J.A. Correa, 2003: Long-distance seed dispersal by tapirs increases seed survival and aggregates tropical trees. *Ecology,* **84(8),** 1998–2006.

Funtowicz, S.O. and J.R. Ravetz, 1993: Science for the post-normal age. *Futures,* **25(7),** 739–755.

Gardner, R.H., B.T. Milne, M.G. Turner, and R.V. O'Neill, 1987: Neutral models for the analysis of broad-scale landscape patterns. *Landscape Ecology,* **1,** 19–28.

Gergel, S.E., E.M. Bennett, B.K. Greenfield, S. King, C.A. Overdevest, and B. Stumborg, 2004: A test of the environmental Kuznets Curve using long-term watershed inputs. *Ecological Applications,* **14(2),** 555–570.

Goldman, A., 1995: Threats to sustainability in African agriculture: searching for appropriate paradigms. *Human Ecology,* **23,** 291–334.

Gunderson, L. and C. Holling (eds.), 2002: *Panarchy: Understanding Transformations in Human and Natural Systems.* Island Press, Washington, DC.

Gunderson, L., C. Holling, and S. Light (eds.), 1995: *Barriers and bridges to the renewal of ecosystems and institutions.* Columbia University Press, New York, 593 pp.

Haberl, H., F. Krausmann, K.H. Erb, and N.B. Schulz, 2002: Human appropriation of net primary production. *Science,* **296(5575),** 1968–1969.

Harremoes, P., D. Gee, M. MacGarvin, A. Stirling, J. Keys, B. Wynne, and S. Guedes Vaz, 2001: *Late Lessons from Early Warning: The Precautionary Principle 1896–2000.* European Environment Agency, Copenhagen, 211 pp.

Heal, G., 2000: Valuing Ecosystem Services. *Ecosystems,* **3(1),** 24–30.

Heck, P., D. Luthi, H. Wernli, and C. Schar, 2001: Climate impacts of European-scale anthropogenic vegetation changes: A sensitivity study using a regional climate model. *Journal of Geophysical Research-Atmospheres,* **106(D8),** 7817–7835.

Heemskerk, M., 2001: Do international commodity prices drive natural resource booms? An empirical analysis of small-scale gold mining in Suriname. *Ecological Economics,* **39(2),** 295–308.

Hellweger, F.L. and A.L. Gordon, 2002: Tracing Amazon River water into the Caribbean Sea. *Journal of Marine Research,* **60(4),** 537–549.

Herwitz, S.R., R.P. Wunderlin, and B.P. Hansen, 1996: Species turnover on a protected subtropical barrier island: A long-term study. *Journal of Biogeography,* **23(5),** 705–715.

Higgins, P.A.T., M.D. Mastrandrea, and S.H. Schneider, 2002: Dynamics of climate and ecosystem coupling: abrupt changes and multiple equilibria. *Philosophical Transactions of the Royal Society: Biological Sciences,* **357,** 647–655.

Higgins, S.I., J.S. Clark, R. Nathan, T. Hovestadt, F. Schurr, J.M.V. Fragoso, M.R. Aguiar, E. Ribbens, and S. Lavorel, 2003: Forecasting plant migration rates: managing uncertainty for risk assessment. *Journal of Ecology,* **91(3),** 341–347.

Hill, J.L. and P.J. Curran, 2003: Area, shape and isolation of tropical forest fragments: effects on tree species diversity and implications for conservation. *Journal of Biogeography,* **30(9),** 1391–1403.

Hillborn, R., 1992: Can fisheries agencies learn from experience? *Fisheries,* **17,** 6–14.

Holling, C.S., 1973: Resilience and stability of ecological systems. *Annual Review of Ecology and Systematics,* **4,** 1–23.

Holling, C.S. (ed.), 1978: *Adaptive Environmental Assessment and Management.* John Wiley & Sons, London, 377 pp.

Holling, C.S., 1986: The resilience of terrestrial ecosystems; local surprise and global change. In: *Sustainable Development of the Biosphere,* W. C.Clark and R. E. Munn. (eds.), Cambridge University Press, Cambridge, 292–317.

Holling, C.S., 1988: Temperate forest insect outbreaks, tropical deforestation and migratory birds. *Memoirs of the Entomological Society of Canada,* **146,** 22–32.

Holling, C.S., 1992: Cross-scale morphology, geometry, and dynamics of ecosystems. *Ecological Monographs,* **62,** 447–502.

Holling, C.S. and G.K. Meffe, 1996: Command and control and the pathology of natural-resource management. *Conservation Biology,* **10(2),** 328–337.

Husband, B.C. and S.C.H. Barrett, 1996: A metapopulation perspective in plant population biology. *Journal of Ecology,* **84(3),** 461–469.

Huston, M.A., 1997: Hidden treatments in ecological experiments: Re-evaluating the ecosystem function of biodiversity. *Oecologia,* **110(4),** 449–460.

IES (Institute for Ecological Science), 2004: Defining ecology: the Institute for Ecological Science's definition of ecology. Available at http://www.ecostudies.org/definition_ecology.html.

Ives, A.R. and J.B. Hughes, 2002: General relationships between species diversity and stability in competitive systems. *American Naturalist,* **159(4),** 388–395.

Ives, A.R., J.L. Klug, and K. Gross, 2000: Stability and species richness in complex communities. *Ecology Letters,* **3(5),** 399–411.

Jackson, J.B.C., M.X. Kirby, W.H. Berger, K.A. Bjorndal, L.W. Botsford, B.J. Bourque, R.H. Bradbury, R. Cooke, J. Erlandson, J.A. Estes, T.P. Hughes, S. Kidwell, C.B. Lange, H.S. Lenihan, J.M. Pandolfi, C.H. Peterson, R.S. Steneck, M.J. Tegner, and R.R. Warner, 2001: Historical overfishing and the recent collapse of coastal ecosystems. *Science,* **293(5530),** 629–638.

John, C.K. and R.S. Nadgauda, 2002: Bamboo flowering and famine. *Current Science,* **82(3),** 261–262.

Jones, K.E., K.E. Barlow, N. Vaughan, A. Rodriguez-Duran, and M.R. Gannon, 2001: Short-term impacts of extreme environmental disturbance on the bats of Puerto Rico. *Animal Conservation,* **4,** 59–66.

Kasischke, E.S., N.L. Christiansen, and Stocks, 1995: Fire, Global Warming, and Carbon Balance of Boreal Forests. *Ecological Applications,* **5,** 437–451.

Keating, B.A., D. Gaydon, N.I. Huth, M.E. Probert, K. Verburg, C.J. Smith, and W. Bond, 2002: Use of modelling to explore the water balance of dryland farming systems in the Murray-Darling Basin, Australia. *European Journal of Agronomy,* **18(1–2),** 159–169.

Kiker, C.F., J.W. Milon, and A.W. Hodges, 2001: Adaptive learning for science-based policy: the Everglades restoration. *Ecological Economics,* **37,** 403–416.

King, J.K., S.M. Harmon, T.T. Fu, and J.B. Gladden, 2002: Mercury removal, methylmercury formation, and sulfate-reducing bacteria profiles in wetland mesocosms. *Chemosphere,* **46(6),** 859–870.

Kinzig, A., S.W. Pacala, and D. Tilman (eds.), 2001: *The Functional Consequences of Biodiversity.* Vol. 33. *Monographs in Population Biology,* Princeton University Press, Princeton and Oxford, 366 pp.

Kinzig, A.P., D. Starrett, K. Arrow, S. Aniyar, B. Bolin, P. Dasgupta, P. Ehrlich, C. Folke, M. Hanemann, G. Heal, M. Hoel, A. Jansson, B.O. Jansson, N. Kautsky, S. Levin, J. Lubchenco, K.G. Maler, S.W. Pacala, S.H. Schneider, D. Siniscalco, and B. Walker, 2003: Coping with uncertainty: A call for a new science-policy forum. *Ambio,* **32(5),** 330–335.

Komdeur, J., 1996: Breeding of the Seychelles magpie robin *Copsychus sechellarum* and implications for its conservation. *Ibis,* **138(3),** 485–498.

Krug, E.C. and D. Winstanley, 2002: The need for comprehensive and consistent treatment of the nitrogen cycle in nitrogen cycling and mass balance studies: I. Terrestrial nitrogen cycle. *Science of the Total Environment,* **293(1–3),** 1–29.

Kuhn, T.S., 1962: *The Structure of Scientific Revolutions.* Vol. 2, The University of Chicago Press, Chicago, 210 pp.

Kuran, T., 1989: Sparks and prairie fires: A theory of unanticipated political revolution. *Public Choice,* **61(1),** 41–74.

Lafleur, P.M. and W.R. Rouse, 1995: Energy partitioning at treeline forest and tundra sites and its sensitivity to climate change. *Atmosphere-Ocean,* **33,** 121–133.

Larkin, P.A., 1977: An epitaph for the concept of maximum sustained yield. *Transactions of the American Fisheries Society,* **106(1),** 1–11.

Lawrence, A., 2003: No forest without timber? *International Forestry Review,* **5(2),** 87–96.

Lawton, J.H. and R.M. May (eds.), 1995: *Extinction Rates.* Oxford University Press, Oxford, UK.

Lee, K., 1993: *Compass and Gyroscope: Integrating Science and Politics for the Environment.* Island Press, Washington, D.C.

Levin, S.A., 1992: The problem of pattern and scale in ecology. *Ecology,* **73(6),** 1943–1967.

Levin, S.A., 1999: *Fragile Dominion: Complexity and the Commons.* Perseus Books, Reading, MA, 431–436 pp.

Levis, S., J.A. Foley, and D. Pollard, 2000: Large scale vegetation feedbacks on a doubled CO2 climate. *Journal of Climate,* **13,** 1313–1325.

Lomolino, M.V. and M.D. Weiser, 2001: Towards a more general species-area relationship: diversity on all islands, great and small. *Journal of Biogeography,* **28(4),** 431–445.

Loreau, M., S. Naeem, P. Inchausti, J. Bengtsson, J.P. Grime, A. Hector, D.U. Hooper, M.A. Huston, D. Raffaelli, B. Schmid, D. Tilman, and D.A. Wardle, 2001: Ecology—Biodiversity and ecosystem functioning: Current knowledge and future challenges. *Science,* **294(5543),** 804–808.

Ludwig, D., 2001: The era of management is over. *Ecosystems,* **4(8),** 758–764.

Ludwig, D., R. Hilborn, and C. Walters, 1993: Uncertainty, resource exploitation, and conservation: Lessons from history. *Science,* **260,** 17 and 36.

MacArthur, R.H. and E.O. Wilson, 1967: *The Theory of Island Biogeography.* Princeton University Press, Princeton, N.J.

Martinez-Alier, J., 2002: *Environmentalism of the Poor: A Study of Ecological Conflicts and Valuation.* Edward Elgar, Northhampton, MA.

Mayr, E., 1991: *One Long Argument: Charles Darwin and the Genesis of Modern Evolutionary Thought.* Penguin, London, UK.

McClanahan, T., N. Polunin, and T. Done, 2002: Ecological states and the resilience of coral reefs. *Conservation Ecology,* **6(2).**

McMichael, A.J., 2004: Environmental and social influences on emerging infectious diseases: past, present and future. *Philosophical Transactions of the Royal Society of London Series B-Biological Sciences,* **359(1447),** 1049–1058.

McNeill, J.R., 2000: *Something New Under the Sun: An Environmental History of the Twentieth-century World.* Norton, New York, NY.

McNeill, W.H., 1976: *Plagues and Peoples.* Anchor Press, Garden City, NY.

Meadows, D., D.L. Meadows, J. Rander, and W.W. Behrens, 1972: *The Limits to Growth.* Universe Books, New York, NY, 311 pp.

Meyer, J.L., M.J. Sale, P.J. Mulholland, and N.L. Poff, 1999: Impacts of climate change on aquatic ecosystem functioning and health. *Journal of the American Water Resources Association,* **35(6),** 1373–1386.

Mols, C.M.M. and M.E. Visser, 2002: Great tits can reduce caterpillar damage in apple orchards. *Journal of Applied Ecology,* **39(6),** 888–899.

Muller-Landau, H.C., S.A. Levin, and J.E. Keymer, 2003: Theoretical perspectives on evolution of long-distance dispersal and the example of specialized pests. *Ecology,* **84(8),** 1957–1967.

Mullin, R.P., 2002: What can be learned from DuPont and the Freon ban: A case study. *Journal of Business Ethics,* **40(3),** 207–218.

Naeem, S., 2002: Biodiversity: Biodiversity equals instability? *Nature,* **416(6876),** 23–24.

Nakićenović, N. and R. Swart (eds.), 2000: *Emissions Scenarios.* London, UK, Cambridge University Press.

Newman, E.I., 1995: Phosphorus inputs to terrestrial ecosystems. *Journal of Ecology,* **83(4),** 713–726.

Ney, S. and M. Thompson, 2000: Cultural discourses in the global climate change debate. In: *Society, behaviour, and climate change mitigation,* E. Jochem, J. Sathaye, and D. Bouille (eds.), Kluwer Academic Publishers, Dodrecht, the Netherlands, 65–92.

Ostrom, E., 2003: How types of goods and property rights jointly affect collective action. *Journal of Theoretical Politics,* **15(3),** 239–270.

Parks, P. and M. Bonifaz, 1994: Nonsustainable use of renewable resources: Mangrove deforestation and mariculture in Ecuador. *Marine Resource Economics,* **9(1),** 1–18.

Parsons, E.A. and W.C. Clark, 1995: Sustainable development as social learning: theoretical perspectives and practical challenges for the design of a research program. In: *Barriers and Bridges to the Renewal of Ecosystems and Institutions,* L.H. Gunderson, C.S. Holling, and S.S. Light (eds.), Columbia University Press, New York, 428–460.

Patz, J.A., P. Daszak, G.M. Tabor, A.A. Aguirre, M. Pearl, J. Epstein, N.D. Wolfe, A.M. Kilpatrick, J. Foufopoulos, D. Molyneux, D.J. Bradley, and Members of the Working Group on Land Use Change and Disease Emergence. 2004: Unhealthy landscapes: policy recommendations on land use change and infectious disease emergence. *Environmental Health Perspectives,* **112(10),** 1092–1098.

Patz, J.A., T.K. Graczyk, N. Geller, and A.Y. Vittor, 2000: Effects of environmental change on emerging parasitic diseases. *International Journal for Parasitology,* **30(12–13),** 1395–1405.

Pauly, D., V. Christensen, J. Dalsgaard, R. Froese, and F. Torres, 1998: Fishing down marine food webs. *Science,* **279(5352),** 860–863.

Peet, R. and M. Watts, 1993: Development Theory and Environment in an Age of Market Triumphalism. *Economic Geography,* **69,** 227–253.

Peterson, G.D., 2000a: Political ecology and ecological resilience: an integration of human and ecological dynamics. *Ecological Economics,* **35,** 323–336.

Peterson, G.D., 2000b: Scaling ecological dynamics: self-organization, hierarchical structure and ecological resilience. *Climatic Change,* **44(3),** 291–309.

Peterson, G.D., C.R. Allen, and C.S. Holling, 1998: Ecological resilience, biodiversity and scale. *Ecosystems,* **1(1),** 6–18.

Peterson, G.D., S.R. Carpenter, and W.A. Brock, 2003a: Uncertainty and the management of multistate ecosystems: an apparently rational route to collapse. *Ecology,* **84(6),** 1403–1411.

Peterson, G.D., G.S. Cumming, and S.R. Carpenter, 2003b: Scenario planning: a tool for conservation in an uncertain world. *Conservation Biology,* **17(2),** pp. 358–366.

Pimm, S.L., 1984: The Complexity and Stability of Ecosystems. *Nature,* **307(5949),** 321–326.

Pinnegar, J.K., N.V.C. Polunin, P. Francour, F. Badalamenti, R. Chemello, M.L. Harmelin-Vivien, B. Hereu, M. Milazzo, M. Zabala, G. D'Anna, and C. Pipitone, 2000: Trophic cascades in benthic marine ecosystems: lessons for fisheries and protected-area management. *Environmental Conservation,* **27(2),** 179–200.

Plotnick, R.E. and R.H. Gardner, 1993: Lattices and landscapes. In: *Lectures on Mathematics in the Life Sciences: Predicting Spatial Effects in Ecological Systems,* R.H. Gardner (ed.), American Mathematical Society, Providence, Rhode Island, 129–157.

Poiani, K.A., B.D. Richter, M.G. Anderson, and H.E. Richter, 2000: Biodiversity conservation at multiple scales: functional sites, landscapes, and networks. *BioScience,* **50,** 133–146.

Powell, R.L., 2002: CFC phase-out: have we met the challenge? *Journal of Fluorine Chemistry,* **114(2),** 237–250.

Prato, T., 2003: Adaptive management of large rivers with special reference to the Missouri River. *Journal of the American Water Resources Association,* **39(4),** 935–946.

Pred, A. and M. Watts, 1992: *Reworking Modernity: Capitalisms and Symbolic Discontent.* Rutgers University Press, New Brunswick, NJ.

Prentice, I.C., G.D. Farquhar, M.J.R. Fasham, M.L. Goulden, M. Heimann, V.J. Jaramillo, H.S. Kheshgi, C.L. Quéré, R.J. Scholes, and D.W.R. Wallace, 2001: The carbon cycle and atmospheric carbon dioxide. In: *Climate Change 2001: The Scientific Basis (Contribution of Working Group I to the Third Assessment Report of the Intergovernmental Panel on Climate Change),* C.A. Johnson (ed.), Cambridge University Press, Cambridge, UK, 183–237.

Quammen, D., 1996: *Song of the Dodo: Island Biogeography in an Age of Extinctions.* Scribner, New York, NY.

Rabin, R.M., S. Stadler, P.J. Wetzel, D.J. Stensrud, and M. Gregory, 1990: Observed effects of landscape variability on convective clouds. *Bulletin of the American Meteorological Association,* **71,** 272–280.

Raskin, P., T. Banuri, G. Gallopin, P. Gutman, A. Hammond, R. Kates, and R. Swart, 2002: *Great Transition: the promise and lure of the times ahead.* Stockholm Environment Institute, Stockholm, 111 pp.

Rial, J.A., R.A. Pielke, M. Beniston, M. Claussen, J. Canadell, P. Cox, H. Held, N. De Noblet-Ducoudre, R. Prinn, J.F. Reynolds, and J.D. Salas, 2004: Nonlinearities, feedbacks and critical thresholds within the Earth's climate system. *Climatic Change,* **65(1–2),** 11–38.

Ricker, W.E., 1975: *Computation and Interpretation of Biological Statistics of Fish Populations.* Bulletin 191, Fisheries Research Board of Canada, Ottawa, Ontario

Robbins, P. 2004. *Political ecology: A critical introduction.* Blackwell Publishers, Oxford.

Robinson, J.G. and E.L. Bennett, 2002: Will alleviating poverty solve the bushmeat crisis? *Oryx,* **36(4),** 332–332.

Rocheleau, D., B. Thomas-Slayter , and E. Wangari (eds.), 1996: *Feminist Political Ecology: Global Issues and Local Experiences.* Routledge, London.

Rojstaczer, S., S.M. Sterling, and N.J. Moore, 2001: Human appropriation of photosynthesis products. *Science,* **294(5551),** 2549–2552.

Roland, J. and P.D. Taylor, 1997: Insect parasitoid species respond to forest structure at different scales. *Nature,* **386,** 710–713.

Roth, A.E., 2002: The economist as engineer: Game theory, experimentation, and computation as tools for design economics. *Econometrica,* **70(4),** 1341–1378.

Sachs, J. and P. Malaney, 2002: The economic and social burden of malaria. *Nature,* **415(6872),** 680–685.

Sala, O.E., F.S. Chapin, J.J. Armesto, E. Berlow, J. Bloomfield, R. Dirzo, E. Huber-Sanwald, L.F. Huenneke, R.B. Jackson, A. Kinzig, R. Leemans, D.M. Lodge, H.A. Mooney, M. Oesterheld, N.L. Poff, M.T. Sykes, B.H. Walker, M. Walker, and D.H. Wall, 2000: Biodiversity—Global biodiversity scenarios for the year 2100. *Science,* **287(5459),** 1770–1774.

Sanchez, B.C. and R.R. Parmenter, 2002: Patterns of shrub-dwelling arthropod diversity across a desert shrubland-grassland ecotone: a test of island biogeographic theory. *Journal of Arid Environments,* **50(2),** 247–265.

Sanderson, E.W., M. Jaiteh, M.A. Levy, K.H. Redford, A.V. Wannebo, and G. Woolmer, 2002: The human footprint and the last of the wild. *Bioscience,* **52(10),** 891–904.

Sarewitz, D., R.A.J. Pielke, and R.J. Byerly, 2000: *Prediction: Science, Decision Making, and the Future of Nature.* Island Press, Washington, DC.

Schaefer, M.B., 1954: Some aspects of the dynamics of populations important to the management of commercial marine fisheries. *Bulletin of the Inter-American tropical tuna commission,* **1,** 25–26.

Scheffer, M., S.R. Carpenter, J.A. Foley, C. Folke, and B. Walker, 2001: Catastrophic shifts in ecosystems. *Nature,* **413,** 591–596.

Schimel, D.S., T.G.F. Kittel, and W.J. Parton, 1991: Terrestrial biogeochemical cycles—global interactions with the atmosphere and hydrology. *Tellus Series a-Dynamic Meteorology and Oceanography,* **43(4),** 188–203.

Schmink, M. and C.H. Wood, 1992: *Contested Frontiers in Amazonia.* Columbia Press, New York, 387 pp.

Schmitz, O.J., 2003: Top predator control of plant biodiversity and productivity in an old field ecosystem. *Ecology Letters,* **6(2),** 156–163.

Scott, J.C., 1998: *Seeing Like a State: How Certain Schemes to Improve the Human Condition Have Failed.* Yale University Press, New Haven, CT.

Selden, T.M. and D.Q. Song, 1994: Environmental-quality and development: Is there a Kuznets curve for air-pollution emissions. *Journal of Environmental Economics and Management,* **27(2),** 147–162.

Simon, K.S. and C.R. Townsend, 2003: Impacts of freshwater invaders at different levels of ecological organisation, with emphasis on salmonids and ecosystem consequences. *Freshwater Biology,* **48(6),** 982–994.

Singh, K.P. and S.K. Tripathi, 2000: Impact of environmental nutrient loading on the structure and functioning of terrestrial ecosystems. *Current Science,* **79(3),** 316–323.

Sinha, P., P.V. Hobbs, R.J. Yokelson, D.R. Blake, S. Gao, and T.W. Kirchstetter, 2003: Distributions of trace gases and aerosols during the dry biomass burning season in southern Africa. *Journal of Geophysical Research-Atmospheres,* **108(D17).**

Snyder, W.E. and D.H. Wise, 2001: Contrasting trophic cascades generated by a community of generalist predators. *Ecology,* **82(6),** 1571–1583.

Spaargaren, G. and A.P.J. Mol, 1992: Sociology, Environment, and Modernity—Ecological Modernization as a Theory of Social-Change. *Society & Natural Resources,* **5(4),** 323–344.

Spencer, C.N., B.R. McClelland, and J.A. Stanford, 1991: Shrimp stocking, salmon collapse and eagle displacement: cascading interactions in the food web of a large aquatic ecosystem. *Bioscience,* **41(1),** 14–21.

Stauffer, D., 1985: *Introduction to Percolation Theory.* Taylor and Francis, London, U.K.

Stauffer, J.R., M.E. Arnegard, M. Cetron, J.J. Sullivan, L.A. Chitsulo, G.F. Turner, S. Chiotha, and K.R. McKaye, 1997: Controlling vectors and hosts of parasitic diseases using fishes. *BioScience,* **47,** 41–49.

Supalla, R., B. Klaus, O. Yeboah, and R. Bruins, 2002: A game theory approach to deciding who will supply instream flow water. *Journal of the American Water Resources Association,* **38(4),** 959–966.

Tansley, A.G., 1935: The use and abuse of vegetational concepts and terms. *Ecology,* **16,** 284–307.

Taylor, L.H., S.M. Latham, and M.E.J. Woolhouse, 2001: Risk factors for human disease emergence. *Philosophical Transactions of the Royal Society of London Series B-Biological Sciences,* **356,** 983–989.

Terborgh, J., L. Lopez, P. Nunez, M. Rao, G. Shahabuddin, G. Orihuela, M. Riveros, R. Ascanio, G.H. Adler, T.D. Lambert, and L. Balbas, 2001: Ecological meltdown in predator-free forest fragments. *Science,* **294(5548),** 1923–1926.

Tilman, D., D. Wedin, and J. Knops, 1996: Productivity and sustainability influenced by biodiversity in grassland ecosystems. *Nature,* **379(6567),** 718–720.

Turner, B.L., W.C. Clark, R.W. Kates, J.F. Richards, J.T. Mathews, and W.B. Meyer (eds.), 1993: *The Earth as Transformed by Human Action.* Cambridge University Press, New York, NY, 713 pp.

Tyagi, B.K., 2004: A review of the emergence of Plasmodium falciparum-dominated malaria in irrigated areas of the Thar Desert, India. *Acta Tropica 89,* **89,** 227–239.

van der Heijden, K., 1996: *Scenarios: The Art of Strategic Conversation.* John Wiley and Sons, New York, NY.

van Notten, P.W.F., J. Rotmans, M.B.A. van Asselt, and D.S. Rothman, 2003: An updated scenario typology. *Futures,* **35(5),** 423–443.

Vanandel, J., J.P. Bakker, and A.P. Grootjans, 1993: Mechanisms of vegetation succession: a review of concepts and perspectives. *Acta Botanica Neerlandica,* **42(4),** 413–433.

Vitousek, P.M., P.R. Ehrlich, A.H. Ehrlich, and P.A. Matson, 1986: Human appropriation of the products of photosynthesis. *BioScience,* **36(6),** 368–373.

Wackernagel, M., N.B. Schulz, D. Deumling, A.C. Linares, M. Jenkins, V. Kapos, C. Monfreda, J. Loh, N. Myers, R. Norgaard, and J. Randers, 2002: Tracking the ecological overshoot of the human economy. *Proceedings of the National Academy of Sciences of the United States of America,* **99(14),** 9266–9271.

Walker, B., A. Kinzig, and J. Langridge, 1999: Plant attribute diversity, resilience, and ecosystem function: The nature and significance of dominant and minor species. *Ecosystems,* **2(2),** 95–113.

Walters, C.Y., 1997: Challenges in adaptive management of riparian and coastal ecosystems. *Conservation Ecology* [online], **1(2),** 1. Available at http://www.consecol.org/vol1/iss2/art1.

Walters, C.J., 1986: *Adaptive Management of Renewable Resources.* McGraw Hill, New York.

Walters, C.J. and R. Hilborn, 1978: Ecological optimization and adaptive management. *Annual Review of Ecology and Systematics,* **9,** 157–188.

Walters, C.J. and C.S. Holling, 1990: Large-scale management experiments and learning by doing. *Ecology,* **71(6),** 2060–2068.

Wang, G.L. and E.A.B. Eltahir, 2000: Modeling the biosphere-atmosphere system: the impact of the subgrid variability in rainfall interception. *Journal of Climate,* **13(16),** 2887–2899.

Watson, R. and C.W. Team (eds.), 2001: *Climate Change 2001: Synthesis Report.* Cambridge University Press, New York, NY.

Watts, M., 1983: *Silent Violence: Food, Famine and Peasantry in Northern Nigeria.* University of California Press, Berkeley.

Weart, S.R., 2003: *The Discovery of Global Warming.* Harvard University Press, Cambridge, MA.

Wennergren, U., M. Ruckelshaus, and P. Kareiva, 1995: The promise and limitations of spatial models in conservation biology. *Oikos,* **74(3),** 349–356.

Westley, F., 1995: Governing design: the management of social systems and ecosystems management. In: *Barriers and Bridges to the Renewal of Ecosystems and Institutions,* L.H. Gunderson, C.S. Holling, and S.S. Light (eds.), Columbia University Press, New York, 489–532.

Westley, F., S.R. Carpenter, W.A. Brock, C.S. Holling, and L.H. Gunderson, 2002: Why systems of people and nature are not just social and ecological systems. In: *Panarchy: Understanding Transformations in Human and Natural Systems,* L.H. Gunderson and C.S. Holling (eds.), Island Press, Washington, DC, 103–119.

Whittaker, R.J., 1995: Disturbed island ecology. *Trends in Ecology & Evolution,* **10(10),** 421–425.

Wilkinson, C.F., 1992: *Crossing the Next Meridian: Land, Water, and the Future of the West.* Island Press, Washington, DC.

Wilson, M.A. and S.R. Carpenter, 1999: Economic valuation of freshwater ecosystem services in the United States: 1971–1997. *Ecological Applications,* **9(3),** 772–783.

Wolf, E. 1972: Ownership and political ecology. *Anthropological Quarterly,* **45,** 201–205.

Wolfenbarger, L.L. and P.R. Phifer, 2000: Biotechnology and ecology—The ecological risks and benefits of genetically engineered plants. *Science,* **290(5499),** 2088–2093.

Zavaleta, E.S., R.J. Hobbs, and H.A. Mooney, 2001: Viewing invasive species removal in a whole-ecosystem context. *Trends in Ecology & Evolution,* **16(8),** 454–459.

Zhang, P.C., G.F. Shao, G. Zhao, D.C. Le Master, G.R. Parker, J.B. Dunning, and Q.L. Li, 2000: Ecology—China's forest policy for the 21st century. *Science,* **288(5474),** 2135–2136.

Zhuang, G.S., J.H. Guo, H. Yuan, and C.Y. Zhao, 2001: The compositions, sources, and size distribution of the dust storm from China in spring of 2000 and its impact on the global environment. *Chinese Science Bulletin,* **46(11),** 895–901.

Chapter 4

State of the Art in Simulating Future Changes in Ecosystem Services

Coordinating Lead Author: Peter Kareiva

Lead Authors: John B. R. Agard, Jacqueline Alder, Elena Bennett, Colin Butler, Steve Carpenter, W. W. L. Cheung, Graeme S. Cumming, Ruth Defries, Bert de Vries, Robert E. Dickinson, Andrew Dobson, Jonathan A. Foley, Jacqueline Geoghegan, Beth Holland, Pavel Kabat, Juan Keymer, Axel Kleidon, David Lodge, Steven M. Manson, Jacquie McGlade, Hal Mooney, Ana M. Parma, Miguel A. Pascual, Henrique M. Pereira, Mark Rosegrant, Claudia Ringler, Osvaldo E. Sala, B. L. Turner II, Detlef van Vuuren, Diana H. Wall, Paul Wilkinson, Volkmar Wolters

Review Editors: Robin Reid, Marten Scheffer, Antonio Alonso

TABLES

Main Messages

Building scenarios and anticipating changes in ecosystem services are modeling exercises. The reliability of models depends on their data inputs and the models themselves. This chapter sketches out the state-of-the-art modeling approaches for critical components of the Millennium Ecosystem Assessment scenarios, examines strengths and weaknesses and alternative approaches, identifies critical uncertainties, and describes high-priority research that could resolve fundamental uncertainty.

In order to evaluate the MA scenarios, readers must understand the capabilities, uncertainties, and frontiers of the models used to project changes in ecosystem services. This chapter provides a rigorous scientific discussion of just how confident we can be about different dimensions of the scenario model analyses and where we need to do a great deal more work. Uncertainty is acceptable as long as it is acknowledged up front. Although many of the models used to inform scenarios have weaknesses, the alternative is to use no models whatsoever. The modeling approaches and the uncertainties vary according to topic. Hence we take up the modeling issues one topic at a time, forecasting land cover change, impacts of land cover changes on local climates, changes in food demand and supply, changes in biodiversity and extinction rates, impacts of changes in phosphorous cycles, impacts of changes in nitrogen cycle and inputs, fisheries and harvest, alterations of coastal ecosystems, and impacts on human health. The final sections evaluate integrated assessment models and look at key gaps in current modeling abilities.

The uncertainties and limitations of models are extensive, and in many cases proven methods do not exist for the forecasting tasks that we face. Recurring limitations and constraints include an absence of models that work well across multiple scales, failure of models to couple interacting processes, and models based on nonrepresentative subsets of Earth's ecosystem services (such as specific and narrow taxonomic groups or geographic regions). These limitations do not mean we should not attempt to make a forecast—only that we should present results with appropriate levels of uncertainty. The act of attempting to make forecasts where apt methods do not exist has already spurred enormous research and innovation, such that in five years our forecasts will become much more reliable. For this reason, we pay particular attention to advances in modeling or data that are likely to greatly enhance our ability to assess alternative ecosystem futures. It is important to recognize that models are not statements of fact but instead are hypotheses to be evaluated in light of coming changes in ecosystem services.

4.1 Introduction

The models used to generate scenarios for the Millennium Ecosystem Assessment are not the only models available. This chapter examines state-of-the-art modeling approaches for critical components of the MA scenarios. It considers the suite of models and modeling approaches that might be drawn on for scenario analyses. The four core models actually used for the global MA were IMAGE for land use change, IMPACT for food demand and agriculture, WaterGAP for water use and availability, and EcoPath and Ecosim for predicting fisheries impacts. (See Chapter 6 for more-detailed descriptions of the models.) In many cases these models were chosen because they are the only ones with global coverage (WaterGAP, for instance).

The state of the art for environmental modeling is changing very rapidly. This chapter describes the key modeling arenas in which we expect major advances over the next 10 years, which in turn could provide improved tools for future global ecosystem assessments. Hence we discuss some models, such as for phosphorous cycling, for which there is no global model but where progress is expected so that future global assessments will have new tools. Climate modeling is not covered, since there have been numerous review papers describing the existing climate models.

In general we seek in this chapter not to advocate or defend the MA's choice of IMAGE, IMPACT, WaterGAP, and Ecosim/Ecopath. Rather, we provide readers with an overview of the modeling field and the variety of approaches being pursued, with pointers to where we expect future research will lead. This venture is so new that there is no commonly accepted suite of models or, as is the case for climate models, some standard approach for testing and contrasting the performance of competing models. In some cases the actual models used by the MA are virtually all that is available. In other cases, such as in models for predicting biodiversity change, there are a variety of options, all under current research development.

The major types of models are:

- statistical models that rely on observed relationships and extrapolate into the future;
- first-principle equations that solve for equilibrium or draw on fundamental laws of transport and mass balance;
- large system (usually simulation) models that mathematically describe relationships among a web of state-variables and attempt to include a somewhat complete representation of the drivers of change;
- expert models and decision support systems that translate qualitative insights or expertise into quantitative assertions; and
- a wide variety of "agent-based" or cellular automata models in which the activities of individual actors are simulated and then aggregated to understand whole-system behavior.

This is not the only taxonomy of models; alternative distinctions include stochastic versus deterministic, simulation versus analytical, spatial versus nonspatial, equilibrium versus nonequilibrium, and so forth. But these categories are most germane to the strategic choices available when attempting to perform a global scenario analysis. We have made an effort to remove as much technical language as possible. We obviously have not succeeded as well as would be ideal. However, in many cases what appears to be "jargon" is necessary for precision and to help other technical experts know exactly what modeling issue is under discussion.

This is not a chapter for light reading. This is a chapter to read with the idea of learning what is going on in the modeling world that might be important for future assessments. The topics chosen are not encyclopedic. We structure discussion around core modeling arenas (such as fisheries or land use or agricultural production). One of the biggest areas of modeling research is coupling models of different processes together and attempting to incorporate feedbacks among processes. Figure 6.3 in Chapter 6 shows how the MA linked together to model different processes.

In the future there may be many more options, and in fact one conclusion of this chapter is that the linking of different processes and scales is probably the biggest research need.

When considering each section, it will be obvious that none of the modeling approaches is ideal. Compromises must be made because of lack of data. Documentation of large-system models is often weak, and transparency is not all it should be. Although many of the models considered for use in the MA are for forecasts, they are also hypotheses, as all models are. This chapter seeks to introduce some alternative approaches that might be selected if existing models fall short. As data are collected, we expect some models to be rejected and new ones to be used. One outcome of the MA is pressure for better modeling practices.

4.2 Forecasting Changes in Land Use and Land Cover

Land use change models attempt to project future changes in land use based on past trends and the drivers thought to determine conversions of land between different categories (forest to agriculture, agriculture to urban, and so forth). One initial motivation behind land use change modeling was the prediction of tropical deforestation, with its many consequences. The field has now broadened geographically and with respect to the type of land cover transitions it examines. Central to understanding the human and ecological aspects of land use and land cover change (or land change) is a movement toward an interdisciplinary perspective of change, where social, ecological, and information sciences are joined (Liverman et al. 1998; Gutman et al. 2004). A core component of integrated land change science is formed by spatially explicit, dynamic land change models that explain and project land cover and land use changes (IGBP-IHDP 1999; Veldkamp and Lambin 2001). Given that land use and land cover are dynamically coupled, land change models provide one of the more powerful ways to combine human and biophysical subsystems, permitting assessments of the consequences and feedbacks between the subsystems. In this sense, these models improve understanding of a broad range of issues critical to the MA, from the resilience of ecosystems to human perturbations to society's responses to changes in ecosystem services.

Land change models are generally classified according to their implementation or scale (e.g., Lambin 1994; Rayner 1994; Kaimowitz and Angelsen 1998; Agarwal et al. 2002). They tend to use a variety of data sources as input. Survey and census data have long been used by land change models and are increasingly joined by spatial data (maps, for instance) on land manager activities and socioeconomic factors. These spatial data are often used in the context of geographic information systems—software systems that store, manipulate, and analyze georeferenced data—and are derived from sources as varied as remotely sensed imagery and global positioning system receivers. Data are also increasingly gathered over several time periods in order to aid understanding of land change trajectories. The chief output of land change models tends to be explanations of past and present use and projections of future land use.

4.2.1 Existing Approaches

Perhaps the simplest land change models use a non-iterative set of equations to seek a single solution where the modeled system can be characterized as static or existing in equilibrium. Gravity models or logistic functions, for example, are used to estimate population-driven land conversion over large areas and coarse resolutions. These models are often based on theories of population growth and diffusion, processes that are thought to determine cumulative land change (Lambin 1994).

System models typically represent stocks and flows of information, material, or energy as sets of differential equations linked through intermediary functions and data (Hannon and Ruth 1994). When differential equations are numerically solved, time advances in discrete steps, which in turn allows dynamic representation of feedbacks so that interacting variables can influence one another's future dynamics. Earlier system models of land change were not spatially explicit, but more recent models are increasingly linked to spatial data (e.g., Voinov et al. 1999; Zhang and Wang 2002).

Another group of land change models relies on statistical methods based on empirical observations (Ludeke et al. 1990; Mertens and Lambin 1997; Geoghegan et al. 2001). For example, econometric models use statistical methods to test theoretical hypotheses concerning the consequences of new road systems (Chomitz and Gray 1996; Nelson and Hellerstein 1997; Pfaff 1999) and of other economic and ecological variables exogenous to the modeled system (Alig 1986; Hardie and Parks 1997). Unless statistical models are tied to a theoretical framework, they may underrate the role of human and institutional choices (Irwin and Geoghegan 2001).

Expert models express qualitative knowledge in a quantitative fashion, often in order to determine where given land uses are likely to occur. Some methods combine expert judgment with Bayesian probability (Bonham-Carter 1991). Symbolic artificial intelligence approaches, in the form of expert systems and rule-based knowledge systems, use logical rules in combination with data to grant models some capacity to address novel situations. Lee et al. (1992), for example, use probabilities to build a set of stochastic branching rules regarding possible land transformations, and then connect those rules with an independent ecological model to assess land change impacts. The probabilities for these branching rules are not estimated in a traditional statistical sense but instead are inferred by interviewing numerous land managers and synthesizing their answers into probabilities.

Land change models that are based on biologically inspired evolutionary computer modeling methods are increasingly common (Whitley 2001). Perhaps the most promising are models based on artificial neural networks, computational analogs of biological neural structures (such as the neuronal structure of the human brain), which are trained to associate outcomes with given stimuli, such as associating spatial land change outcomes with inputs like population density or distance to water bodies (e.g., Shellito and Pijanowski 2003). Another body of research applies computational models of Darwinian evolution, such as genetic programming or classifier systems, to tease out causal linkages between various factors and land change (Xiao et al. 2002; Manson 2004).

A growing number of land use and land cover change models are based on cellular modeling methods, which use models that conduct operations on a lattice of congruent cells, such as a grid. In the common cellular automata approach, cells in a regular two-dimensional grid exist in one of a finite set of states, each state representing a kind of land use, for instance. Time advances in discrete steps, and future states depend on transition rules based on a the condition (or state) of the surrounding immediate neighborhood (Hegelsmann 1998). In another common cellular modeling method—Markov modeling—the states of cells arrayed in a lattice depend probabilistically but simply on previous cell states. Cellular models have proved their utility for modeling land change in linked human-environment systems (Li and Reynolds 1997; White and Engelen 2000).

Agent-based models are a relatively recent development in land change modeling. They are collections of agents, or software programs, which represent adaptive autonomous entities (like farmers, or institutions that build roads, or local elders) that extract information from their environment and apply it to behavior such as perception, planning, and learning (Conte et al. 1997). Agent-based models have been used in particular to model small-scale decision-making of actors in land change (Gimblett 2002; Janssen 2003; Parker et al. 2003).

4.2.2 Critical Evaluation of Approaches

Equation models have the advantage of being simple and elegant and relatively transparent. These models can provide a good first estimate of land change at broad scales, which can be tied to driving forces such as population or economics. Their chief limitation is the degree of simplification necessary to create an analytically tractable system of equations, which often results in a highly abstract model that does not reflect many aspects of reality (Baker 1989). Cross-scale relationships (interactions between different spatial scales) and time-dependent relationships can be difficult to model with simultaneous equations, given the need for common parameters across scales or time and equilibrium assumptions.

System models are a powerful means of representing dynamic systems. They are widespread across many academic disciplines. System models can face limitations, however, because the complexity of real-world parameters can be difficult to convey in the form of linked equations. Equations can also limit system models to statistically idealized flows and stocks, which means that discrete actions or the decision-making that led to them are not included unless the system in question is modeled at a small scale or in detail (Vanclay 2003). In order to focus on key dynamics, the modeler typically makes assumptions about the aggregate results of behavior at the potential cost of glossing over or assuming away key system behavior.

Statistical models are widely accepted and well understood. Despite this, considerable care must be taken to ensure assumptions such as independence of observations, especially in light of spatial or temporal autocorrelation (where observations are correlated in space or time) (Griffith 1987) and issues of data aggregation (combining data from multiple sources or scales) (Rudel 1989). Researchers have begun to devote attention to the statistical problems that arise from using spatial data, thereby decreasing the bias and inefficiencies in parameter estimates and using spatial and temporal autocorrelation to inform model construction (Kaufmann and Seto 2001; Overmars 2003). Another complication is that the choice of a land change descriptor can dramatically influence the result, such as when tracking change in areal extent of land cover, as opposed to the rate of change in land cover (Kummer and Sham 1994). Either variable is correct in the sense that it measures phenomena of interest, but the variables can give different statistical results.

Expert models are useful for rendering qualitative expert domain knowledge into formats traditionally considered quantitative knowledge. The underlying logical basis that allows these models to function can create difficulties, however, since it is challenging to include all aspects of the problem domain, which can lead in turn to inconsistencies in model results. Application of expert systems to land change has remained underexplored due to the difficulty of logically encoding knowledge that adequately maps onto the complex spatiotemporal nature of most land use and land cover change situations. Similarly, it can be difficult to find experts sufficiently versed in a given situation; when they are found, it is also difficult to parse their knowledge into the logical rules and structures necessary to create expert models (Skidmore et al. 1996).

Evolutionary models have been used with success to project land use and land cover change. They are in essence powerful directed-search methods that excel in identifying patterns and relationships in highly dimensional, noisy, stochastic environments (Kaboudan 2003). At the same time, theories on how and why evolutionary methods work are subject to ongoing debate, which serves to blunt the edge of any analysis based on them (Whitley 2001). One side effect is that identification of structures of causality and correlation is more straightforward with statistical methods, for example, than with evolutionary methods because the latter can too easily be used to create convoluted computer programs that produce seemingly good results but at the expense of understanding how or why.

Cellular models are appealing for their capacity to use relatively simple rules to represent local interactions that can in turn lead to complex outcomes (Phipps 1989). Cellular models can suffer from "spatial orientedness" (Hogeweg 1988), however, where the simple cellular neighborhood relationships do not reflect actual spatial relationships. As such, these methods may not be suited to model land change where there are non-uniform or non-local interactions. As a result, they must be buttressed with complex rule sets to differentiate between the kinds of decision-making that apply to groups of cells, such as local land tenure structure (e.g., Li and Gar-on Yeh 2000). While effective, these deviations from classic cellular models come at the potential cost of moving models away from their key advantage of simplicity (Torrens and O'Sullivan 2001).

Agent-based models typically complement other approaches to modeling land change. Their strength lies in the ability to represent heterogeneous agents (Huston et al. 1988) and to incorporate interaction and communication among agents (Judson 1994) in a manner unlike that of other modeling methods. Agent-based models can be difficult to use, however, since they are often tailored to a particular setting and create results that are often not generalizable (Durlauf 1997). Much work remains to be done on establishing common modeling platforms and devising means of validating agent-based models, particularly when distinguishing legitimate results from modeling artifacts, as many of these models remain underevaluated. Similarly, they are often used at small spatial scales; they need to be scaled up to larger ones useful for full ecological assessment (Veldkamp and Verburg 2004; Manson 2003)

In sum, modeling land use change runs the gamut from relatively straightforward equation-based models to complicated and computationally intensive models. There is a movement toward greater integration and hybridization of these approaches in order to compensate for shortcomings of individual methods and to address outstanding issues in land change, such as interdisciplinary integration, spatio-temporal scale issues, and the complexity of land change (Brown et al. 2004).

Dynamic spatial simulation modeling, for example, incorporates cellular modeling to address spatial heterogeneity and uses system models to represent social and economic mechanisms in addition to ecological processes such as secondary succession (Lambin 1997). GEOMOD2 combines statistical modeling, systems approaches, and expert decision rules to project land change at the regional scale (Pontius et al. 2001). Another example is given by the CLUE family of models, which use a combination of approaches to model land change and associated phenomena at regional scales (Veldkamp and Fresco 1996; Verburg et al. 2002). A final example is found in Integrated Assessment Models that incorporate links to the terrestrial environment at continental and global scales. The IMAGE 2.0 model, for example, incorporates land use, land use change, soil information, and element fluxes at the global scale at half-degree resolution (Alcamo 1994). Integrated Assessment Models are discussed at the end of this chapter.

4.2.3 Research Needs

The future of land change modeling is defined in part by three themes: interdisciplinary research, better integration of theory and method, and refinement of modeling techniques, including establishing standard rules, measures, and metrics that provide the rigor found in less expansive modeling approaches common to established subfields (such as demographic or econometric models) (Rindfuss et al. 2004). First, integration across disciplines appears increasingly necessary with respect to understanding the webs of causality underlying land change. The land change research community has identified three conceptual foci: social systems, ecological systems, and land managers or decision-makers (IGBP-IHDP 1995). Heretofore, social systems such as institutions (rules) have not been incorporated well into human-environment models. Land change models, however, increasingly account for institutional settings, thereby increasing their robustness for use at local to regional scales of analysis (Gutman et al. 2004).

Second, it is increasingly apparent that land change modeling can be improved and can create greater interest among the core social and environmental sciences if it is informed by critical concepts and theory relevant to both sciences and their coupling. Irwin and Geoghegan (2001), for example, argue that land change models often claim to represent human behavior while not explicitly using theories of human behavior. Similarly, other models proclaim to address the environment, but it is reduced to nature as a resource stock for human use, not as part of a functioning ecosystem. There is therefore an increasing focus on incorporating ecosystem models, such as landscape-scale forest models, with models of land use (He 1999). Greater engagement across disciplines will likely (or it is hoped will) accelerate the trend of better integration of theory and method.

Third, the greater integration and hybridization of the approaches noted earlier speaks to ongoing development of new methods and metrics of performance. It is important to note that models are increasingly oriented toward pursuing the fine spatial and temporal resolution necessary to assess human dynamics, such as individual decision-making, and ecological phenomena, such as biodiversity. Having both temporal and spatial explicitness is a key need, and therefore a goal, of land change modeling (Agarwal et al. 2002). Getting the magnitude of land change right is only part of the goal; getting its location right is the other. Model performance regarding both needs requires new metrics (Pontius 2000).

Also important is the extent to which models can also serve as vehicles to integrate disciplines in a manner that captures interactions across various real-world human and environment systems. Most of these methods, for example, can incorporate some degree of feedback between human and biogeophysical systems. Some do so explicitly, such as system models or cellular models, while others, such as expert models or agent-based models, have been adapted to accommodate dynamic interactions.

Finally, there is a movement toward increasing model transparency through mechanisms, such as better user interfaces and communication of model results both in and out of the research community (Parker et al. 2003). The key stumbling block to incorporating these changes resides less in the models themselves and more in the disciplinary contexts in which they are used. This caution notwithstanding, land change models promise to provide the foundational basis for understanding and projecting human-environment interactions for terrestrial ecosystems.

4.3 Forecasting Impacts of Land Cover Change on Local and Regional Climates

Terrestrial ecosystems both influence the climate and in turn are themselves influenced by the climate (Foley et al. 2003). Scenarios of the future paths of the biosphere (e.g., DeFries et al. 2002a) must therefore be viewed as interactive with the climate system. A detailed analysis of this issue would require intertwining Intergovernmental Panel on Climate Change predictions and dynamical representations of future greenhouse gas emissions and their impacts on climate with a MA-type model of vegetation and biotic responses that in turn fed back on the greenhouse gas scenarios. Such an analysis is currently not feasible. However, a general awareness of the techniques that might be used should promote improved treatments in future assessments. (Various anthropogenic processes that drive climate change and can feed back on the biosphere are discussed extensively in Chapter 13 of MA *Current State and Trends*.) An earlier review from a climate modeling perspective is given in Dickinson (1992).

What are the changes of land use that may significantly affect climate? They include conversion of natural forest to other uses, including agroforestry, grazing, and crops; conversion of grasslands by natural or human factors to other covers, including shrubs (e.g., Hoffman and Jackson 2000) and croplands; desertification; initiation or cessation of irrigated agriculture; urbanization; draining or creation of seasonal or permanent wetlands; and, in general, anything changing the overall vegetation density, commonly expressed in models by its leaf-area index, or changing the hydraulic or nutrient properties of the soil (such as compaction or salinization).

Discussion here is limited to the question of the current modeling basis for describing how changes in human land use can modify climate. Any quantification of the future impacts on climate of land use change must start with predictions or scenarios of land use change. These changes must be described in terms of the parameters that characterize the impacts of land on climate through biophysical (that is, energy and water balance) or biogeochemical effects (modifying the atmospheric gaseous or particulate composition). Chapter 13 in the MA *Current State and Trends* volume addresses current knowledge as to land use change modification of climate drivers. Here, we focus on the assessment of the capabilities for future prediction, assuming we have appropriate information on current conditions (which in reality we may not always have).

Describing the biophysical impacts of land use and land cover change on local and regional climate is an area where the general modeling strategy and methodology is relatively free of controversy. A single modeling approach—based on global and regional climate models—is accepted as credible, so the assessment task is a judgment as to which implementations of this method are likely to be most successful. However, application of these models to the question of the impacts of land cover change on climate is sufficiently immature that their evaluation has been limited. In other words, any such application may provide useful guidance, but details would have to be presented with ample caveats and with considerable uncertainties. The most serious bottlenecks for progress are quantification of scenarios of future land use and land cover change in terms that provide parameters needed by the models, the lack of test cases with which modeled impacts of land use/land cover change have been compared to observations, the absence of a full incorporation of feedbacks from changes of vegetation cover in climate simulations, and uncertainties in the treatment of the coupling between land cover change and atmospheric boundary layer processes connected to rainfall.

Biogeochemical connections to terrestrial ecosystems are detailed in Chapter 13 of the MA *Current State and Trends* volume and various management strategies are indicated in Chapter 12 of the MA *Policy Responses* volume. This brief overview of the biogeochemical consequences of land use change indicates what should be included as output of such modeling: The terrestrial system includes important stores of carbon that through land use change can become sources for greenhouse gases or other significant atmospheric constituents. In addition, terrestrial processes can sequester carbon dioxide from the atmosphere and so reduce the impact of that added by fossil fuel combustion. The release of methane to the atmosphere by carbon cycling in anoxic soils can be modified by land use change. Land use change that extends livestock grazing may increase methane emissions. Changes may also occur in the release of volatile organic compounds and so affect air quality or the formation of aerosols, and the latter may have impacts on cloud formation and precipitation.

4.3.1 Existing Approaches

4.3.1.1 Climate Models as Used to Address Biophysical Impacts of Land Cover Change

The only general approach to assessing impacts of land use/land cover change on climate that has a good likelihood of providing useful information from a policy viewpoint is a comprehensive approach that integrates global or regional climate models. Such integration is only likely to be credible if its starting point is current state-of-the-art climate models. There are perhaps a dozen state-of-the-art models worldwide as developed and maintained by large groups of scientists. These models are extensively evaluated by the IPCC Working Group I (e.g., IPCC 2001), and so their summary here can be brief. Such models serve as national or international resources that are generally available to appropriate collaborators and in some cases freely distributed from the Internet in a form suitable for use on multiple computer platforms with documentation to facilitate their use by independent scientists (e.g., the Community Climate System Model; Blackmon et al. 2000, 2001). However, meaningful use of these models still requires adequate scientific background, considerable individual commitment, and adequate computational resources. Such models have been under development for several decades by various groups and have many applications (e.g., Manabe and Stouffer 1996; Osborne et al. 2004).

Climate is modeled by simulation of the atmospheric weather and coupled surface processes on an hour-by-hour and day-by-day basis. The surface processes include an ocean model, a sea ice model, and a land surface model. The fluid behavior of the atmosphere and oceans are described by partial differential equations that are numerically integrated. The most advanced such "Earth System" models have been developed by large teams of scientists with considerable institutional support. The World Climate Research Programme of the World Meteorological Organization is largely devoted to coordination between the various modeling groups and sponsorship of multiple evaluation activities of model components and complete model simulations to improve these models.

Climate is established from the simulations of these models through various kinds of averaging in time and space and other such analyses that are commonly used with observational data or to help diagnose the functioning of the system. A variety of model outputs are produced, each of which may be of interest to a different community of scientists.

The overall strategy for use of these models to address impacts of some imposed change is relatively simple. The equations are integrated over a sufficiently long period, often many simulated years, to establish their climatology. Such a simulation is then repeated, except for the assumed change, such as land use (e.g., Maynard and Royer 2004), and the consequent climate impact and its statistical confidence level is assessed by the difference between the two states and through its comparison with natural fluctuation statistics.

The production of a sufficient number of independent samples from this naturally chaotic system is achieved by some combination of simulating for a sufficiently long period, carrying out the integrations multiple times (ensemble approach, cf. Boer 2004), and using statistical methods of pattern analysis to optimize the signal (climate change, for example) relative to the natural variability (that is, noise) of the system. Available computational resources, and more important, questions as to the correctness of model details may limit such analyses.

The latter issue is addressed and more robust conclusions are obtained by carrying out the same integration with multiple independent models and identifying and analyzing any significantly different results that arise or by assessing the model components most important for the answer being sought in terms of how they may contribute to the uncertainty of the result. The time required to equilibrate is necessarily at least as long as the longest time scales of the individual systems. For the oceans, this is determined by depth included and can be many centuries for a full ocean. For prescribed vegetation and soil properties (that is, fixed soil carbon and nitrogen), soil moisture takes longest to equilibrate. Models that include the development of forests or some of the slower soil processes may also require centuries. Large ice sheets may require millennia.

4.3.1.2 Model Components Most Closely Connected to Biophysical Impacts of Land Use Change

Land surface models couple atmospheric processes with the conservation of energy and water, which may depend on land cover. These models initially simply tracked reservoirs of water whose temperature was adjusted to conserve energy by turbulent exchanges with the atmosphere (e.g., Manabe et al. 1965; Manabe and Bryan 1969; Manabe and Wetherald 1975). Later authors (e.g., Dickinson et al. 1981; Dickinson 1984; Sellers et al. 1986) addressed the need to include land cover elements that varied geographically and included greater complexity, such as a vegetation component and multiple temperature and soil water variables.

The largest modifications in simulations from those of earlier efforts resulted from the stomatal controls in plants on transpiration. This aspect is now modeled from carbon assimilation. Its inclusion (e.g., Sellers et al. 1997) and that of snow cover and micrometeorology is now relatively advanced and well understood, but implementations may differ in details because of different objectives and institutional histories (e.g., Dai et al. 2003). Improvements in the mapping of different land covers and their correlations with leaf area and albedo are being implemented with use of new global remote sensing data (e.g., Buermann et al. 2002; Tian et al. 2004a, 2004b). Many of the more important impacts of land use/land cover change, such as impacts on the hydrological cycle, require not only these components but also the coupling of the surface micrometeorology to atmospheric boundary layer processes.

What attributes of land use changes need to be incorporated in climate models? The answer to this question entails all key attributes that current climate models use as inputs when calculating energy fluxes. These include the LAI and hydraulic properties mentioned earlier, albedo and surface roughness, stomatal functioning as an element of evapotranspiration and carbon cycling, and the ways energy fluxes might be changed independent of the above—for example, through nutrient changes. These properties are included either from prescribed vegetation cover with seasonal phenologies or through models of the vegetation dynamics.

It is currently not practical to include a wide variety of plant species, so that the climate role of vegetation is represented by 10–20 "plant functional types" (e.g., Bonan et al. 2002). The extra detail of subtle species-differences within the same functional group may never yield sufficient additional predictive power to make species-specific models desirable. The dynamics of vegetation as it interacts with soil moisture and its climatic environment can be formulated at various levels depending on the time scales involved. For changes over a few years or less, only leaf properties need to be included. On longer time scales, growth, competition, and hence initiation and survival of individual plant types may have to be included to characterize the terrestrial feedbacks adequately.

In general, the vegetation cover can have strong influences on the surface exchanges of energy and water. Past sensitivity studies (e.g., Bonan et al. 1992; Foley et al. 1994) have revealed that a transition from systems shaded by trees to short vegetation covered by snow can have a large positive feedback on climate in high latitudes. Raupach (1998) has shown these sorts of response to gradual temperature changes are not always incremental but can instead involve abrupt transitions or thresholds. In addition, spatial scales of surface heterogeneity in terms of wind, soils (e.g., Zender and Newman 2003), soil moisture, and vegetation will interact with the underlying turbulent convection to structure its spatial scale and consequently the probability and amounts of precipitation.

4.3.1.3 Examples of Land Use Change Addressed in Past Literature

Various idealized scenarios have been studied. One popular question has been the possible impacts of complete conversion of the Amazon forest to degraded pasture (e.g. Dickinson and Henderson-Sellers 1988; Lean and Warrilow 1989; Shukla et al. 1990; Nobre et al. 1991; Zhang et al. 1996; Voldoire and Royer 2004). Early analyses of this extreme scenario differed fairly widely not only in results but also in how the scenario was translated into model parameters. The controlling parameters are the surface albedo, surface roughness, soil hydrological properties, and possibly the capacity of the vegetation to transpire through stomatal functioning.

The most obvious changes when the Amazon forest is converted to pasture are that its albedo increases and it becomes a much smoother surface. The increased albedo reduces surface sensible and latent fluxes and ultimately alters precipitation in almost all models; the decreased surface roughness tends to make the surface warmer, and the increased upward infrared radiation leads to further reduction of boundary layer buoyancy generation. Substantial feedbacks occur with atmospheric cloud cover, with less precipitation being accompanied by less cloud cover and more surface solar heating. The need for more realistic scenarios that address the consequences of conversion of smaller areas and forest fragmentation is recognized but has not yet been adequately addressed.

A few studies have addressed conversions in the United States between forest and cropland (e.g., Bonan1999; Pan et al. 1999; DeFries et al. 2002a). Although such conversions superficially appear similar to Amazon deforestation, results have been remarkably different, and, in particular, the surface temperatures for cropland have declined rather than increased. This effect appears to be a result of increases in evaporative cooling and suggests that plant nutrition effects, especially nitrogen levels, may provide strong coupling to surface temperatures and precipitation.

4.3.1.4 Models of Biogeochemical Impacts of Land Cover Change on Climate

Biogeochemical models to estimate greenhouse gas emissions from land cover change are much simpler than the climate models used to estimate the biophysical impacts. Most efforts have focused on the release and uptake of atmospheric carbon dioxide from land cover change, particularly deforestation. The most widely used approach is a "bookkeeping" model, in which estimates of the areas of each type of land use change are combined with prescribed response curves for decay and regrowth (Fearnside 2000; Houghton and Hackler 2001).

The land use changes in the bookkeeping model include the conversion of natural ecosystems to croplands and pastures, the abandonment of agricultural lands with subsequent recovery of natural vegetation, shifting cultivation, harvest of wood (forestry), plantation establishment, and, in some instances, fire management (exclusion and suppression of fire). The bookkeeping approach requires three basic inputs: rates of clearing, biomass at time of initial clearing, and decay and regrowth rates following clearing. Houghton and Hackler (2001) have applied the model at a very coarse, continental scale. More recently, remote sensing analysis to determine rates of deforestation have been combined with the bookkeeping model for more spatially explicit estimates of carbon emissions (Achard et al. 2002; DeFries et al. 2002b).

In addition to the bookkeeping approach, process models of the terrestrial carbon cycle have been used to estimate carbon fluxes from land use change (DeFries et al. 1999; McGuire et al. 2001). Such models simulate carbon stocks in vegetation and soil and the uptake and release of carbon through photosynthesis and respiration, based on variables such as climate, incoming solar radiation, and soil type. The most recent developments are dynamic models that simulate the response of vegetation to climate change and enhanced growth from elevated atmospheric carbon dioxide concentrations, as well as the resulting feedbacks to the atmosphere through changes in the vegetation's uptake and release of carbon dioxide (Cox et al. 2000). Anthropogenic land cover changes have not yet been incorporated in this framework.

Most efforts to model greenhouse gas emissions from land cover change have focused on carbon dioxide. Models to estimate emissions of other greenhouse gases from land use change, including methane from landfills, rice paddies, and cattle, and nitrous oxide from agricultural soils, are hampered by incomplete understanding of the biological processes.

4.3.2 Critical Evaluation of Approaches

4.3.2.1 Modeling of Biophysical Impacts of Land Cover Change

This section assesses how successful we judge models to be in attempting to model the impacts of land cover change on local climates. The local and regional climate variables that are modeled are primarily surface temperature and humidity and rainfall. We address how well this is done from the viewpoint of somebody who might want to use these models as a tool. A description of the details of any one model is far too complex to be presented here. Rather, in order to assess their success, we describe in broad-brush terms what the models are trying to do. Details are omitted, such as the fact that surface temperature is not a single variable but has several important elements that must be individually modeled. The type of rainfall believed to be most affected by land use change is of the "convective" (thunderstorm) variety. Because it involves important processes that occur on scales that are not incorporated in current models, we judge that changes of this variable are not yet reliably modeled.

4.3.2.1.1 Modeling of temperature change

Current modeling can be judged to provide, in principle, adequate results for temperature changes on a large scale.

Temperatures change with land cover change because of changes in absorbed solar radiation (controlled by cloud cover and albedo), because of changes in the fraction of energy going into evapotranspiration, and because of changes in roughness elements alter turbulence patterns. The land surface modeling of these relationships is limited primarily by uncertainties as to how albedo, roughness elements, clouds, and precipitation will change with land use change. Mathews et al. (2004) estimate that past land cover change has cooled the world by between 0.1 and 0.2 K and that the carbon released by this land cover change has warmed the world by a comparable amount.

4.3.2.1.2 Representation of heterogeneous land surface in models

Land cover is heterogeneous on a spatial scale that is much finer than the coarse grid cell size of climate models; these models treat all vegetation within grids the size of many thousands of square kilometers as essentially homogeneous. Because the resolution is so coarse, it is not possible to simulate with confidence the possible effects of forest fragmentation and inclusion of patches of other cover, such as crops or pasture. Local micrometeorological factors that will change with land cover exert considerable controls on local and regional temperatures. Since such changes are confined to the fraction of land whose use/cover has been changed and can go in both directions, their contribution to global temperature changes is usually thought to be relatively small. However, they can become considerably more significant if temperature changes are weighted by various risk factors such as proximity to human populations. The micrometeorological effects for particular local or regional systems may have important consequences for precipitation, as discussed later, but we do not understand how to include such effects in climate models.

4.3.2.1.3 Difficulties in modeling rainfall

Modification of rainfall is potentially one of the most important climatic impacts of land use change (e.g., Pielke 2001). This interesting issue is not well developed because it involves scaling aspects of modeling that are not sufficiently advanced. However, it is possible to clarify what is most important. Precipitation in summertime and tropical systems (that is, rainfall) is largely or entirely convective. It is this type of precipitation that is most sensitive to the atmosphere's lower boundary and hence land use change. Wintertime precipitation is largely generated by large-scale storm systems that are less connected to the surface and often originate over the ocean.

Convective rainfall is initiated primarily because of the instability (positive buoyancy) of near surface air that acts in two ways: it allows convective plumes to penetrate from the boundary layer up to the level of free convection, where moist instability carries it further, as high as the top of the tropopause; and through horizontal gradients it creates pressure forces that drive horizontal convergence, hence further uplift. The drivers of these mechanisms are land heterogeneities, which are currently lost in the processes used to scale the effects of motions on these scales to the scales resolved by climate models. In principle, scaling should include all the statistical properties given by distributions of the small-scale systems that couple back to the large scale. However, the formulations currently used assume an underlying homogeneous surface and may be intrinsically incapable of determining the changes of rainfall from land use change.

4.3.2.1.4 Difficulties assessing the significance of modeled impacts of land use change

Modeling studies test the sensitivity of climate to land cover changes by varying only the land cover in the model. In reality, land cover is only one of many factors that determine climate, including winds, incoming radiation, and clouds. These confounding factors make it difficult to identify a "land cover signal" from natural variability either in a model or in observations. Although sound statistical procedures are available for determining the "signal to noise" ratio of systems with spatially and temporally correlated randomness (e.g., Von Storch and Zwiers 1999), these have commonly not been used in studies of climate change from land use change. This limitation hampers interpretation of impacts reported in the literature.

4.3.2.1.5 Other complexity issues related to inadequacies of scaling methodology

More detailed models traditionally calculate evapotranspiration in terms of three components: transpiration, soil evaporation, and canopy evaporation (interception loss). Model results depend on how these are apportioned, which depends strongly on precipitation intensities. One difficulty with many models is that they apply their calculated precipitation and radiation from the atmospheric model uniformly over their grid-squares. These resolution elements are generally of much larger scale than the occurrence of individual convective systems, however, and hence poorly match actual local precipitation intensities and radiation. Appropriate precipitation and radiation downscaling must be used in the model to obtain better results. Because of the use of faulty satellite-derived data, some models have underestimated the LAI of tropical forests and in doing so have exaggerated the losses of soil water to bare soil evaporation.

Current parameterizations of runoff do not provide very plausible schemes for the downscaling to the scales on which precipitation and runoff occur. Because runoff provides a major feedback on soil moisture, inadequacies in its treatment introduce uncertainty into the issue of soil-moisture/vegetation interaction (e.g., Koster and Milly 1997).

4.3.2.2 Modeling of Biogeochemical Impacts of Land Cover Change

Model estimates of greenhouse gas fluxes from land cover change have a large range of uncertainty. Carbon dioxide emissions from deforestation and uptake from regrowth are the most uncertain components of the global carbon budget reported by the IPCC. The uncertainties arise from imprecise data on the required model inputs. Estimates of the rates of deforestation vary, and spatially explicit data covering the entire tropical belt are not available. These estimates are improving with the use of satellite data, but there is currently no pan-tropical observational system for monitoring deforestation. A second source of uncertainty arises from lack of spatial data on biomass distributions prior to clearing. Field measurements of biomass are based on point-samples, which are difficult to extrapolate over larger areas. Satellite capabilities to assess biomass distributions over large areas are not in place.

Estimates of other greenhouse gas fluxes—methane and nitrous oxide—from land cover change are even more uncertain than estimates of carbon dioxide fluxes. For these gases, inadequate understanding of the biological processes limits the modeling capabilities.

4.3.3 Research Needs

Future scenarios need to provide data quantified for input to climate models. Specifically, they need to describe in quantitative terms how the surface structure and its radiative properties have been modified, using parameters used by the climate models.

Test cases are needed with simultaneous observations of land use change and climate change to test modeling predictions; some areas expected to undergo large land use change in the future should be equipped with an adequate observational system to measure the consequent climate change.

An interactive vegetation-climate dynamical system needs to be a component of future scenarios. That is, quantitative trajectories of land use on a global basis should be prescribed in terms of quantities that can be used as boundary conditions for climate models. In this way, it would become possible to address the synergies between land use-driven climate change and greenhouse warming.

For the biogeochemical fluxes, improved monitoring systems for deforestation and biomass distributions are needed to reduce uncertainties of carbon emitted to the atmosphere as a result of land cover change. In addition, dynamic models that estimate changes in carbon fluxes as a result of vegetation being altered by climate change need to also include vegetation changes expected because of human activities, which will also be altering the landscape.

4.4 Forecasting Change in Food Demand and Supply

Over the past 50 years, there have been at least 30 quantitative projections of global food prospects (supply and demand balances), as well as numerous qualitative predictions, with the latter often tied to short-term spikes in global food prices. Global simulation models that simulate the interrelationships among population growth, food demand, natural resource degradation, and food supply are yet another class of forecasting exercises (Meadows et al. 1972, 1992; Mesarovic and Pestel 1974; Herrera 1976); but they are not commonly used today.

The number of players engaging in projections of future food demand, supply, and related variables at the global level has been declining over time. Important organizations conducting projections on the global scale include the Food and Agriculture Organization of the United Nations, the Food and Agriculture Policy Research Institute, the International Food Policy Research Institute, the Organisation for Economic Co-operation and Development, and the U.S. Department of Agriculture. Other food projection exercises focus on particular regions, like the European Union. Finally, many individual analyses and projections are carried out at the national level by agriculture departments and national-level agricultural research institutions. Results from some of these models are published periodically with updated projections. In addition to their differing coverage and regional focus, existing approaches also vary in the length of the projections period, the approach to modeling, and in the primary assumptions made in each model. The focus in this section will be on global food projection models.

4.4.1 Existing Approaches

This section examines the evolution of food supply and demand projections and examines current food projection models based on various criteria, following McCalla and Revoredo (2001), who carried out a critical review of food projection models.

4.4.1.1 Evolution of Food Supply and Demand Projections

Early food models (for example, the mathematical model of population growth posited by Thomas R. Malthus) focused on potential food gaps by comparing fixed land resources with rates of growth of population. This was followed by a requirements approach, where minimum nutritional needs were multiplied by population to produce projected food needs; on the supply side, yield increases were added to supply projections.

By the 1960s, income and Engel curves (statistical relationship between consumption and income) were added to food demand projections, while Green Revolution changes in food production, as well as resource limits, were added on the supply side. Food price instabilities in the early 1970s, making the assumption of constant prices illusionary, spurred the disaggregation of global models to the national level, with domestic supply and demand, country by country, and appropriate cross-commodity (maize and wheat, for instance) relationships embedded and with explicit recognition of policy built in. Disaggregation also allowed for a more detailed representation of changes in food preferences, including the diversification of diets with changing income levels. Models thus graduated from supply-and-demand gap projections into global price equilibrium trade models, which are more sophisticated, much larger, and more expensive to maintain (McCalla and Revoredo 2001).

4.4.1.2 Approach to Food Projections Modeling

The main types of global food projection models fall into two categories, trend projection models and world trade models.

Trend projection models project supply and demand separately based on historical trends. Relative prices are assumed to be constant over time. In pure trend projection models, which include most of the existing trend models, the difference between projected consumption and projected production creates a gap, indicating food surpluses and shortages at the regional or global level, which can be bridged through trade. FAO's projections during the 1960s to the 1980s, IFPRI's 1977 and 1986 projections (IFPRI 1977; Paulino 1986), OECD's 1960s projections, and USDA's 1960s projections fall into this category. In extended trend projection models, a spatial trade model is used to distribute the projected surpluses or deficits among regions and countries (Blakeslee et al. 1973). This can be done in the form of transportation models, which minimize the cost of moving surpluses to shortage locations by estimating food flows over geographical regions.

The simplest world trade models assume a global supply and demand equilibrium. In particular, these models estimate supply and demand functions at the country/regional levels; projections are aggregated at the world market, where prices adjust until global supply equals global demand. These models are also called price endogenous models. All major food projection models in use fall into the class of global non-spatial trade models. They include the different versions of FAO's World Food Model (FAO 1993), partly used in FAO's *World Agriculture: Towards 2015/30* study (Bruinsma 2003), IFPRI's IMPACT model (Rosegrant et al. 2001), FAPRI's commodity models (Meyers et al. 1986), the World Agricultural Model from the International Institute for Applied Systems Analysis (Parikh and Rabar 1981), the Free University of Amsterdam's Model of International Relations in Agriculture (Linnemann et al. 1979), and the World Bank model (Mitchell et al. 1997).

An alternative form of world trade models starts by assuming the costs of trade among regions have been minimized (using output from transportation models), subject to constraints that represent the characteristics of the different regions (Thompson 1981). These models depict spatially varying patterns of trade between different regions. Unfortunately, spatial models do not predict trade flows well, due to the reality of quantitative trade barriers, the heterogeneity of commodities in terms of characteristics and seasonality, and risk diversification strategies being pursued by importers (McCalla and Revoredo 2001).

Earlier models used linear equations, while more recent versions are based on nonlinear elasticity equations, which can better handle sharp perturbations (or "shocks"). Early projection models were static, with point estimates for future years dependent on projected rates of change in population and other key variables. The alternative approach entails recursive models that estimate all variables annually, moving repeatedly toward the final year, allowing for the observation of the path of adjustment. Most projection models are partial equilibrium models—that is, they focus on the agricultural sector instead of representing the entire economy.

4.4.1.3 Coverage

Country and commodity coverage differ by model. Whereas some models explicitly focus on developing countries (for example, the IMPACT and FAO models) and therefore tend to aggregate industrial countries and regions, others have been specifically developed to analyze the food supply, demand, and trade projections of industrial countries, in general, or the European Union, in particular, like CAPRI of Bonn University (CAPRI 2004).

The European Commission, for example, recently commissioned a study on the impact of its Mid-Term Review proposals for the year 2009 with reference to a status quo policy situation, involving six different agricultural projection models: the EU-15 agricultural markets model and the ESIM model, both under the Directorate General for Agriculture of the EU (DG-AGRI 2003a); the FAPRI model (FAPRI 2002a, 2002b); the CAPRI model of the University of Bonn, operating at the regional level (DG-AGRI 2003b); the CAPMAT model of the Centre for World Food Studies of the University of Amsterdam and the Netherlands Bureau for Economic Policy Analysis in The Hague (DG-AGRI 2003a); and the CAPSIM model, operating at the national level, also from the University of Bonn (DG-AGRI 2003a, 2003c). Similarly, commodity coverage differs among models, depending on the region or issue of concern. IMPACT, for example, started out with a focus on rice, followed by other staple crops of importance to the food security situation of poor countries, before adding higher-value commodities.

4.4.1.4 Projections Period

Long-term projections include those by IFPRI and FAO (as presented in *World Agriculture: Towards 2015/30*). Short-term projections have been developed by FAO, FAPRI, USDA, and OECD.

FAO has produced a series of long-term projections, beginning with the *Indicative World Plan for Agricultural Development* (FAO 1970), followed by *World Agriculture: Towards 2000* (Alexandratos 1988), *World Agriculture: Towards 2010* (Alexandratos 1995), and *World Agriculture: Towards 2015/2030* (Bruinsma 2003). These are recursive global non-spatial trade models. The most recent study has a base year of 1997–99, incorporates the medium-variant U.N. population projections (2001), GDP data from the World Bank, and agricultural data from its own databases to project food supply, demand, and net trade for crops and livestock products for 2015 and 2030.

FAO develops projections through many iterations and adjustments in key variables based on extensive consultations with experts in different fields, particularly during analysis of the scope for production growth and trade. The end product may be described as a set of projections that meet conditions of accounting consistency and to a large extent respect constraints and views expressed by the specialists in the different disciplines and countries (Bruinsma 2003, p. 379). The FAO study only uses one scenario: a baseline that projects the future that the authors anticipate to be most likely.

The International Model for Policy Analysis of Agricultural Commodities and Trade was developed at IFPRI in the early 1990s (Rosegrant et al. 1995). IMPACT is a representation of a competitive world agricultural market for 32 crop and livestock commodities and is specified as a set of

43 country or regional sub-models; supply, demand, and prices for agricultural commodities are determined within each of these. World agricultural commodity prices are determined annually at levels that clear international markets.

IMPACT generates annual projections for crop area, yield, and production; demand for food, feed, and other uses; crop prices and trade; and livestock numbers, yield, production, demand, prices, and trade. The current base year is 1997 (average of 1996–98). The model uses FAOSTAT agricultural data (FAO 2000); income and population data and projections from the World Bank (World Bank 1998, 2000) and the United Nations (UN 1998); a system of supply and demand elasticities from literature reviews and expert estimates; rates of malnutrition from the U.N. Administrative Committee on Coordination–Subcommittee on Nutrition (ACC/SCN 1996) and the World Health Organization (WHO 1997); and calorie-malnutrition relationships developed by Smith and Haddad (2000).

The Food and Agricultural Policy Research Institute publishes short-term projections of the U.S. as well as an annual world agricultural outlook (FAPRI 2003). This consists of an integrated set of non-spatial partial equilibrium models for major agricultural markets, including world markets for cereals, oilseeds, meats, dairy products, cotton, and sugar. For each commodity, the largest exporting and importing countries are treated separately, with other countries included in regional groupings or a "rest of world" aggregate. For most countries and commodities, the model estimates production, consumption, and trade; in many cases the model also estimates domestic market prices, stocks, and other variables of interest. Parameters are estimated based on econometric techniques, expert opinions, or a synthesis of the literature. Similar to IFPRI's IMPACT model, area is generally a function of output and input prices and government policies, while yield equations incorporate technical progress and price responses. The projection horizon is 10 years (DG-AGRI 2003a).

The OECD *Agricultural Outlook* provides a short-term assessment (five years ahead) of prospects for the markets of the major temperate-zone agricultural products of OECD members (OECD 2003). The projections to 2008, presented in the latest *Outlook* based on the recursive AGLINK model, are considered a plausible medium-term future for the markets of key commodities. Projections are developed by the OECD Secretariat together with experts in individual countries from annual questionnaires supplemented with data from FAO, the United Nations, the World Bank, and the IMF to determine market developments in the non-OECD area. National market projections are then developed with AGLINK and linked with one another through trade in agricultural products. Final results are presented following a series of meetings of the various commodity groups (cereals, animal feeds and sugar, meat and dairy products) of the OECD Committee for Agriculture (Uebayashi 2004).

The U.S. Department of Agriculture also produces short-term baseline projections of the agricultural situation 12 years into the future (USDA 2003). These projections cover agricultural commodities, trade, and other indicators, including farm income and food prices. The USDA presents a baseline scenario that projects a future with no shocks, a continuation of U.S. policies, and other specific assumptions related to agricultural policies, the macro economy, weather, and international development. Crops included are corn, sorghum, barley, oats, wheat, rice, upland cotton, and soybeans, as well as some fruit, vegetable, and greenhouse/nursery products. The model also produces projections for livestock, including beef, poultry, and pork. The projections that are presented tend to focus on the situation in the United States.

4.4.2 Critical Evaluation of Approaches

In their assessment of global projection models, McCalla and Revoredo (2001 p. 39) conclude that projections with shorter time horizons are more accurate than those with longer horizons; that projections are more accurate for aggregations of components—regions, commodities—than for the component parts themselves; and that projections for larger countries tend to be more accurate than for smaller ones. Data problems are a major cause of error, especially in developing countries. Moreover, data deficiencies are most frequently encountered in countries with severe food security problems, leading to erroneous conclusions and making it particularly difficult to develop adequate policy interventions.

For industrial countries, modeling rapidly changing, complex domestic policies, including quantitative border restrictions, is a major issue of concern. Rosegrant and Meijer (2001) point out that for IMPACT, the main discrepancies in the projections are due to short-term variability, such as the collapse of the Soviet Union or major weather events, which cause large departures from fundamental production and demand trends—variability that long-term projection models are not intended to capture.

Most of the currently used projection models are equilibrium models, which by construction require continuous adjustment to produce consistent, stable conclusions (McCalla and Revoredo 2001).

Food projection models typically produce alternative scenario results, which can then be used to alert policymakers and citizens to major issues that need attention. A test for the usefulness of these models may therefore be whether the results of these models (for example, the different scenarios) enrich the policy debate (McCalla and Revoredo 2001).

While models can make important contributions at the global and regional levels, food insecurity will be increasingly concentrated in individual countries with high population growth, high economic dependence on agriculture, poor agricultural resources, and few alternative development opportunities. These countries continue to be overlooked in regional and global studies because, overall, resources are sufficient to meet future food demands.

Each of the models described includes several critical assumptions in their approaches. Although the general methodology and underlying supply and demand functional forms are well established in the literature and have been widely validated, the details of how to implement these principles in specific models are not agreed on. For example, the elasticity of supply and demand functions is often unknown. Moreover, the supply and demand functions must be adjusted by growing incomes or population growth, which are not easily predicted and thus introduce an exogenous layer of uncertainty.

4.4.3 Research Needs

Future research should include the integration of poverty projections with global food supply and demand projections. In addition, distributional consequences of such projections need examination. Moreover, research aimed at generating future food security–environment scenarios must sufficiently disaggregate agroecologies and commodities so that chronically food-insecure countries do not get overlooked in regional/global modeling exercises. To represent the nexus of poverty, food insecurity, and land degradation, we will need models that better treat the way changes in ecosystems influence these factors and in turn are driven by them. Last, there is clearly the possibility that new technologies, most notably genetically modified crops, could alter food production systems in ways that have implications for human well-being, local economies, and land practices. A forthcoming study by the National Research Council of the U.S. National Academy of Sciences is focusing on alternative futures due to biotechnology, and that study could be a foundation for better models of food production.

4.5 Forecasting Changes in Biodiversity and Extinction

Forecasting changes in biodiversity is key to developing plausible scenarios of the future. Unfortunately, biodiversity does not mean the same thing to all ecologists and cannot be assigned one unambiguous metric. For practical reasons, the MA scenarios focus on species richness. Changes in species richness include both gains and losses of species. This section briefly outlines several approaches to predicting changes in species richness on the time scale of 100 to 1,000 years. Global extinction is considered by some to be the most serious of all the anthropogenic global changes because it is the only one that will never be reversed.

4.5.1 Existing Approaches

4.5.1.1 Qualitative Approach

One of the earliest attempts to develop global biodiversity scenarios used a qualitative approach (Sala et al. 2000; Chapin III et al. 2001). The exercise focused on terrestrial biomes and freshwater ecosystems, but the qualitative approach could be used similarly to evaluate patterns of biodiversity change in the oceans. The globe was divided into 11 biomes and two types of freshwater ecosystems, and scenarios were developed for the year 2100. The first step was to identify, based on expert opinion, the major drivers of global biodiversity change. The major drivers were determined to be changes in land use, climate, nitrogen deposition, biotic exchange, and atmospheric concentration of carbon dioxide. Biotic exchange referred to the accidental or deliberate introduction of non-native species into an ecosystem.

The second step broke the analysis into two components: assessing the drivers of biodiversity change and characterizing biome-specific sensitivity to changes in those drivers. Patterns of change for drivers were described from a series of existent models. For example, the IMAGE 2 model (Alcamo 1994) provided patterns of global land use change, and Biome 3 (Haxeltine and Prentice 1996) yielded estimates of climate change and potential vegetation. More qualitative models were used to estimate the global patterns of the other drivers. Drivers were not expected to change uniformly across biomes. Although there was agreement that biomes ought to react differently to changes in drivers, there was no quantitative assessment of biome sensitivity to each driver. Instead, the exercise developed a ranking of sensitivity for each driver and biome based on the opinion of experts representing each biome. These experts have all worked in the biome that they were representing; their appreciations of biome sensitivity were based on their understanding of the ecology of each biome and were calibrated across biomes at a workshop at the National Center for Ecological Analysis and Synthesis, University of California, Santa Barbara. Sensitivity estimates were ranked on a scale from 1 to 5.

The relative expected change in biodiversity from each biome and freshwater ecosystem resulting from each driver was calculated as the product of the expected change in the driver and the biome sensitivity. Finally, the total biodiversity change per biome depended on the interactions among drivers of biodiversity change. The exercise developed three alternative scenarios by assuming that there were no interactions among drivers, that the interactions were synergistic, or that the interactions were antagonistic. The three scenarios would encompass a range of potential outcomes. Information was not available to assign a higher probability for any of the alternatives. Computationally, the no-interaction scenario calculated total biodiversity change as the sum of the effects of each driver, the synergistic scenario used the product of the change resulting from each driver, and the antagonistic scenario used the change resulting from the single most influential driver.

4.5.1.2 Correlation to Environmental Variables

Climate has been identified as one of the most important correlates of species richness for a wide range of taxa (Rosenzweig 1995; Whittaker et al. 2001; Brown 2001). In combination with changing topography, climate change has been invoked to explain speciation and extinction events in such diverse taxa as hominids (Foley 1994), trees (Ricklefs et al. 1999), carabid beetles (Ashworth 1996), and birds (Rahbek and Graves 2001). Research on elevational clines in species richness (e.g., Rosenzweig 1995; Lomolino 2001; Brown 2001) further supports a fundamental role for climate as a determinant of patterns of biodiversity.

Many species exhibit a latitudinal gradient in species richness (e.g., Turpie and Crowe 1994; Cumming 2000) that is strongly linked to climate (Gaston 2000; Whittaker et al. 2001), available energy, and primary production (e.g., O'Brien 1998). These gradients have been used as one way of predicting species richness over large areas, based on the correlation between species occurrences and environmental conditions. Correlative approaches typically use general linear models to estimate species richness in unsampled areas. Species richness is frequently considered as a response variable in its own right. An alternative but more time-consuming approach is to model individual species occurrences and then to stack species models to produce estimates of species richness.

A special class of "correlative models" that has been used to predict biodiversity impacts is the so-called bioclimatic envelope approach (Midgley et al. 2002; Erasmus et al. 2002). These models identify the current distribution of species in terms of climatic and other environmental variables (topography, soils, etc.) and then infer local disappearance of species because new conditions are outside the species "bioclimatic envelope." This approach cannot really predict whether a species will become extinct; instead, it predicts changes in where species should occur. If no appropriate climate zone exists, then it might be concluded that a species will go extinct, but this is not certain.

4.5.1.3 Species-Area Relationship

The most widely used approach for predicting species loss entails the application of the species-area relationship (Pimm et al. 1995; May et al. 1995; Reid 1992). This describes one of the most general patterns in ecology (Rosenzweig 1995; Brown and Lomolino 1998; Begon et al. 1998): the relationship between the area of sampling and the number of species in the sample follows the power law,

$$S = cA^z$$

where S is the number of species, A is the sampled area, z is a constant that typically depends on the type of sampling, and c is a constant that typically depends on the region and taxa sampled. This was the approach adopted for the terrestrial biodiversity scenarios, and its assumptions and uncertainties are discussed at length in Chapter 10. Here we simply revisit the high and low points of this approach.

The idea behind using the SAR to estimate extinction rates is relatively straightforward; it simply assumes that the number of species remaining after native habitat loss follows a species-area curve, where A is the area of native habitat left. A typical value of z for islands of an oceanic archipelago or other types of habitat isolates (mountaintops, forest fragments, etc.) is 0.3 (Rosenzweig 1995). It takes from a few decades to several centuries for the species number in the remnant habitat to reach the equilibrium predicted by the SAR (Brooks et al. 1999; Ferraz et al. 2003; Leach and Givnish 1996; see also Chapter 10). If habitat restoration takes place during this time, the extinctions will be fewer than predicted by the SAR.

4.5.1.4 Threat Analyses Approaches

In many cases species are at risk because of habitat degradation and environmental threats that do not lend themselves easily to a strict species-area curve approach. This is especially true for freshwater and marine biodiversity. In freshwater systems, water withdrawals and dewatering of streams obviously can make it impossible for fish to survive. One approach is to link water discharge to fish diversity (e.g., Oberdorff et al. 1995). Similarly, in the marine environment extinctions are very hard to observe, and the primary impact that has actually been measured is a change in the biomass occupying different trophic levels. Chapter 10 presents a method for relating changes in biomass at trophic levels to changes in biodiversity. These threat analyses approaches typically start with a statistical relationship between some measured stress and a measured response in a particular taxa (like fish in response to reduced water discharge).

4.5.1.5 Population Viability Analyses

In order to categorize species according to their risk of extinction, modelers routinely conduct population viability analyses. These range from simple diffusion approximations for population fluctuations to detailed stochastic demographic matrix models. In all cases extinction is primarily a function of the current population size, environmental variability, and rate of population growth rate (Morris and Doak 2002). Changes in extinction risk will result if any of these key factors is altered. To date, most models that attempt to predict changing extinction risk per species tend to focus on what happens because species abundance is reduced. There is no reason, however, that changes in environmental variability (which is expected to be affected by climate change) could not be used to project an altered extinction risk for any given species. In theory, someone could sum PVA models over many species and then generate predictions about aggregate extinction risks.

4.5.2 Critical Evaluation of Approaches

4.5.2.1 Qualitative Approach

The strengths of the qualitative approach are that it is simple, tractable, and easy to communicate. It made global biodiversity scenarios possible before global species richness was fully described. The major weaknesses of the approach are associated with its scale and qualitative nature. The scale at which the qualitative biodiversity scenarios were run was very coarse, with only 11 terrestrial units and two freshwater ecosystem types. The coarse scale resulted in large errors in driver patterns and also yielded results at a scale that was too coarse to be used in management and decision-making. Most decisions about biodiversity occur at finer scales—from paddocks to nations—and never reach the level of biomes. The second weakness is the qualitative nature of the exercise that is related with the scale. This type of exercise may only be doable at a coarse scale, where differences among biomes are large enough to be captured without more sophisticated calculations.

4.5.2.2 Correlation to Environmental Variables

In general, correlative approaches offer a reasonable alternative to mechanistic methods. Their main weaknesses are the same as those of any statistical analysis. The results may be influenced by biases in sampling regime, they rely on large sample sizes for accurate prediction, and they may fail to take adequate account of complex system dynamics and nonlinearity. Correlative approaches also make the assumption that species will occur wherever habitat is favorable, ignoring the potential for dispersal limitation and other confounding biotic interactions (such as competition and predation) unless they are explicitly included in the model.

Ideally, linear models of species richness should be based on a small number of variables with well-demonstrated relevance to the occurrence of the study taxon. A number of studies have reported poor out-of-sample prediction, which can often result from over fitting; a variety of statistical methods (bootstrapping, jack-knifing, and model averaging) can be used to overcome this problem (Raftery et al. 1997; Fielding and Bell 1997; Hoeting et al. 1999). Several authors have used the predictive power of correlative methods to consider the likely implications of climate change for species distributions and population processes (e.g., Schwartz et al. 2001; Thomas et al. 2001; Kerr 2001; Peterson et al. 2001), although a failure to take account of covariance between climatic variables may result in oversimplification of the problem (Rogers and Randolph 2000).

A further weakness in correlative approaches is that species will not exploit their full potential geographic range if individuals are unable to reach areas where the habitat is suitable. Island biogeography (MacArthur and Wilson 1967; Hubbell 2001) has been a widely used framework for thinking about changes in species distributions. Recent studies of invasive species (e.g., Parker et al. 1999) also shed light on the dispersal ability of organisms and the ways in which physical and biological variables interact to change the extents of species ranges.

4.5.2.3 Species-Area Relationship

One strength of the SAR approach is its simplicity: it is very straightforward way to explain how the calculation of biodiversity loss is made. Furthermore, the SAR is an ubiquitous pattern in nature, with more than 150 studies documenting SARs for different taxa and regions (see, e.g., the review in Lomolino and Weiser 2001). One weakness of the SAR approach is that it does not distinguish among species and hence does not tell us where to direct conservation efforts. Second, it does not predict when the loss of ecosystem services associated with a species or a group of species is going to occur, which may be before a species is locally extirpated or may not occur even after that event. Finally, the SAR approach accounts only for the impacts of habitat loss. While habitat loss is the major driver of biodiversity loss (Sala et al. 2000; Hilton-Taylor 2000), other drivers such as hunting, trade, invasive species, and climate change are also important. Nevertheless, in the case of climate change the SAR has been used to predict biodiversity loss based on predictions of habitat loss induced by climate (Thomas et al. 2004).

A major assumption of the SAR approach is that no species survive outside native habitat. Put another way, instead of stating that habitat is lost, it is more accurate to say that habitat is changed. The actual changes in habitat, even if from a forest to a plantation, do not correspond to total habitat degradation. Some species will remain, even as habitat is altered. In Costa Rica, for instance, about 20% of the bird species are exclusively associated with human-altered habitats, and the majority of species (about 66%) use both natural and human-altered habitats (Pereira et al. 2004). Furthermore, when the SAR is applied to a given region, it will project only regional extirpations. Global extinctions will depend on how many of the species going extinct are endemic to the region being considered. (See Chapter 10 for a thorough discussion of this issue.) Also, the SAR approach applies only to the native species of a region.

One large uncertainty associated with the SAR projections of biodiversity loss is the choice of the z-value (see Equation above). SARs can be classified according to the type of sampling in three categories: continental, where nested areas are sampled within a biogeographic unit, and typical z-values are in the range 0.12–0.18 (Rosenzweig 1995, but see Crawley and Harral 2001); island, where islands of an archipelago or habitat islands such as forest patches are sampled, and typical z-values are in the range 0.25–0.35 (MacArthur and Wilson 1967; Rosenzweig 1995); and inter-province, where areas belonging to different biogeographic provinces are sampled, and z-values cluster around 1 (Rosenzweig 1995). A review of the literature of SARs in vascular plants (see Chapter 10) suggests that typical intervals for the z-values can be even wider, and the mean values may differ from the ranges discussed above. It has been standard practice to use z-values of island SARs to estimate biodiversity loss (Pimm et al. 1995; May et al. 1995; Reid 1992), but arguments could be made for using the z-values of the continental or even the inter-province SARs, depending on the time scale of interest (Rosenzweig 2001).

4.5.2.4 Threat Analyses Approaches

Models that link threats like reduced water discharge to reduced diversity are in some sense a special class of correlative models, and hence have the same weaknesses as described for correlative approaches. This is a very new branch of modeling, and it is plagued by highly norepresentative taxonomic treatments. For instance, predictions of freshwater extinctions or changes in diversity are made only for fish. With the exception of coral reefs, predictions about marine diversity are also made only for fish. In fact, the data on trophic level and biomass comes from fish landed commercially; hence the relationships are based on harvested species.

4.5.2.5 PVA Models

The biggest limitation of population viability analyses models for predicting changes in biodiversity is that they are impractical because of their huge appetite for data and species-specific analyses. PVA models are best used to examine specific species. Even when applied to single species, there is extensive debate about their value because of huge uncertainty that often leads to a probability of extinction between 0 and 1 (Ludwig 1999). In management, PVAs are used for comparing relative risks of alternative options, but they do not lend themselves so well to point estimates of any given extinction probability.

4.5.3 Research Needs

There is a need for research that examines which estimation procedures for z-values and also which z-values are most appropriate for describing biodiversity loss as a consequence of habitat loss. Retrospective studies (e.g., Pimm and Askins 1995) that analyze past extinctions associated with habitat loss as well as long-term experimental studies concerning the effect of habitat manipulations on local species extinctions are needed. Also needed are approaches to validate diversity models, which includes the generation of suitable data sets of biodiversity that are publicly available.

A generalization of the SAR to multiple habitats is needed. Recently, Tjorve (2002) and Pereira and Daily (in review) have discussed the implications for biodiversity of changing the proportion of cover of different habitats in a landscape. However, more empirical and theoretical research is needed to determine how the SAR can be extended to complex landscapes with several habitats. Open questions include: Which z values should be used for different habitats? How do we estimate species loss caused by habitat conversion for species that prefer native habitat but can also survive in the agricultural landscape?

The development of models of biodiversity change that do not conveniently fit the species-area paradigm is very much in its infancy. For aquatic systems, for example, currently available data tend to represent a highly biased taxonomic sample (such as fish, while invertebrates or aquatic plants are neglected). In most cases the threat-models are extrapolated well beyond their original data range in order to obtain global predictions. These extrapolations need scrutiny. We expect that within 10 years there could be huge advances in tools for predicting aquatic biodiversity and how it responds under different scenarios.

There will probably never be "a biodiversity model" that enables us to predict changes of species richness at all of the geographic and temporal scales or for all ecosystems and taxa. The multiscale approach of the MA was met by a combined use of different quantitative and qualitative tools for estimating the future development of biodiversity in a hierarchical approach. This approach made it possible for us to answer such diverse questions as: What will the future pattern of species diversity of known groups be and what trends are to be expected in the future? Is there any scenario for significantly reducing biodiversity loss? How well can other, easy-to-obtain abiotic data be used to make predictions about biodiversity patterns? Judgment on the most appropriate methods to use should be based on the following criteria: spatio-temporal resolution, taxa of interest and data availability, ease of application, and ability to cope with unexpected events (such as increased extinction risk due to sudden population decline).

4.6 Forecasting Changes in Phosphorus Cycling and Impacts on Water Quality

Phosphorus (P) is frequently the limiting nutrient for primary production in freshwater ecosystems (Schindler 1977) and is recognized as a critical nutrient in marine ecosystems (Van Capellen and Ingall 1994; Tyrell 1999). Excess P input causes harmful algal blooms (including blooms of toxic species) as well as excessive growth of attached algae and macrophytes. Excess plant growth can damage benthic habitats such as coral reefs. Harmful algal blooms cause deoxygenation and foul odors, fish mortality, and economic losses. Impacts on human well-being include health problems caused by toxic algae blooms and waterborne diseases, as well as loss of aquatic resources. Economic costs derive from health impacts as well as increased costs of water purification and impairment of water supply for agriculture, industry, and municipal consumption (Carpenter et al. 1998; Postel and Carpenter 1997; Smith 1998). Because of the role of P in eutrophication and water quality, it was important that the MA Scenarios Working Group considered potential future changes in P flow to freshwater and marine ecosystems.

In this section, we assess the available models for P transport to water bodies, eutrophication, and impact on ecosystem services. We discuss process-based and export-coefficient models of P transport and many types of in-lake eutrophication models, from simple empirical models to more complex ones that include recycling and biotic effects. We also briefly discuss models that include interactions of policy with water quality.

4.6.1 Existing Approaches

4.6.1.1 Phosphorus Transport Models

4.6.1.1.1 Process-based models

Before discussing phosphorus models, it is important to understand how P moves through the environment. P arrives in surface waters primarily via runoff. P runoff can be dissolved in water, which moves on the surface and in subsurface flows. More commonly, P is delivered in the form of soil particles. Most P runoff is absorbed to soil particles and moves during major storms with heavy erosion (Pionke et al.1997). Once it enters the aquatic environment, P can be released in forms available for plant growth. Measures of P availability used in terrestrial ecology tend to underestimate the amount of P that can be released after soil is eroded into aquatic ecosystems (Sharpley et al. 2002).

Process-based models simulate P transport across watersheds to surface water. They have been used for small and large watersheds, in diverse soils and topographies. Such models are often used to estimate the impacts of different types of land use and management on P transport. For example, several have been used to highlight best management practices (Sharpley et al. 2002). Some of these models are based on or use the Universal Soil Loss Equation, first developed by Wischmeier (1958).

Some, like AGNPS–Agricultural Nonpoint Pollution Source (Young et al. 1989), estimate runoff in large watersheds (up to 20,000 ha) but can be used to analyze runoff from individual fields, and the impact of specific best management practices, within the overall watershed.

Other process-based models for simulating P transport include ANSWERS–Areal Nonpoint Source Watershed Environment Response Simulation (Beasley et al. 1985), GAMES–Guelph Model for Evaluating Effects of Agricultural Management Systems on Erosion and Sedimentation (Cook et al. 1985), ARM–Agricultural Runoff Model (Donnigan et al. 1977), and EPIC–Erosion-Productivity Impact Calculator (Sharpley and Williams 1990). Recent developments include models like INCA-P, a dynamic mass-balance model that investigates transport and retention of P in the terrestrial and aquatic environments (Wade et al. 2002). These models were largely developed in order to estimate the benefit or drawback of specific land management techniques and estimated P transport at the edges of agricultural fields.

4.6.1.1.2 Export coefficient models

Export coefficient models are steady-state models used to estimate P load based on the sum of P loads from various land types in a watershed (Wade et al. 2001). These models are generally simple, empirically driven, and often not spatially explicit. The name comes from the fact that each type of land in the model has associated with it an export coefficient that is an empirically determined estimate of P runoff from that type of land. Land may be defined by soil properties, land use, land management, or some combination of factors.

A few spatially explicit export coefficient models for P have been developed, such as those developed by Soranno et al. (1996), who included distance and routing to the lake, and Gburek and Sharpley (1998), who routed export to the stream from source areas in the watershed. These models tend to be highly data-intensive and are thus limited in their applicability.

Data generally come from measurements of edge-of-field P transport and field data such as the physical and chemical properties of the soil, land use, and land management properties. Export coefficient models can be linked to a GIS to estimate a runoff over a given watershed or watershed area. They often require plentiful data for parameterization and calibration (Sharpley et al. 2002).

A few recent export coefficient models have been developed to model the impact of land use change on P transport (Wickham et al. 2002). Wickham et al. (2000) use an export coefficient model to estimate total P transport with one land use map and then again with another land use map and compare the difference between the two to understand how land use change might affect P runoff.

4.6.1.2 Freshwater Eutrophication Models That Predict Important Ecosystem and Policy-relevant Impacts

4.6.1.2.1 Simple empirical models

The simplest, and in many ways most widely applicable, models of eutrophication are the empirical relationships between P input, or loading, to a water body and the biomass of primary producers in that water body (Rigler and Peters 1995). We will refer to these as simple empirical models for eutrophication. The general form of SEMEs is

$$A = f(\text{P input, covariates, parameters}) + \epsilon$$

In this equation, A is a measure of algal abundance such as cell concentration, chlorophyll a concentration, or primary production; P input is expressed as a rate or annual load; covariates (if present) include variates such as lake morphometry or hydrology; and ϵ are errors with a specified probability distribution (usually normal or lognormal, with moments estimated from the data). The parameters of the function f and the distribution of ϵ are fitted by regression methods.

SEMEs have a long history in limnology. An early and frequently adopted model was introduced by Vollenweider (1968). His model predicts chlorophyll concentration from P input rate, adjusted by simple corrections for water depth and hydraulic retention time. P input has also been used to predict other biotic variates, such as biomass of consumers, in lakes (Håkanson and Peters 1995). Of particular interest to the MA, P input has been used in conjunction with N:P ratios to predict concentration of cyanobacteria, an important type of toxic algae (Smith 1983; Stow et al. 1997). Harmful algal blooms are highly variable in space and time (Hallegraeff 1993; Soranno 1997). In general, predictions of chlorophyll have lower uncertainties than predictions of the timing and spatial pattern of harmful algal blooms.

4.6.1.2.2 Recycling and biotic effects models

The wide confidence intervals around predictions of empirical models and the desire to extrapolate beyond the calibration data have prompted considerable research on more complicated models of eutrophication. Much of this work has focused on P recycling from sediments and food web processes.

P recycling from sediments can cause lakes to have higher biomass of algae than expected from typical P input–chlorophyll relationships. P recycling from sediments is caused by anoxia (which increases solubility of iron-P complexes found in sediments) or turbulent mixing of sediments into the water. P recycling due to anoxia has been modeled empirically, using its correlation with other limnological drivers (Nürnberg 1984, 1995). More mechanistic models of recycling have also been developed (Tyrrell 1999).

Under conditions of excess P input, a positive feedback can maintain a quasi-stable eutrophic state. High production of algae leads to rapid depletion of oxygen in deeper water, as decaying algae sink to the bottom. Anoxia promotes recycling of P from sediments, leading to more production of algae, thereby creating a self-sustaining feedback. A model of this phenomenon exists for lakes (Carpenter et al. 1999b; Ludwig et al. 2003), though for many applications at least one parameter of the model has a large standard deviation (Carpenter 2003). This mechanism is exacerbated by sulfate deposition caused by coal-burning industrial processes. In anoxic waters or sediments of lakes and reservoirs, sulfate reduction leads to formation of iron sulfide, decreasing the availability of iron to bind P in sediments (Caraco et al. 1991). Sulfate is abundant in sea salt. A similar feedback among algal production, anoxic events, and P recycling has been modeled for marine systems (Van Capellen and Ingall 1994; Tyrrell 1999).

In shallow lakes or seas, P can be recycled by physical mixing of sediments by waves or bottom-feeding fishes (Scheffer et al. 1993; Jeppesen et al. 1998; Scheffer 1998). Feedbacks among water clarity, macrophytes, and bottom-feeding fishes result in two quasi-stable states—one with macrophytes, clear water, and few bottom-feeding fishes and the other with turbid water, no macrophytes, and abundant bottom-feeders. Models of shallow lakes are well understood (Scheffer 1998; Scheffer et al. 2001a, 2001b). The extent to which these models can predict lake dynamics is currently an area of active research.

Fish predation can change grazer communities and thereby change chlorophyll concentrations or primary production of pelagic systems (Carpenter and Kitchell 1993). The grazer community affects phytoplankton through direct consumption as well as excretion of P. Examples of food web impacts on phytoplankton are known from both lakes and oceans (Carpenter 2003). In lakes, the impact of grazers on chlorophyll or primary production can often be predicted from measurements of the body length of crustacean zooplankton (Pace 1984; Carpenter and Kitchell 1993). A simple empirical model of grazer effects substantially reduced the variance of predicted chlorophyll when the model was applied to a cross-section of North American lakes (Carpenter 2002).

Mechanistically rich simulation models have been used to understand or manage eutrophication in many situations (Chapra 1997; Xu et al. 2002). Such models address a diversity of climatic, biogeochemical, and biological factors that may affect eutrophication. Although these models have a number of fundamental similarities, such as the central role of nutrient supply, they also include a number of site-specific features (e.g., Bartell et al. 1999; Drago et al. 2001; Everbecq et al. 2001; Gin et al. 2001; Håkanson and Boulion 2003; Karim et al. 2002; Pei and Wang 2003). Because of the diversity and site-specificity of this family of models, it was not possible to recommend one particular mechanistic model for use by the MA.

Recycling and food web dynamics create threshold behaviors in the P cycle (Carpenter 2003). For the MA, the most important thresholds are those that, if crossed, create self-perpetuating eutrophication of a water body. These thresholds are difficult to discern before they are crossed (Carpenter 2003). For example, in a region of Wisconsin in the United States with generally high water quality, nearly half of the lakes were judged susceptible to self-perpetuating eutrophication (Beisner et al. 2003). Extensive regional data bases and whole-lake experiments that deliberately eutrophied three experimental lakes were necessary to make this calculation. Comparable data are available for few other regions of the world.

At present, validated and generally applicable models for the prediction of eutrophication thresholds do not exist. Certain key elements of such models are present in the research summarized here, but these elements have not been aggregated in a globally applicable modeling framework. Statistical studies of eutrophication thresholds reveal broad confidence intervals, indicating that uncertainties about the location of thresholds are high (Carpenter 2003).

4.6.1.2.3 Models for interaction of policy and water quality

Some policy-related models have addressed water quality, eutrophication, or harmful algal blooms. For example, the global scenario model PoleStar (Raskin et al. 1999) presents water quality indicators. The water quality module of PoleStar is relatively simple (and therefore readily testable with data where these exist) and transparent and has been used in a number of global scenario exercises.

For single lakes, stochastic dynamic optimization models have been used to determine optimal P input in the presence of thresholds and uncertainty about parameters (Carpenter et al. 1999b; Ludwig et al. 2003). Variants of these models have been used to study the possibility of estimating the thresholds for eutrophication by active adaptive management (Carpenter 2003; Peterson et al. 2003). The general finding is that someone is unlikely to learn the threshold without crossing it and thereby eutrophying the lake (Carpenter 2003).

Various other models have been used to study economic, social, or political processes that interact with ecosystem dynamics to determine ecosystem services derived from fresh water (Brock and de Zeeuw 2002; Guneralp and Barlas 2003; Janssen 2001; Scheffer et al. 2003; Tundisi and Matsumura-Tundisi 2003). These include comparisons of policies that maximize a measure of expected net utility over long time horizons as well as game theory models of stakeholder interactions. Other models have considered the dynamics of uncertainty about thresholds as managers attempt to maximize expected net utility for lake ecosystem services (Carpenter et al. 1999a; Janssen and Carpenter 1999; Carpenter 2003; Peterson et al. 2003). Extensions of such models in the form of computer games can be used to help stakeholders understand the vulnerabilities of water quality and form expectations that are consistent with sustainable use of fresh waters (Carpenter et al. 1999b; Peterson et al. 2003).

4.6.2 Critical Evaluation of Approaches

4.6.2.1 Phosphorus Transport Models

Phosphorus transport models can provide reasonably accurate estimates of P transport, especially in small, agricultural watersheds, for which most of them were developed. A recent review by Sharpley et al. (2002) provides an excellent overview of data and relationships available to update P transport models.

Applications of transport models are often limited by lack of data available for the detailed parameterization that is necessary. The more realistic these models attempt to be in terms of mechanisms of P transport, the more data they require. Predicting P transport may require highly accurate land use maps, digital elevation models, and data about fertilizer and manure use, including the P content of the manure. Increasing mechanistic detail, watershed size, or spatial resolution can quickly cause run times to become extremely long and the models themselves to be difficult to apply (Sharpley et al. 2002).

In addition, many models of P transport are designed to simulate P transport only in small-scale agricultural systems (Gburek and Sharpley 1998). The accuracy of scaling-up is dependent on how processes at finer scales relate to processes that govern P transport at larger, watershed scales (Sharpley et al. 2002). These models were unlikely to be useful for quantifying distinctions in the MA scenarios at global scales due to difficulties in generalizing across differences in the processes that drive eutrophication around the world.

Most P transport models are not linked to models that simulate impact of P on the aquatic ecology (Wade et al. 2002). Linking these two models in order to understand the impact of land use and management strategies on aquatic ecosystems will be an important next step for researchers and watershed managers.

4.6.2.2 Freshwater Eutrophication Models

Empirical eutrophication models of the type introduced by Vollenweider (1968) could provide a robust foundation for eutrophication estimates. Advantages of the empirical models include a long history of usage leading to considerable information about strength and limitations of the models, simple transparent mathematical structure, and the possibility of rigorous uncertainty analyses. Limitations of empirical models include uncertainty of extrapolation beyond the conditions of the data used to fit the regressions and the rather wide confidence intervals of prediction. Cole et al. (1991) and Pace (2001) discuss the strengths and limitations of empirical models in ecosystem science.

The simple empirical models omit a number of important effects. For the quantification of the global scenarios of MA, however, it was impractical to obtain data on these other factors at the necessary scales.

Recycling and food web effects are two important omissions from the simple P input models to predict chlorophyll. For the purposes of MA, these omissions probably caused simple empirical models to underestimate chlorophyll and the harmful effects of eutrophication under conditions when P recycling was likely to be high. These include situations in which there has been a long history of high P input, causing sediments to become enriched with P; water temperatures were warm (Nürnberg 1995); or ecosystems have received high inputs of sulfate (such as emissions from burning coal). Overfishing of top trophic levels may cause cascading effects that exacerbate eutrophication. In summary, then, the simple P input models could underestimate the severity of eutrophication in warmer regions of the world or under conditions of chronic heavy loading, climate warming, or food web transformation by fishing. These are the future conditions that needed to be addressed in some MA scenarios.

In situations where extensive data on lake morphometry, biogeochemistry, hydrodynamics, and food web structure are available, it may be advisable to use more detailed models to predict eutrophication. Pragmatically speaking, however, the current global data bases are not likely to provide enough detail to warrant one of these more sophisticated modeling approaches.

4.6.3 Research Needs

Many of the P transport models presented here require a large amount of data to be parameterized and calibrated for the particular watershed or region in question. Yet the underlying principles are similar, and it should be possible to develop more generally applicable P loading models. Improved spatial data sets for soils and topography (digital elevation maps) will also advance our ability to predict P loads. Better information is needed about how soil P concentrations and pools interact with land use and change in land use to affect P transport. Because P transport is affected by soil type, vegetation type, climate, and many other local variables, it has been difficult for researchers to generalize about P transport while attempting to model transport in a specific watershed. Sharpley et al. (2002) present an overview of the generalizations for which data exist.

While eutrophication may appear to be a regional problem, it is related to global changes that people are making to the P cycle through mining and widespread use of fertilizers (Bennett et al. 2001). Eutrophication of a given lake is not independent of eutrophication happening elsewhere; it is a global pattern. Developing simple large-scale or even global models that can be used to indicate P use, transport, and eutrophication based on land use should be a priority.

Developing models to understand the impact of land use change on P transport is also important. At present, the available models assume that the process of land use change does not release P (Wickham et al. 2000, 2002). That is, they figure that if agriculture exports equal X g P/ha and urban exports equal Y g P/ha, converting agricultural area to urban decreases P export by X-Y g P/ha. However, several studies have indicated that the period of transition to urbanized use is a critical period of high P transport (Kaufman 2000; Owens et al. 2000). Models that address the impact of this period of transition are important. Development of these models will require land use change data as well as a better quantitative data about the impact of periods of land use transition on P transport.

A priority for freshwater eutrophication models is developing a better understanding and predictive capability for internal recycling and sequestration mechanisms (including those mediated by organisms, especially invasive species). The interaction of internal recycling and sequestration with loading processes will be a critical aspect of this understanding. Although some models have been developed for integrating management with transport and eutrophication models, further development of models that integrate human interventions (including management actions) to water quality, aquatic ecosystem services, and human well-being is needed.

4.7 Forecasting Changes in the Nitrogen Cycle and Their Consequences

The nitrogen (N) cycle is a key regulator of the Earth system, linking terrestrial, marine, photochemical, and industrial processes. Biodiversity, carbon storage, and atmospheric chemistry are all regulated in part by the cycling of reactive nitrogen compounds. Over the last century and a half, expansion and intensification of agriculture together with fossil fuel combustion have led to an acceleration of natural microbial N cycling and have more than doubled N inputs to terrestrial ecosystems. Measurable results of these perturbations include a 17% increase in the atmospheric concentration of nitrous oxide (N_2O), a potent greenhouse gas, and a doubling of dissolved nitrogen export from rivers to coastal zones and of natural reactive nitrogen emissions to the atmosphere (Galloway et al. 2004). Since many of the effects of human N cycle perturbations are difficult to measure directly, ecosystem models play an important role in assessing and quantifying past and present impacts and in making future predictions.

4.7.1 Existing Approaches

4.7.1.1 Transport Models

4.7.1.1.1 Process-based models of the terrestrial nitrogen cycle

A number of models have been developed to simulate nitrogen biogeochemistry. The simulation of net primary productivity is central to all of these models. NPP is either derived from satellite normalized difference vegetation index data (Potter et al. 1996; Asner et al. 2001) or calculated as a function of climatological inputs like temperature, solar insolation, and precipitation (which drives the soil water balance component of the models). In terrestrial models that simulate the coupled carbon-nitrogen cycle, including CENTURY, TEM, BIOME-BGC, pNET, and the NCAR CLM2, the calculation of NPP is modulated by nutrient limitation—that is, soil nitrogen availability (McGuire et al. 1992; Aber et al. 1997; White et al. 2000; Parton et al. 2001; Bonan et al. 2002).

The availability of mineral nitrogen in the soil is controlled at short time scales by soil temperature and moisture conditions, at intermediate time scales by the supply of new organic matter from plant litter and decomposition of existing organic matter, and at longer time scales by changes in litter quality due to changing plant community composition and soil texture. In the absence of anthropogenic inputs, on very long time scales (decades to centuries), the composition of natural plant communities depends on the balance between accumulation and loss of fixed nitrogen, with accumulation from N deposition and from the symbiotic and asymbiotic fixation of atmospheric N_2 and with loss due to leaching and transport in outflow, microbial dentrification, and denitrification during biomass burning.

The different processes that determine soil N availability are simulated by current models with varying degrees of sophistication. Many terrestrial models have multiple compartments describing woody and herbaceous litter and rapidly and slowly degrading soil organic matter pools (Parton et al. 1987). These models contain detailed algorithms for soil organic matter dynamics that include competition between plants and soil biota for soil mineral nitrogen resources and nitrogen constraints for carbon assimilation and allocation to different plant tissues. Some models, like the NCAR CLM2, are beginning to incorporate dynamic vegetation algorithms, with explicit competition between multiple plant functional types for common soil water and mineral nitrogen resources, which allows for evolution of plant communities in the face of human perturbations and global change (Bonan et al. 2002).

In contrast to these relatively sophisticated algorithms, other key processes like N_2 fixation are still parameterized rather crudely—for example, as simple functions of precipitation or based on biome type (Parton et al. 1987). Furthermore, a recent assessment of terrestrial N_2 fixation, based on a compilation of measurements from different ecosystems, revealed a large uncertainty (a range of 100–300 Tg N/yr) in the global terrestrial fixation rate (Cleveland et al. 1999).

4.7.1.1.2 Modeling N export in rivers

A major consequence of N cycle perturbation is the increased transport of leached N in rivers to coastal regions. Increased N loading in coastal areas can stimulate harmful algal blooms and associated heavy loads of decaying organic matter, which can lead to hypoxic or anoxic conditions. This phenomenon, known as eutrophication, is often accompanied by changes in plant and algal species composition, fish death, coral reef degradation, and decreases in species diversity (NRC 2000). In extreme cases, such as the outlet of the Mississippi River into the Gulf of Mexico, eutrophication can turn coastal waters into "dead zones" (Rabalais et al. 2002). Increased N delivery to coastal areas also can lead to enhanced microbial production of nitrogen trace gases, including NH_3 and the greenhouse gas N_2O (Naqvi et al. 2000).

In recent years, a number of studies have attempted to quantify and identify the origin of N exported in rivers to coastal regions. These studies generally have used empirical models to relate N export in rivers to various independent variables, including basin runoff , land cover type, soil texture, human population, and N inputs from fertilizer, sewage, and atmospheric deposition (Seitzinger and Kroeze 1998; Caraco and Cole 1999; Lewis et al. 1999; Alexander et al. 2000; Lewis 2002; R. A. Smith et al. 2003; S. V. Smith et al. 2003). The statistical models range from simple linear regressions to complex nonlinear matrix inversions.

Recently, Donner et al. (2002) published the first process-based simulation of N transport in rivers on a regional scale. Their study coupled water runoff rates from a carbon-only terrestrial ecosystem model to prescribed N leaching fluxes and a hydrological routing model for the Mississippi River basin. The study did not account for increases in fertilizer and other anthropogenic N inputs over time, since its main purpose was to isolate the effect of changes in hydrology (that is, increased runoff) on N export in the Mississippi River from 1955 to 1996.

Although the terrestrial ecosystem model used in Donner et al. (2002) was a carbon-only model, most of the terrestrial coupled carbon-nitrogen biogeochemistry models just discussed explicitly calculate NO_3-leaching rates as well as water runoff rates. These models commonly calculate an N leaching term at each individual grid cell, which goes into a global accounting pool but otherwise effectively leaves the grid cell and disappears from the model. Future improvements in N cycle modeling will involve coupling terrestrial N leaching rates to hydrological and river routing models, permitting evaluation of downstream and long-term impacts of N leaching.

4.7.1.1.3 Ocean and coastal nitrogen biogeochemistry models

Human N cycle perturbations have impacts not only on terrestrial ecosystems but also on coastal regions and, potentially, on the open ocean. The transport of leached N in rivers to coastal regions and its stimulation of coastal eutrophication have already been described. In addition, a significant fraction of the reactive NH_3 and NO_x volatilized from soil and produced by fossil fuel combustion eventually deposits on coastal or open ocean waters (Holland et al. 1997; Paerl 2002). Global change may also have profound impacts on oceanic N cycling through other physical and chemical mechanisms, such as decreases in ocean pH associated with increasing atmospheric CO_2 and enhancement of thermal stratification due to global warming. The latter may increase water-column O_2 depletion, thus promoting denitrification (Altabet et al. 1995).

Three-dimensional ocean biogeochemistry-circulation models have historically not been well designed to simulate the impact of anthropogenic N inputs on the oceanic nitrogen cycle. Most early ocean carbon models were constructed around phosphorus as the limiting nutrient, in large part to avoid the complications of simulating biological sources and sinks of oceanic fixed N. Furthermore, rather than being computed mechanistically, oceanic primary production and associated organic matter remineralization were estimated as the fluxes needed to produce dissolved phosphate concentrations that match observed climatologies (Najjar and Orr 1998).

An additional complication is that Fe rather than N appears to be the limiting nutrient in high nitrate–low chlorophyll regions of the sub-tropical North Pacific and the Southern Ocean, where enhancements in primary production could lead to significant increases in oceanic sequestration of fossil CO_2. As a result, most recent model and empirical studies of oceanic nutrient limitation have focused on Fe (Fuhrman and Capone 1991; Moore et al. 2002b; Jin et al. 2002). Some of these studies have predicted that Fe fertilization may stimulate oceanic primary production and accompanying remineralization and nitrification, with a resulting increase in oceanic N_2O production. The resulting increase in atmospheric N_2O could offset or even outweigh the gains in greenhouse gas reduction associated with CO_2 sequestration. However, a better understanding of oceanic N_2O production and an improvement in its parameterization in ocean models are needed before such simulations can be fully credible.

As in terrestrial ecosystems, the importance of oceanic biological N_2 fixation is not well understood. Current estimates of oceanic N_2 fixation range over an order of magnitude (Gruber and Sarmiento 1997; Codispoti et al. 2001). The rate of oceanic denitrification is also uncertain, although current global estimates are somewhat better constrained, thanks to geochemical tracer studies (analyses based on observed nitrate:phosphate ratios) (Howell et al. 1997; Deutsch et al. 2001). The balance between oceanic N_2 fixation and denitrification, together with riverine N inputs, ultimately determines the availability of oceanic fixed N to primary producers and has been hypothesized in simple box model studies to regulate atmospheric CO_2 levels on millennial time scales (Falkowski 1997). Progress is being made in reconciling estimates of ocean N_2 fixation and denitrification derived from biological extrapolations versus geochemical tracers, thereby providing improved constraints on these important N cycle fluxes (Hansell et al. 2004).

Other promising developments in coastal and oceanic N cycle modeling include the development of regional approaches to estuarine and coastal nutrient biogeochemistry, which involve embedding higher resolution regional submodules and/or off-line regional and global compartment simulations and databases (Jickells 2002; Mackenzie et al. 1998). In addition, open ocean ecosystem models that move away from climatological "nutrient-restoring" approaches and toward more process-based simulations are under development. Such models include multinutrient (NO_3, NH_4, PO_4, SiO_3, Fe) limitation and explicitly resolve community structure (picoplankton, diatoms, calcifiers, diazotrophs) in the upper ocean (Moore et al. 2002a, 2002b; Le Fevre et al. 2003).

4.7.1.2 Models Emphasizing Feedback between Nitrogen and Key Ecosystem Processes

4.7.1.2.1 Modeling N regulation of NPP

Models such as the Terrestrial Ecosystem Model and Biome BGC have examined the influence of vegetation C:N ratio, and its consequent feedbacks on soil N availability, in regulating NPP. These models have demonstrated why temperate forest ecosystems, which have large, carbon-rich woody vegetation fractions and wide C:N leaf ratios, tend to be chronically N-limited (McGuire et al. 1992; White et al. 2000). In contrast, humid and dry tropical forests, which contain N-rich vegetation, are more likely to be limited by phosphorus (Vitousek 1994). The TEM model has also demonstrated the importance of considering N availability in predicting how climate change will affect NPP (Rastetter et al. 1992). Models that consider carbon biogeochemistry alone tend to predict an increase in soil respiration at warmer temperatures and therefore net CO_2 loss to the atmosphere from soil organic carbon. However, models that consider coupled carbon-nitrogen dynamics predict an increase in soil N mineralization rates and therefore soil N availability that can help increase NPP and thus offset soil carbon losses.

4.7.1.2.2 Modeling N trace gas emissions from terrestrial ecosystems

The "leaky pipe" model of Firestone and Davidson (1989) provides the conceptual framework for estimating microbial NO_x, N_2O, and N_2 emissions from soils. Soil gas diffusivity, a function of soil type and soil water content, regulates the partitioning of N trace gas in this conceptual model. Low soil gas diffusivity favors the emission of N_2 over N_2O and N_2O over NO_x.

The leaky pipe model has been incorporated into some process-based model algorithms. For example, the CASA model estimates N trace gas emissions as a fixed fraction (say, 1–2%) of the rate of soil N mineralization, with a soil-moisture dependent partitioning between NO_x and N_2O (Potter et al. 1996). The N gas sub-model of CENTURY uses calculated soil water content, temperature, and microbial N cycling rates to simulate daily N_2, N_2O, and NO_x emissions from nitrification and denitrification. A daily time step is used because this degree of resolution is needed to reproduce the short-term events that are often responsible for the majority of N gas emissions from soils (Parton et al. 2001). The daily time step is also more appropriate to the needed coupling to atmospheric chemistry transport models.

NH_3 emissions are largely associated with volatilization from livestock manure and ammonium fertilizers and are equal to or exceed NO_x emissions on a global scale. Modeling of NH_3 emissions is generally based on simple empirical regression models (which are also used to estimate NO_x and N_2O losses from fertilizer) (Bouwman et al. 2002). Alternatively, some process-based models like CENTURY estimate immediate NH_3 volatilization losses as a product of the livestock manure or fertilizer input and a soil texture-dependent emission coefficient.

Soil N trace gas models have been successfully evaluated at specific test sites (Parton et al. 2001), although the extrapolation of model results to the regional and global scales is still highly uncertain. An additional shortcoming of current trace gas emission models is that they do not track the atmospheric transport, chemical transformation, and deposition of NH_3 and NO_x, and they thus neglect additional emissions and other effects that may occur downwind.

4.7.1.2.3 Feedbacks between N fluxes and ecosystem response

Much of the anthropogenic NO_x and NH_x emitted to the atmosphere as a result of agriculture, fossil fuel combustion, and other human activities deposits close to its point origin. However, a portion may be transported long distances, crossing national boundaries and even oceans before depositing. The lifetime of NO_x and NH_x in the atmosphere is short (hours to days), but these reactive species can be transformed to longer-lived species such as HNO_3 and PAN or may simply escape the boundary layer and be rapidly transported in strong upper tropospheric winds.

Long-distance transport of NH_x and NO_y can have profound impacts on downwind ecosystems. N deposition on formerly pristine natural ecosystem, such as N-limited grasslands, forests, and aquatic systems, in some cases can stimulate productivity, leading to increased carbon uptake and storage (Galloway et al. 1995; Holland et al. 1997). In other cases, excessive N deposition can cause acidification, forest decline, a decrease in plant species diversity, declining production, and C storage and accelerated N losses (Galloway et al. 1995; Vitousek et al. 1997; Aber et al. 1998). N deposition onto formerly pristine areas can also alter the emission and uptake of other trace species. For example, N deposition can lead to increased N_2O emissions (Mosier et al. 1998) and decreased soil consumption of atmospheric CH_4, which is also important greenhouse gas. Both increased N_2O emissions and decreased CH_4 consumption contribute to increased radiative heating of Earth's atmosphere.

Northern temperate forest ecosystems, which are often located downwind from centers of fossil fuel combustion, have seen the greatest changes in N inputs from the atmosphere and have been the focus of most terrestrial biogeochemical modeling studies. Models have predicted an initial stimulation of NPP in these generally N-limited ecosystems, followed in some cases by eventual N saturation, which is characterized by increased nitrate leaching rates and ultimately declining forest productivity (Aber et al. 1997). The TerraFlux model has been applied to biomes other than temperate forests, notably semiarid and tropical regions, where much of the future growth in atmospheric N deposition is projected to occur. These ecosystems may respond to excess N in markedly different ways than temperate forests and may be more likely to suffer deleterious effects. TerraFlux results suggest that N-rich tropical forests may have reduced productivity following excess N deposition, associated with increased leaching of NO_3 and the related loss of important, potentially nutrient-limiting cations like Ca+2, Mg+2 (Asner et al. 2001). However, TerraFlux predicts increases in productivity in semiarid systems following N input if water availability is sufficient and water losses are moderate.

4.7.2 Critical Evaluation of Approaches

One of the major weaknesses in current N cycle models is that they fail to account for the significant fraction of anthropogenic N inputs to terrestrial ecosystems that is denitrified (lost to gaseous N_2 or N_2O), reassimilated into biomass, or stored within groundwater or wetlands before reaching rivers. Detailed N budget studies in individual watersheds have found that only approximately 15–25% of N fertilizer and other inputs to watersheds ends up in rivers (Howarth et al. 1996; Caraco and Cole 1999). Typically, ~40% of N inputs to watersheds cannot be accounted for (Howarth et al. 1996). This missing N is assumed to be denitrified or stored in the landscape. A relatively smaller fraction of N inputs (~10%) is observed to be denitrified within rivers. Improved accounting for N losses that occur in between the soil leaching and coastal delivery stages is necessary for a credible simulation of the impact of human N cycle perturbations in a comprehensive Earth System model.

4.7.3 Research Needs

Addressing the changing nitrogen cycle and ecosystem services requires a variety of models at a range of temporal and spatial scales, from local to global. One of the clear gaps in our knowledge is in modeling the dynamics of coupled systems, in which terrestrial and atmospheric systems interact with economic trends or cycles. Coupled models, including coupling among the biogeochemical cycles, will need to be improved so that they can anticipate the crossings of thresholds that yield entirely new ecosystem states. More specifically, we must move toward an integrated model of the terrestrial, aquatic, and atmospheric components of the nitrogen cycle, which encompasses the complex feedback-response relationships and key nonlinearities of the cycle. Such a whole Earth system model should include terrestrial biogeochemistry models coupled with atmospheric chemistry/dynamics models, river transport models, and coastal and open ocean biogeochemistry/circulation models.

4.8 Forecasting Fish Populations and Harvest

The goal of fisheries assessment is to predict the consequences of fishing and other environmental interventions and, on that basis, evaluate how different management schemes fare at achieving various management goals. Forecasting the state and harvest of exploited populations and communities is thus central to fishery science.

There are two broad approaches that can be used to forecast fisheries population and harvest. On the one hand, there are short-term forecasts aimed at predicting the size of the exploitable stock for the upcoming fishing season in order to implement a predetermined feedback harvest rule. In this case, the forecast is part of the tactic used to define regulatory measures for the fishing season, such as the total allowable catch or the number of allowable effort units. This type of forecast is critical for fisheries based on short-lived or semelparous species (species that reproduce once and then die), where the bulk or all of the annual catch is made up of new recruits.

Mid- and long-term forecasts of populations, on the other hand, are used in policy design to examine likely consequences of different management options and thus guide strategic decision-making. In contrast to short-term tactical forecasts, mid- and long-term forecasts are not meant to actually predict the future of the system under exploitation; rather, they attempt to represent a full range of scenarios that are deemed possible based on historical experience. Because our ability to actually predict the responses of natural systems to harvest is admittedly limited, the emphasis in policy design is on feedback and robustness of performance across scenarios. Mid- and long-term forecasts aimed at guiding general management approaches are difficult because they require more information than the most recent harvest rates and data on catch per unit of effort, but it is not clear which of many possible auxiliary data will be most useful or how much history to consider.

4.8.1 Existing Approaches

The basic approach to fisheries forecasting has three components: a mathematical model used to describe the dynamics of the system under study as it is impacted by fishing, an approach used to condition the model on available information, and numerical tools used to implement forecasts under various management regimes.

4.8.1.1 Single-Species Approaches

Fisheries assessment and management have been dominated by single-species approaches aimed at controlling fishing impacts on unitary stocks by considering them in isolation from the ecosystem of which they are part. As a result, quantitative methods used for fisheries forecasting and policy evaluation have emphasized single-species modeling.

4.8.1.1.1 Models

Models used to represent single-stock dynamics range widely in complexity. The simplest models correspond to biomass-aggregated stock-production models, such as the Schaefer or Pella Tomlinson models (Quinn and Deriso 1999), which specify production as a simple nonlinear function of aggregate stock biomass. Surplus production is zero when the stock is at carrying capacity, and it increases to some maximum at some intermediate stock size. An increase in realism relative to simple stock-production models is achieved in the so-called delay-difference models (Deriso 1980; Quinn and Deriso 1999) by explicitly modeling the separate contributions of growth and births (actually recruitment of new-year classes to the exploited stock) to stock production. In these models, the stock is represented by the aggregate biomass of the exploited, mature component, animals recruit to the stock at some age r, and annual recruitment is a stochastic function of the mature biomass r years earlier. Generalized versions of this model include equations to predict the changes in size composition of the exploited stock (Hilborn and Walters 1992).

The models most widely used for fisheries forecasting are substantially more complex, including a representation of the age and size structure of the stock as well as age/size-specific fishing mortalities. As in delay-difference models, stochastic, density-dependent stock-recruitment relationships of various types are used to generate recruitment as a function of mature biomass. The standard Virtual Population Analysis and statistical catch-at-age models (Hilborn and Walters 1992; Quinn and Deriso 1999) used commonly for fish stock assessment and forecasting belong in this class. Finally, even more complex are models that incorporate spatial structure in addition to age or size structure, such as MULTIFAN CL (Fournier et al. 1998; Hampton and Fournier 2001). Each increase in realism is achieved by an increase in the number of parameters. For example, while a single fishing mortality rate per year is used in stock-production and delay-difference models, a vector of age/size-specific mortality parameters per year is used in standard age/size-structured models.

Forward projections constructed with these models always include stochasticity in at least some of the key processes. Recruitment variability induced by environmental forces is usually the dominant source. Typically, this variability is captured using a probabilistic distribution (such as log-normal with independent or autocorrelated random-year effects) as an empirical descriptor without attempting to model the actual environmental factors and processes underlying the variability. The inclusion of regime shifts in some of the scenarios (e.g., MacCall 2002; Parma 2002a, 2002b) is becoming more common, as empirical evidence is gained in their support (Francis and Hare 1994, 1998).

4.8.1.1.2 Conditioning approaches

Whichever the structure of the population model, its parameters are estimated by fitting time series of data on the stock and its fishery or fisheries, often making use of other sources of relevant information, such as information "borrowed" from other similar stocks. Fishery models are conditioned using formal statistical methods based on maximum likelihood or, increasingly, Bayesian techniques (Hilborn and Mangel 1997; Punt and Hilborn 1997, 2002). Maximum likelihood methods aim at providing best point estimates of abundance and fishing mortality rates over time and their associated estimation error. By contrast, Bayesian methods are used to derive joint probability posterior distributions of model parameters (and functions of them), conditioned on all observations and prior information.

Models are fitted to different types of data, depending on model complexity. Most critical for the estimation of the level of stock depletion is the availability of indices of stock abundance. These are derived from research surveys or commercial catch per unit of effort. Tagging data can also be used to provide information on abundance or exploitation rates. In addition, age/size-structured models use information on the age/size composition of the commercial and survey catches to help estimate trends in year-class strength. All these different sources of information are generally analyzed using an integrated statistical approach, where the likelihood function has several components, one for each type of data, with each based on a probability model deemed appropriate for the data in question. When estimation is done using Bayesian methods, prior information other than hard data may also be incorporated (e.g., McAllister et al. 2001).

Advances in computer technology and development of efficient methods of nonlinear estimation (such as use of automatic differentiation in AD Model Builder; available at www.otter-rsch.com/admodel.htm) have made it possible to build very complex models that incorporate process variability in many parameters assumed to be constant in simpler models. For example, fishing catchability and selectivity may be assumed to vary over time according to some specified random process. While in the past, estimation was done assuming that either all the noise was due to measurement error or process error, the new generation of fishery models incorporate both process and measurement error in the estimation.

4.8.1.1.3 Numerical tools

Monte Carlo techniques are used to simulate future stock trajectories incorporating different sources of uncertainty, as discussed below. Bayesian Markov Chain Monte Carlo methods (Punt and Hilborn 1997, 2002) are increasingly used to approximate posterior distributions of model parameters and then sample from them to project populations under various candidate fishing policies (e.g., Patterson 1999; Parma 2002a, 2002b).

4.8.1.2 Multispecies Approaches

Concerns about the impacts of fisheries on non-target species, habitats, and marine communities have increased over the last decade, leading to strong pressure to move from single-species management to ecosystem management. This has encouraged further developments of ecosystem models, which are needed to address the type of questions now being posed to fisheries assessment scientists.

4.8.1.2.1 Generalizations of single-species models to include features of multispecies systems

The simplest way in which multispecies effects have been incorporated into single-species fishery models is by adding mortality terms to represent the effects of predation on a target species. The dynamics of the predator is not explicitly modeled but instead is used as a driving variable in modeling the dynamics of the target species. Walters et al. (1986), for example, modeled the stock-recruitment relationship of Pacific herring (*Clupea harengus*) as affected by the abundance of its main predator, Pacific cod (*Gadus macrocephalus*). Similar models have been developed for pollock (*Theragra calcograma*) in the eastern Bering Sea (Livingston and Methot 1998) to assess the influence of predation and climate effects on recruitment. Also, Punt and Butterworth (1995) used a three-species model to evaluate the impact of culling the predator fur seals (*Arctocephalus pusillus pusillus*) on the abundance and catches of the Cape hakes *Merluccius capensis* and *M. paradoxus*. They considered this to be the "minimal realistic model" needed to examine their question and emphasized that great care needs to be taken when designing such models to ensure that all the important predator-prey interactions are incorporated.

A coarse approach for applying single-species models to multispecies systems are the so-called aggregated production models (Hilborn and Walters 1992), which simply apply stock-production models (biomass logistic models with harvest) to aggregates of species. These models have been tuned to time series of catch rate and fishing effort. Ralston and Polovina (1982) found that in several tropical fisheries, trends in catch rate and yield for mixed-species assemblages were consistent, while results from production models applied to single-species were erratic.

4.8.1.2.2 Multispecies "top-down" models based on the mass action principle

Most of the early multispecies fishery models rely on the mass action principle to represent predator-prey interactions (Walters and Martell in review). Under this principle, the number of encounters between species is proportional to the product of their densities. Predation rates, whether or not they are affected by predator satiation and handling time (so-called type II functional response by predators), are directly predicted from such encounter rates. These models generally predict very strong "top-down" control of abundances by predators.

The simplest models based on the mass-action principle are generalizations of single-species stock-production models. They depict the biomass dynamics of multiple species using logistic models linked by Lotka-Volterra predator-prey equations (Larkin and Gazey 1982).

A second approach is multispecies virtual population analysis (Sparre 1991), a detailed age-structured model with age-specific harvest and predation rates, originated in the North Sea model of Andersen and Ursin (1977). MSVPA focuses on the interactions between commercially exploited fish stocks for which catch-at-age data are available. It assumes that individual food intake and growth are constant, and it uses data on stomach contents of all modeled species to estimate prey suitabilities. Historical trends in abundance are estimated from historical catches. A forecasting version of MSVPA, MSFOR, is being applied to the analysis of exploited ecosystems of the North Sea (Rice et al. 1991; Vinther et al. 2002) and eastern Bering Sea (Jurado-Molina and Livingston 2002).

4.8.1.2.3 Mass balance multispecies approaches

While all the previous approaches are conditioned on past data on the dynamic states of the populations represented, mass-balance methods are founded on a static description of the ecosystem, represented by biomasses aggregated into ecologically functional groups. The basic idea behind the mass-balance assumption is that for the collection of functional groups considered, production ought to be balanced by predation, harvest, migration, and biomass change. The most widely used mass-balance model is Ecopath, which is based on static flow models (Polovina 1984; Christensen and Pauly 1992), defined by a series of simultaneous linear equations that represent trophic interactions and fishing. The essential parameters required for each functional group are generally the same as those of other multispecies models—namely biomass, production rate, consumption rate, diet composition, and fisheries catch. One extra parameter per group controls the fraction of the production that is accounted for in the model. The diet composition matrix plus four of the five group-specific parameters need to be "known," and the mass-balance equations are solved for the remaining parameters.

Unlike MSVPA, Ecopath uses data on the production/biomass ratio as input (Christensen and Walters 2000). Ecopath does not require a representation of individual species or their age structure. Another difference between the two approaches is that while MSVPA considers the subset of commercially important species and their key preys and predators, Ecopath attempts to portray ecosystem-wide dynamics, including primary production.

Ecosim is a dynamic extension of Ecopath that simulates time trajectories of the different functional groups modeled and thus can be used to examine influences on ecosystem dynamics resulting from any given harvest policy (Walters et al. 1997). Ecosim replaces the static biomass flow of Ecopath by a system of differential equations but it retains the mass-balance assumption of Ecopath by tuning the model to the baseline observations on biomasses and consumption rates of the functional groups at a given reference time. We should note that this does not imply equilibrium; known changes in biomasses can be incorporated in the Ecopath biomass flow equations.

A fundamental difference with mass-action models is the introduction in Ecosim of the foraging arena concept (Walters and Juanes 1993; Walters and Martell in review), by which only a dynamic fraction of each ecosystem component is vulnerable to predators. The fact that parts of the prey populations are not vulnerable effectively augments bottom-up effects compared with typical Lotka-Volterra-based models. Through alternative parameterizations, users can represent a variety of assumptions about the nature of predator-prey interactions.

Ecopath with Ecosim software is a widely used tool for the quantitative analysis of food webs and ecosystem dynamics (e.g., Pauly et al. 2000), and new capabilities are being constantly developed. In particular, the software can input historical trends in fishing mortality or effort, productivity indices (such as upwelling), recruitment indices, and biomass of other, nonmodeled species to drive the dynamics. Also, predicted trends can be fitted to observed trends in relative or absolute abundances, to direct estimates of total mortality rate, and to historical catches. Advanced users are beginning to experiment with fitting the model to time series data using formal statistical methods, but this is in its early stages of development compared with single-species approaches.

4.8.2 Critical Evaluation of Approaches

4.8.2.1 Uncertainty Analysis

Our limited ability to forecast population abundances and catches has several roots: observation uncertainty, process uncertainty, model uncertainty, and institutional uncertainty.

First, we do not "observe" marine populations directly, nor do we observe all the relevant variables to be able to estimate population abundance confidently and understand the relationships that govern their interactions. Errors in estimates of current exploited stock sizes obtained by modern assessment methods commonly exceed 30% (NRC 1998). Much larger errors, as large as 200%, have resulted from the use of flawed assessment models (Walters and Maguire 1996). The abundance of other unexploited ecosystem components may be even less known. This means that there is substantial uncertainty about the initial conditions of the variables involved in running any forecast model. Imprecision and biases in diet composition data used to parameterize predator-prey relationships are also a problem. In particular, Walters and Martell (in review) have cautioned about the risk of missing small prey items infrequently eaten by abundant predators, a phenomenon that may strongly affect prey dynamics.

Second, natural processes are inherently variable, and no matter how good our models may be, they cannot predict the exact state of the system at any given time in the future. In marine populations, most life histories involve an early larval stage that is subject to the vagaries of the planktonic environment. As a result, variability in recruitment contributes substantial process uncertainty (Botsford and Parma in press). Oceanic processes are subject to large-scale decadal oscillations, as well as episodic events like El Niño that alter primary production and hence fish productivity. It has proved very hard to build these environmental drivers into fishery models.

Third, models consist of relationships between variables (functional or probabilistic) and parameters, and there is uncertainty in both: structural uncertainty and parameter uncertainty. As a rule, many alternative models are consistent with experience and historical data, but some model uncertainties are more critical than others. For example, the productivity of a stock at low biomass is critical for the estimation of sustainable harvest rates; uncertainty about the stock structure and the proper spatial resolution to consider may affect interpretation of most observed patterns; different parameterizations of predator-prey relationships may completely change the behavior of a multispecies model, from top-down to bottom-up (Walters and Martell in review).

Fourth, forecasts usually assume that some management scheme will be in place in the future. In reality, there is substantial uncertainty about future management decisions and the degree of compliance with management regulations. Although this could be considered part of model uncertainty, institutional uncertainty brings into play a higher order of complexity associated with forecasting how society and its institutions will behave in response not only to the vagaries of natural systems but also to economic, social, and political forces.

Different forecasting approaches have different capacities for dealing with uncertainty. Modern single-species forecasts usually incorporate uncertainty in initial conditions, process uncertainty, and parameter uncertainty using Bayesian techniques. Structural uncertainty is commonly treated in a more ad hoc way, by conducting forecasts using a small number of alternative model structures when searching for robustness in policy performance (e.g., Butterworth and Punt 1999). Less frequently, formal Bayesian techniques are used to estimate the plausibility of alternative model structures (e.g., Patterson 1999; McAllister and Kirchner 2002; Parma 2002a).

Multispecies models of intermediate complexity, like those derived by extending single-species models to account for predation effects, facilitate the incorporation of current state-of-the-art tools used in single-species models. When it comes to large multispecies approaches, uncertainties that surround model forecasts are much less frequently conveyed. For example, uncertainties about input parameters to Ecopath can be explored using the ECORANGE routine, but the sensitivity of Ecosim predictions to these uncertainties is often overlooked. Walters and Martell (in review) recommend the use of alternative values to partition predation mortalities, in addition to those implied by diet composition data, to evaluate sensitivity with respect to uncertainty in the data. Despite similar warnings repeatedly made by modelers (Aydin and Friday 2000; Walters et al. 1997), the software has often been often used as a black box.

4.8.2.2 Strengths and Weaknesses of Different Approaches

The merits and limitations of the different forecasting approaches need to be viewed in the context of the purpose of the forecasting exercise—that is, what question is being addressed. Single-species approaches continue to be the preferred approach for evaluating the performance of single-species management procedures. Their main strength is that they are formally conditioned on past data (which means management draws its lessons from history). The methods for doing this are supported by a significant development of statistical techniques (and specialized software) for estimating model parameters efficiently, quantifying uncertainty based on modern Bayesian techniques, and incorporating this uncertainty in decision analysis. Their limitations are mainly a function of the quality of the data used for model fitting and the peculiar history of exploitation of the stock in question.

In many cases, stock assessments, and in turn population forecasts, rely solely on fishery-dependent data. This is problematic because there are many ways in which catch per unit of effort may fail badly as an index of stock abundance (Hilborn and Walters 1992). Also, many exploitation histories correspond to depletion trajectories, the so-called one-way trips (Hilborn and Walters 1992), which do not provide needed contrast in abundance and effort, making it hard to distinguish productivity and mortality parameters when conditioning on past data. Conditioning is so critical because by fitting a model to historical data before attempting a forecast, we demonstrate that the model is able to reproduce historical trends and we constrain the universe of possible models. This, however, is no guarantee that the model will be able to extrapolate system responses correctly to novel perturbations in the future.

Aside from these limitations, single-species approaches obviously cannot be used to address ecosystem questions, such as forecasting the impact of large-scale perturbations or determining how human interventions propagate from particular components to the rest of the ecosystem. Some multispecies concerns, such as the impact of a given single-species management procedure on other species, may best be addressed by modeling only the linked dynamics of the key species involved, as in the "minimal realistic model" of the hake-seal system developed by Punt and Butterworth (1995). The uncertainty present in single-species models is here compounded by uncertainty in the parameters and relationships that govern species interactions. This increase in model uncertainty was the main reason why the large multispecies approaches developed in the late 1970s and early 1980s fell out of favor (Quinn in press).

The resurgence of multispecies models, such as MSVPA in the North Sea and the Bering Sea, was accompanied by major investments in field programs of stomach data collection to help fill some of the information gaps. Unfortunately, such investments are seldom possible, so lack of basic data to support multispecies models will be a major limitation for their widespread use.

Beyond observation uncertainty, other problems detected when using top-down multispecies models to forecast ecosystem changes seem to reflect limitations of the basic core assumptions used to represent predator-prey interactions. MSVPA, for example, assumes that the predator is always able to consume a fixed ration of food. According to Walters and Martell (in review), forecasts done using top-down multispecies models based on the mass-action principle have produced some unrealistic results (such as strong trophic cascades, loss of biodiversity, and dynamic instability) more drastically and more frequently than is supported by field evidence.

The development of the foraging arena concept of Ecosim as an alternative to mass-action assumptions was a major step forward, allowing control of the degree to which the model behaves as top down or bottom up. However, choosing appropriate values for prey vulnerability parameters is difficult due to the lack of information to quantify these processes (Plagányi and Butterworth in review). The alternative of using default values followed in many applications is unsatisfactory, as Ecosim predictions are highly sensitive to the choice of vulnerability settings (Shannon et al. 2000). Further development of time series–fitting approaches and alternative estimation schemes will be needed to help parameterize these processes. Some alternatives are discussed by Walters and Martell (in review).

Like the foraging arena formulation, the implications of several other assumptions in Ecosim are just beginning to be explored (see detailed discussions in Aydin and Friday 2000; Christensen and Walters 2000; Plagányi and Butterworth in review; Walters and Martell in review). In particular, the assumption of mass balance and the use of Ecopath as a starting point have raised criticism. On the one hand, the mass balance condition is a strength in that information is added by forcing the productivity of all consumers to be supported by primary productivity in the ecosystem. This is in sharp contrast to Lotka-Volterra systems, where the overall productivity is unconstrained, leading to instability.

This information, however, is not without a cost. First, it augments the requirements for ecosystem-wide data, including primary production, and specification of the form of functional responses in predator-prey relationships among all functional groups represented. Second, reliance on Ecopath biomass flows balanced for some reference time implies that predator-prey parameters are time-invariant, which may not hold when the ecosystem changes substantially from the reference situation. The assumption of equilibrium commonly made when balancing Ecopath equations may be even more problematic when biomass of some of the key groups is changing during the reference time (Walters et al. 1997; Plagányi and Butterworth in review).

To conclude, single-species approaches will be hard to beat as tools for describing the dynamics of exploited stocks. Improvements on predation mortalities may result from incorporation of multispecies interactions but only when the data required to parameterize them are available. Large multispecies models such as MSFOR and Ecosim, on the other hand, have been and will be useful to explore ecosystem function and to postulate alternative scenarios about plausible ecosystem responses to environmental change and human interventions, when applied taking due care of uncertainties and potential pitfalls. To date, calibration and diagnostic methods and expertise have not yet advanced to the standards common in single-species fishery models. The value of these approaches to guide management decisions in the future will be a direct function of the expertise gained in that area.

None of the forecasting tools should ever be used without at the very least exploring the sensitivity of predictions to alternative model parameterizations and observation uncertainties. This is less of a problem in single-species fishery forecasting because assessment scientists have learned (the hard way) not to trust model predictions, and growing emphasis is placed on robustness and precaution. The risk is more serious with multispecies powerful packages like Ecopath with Ecosim, which are relatively easy to set to run without careful consideration of the implications of default choices.

4.8.3 Research Needs

In general, it can be argued that fishery models must be unsuccessful because such a large portion of fishery stocks have been overexploited, and some have even collapsed. In many of these cases, however, the collapse of the stocks was anticipated by the models (of, for example, Georges Bank cod). Thus it is not clear what is a failure of modeling and what is a failure of policy-making.

Improved understanding of ecosystem function and improved forecasting capabilities will require research and developments in several fronts, including:

- research aimed at improving understanding of spatial structure and relevant scales at which different natural processes operate and at evaluating when is it worth the effort to incorporate the space dimension into single or multispecies models and collect the required data;
- development of approaches to bridge the gap between detailed process-oriented studies and simpler empirical models useful for fisheries forecasting (can we use the results of process-oriented studies to build alternative scenarios, represented as simple model prototypes, and assign relative plausibility?);
- further development and testing of formal approaches for conditioning ecosystem models to different sources of information, including times series data;
- development of new methods for estimating predator-prey vulnerability parameters; and
- further exploration of the implications of core assumptions made in ecosystem models and consideration of alternative formulations.

4.9 Forecasting Impacts on Coastal Ecosystems

Ecological forecasts predict the effects of biological, chemical, physical, and human-induced changes on ecosystems and their components. Short-term coastal ecosystem forecasts, such as predicting landfall of toxic algal blooms, are similar to those done for weather and hurricane prediction, which also affect human well-being. On the other hand, forecasting large-scale, long-term ecosystem changes has similarities with macroeconomic forecasts that rely heavily on expert judgment, analysis, and assessment, in addition to numerical simulation and prediction. Forecasts of such broad-based, long-term effects are particularly important because some of the most severe and long-lasting effects on ecosystems may result from chronic influences that are subtle over short time frames. In this sense, one aspect of coastal forecasting has a longer history because of the well-developed use of modeling in fisheries stock assessments, as described in the preceding section.

Coastal ecosystem models should be able to predict long-term changes in ecosystem function based on past and present environmental and societal change and on coastal governance processes. The results of such a model will depend not only on the modeling of physical, chemical, and biological spatial and temporal data but also on the value judgments involved in the decision-making process. The key direct drivers of coastal ecosystem change that may affect human well-being and that are a priority for coastal forecasting are eutrophication, habitat modification, hydrologic and hydrodynamic disruption, exploitation of resources, toxic effects, introduction of non-native species, global climate change and variability, shoreline erosion and hazardous storms, and pathogens and toxins that affect human health.

4.9.1 Existing Approaches

The basic structure of a simple coastal forecasting model consists of a meta-data portal linked to an analysis system. The meta-data portal is an algorithm that integrates and manages data from many disparate sources and organizes them into a form that can be used by the analysis system. Data can be from archived sources or arrive in real time from satellites and onsite measuring instruments. The analysis component that processes the physical, chemical, and biological data required to monitor the direct drivers is normally handled by a set of interactive deterministic multimodels, the choice of which depends on the questions being asked. Interactive or coupled multimodels consist of a set of stand-alone models that handle different kinds of data (such as in situ optics, chemical analyses, and remotely sensed data) and that communicate with each other in order to predict the likelihood of some outcome (such as a harmful algal bloom). Final output is usually in the form of a visual display in a GIS or mapping module.

Deterministic models are commonly of two types: empirical models based on observation or experience in particular places and mechanistic models based on theories, which explain phenomena in purely physical terms. Empirical models are very common but of limited general use and are not discussed further. What we need instead are models that can accommodate the value judgments involved in the decision-making process of humans. For this, some sort of decision support system is required. The current state of the art is still a long way from integrating this latter aspect into existing coastal forecasting systems, which at present do not allow sufficient lead time for coastal managers to intervene in order to avert a potentially undesirable long-term outcome. This assessment is not meant to review the many excellent scientific publications on coastal models. It aims rather to give a sense of the state of the art in relation to the aspiration of building scenarios that inform the decisions of coastal managers for long-term planning.

4.9.1.1 Nowcast/Forecast Modeling Approach

A nowcast/forecast type system results when data from multiple sources is fed to a meta-data portal in real time. The term "nowcast" is used because it refers to the fact there is no lag time between events and analyses—data are immediately fed into the decision analyses. Nowcast models are of interest because they promise the possibility of allowing scenarios over long time scales to respond to discontinuities and surprises as they arise. The Global Ocean Observing System has scientific details of the operation of a nested set of regional coastal and ocean forecasting systems (see http//:ioc.unesco.org/goos).

An illustrative state-of-the-art example of a regional nowcast/forecast system is the New Jersey Shelf Observing System being developed by the Coastal Ocean Observatory Laboratory. The aim is to provide a synoptic 3-D picture of the biogeochemical cycling of elements and physical forcing of continental shelf primary productivity in the New York Bight (Schofield et al. 2002). The system under development consists of an array of surface current radar systems, color satellites, and autonomous underwater vehicles. These will input mainly bio-optical data into a new generation of physical-biological ocean models for hindcast and real-time continental shelf predictive experiments. Results will be disseminated in real time to both field scientists and water quality forecasters over the Internet (marine.rutgers.edu/cool) as well as to the general public (www.coolroom.org).

The ensemble of forecasts is generated by an extensive suite of atmospheric, ocean, and biological models including ROMS–the Regional Ocean Modeling System and TOMS–the generalized Terrain-following Ocean Modeling System. ROMS is interfaced with a suite of atmospheric forecast models. TOMS is coupled with a bio-optical ecosystem model, which uses the spectral distribution of light energy along with temperature and nutrients to estimate the growth of phytoplankton functional groups representing broad classes of the phytoplankton species. The biological forecast can then be validated in real time from the field measurements that can guide the evolution of the model. Errors between the model prediction and the field measurements are used to direct autonomous underwater vehicles into regions where more data are needed.

4.9.1.2 Decision-Support Approach

Another approach is to supplement deterministic models by decision-support techniques. Decision analysis is a step-by-step analysis of the consequences of choices under uncertainty. Decision-support techniques include cost-benefit analysis, cost-effectiveness analysis, multicriteria analysis, risk-benefit analysis, decision analysis, environmental impact assessment, and trade-off analysis. Of the various decision-support techniques in use, only multicriteria analysis uses mathematical programming techniques to select options based on objective functions with explicit weights, which stakeholders can then apply. The other approaches are not easily adaptable to mathematical programming techniques because of lack of clear techniques for incorporating information in the decision-making process.

SimCoast provides an illustrative example of how a fuzzy logic expert system can be objectively programmed into a coastal ecosystem decision-support model. According to McGlade and Price (1993), three key intelligent systems techniques potentially useful for sustainable coastal zone management are neural networks, expert systems, and genetic algorithms. However, interactions between these methods and other approaches such as fuzzy logic and issue analysis also give users the ability to assess the combined uncertainty and imprecision in their knowledge and data.

SimCoast is a fuzzy logic, rule-based, expert system in which a combination of a fuzzy logic and issue analysis has been used to produce a soft intelligence system for multi-objective decision-making. It is designed to enable researchers, managers, and decision-makers to create and evaluate different policy scenarios for coastal zone management. The conceptual basis is a two-dimensional multi-zoned transect onto which key features such as ports, laws, mangroves, and activities such as fisheries, aquaculture, shipping, and tourism are mapped. These activities are associated with different zones and with the process to which they are linked (such as land tenure, erosion, organic loading). The effects of activities on the features are evaluated in relation to defined policy targets (water quality, system productivity, ecosystem integrity, for instance) measured in particular units (oxygen level, turbidity, *E. coli* concentrations, number of species, or biomass). This evaluation is the result of consensual expert rules defined during workshops. Fuzzy logic and certainty factors are used to combine new data and build scenarios based on the ideas or even alternative hypotheses of experts.

4.9.2 Critical Evaluation of Approaches

4.9.2.1 Nowcast/Forecast Modeling

There are numerous coastal ecosystem nowcast/forecast systems based on deterministic models coupled to a GIS to predict harmful algal blooms, oxygen depletion effects, oil spills, climate change effects, and the like. In spite of their great scientific interest, these approaches require considerable expertise and resources that are not widely available to coastal ecosystem managers around the world. Further, each coastal area has different conditions and priority problems, so that no single system of deterministic models will be useful in most situations.

Coastal nowcast/forecast systems are typically dominated by uncertainties in model initialization largely attributable to under-sampling. To deal with this, an ensemble of forecasts with differing initial conditions is used to identify regions in which additional data are required (Schofield et al. 2002). Hence the models provide insight into what has not been sampled and guidance for further real time observational updates using multiple platforms, including remote (satellites, aircraft, and shore-based), stationary (surface and subsurface), movable (ships and autonomous underwater vehicles), and drifting (surface or vertically mobile) systems.

This rapid environmental assessment capability changes the entire paradigm for adaptive sampling and nowcast/forecast modeling. Forecast errors or misfits may now be dominated by uncertainties in the model formulations or boundary conditions, and ensemble forecasts with differing model parameterizations identify regions in which additional data are needed to keep a model on track. In the time it takes to prepare the ensemble of forecasts for the well-sampled ocean, additional data have arrived, and on-the-fly model-data metrics can be used to quantify which forecast in the ensemble is the least uncertain (Schofield et al. 2002). This approach offered the MA the potential to have scenarios and long-term forecasts dynamically updated to adapt to discontinuities and surprises, which are an increasingly common feature of the modern world.

4.9.2.2 Decision Support Systems

Decision support techniques are not really forecasting methods. They do, however, provide a structured framework through which a choice between alternatives can be made with regard to a given set of criteria. This more widely accessible approach can be criticized for theoretical difficulties associated with aggregating preferences for use as weights in the models.

Expert systems, by their very nature, deal with a good deal of uncertain data, information, and knowledge. Decision support systems use many methods to integrate uncertain information for inference: these most commonly include Bayes theorem or the Dempster-Shafer theory of evidence, but certainty factors and fuzzy sets are sometimes also used. The commonly used Dempster-Shafer theory of evidence (Dempster 1968; Shafer 1976), which is an extension of Bayes theorem, appears to be a robust approach. According to Moore et al. (2001), this theory does not require exhaustive prior or conditional probabilities before calculation can take place, and it can be used where evidence is based on vague perceptions or entirely lacking.

Normally, where probabilities are not known, equal prior probabilities are unrealistically assigned to each competing piece of evidence, and the sum of all assigned probabilities must equal one. With the Dempster-Shafer theory, an ignorance value close to zero (ignorance = 0 represents complete ignorance) can be used to represent the lack of information, rectifying what would be erroneous with probability. Related to this is the fact that when belief is assigned to a particular hypothesis, the remaining belief does not necessarily support the hypothesis' negation. Other advantages of using the Dempster-Shafer theory include the ability to use evidence supporting more than one hypothesis (a subset of the total number of hypotheses). Finally, this approach models the narrowing of the hypothesis set with the accumulation of evidence, which is exactly how experts reason. It held the possibility for the MA to allow fully specified uncertainties being attached to scenarios. It creates a feedback loop, which stimulates human beings to take decisions that they think may change the scenario outcome.

4.9.3 Research Needs

Traditional ecological forecasting and decision-support methods are no longer adequate because they have a limited ability to predict discontinuities—significant nonlinear changes in the direct or indirect drivers of change that force a fundamental re-evaluation of strategy or goals. We therefore need to look for new models that are able to accommodate discontinuities and facilitate proactive scenario planning where the past is an increasingly unreliable guide to the future.

Perhaps the deliberative process that attempts to look at the long-term future of ecosystems advocated by Clark et al. (2001) and adopted by the MA should advance beyond conventional ecological forecasting and be more accurately referred to as ecological foresighting. This is taken to mean not the identification of the most likely scenario but the evaluation of many possible, feasible, or even desirable scenarios. This helps develop a deeper understanding of the options and promotes better planning from a backcasting standpoint. Since long-term ecological foresighting must have as a base significant hindcast and nowcast information, both nowcast/forecast deterministic models and scenario-based decision support systems need to be linked so that relevant up-to-date alternatives are presented to coastal managers.

We need to also develop ways of modeling and of estimating how ecosystem services respond to combinations of stresses at local and regional scales. The combination of complex interactions among a large number of components with the variable nature of ecosystems and their driving forces makes the development of such tools a significant challenge. Potential techniques include neural nets, artificial intelligence, fuzzy sets, and massively parallel algorithms.

4.10 Forecasting Impacts on Human Health

Forecasting the impacts of future ecosystem change on human health (McMichael et al. 2003) at a global scale and over the next century is daunting. This task is subject to such large uncertainty that it might reasonably be considered impossible, or at least not scientific. Nevertheless, uncertainty is not infinite, and many boundary conditions can be identified. These include not only the consistency of anatomy, physiology, and pathophysiology (disease processes), but also the fact that social and technological factors will continue to influence and modify human health, as they have for millennia. Though substantial uncertainty remains about the characteristics of even present health at the global scale, a great deal is known.

4.10.1 Existing Approaches

One dominant approach to modeling human health and disease is embodied by classical epidemiological and bioclimatic treatments of malaria, and even wildlife disease (e.g., Dobson 2000; Rogers and Randolph 2000). These approaches typically start with a detailed look at one disease and one host, and then build toward predictions of critical thresholds for epidemics and the implications for public policy. Land changes or climatic changes are especially important when the diseases are transmitted by vectors, such as malaria being transmitted by mosquitoes; this is because temperatures alter vector survival and behavior, and land alterations can create or destroy habitats for vectors. Because they are host-pathogen specific, these approaches do not lend themselves easily to global assessments; nor are they motivated toward global assessment, since health policy is often formulated one disease at a time. Currently, there is a move away from modeling one species at a time—even from the perspective of classical epidemiology. For instance, there is a growing recognition that climatic and habitat perturbation may underlie the emergence of many new human diseases (Patz et al. 1996), but no models of this process are yet available.

Because of the limitations of single-species disease models, an alternative emphasis for global assessments has been on a more phenomenological or aggregate approach that emphasizes connections between population demography, social behavior, and poverty. This topic can appear overwhelmingly complex. It may be useful to consider that uncertainty applies to three key "black box" determinants relevant to this task. These boxes can be conceptualized as input, output, and modifying determinants, where input represents the state of ecosystems, output the state of health, and modifiers the social, technological, and political co-factors that can either dampen or exacerbate the importance of ecosystem change upon health. Of course, in reality these categories are not clearly separate. At all times, a continuous interaction exists between these three categories, though often the causal links are subtle, poorly identified, and at least in some cases—away from thresholds—unimportant.

Clearly, myriad possible interactions and cascading responses between ecosystem services, human health, and the societies and institutions that modify and influence both health and ecosystems are possible. The outcomes can be positive or negative. One method has been proposed to classify potential adverse responses into one of four broad groups, called "direct," "mediated," "modulated," and "systems failure" effects (Butler et al. in press).

In this categorization, direct health effects manifest through the loss of a useful ecosystem service, such as the provision of sufficient food, clean water, or fertile soil or the restriction of erosion and flooding. Direct effects occur as the result of physical actors but do not include pathogens per se. They are probably the most easily understood of the four effects.

Mediated effects, as opposed to direct ones, have increased causal complexity and in some cases involve pathogens. Some have potentially high morbidity and mortality. There is also often a longer lag between the ecosystem change and the health outcome than for direct effects. Many infectious and some chronic diseases fall in this category. In many of these cases disease have emerged as a result of the increased food-producing capacity of ecosystems (a provisioning ecosystem service)—for example, by animal domestication, irrigation, dams, and other intensive farming practices. A trade-off has been the unforeseen increase in the incidence and prevalence of many of these communicable diseases.

Effects called "modulated" and "systems failure" have also been identified as larger-scale, more lagged, and more causally complex adverse consequences of ecosystem change. Modulated effects refer to episodes of state failure or of nascent or realized large-scale social and economic collapse. Systems failure refers to economic and social collapse at a supra-national scale as a result of coalescing, interacting modulated effects.

This classification can be mentally fitted to different ecosystem futures. For example, if a region experiences marked loss of provisioning services, then direct adverse effects, such as the disruption of flooding or increased hunger, are likely. Loss of biodiversity, and more contact between humans and undomesticated species may mean the occurrence of more mediated effects, including emerging infectious diseases. Greater intensification of animal husbandry may facilitate the spread of recombinant forms of known diseases, such as influenza. However, more realistic and useful predictions of the health impact of ecosystem change require greater understanding and at least attempted predictions of the social, institutional, and technological factors that modify health—the third of the black boxes identified earlier.

In some cases, such as modulated health effects, health itself may critically affect the quality of these human services. It is true that modulated effects are far less likely to occur in societies that have strong institutions, reasonable governance, and high technology. Adverse health effects consequent to ecosystem change and reduced ecosystem services may in some cases overwhelm societies that are already fragile, however, causing them to exceed a "tipping point" beyond which decline is highly likely.

Although far less likely, it is also possible to envisage pathways and causal webs that test the social, economic, and health fabrics of societies that currently appear almost invulnerable to adverse ecosystem change. For example, an interlocking cascade of adverse events and erroneous decisions led in the last century to the Great Depression and World War II. Although perceived resource scarcity was a factor in this cascade (for example, in both the German and Japanese peoples' desire for expansion), ecosystem buffers—especially on a per capita basis—were far higher then than they are now.

4.10.2 Critical Evaluation of Approaches

Conventionally, scenario theorists accentuate and extrapolate existing trends into the future, imagining and modeling how different futures will unfold. One problem with the scenario approach is that assumptions about the future are typically "bundled together," as in "TechnoGarden," whereby across-the-board technical innovation is expected to meet global challenges. But an imaginary future marked by rapid technological progress may adversely affect human health if it is not also accompanied by other forms of progress. That is to say, affordable vaccines, surgery, and pharmaceuticals may not be fully able to compensate for poor health if high technology is also accompanied by a dehumanization of society.

Second, human health is a mix of technology, environment, and social systems. Predicting how these interlink is almost impossible, and we lack both theory and empirical relationships as guidance. We have studies that predict how climate variability might directly affect human health through air pollution and diseases (McMichael et al. 2003), but the fact that climate change will also alter distribution of wealth and disrupt societies is not factored into these health predictions.

Reductionist approaches to science have sought to simplify and analyze reality by considering isolated elements, and at ever finer resolution. This approach has been very successful in many fields and has contributed substantially to the enormous scientific, technological, and material progress that marks our time. And although material progress is not sufficient for either human health or human well-being, it is clearly an important contributor to both. Yet reductionist methods have limits that are increasingly recognized. A major drawback is that they mask the significance of threshold effects and, even more important, they can actually hide the concept of thresholds.

In a linear, reductionist conceptualization of reality, an increment of change is not of itself very important. If the experimenter (including the unwitting social and ecological experimenter) concludes that an increment of change causes an undesirable increment of response, then linear thinking suggests that the remedy is simply to subtract that increment. In many cases this is possible. In ecosystems, society, and human health, however, innumerable though still poorly defined thresholds exist beyond which reversal is either impossible or prohibitively expensive. This phenomenon of costly or impossible reversibility, also known as hysteresis, is increasingly recognized in ecology (Scheffer et al. 2001a), but has as yet been little studied in relation to health and society (Butler et al. in press).

4.10.3 Research Needs

The critical areas of research entail linking together the many threads that affect human health. We still lack some of the most basic information on how the epidemiology of disease is altered by the environment. Inevitably, human systems respond to health threats, and any sort of prediction regarding possible adaptive responses is totally lacking in global health projections. We know historically that the collapse of social systems can drastically affect human health. Systems failure, were it to occur, is possible as the century unfolds. Yet we have very little ability to predict or anticipate these system failures, which in turn put so much stress on health care systems.

There may be many opportunities for translating the successes of specific epidemiological models into true global assessments. The situation now is very much like that of ecology 30 years ago, when the focus was on single-species models or models of pairs of interacting species. The rise of ecosystem science and attention to biodiversity prompted a whole new generation of models, but still with critical ecological underpinnings. Global health assessments would benefit from attempts at using epidemiological models to scale up and aggregate over many diseases toward summary predictions.

4.11 Integrated Assessment Models

Many environmental problems are caused by a complex web of causes and effects that have environmental, social, and economic dimensions. The fact that these webs cannot be well described by disciplinary approaches has led to an increasing interest in integrated assessment models. Weyant et al. (1996), Van der Sluijs (1997), Rotmans and Dowlatabadi (1998), and Toth and Hizsnyik (1998) provide interesting reviews of integrated assessment definitions and methods. In general, they describe IAMs as modeling frameworks to organize and structure various pieces of (disciplinary) scientific knowledge in order to analyze the cause-effect relationships of a specific problem. The analyses should have a wide coverage and include cross-linkages and interactions with other problems.

The term IAMs is applied in particular to models that include some description of the socioeconomic system (including economic activities and human behavior, population dynamics, and resource use) and its interaction with the environmental system (regional air pollution, the climate system, land cover/land use, and so on). They can be qualitative (conceptual models) or quantitative (formal computer models). IAMs are used to synthesize available scientific information from different disciplines, and their specific approaches and assumptions, in an organized way. An essential feature of IAMs is their focus on application for policy support and on assessment rather than scientific research per se. In such policy-oriented applications, IAMs have different functions. They serve as an early warning system and an exploration of possible futures, they are used for policy evaluation, and they provide tools that directly support public decision-making and negotiations.

IAMs can be categorized in various ways. A first classification is the dominant modeling paradigm. IAMs can be calculated as either simulation or optimization models. Most simulation IAMs are based on differential equation descriptions. Sometimes partial optimization is used. The IAMs derived from an economic problem setting tend to use optimization techniques to evaluate the minimum cost or other objective function of certain trajectories. These types of IAMs are like dynamic cost-benefit analyses. Within both modeling paradigms, deterministic as well as stochastic approaches are used, although the former dominate. Within optimization IAMs, a further distinction can be made according to the degree to which the objective is to satisfy exogenous constraints or targets.

A second classification concerns horizontal and vertical integration. Vertical integration refers to the degree to which a model covers the full cause-effect chain of the relevant issue—from driving forces to pressures, to changes in state, to impacts, and finally to possible responses. Vertical integration has emerged from the pressure-state-impact-response framework in environmental policy. Horizontal integration refers to the integration between different aspects of the object of study. This can be in a rather narrow sense, such as integrating the interactions between water and land cover/land use, or much wider, as in the integration of demographic/health issues and the state of the environment. A related classification is in terms of the topic they focus on. This is often a reflection of the intended policy application. Many IAMs have been used to explore the dynamics of acidifying pollutants and greenhouse gases and their impacts. To make IAMs policy-relevant, cost modules and allocation algorithms can be added.

As with other models, IAM outcomes depend strongly on the assumptions made. As IAMs often cover a broad range of different topics and focus on integration of disciplinary knowledge (scientific information on the linkages among different disciplines is often less "strong" and involves more expert assessment), these assumption play a more important role than in other areas of modeling. In IAMs dealing with climate change, for instance, key uncertainties include developments of population and economy, sociopolitical choices with respect to human development (such as environmental policies), technology development, and discount rates. Several tools have been developed to deal with uncertainties. An important tool includes the use of storyline-based scenarios, defining consistent sets of assumptions. Others include more traditional uncertainty analysis and the assignment of qualifications to uncertain model outcomes (see, for example, www.nusap.net).

Almost by definition, the field of IA modeling is rather broad and vaguely defined. One reason for this is that it is relatively new, and universal rules and principles have not yet crystallized. This makes the overview in this section rather eclectic and limited, focusing on specific examples. We do not intend to give an extensive overview of all available models in the field. For many specific applications, such as integrated assessment of climate change, comprehensive overviews already exist (Weyant et al. 1996; Rotmans and Dowlatabadi 1998). We confine ourselves to models with a relatively large level of integration that have not yet been covered in other sections of this chapter.

4.11.1 Existing Approaches

IAMs first became popular in the fields of air pollution control and climate change. The work in these areas in the late 1980s and during the 1990s generated several models that are useful for carrying out assessments in global environmental change or sustainable development because they have typically both a long time horizon and a global perspective. For example, UNEP's *Global Environment Outlook* (UNEP 2002) and IPCC's *Special Report on Emissions Scenarios* (Nakićenović and Swart 2000) have all built on these models.

The first well-known IAM was built in the early 1970s in response to the concerns about world trends of a group of industrialists and civil servants, the Club of Rome (Meadows et al. 1972). The computer simulation model World3 described couplings between the major demographic, resource, and economic components of the world system at the global level. It used the system dynamics method developed at the Massachusetts Institute of Technology from electrical engineering science. Its main purpose was to raise awareness about the nature of exponential growth in a finite world and the systemic nature of the observed and anticipated trends due to the various linkages between what were at the time largely seen as separate processes. It showed the risks in continuing business-as-usual development paths—so much so that only its doomsday message came across. This model, and subsequent efforts at more disaggregated models (e.g., Mesarovic and Pestel 1974) inspired the development of a whole set of derived models. It also started a vigorous debate about the nature of market processes, which according to most economists would solve most problems long before a catastrophe would unfold.

A second generation of IAMs had a more narrow focus on a particular environmental problem and the ways in which policy could deal with it. This was partly in response to the aggregate nature of the first generation models that made it hard to perform meaningful validation and policy support. An outstanding and well-known example in this respect is RAINS—the Regional Acidification Information and Simulation model of acidification in Europe developed in the 1980s (Alcamo et al. 1990). It played and still plays a major role in the international air pollution negotiations in Europe.

The first steps of IAMs dealing with the causes and consequences of climate change were taken in the late 1970s. Examples include models by Nordhaus (1979) and Häfele et al. (1981), although the environmental part in these examples was extremely simple, including only atmospheric CO_2 concentration. Mintzer (1987), Lashof and Tirpack (1989), and Rotmans et al. (1990) extended these models by including more physical and chemical aspects of the climate system. Since then, a large number of such models have been developed and currently more than 50 climate change-oriented IAMs coexist (Van der Sluijs 2002). More recently, IAMs have been developed with an expanded or different emphasis, such as water (e.g., Döll et al. 1999) and human health (Martens 1997). In the community of economists, the emphasis has been on the merging of a neoclassical economic growth model with a simple climate system model and on using it in the search for cost-effective abatement strategies. Here, too, a series of additions has followed, such as a more elaborate energy system, more in-depth treatment of technological dynamics, and integration with impact modules.

As these third-generation IAMs are expanding their scope and level of integration, they are slowly developing from environmental or climate change models into global change or sustainable development models. (See Table 4.1.)

4.11.2 Critical Evaluation of Approaches

Three evaluation criteria were developed for discussing IAMs in the context of quantifying the MA scenarios:

- Is there any integration between ecosystems and other parts of the (world) system, such as land, water, atmosphere, population, and economy, and if so, how is it done?
- At what spatial and time scale(s) are the ecosystem and the interactions with other parts modeled, ranging from short-term local dynamics to large-scale and global long-term dynamics?
- To what extent is the model used, or has it been used, for policy applications, and if so, how is the interface with decision-makers or analysts constructed and applied?

For our purpose, we clustered three groups of IAM models. The first group contains models that have been built with the explicit objective of providing an integrated insight into a broad range of environmental, economic, and social aspects of sustainable development. The second group contains models that have been mostly built around the link between economic development, the energy sector, and the climate system. The third group of models is a subclass of the previous one. These models started out as energy-environment models but have evolved to a point where the newly developed ones are now better characterized as a third category: global change models. (We have not included the IAM models on regional air pollution control here because of their narrow focus. It should be noted, however, that the RAINS modeling team is currently extending its framework to cover not only acidification, eutrophication, and ground-level ozone but also greenhouse gas emissions.)

4.11.2.1 Sustainable Development Models

This group of IAMs has the highest level of integration in terms of social, economic, and environmental issues. In order to avoid levels of complexity that are too large, they use a rather high level of aggregation. They use expert-model derived meta-level descriptions of underlying processes—often correlations or "stylized facts"—and a low level of spatial or regional disaggregation. This group includes the system-dynamics World3 model and the more recent related models such as International Futures (Hughes 1999), TARGETS (Rotmans and De Vries 1997), Threshold 21 (Barney 2000), and GUMBO (Bouwmans et al. 2002). It also includes the Polestar system, which systematically links scenario assumptions and scenario outputs for a wide range of issues (Raskin et al. 1999).

Consistent with their high level of aggregation, most of these models try to answer rather broadly formulated questions, identifying possible trade-offs between economic developments and ecological functioning without providing detailed and concrete strategies or policy advice on how to deal with the trade-offs. From the perspective of developing scenarios for ecosystem services, the TARGETS model, the Polestar model, and the GUMBO model have provided interesting insights. In the context of the MA, these models had major advantages of allowing for a high level of integration and of including several feedbacks. Their description of detailed (ecological) processes is often simple, however.

The TARGETS model (Rotmans and de Vries 1997; De Vries 2001) has been developed at the Dutch National Institute of Public Health and the Environment and applied to work out three consistent perspectives on sustainable development. It includes five submodels, one of which simulates key biogeochemical cycles. The land cover/land use and the food, water, and energy supply-and-demand dynamics are simulated at an aggregate level. Scenarios were built around the framework of the Cultural Theory (Thompson et al. 1990), with an evaluation of parameter assumptions and model outcomes in terms of "utopian" and "dystopian" courses of events. The model allows a clear linkage with ecosystem services at the high aggregation

Table 4.1. Examples of Integrated Assessment Models (Adapted from Bakkes et al. 2000)

Model	Analytical Technique	Horizontal Integration[a]	Vertical Integration[b]	Key References	Key Existing Scenarios	Ease of Use by Non-Developers[c]
World 3	system dynamics	present	limited	Meadows et al. 1972; Meadows et al. 1992	13 explorative scenarios	limited
Int. Futures	system dynamics	advanced	limited	Hughes 1999	base scenario	very high
TARGETS	system dynamics	present	present	Rotmans and de Vries 1997	reference case with varieties	high
Threshold 21	system dynamics	advanced	present		several explorative scenarios	very high
Polestar	accounting	present	limited	Raskin et al. 1999; SEI Boston Center 1999	SGS-scenarios	very high
MESSAGE	dynamic linear programming	limited	limited	Messner and Strubegger 1995; Riahi and Roehrl 2000	SRES, WEC	limited
MiniCAM	partial equilibrium	limited	limited	Edmonds et al. 1996	SRES	high
AIM	general equilibrium	limited	limited	Morita et al. 1994	SRES, GEO	limited
IMAGE	system dynamics/ simulation	limited	advanced	Rotmans 1999; Alcamo 1994; IMAGE-team 2001	SRES, GEO	limited

[a] Very limited indicates a lack of integration between domains as well as within a domain. Limited refers to a lack in one of the two. Present indicates several domains are covered in an integrated manner. Advanced is used for models that include environmental, economic, and sociocultural aspects.

[b] Limited refers to models where several parts of the cause–effect chains modeled are missing or not explicit. Present refers to the models where the casual chain is modeled, but there is a lack of feedback from the output of the model to the input. The term advanced is reserved for models where this final loop is also closed.

[c] Very limited refers to models that are not accessible to non-developers. Limited refers to models where the models can be used by outsiders after considerable training. The term high classifies models that exhibit an interface and a level transparency that makes it very easy for non-developers to apply the model and to adjust it to their own needs.

level used. In contrast to most other global models, the TARGETS model included a full link back from environmental change into demographic developments (including health) and a partial evaluation of feedbacks upon the economy. The main disadvantages of the TARGETS model are the lack of regional disaggregation and the fact that, like most IAMs in this group, it was not related to a specific decision-making process.

Polestar is an integrated accounting framework developed by the Stockholm Environment Institute's Boston Center (SEI Boston Center 1999). Its best known application has been in conjunction with the Global Scenarios Group (Gallopin et al. 1997). The backbone of the model is an extensive data set containing a wide range of social, economic, and environmental variables. Polestar has relatively little relationships between each of the variables in the system. In that sense, it is not so much of a model as an accounting framework to explore various assumptions. It is more suitable for exploring the range of possible futures and the possible impact of certain policy interventions than for deriving integrated and balanced answers for issues related to political decision-making. Polestar has been very successfully applied in supporting various scenario development processes with strong user involvement, including scenarios that describe some elements of ecological functioning.

Finally, the GUMBO (Global Unified Metamodel of the Biosphere) model was developed explicitly to deal with a description of ecological services and their possible future development under various assumptions (Bouwmans et al. 2002). It is a "metamodel" in that it represents a synthesis and a simplification of several existing dynamic global models in both the natural and social sciences at an intermediate level of complexity. GUMBO includes dynamic feedbacks among human technology, economic production, and welfare and ecosystem goods and services within the dynamic Earth system. It includes modules to simulate carbon, water, and nutrient fluxes through the environmental and ecological systems. GUMBO links these elements across eleven biomes that together cover the entire surface of Earth (open ocean, coastal ocean, forests, grasslands, wetlands, lakes/rivers, deserts, tundra, ice/rock, cropland, and urban). The model also nicely links to several socioeconomic elements of sustainable development, such as the different types of capital (human, built, social, and natural) that form an essential element of the World Bank's approach to sustainable development and measures of sustainable social welfare.

4.11.2.2 Models Concentrating on Economy, Energy, and Climate Relationships

IAMs have been very successful in the field of climate change. A large number of IAMs have been developed, and their results are regularly presented to and discussed with decision-makers. In this way, they have clearly influenced policy-making. Examples of interaction between outcomes of IAMs and decision-makers include use and development of the IMAGE model in dialogue with policy-makers in the Netherlands (Alcamo et al. 1996) or the contribution of IAMs to several chapters of IPCC's Third Assessment Report, in particular with respect to mitigation strategies (Metz et al. 2001). Reasons for this include that the issue of climate change refers to a rather complex web of causes, environmental processes, and impacts (which can only be understood well in an integrated way), that many crucial relationships are well known, and, more recently, that an institutionalized policy process exists. As a result, currently more than 50 IAMs exist that cluster around the relationship of economic development, energy use, and climate change. Some of these also describe related issues such as other atmospheric pollutants and depletion of resources. In some publications, these models are referred to as 3E models: Economy, Energy, and Environment.

A clear difference within this group can be made between basic macroeconomic models, with little technological and climate detail, and models that include a detailed description of the energy sector. The first category includes, among others, rather aggregated meta-models that aim to link both economic causes of climate change and the resulting impacts in order to perform cost-benefit analysis, such as the DICE model (Nordhaus 1994) and the FUND model (e.g., Tol 1997, 2003). Both of these have been applied in a large number of studies. A strong point of these models is that they fully describe the cycle—from economic development to climate change, possible damage from climate change, and its feedback on the economy. However, the detail in describing ecological functions in these models is rather low and abstract. In particular, DICE describes climate change in terms of a limited set of equations for global temperature increase with a couple of damage functions. The FUND model, in time, has become more comprehensive in its description of climate change impacts; now, for instance, it also deals with spread of diseases.

Another group of models includes a more detailed description of the energy and the climate system but, in turn, sometimes lacks a feedback on economic development. This includes most of the models used in the recent IPCC *Special Report on Emission Scenarios* study (ASF, MESSAGE-MACRO, MARIA, MiniCAM, AIM, and IMAGE). (Descriptions and references of these models are provided by Nakićenović and Swart, 2000; see also www.grida.no/climate/ipcc/emission/index.htm.) Other influential IAM models that fall into this category are the ICAM model (Dowlatabadi 1995) and MIT's Integrated Assessment Model (Prinn et al. 1998). The ICAM model emphasizes the role of uncertainty. The MIT model consists of a framework of underlying state-of-the-art disciplinary models and probably represents currently one of the most well developed models in the field.

4.11.2.3 Global Change Models

The models in this group are similar to those just described, in that the emphasis is placed on the relationships between energy and the environment. Reflecting a trend of the past 10 years and taking advantage of new computer tools and satellite data, the developers of these models have widened their scope to include other aspects as well, in particular land cover/land use change and the relationships between land use change and climate change. For this reason, we have chosen to characterize them as global change models. This section focuses on the IMAGE and AIM models.

The IMAGE model is one of the well-known integrated assessment models in this category (IMAGE-team 2001). It includes a description of the energy/industry system, of land use change, and of the climate system, partly based on 19 global regions and partly on a 0.5x0.5 grid. In more recent years, it has both aimed at applications within the climate change community (for example, in the IPCC-SRES) and at much broader applications, such as UNEP's *Global Environment Outlook* (UNEP 2002). Ecological services that are described within the model include the role of ecosystems within carbon and nutrient cycling, provision of food and energy, and some more abstract indicators of ecological functioning. The number of linkages in IMAGE between the environmental system and socioeconomic system, however, are limited. Within its more narrow focus of environmental problems, nevertheless, horizontal integration is exemplary. Vertical integration is more limited, as there is a lack of feedback from the environmental impacts calculated by the model on the macroeconomic trajectories that the model assumes.

AIM is a set of models developed by the National Institute of Environmental Studies in Japan, including a general equilibrium model but also some fuller integrated assessment models (Kainuma et al. 2002). In terms of its coverage and applications it is quite similar to the IMAGE model, although the emphasis is more on East and South Asia.

4.11.3 Conclusions

Over the years several IAMs have provided relevant information on the future of ecological services. This is in particular the case for regulating and providing ecological services such as food and water provision and those related to biogeochemical cycles. Only a few models have been specifically developed to provide information on "sustainable development" issues, however. (That term is used here to indicate the central focus of the MA—ecological functions and their relevance for human well-being.) Existing models only cover trends for a few selected services. Evaluating IAMs against the three criteria described earlier leads to several conclusions.

4.11.3.1 Integration and Feedbacks

Although the integration in most models is high from the perspective of the limited (environmental) problems they were developed for, their integration from a perspective of the MA's objective is still rather low. In particular, the number of feedbacks that are included from ecological changes on socioeconomic drivers are scarce. (Some exceptions are the impacts of food production and climate policy on socioeconomic drivers.)

We believe that a better description of the linkages and feedbacks from an overall sustainable development perspective (instead of a single issue perspective) could improve the relevance of a selected set of IAMs for broad global assessments. An international attempt to further specify these linkages and establish priorities for them against an overall "sustainability science" context could be helpful.

At the same time, it is clearly not realistic to expect models to be both comprehensive and detailed. Therefore, nested approaches could provide a significant improvement over existing work. Here, comprehensive but more aggregated models provide drivers for more dedicated (and thus more detailed) models. Such models could focus on particular issues or regions.

4.11.3.2 Level of Geographical Aggregation

The processes in ecosystems and hence the provision of ecosystem services are most adequately considered as nested dynamical processes occurring at various scales (Gunderson et al. 1995). A proper understanding of their response under human-induced direct and indirect (such as climate change) perturbations demands models that cover various scales in both space and time.

Given the purpose of the MA, models needed to acknowledge heterogeneity and include a sufficiently detailed regional/local specificity. With the new tools of geographic information systems and an ever-growing amount of satellite and field data, there is a clear tendency to invest in ever-higher spatial resolution in models. This has a price, too, as it usually implies less than global or even regional coverage. At the same time, given ongoing globalization, many economic processes potentially have ever-wider consequences for ecosystem perturbations. Thus, at the economic level as well, understanding the nature and dynamics of regional differences and interregional links, in particular trade, will be an important research issue for integrated modeling in the coming years. Again, a nested approach to integrated assessment modeling could be a helpful way forward, in which global models provide context for detailed, regional (ecological) models.

4.11.3.3 Areas That Are Poorly Covered

Ideally, IAMs should cover a wide range of different aspects of sustainable development if they are to be used for policy-relevant assessment with as broad a scope as the MA. Clearly, some areas of ecosystem change are poorly covered. In particular, the linkages between ecosystem change and human development, in a broad economic and sociopolitical sense, are weak or absent in most IAMs. The institutional components among them are notoriously weak. This reflects the large and growing complexity of the economic and social processes in an ever more integrated world.

There is as yet limited experience with and agreement on how to connect the various layers of a vertically integrated model. The emerging science of complexity, especially when it entails detailed simulation models of socioecological systems, may indicate a way forward (Janssen 2003; de Vries and Goudsblom 2002). Another strategy is to include qualitative narratives and then try to support certain parts of the narratives with the rigor and consistency of quantitative models. Some parts of the environment are almost systematically lacking in IAMs, such as marine ecosystems and coastal zones. This is also the case for a large number of ecosystem services, in particular regulating and cultural services. Our representation of the parts that are included suffers from an incomplete understanding of the underlying processes.

4.11.3.4 Application of IAMs

A major area of use of IAMs is in developing and analyzing scenarios. Models should be only one tool in this process, their main role being to generate and organize quantitative projections. Descriptive narratives are powerful tools to convey the broader significance of scenarios, as indicated, for instance, in IPCC's SRES scenario work. Among other things, narratives bring in qualitative elements that quantitative models cannot handle and convey that different scenarios constitute very different worlds and, therefore, strategies that will work in one future world may very well be out of place in another.

Finally, we need to note that uncertainties are a key element in IAMs, given their high complexity and focus on decision-making. These uncertainties include, for example, variability of parameters, inaccuracy of model specification, and lack of knowledge with regard to model boundaries. Although the existence of uncertainties has been recognized early in the process of developing IAMs, in many of them uncertainty analysis is included only partially or not at all. Several new projects have been set up to work on uncertainties in a more specific way (see, e.g., www.nusap.net).

4.12 Key Gaps in Our Modeling Abilities

Many of the shortcomings of models pertain to data limitations or limitations of the models themselves. For example, our ability to incorporate spatially explicit data is often lacking. Two fundamental conceptual gaps stand out as especially important.

The first gap concerns the absence of critical feedbacks in many cases. Examples can be found in virtually every arena of forecasts. As food supply changes, so will patterns of land use, which will then feedback on ecosystem services and climate alteration and future food supplies. Land use changes modify the climate, but the climate then alters the vegetation possible on any parcel of land, which in turn constrains the types of land cover possible. These types of feedbacks are lacking throughout. This means the forecasts are best over shorter time scales, before the feedbacks are given time to resonate back through systems. It may be that 50 years, which is the timeframe of the MA projections, is sufficiently short that the absence of critical feedbacks is not as much of a liability as it might seem at first glance. If these feedbacks are important, our models may be seriously wrong in their predictions.

The second major gap is the absence of theories and models that anticipate thresholds, which once passed yield fundamental system changes or even system collapse. We know it is possible to hunt or fish a species to extinction—to total collapse, in other words. A short time frame for forecasts does not confer any immunity to thresholds, since we may be very close to certain thresholds at this point in time. For this reason, the greatest priority for advancing MA models is more explicit attention to anticipation of thresholds.

Finally, there is the issue of model transparency and the use of models by decision-makers. Much of what makes models fail as useful assessment tools is that modelers often get the technical process of modeling right but do not account for the fact that assessment tools need to be part of existing social and political processes. There needs to be much more work aimed at: how to ask modeling questions so that they are relevant to policy and other processes; how to find new ways to communicate complexity to nonspecialists because of the abundance of nonlinearities, feedbacks, and time lags in most global ecosystems; how to elicit knowledge with and from stakeholders at different levels of organization (local, regional, national, international), how to understand the way models fit or do not fit into social and political processes; and how to communicate model uncertainty to non-specialists.

References

ACC/SCN (United Nations Administrative Committee on Coordination–Subcommittee on Nutrition), 1996: *Update on the Nutrition Situation.* ACC/SCN, Geneva.

Aber, J.D., S.V. Ollinger, and C.T. Driscoll, 1997: Modeling nitrogen saturation in forest ecosystems in response to land use and atmospheric deposition. *Ecological Modelling,* **101,** 61–78.

Aber, J., W. McDowell, K. Nadelhoffer, A. Magill, G. Bernston, M. Kamakea, S. McNulty, W. Currie, L. Rustad, and I. Fernandez, 1998: Nitrogen saturation in temperate forest ecosystems. *Bioscience,* **48,** 921–934.

Achard, F., H. Eva, H.J. Stibig, P. Mayaux, J. Gallego, and T. Richards, 2002: Determination of deforestation rates of the world's humid tropical forests. *Science,* **297,** 999–1002.

Agarwal, C., G.L. Green, J.M. Grove, T. Evans, and C. Schweik, 2002: *A Review and Assessment of Land-Use Change Models: Dynamics of Space, Time, and Human Choice.* Bloomington, Indiana, Center for the Study of Institutions, Population, and Environmental Change at Indiana University-Bloomington and the USDA Forest Service Northeastern Forest Research Station, Burlington, Vermont, 61 pp.

Alcamo, J., Shaw, R., and Hordijk, L., 1990: *The RAINS model of acidification: Science and Strategies in Europe,* Kluwer Academic Publishers, Dordrecht, Netherlands.

Alcamo, J., 1994: *IMAGE 2.0—Integrated Modeling of Global Climate Change.* Kluwer Academic Publishers, Dordrecht.

Alcamo, J., Kreileman, E. and Leemans, R., 1996: Global models meet global policy. How can global and regional modelers connect with environmental policy makers? What has hindered them? What has helped? *Global Environmental Change.* **6(4),** 255–259.

Alexander, R.B., R.A. Smith, and G.E. Schwarz, 2000: Effect of stream channel size on the delivery of nitrogen to the Gulf of Mexico. *Nature,* **403,** 758–761.

Alexandratos, N., (ed.), 1995: *World Agriculture: Towards 2010: An FAO Study.* Food and Agriculture Organization of the United Nations, Rome and John Wiley and Sons, Chichester, England, 488 pp.

Alexandratos, N., (ed.), 1988: *World Agriculture: Toward 2000: An FAO Study.* Belhaven Press, London, 338 pp.

Alig, R.J., 1986: Econometric analysis of the factors influencing forest acreage in the Southeast. *Forest Science,* **32(1),** 19–134.

Altabet, M.A., R. Francois, D.W. Murray, and W.L. Prell, 1995: Climate-related variations in denitrification in the Arabian Sea from sediment $^{15}N/^{14}N$ ratios. *Nature,* **373,** 506–509.

Andersen, K.P., and E. Ursin, 1977: A multispecies extension to the Beverton and Holt theory of fishing, with accounts of phosphorous circulation and primary productivity. *Meddelelser,* **7,** 319–435.

Ashworth, A.C., 1996: The response of arctic Carabidae (Coleoptera) to climate change based on the fossil record of the Quaternary Period. *Annales Zoologici Fennici,* **33,** 125–131.

Asner, G.P., A.R. Townsend, W. Riley, P.A. Matson, J.C. Neff, and C.C. Cleveland, 2001: Physical and biogeochemical controls of terrestrial ecosystems responses to nitrogen deposition. *Biogeochemistry,* **54,** 1–39.

Aydin, K., and N. Friday, 2000: The early development of ECOSIM as a predictive multispecies fisheries management tool. IWC SC/53/E WP2.

Baker, W.L., 1989: A review of models of landscape change. *Landscape Ecology,* **2(2),** 111–133.

Bakkes, J.A., J. Grosskurth, A.M. Idenburg, D.S. Rothman and D.P. van Vuuren, 2000: Descriptions of selected global models for scenario studies on environmentally sustainable development. RIVM report no. 402001018. National Institute of Public Health and the Environment, Bilthoven, Netherlands.

Barney, G., 2000: Threshold 21 (T21) Overview. Millennium Institute. Washington D.C.

Bartell, S.M., G. Lefebvre, G. Kaminski, M. Carreau, and K.R. Campbell, 1999: An ecosystem model for assessing ecological risks in Quebec rivers, lakes, and reservoirs. *Ecological Modelling,* **124,** 43–67.

Beasley, D.B., E.J. Monike, E.R. Miller, and L.F. Huggins, 1985: Using simulation to assess the impacts of conservation tillage on movement of sediment and phosphorus into Lake Erie. *Journal of Soil and Water Conservation,* **40,** 233–237.

Begon, M., C.R. Townsend, and J.L. Harper, 1998: *Ecology: Individuals, Populations and Communities.* 3rd ed. Blackwell Science Inc., Oxford.

Beisner, B.E., C.L. Dent, and S.R. Carpenter, 2003: Variability of lakes on the landscape: roles of phosphorus, food webs and dissolved organic carbon. *Ecology,* **84,** 1563–1575.

Bennett, E.M., S.R. Carpenter, and N.F. Caraco, 2001: Human impact on erodable phosphorus and eutrophication: A global perspective. *BioScience,* **51,** 227–234.

Blackmon, M.B., B. Boville, F. Bryan, R. Dickinson, P. Gent, J. Kiehl, R. Moritz, D. Randall, J. Shukla, S. Solomon, G. Bonan, S. Doney, I. Fung, J. Hack, E. Hunke, J. Hurrell, et al., 2001: The Community Climate System Model. *BAMS,* **82(11),** 2357–2376.

Blackmon, M.B., B. Boville, F. Bryan, R. Dickinson, P. Gent, J. Kiehl, R. Moritz D. Randall, J. Shukla, S. Solomon, J. Fein and CCSM Working Group Co-Chairs, cited 2000: *Community Climate System Model Plan 2000–2005.* Available on-line from http://www.ccsm.ucar.edu/management/plan2000/index.html.

Blakeslee, L.L., E.O. Heady, and C. Framingham, 1973: *World food production, demand and trade.* Iowa State University Press, Ames, IA.

Boer, G.J., 2004: Long-timescale potential predictability in an ensemble of coupled climate models. *Climate Dynamics,* in press.

Bonan, G.B., 1999: Frost followed the plow: Impacts of deforestation on the climate of the United States. *Ecological Applications,* **9(4),** 1305–1315

Bonan, G.B., D. Pollard, and S.L. Thompson, 1992: Effects of boreal forest vegetation on global climate. *Nature,* **359,** 716–717

Bonan, G.B, S. Levis, L. Kergoat, and K.W. Oleson, 2002. Landscapes as patches of plant functional types: an integrating concept for climate and ecosystem models. *Global Biogeochem. Cycles,* **16,** 10 1029/2000GB001360.

Bonham-Carter, G.F., 1991: Integration of geoscientific data using GIS. In: *Geographical Information Systems. Vol. I: Principles and Applications,* D.J. Maguire, M.J. Goodchild and D.W. Rhind (eds.), Longman Scientific and Technical, London, pp. 171–84.

Botsford, L.W., and A.M. Parma, 2005: Uncertainty in marine management. In: *Marine Conservation Biology: the science of maintaining the sea's diversity,* E.A. Norse and L.B. Crowder (eds.). Island Press, Washington, DC.

Bouwman, A.F., L.J.M. Boumans, and N.H. Batjes, 2002: Emissions of N_2O and NO from fertilized fields: Summary of available measurement data. *Global Biogeochemical Cycles,* **16,** doi:1-/1-202—1GB001811.

Bouwmans, R., Costanza, R., Farley, J., Wilson, M.A., Portela, R., Rotmans, J., Villa, F. and Grasso, M., 2002: Modeling the dynamics of the integrated earth system and the value of global ecosystem services using the GUMBO model. *Ecological Economics,* **41,** 529–560.

Brock, W.A. and A. de Zeeuw, 2002: The repeated lake game. *Economics Letters,* **76,** 109–114.

Brooks, T.M., S.L. Pimm, and J.O. Oyugi, 1999: Time lag between deforestation and bird extinction in tropical forest fragments. *Conservation Biology,* **13,** 1140–1150.

Brown, D.G., R. Walker, S. Manson, and K. Seto, 2004: Modeling land use and land cover change. In: *Land Change Science: Observing, Monitoring, and Understanding Trajectories of Change on the Earth's Surface,* G. Gutman, A. Janetos, C. Justice, E. Moran, J. Mustard, R. Rindfuss, D. Skole and B.L. Turner, II., (eds.), Kluwer Academic Publishers, Dordrecht, Netherlands, (Forthcoming).

Brown, J.H., 2001: Mammals on mountainsides: elevational patterns of diversity. *Global Ecology and Biodiversity,* **10,** 101–109.

Brown, J.H., and M.V. Lomolino, 1998: *Biogeography.* 2nd ed. Sinauer Associates, Sunderland, MA., 560 pp.

Bruinsma, J., (ed.), 2003: *World Agriculture: Towards 2015/2030: An FAO Study.* Food and Agriculture Organization of the United Nations, Rome and Earthscan Publications Ltd, London, 520 pp.

Buermann, W., Y. Wang, J. Dong, L. Zhou, Z. Zeng, R. E. Dickinson, C. S. Potter and R. B. Myneni, 2002. Analysis of a multiyear global vegetation leaf area index data set. *Journal of Geophysical Research,* Vol. **107,** No. D22, 4646, doi:10.1029/2001JD000975.

Butler, C.D., C.F. Corvalan, and H.S. Koren, in press: Human health, well-being and global ecological scenarios. *Ecosystems.*

Butterworth, D.S., and A.E. Punt, 1999: Experiences in the evaluation and implementation of management procedures. *ICES J. Mar. Sci.* **56,** 985–998.

CAPRI (Common Agricultural Policy Regional Impact Analysis), 2004: http://www.agp.uni-bonn.de/agpo/rsrch/capri/caprifp4_e.htm [project website, accessed May 2004].

Caraco, N.F., J.J. Cole, and G.E. Likens, 1991: A cross-system study of phosphorus release from lake sediments. In: *Comparative Analysis of Ecosystems,* J. Cole, G. Lovett and S. Findlay (eds.), Springer-Verlag, New York, pp. 241–258.

Caraco, N.F., and J.J. Cole, 1999: Human impact on nitrate export: An analysis using major world rivers. *Ambio,* **28(2),** 167–170.

Carpenter, S.R., 2002: Ecological futures: building an ecology of the long now. *Ecology,* **83,** 2069–2083.

Carpenter, S.R., 2003: *Regime Shifts in Lake Ecosystems: Pattern and Variation. Vol. 15,* In: the Excellence in Ecology Series, Ecology Institute, Oldendorf/Luhe, Germany.

Carpenter, S.R., and J.F. Kitchell, (eds.), 1993: *The Trophic Cascade in Lakes.* Cambridge University Press, Cambridge, UK., 399 pp.

Carpenter, S.R., N.F. Caraco, D.L. Correll, R.W. Howarth, A.N. Sharpley, and V.H. Smith, 1998: Nonpoint pollution of surface waters with phosphorus and nitrogen. *Ecological Applications,* **8,** 559–568.

Carpenter, S.R., W.A. Brock, and P.C. Hanson, 1999a: Ecological and social dynamics in simple models of ecosystem management. *Conservation Ecology* **3(2),** 4.

Carpenter, S.R., D. Ludwig, and W.A. Brock, 1999b: Management of eutrophication for lakes subject to potentially irreversible change. *Ecological Applications,* **9,** 751–771.

Chapin, III., F.S., O.E. Sala, and E. Huber-Sannwald, (eds.), 2001: *Global Biodiversity in a Changing Environment: Scenarios for the 21st Century.* Springer-Verlag, New York, 376 pp.

Chapra, S.C., 1997: *Surface water quality modeling.* McGraw-Hill, New York, 784 pp.

Chomitz, K.M., and D.A. Gray, 1996: Roads, land use, and deforestation: a spatial model applied to Belize. *The World Bank Economic Review,* **10(3),** 487–512.

Christensen, V., and D. Pauly, 1992. ECOPATH II—A software for balancing steady-state models and calculating network characteristics. *Ecol. Modelling,* **61,** 169–185.

Christensen, V., and C. Walters, 2000: ECOPATH with ECOSIM: Methods, capabilities and limitations. In: Methods for evaluating the impacts of fisheries on North Atlantic ecosystems, D. Pauly and T.J. Pitcher (eds.), Fisheries Centre Research Reports **8,** 79–105.

Clark, J.S., S.R. Carpenter, M. Barber, S. Collins, A. Dobson, J.A. Foley, D.M. Lodge, M. Pascual, R. Pielke, Jr., W. Pizer, C. Pringle, W.V. Reid, K.A. Rose, O. Sala, W.H. Schlesinger, D.H. Wall, and D. Wear, 2001: Ecological forecasts: An emerging imperative. *Science,* **293(5530),** 657–660.

Cleveland, C.C., A.R. Townsend, D.S. Schimel, H. Fisher, R.W. Howarth, L.O. Hedin, S.S. Perakis, E.F. Latty, J.C. Von Fischer, A. Elseroad, and M.F. Wasson, 1999: Global patterns of terrestrial biological nitrogen (N2) fixation in natural ecosystems. *Global Biogeochemical Cycles,* **13,** 623–645.

Codispoti, L.A., J.A. Brandes, J.P. Christensen, A.H. Devol, S.W.A. Naqvi, H.W. Paerl, and T. Yoshinari, 2001: The oceanic fixed nitrogen and nitrous oxide budgets: Moving targets as we enter the anthropocene? *Scientia Marina,* **65,** 85–105.

Cole, J., G. Lovett, and S. Findlay, (eds.), 1991: *Comparative Analyses of Ecosystems: Patterns, Mechanisms, and Theories.* Springer-Verlag, New York, 375 pp.

Conte, R., R. Hegselmann, and P. Terna, (eds.), 1997: *Simulating Social Phenomena.* Springer-Verlag, Berlin, 536 pp.

Cook, D.J., W.T. Dickinson, and R.P. Rudra, 1985: *GAMES—The Guelph Model for Evaluating the Effects of Agricultural Management Systems in Erosion and Sedimentation: User's Manual.* Technical Report. School of Engineering, University of Guelph, Ontario, pp. 126–7.

Cox, P.M., R.A. Betts, C.D. Jones, S.A. Spall, and I.J. Totterdell, 2000: Acceleration of global warming due to carbon-cycle feedbacks in a coupled climate model. *Nature,* **408,** 184–187.

Crawley, M.J., and J.E. Harral, 2001: Scale dependence in plant biodiversity. *Science,* **291(5505),** 864–868.

Cumming, G.S., 2000: Using habitat models to map diversity: pan-African species richness of ticks (Acari: Ixodida). *Journal of Biogeography,* **27,** 425–440.

Dai, Y., X. Zeng, R.E. Dickinson, I. Baker, G.B. Bonan, M.G. Bosilovich, A.S. Denning, P.A. Dirmeyer, P.R. Houser, G-Y. Niu, K.W. Oleson, C. A. Schlosser, and Z.L. Yang, 2003: The Common Land Model (CLM) *Bull. Amer. Meteor. Soc.,* **84(8),** 1013–1023.

DeFries, R.S., C. Field, I. Fung, G. Collatz, and L. Bounoua, 1999: Combining satellite data and biogeochemical models to estimate global effects of human-induced land cover change on carbon emissions and primary productivity. *Global Biogeochemical Cycles,* **13(3),** 803–815.

DeFries, R.S., L. Bounoua, G.J. Collatz, 2002a: Human modification of the landscape and surface climate in the next fifty years. *Global Change Biology,* **8(5),** 438–458.

DeFries, R.A., R.A. Houghton, M. Hansen, C. Field, D.L. Skole, and J. Townshend, 2002b: Carbon emissions from tropical deforestation and regrowth based on satellite observations for the 1980s and 90s. *Proceedings of the National Academy of Sciences,* **99(22),** 14256–14261.

Dempster, A.P., 1968: A generalization of Bayesian inference. J. *Roy. Statist. Soc. B.,* **30,** 205–247.

Deriso, R.B., 1980: Harvesting strategies and parameter estimation for an age-structured model. *Can. J. Fish. Aquat. Sci.,* **37,** 268–282.

Deutsch, C., Gruber, N., R.M. Key, and J.L. Sarmiento, 2001: Denitrification and N_2 fixation in the Pacific Ocean. *Global Biogeochemical Cycles,* **15,** 483–506.

de Vries B. and J. Goudsblom (eds.), 2002: Mappae Mundi: Humans and their Habitats in a Long-Term Socio-Ecological Perspective: Myths, Maps and Models. Amsterdam University Press. Amsterdam.

de Vries, H.J.M., 2001: Perceptions and risks in the search for a sustainable world. A model based approach. *International Journal for Sustainable Development,* **4(4).**

DG-AGRI (Directorate-General for Agriculture of the European Commission), 2003a: *Mid-term review of the common agricultural policy. July 2002 proposals.* Impact analyses. European Commission, Brussels.

DG-AGRI, 2003b: Impact Analysis of the Medium Term Review Proposals with the CAPRI Modelling System. Prepared by University of Bonn. Department for Economics and Agricultural Policy. European Commission, Brussels.

DG-AGRI, 2003c: Reform of the common agricultural policy medium-term prospects for agricultural markets and income in the European Union 2003–2010. European Commission, Brussels.

Dickinson, R.E., 1984: Modeling evapotranspiration for three-dimensional global climate models. *Climate Processes and Climate Sensitivity.* J.E. Hansen and K. Takahashi, eds., American Geophysical Union Geophysical Monograph 29, Maurice Ewing **5,** 58–72.

Dickinson, R.E., 1992: Changes in Land Use, Chapter 22, 689–701, Climate Systems Modeling, K. Trenberth, ed., Cambridge University Press.

Dickinson, R.E., and A. Henderson-Sellers, 1988: Modelling tropical deforestation: A study of GCM land-surface parameterizations. *Quart. J. Roy. Meteor. Soc.,* **114,** 439–462.

Dickinson, R.E., E. Ridley, and R. Roble, 1981: A three-dimensional general circulation model of the thermosphere. *Journal of Geophysical Research,* **86,** 1499–1512.

Dobson, A.P., 2000: Raccoon rabies in space and time. *Proceedings of the National Academy of Science,* **97,** 14041–14043.

Döll, P., F. Kaspar, and J. Alcamo, 1999: Computation of global water availability and water use at the scale of large drainage Basins. *Mathematische Geologie,* **4,** 111–118.

Donner, S.D., M.T. Coe, J.D. Lenters, T.E. Twine and J.A. Foley, 2002: Modeling the impact of hydrological changes on nitrate transport in the Mississippi River basin from 1955 to 1994. *Global Biogeochem. Cycles,* **16(3),** 1043, 10.1029/2001GB001396.

Donnigan, Jr., A.S., D.C. Beverlein, H.H. David, Jr., and N.H. Crawford, 1977: Agricultural Runoff Management (ARM) Model Version II: Refinement and Testing. EPA 600/3-77-098. Environmental Research Library, Athens, GA, 294 pp.

Dowlatabadi, H., 1995: *Integrated assessment climate assessment model 2.0.* Technical documentation. Department of Engineering and Public Policy, Carnegie-Mellon University, Pittsburgh.

Drago, M., B. Cescon, and L. Iovenitti, 2001: A three-dimensional numerical model for eutrophication and pollutant transport. *Ecological Modelling,* **145,** 17–34.

Durlauf, S.N., 1997: Insights for socioeconomic modeling. *Complexity,* **2**(3), 47–49.

Edmonds, J. M., M. Wise, et al, 1996: An integrated assessment of climate change and the accelerated introduction of advanced energy technologies: An application of MiniCAM 1.0. *Mitigation and Adaptation Strategies for Global Change,* **1(4),** 311–339.

Erasmus, B.F.N., A.S. Van Jaarsveld, S.L. Chown, M. Kshatriya, and K.J. Wessels, 2002: Vulnerability of South African Animal Taxa to Climate Change, *Global Change Biology,* **8(7),** 679–693.

Everbecq, E.V. Gosselain, L. Viroux and J.P. Descy, 2001: POTAMON: a dynamic model for predicting phytoplankton biomass and composition in lowland rivers. *Water Research,* **35,** 901–912.

Falkowski, P.G., 1997. Evolution of the nitrogen cycle and its influence on the biological sequestration of CO_2 in the ocean. *Nature,* **387,** 272–275.

FAO (Food and Agriculture Organization of the United Nations) 1970: Provisional Indicative World Plan for Agricultural Development. Rome.

FAO, 1993: The World Food Model. Model Specification. Document ESC/M/93/1. Food and Agriculture Organization of the United Nations, 1970: Provisional Indicative World Plan for Agricultural Development. Rome.

FAO, 2000: FAOSTAT database. [online] Food and Agriculture Organization of the United Nations, Rome. Available at http://apps.fao.org/.

FAPRI (Food and Agricultural Policy Research Institute) 2003: *U.S. and World Agricultural Outlook 2003.* FAPRI Staff Report #1–03. University of Iowa and University of Missouri-Columbia.

FAPRI, 2002a: *FAPRI 2002 World Agricultural Outlook.* FAPRI Staff Report #1–02. FAPRI, Iowa State University, Ames, IA.

FAPRI, 2002b: *FAPRI 2002 U.S. Baseline Briefing Book.* FAPRI-UMC Technical Data Report #2-02. Food and Agricultural Policy Research Institute, University of Missouri, Columbia, Missouri.

Fearnside, P.M., 2000: Global warming and tropical land use change: Greenhouse gas emissions from biomass burning, decomposition and soils in forest conversion, shifting cultivation and secondary vegetation. *Climatic Change,* **46,** 115–158.

Ferraz, G., G.J. Russell, P.C. Stouffer, R.O.J. Bierregaard, P.S. L., and L.T. E., 2003: Rates of species loss from Amazonian forest fragments. *Proceedings of the National Academy of Sciences USA,* **100,** 14069–73.

Fielding, A.H., and J.F. Bell, 1997: A review of methods for the assessment of prediction errors in conservation presence/absence models. *Environmental Conservation,* **24,** 38–49.

Firestone, M.K., and E.A. Davidson, 1989: Microbial basis of NO and N_2O production and consumption. In: *Exchange of trace gases between terrestrial ecosystems and the atmosphere,* M.O. Andreae and D.S. Schimel (eds.), John Wiley and Sons Ltd., Dahlem Konferenzen, Chicheste.

Foley, R.A., 1994: Speciation, extinction and climatic change in hominid evolution. *Journal of Human Evolution,* **26,** 275–289.

Foley, J.A., J.E. Kutzbach, M.T. Coe and S. Levis, 1994: Feedbacks between climate and boreal forests during the Holocene epoch. *Nature,* **371,** 52–54.

Foley, J.A., M.H. Costa, C. Delire, N. Ramankutty, and P. Snyder, 2003: Green Surprise? How terrestrial ecosystems could affect earth's climate. *Frontiers in Ecology and the Environment,* **1(1),** 38–44.

Fournier, D.A., J. Hampton, and J.R. Sibert, 1998: MULTIFAN-CL: a length-based, age-structured model for fisheries stock assessment, with application to South Pacific albacore, *Thunnus alalunga. Can. J. Fish. Aquat. Sci.,* **55,** 2105–2116.

Francis, R.C., and S.R. Hare, 1994: Decadal-scale regime shifts in the large marine ecosystems of the North-east Pacific: a case for historical science. *Fish. Oceanogr.,* **3,** 279–291.

Francis, R.C., S.R. Hare, A.B. Hollowed, and W.S. Wooster, 1998: Effects of interdecadal climate variability on the oceanic ecosystems of the NE Pacific. *Fish. Oceanogr.,* **7,** 1–21.

Fuhrman, J.A. and D.G. Capone, 1991: Possible biogeochemical consequences of ocean fertilization. *Limnol. Oceanogr.,* **36,** 1951–1959.

Gallopin, G., Hammond, A., Raskin, P. and Swart, R., 1997: Branch points: global scenarios and human choice. PoleStar Series Report no. 7. Stockholm, pp. 47.

Galloway, J.N., W.H. Schlesinger, H. Levy II, A. Michaels, and J.L. Schnoor, 1995: Nitrogen fixation: Anthropogenic enhancement-environmental response. *Global Biogeochemical Cycles,* **9,** 235–252.

Galloway, J.N., et al., 2004: *Nitrogen Cycles: Past, Present and Future.* in preparation.

Gaston, K.J., 2000: Global patterns in biodiversity. *Nature,* **405,** 220–227.

Gburek, W.J., and A.N. Sharpley, 1998: Hydrological controls on phosphorus loss from upland agricultural watersheds. *Journal of Environmental Quality,* **27,** 267–277.

Geoghegan, J., S. Cortina Villar, P. Klepeis, P. Macario Mendoza, Y. Ogneva-Himmelberger, R.R. Chowdhury, B.L. Turner, II and, C. Vance, 2001: Modeling tropical deforestation in the southern Yucatán peninsular region: comparing survey and satellite data. *Agriculture, Ecosystems and Environment,* **85,** 25–46.

Gimblett, H.R., (ed.), 2002. *Integrating Geographic Information Systems and Agent-Based Modeling Techniques for Simulation of Social and Ecological Processes.* Santa Fe Institute Studies on the Sciences of Complexity, Oxford University Press, New York.

Gin, K.Y.H., Q.Y. Zhang, E.S. Chan, and L.M. Chou, 2001: Three-dimensional ecological eutrophication model for Singapore. *Journal of Environmental Engineering ASCE,* **127,** 928–937.

Griffith, D.A., 1987: *Spatial Autocorrelation: A Primer.* American Association of Geographers, Washington, DC.

Gruber, N., and J.L. Sarmiento, 1997: Global patterns of marine nitrogen fixation and denitrification. *Global Biogeochemical Cycles,* **11,** 235–266.

Gunderson, L., C.S. Holling, and S.S. Light, 1995: *Barriers and bridges to the renewal of ecosystems and institutions.*

Guneralp, B., and Y. Barlas, 2003: Dynamic modeling of a shallow freshwater lake for ecological and economic sustainability. *Ecological Modelling,* **167,** 115–138.

Gutman, G., A. Janetos, C. Justice, E. Moran, J. Mustard, R. Rindfuss, D. Skole, and I.B.L. Turner (eds.), 2004: *Land Change Science: Observing, Monitoring, and Understanding Trajectories of Change on the Earth's Surface.* Kluwer Academic Publishers, Dordrecht, Netherlands.

Häfele, W., J. Anderer., A. McDonald and N. Nakićenović, 1981: *Energy in a finite world.* Ballinger. Cambridge. MA.

Håkanson, L., and V.V. Boulion, 2003: A general dynamic model to predict biomass and production of phytoplankton in lakes. *Ecological Modelling,* **165,** 285–301.

Håkanson, L., and R.H. Peters, 1995: *Predictive Limnology: Methods for Predictive Modeling.* SPB Academic, Amsterdam, Netherlands.

Hallegraeff, G.M., 1993: A review of harmful algal blooms and their apparent global increase. *Phycologia,* **32,** 79–99.

Hampton, J., and D.A. Fournier, 2001: A spatially disaggregated, length-based, age-structured population model of yellowfin tuna (*Thunnus albacares*) in the western and central Pacific *Ocean. Mar. Fresh. Res.,* **52,** 937–963.

Hannon, B., and M. Ruth, 1994: *Dynamic Modeling.* Springer-Verlag, New York.

Hansell, D.A., N.R. Bates, and D.B. Olson, 2004: Excess nitrate and nitrogen fixation in the North Atlantic. *Marine Chemistry,* in press.

Hardie, I.W., and P.J. Parks, 1997: Land use with heterogeneous land quality: an application of an area base model. *American Journal of Agricultural Economics,* **79(2),** 299–310.

Haxeltine, A. and I.C. Prentice, 1996: BIOME 3: an equilibrium terrestrial bioshpere model based on ecophysiological constraints, resource availability, and competition among plant functional types. *Global Biogeochemical Cycles,* **10(4),** 693–710.

He, H.S., 1999: The effects of seed dispersal on the simulation of long-term forest landscape change. *Ecosystems,* **2,** 308–319.

Hegelsmann, R., 1998: Modeling social dynamics by cellular automata. In: *Computer modeling of social processes,* W.B.G. Liebrand, A. Nowak, and R. Hegselmann (eds.), Sage, Thousand Oaks, CA., pp. 37–64.

Herrera, A.O, 1976: *Catastrophe or New Society? A Latin American World Model.* International Development Research Centre, Ottawa, Canada.

Hilborn, R., and C.J. Walters, 1992: *Quantitative Fisheries Stock Assessment: Choice, Dynamics, & Uncertainty.* Chapman and Hall, New York,

Hilborn, R., and M. Mangel, 1997: *The Ecological Detective: Confronting Models with Data.* Princeton University Press, Princeton, 315 pp.

Hilton-Taylor, C., (compiler), 2000: *2000 IUCN Red List of Threatened Species.* IUCN/SSC, Gland, Switzerland and Cambridge, UK.

Hoeting, J.A., D. Madigan, and A.E. Raftery, 1999: Bayesian model averaging: a tutorial. *Statistical Science,* **14,** 382–401.

Hoffman, W., and Jackson, R., 2000: Vegetation-climate feedbacks in the conversion of tropical sananna to grassland. *J. Cli.* **13,** 1593–1602.

Hogeweg, P., 1988: Cellular automata as a paradigm for ecological modelling. *Applied Mathematics and Computation,* **27(1),** 81–100.

Holland, E.A., B.H. Braswell, J.F. Lamarque, A. Townsend, J.M. Sulzman, J.F. Müller , F. Dentener, G. Brasseur, H. Levy, II, J.E. Penner, and G. Roelofs, 1997: Variations in the predicted spatial distribution of atmospheric nitrogen deposition and their impact on carbon uptake by terrestrial ecosystems. *Journal of Geophysical Research,* **102(D13),** 15,849.

Houghton, R.A. and J.L. Hackler, 2001: *Carbon Flux to the Atmosphere from Land Use Changes: 1850 to 1990.* ORNL/CDIAC-131 NDP-050/R1, Carbon Dioxide Information Analysis Center, Environmental Sciences Division, Oak Ridge National Laboratory, Oak Ridge, Tenn., 86 pp. pp.

Howarth, R.W., et al., 1996: Regional nitrogen budgets and riverine N and P fluxes for the drainages to the North Atlantic Ocean: Natural and human influences. *Biogeochemistry,* **35,** 75–79.

Howell, E.A., S.C. Doney, R.A. Fine, and D.B. Olson, 1997: Geochemical estimates of denitrification in the Arabian Sea and the Bay of Bengal during WOCE. *Geophys. Res. Lett.,* **24,** 2549–2552.

Hubbell, S.P., 2001: *The Unified Neutral Theory of Biodiversity and Biogeography.* Princeton Monographs in Population biology, Princeton University Press. Princeton, NJ., 375 pp.

Hughes, B.B., 1999: International Futures: *Choices in the face of uncertainty.* Oxford, UK, Westview Press.

Huston, M., D. DeAngelis, and W. Post, 1988: New computer models unify ecological theory. *Bioscience,* **38(10),** 682–91.

IFPRI (International Food Policy Research Institute), 1977: *Food Needs of Developing Countries: Projections of Production and Consumption to 1990.* Research Report 3, IFPRI. Washington, DC.

IGBP-IHDP (International Geosphere-Biosphere Programme and International Human Dimensions Programme), 1995: *Land-Use and Land-Cover Change: Science/Research Plan.* Stockholm, International Geosphere-Biosphere Programme and International Human Dimensions Programme.

IGBP-IHDP, 1999: *Land-Use and Land-Cover Change Implementation Strategy.* International Geosphere-Biosphere Programme and International Human Dimensions Programme, Stockholm,

IMAGE-team, 2001: *The IMAGE 2.2 implementation of the SRES scenarios. A comprehensive analysis of emissions, climate change and impacts in the 21st century.* CD-ROM publication 481508018, Bilthoven, Netherlands.

IPCC (Intergovernmental Panel on Climate Change), 2001: *Climate Change 2001: The Scientific Basis. Contribution of Working Group I to the Third Assessment Report of the Intergovernmental Panel on Climate Change.* Houghton, J.T., Y. Ding, D.J. Griggs, M. Noguer, P.J. van der Linden, X. Dai, K. Maskell, and C.A. Johnson (eds.), Cambridge University Press, United Kingdom and New York, NY, 881 pp.

Irwin, E., and J. Geoghegan, 2001: Theory, data, methods: developing spatially explicit economic models of land use change. *Agriculture, Ecosystems, and Environment,* **85,** 7–23.

Janssen, M.A., 2001: An exploratory integrated model to assess management of lake eutrophication. *Ecological Modelling,* **140,** 111–124.

Janssen, M., (ed.), 2003: *Complexity and Ecosystem Management: The Theory and Practice of Multi-Agent Approaches.* Edward Elgar Publishers, Northampton, MA.

Janssen, M.A,. and S.R. Carpenter, 1999: Managing the Resilience of Lakes: A multi-agent modeling approach. *Conservation Ecology* **3**(2), 15. [online] URL: http://www.consecol.org/vol3/iss2/art15.

Jeppeson, E., M. Sondergaard, M. Sondergaard, and K. Christofferson (eds.), 1998: *The Structuring Role of Submerged Macrophytes in Lakes.* Springer-Verlag, Berlin.

Jickells, T.D., 2002: Nutrient biochemistry of the coastal zone. Science, **281,** 217–222.

Jin, X., C. Deutsch, N. Gruber, and K. Keller, 2002: Assessing the Consequences of Iron Fertilization on Oceanic N_2O Emissions and Net Radiative Forcing. *Eos. Trans. AGU,* **83(4),** Ocean Sciences Meet. Suppl., Abstract OS51F-10.

Judson, O.P., 1994: The rise of the individual-based model in ecology. *Trends in Ecology and Evolution,* **9(1),** 9–14.

Jurado-Molina, J., and P.A. Livingston, 2002: Multispecies perspectives on the Bering Sea Groundfish fisheries management regime. *North American Journal of Fisheries Management,* **22,** 1164–1175.

Kaboudan, M.A., 2003: Forecasting with computer-evolved model specifications: a genetic programming application. *Computers & Operations Research,* **30(11),** 1661–81.

Kaimowitz, D., and A. Angelsen, 1998: *Economic Models of Tropical Deforestation: A Review.* Centre for International Forestry Research, Jakarta.

Kainuma, M., Matsuoka, Y. and Morita, T., 2002: *Climate policy assessment.* Springer.

Karim, M.R., M. Sekine, and M. Ukita, 2002: Simulation of eutrophication and associated occurrence of hypoxic and anoxic conditions in a coastal bay in Japan. *Marine Pollution Bulletin,* **45,** 280–285.

Kaufman, M.M., 2000: Erosion Control at Construction Sites: The Science-Policy Gap. *Environmental Management,* **26(1),** 89–98.

Kaufmann, R.K. and K.C. Seto, 2001: Change detection, accuracy, and bias in a sequential analysis of Landsat imagery of the Pearl River Delta, China: econometric techniques. *Agriculture Ecosystems and Environment,* **85,** 95–105.

Kerr, J.T., 2001: Butterfly species richness patterns in Canada: Energy, heterogeneity, and the potential consequences of climate change. *Conservation Ecology,* **5,** 1–17.

Koster, R.D., and P.C. Milly., 1997: The interplay between transpiration and runoff formulation in land surface schemes used with atmospheric models. *J. Climate,* **10,** 1578–1591.

Kummer, D., and C.H. Sham, 1994: The causes of tropical deforestation: A quantitative analysis and case study from the Philippines. In: *The Causes of Tropical Deforestation,* K. Brown, and D.W. Pearce (eds.), University College of London Press, London, pp. 146–58.

Lambin, E.F., 1994: *Modelling Deforestation Processes: A Review.* European Commission, Luxemburg.

Lambin, E.F., 1997: Modelling and monitoring land-cover change processes in tropical regions. *Progress in Physical Geography,* **21(3),** 375–93.

Larkin, P.A., and W. Gazey, 1982: Applications of ecological simulation models to management of tropical multispecies fisheries. In: *Management of tropical fisheries,* D. Pauly, and G.I. Murphy (eds.), ICLARM Conf. Proc. 9, Manila, pp. 123–140.

Lashof, D.A. and D.A. Tirpack, 1989: Policy options for stabilizing global climate, draft report to Congress, U.S. Environmental Protection Agency, Washington, DC.

Leach, M.K. and T.J. Givnish, 1996: Ecological Determinants of Species Loss in Remnant Prairies. *Science,* **273,** 1555–1558.

Lean, J. and D.A. Warrilow, 1989: Simulation of the regional climatic impact of Amazon deforestation. *Nature,* **342,** 411–413.

Lee, R.G., R. Flamm, M.G. Turner, C. Bledsoe, P. Chandler, C. DeFerrari, R. Gottfried, R.J. Naiman, N. Schumaker, and D. Wear, 1992: Integrating sustainable development and environmental vitality: A landscape ecology approach. In: *Watershed Management: Balancing Sustainability and Environmental Change,* R.J. Naiman (ed.), Springer-Verlag, New York, pp. 499–521.

LeFevre, N. A.H. Taylor, F.J. Gilbert and R.J. Geider, 2003: Modeling carbon to nitrogen and carbon to chlorophyll a ratios in the ocean at low latitudes: Evaluation of the role of physiological plasticity. *Limnology and Oceanography,* **48(5),** 1796–1807.

Lewis, W.M., 2002: Yield of nitrogen from minimally disturbed watersheds of the United States. *Biogeochemistry,* **57/58,** 375–385.

Lewis, W.M., J.M. Melack, W.H. McDowell, M. McClain, and J.E. Richey, 1999: Nitrogen yields from undisturbed watersheds in the Americas. *Biogeochemistry,* **46,** 149–162.

Li, H., and J.F. Reynolds, 1997: Modeling effects of spatial pattern, drought, and grazing on rates of rangeland degradation: A combined Markov and cellular automaton approach. In: *Scale in Remote Sensing and GIS,* D.A. Quattrochi and M.F. Goodchild (eds.), Lewis Publishers, New York, pp. 211–30.

Li, X., and A. Gar-on Yeh, 2000: Modelling sustainable urban development by the integration of constrained cellular automata. *International Journal of Geographical Information Science,* **14(2),** 131–52.

Linnemann, H.J. de Hoogh, M.A. Keyzer, and H.D.J. van Heemst, 1979: *MOIRA: Model of International Relations in Agriculture.* North Holland, Amsterdam.

Liverman, D., E.F. Moran, R.R. Rindfuss, and P.C. Stern, (eds.), 1998: *People and Pixels: Linking Remote Sensing and Social Science.* National Academy Press, Washington, DC.

Livingston, P.A., and R.D. Methot, 1998: Incorporation of predation into a population assessment model of eastern Bering Sea walleye pollock. In: *Fishery Stock Assessment Models,* Alaska Sea Grant College Program Publication AK-SG-98–01, pp. 663–678.

Lomolino, M.V., 2001: Elevation gradients of species-density: historical and prospective views. *Global Ecology and Biogeography,* **10,** 3–13.

Lomolino, M.V., and M.D. Weiser, 2001: Toward a more general species-area relationship: diversity on all islands, great and small. *Journal of Biogeography,* **28,** 431–435.

Ludeke, A.K., R.C. Maggio, and L.M. Reid 1990: An analysis of anthropogenic deforestation using logistic regression and GIS. *Journal of Environmental Management,* **31,** 247–59.

Ludwig, D., 1999: Is it meaningful to estimate a probability of extinction? *Ecology,* **80,** 298–310.

Ludwig, D., S. Carpenter, and W. Brock, 2003: Optimal phosphorus loading for a potentially eutrophic lake. *Ecological Applications,* **13,** 1135–1152.

MacArthur, R.H., and E.O. Wilson, 1967: *The theory of island biogeography*: Monographs in population biology, No. 1. Princeton University Press, Princeton, NJ.

MacCall, A.D., 2002: Fishery-management and stock-rebuilding prospects under conditions of low-frequency environmental variability and species interactions. In: *Targets, Thresholds, and the Burden of Proof in Fisheries Management,* M. Mangel (ed.), *Bulletin of Marine Science,* **70,** 613–628.

Mackenzie, F.T., L.M. Ver, and A. Lerman, 1998: Coupled biogeochemical cycles of carbon, nitrogen, phosphorus and sulfur in the land-ocean-atmosphere system. In: *Asian Change in the Context of Global Climate Change,* J.N. Galloway and J.M. Melillo (eds.), Cambridge University Press, 1998.

Manabe, S., and K. Bryan, 1969: Climate calculations with a combined ocean-atmosphere model. J. Atmos. Sci., **26,** 786–789.

Manabe , S., and R.J. Stouffer, 1996: Low frequency variability of surface air temperature in a 1000-year integration of a coupled atmosphree-ocean-land model. J. *Climate* **9**, 376–393.

Manabe, S., and Wetherald, 1975: The effects of doubling the CO2 concentration on the climate of a general circulation model. J. Atmos. Sci., 37, 3–15.

Manabe, S., J. Smagorinsky, and R.F. Strickler, 1965: Simulated climatology of a general circuilation model with a hydrological cycle,. Monthly Weather Review, **93,** 769–798.

Manson, S.M., 2003: Validation and verification of multi-agent models for ecosystem management. In: *Complexity and Ecosystem Management: The Theory and Practice of Multi-Agent Approaches,* M. Janssen (ed.), Edward Elgar Publishers, Northampton, MA., pp. 63–74.

Manson, S.M., 2004: The SYPR integrative assessment model: complexity in development. In: *Integrated Land-Change Science and Tropical Deforestation in the Southern Yucatán: Final Frontiers,* B.L. Turner, II, D. Foster, and J. Geoghegan (eds.), Clarendon Press, Oxford, UK, 271–291.

Martens, P., 1997: *Health impacts of climate change and ozone depletion: An eco-epidemiological modelling approach.* (English translation from Dutch) PhD Thesis, University of Maastricht, Netherlands.

Matthews, H. D., Weaver, A.J., Meissner, K.J., Gillett, N.P., and Eby, M., 2004 : Natural and anthropogenic climate change: incorporating historical land cover change, vegetation dynamics and the global carbon cycle. *Climate Dynamics* **22,** 461–479.

May, R.M., J.H. Lawton, and N.E. Stork, 1995: Assessing extinction rates. In: *Extinction Rates,* J.H. Lawton and R.M. May (eds.), Oxford University Press, Oxford, UK, 1–24.

Maynard, K. and J.F.Royer., 2004: Sensitivity of a general circulation model to land surface parameters in African tropical deforestation experiments. *Climate Dynamics,* **22,** 555–572.

McAllister, M.K., E.K Pikitch, and E.A. Babcock, 2001: Using demographic methods to construct Bayesian priors for the intrinsic rate of increase in the Schaefer model and implications for stock rebuilding. *Canadian Journal of Fisheries and Aquatic Sciences,* **58,** 1871–1890.

McAllister, M.K. and C.H. Kirchner, 2002: Accounting for structural uncertainty to facilitate precautionary fishery management: illustration with Namibian orange roughy. In: *Targets, Thresholds, and the Burden of Proof in Fisheries Management,* M. Mangel (ed.), *Bulletin of Marine Science,* **70,** pp. 499–540.

McCalla, A.F., and C.L. Revoredo, 2001: *Prospects for Global Food Security: A Critical Appraisal of Past Projections and Predictions.* 2020 Vision for Food, Agriculture, and the Environment Discussion Paper 35, IFPRI, Washington, DC.

McGlade J.M. and A.R.G. Price, 1993: Multi-disciplinary modeling: an overview and practical implications for the governance of the Gulf region, Mar. Pollut. Bull., **27,** 361–377.

McGuire, A.D., J.M. Melillo, L.A. Joyce, D.W. Kicklighter, A.L. Grace, B. Moore III, and C.J. Vörösmarty, 1992: Interactions between carbon and nitrogen dynamics in estimating net primary productivity for potential vegetation in North America. *Global Biogeochem. Cycles,* **6(2),** 101–124.

McGuire, A.D., S. Sitch, J.S. Clein, R. Dargaville, G. Esser, J. Foley, M. Heimann, F. Joos, J. Kaplan, D.W. Kicklighter, R.A. Meier, J.M. Melillo, B.I. Moore, I.C. Prentice, N. Ramankutty, T. Reichenau, A. Schloss, H. Tian, L.J. Williams, and U. Wittenberg, 2001: Carbon balance of the terrestrial biosphere in the twentieth century: Analyses of CO_2, climate and land use effects with four process-based ecosystem models. *Global Biogeochemical Cycles,* **15(1),** 183–206.

McMichael, A.J., D.H. Campbell-Lendrum, C.F. Corvalán, K.L. Ebi, A. Githeko, J.D. Scheraga, and A. Woodward, 2003: *Climate change and human health: risks and responses.* World Health Organization, Geneva. 250pp.

McMichael A.J., R. Chamber, K. Chopra, P. Dasgupta, A. Duraiappah, W. Niu, and C.D. Butler, 2003: Ecosystems and Human Well-being. In: *People and Ecosystems: A Framework for Assessment and Action,* Millennium Ecosystem Assessment. Island Press, Washington, DC.

Meadows, D.H., D.L. Meadows, J. Randers, and W.W. Behrens, 1972: *The Limits to Growth: A Report for the Club of Rome's Project on the Predicament of Mankind.* Universe Books, New York.

Meadows, D.H., D.L. Meadows, and J. Randers, 1992: *Beyond the limits: Confronting global collapse, envisioning a sustainable future.*

Mertens, B., and E. F. Lambin, 1997: Spatial modeling of deforestation in southern Cameroon. *Applied* Geography, **17(2),** 143–62.

Mesarovic, M., and E. Pestel, 1974: *Mankind at the Turning Point: The Second Report of the Club of Rome.* Dutton, New York.

Messner, S. and M. Strubegger, 1995: *User's Guide for MESSAGE III.* WP-95-69, International Institute for Applied Systems Analysis, Laxenburg, Austria, 155 pp.

Metz, B. Davidson, O., Swart, R., and Pan, J. (2001). Climate Change 2001: Mitigation. Cambridge University Press. Cambridge.

Meyers, W.H., S. Devadoss, and M. Helmar, 1986: *Baseline projections, yields impacts and trade liberalization: Impacts for soybeans, wheat, and feed* grains: A FAPRI trade model analysis. Center for Agricultural and Rural Development (CARD), Iowa State University, Ames, IA.

Midgley, G.F., L. Hannah, D. Millar, M.C. Rutherford, and L.W. Powrie, 2002: Assessing the vulnerability of species richness to anthropogenic climate change in a biodiversity hotspot. *Global Ecology and Biogeography,* **11,** 445451.

Mintzer, I., 1987: A matter of degrees: The potential for controlling the greenhouse effect. Research Report 5, World Resources Institute. Washington D.C.

Mitchell, D.O., M.D. Ingco, and R.C. Duncan, 1997: *The World Food Outlook.* Cambridge University Press, New York.

Moore A.B., A.R. James, P.C. Sims, and G.K. Blackwell, 2001: Intelligent Metadata Extraction for Integrated Coastal Zone Management. Proceedings of GeoComputation 2001, University of Queensland, Brisbane, Australia, ISBN 18664995637 (available at http://www.geocomputation.org/2001/papers/moore.pdf)

Moore, J.K., S.C. Doney, J.A. Kleypas, D.M. Glover and I.Y. Fung, 2002a: An intermediate complexity marine ecosystem model for the global domain. *Deep-Sea Research II,* in press.

Moore, J.K., S.C. Doney, J.A. Kleypas, D.M. Glover and I.Y. Fung, 2002b: Iron cycling and nutrient limitation patterns in surface waters of the world ocean, *Deap-Sea Research II,* in press.

Morita, T., Y. Matsuoka, M. Kainuma, and H. Harasawa, 1994: AIM—Asian Pacific integrated model for evaluating policy options to reduce GHG emissions and global warming impacts. In *Global Warming Issues in Asia,* S. Bhattacharya et al. (eds.), AIT, Bangkok, pp. 254–273

Morris, W. and D. Doak, 2002: *Quantitative Conservation Biology.* Sinauer Publishers, MA, 480 pp.

Mosier, A., C. Kroeze, C. Nevison, O. Oenema, S. Seitzinger, and O. van Cleemput, 1998: Closing the global atmospheric N_2O budget: Nitrous oxide emissions through the agricultural nitrogen cycle. *Nutrient Cycling in Agroecosystems,* **52,** 225–248.

Najjar, R.G., and J. Orr, 1998: Design of OCMIP-2 simulations of chlorofluorocarbons, the solubility pump and common biogeochemistry, http://www.ipsl.jussieu.fr/OCMIP/phase2/simulations/design.ps, 19 pp.

Nakićenović, N., and Swart (eds.), 2000: *Special Report on Emission Scenarios.* Cambridge University Press, Cambridge.

Naqvi, S.W.A., D.A. Jayakumar, P.V. Narvekar, H. Naik, V.V.S.S. Sarma, W. D'Souza, S. Joseph, and M.D. George, 2000: Increased marine production of N_2O due to intensifying anoxia on the Indian continental shelf. *Nature,* **408,** 346–349.

Nelson, G.C., and D. Hellerstein, 1997: Do roads cause deforestation? Using satellite images in econometric analysis of land use. *American Journal of Agricultural Economics,* **79**(5), 80–88.

Nobre, C.A., P.J. Sellers, and J. Shukla, 1991: Amazonian deforestation and regional climate change. *Journal of Climate,* **4,** 957–987.

Nordhaus, W.D., 1979: *The Efficient Use of Energy Resources.* Yale University Press, New Haven. CT.

Nordhaus, W.D., 1994: *Managing the Global Commons.* MIT Press, Cambridge, MA.

NRC (National Research Council), 1998: *Improving Fish Stock Assessments.* National Academy Press, Washington, DC.

NRC (Committee on the Causes and Management of Coastal Eutrophication, Ocean Studies Board and Water Science and Technology Board), 2000: *Clean Coastal Waters. Understanding and Reducing the Effects of Nutrient Pollution.* National Academy Press, Washington, DC., 393 pp.

Nürnberg, G.K., 1984: The prediction of internal phosphorus load in lakes with anoxic hypolimnia. *Limnology and Oceanography,* **29,** 111–124.

Nürnberg, G.K., 1995: Quantifying anoxia in lakes. *Limnology and Oceanography,* **40,** 1100–1111.

Oberdorff, T., J.F. Guegan, and B. Hugueny, 1995: Global scale patterns of fish species richness in rivers, *Ecography,* **18,** 345–352.

O'Brien, E.M., 1998: Water-energy dynamics, climate, and prediction of plant species richness: an interim general model. *Journal of Biogeography,* **25,** 379–398.

OECD (Organisation for Economic Co-operation and Development), 2003: Agricultural Outlook 2003–2008. Paris.

Osborne, T.M., D.M. Lawrence, J.M. Slingo, A.J.Challinor., and T.R. Wheeler., 2004: Influence of vegetation on the local climateand hydrology in the tropics: sensitivity to soil parameters. *Climate Dynamics* **23,** 45–61.

Overmars, K.P., G.H.J. de Koning, and A. Veldkamp, 2003: Spatial autocorrelation in multi-scale land use models. *Ecological Modelling,* **164,** 257–270.

Owens, D.W., P. Jopke, D.W. Hall, J. Balousek, and A. Roa, 2000: Soil erosion from two small construction sites, Dane County, Wisconsin. U.S.G.S. Fact Sheet, U.S.G.S., Madison, WI.

Pace, M.L., 1984: Zooplankton community structure, but not biomass, influences the phosphorus-chlorophyll *a* relationship. *Canadian Journal of Fisheries and Aquatic Sciences,* **41,** 1089–96

Pace, M.L., 2001: Prediction and the aquatic sciences. *Canadian Journal of Fisheries and Aquatic Sciences,* **58,** 63–72.

Paerl, H.W., 2002: Connection atmospheric nitrogen deposition to coastal eutrophication, *Environ. Science and Technology,* 323–326.

Pan, Z., E. Takle, S. M., and R. Arritt, 1999: Simulation of potential impacts of man-made land use changes on U.S. summer climate under various synoptic regimes. *Journal of Geophysical Research,* **104(D6),** 6515–6528.

Parikh, K., and Rabar, F., 1981: *Food for all in a sustainable world: The IIASA Food and Agriculture Program.* International Institute for Applied System Analysis, Laxenburg, Switzerland.

Parker, D.C., S.M. Manson, M. Janssen, M.J. Hoffmann, and P.J. Deadman, 2003: Multi-agent systems for the simulation of land use and land cover change: a review. *Annals of the Association of American Geographers,* **93(2),** 316–40.

Parker, I.M., D. Simberloff, W.M. Lonsdale, K. Goodell, M. Wonham, P.M. Kareiva, M. Williamson, B. von Holle, P.B. Moyle, J.E. Byers, and L. Godwasser, 1999: Impact: Toward a framework for understanding the ecological effect of invaders. *Biological Invasions,* **1,** 3–19.

Parma, A.M., 2002a: Bayesian approaches to the analysis of uncertainty in the stock assessment of Pacific halibut. In: *Incorporating Uncertainty into Fisheries Models,* J.M. Berkson, L.L. Kline and D.J. Orth (eds.), American Fisheries Society, Bethesda, MD., **27,** 113–136.

Parma, A.M., 2002b: In search of robust harvest rules for Pacific halibut in the face of uncertain assessments and decadal changes in productivity. In: *Targets, Thresholds, and the Burden of Proof in Fisheries Management,* Mangel, M. (ed.), *Bulletin of Marine Science,* **70,** 455–472.

Parton, W.J., D.S. Schimel, C.V. Cole, and D.S. Ojima, 1987: Analysis of factors controlling soil organic matter levels in Great Plains grasslands. *Soil Science Society of America Journal,* **51,** 1173–1179.

Parton, W.J., E.A. Holland, S.J. Del Grosso, M.G. Hartman, R.E. Martin, A.R. Mosier, D.A. Ojima, and D.S. Schimel, 2001: Generalized model for NO_x and N_2O emissions from soils. *J. Geophys. Res.,* **106,** 17,403–17,419.

Patterson, K.R., 1999: Evaluating uncertainty in harvest control law catches using Bayesian Markov Chain Monte Carlo virtual population analysis with apadtive rejection sampling and including structural uncertainty. *Canadian Journal of Fisheries and Aquatic Sciences,* **56,** 208–221.

Patz, J.A., P.R. Epstein, T.A. Burke, and J.M. Balbus, 1996: Global climate change and emerging infectious diseases. *Journal of the American Medical Association,* **275(3),** 217–223.

Paulino, L.A, 1986: *Food in the Third World: Past Trends and Projections to 2000.* Research Report 52. International Food Policy Research Institute, Washington, DC.

Pauly, D., V. Christensen, and C.J. Walters, 2000: Ecopath, Ecosim, and Ecospace as tools for evaluating ecosystem impact of fisheries. *ICES Journal of Marine Science,* **57,** 697–706.

Pei, H.P. and Y. Wang, 2003: Eutrophication research of West Lake, Hangzhou, China: Modeling under uncertainty. *Water Research,* **37,** 416–428.

Pereira, H., and G.C. Daily, in review: Modeling Frameworks for Biodiversity Dynamics: Species-Area and Reaction-Diffusion Approaches. In: *Determinants of Biodiversity Change: Ecological Tools for Building Scenarios*

Pereira, H., G.C. Daily, and J. Roughgarden, 2004: A framework for assessing the relative vulnerability of species to land-use change. *Ecological Applications,* **14,** 730–742.

Peterson, A.T., C.V. Sanchez, J. Soberon, J. Bartley, R.W. Buddemeier, and S.A. Navarro, 2001: Effects of global climate change on geographic distributions of Mexican Cracidae. *Ecological Modeling,* **144,** 21–30.

Peterson, G.D., S.R. Carpenter, and W.A. Brock, 2003: Uncertainty and the management of multistate ecosystems: An apparently rational route to collapse. *Ecology,* **84,** 1403–1411.

Pfaff, A., 1999: What drives deforestation in the Brazilian Amazon? *Journal Environmental and Economic Management,* **37,** 26–43.

Phipps, M., 1989: Dynamical behavior of cellular automata under the constraint of neighborhood coherence. *Geographical Analysis,* **21(3),** 197–215.

Pielke, R.A., 2001: Influence of the spatial distribution of vegetation and soils on the prediction of cumulus convective rainfall. *Reviews of Geophysics,* **39(2):** 151–177.

Pimm, S.L., G.J. Russell, J.L. Gittleman, and T.M. Brooks, 1995: The future of biodiversity. *Science,* **269,** 347–350.

Pimm, S.L., and R.A. Askins, 1995: Forest losses predict bird extinctions in eastern North America. *Proceedings of the National Academy of Sciences,* **92,** 9347–9347.

Pionke, H. B., W. J. Gburek, A .N. Sharpley, and J. A. Zollweg. 1997. Hydrologic and chemical controls on phosphorus loss from catchments. in Tunney, editor. Phosphorus loss from soil to water. CAB International, New York.

Plaganyi, E.E., and D.S. Butterworth. in review: The global eco-epidemic: a critical look at what Ecopath with Ecosim can and cannot achieve in Practical fisheries management.

Polovina, J.J., 1984: Model of a coral reef ecosystem. I. The Ecopath model and its application to French Frigate Shoals. *Coral Reefs,* **3,** 1–11.

Pontius, R.G., Jr., 2000: Quantification error versus location error in comparison of categorical maps. *Photogrammetric Engineering and Remote Sensing,* **66**(8), 1011–16.

Pontius, R.G., Jr., J. Cornell and C. Hall, 2001: Modeling the spatial pattern of land-use change with GEOMOD2: application and validation for Costa Rica. *Agriculture, Ecosystems & Environment,* **85**(1–3), 191–203.

Postel, S., and S.R. Carpenter, 1997: Freshwater ecosystem services. In: *Nature's Services,* G. Daily (ed.), Island Press, Washington, DC, pp. 195–214.

Potter, C.S., P.A. Matson, P.M. Vitousek, and E.A. Davidson, 1996: Process modeling of controls on nitrogen trace gas emissions from soils worldwide. *J. Geophy. Res.,* **101(D1),** 1361–1377.

Prinn, R., H. Jacoby, A. Sokolov, C. Wang, X. Xiao, Z. Yang, R. Eckaus, P. Stone, D. Ellerman, J. Melillo, J. Fitzmaurice, D. Kicklighter, G. Holian, and Y. Liu, 1998: Integrated Global System Model for climate policy assessment: Feedbacks and sensitivity studies. *Climatic Change,* **41(3/4),** 469–546.

Punt, A.E., and D.S. Butterworth, 1995: The effects of future consumption by the Cape fur seal on catches and catch rates of the Cape hakes. 4. Modeling the biological interaction between Cape fur seals *Arctocephalus pusillus pusillus* and the Cape hakes *Merluccius capensis* and *M. paradoxus. South African Journal of Marine Science,* **16,** 255–285.

Punt, A.E., and R. Hilborn, 1997: Fisheries stock assessment and Bayesian analysis: the Bayesian approach. *Reviews in Fish Biology and Fisheries,* **7,** 35–63.

Punt, A.E., and R. Hilborn, 2002: Bayesian stock assessment methods in fisheries. FAO Computerized Information Series, *Fisheries,* **12,** 56 pp.

Quinn, II, T.J., in press: Ruminations on the development and future of population dynamics models in fisheries.

Quinn, II, T.J., and R.B. Deriso, 1999: *Quantitative Fish Dynamics,* Oxford University Press, New York.

Rabalais, N.N., R.E. Turner, and D. Scavia, 2002: Beyond science and into policy: Gulf of Mexico hypoxia and the Mississippi River. *BioScience,* **52,** 129–142.

Raftery, A.E., D. Madigan, and J.A. Hoeting, 1997: Bayesian model averaging for linear regression models. *Journal of the American Statistical Association,* **92,** 179–191.

Rahbek, C., and G.R. Graves, 2001: Multiscale assessment of patterns of avian species richness. *Proceedings of the National Academy of Sciences of the United States of America,* **98,** 4534–4539.

Ralston, S. and J.J. Polovina, 1982: A multispecies analysis of the commercial deep-sea handline fishery in Hawaii. *Fisheries Bulletin,* **80(3),** 435–448.

Raskin, P., C. Heaps, J. Sieber, and E. Kemp-Benedict, 1999: *PoleStar System Manual.* Stockholm Environment Institute, Boston, Available on the Internet. URL: http://www.seib.org/polestar.

Rastetter, E.B., R.B. McKane, G.R. Shaver, and J.M. Melillo, 1992: Changes in C storage by terrestrial ecosystems: How C-N interactions restrict responses to CO_2 and temperature. *Water, Air and Soil Pollution,* **64,** 327–344.

Raupach, M.R., 1998: Influences of local feedbacks on land-air exchanges of energy and carbon. *Global Change Biology,* **4,** 477–494.

Rayner, S., 1994: A wiring diagram for the study of land-use/cover change. In: *Changes in Land-Use and Land-Cover: A global perspective,* W.B. Meyer, and B.L. Turner, II (eds.), Cambridge University Press, Cambridge, UK, pp. 13–53.

Reid, W.V., 1992: How Many Species Will There Be? In: *Tropical deforestation and species extinction,* T.C. Whitmore and J.A. Sayer (eds.), Chapman and Hall, London, UK, pp. 53–73.

Riahi, K. and R.A. Roehrl, 2000: Greenhouse gas emissions in a dynamics as usual scenario of economic and energy development. *Technological Forecasting & Social Change,* **63(2–3).**

Rice, J.C., N. Daan, J.G. Pope, and H. Gislason, 1991: The stability of estimates of suitabilities in MSVPA over four years of data from predator stomachs. *ICES Marine Science Symposia,* **193,** 34–45.

Ricklefs, R.E., R.E. Latham, and Q. Hong, 1999: Global patterns of tree species richness in moist forests: Distinguishing ecological influences and historical contingency. *Oikos,* **86,** 369–373.

Rigler, F.H., and R.H. Peters, 1995: *Science and Limnology.* Ecology Institute, Oldendorf/Luhe, Germany.

Rindfuss, R.R., S.J. Walsh, B.L. Turner, II, J. Fox and V. Mishra, 2004: Developing a science of land change: challenges and methodological Issues. *Proceedings of the National Academy of Sciences,* **101(39),** 13976–81.

Rogers, D.J., and S.E. Randolph, 2000: The global spread of malaria in a future, warmer world. *Science,* **289,** 1763–1766.

Rosegrant, M.W., and S. Meijer, 2001: Does IMPACT Work? Comparison of Model Projections with Actual Outcomes. Mimeo, August, International Food Policy Research Institute, Washington, DC.

Rosegrant, M. W., M. S. Paisner, S. Meijer, and J. Witcover, 2001: *Global Food Projections to 2020: Emerging Trends and Alternative Futures.* International Food Policy Research Institute, Washington, DC.

Rosegrant, M.W., M. Agcaoili-Sombilla, and N.D. Perez, 1995: *Global Food Projections to 2020: Implications for Investment.* 2020 Discussion Paper No. 5. International Food Policy Research Institute, Washington, DC.

Rosenzweig, M.L., 1995: *Species diversity in space and time.* Cambridge University Press, Cambridge, UK.

Rosenzweig, M.L., 2001: Loss of speciation rate will impoverish future diversity. *Proceedings of the National Academy of Sciences of the United States of America,* **98,** 5404–5410.

Rotmans, J. and Dowlatabadi, H., 1998: Integrated assessment modelling. In: *Human choice and climate change. Volume 3: The tools for policy analysis,* S. Rayner and E.L. Malone (eds.), Batelle Press, Columbus, OH, pp. 291–377.

Rotmans, J. and H.J.M. de Vries, (eds.),1997: *Perspectives on Global Change: The TARGETS approach.* Cambridge University Press, Cambridge, UK.

Rotmans, J., H. de Boois, and R.J. Swart, 1990: An integrated model for the assessment of the greenhouse effect: the Dutch approach. *Climatic Change,* **16,** 331–356.

Rudel, T.K. 1989: Population, development, and tropical deforestation: A cross-national study. *Rural Sociology,* **54,** 327–38.

Sala, O.E., F.S. Chapin, J.J. Armesto, E. Berlow, J. Bloomfield, R. Dirzo, E. HuberSanwald, L. F. Huenneke, R.B. Jackson, A. Kinzig, R. Leemans, D.M. Lodge, H.A. Mooney, M. Oesterheld, N.L. Poff, M.T. Sykes, B.H. Walker, M. Walker, and D.H. Wall, 2000: Biodiversity: Global biodiversity scenarios for the year 2100. *Science,* **287,** 1770–1774.

Scheffer, M., 1998: *Ecology of Shallow Lakes.* Chapman and Hall, London.

Scheffer, M., S.H. Hosper, M.L. Meijer, B. Moss, and E. Jeppesen, 1993: Alternative equilibria in shallow lakes. *Trends in Ecology and Evolution,* **8,** 275–279.

Scheffer, M., S. Carpenter, J.A. Foley, C. Folke, B. Walker, 2001a: Catastrophic shifts in ecosystems. *Nature,* **413,** 591–596.

Scheffer, M., D. Straile, E.H. van Nes, and H. Hosper, 2001b: Climatic warming causes regime shifts in lake food webs. *Limnology and Oceanography,* **46,** 1780–1783.

Scheffer, M., F. Westley, and W. Brock, 2003: Slow response of societies to new problems: causes and costs. *Ecosystems,* **6,** 493–502.

Schindler, D.W., 1977: The evolution of phosphorus limitation in lakes. *Science,* **195,** 260–262.

Schofield O., T. Bergmann, W.P. Bissett, F. Grassle, J. Kohut, M.A. Moline, and S. Glenn, 2002: Linking regional coastal observatories to provide the foundation for a regional ocean observatory network. *J. Ocean. Eng.* **27(2),** 146–154.

Schwartz, M.W., L.R. Iverson, and A.M. Prasad, 2001: Predicting the potential future distribution of four tree species in Ohio using current habitat availability and climatic forcing. *Ecosystems,* **4,** 568–581.

SEI Boston Center (Stockholm Environment Institute), 1999: *PoleStar 2000.* SEI Boston Center, Tellus.

Seitzinger, S.P., and C. Kroeze, 1998: Global distribution of nitrous oxide production and N inputs in freshwater and coastal marine ecosystems. *Global Biogeochem. Cycles,* **12(1),** 93–113.

Sellers, P.J., Y. Mintz, Y.C. Sud, and A. Dalcher, 1986: A simple biosphere model (SiB) for use with general circulation models. *Journal of Atmospheric Science,* **43(6),** 505–531.

Sellers, P.J., R.E. Dickinson, D.A. Randall, A.K. Betts, F.G. Hall, H.A. Mooney, C.A. Nobre, N. Sato, C.B. Field, and A. Henderson-Sellers, 1997: Modeling the exchanges of energy, water, and carbon between continents and the atmosphere. *Science,* **275,** 502–509.

Shafer G., 1976: *A Mathematical Theory of Evidence.* Princeton Univ. Press, NJ.

Shannon, L.J., P.M. Cury, and A. Jarre, 2000: Modelling effects of fishing in the Southern Benguela ecosystem. *ICES Journal of Marine Science,* **57,** 720–722.

Sharpley, A.N., P.J.A. Kleinman, R.W. McDowell, M. Gitau, and R.B. Bryant, 2002: Modeling phosphorus transport in agricultural watersheds: Processes and possibilities. *Journal of Soil and Water Conservation,* **57,** 425–439.

Sharpley, A.N. and J.R. Williams (eds.), 1990: EPIC—Erosion/Productivity Impact Calculator. Model Documentation USDA Technical Bulletin 1768.

Shellito, B.A. and B.C. Pijanowski, 2003: Using Neural Nets to Model the Spatial Distribution of Seasonal Homes. *Cartography and Geographic Information Science,* **30,** 281–290.

Shukla, J., C. Nobre, and P. Sellers, 1990: Amazon deforestation and climate change. *Science,* **247,** 1322–1325.

Skidmore, A.K., F. Watford, P. Luckananurag, and P.J. Ryan, 1996: An operational GIS expert system for mapping forest soils. *Photogrammetric Engineering and Remote Sensing,* **62(5),** 501–11.

Smith, L., and L. Haddad, 2000: *Explaining Child Malnutrition in Developing Countries: A Cross-Country Analysis.* International Food Policy Research Institute, Washington, DC.

Smith, R.A., R.B. Alexander, and G.E. Schwarz, 2003: Natural background concentrations of nutrients in streams and rivers of the conterminous United States. *Environ. Science and Technology.* **37(14),** 3039–3047.

Smith, S.V., D.P. Swaney, L. Talaue-McManus, J.D. Bartley, P.T. Sandhei, C.J. McLaughlin, V.C. Dupra, C.J. Crossland, R.W. Buddemeier, B.A. Maxwell, and F. Wulff, 2003: Humans, hydrology, and the distribution of inorganic nutrient loading to the ocean, *Bioscience,* **53(3),** 235–245.

Smith, V.H., 1983: Low nitrogen to phosphorus ratios favor dominance by blue-green algae in lake phytoplankton. *Science,* **221,** 669–671.

Smith, V.H., 1998: Cultural eutrophication of inland, estuarine and coastal waters. In: *Successes, Limitations and Frontiers of Ecosystem Science,* M.L. Pace and P.M. Groffman (eds.), Springer-Verlag, New York, pp. 7–49.

Soranno, P.A., 1997: Factors affecting the timing of surface scums and epilimnetic blooms of blue-green algae in a eutrophic lake. *Canadian Journal of Fisheries and Aquatic Sciences,* **54,** 1965–1975.

Soranno, P.A., S.L. Hubler, S.R. Carpenter, and R.C. Lathrop, 1996: Phosphorus loads to surface waters: a simple model to account for spatial pattern of land use. *Ecological Applications,* **6,** 865–878.

Sparre, P., 1991: Introduction to multispecies virtual population analysis. *ICES Mar. Sci. Symp.,* **193,** 12–21.

Stow, C.A., S.R. Carpenter, and R.C. Lathrop, 1997: A Bayesian observation error model to predict cyanobacterial biovolume from spring total phosphorus in Lake Mendota, Wisconsin. *Canadian Journal of Fisheries and Aquatic Sciences,* **54,** 464–473.

Thomas, C.D., E.J. Bodsworth, R.J. Wilson, A.D. Simmons, Z.G. Davies, M. Musche, and L. Conradt, 2001: Ecological and evolutionary processes at expanding range margins. *Nature,* **411,** 577–581

Thomas, C.D., A. Cameron, R.E. Green, M. Bakkenes, L.J. Beaumont, Y.C. Collingham, B.F. N. Erasmus, M. Ferreira de Siqueira, A. Grainger, L. Hannah, L. Hughes, B. Huntley, A.S. van Jaarsveld, G.F. Midgley, L. Miles, M. Ortega-Huerta, A. Townsend Peterson, O.L. Phillips, and S. E. Williams, 2004: Extinction risk from climate change. *Nature,* **427,** 145–148.

Thompson, M., R. Ellis, et al., 1990: *Cultural Theory.* Westview Press, Boulder, CO.

Thompson, R.L, 1981: *A survey of recent U.S. developments in international agricultural trade models.* Bibliographies and Literature of Agriculture Number 21. U.S. Department of Agriculture, Washington, DC.

Tian, Y., R.E. Dickinson, L. Zhou, X. Zeng, Y. Dai, R.B. Myneni, Y. Knyazikhin, X., Zhang, M. Friedl, H. Yu, W. Wu, and M. Shaikh, 2004a: Comparison of seasonal and spatial variations of leaf area index and fraction of absorbed photosynthetically active radiation from Moderate Resolution Imaging Spectroradiometer (MODIS) and Common Land Model. *J. Geophys. Res.,* **109,** doi:10.1029/2003JD003777.

Tian, Y., R.E. Dickinson, L. Zhou, X. Zeng, Y. Dai, R.B. Myneni, M. Friedl, C.B. Schaaf, M. Carroll, and F. Gao, , 2004b Land boundary conditions from MODIS data and consequences for the albedo of a climate model. *Geophys. Res. Lett.,* 31, doi:10.1029/2003GL019104.

Tjorve, E., 2002: Habitat size and number in multi-habitat landscapes: A model approach based on species-area curves. *Ecography,* **25,** 17–24.

Tol, R.S.J., 1997: On the Optimal Control of Carbon Dioxide Emissions: An Application of *FUND. Environmental Modelling and Assessment,* **2,** 151–163.

Tol, R.S.J., 2003: Is the Uncertainty about Climate Change Too Large for Expected Cost-Benefit Analysis? *Climatic Change,* **56 (3),** 265–289

Torrens, P.M., and D. O'Sullivan, 2001: Cellular automata and urban simulation: where do we go from here? *Environment and Planning B,* **28(2),** 163–68.

Toth, F.L. and E. Hizsnyik, 1998: Integrated environmental assessment methods: Evolution and applications. *Environmental Modeling and Assessment,* **3,** 193–207.

Tundisi, J.G., and T. Matsumura-Tundisi, 2003: Integration of research and management in optimizing multiple uses of reservoirs: the experience in South America and Brazilian case studies. *Hydrobiologia,* **500,** 231–242.

Turpie, J.K., and T.M. Crowe, 1994: Patterns of distribution, diversity and endemism of larger African mammals. *South African Journal of Zoology,* **29,** 19–32.

Tyrrell, T., 1999: The relative influences of nitrogen and phosphorus on oceanic primary production. *Nature,* **400,** 525–531.

Uebayashi, A. 2004. OECD Agricultural Outlook and Its Baseline Process using AGLINK model. OECD, Paris (Mimeo). Accessed at http://www.unece.org/stats/documents/ces/sem.44/wp.7.e.pdf.

UN (United Nations), 2001: *World Population Prospects, the 2000 Revision-Highlights.* Doc. No. ESA/P/WP.165. United Nations, New York.

UN, 1998: *World Population Prospects: 1998 Revisions.* United Nations, New York.

UNEP (United Nations Environment Programme), 2002: *Global Environmental Outlook 3.* EarthScan Publications, London.UNEP (2002). Global Environment Outlook. Oxford University Press. Oxford.

USDA (United States Department of Agriculture), 2003: *USDA Agricultural Baseline Projections to 2012.* Staff Report WAOB-2003–1. Office of the Chief Economist, Washington, D.C.

Van Cappellen, P., and E.D. Ingall, 1994: Benthic phosphorus regeneration, net primary production, and ocean anoxia: A model of the coupled marine biogeochemical cycles of carbon and phosphorus. *Paleoceanography,* **9,** 677–692.

Vanclay, J. K., 2003: Why model landscapes at the level of households and fields? *Small-scale Forest Economics, Management and Policy,* **2(2),** 121–34.

Van der Sluijs, J.P., 2002: Definition of Integrated Assessment. In: *Encyclopedia of Global Environmental Change,* Munn (ed.), John Wiley and Sons. Chichester.

Van der Sluijs, J.P., 1997: Anchoring amid uncertainty; On the management of uncertainties in risk assessment of anthropogenic climate change, PhD-thesis, Universiteit Utrecht.

Veldkamp, A. and L.O. Fresco, 1996: CLUE: a conceptual framework to study the Conversion of Land Use and its Effects. *Ecological modelling,* **85,** 253–270.

Veldkamp, A., and E. Lambin, 2001: Special Issue: Predicting Land-Use Change. *Agriculture, Ecosystems and Environment,* **85.**

Veldkamp, A. and P.H. Verburg (eds.), 2004: Modelling land use change and environmental impact. *Journal of Environmental Management,* **72,** 1–115.

Verburg, P.H., W. Soepboer, R. Limpiada, M.V.O. Espaldon, M. Sharifa, and A. Veldkamp, 2002: Land use change modelling at the regional scale: the CLUE-S model. *Environmental Management,* **30,** 391–405.

Vinther, M., P. Lewy, and L. Thomsen, 2002: *Specification and documentation of the 4M package containing species, multi-fleet and multi-area models.* Danish Institute for Fisheries and Marine Research, Charlotte Castle, DK-2920 Charlottenlund, Denmark.

Vitousek, P.M., 1994: Beyond Global Warming: Ecology and Global Change. *Ecology,* **75,** 1861–1876.

Vitousek, P.M., J. Aber, R.W. Howarth, G.E. Likens, P.A. Matson, D.W. Schindler, W.H. Schlesinger, and G.D. Tilman, 1997: Human alterations of the global nitrogen cycle: Causes and consequences. *Ecological Issues,* **1,** 1–15.

Voinov, A., H. Voinov, and R. Costanza, 1999: Surface Water Flow in Landscape Models: 2. Patuxent Watershed Case Study. *Ecological Modelling,* **119**(2–3), 211–30.

Voldoire, a., and J.F. Royer, 2004: Tropical deforestation and climate variability. Climate Dynamics 22, 857–874

Vollenweider, R.A., 1968: The scientific basis of lake and stream eutrophication with particular reference to phosphorus and nitrogen as eutrophication factors. In: *Technical report OECD,* DAS/C81/68, Paris, France, pp. 1–182.

Von Storch, H., and Zwiers F., 1999: Statistical analyses in climate research. Cambridge University Press, Cambridge UK.

Wade, A.J., G.M. Hornberger, P.G. Whitehead, H.P. Jarvie, and N. Flynn, 2001: On modeling the mechanisms that control in-stream phosphorus, macrophyte, and epiphyte dynamics: An assessment of a new model using general sensitivity analysis. *Water Resources Research,* **37,** 2777–2792.

Wade, A.J., P.G. Whitehead, and D. Butterfield, 2002: The Integrated Catchments model of Phosphorus dynamics (INCA-P), a new approach for multiple source assessment in heterogeneous river systems: model structure and equations. *Hydrology and Earth System Sciences,* **6,** 583–606.

Walters, C., and F. Juanes, 1993: Recruitment limitations as a consequence of natural selection for use of restricted habitats and predation risk taking by juvenile fishes. *Canadian Journal of Fisheries and Aquatic Science,* **50,** 2058–2070.

Walters, C., V. Christensen, and D. Pauly, 1997: Structuring dynamic models of exploited ecosystems from trophic mass-balance assessments. *Reviews in Fish Biology and Fisheries,* **7,** 139–172.

Walters, C., and J.J. Maguire, 1996: Lessons for stock assessment from the northern cod collapse. *Reviews in Fish Biology and Fisheries,* **33,** 145–159.

Walters, C., and S. Martell, in review: Harvest Management of Aquatic Ecosystem.

Walters, C.J., M. Stocker, A.V. Tyler, and S.J. Westrheim, 1986: Interaction between Pacific cod (*Gadus macrocephalus*) and herring (*Clupea harengus pallasi*) in the Hecate Strait, British Columbia. *Canadian Journal of Fisheries and Aquatic Sciences,* **43,** 830–837.

Weyant, J., O. Davidson, H. Dowlabathi, J. Edmonds, M. Grubb, E.A. Parson, R. Richels, J. Rotmans, P.R. Shukla, R.S.J. Tol, W. Cline, and S. Fankhauser, 1996: Integrated assessment of climate change: an overview and comparison of approaches and results. In: *Climate Change 1995: Economic and social dimensions of climate change,* J.P. Bruce, H. Lee and E.F. Haites (eds.), Cambridge University Press, Cambridge, pp. 367–396.

White, M.A., P.E. Thornton, S.W. Running and R.R. Nemani, 2000: Parameterization and sensitivity analysis of the BIOME-BGC terrestrial ecosystem model: net primary production controls, *Earth Interactions, 4,* Paper no. 3.

White, R., and G. Engelen, 2000: High-resolution integrated modelling of spatial dynamics of urban and regional systems. *Computers, Environment, and Urban Systems,* **24,** 383–400.

Whitley, D., 2001: An overview of evolutionary algorithms: practical issues and common pitfalls. *Information and Software Technology,* **43**(14), 817–31.

Whittaker, R.J., K.J. Willis, and R. Field, 2001: Scale and species richness: towards a general, hierarchical theory of species diversity. *Journal of Biogeography* 28: 453–470.

WHO (World Health Organization), 1997: *WHO Global Database on Child Growth and Malnutrition.* Programme of Nutrition, WHO Document #WHO/NUT/97.4, World Health Organization, Geneva.

Wickham, J.D., R.V. O'Neill, K.H. Riitters, E.R. Smith, T.G. Wade, and K.B. Jones, 2002: Geographic targeting of increases in nutrient export due to future urbanization. *Ecological Applications,* **12,** 93–106.

Wickham, J.D., K.H. Riitters, R.V. O'Neill, E.R. Smith, T.G. Wade, and K.B. Jones, 2000: Land cover as a framework for assessing risk of water pollution. *Journal of the American Water Resources Association,* **36,** 1417–1422.

Wischmeier, W.H., 1958: USDA Agriculture Handbook 282.

World Bank, 2000: *World Development Indicators on CD-Rom.* The World Bank, Washington, DC.

World Bank, 1998: *World Development Indicators on CD-Rom.* The World Bank, Washington, D.C.

Xiao, N.C., D.A. Bennett, and M.P. Armstrong, 2002: Using evolutionary algorithms to generate alternatives for multiobjective site-search problems. *Environment and Planning A,* **34**(4), 639–56.

Xu, F.L., S. Tao, R.W. Dawson, and X.Y. Lu, 2002: History, development and characteristics of lake ecological models. *Journal of Environmental Sciences—China,* **14,** 255–263.

Young, R.A., C.A. Onstad, D.D. Bosch, and W.P. Anderson, 1989: AGNPS: A nonpoint-source pollution model for evaluating agricultural watersheds. *Journal of Soil and Water Conservation,* **44,** 168–173.

Zender, C.S., and D. Newman, 2003: Spatial heterogeneity in aeolian erodibility: Uniform, topographic, geomorphic, and hydrologic hypotheses. *J. Geophysical Res.,* **108(D17),** 4543, doi:10.1029/2002JD003039.

Zhang, X., and Y. Wang, 2002: Spatial dynamic modeling for urban development. *Photogrammetric Engineering and Remote Sensing,* **67(9),** 1049–57.

Zhang, H., K. McGuffie, and A. Henderson-Sellers., 1996: Impacts of tropical deforestation Part I: Process analysis of local climatic change. *J. Clim.* **9,** 1497–1516.

Part II

The Millennium Ecosystem Assessment Scenarios

Chapter 5

Scenarios for Ecosystem Services: Rationale and Overview

Coordinating Lead Authors: Elena Bennett, Steve Carpenter
Lead Authors: Steve Cork, Garry Peterson, Gerhardt Petschel-Held, Teresa Ribeiro, Monika Zurek
Review Editors: Antonio Alonso Concheiro, Yuzuru Matsuoka, Allen Hammond

BOXES

FIGURES

TABLES

Main Messages

The Millennium Ecosystem Assessment scenarios address plausible future changes in ecosystems, in the supply of and demand for ecosystem services, and in the consequent changes in human well-being.

A survey of user needs and a set of interviews with decision-makers and other leaders identified a set of key concerns to be addressed by the MA scenarios. These concerns included globalization, leadership, poverty and inequality, technology, local flexibility, and surprises. Uncertainties about these factors have large implications for future ecosystem services. The uncertainties are related to ecosystem management dilemmas—situations in which significant risks are associated with each possible decision. Two dilemmas identified by respondents were: What degree of ecological complexity is needed to provide reliable ecological services? And to what degree can people use technology to substitute for the role of relatively undisturbed ecosystems in the provision of services? Exploring the consequences of different outcomes for the key concerns and different decisions made about the dilemmas form the underlying basis for the differences in the four scenarios.

The MA scenarios were designed to explore a wide range of contexts under which sustainable development will be pursued, as well as a wide range of approaches to sustainable development. With respect to context, we explore two basic futures—one that becomes increasingly globalized and one that becomes increasingly regionalized. In terms of approaches, we focus on futures that emphasize economic growth and promotion of public goods and futures that emphasize proactive management of ecosystems and their services. Framed in terms of contexts and approaches, the scenarios are:

- Global Orchestration (globalized, with emphasis on economic growth and public goods),
- Order from Strength (regionalized, with emphasis on national security and economic growth),
- Adapting Mosaic (regionalized, with emphasis on local adaptation and flexible governance), and
- TechnoGarden (globalized, with emphasis on green technology).

The focus on alternative approaches to sustaining ecosystem services distinguishes the MA scenarios from previous global scenario exercises. For each of the four scenarios, we analyzed a set of plausible socioeconomic changes consistent with the contrasting approaches to ecosystem management.

The purpose of the scenarios is to explore the consequences of the four futures for ecosystem services and human well-being. The four futures that we examine were developed based on interviews with leaders in nongovernmental organizations, governments, and business from five continents, on the literature, and on policy documents addressing linkages between ecosystem change and human well-being. No scenario will match the future as it actually occurs. No scenario represents business as usual, though all are based on current conditions and trends. None of the scenarios represents a "best" path or a "worst" path. There could be combinations of policies and practices that produce significantly better, or worse, outcomes than any of the scenarios. The future will represent a mix of approaches and consequences described in the scenarios, as well as events and innovations that were not imagined at the time of writing. The scenarios explore a wide variety of choices and their consequences.

The Global Orchestration scenario explores the possibilities of a world in which global economic and social policies are the primary approach to sustainability. The recognition that many of the most pressing global problems seem to have roots in poverty and inequality evokes fair policies to improve the well-being of those in poorer countries by removing trade barriers and subsidies. Environmental problems are dealt with in an ad-hoc manner since people generally assume that improved economic well-being will create both the demand for and the means to achieve a well-functioning environment. Nations also make progress on global environmental problems, such as greenhouse gas emissions and depletion of pelagic marine fisheries. However, some local and regional environmental problems are exacerbated. The results for ecosystem services are mixed. While human well-being is improved in many of the poorest countries (and in some rich countries), a number of ecosystem services deteriorate by 2050.

The Order from Strength scenario examines the outcomes of a world in which protection through boundaries becomes paramount. The policies enacted in this scenario lead to a world in which the rich protect their borders, attempting to confine poverty, conflict, environmental degradation, and deterioration of ecosystem services to areas outside those borders. Poverty, conflict, and environmental problems often cross the borders, however, impinging on the well-being of those within. Protected natural areas are not sufficient for nature preservation or the maintenance of ecosystem services.

The Adapting Mosaic scenario explores the benefits and risks of local and regional management as the primary approach to sustainability. In this scenario, lack of faith in global institutions, combined with increased understanding of the importance of resilience and local flexibility lead to approaches that favor experimentation and local control of ecosystem management. The results are mixed, as some regions do a good job managing ecosystems and others do not. High levels of communication and interest in learning leads regions to compare experiences and learn from one another. Gradually the number of successful experiments begins to grow. While global problems are ignored initially, later in the scenario they are approached with flexible strategies based on successful experiences with locally adaptive management. However, some systems suffer long-lasting degradation.

The TechnoGarden scenario explores the potential role of technology in providing or improving the provision of ecosystem services. The use of technology and the focus on ecosystem services is driven by a system of property rights and valuation of ecosystem services. In this scenario, people push ecosystems to their limits of producing the optimum amount of ecosystem services for humans through the use of technology. Often, the technologies they use are more flexible than today's environmental engineering and they allow multiple needs to be met from the same ecosystem. Provision of ecosystem services in this scenario is high worldwide, but flexibility is low due to high dependence on a narrow set of optimal approaches. In some cases, unexpected problems created by technology and the erosion of ecological resilience lead to vulnerable ecosystem services, which are subject to interruption or breakdown. In addition, success in increasing the production of ecosystem services often undercuts the ability of ecosystems to support themselves, leading to surprising interruptions of service provision and collapse of some ecosystem services. These interruptions and collapses sometimes have serious consequences for human well-being.

Different modes of governance and management of ecosystem services have complementary advantages and disadvantages:

- Economic growth and expansion of public goods (such as education and accessible technologies) enables society to respond effectively when environmental problems emerge. However, if the focus on public goods overwhelms attention to the environment and proactive environmental policies are not pursued, there is increased risk of regional interruptions in provision of ecosystem services.

- A focus on strong national security creates some opportunities for ecosystem preserves, but if this is not coupled with active ecosystem management outside the reserves, then pressure on the environment increases and there is greater risk of large disturbances of ecosystem services and vulnerability to interruptions in provision of ecosystem service.

- When regional ecosystem management is proactive and oriented around adapting to change, ecosystem services become more resilient and society becomes less vulnerable to disturbances of these services. However, a regional focus diminishes attention to the global commons and exacerbates global environmental problems such as climate change and declining oceanic fisheries.

- Technological innovations and ecosystem engineering, coupled with economic incentive measures to facilitate their uptake, lead to highly efficient provision of ecosystem services. However, novel technologies can create novel environmental problems, and in some cases the resulting disruptions of ecosystem services affect large numbers of people.

The scenarios differ in the frequency and magnitude of surprising changes in ecosystem services. In Order from Strength, extreme disturbances of ecosystem services have a moderately wide range with a relatively high mode. Most of the human population is in relatively impoverished regions with deteriorating ecosystem services, and this situation is reflected in breakdowns that affect a relatively large number of people. Global Orchestration has a comparable range but a lower mode. Some severe breakdowns of ecosystem services still occur, but these tend to affect fewer people than in Order from Strength. In Adapting Mosaic, the distribution of extreme events is bimodal. The bimodality results from local vulnerability in some regions, underlying events that affect smaller numbers of people, and from diminished attention to the global commons, which underlies some events that affect large numbers of people. TechnoGarden leads to the widest distribution of large-scale breakdown event magnitudes. The mode is moderate, but the range is wide and some breakdowns affect large numbers of people.

The future of ecosystem services will likely have elements of each of the four scenarios. Changes in global trends could cause any of the scenarios to branch into one of the other ones.

5.1. Introduction

An infinite number of imaginable futures might be explored with the Millennium Ecosystem Assessment scenarios. However, scenarios are most powerful when presented as a small set with clear and striking differences (Van der Heijden 1996). Thus, the Scenarios Working Group had to decide how to compress an infinity of dimensions into a few comprehensible ones. In this chapter, we explain why we chose the four storylines that we develop, describe the key differences among them, and provide a brief sketch of each scenario. We summarize the potential benefits and inadvertent negative consequences of each scenario and describe how each scenario could potentially branch into one of the others. This chapter sets the stage for more detailed presentation of the scenarios in Chapter 8. While the scenarios are both qualitative and quantitative, in this chapter we focus on the qualitative. The quantitative material can be found primarily in Chapter 9.

5.2. Why Think about the Future of Ecosystem Services?

In order to make sound choices, people need to understand what the consequences of their actions, or inaction, will be. We have means of estimating how ecosystems and their services may change in coming decades given specific changes in driving forces such as population, economic growth, trade policies, resource management policies, and so forth, but the potential outcomes are both complex and variable. How can a decision-maker weigh different policy options in the face of such complexity and uncertainty?

The MA scenarios are designed to highlight key comparisons among approaches to development and to inform decision-makers about the consequences for ecosystem services of contrasting development paths. The central idea behind scenarios is to examine multiple possible futures and to let differences between them illuminate cause and effect and probable outcomes of certain approaches or decisions. While predictions and forecasts, more common approaches in ecology, focus on the single best or optimal approach, scenarios explicitly consider uncertainties and unknowns.

5.3. What Issues Should the Scenarios Address?

The goal of the MA scenarios is to inform diverse decision-makers about the potential futures of ecosystems and ecosystem services and how decisions can affect them. For this purpose, the scenarios needed to address the concerns of decision-makers and represent key aspects of the ecosystem dynamics behind those concerns. To identify focal issues for the MA scenarios, we used interviews with individual decision-makers and leading environmental thinkers, a survey of the needs of the MA's designated user community, and expert understanding of global ecosystem services and their connections to human well-being. Here, we present findings of each of these efforts and explain how they are represented in our four scenarios.

5.3.1 User Needs

Scientific assessments are most helpful to decision-makers when the intended users are active stakeholders in the assessment process and, in particular, when the users directly help shape the questions that the assessment will answer. An extensive effort was made to identify the needs of various MA audiences for information from the assessment and to engage those audiences in the governance and design of the MA process. This effort included directly asking various users what questions they wanted the MA to address. Users who responded included representatives from the Convention on Biological Diversity, the Convention to Combat Desertification, Ramsar, and other national government representatives; individuals from the private sector; and members of international nongovernmental organizations, civil society, and indigenous groups. This effort led to a greater understanding of what the active stakeholders hoped to gain from the MA scenarios.

Core questions for scenarios were derived from the user needs identified through these questions:

- What are the **plausible future changes in ecosystems** and in the supply of and demand for ecosystem services and the consequent changes in the constituents of well-being?
- What are the **costs, benefits, and risks** of plausible future changes in ecosystems and how will these costs, benefits, and risks affect different sectors of society and different regions of the world?
- What are the **inadvertent negative consequences** associated with various futures?
- What response options can lessen the **vulnerability** of people/communities?
- Under what circumstances are **thresholds, regime shifts, or irreversible changes** likely to occur?

There were also questions about specific drivers and responses:

- What policies and actions concerning ecosystems can best contribute to **reducing poverty**?
- What will be the positive and negative consequences of a further increase in flows of **nitrogen and phosphorus** in the next several decades?
- What will be the consequence of **biodiversity** loss for ecosystem services and human well-being?
- What will be the impact of changes in **desertification** on provision of ecosystem services and how will this vary across regions? How will demand for ecosystem services increase or decrease the rate of desertification?
- What are the impacts of changes in **wetlands** on provision of ecosystem services? How will demand for ecosystem services increase or decrease the rate of loss of wetlands?

The scenarios address these core questions. They explore the potential futures of ecosystems and the services they provide, including the possible benefits and inadvertent consequences that could emerge in each future. Each scenario also considers vulnerability, resilience, and possibilities for thresholds and regime shifts in socioecological systems given the specific details of how the scenario unfolds.

5.3.2 Interviews

Insights from leaders helped focus the MA scenarios directly on the most pressing interests of decision-makers and other scenario users. In addition to the user needs survey described above, we interviewed 59 leaders in NGOs, governments, and business from five continents. (See Figure 5.1.) The leaders were chosen based on recommendations from the MA Board (who were themselves selected from MA users to guide the MA process). The selection process was not random, but it aimed for diversification. We intentionally chose leaders from many sectors and nations in order to gain access to a wide range of concerns and responses. While it would have been interesting to get a broader view by interviewing many additional people, including people who are not leaders, this was not possible due to time constraints.

Based on previous scenario work (Van der Heijden 1996), we designed open-ended, general questions that would elicit a wide range of conversations about issues that interviewees thought were critical determinants of the current and future states of the world. (See Box 5.1.) Interviewees received the questions by e-mail in advance of the interview. Most interviews were conducted by telephone and typically lasted about one hour. In some cases, respondents followed up the interview with further comments by e-mail. A few interviews were conducted entirely by e-mail. Further information about the interview process can be found in Chapter 6.

Most of the interviewees were concerned that ecosystems are changing for the worse, reducing the quality and quantity of many important ecosystem services. That is, the respondents were concerned about the sustainability of ecosystem services. The interviewees disagreed, however, about the main causes of ecosystem degradation. Poverty, inequality, overconsumption, and mismanagement were a few of the factors that interviewees listed as factors in ecosystem degradation. The MA scenarios should therefore elucidate the links between interviewee concerns about ecosystem services and the types of problems that may be caused by each of the key sources they mentioned.

There were also important differences in views about how to address the challenges of sustainable provision of ecosystem services. Generally, these unfold from a basic disagreement about whether the world is generally vulnerable and fragile or generally resilient and recoverable. While one respondent said, "What gives me the most hope for the future is the tremendous resilience that nature has demonstrated in responding to opportunities," another said, "The environment . . . is resilient at the moment, but we cannot treat it with impunity forever." Stances on the resilience or vulnerability of ecosystems were associated with beliefs about the effort that should be placed on environmental problems. Some felt it is imperative to make the environment the key focus of society immediately, while others felt that society should focus first on improving human well-being, with a hope that this would lead to environmental improvements. Useful scenarios will attempt to embrace such divergent views and provide a framework in which these viewpoints can be debated.

Interviewees also talked about the variables that will determine how the future unfolds. Many respondents cited the same variables, but there were diverse views about how those affect the future—even to the point of disagreeing over whether the outcomes would be positive or negative. The factors identified by many respondents were globalization or global connectedness, human values, inequality, leadership, urbanization, technology, and energy sources.

While some respondents said that increased global connectivity would increase communication, trade, and the range of opportunities available to people, other decision-makers expressed concerns about homogenization of Earth's biological and social systems due to globalization. While some respondents were excited about using technology to solve problems and enhance the provision of ecosystem services, others feared that technology might cause more problems than it would solve. Nearly all the decision-makers we interviewed mentioned concern about energy sources. There was consensus that the ways in which society obtains

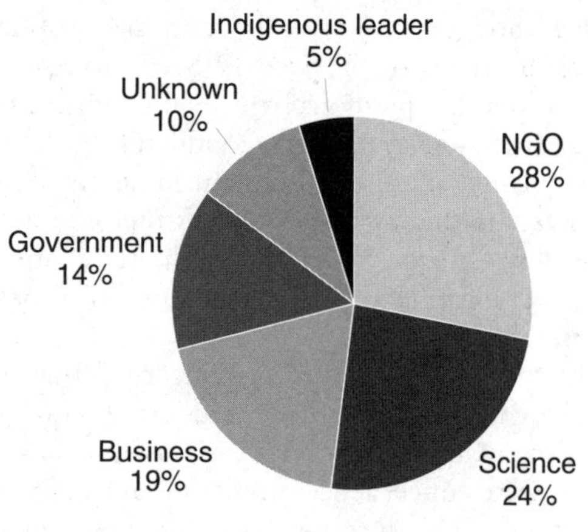

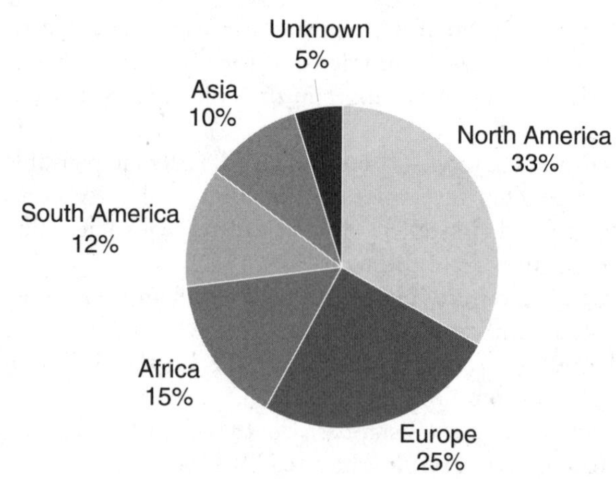

Figure 5.1. Interviewees by Sector and by Region

BOX 5.1

Interview Questions

What words would you use to describe the current state of the Earth's natural and human systems?

What words would you use to describe the ideal state of the Earth's natural and human systems in 2050?

What obstacles do you envision to achieving this ideal world?

If you could talk to someone who visited the world in 2050, what would you need to know to understand what the world really looks like in 2050?

Who or what will be most influential in determining the pathway of change into the future?

What is the biggest change you expect between 2003 and 2050?

What surprises might you envision between now and 2050?

What gives you the most hope for the future?

and uses energy will affect the future. Of course, there was disagreement about the best sources of energy. Finally, human values, leadership, and inequality were all considered to be important drivers of the way the future would unfold. However, whether strong leaders and closed borders were better than free trade and cooperation was debated. Inequality was seen as an important driver of many of today's problems, but interviewees did not agree on whether inequality could or even should truly be reduced or on the best way to reduce inequality, nationally, or globally.

In general, interviewees agreed that the current situation could be improved, but there was little agreement on how to do it. There were diverse beliefs about how to put the world on a sustainable path. One interviewee said, "We sorely need inspirational leaders. We don't know how to breed or train them. They appear to be almost accidental. But we sorely need them now." Another argued that mechanisms for rewarding people for "good environmental living" were the most critical need. There was great disagreement about the role of governments. One respondent said that "governments and heads of state have proven themselves irrelevant," while the next said that "governments must work together." (See Box 5.2.)

The scenarios should address the factors that the interviewees found to be important. The scenarios should also attempt to embrace the diversity of viewpoints held by the interviewees. By organizing diverse viewpoints in scenarios, we hope to facilitate debate and discussion. Clarification of terms is one way in which scenarios could facilitate discussion. In the interviews, it was often difficult to determine whether apparent disagreements were actual disagreements about the facts or simply misunderstandings derived from different interpretations of the same words. For example, interviewees disagreed about whether globalization was a positive or a negative factor. This disagreement could reflect different beliefs about the future, or it could reflect different definitions of globalization (for instance, globalization as trade dominated by policies that favor wealthy nations versus globalization with policies that open international markets for all nations versus globalization as something larger than just trade).

Interviewees agree that sustainable development is needed, but disagree about how best to achieve it. There are diverse views about which actions to take and about the sequences in which actions should be taken. The message for scenario building is clear. Useful scenarios will help decision-makers understand the possible and likely effects of key actions or paths that we might choose to take: fair global economic policies, use of technology to provide or improve provision of ecosystem services, the role of top-down control and leadership, the effects of multiscale decision-making and local flexibility, and the role of ecosystem dynamics in determining the end result of decisions. Useful scenarios will also embrace the diversity of views about the importance of these factors.

5.3.3 Ecological Management Dilemmas

Ecological dynamics underlie the concerns of decision-makers and MA users. These ecological dynamics influence the results of management actions, but important aspects of

BOX 5.2

Selected Quotes from the Interviews

Sustainable development is needed, and managing ecosystem services and human well-being is a key aspect of that:

"There is tangible evidence that natural systems are stressed to the limits of tolerance."

"Natural systems are fragile, threatened, degraded, and overburdened by human demand. At the same time, human systems are unequal in access to resources."

Globalization is an important player, but there is disagreement about whether it is a major problem or a significant solution:

"Governments and heads of state have proven themselves irrelevant when it comes to solving real problems. They are more successful when acting in their homes but not when coming together to face global issues."

"Governments must work together – we can't save half a planet."

Considering poverty is important:

"There is unequal distribution of resources, population, and trade, leading to a vicious circle of environmental degradation in the most vulnerable parts of the world, which will ultimately negatively impact the whole globe's security."

Considering technology is important:

"We might also see some astonishing technological breakthroughs involving biotechnology, not just in genetically modified organisms but also in fields such as organic computers that may be self-reproducing."

On the importance of local and regional flexibility:

"The ideal state of the world is when there is respect for the ecosystem and living within its limitations avoiding experimentation with changing it, where everyone has enough to cover the basic needs of water, food, and shelter and conserve the natural resources, where everyone tries to live with the seasonal changes without the need to modify the surroundings (e.g., temperature and humidity) artificially within the limits of our body adaptability (which we should use to its maximum capacity)."

"Business leaders understand that surprise is the rule and flexibility is key to surviving the surprises."

On the importance of surprises:

"The next 50 years will tell us whether that self-proclaimed marvel of evolution, the human mind, can surprise us even as we are surprised by chaotic events."

the dynamics are unknown and uncontrollable. Thus, far-reaching ecosystem management decisions are often made in situations where the ecological responses are unknown. (See Box 5.3.) In these cases, all options appear to have potentially severe negative outcomes, and the outcomes are highly ambiguous (Ludwig 2001). These situations are termed ecological management dilemmas (Bennett et al. in press). Managers generally transform or manage an ecosystem with the aim of obtaining a set of desired ecosystem services. A number of perverse consequences are possible, however, such as reductions in future ecosystem services, increases in vulnerability of ecosystems to disturbance, or unforeseen trade-offs in other ecosystem services. (See Chapter 3.) The prospect of perverse consequences creates dilemmas for ecological management.

Ecological management dilemmas challenge decision-makers to seek policies that are robust to uncertainty, surprise, and failure of actions to evoke expected responses. That is, the policy should achieve acceptable outcomes even under unexpected conditions. Flexibility and learning mechanisms become an essential part of the management process to cope with the fact that management actions need to continually adapt to evolving ecological dynamics (Walters 1986; Gunderson et al. 1995; Carpenter 2003). Ecological dilemmas are not susceptible to the routine approaches of ecosystem management because they involve complex ecological dynamics and uncertainties (Holling and Meffe 1996; Ludwig 2001). Instead, they require approaches that are more flexible, more attentive to change, and more innovative (Gunderson and Holling 2002). The construction of institutions that address ongoing change in ecosystems, emerging ecological dilemmas, and sustainable management of ecosystem services is currently an active area of uncertainty, debate, and research (National Research Council 1999, 2002; Berkes et al. 2003). The scenarios should reflect the current diversity of viewpoints about how ecological management dilemmas should best be addressed.

Two ecological dilemmas were frequently raised by the interviewees and MA user community: What degree of ecological complexity is needed to provide reliable ecological services? (See Box 5.4.) And to what degree can people use technology to substitute for the role of relatively undisturbed ecosystems in provision of services? These unknowns are critical because the answers provide a clue about the best approaches to managing for ecosystem services in any particular situation. The answers will affect the resolution of many of the questions asked by the MA user community and the concerns of the interviewees. Since we currently do not know how much ecological complexity is enough, the costs and benefits of future complexity are hard to evaluate. We also do not fully understand when technology can be used to substitute for an ecosystem's role in provision of ecosystem services and when technology might lead to deleterious side effects. We sought scenarios that address these ecological dilemmas in a useful way with respect to the concerns that decision-makers presented in the interviews.

5.3.4 Drivers and Current Conditions

The scenarios should also be rooted in the present. Transformations described in the scenarios should emerge from the important drivers and current condition of socioecological systems. These are presented in Chapter 7. Working

BOX 5.3
Catastrophic Change in Ecosystems

Most of the time, changes in ecosystems and their services are gradual and incremental. Most of these gradual changes are detectable, and many are predictable. However, some changes in ecosystems and their services are large in magnitude as well as difficult, expensive or impossible to reverse (*high certainty*) (Scheffer et al. 2001). These changes are important, massive, and hard to predict, so they may come as surprises. Some systems that are known to exhibit large, hard-to-reverse changes (adverse changes indicated in parentheses here) include pelagic fisheries (economic collapse), freshwater lakes and reservoirs (toxic blooms, fish kills), pastoral lands (conversion to woodland), and dryland agriculture (salinization, desertification) (Carpenter 2003; Folke et al. 2005; Walker and Meyers 2004).

Slow losses of resilience set the stage for large changes that occur after the ecosystem crosses a threshold or is subjected to a random event such as a climate fluctuation (*established but incomplete*) (Folke et al. 2005; Groffman et al. 2005). For example, slow buildup of phosphorus in soils gradually increases the vulnerability of lakes and reservoirs to runoff events that trigger oxygen depletion, toxic algae blooms, and fish kills (Carpenter 2003). Gradual overfishing and nutrient runoff make coral reefs susceptible to severe deterioration triggered by storms, invasive species, or disease (Bellwood et al. 2004; Hughes et al. 2003). Slow decrease in grass cover crosses a threshold so that grasslands can no longer carry a fire, allowing woody vegetation to dominate and severely decreasing forage for livestock (Walker 1993). In the Sahel, decades-long droughts are caused by strong feedbacks between vegetation and the atmosphere and may be triggered by slow changes in land degradation (Foley et al. 2003).

Because multiple, interacting stresses on ecosystems are increasing, it is likely that harmful large ecosystem shifts will become more common in the future (*established but incomplete*). On the other hand, proactive ecosystem management and wise use of ecological technology can reduce the impact of harmful shifts in ecosystems and assist people in adapting to unexpected change (*established but incomplete*).

from these initial conditions and drivers, the Scenarios Working Group developed plausible pathways to four very different futures by 2050. The year 2050 was chosen to be far enough in the future to reveal the effects of important ecological feedbacks and to consider long-term futures and yet near enough that the causal chain between current decisions and eventual outcomes could be reasonably traced.

5.4. Introduction to and Overview of the Scenarios

The interviewee concerns and user needs, and the ecological uncertainties that underlie them, are the factors that the scenarios should address. We identified four clusters of beliefs that embrace most of the fears, hopes, and expectations for the future that were encountered in the interviews and the statements of user needs.

Many leaders felt that the future would bring increased emphasis on national security, leading to greater protection of borders with associated consequences for economic development and changes in direct drivers of ecosystem services. Other respondents felt that the future could, or should, bring greater emphasis on fair, globally accepted economic and environmental policies, as well as greater attention by governments to public goods. Some interviewees pointed to the prospect of technology for managing ecosystem services with greater efficiency. Still others found hope in local adaptive capacity for flexible, innovative management of socioecological systems. The future may well involve a mix of these perspectives.

Our approach to the four clusters of beliefs was informed by previous explorations of sustainability concepts. Among these are ideas about investment in manufactured, human, and natural capital (Dasgupta and Mäler 2000, 2001); objectives of business development, community empowerment, and environmental conservation (Munasinghe and Shearer 1995); trade-offs among individualist, hierarchist, and egalitarian social perspectives (Janssen and DeVries 1998); and integrated theories for ecosystems, social systems, and management systems (Gunderson and Holling 2002).

Economic development is sometimes viewed as the key to sustainable development. The Environmental Kuznets Curve suggests that as economic growth occurs, environmental quality is first degraded and later improved (Stern 1998). The conclusion that many have drawn from this theory and the evidence supporting it is that economic growth should lead to improvements in the environment. Other evidence also indicates that poverty alleviation may lead to improvements in ecosystems. For example, the poorest people are often directly dependent on ecosystems for services such as food, fuel, and water. In times of scarcity or high population, these groups may overharvest from local ecosystems. By diversifying economic opportunity, both human well-being and direct impacts on ecosystems may be reduced. On the other hand, greater consumption is often associated with greater impact on the environment (Wackernagel and Rees 1995). The disparity in income among nations leads to enormous disparity in political and economic power as well as a much greater impact on global life-support systems by rich countries than poor (Ehrlich and Ehrlich 2004). The connections between economic policies and the status of ecosystem services are multiple and complex. All the scenarios explore these connections to some degree.

In the Global Orchestration scenario, we explore the possibilities of a world in which global economic policies are the primary approach to sustainability. The recognition that many of the most pressing problems of the time seem to have roots in poverty and inequality leads many leaders toward a strategy of globally orchestrating fair policies to improve well-being of those in poorer countries by removing trade barriers and subsidies. Nations also make progress on global environmental problems, such as greenhouse gas emissions and depletion of pelagic marine fisheries. The results for ecosystem services are mixed. While human well-being is improved in many of the poorest countries, it is still not clear in 2050 whether the net impact on ecosystems will be positive or negative.

Some respondents believe that national security will become an overarching concern in the future. Should this

BOX 5.4

Biodiversity, Disturbance, and Resilience of Ecosystem Services

Ecosystem resilience is maintained by genetic and species diversity as well as by spatial patterns of landscapes, environmental fluctuations, and temporal cycles with which species evolved. Management for resilience recognizes the importance of heterogeneity and change, including the natural processes of species turnover, extinction, and evolution. Ecosystem resilience is the amount of disturbance that an ecosystem can withstand and still maintain essentially the same structure, processes, and flow of ecosystem services (Holling 1973). As described here, the renewal and reorganization of ecosystems after disturbance depends on the functional groups of species within ecosystems and the diversity of responses to environmental fluctuations within those functional groups.

Disturbance is routine in ecosystem dynamics (White and Pickett 1985). All species evolved in the presence of certain types, magnitudes, frequencies, and spatial patterns of disturbance and are thus adapted to these disturbances (Paine et al. 1998). Disturbances within the typical range usually result in little long-term change in ecosystem characteristics, processes, or services, even though species turnover may be extensive (Turner et al. 1997). Moreover, the typical disturbance regime is often necessary for maintaining ecosystem resilience. Without disturbance, critical groups of species or processes disappear over time (White and Pickett 1985).

Events outside the range of typical disturbances can transform ecosystems, creating new and surprising ecosystem structures and processes. Disturbances that cause surprising transformations often involve compounded perturbations, with multiple events within the normal recovery interval of the ecosystem or unusual combinations of drivers (Paine et al. 1998). Ecosystem transformations can also result from anthropogenic disturbances, which are often chronic (instead of pulsed) and may be unlike anything experienced before in the evolutionary history of the species (Bengtsson et al. 2003).

The biotic structure of an ecosystem also affects the outcome of disturbance. Population attributes such as dispersal ability or generation time affect the response of particular species to disturbance. Aspects of community structure, including biodiversity, play a critical role in the responses of ecosystems to disturbance (Chapin et al. 2000; Kinzig et al. 2002; Loreau et al. 2002).

Functional groups are sets of species that perform similar ecosystem processes. Ecologists have identified functional groups by clustering microbes, plants, or animals according to biological similarities (Holling 1992; Frost et al. 1995; Walker et al. 1999; Havlicek and Carpenter 2001). At least two different effects of functional groups on ecosystem processes have been recognized (Yachi and Loreau 1999; Ives et al. 1999, 2000). First, if several functional groups are complementary in their use of resources, the diversity among functional groups tends to increase the total flow of ecosystem services (Yachi and Loreau 1999; Hulot et al. 2000; Reich et al. 2004; Petchey et al. 2004). For example, functional groups of plants that root at different depths, that grow or flower at different times of the year, and that differ in seed dispersal and dormancy act together to increase ecosystem productivity.

Second, diversity within functional groups maintains the rate of ecosystem processes despite environmental fluctuations if the individual species respond differently to such fluctuations (Yachi and Loreau 1999; Ives et al. 1999, 2000; Walker et al. 1999; Norberg 2004; Bai et al. 2004). This phenomenon is called response diversity. When the environment changes, a formerly rare species with different characteristics can become dominant (Frost et al. 1995). Response diversity is the key to the insurance effect of biodiversity on ecosystem services (Elmqvist et al. 2003). In the face of uncertain and often novel anthropogenic changes in the environment, preserving the diversity of species and functional groups increase the chance that species are retained that later play a crucial role in the ecosystem. In this sense, species and functional diversity provide "insurance" against future environmental change. In contrast to monetary insurance against unexpected accidents, however, the insurance provided by diversity is not guaranteed, and the environmental change for which diversity may provide insurance is not unexpected. Preserving biodiversity is not a substitute for reducing other kinds of anthropogenic stresses on ecosystems.

It is an oversimplification to equate species richness with resilience of ecosystem services. Instead, the effect of diversity on resilience depends on organization of species among functional groups, spatial pattern, and scaling of ecosystem processes in time and space (Elmqvist et al. 2003; Folke et al. 2004). A species invasion that adds to species richness can decrease the resilience of ecosystem services if it reduces response diversity.

When chronic, progressively worsening stress to an ecosystem removes species in order of their susceptibility to the stress, the surviving species are more tolerant of this specific stress (Ives and Cardinale 2004). These species, however, may provide little insurance against other types of environmental changes. If the ecosystem is subjected to a different kind of stress or disturbance, these few species may be eliminated, thereby causing a greater loss of ecosystem services. Thus the maintenance of ecosystem services requires the maintenance of diversity during multiple, successive environmental changes.

Diversity of spatial pattern creates a kind of response diversity (Elmqvist et al. 2003). Dispersal of species among patches in heterogeneous landscapes confers resilience to disturbances that affect only part of the landscape or seascape (Peterson et al. 1998; Nyström and Folke 2001; Loreau et al. 2003; Cardinale et al. 2004). If a process is eliminated from part of the landscape or seascape but is present in other patches within dispersal range of the affected patch, then the missing process can be reestablished. Furthermore, the pattern of local elimination and recolonization through dispersal may establish numerous ecosystem configurations, thereby creating local ecosystem diversity throughout a landscape or seascape.

Response diversity acts across scales through interspecific differences in the use of space (such as dispersal ability, patch size, and home range size) and time (such as generation time, dormancy period, and seasonality). Ecological disturbances usually occur in a specific range of time-space scales, allowing persistence of species, structures, or processes that occur at the scales that were not affected (Elmqvist et al. 2003). Therefore, replication of ecological processes across a wide range of scales confers resilience (Peterson et al. 1998). Species that act across a wide range of space scales (such as highly mobile species) or time scales (such as long-lived species or large-bodied generalist predators) are an important element of ecosystem response diversity (Peterson et al. 1998). Regional losses of such species increase the risk of catastrophic ecosystem changes that cause large reductions in ecosystem services (Elmqvist et al. 2003).

Traditional societies may have known about response diversity for a long time. Berkes et al. (2003) describe several societies that appear to manage for response diversity and may thereby build resilience of ecosystem services.

In summary, proactive ecosystem management builds ecosystem resilience through maintenance of genetic and species diversity, as well as spatial patterns of landscapes and temporal cycles of environmental fluctuations and disturbance with which species evolved. In contrast, ecosystem management practices that reduce response diversity, remove whole functional groups or trophic levels, expose ecosystems to chronic novel stress or novel disturbances, or create compounded perturbations (unusual combinations of disturbances at intervals shorter than the normal recovery cycle of the ecosystem) increase the risk of large-scale breakdowns in ecosystems and losses of ecosystem services (Folke et al. 2004).

occur, nations or blocs may concentrate on inward-looking economic development, so that the globalization of the economy may proceed more slowly than in Global Orchestration. While some regions would remain well endowed with ecosystem services, other regions that now have fewer ecosystem services may remain impoverished.

The Order from Strength scenario examines the outcomes of a world in which protection through boundaries becomes paramount. The policies enacted in this scenario lead to a world in which the rich protect their borders, attempting to confine poverty, conflict, environmental degradation, and deterioration of ecosystem services to areas outside the borders. Poverty, conflict, and environmental problems often cross the borders, however, impinging on the well-being of those within. Protected natural areas are not sufficient for nature preservation or the maintenance of ecosystem services. In addition to losses of ecosystem services in poor regions, global ecosystem services are degraded due to lack of attention to the global commons.

The survey and interview results indicated that many of those interviewed think that complexity and local flexibility are a critical component of the path to sustainability. Social and ecological scientists have addressed the conditions in which disaggregated management systems outperform centralized ones (Grossman 1989; Scott 1998; Gunderson et al. 1995; National Research Council 2002). Because ecosystems are subject to large and potentially irreversible changes (Chapter 3), certain types of centralized ecosystem management schemes are subject to catastrophic failure (Holling and Meffe 1996). According to a large number of case studies, enabling conditions for successful ecosystem management include small size, well-defined boundaries, shared norms, social capital, appropriate leadership, fairness in allocation of ecosystem services, and locally devised, easily enforceable access and management rules (National Research Council 2002). It is important to note that central governments do not undermine local authority for ecosystem management (Ostrom 1990; Wade 1988; National Research Council 2002).

While there is clear evidence of success in local ecosystem management, the multiscalar nature of ecosystems poses challenges for this approach. For example, local management of thousands of subwatersheds may not lead to sustainable management of a continental river system if there are significant externalities that are not properly included in local accounting. Some of our respondents feared that a disaggregated world would exacerbate global problems or benefit ecological services only in regions that were relatively wealthy, well educated, and well endowed with natural capital. These trade-offs in the scales of ecosystem management are addressed in the scenarios.

The Adapting Mosaic scenario explores the benefits and risks of disaggregation. In this scenario, lack of faith in global financial and environmental institutions, combined with increasing understanding of the importance of resilience and local flexibility, leads to diminishing power and influence of these institutions compared with local and regional ones. Eventually, this leads to diverse local practices for ecosystem management. The results are mixed, as some regions do a good job managing ecosystems and others do not. High levels of communication enable regions to compare experiences and learn from one another. Gradually, the number of successful experiments begins to grow. While global problems are ignored initially, later in the scenario they are approached with flexible strategies based on successful experiences with locally adaptive management.

Still others are optimistic about the use of technology to sustain ecosystem services. Technology has led to great improvements in agricultural production efficiency, in medicine, and in the provision of other ecosystem services. Advances in technology have the potential to build human well-being through more efficient use of ecosystem services as well as through better understanding of ecosystem conditions and trends. Greater efficiency could reduce the overall impact on ecosystems and thereby increase opportunity for sustainability of ecosystem services.

On the other hand, technological solutions sometimes lead to unexpected problems (Tenner 1997). Acceleration of technology may be a factor in the increased incidence of environmental problems, demanding more and more ingenious responses (Homer-Dixon 2000). For example, increased use of pesticides in agriculture may lead to pests that are resistant, requiring a newer and better technology to remove them. In addition, efficiency gains are often focused on a single service, rather than a bundle of services; in fact, increased efficiency in provision of one service may cause declines in provision of other services. Highly efficient environmental management systems often rely on predictions, but ecosystem changes are often unpredictable and errors in prediction lead to costly mistakes (Oreskes 2003; Pielke 2003). Other problems derive from the complexity of the decision systems in which environmental predictions are used (Dörner 1996; Sarewitz et al. 2000). For these reasons, some experts are cautious about the use of technology to manage ecosystem services more efficiently. Increasing reliance on technology could increase the frequency and severity of unexpected problems, erode the resilience of ecosystems, and over time cause ecosystem services to become more vulnerable.

The TechnoGarden scenario explores the potential role of technology in providing or improving the provision of ecosystem services. In this scenario, people push ecosystems to their limits of producing the optimum amount of ecosystem services through the use of technology. Often, the technologies they use are more flexible than today's environmental engineering and they allow multiple needs to be met from the same ecosystem. In the beginning of the scenario, these technologies are primarily developed in wealthier countries and slowly dispersed to poorer places, but later—promoted by a global focus on education—they are developed everywhere. Provision of ecosystem services in this scenario is high worldwide, but flexibility is low due to high dependence on a consistent provision of services. In some cases, unexpected problems and secondary effects created by technology and erosion of ecological resilience lead to vulnerable ecosystem services that are subject to interruption or breakdown.

In summary, the four MA scenarios represent diverse views of the future of ecosystem services. While advocates for particular viewpoints may state them as assertions, the Scenarios Working Group regards them as questions to be addressed. Each of the four scenarios addresses different sets of beliefs about how the global system might change in directions that could sustain ecosystem services. (See Figure 5.2.) Global Orchestration describes a world in which policy initiatives attempt to establish fair global markets and organize transnational responses to certain global environmental problems. Order from Strength addresses the beliefs of those who hold that the future will, or should, bring security, including protection of natural resources and ecosystem services. In the world of Adapting Mosaic, the focus of economics and politics shifts to local or regional scales. TechnoGarden presents a future in which great emphasis is placed on the development of technology for efficient management of ecosystem services.

Some key characteristics of the global system during each scenario are presented and compared in Table 5.1. As we shall see, the contrasting conditions of these scenarios lead to different bundles of benefits and risks for ecosystem services and human well-being. In the remainder of the chapter, we present short sketches of each scenario, compare their benefits and risks, and describe situations in which the conditions of one scenario could branch toward the conditions of a different scenario.

5.5 Sketches of the Scenarios

This section presents short synopses of the four scenarios. Each scenario is told by an observer looking back at 2000 from 2050. These brief descriptions are intended to provide an overview of the dynamics in each scenario. Longer, more detailed narratives are presented in Chapter 8. Some quantitative model results are presented in Chapter 8, and full model results are presented in Chapter 9.

5.5.1 Global Orchestration

Summary: *The past 50 years have shown that some ecosystem services can be maintained or improved by appropriate macroscale policies. Notable successes occurred in reducing or controlling many global pollutants and in slowing, or in some cases reversing, loss of marine fish stocks. In some situations, it turned out that ecosystem services improved as economies developed. On the other hand, it appears that global action focused primarily on the economic aspects of environmental problems is not enough. In some regions and nations, ecosystem services have deteriorated despite economic advancement. Also, it was sometimes difficult to adjust large-scale environmental policies for local and regional issues. Despite some significant environmental disasters, this lesson has not yet been learned. As we look to 2100 and beyond, multiscale management of ecosystem services is a top challenge for environmental policy.*

At the beginning of the twenty-first century, poverty and inequality, together with environmental degradation and climate change, were pressing problems on the agendas of global and national decision-makers. Concerns about social tensions arising from inequalities in and uneven access to global markets were growing, as these tensions were often seen as the un-

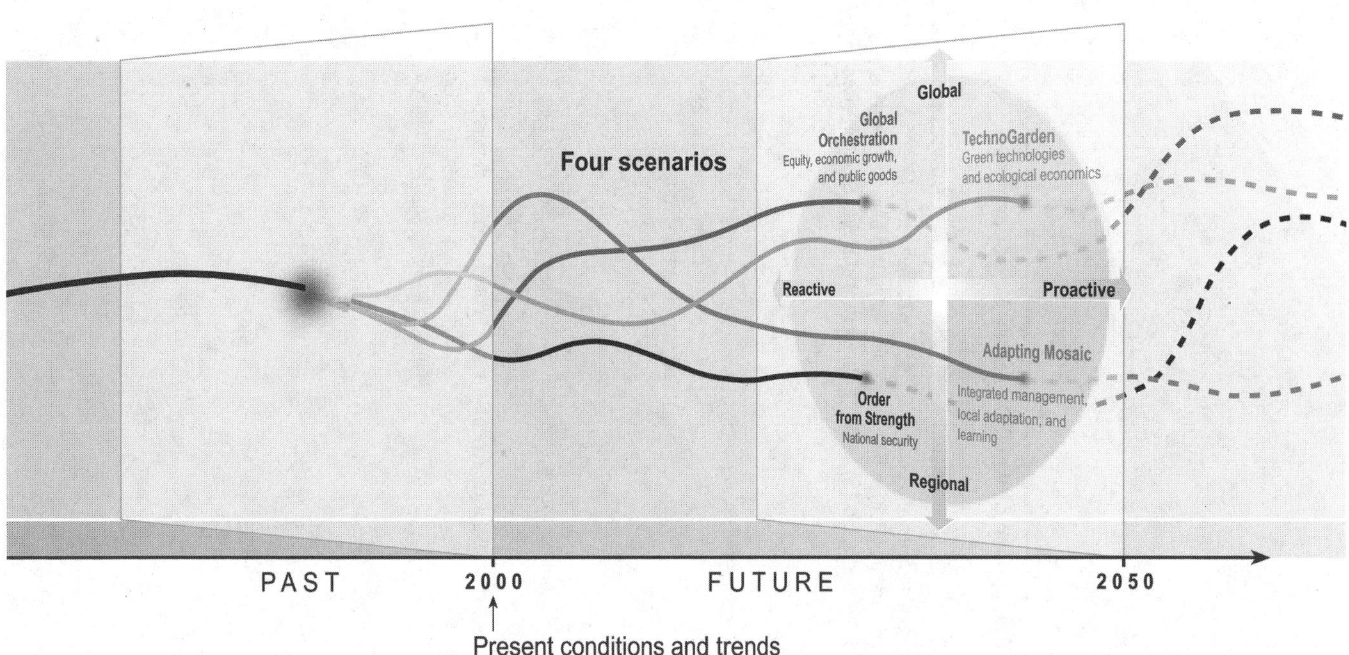

Figure 5.2. Contrasting Approaches among MA Scenarios. The scenario differences are based on the approaches pursued toward governance and economic development (regionalized versus globalized) and ecosystem service management (reactive versus proactive).

Table 5.1. Comparison of Variables across Scenarios

Variables	Order from Strength	Global Orchestration	TechnoGarden	Adapting Mosaic
Priorities for investment in (and consumption of) natural (N), human (H), manufactured (M), and social (S) capitals	high investment in H, M, and S among elites within and between societies; low investment in H and S of non-elites by elites	high investments in M, H, and S, but not with respect to ecosystem goods and services; investment in N or in environmental research or education only to the extent it matters for development	balanced investment in all forms of capital	strong investment in S, through encouragement of local institutions, and H, through investment in environmental research and education; these lead to investments in N, through both protection and repair
Perspective on ecosystem services and socioecological systems	ecosystem services will be maintained by local or national actions to conserve representative ecosystems	ecosystem services can be sustained by global social economic policies; ecological systems are either resilient (and will remain so) or resilience will be enhanced as a consequence of social policies	ecosystem services can be sustained by technologies, including methods of ecological engineering and ecological economics	ecosystem services are an evolving aspect of socioecological interactions; ability to control and predict ecosystem services is limited, so continual learning and adaptation are necessary to sustain ecosystem services
Changes in connectivity of socioecological systems	variability in socioecological links among nations	loosening of socioecological links	tightening of socioecological links driven by social and economic objectives	exploring and adjusting socioecological links
Root of unexpected breakdowns in ecosystem services	lack of consideration of the impact of slowly changing variables; feedbacks emerge as surprising due to lack of monitoring	lack of consideration of the impact of slowly changing variables; feedbacks emerge as surprises due to lack of monitoring	over-engineering; some ecological systems are more complex than expected; lack of understanding of how technological innovation interacts with ecological complexity, including long-term dynamics and spatial connectedness of ecosystem processes	risks of experimentation 9or inaction0 lead to some localized breakdowns, while global ecosystem services diminish due to limited capacity to address global ecosystem issues

Response to unexpected breakdowns of ecosystem services	elites assume that local and national management of ecosystem services is sustainable	it is assumed that global policies, markets, substitution, and adaptation are sufficient to maintain ecosystem services; little strategic planning as there is little investment in research or monitoring	proactive in the sense that ecosystems are engineered to provide specific ecosystem goods and services and avoid cross-scale ecological feedbacks; assumption that technologies will prepare managers to deal with unexpected feedbacks	adaptive, innovative, and proactive for local, national, or regional ecosystem services, but global coordination of ecosystem management is lacking
Investment in learning about the environment	lowest of the four scenarios	low, motivated by breakdowns in ecosystem services	high investment in ecosystem research motivated by technological development	variable; high in some locales; learning by comparing regional experiences, but little investment in understanding of global ecosystem services
Approach to global commons	action on global issues that affect elites	strong global action on global issues if and when they affect social and economic goals	coordinated and proactive action on global issues, emphasizing technological solutions	little attention to global issues that affect ecosystem services
Programs for multiscale resources management	sometimes exist in wealthier regions; in other regions, institutions for multiscale resource management are slow to develop	focus on global emissions with little emphasis on multiscale coordination; global practices sometimes overlook regional variation in resource issues	scale of intervention tends to be driven by technological capability and opportunity	variable among regions; growing attention to multiscale management of ecosystem services emerges by ~2050
Institutions, incentives, property rights for ecosystems management	regulatory control by elites, who have greater access to ecosystem services	weak institutions for ecosystem services, except for global commons issues; ecosystem services not seen as fundamental to wealth creation; property rights often discounted from benefits of ecosystem services	property rights and market mechanisms are emphasized to align private incentives with sustainability of ecosystem services	high variability of institutions among regions; exploration of practices to build resilience of ecosystem services
Emphasis on development of environmental technology	low; in wealthy regions, some progress on environmental technology in response to recognition of novel environmental problems	moderate, in response to emergence of novel environmental problems	rapid development of advanced environmental technologies that support ecological design; technologies are used to decrease the dependency of economic growth and consumption	moderate; variable among regions

derlying causes of uncontrolled migration, conflicts, and even terrorism. Leaders were also concerned about inequalities among people, including differential access to technology and education and other drivers of inequality. There were great debates about the best approach to solving these problems.

Eventually, globally orchestrated policy reforms took hold as the dominant strategy. Policy reforms were used to reshape the world's economic and governance systems. The emphasis of these reforms was on creating markets that allowed equal participation and provided equal access to goods and services. The reforms also targeted the creation of more transparent governance systems worldwide as the necessary foundation of economic growth. As the world became increasingly connected financially, it was necessary to create global policies to deal with problems arising from the connections. Thus, one result of globalized economic systems was a strengthening of global and regional standard-setting bodies such as the World Trade Organization. The focus on policy reforms and faith in global institutions also led to strengthening of the United Nations and some other multinational alliances.

At about this same time, governments found themselves making decisions about how to handle terrorism and conflicts among nations. Should rich countries focus on borders and protection or should they assist with development in poorer countries to spread goodwill? Generally, rich nations leaned toward helping poor nations meet their basic needs, as this was thought to be the better long-term solution. Trade practices that had hindered economic development in poor countries were discontinued. These reforms were followed by increased wealth in many poor countries, which led to secondary improvements in governance and democracy. In most regions of the world, governments invested more heavily in public goods, such as education and public transportation.

Trade expanded globally, driven by removal of subsidies and increasing demand for goods and services around the planet. Economies in China, India, and Southeast Asia began to grow rapidly again. A focus on education and, in some cases, political reform helped civil society grow in poorer countries. In countries that profited from increased market access and production opportunities, a wealthier middle class began to develop. Civil society and the growing middle class, in turn, brought about further reforms.

By the 2020s, a growing middle class was demanding cleaner cities, less pollution, and a more beautiful environment. This was particularly true for problems that occurred in and around urban settings and those that directly affected human health. Nevertheless, problems of intensified agricultural systems and the slow loss of wildlands received only limited attention. Environmental problems that were difficult to reverse, such as biodiversity loss, were more or less ignored by the general population because so many other things were going well.

Driven by policies aimed at increasing gross domestic product and human well-being, agricultural area expanded in poor countries, leading to increased human impacts on terrestrial ecosystems. Agricultural specialization increased, driven by the selection of high-yield and commercially valuable crops and livestock. Local ecological knowledge was often replaced by uniform industrial methods. Consequently, by the 2020s, wild varieties of agricultural species existed primarily in gene banks, and the number of domestic varieties in use was greatly reduced. Diverse landraces persisted mostly in marginal areas. By 2025, many small farms had consolidated into large agricultural operations. All farms, small and large, had become more highly mechanized and industrial. By sometime in the 2030s, the rate of increase in agricultural area had begun to slow down due to replacement of traditional agriculture with more-efficient industrial systems.

As the rate of agricultural expansion declined, particularly in rich countries, and as people moved from the countryside into cities, many terrestrial ecosystems began to recover from intensive human use. This recovery was aided by increased productivity of farms, which allowed some reduction in agricultural land area. Recovery of ecosystem function in these areas was aided by replanting and some restoration. Ecosystem restoration was driven by people's interest in increasing the supply of fuelwood and other biomass products, in addition to the expansion of intensively managed spaces for recreation. In contrast to the agricultural land recovery, coastal marine ecosystems and wetlands declined significantly because the increased urban growth was mostly concentrated in a 100-kilometer band along the coastline.

Increases in wealth and in the availability of technology resulted in the continuing improvement of health around the planet. Regional inequalities in health were prevalent until the mid 2020s. Obesity-related diseases remained a threat, particularly in rapidly developing areas, as new food choices became available and societies shifted their eating habits to less healthy diets. Emerging infectious diseases were also a risk. The potential for the origination and spread of novel pathogens was high in areas where ecosystem function was disregarded. It turned out that disruption of ecosystem regulation processes increased the likelihood of exposure to pathogens originating from wild animals and plants, and the movement of exotic species around the world through widespread trade further facilitated the spread of pathogens. While these surprises occurred in rich and poor countries, the capacity to respond was higher in rich countries, and hence the impact was much higher in poorer countries. Positive surprises, such as the success of genetically modified organisms in reducing the agricultural expansion, also occurred.

Despite economic policies designed ultimately to lead to a better environment, the simplification of ecosystems eventually led to a decrease of environmental security as ecological surprises became more common. One surprise of the past 50 years was the high impact that widespread trade had on hastening the spread of invasive species. It seems that reduced diversity limited the options of ecosystems to respond to ever increasing ecological surprises, although it is hard to tell if the problem was this or simply increased population pressure. People in poor countries are generally doing better than they were in 2000, but, looking to 2100, we wonder whether the early policies to increase economic growth will provide the necessary resilience to cope with future surprises.

5.5.2 Order from Strength

Summary: *Since 2000, the availability of ecosystem services has fallen below minimal needs for human well-being in some regions of the world while being maintained or even improved in other regions. Widespread loss of faith in global institutions and fear of terrorism led rich countries to favor policies that ensured security and erected boundaries against outsiders. Even in better-off areas, though, there have been some breakdowns of ecosystem services. It turned out that climate change was often more rapid than response capacity, leading to local degradation of ecosystem services in some places, even in rich nations. Overall, the current global condition of ecosystem services is highly variable and declining on average. Even the places in the best condition are at risk, although citizens of wealthy nations enjoy a tolerable level of ecosystem services and human well-being. As we look to 2100 and beyond, Earth's ecosystem services seem fragmented and imperiled. Problems exist at all scales, from global fisheries collapses to regions of the world where ecosystem services are sorely in need of restoration and other regions where ecosystem services are currently fine but threatened. We have learned that it is impossible to build walls that are high enough to keep out all the world's ills, but also that it is sometimes a reasonable policy to focus minimal resources on carefully protecting a few areas rather than only partially protecting everywhere.*

At the beginning of the twenty-first century, terrorism, war, and loss of trust in global institutions led many people to believe that there was a need for powerful nations to maintain peace and achieve equity. Governments of the industrial world reluctantly accepted that militarily and economically strong democratic nations could maintain global order, protect lifestyles in the industrial world, and provide some benefits for any developing countries that elected to become allies. Countries were often unwilling to participate in international and global institutions as they concentrated on building strength as nations. As a result, global institutions began to stagnate as people lost confidence in them and their power eroded.

The EU and the United States turned inward, striving to preserve national security. Trade policies veered toward increasing protectionism. Religious fundamentalism and nationalism were mutually reinforcing in some nations. In some cases, parts of civil society saw this inward focus as dangerous and tried to oppose it, but they were mostly silenced by already strong national governments. Just as the focus of nations was turned to protecting borders, environmental policies concentrated on securing resources for human consumption. Building strong nations was a priority, as many felt that environmental challenges could not be adequately addressed without first strengthening nations and economies. Conservation focused on parks and preserves.

By sometime around 2018, this had increased the separation between the rich, powerful countries and the poverty-stricken ones, with very few countries left in between. Societies were also stratified within nations: rich and powerful people and poor people existed within both rich and poor nations. Within nations, rich and powerful people increasingly turned to gated communities as a way to protect themselves from outsiders.

In the rich world, the drive for security and protection led to privatization of access to many natural resources, as businesses stepped in to help governments assure consistent access to resources. In turn, governments protected the economic interests of these businesses. This led to increasingly tighter connections between governments and business at all scales. There was also very little trade with poor countries.

The world outside the rich people's walls experienced a lot of conflict during this period. The disputes were largely over access to natural resources like water, oil, and fuelwood. Many in poorer countries felt that the way out was to immigrate to a rich country or become part of the elite in their own country, which historians believe entrenched the compartmentalization. With most poor people spending all their time and energy trying to become one of the elites, there were few left to argue for other priorities. Some elites did demand better treatment of the poor and were sometimes able to effect change. Significant economic problems persisted in the poor world due to corruption, disease, and pollution. As poor countries spent most of their time attending to crises of disease and other problems, widespread improvements in economic well-being became rare. Although fertility had been starting to drop in poor countries at the beginning of the twenty-first century, the collapse of nascent social safety nets resulted in increases in fertility; population growth rates reversed course and began to increase.

Powerful countries often coped with problems by shifting the burdens to other, less powerful countries, increasing the gap between the rich and poor. In particular, resource-intensive industries were moved to poorer countries or to poorer parts of wealthy countries. This taxed poor people's environment further, leading to widespread migration from collapsed places to new parts of poorer countries. This migration created stresses that sometimes led to environmental degradation in the new places. For example, refugees who left one place for another increased the pressure on the new area's environment until it collapsed. Disease, particularly contagious diseases, became rampant in poor areas.

Rich nations also attempted to make their lands more livable by moving food production to poor countries. The price of food rose as conflict in poor areas affected their ability to produce food. In some cases, this led rich nations to attempt to stabilize poorer ones through a combination of military and economic intervention. In other cases, rich nations simply produced more of their own food.

The inward focus of wealthier nations did lead to some benefits, including high levels of protection, easy access to goods and services inside the wealthy areas, and pockets of very well preserved wilderness in rich countries and in places that wealthy people wanted to visit on holiday. The spread of invasive species was also a lot lower than researchers had predicted in 2000, a surprise attributed to the decrease in trade among countries. The rate of successful invasions was higher than in 2000, since degraded ecosystems were more susceptible to successful invasion when exotic species were present.

During times when powerful countries became more assured of their security, they did turn somewhat to global issues, particularly those that would obviously affect themselves. Sometimes funding was made available to help poor countries with particularly pressing problems. The focus for this funding was often on conflicts or refugee problems (which were seen as having secondary impacts on rich countries). Generally, when funding was available for poorer areas, the focus was on physical safety rather than social welfare issues. Some global environmental issues that affected rich countries were addressed in the same way, through cautious agreements among rich nations, and this led to some improvements on global environmental issues. However, progress has been slow on those issues that are not of direct concern to the powerful.

As the attention of governments was on economic and military strength, there was less focus on the environment. Global issues (such as climate change) and international issues (such as large river management) were almost always impossible to address as at least one key nation was unwilling to cooperate. Ironically, global climate change increased less than had been expected at the turn of the century, due to a larger than expected proportion of the world's population being forced to live a simpler and less materialistic existence.

Now, in 2050, some poorer regions have finally gained a reasonable amount of stability, and are finding themselves able to form coalitions and trade agreements to better their situation. Generally, these coalitions have worked well to lift some poor areas out of totally abject poverty. This was especially true for nations that had crossed the digital divide. Some Asian, South American, and African nations had established digital networks, which gave their people an advantage in terms of access to global markets and information. These countries in particular were able to gain more stability. As soon as things start getting better, many people want to immigrate to these areas. Thus, countries often are forced to create strong laws against immigration in order to keep their society safe and orderly. The future of these regions is uncertain.

Today, it is apparent that there was not a linear trend toward higher and higher walls, even though it sometimes felt that way. Instead, we saw episodes of rapid change and periods of relative stability. There were some fluctuations of increasing and decreasing compartmentalization as the powerful countries periodically invested in keeping conditions tolerable for the poor in order to reduce illegal immigration and other problems. There were also activist groups and intellectual dissidents in wealthy nations that tried to support the poor and poor nations. Looking forward to 2100, these activist groups are one of the main sources of hope in an otherwise bleak situation. People and ecosystems are generally doing worse than in 2000, but some hope can be found in the activists working to support the poor and improve management.

5.5.3 Adapting Mosaic

Summary: *The past 50 years have brought a mix of successes and failures in managing ecosystem services. Approaches to management have been heterogeneous. Some regions strengthened the centralized environmental agencies that emerged late in the twentieth century, while others embarked on novel institutional arrangements. Some approaches turned out to be disastrous, but others proved able to maintain or improve ecosystem services. Many nations have emulated the successes of other nations, and the number of successes has begun to climb by 2050. As a result, the world in 2050 is a diverse mosaic with respect to ecosystem services and human well-being. A considerable variety of approaches still exists, and regrettably some regions still cannot provide adequate ecosystem services for their people. Other regions are doing well, and remarkable successes have occurred on every continent. With respect to global-scale environmental problems, progress has been slow. As we look to 2100 and beyond, policy and ecological science face a twin challenge: to rebuild ecosystem services in the regions where they have collapsed and to transfer the lessons of regional success to problems of the global commons.*

Opportunities for, and interest in, learning about socioecological systems were a defining feature of the early twenty-first century. People had great optimism that they could learn to manage socioecological systems better, but they also retained humility about limits to human control and foresight and the prospects for surprise. Learning to improve socioecological systems came at a great cost. There were failures as well as successes, and learning diverted some of society's resources. Economic growth was probably lower than it could have been had decision-makers put all our investments toward manufactured capital, but economic growth has begun to improve recently as the benefits of better socioecological systems are now slowly being realized.

At the turn of the century, some people in the rich world held beliefs that promoted regionalization of trade, nationalism, and local or regional management of natural resources. Global trade barriers for goods and products were increased, but trade barriers decreased within regional blocs such as ASEAN, NAFTA, and the EU. In contrast, global barriers for information flow nearly disappeared due to improving communication technologies and the rapidly decreasing cost of information access. Political focus followed the economic emphasis on regional or national trade.

The regionalization of markets and politics was associated with a decline in the relative power of global international institutions. The decline was partly linked to loss of confidence in the effectiveness of global governance and dissatisfaction with distortions of global markets. But the strengthening of interactions within nations and within regional blocs was also an important factor in the relative deemphasis of global institutions. Dissatisfaction with the results of global environmental summits and other global approaches led many people to perceive global institutions to be ineffective at environmental management. Climate change negotiations had broken down by 2010. International agreements failed to prevent the depletion of most marine fisheries, and regulation of transboundary pollutants proved ineffective.

Within some nations, power devolved to local authorities. There was variation among nations and regions in styles of management, including natural resource management. Some managed with rigid centralized bureaucracies. Others focused on market incentives or other economic measures. Still others attempted some form of adaptive management for the nation or region as a whole. Some local areas explored actively adaptive management, investigating alternatives through experimentation. Some were passively adaptive, investing in a certain amount of monitoring but dealing with change in a reactive way. Still other locales largely ignored the environment, dealing with crises only as they arose.

There was great diversity in the outcome of these varied approaches to managing socioecological systems. Some notable disasters were poorly handled. Sometimes, methods that succeeded in one region failed when imported to another region because of unforeseen differences in social practices, politics, or ecosystems. Reactions to resource breakdowns were also diverse. Perversely, failed practices were sometimes sustained by subsidies from other regions or other sectors of the economy. In other cases, breakdowns were followed by innovations that eventually made things better.

Groups began to experiment with innovative local and regional management practices that put special emphasis on investments into human and social capital, such as education and training. Information about success stories was shared among locations. Information sharing was facilitated by cheap communication tools such as the Internet. The experiments varied in their success. As more and more experience and knowledge were collected, the conditions for success were better understood and experiments became more successful on average. Food production became more localized, feeding into national or regional markets that valued clean, green production processes. Environmental technologies were developed based on local needs and conditions, leading to a gradual improvement in management of socioecological systems and natural resources.

By the 2020s, global tourism had begun to encourage development and application of local learning as a celebration of diversity in reaction against global homogenization and the sameness of products. Traveling was seen as a means to experience heterogeneity, but, in the end, had negative feedbacks due to increased transportation and human impact on poorer regions.

Throughout this period of varied learning, there was relatively little focus on global commons problems such as climate change, marine fisheries, and transboundary pollution. Crucial ecological feedbacks were acting over spatial extents that were too large to be noticed by local institutions. As a result, large-scale environmental crises eventually became more frequent. Technological disasters occurred in some natural resource systems. Climate shifts led to more storm surges in coastal areas. Top predators vanished from most marine ecosystems, leaving jellyfish as the apex predator for vast areas of the world. Coastal pollution increased drastically, which led to further degradation of coastal fisheries and severe health risks to humans from eating shellfish, shrimp, and other filter feeders. There were also outbreaks of new diseases, such as rapidly evolving bacteria resistant to antibiotics. Luckily, climate change was not as bad as it could have been because people were trying to curtail local pollutants like nitrogen oxides and sulfur dioxide, which also act as agents of climate change. But sometimes the global phenomena affected local socioecological systems in severe ways.

At about the same time, businesses became more interested in finding new markets in other parts of the world and consumers began to demand a greater diversity of choices. The renaissance of global business led to greater internationalization of governance and negotiation of new international trade agreements. Some global barriers to trade started to erode, and the economy gradually became more globalized.

The negative large-scale environmental events were largely seen as being caused by inadequate management of the global environmental commons. The growing international framework of trade and political institutions provided a foundation on which global environmental management institutions could be rebuilt. The rebuilding was slow and tenuous, due to slowly changing institutions that often needed disaster as a goad to action. Nevertheless, renewal began. The emerging institutions for international environmental management drew on decades of local and regional experience, including a rich history of successes and failures. The emerging institutions were more focused on ecosystem units than in the early decades of the century. Watersheds, air basins, and coastal regions, rather than states or nations, became the basis for management. New large-scale management was also more cautious, focused on learning while managing, based on the successes that learning had brought to many locales earlier. When two or more regions came together to manage a jointly shared problem, they often participated in deliberate small-scale trials to determine the best management practices.

In the year 2050, Earth's socioecological systems seem poised at a branch point. Local ecosystem management is varied and improving in many regions. While problems exist, the situation is better than in 2000. On the other hand, global environmental problems have become more pressing. It seems possible that new approaches will emerge for addressing them, built in part on the varied experiments of preceding decades. This hope beckons at the dawn of the second half of the twenty-first century.

5.5.4 TechnoGarden

Summary: *Significant investments in environmental technology seem to be paying off. At the beginning of the century, doomsayers felt that Earth's ecosystem services were breaking* *down. As we look back over the past 50 years, however, we see many successes in managing ecosystem services through continually improving technology. Investment in technology was accompanied by significant economic development and education, improving people's lives and helping them understand the ecosystems that make their lives possible. On the other hand, not every problem has succumbed to technological innovation. In some cases, we seem to be barely ahead of the next threat to global life support. Even worse, new environmental problems often seem to emerge from the most recent technological solution, and the costs of managing the environment are continually rising. Many wonder if we are in fact on a downward spiral, where new problems arise before the last one is really solved. As we look to 2100 and beyond, we need to cope with a situation in which problems are multiplying faster than solutions. The science and policy challenge for the next 50 years is to learn how to organize socioecological systems so that ecosystem services are maintained without taxing society's ability to invent and pay for solutions to novel, emergent problems.*

Early in the twenty-first century, increased recognition of the importance of ecosystem services led to increasingly formalized patterns of human/ecological interactions. The trend to formalization led to definition of a wide variety of ecological property rights, which were assigned to a variety of communal groups, states, individuals, and corporations. These rights often prompted ecosystem engineering to maintain provision of the desired ecosystem services. Investment in ecological understanding and natural capital meant that environmental problems were often identified before they became severe.

Such property rights systems eased industrial countries away from protective subsidies and improved income opportunities for developing countries. They also led to increasing government control through "green" taxes and subsidies of research and development. Policies emphasizing research and development led to significant scientific efforts, particularly in the use of technological control to maintain consistent resource flows. There was also a strong belief that "natural capitalism"—a focus on looking for profits in working with nature—could be profitable for both individuals and society. Big business became interested in research and development of new technologies to produce or enhance production of ecosystem services. The impossibility of maintaining exclusive access to information drove ever more rapid innovation during the early period. It was a time of rapid gain and spread of knowledge around the globe. Global communication, combined with open trade policies, allowed the developing world to apply some of the new technologies and start developing their own.

As population continued to grow and demand for resources intensified, people increasingly pushed ecosystems to their limits of production. This ecological engineering was done privately at local, small, or regional scales by a variety of private, public, and community and individual actors and was done within different types of property rights schemes at different locations. Some areas established property rights schemes based on command and control, common property, or market-based schemes, while others remained open access. This engineering was far more sophisticated, subtle, and adaptive than many traditional attempts at ecological engineering. The new ecological engineers were schooled in the engineering approach of "fast, cheap, and out of control" and used advances in computer, communication, and materials sciences to permit human infrastructure to be increasingly flexible, dynamic, and adaptive, like wild ecosystems. Innovations such as pop-up infrastructure allowed people to intervene in ecological dynamics rapidly and flexibly.

In response to negative consequences of intensive agriculture in the industrial world—including land degradation, eutrophication of lakes and estuaries, and disease outbreaks—demand for ecological agriculture began to increase. In the 1990s, governments in several European countries had already begun to change or remove agricultural subsidies following a series of agricultural crises in Europe (mad cow disease, foot-and-mouth disease, swine fever, contamination of food with halogenated organic compounds).

Ecological agriculture unfolded in two intertwined planes. Due to the increasing focus on ecosystem services, people began to realize that agricultural systems were embedded within landscapes and that agriculture could not just produce food or fiber at the expense of all other potential services. This led to policies that encouraged farmers to create a landscape that produced a variety of ecosystem services rather than focusing on food as a single service. The goal of multifunctionality moved government agricultural policy away from a focus on the volume of agriculture production to a focus on agricultural profitability. Despite initial concerns that multifunctional agriculture would destroy farming as a way of life and reduce yields, its profitability and lowered risk encouraged many farmers in Europe and North American to convert their operations. This trend began in the 1990s, and its expansion first in Europe and then North America meant that by 2010 nearly half of European and 10% of North American farms were focusing on a multifunctional existence. By 2025, these numbers had jumped to nearly 90% in Europe and 60% in North America. The diversification of agricultural production and lower yields increased the profitability of farming—particularly smaller-scale farming—and reduced the power of large-scale agribusiness.

Ecological agriculture and the end of widespread subsidies opened the rich world to agricultural inputs from poor countries, and this spurred radical changes in agriculture in Eastern Europe and later in Africa and Latin America. Increased ability of developing countries to export agricultural production encouraged investment in intensification. The demand of industrial countries for at least nominally safe and ecologically friendly production helped stimulate intensification efforts to increase production in environmentally friendly ways. Some of these developments came from the use of genetically modified crops. Despite initial opposition in the EU, the absence of all but a few minor ecological problems led to their widespread use. As crop production for the developing world remained somewhat less sensitive to ecological issues, some local ecological degradation resulted from the agricultural intensification. Water pollution, eutrophication, deforestation, and erosion became significant problems in some locations.

These changes did not happen evenly across the entire world. For example, development of green agriculture spread most rapidly in North European countries. East European countries were well positioned to export agricultural products to the EU and were the first to intensify. In Africa the situation was quite heterogeneous; some countries in southern Africa intensified their agricultural production rapidly, while other African countries were unable to respond to these opportunities due to local problems in governance, lack of infrastructure, or water shortages and droughts.

The engineering approach took hold in urban and suburban areas, too. The best urban management focused on creating low or positive impact on ecosystems using green architecture and on diverse transportation strategies and urban parks as functional ecosystems. In rich countries, new housing developments begin to include rain gardens and wetland areas to clarify runoff and provide wildlife habitat. The specific activities that people engaged in varied by location, based on the ecosystem services they desired and the difficulty of providing those services. In general, rich countries focused on providing water regulation services and cultural services, while developing countries focused more on the production and regulation of water and the production of provisioning services. Regional differences within rich and poor worlds continued to exist due to culture, governance, environmental factors, and the way that property rights were organized.

The highly managed urban garden approach sometimes led to destruction of local, rural, and indigenous cultures. Since the dominant values tended to be functional, culture for culture's sake was not highly valued. The degree of this loss was variable across regions, but some cultural loss was inevitable everywhere. This lowered the adaptive capacity of local ecosystem management by diminishing society's capability to detect subtle changes in local ecological processes, particularly in terms of detecting gradual changes in slow processes. On the other hand, sensitive and cheap ecological monitoring did allow for the rapid accumulation of short-term ecological knowledge.

Highly engineered systems turned out to be very vulnerable to disruptions, however. Even successful management was at risk from loss of process diversity, loss of local knowledge, and people's dependence on stable, consistent supplies of ecosystem services. Ecosystems tended to be simplified because the more obscure and apparently unimportant processes were not supported or maintained. At the same time, increasing social reliance on the provision of ecosystem services led to declines in alternative mechanisms of supplying them. These factors combined to greatly increase the risk of a major breakdown in provision of ecosystem services. The problems were especially severe at the boundaries between ecosystems and across scales, where local effects of management interacted with large-scale fluctuations in ecosystem conditions and function.

Looking back from the year 2050, it seems that we did a pretty good job managing and understanding a rapidly changing world. There are some persistent or growing social and ecological problems, like the loss of local knowledge about ecosystem services and eutrophication of fresh waters and coastal oceans. But in general people around the world have better access to resources and we seem to be thinking more about multifunctionality and systems approaches rather than single goals. Looking forward to 2100, there is great hope for continuing improvement in ecosystem management. We will need to cope with a situation in which problems (caused by new technologies) are sometimes multiplying faster than solutions. The science and policy challenge for the next 50 years is to learn how to organize socioecological systems so that ecosystem services are maintained without taxing society's ability to invent and pay for solutions to novel, emergent problems.

5.6 Potential Benefits and Inadvertent Negative Consequences of the Scenarios

Each scenario illustrates the potential benefits and potential risks inherent in the path of each particular storyline. (See Table 5.2.) It is important to note that each scenario emerges from the complex interactions of billions of people and millions of institutions, not from the action of a centralized global controller. The world cannot be directed in one of these four ways, but it could self-organize in one of the ways envisioned by the scenarios or in some hybrid of the four scenarios. At the level of individuals and nations, decisions by people will affect this self-organization of the Earth system.

It is reasonable to consider the relative benefits and negative consequences of the scenarios. These are important for those who are considering their own decisions in the context of the scenarios. Also, there are decisions that could tip the world incrementally toward one scenario or another, and decision-makers may wish to take this into account. Finally, we found that individuals hold contrasting views about the desirability of different paths toward sustainability, and by considering benefits and risks we contribute to the dialogue among contrasting points of view.

Global Orchestration shows some obvious positives. Economic prosperity, global economic growth, and increased equity may lead to higher human well-being around the world. If this wealth leads to increased demand for a better environment or to higher capacity to create a better environment, ecosystems may be restored or better protected. As with all paths to the future, there is the potential for inadvertent negative consequences. Increased wealth may not lead to increased demand for a better environment, but only to increased demand for ecosystem services, which could degrade ecosystems through overuse. The focus on global issues in this scenario and the top-down delivery of globally orchestrated policies comes at the expense of local

Table 5.2. Benefits and Inadvertent Consequences of Four Scenarios

Scenario	Potential Benefits	Inadvertent Consequences
Order from Strength	increased security for those who can afford it political and trade barriers slow the spread of invasive species and some diseases some regions remain well-endowed with ecosystem services	lower economic growth because of fragmentation, inequality, conflict, and lost human potential risk of security breaches (from middle well-off countries) environmental degradation of the global commons, and losses of ecosystem services in poor regions vulnerability due to fragmentation of ecosystem services
Global Orchestration	economic prosperity and increased equality due to more efficient global markets wealth increases demand for a better environment and the capacity to create a better environment	progress on global environmental problems may be insufficient to sustain local and regional ecosystem services breakdowns of ecosystem services create inequality (disproportionate impacts on the poor reactive management may be more costly than preventive or proactive management
TechnoGarden	highly efficient management and utilization of ecosystems technological enhancement of ecosystem services forward-looking market mechanisms efficiently allocate ecosystem services	increasing reliance on particular technologies may decrease the diversity of systems for providing ecosystem services, thereby increasing vulnerability to surprising breakdowns some technological innovations create the need for new technological innovations wilderness disappears as "gardening" of nature increases, and people have fewer experiences of nature less economic growth because of diversion of resources to environmental technology
Adapting Mosaic	integration of management institutions with ecological processes to improve the resilience of ecosystem services growth of adaptive capacity to sustain ecosystem services in a changing world	little progress on global ecosystem problems less economic growth than maximum possible because of regionalization of economies and inefficiencies of experimentation

and regional flexibility. Progress on global environmental problems may not be enough to sustain local ecosystem services, and without flexibility, these local issues may not be appropriately addressed. Finally, people and institutions in this scenario are generally reactive to environmental problems rather than proactive. Such reactive management may be more costly than preventive management and may experience costly failures in some cases.

Order from Strength has some adverse outcomes for ecosystem services and human well-being. But there are also some possible positive outcomes for ecosystem services. Lower international trade may mean that fewer invasive species are transported. It may also mean that fewer diseases are spread or that diseases are not spread as quickly or as far as they might be in a more globally connected world. The scenario implies that some wealthy people might have high levels of security and that some ecosystems in wealthy areas might be well protected. The potential inadvertent adverse outcomes are more obvious. Fragmentation, inequality, and conflict may lead to lower economic growth and lost human potential. Security may not be high because pressures from the dispossessed will be extremely high. A globally fragmented world also risks degradation of the global commons and problems caused by fragmentation of ecosystems. Severe losses of ecosystems and their services could occur in some areas.

Adapting Mosaic focuses on flexibility locally and regionally. Local empowerment allows management to be proactive with respect to addressing ecosystem management and to integrate management institutions with ecological processes to improve the resilience of ecosystem services. Because the benefits of ecosystem services are allocated fairly, management institutions tend to focus on the current and future provision of ecosystem services. Also, the focus of most people in the scenario is on adaptive capacity, which may help management institutions approach change more flexibly and better sustain provision of services in a rapidly changing world. However, as with all scenarios, this one has potential for unintentional negatives. The high degree of focus on local and regional management leads to less progress on global problems than in a more globalized world. Also, there is less than the maximum possible economic growth because of the regionalization of economics and the inefficiencies of experimentation.

TechnoGarden uses technology to maintain and improve the provision of ecosystem services. The benefits are a highly efficient utilization of ecosystems for service provision of targeted services and actual enhancement of the services provided. This scenario also includes forward-looking market mechanisms, such as futures markets for ecosystem services and appropriate systems of property rights to allocate and manage ecosystem services efficiently. When conditions are stable and predictable, the provision of services is high and extremely reliable. However, increasing reliance on technologies decreases the diversity of systems that provide any one service and increases the number of connections, thus increasing the vulnerability to unexpected environmental or social changes. Some of these technological innovations will create problems that lead to the need for new technological innovations to solve the problems created by the previous innovation. Finally, the "gardening" approach to natural resource management and provision of ecosystem services may cause wilderness to disappear and people to have fewer experiences of nature and wildness. People who are less familiar with ecosystems may be less likely to understand the processes that build resilient and sustainable ecosystem services.

5.7 Breakdowns of Ecosystem Services in the Four Scenarios

Interruptions, breakdowns, and surprising changes in ecosystem services have occurred throughout human history and occur in all plausible scenarios of ecosystem futures. Rapid, potentially irreversible changes are an important feature of ecosystems that can confound human capacity for prediction and control. (See Chapter 3.) Surprises related to ecosystem dynamics were identified by MA interviewees as an area of concern. The different scenarios are associated with varying patterns of disturbance to ecosystem services.

Probability distributions of extreme events are one way of describing the differences among scenarios with respect to surprises. Suppose that all disturbances of ecosystem services were documented during each year for Earth as a whole and ranked in magnitude by the number of people affected by the disturbance. Given such data from many years, a distribution could be constructed showing the likelihood of extreme ecosystem events as a function of their magnitude. Distributions of extreme ecosystem events consistent with each scenario are presented in Figure 5.3. These distributions illustrate our qualitative inferences about extreme events in the scenarios. They are not based on data, because no appropriate data or global models exist. These distributions are integrated to produce the cumulative probability diagrams shown in Figure 5.4.

The scenarios are expected to be different in the frequency and magnitude of surprising changes in ecosystem services. In Figure 5.3A, the magnitude of an ecosystem disturbance is measured by the number of people it affects (x-axis). Because of the great range in event severity, we use a logarithmic (base-10) scale for the x-axis. The likelihood of a disturbance of a given size is given by the corresponding y-axis value.

In Order from Strength, in which people have a reactive and geographically limited approach to sustaining ecosystem services, there is a high chance of extreme disturbances. That is, extreme disturbances of ecosystem services have a moderately wide range with a rather high modal value (see Figure 5.3A). In Figure 5.4, the Order from Strength line is far to the right of all other lines, indicating that there is a high probability of a large disturbance event. Most of the human population inhabits relatively impoverished regions with deteriorating ecosystem services, and this situation is

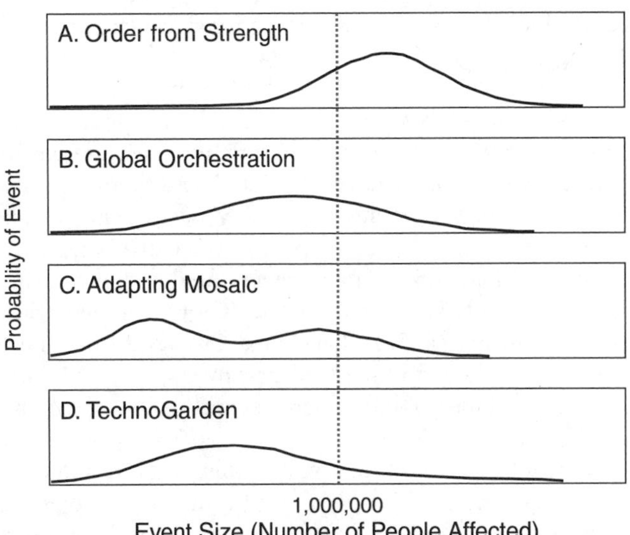

Figure 5.3. Distributions of Extreme Events during MA Scenarios. The x-axis is the magnitude of the disturbance of ecosystem services, measured by the number of people affected. The y-axis is the likelihood of an extreme ecosystem event of a given magnitude. The total area under each curve is the same, because for each scenario the probabilities of all event magnitudes must sum to 1. Order from Strength has a very high probability of extreme events affecting just over one million people. Global Orchestration has a moderate probability of extreme events affecting a small number of people due to regional breakdowns in ecosystem services. It has a somewhat lower, but still significant, probability of larger, multi-region breakdowns. TechnoGarden has a moderate to high probability of relatively small events and a low but significant probability of breakdowns that affect extremely large numbers of people.

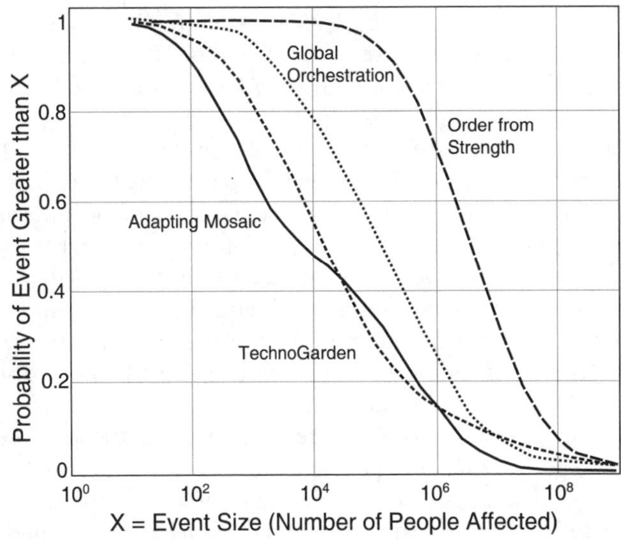

Figure 5.4. Cumulative Probability Distributions of Extreme Events. These distributions are derived from the distributions in Figure 5.3. The x-axis is the number of people affected by a given event, and the y-axis is the probability of an event in which more people are affected.

reflected in breakdowns that affect a relatively large number of people.

Global Orchestration, in which the primary approach is fair trade and global policies to ameliorate poverty, has slightly less chance of extreme disturbances because wealth is greater, the world is more connected, and there is greater capacity to react to events when they occur. Also, breakdowns tend to affect fewer people than in Order from Strength. In Figure 5.3B, this is represented in the comparable range but a lower modal value for Global Orchestration. Similarly, in Figure 5.4, the Global Orchestration line is somewhat to the left of the Order from Strength line, indicating that the probability of large events will be somewhat smaller.

In Adapting Mosaic, local vulnerability leads to some extreme events that affect only a small number of people. At the same time, diminished attention to the global commons underlies a small number of extreme events that affect large numbers of people. These large breakdowns are less common than in Order from Strength or Global Orchestration. This is represented as a bimodal distribution of extreme events (see Figure 5.3C). Local adaptation reduces the number of ecosystem service breakdowns that affect large numbers of people. Some regions become vulnerable, and in many years the most extreme breakdowns affect modest numbers of people in these vulnerable regions. At the same time, management of global commons problems, such as the atmosphere and marine pelagic fisheries, tends to be neglected in Adapting Mosaic. Consequently, in some years breakdowns of ecosystem services affect relatively large regions and relatively large numbers of people, thereby creating the second mode in the curve of Figure 5.3C. This bimodal distribution can be seen in the changing slope of the Adapting Mosaic line in Figure 5.4. Also note that the line is to the left of the graph, indicating that most disturbance events affect only a small number of people.

TechnoGarden has the widest distribution of ecosystem event magnitudes (Figure 5.3D). The typical extreme event affects fewer people than Global Orchestration or Order from Strength, but there are many more of these events than in any other scenario. This is also shown in Figure 5.4: the right-side tail of the TechnoGarden line is higher than the lines for all other scenarios. The modal value of extreme breakdowns is lower than the mode for Global Orchestration and lies between the modes for Adapting Mosaic. However, the distribution is widely dispersed; in many years, the most extreme breakdowns of ecosystem services are as large as the upper mode of Adapting Mosaic or the mode in Order from Strength.

The probability of an extreme event that affects more than a given number of people is the area of the curve to the right of that number of people. The vertical dotted line in Figure 5.3 indicates extreme events that affect 1 million people. Thus the area of each curve to the right of the line is the probability of extreme ecosystem events that affect at least a million people. These areas are collected in Figure 5.5. Extreme events that affect at least 1 million people are most common in Order from Strength. They are less common in Global Orchestration. Extreme events are least common in Adapting Mosaic and TechnoGarden, but for different reasons. In Adapting Mosaic, the emphasis of local, not global, commons problems means that there are some large-scale breakdowns of ecosystem services. In TechnoGarden, the emphasis of high efficiency and rigid control makes ecosystem management vulnerable to unexpected events.

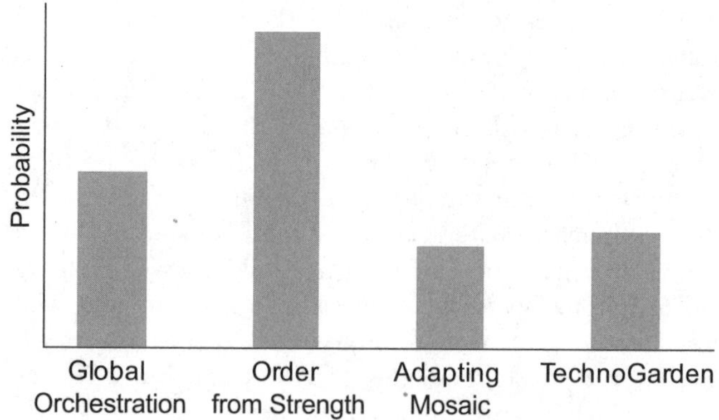

Figure 5.5. Probabilities of Extreme Ecosystem Events Affecting at Least 1 Million People. Derived from Figure 5.3.

The impact of an extreme disturbance of ecosystem services will depend on society's capacity to respond, compensate, and adapt to the disturbance. These capacities are expected to differ among the scenarios. In Global Orchestration, there is good capacity to respond to disturbances after the fact, but little attention to addressing underlying causes of ecosystem disturbances. In Order from Strength, rich nations may have considerable capacity to respond to internal disturbances, but the capacity to respond to disturbances in poor nations may be much less. In Adapting Mosaic, local and regional institutions create considerable capacity to address disturbances at those spatial scales, but the Earth system is more vulnerable to global disturbances that affect a relatively large number of people. In TechnoGarden, technology provides a capacity to address some kinds of disturbances, but it also creates new vulnerabilities to the possibility of novel disturbances. The complex interactions of disturbance regimes and capacities to respond or adapt give rise to many of the complex dynamics that are thought to occur in the scenarios.

5.8 Transitions among the Scenarios

The scenarios are not predictions. The future of ecosystem services will likely have elements from each of the four scenarios. Indeed, the roots of all four scenarios are evident in the present. Some of our interviewees see tendencies toward Order from Strength in current events. Others see the potential to change the world now through global policies, adaptive local management, and technological innovations. Each scenario proceeds like a river in its own unique channel, but in actuality global dynamics will be more like a braided river, with different channels connecting at some times and diverging at other times. Table 5.3 presents some events that could cause one scenario to branch into another one.

Global Orchestration could branch into Order from Strength if global economic agreements break down, if

Table 5.3. Potential for Each Scenario to Branch into Another Scenario

Scenario	Order from Strength	Global Orchestration	Adapting Mosaic	TechnoGarden
Order from Strength		emergence of strong global institutions; these institutions are recognized as legitimate by most nations, while the derive capacity from the economic, political, and military power of wealthy nations; wealthy nations recognize that their societies cannot be sustained in isolation, and that global reform is necessary	strong regional economic blocs develop; in some regions, there is growing recognition of the importance of ecosystem services and the will to invest in learning to sustain ecosystem services; this recognition spreads, slowly and patchily	same as for Global Orchestration, but more emphasis on environmental technology as the key to building ecosystem services for human well-being
Global Orchestration	globalization of the economy stalls; global agreements break down, including those related to the environment; conflict and nationalism spread; wealthy nations look inward		globalization of the economy gives way to stronger regional blocs; recognition that local ecosystem services are critical for human well-being; devolution of property rights and responsibility for ecosystem services to local authorities	recognition that human well-being depends on ecosystem services and that technology can be used to manage ecosystem services more efficiently; rapid growth of investment in the environmental technology sector
Adapting Mosaic	spreading conflict overtakes the collective problem-solving necessary for adaptive management of ecosystem services	increased connectivity of the global economy and expansion of global institutions are driven by growing recognition of the economic opportunities from expanded international trade and by appreciation of common interest in solving global problems of inequity, hunger, disease, and breakdown of global environmental commons		same as for Global Orchestration, but more emphasis on environmental technology as the key to building ecosystem services for human well-being
TechnoGarden	globalization of the economy stalls; global agreements break down, including those related to the environment; conflict and nationalism spread; wealthy nations look inward	environmental technology sector does not compete well economically, so it does not expand to the level envisioned in TechnoGarden	globalization of the economy gives way to stronger regional blocs; recognition that low controllability and low predictability of ecosystem services favor experimental management with multiple approaches and diversified ecosystems; devolution of property rights and responsibility for ecosystem services to local authorities; loss of economies of scale for technological solutions, and loss of confidence in large-scale technological fixes	

conflict, fundamentalism, and nationalism spread, and if rich nations look inward. Transitions of this type have been considered in previous global scenario exercises. (See Chapter 2.) On the other hand, if globalization gave way to regionalization of economic activity combined with devolution of authority for ecosystem services to institutions at appropriate scales, the resulting system would resemble Adapting Mosaic more than Order from Strength. If in Global Orchestration a strong technological sector emerged, and if society were generally enthusiastic about technological approaches to environmental needs, the system could branch toward TechnoGarden.

It is more difficult to imagine transitions away from Order from Strength, because low economic growth, social breakdown, and environmental degradation would reduce the store of capital necessary for global transformation. If wealthy societies recognized that isolation were no longer sustainable, perhaps they would have the capacity to build global institutions that could move the system toward Global Orchestration. Alternatively, stronger regional economic blocs could develop, at least in some parts of the world. If this economic growth were coupled with investment in human, social, and natural capital and with the development of appropriate institutions for ecosystem management, the system could move toward Adapting Mosaic. This transformation, however, would probably encompass less of the world than envisioned in the global scenario for Adapting Mosaic. In wealthy parts of the world, investments in environmental technology might lead to a sort of TechnoGarden. However, the dispersion of technological innovations globally would probably not occur unless global institutions were expanded. Thus in this family of scenarios, Order from Strength acts like a basin of attraction—it is easier to understand how the global system might move into Order from Strength than it is to understand how the system might move out of it.

The Adapting Mosaic scenario could branch toward Global Orchestration if there were sufficient impetus from transnational economic activity or if global commons problems were perceived as more pressing and urgent. Indeed, a movement toward a more multiscaled sort of Global Orchestration is envisioned near the end of the Adapting Mosaic scenario. On the other hand, Adapting Mosaic could shift toward Order from Strength if slow economic growth exacerbated conflict, fundamentalism, or nationalism. If the diverse approaches to ecosystem management led to successful technological innovations, technology could become an important part of Adapting Mosaic. This would move the system toward TechnoGarden, although the focus would be on local ecosystem management instead of the global focus envisioned in the TechnoGarden scenario.

The TechnoGarden scenario could branch toward Global Orchestration if the environmental technology sector of the economy does not compete well and fails to expand to the level envisioned in the TechnoGarden scenario. The events that could cause TechnoGarden to branch toward Order from Strength are similar to those for Global Orchestration. If globalization stalls, if conflict, fundamentalism, and nationalism expand, and if wealthy nations look inward, the system could move toward Order from Strength. TechnoGarden could move toward Adapting Mosaic if regional trading blocs became stronger. Also, if technological failures led people to think that ecosystems were not predictable and controllable, ecosystem management could move toward diversified adaptive approaches. This would involve devolution of property rights and authority to appropriately scaled institutions. If such changes occurred, the world of TechnoGarden could branch toward that of Adapting Mosaic.

References

Bai, Y.F., X.G. Han, J.G. Wu, Z.Z. Chen, and L.H. Li, 2004: Ecosystem stability and compensatory effects in the Inner Mongolia grassland, *Nature,* **431,** pp. 181–84.

Bellwood, D.R., T.P. Hughes, C. Folke, and M. Nystrom, 2004: Confronting the coral reef crisis, *Nature,* **429,** pp. 827–33.

Bengtsson, J., P. Angelstam, T. Elmqvist, U. Emanuelsson, C. Folke, et al., 2003: Reserves, resilience and dynamic landscapes, *Ambio,* **32,** pp. 389–96.

Bennett, E.M., G.D. Peterson, and E.A. Levitt, Looking to the future of ecosystem services, *Ecosystems.* In press.

Berkes, F., J. Colding and C. Folke (eds.), 2003: *Navigating Social-Ecological Systems,* Cambridge University Press, Cambridge, UK.

Cardinale, B.J., A.R. Ives, and P. Inchausti, 2004: Effects of species diversity on the primary productivity of ecosystems: Extending our spatial and temporal scales of inference, *Oikos,* **104,** pp. 437–50.

Carpenter, S.R., 2003: *Regime Shifts in Lake Ecosystems,* Excellence in Ecology, Volume 15, Ecology Institute, Oldendorf/Luhe, Germany.

Chapin, F.S. III, E.S. Zavaleta, V.T. Eviner, R.L.Naylor, P. M. Vitousek, et al., 2000: Consequences of changing biodiversity, *Nature,* **405,** pp. 234–42.

Dasgupta, P. and K.G. Mäler, 2000: Net national product, wealth and social well-being, *Environment and Development Economics,* **5,** pp. 69–93.

Dasgupta, P. and K.G. Mäler, 2001: Wealth as a criterion for sustainable development, *World Economics,* **2,** pp. 19–44.

Dörner, D., 1996: *The Logic of Failure,* Metropolitan Books, New York, NY.

Ehrlich, P.R. and A.H. Ehrlich, 2004: *One with Nineveh: Politics, Consumption, and the Human Future,* Island Press, Washington, DC.

Elmqvist, T., C. Folke, M. Nyström, G. Peterson, J. Bengtsson, et al., 2003: Response diversity, ecosystem change and resilience, *Frontiers in Ecology and the Environment,* **1,** pp. 488–94.

Foley J.A., M.T. Coe, M. Scheffer, and G. Wang, 2003: Regime shifts in the Sahara and Sahel: Interactions between ecological and climatic systems in Northern Africa, *Ecosystems,* 6, pp. 524–39.

Folke, C., S. Carpenter, B. Walker, M. Scheffer, T. Elmqvist, et al., 2004: Regime shifts, resilience, and biodiversity in ecosystem management, *Annual Review of Ecology, Evolution, and Systematics,* **35,** pp. 557–81.

Folke, C., S. Carpenter, B. Walker, M. Scheffer, T. Elmqvist, et al., 2005: Regime shifts, resilience and biodiversity in ecosystem management, *Annual Review of Ecology Evolution and Systematics,* **35,** pp. 557–81.

Frost, T.M., S.R. Carpenter, A.R. Ives, and T.K. Kratz, 1995: Species compensation and complementarity in ecosystem function. In: *Linking Species and Ecosystems,* C. Jones and J. Lawton (eds.), Chapman and Hall, New York, NY, pp. 224–39.

Groffman, P.M., J.S. Baron, T. Blett, A.J. Gold, I. Goodman, et al., 2005: Ecological thresholds: The key to successful environmental management or an important concept with no practical application? *Ecosystems.* In press.

Grossman, S., 1989: *The Informational Role of Prices,* MIT Press, Cambridge, MA.

Gunderson, L.H. and C.S. Holling (eds.), 2002: *Panarchy,* Island Press, Washington DC.

Gunderson, L.H., C.S. Holling and S.S. Light (eds.), 1995: *Barriers and Bridges to the Renewal of Ecosystems and Institutions,* Columbia University Press, New York, NY.

Havlicek, T. and S.R. Carpenter, 2001: Pelagic species size distributions in lakes: Are they discontinuous? *Limnology and Oceanography,* **46,** pp. 1021–33.

Holling, C.S., 1973: Resilience and stability of ecological systems, *Annual Review of Ecology and Systematics,* **4,** pp. 1–23.

Holling, C.S., 1992: Cross-scale morphology, geometry and dynamics of ecosystems, *Ecological Monographs,* **62,** pp. 447–502.

Holling, C.S. and G.K. Meffe, 1996: Command and control and the pathology of natural resource management, *Conservation Biology,* **10,** pp. 328–337.

Homer-Dixon, T.F., 2000: *The Ingenuity Gap,* Knopf, New York, NY.

Hughes, T.P., A.H. Baird, D.R. Bellwood, M. Card, S.R. Connolly, et al., 2003: Climate change, human impacts, and the resilience of coral reefs, *Science,* **301,** pp. 929–33.

Hulot, F.D., G. Lacroix, F.O. Lescher-Moutoue, and M. Loreau, 2000: Functional diversity governs ecosystem response to nutrient enrichment, *Nature,* **405,** pp. 340–44.

Ives, A.R. and B.J. Cardinale, 2004: Food-web interactions govern the resistance of communities after non-random extinctions, *Nature,* **429,** pp. 174–77.

Ives, A.R., J.L. Klug, and K. Gross, 1999: Stability and variability in competitive communities, *Science,* **28,** pp. 542–44.

Ives, A.R., J.L. Klug, and K. Gross, 2000: Stability and species richness in complex communities, *Ecology Letters,* **3,** pp. 399–411.

Janssen, M. and H.J.M. deVries, 1998: The battle of perspectives: A multi-agents model with adaptive responses to climate change, *Ecological Economics,* **26,** pp. 43–65.

Kinzig, A.P., S.W. Pacala, and D. Tilman (eds.), 2002: *The Functional Consequences of Biodiversity,* Princeton University Press, Princeton, NJ.

Loreau, M., N. Mouquet, and A. Gonzalez, 2003: Biodiversity as spatial insurance in heterogeneous landscapes, *Proceedings of the National Academy of Sciences,* **100,** pp. 12765–70.

Loreau, M., S. Naeem, and P. Inchausti (eds.), 2002: *Biodiversity and Ecosystem Functioning: Synthesis and Perspectives,* Oxford University Press, London, UK.

Ludwig, D., 2001: The era of management is over, *Ecosystems,* **4,** pp. 758–64.

Munasinghe, M. and W. Shearer (eds.), 1995: Defining and measuring sustainability: The biogeophysical foundations, Distributed for the United Nations University by the World Bank, Washington, DC.

National Research Council, 1999: *Our Common Journey,* National Academy Press, Washington, DC.

National Research Council, 2002: *The Drama of the Commons,* National Academy Press, Washington, DC.

Norberg, J., 2004: Biodiversity and ecosystem functioning: A complex adaptive systems approach, *Limnology and Oceanography,* **49,** pp. 1269–77.

Nyström, M. and C. Folke, 2001: Spatial resilience of coral reefs, *Ecosystems,* **4,** 406–17.

Oreskes, N., 2003: The role of quantitative models in science. In: *Models in Ecosystem Science,* C. Canham, J. Cole, and W. Lauenroth (eds.), Princeton University Press, Princeton, NJ, pp. 13–31.

Ostrom, E., 1990: *Governing the Commons: The Evolution of Institutions for Collective Action,* Cambridge University Press, Cambridge, UK.

Paine, R.T., M.J. Tegner, and E.A. Johnson, 1998: Compounded perturbations yield ecological surprises, *Ecosystems,* **1,** pp. 535–45.

Peterson, G., C.R. Allen, and C.S. Holling, 1998: Ecological resilience, biodiversity, and scale, *Ecosystems,* **1,** pp. 6–18.

Pielke, R.A., Jr., 2003: The role of models in predictions for decision. In: *Models in Ecosystem Science,* C. Canham, J. Cole, and W. Lauenroth (eds.), Princeton University Press, Princeton, NJ, pp.111–38.

Sarewitz, D., R.A. Pielke Jr., and R. Byerly, Jr., 2000: *Prediction: Science, Decision Making, and the Future of Nature,* Island Press, Washington, DC.

Scheffer, M., S. Carpenter, J. Foley, C. Folke, and B. Walker, 2001: Catastrophic shifts in ecosystems, *Nature,* **413,** pp. 591–6.

Scott, J.C., 1998: *Seeing Like a State: How Certain Schemes to Improve the Human Condition Have Failed,* Yale University Press, New Haven, CT.

Stern, D.I., 1998: Progress on the environmental Kuznets curve? *Environment and Development Economics,* **3,** pp. 173–96.

Tenner, E., 1997: *Why Things Bite Back: Technology and the Revenge of Unintended Consequences,* Vintage, NY.

Turner, M.G., V.H. Dale, and E.H. Everham III, 1997: Crown fires, hurricanes and volcanoes: A comparison among large-scale disturbances, *BioScience,* **47,** pp. 758–68.

Van der Heijden, K., 1996: *Scenarios: The Art of Strategic Conversation,* John Wiley and Sons, Inc., New York, NY.

Wackernagel, M., and W.E. Rees, 1995: *Our Ecological Footprint: Reducing Human Impact on the Earth,* New Society Publishers, Gabriola Island, BC.

Wade, R., 1988: *Village Republics: Economic Conditions for Collective Action in South India,* ICS Press, San Francisco, CA.

Walker, B., 1993: Rangeland ecology: Understanding and managing change, *Ambio,* **22,** pp. 2–3.

Walker, B.H., A.P. Kinzig, and J. Langridge, 1999: Plant attribute diversity, resilience, and ecosystem function: The nature and significance of dominant and minor species, *Ecosystems,* **2,** pp. 95–113.

Walker, B. and J.A. Meyers, 2004: Thresholds in social-ecological systems: A developing data base, *Ecology and Society,* **9** [Online]. Available at http://www.ecologyandsociety.org/vol9/iss2/art3.

Walters, C.J., 1986: *Adaptive Management of Renewable Resources,* MacMillan, New York, NY.

Yachi, S. and M. Loreau, 1999: Biodiversity and ecosystem productivity in a fluctuating environment: The insurance hypothesis, *Proceedings of the National Academy of Sciences,* **96,** pp. 1463–8.

Chapter 6

Methodology for Developing the MA Scenarios

Coordinating Lead Authors: Joseph Alcamo, Detlef van Vuuren, Claudia Ringler
Lead Authors: Jacqueline Alder, Elena Bennett, David Lodge, Toshihiko Masui, Tsuneyuki Morita,* Mark Rosegrant, Osvaldo Sala, Kerstin Schulze, Monika Zurek
Contributing Authors: Bas Eickhout, Michael Maerker, Kasper Kok
Review Editors: Antonio Alonso Concheiro, Yuzuru Matsuoka, Allen Hammond

---

---

*Deceased.

BOXES

FIGURES

TABLES

*This appears in Appendix A at the end of this volume.

Main Messages

The Millennium Ecosystem Assessment scenarios break new ground in global environmental scenarios by explicitly incorporating both ecosystem dynamics and feedbacks. The goal of the MA is to provide decision-makers and stakeholders with scientific information on the links between ecosystem change and human well-being. Scenarios are used in this context to explore alternative futures on the basis of coherent and internally consistent sets of assumptions. The scenarios are novel in that they incorporate feedbacks between social and ecological systems and consider the connections between global and local socioecological processes.

The approach to scenario development used in the MA combines qualitative storyline development and quantitative modeling. In this way, the scenarios capture the aspects of ecosystem services that are possible to quantify, but also those that are difficult or even impossible to express in quantitative terms. The MA developed scenarios of ecosystem services and human well-being to 2050, with selected results up to 2100. Scenarios were developed in an iterative process of storyline development and modeling. The storylines covered many complex aspects of society and ecosystems that are difficult to quantify, while the models helped ensure the consistency of the storylines and provided important numerical information where quantification was possible.

In the MA, scenarios were partly quantified by using linked global models to ensure integration across future changes in ecosystem services. Available global models do not allow for a comprehensive assessment of the linkages among ecosystem change, ecosystem services, human well-being, and social responses to ecosystem change. To assess ecosystem change for a larger set of services, several global models were linked and run based on a consistent set of scenario drivers to ensure integration across future changes in ecosystem services.

While advances have been made by the MA in scenario development to explore possible futures of the linkages between ecosystem change and society, still further progress is possible. Using quantitative and qualitative tools, the MA scenarios cover a large number of ecological services and drivers of ecosystem change. In the course of developing these scenarios, we also identified areas where analytical tools are relatively weak. For quantification of ecosystem service scenarios, we particularly need models that further disaggregate services to local scales, address cultural and supporting ecosystem services, and consider feedbacks between ecosystem change and human development.

6.1 Introduction

The goal of the Millennium Ecosystem Assessment is to provide decision-makers and stakeholders with scientific information on the links between ecosystem change and human well-being. The MA focuses on ecosystem services (such as food, water, and biodiversity) and on the consequences of changes in ecosystems for human well-being and for other life on Earth. Ecosystem change, on the other hand, is significantly affected by human decisions, often over long time horizons (Carpenter 2002). For example, changes in soils or biodiversity of long-lived organisms can have legacy effects that last for decades or longer. Thus it is crucially important to consider the future when making decisions about the current management of ecosystem services.

The MA developed scenarios to provide decision-makers and stakeholders with scientific information on the links between ecosystem change and human well-being. This chapter describes the methodology used to develop the scenarios. It first provides background information on scenarios in general, followed by an overview of the methodology used in the MA scenario development. The development of the qualitative storylines and the global modeling exercise are then described in detail. Finally, we briefly describe how uncertainty and scale issues were handled in the scenarios. Eight Appendixes provide detailed descriptions of the models we relied on.

6.2 Background to the MA Scenarios

Applied (natural and social) sciences have used many methods for devising an understanding of the future, including predictions, projections, and scenario development. (For an overview of methods, see, e.g., Glenn and Gordon 2005.) Each approach has its own methodology, levels of uncertainty, and tools for estimating probabilities. It should be noted that these terms are often not strictly separated in the literature. The conventional difference, however, is that a prediction is an attempt to produce a most likely description or estimate of the actual evolution of a variable or system in the future. (The term "forecasting" is also often used; it is used interchangeably with prediction in this chapter.) Projections differ from predictions in that they involve assumptions concerning, for example, future socioeconomic and technological developments that may not be realized. They are therefore subject to substantial uncertainty. Scenarios are neither predictions nor projections and may be based on a narrative storyline. Scenarios may be derived from projections but often include additional information from other sources.

Over the years, experience with global assessment projects in the ecological and environmental realm has shown that prediction over large time periods is difficult if not impossible, given the complexity of the systems examined and the large uncertainties associated with them—particularly for time horizons beyond 10–20 years. In the case of ecosystems, heterogeneity, non-linear dynamics, and cross-scale interactions of ecosystems contribute to system complexity (Holling 1978; Levin 2000). Furthermore, ecological predictions are contingent on drivers that may be even more difficult to predict, such as human behavior. As a result, people rarely have enough information to produce reliable predictions of ecosystem behavior or environmental change (Sarewitz et al. 2000; Funtowicz and Ravetz 1993). Despite these problems with predicting the future, people need to take decisions with implications for the future. Scenario development offers one approach to dealing with uncertainty.

The MA uses the IPCC definition of scenarios as "plausible descriptions of how the future may develop, based on a coherent and internally consistent set of assumptions about key relationships and driving forces" (such as rate of technology changes and prices) (IPCC 2000b). As such, scenarios are used as a systematic method for thinking creatively about complex, uncertain futures for both qualitative and quantitative aspects. Figure 6.1 presents an overview of the additional value created by scenario analysis compared with more deterministic approaches such as predictions.

Scenarios can serve different purposes (see, e.g., Alcamo 2001; van der Heijden 1997). They can be used in an explorative manner or for scientific assessment in order to understand the functioning of an investigated system. For the MA, researchers are interested in exploring hypothesized interactions and linkages between key variables related to ecosystems and human well-being. Scenario outcomes can then form part of planning and decision-making processes and help bridge the gap between the scientific and the policy-making communities. The MA scenarios can also be used in an informative or educational way. Depending on the process used, scenarios can also challenge the assumptions that people have about the future and can illustrate the different views on their outcomes held by participants of the scenario-building exercise.

In general, scenarios contain a description of step-wise changes or a storyline, driving forces, base year, and time steps and horizon (Alcamo 2001). They are often classified by the method used to develop them, their goals and objectives, or their output. One classification of scenarios distinguishes between "exploratory" and "anticipatory" scenarios. Exploratory scenarios are descriptive and explore trends into the future. Anticipatory scenarios start with a vision of the future that could be optimistic, pessimistic, or neutral and work backwards in time to discern how that particular future might be reached. This type of scenario is sometimes also referred to as a "normative" scenario. The MA scenarios were developed using mostly exploratory approaches.

Finally, scenarios can consist of qualitative information, quantitative information, or both. Chapter 2 contains some examples of each of these types of scenarios. Qualitative scenarios, using a narrative text to convey the main scenario messages, can be very helpful when presenting information

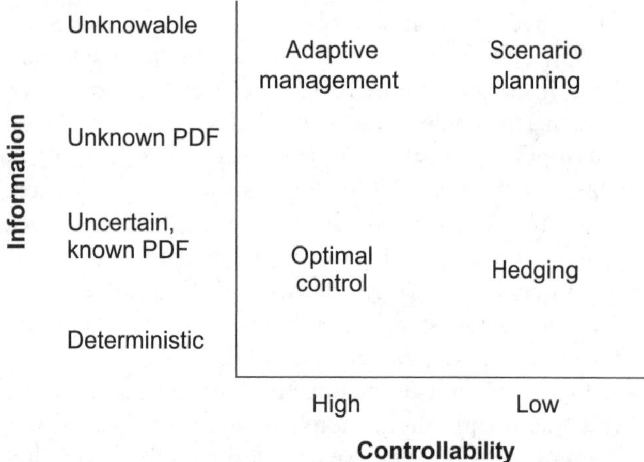

Figure 6.1. Status of Information, from Low to High, versus Degree of Controllability. PDF = probability distribution function. The figure shows the domain of traditional decision tools such as utility optimization and the domain where scenarios may be helpful. (Adapted from Peterson et al. 2003)

to a nonscientific audience. Quantitative scenarios usually rely on modeling tools incorporating quantified information to calculate future developments and changes and are presented in the form of graphs and tables.

Both scenario types can be combined to develop internally consistent storylines assessed through quantification and models, which are then disseminated in a narrative form. (A "storyline" is a scenario in written form, and usually takes the form of a story with a very definite message or "line" running through it.) This approach was used to develop the MA scenarios. The qualitative scenarios (storylines) provide an understandable way to communicate complex information, have considerable depth, describe comprehensive feedback effects, and incorporate a wide range of views about the future. The quantitative scenarios are used to check the consistency of the qualitative scenarios, to provide relevant numerical information, and to "enrich" the qualitative scenarios by showing trends and dynamics not anticipated by the storylines. By "consistency" we mean that the storylines do not contain elements that are contradictory according to current knowledge. On the other hand, the goal of developing consistent scenarios should not lead us to omit elements that may look contradictory but in reality are only surprising new connections or results. Often, uncovering these unanticipated connections that challenge current beliefs and assumptions is one of the most powerful results of the scenarios analysis. For example, is a scenario about climate impacts only consistent when it assumes that climate change will lead to global warming? Indeed, there are "surprising" yet plausible and consistent scenarios that postulate that climate change will lead to cooling of parts of the lower atmosphere.

Together, the qualitative and quantitative scenarios provide a powerful combination that compensates for some of the deficits of either one on its own. The combination of qualitative with quantitative scenarios has been used in many recent global environmental assessments, such as IPCC's Special Report on Emissions Scenarios (IPCC 2000b), UNEP's Global Environment Outlook (UNEP 2002), the scenarios of the Global Scenario Group (Raskin et al. 1998), and the World Water Vision scenarios (Cosgrove and Rijsberman 2000; Alcamo et al. 2000).

The distinction between qualitative and quantitative scenarios is sometimes blurred, however. Qualitative scenarios can be derived by formalized, almost quantitative methods, while quantitative scenarios can be developed by soliciting numerical estimates from experts or by using semi-quantitative techniques such as fuzzy set theory. Storylines can also be interspersed with numerical data and thereby be viewed as both qualitative and quantitative.

As noted in earlier chapters, the main objective of the MA scenarios is to explore links between future changes in world ecosystems and their services and human well-being. The scenario analysis focuses on the period up to 2050, with selected prospects for 2100.

6.3 Overview of Procedure for Developing the MA Scenarios

This section describes the process used to develop the MA scenarios. The procedure consists of 14 steps organized into three phases (see Box 6.1); the details of the storyline development and the modeling exercise are explained in later sections. In the first phase, the scenario exercise was organized and the main questions and focus of the alternative scenarios were identified. In the second phase, the storylines were written and the scenarios were quantified using an iterative procedure. During the third phase, the results of the scenario analysis were synthesized, and scenarios and their outcomes were reviewed by the stakeholders of the MA, revised, and disseminated. These elements are also indicated in Figure 6.2. While Figure 6.2 suggests that activities were completed once processed, in reality earlier activities were often revised during an iterative process.

Two essential activities within the overall scenario development framework were the formulation of alternative scenario storylines and their quantification. These two elements were designed to be mutually reinforcing. The development of scenario storylines facilitates internal consistency of different assumptions and takes into account a broad range of elements and feedback effects that are either difficult to quantify or for which no modeling capability exists, or both. Based on initial storylines, the quantification process helps to provide insights into those processes where sufficient knowledge exists to allow modeling, and to take into account the interactions among the various drivers and services. During scenario development, several interactions were organized between the storyline development and the modeling exercise in order to increase the consistency of the two approaches.

6.3.1 Organizational Steps

The first phase in the MA scenario development consisted of establishing a scenario guidance team, composed of the chairpersons and secretariat of the Scenarios Working Group, to lead and coordinate the scenario-building process. In addition, a larger panel, composed mainly of scientific experts, was assembled to build the scenarios.

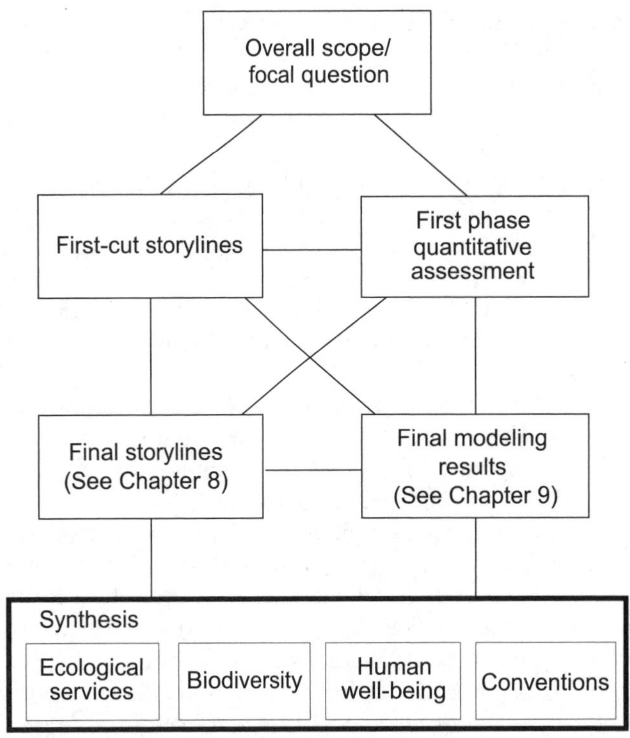

Figure 6.2. Overall Methodology of MA Scenario Development

The scenario guidance team conducted a series of interviews with potential users of the scenarios to obtain their input for developing the goals and focus of the scenarios. This was especially important for the MA because the number of potential users is very large and diverse. These interviews also ensured input from stakeholders and users early on in the study. Understanding the needs and desires of users and their outlook on future development helped the team to devise the main focal questions of the scenarios.

Based on the results of the user interviews and discussions with the scenario panel, the objectives, focus, leading themes, and hypotheses of the scenarios were derived by the scenario guidance team and panel (and later confirmed by the MA Assessment Panel). For the MA, the main objective of scenario development was to explore alternative development paths for world ecosystems and their services over the next 50 years and the consequences of these paths for human well-being. Based on these results, the scenario team clarified the focal questions to be addressed by the scenarios (for the rationale behind the choice of questions, see Chapter 5). The main question was:

What are the consequences of plausible changes in development paths for ecosystems and their services over the next 50 years and what will be the consequences of those changes for human well-being?

The key focal question was then defined through a series of more specific questions—that is, what are the consequences

BOX 6.1

MA Procedure for Developing Scenarios

Phase I: Organizational steps

1. Establish a scenario guidance team.
2. Establish a scenario panel.
3. Conduct interviews with scenario end users.
4. Determine the objectives and focus of the scenarios.
5. Devise the focal questions of the scenarios.

Phase II: Scenario storyline development and quantification

6. Construct a zero-order draft of scenario storylines.
7. Organize modeling analyses and begin quantification.
8. Revise zero-order storylines and construct first-order storylines.
9. Quantify scenario elements.
10. Revise storylines based on results of quantifications.
11. Revise model inputs for drivers and re-run the models.

Phase III: Synthesis, review, and dissemination

12. Distribute draft scenarios for general review.
13. Develop final version of the scenarios by incorporating user feedback.
14. Publish and disseminate the scenarios.

for ecosystem services and human well-being of strategies that emphasize:

- economic and human development (e.g., poverty eradication, market liberalization) as the primary means of management?
- local and regional safety and protection, giving far less emphasis to cross-border and global issues?
- development and use of technologies, allowing greater eco-efficiency and adaptive control?
- adaptive management and local learning about the consequences of management interventions for ecosystem services?

Ecosystem services are defined as "the conditions and processes supported by biodiversity through which ecosystems sustain and fulfill human life, including the provision of goods" (MA 2003; see also Chapter 1). Ecosystem processes are seldom traded in markets, typically have no market price, and therefore usually do not enter in economic decision-making or cost-benefit analyses even though they are essential for human well-being. The MA considered the following interlinked categories of ecosystem services: provisioning services (food, fresh water, and other biological products), supporting and regulating services (including soil formation, nutrient cycling, waste treatment, and climate regulation), and cultural services.

6.3.2 Scenario Storyline Development and Quantification

Following a review and evaluation of current and past scenario efforts, scenario building blocks for driving forces, ecological management dilemmas, branch points, and so on were mapped out. Storyline outlines were then developed around these building blocks. Additional details on the storyline development are provided later in this chapter.

While the initial storylines were being developed, a team of modelers representing several global models was organized to quantify the scenarios. Five global models covering global change processes or ecosystem provisioning services and two global models describing changes in biodiversity were chosen. Criteria that were used to select these models included global coverage, publications of model structure and/or model application in peer-reviewed literature, relevance in describing the future of ecosystem services, and ability to be adapted to the storylines of the MA. (The Ecopath with Ecosim models used to describe marine ecosystems and their service to global fisheries forms an exception to the rule of global coverage, as no global model for this issue was available.) Although all the models had been developed previously, linkages among models and projections out to 2050 and 2100 did require adjustments for several of them. Test calculations were carried out using preliminary driving force assumptions. These test calculations were helpful in identifying the potential contribution of different models to the analysis and in clarifying the procedures of linking the different models.

After a series of iterations, the zero-order storylines were revised and cross-checked for internal consistency. One measure used to accomplish this was the development of timelines and milestones for the various scenarios.

In the next step, the modeling team, in consultation with the storyline team, developed quantitative driving forces that were considered to be consistent with the storylines. Obviously, there is room for interpretation regarding the consistency of driving forces and the storylines, and multiple sets of driving forces are possible. The driving forces and quantified drivers for the modeling exercise chosen for the MA scenarios are discussed in detail in Chapters 7, 8, and 9. Based on the model outcomes of the quantitative scenarios, the scenario team further elaborated or adapted the storylines. A number of feedback workshops with the MA Board and stakeholder groups were held to improve the focus and details of the storylines.

Based on the results of the first round of quantified scenarios, small adjustments in the specification of drivers and linkages among models were made, new model calculations were carried out with the modeling framework, and the storylines were revised (in other words, there was one iteration between storyline development and quantification). Ideally, a series of iterations between storyline improvement, quantification, and stakeholder feedback sessions would have helped to better harmonize the quantitative and qualitative scenarios, but time constraints limited the number of iterations for the MA. The quantified scenario results are described in detail in Chapter 9, while the scenario storylines can be found in Chapter 8.

6.3.3 Synthesis, Review, and Dissemination

The scenario outcomes were assessed in the context of the focal questions and user needs of the various MA user groups. These results are described in Chapters 11, 12, 13, and 14, based on an analysis of both qualitative and quantitative scenario outcomes. Feedback from the assessment component of the scenario team led to further refinement of the storylines and the provision of additional model details.

The scenarios, consisting of the qualitative storylines and quantitative model calculations, were disseminated for review by interested user groups. This was accomplished through presentations, workshops, the MA review process, and Internet communications. Reviewer comments were then incorporated into the scenarios. Both review and dissemination are considered important elements for the success of the scenario exercise.

6.3.4 Linkages between Different Spatial and Temporal Scales

In order to deal with the multiscale aspects of the relationships between ecosystem services and human well-being, the MA called for a large number of sub-global assessments in addition to the global assessment. Several of the sub-global assessments also developed scenarios. As these were often targeted at specific user groups or addressed very specific questions, it was not always possible to directly link the sub-global assessments to the global assessment. (See also Box 6.2.) Nevertheless, to harmonize the global and sub-global scenario exercises as much as possible, the following steps were taken:

BOX 6.2
A Comparison of Global and Sub-global Scenario Development

One of the goals of both the global and sub-global scenarios was to provide foresight about potential futures for ecosystems, including the provision of ecosystem services and human well-being.

Despite similar goals, sub-global scenario development was somewhat different from global scenario development. The key differences were the extensive use of quantitative models in the global scenarios, greater involvement of decision-makers in the sub-global scenarios, and use of sub-global scenarios directly as a tool for decision-making (versus a broader learning-focused global scenario set).

Because the group of decision-makers at the global scale is more diffuse, involvement of decision-makers in the global scenario development was less intense than it was for the sub-global scenarios. Representatives from the business community, the public sector, and the international conventions were periodically informed of progress in the development of the global scenarios and asked for feedback. The sub-global assessments, because they often focused on issues for which key decision-makers could be identified, had closer contact with their primary intended users. The ultimate result of having decision-makers more involved in scenario development was that the scenarios themselves were built more as a direct tool for engaging people in decision-finding processes. Thus, in most sub-global assessments, scenario development focused on futures over which local decision-makers have at least some direct control.

The global scenarios provide four global storylines from where a look down enriches these stories with regional and local details. The sub-global scenarios provide a large number of local stories from where a look up enriches the stories with regional and global "details." A more complete description of the sub-global scenarios can be found in Chapter 9 of the MA *Multiscale Assessments* volume.

- Representatives of some sub-global assessments participated in the global scenario team and contributed to the scenario development.
- Members of the global scenario guidance team participated at various occasions in meetings of the sub-global scenario assessments, explaining both the preliminary global scenario results and the procedure followed in developing the global scenarios.
- Some of the sub-global assessments used the storylines of the global assessment as background for their work or otherwise linked their scenarios to the global assessment.
- After the storylines and the model runs of the global scenarios were finalized, results and findings of the sub-global assessment were used to illustrate how the scenarios could play out at the local scale. (See Boxes in Chapter 8.)

As well as addressing changes in ecosystems and their services at several spatial scales, the MA also considered different temporal scales. For a more detailed discussion on the general issue of scales in the MA, see the MA conceptual framework report (MA 2003). (See also Chapter 7 of this volume and Chapter 4 in the MA *Multiscale Assessments* volume).

The question of temporal scale was important for the construction of the MA scenarios. Although the global scenarios were primarily developed to the year 2050, scenario results of the quantitative scenario elaboration were also reported for 2020, 2050, and 2100. The 2020 report provided a link between the scenarios and medium-term policy objectives, such as the 2015 Millennium Development Goals. It also linked the global and sub-global scenarios, many of which extend only to approximately 2025. Meanwhile, the results for 2100 impart insight into longer-term trends in ecosystem services. Results for 2100 were only reported for parameters that are determined by strong inertia within the natural system, such as climate change and sea level rise.

While several of the models used within the modeling exercise perform their calculations for 10–40 global regions or countries, a much lower resolution was chosen for reporting. Quantitative results (Chapter 9) are mainly presented for six reporting regions: sub-Saharan Africa, Middle East and North Africa, the Organisation of Economic Cooperation and Development, the former Soviet Union, Latin America, and Asia. (See Figure 6.3 in Appendix A.) These are sometimes aggregated into "rich" or "wealthy" countries and "poorer" countries.

The reasons for using this lower resolution include the amount of information that could be presented within this volume and checked for internal consistency. In addition, some models use a global grid of half-degree latitude and longitude to calculate changes in environmental and ecological parameters. The latter are presented in case they are relevant. Grid-level results should be interpreted as broad-brush visualizations of the geographic patterns underlying the scenarios, not as specific predictions for small regions or even grid cells.

6.4 Building the Qualitative Scenarios: Developing Storylines

Significant emphasis was placed on storyline development. Storylines can be provocative because they challenge the tendency of people to extrapolate from the present into the future. They can be used to highlight key uncertainties and surprises about the future. They can consider nonlinearities and complicated causal links more easily than global models can. Moreover, they can incorporate important ecological processes, which so far have not been satisfactorily considered in existing global models. (See Chapter 3.) Since the MA's goal for scenarios development was to specifically consider the future of ecosystems and their services, storyline development was used to incorporate processes that the models could not fully address. Moreover, the qualitative stories provided the input variables for the global models.

The qualitative storylines were developed through a series of discussions among the scenario development panel alternating with feedback from MA user groups and outside experts. The storyline development followed six steps:

- identification of what the MA user groups wanted to learn from the scenarios,
- development of a set of scenario building blocks,
- determination of a set of basic storylines that reflected the MA goal and responded to user needs,

- development of rich details for the storylines,
- harmonization of the storylines with modeling results, and
- feedback from experts and user groups and its incorporation into the final storylines.

Although these steps are presented in order, the process in reality cycled through some of the steps many times until it was felt that consensus was reached on the storylines.

Key questions about the future and main uncertainties of MA user groups were identified through a series of feedback techniques. Approximately 70 leaders and decision-makers from around the world and in many different decision-making positions were interviewed about their hopes and fears for the future. A formal User Needs survey developed by the MA at the beginning of the assessment was used as additional input. This survey was sent to representatives of the MA user community and contained questions on expected outcomes of the MA process. Synthesis of these surveys led to the formulation of the key questions listed earlier.

In the second step, the scenario team developed a number of scenario building blocks, including the factors differentiating the scenarios. In addition, the scenario team identified possible driving forces of socioecological systems into the future, as well as the main uncertainties of these driving forces and the prospects for being able to steer them. Other scenario building blocks included discussions on ecological dilemmas that decision-makers are likely to face in the near future, possible branching points of scenarios, the occurrence of cross-scale ecological feedback loops, and assumptions that decision-makers hold about the functioning of ecological systems (such as whether ecological systems are fragile or resilient).

A first set of scenario storylines was developed using a number of different development paths to distinguish among them. This was done through a combination of writing, presentations, and discussions within the scenario team and feedback from other working groups within the MA. Once these storylines were developed, they were presented to a wider group of experts, including the MA Board, members of the World Business Council for Sustainable Development, scenario experts from other scenario exercises, and several decision-maker communities. Feedback from this exercise led to further refinement of the storylines.

As the results of the quantified scenarios became available, they were compared with the qualitative storylines. This led to further discussions about the logical pathways to the final sequence of events in the scenarios. These discussions were encouraged by the structure of the scenario team, which included both storyline-writers and members of the modeling teams. As a result of these discussions, storylines and model driving forces were adjusted. These discussions also led to new interpretations of the storylines into model parameters.

6.5 Building the Quantitative Scenarios: The Global Modeling Exercise

6.5.1 Organization of the Global Modeling Exercise

As noted in previous sections, the storyline development was complemented by building quantitative scenarios using a linked set of global models. The purpose of the modeling exercise was both to test the consistency of the storylines as developed in the first round and to elaborate and illustrate the scenarios in numerical form. This "quantification of the scenarios" had five main steps:

- Assembling several global models to assess possible future changes in the world's ecosystems and their services. These models are briefly described in Box 6.3 and in the Appendixes. In addition, several models were used to describe certain aspects of changes in biodiversity.
- Specifying a consistent set of model inputs based on the scenario storylines.
- Running the models with the specified model inputs.
- Soft-linking the models by using the output from one model as input to another (we use the term soft-link as the models were not run simultaneously).
- Compiling and analyzing model outputs about changes in future ecosystem services and implications for human well-being. The models were used to analyze the future state of indicators for "provisioning," "regulating," and "supporting" ecosystem services. These indicators are listed in Table 6.1. The analysis of modeling results is presented in Chapter 9.

6.5.2 Specifying a Consistent Set of Model Inputs

The first version of the storylines of the MA scenarios (and in particular, tables containing their main characteristics) formed the basis of the main model assumptions for the quantitative exercise. Over several workshops, the storylines were translated into a consistent set of model assumptions that closely corresponded to the "indirect drivers" of ecosystem services. These included:

- *population development,* including total population and age distribution in different regions;
- *economic development* as represented by assumed growth in per capita GDP per region and changes in economic structure;
- *technology development,* covering many model inputs such as the rate of improvement in the efficiency of domestic water use or the rate of increase in crop yields;
- *human behavior,* covering model parameters such as the willingness of people to invest time or money in energy conservation or water conservation; and
- *institutional factors,* such as the existence and strength of institutions to promote education, international trade, and international technology transfer. The latter are represented directly (trade barriers, for instance, and import tariffs) or indirectly (income elasticity for education) in the models, based on the storylines.

For each of these factors, trends were developed for model inputs that corresponded to the qualitative statements of the storylines. For example, statements in the storylines about "high" or "low" mortality were interpreted such that the trend in mortality would be in the upper or lower 20% of the probabilistic demographic projections. (See also Chapter 9.) Another example is that the scenario with the highest level of agricultural intensification was assumed to have the fastest rate of improvement of crop yield

BOX 6.3

Models Used in MA Global Modeling Exercise

The models in the global modeling exercise include:

- The IMPACT model of the International Food Policy Research Institute in the United States, which computes food supply, demand, trade, and international food prices for countries and regions (Rosegrant et al. 2002).
- The WaterGAP model of the University of Kassel in Germany, which computes global water use and availability on a watershed scale (Alcamo et al. 2003a, 2003b).
- The AIM global change integrated model of the National Institute for Environment Studies in Japan, which computes land cover and other indicators of global change worldwide, with an emphasis on Asia (Kainuma et al. 2002).
- The IMAGE 2.2 global change model of the National Institute of Public Health and the Environment in the Netherlands, which computes climate and global land cover on a grid scale and several other indicators of global change (IMAGE-team 2001).
- The Ecopath with Ecosim model of the University of British Columbia in Canada, which computes dynamic changes in selected marine ecosystems as a function of fishing efforts (Pauly et al. 2000).

and largest expansion of irrigation development. An overview of model inputs is provided in Chapter 9.

6.5.3 Soft-linking the Models

To achieve greater consistency between the calculations of the different models, they were "soft-linked" in the sense that output files from one model were used as inputs to other models. (See Figure 6.4.) The time interval of data that were exchanged between the models was usually one year. The following model linkages were included:

- Computations of regional food supply, demand, and trade from the IMPACT model were aggregated to the 17 IMAGE world regions and the 12 IMAGE animal and crop types. These data were then used as input to the IMAGE land cover model that computed on a global grid the changes in agricultural land that are consistent with the agricultural production computed in IMPACT. In addition, IMAGE was used to calculate the amount of grassland needed for the livestock production computed in IMPACT. Two iterations were done between IMPACT and IMAGE to increase the consistency of the information on agricultural production, availability of land, and climate change. (First, preliminary runs were used as a basis for discussion between the two groups on the consistency of trends under each of the four scenarios for agricultural production, yields changes, and impacts on total land use in each region; for the final runs, an additional iteration was done, providing information from IMAGE to IMPACT on yield changes resulting from climate change and use of marginal lands.) The linkage between IMPACT and IMAGE was done for 2000, 2020, 2050, and 2100.

Table 6.1. Global Modeling Output

Ecosystem Service	Indicator	Model Used to Calculate Indicator
Direct drivers of ecosystem change		
Climate change	temperature change, precipitation change	IMAGE
Changes in land use and land cover	areas per land cover and land use type	IMAGE
Technology adaptation and use	water use efficiency, energy efficiency, area and numbers growth, crop yield growth, and changes in livestock carcass weight	IMPACT, IMAGE, WaterGAP
Exogenous inputs	fertilizer use, irrigation, wage rates	IMPACT, IMAGE
Air pollution emissions	sulfur and NO_x emissions	AIM, IMAGE
Provisioning services		
Food	total meat, fish, and crop production; consumption; trade, food prices	IMPACT
Food	potential food production, crop area, pasture area, and area for biofuels	IMAGE, AIM
Fish	stock	Ecopath/Ecosim
Fuelwood	biofuel supply	IMAGE, AIM
Fresh water	annual renewable water resources, water withdrawals and consumption, return flows	WaterGAP, AIM
Regulating services		
Climate regulation	net carbon flux	IMAGE
Erosion	erosion risk	IMAGE
Supporting services		
Primary production	primary production	IMAGE, AIM
Food security	calorie availability, food prices, share of malnourished pre-school children in developing countries	IMPACT
Water security	water stress	WaterGAP, AIM
Input to biodiversity calculations		
Terrestrial biodiversity	land use area, climate change, nitrogen and sulfur deposition	IMAGE
Aquatic biodiversity	river discharge	WaterGAP
	climate change	IMAGE

- Changes in irrigated areas computed in IMPACT were allocated to a global grid in the WaterGAP model and then used to compute regional irrigation water requirements. These irrigation water requirements were then added to water withdrawals from the domestic and in-

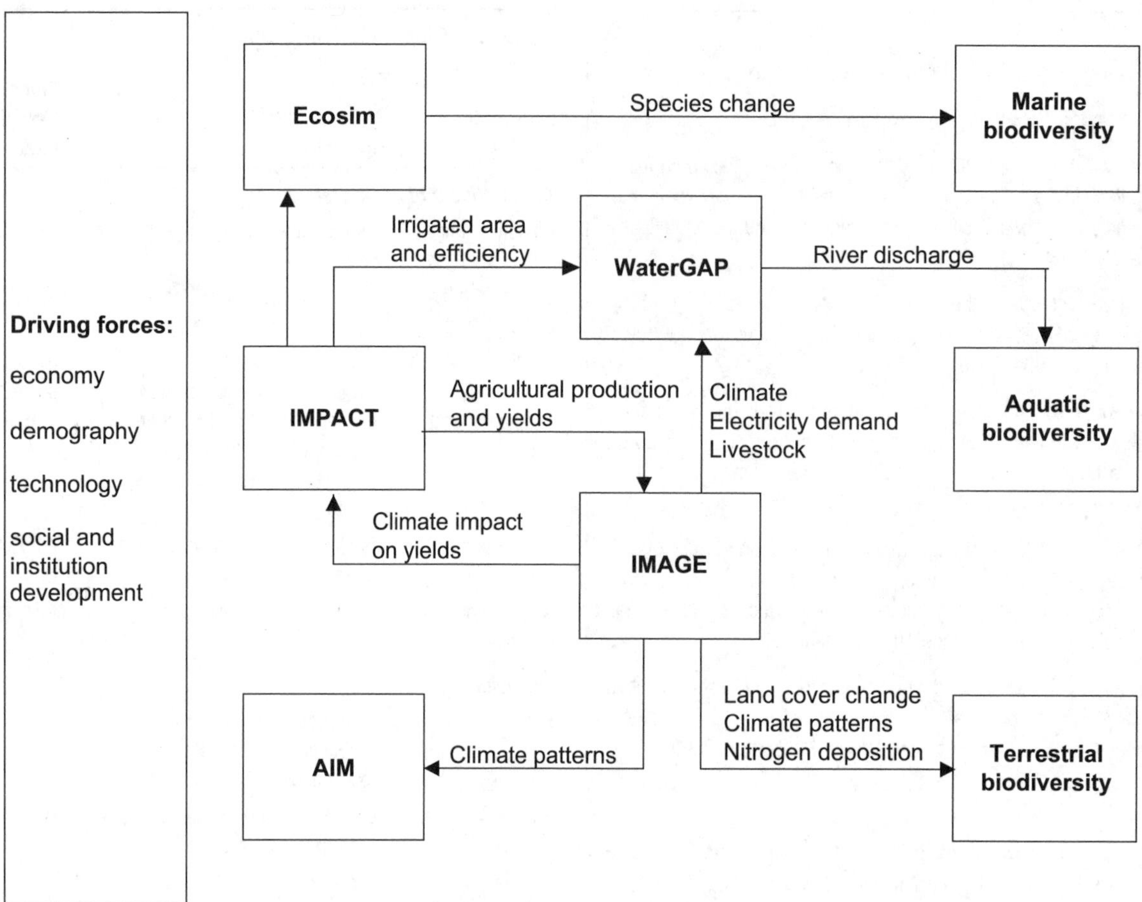

Figure 6.4. Linkages between Models

dustrial sectors (changes for these sectors were calculated by WaterGAP) and compared with local water availability. From this comparison, WaterGAP estimated water stress on a global grid.

- Changes in temperature and precipitation were calculated in IMAGE based on trends in greenhouse gas emissions from energy and land use. These climate change calculations were used in WaterGAP to compute changes in water availability and in IMPACT to estimate changes in crop yield.
- The IMAGE model was used to compute changes in electricity use and livestock production, the latter obtained from IMPACT, which were used by WaterGAP to estimate future water requirements in the electricity and livestock sectors.
- The model for Freshwater Biodiversity used inputs of river discharge from WaterGAP and climate from IMAGE.
- The calculations of changes in terrestrial biodiversity relied on calculations of land cover changes, nitrogen deposition, and climate from IMAGE. Calculations were done on the basis of annual IMAGE output data.
- The AIM modeling team used the same drivers in terms of population, economic growth, and technology development, but no linkages were made between the AIM model and outputs of other models.

Linkages between models are seldom straightforward, and usually require the upscaling (aggregation) or downscaling of various data. The IMPACT model, for instance, uses a more detailed regional disaggregation level than the IMAGE model. In almost all cases, however, regional scaling was possible by combining regions (upscaling) or by assuming proportional changes (downscaling). The models were not recalibrated on the basis of the new input parameters provided by the other models, but in most cases the models had been calibrated using comparable international databases. The new linkages therefore did not lead to major inconsistencies in assumptions between the models.

6.5.4 Modeling Changes in Biodiversity

Currently, there are no global models that describe changes in biodiversity on the basis of global scenarios. For the MA analysis, several smaller models or algorithms were developed on the basis of linkages between global change parameters and species diversity to describe elements of biodiversity change. Outcomes from these tools were then used for a more elaborate discussion of the possible impacts of the different scenarios for global biodiversity. The methods used are discussed in detail in Chapter 10.

6.5.4.1 Terrestrial Biodiversity

The possible future changes in terrestrial biodiversity under the four MA scenarios were explored on the basis of the assessment of changes in native habitat cover over time, climate change, and changes in nitrogen deposition. A review of global threats to biodiversity identified land use change, climate change, the introduction of alien species, nitrogen deposition, and carbon fertilization as major driving forces for extinction (Sala et al. 2000). Simple and well-established existing relationships between these threats and species diversity were used to explore the possible impacts under the MA scenarios. For land use change, for instance, the species-area relationship was used to describe potential loss of plant diversity.

A second important cause of potential change in future global biodiversity is climate change. The potential impacts of climate change were explored using several tools that provide insight into changes in biomes and plant diversity as a result of climate change. More qualitative assessments were made for the impacts of nitrogen and invasive species. Finally, changes in local biodiversity were directly estimated on the basis of the land cover change scenarios.

6.5.4.2 Aquatic and Marine Biodiversity

Oberdorff et al. (1995) developed a global model to describe the diversity of fresh waters as a function of river discharge, net primary production per area of the watershed, watershed area, and fish species richness of the continent. As river discharge can be used as a measure of the size of the freshwater habitat, Oberdorff's model is, in fact, an approximation or expression of the species-area curve for freshwater biodiversity. In our exercise, we updated the fish species numbers and discharge data in Oberdorff's model and used it to describe changes in fish biodiversity as a function of changes in drivers that affect river discharge.

For marine biodiversity, several studies are available on the impacts of fisheries on the marine diversity for different trophic levels. However, to date no global model has been developed. Instead, we applied several regional models to different seas around the world using the MA storylines to specify their main assumptions. The results of these localized case studies served to indicate possible changes in global biodiversity. In addition, a qualitative discussion based on expert knowledge of the impacts of other pressures, including climate change, on marine biodiversity added to the interpretation of model results.

6.5.5 Comments on the Modeling Approach

The approach of using global models to quantify the scenarios has certain disadvantages that should be made explicit. The outcomes of global modeling exercises are highly uncertain, and their assumptions, drivers, and equation systems are often difficult to explain to a nontechnical audience. Furthermore, the models brought together in the MA global modeling exercise were developed independently and were therefore not fully compatible. For example, they have different spatial and temporal resolutions.

The advantages of using global models as part of the scenario building process outweigh the disadvantages, however. Although the model results are uncertain, the models themselves have been published in the scientific literature and have undergone peer review. They have also been found useful for linking science with policy issues in earlier international policy-relevant applications. The models used in the MA exercise provide insights into the trends of many different types of ecosystem services, including global food production, the status of global freshwater resources, and global land cover. In combination, they provide a unique opportunity to generate globally comprehensive, rich, and detailed information for enriching the MA storylines.

While more than 20 indicators were computed in the linked modeling system, coverage of global ecosystem services and feedback effects remained limited. As noted, we tried to make up for this deficit by developing qualitative storylines, which in text form can describe additional indicators and aspects of ecosystem services. At the same time, the modeling exercise addressed some of the deficits of the storylines. For example, model calculations can be used to interlink outcomes for various ecosystem services and to explore the consistency of the storyline assumptions. As part of an overall approach, the storyline-writing and modeling exercise complement each other.

6.6 Discussion of Uncertainty and Scenarios

6.6.1 Using the Scenario Approach to Explore Uncertainty

As explained in the introduction, the main reason to use a scenario approach to explore the future development of ecosystem services is that the systems under study are too complex and the uncertainties too large to use alternative approaches, such as prediction. (See also Chapter 3.) Therefore scenario analysis is used as a tool to address the uncertainty of the future. The MA scenario analysis provides concrete information about plausible future development paths of ecosystem services and their relation to human well-being. The range of scenarios exemplifies the range of possible futures and in so doing helps stakeholders and decision-makers to design robust strategies to preserve ecosystem services for human well-being.

The high level of uncertainty about the future of ecosystem services also implies that is not possible to distinguish between the probability of one scenario versus another. In scenario analysis we sometimes have an intuitive sense that one scenario is more probable than another, but for the MA and most scenario exercises it is not fruitful to dwell on their relative probabilities. With regards to the MA scenarios, other scenarios are also possible, and it is highly unlikely that any of the four scenarios developed for the MA would materialize as described. In other words, the four scenarios are only a small subset of limitless plausible futures. They were selected because they sampled broadly over the space of possible futures, they illustrated points about ecosystem services and human well-being that the MA was charged to address, and they enabled us to answer the focal questions posed by the MA Scenarios Working Group.

6.6.2 Communicating Uncertainties of the Scenarios

Despite the uncertainty of the future, the scenarios contain statements that we intuitively judge as more likely than others. To communicate this certainty/uncertainty we use the expressions shown in Figure 6.5. This scheme was developed for handling uncertainty in assessments of the Intergovernmental Panel on Climate Change (Moss and Schneider 2000).

Associated with each statement of confidence is a quantitative confidence level or range of probability. According to this scale, a confidence level of 1.0 implies that we are absolutely certain that a statement is true, whereas a level of 0.0 implies that we are absolutely certain that the statement is false. It should be noted, however, that in this volume confidence levels are typically not estimated numerically. Instead, they are based on the subjective judgments of the scientists. Also, it is unusual to make statements that do not have at least *medium certainty* (unless they are high-risk events).

Another way to communicate uncertainty is shown in Figure 6.6, which describes a set of expressions for describing the state of knowledge about models and parameters used for constructing the MA scenarios. These expressions can be used to supplement the five-point scale of Figure 6.5 in order to explain why a model outcome is associated with high, medium, or low confidence. These expressions are used, for example, in the Appendix of this chapter to describe the uncertainties of different aspects of the models used in the global modeling exercise. They are also used extensively in Chapter 9 to explain estimates of future changes in ecosystem services.

6.6.3 Sensitivity Analysis

In some cases, formal sensitivity analysis was used as part of the global modeling exercise to estimate the uncertainty of calculations. For example, the MA population scenarios were selected from a stochastic calculation of population projections. (See Chapter 7.) Another example is the assessment of the uncertainty of climate change on water availability. (See Chapter 9.) A third example is the use of Monte-Carlo analysis as part of the calculation of changes in terrestrial biodiversity. (See Chapter 10.)

(1.00) "Very certain" (0.975)
(0.975) "High certainty" (0.83)
(0.83) "Medium certainty" (0.67)
(0.67) "Low certainty" (0.525)
(0.525) "Very uncertain" (0.5)

Figure 6.5. Scale for Assessing State of Knowledge and Statement Confidence (Moss and Schneider 2000)

Level of Agreement			
	High	Established but incomplete	Well-established
	Low	Speculative	Competing explanations
		Low	**High**
		Amount of evidence (theory, observations, model outputs)	

Figure 6.6. Scheme for Describing "State of Knowledge" or Uncertainty of Statements from Models or Theories (Moss and Schneider 2003)

6.7 Summary

The goal of the MA is to provide decision-makers and stakeholders with scientific information about linkages between ecosystem change and human well-being. Several MA scenarios were developed to explore alternative futures on the basis of coherent and internally consistent sets of assumptions. Scenario development was chosen instead of other approaches, such as predictions, as scenarios are better suited to deal with the large inherent uncertainties of the complex relationships between ecological and human systems and within each of these systems.

An important aspect of the MA scenarios is that they need to take into account ecosystem dynamics and ecosystem feedbacks. As earlier global scenarios have been generated for other purposes, incorporation of realistic ecosystem dynamics is a novel aspect of the MA scenarios.

The MA developed scenarios of ecosystems services and human well-being from 2000–50 with selected outlooks to 2100. The MA scenarios were developed by first defining qualitative storylines, followed by quantification of selected storyline drivers and parameters in an iterative process. The development of scenario storylines allowed the process to focus on internal consistency of different assumptions and also to take into account a broad range of elements and feedback effects that often cannot be quantified. Based on initial storylines, the quantification process helped to provide insights into processes where sufficient knowledge exists to allow for modeling and to take into account the interactions among the various drivers and ecosystem services.

APPENDIXES: Descriptions of Models

Appendix 6.1 The IMAGE 2.2 Model

The IMAGE modeling framework—the Integrated Model to Assess the Global Environment—was originally developed to study the causes and impacts of climate change within an integrated context. At the moment, however, IMAGE 2.2 is used to study a whole range of environmental and global change problems, in particular in the realm of land use change, atmospheric pollution, and climate change. The model and its sub-models have been described in detail in several publications (see, in particular, Alcamo et al. 1998; IMAGE-team 2001).

Model Structure and Data

In general terms, the IMAGE 2.2 framework describes global environmental change in terms of its cause-response chain. Appendix Figure 6.1 provides an overview of the different parts of the model.

The cause-response chains start with the main driving forces—population and macroeconomic changes—that determine energy and food consumption and production. Cooperation with a macroeconomic modeling team (CPB 1999) working on a general equilibrium model ensures in several cases an economic underpinning of assumptions made. Next, a detailed description of the energy and food consumption and production are developed using the TIMER Global Energy Model and the AEM Food Demand and Trade model (for the MA, the latter was replaced by a link to the IMPACT model). Both models account for various substitution processes, technology development, and trade.

The changes in production and demand for food and biofuels (the latter are calculated in the energy model) have implications for land use, which is modeled in IMAGE on a 0.5 by 0.5 degree grid. Changes in both energy consumption and land use patterns give rise to emissions that are used to calculate changes in atmospheric concentration of greenhouse gases and some atmospheric pollutants such as nitrogen and sulfur oxides. Changes in concentration of greenhouse gases, ozone precursors, and species involved in aerosol formation comprise the basis for calculating climatic change. Next, changes in climate are calculated as global mean changes that are downscaled to the 0.5 by 0.5 degree grids using patterns generated by general circulation models.

The Land-Cover Model of IMAGE simulates the change in land use and land cover in each region driven by demands for food (including crops, feed, and grass for animal agriculture), timber, and biofuels, in addition to changes in climate. The model distinguishes 14 natural and forest land cover types and 5 humanmade types. A crop module based on the FAO agro-ecological zones approach computes the spatially explicit yields of the different crop groups and grass and the areas used for their production, as determined by climate and soil quality (Alcamo et al. 1998). In case expansion of agricultural land is required, a rule-based "suitability map" determines which grid cells are selected. Conditions that enhance the suitability of a grid cell for agricultural expansion are its potential crop yield (which changes over time as a result of climate change and technology development), its proximity to other agricultural areas, and its proximity to water bodies.

The Land-Cover Model also includes a modified version of the BIOME model (Prentice et al. 1992) to compute changes in potential vegetation (the equilibrium vegetation that should eventually develop under a given climate). The shifts in vegetation zones, however, do not occur instantaneously. In IMAGE 2.2, such dynamic adaptation is modeled explicitly according to the algorithms developed by Van Minnen et al. (2000). An important aspect of IMAGE is that it accounts for significant feedbacks within the system, such as temperature, precipitation, and atmospheric CO_2 feedbacks on the selection of crop types and the migration of ecosystems. This allows for calculating changes in crop and grass yields and, as a consequence, the location of different types of agriculture, changes in net primary productivity, and migration of natural ecosystems.

Application

The IMAGE model has been applied in several assessment studies worldwide, including work for IPCC and analyses for UNEP's Global Environment Outlook (UNEP 2002). For instance, the IMAGE team was one of the six models that took part in the development of the scenarios of IPCC's Special Report on Emission Scenarios (IPCC 2000a; de Vries et al. 2000; Kram et al. 2000). The model has also been used for a large number of studies that aim to identify strategies that could mitigate climate change, mostly focusing on the role of technology or relevant timing of action (e.g., van Vuuren and de Vries 2001). IMAGE also contributed to European projects, including the regularly published *State of the Environment* report on Europe of the EU/European Environment Agency and work for the Directorate-General for the Environment of the European Commission. Recently, the geographic scale of IMAGE was further disaggregated to the country level in Europe, using the model in a large project for land use change scenarios in Europe. In addition, on a project basis the capabilities of the model to describe the nitrogen cycle are improved. In recent years, the links to biodiversity modeling have also been improving.

Uncertainty

As a global Integrated Assessment Model, the focus of IMAGE is on large-scale, mostly first-order drivers of global environmental change. This obviously introduces some important limitations to its results, and in particular how to interpret their accuracy and uncertainty. An important method to handle some of the uncertainties is by using a scenario approach. A large number of relationships and model drivers that are currently not known or that depend on human decisions are varied in these scenarios to explore

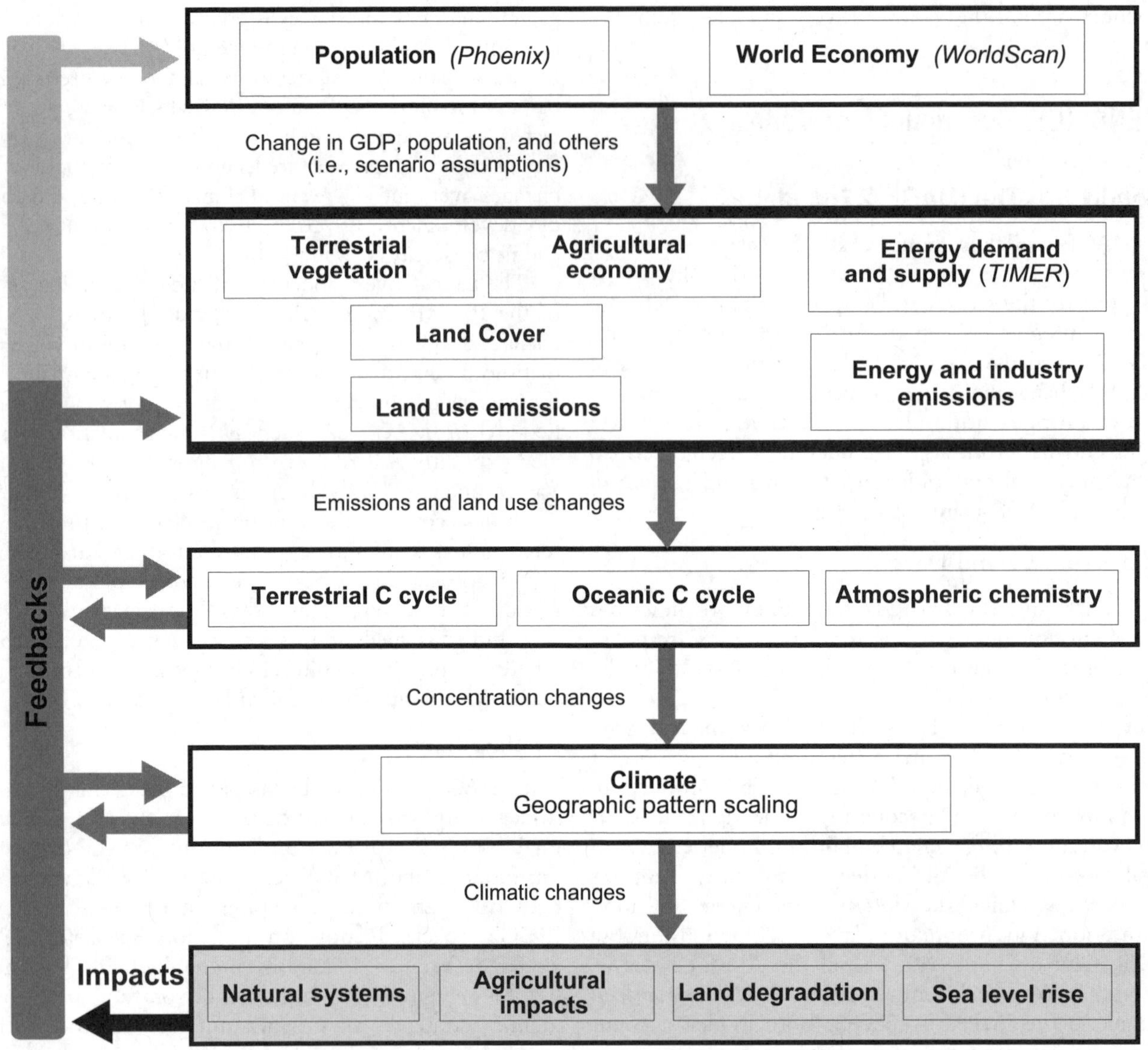

Appendix Figure 6.1. Structure of IMAGE 2.2 Model. Phoenix, WorldScan, and TIMER are submodels of the IMAGE 2.2 model.

the uncertainties involved in them (see IMAGE-team 2001). For the energy model, in 2001 a separate project was performed to evaluate the uncertainties in the energy model using both quantitative and qualitative techniques. Through this analysis we identified that the model's most important uncertainties had to do with assumptions for technological improvement in the energy system and how human activities are translated into a demand for energy (including human lifestyles, economic sector change, and energy efficiency).

The carbon cycle model has also been used in a sensitivity analysis to assess uncertainties in carbon cycle modeling in general (Leemans et al. 2002). Finally, a main uncertainty in IMAGE's climate model has to do with the "climate sensitivity" (that is, the response of global temperature computed by the model to changes in atmospheric greenhouse gas concentrations) and the regional patterns of changed temperature and precipitation. IMAGE 2.2 has actually been set up in such a way that these variables can be easily varied on the basis of more scientifically detailed models, and a separate CD-ROM has been published indicating the uncertainties in these relationships in detail. To summarize, in terms of the scheme discussed in this chapter on the certainty of different theories, most of IMAGE would need to go into the category of established but incomplete knowledge.

Appendix Tables 6.1 and 6.2 give an overview of the main sources of uncertainty in the IMAGE model.

Appendix 6.2 The IMPACT Model

IMPACT—the International Model for Policy Analysis of Agricultural Commodities and Trade—was developed in the early 1990s as a response to concerns about a lack of vision and consensus regarding the actions required to feed the world in the future, reduce poverty, and protect the natural resource base.

Model Structure and Data

IMPACT is a representation of a competitive world agricultural market for 32 crop and livestock commodities, includ-

Appendix Table 6.1. Overview of Major Uncertainties in the IMAGE Model

	Model Component		
Uncertainty	Energy and Related Emissions	Land Use and Land Cover	Environmental System and Climate Change
Model structure	integration in larger economy, feedbacks dynamic formulation in energy model (learning by doing, multinomial logit)	rule-based algorithm for allocating land use	scheme for allocating carbon pools in the carbon cycle model
Parameter	resource assumptions learning parameters	biome model parameter setting CO_2 fertilization	climate sensitivity climate change patterns multipliers in carbon model (impact of climate and carbon cycle)
Driving force	population assumptions economic assumptions assumptions on technology change lifestyle, material intensity, diets environmental policies agricultural production levels (from IMPACT)		
Initial condition	emissions in base year (1995) historic energy use	initial land use / land cover maps historic land use data (FAO)	climate in base year (average global values and maps)
Model operation			downscaling method

Appendix Table 6.2. Level of Confidence in Different Types of Scenario Calculations from IMAGE

Level of Agreement		Amount of Evidence (theory, observation, model outputs): Low	Amount of Evidence (theory, observation, model outputs): High
	High	**Established but incomplete** climate impacts on agriculture and biomes carbon cycle	**Well-established** energy modeling and scenarios
	Low	**Speculative** grid-level changes in driving forces impacts of land degradation	**Competing explanations** global climate change—including estimates of uncertainty local climate change land use change

ing all cereals, soybeans, roots and tubers, meats, milk, eggs, oils, oilcakes and meals, sugar and sweeteners, fruits and vegetables, and fish. It is specified as a set of 43 country or regional sub-models, within each of which supply, demand, and prices for agricultural commodities are determined. The country and regional agricultural sub-models are linked through trade, a specification that highlights the interdependence of countries and commodities in global agricultural markets.

The model uses a system of supply and demand elasticities incorporated into a series of linear and nonlinear equations to approximate the underlying production and demand functions. World agricultural commodity prices are determined annually at levels that clear international markets. Demand is a function of prices, income, and population growth. Growth in crop production in each country is determined by crop prices and the rate of productivity growth. (See Appendix Figure 6.2.) The model is written in the General Algebraic Modeling System programming language and makes use of the Gauss-Seidel algorithm. This procedure minimizes the sum of net trade at the international level and seeks a world market price for a commodity that satisfies market-clearing conditions.

IMPACT generates annual projections for crop area, yield, and production; demand for food, feed, and other uses; and prices and trade. It also generates projections for livestock numbers, yield, production, demand, prices, and trade. The current base year is 1997 (three-year average of 1996–98) and the model incorporates data from FAOSTAT (FAO 2000); commodity, income, and population data and projections from the World Bank (World Bank 1998, 2000a, 2000b) and the UN (UN 1998); a system of supply and demand elasticities from literature reviews and expert estimates; and rates for malnutrition from ACC/SCN (1996)/WHO (1997) and calorie-child malnutrition relationships developed by Smith and Haddad (2000). For MA purposes, the projections period was updated from 1997–2025 to 2100. Additional updates on drivers and parameters are described in Chapter 9.

Application

IMPACT has been applied to a wide variety of contexts for medium- and long-term policy analysis of global food markets. Applications include commodity-specific analyses (for example, for roots and tubers (Scott et al. 2000), for livestock (Delgado et al. 1999), and for fisheries (Delgado et al. 2003)) and regional analyses (for example, on the consequences of the Asian financial crisis (Rosegrant and Ringler 2000)). In 2002 a separate IMPACT-WATER model was developed that incorporates the implications of water availability and nonagricultural water demands on food security and global food markets (Rosegrant et al. 2002).

Uncertainty

As IMPACT does not contain equations with known statistical properties, formal uncertainty tests cannot be carried

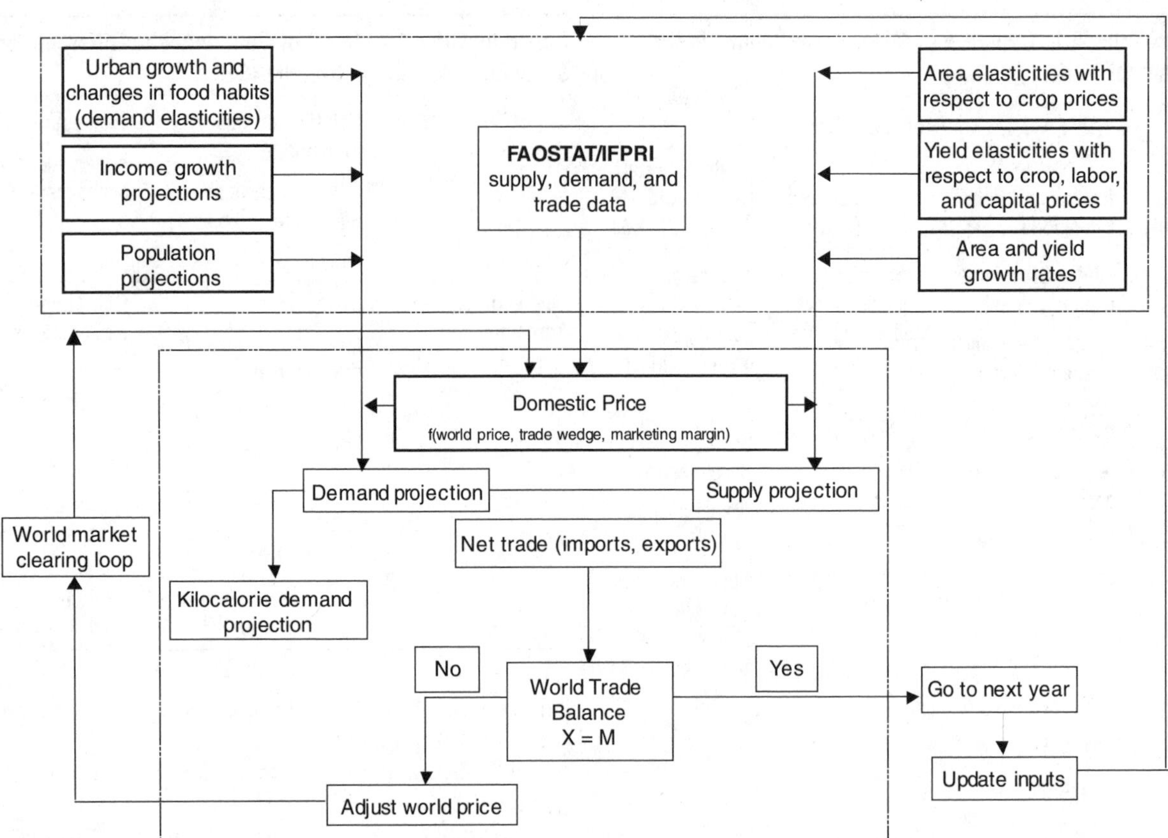

Appendix Figure 6.2. Structure of IMPACT Model

out. However, the robustness of results has been tested and the sensitivity of results with respect to the various drivers in the model have been carried out through numerous scenario analyses, as described in Rosegrant et al. (2001) and the other IMPACT publications cited in this assessment. The drivers associated with uncertainty that have important implications for model outcomes include population and income growth, as well as drivers affecting area and yield and livestock numbers and slaughtered weight growth.

Appendix Tables 6.3 and 6.4 summarize the points related to uncertainty in the model, based on the level of agreement and amount of evidence.

Appendix 6.3 The WaterGAP Model

The principal instrument used for the global analysis of water withdrawals, availability, and stress is the WaterGAP model—Water Global Assessment and Prognosis—developed at the Center for Environmental Systems Research of the University of Kassel in cooperation with the National Institute of Public Health and the Environment of the Netherlands. WaterGAP is currently the only model with global coverage that computes both water use and availability on the river basin scale.

Model Structure and Data

The aim of WaterGAP is to provide a basis both for an assessment of current water resources and water use and for an integrated perspective of the impacts of global change on the water sector. WaterGAP has two main components, a global hydrology model and a global water use model. (See Appendix Figure 6.3.)

The global hydrology model simulates the characteristic macroscale behavior of the terrestrial water cycle to estimate water availability; in this context, we define "water availability" as the total river discharge, which is the combined surface runoff and groundwater recharge. The model calculates discharge based on the computation of daily water balances of the soil and canopy. A water balance is also performed for open waters, and river flow is routed via a global flow routing scheme. In a standard global run, the discharge of approximately 10,500 rivers is computed. The global hydrology model provides a testable method for taking into account the effects of climate and land cover on runoff.

The global water use model consists of three main submodels that compute water use for households, industry, and irrigation in 150 countries. Beside the water withdrawal (total volume of water that is abstracted from surface or groundwater sources), the consumptive water use (water that is actually used and not returned to the water cycle as return flow) is quantified. Both water availability and water use computations cover the entire land surface of the globe except Antarctica (spatial resolution 0.5°—that is, 66,896 grid cells). A global drainage direction map with a 0.5° spatial resolution allows for drainage basins to be chosen flexibly; this permits the analysis of the water resources situation in all large drainage basins worldwide. For a more detailed

Appendix Table 6.3. Overview of Major Uncertainties in the IMPACT Model

Model Component	Uncertainty
Model structure	based on partial equilibrium theory (equilibrium between demand and supply of all commodities and production factors) underlying sources of growth in area/numbers and productivity structure of supply and demand functions and underlying elasticities, complementary and substitution of factor inputs.
Parameters	**Input parameters** base year (three-year centered moving averages for) area, yield, production, numbers for 32 agricultural commodities and 43 countries and regions elasticities underlying the country and regional demand and supply functions commodity prices driving forces **Output parameters** annual levels of food supply, demand, trade, international food prices, calorie availability, and share and number of malnourished children
Driving force	**Economic and demographic drivers** income growth (GDP) population growth **Technological, management, and infrastructural drivers** productivity growth (including management research, conventional plant breeding) area and irrigated area growth livestock feed ratios **Policy drivers** including commodity price policy as defined by taxes and subsidies on commodities, drivers affecting child malnutrition, and food demand preferences
Initial condition	baseline—three-year average centered on 1997 of all input parameters and assumptions for driving forces
Model operation	—

description of the model, see Alcamo et al. (2000, 2003a, 2003b), Alcamo (2001), and Döll et al. (2003).

Application

Results from the model have been used in many national and international studies, including the World Water Assessment, the International Dialogue on Climate and Water, UNEP's Global Environmental Outlook, and the World Water Vision scenarios disseminated by the World Water Commission.

Uncertainty

Appendix Tables 6.5 and 6.6 summarize some of the most important sources of uncertainty in WaterGAP calculations.

Appendix Table 6.4. Level of Confidence in Different Types of Scenario Calculations from IMPACT

		Low (Amount of Evidence)	High (Amount of Evidence)
Level of Agreement/ Assessment	High	**Established but incomplete** projections of area projections of irrigated area, yield projections of livestock numbers, production number of malnourished children calorie availability	**Well-established** changes in consumption patterns and food demand
	Low	**Speculative**	**Competing explanations** projections of commodity prices commodity trade

Amount of Evidence (theory, observations, model outputs)

Alcamo et al. (2003b) found that the magnitude of uncertainty of model calculations was very spatially dependent.

Appendix 6.4 The Asia-Pacific Integrated Model

AIM—the Asia-Pacific Integrated Model—is a large-scale computer simulation model developed by the National Institute for Environmental Studies in collaboration with Kyoto University and several research institutes in the Asia-Pacific region. AIM assesses policy options for stabilizing global climate, especially in the Asia-Pacific region, with objectives of reducing greenhouse gas emissions and avoiding the impacts of climate change. Modelers and policymakers have recognized that climate change problems have to be solved in conjunction with other policy objectives, such as economic development and environmental conservation. The AIM model has thus been extended to take into account a range of environmental problems, such as ecosystem degradation and waste disposal, in a comprehensive way.

Model Structure and Data

AIM/Water estimates country-wise water use (withdrawal and consumption in agricultural, industrial, and domestic sector), country-wise renewable water resource, spatial distribution of water use and renewable water resources with resolution of $2.5° \times 2.5°$, and basin-wise water stress index. Appendix Figure 6.4 presents an overview over the main components of the AIM/Water sub-model. Future scenarios of population, GDP, technological improvements, and historical trends of population with access to water supply are the basic inputs used in the estimation of water use. The country-wise water use is then disaggregated to grid cells in proportion to the spatial densities of population and crop-

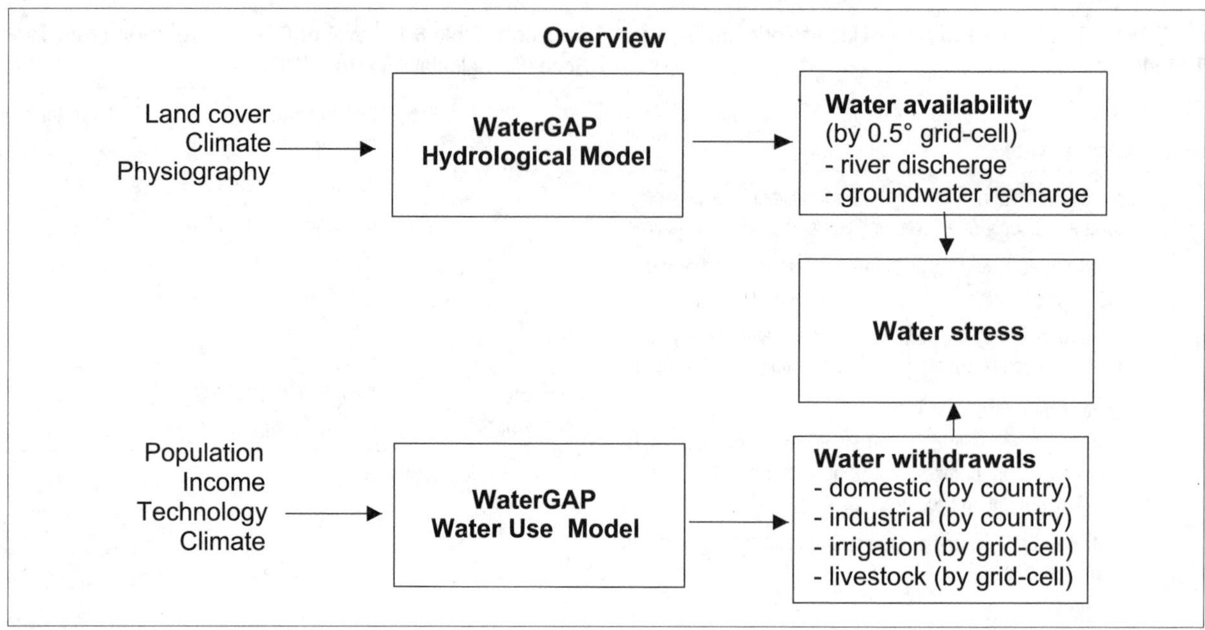

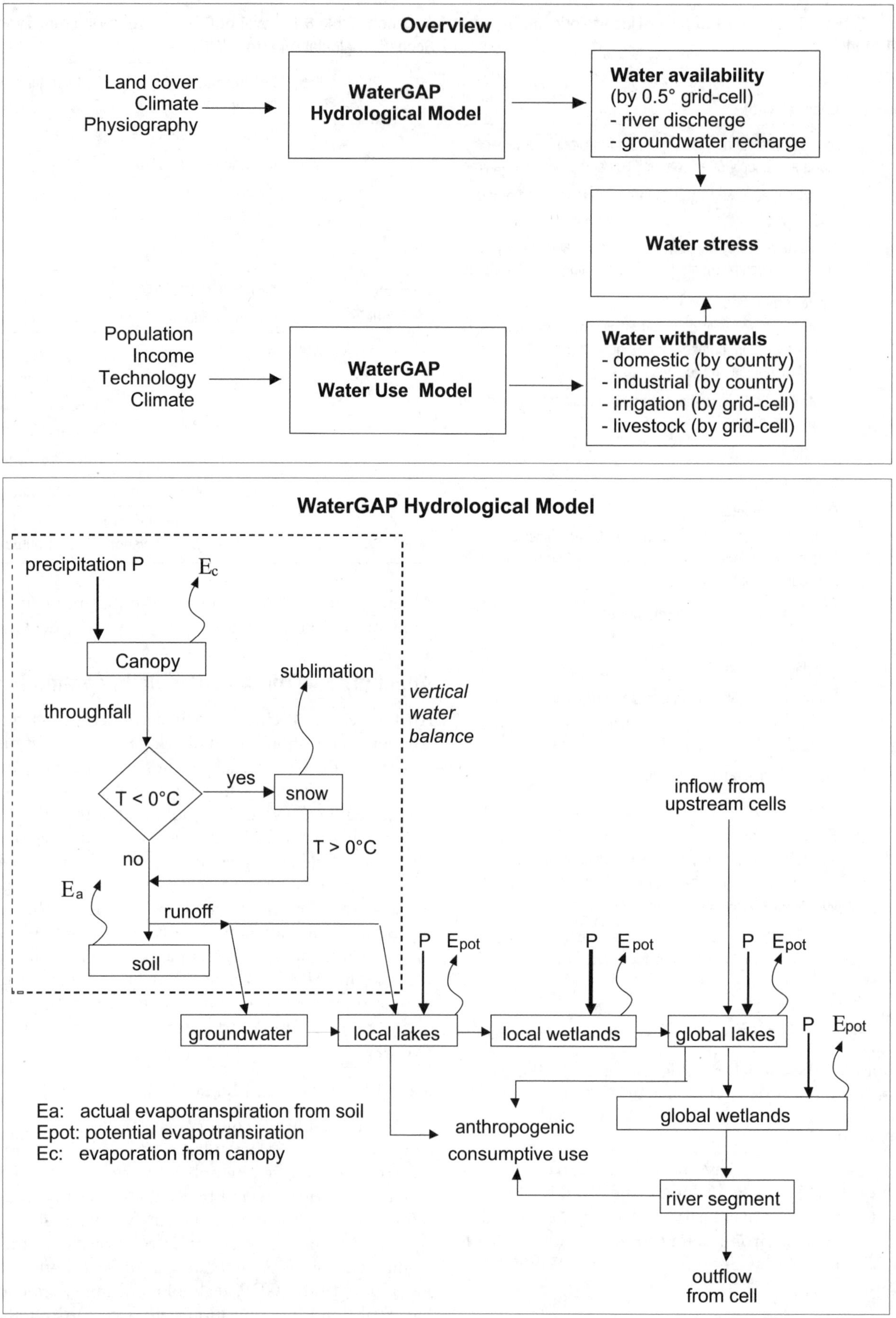

WaterGAP Hydrological Model

precipitation P
E_c
Canopy
throughfall
sublimation
vertical water balance
T < 0°C
yes
snow
no
T > 0°C
E_a
runoff
soil
inflow from upstream cells
P E_{pot}
P E_{pot}
P E_{pot}
groundwater
local lakes
local wetlands
global lakes
P E_{pot}
global wetlands
anthropogenic consumptive use
river segment
outflow from cell

Ea: actual evapotranspiration from soil
Epot: potential evapotransiration
Ec: evaporation from canopy

Appendix Figure 6.3. Structure of WaterGAP Model. Top: overview of main components. Bottom: WaterGAP hydrological model. (Döll et al. 2003)

Appendix Table 6.5. Overview of Major Uncertainties in WaterGAP Model

Model Component	Uncertainty
Model structure	evapotranspiration river transport time snowmelt mechanism
Parameters	watershed calibration parameter parameter for allocating total discharge to surface and subsurface flow
Driving force	local precipitation inputs, frequency of rain days
Initial condition	current direction of flows in flat and wetland areas grid resolution of current water withdrawals
Model operation	downscaling method for country-scale domestic and industrial withdrawals; interpolation of climate data

Appendix Table 6.6. Level of Confidence in Different Types of Scenario Calculations from WaterGAP

Level of Agreement	High	**Established but incomplete** water stress annual withdrawals in developing countries annual water availability in areas without long-term hydrologic gauges	**Well-established** annual withdrawals in industrialized countries annual water availability where there are long-term hydrologic gauges (about 50% of the area of Earth)
	Low	**Speculative** return flows	**Competing explanations** water quality freshwater biodiversity (WaterGAP contributes to these calculations)
		Low	**High**
		Amount of Evidence (theory, observations, model outputs)	

land. The change in renewable water resource is estimated by considering future climate change as input data. In order to obtain a water stress index, water withdrawal and renewable water resources are compared in each river basin. See also Harasawa et al. (2002) for a more detailed description.

AIM/Agriculture estimates potential crop productivity of rice, wheat, and maize with the spatial resolution of 0.5° × 0.5°. Climatic factors are then taken as inputs to simulate net accumulation of biomass through photosynthesis and respiration. They include monthly temperature, cloudiness, precipitation, vapor pressure, and wind speed. The physical and chemical properties of soil such as soil texture and soil slope are also considered in estimating suitability for agriculture (Takahashi et al. 1997).

AIM/Ecosystem is a global computable general equilibrium model. The model structure is shown in Appendix Figure 6.5. It is an economic model with 15 regions and 15 sectors. The model has been developed for the period 1997–2100 with recursive dynamics. Prices and activities are calculated in order to balance demand and supply for all commodities and production factors. AIM/Ecosystem model is linked to AIM/Agriculture model in terms of land productivity changes resulting from climate change. The main drivers of these dynamics are population, production investment, and technology improvement. In this model, various environmental issues such as deforestation and air pollution are included. These interact with the economy through provision of resources and maintenance and degradation of the environment. This model thus consistently estimates economic activities such as GDP and primary energy supply, the related environmental load such as air pollution, and environmental protection activities such as investments in desulfurization technologies (Masui et al. forthcoming).

Application

The AIM model has been used in the development of one of the marker scenarios for IPCC/SRES. The extended version was used for UNEP's GEO3 report. Long-term scenarios of environmental factors quantified using AIM/Water, AIM/Agriculture, and AIM/Ecosystem have been used for the MA.

Uncertainty

The AIM models are based on a deterministic framework. The time scale of each model is more than 100 years. This long-term framework stresses theoretical consistency. It is preferable to models that use past trends because those are not suitable for studying long-term dynamics. Our approach to uncertainty is not to evaluate each parameter or function individually but to assess the robust options or policies derived from the various simulation results.

Appendix Tables 6.7 and 6.8 summarize the uncertainties in each model.

An option or sets of options related to the elements in this table are introduced to the models and simulated under the different scenarios. When the options always produce similar results even in different scenarios, they are regarded as robust.

Appendix 6.5 Ecopath with Ecosim

EwE—Ecopath with Ecosim—is an ecological modeling software suite for personal computers; some components of EwE have been under development for nearly two decades. The approach is thoroughly documented in the scientific literature, with over 100 ecosystems models developed to date (see www.ecopath.org). EwE uses two main components: Ecopath, a static, mass-balanced snapshot of the system, and Ecosim, a time dynamic simulation module for policy exploration that is based on an Ecopath model.

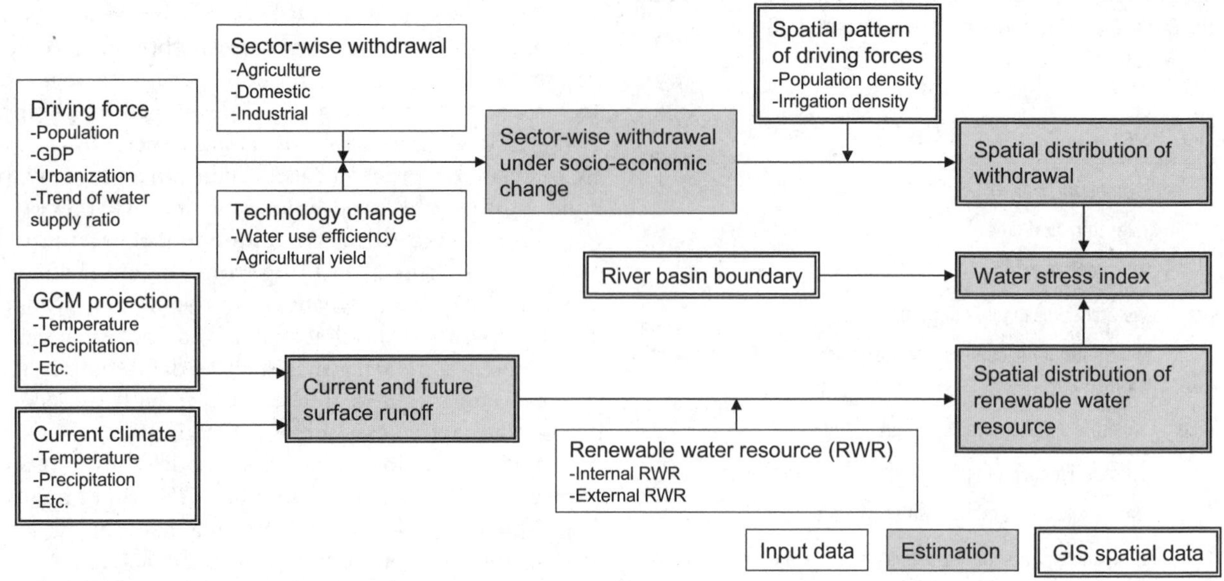

Appendix Figure 6.4. Structure of AIM/Water Model

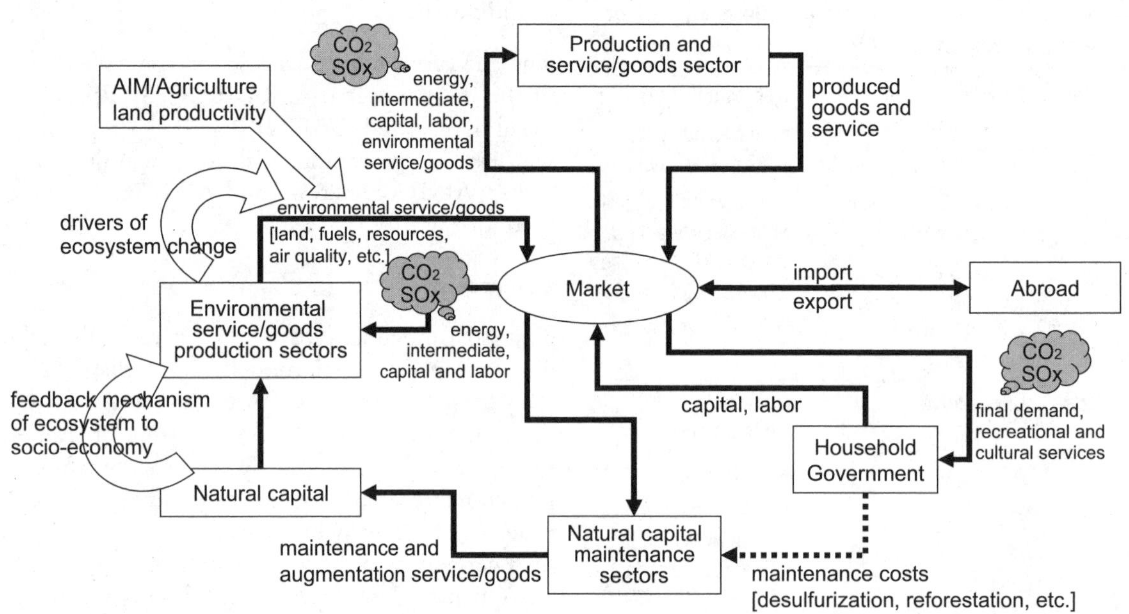

Appendix Figure 6.5. Structure of AIM/Ecosystem Model

Model Structure and Data

The foundation of the marine fisheries calculations is an Ecopath model (Christensen and Pauly 1992; Pauly et al. 2000). The model creates a static, mass-balanced snapshot of the resources in an ecosystem and their trophic interactions, represented by trophically linked biomass "pools." The biomass pools consist of a single species or of species groups representing ecological guilds. Pools may be further split into ontogenetic (juvenile/adult) groups that can then be linked together in Ecosim.

Ecopath data requirements of biomass estimates, total mortality estimates, consumption estimates, diet compositions, and fishery catches are relatively simple and are generally available from stock assessment, ecological studies, or the literature. The parameterization of Ecopath is based on satisfying two key equations: the production of fish and the conservation of matter. In general, Ecopath requires input of three of the following four parameters: biomass, production/biomass ratio (or total mortality), consumption/biomass ratio, and ecotrophic efficiency for each of the functional groups in a model. Christensen and Walters (2004) detail the methods used and the capabilities and pitfalls of this approach.

Ecosim has a dynamic simulation capability at the ecosystem level, with key initial parameters from the base Ecopath model. (See Appendix Figures 6.6 and 6.7.) The key computational aspects are:

- use of mass-balance results (from Ecopath) for parameter estimation;

Appendix Table 6.7. Overview of Major Uncertainties in AIM Model

Model Component	Uncertainty	
	AIM/Water	
	Water Withdrawal Model	**Renewable Water Resource Model**
Model structure	assumption that the spatial pattern of population and land use will not change in future	choice of method for estimating potential evapotranspiration
Parameter	assumption regarding water use efficiency improvement in each sector	assumption of the model parameter for relating actual and potential evapotranspiration
	assumption of the model parameter for estimating urbanization ratio	
Driving force	population	climate projected by GCM
	increasing trend in population with access to water	
	degree of economic activity	
Initial condition	error in the estimated sectoral water withdrawal in the base year	error in the estimated renewable water resource in the base year
		error in the observed climate data
Model operation		procedure to develop climate scenario from GCM result
	AIM/Agriculture	
Model structure	choice of method for estimating photosynthesis ratio	
Parameter	assumption of the model parameter which describes crop growth characteristics	
Driving force	future climate projected by GCM	
Initial condition	error in the observed climate data	
	error in the soil data	
Model operation	procedure to develop climate scenario from GCM result	
	AIM/Ecosystem	
Model structure	based on the general equilibrium theory (equilibrium between demand and supply of all commodities and production factors)	
	investment function in each period	
	structures of production function and demand function: especially elasticity of substitution among the inputs	
Parameter	change of preference	
	relationship between cost and performance in pollution reduction	
Driving force	technology assumption and population projection	
Initial condition	disaggregation of economic data into more detailed subsectors (inputs to each power generation such as thermal power, nuclear, and hydro power)	
	environmental investment and stock of environmental equipment besides the stock of production equipment	
Model operation	nonlinearity in demand and production functions	

- variable speed splitting, which enables efficient modeling of the dynamics of both "fast" (phytoplankton) and "slow" groups (whales);
- the effects of micro-scale behaviors on macro-scale rates: top-down versus bottom-up control explicitly incorporated; and
- inclusion of biomass and size structure dynamics for key ecosystem groups, using a mix of differential and difference equations.

Ecosim uses a system of differential equations that express biomass flux rates among pools as a function of time varying biomass and harvest rates (Walters et al. 1997, 2000). Predator-prey interactions are moderated by prey behavior to limit exposure to predation, such that biomass flux patterns incorporate bottom-up as well as top-down control (Walters 2000). Repeated simulations in Ecosim allow for the fitting of predicted biomasses to time series data. Ecosim can thus incorporate time series data on: relative abundance indices (such as survey data or catch per unit effort data), absolute abundance estimates, catches, fleet effort, fishing rates, and total mortality estimates.

Ecosim can be used in optimization and gaming modes. In the latter, it can explore policy options by "sketching" fishing rates over time, with the results (catches, economic

Appendix Table 6.8. Level of Confidence in Different Types of Scenario Calculations from AIM

Level of Agreement	**High**	**Established but incomplete** pollution reduction water demand water stress	**Well-established** economic activity (based on general equilibrium theory) production function
	Low	**Speculative** —	**Competing explanations** impact on agricultural products (from economic model) renewable water resource
		Low	**High**
		Amount of Evidence (theory, observation, model outputs)	

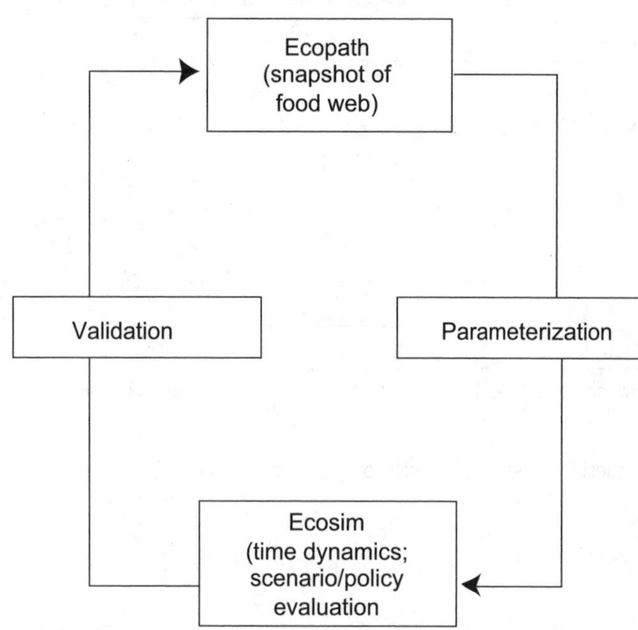

Appendix Figure 6.6. Structure of Ecopath with Ecosim (EwE) Model

performance indicators, biomass changes) examined for each sketch. Formal optimization methods can be used to search for fishing policies that would maximize a particular policy goal or "objective function" for management. The objective function represents a weighted sum of the four objectives: economic, social, legal, and ecological. Assigning alternative weights to these components is a way to look at conflict or trade-off with one another in terms of policy choice. The goal function for policy optimization is defined by the user in Ecosim, based on an evaluation of four weighted policy objectives:

- maximize fisheries rent,
- maximize social benefits,
- maximize mandated rebuilding of species, and
- maximize ecosystem structure or "health."

The fishing policy search routine described estimates time series of relative fleet sizes that would maximize a multi-criterion objective function. In Ecosim, the relative fleet sizes are used to calculate relative fishing mortality rates by each fleet type, assuming the mix of fishing rates over biomass groups remains constant for each fleet type (that is, reducing a fleet type by some percentage results in the same percentage decrease in the fishing rates that it causes on all the groups that it catches). However, density-dependent catchability effects can be entered, and if so reductions in biomass for a group may result in the fishing rate remaining high despite reductions in total effort by any or all fleets that harvest it. Despite this caveat, the basic philosophy in the fishing policy search is that future management will be based on control of relative fishing efforts by fleet type rather than on multispecies quota systems.

Application

Ecopath with Ecosim has been applied to a number of marine ecosystems throughout the world and at varying spatial scales—from small estuary and coral reef systems to large regional studies such as the North Atlantic. For the MA, three well-documented and peer-reviewed EwE models were used: Gulf of Thailand, Central North Pacific, and North Benguela. (See Appendix Box 6.1.) For each one, the narrative storylines of the MA scenarios were interpreted in terms of specific model parameters (mostly the objective function specifying focus on profits, conservation of jobs, or ecosystem management). The landings, value of the landings (see Chapter 9), and the diversity of the landings (see Chapter 10) were used to investigate the differences between the various scenarios for each ecosystem.

Uncertainty

EwE models include routines to explicitly deal with uncertainty in input parameters and with the way this uncertainty may affect results from the simulation modeling. Parameterization for this is as a rule straightforward. The biggest problem in the analysis is usually centered on the state of the knowledge of how exploitation has affected the ecosystem resources over time. It is, for instance, difficult to evaluate if a certain catch history is caused by light exploitation of a large stock or heavy exploitation of a small stock. In order to evaluate this, it is necessary to have information about the population histories in a given ecosystem, and such information is often not accessible, especially for tropical areas.

Appendix Tables 6.9 and 6.10 sum up the uncertainties of the EwE models.

Appendix 6.6 Terrestrial Biodiversity Model

The concept of biodiversity has several dimensions. First of all, it is used for different conceptual levels—genetic diversity, species diversity, and ecosystem diversity. In addition, it refers to both richness and levels of abundance. And finally, the term can be applied both at the local and the

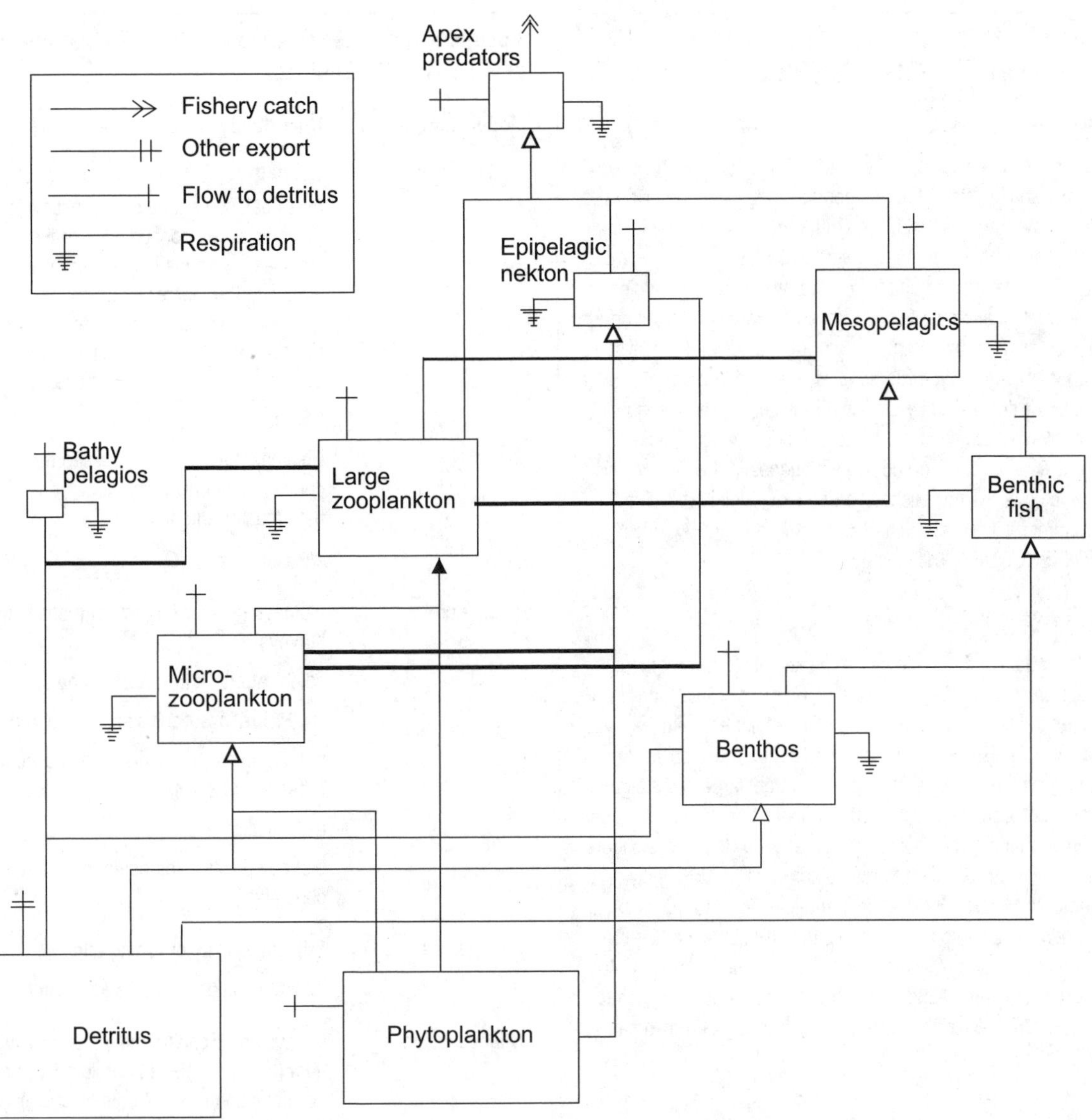

Appendix Figure 6.7. Structure of Ecopath Model

global level. These aspects relate in a different way to ecological services, as discussed in Chapter 10. In the context of the MA, the main focus was on the analysis of species diversity. Chapter 4 provides an overview of the different available methods to assess changes in biodiversity and their strengths and weaknesses.

Model Structure and Data

The assessment applied in Chapter 10 covers four different causes of biodiversity loss: loss of habitat, climate change, nitrogen deposition, and introduction of alien species. For the quantitative analysis, the basis of the analysis is formed by the species-area relationship, defined as $S = c\ A^{z}$, where S is the number of species in an area, A is the habitat area, and c and z are constants. This SAR is applied to about 60 biomes defined by the combination of the biomes defined in IMAGE 2.2 and the realms defined in the biodiversity ecoregion map of Olson et al. (2001). Within the analysis, data on vascular plants were used as an example of possible changes in biodiversity. Data on the c value of the SAR (which represents the intrinsic diversity of each system for a unity size) for vascular plants were obtained by comparing the global land cover map of the IMAGE 2.2 model to a map on diversity of vascular plants (Barthlott 1999). For the z-value of the SAR, a large range of different values was used, as explained in detail in Chapter 10.

The analysis focuses on both global and local changes in biodiversity. For the three different drivers of loss used in the quantitative analysis, the following analysis was performed:

- biodiversity loss from land use change—the SAR was applied directly on the basis of the changes in the IMAGE land use maps;
- biodiversity loss from climate change—three different methods to describe impacts from climate change were applied (a process-oriented model, a method related to species ranges, and the biome description applied in IMAGE 2.2); and
- biodiversity loss from nitrogen deposition—the method describing risks of nitrogen deposition based on critical loads as described by Bouwman et al. (2002) was used.

APPENDIX BOX 6.1

The Three Ecopath with Ecosim Models Used

Gulf of Thailand

The Gulf of Thailand is located in the South China Sea. It is a shallow, tropical coastal shelf system that has been heavily exploited since the 1960s. Prior to the early 1960s, fishing in the area was primarily small scale, with minimal impact on the ecosystem. A trawl fishery was introduced in 1963, however, and since then the area has been subjected to intense, steadily increasing fishing pressure (Pauly and Chuenpagdee 2003). The system has changed from a highly diverse ecosystem with a number of large, long-lived species (such as sharks and rays) to one that is now dominated by small, short-lived species that support a high-value invertebrate fishery. Shrimp and squid caught primarily by trawl gear are economically the dominant fisheries in the Gulf of Thailand. The bycatch of the trawl fishery is used for animal feed. The Gulf of Thailand model is well established and detailed in an FAO technical report (FAO/FISHCODE 2001).

Central North Pacific

The area of Central North Pacific that is modeled is focused on epipelagic waters from 0N to 40N latitude and between 150W and 130E longitude (Cox et al. 2002). Tuna fishing is the major economic activity in the area after tourism in the Hawaiian Islands. The tuna fishery is divided into deepwater longline fisheries that target large-sized bigeye, yellowfin, and albacore tuna and surface fleets that target all ages/sizes of skipjack tuna, small sizes of bigeye, yellowfin, and albacore using a range of gear including purse seine, large-mesh gillnet (driftnet), small-mesh gillnet, handline, pole-and-line, and troll (Cox et al. 2000). Recent assessments of the tuna fisheries indicate that top predators such as blue marlin (*Makaira spp.)* and swordfish (*Xiphias gladius*) declined since the 1950s while small tunas, their prey, have increased. The Central North Pacific model is described in detail in Cox et al. (2000).

North Benguela

The North Benguela Current is an upwelling system off the west coast of Southern Africa. This system is highly productive, resulting in a rich living marine resource system that supports small, medium, and large pelagic fisheries (Heymans et al 2004). The system undergoes dramatic changes due to climatic and physical changes and therefore the marine life production can be quite variable. Sardine or anchovy used to be the dominant small pelagics; both species, however, have been at very low abundance for years, as indicated by surveys in the late 1990s (Boyer and Hampton 2001). The North Benguela ecosystem model is now used by the Namibian Fisheries Research Institute and is described in detail in Heymans et al. (2004).

Appendix Table 6.9. Overview of Major Uncertainties in Ecopath with Ecosim (EwE) Model

Model Component	Uncertainty
Model structure	**Ecopath:** mass balance based on the estimation of the biomass and food consumption of the state variables (species or groups of species) of an aquatic ecosystem, with the master equation: Production = catches + predation mortality + biomass accumulation + net migration + other mortality; with *P*roduction = *B*iomass * P/B ratio.
	Ecosim:
	Ebiomass dynamics expressed through a series of coupled differential equations derived from the Ecopath master equation: $dB_i / dt = g_i \sum_j Q_{ji} - \sum_j Q_{ij} I_i - (M_i + F_i + e_i)B_i$
Parameter	biomass by species (group) (usually from fisheries surveys)
	catch rate ($t \cdot km^{-2}$ $year^{-1}$) by species group
	the P/B ratio is equivalent to total mortality
	net migration rate and biomass accumulation rate (often set to zero)
	assimilation rate
	diet composition (obtained from predators' stomach contents)
Driving force	fishing mortality of fishing effort
	forcing functions expressing environmental variables
Initial condition	the system (Ecopath model) is initially set to equilibrium (but can include net migration rate and biomass accumulation rate not equal to zero)
Model operation	**Ecopath:**
	used to obtain "snapshot" representation of the food web and biomass in an ecosystem
	Ecosim:
	optimization or gaming modes, policies explored by "sketching" fishing rates over time and with the results (catches, economic performance indicators, biomass changes) examined for each simulation (with or without previous fitting of biomass or the time series)

Uncertainty and Limitations

In the overall methodology, several major assumptions had to be made:

- We assume that the SAR can be applied independently for each ecoregion-biome combination, thus assuming that the overlap in numbers of species is minimum relative to the number of species that are endemic to each ecoregion-biome combination.
- We assume that diversity loss will occur as a result of the transformation of natural vegetation into a human-dominated land cover unit, so we assume that human-dominated vegetation has a diversity of zero endemic species for the purposes of the SAR calculations.
- In our calculations, we have not assumed any extinction rate with time, but simply assume that at some point the number of species will reach the level as indicated by the SAR. This means that our results should not be interpreted in terms of a direct loss of number of species, but in terms of species that are "committed" to extinction.

The method to estimate impacts from climate change is a simplification of the response at the level of individual species that will occur in reality.

Appendix Table 6.10. Level of Confidence in Different Types of Scenario Calculations from Ecopath with Ecosim

Level of Agreement	**High**	**Established but incomplete** Ecopath: diet composition and biomass Ecosim: dynamics of depleted groups	**Well-established** Ecopath: food consumption rate, P/B ratios Ecosim: dynamics of abundant groups
	Low	**Speculative** Ecopath: treatment of lower trophic levels, especially microbial groups Ecosim: trophic versus environment forcing	**Competing explanations** Ecopath: n.a. Ecosim: trophic mediation (control of interaction of two species by third species)
		Low	**High**
		Amount of Evidence (theory, observation, model outputs)	

The method is dependent on several uncertainties in data, including the value of z and the number of ecoregions that is defined. In Chapter 10, an extensive uncertainty analysis is performed with regard to these aspects. Appendix Table 6.11 gives a brief overview of the uncertainties of the model.

Appendix Table 6.11. Overview of Major Uncertainties in the Terrestrial Biodiversity Model

Model Component	Uncertainty
Model structure	use of species-area curve
	assumption of irreversibility of species loss
	fraction of species remaining after conversion of natural area into agricultural land
	BIOME model to describe impacts of climate change
	use of critical loads to describe loss of biodiversity for nitrogen deposition
Parameter	value of z (determining biodiversity loss for reduction of habitat), scale of analysis (provincial, island, continental)
	biodiversity loss multiplier for areas affected by climate change
	biodiversity loss multiplier as a function of nitrogen deposition (excess of critical load)
Driving force	land use change (from IMAGE)
	climate change and biome response (from IMAGE)
	nitrogen deposition (from IMAGE)
Initial condition	number of species per habitat type
	land use maps and biome map for start year
Model operation	number of separate biomes and the amount of overlap in species numbers between biomes

Appendix 6.7 Freshwater Biodiversity Model

Freshwater ecosystems have been underrepresented in past global studies of biodiversity. No global quantitative models to forecast the response of freshwater biodiversity to environmental drivers exist. Thus we adapted a previously published descriptive model (Oberdorff et al. 1995) on the relationship between the number of fish species to river size (as measured by volume of water discharged at the river mouth). While fishes are only one component of freshwater biodiversity, they are frequently of great ecological importance and of great value to humans in fisheries. They were the only group of freshwater biota for which near-global data exist. Likewise, because analogous statistical relationships are not quantified for other freshwater habitats (such as lakes and wetlands), we were limited in our quantitative effort to rivers.

Model Structure and Data

Using the statistical approach of Oberdorff et al. (1995), we constructed a regression model relating the number of fish species (taken from Oberdorff et al. 1995 and FishBase) to river discharge for 237 rivers worldwide. Baseline river discharge data from the WaterGAP model correlated strongly with data used by Oberdorff et al. (1995). Thus, for our scenarios we used river discharge output from WaterGAP for both baseline and future conditions. In brief, WaterGAP provided future river discharge that was then used as the dependent variable in a simple regression model to predict the future number of fish species.

Applications

While variations on this statistical model have been used successfully to predict current patterns of riverine fishes among rivers (Oberdorff et al. 1995; Guegan et al. 1998; Hawkins et al. 2003), there has been no previous application of the approach to future scenarios.

Uncertainty and Limitations

For reasons of model structure and data limitation, output applies only to rivers and fishes; lakes and wetlands and other aquatic taxa are not addressed with this quantitative model. Because the data for fish species in each river do not distinguish endemic species from species that also occur in other rivers, the scenario estimates are for river-specific losses of species, not global extinctions. Because other considerations (for example, the pace of evolution, the rate at which species might migrate, and the prevalence of human introductions of species) dictate that the model should not be used to make quantitative forecasts of increases in fish species number, quantitative scenarios are only possible where river discharge declines in scenarios.

Because the independent variable in the regression model (river discharge) is an output from WaterGAP, values of future discharge are subject to all the uncertainties described for WaterGAP. Appendix Table 6.12 summarizes the main sources of uncertainty.

Appendix Table 6.12. Overview of Major Uncertainties in the Freshwater Biodiversity Model

Model Component	Uncertainty
Model structure	assumption of log-linearity between number of fish species and river discharge
Parameter	uncertain coefficients of statistical model
Driving force	uncertainty of river discharge (see WaterGAP description)
Initial condition	uncertain current relationship between number of fish species and river discharge
Model operation	inclusion of only river discharge as independent variable

The model also assumes that all other features of the riverine habitat that are important biologically remain constant. This assumption is most certainly violated, but the magnitude of the consequences of any violation is impossible to ascertain. Violations would include changes in other aspects of river flow besides mean annual discharge (the timing and duration of low and high flows are also important to fishes), and the variety of other drivers (eutrophication, acidification, temperature, xenochemicals, habitat structure, other species in the food web, and so on) that would interact strongly with discharge. Finally, lag times of unknown duration would characterize the pace at which fish species numbers would equilibrate to lower discharge.

While the magnitude of error in scenarios is impossible to quantify, there are strong reasons to expect that the directions of errors are likely to produce underestimates of species loss (once species number equilibrates with reduced discharge levels). Reductions in species number are likely to be greater then predicted by the species-discharge model because interactions between discharge and other habitat features will change conditions away from those to which local species are adapted. Thus the species-discharge model will provide a conservative index of river-specific extirpation of fish species, as a function of the drivers that affect discharge (climate and water withdrawals).

Appendix 6.8 Multiscale Scenario Development in the Sub-global Assessments

As part of the MA's sub-global assessments, a number of scenario exercises were carried out in order to develop scenarios at regional scales. The number of sub-global assessments was limited by the available human and financial resources. Assessments were carried out in over 15 locations. (See Appendix Figure 6.8.) They have yielded scientific insights and policy-relevant information and prove the potential of the multiscale design of the MA.

Multiscale Design

Three of the sub-global assessments are themselves multiscalar. The SAfMA, PtMA, and CARSEA sub-global assessments contained nested assessments. Appendix Figure 6.8 depicts the designs of these three nested assessments. Each has opted for a slightly different design. In SAfMA, scenarios were developed for the Southern African Development Community region, for two large basins (particularly the Gariep basin) within SADC, and for a number of local small watersheds within these basins. The design was strictly hierarchical, although scenarios were developed independently. In Portugal, the design is not completely hierarchical and an attempt will be made to scale the national scenarios down to a very small community and to a single farm directly. National scenarios for PtMA are themselves downscaled from the global scenarios. In the Caribbean Sea assessment, scenarios were developed independently in two separate nested assessments. A link will be attempted after the scenarios have been fully developed.

Advantages of Multiscale Scenarios

The advantages of generating multiscale scenarios are:

- *Global and local scenarios are linked.* A single approach for developing both global and local scenarios simultaneously produces a higher level of consistency and integration than if they were independently developed.
- *Different purposes to develop scenarios can be "merged."* Local adaptive management strategies can be compared with regional and global explorations of future changes in ecosystems and human well-being.
- *The audience for scenario results can be increased.* Scenarios can be effective tools for integrating and communicating complex information about ecosystem services and other subjects. Producing both global and local scenarios at the same time can in principle broaden the audience of MA results to include local indigenous people (through theater plays), local decision-makers (through models), and the national or international public (through newspapers) or policy-makers (through combined stories and models).

When to Focus on Single-Scale Scenarios?

Large amounts of resources—time and money—are required for multiscale scenario development, especially if it involves a high degree of stakeholder participation or an iterative process between stakeholders, scenarios writers, and modelers. When adequate resources are not available, it might be sufficient to develop scenarios at a single scale. Despite the advantages of multiscale scenarios, there are a number of situations in which the full development of multiscale scenarios might not prove to have a large added value. For example:

- *The importance of a local issue can be decoupled from issues at the global scale.* In the sub-global assessment in the Kristianstad wetlands in Sweden, recent flooding events put the question of coastal protection high on the local agenda. The main issue was whether dikes should be raised or natural area put aside as a flooding area.
- *The national government is dominant in the management of national resources.* Because China is very large and has many nationalities, the national government is dominant in the organization of national resources. Hence the sub-

Southern Africa MA (SAfMA) | Portugal MA (PtMA)

Global storylines | Global | Global storylines

SADC | Regional/National | Portugal

Gariep, Zambezi | Basin | Mondego, Mira

Local

Appendix Figure 6.8. Multiscale Design of the Sub-global Assessments in Southern Africa and Portugal

global assessment in western China accounted mostly for national rather than global actors in its analyses.

References

ACC/SCN (United Nations Administrative Committee on Coordination–Subcommittee on Nutrition), 1996: *Update on the Nutrition Situation*. ACC/SCN, Geneva.

Alcamo, J., P. Döll, T. Henrichs, F. Kaspar, B. Lehner, T. Rösch, and S. Siebert, 2003a: Development and testing of the WaterGAP 2 global model of water use and availability. *Hydrological Sciences*. **48(3),** 317–337.

Alcamo, J., P. Döll, T. Henrichs, F. Kaspar, B. Lehner, T. Rösch, and S. Siebert, 2003b: Global estimation of water withdrawals and availability under current and "business as usual" conditions. *Hydrological Sciences*. **48(3),** 339–348.

Alcamo, J., 2001: *Scenarios as tools for international assessments,* Expert's corner report Prospects and Scenarios No. 5, European Environment Agency. Copenhagen, 31 pp.

Alcamo, J., T. Henrichs, and T. Rösch, 2000: *World Water in 2025—Global modeling and scenario analysis for the World Water Commission on Water for the 21st Century*. Report A0002, Center for Environmental Systems Research, University of Kassel, 48pp.

Alcamo, J., R. Leemans, and E. Kreileman (eds.), 1998: *Global change scenarios of the 21st century. Results from the IMAGE 2.1 model*. Pergamon & Elseviers Science, London, 296 pp.

Barthlott, W., N. Biedinger, G. Braun, F. Feig, G. Kier, and J. Mutke, 1999: Terminological and methodological aspects of the mapping and analysis of global biodiversity. Acta Botatica Fennica, **162,** 103–110.

Bouwman, A.F., D.P. van Vuuren, R.G. Derwent, and M. Posch, 2002: A global analysis of acidification and eutrophication of terrestrial ecosystems. *Water, Air and Soil Pollution*. **141,** 349–382.

Boyer, D. C. and I. Hampton, 2001: An overview of the living marine resources of Namibia. *South African Journal of Marine Science,* **23,** 5–35

Carpenter, S.R., 2002: Ecological futures: building an ecology of the long now. *Ecology,* **83(8),** 2069–2083.

Christensen, V. and C. Walters, 2004: Ecopath and Ecosim: methods, capabilities and limitations. In: *Methods for assessing the impact of fisheries on marine ecosystems of the North Atlantic*. Pauly, D. and T.J. Pitcher (eds.), Fisheries Centre Research Report, **8(2),** 79–105.

Christensen, V. and D. Pauly, 1992: Ecopath II—a software for balancing steady-state ecosystem models and calculating network characteristics. *Ecological Modelling,* **61(3/4),** 169–185.

Cosgrove, W., and F. Rijsberman, 2000: *World Water Vision: Making water everybody's business*. Earthscan Publications, Ltd., London, 108 pp.

Cox, S. P., S. J. D. Martell, C. J. Walters, T. E. Essington, J. F. Kitchell, C. Boggs, and I. Kaplan, 2002: Reconstructing ecosystem dynamics in the central Pacific Ocean, 1952–1998. 1. Estimating population biomass and recruitment of tunas and billfishes. *Canadian Journal of Fisheries and Aquatic Sciences,* **59,** 1724–1735.

CPB (Centraal Planbureau), 1999: *WorldScan: The Core Version*. CPB Netherlands Bureau for Economic Policy Analysis, The Hague, 137 pp.

de Vries, B., J. Bollen, L. Bouwman, M. den Elzen, M. Janssen, and E. Kreileman, 2000: Greenhouse gas emissions in an equity-, environment- and service-oriented world: an IMAGE-based scenario for the 21st century. *Technological Forecasting and Social Change,* **63,** 137–174.

Delgado, C., N. Wada, M. Rosegrant, S. Meijer, and M. Ahmed, 2003: *Fish to2020: Supply and demand in changing global markets*. International Food Policy Research Institute and WorldFish Center, Malaysia, 226 pp.

Delgado C., M. Rosegrant, H. Steinfeld, S. Ehui, and C. Courbois, 1999: *Livestock to 2020 the next revolution*. Food, Agriculture and the Environment Discussion Paper 28, IFPRI, Washington DC.

Döll, P., F. Kaspar, and B. Lehner, 2003: A global hydrological model for deriving water availability indicators: model tuning and validation. *Journal of Hydrology,* **270,** 105–134.

FAO (Food and Agriculture Organization of the United Nations), 2000: FAOSTAT Statistics Database. [online] Rome. Cited May, 2004. Available at http//www.fao.org/FAOSTAT.

FAO/FISHCODE, 2001: *Report of a bio-economic modeling workshop and a policy dialogue meeting on the Thai demersal fisheries in the Gulf of Thailand,* 31 May–9 June 2000, Hua Hin, Thailand, FI:GCP/INT/648/NOR: Field Report F-16 (En), Rome, FAO, 104 pp.

Funtowicz, S. O. and J. R. Ravetz, 1993: Science for the post-normal age. *Futures,* **25(7),** 739–755.

Glenn, J., T. Gordon (eds.) 2005: Futures research methodology. Version 2.0. AC/UNU Millennium Project. United Nations University Press.

Guegan, J.F., S. Lek., T. Oberdorff, 1998: Energy availability and habitat heterogeneity predict global riverine fish diversity. *Nature,* **391(6665),** 382–384

Harasawa, H., Y. Matsuoka, K. Takahashi, Y. Hijioka, Y. Shimada, Y. Munesue, and M. Lal, 2002: Potential impacts of global climate change, In: *Climate Policy Assessment—Asia-Pacific Integrated Modeling,* Kainuma, M., Matsuoka, Y., and Morita, T. (eds.). Springer, Tokyo, pp. 37–54.

Hawkins, B.A., Field, R., Cornell, H.V., 2003: Energy, water, and broad-scale geographic patterns of species richness. *Ecology,* **84(12),** 3105–3117.

Heymans, J.J., L.T. Shannon, and A. Jarre, 2004: Changes in the Northern Benguela ecosystem over three decades: 1970s, 1980s and 1990s. *Ecological Modelling,* **172(2–4),** 175–196.

Holling, C. S., (ed.), 1978: *Adaptive Environmental Assessment and Management*. John Wiley & Sons, London, 377 pp.

IMAGE-team, 2001: The IMAGE 2.2 implementation of the SRES scenarios. A comprehensive analysis of emissions, climate change and impacts in the 21st century [CD-ROM], RIVM CD-ROM publication 481508018, National Institute for Public Health and the Environment, Bilthoven, the Netherlands.

IPCC (Intergovernmental Panel on Climate Change), 2000a: Special Report on Emission Scenarios. A Special Report of IPCC Working Group III. Edited by N. Nakićenović and R. Swart. *Cambridge University Press,* Cambridge, United Kingdom.

IPCC, 2000b: *Special Report on Emissions Scenarios*. Intergovernmental Panel on Climate Change, Cambridge University Press, Cambridge.

Kainuma, M., Y. Matsuoka, and T. Morita, 2002: Climate Policy Assessment. Springer, Tokyo, 432 pp.

Kram, T., T. Morita, K. Riahi, R.A. Roehrl, S, van Rooien, A. Sankowski, and B. De Vries, 2000: Global and Regional Greenhouse Gas Emissions Scenarios. *Technological Forecasting and Social Change,* **63(2–3),** 335–372.

Leemans, R., B. Eickhout, B. Strengers, L. Bouwman, and M. Schaeffer, 2002: The consequences of uncertainties in land use, climate and vegetation responses on the terrestrial carbon. In: *Land Use/Cover Change Effects on the Terrestrial Carbon Cycle in the Asia-Pacific Region,* Science in China, Vol. 45, J.G. Canadell, G. Zhou, I. Noble (eds.), pp. 126–141.

Levin, S. A., 2000: *Fragile dominion: complexity and the commons,* Perseus Publishing, Reading, MA, 256 pp.

MA (Millennium Ecosystem Assessment), 2003: *Ecosystems and Human Well-Being: A Framework for Assessment.* Island Press, Washington, DC.

Masui, T., Y. Hijioka, K. Takahashi, J. Fujino, Y. Matsuoka, H. Harasawa, and T. Morita, Analysis of economic instruments for global natural conservation, forthcoming.

Moss, R.H. and S.H. Schneider, 2000: Uncertainties in IPCC TAR. Recommendation to Lead Authors for more consistent Assessment and Reporting. In: *Guidance papers on the cross cutting issues of the Third Assessment Report of IPCC,* R. Pachuari, T. Taniguchi, and K. Tanaka (eds.), Geneva, Switzerland, World Meteorological Organization, 33–51.

Oberdorff T., J.F. Guégan, and B. Hugueny, 1995: Global scale patterns of fish species richness in rivers. *Ecography,* **18,** 345–352.

Olson, D. M., E. Dinerstein, E.D. Wikramanayake, N.D. Burgess, G.V.N. Powelle, E.C. Underwood, J.A. D'amico, I. Itoua, H.E. Strand, J.C. Morrison, C.J. Loucks, T.F. Allnutt, T.H. Ricketts, Y. Kura, J.F. Lamoreux, W.W. Wettengel, P. Hedao, K.R. Kassem, 2001: Terrestrial ecoregions of the world: A new map of life on Earth. *BioScience,* **51(11),** 933–938.

Pauly, D. and R. Chuenpagdee, 2003: Development of fisheries in the Gulf of Thailand large marine ecosystem: Analysis of an unplanned experiment. In: *Large Marine Ecosystems of the World: Change and Sustainability,* G. Hempel and K. Sherman (eds.) Elsevier Science.

Pauly, D., V. Christensen, and C. Walters, 2000: Ecopath, Ecosim, and Ecospace as tools for evaluating ecosystem impact of fisheries. *ICES Journal of Marine Science,* **57(3),** 697–706.

Peterson, G.D., Cumming, G.C., and S.R. Carpenter, 2003: Scenario Planning, a tool for conservation in an uncertain world. *Conservation Biology,* **17,** 358–366

Prentice, I.C., W. Cramer, S.P. Harrison, R. Leemans, R. A. Monserud, and A. M. Solomon, 1992: A global biome model based on plant physiology and dominance, soil properties and climate. *Journal of Biogeography,* **19,** 117–134

Raskin, P., G. Gallopin, P. Gutman, A. Hammond, and R. Swart, 1998: Bending the curve: toward global sustainability. Stockholm Environment Institute. Lilla Nygatan 1. Box 2142, S-103, Stockholm, Sweden.

Rosegrant, M.W., X. Cai, and S. Cline, 2002: *World Water and Food to 2025: Dealing with Scarcity.* International Food Policy Research Institute, Washington, DC, 310 pp.

Rosegrant, M.W., M. Paisner, S. Meijer, and J. Witcover, 2001: *Global Food Projections to 2020: Emerging Trends and Alternative Futures.* International Food Policy Research Institute. Washington, DC, 206 pp.

Rosegrant, M.W. and C. Ringler, 2000: Asian economic crisis and the long-term global food situation. *Food Policy* **25(3),** 243–254.

Sala, O.E., F.S. Chapin III, J.J. Armesto, R. Berlow, J. Bloomfield, R. Dirzo, E. Huber-Sanwald, L.F. Huenneke, R.B. Jackson, A. Kinzig, R. Leemans, D. Lodge, H.A. Mooney, M. Oesterheld, N.L. Poff, M.T. Sykes, B.H. Walker, M. Walker, and D.H. Wall, 2000: Global biodiversity scenarios for the year 2100. *Science,* **287,** 1770–1774.

Sarewitz, D., R.A. Pielke Jr., and R. Byerly Jr., 2000: *Prediction—Science, Decision Making, and the Future of Nature.* Island Press, Washington DC, 405 pp.

Scott, G.J., M.W. Rosegrant, and C. Ringler, 2000: *Roots and Tubers for the 21st century: Trends, projections, and policy options.* 2020 Vision for Food, Agriculture, and the Environment Discussion Paper No. 31. International Food Policy Research Institute, Washington, DC, 2pp.

Smith, L. and L. Haddad, 2000: *Explaining Child Malnutrition in Developing Countries: A Cross-Country Analysis.* Research Report 111. International Food Policy Research Institute, Washington, DC, 112 pp.

Takahashi, K., H. Harasawa, Y. and Matsuoka, 1997: Climate change impact on global crop production, *Journal of global environmental engineering,* **3,** 145–161.

UN (United Nations), 1998: *World Population Prospects: 1998 Revisions.* United Nations, New York.

UNEP (United Nations Environment Programme), 2002: *Global Environmental Outlook 3, Past, present and future perspectives.* UNEP and Earthscan Publications Ltd, 446 pp.

Van der Heijden, K., 1996: *Scenarios: The Art of Strategic Conversation.* Wiley, NY.

Van Minnen, J., R. Leemans, and F. Ihle, 2000: Defining the importance of including transient ecosystem responses to simulate C-cycle dynamics in a global change model. *Global Change Biology,* **6,** 595–612.

Van Vuuren, D.P. and H.J.M. De Vries, 2001: Mitigation scenarios in a world oriented at sustainable development: the role of technology efficiency and timing. *Climate Policy,* **1,** 189–210.

Walters, C. J., J.F. Kitchell, V. Christensen, and D. Pauly, 2000: Representing density dependent consequences of life history strategies in aquatic ecosystems: EcoSim II. *Ecosystems,* **3,** 70–83.

Walters, C., V. Christensen, and D. Pauly, 1997: Structuring dynamic models of exploited ecosystems from trophic mass-balance models of marine ecosystems. *Reviews in Fish Biology and Fisheries,* **7(2),** 139–172.

WHO (World Health Organization), 1997: *WHO Global Database on Child Growth and Malnutrition.* Program of Nutrition. WHO Document WHO/NUT/97.4. World Health Organization, Geneva.

World Bank, 2000a: World Development Indicators [CD-Rom] The World Bank, Washington, DC.

World Bank, 2000b: *Global Commodity Markets: A Comprehensive Review and Price Forecast.* Developments Prospects Group, Commodities Team. The World Bank, Washington DC.

World Bank, 1998: World Development Indicators [CD-Rom] The World Bank, Washington, DC.

Chapter 7

Drivers of Change in Ecosystem Condition and Services

Coordinating Lead Author: Gerald C. Nelson

Lead Authors: Elena Bennett, Asmeret Asefaw Berhe, Kenneth G. Cassman, Ruth DeFries, Thomas Dietz, Andrew Dobson, Achim Dobermann, Anthony Janetos, Marc Levy, Diana Marco, Nebojsa Nakićenović, Brian O'Neill, Richard Norgaard, Gerhard Petschel-Held, Dennis Ojima, Prabhu Pingali, Robert Watson, Monika Zurek

Review Editors: Agnes Rola, Ortwin Renn, Wolfgang Weimer-Jehle

BOXES

FIGURES

TABLES

*This appears in Appendix A at the end of this volume.

Main Messages

A driver is any natural or human-induced factor that directly or indirectly causes a change in an ecosystem. A *direct driver* unequivocally influences ecosystem processes. An *indirect driver* operates more diffusely, by altering one or more direct drivers. Millennium Ecosystem Assessment categories of indirect drivers of change are demographic, economic, sociopolitical, scientific and technological, and cultural and religious. Important direct drivers include changes in climate, plant nutrient use, land conversion, and diseases and invasive species.

World population, a key indirect driver, will likely peak before the end of the twenty-first century at less than 10 billion people. The global population growth rate peaked at 2.1% per year in the late 1960s and fell to 1.35% per year by 2000, when global population reached 6 billion. Population growth over the next several decades is expected to be concentrated in the poorest urban communities in sub-Saharan Africa, South Asia, and the Middle East. Populations in all parts of the world are expected to experience substantial aging during the next century. While industrial countries will have the oldest populations, the rate of aging could be extremely fast in some developing countries.

Between 1950 and 2000, world GDP grew by 3.85% per year on average, resulting in an average per capita income growth rate of 2.09%. In the MA scenarios, per capita income grows two to four times between 2000 and 2050, depending on scenario. Total economic output grows three to six times during that period. With rising per capita income, the structure of consumption changes, with wide-ranging potential for effects on ecosystem condition and services. At low incomes, demand for food quantity initially increases and then stabilizes. Food expenditures become more diverse, and consumption of industrial goods and services rises. These consumption changes drive area expansion for agriculture and energy and materials use. In the MA scenarios, land used for agriculture and biofuels expands from 4.9 million square kilometers in 2000 to 5.3–5.9 million square kilometers in 2050. Water withdrawals expand by 20–80% during the same period.

In the 200 years for which reliable data exist, the overall growth of consumption has outpaced increases in materials and energy efficiency, leading to absolute increases of materials and energy use.

Nations with lower trade barriers, more open economies, and transparent government processes tend to have higher per capita income growth rates. International trade is an important source of economic gains, as it enables comparative advantage to be exploited and accelerates the diffusion of more efficient technologies and practices. Where inadequate property rights exist, trade can accelerate exploitation of ecosystem services.

Economic policy distortions such as taxes and subsidies can have serious environmental consequences, both in the country where they are implemented and abroad. Subsidies to conventional energy are estimated to have been $250–300 billion a year in the mid-1990s. Changes in greenhouse gas emissions in the MA scenarios range from negative (a decline of 25%) to positive (160%), depending on overall economic growth and subsidy reductions. The 2001–03 average subsidies paid to the agricultural sectors of OECD countries were over $324 billion annually. OECD protectionism and subsidies cost developing countries over $20 billion annually in lost agricultural income.

Since the mid-twentieth century, public sector investments in crop research and infrastructure development have resulted in substantial yield increases worldwide in some major food crops. These yield increases have reduced the demand for crop area expansion arising from population and income growth.

Among the main direct drivers, Earth's climate system has changed since the pre-industrial era, in part due to human activities, and is projected to continue to change throughout the twenty-first century. During the last 100 years, the global mean surface temperature has increased by about 0.6° Celsius, precipitation patterns have changed spatially and temporally, and global average sea level rose 0.1–0.2 meters. The global mean surface temperature is projected to increase 1.4–5.8° Celsius between 1990 and 2100, accompanied by more heat waves. Precipitation patterns are projected to change, with most arid and semiarid areas becoming drier and with an increase in heavy precipitation events, leading to an increased incidence in floods and drought. Global mean sea level is projected to increase by 0.05–0.32 meters in the 1990–2050 period under the MA scenarios (0.09–0.88 meters between 1990 and 2100).

Plant nutrient application is essential to food production, but current methods of use contribute to environmental and socioeconomic problems caused by greenhouse gas emissions, eutrophication, and off-farm hypoxia. Nitrogen application has increased eightfold since 1960, but 50% of the nitrogen fertilizer applied is often lost to the environment. Improvements in nitrogen use efficiency require more investment in technologies that achieve greater congruence between crop nitrogen demand and nitrogen supply from all sources and that do not reduce farmer income. Phosphorus application has increased threefold since 1960, with steady increase until 1990, followed by leveling off at a level approximately equal to 1980s applications. These changes are mirrored by phosphorus accumulation in soils, which can serve as an indicator of eutrophication potential for freshwater lakes and P-sensitive estuaries.

Land cover change is a major driver of ecosystem condition and services. Deforestation and forest degradation affect 8.5% of the world's remaining forests, nearly half of which are in South America. Deforestation and forest degradation have been more extensive in the tropics over the past few decades than in the rest of the world, although data on boreal forests are especially limited, and the extent of change in this region is less well known. Approximately 10% of the drylands and hyper-arid zones of the world are considered degraded, with the majority of these areas in Asia. Cropped areas currently cover approximately 30% of Earth's surface. In the MA scenarios, cropped areas (including pastures) increase 9–21% between 1995 and 2050.

Human-driven movement of organisms, deliberate and accidental, is causing a massive alteration of species ranges and contributing to changes in ecosystem function. In some ecosystems, invasions by alien organisms and diseases result in the extinction of native species or a huge loss in ecosystem services. However, introductions of alien species can also be beneficial in terms of human population; most food is produced from introduced plants and animals.

7.1 Introduction

This chapter examines indirect and direct drivers of change in ecosystem services (the two right boxes in the MA conceptual framework; see Chapter 1 for the diagram and description of the conceptual framework). The goal is to provide an overview at the global level of important drivers of ecosystem condition and the ability to deliver services that improve human well-being.

It is important to recognize that this chapter does not cover the remaining two boxes of the framework—the mechanisms by which the drivers interact with specific ecosystems to alter their condition and ability to deliver services and the effects on human well-being. That discussion is left to the individual condition and services chapters in the *Current State and Trends* volume and to later chapters in this volume. For selected driver categories, we do provide a brief overview of some general ecosystem consequences and some review of the range of values in the scenarios. The MA conceptual framework is not the only way to organize an assessment of ecosystems. Other popular frameworks that examine human-environment interactions include the ecological footprint (Wackernagel and Rees 1996), IPAT and its derivatives (Ehrlich and Holden 1971; York et al. 2003b), and consumption analysis (Arrow et al. 2004).

The MA definition of a driver is any natural or human-induced factor that directly or indirectly causes a change in an ecosystem. A *direct driver* unequivocally influences ecosystem processes. An *indirect driver* operates more diffusely, by altering one or more direct drivers. The categories of global driving forces used in the MA are: *demographic, economic, sociopolitical, cultural and religious, science and technology,* and *physical and biological.* Drivers in all categories other than physical and biological are considered indirect. Important direct (physical and biological) drivers include changes in climate, plant nutrient use, land conversion, and diseases and invasive species.

This chapter does not include natural drivers such as solar radiation, natural climate variability and extreme weather events, or volcanic eruptions and earthquakes. Although some of these can have significant effects on ecosystem services (such as the explosion of the Krakatoa volcano in 1883, which resulted in lower temperatures globally for several years, with negative impacts on agriculture worldwide), space limitations preclude their inclusion. The focus here is on anthropogenic drivers.

7.2 Indirect Drivers

We begin this chapter with a discussion of indirect driver categories—demographic, economic, sociopolitical, cultural and religious, and science and technology.

7.2.1 Demographic Drivers

The number of people currently residing on Earth is widely acknowledged to be an important variable in influencing ecosystem condition. There is also a growing recognition that how population is distributed across age groups, urban and rural regions, living arrangements, and geographic regions affects consumption patterns and therefore ecosystem impacts. These influences are also moderated by resources available to individuals, how they choose to allocate them (economic, sociopolitical, and cultural drivers), and the changing technical relationships needed to convert raw materials provided by ecosystems into services of value to humans (science and technology drivers). In this section, we address population dynamics, focusing on current conditions, projections of the future, and the primary determinants of population change: fertility, mortality, and migration.

7.2.1.1 Current Conditions

Global population increased by 2 billion during the last quarter of the twentieth century, reaching 6 billion in 2000. During that time, birth rates in many parts of the world fell far more quickly than anticipated, and life expectancies—with some notable exceptions—improved steadily. Now population growth rates are declining nearly everywhere. The global growth rate peaked at 2.1% per year in the late 1960s and has since fallen to 1.35% (see Table 7.1), and the annual absolute increment of global population peaked at about 87 million per year in the late 1980s and is now about 78 million (United Nations 2003a). This does not mean that little additional population growth is to be expected; global population is likely to increase by another 2 billion over the next few decades. (See Figure 7.1.) Nonetheless, the end of world population growth, while not imminent, is now on the horizon (Lutz et al. 2001).

Within the boundaries of the MA ecosystems, the cultivated system contains the greatest number of people (4.1 billion in 2000). The coastal system has the highest population density, at 170 people per square kilometer (see MA *Current State and Trends,* Table 5.1). Population growth over the period 1990–2000 has varied across ecosystems. The highest growth rate (18.5% net over the decade) occurred in drylands. The greatest increase in population density was found in the coastal zone, where population growth totaled 23.3 people per square kilometer over the decade. The cultivated systems witnessed the greatest total increase in population during the period—506 million additional people. (See MA *Current State and Trends,* Chapter 5.)

The recent decades of great demographic change have produced unprecedented demographic diversity across regions and countries (Cohen 2003). Substantial population increases are still expected in sub-Saharan Africa, South Asia, and the Middle East. In Europe and East Asia, growth has slowed or even stopped, and rapid aging has become a serious concern.

Traditional demographic groupings of countries are breaking down. In the United States, a high-income country, a doubling of population in the future is anticipated. Many developing countries, including China, Thailand, and North and South Korea, now have low fertility rates that until recently were found only in high-income countries. In 24 countries, mainly in Europe, fertility has fallen to very low levels—below an average of 1.5 births per woman—prompting serious concerns not only about aging but also about population decline.

Mortality and urbanization rates vary widely across countries as well. Life expectancies are substantially higher in high-income countries (about 75 years) than in developing ones (63 years), although the gap has closed from over 17 years of difference in the early 1970s to about 12 years today. Particular countries or regions have done much better, or worse, than the average: life expectancy in East Asia grew impressively over the past few decades, while improvements stalled in Russia and have been reversed in some parts of Africa (due primarily to the impact of HIV/AIDS).

Table 7.1 Global Population Trends since 1950 (United Nations 2003d)

Total population size

	1950	1975	2000
		(million)	
Higher-income countries	813	1,047	1,194
Lower-income countries	1,706	3,021	4,877
Africa	221	408	796
Asia	1,398	2,398	3,680
Europe	547	676	728
Latin America and Caribbean	167	322	520
North America	172	243	316
Oceania	13	22	31
World	**2,519**	**4,068**	**6,071**

Life expectancy, both sexes combined

	1950–55	1970–75	1995–2000
		(years)	
Higher-income countries	66.1	71.4	74.8
Lower-income countries	41.0	54.7	62.5
Africa	37.8	46.2	50.0
Asia	41.4	56.3	65.7
Europe	65.6	71.0	73.2
Latin America and Caribbean	51.4	60.9	69.4
North America	68.8	71.6	76.4
Oceania	60.3	65.8	73.2
World	**46.5**	**58.0**	**64.6**

Population growth rate per year

	1950–55	1970–75	1995–2000
		(percent)	
Higher-income countries	1.20	0.78	0.34
Lower-income countries	2.08	2.36	1.61
Africa	2.19	2.66	2.35
Asia	1.95	2.24	1.41
Europe	0.99	0.59	0.02
Latin America and Caribbean	2.65	2.45	1.56
North America	1.71	0.97	1.07
Oceania	2.15	2.07	1.41
World	**1.80**	**1.94**	**1.35**

Total fertility rate

	1950–55	1970–75	1995–2000
		(fertility rate)	
Higher-income countries	2.84	2.13	1.58
Lower-income countries	6.16	5.42	3.11
Africa	6.74	6.71	5.22
Asia	5.89	5.06	2.72
Europe	2.66	2.16	1.42
Latin America and Caribbean	5.89	5.03	2.72
North America	3.47	2.01	2.01
Oceania	3.90	3.25	2.45
World	**5.02**	**4.48**	**2.83**

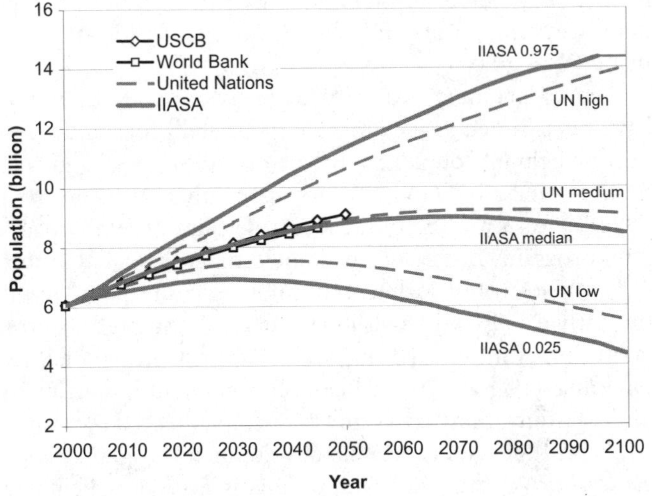

Figure 7.1. Global Population Projections, 2005–2100 (Lutz et al. 2001; United Nations 2003c; United States Census Bureau 2003; World Bank 2003)

Regions are also distinguished by different urbanization levels. High-income countries typically have populations that are 70–80% urban. Some developing-country regions, such as parts of Asia, are still largely rural, while Latin America, at 75% urban, is indistinguishable from high-income countries in this regard. Within the MA system boundaries, the coastal zone has the highest urbanization rate, at 64% (see MA *Current State and Trends,* Chapter 27; see also the discussion of urbanization in the land conversion section later in this chapter).

The diversity of current conditions means that the outlook for future demographic change is highly variable as well. Current age structures are a key determinant of population growth over the next few decades, due to the "momentum" inherent in young populations. Thus, currently high-fertility regions with young age structures (sub-Saharan Africa, Middle East, and South Asia) have some population growth built in. In contrast, the very low fertility and relatively old age structures of some European countries have generated "negative momentum," a built-in tendency for the population to decline even if fertility rises in the future (Lutz et al. 2003).

Differences in urbanization, education, economic and social conditions (especially for women), and other determinants of fertility change are also major factors in the outlook for future demographic changes, and vary widely. As a result, growth over the next several decades is expected to be concentrated in particular sub-populations, especially the poorest, urban communities in developing countries. In addition, while substantial population aging is expected in all regions during the century, there will be strong differences in timing and degree. The lowest fertility high-income countries are already aging and will lead the transition to relatively old age structures, but in some developing countries such as China, which saw sharp declines in fertility over the past few decades, aging is expected to proceed more quickly.

7.2.1.2 Determinants of Fertility, Mortality, and Migration

Since the 1940s, demographers have projected future changes in population size and composition by age and sex using the cohort component method, which involves first projecting trends in the components of population change: fertility, mortality, and migration. Trends in these variables are themselves influenced by a wide range of social, economic, and cultural factors. This section briefly sketches the principal determinants for each component.

7.2.1.2.1 Fertility

The conceptual basis for projecting future fertility changes in a country differs, depending on its current level of fertility. Unless otherwise indicated, what is meant by "fertility" is the period total fertility rate. The TFR is defined as the number of children a woman would give birth to if through her lifetime she experienced the set of age-specific fertility rates currently observed. Since age-specific rates generally change over time, TFR does not in general give the actual number of births a woman alive today can be expected to have. Rather, it is a synthetic index meant to measure age-specific birth rates in a given year.

In those countries with high fertility—defined here as fertility above the "replacement level" of slightly more than two births per woman—demographic transition theory provides the primary basis for forecasting fertility trends. The concept of demographic transition is a generalization of events observed over the past two centuries in those countries with the highest incomes today. While different societies experienced the transition in different ways, in general these societies have gradually shifted from small, slowly growing populations with high mortality and high fertility to large, slowly growing populations with low mortality and low fertility (Knodel and Walle 1979; van de Kaa 1996; Lee 2003). During the transition itself, population growth accelerates because the decline in death rates precedes the decline in birth rates, creating a sudden "surplus" of births over deaths.

Evidence from all parts of the world overwhelmingly confirms the relevance of the concept of demographic transition. The transition is well advanced in all developing countries, except in sub-Saharan Africa, and even there the beginnings of a fertility decline have become apparent (Cohen 1998a; Garenne and Joseph 2002). Fertility is already at or below replacement level in several developing countries, including China.

The idea that reduced demand for children drives fertility decline was given theoretical rigor in the 1960s with the development of a theory based on changes in determinants of parents' demand for children (Becker 1960; Becker and Lewis 1973; Becker and Barro 1988). The model assumes that fertility falls because, as economic development proceeds, parents' preferences shift toward higher "quality" children requiring greater investments in education and health, while increases in women's labor force participation and wages increase the opportunity costs of raising children. At the same time, development leads to a decline in some of the economic benefits parents may derive from children, such as household labor, income, and old age security. Thus, as the net cost of children rises, demand falls.

While an important advance in understanding fertility behavior, this framework by itself is insufficient to explain the diversity of observed fertility change. It has been extended and made more flexible by taking into account sociological aspects such as the effects of development on attitudes toward fertility regulation (Easterlin 1969, 1975), the importance of education and social change in changing women's childbearing desires and their ability to achieve them, changes in cultural contexts such as a decline in religious beliefs and increased materialism (Lesthaeghe 1983; Ryder 1983), and shifts away from extended family structures toward the child-centered nuclear family (Caldwell 1982). Many researchers have also emphasized the importance of the spread of new ideas in general, and in particular those regarding the feasibility and acceptability of birth control (Cleland and Wilson 1987; Bongaarts and Watkins 1996).

Population-related policies have clearly played a role in the decline of fertility in developing countries over the past several decades and are likely to be one determinant of future fertility. For example, it is widely acknowledged that China's population policy played a central role in its very rapid decline in fertility in the 1970s and in maintaining China's currently low fertility. Future changes to its one-child policy could affect fertility trends over the coming decades, although this link should not be considered automatic (Wong 2001).

Family planning programs in many parts of the world have been aimed at meeting "unmet need" for contraception by helping couples overcome obstacles (social and cultural, as well as economic) to contraceptive use (Bongaarts 1994). Measuring the influence on fertility of family planning programs is difficult, but one estimate concluded that 43% of the fertility decline in developing countries between the early 1960s and late 1980s could be attributed to program interventions (Bongaarts 1997). Although family planning programs have been the main intervention to lower fertility, they have often been implemented concomitantly with general economic development programs, educational programs, and health programs, each of which may have an indirect effect on changing fertility. Future change in fertility may also be affected by public policies that address such social and economic factors as women's status, educational and employment opportunities, and public health, as called for by the Cairo Program of Action, an outcome of the 1994 International Conference on Population and Development in Cairo. Although these policies are not primarily motivated by their potential effect on demographic trends, achieving them would likely lead to lower fertility (and lower mortality).

While each perspective on the determinants of fertility and mortality change offers important insights, no single, simple theory explains the multifaceted history of demographic transition around the world. Each explanation suffers from its own shortcomings, and for each, exceptions can be found (Oppenheim-Mason 1997; Robinson 1997). It is probably best to think of fertility and mortality transitions as being driven by a combination of factors rather than a single cause, but determining the precise mix of factors at work in a particular population at a given time remains an elusive goal (Hirschman 1994; National Research Council 2000).

Demographic transition theory provides little guidance on future fertility trends in countries that have already completed the transition to low fertility. Traditionally, many population forecasters assumed that fertility in all countries would eventually stabilize at replacement level, leading to stabilization of population growth. This approach has been strongly criticized as assigning a magnetic force to "replacement-level" fertility, without any empirical evidence that total fertility rates will naturally drift to that level (Demeny 1997). Total fertility has been below replacement level in 20 European countries for at least two decades, and it is currently below 1.5 children per woman in 21 European countries (United Nations 2003a). Fertility has also fallen below replacement level in several developing countries. By 2000, 59 countries were below replacement level fertility, accounting for 45% of the world population (United Nations 2003a).

Many arguments support the idea that fertility will decline below replacement level in more populations in the future. These arguments can be grouped together under the term "individuation," which encompasses the weakening of family ties, characterized by declining marriage rates and high divorce rates, the increasing independence and career orientation of women, and value shifts toward materialism and consumerism (Bumpass 1990). Individuation, together with increasing demands and personal expectations for the amount of attention, time, and money devoted to children, is likely to result in fewer couples who have more than one or two children and an increasing number of childless women.

While current trends and some plausible explanations may suggest that low fertility will continue, there is no compelling theory that can predict reproductive behavior in low-fertility societies. Although fertility typically continues to fall after reaching replacement level, there is no clear pattern to subsequent fertility trends. In some countries, fertility falls quickly to very low levels, while in others it has followed a more gradual slide. In the United States, Sweden, and some other countries, fertility declined well below replacement level and then rose nearly to replacement level again (and in Sweden, then returned again to low levels).

Some of these changes have been due to changes in the timing of births, even if the actual number of births women have over their lifetimes has changed little. Since the mean age of childbearing has been increasing in many high-income countries over the past several decades, part of the decline in TFR has been due to this timing effect and not to a change in the completed fertility of women (Bongaarts and Feeney 1998). Proposed explanations for the trend toward later childbearing include economic uncertainty for young adults, lack of affordable housing, increases in higher education enrollment rates, and difficulties women face in combining child raising with careers, including cultural factors and inflexible labor markets (Kohler et al. 2002).

Some demographers argue that the TFR is likely to increase in the future once the mean age of childbearing stops rising, as happened in the 1980s in the United States when fertility rose to its current, near-replacement level. An additional argument against continued very low fertility is that in surveys conducted in much of Europe, women consistently say they want about two children (Bongaarts 1999). There are many reasons why women may fail to reach this target (career plans, divorce, or infertility, for example), but this finding suggests that fertility is unlikely to remain extremely low, especially if societies make it easier for women to combine careers and childbearing. Nonetheless, it is unclear whether the younger women who are currently postponing births will recuperate this delayed fertility at older ages (Lesthaeghe and Willems 1999; Frejka and Calot 2001).

7.2.1.2.2 Mortality

Mortality decline as a component of the demographic transition has typically begun with reductions in infectious disease driven by improvements in public health and hygiene along with better nutrition as incomes rose and the impacts of famines reduced (Lee 2003). Reduced infant mortality, in particular, is a relatively straightforward consequence of public health and sanitation expenditures. Later, reductions in mortality are driven by reductions in chronic and degenerative diseases such as heart disease and cancer.

Mortality projections are based on projecting future life expectancy at birth—that is, the average number of years a child born today can expect to live if current age-specific mortality levels continued in the future. (Life expectancy—like the total fertility rate—measures the situation at a given period of time; it does not reflect the actual experience of an individual. Nonetheless, life expectancy provides a useful summary of the mortality rates for each age and sex group in a population at a particular time.)

Uncertainties about future changes in life expectancy are quite different in high- and low-mortality countries. In the latter, primarily in industrial regions, mortality is concentrated at old ages. The long-term outlook for life expectancy improvements depends mainly on whether or not a biological upper limit to life expectancy exists and, if it does, how soon it might be reached. Death rates have been declining steadily at old ages, but there is a range of opinions on how long this trend can continue.

One point of view is that life expectancy in higher-income countries is unlikely to increase much beyond 85 years from its current level of about 75 years. Some have argued that this age represents an intrinsic (genetically determined) limit to the human life span (DeFries et al. 2002). Reductions in mortality that do occur are likely to increase an individual's chances of surviving to the maximum life span but not extend the maximum itself. Others argue that while the intrinsic limit may be modifiable, in practical terms it is unlikely to be exceeded without unforeseeable medical breakthroughs (Olshansky et al. 1990; Olshansky et al. 2001). This view is based on calculations showing that increasing life expectancy to 85 years would require dramatic reductions in mortality rates, particularly among the elderly. Olshansky et al. point out that complete elimination of deaths from diseases such as heart disease, cancer, and diabetes—which account for a large proportion of deaths among the elderly—would not extend average life expectancy beyond 90. Only breakthroughs in controlling the fundamental rate of aging could do that.

Other researchers hold that reduced mortality among the oldest ages could produce substantial improvements in life expectancy. Data from several higher-income countries shows that death rates at old ages have been falling over the past several decades, and this improvement has been accelerating, not decelerating, as would be expected if a limit were being approached (Vaupel 1997; Oeppen and Vaupel 2002).

In most developing countries, possible limits to the life span are not as relevant to projections because life expectancies are lower and mortality is not as concentrated at the oldest ages. Future life expectancy will be determined by the efficiency of local health services, the spread of traditional diseases such as malaria and tuberculosis and of new diseases such as HIV/AIDS, as well as living standards and educational levels. Projecting mortality in developing countries is difficult because of the relative scarcity and poor quality of data on current and past trends. In addition, the future course of the HIV/AIDS epidemic could substantially affect mortality in many countries, especially in sub-Saharan Africa where HIV prevalence rates are especially high. HIV/AIDS has slowed, and in some cases reversed, the impressive gains in life expectancy in developing countries over the past several decades. In Botswana, for example, life expectancy has dropped from about 65 years in the early 1990s to 56 years in the late 1990s and is expected to decline to below 40 years by 2005 (United Nations 2003e).

The possibility of environmental feedbacks is sometimes suggested as important when considering future mortality rates. The most frequently discussed possibilities for future effects center around the idea of carrying capacity (the maximum number of people that Earth can support) and the potential health impacts of climate change. Currently, however, population projections do not take explicit account of possible large-scale environmental feedbacks on mortality that have not yet occurred, although they do implicitly consider smaller effects, since they have affected average trends in the past, and since past trends serve as an important component of projections (National Research Council 2000).

There are at least three reasons that carrying capacity is not considered in long-term population projections. First, there is no agreement on what the limiting factors to population growth might be. Carrying capacity is contingent on economic structure, consumption patterns, preferences (including those regarding the environment), and their evolution over time (Arrow et al. 1995), defying attempts to attach a single number to the concept. Any proposed limit relevant to projections over the next century or two would depend primarily on which factor or factors were assumed to be limiting, as well as on how thinly any one factor had to be spread to begin to exert its limiting influence. Proposed limits have been based on a wide range of factors, including supplies of energy, food, water, and mineral resources, as well as disease and biological diversity. No consensus on the human carrying capacity has emerged; on the contrary, the range of estimates has widened over time (Cohen 1995). Second, even if the relevant factors could be agreed on, it would be difficult to project the future evolution of those factors for use in population projections (Keyfitz 1982). Future agricultural systems, energy supplies, and water availability are difficult to foresee in their own right, and there is no consensus in these areas to which demographers might turn. Third, even if these factors could be reliably predicted, their effects are mediated through economic, political, and cultural systems in ways that are not possible to quantify with confidence (Cohen 1998b).

Environmental effects on mortality short of a large-scale catastrophe have received increasing attention, especially those that might be driven by future climate change, particularly in combination with other trends such as changing spatial distributions of population, land use change, and agricultural intensification. The ultimate mortality impact of these environmental health risks is uncertain (Daily and Ehrlich 1996). (See Chapter 9 of IPCC (2002) for a discussion of climate change and human health, and Chapter 11 in this volume.) Yet even the most pessimistic forecasts for additional deaths, when spread over large populations, do not significantly change the general outlook for mortality globally.

7.2.1.2.3 International migration

Future international migration is more difficult to project than fertility or mortality. Migration flows often reflect short-term changes in economic, social, or political factors, which are impossible to predict. And since no single, compelling theory of migration exists, projections are generally based on past trends and current policies, which may not be relevant in the future. Even past migration flows provide minimal guidance because there is often little information about them.

Projections of international migration generally begin with consideration of current and historical trends (Zlotnik 1998; United Nations 2003b). For example, most projections foresee continued net migration into traditional receiving countries such as the United States, Canada, and Australia. These trends may then be modified based on potential changes in underlying forces affecting migration. The forces are complex, and no single factor can explain the history of observed migration trends. For example, population growth rates in sending regions are not a good indicator of emigration flows. In general, correlations between rates of natural increase in developing countries and levels of emigration to higher-income countries have been weak or nonexistent.

A number of theories from different disciplines have attempted to explain migration flows (Massey et al. 1998). International migration is often viewed mainly as a mechanism for redistributing labor to where it is most productive (Todaro 1976), driven by differences in wages among areas. Individuals decide whether to migrate by weighing the estimated benefits of higher wages in a new location against the costs of moving. The choice of destination will depend on where migrants perceive their skills to be most valuable.

This basic model emphasizing the labor market has been extended to address recognized shortcomings. The newer approach, sometimes called new economics models, assumes that migration decisions are not strictly individual but are affected by the preferences and constraints of families. Decisions are made not only to maximize income, for example, but also to meet family or household demands for insurance. By diversifying family labor, households can minimize risks to their well-being (Stark 1991).

Migration theory also requires consideration of political factors, especially to explain why international flows are much lower than would be predicted based solely on economic costs and benefits. Since a fundamental function of the state is to preserve the integrity of a society by controlling entry of foreigners, explanations must balance the interests of the individual with those of society as expressed through migration policies.

The various factors influencing migration decisions are often categorized according to whether they attract migrants to a region of destination ("pull" factors), drive migrants out of regions of origin ("push" factors), or facilitate the process of migration ("network" factors) (Martin and Widgren 1996). In addition to the factors evoked by these theories, others might include the need to flee life-threatening situations, the existence of kin or other social networks in destination countries, the existence of an underground market in migration, and income inequality and changes in cultural perceptions of migration in sending countries that are induced by migration itself (United Nations 1998).

An additional factor particularly relevant to the MA is the potential for growing numbers of "environmental refugees"—people driven to migrate by environmental factors (El-Hinnawi 1985). There is considerable debate on the relevance of environmental change to migration (Suhrke 1994; Hugo 1996). At one end is the view that environmental conditions are just one of many "push" factors influencing migration decisions (MacKellar et al. 1998). Environmental change, in this view, primarily acts indirectly by reducing income (by, for example, reducing agricultural productivity), making income less stable or negatively affecting health or environmental amenities. It also acts in concert with other factors, and therefore its relative role is difficult to isolate. At the other end lies the view that deteriorating environmental conditions are a key cause of a significant number of migrants in developing countries (Jacobson 1989; Myers 2002). While other factors such as poverty and population growth may interact with environmental change, environmental degradation is assumed to play a principal role.

The degree to which environmental migration is relevant to long-term population projections depends in part on the anticipated magnitude of the population movements. One estimate puts the number of environmental refugees in the mid-1990s at 25 million (over half of them in sub-Saharan Africa) (Myers and Kent 1995; Myers 2002). In comparison, there were about 26 million refugees as traditionally defined and an estimated 125 million international migrants—that is, people living in a country other than the one in which they were born (Martin and Widgren 1996; United Nations 1998). According to the United Nations High Commissioner on Refugees, there were about 13 million refugees in 1995, and an additional 13 million "persons of concern to the UNHCR," a group that includes people forced from their homes or communities but still residing in their own countries. Since Myers includes displaced persons who have not crossed international borders in his definition of "environmental refugees," the total figure—26 million—is the most relevant for comparison. Myers (2002) predicts that the number of environmental refugees is likely to double by 2010, and could swell to 200 million by 2025 due to the impacts of climate change and other sources of environmental pressure.

The potential relevance of these figures to population projections also depends on the level of aggregation. Most environmental migration occurs within national boundaries and therefore would not affect any of the long-term population projections. Additionally, some long-term projections are made at the level of world regions, so that much of the international migration would be masked as well. Finally, if environmental migration occurs in the future, its relevance compared with other factors driving migration, such as economic imbalances, must be weighed before concluding it is important to long-range population projections.

7.2.1.3 Demographic Change and Ecosystem Consequences

The ways in which population change influences ecosystems are complex. The basic pathway is from growing consumption driven by population to production processes that rely in part on ecosystem services to meet that consumption. The ultimate effects on ecosystems of an additional person are influenced by the entire range of indirect drivers. Research on direct population-environment interactions has focused increasingly on demographic characteristics that go beyond population size, and on a wide range of mediating factors.

For example, the energy studies literature has identified household characteristics such as size, age, and composition as key determinants of residential energy demand (Schipper 1996a). Household size appears to have an important effect on per capita consumption, most likely due to the existence of substantial economies of scale in energy use at the household level (Ironmonger et al. 1995; Vringer and Blok 1995; O'Neill and Chen 2002). Smaller household size is often accompanied by a larger number of households, and there is some evidence that the number of households, rather than population size per se, drives environmentally significant consumption (Cramer 1997, 1998). Research on ecological impacts of changes in household size and numbers has a shorter history but has also suggested a link between the two (Liu et al. 2003). Much of this literature is based on case studies, which suggest the importance of local context in determining population and land use interactions (National Academy of Sciences 2001). It has also been suggested that larger populations create difficulties for democratic institutions and thus environmental governance (Dahl and Tufte 1973; Frey and Al-Mansour 1995; Dietz 1996/1997).

These relationships may be important determinants of aggregate consumption in the future, since the proportion of the population living in various household types may shift substantially. The aging process itself leads to a shift toward smaller households, as children become a smaller fraction of the population. In addition, changes in preferences for nuclear over extended households may lead to even greater shifts in developing countries.

7.2.1.4 Overview of Demographic Drivers in the MA Scenarios

Four population projections were developed for the MA Scenarios Working Group to use in the quantification of storylines. All four projections are based on the IIASA 2001 probabilistic projections for the world (Lutz et al. 2001), but they are designed to be consistent with the four MA storylines, as judged by the Working Group with additional input from IIASA demographers (O'Neill 2005).

Table 7.2 lists the qualitative assumptions about fertility, mortality, and migration for each storyline. These assumptions are expressed in terms of high, medium, and low (H/M/L) categories, defined not in absolute but in relative terms—that is, a high fertility assumption for a given region means that fertility is assumed to be high relative to the median of the probability distribution for future fertility in the IIASA projections. Since the storylines describe events unfolding through 2050, the demographic assumptions specified here apply through 2050 as well. For the period 2050–2100, the demographic assumptions were assumed to remain the same, in order to gauge the consequences of trends through 2050 for the longer term. This is not intended to reflect any judgment regarding the plausibility of trends beyond 2050.

Figure 7.2 shows results for global population size. The range of population values in the scenarios is 8.1–9.6 billion in 2050 and 6.8–10.5 billion in 2100. These ranges cover 50–60% of the full uncertainty distribution for population

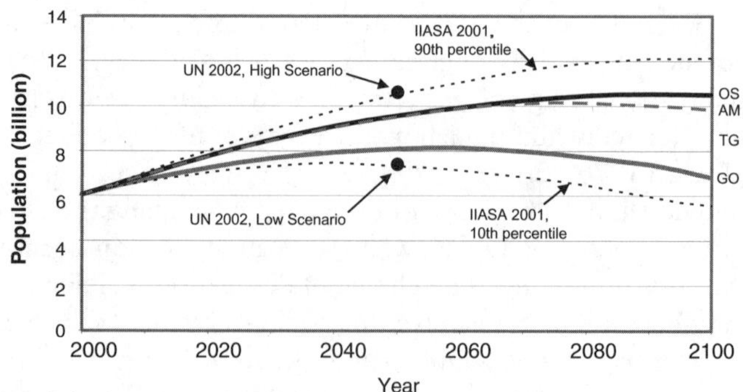

Figure 7.2. MA World Population Scenarios, 2000–2100. OS: Order from Strength, TG: TechnoGarden, AM: Adapting Mosaic, GO: Global Orchestration. An overview of each scenario can be found in the Summary.

size in the IIASA projections. The primary reason that these scenarios do not fall closer to the extremes of the full uncertainty distribution is that they correlate fertility and mortality: the Order from Strength scenario generally assumes high fertility and high mortality, and the Global Orchestration scenario generally assumes low fertility and low mortality. Both of these pairs of assumptions lead to more moderate population size outcomes.

7.2.2 *Economic Drivers: Consumption, Production, and Globalization*

Economic activity is a consequence of humans striving to improve their well-being. It is the myriad of technological processes that combine physical inputs, many derived from ecosystems, with human effort to generate goods and services that can improve human well-being. This activity is influenced by the endowment of natural resources, including ecosystem services (natural capital), the number and skills of humans (human capital), the stock of built resources (manufactured capital), and the nature of human institutions, both formal and informal (social capital). An early definition of social capital was "features of social organization, such as trust, norms, and networks that can improve the efficiency of society by facilitating coordinated actions" (Putnam et al. 1993, p 167). We expand the definition here to include formal institutions, such as the various levels of

Table 7.2 Fertility, Mortality, and Migration Assumptions, by Scenario[a]

Variable	Global Orchestration	TechnoGarden	Order from Strength	Adapting Mosaic
Fertility	HF: low LF: low VLF: medium	HF: medium LF: medium VLF: medium	HF: high LF: high VLF: low	Order from Strength until 2010, deviate to medium by 2050
Mortality	D: low I: low	D: medium I: medium	D: high I: high	Order from Strength until 2010, deviate to medium by 2050
Migration	high	medium	low	low

Key: I = more-developed country regions, D = less-developed country regions, HF = high-fertility regions (TFR > 2.1 in year 2000), LF = low-fertility regions (1.7 < TFR < 2.1), VLF = very-low-fertility regions (TFR < 1.7).

Notes: [a]In the ITASA projections, migration is assumed to be zero beyond 2070, so all scenarios have zero migration in the long run.

governments and their policies and regulations, and cultural and religious aspects of social organization.

Economic activity is also strongly influenced by available technologies. Local resource endowments of all kinds, and technologies, can be enhanced by access to other markets. International flows of goods and services, capital, labor, and ideas change the mix of economic activities that can be undertaken at home and the variety of items available for consumption.

7.2.2.1 Economic Growth, Changing Consumption Patterns, and Structural Transformation

Human well-being is clearly affected by economic growth and its distribution. Income received by individuals and families determines their level and nature of consumption. As per capita income grows, the nature of consumption changes, shifting from basic needs to goods and services that improve the quality of life. Businesses respond to these changing demands by producing an evolving mix of products.

As income increases, the mix of economic activities changes. This process, sometimes referred to as structural transformation, is driven by human behavior summarized in the form of two related economic "laws" with important consequences for ecosystems—Engel's law and Bennett's law. Engel's law, named after the German statistician who first observed the resulting statistical regularity, states that as income grows, the share of additional income spent on food declines. This relationship follows from basic human behavior. After basic food needs are met, the demand for an additional quantity of food drops off rapidly. Bennett's law states that as incomes rise, the source of calories changes. (A monograph by M. K. Bennett on Wheat in National Diets in a 1941 issue of *Wheat Studies* (Bennett 1941) and related comparative studies of the consumption of staple foods led to the empirical generalization that there is an inverse relationship between the percentage of total calories derived from cereals and other staple foods and per capita income.) The importance of starchy staples (e.g. rice, wheat, potatoes) declines, and diets include more fat, meat and fish, and fruits and vegetables. This behavior is the result of a general human desire for more dietary diversity and the ability to afford it as income rises.

These laws have several consequences for ecosystem condition and demand for ecosystem services. As income grows, the demand for nonagricultural goods and services increases more than proportionally. Producers respond by devoting relatively more resources to industry and service activities than agriculture. The share of agricultural output in total economic activity falls. The shift to a more diverse diet, in particular to more animal- and fish-based protein intake, slows the shift away from agriculture. Total consumption of starchy staples rises over some range of incomes as animal consumption that relies on feed grains gradually replaces direct human consumption of those grains. Eventually the demand for more diverse diets is satisfied, and further income growth is spent almost entirely on nonagricultural goods and services.

Industrial output share rises initially but then falls. Throughout the process of economic development, the importance of services in economic output rises continuously. A consequence of this shift toward services is that by the late 1990s, services provided more than 60% of global output, and in many countries an even larger share of employment (World Trade Organization 2004). In 2000, agriculture accounted for 5% of world GDP, industry accounted for 31%, and service industries, 64% (World Resources Institute et al. 2002).

Figure 7.3 documents the shift in economic structure in the past two centuries of the world's largest economies from agricultural production to industry and, to a greater extent, services. In developing regions a marked decline in the contribution of agriculture to GDP has occurred in recent years, but the contribution of services is larger than it was historically in industrial countries at the same level of income.

The shift away from agriculture and toward nonagricultural goods and then services is sometimes viewed as a process that ultimately reduces pressures on ecosystems, since services are assumed to be the least demanding of ecosystem products. But this outcome must be interpreted carefully. It is important to distinguish between absolute and relative changes. High-income countries almost always produce more agricultural output than when they were poor, but industrial and services output grows much faster, so the relative contribution of agriculture declines. Technological change further replaces most of the labor force in agriculture, potentially altering the demands for ecosystem services. The way agricultural statistics are reported also tends to overemphasize the extent of the decline in the economic importance of agriculture. In developing countries, the agricultural sector is "vertically integrated"—farmers produce everything from seeds and agricultural infrastructure to food services. In high-income countries, "agricultural" statistics focuses only on production of primary products. "Food" production is reported in statistics on industry, transport, and services (fast food, restaurants, and so on).

Urbanization also influences the structure of food consumption, increasing the service content dramatically. Rural consumers are more likely to consume food produced at home. Urban consumers are more likely to demand easily prepared, quick meals and to purchase them from restaurants. Supermarkets replace neighborhood stores and street vendors.

A few examples from Asian countries with rapidly rising incomes illustrate these phenomena, although similar changes are occurring throughout the developing world today and happened in the industrial world in the twentieth century. Data from China show that the human intake of cereals and the consumption of coarse grains decreased during the past two decades in both urban and rural populations, and there was a dramatic increase in the consumption of animal foods. A similar, but less dramatic change is also observed in India, with figures that suggest a doubling in the intake of fat calories over a 20-year period. Although the Indian consumption of rice and wheat has been increasing, the percentage of all cereals in household expenditure

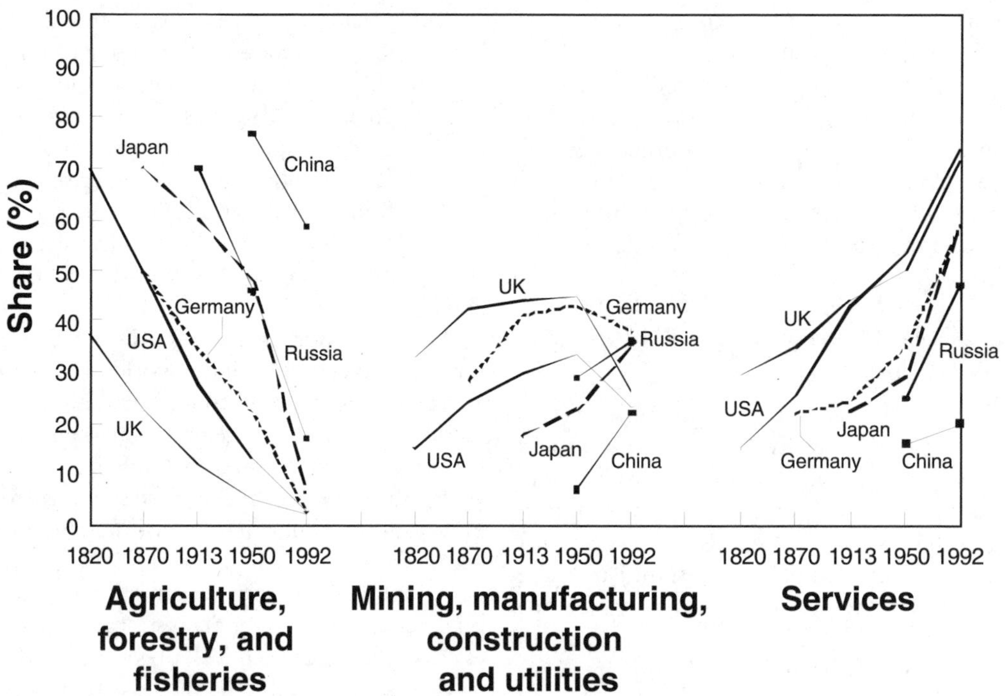

Figure 7.3. Changes in Economic Structure for Selected Countries, 1820–1992 (Nakićenović et al. 2000; Maddison 1995)

has been declining. Meat consumption in India has been growing, although not as fast as in China (FAO 2004). The major increases in food consumption in India are in milk, eggs, fruit, and vegetables. Vegetable oil demand is also growing (USDA 2001).

By 2002, the share of supermarkets in the processed and packaged food retail market was 33% in Southeast Asia and 63% in East Asia. The share of supermarkets in fresh foods was roughly 15–20% in Southeast Asia and 30% in East Asia outside China. The 2001 supermarket share of Chinese urban food markets was 48%, up from 30% in 1999. Supermarkets are also becoming an emerging force in South Asia, particularly in urban India since the mid-1990s (Pingali and Khwaja in press).

7.2.2.2 Economic Growth, Distribution, and Globalization

The rate of growth and its sectoral composition depend on resource endowments, including ecosystem condition, on the technologies available, and on the extent of market reach. Hence the effects of global economic performance on ecosystems are more than straightforward changes in national income. International trade, capital flows, technology transfer, and technical change are crucial elements in global growth.

Perhaps the most comprehensive compilation of data on historical economic development is that of Maddison (1995). Table 7.3 shows Maddison's per capita GDP growth rate estimates for selected regions and time periods. Since 1820, global GDP has increased by a factor of 40, or at a rate of about 2.2% per year. In the past 110 years, global per capita GDP grew by a factor of more than five, or at a rate of 1.5% per year. Between 1950 and 2000, world GDP grew by 3.85%, resulting in an average per capita income growth rate of 2.09% for that period (Maddison 2003).

In the late twentieth century, income was distributed unevenly both within countries and around the world. (See Figure 7.4.) The level of per capita income was highest in North America, Western Europe, Australasia, and Northeast Asia (see Figure 7.5 in Appendix A), but growth rates were highest in South Asia, China, and parts of South America (see Figures 7.6a and 7.6b in Appendix A). If these trends continue, global income disparities will be reduced, although national disparities might increase. Africa is a conspicuous exception to the trend of growing incomes.

Economic growth is facilitated by trade. Growth in international trade flows has exceeded growth in global production for many years, and the differential may be growing. (See Figure 7.7.) In 2001, international trade in goods was equal to 40% of gross world product (World Bank 2003). Growth in trade of manufactured goods has been much more rapid than trade in agricultural or mining products. (See Figure 7.8.)

High incomes in OECD countries and rapid growth in income in some lower-income countries, combined with unprecedented growth of global interconnectedness, is leading to dramatic changes in lifestyles and consumption patterns. For example, tourism is one of the most rapidly growing industries, and growth in trade of processed food products and fresh fruits and vegetables is much more rapid than growth in trade of raw agricultural commodities.

Economic growth requires an expansion of physical and institutional infrastructure. The development of this infrastructure can play a major role in the impacts on ecosystems. In a review of 152 studies of tropical deforestation, Geist and Lambin (2002) found that 72 studies cited infrastructure extension (including transportation, markets, settlements, public services, and private-sector activities) as an important direct driver. In a similar study evaluating 132

Table 7.3 Per Capita GDP Growth Rates for Selected Regions and Time Periods (Nakićenović et al. 2000, based on Maddison 1995, with 1990–2000 data from Maddison 2003)

	1870–1913	1913–50	1950–80	1980–92	1990–2000
			(percent per year)		
Western Europe	1.3	0.9	3.5	1.7	1.7
Australia, Canada, New Zealand, United States	1.8	1.6	2.2	1.3	1.9
Eastern Europe	1.0	1.2	2.9	–2.4	0.6
Latin America	1.5	1.5	2.5	–0.6	1.4
Asia	0.6	0.1	3.5	3.6	3.2
Africa	0.5	1.0	1.8	–0.8	0.1
World (sample of 199 countries)	**1.3**	**0.9**	**2.5**	**1.1**	**1.5**

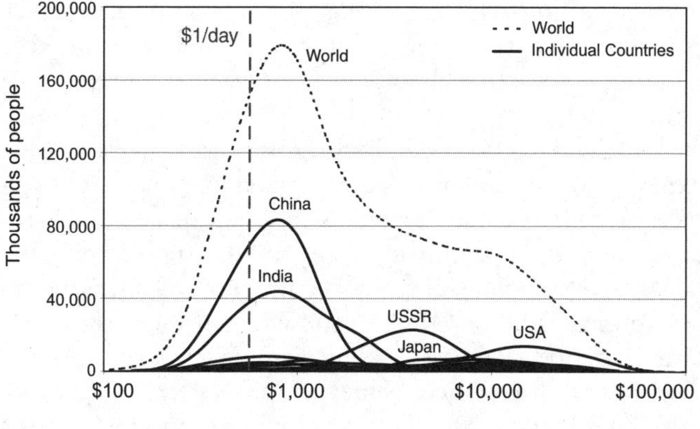

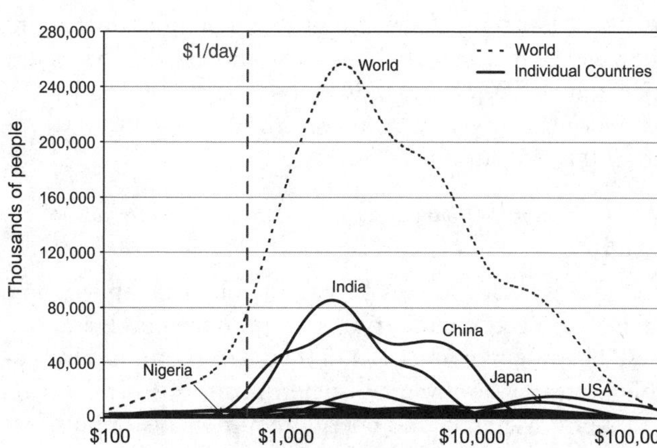

Figure 7.4. Income Level and Distribution, 1970 and 2000. Note: the data used are adjusted to 1985 prices and are PPP adjusted, drawing on various Summers and Heston/Penn-World Tables work. (Sala-i-Martin 2003, as reproduced in Barro and Sala-i-Martin 2003)

studies on desertification, they found infrastructure extension cited 73 times (Geist and Lambin 2004).

7.2.2.3 Economic Distortions

Government policies can alter market outcomes, increasing or reducing prices and changing production and consumption levels. By some estimates, distortions in agricultural markets are the largest. Total support to agriculture in OECD countries averaged over $324 billion per year in 2001–03; about three quarters of this amount was used to

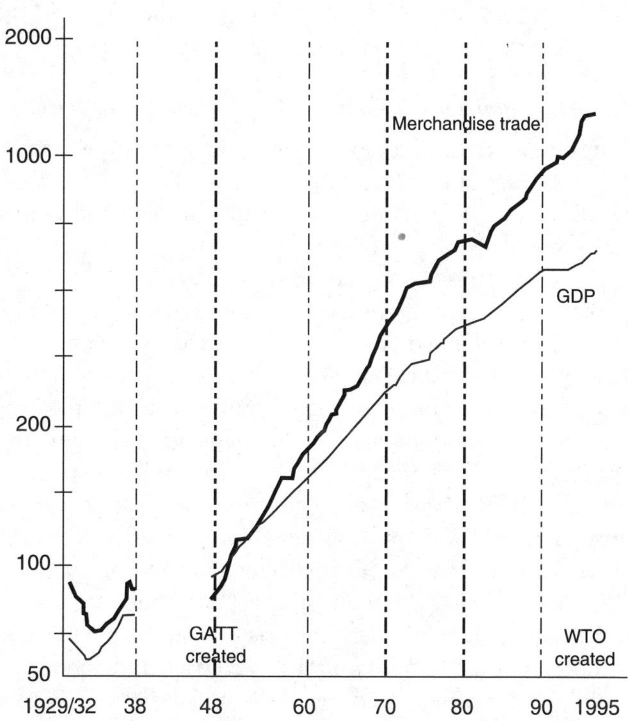

Figure 7.7. World Trade and GDP Growth, 1930–95 (World Trade Organization 2003)

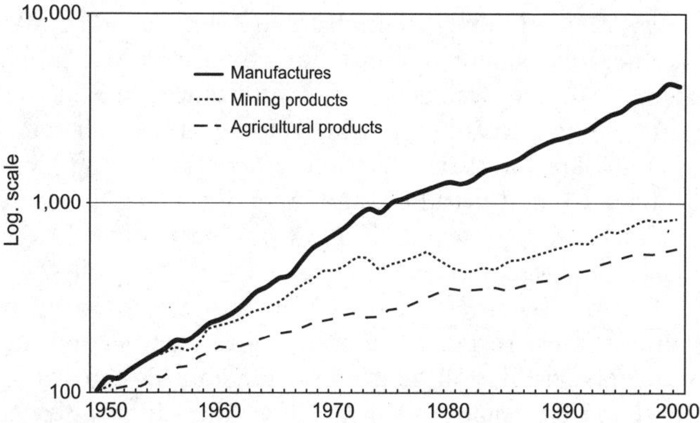

Figure 7.8. World Merchandise Trade by Major Trade Groups, 1950–2001 (http://www.wto.org/English/res_e/statis_e/its2002_e/its02_longterm_e.htm)

support farm income directly, while the remainder went into general infrastructure improvements, research, marketing, and so on (OECD 2004). Many of these subsidies were in the form of higher prices to farmers, providing direct incentive to increase agricultural production. In low-income countries, on the other hand, governments sometimes tax agriculture directly or indirectly and do not provide support systems of research, marketing, and transportation infrastructure (Anderson and Hayami 1986).

According to the U.N. Environment Programme (UNEP 1999), global energy subsidies currently total $200 billion a year (de Moor 2002). OECD countries spend some $82 billion a year subsidizing energy production, mostly through tax breaks, cheap provision of public infrastructure and services, subsidized capital, and price support (OECD 1997). Globally, more than 80% of the subsidies are for fossil fuels use, among the most polluting energy sources.

7.2.2.4 Determinants of Economic Growth and Development

Economic growth and development depend on growth in the availability of resources, the mobility of those resources, the efficiency of their use, and the institutional and policy environment. Growth in per capita output depends on total output growing more rapidly than population.[1]

Numerous preconditions must be met before any "take off" into accelerated rates of productivity and economic growth can materialize. Research based on historical evidence allows a number of generalizations as to the patterns of advances in productivity and economic growth—the importance of economic openness to trade and capital flows and the contribution of technological change either through innovation or adoption. Little research has been undertaken on the role of ecosystems in economic growth.

Chenery et al. (1986) and Barro (1997) indicate strong empirical evidence of a positive relationship between trade openness and productivity, industrialization, and economic growth. For example, between 1990 and 1998 the 12 fastest-growing developing countries saw their exports of goods and services increase 14% and their output 8% (World Bank 2002). Dosi et al. (1990) highlight the critical roles of policies and institutions in realizing economic gains from the international division of labor.

It is theoretically possible that trade liberalization could have negative economic impacts on countries where property rights are not well defined, but little empirical evidence exists of this actually occurring. However, not all trade flows are equal in their effects on growth. Dollar and Collier (2001) found that the countries experiencing the most rapid trade-driven economic growth were trading a large share of high-technology products.

The late twentieth century trend toward more open economies led to greater uniformity in macroeconomic (monetary, fiscal, and exchange rate) policies across the world and facilitated capital mobility. International capital flows are critical to economic growth because they relieve resource constraints and often facilitate technology transfers that enhance productivity of existing resources. But not all developing countries participated equally. For instance, the vast majority of private-sector capital flows is concentrated in the 10 largest developing countries (World Bank 2002).

Adoption of technical improvements leads to the productivity growth essential to improvements in per capita income. Expenditures on research and development typically have high returns. Late-developing countries can, for a time, rely on borrowing technologies to improve productivity. Growth rates tend to be lower for economies at the technology and productivity frontier. For instance, nineteenth-century productivity and per capita GDP growth rates in the rapidly industrializing United States far exceeded those of England, then at the technology and productivity frontier. Likewise, in the post–World War II period, growth rates in Japan and most of Western Europe exceeded U.S. rates (by then at the technology and productivity frontier) (Maddison 1991, 1995). High human capital (education), a favorable institutional environment, free trade, and access to technology are key factors for rapid economic catch-up.

Sustained high-productivity growth and in some cases exploitation of natural resources (including ecosystem services) resulted in the current high levels of per capita income in OECD countries. Latecomers (such as Austria, Japan, and Scandinavia) rapidly caught up to the productivity frontier of other OECD economies in the post–World War II period. Per capita GDP growth rates of 3.5% per year were, for instance, achieved in Western Europe between 1950 and 1980. The developing economies of Asia achieved high per capita GDP growth rates beginning in the 1960s. Per capita GDP growth rates of individual countries have been extremely high for short periods—8% a year in Japan in 1950–73, 7% in Korea between 1965 and 1992, and over 6% per year in China from the early 1980s to the mid-2000s (Maddison 1995).

7.2.2.5 Economic Productivity and Energy and Materials Intensity

Economy activity requires energy and physical inputs, some of which are ecosystem services, to produce goods and services. The rate of conversion of inputs to economically valuable outputs is an important determinant of the impact on ecosystems. Materials and energy requirements (inputs) per unit of economic activity (often measured by GDP) are referred to as materials and energy intensity, respectively. Some evidence suggests that materials and energy intensity follow an inverted U-curve (IU hypothesis) as income grows—that is, the requirements per unit of economic activity rise for some earlier increases in economic activity and then decline. For some materials, the IU hypothesis (Moll 1989; Tilton 1990) holds quite well. The underlying explanatory factors are a mixture of structural change in the economy along with technology and resource substitution and innovation processes. Recent literature suggests that an N-shaped curve better describes the relationship of material intensity in high-income countries (De Bruyn and Opschoor 1994; De Bruyn et al. 1995; Suri and Chapman 1996; Ansuategi et al. 1997).[2]

Figure 7.9 in Appendix A shows material intensity versus per capita income data for 13 world regions for some metals (van Vuuren et al. 2000; see also the discussion in De Vries et al. 1994). Figure 7.10 in Appendix A shows a similar curve for total energy intensity (including traditional noncommercial energy forms) for 11 world regions, again as a function of per capita income (Nakićenović et al. 1998).

Commercial energy intensity of GDP generally follows the IU hypothesis, although the initially rising part of commercial energy intensity stems from replacing traditional energy forms and technologies with modern commercial energy forms. Traditional energy sources such as fuelwood; agricultural wastes, including dung; work of animals; wind mills; and water wheels have low energy intensity compared with modern energy sources such as oil products and electricity. The traditional methods of biomass combustion are not only inefficient but lead to a wide range of health hazards, such as indoor air pollution (Smith and Mehta 2003). Replacing traditional energy sources and carriers with modern sources increases the conversion efficiencies, especially at the point of end use. Consequently, the resulting aggregate total (commercial plus noncommercial) energy intensity shows a persistent declining trend over time, especially with rising incomes (Watson et al. 1996; Nakićenović et al. 1998). (See Figure 7.11 in Appendix A.)

There are two important points to retain from Figures 7.9–7.11 in Appendix A. First, energy and materials intensity (that is, energy use per unit of economic output) tend to decline with rising levels of GDP per capita. In other words, energy and material productivity—the inverse of energy intensity—improve in line with overall macroeconomic productivity.

Second, growth in productivity and intensity improvement has historically been outpaced by economic output growth. Hence, materials and energy use have risen in absolute terms over time (Nriagu 1996; Watson et al. 1996; Grübler 1998). An important issue for the future is whether technological advancement can outpace economic growth and lead to reductions in materials and energy use.

It is also important to emphasize that energy and material intensity are affected by many factors other than macroeconomic productivity growth. OECD (1998) notes that high rates of productivity increase have been associated in the past with new competitive pressures, strong price or regulatory incentives, catching up or recovery, and a good "climate for innovation." Table 7.4 summarizes selected macroeconomic, labor, energy, and material productivity increases that have been achieved in a range of economies and sectors at different times. (See also the discussion later on science and technology drivers.) This historical evidence suggests that continued productivity growth is a reasonable assumption for the future.

For instance, low historical rates of energy intensity improvement reflect the low priority placed on energy efficiency by most producers and users of technology. On average, energy costs account for only about 5% of GDP. Energy intensity reductions average about 1% per year, in contrast to improvements in labor productivity above 2% per year over the period 1870 to1992. Over shorter time periods, and given appropriate incentives, energy intensity improvement rates can be substantially higher, as in the OECD countries after 1973 or in China since 1977, where energy intensity improvement rates of 5% have been observed.

Rapid productivity growth can also occur during periods of successful economic catch-up; for instance, Japanese labor productivity grew at 7.7% annually during 1950–73 (Maddison 1995). Similar high-productivity growth was also achieved in industrial oil usage in the OECD or U.S. car fuel economies after 1973. Of the examples given in Table 7.4, productivity increases are the highest for communication. Many observers consider that communication may become as an important a driver of economic growth in the future as traditional resource- and energy-intensive industries have been in the past.

7.2.2.6 Economic Drivers and Ecosystem Consequences

The twin questions of the sustainability of economic growth and its impact on the environment were given high visibility by the 1987 report of the World Commission on Environment and Development (World Commission on Environment and Development 1987). Researchers have started to look for empirical evidence to answer these questions and provide a theoretical basis for understanding the interactions between economic development and environmental quality (Grossman 1995; Dasgupta et al. 1997).

Some argue that continuous economic growth requires an ever-increasing amount of resources and energy and produces rising pollution and waste levels. As Earth's natural resources and its capacity to absorb waste are finite, continuous economic growth will eventually overwhelm the carrying capacity of the planet (Georgescu-Roegen 1971 cited in Meadows et al. 1972; Panayotou 2000). Therefore, they argue, economic systems must eventually be transformed to steady-state economies, in which economic growth ceases (Daly 1991).

Others argue that economic growth results eventually in a strengthening of environmental protection measures and hence an increase in environmental quality. Higher per capita income levels spur demand for a better environment, which induces development of policies and regulations to address environmental quality problems. As the IU hypothesis suggests, initial empirical research found that at lower levels of income, economic growth is connected with increasing environmental damage. But after reaching a certain level of per capita income, the impact on at least some elements of environmental quality reverses.

Pollution abatement efforts appear to increase with income, a growing willingness to pay for a clean environment, and progress in the development of clean technology. This process seems well established for traditional pollutants, such as particulates and sulfur (e.g., World Bank 1992; Kato 1996; Viguier 1999), and there have been some claims that it might apply to greenhouse gas emissions. Schmalensee et al. (1998) found that CO_2 emissions have flattened and may have reversed for highly developed economies such as the United States and Japan. This IU relationship is sometimes referred to as the Environmental Kuznets Curve, named for a similar-looking relationship between income and inequality identified by Simon Kuznets (Kuznets 1955).

Table 7.4 Examples of Productivity Growth for the Entire Economy and for Selected Sectors and Countries (Nakićenović et al. 2000)

Sector/Technology	Region	Productivity Indicator	Period	Annual Productivity Change (percent)
Whole economy [a]	12 countries Europe	GDP/capita	1870–1992	1.7
Whole economy [a]	12 countries Europe	GDP/hour worked	1870–1992	2.2
Whole economy [a]	U.S.	GDP/hour worked	1870–1973	2.3
Whole economy [a]	U.S.	GDP/hour worked	1973–92	1.1
Whole economy [a]	Japan	GDP/hour worked	1950–73	7.7
Whole economy [a]	South Korea	GDP/hour worked	1950–92	4.6
Whole economy [b]	World	GDP/primary energy	1971–95	1.0
Whole economy [b]	OECD	GDP/primary energy	1971–95	1.3
Whole economy [b]	U.S.	GDP/primary energy	1800–1995	0.9
Whole economy [b]	United Kingdom	GDP/primary energy	1890–1995	0.9
Whole economy [b]	China	GDP/primary energy	1977–95	4.9
Whole economy [c]	Japan	GDP/material use	1975–94	2.0
Whole economy [c]	U.S.	GDP/material use	1975–94	2.5
Agriculture [d,e]	Ireland	tons wheat/hectare	1950–90	5.3
Agriculture [e]	Japan	tons rice/hectare	1950–96	2.2
Agriculture [e]	India	tons rice/hectare	1950–96	2.0
Industry [a]	OECD (6 countries)	value added/hour worked	1950-84	5.3
Industry [a]	Japan	value added/hour worked	1950–73	7.3
Industry [b]	OECD	industrial production/energy	1971–95	2.5
Industry [b]	OECD	industrial production/energy	1974–86	8.0
New cars [f]	U.S.	vehicle fuel economy	1972–82	7.0
New cars [f]	U.S.	vehicle fuel economy	1982–92	0.0
Commercial aviation [g]	World	ton-km/energy	1974–88	3.8
Commercial aviation [g]	World	ton-km/energy	1988–95	0.3
Commercial aviation [g]	World	ton-km/labor	1974–95	5.6
Telephone call costs [d]	Transatlantic	London–NY, costs for 3 minutes	1925–95	8.5
Telephone cables [d]	Transatlantic	telephone calls/unit cable mass	1914–94	25.0

Data sources: [a] Maddison 1995. [b] OECD and IEA statistics. [c] WRI 1997. [d] OECD 1998b; Waggoner 1996; Hayami and Ruttan 1985.
[e] FAO (various years 1963–96) Production Statistics. [f] Includes light trucks; Schipper 1996. [g] International Civil Aviation Organization statistics.

(See Grossman and Krueger 1992; Shafik and Bandyopadhay 1992; Panayotou 1993.)

The principal explanation for this relationship is changes in economic structure, from more industrial to more services, which occur with development and technical innovation that provides more resource-efficient technologies (Stern 2004). (While it often assumed that the shift toward services will reduce environmental impact, this assertion has been challenged (Salzman and Rejeski 2002; York et al. 2003b). In addition, a number of more indirect factors, such as the growing awareness of environmental problems, education, and improved environmental regulations, are now also seen as affecting the shape of the curve (Stern 2004).

The EKC hypothesis has generated many studies (for a thorough literature review, see Panayotou 2000 and Stern 2004). Stern (2004) provides one of the most recent reviews of the criticism and states that "there is little evidence for a common inverted U-shaped pathway which countries follow as their income rises." There are three main criticisms of the EKC results: First, its econometric foundations have been challenged (Harbaugh et al. 2000; Stern 2004). A second issue is the choice of indicators selected to represent environmental quality (Grossman 1995). A third criticism highlights omitted variables. Panayotou (2000) describes the use of per capita income in the EKC analysis as an "omnibus variable representing a variety of underlying influences." Urbanization, infrastructure, poverty, and income distribution are other factors in the complex interplay between economic growth, population, and environment (see, e.g., Rotmans and de Vries 1997; DeVries et al. 1999; O'Neill et al. 2001).

In summary, the EKC is at best a reduced form description, not a precise formulation of cause and effect, which captures a few of the complex interactions among economic activity and ecosystem condition and service. Other attempts at describing, identifying, and explaining aggregate relationships between economic activity and the environment, such as the ecological footprint (York et al. 2003a), are subject to similar criticisms.

In conclusion, there is controversy about whether the current rate of global economic growth is sustainable. There is little question that some, perhaps many, of the world's ecosystems have experienced unsustainable pressure as resources are extracted and ecosystem services are used to produce this growth. However, given the complexity of the interactions among economy and environment, simple formulations of the relationships between ecosystem sustainability and economic growth are not possible.

7.2.2.7 Tourism as an Example of Economic Drivers and the Environment

Tourism provides a good example of the complexities of the interactions among economic growth and ecosystems. World tourism spending is expected to grow at over 6% per year (World Tourism Organization as referenced in Hawkins and Lamoureux 2001), making it one of the world's fastest growing industries and a major source of foreign exchange earning and employment for many developing countries.[3]

Tourism is increasingly focusing on natural environments. Specialty tourism, including ecotourism, accounted for about 20% of total international travel in the late 1990s (World Tourism Organization (1998) as cited in Hawkins and Lamoureux 2001). Ecotourism has the potential to contribute in a positive manner to socioeconomic well-being, but fast and uncontrolled growth can be the major cause of ecosystem degradation and loss of local identity and traditional cultures. Paradoxically, the very success of tourism can lead to the degradation of the natural environment. By drawing on local natural resources, tourism can reduce a location's attractiveness to tourists. The discussions in Weaver (2001) on tourism in rainforests, mountain ecosystems, polar environments, islands and coasts, deserts, and marine environments highlight these trade-offs.

Tourism is sometimes seen as an opportunity for economic development, economic diversification, and the growth of related activities, especially in developing countries. Among the benefits are direct revenues, generated by fees and taxes incurred and by voluntary payments for the use of biological resources. These revenues can be used for the maintenance of natural areas and a contribution of tourism to economic development, including linkage effects to other related sectors and job creation. However, it is well known that in developing countries a significant share of the initial tourist expenditures leave the destination country to pay for imported goods and services (Lindberg 2001).

Sustainable tourism can make positive improvements to biological diversity conservation, especially when local communities are directly involved with operators. If local communities receive income directly from a tourist enterprise, they are more likely to provide greater protection and conservation of local resources. Moreover, sustainable tourism can serve as an educational opportunity, increasing knowledge of and respect for natural ecosystems and biological resources. Other benefits include providing incentives to maintain traditional arts and crafts and traditional knowledge, plus innovations and practices that contribute to the sustainable use of biological diversity.

The impacts of tourism on ecosystems in general and on biodiversity in particular can be positive or negative, direct or indirect, and temporary or lasting, and they vary in scale from global to local (van der Duim and Caalders 2002). The different pathways these effects take are depicted in Figure 7.12.

Clearly, tourism acts on various direct drivers of the MA conceptual framework. Gössling (2002a) reports:

- Changes in land cover and land use due to tourism-related investments, including accommodation and golf courses, amount to more than 500,000 square kilometers, with the largest contribution coming from traffic infrastructure. Yet these numbers must be interpreted with caution, as roads, airports, and railways are used for nontourism activities also.
- Energy use and related carbon emissions occur due to transportation, but also due to accommodation and on-site activities. The overall contribution to carbon emissions is estimated to be on the order of 5%, of which transportation contributes about 90%.
- Biotic exchange is difficult to assess on a global scale, but various processes for exchange do exist. This includes the intentional or unintentional introduction of new species, both in the home and the hosting region. Extinction due to collection, hunting, or gathering of threatened species is another direct effect.
- Of indirect effect, yet from the human ecology perspective of high relevance, is the change of the human-nature relationships due to intercultural exchange during travel. Though it is believed that traveling can contribute to the increase in environmental consciousness of the guest, it is also possible that the people in the host country change their perception and understanding of the environment, for example by introducing the modern separation of "culture" and "nature" into indigenous societies (Gössling 2002b).

The potential adverse impacts of tourism can be roughly divided into environmental and socioeconomic, the latter often imposed on local and indigenous communities.

Direct use of natural resources, both renewable and nonrenewable, in the provision of tourist facilities is one of the most significant direct impacts of tourism in a given area. Land conversion for accommodation and infrastructure provision, the choice of the site, and the use of building materials are mechanisms by which ecosystems can be altered (Buckley 2001). Negative impacts on species composition and on wildlife can be caused by even such benign behavior as bird-watching (Sekercioglu 2002) and exacerbated by inappropriate behavior and unregulated tourism activities (such as off-road driving, plant-picking, hunting, shooting, fishing, and scuba diving). Tourist transportation can increase the risk of introducing alien species (water-borne pathogenic bacteria and protozoa, for instance) (Buckley 1998 as cited in Buckley 2001). And the manner and frequency of human presence can disturb the behavior of animals, as in the collapse of the feeding and mating systems of the Galapagos land iguana caused by tourist disturbance (Edington and Edington 1986 as cited in Buckley 2001).

Tourism has for many years been focused on mountain and coastal areas. In the mountains, sources of damage include erosion and pollution from the construction of hiking

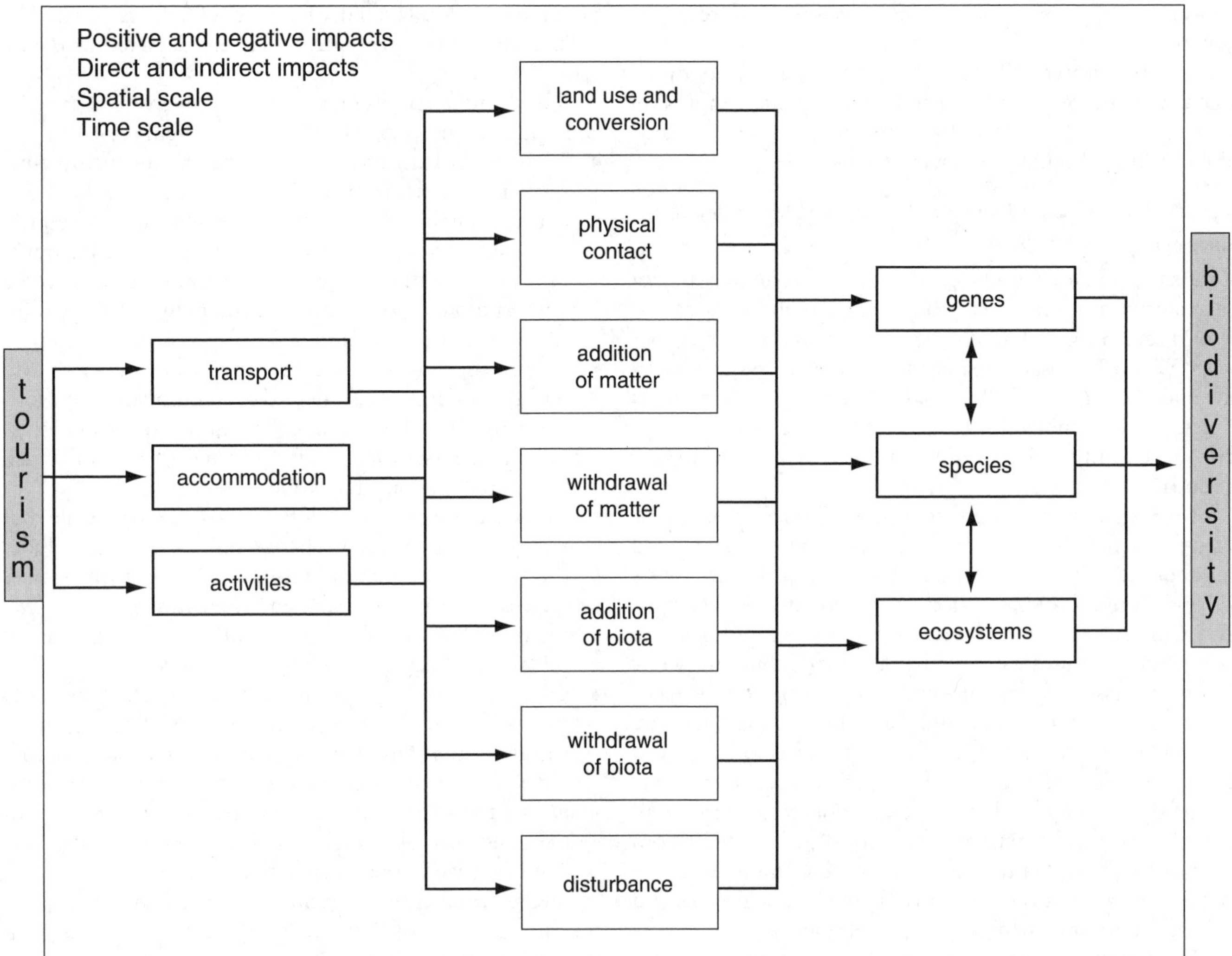

Figure 7.12. Pathways of the Ecological Impacts of Tourism on Biodiversity (van der Duim and Caalders 2002)

trails, bridges in high mountains, camp sites, chalets, and hotels (Frost 2001). In marine and coastal environments, impacts arise from inappropriate planning, irresponsible behavior by tourists and operators, and lack of education on and awareness of the impacts of, for example, tourist resorts along the coastal zones (Cater and Cater 2001; Halpenny 2001).

Tourism is a water-intensive activity with a large production of waste. The extraction of groundwater by some tourism activities can cause desiccation, resulting in loss of biological diversity. Moreover, the disposal of untreated effluents into surrounding rivers and seas can cause eutrophication, and it can also introduce pathogens into water bodies. Disposal of waste produced by the tourism industry may cause major environmental problems.

Negative socioeconomic and cultural consequences of tourism include impacts on local communities and cultural values (Wearing 2001). Increased tourism activities can cause an influx of people seeking employment or entrepreneurial opportunities who may not be able to find suitable employment. If an economy relies heavily on tourism, a recession elsewhere can result in a sudden loss of income and jobs. When tourism development occurs, economic benefits are sometimes unequally distributed among members of local communities. Lindberg (2001) points out that in some circumstances, tourism can actually increase inequalities, and thus relative poverty, in communities. However, there appears to be no evidence about whether this is a systematic problem or relatively infrequent. Negative cultural outcomes include intergenerational and gender conflicts, changes in traditional practices, and loss of access by indigenous and local communities to their land and resources as well as sacred sites. Positive outcomes include a renewed interest in maintaining local cultural practices.

Within the tourism industry and also within research on tourism, the notion of "sustainable tourism" has emerged to promote traveling with fewer negative impacts on sustainability. According to the World Tourism Organization, sustainable tourism development meets the needs of present tourists and host regions while protecting and enhancing opportunities for the future. It is envisaged as leading to management of all resources in such a way that economic, social, and aesthetic needs can be fulfilled while maintaining cultural integrity, essential ecological processes, biological diversity, and life support systems. Projects with headlines like "ecotourism" or "alternative tourism" have emerged under the overall premise of sustainable tourism. Though success stories do exist, there is evidence that the majority of projects cannot be genuinely conceived of as being sustainable (Collings 1999).

7.2.2.8 Overview of Economic Drivers in the MA Scenarios

The MA scenarios include a rich set of economic drivers, and Chapter 9 provides a more-detailed discussion of these. A useful summary statistic is the differences in per capita income. In the MA scenarios, per capita income grows two to four times between 2000 and 2050, depending on scenario. Total economic output grows three to six times during that period.

7.2.3 Sociopolitical Drivers

"Sociopolitical" drivers encompass the forces influencing decision-making in the large conceptual space between economics and culture. The boundaries among economic, sociopolitical, and cultural driver categories are fluid, changing with time, level of analysis, and observer (Young 2002).

Sociopolitical driving forces have been seen as important in past environmental change (see, e.g., Redman 1999; de Vries and Goudsblom 2002). However, these drivers have been more the subject of taken-for-granted assumptions and speculation than sound theoretical and empirical research, so the basis for strong conclusions about how these drivers work is limited. This is an active area of research across many disciplines, so the state of knowledge is improving rapidly. In this regard it is important to remember the adage that the lack of evidence of an effect is not evidence for a lack of effect. Sociopolitical drivers may be some of the most fundamental influences on how humans influence the environment. They should always be given careful consideration in understanding environmental change, and they deserve high priority in future research agendas.

Some topics under the general theme of sociopolitical drivers have been well researched. For example, a strong literature examines governance of the commons at scales ranging from the local to the global.

Many writers have argued for the importance of democracy and equitable distribution of power for protecting the environment, and there are numerous accounts of how arbitrary uses of power harm the environment (see, for example, Ehrlich and Ehrlich 2004). When we turn to the scientific literature investigating the effects of democratic institutions and other political forms as drivers of environmental impact, however, systematic research is still in its early stages. There is some theoretical work of long standing (e.g., Beck 1992; Buttel 2003; Mol and Sonnenfeld 2000; Schnaiberg 1980; York and Rosa 2003). But once we move from the well-developed literature on institutions for commons governance, empirical and analytical research is still in the early stages of development. Thus we can characterize the directions current research is taking but cannot draw strong conclusions.

For convenience in discussing this general literature, we have divided sociopolitical drivers into four categories related to governance:

- the quantity of public participation in public decision-making,
- the makeup of participants in public decision-making,
- the mechanisms of dispute resolution, and
- the role of the state relative to the private sector.

7.2.3.1 The Quantity of Public Participation in Public Decision-making

The general role of the public in decision-making appears to be expanding, as evidenced by the extent of democratization. Despite some backsliding, there has been a trend away from centralized authoritarian governments and a rise of elected democracies. As well, there is some evidence of improving administrative competence across the developing world (Kaufmann et al. 2003). It is generally assumed that democratization leads to government actions that are friendlier to the environment, but the evidence in support of this assertion is limited (Congleton 1996; York et al. 2003a).

The literature on public participation in environmental assessment and decision-making is much more robust and indicates that such involvement at the local and regional level generally leads to more-sustainable approaches to managing resources (Stern and Fineberg 1996; Dietz and Stern 1998; Tuler and Webler 1999; Lubell 2000; Beierle and Cayford 2002; Dietz et al. 2003).

Finally, there is a substantial and robust literature on a key category of environmental problem—the governance of commons (Ostrom et al. 2002). Some strong generalizations have emerged from this literature that contrast with Hardin's original stark conclusions (Hardin 1968). We are more likely to govern commons sustainably when:

- "Resources and use of the resources by humans can be monitored, and the information can be verified and understood at relatively low cost, . . .
- rates of change in resources, resource-user populations, technology and economic and social conditions are moderate,
- communities maintain frequent face-to-face communication and dense social networks—sometimes called social capital—that increase the potential for trust, allow people to express and see emotional reactions to distrust and lower the cost of monitoring behavior and inducing rule compliance,
- outsiders can be excluded at relatively low cost from using the resource . . . , and
- users support effect monitoring and rule enforcement" (Dietz et al. 2003, p. 1908).

The challenge is to find ways of structuring decision-making processes that support the emergence of these conditions or that adapt when they do not obtain.

7.2.3.2 The Makeup of Participants in Public Decision-making

The voices heard in public decision-making and how they are expressed have changed, as evidenced in the changing role of women, the rise of civil society, and the growth of engaged fundamentalism. Democratic institutions have also encouraged decentralized decision-making, with the intended beneficiaries having a greater say in the decisions made. This trend has helped empower local communities, especially rural women and resource-poor households. Decentralization trends have also had an impact on decisions made by regional and international institutions, with the increasing involvement of NGOs and grassroots organizations, such as traditional peoples groups.

The power of NGOs arises in part from their ability to mobilize voters in societies where the average citizen does not participate actively in the political process. Hence, more openness and transparency in public decision-making enhances the influence of NGOs (Princen and Finger 1994).

7.2.3.3 The Mechanisms of Dispute Resolution

The mechanisms by which nations solve their disputes, peaceful and otherwise, are changing. Although the cold war has ended, the persistence of regional and civil wars and other international conflicts in some parts of the world continues to be a matter of concern. There is an urgent need to understand the driving forces behind such conflicts and their impact on sustainable livelihoods and the natural resource base. (See Box 7.1.)

Numerous mechanisms have been proposed to help incorporate the views of diverse stakeholders into environmental decision-making (Renn et al. 1995). While it is not yet clear how these function in practice, there is a growing body of research on public participation in environmental assessment and decision-making, and several synthetic efforts to understand these processes are under way (U.S. National Research Council; Kasemir et al. 2003).

7.2.3.4 The Role of the State Relative to the Private Sector

The declining importance of the state relative to the private sector—as a supplier of goods and services, as a source of employment, and as a source of innovation—seems likely to continue. The future functions of the state in provisioning public goods, security, and regulation are still evolving, particularly in the developing world. In all countries the implications of privatization trends on the sustainable management of the local and global resource base are still not clear. The old two-way relationships between governments and private firms have been radically changed with the emergence of large numbers of NGOs, which have become important actors in the political and social scene. For example, some environmental NGOs employ scientists who play important roles in bridging scientific communities to assess environmental problems, educating the public, and influencing the political process.

An important driver of the new role of NGOs has been improved communications technologies that make it easier for large numbers of like-minded people to work together. Clearly, there is a rapid transition under way in how we organize and communicate. A major intellectual effort to come to grips with this driver helps identify the issues but by no means provides a clear picture (Castells 1996).

7.2.3.5 Education, Knowledge, and Sociopolitical Drivers

Democracy allows more participants into the political process, but what if some people know more than others, or if people know different things? Formal schooling increases economic productivity and the ability to comprehend and participate in the political process, but it is also likely to contribute to the loss of some indigenous and experiential knowledge. Formal education is often lowest among the poor who interact most directly with ecosystems.

Average levels of formal education are increasing around the globe, but with great differences in rates, especially within developing regions (United Nations 2003a). Some developing countries, such as China, have invested heavily in primary and secondary education over the past several decades and can anticipate an increasingly well educated population as educated children become educated adults. In contrast, other regions, such as South Asia and sub-Saharan Africa, have very low average education levels that will take decades to increase substantially (Lutz and Goujon 2001).

Formal education beyond secondary school becomes increasingly specialized. Knowledge narrowing, complemented by the economic gains of specialization provided by expanding markets, means the attention of any particular individual is increasingly focused on a small subset of the human enterprise. Thus knowledge and employment specialization, along with the shift to urban life, have separated people from traditional understandings of nature as a whole and divided modern understanding into multiple parts and disbursed it among people (Giddens 1990). Another consequence is that while the "volume" of knowledge is far greater today than at any time in the past, it is widely dispersed. The deep knowledge held by one type of experts is very difficult to link with that of other experts. And syntheses of knowledge into general information that can be widely conveyed to inform collective action are also difficult. The modern human dilemma can be characterized as the challenge of rallying disparate human knowledge of complex systems to inform action (Norgaard 2004).

Incorporating the knowledge of separate experts in democratic and bureaucratic systems presents its own challenges. The primary response, generally referred to as "progressive governance," has been to argue that the voice of experts should be heard before legislatures vote or administrators decide. Progressive governance is not without its own problems. The separate voices of different experts rarely speak coherently to systemic social and ecological problems, nor do they speak to the concerns of people, especially local people. Facts and values are not always separable, and experts end up speaking for particular interests even when they think they are speaking for the public. Experts also become captured by special interest groups, and government agencies can become an interest of their own. Interagency task forces, stricter peer review, and increasing public participation through hearings and other mechanisms offset some of these problems (Dupre 1986; Jasanoff 1990; Irwin 1995; Fischer 2000).

7.2.3.6 Sociopolitical Drivers and Ecosystem Consequences

Widely accepted generalizations about the effects of sociopolitical drivers on ecosystems do not exist. Some examples

BOX 7.1

War as a Driver of Change in Ecosystem Services and Human Well-being

War acts as both a direct and an indirect driver of change in ecosystem services and human well-being, as nature becomes the intended victim or recipient of "collateral damage." The number of wars waged reached a maximum of 187 in the mid-1980s. That number was reduced by half by 2000 (Marshall et al. 2003). Most of these wars were internal conflicts. They were distributed highly unevenly across the world, predominantly in poor countries. (See MA *Current State and Trends,* Chapter 5, for a map of the distribution of internal wars from 1975 to 2003.)

During the twentieth century, 191 million people lost their lives in conflict-related incidents; 60% of these were non-military casualties (Krug 2002). Since World War II, more than 24 million people have been killed and another 50 million injured in state-sponsored wars and armed conflicts globally. Civilian casualties accounted for 90% of the dead. Between 1985 and 1995 alone, UNICEF estimated that wars have claimed the lives of 2 million children, wounded or disabled 4–6 million, orphaned 1 million, and left 12 million children homeless (UNICEF 1996; Pendersen 2002). Globally, armed conflicts around the world were responsible for the internal displacement of 11.7 million people during 1998 alone. In addition, armed conflicts forced 23 million people to seek refuge outside their countries in 1997 alone, while an additional 2.5 million abandoned their livelihoods and crossed national boundaries as a result of war-related violence (Krug 2002).

Environmental effects of warfare include damage to animals, defoliation, destruction of flora, degradation of soil, and loss of biodiversity (Pendersen 2002). The impacts of war on nature depend on the magnitude and duration of the conflict and its effects, the type of ecosystem involved (its resistance and resilience), the type of weapons used, the process of weapons production, and the cumulative effects sustained from the military campaigns. The main activities that result in protracted and persistent effects on ecosystem health, services, and human well-being are the manufacturing and testing of nuclear weapons, aerial and naval bombardment of landscapes, dispersal of landmines and de-mining, and the use and storage of military toxins and waste (Leaning 2000).

Ecological disturbance from armed conflict is not solely a contemporary issue. Scorched earth tactics (meant to inflict maximum damage to resources and facilities in an area in order to deny their use by the enemy) were used by the Greeks in the Peloponnesian wars (431–04 BC) (Thucydides 1989). Romans applied salts to soils in Carthage after their victory in the Punic wars (264–146 BC), and dykes were destroyed in the Netherlands to prevent a French invasion in 1792 (Bodansky 2003).

Modern warfare, however, has had particularly severe impacts. Chemical defoliants, bombs, and physical disruption by the United States army destroyed 14% of the Vietnamese total forest cover between 1961 and 1970, including 54% of the mangrove forests (Bodansky 2003).

In the first Gulf War in 1991, Iraq released about 2 million barrels of oil into the Persian Gulf and ignited 736 Kuwaiti oil wells, spreading clouds of black soot throughout the Middle East and the surrounding regions. The soot emission changed the local temperature, and the sulfur released with it contributed to acid rain. The effects suffered due to oil contamination included lesions and cancerous growths, sublethal effects of lower tolerance to environmental stress, and loss of insulation and motility by mammals from oiling of feathers and fur (World Conservation Monitoring Center 1991).

During the Eritrean war for independence from Ethiopia (1961–91), 30% forest cover of the country was reduced to less than 1%. The near-complete elimination of vegetation of some areas resulted in high rates of soil erosion with consequent problems of reduced crop yield, reduced wildlife populations, and sedimentation of rivers and reservoirs (Berhe in press).

Indirect impacts of wars include changes in land use as cultivators change crops, abandon fields, or retreat to more secure areas that are sometimes environmentally sensitive. Infrastructure damage or construction of new roads alters the incentives of an ecosystem's inhabitants (Westing 1988; W. Ghiorghis 1993; Berhe 2000). War-driven environmental degradation can initiate social degradation and protracted cycles of social and environmental decline by creating poverty, overexploitation of marginal resources, underdevelopment, and in extreme cases famine and social destruction (Lanz 1996; Berhe 2000, in press).

Since the fall of the Soviet Union, the danger of nuclear weapons has become a serious threat. It is estimated that a large-scale nuclear war could result in more than 1 billion deaths and injury of an equal number of people due to the combined effects of blast, fire, and radiation. Among the most serious environmental consequences of a nuclear war of this magnitude are subfreezing temperatures; a reduction in solar radiation at Earth's surface by half; triple the amount of UV-B radiation; increased doses of ionizing radiation from gamma and neuron flux of the fireball and radioactive debris fallout; chemical pollution of surface waters with pyrotoxins released from chemical storage areas; atmospheric pollution from the release of nitric oxide, ozone, and pyrogenic pollutants from the detonation; and release of toxic chemicals from secondary fires and storage areas (Ehrlich et al. 1983).

Landmines are not usually considered weapons of mass destruction, but more people have been killed and injured by anti-personnel mines than by nuclear and chemical weapons combined. Land mines maim or kill an estimated 400 people a week (International Campaign to Ban Landmines 2002). It is believed that there are about 80–110 million landmines spread in 82 countries, with another 230 million waiting to be deployed in 94 countries (Berhe 2000; International Campaign to Ban Landmines 2002). The distribution is not uniform. In Cambodia and Bosnia, there is approximately one mine in the ground for each citizen; in Afghanistan, Iraq, Croatia, Eritrea, and Sudan, the ratio is one mine for every two persons (Berhe 2000).

A landmine costs $3 to manufacture but $300–1,000 to clear (Nachón 2000). Landmines introduce nonbiodegradable and toxic waste of depleted uranium and 2,4,6-trinitrotoulene (TNT), hexahydro-1,3,5-trinitro-1,3,5-triazide (RDX, or Cyclonite), or tetryl as high-explosive filters. These compounds have been known to leach into soil and underground water as the metal or timber casing of the mine disintegrates (Gray 1997).

The impacts on ecosystem services and human well-being include access denial (which can lead to ecosystem recovery), disruption of land stability, pollution, and loss of biodiversity in the short run. These effects can be manifested as being biophysical, chemical, socioeconomic, or political in nature. The biophysical or chemical effects include destruction of soil structure: an increased rate of sediment transport by erosion and loads into streams; contamination with toxic pollutants; and a changing proportion, diversity, and productivity of flora and fauna coupled with habitat destruction. The socioeconomic or political effects include loss of income, food shortage, poverty, vulnerability, change in population per unit area, increased social polarization, declining health care, migration or displacement, destruction of essential infrastructures, arrested development of regions, and sociopolitical instability (Berhe 2000).

can be found in the final section of this chapter, on interactions among drivers and ecosystems.

7.2.3.7 Overview of Sociopolitical Drivers in the MA Scenarios

Sociopolitical drivers enter the MA scenarios in a number of ways—human capital and research investments, extent of international cooperation and attitudes toward environmental policies, and lifestyle choices affecting energy demand. (See Chapter 9 for further information on the different assumptions made about these in the four MA scenarios.)

7.2.4 Cultural and Religious Drivers

The word "culture" has many definitions in both the social sciences and in ordinary language. To understand culture as a driver of ecosystem change, it may be most useful to focus on the values, beliefs, and norms that a group of people share and that have the most influence on decision-making about the environment. (Of course, other aspects of culture, such as risk perception and willingness to accept risk, or preference for present versus future benefits, also influence environmental decision-making, but they are less salient to this discussion.) In this sense, culture conditions individuals' perceptions of the world, influences what they consider important, and suggests courses of action that are appropriate and inappropriate. And while culture is most often thought of as a characteristic of national or ethnic groups, this definition also acknowledges the emergence of cultures within professions and organizations, along with the possibility that an individual may be able to draw on or reconcile more than one culture.

There is a substantial literature examining the role of culture in shaping human environmental behavior. Productive work on this has focused on approaches that link directly with the social psychology of environmental concern and the influence of values, beliefs, and norms on individual decisions. Broad comparisons of whole cultures have not proved useful because they ignore vast variations in values, beliefs, and norms within cultures. But a growing number of studies are conducted in parallel in multiple societies and allow for systematic examination of the role of culture without overgeneralizing (Dunlap and Mertig 1995; Rosa et al. 2000; Hanada 2003). This work builds on social psychological analyses conducted within nations and cultures and then systematically compares them across nations.

Consumption behavior, especially in affluent nations and within affluent groups in developing nations, may be a particularly important driver of environmental change, but its cultural elements have not been extensively studied (Stern et al. 1997). It is clear that there is broad concern with the environment throughout the globe (Dunlap et al. 1993), but it is less clear how that concern translates into either changes in consumer behavior or political action. This is because consumption has many determinants beyond satisfying basic human needs. Values and beliefs regarding the environmental impacts of consumption may play a role, but consumption is also driven by peer group expectations and efforts to establish a personal identity, both of which are the targets of advertising. Further, some consumption is relatively easy to change (such as turning off a light), while other aspects require substantial investments (such as buying a high-efficiency refrigerator) (Dietz and Stern 2002).

A substantial body of literature provides lessons on how policies and programs can most effectively produce cultural change around environmental behavior (Dietz and Stern 2002). Obviously, the relationship between culture and behavior is context-specific. Indeed, one important lesson of research on this topic is that overarching generalizations are seldom correct.

The ability of culture to shape behavior depends on the constraints faced by individuals, and the effects of changing constraints on behavior depend on the culture of the individuals encountering the changes (Gardner and Stern 1995; Guagnano et al. 1995). In many circumstances changing values, beliefs, and norms will have no effect on behavior because individuals face structural constraints to pro-environmental behavior. (For example, public education about the problems of using tropical hardwoods will have little impact if people have no way of knowing the origins of lumber they may purchase.) In other circumstances changing constraints on behavior will have little effect because the mix of values, beliefs, and norms provides little reason to behave in an environmentally protective way. (For instance, making environmentally friendly food choices available may not influence consumptions if people are not aware of the environmental impacts of their consumption patterns).

Just as it is important to differentiate structural constraints (how hard it is to behave in an environmentally protective way) from culture (values, norms, and beliefs), it is also important to realize that there are multiple forms of pro-environmental behavior, and different factors may drive them. For example, Stern (Stern et al. 1999; Stern 2000) notes that environmental consumer behavior, environmental political behavior, and a willingness to make sacrifices to protect the environment, while positively correlated, are somewhat distinct and are influenced by different elements of culture.

There has been extensive and diverse research on the influence of values on pro-environmental behavior. A tradition stretching from Kluckholm (1952) through Rokeach (1968, 1973) to Schwartz (1987, 1990, 1992) has provided theoretical and empirical arguments to support the idea that values—things that people consider important in their life—are important in shaping behavior and are relatively stable over the life course (Schwartz 1996). Two strains of research have applied this logic to environmental concerns. Inglehart (1995) has suggested that a set of values he labels "post-materialist" predict environmental concern. His argument is that only once people have achieved a reasonable degree of material security can they assign priority to issues such as the environment. This argument has some strong parallels to the Environmental Kuznets Curve and to the ecological modernization theory in sociology (York et al. 2003b). However, as indicated, there is considerable controversy regarding the empirical support for this argument at either the individual or the national level (e.g., Brechin and Kempton 1994; Dunlap and Mertig 1995; Brechin and Kempton 1997; Dunlap and Mertig 1997; Brechin 1999; Stern et al. 1999; York et al. 2003b).

A number of researchers have suggested that altruism is a key value underpinning environmental concern, and that the scope of altruism may be limited to other humans or may extend to other species and the biosphere itself (Dunlap et al. 1983; Merchant 1992; Stern et al. 1993; Karp 1996). Empirical work in this tradition has deployed Schwartz's cross-cultural measures of values, and finds fairly consistent support for the idea that altruism predicts environmental concern, as well as evidence that traditional values—which might be termed conservatism and appear to be related to fundamentalism in many faiths (Schwartz and Huismans 1995)—lead to less concern with the environment (e.g., Kempton et al. 1995; Karp 1996; Schultz and Zelezny 1998; Schultz and Zelezny 1999; Stern et al. 1999).

The social psychological literature places considerable emphasis on beliefs as predictors of behaviors. At one level, this literature emphasizes very specific beliefs about a behavior—its consequences, the difficulty in conducting it, and so on (Ajzen and Fishbein 1980; Ajzen 1991). While this approach is helpful in designing interventions to change environmentally significant behavior, such beliefs are generally not what is thought of as culture. However, it has been argued that broad general beliefs about the environment and human interactions with it provide a backdrop that influence the acceptability of or resistance to more specific beliefs (Stern et al. 1995).

One of the most widely used measures of environmental concern, the "new ecological paradigm," can be interpreted as a measure of such general beliefs (Dunlap et al. 2002). The idea is that people vary in the degree to which they think that human interventions can cause significant harm to the environment. Variation in this general view makes individuals more or less accepting of new information about specific environmental problems. There is evidence that this new paradigm as well as values influence risk perception (Whitfield et al. 1999), including perception of ecological risk (Slimak 2003).

Norms are also a key part of culture related to environmental behavior. Norms signal to the individual what is appropriate behavior. Substantial work has been done on what activates norms regarding pro-environmental behavior (Van Liere and Dunlap 1978; Stern et al. 1986). Norms have been shown to be particularly important in local commons management, where they often guide behaviors that prevent overexploitation of a common pool resource (Ostrom et al. 2002).

Flowing from White's essay (White 1967), there has been substantial interest in the possible influences of religion on environmental concern and pro-environmental behavior. Some work on values has shown that traditional values are negatively related to environmental concern. There is a substantial literature in the United States that seems to indicate that adherents to fundamentalist Christian beliefs have less environmental concern than those with more liberal or moderate beliefs (Eckberg and Blocker 1989, 1996; Kanagy and Willits 1993).

While the social psychological literature is quite rich, as noted, the literature on cross-national comparisons is just emerging. The implications of that literature for our understanding of how environmental policy might be developed are unclear. And work on how environmentally relevant values, beliefs, and norms change is still in its earliest stages (but see Richerson et al. 2002).

7.2.4.1 Cultural and Religious Drivers and Ecosystem Consequences

It is common to hear arguments that we can best protect ecosystems, and by implication the services they provide, by changing values, beliefs, and norms. Yet as this section has described, values, beliefs, and norms—culture—are complex in themselves and their links to environmentally consequential behavior add further complexity. Changing values or beliefs will have little effect if changes in behavior to benefit the environment have high costs in time or money. In other cases, simply making people aware of the ecosystem consequences of their behavior can bring about substantial change, such as the growing demand for "ecosystem-friendly" consumer goods (Thogersen 2002). While cultural and religious factors may have a substantial influence on ecosystems, research has shown that broad generalizations are unwarranted and that analyses must always be context-specific.

7.2.4.2 Overview of Cultural and Religious Drivers in the MA Scenarios

Few cultural and religious drivers are built explicitly into the MA scenario quantitative modeling. However, changes in culture are an important part of the qualitative elements of some of the scenarios, particularly in Adapting Mosaic and to a certain extent in TechnoGarden. Both scenarios are built on the assumption that a general shift will occur in the way ecosystems and their services are valued. In both cases decision-makers at various scales develop a proactive approach to ecosystem management, but they pursue different management strategies to reach this goal. In TechnoGarden, the supply of ecosystem services is maintained by controlling ecosystem functions via technology. In Adapting Mosaic, the aim is to create a set of flexible, adaptive management options through a learning approach. Culturally diverse forms of learning about and adapting to ecosystem changes are fostered. Devising ways of incorporating traditional ecological and local knowledge into management processes and protecting the cultural and spiritual values assigned to nature in various cultures become part of the developed strategies.

7.2.5 Science and Technology Drivers

The development and diffusion of scientific knowledge and technologies that exploit knowledge have profound implications for ecological systems and human well-being. The twentieth century saw tremendous advances in understanding how the world works physically, chemically, biologically, and socially and in the applications of that knowledge to human endeavors. Earlier sections have documented examples of tremendous productivity gains in many industries.

From the introduction of the automobile in the early years to commercialization of genetically modified crops and widespread use of information technology more recently, many new products drew both praise and damnation regarding their effects on ecosystems. The twenty-first century is likely to see continued breathtaking advances in applications of materials science, molecular biology, and information technology—with real potential to improve human well-being and uncertain consequences for ecosystems.

Humans have been extremely successful in institutionalizing the process of scientific and technical change. Organizational structures that encourage researchers to make breakthroughs and to use them to develop potentially valuable products—such as research universities, publicly funded research centers, public-private collaborations for research and development, private research programs, regulatory institutions, and international agreements that collectively determine intellectual property rules (patents and copyrights, for instance)—are either in place or being implemented in the industrial world. However, they are not in place in most developing countries. Furthermore, institutions to use and reward development of indigenous knowledge are not well developed.

Society's ability to manage the process of product dissemination—identifying the potential for adverse consequences and finding ways to minimize them—has not always kept pace with our ability to develop new products and services. This disparity became especially obvious as the introduction of genetically modified crops met widespread opposition in some parts of the world. The protests in part resulted from the speed of advancement, as the rate of commercial adoptions of the first products of this new technology was unprecedented. At least 30 years passed between the development and widespread use of hybrid maize in industrial countries. For semi-dwarf rice and wheat in developing countries, a similar rate of use was reached only 15 years after development began. But use of genetically modified soybeans reached similar levels of use after only 5 years in Argentina and the United States. The use of the Internet accelerated worldwide communication and the organization of protests.

The state of scientific and technical knowledge at any moment depends on the accumulation of knowledge over time. Decision-makers can affect the rate of change in scientific and technical knowledge through setting research priorities and changing levels of funding. Domestic government funding for science and technology is driven by objectives such as scientific education, technology development, export markets, commercialization and privatization, and military power. The private sector responds to the perceived future for their products, looking for those that will be the most acceptable and profitable.

7.2.5.1 Innovation and Technological Change

The importance of "advances in knowledge" and technology in explaining the historical record of productivity growth has already been mentioned. In the original study on contributions of productivity growth to overall growth by Solow (1957), productivity enhancements were estimated to account for 87% of per capita growth (the remainder was attributed to increases in capital inputs). Since then, further methodological and statistical refinements have reduced the unexplained "residual" of productivity growth that is equated to advances in knowledge and technology, but they remain the largest single source of long-run productivity and economic growth. Science and technology are estimated to have accounted for more than one third of total GDP growth in the United States from 1929 to the early 1980s (Denison 1985), and for 16–47% of GDP growth in selected OECD countries for the period 1960 to 1995 (Table 10.1 in Barro and Sala-i-Martin 2004).[4]

The observed slowdown in productivity growth rates from the early 1970s to the mid- 1990s is generally interpreted as a weakening of the technological frontier in the OECD countries (Maddison 1995; Barro 1997), although quantitative statistics (and even everyday experience) do not corroborate the perception of a slowdown in technological innovation and change. An alternate interpretation is that the OECD countries have moved out of a long period of industrialization and into post-industrial development as service economies. In such economies, productivity is hard to measure, partly because services include a mixture of government, nonmarket, and market activities, partly because economic accounts measure services primarily via inputs (such as the cost of labor) rather than outputs, and partly because it is difficult to define service quality. Nevertheless, labor productivity in the service sector appears to grow more slowly than in the agricultural and industrial sectors (Millward 1990; Baumol 1993).

Finally, another interpretation is that productivity growth lags behind technological change because institutional and social adjustment processes take time to be implemented (Freeman and Perez 1988; David 1990). Once an appropriate "match" (Freeman and Perez 1988) between institutional and technological change is achieved, productivity growth accelerates. Maddison (1995) observed that the nineteenth century productivity surge in the United States was preceded by a long period of investment in infrastructure. Landes (1969) notes that both the German and Japanese economic accelerations were preceded by long periods of investment in education. Maddison (1995) has further suggested that recent developments in information technology involve considerable investment, both in hardware and in human learning. There is some preliminary evidence that the payoff from this investment is beginning to be important in the early twenty-first century.

7.2.5.2 Agricultural Science and Technology

We focus here on agricultural science and technology because of its obvious implications for land conversion and widespread consequences for many ecosystems. The ground-breaking research of Gregor Mendel in the 1860s on the heritability of phenotypical characteristics in garden peas laid the foundation for plant and livestock breeding research and the improvement of food crops in the twentieth century (Huffman and Evenson 1993). Since the beginning of that century, there have been three waves of agricultural biological technology development and diffusion.

The first wave took place mainly in North America and Europe in the 1930s and focused on important temperate climate crops. The discovery of hybridization—in which a cross of two genetically very different parent plants can produce a plant of greater vigor and higher yield than the parents—set the stage for major yield improvements in some of the most important food crops, especially maize. The first maize hybrids were commercially available for farmers in the U.S. Corn Belt in the 1930s. Average U.S. maize yields improved from 24.4 bushels per acre (1.53 tons per hectare) in 1860 to 116.2 bushels per acre (7.31 tons per hectare) in 1989, a more than fourfold increase over about 130 years. But even crops such as wheat, for which hybridization has not been commercially viable, saw similar improvements. Wheat yields grew from 11.0 bushels per acre (0.74 tons per hectare) in 1860 to 32.8 bushels per acre (2.21 tons per hectare) in 1989 (Huffman and Evenson 1993).

Some of the key developments were agricultural research systems that included universities, agricultural field stations, agricultural input companies, and extension services covering the chain from basic crop improvement research via field trials to disseminating information and new seed material to farmers (Ruttan 2001).

Agronomic research on improved inputs for crop production like fertilizer or pesticides and new agricultural management practices emerged, which further enhanced crop yields in farmers' fields. Funding for organized agricultural research was mainly provided by the public sector in the first half of the century, while the importance of private-sector research grew substantially in the second half as the private sector gained legal rights to protect genetic modifications (Huffman and Evenson 1993). In labor-scarce countries, particularly the United States, the private sector played a central role from the beginning in the development of agricultural machinery.

The second wave of agricultural technology development was particularly important in the developing world because it extended plant breeding and nutrient management techniques to important food crops in low-income countries. Since the early 1960s, productivity growth has been significant for rice in Asia, wheat in irrigated and favorable production environments worldwide, and maize in Mesoamerica and selected parts of Africa and Asia. High rates of investment in crop research, infrastructure, and market development combined with appropriate policy support fueled this land productivity (Pingali and Heisey 2001).

The Green Revolution strategy for food crop productivity growth was explicitly based on the premise that, given appropriate institutional mechanisms, technology spillovers across political and agro-climatic boundaries can be captured. Hence the Consultative Group on International Agricultural Research was established specifically to generate spillovers, particularly for nations that are unable to capture all the benefits of their research investments.

The major breakthroughs in yield potential that epitomize the Green Revolution came from conventional plant breeding approaches, characterized as crossing plants with different genetic backgrounds and selecting from among the progeny individual plants with desirable characteristics. Repeating the process over several generations leads to plant varieties with improved characteristics such as higher yields, improved disease resistance, and improved nutritional quality. The yields for the major cereals, especially with increased use of inorganic fertilizers with high nitrogen content, have continued to rise at a steady rate after the initial dramatic shifts in the 1960s for rice and wheat. For example, Table 7.5 shows that yields in irrigated spring wheat rose at the rate of 1% per year over the past three decades, an increase of around 100 kilograms per hectare per year (Pingali and Rajaram 1999).

In addition to work on shifting the yield frontier of cereal crops, plant breeders continue to have successes in the less glamorous areas of maintenance research. These include development of plants with durable resistance to a wide spectrum of insects and diseases, plants that are better able to tolerate a variety of physical stresses, crops that require significantly fewer days from planting to harvest, and cereal grain with enhanced taste and nutritional qualities.

The third, ongoing wave of agricultural technology development has been called the Gene Revolution and is based on techniques of transferring genetic material from one organism to another that would not occur via normal reproductive methods. The early phase of this wave was characterized by the development of a few commercial products (glyphosate-resistant soybeans, and maize and cotton that are resistant to lepidopteran pests), principally by private research firms (Nelson 2001). These early products of genetic engineering do not have higher potential yields than traditional varieties, but they often have higher effective yields because they reduce the cost of pest control. Varieties of food crops with other desirable characteristics such as increased beta-carotene content (the so-called golden

Table 7.5 Rate of Growth of Wheat Yield by Mega-environment, Elite Spring Wheat Yield Trial, 1964–95. Wheat breeders classify the developing world's spring wheat-growing areas into six distinct mega-environments: irrigated (ME1); high rainfall (ME2); acid soil (ME3); drought-prone (ME4); high temperature (ME5); and high latitude (ME6). A mega-environment is a broad, frequently transcontinental but not necessarily contiguous area occurring in more than one country, with similar biotic and abiotic stresses, cropping system requirements, volume of production, and, possibly, consumer preferences (Pingali and Rajaram 1999; Lantican et al. 2003).

Period	ME1–Irrigated	ME2–High Rainfall	ME4–Drought-prone	ME5–High Temperature
	(percent per year/kilograms per year)			
1964–78	1.22	1.72	1.54	1.41
	71.6	81.5	32.4	34.9
1979–99	0.82	1.16	3.48	2.10
	53.5	62.5	87.7	46.1

rice), increased drought tolerance (wheat), and virus resistance (papaya) are in various stages of development.

Substantial empirical evidence exists on the production, productivity, income, and human welfare impacts of modern agricultural science and the international flow of modern varieties of food crops. Evenson and Gollin (2003) provide detailed information for all the major food crops on the extent of adoption and impact of improved variety use. The adoption of modern varieties during the first 20 years of the Green Revolution—aggregated across all crops—reached 9% in 1970 and rose to 29% in 1980. In the subsequent 20 years, far more adoption has occurred than in the first two decades. By 1990, adoption of improved varieties had reached 46%, and it was 63% by 1998. Moreover, in many areas and in many crops, first-generation modern varieties have been replaced by second and third generations of improved varieties (Evenson and Gollin 2003).

Much of the increase in agricultural output over the past 40 years has come from an increase in yields per hectare rather than an expansion of area under cultivation. For instance, FAO data indicate that for all developing countries, wheat yields rose by 208% from 1960 to 2000, rice yields rose 109%, maize yields rose 157%, potato yields rose 78%, and cassava yields rose 36% (FAOSTAT). (See MA *Current State and Trends,* Chapter 26, for information on crop area expansion.) Trends in total factor productivity are consistent with partial productivity measures, such as rate of yield growth. Pingali and Heisey (2001) provide a comprehensive compilation of total factor productivity evidence for several countries and crops.

Widespread adoption of improved seed-fertilizer technology led to a significant growth in food supply, contributing to a fall in real food prices. The primary effect of agricultural technology on the non-farm poor, as well as on the rural poor who are net purchasers of food, is through lower food prices.

The effect of agricultural research on improving the purchasing power of the poor—both by raising their incomes and by lowering the prices of staple food products—is probably the major source of nutritional gains associated with agricultural research. Only the poor go hungry. Because a relatively high proportion of any income gains made by the poor is spent on food, the income effects of research-induced supply shifts can have major nutritional implications, particularly if those shifts result from technologies aimed at the poorest producers (Pinstrup-Andersen et al. 1976; Hayami and Herdt 1977; Scobie and Posada 1978; Binswanger 1980; Alston et al. 1995).

Several studies have provided empirical support to the proposition that growth in the agricultural sector has economy-wide effects. One of the earliest studies showing the linkages between the agricultural and nonagricultural sectors was done at the village level by Hayami et al. (1978). Hayami provided an excellent micro-level illustration of the impacts of rapid growth in rice production on land and labor markets and the nonagricultural sector.

More recent assessments on the impacts of productivity growth on land and labor markets have been done by Pinstrup-Andersen and Hazell (1985) and by David and Otsuka (1994). Pinstrup-Andersen and Hazell (1985) argued that landless labor did not adequately share in the benefits of the Green Revolution because of depressed wage rates attributable to migrants from other regions. David and Otsuka (1994), on the other hand, found that migrants shared in the benefits of the Green Revolution through increased employment opportunities and wage income. The latter study also documented that rising productivity caused land prices to rise in the high-potential environments where the Green Revolution took off. For sector-level validation of the proposition that agriculture does indeed act as an engine of overall economic growth, see Hazell and Haggblade (1993), Delgado et al. (1998), and Fan et al. (1998).

The profitability of farming systems using improved varieties has been maintained despite falling food prices (in real terms), owing to a steady decline in the cost per ton of production (Pingali and Traxler 2002). Savings in production costs have come about from technical change in crop management and increased input-use efficiencies. Once improved varieties have been adopted, the next set of technologies that makes a significant difference in reducing production costs includes machinery, land management practices (often in association with herbicide use), fertilizer use, integrated pest management, and (most recently) improved water management practices.

Although many Green Revolution technologies were developed and extended in package form (such as new plant varieties plus recommended fertilizer, pesticide, and herbicide rates, along with water control measures), many components of these technologies were taken up in a piecemeal, often stepwise manner (Byerlee and Hesse de Polanco 1986). The sequence of adoption is determined by factor scarcities and the potential cost savings achieved. Herdt (1987) provided a detailed assessment of the sequential adoption of crop management technologies for rice in the Philippines. Traxler and Byerlee (1992) provided similar evidence on the sequential adoption of crop management technologies for wheat in Sonora, northwestern Mexico.

Although high-potential environments gained the most in terms of productivity growth from the Green Revolution varieties, the less favorable environments benefited as well through technology spillovers and through labor migration to more productive environments. According to David and Otsuka (1994), wage equalization across favorable and unfavorable environments was one of the primary means of redistributing the gains of technological change. (Wages of workers in unfavorable environments are pulled up by demand for additional labor in the favorable environments.) Renkow (1993) found similar results for wheat grown in high- and low-potential environments in Pakistan.

Byerlee and Moya (1993), in their global assessment of the adoption of improved wheat varieties, found that over time the adoption of modern varieties in unfavorable environments caught up to levels of adoption in more favorable environments, particularly when germplasm developed for high-potential environments was further adapted to the more marginal environments. In the case of wheat, the rate of growth in yield potential in drought-prone environments was around 2.5% per year during the 1980s and 1990s (Lantican et al. 2003). Initially the growth in yield potential for the marginal environments came from technological spillovers as varieties bred for the high-potential environments were adapted to the marginal environments. During the 1990s, however, further gains in yield potential came from breeding efforts targeted specifically at the marginal environments.

Since the 1990s, the locus of agricultural research and development has shifted dramatically from the public to the private multinational sector. Three interrelated forces in this latter wave of globalization are transforming the system for supplying improved agricultural technologies to the world's farmers. The first is the evolving environment for protecting intellectual property in plant innovations. The second is the rapid pace of discovery and the growth in importance of molecular biology and genetic engineering. Finally, agricultural input and output trade is becoming more open in nearly all countries. These developments have created a powerful new set of incentives for private research investment, altering the structure of the public/private agricultural research endeavor, particularly with respect to crop improvement (Falcon and Fowler 2002; Pingali and Traxler 2002).

7.2.5.3 Science and Technology Drivers and Ecosystem Consequences

The consequences of technical change for ecosystems are as diverse as those for economic drivers, because technical change alters the interplay among inputs, resource use, and outputs. Two examples—in agriculture and fisheries—illustrate this complexity. The development of high-yielding crop varieties meant that less land could be used to produce the same amount of food with positive effects on the condition of the unconverted ecosystems. The invention of the Haber-Bosch process to convert atmospheric nitrogen cheaply into nitrogenous fertilizer meant that plants with high yield response to this fertilizer were favored in the marketplace. More nitrogen was applied to fields, altering dramatically the natural nitrogen cycle, with negative consequences in coastal ecosystems. (See later discussion on nitrogen use and ecosystem consequences.) Use of pesticides also had unintended consequences, such as the reduction in availability of by-product protein from irrigated fields (fish and amphibians).

Improved marine fishing technologies have made it possible to extract considerable fish biomass from the marine system. In fact, humankind has probably reached the maximum (and in some places exceeded) levels of fish biomass removal before significant ecosystem changes are induced. (Since fish biomass is a small fraction of the marine standing biomass, current fish extraction probably removes only a small portion of total marine biomass; recent initiatives such as the Scientific Committee on Ocean Research workshop on the impacts of fishing on marine ecosystems are trying to address the question of impacts.) For example, in the Gulf of Thailand higher trophic animals are no longer present and the system is dominated by lower trophic species with a high biomass turnover (Christianson 1998). Research in West Africa (Alder and Sumaila 2004) and the North Atlantic (Christianson et al. 2003) indicates similar changes.

A less diverse system is in principle more vulnerable (less resistant) to perturbations such as disease and climate change, but there is little documented evidence that the loss of fish biomass is driving changes in the carbon and nitrogen cycle as well as diseases. In parts of the Caribbean, removal of the top predators ultimately resulted in the areas shifting from coral-dominated reefs to algal-dominated ones. In this case, the removal of top predators resulted in reefs dominated by sea urchins, but a disease wiped them out so there were few animals left to clean algae from the corals, resulting in a rapid shift from coral to algal-dominated reefs. (This process is described in detail in Chapter 19 of the *Current State and Trends* volume.)

7.2.5.4 Overview of Science and Technology Drivers in the MA Scenarios

The MA scenarios include a number of different science and technology drivers. A qualitative assessment about changes in aggregate technology development is implemented in a variety of assumptions about crop yield changes, energy and water efficiency, and materials intensity. The assumptions about the development of science and technology in each of the scenarios can be found in Chapter 9.

7.3 Direct Drivers

This section reviews some of the most important direct drivers of ecosystem condition: climate variability and change, plant nutrient use, land conversion, invasive species, and diseases.

7.3.1 Climate Variability and Change

Key climatic parameters that affect ecological systems include mean temperature and precipitation and their variability and extremes, and in the case of marine systems, sea level. In this section we review the available record on climate and the drivers of climate variability and change.

This section is primarily based on the expert and government peer-reviewed reports from the Intergovernmental Panel on Climate Change—especially on the Working Group I Report of the Third Assessment Report (Houghton et al. 2001; IPCC 2002) and the Special Report on Emissions Scenarios (Nakićenović et al. 2000)—and on the Convention on Biological Diversity Technical Series No. 10 (Secretariat of the Convention on Biological Diversity 2003). This section highlights the key conclusions of these assessments, which, given their comprehensive nature and recent publication, are still valid.

7.3.1.1 Observed Changes in Climate

The global-average surface air temperature has increased by 0.6 ± 0.2° Celsius since about 1860. (See Figure 7.13 in Appendix A.) The record shows a great deal of spatial and temporal variability; for example, most of the warming occurred during two periods (1910–45 and since 1976). It is very likely that the 1990s was the warmest decade, and 1998 the warmest year, of the instrumental record. Extending the instrumental record with proxy data for the Northern Hemisphere indicates that over the past 1,000 years, the twentieth century's increase in temperature is likely to have been the largest of any century, and the 1990s was likely the warmest decade (IPCC 2002, p. 42). On average, night-time daily minimum temperatures over land have increased at about twice the rate of daytime daily maximum temperatures since about 1950 (approximately 0.2° versus 0.1° Celsius per decade). This has lengthened the freeze-free season in many mid- and high-latitude regions.

Precipitation increased by 0.5–1% per decade in the twentieth century over most mid- and high-latitudes of the Northern Hemisphere continents. Rainfall has decreased over much of the sub-tropical land areas (−0.3% per decade), although it appeared to recover in the 1990s. It is likely that there has been an increase in heavy and extreme precipitation events, on average, in the mid- and high-latitudes of the Northern Hemisphere.

There has been a widespread retreat of mountain glaciers in nonpolar regions during the twentieth century, decreases of about 10% in the extent of snow cover since the late 1960s, and a reduction of about two weeks in the annual duration of lake- and river-ice cover in the mid- and high latitudes of the Northern Hemisphere over the twentieth century. Northern Hemisphere spring and summer sea-ice extent has decreased by about 10–15% since the 1950s. It is likely that there has been about a 40% decline in Arctic sea-ice thickness during late summer to early autumn in recent decades and a considerably slower decline in winter sea-ice thickness.

Global-average sea level rose 10–20 centimeters during the twentieth century. Warm episodes of the El Niño/Southern Oscillation phenomenon have been more frequent, persistent, and intense since the mid-1970s. This recent behavior has been reflected in regional variations of precipitation and temperature over much of the tropics and sub-tropics.

7.3.1.2 Observed Changes in Atmospheric Composition: Greenhouse Gases and Aerosol Precursors

Since 1750, the atmospheric concentration of carbon dioxide has increased by about 32% (from about 280 to 376 parts per million in 2003. Approximately 60% of that increase (60 ppm) has taken place since 1959 (Keeling and Whorf, at cdiac.esd.ornl.gov/trends/co2/sio-mlo.htm). (See Figure 7.14 in Appendix A.) Nearly 80% of the increase during the past 20 years is due to fossil fuel burning, with the rest being due to land use changes, especially deforestation and, to a lesser extent, cement production. The rate of increase of atmospheric CO_2 concentration has been about 0.4% per year over the past two decades. During the 1990s, the annual increase varied by a factor of three, with a large part of this variability being due to the effect of climate variability on CO_2 uptake and release by land and oceans.

Atmospheric methane concentrations have increased by a factor of 2.5 since 1750 (from about 700 to 1,750 parts per billion), and they continue to increase. The annual increase in CH_4 atmospheric concentrations became slower and more variable in the 1990s compared with the 1980s. The atmospheric concentration of nitrous oxide has increased by about 17% since 1750 (from about 270 to 315 parts per billion).

The atmospheric concentrations of many of the halocarbon gases that are both ozone-depleting and greenhouse gases are either decreasing or increasing more slowly in response to reduced emissions under the regulations of the Montreal Protocol on the ozone layer and its amendments. Their substitute compounds and some other synthetic compounds (such as perfluorocarbons and sulfur hexafluoride) are increasing rapidly in the atmosphere, but from recent near-zero concentrations.

Tropospheric ozone has increased by about 35% since 1750 due to anthropogenic emissions of several ozone-forming gases (non-methane hydrocarbons, NO_x, and carbon monoxide). Ozone varies considerably by region, and because of its short atmospheric lifetime it responds much more quickly to changes in precursor emissions than the long-lived greenhouse gases, such as CO_2.

7.3.1.3 Projections of Changes in Atmospheric Composition: Greenhouse Gases and Aerosol Precursors

Carbon emissions due to fossil-fuel burning are virtually certain to be the dominant influence on the trends in atmospheric CO_2 concentration during the twenty-first century. As the CO_2 concentration increases and climate changes, the oceans and land will take up a progressively decreasing proportion of anthropogenic carbon emissions. By the end of the twenty-first century, models project atmospheric concentrations of CO_2 of 540–970 parts per million. (See Figure 7.15 in Appendix A.) This range of projected concentrations is primarily due to differences among the emissions scenarios (the IPCC-SRES scenarios show a range of 5–28 gigatons of carbon per year in 2100, compared with 7.1 gigatons in 1990); different carbon model assumptions would add at least ±10% uncertainty to these projections. Sequestration of carbon by changing land use could influence atmospheric CO_2 concentration. However, even if all of the carbon so far released by land use changes could be restored to the terrestrial biosphere (through reforestation, for example), projected levels of CO_2 concentration would be reduced by only 40–70 parts per million.

Model projections of the emissions of the non-CO_2 greenhouse gases vary considerably by 2100 across the IPCC-SRES emissions scenarios. Annual anthropogenic CH_4 and N_2O emissions are projected to be 270–890 teragram CH_4 and 5–17 teragram N in 2100, compared with 310 teragram CH_4 and 6.7 teragram N in 1990. By the end of the century, models project the atmospheric concentrations of CH_4 to be 1,550–3,750 parts per billion and of N_2O to be 340–460 parts per billion.

Model projections of the emissions of the precursors of tropospheric ozone—that is, CO, NMHCs, and NO_x—also vary considerably by 2100 across the IPCC-SRES emissions scenarios. Annual anthropogenic emissions are projected to be 360–2,600 teragram CO, 90–420 teragram NMHCs, and 19–110 teragram N in 2100, compared with about 880 teragram CO, 140 teragram NMHCs, and 31 teragram N in 1990. In some scenarios, tropospheric ozone would become as important a radiative forcing agent as CH_4 and would threaten the attainment of air quality targets over much of the Northern Hemisphere.

The IPCC-SRES scenarios primarily project decreases in anthropogenic sulfur dioxide emissions, leading to projected decreases in the atmospheric concentrations of sulfate aerosols. Model projections of the emissions of SO_2 vary considerably by 2100 across the IPCC-SRES emissions scenarios. Annual anthropogenic SO_2 emissions are projected to be 20–60 teragram S in 2100, compared with about 71 teragram S in 1990. In addition, natural aerosols (such as sea salt, dust, and emissions leading to the production of sulfate and carbon aerosols) may increase as a result of changes in climate and atmospheric chemistry.

7.3.1.4 Existing Projections of Changes in Climate

The global climate of the twenty-first century will depend on natural changes and the response of the climate system to human activities. Climate models can simulate the response of many climate variables, such as increases in global surface temperature and sea level, in various scenarios of greenhouse gas and other human-related emissions. (See Table 7.6.) The globally averaged surface air temperature increase from 1990 to 2100 for the range of IPCC-SRES scenarios is projected to be 1.4–5.8° Celsius. This increase would be without precedent during the last thousand years.

Table 7.6 Observed and Modeled Changes in Extremes of Weather and Climate *(IPCC 2002)*

Change	Observed (1950–2000)	Projections from Models (2050–2100)
Higher maximum temperatures and more hot days	nearly all land areas	most models
Increase of heat index	many land areas	most models
More intense precipitation events	many northern hemisphere mid- to high-latitude land areas	most models
Higher minimum temperatures and fewer cold days	virtually all land areas	most models
Fewer frost days	virtually all land areas	physically plausible based on increased minimum temperatures
Reduced diurnal temperature range	most land areas	most models
Summer continental drying	few areas	most models
Increase in tropical cyclone peak wind intensities	not observed, but very few analyses	some models
Increase in tropical cyclone mean and peak precipitation intensities	insufficient data	some models

A coherent picture of regional climate change using regionalization techniques is not yet possible. However, based on recent global model simulations, it is likely that nearly all land areas will warm more rapidly than the global average, particularly those at high latitudes in the cold season. Most notable is the warming in the northern regions of North America and in northern and central Asia, which is in excess of 40% above the global-mean change. In contrast, the warming is less than the global-mean change in South and Southeast Asia in summer and southern South America in winter.

Globally averaged precipitation is projected to increase. Based on recent global model simulations, it is likely that precipitation will increase over northern mid- and high latitudes and over Antarctica in winter. At low latitudes, there are both regional increases and decreases, which are likely to depend on the emissions scenario, but in general most arid and semiarid areas are projected to become drier.

Analyses of past data and improvements in climate models have enabled changes in extreme events observed to date (such as heat waves, heavy precipitation events, and droughts) to be compared to similar changes in model simulations for future climate.

For some other extreme phenomena, many of which may have important impacts on ecosystems and society, there is currently insufficient information to assess recent trends, and the confidence in models and understanding is inadequate to make firm projections on, for instance, the intensity of mid-latitude storms. Further, very small-scale phenomena, such as thunderstorms, tornadoes, hail, and lightning, are not simulated in global models. Recent trends for conditions to become more El Niño-like in the tropical Pacific are projected to continue in many models, although confidence in such projections is tempered by some shortcomings in how well El Niño is simulated in global climate models.

Northern Hemisphere snow cover and sea-ice extent are projected to decrease further. Glaciers and icecaps (excluding the ice sheets of Greenland and Antarctica) will continue their widespread retreat during the twenty-first century.

For the range of IPCC-SRES scenarios, a sea level rise of 9–88 centimeters is projected for 1990 to 2100, with a central value of 0.47 meters, which is about two to four times the rate over the twentieth century.

7.3.1.5 Climate Drivers and Ecosystem Consequences

Climate change and elevated atmospheric concentrations of CO_2 are projected to affect individuals, populations, species, and ecosystem composition and function both directly (through increases in temperature and changes in precipitation, changes in extreme climatic events and in the case of aquatic systems changes in water temperature, sea level, and so on) and indirectly (through climate changing the intensity and frequency of disturbances such as wildfires and major storms). The magnitude of the impacts will, however, depend on other anthropogenic pressures, particularly increased land use intensity and the associated modification, fragmentation, and loss of habitats (or habitat unification, especially in the case of freshwater bodies); the introduction of invasive species; and direct effects on reproduction, dominance, and survival through chemical and mechanical treatments.

No realistic projection of the future state of Earth's ecosystems can be made without taking into account all of these pressures—past, present, and future. Independent of climate change, biodiversity is forecast to decrease in the future due to the multiple pressures from human activities—climate change constitutes an additional pressure. Quantification of the impacts of climate change alone, given the multiple and interactive pressures acting on Earth's ecosystems, is difficult and likely to vary regionally.

The general impact of climate change is that the habitats of many species will move poleward or to higher elevations from their current locations, with the most rapid changes being where the general tendency is accelerated by changes in natural and anthropogenic disturbance patterns. For example, the climatic zones suitable for temperate and boreal plant species may be displaced by 200–1,200 kilometers poleward over the next 100 years. Weedy species (those that are highly mobile and can establish quickly) and invasive species will have advantage over others. Drought and desertification processes will result in movements of habitats of many species toward areas of higher rainfall from their current locations.

Species and ecosystems are projected to be affected by extreme climatic events—for example, higher maximum temperatures, more hot days, and heat waves are projected to increase heat stress in plants and animals and reduce plant productivity. Higher minimum temperatures, fewer cold days, frost days, and cold waves could result in an extended range and activity of some pest and disease vectors and increased productivity in some plant species and ecosystems. More-intense precipitation events are projected to result in increased soil erosion, increased flood runoff. Increased summer drying over most mid-latitude continental interiors and associated risk of drought are projected to result in decreased plant productivity, increased risk of wild fires and diseases, and pest outbreaks. Increased Asian summer monsoon precipitation variability and increased intensity of mid-latitude storms could lead to increased frequency and intensity of floods and damage to coastal areas.

7.3.1.6 Overview of Climate Drivers in the MA Scenarios

The MA scenarios use the IPCC-SRES scenarios as a basis for their assumptions about the energy and climate developments. The range of climate drivers in the MA scenarios can be found in Chapter 9.

7.3.2 Plant Nutrient Use

All plants require three macronutrients—nitrogen, phosphorus, and potassium—and numerous micronutrients for growth. Crop production often requires supplementation of natural sources. Nitrogen and phosphorus can move beyond the bounds of the field to which they are applied, potentially affecting ecosystems offsite. In addition, phosphorus used in detergents and output from sewer systems has been an important contributor to aquatic plant growth in water bodies near population centers in some parts of the world. In this section, we focus on nitrogen and phosphorus as drivers of ecosystem changes. Potassium is relatively immobile and mostly benign offsite, so it is not discussed. Other nutrients, particularly micronutrients, are of great importance in many parts of the world as drivers for sustainable crop production and human health (Welch and Graham 1999). Finally, carbon is itself a fertilizer, and rising carbon concentrations, especially the upper range across IPCC-SRES emissions scenarios, would also affect growth of plants.

7.3.2.1 Nitrogen Use and Trends

Atmospheric N is mostly inert N_2 gas, which is fixed into reactive, biologically available forms through both natural and anthropogenic fixation processes. Human activities are dramatically changing the rate of N_2 fixation and global atmospheric deposition of reactive N (Galloway and Cowling 2002). Reactive N is defined as all biologically, photochemically, or radiatively active forms of N, a diverse pool that includes mineral N forms such as NO_3^- and NH_4^+, gases that are chemically active in the troposphere (NO_x and NH_3), and gases such as N_2O that contribute to the greenhouse effect (Galloway et al. 1995). In 1990, the total amount of reactive N created by human activities was about 141 teragram N per year (see Table 7.7), which represents a ninefold increase over 1890, compared with a 3.5-fold increase in global population (Galloway and Cowling 2002). Between 1960 and 2002, use of nitrogenous fertilizers increased eightfold, from 10.83 million to 85.11 million tons of nitrogen (IFA 2004).

In the past, creation of reactive N was dominated by natural processes, which also increased forest biomass production and storage of atmospheric CO_2 in plant and soils (Mosier et al. 2001) At present, biological N_2 fixation occurring in cultivated crops, synthetic N production through the Haber-Bosch process, and fossil-fuel combustion have become major sources of reactive N (Galloway and Cowling 2002). Vitousek et al. (1997) estimate that humans have approximately doubled the rate of N input to the terrestrial N cycle.

Table 7.7 Regional Creation of Reactive Nitrogen in the Mid-1990s (Galloway and Cowling 2002)

World Regions	Reactive Nitrogen Creation	
	Total (teragram per year)	Per Person (kilogram per year)
Africa	5.3	7
Asia	68.9	17
Europe and former Soviet Union	26.5	44
Latin America	9.4	19
North America	28.4	100
Oceania	2.2	63
World	140.7	24

Large regional differences in reactive N creation occur. Whereas Asia accounts for nearly 50% of the net global creation of reactive N, per capita creation is by far largest in North America, followed by Oceania and Europe. Inorganic fertilizers contribute about 82 teragram N per year reactive N, whereas managed biofixation adds about 20 teragram N per year and recycling of organic wastes 28–36 teragram N per year (Smil 1999).

It is generally believed that organic nutrient sources offer environmental and other benefits, but the potential benefits from relying on organic sources must be weighed against potential limitations (Cassman et al. 2003). Although demand for organic food is predicted to grow, especially in higher-income countries, only 1% of the world's cropland (about 16 million hectares) is currently under certified organic production (FAO 2002). Globally, organic N sources are not available in amounts that would be large enough to meet nutrient needs for food production. On a global basis, manure and legumes can contribute about 25% of crop N requirements (Roy et al. 2002). Except for soybeans, however, legume area is declining worldwide, and this trend is unlikely to be reversed in the near future (Smil 1997). Moreover, controlling the fate of N from organic sources is as difficult as managing the fate of N from mineral fertilizer. Organic or low-input agriculture alone cannot secure the future food supply in the developing world, where maintaining low food prices contributes to reducing poverty and increasing economic wealth (Senauer and Sur 2001).

Crop production at a global scale will largely depend on mineral N fertilizer to meet current and future food demand (Cassman et al. 2003). Nitrogen use in developing countries has increased exponentially during the course of the Green Revolution (see Figure 7.16 in Appendix A) due to rapid adoption of improved high-yielding varieties that could respond to the increased N supply and greater cropping intensity (Cassman and Pingali 1995). In industrial countries, excluding Eastern Europe and the former Soviet Union, N fertilizer use has remained relatively constant during the past 25 years, although yields of many crops continue to rise slowly. In Eastern Europe and the countries of the former Soviet Union, N consumption dropped in the 1990s as a result of political and economic turmoil.

The three major cereals—maize, rice, and wheat—account for about 56% of global N fertilizer use (IFA 2002). At a global scale, cereal yields and fertilizer N consumption have increased in a near-linear fashion during the past 40 years and are highly correlated with one another. The ratio of global cereal production to global fertilizer N consumption shows a curvilinear decline in the past 40 years, raising concerns that future increases in N fertilizer use are unlikely to be as effective in raising yields as in the past (Tilman et al. 2002). Because the relationship between crop yield and N uptake is tightly conserved (Cassman et al. 2002), achieving further increases in food production will require greater N uptake by crops and, consequently, either more external N inputs or more efficient use of N. Recent estimates for major cereals suggest that increases in N consumption by about 20–60% during the next 25 years will be required to keep pace with the expected demand (Cassman et al. 2003).

Agricultural lands lose a substantial fraction of the fertilizer N applied, often 40–60%. Therefore, enhancing the efficiency of N use from fertilizer (nitrogen use efficiency) is a key measure for reducing the amount of reactive N released into ecosystems (Galloway et al. 2002). The simplest measure of NUE is the amount of grain (or other harvest product) produced per unit N input, which is an aggregate efficiency index that incorporates the contributions from indigenous soil N, fertilizer uptake efficiency, and the efficiency with which N acquired by the plant is converted into grain yield.

Large differences in NUE exist among countries, regions, farms, fields within a farm, and crop species, because crop yield response functions to N vary widely among different environments (Cassman et al. 2002). Therefore, policies that promote an increase or decrease in N fertilizer use at state, national, or regional levels would have a widely varying impact on NUE, yields, farm profitability, and environmental quality.

Large-scale NUE can be increased with sufficient investments in research and policies that favor increases in NUE at the field scale. Figure 7.17 shows that in U.S. maize systems, NUE increased from 42 kilograms of crop per kilogram of N in 1980 to 57 kilograms of crop in 2000 (Cassman et al. 2002). Three factors contributed to this improvement: increased yields and more vigorous crop growth associated with greater stress tolerance of hybrids (Duvick and Cassman 1999); improved management of production factors other than N (conservation tillage, seed quality and higher plant densities); and improved N fertilizer management (Dobermann and Cassman 2002).

In Japan, NUE of rice remained virtually constant at about 57 kilograms of crop per kilogram N from 1961 to 1985, but this has increased to more than 75 kilograms of

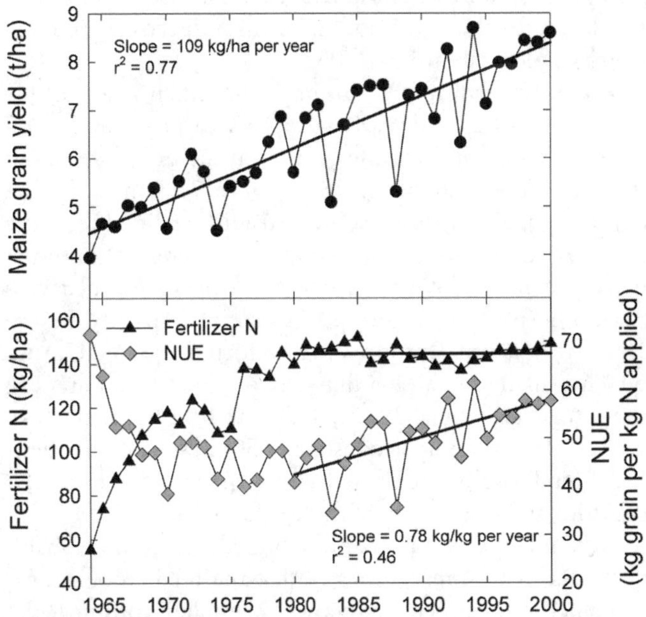

Figure 7.17. Trends in Maize Grain Yield, Use of Nitrogen Fertilizer, and Nitrogen Use Efficiency in United States, 1965–2000 (modified from Cassman et al. 2002)

crop in recent years (Mishima 2001). Key factors contributing to this increase were a shift to rice varieties with better grain quality, which also had lower yield potential and nitrogen concentrations, and the adoption of more knowledge-intensive N management technologies (Suzuki 1998).

Increasing NUE in the developing world presents a greater challenge. Nitrogen use efficiency is particularly low in intensive irrigated rice systems of sub-tropical and tropical Asia, where it has remained virtually unchanged during the past 20–30 years (Dobermann and Cassman 2002). Research has demonstrated, however, that rice is capable of taking up fertilizer N very efficiently provided the timing of N applications is congruent with the dynamics of soil N supply and crop N demand (Peng and Cassman 1998). These principles have recently become embedded in a new approach for site-specific nutrient management, resulting in significant increases in yields and NUE at numerous farms sites across Asia (Dobermann et al. 2002). Improving the congruence between crop N demand and N supply also was found to substantially increase N fertilizer efficiency of irrigated wheat in Mexico (Matson et al. 1998; Riley et al. 2003).

7.3.2.2 Phosphorus Use and Trends

Phosphorus is widely used in fertilizers for agricultural crops, as well as on lawns in high-income countries. It is also used as a nutrient in supplements for dairy cattle in some parts of the world. More than 99% of all phosphate fertilizers are derived from mined phosphate rock. Just six countries account for 80% of the world phosphate rock production (United States Geological Survey 2003): the United States (27% of world total), Morocco and Western Sahara (18%), China (16%), Russia (8%), Tunisia (6%), and Jordan (5%). At current annual mining rates of 133 million tons per year, known phosphate rock reserves and resources would last for about 125 and 375 years, respectively (United States Geological Survey 2003).

Globally, fertilizer-P consumption, which rose steadily from 1961 to the late 1980s, appears to have leveled off around 33 million tons after a drop in the early 1990s, but regional differences occur. (See Figure 7.18 in Appendix A.) Phosphorus consumption has steadily declined since the late 1970s in higher-income countries, whereas it continues to rise in many lower-income countries, particularly in Asia. Phosphorus consumption dropped sharply in the 1990s in Eastern Europe and the former Soviet Union, which is likely to cause a drawdown of soil P resources in these regions. China, Brazil, India, and the United States are currently the major consumers of phosphate fertilizers. Fertilizer P use has remained low in many poor countries of Africa and other parts of the world.

Livestock excreta play an important role in the global P cycle. Recent estimates suggest that the total amount of P contained in livestock excreta was 21 million tons in 1996, of which 8.8 million tons were recovered in manure (Sheldrick et al. 2003a). However, manure P input has declined from 50% of the fertilizer plus manure P input in 1961 to 38% in 1996 (Sheldrick et al. 2003a).

Global agricultural P budgets (inputs are fertilizers and manures and outputs are agricultural products and runoff) indicate that average P accumulation in agricultural areas of the world is approximately 8 million tons P per year. While P is still accumulating, the rate of annual accumulation has begun to plateau, possibly as early as the 1980s. (See Figure 7.19.) Slowing rates of annual accumulation is causing the rate of increase in cumulative accumulation to decline, but this decline (2–4 million tons P per year) is minimal compared with total cumulative accumulation (over 300 million tons P per year).

P accretion is occurring in many countries, but rates of P accumulation over time vary. (See Figure 7.20.) On average, rates of P accumulation on agricultural land have

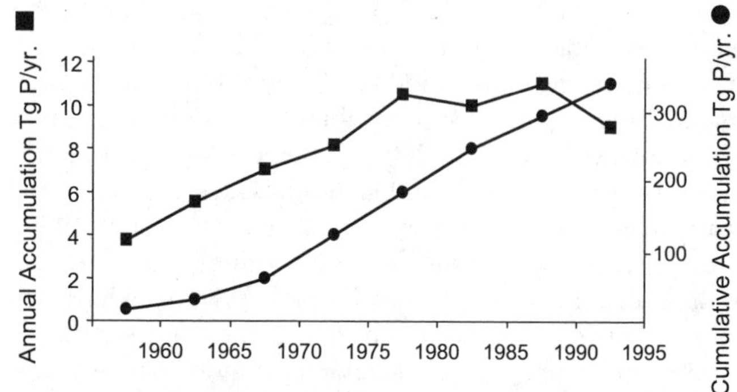

Figure 7.19. Phosphorus Accumulation in Agricultural Soils, 1960–95, as Determined by Global Budget. Squares indicate annual accumulation based on five-year averages. Circles represent cumulative P accumulation. Inputs are fertilizer and manure, based on global estimates of fertilizer use from the FAO 1950–97 and estimates of manure production based on animal densities. Outputs are agricultural products such as meat, eggs, and grains based on agricultural production data from FAO 1950–98, and the percentage P of those products and P in runoff. (Bennett et al. 2001)

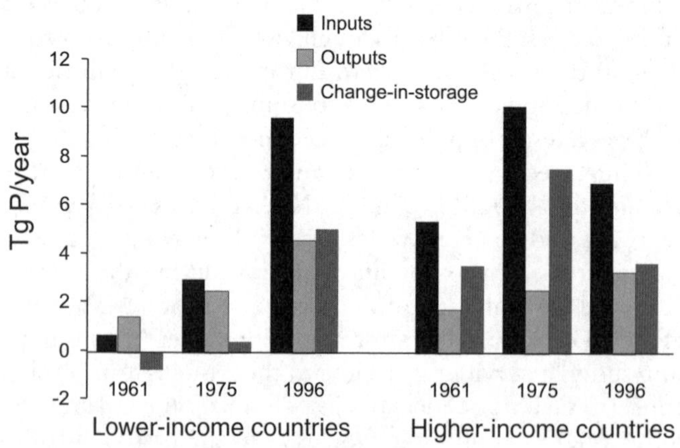

Figure 7.20. Estimated Inputs, Outputs, and Change in Storage of Phosphorus on Agricultural Lands in Lower- and Higher-income Countries, 1961–96. Inputs include fertilizers and manure; outputs are runoff and crops harvested. Note that the drop in fertilizer use in the industrial countries in 1996 may be due to greatly reduced fertilizer use in the Former Soviet Union and Eastern Europe. (Bennett et al. 2001)

started to decline in higher-income countries but are still rising in lower-income ones. While agricultural lands in the latter had a net loss of P in 1961, they now accumulate more phosphorus per year than do higher-income countries' agricultural areas, making up 5 million of the 8 million tons per year total global P accumulation on agricultural lands. This accumulation is a global average; actual accumulation will vary at regional, national, and local scales.

Calculating nutrient budgets is associated with many uncertainties that are seldom taken into account (Oenema et al. 2003). Many estimates are partial budgets or include numerous assumptions about components that are difficult to quantify at the scales of interest. Moreover, great diversity exists in P budgets among countries, within a country, or even between fields in the same farm. Nutrient audits for China suggest average annual P losses of 5 kilograms per hectare of agricultural land (Sheldrick et al. 2003b). Similarly, an annual loss of 3 kilograms of phosphorus per hectare was estimated for 38 countries of sub-Saharan Africa (Stoorvogel et al. 1993).

In contrast, on-farm studies conducted at 207 sites in China, India, Indonesia, Thailand, and the Philippines showed an average annual surplus of 12 kilograms P per hectare under double-cropping of irrigated rice (Dobermann and Cassman 2002). A study in Uganda demonstrated that the P balance was near neutral for a whole district, but it varied from positive in fields near homesteads to negative in outfields (Bekunda and Manzi 2003). In Southern Mali, P was at balance for the whole region, but large local variations occurred (Van der Pol and Traore 1993). Soil surface P balances studied in nine organic farms in the United Kingdom were positive in six cases, resulting from supplementary P fertilizer (rock phosphate) and additional feed for non-ruminant livestock, whereas a stockless system without P fertilizer resulted in a large P deficit (Berry et al. 2003). In slash-and-burn agriculture, the P balance is only positive when fertilizer is applied (Hoelscher et al. 1996).

Global differences in soil quality also affect the relationships between fertilizer P use, crop productivity, and risk of P pollution. In many tropical regions, small farmers operate on acid upland soils with high P-fixation potential. Reclamation of P-deficient land requires larger phosphate applications (Uexkuell and Mutert 1995) and good soil conservation to prevent P losses by erosion.

7.3.2.3 Potassium Use and Trends

Annual global potash production amounts to about 26 million tons, of which 93% is used in agriculture. Roughly 96% of all potash is produced in North America, Western and Eastern Europe, and the Middle East. Nine companies in Canada, Russia, Belarus, Germany, the United States, and Israel account for 87% of the global potash production. Large amounts of potash fertilizers are shipped around the globe to satisfy the needs of crop production for this important macronutrient.

Trends in global potash fertilizer use are similar to those observed for N and P: stagnating or slightly declining use in many higher-income countries, significant increases in selected lower-income countries, and a sharp drop in Eastern Europe and the former Soviet Union in recent years. (See Figure 7.21.)

Negative potassium (or K) budgets in many parts of the developing world raise concern about the long-term sustainability of crop production, particularly under intensive cultivation. For example, although K use has increased on agricultural land in China during the past 20 years, the nation's overall annual K budget remains highly negative at about negative 60 kilograms per hectare (Sheldrick et al. 2003b). Similar estimates for India and Indonesia suggest annual K losses of about 20–40 kilograms of K per hectare, and these losses have been increasing steadily during the past 40 years. An average annual K loss of nearly 20 kilograms per hectare was estimated for all of sub-Saharan Africa (Stoorvogel et al. 1993). Average annual K losses of about 50 kilograms per hectare are common in double cropping rice systems of Asia (Dobermann et al. 1998).

Overall, potassium deficiency in agriculture is not yet widespread, but if present trends continue, it is only a matter of time until depletion of soil K resources will reach levels that could cause significant limitations for crop production. Because most of the K taken up by plants is contained in vegetative plant parts, recycling of crop residue and organic wastes is a key issue for sustaining K input-output balances. Mineral fertilizer use must be balanced with regard to K, which is an important measure to increase yields and nitrogen use efficiency. To achieve this, annual rates of K use in the developing world may have to rise much faster than those of N use.

7.3.2.4 Plant Nutrient Drivers and Ecosystem Consequences

7.3.2.4.1 *Nitrogen and ecosystem consequences*

Synthetic production of fertilizer N has been a key driver of the remarkable increase in food production that has occurred during the past 50 years (Smil 2001), but crop production also contributes much to global accumulation of reactive N. Only about half of all anthropogenic N inputs on croplands are taken up by harvested crops and their resi-

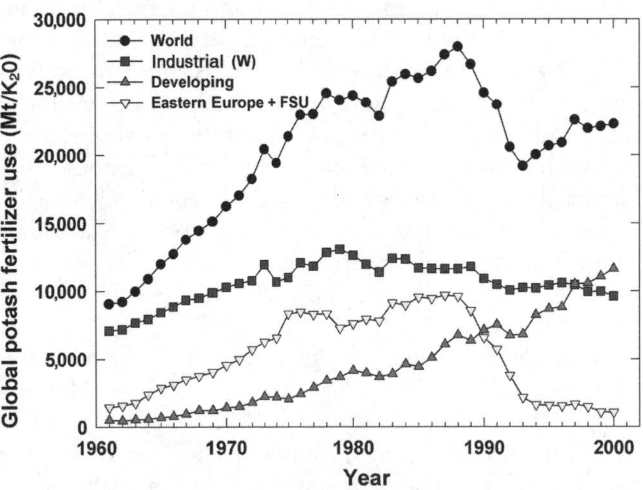

Figure 7.21. Trends in Global Consumption of Potash Fertilizer, 1961–2000 (IFA 2003)

dues. Losses to the atmosphere amount to 26–60 million tons N per year, while waters receive 32–45 million tons N per year from leaching and erosion (Smil 1999). Many uncertainties are associated with these estimates, but there is increasing concern about the enrichment of the biosphere with reactive N forms and significant changes in reactive N distribution (Galloway and Cowling 2002). The key challenge is to meet the greater N requirements of higher-yielding crops while concurrently increasing nitrogen use efficiency and reducing the reactive N load attributable to agriculture.

If left unchecked, significant costs to society may arise through both direct and indirect effects on the environment and human health (Wolfe and Patz 2002) caused by increased emissions of NO_x, NH_3, N_2O, NO_3, and dissolved organic N compounds or by deposition of NO_y and NH_x compounds (Mosier et al. 2001; Townsend et al. 2003). Consequences include decreased drinking water quality (Spalding and Exner 1993), eutrophication of freshwater ecosystems (McIsaac et al. 2001), hypoxia in coastal marine ecosystems (Rabalais 2002), nitrous oxide emissions contributing to global climate change (Smith et al. 1997; Bouwman et al. 2002), and air pollution by NO_x in urban areas (Townsend et al. 2003). Occurrence of such problems varies widely in different world regions. However, preliminary estimates for the United Kingdom (Pretty et al. 2000) and Germany (Schweigert and van der Ploeg 2000) suggest that the environmental costs of excessive N fertilizer use may be a substantial proportion of the value of all farm goods produced.

7.3.2.4.2 Phosphorus and ecosystem consequences

Approximately 20–30% of P in fertilizer is taken up by agricultural plants when they grow (Sharpley et al. 1993). The remainder accumulates in the soil or runs off. Phosphorus moves downhill in particulate or dissolved form, where it has secondary impacts on other ecosystems.

Phosphorus is the limiting nutrient in most freshwater lakes (Vollenweider 1968; Schindler 1977) and is a critical nutrient in the productivity of marine ecosystems (Tyrell 1999). Excess P degrades lakes through a syndrome of increased productivity called eutrophication (Vollenweider 1968; Schindler et al. 1971; Dillon and Rigler 1974). Phosphorus is the primary cause of most freshwater eutrophication and a component of estuarine eutrophication. Eutrophic lakes frequently experience noxious algal blooms, increased aquatic plant growth, and oxygen depletion, leading to degradation of their ecological, economic, and aesthetic value by restricting use for fisheries, drinking water, industry, and recreation (National Research Council 1992; Sharpley et al. 1994).

When lakes become eutrophic, many ecosystem services are reduced. Water from lakes that experience algal blooms is more expensive to purify for drinking or other industrial uses (Carpenter et al. 1998). Eutrophication can also restrict use of lakes for fisheries because low oxygen levels reduce fish populations (Smith 1998). While habitat for some aquatic plants, such as noxious algae, are increased by eutrophication, habitat for other plants, such as aquatic macrophytes, may decrease. Underlying all these changes are reductions in some of the supporting services provided by lakes, such as nutrient cycling. Possibly the most striking loss is that of many of the cultural services provided by lakes. Foul odors of rotting algae, slime-covered lakes, and toxic chemicals produced by some blue-green algae during blooms keep people from swimming, boating, and otherwise enjoying the aesthetic value of lakes.

In marine systems, excess P input causes harmful algal blooms (including blooms of toxic species); growth of attached algae and macrophytes that can damage benthic habitats, including coral reefs; problems with deoxygenation and foul odors; fish mortality; and economic losses associated with increased costs of water purification and impairment of water supply for agriculture, industry, and municipal consumption (Postel 1997; Carpenter et al. 1998; Smith 1998). The U.S. Environmental Protection Agency called eutrophication the most ubiquitous water quality problem in the United States (U.S. Environmental Protection Agency 1996). Although it is currently a key problem primarily in industrial countries with high rates of fertilizer use, recent increases in phosphate fertilizer use in developing countries indicate that it may become of increasing importance there as well.

It is extremely difficult to control P runoff because it comes from widely dispersed sources and can vary with weather. Studies indicate that agriculture is a major source of P (Sharpley 1995) and that most of the runoff happens during spring storms that carry soils and their P into bodies of water (Pionke et al. 1997), but it has proved very difficult to locate particular fields as important sources in ways other than through proxy measurements, such as high concentrations of P in the soil.

Accumulation of P uphill from bodies of water has been termed a chemical time bomb (Stigliani et al. 1991) because it is simply waiting to be transported to downstream water bodies through erosion and other processes. As a global average, if we completely ceased all P fertilizer use it would take approximately 40 years for soil P concentrations to return to 1958 levels (calculated based on Bennett et al. 2001). It is important to keep in mind that turnover rates for P in soils are highly variable in space. Some areas with high soil P may have very slow turnover. For example, a study of the Lake Mendota watershed in Wisconsin USA, suggests that if over fertilizing were to stop, it would still take over 260 years for P in the soil to drop to 1974 levels (Bennett et al. 1999). Other budgets for high P soils indicate similarly long turnover times. However, many regions, especially in highly weathered tropical soils, have soils that are very impoverished in P and would have a shorter drawdown time.

7.3.2.4.3 Potassium and ecosystem consequences

Negative environmental impacts of potassium are negligible, so that its major role as an ecosystem driver is that of increasing crop productivity. Potassium has no known negative consequences for ecosystems, and its excessive accumulation in soils and waters is rare.

7.3.2.5 Summary of Plant Nutrient Drivers

Human activities have dramatically altered plant nutrient cycles. Human interventions have had a positive impact on crucial ecosystem services such as food supply but have also caused numerous negative consequences for other ecosystem services. These impacts (negative and positive) vary by location.

Changes in the global atmospheric deposition of reactive N and in the concentrations of soil P have had a positive impact on increasing food production at affordable prices but have also had negative consequences for some ecosystem services. Traditional cropping practiced before synthetic N fertilizers became available could provide today's average diets for only about 40% of the existing population (Smil 2001). As the human population continues to increase, fertilizer N will continue to play a dominant role in the global N cycle. Therefore, increases in nitrogen use efficiency are necessary to control the cycling of the most potent reactive N compounds. There is much potential for fine-tuning N management in agricultural crops through technologies that achieve greater congruence between crop N demand and N supply from all sources—including fertilizer, organic inputs, and indigenous soil N (Cassman et al. 2002). Such improvements are likely to have a large impact on the global N cycle, but they require collaboration among agronomists, soil scientists, agricultural economists, ecologists, and politicians (Galloway et al. 2002), as well as significant long-term investments in research and education.

Patterns of P supply, consumption, and waste production have also become decoupled from natural P cycles (Tiessen 1995). Changing the P cycle has had a positive impact on production of food from agricultural areas, but a negative impact on provision of services from aquatic ecosystems downstream from those areas. Phosphorus surpluses due to fertilizer use, livestock industry, and imports of feed and food have become widespread in high-income countries. In contrast, both P surpluses and deficits are found in developing countries, including large areas that are naturally P-deficient. Finding ways to maintain or increase agricultural production without sacrificing water quality will be a major issue in the coming years.

7.3.2.6 Overview of Plant Nutrient Drivers in the MA Scenarios

Changes in nutrient use vary among the scenarios, depending on the key indirect drivers of food demand and technological change. Increased food demand generally leads to increases in the amount of fertilizer used; however, this can be tempered by technological change that results in increased food production without additional fertilizers.

All scenarios will experience some increase in the use of N and P fertilizers due to growing human population and increased demand for food. Global Orchestration is projected to have the highest increase in fertilizer use. Increasing economic openness in this scenario makes fertilizers available in locations where they currently are not commonly used. Much of the increase in fertilizer use will be in currently poor countries. Order from Strength will also have a large increase in fertilizer use. In this scenario, however, most of the increase comes from higher use in wealthier countries.

Increased agricultural production in wealthier countries requires increased efficiency of production from lands already in cultivation in addition to a greater amount of cultivated land. Without additional technological advances, production is likely to be improved through heavier use of fertilizers. Food production in the TechnoGarden scenario also increases, but technical change is the primary driver, so fertilizer use increase is the lowest in this scenario. Nutrient use in the Adapting Mosaic scenario falls in the range of the other scenarios. The environmentally proactive approach should lead to decreased use of fertilizers, but low amounts of trade will force people in some areas to use fertilizers heavily simply to feed people locally.

7.3.3 Land Conversion

Humans change land use to alter the mix of ecosystem services provided by that land. Sometimes the land conversion effort is intentional, such as plowing grassland to grow crops. In other cases, land conversion is a consequence of other activities. For example, salinization is the unintended consequence of irrigation that does not have adequate drainage.

The Millennium Ecosystem Assessment sponsored an international effort to document regions around the world in which rapid and recent change (since the 1970s) in land cover can be shown to have occurred (Lepers et al. in press). In this section, we summarize its results, focusing on four types of land conversion—deforestation, dryland degradation, agricultural expansion and abandonment, and urban expansion.

7.3.3.1 Deforestation

Deforestation is the single most measured process of land cover change at a global scale (FAO 2001a; Achard et al. 2002; DeFries et al. 2002). According to most definitions, a forest is defined as land with tree canopy cover above some minimum threshold or as land that is intended to be used as forested land. (Different thresholds have been used to define forested land; for instance FAO (2001a) uses a 10% tree canopy cover threshold, whereas the IGBP land-cover classification uses a 60% threshold (Scepan 1999).) Deforestation occurs when forest is converted to another land cover and intended land use. Not all disturbances in a forest ecosystem lead to deforestation. Forested land that has been harvested and has not yet regrown to forest is often still categorized as forest in many databases. Forest degradation—"changes within the forest that negatively affect the structure or function of the stand or site and thereby lower the capacity to supply products and/or services"—can but does not necessarily lead to deforestation (FAO 2001a).

Afforestation is the conversion from some other land use to forest; reforestation is afforestation on land that at some point in the past was forested. Data on afforestation and reforestation are much less readily available than on deforestation. Few spatially explicit data sets on afforestation, forest expansion, or reforestation were found in the literature at a regional scale, even though these positive changes in forest represent a major ecosystem change and should not be ignored (see Munroe et al. 2002 for an example of a region of Honduras where both deforestation and reforestation have occurred, and additional references in Turner II et al. 2004). According to FAO (2001a), global forest area underwent a net change of −9.4% between 1990 and 2000. Losses of 16.1% were balanced by a 6.7% gain in forest area (mainly forest plantations or regrowth). The latter figures suggest that the extent of afforestation is considerable.

Remotely sensed data can often be used to distinguish between forested and nonforested land. Three quarters of the world's forests are covered by at least one remote sensing–based or expert-opinion data set on the occurrence of deforestation or degradation. Approximately 8% of the forests are covered by five data sets, with coverage in the Amazon Basin being especially extensive. On the other hand, one quarter of the world's forests only have statistical information on deforestation at national or sub-national scales. This is the case, for instance, for Europe and Canada.

Using the forest classes of the GLC 2000 and IGBP DIScover data sets, forests covered 39% of Earth's surface in 2000. Around 30% of the world's forests were located in Asia. Africa, South America, and North America (Canada, the United States, and Mexico) each accounted for less than 20% in 2000.

As noted in Chapter 21 of the MA *Current State and Trends* volume, during the industrial era, global forest area was reduced by 40%, with three quarters of this loss occurring during the last two centuries. Forests have completely disappeared in 25 countries, and another 29 have lost more than 90%. Although forest cover and biomass in Europe and North America is currently increasing following radical declines in the past, deforestation of natural forests in the tropics continues at an annual rate of over 10 million hectares a year—an area larger than Greece, Nicaragua, or Nepal, and more than four times the size of Belgium. Moreover, degradation and fragmentation of remaining forests are further impairing ecosystem functioning. The area of planted forests is, however, growing, decreasing the likelihood of a global shortage of wood.

Existing data suggest that deforestation and forest degradation have been more extensive in the tropics than in the rest of the world, but this is possibly due to the fact that most data sets cover the tropics and not the boreal zones. Only 20% of the forests of nontropical areas are covered by at least one remote sensing–based or expert-opinion data set, while the entire tropical region is covered. Areas of deforestation or degradation may thus go uncounted in the boreal or temperate regions. Some locations in Canada and Europe are also subject to intensive forest exploitation and could also appear as degraded as Siberia.

7.3.3.2 Dryland Degradation

Drylands are defined in the World Atlas of Desertification as zones having a ratio of average annual precipitation to potential evapotranspiration lower than 0.65 (Middleton and Thomas 1997). Dryland degradation, also called desertification, has affected parts of Africa, Asia, and Mediterranean Europe for centuries, parts of America for one or two centuries, and parts of Australia for 100 years or less (Dregne 2002).

There is little agreement on the definition and indicators of desertification, so this category of land cover change is not easy to map at a global scale. The most commonly accepted definition of desertification is "land degradation in arid, semi-arid and dry sub-humid areas resulting from various factors, including climatic variations and human activities" (UNCED 1992). In the UNCED definition from *Agenda 21,* desertification only applies to the drylands of the world. Hyper-arid zones are not part of the definition because they are presumed to be so dry that human degradation is severely limited unless irrigation is practiced. But they are included in the statistics reported here as well as in Chapter 22 of the *Current State and Trends* volume.

Despite the global importance of desertification, available data on the extent of land degradation in drylands are extremely limited. To date, only two studies are available with global coverage. Both have considerable weakness, but for lack of anything better they are widely used as a basis for many national, regional, and global environmental assessments. The most well known study is the Global Assessment of Soil Degradation, or GLASOD (Oldeman et al. 1991). This study found that 20% of the world's drylands (excluding hyper-arid areas) suffered from soil degradation. Water and wind erosion was reported as the prime cause for 87% of the degraded land (Thomas and Middleton 1997; Oldeman and van Lynden 1997).

The second global study is that of Dregne and Chou (1992), which covered both soil and vegetation degradation. It was based on secondary sources, which they qualified as follows: "The information base upon which the estimates in this report were made is poor. Anecdotal accounts, research reports, travelers' descriptions, personal opinions, and local experience provided most of the evidence for the various estimates." This study reported that some 70% of the world's drylands (excluding hyper-arid areas) were suffering from desertification (soil and vegetation degradation).

The MA-commissioned study by Lepers et al. (in press) compiled more detailed (sometimes overlapping) regional data sets (including hyper-arid drylands) derived from literature review, erosion models, field assessments, and remote sensing. The data sets used to assess degradation covered 62% of the drylands of the world. Major gaps in coverage occurred around the Mediterranean Basin, in the Sahel, in the east of Africa, in parts of South America (north of Argentina, Paraguay, Bolivia, Peru, and Ecuador), and in the federal lands of the United States. This study found less alarming levels of land degradation (soil plus vegetation) in the drylands and hyper-arid regions of the world. With only partial coverage, and in some areas relying on a single data set, it estimated that 10% of global drylands were degraded.

7.3.3.3 Agricultural Conversion

Most of the studies and data sets related to changes in agricultural land focus on changes in arable land and permanent crops. As defined by FAO, arable land refers to land under temporary crops, temporary meadows for mowing or pasture, land under market and kitchen gardens, and land temporarily fallow (less than five years). Permanent crop represents land cultivated with crops that occupy the land for long periods and need not be replanted after each harvest, such as cocoa, coffee, and rubber. This category includes land under flowering shrubs, fruit trees, nut trees, and vines but excludes land under trees grown for wood or timber (FAO 2001b). As defined by FAO, agricultural land also includes permanent pasture. However, this category of land use was not included here due to lack of data and the difficulty of differentiating pastures from natural unmanaged grasslands in many parts of the world. So the term "cropland class," as used here, refers to arable land and permanent crops.

The cropland class, defined as areas with at least 10% of croplands within each pixel, covered 30% of Earth's surface in 1990. The exact proportion was between 12% and 14%, depending on whether Antarctica and Greenland were included (Ramankutty and Foley 1998). Around 40% of the cropland class was located in Asia; Europe accounted for 16% and Africa, North America, and South America each accounted for 13%.

Lepers et al. (in press) attempted to assess the change in cropland area between 1980 and 1990 by combining data from a variety of sources and time periods. This effort resulted in unexpected outcomes, including extensive cropland increase in the central United States, southern Ireland, and the island of Java. It found conversely that cropland area decreased in China and that there no identifiable hotspots of cropland increase in the central part of Africa. Clearly, much more analysis is needed to identify changes in this critical driver of ecosystem change. (See Box 7.2 for a profile of land use change in one region.)

7.3.3.4 Urbanization

In 2000, towns and cities sheltered more than 2.7 billion of the world's 6 billion people (UNDP 2000). Growth in urban population usually results in conversion of agricultural, often highly productive land and other land types for residential, infrastructure, and amenity uses. These growing cities have an impact on their surrounding ecosystems through their demand for food, fuel, water, and other natural resources (Lambin et al. 2003).

So far no global data sets or direct measurements have been developed to delineate the changes in extent and shape of urban areas. Only indirect indicators such as human population or the evolution through time of the night-time lights viewed from space can be used as a proxy to measure the change in urban extent. The data presented here are based on such indirect indicators and should be considered tentative.

Global population is strongly clustered, with 50% of the world's population inhabiting less than 3% of the available land area (excluding Antarctica), at average densities greater than 300 people per square kilometer (Small 2001). Today's largest cities are mainly located on the eastern coast of the United States, in western Europe, in India, and in East Asia; the most rapidly growing cities are all located in the tropics.

7.3.4 Biological Invasions and Diseases

7.3.4.1 Biological Invasions and Ecosystem Effects

At present, no definition of biological invasion has been unanimously accepted by the scientific community, due in part to the proliferation of terms to describe various concepts used by different authors (Richardson et al. 2000). However, consensus is growing around at least two main aspects of the invasion process: the traits that enable a species to invade a habitat, called invasiveness, and the habitat characteristics that determine its susceptibility to the establishment and spread of an invasive species, its "invasibility" (Lonsdale 1999; Alpert et al. 2000).

This conceptual framework allows for an operational definition of invasive species as a species that spreads in space, either occupying new habitats or increasing its cover in areas previously occupied. This approach allows for a more general treatment of the invasion problem, since cases in which native species become invaders after some habitat or climatic change (for example, shrub encroachment, crayfish, and lampreys (Jeltsch et al.1997; Lodge et al. 2000; van Auken 2000)) can also be considered. Thus, although most of invasions are thought to be caused by non-native species, the key distinction becomes between invasive and noninvasive species rather than between native and non-native species (Crawley 1986; Alpert et al. 2000; Sakai et al. 2001).

Biological invasions are a global phenomenon, affecting ecosystems in most biomes (Mack et al. 2000). Human-driven movement of organisms, deliberate or accidental, has caused a massive alteration of species ranges, overwhelming the changes that occurred after the retreat of the last Ice Age (Semken 1983). Ecosystem changes brought about by invasions can have both short-term (ecological) and long-term (evolutionary) consequences. For example, the tropical alga *Caulerpa taxifolia* evolved tolerance for colder temperatures while it was growing in aquaria. After escaping, it is invading vast portions of the northwest Mediterranean Sea (Meisnesz 1999).

Ecological interactions between invaders and native species can be complex. Invaders can affect native species, but natives also affect performance of the invaders during the invasion process.

Acceleration of extinction rates as a result of negative interactions is one of the most important consequences of biological invasions (Ehrlich and Daily 1993; Daily and Ehrlich 1997; Hughes et al. 1998). In the United States, invasions of non-native plants, animals, and microbes are thought to be responsible for 42% of the decline of native species now listed as endangered or threatened (Pimentel 2002).

In some cases, the invader causes inhibition of the establishment of native species (a kind of ammensalism) (Walker and Vitousek 1991). Cases of commensalism and mutualism are less well documented. (Symbiosis is a mutually beneficial relationship between two organisms; commensalism is a relationship in which only one of the two organisms ob-

BOX 7.2

Drivers of Land Use Change in Central Asia

The Central Asian region consists of a broad range of ecosystems and associated land uses, spanning an area from west of the Urals to the Mongolian Plateau and the forests of the Far East of Russia. Historically, this region has been an important supplier of meat, milk, wool, forest products, and grains. Low population densities of nomadic pastoralists in most of this region used the rich grasslands to graze mixed herds of cattle, sheep, goats, horses, and camels. Grazing patterns were dictated more by intra- and interannual climate variability than political or economic factors.

Political and social changes have lead to dramatic adjustments in land use. During the early 1900s, most croplands and livestock systems were placed under collective management and remained there until the dissolution of the Soviet Union. These land use policies led to major changes in land management throughout the region and altered the transhumance patterns of livestock and rangeland management. Many aspects of the traditional nomadic culture were replaced. The largest decline in world forests and grassland areas in the past two centuries took place in the Eurasian region. Large-scale agricultural intensification took place during the Soviet era, and livestock production was made more sedentary. Cropland conversion peaked during the late 1980s.

Land degradation due to overgrazing, sowing of monocultures, poor irrigation management, cropping marginal lands, and increased frequency of fires has become a serious environmental concern. UNEP estimates that 60–70% of the grasslands in China, Mongolia, Central Asia, and Russia are overgrazed or degraded due to inappropriate cropping.

The dissolution of the Soviet Union led to changing pressures on land use once again. Since 1990, land managers have been adjusting to life in transition economies, characterized by a combination of volatile markets, policy changes, reductions in public services, and unclear and uncertain land tenure systems. It is not any one of the factors but rather the interactions among them that makes these systems and peoples vulnerable.

The global factors leading to increased vulnerability in this region stem from the socioeconomic adjustments to globalization and resource use adjustments resulting from climate warming. In addition, regional adjustments have triggered demographic changes and, in some cases, growth to take advantage of opportunities associated with globalization of the resource base.

The consequences of the latest changes include cropland abandonment, destocking of certain rangelands and increased stocking of others, degradation of soils due to salinization and desertification, and damage to wetlands due to modifications of water regimes and industrial development. A major shift in livestock rates has been observed in the Eurasian steppe, with increases in China and Mongolia, where opportunities arose with free market access, whereas in Central Asian countries major destocking occurred due to loss of social infrastructure. During spring 2003, the region was the source of large dust plumes that traveled as far the North American continent and led to health concerns for the Asian populations in their path.

tains some sort of advantage, such as between remoras and sharks; mutualism refers to a relationship in which both organisms benefit; ammensalism is an interaction in which one organism is harmed and the other is unaffected.) Commensalism is related to facilitation processes like nurse effects and mutualism. For example, the native *Turdus* bird disperses seeds of introduced *Ligustrum lucidum* in Argentina (Marco et al. 2002). An example of mutualism is the introduction of exotic N-fixing bacteria or mycorrhizal fungi with crops (Read et al. 1992).

In many cases, disruption of an interaction in the native habitat, like the absence of pathogens, plays an important role in contributing to the invasive success of exotic plant and animal species. On average, 84% fewer fungi and 24% fewer viruses infect plant species in the naturalized range than in the native range (Mitchell and Power 2003). Torchin et al. (2003) show that invasive animal species lose a significant proportion of the parasite diversity they support in their native habitat. In a survey of 24 invasive animal species, Torchin et al. (2003) show that whereas these species had an average of 16 species of parasites in their native range, only two of the parasites successfully colonized with the host in the new range. The original parasites were supplemented by four parasites native to the host's exotic habitat. This considerable reduction in parasite burden might give introduced species an advantage over native species with a larger diversity of parasites. Invasive plant species that are more completely released from pathogens are more widely reported as harmful invaders of both agricultural and natural ecosystems.

The threats that biological invasions pose to biodiversity and to ecosystem-level processes translate directly into economic consequences such as losses in crops, fisheries, forestry, and grazing capacity (Mack et al. 2000). Mismanagement of semiarid grasslands combined with climatic changes has caused woody plant invasion by native bushes and the loss of grazing lands in North and South America (van Auken 2000). Pimentel (2002) estimated that exotic weeds in crops cost U.S. agriculture $27 billion a year. Marine and estuarine waters in North America are heavily invaded, mainly by crustaceans and mollusks in a pattern corresponding to trade routes (Ruiz et al. 2000). U.S. lakes and watersheds are invaded as well. The costs associated with zebra mussel (*Dreissena polymorpha*) invasion were estimated at $100 million a year by Pimentel et al. (1999), although the $5 billion a year estimated by the Department of Environmental Quality of Michigan Government is perhaps a more realistic figure (Michigan Department of Environmental Quality 2004). Leung et al. (2002) found an expenditure of up to $324,000 per year to prevent invasion by zebra mussels into a single lake with a power plant.

In spite of these alarming figures, can introductions of alien species be beneficial? Some 98% of the U.S. food supply comes from introduced species, such as corn, wheat, rice, and other crops, as well as cattle, poultry, and other livestock (Pimentel 2002). The same proportion of introduced to native food species is likely to be found in most countries. Some species are introduced to control other introduced species, sometimes successfully (*Galerucella calmariensis* and *G. pusilla* to control purple loosestrife in North

America) and sometimes unsuccessfully (*Bufo marinus* (cane toad) to control greyback cane beetle (*Dermolepida albohirtum* (Waterhouse)) and French's Cane Beetle (*Lepidiota frenchi* (Blackburn)) in Australia and other sugar-growing countries).

Many introduced species are alternative resources for native species (for example, alien plants and fruit-eating birds already mentioned). But how many of them have become invasive? In Britain, 71 out of 75 non-native crop plant species are naturalized, in part because they are strongly selected to growth where they are cultivated (Williamson and Fitter 1996). However, not all habitats are equally invasible. In an extensive study of plant invasions in the British Isles, Crawley has shown that lowland disturbed habitats are host to a much larger diversity of alien plant species than are upland habitats, which remain relatively uninvaded (Crawley and May 1987; Crawley 1989).

7.3.4.2 Diseases and Ecosystem Effects

Many biological invasions are regarded as affecting human well-being, either by affecting public health or by impairing economic activities such as agriculture or fisheries. Among the most common sanitary problems are the invasions by disease-causing organisms or by vectors of disease-causing parasites. Tuberculosis, AIDS, influenza, cholera, bubonic plague, bovine spongiform encephalopathy, and severe acute respiratory syndrome are among the diseases caused by introduced pathogenic biological agents in many countries. For example, more than 100 million people are infected with the influenza virus each year in just the United States, Europe, and Japan. In the United States alone, influenza represents on average $14.6 billion in direct and indirect costs each year. The influenza virus is originally Asiatic, and aquatic birds provide the natural reservoir from which occasional genetic mutations can spread to domestic poultry and humans (Zambon 2001).

While the term "invasive alien species" is commonly associated with plants and insects, parasites and pathogens possess a considerable potential to significantly modify ecosystem function. This potential stems from both their ability to multiply very rapidly and also their diversity. Arguably more than half of biodiversity consists of species that are parasitic on more conspicuous free-living species (Dobson et al. 1992). In the last 20 years, studies demonstrating how pathogens modify and regulate free-living hosts have completely modified our understanding of the role that parasites play in natural and human-modified systems (Dobson and Hudson 1986; Dobson and Crawley 1994; Grenfell and Dobson 1995). Examples from a range of ecosystems and a variety of key processes are reviewed in this section.

Jim Porter and his colleagues have been intensely monitoring the coral communities of the Florida Keys since the early 1990s. The work uses regular digital photographic censuses along fixed transects throughout the Keys, allowing detailed analysis of the growth and structure of the coral community. In the late 1990s the scientists noted dramatic changes in the structure of the coral community at some transects. The dominant structural species, Elkhorn coral, was first infected with a new disease syndrome, white pox disease (the most severely affected Elkhorn species was *Acropora cervicornis*). This infection affected several Elkhorn coral while also rapidly spreading to other coral species (Patterson et al. 2002). Within a few months, once-vibrant coral reefs were reduced to bare rock. The work provides a vivid example of how quickly a pathogen can modify and destroy the structure of a complex community. More disconcertingly, it is but one of an increasing number of examples that suggest pathogen outbreaks are increasing in the oceans (Harvell et al. 1999).

The Florida Keys example echoes a dramatic terrestrial example that occurred almost a century earlier, when rinderpest virus was accidentally introduced into sub-Saharan Africa. Rinderpest virus causes high fatality rates in the hoofed mammals it infects (cattle, wildebeest, and buffalo), is a member of the Morbillivirus family, and is a recent ancestor of human measles. It took 10 years for rinderpest to spread from the Horn of Africa to the Cape. The pandemic it caused reduced the abundance of susceptible species by up to 90% in many areas (Spinage 1962; Plowright 1982).

The huge reduction in the abundance of herbivores had impacts that spread throughout the ecosystems of sub-Saharan Africa (Prins and Weyerhaeuser 1987; McNaughton 1992). Fire frequency increased as uneaten grasses accumulated and ignited during lightning storms in the dry season. This combined with a reduction in browsers to allow miombo bush to spread across areas that were previously grass savannas. The absence of prey caused a decline in the abundance of lions, although other carnivores, such as cheetahs and hunting dogs, may have increased as habitats appeared that were better suited to their hunting methods. More insidiously, the absence of buffalo, wildebeest, and cattle caused tsetse flies to increase their feeding rates on humans, which led to a significant epidemic of sleeping sickness (Simon 1962).

The development of a vaccine for rinderpest in the late 1950s allowed a reversal of many of these trends (Plowright 1982; Dobson and Crawley 1994). Although only cattle have been vaccinated, their role as reservoir hosts is sharply illustrated by the speed at which the pathogen disappeared from wildlife. This in turn led to a reversal of the balance between grassland and miombo bushland and to a significant increase of, first, herbivores and then the carnivores that feed upon them.

In a third example, the fungal disease amphibian chytridiomycosis and other pathogens are implicated in amphibian population declines worldwide (Daszak et al. 2003). Chytridiomycosis appears to be locally increasing in impact or moving into new regions. In common with many emerging infectious diseases of humans, domestic animals, and other wildlife species, the emergence of chytridiomycosis may be driven by anthropogenic introduction. Emerging infectious diseases are linked to anthropogenic environmental changes that foster increased transmission within or between hosts.

These three examples illustrate how the introduction of a novel pathogen has caused ecosystem-wide changes in species abundance. If pathogens were considered as keystone species, they would plainly provide some of the most dramatic examples of individual species that cause dramatic changes in the abundance of all others in the ecosystem (Burdon 1991; Power et al. 1996). As ecologists focus more attention on the role of parasites in natural systems, more examples will be found where highly specific pathogens have significantly affected a single species in a community and caused cascading effects on other species that use the infected species as a resource (Dobson and Hudson 1986; Dobson and Crawley 1994). Nevertheless, this emphasis on dramatic examples of ecosystem change should not distract attention from the fundamental role that endemic parasites play in driving natural processes.

7.4 Examples of Interactions among Drivers and Ecosystems

Changes in ecosystem services are almost always caused by multiple, interacting drivers. Changes are driven by combinations of drivers that work over time (such as population and income growth interacting with technological advances that lead to climate change), over level of organization (such as local zoning laws versus international environmental treaties), and that happen intermittently (such as droughts, wars, and economic crises).

Changes in ecosystem services feed back to the drivers of changes. For example, they create new opportunities and constraints on land use, induce institutional changes in response to perceived and anticipated resource degradation, and give rise to social effects such as changes in income inequality (as there are winners and losers in environmental change). Reviews of case studies of deforestation and desertification (Geist and Lambin 2002, 2004) reveal that the most common type of interaction is synergetic factor combinations—combined effects of multiple drivers that are amplified by reciprocal action and feedbacks.

Drivers interact across spatial, temporal, and organizational scales. Global trends like climate change or globalization can influence regional contexts of local ecosystem management. Research on the combined effects of these two major global trends on the local farmers, other agents, and finally on ecosystems is sparse but growing. For example, a study in South Africa found that changes in export prices of cash crops can trigger land use changes on the local level and that removal of national credits and subsidies can make some farmers more vulnerable to environmental changes while others profit from easier access to markets and are less vulnerable to climate change (Leichenko and O'Brien 2002).

Any specific ecosystem change is driven by a network of interactions among individual drivers. Though some of the elements of these networks are global, the actual set of interactions that brings about an ecosystem change is more or less specific to a particular place. For example, a link between increasing producer prices and the extension of production can be found in many places throughout the world. The strength of this effect, however, is determined by a range of location-specific factors including production conditions, the availability of resources and knowledge, and the economic situation of the farmer (Jones 2002). No single conceptual framework exists that captures the broad range of case study evidence (Lambin et al. 2001).

To generalize, simplifications are needed, but these simplifications are often sufficient to explain general trends in ecosystem change. There exists a broad range of scientific schools of generalizing simplifications, each one entering the arena from a different perspective and thus adding different pieces to the broadening knowledge base on causes for ecosystem change. Many of these approaches emerge from disciplinary perspectives, such as spatial (agro-)economic models of land use change (Alcamo et al. 1998; Nelson and Geoghegan 2002). Others use generalizing schemes on a highly aggregated basis like the IPAT approach, where environmental impact is seen as the product of population, affluence, and technology (Ehrlich and Holden 1971; Ehrlich and Daily 1993; Waggoner and Asubel 2002). Others analyze case studies in order to identify common threads and processes (Scherr 1997; Geist and Lambin 2002), sometimes attempting to identify trajectories of environmental criticality (Kasperson et al. 1995) or to formulate qualitative models (Petschel-Held et al. 1999; Petschel-Held and Matthias 2001).

Based on the findings of the Multiscale Assessments of the MA and recent literature, examples of causal complexes for ecosystem change can be given. This section reflects on some of these patterns, organized along major direct drivers.

7.4.1 Land Use Change

Ten years of research within the international program on land use and land cover change of IGPB concluded that neither population nor poverty alone constituted the sole and major underlying causes of land cover change worldwide. (See Figure 7.22 in Appendix A.) Rather, responses to economic opportunities, mediated by institutional factors, drive land cover changes. Opportunities and constraints for new land uses are created by local as well as national markets and policies. Global forces become the main determinants of land use change, as they amplify or attenuate local factors (Lambin et al. 2001).

7.4.1.1 Tropical Deforestation

Case study reviews (Geist and Lambin 2002) as well as findings within the MA's alternatives to slash-and-burn crosscutting Sub-global Assessment reveal regional patterns for tropical deforestation. In all regions of the humid tropics—Latin America, Southeast Asia, and Central Africa—deforestation is primarily the result of a combination of commercial wood extraction, permanent cultivation, livestock development, and the extension of overland transport infrastructure.

However, many regional variations on this general pattern are found. Deforestation driven by swidden agriculture is more widespread in upland and foothill zones of Southeast Asia than in other regions. Road construction by the state followed by colonizing migrant settlers, who in turn practice slash-and-burn agriculture, is most frequent in lowland areas of Latin America, especially in the Amazon Basin.

Pasture creation for cattle ranching is causing deforestation almost exclusively in the humid lowland regions of mainland South America. The spontaneous expansion of smallholder agriculture and fuelwood extraction for domestic uses are important causes of deforestation in Africa. These regional differences mostly come from varying mixes of economic, institutional, technological, cultural, and demographic factors underlying the direct causes of deforestation.

7.4.1.2 Agricultural Intensification

Agricultural intensification is usually defined as substantial use of purchased inputs, especially fertilizer, in combination with new plant varieties that respond well to the increased inputs. Globally, intensification has been a major contributor to the doubling of food production over the last 40 years (MA *Current State and Trends,* Chapter 8). In higher-income countries, this increase was primarily achieved via intensification. In lower-income countries, 71% of the increase in crop production came from yield growth; the remaining 29% from area expansion (Bruinsma 2003). These successes are tempered by concerns about whether they can be maintained in current locations and about the consequences of the driving forces in regions where intensification has not yet occurred.

Lambin et al. (2001) identify three major pathways toward agricultural intensification that are supported by a number of sub-global assessments:

- Land scarcity in economies weakly integrated into the world markets can be caused by interactions among population growth and institutional changes, such as implementation of legal property rights (Ostrom et al. 1999). For example, the Western Ghat sub-global assessment reports about land reform acts in this respect, by growth in population and its density, where it is often related to land division (Kates and Haarmann 1992). In some instances land scarcity can contribute to deforestation (see the MA sub-global assessments on the Darjeeling Region in India and on Papua New Guinea).
- Market integration (called commodification by Lambin et al. 2001) is brought about by interactions among general infrastructure development (such as road building projects and creation of a national food safety inspection service), changes in macroeconomic policies, and technological change. For an individual farmer, it can mean changes in input prices and availability and changes in sales opportunities and prices. Economic specialization, wage labor, contract farming, and other adjustments follow.
- Region-specific interventions, in the form of state-, donor-, or NGO-sponsored projects, are done to foster development through commercial agriculture. Many of these projects are planned and managed from afar, creating the potential for inefficient management and inadequate attention to specific local constraints. These projects are also vulnerable to changes in government or donor policy and to public-sector financial constraints (Altieri 1999). The Sinai sub-global assessment (Egypt) demonstrates this possibility. Agricultural development projects are designed to attract as many as 6 million people to the Sinai region in order to relieve population pressure from the Nile valley.

7.4.1.3 Urban Growth and Urbanization

Though only about 2% of Earth's land surface is covered by built-up area (Grübler 1994), the effect of urban systems on ecosystems extends well beyond urban boundaries. Three processes of urban change appear to be of relevance for ecosystem change: the growth of urban population (urbanization), the growth of built-up area (urban growth), and the spreading of urban functions into the urban hinterland connected with a decrease in the urban-rural gradient in population density, land prices, and so on (urban sprawl). Each of these processes has its own pathways in which it affects ecosystems and human well-being:

- Urbanization: In developing countries, urban administration often is not capable of developing needed infrastructure as fast as urban populations grow (Kropp et al. 2001). Insufficient housing conditions, a lack of access to safe water and sanitation, and major health effects follow. Rural-urban migration is an important source of urbanization in developing countries. This arises when opportunities for gainful employment in agriculture cannot keep up with growth in the rural workforce and when urban areas provide the potential of gainful employment. Degradation of rural environments can contribute to the pressure. There is, however, also a reverse effect, with remittances possibly allowing the use of cultural practices that reduce pressure on rural ecosystems.
- Urban growth and sprawl: In industrial countries, urban growth and in particular urban sprawl play a much more significant role: landscapes become fragmented, traffic and energy use increase, and land sealing can change flood regimes. The latter has been observed in the sub-global assessment of the Kristianstad Vattenrike in Sweden as a major driving process of ecosystem change.

7.4.2 Tourism

Several of the sub-global assessments report the importance of tourism, both as a driver of ecosystem change and as a potential response option for income generation and reduction of pressure on ecosystems (e.g., the Southern African, Portugal, Caribbean, Trinidad Northern Range, and Costa Rica assessments).

The Southern African Sub-global Assessment provides a good example of how biodiversity serves as an important basis for income generation, thus representing a major asset of a region or community (Biggs et al. 2004). In many areas in Southern Africa, nature-based tourism represents a major source of income, contributing about 9% of GDP in the Southern Africa Development Community region. Though no quantitative estimates on faunal biodiversity exist for the region, it is evident that animal biodiversity is one of the most important motives for tourists. Also, nature-based tourism, together with hunting, contributes to the privatization of the nature conservation efforts in the region, as in private nature parks. This privatization represents an important pillar in the overall conservation efforts in the region.

From an integrated point of view—that is, by considering the region as a whole—tourism appears to be more a driver for ecosystem conservation than for change.

In contrast, the Caribbean Sea Sub-global Assessment shows the other side of tourism—its negative effects on ecosystems and finally on human well-being. Tourism is seen as an indirect and endogenous driver, which together with exogenous drivers acts synergistically on coastal and marine ecosystems in the region. One major exogenous driver in the Caribbean is climate change, with a potential increase in intensity and frequency of tropical cyclones. The evidence of effects of climate change in the recent past is mixed, but it is widely seen as a threat for the future. Other exogenous drivers described by the sub-global assessment include increased river discharge of the Amazon and the introduction of alien bacterial species carried by dust blown from the Sahara. Both bring about changes to marine and coastal ecosystems.

The exogenous drivers in the Caribbean are believed to lower the resilience of the coastal ecosystem to perturbations from infrastructure development along the coast for tourism. These interactions are especially important, as tourism constitutes the single most important income source for many of the Caribbean states (98% of GDP in Barbados, for instance). This decrease in resilience increases the likelihood of major damage by tropical storms, with a direct impact on human well-being.

7.5 Concluding Remarks

This chapter has presented an overview of the important drivers of ecosystem change at the global level, with occasional forays into regional effects. It has also sketched out a few of the important links between direct and indirect drivers and the consequences for ecosystems. Readers interested in how these interactions might play out in the future are advised to proceed to later chapters in this volume. Readers interested in more details about how drivers affect individual ecosystems or services should turn to the *Current State and Trends* volume and its chapters on ecosystems and services. Finally, for a review of the plethora of options for human response to undesirable changes, readers are directed to the *Policy Responses* volume.

Notes

1. This section borrows heavily from Nakićenović 2000, Section 3.3.2.
2. This section borrows heavily from Nakićenović 2000.
3. This section borrows heavily from Convention on Biological Diversity 2003.
4. This section borrows extensively from Nakićenović 2000.

References

Achard, F., H.D. Eva, and H.J. Stibbig, 2002: Determination of deforestation rates of the world's humid tropical forests. *Science,* **297,** 999–1002.

Ajzen, I., 1991: The Theory of Planned Behavior. *Organizational Behavior and Human Decision Processes,* **50,** 172–211.

Ajzen, I. and M. Fishbein, 1980: *Understanding Attitudes and Predicting Behavior.* Prentice-Hall, Englewood Cliffs, New Jersey.

Alcamo, J., E. Kreileman, M. Krol, R. Leemans, J. Bollen, J.v. Minnen, M. Schaefer, S. Toet, and B. deVries, 1998: Global modelling of environment change: an overview of IMAGE2.1. In: *Global Change Scenarios of the 21st Century—Results from the IMAGE2.1 Model,* J. Alcamo, R. Leemans, and E. Kreileman (eds.)Oxford.

Alder, J. and R.U. Sumaila, 2004: West Africa: fish basket of Europe past and present. *Journal of Environment and Development.*

Alpert, P., E. Bone, and Holzapfel, 2000: Invasiveness, invasibility and the role of environmental stress in the spread of non-native plants. *Perspectives in Plant Ecology, Evolution and Systematics,* **31,** 52–66.

Alston, J.M., G.W. Norton, and P.G. Pardey, 1995: *Science under Scarcity: Principles and Practices for Agricultural Research Evaluation and Priority Setting.* Cornell University Press, Ithaca.

Altieri, M.A., 1999: The ecological role of biodiversity in agroecosystems. *Agriculture, Ecosystems and Environment,* **74,** 19–31.

Anderson, K. and Y. Hayami, 1986: *The Political Economy of Agricultural Protection: East Asia in International Perspective.* Allen and Unwin, Boston, London and Sydney.

Ansuategi, A., E. Barbier, and C. Perrings, 1997: *The Environmental Kuznets Curve.* Amsterdam, USF Workshop on Economic Modelling of Sustainable Development: Between Theory and Practice, Tinbergen Institute.

Arrow, K., B. Bolin, R. Costanza, P. Dasgupta, and G.C. Daily, 1995: Economic growth, carrying capacity, and the environment. *Science,* **268,** 520–521.

Arrow, K., P. Dasgupta, L. Goulder, G. Daily, P. Ehrlich, G. Heal, S. Levin, K.-G. Maler, S. Schneider, D. Starrett, and B. Walker, 2004: Are We Consuming Too Much? *Journal of Economic Perspectives,* **18(3),** 147–172.

Barro, R.J., 1997: *Determinants of Economic Growth.* The MIT Press, Cambridge.

Barro, R.J. and X. Sala-i-Martin, 2004: *Economic Growth.* 2 ed. MIT Press, New York, 654 pp.

Baumol, W.J., 1993: *Social Wants and Dismal Science: the Curious Case of the Climbing Costs of Health and Teaching.* WP-64–93, Fondazione Enrico Mattei, Milan.

Beck, U., 1992: *Risk Society: Toward a New Modernity.* Sage Publications, London.

Becker, G., 1960: An Economic Analysis of Fertility. *Demographic and Economic Change in Developed Countries, Universities-National Bureau Conference Series,* **11** (Princeton: Princeton University Press).

Becker, G.S. and H.G. Lewis, 1973: On the Interaction Between Quantity and Quality of Children. *Journal of Political Economy,* **82,** 279–88.

Becker, G.S. and R. Barro, 1988: A Reformulation of the Economic Theory of Fertility. *Quarterly Journal of Economics,* **103,** 1–25.

Beierle, T.C. and J. Cayford, 2002: *Democracy in Practice: Public Participation in Environmental Decisions.* Resources for the Future, Washington, D.C.

Bekunda, M. and G. Manzi, 2003: Use of the partial nutrient budget as an indicator of nutrient depletion in the highlands of southwestern Uganda. *Nutrient Cycling in Agroecosystems,* **67,** 187–195.

Bennett, E.M., S.R. Carpenter, and N.F. Caraco, 2001: Human Impact on Erodable Phosphorus and Eutrophication: A Global Perspective. *BioScience,* **51(3),** 227–234.

Bennett, E.M., T. Reed-Andersen, J.N. Houser, J.G. Gabriel, and S.R. Carpenter, 1999: A phosphorus budget for the Lake Mendota watershed. *Ecosystems,* **2,** 69–75.

Bennett, M.K., 1941: Wheat in National Diets. *Wheat Studies,* **18(2),** 37–76.

Berhe, A.A., 2000: *Landmines and Land Degradation: A Regional Political Ecology Perspective on the Impacts of Landmines on Environment and Development in the Developing World. Department of Resource Development,* Michigan State University, East Lansing, Michigan, 157 pp.

Berhe, A.A., In Press: The Impact of Erosion on the Terrestrial Carbon Reservoir: Ecological and Socioeconomic Implications of Soil Erosion in Eritrea. In: *Proceedings of the International Conference Commemorating the 10th Anniversary of the Independence of Eritrea: Lessons and Prospects. July 22–26, 2001,* Asmara, Eritrea.

Berry, P.M., E.A. Stockdale, R. Sylvester-Bradley, L. Philipps, K.A. Smith, E.I. Lord, C.A. Watson, and S. Fortune, 2003: N, P and K budgets for crop rotations on nine organic farms in the UK. *Soil Use and Management,* **19,** 112–118.

Biggs, R., E. Bohensky, P.V. Desanker, C. Fabricius, T. Lynam, A.A. Misselhorn, C. Musvoto, M. Mutale, B. Reyers, R.J. Scholes, S. Shikongo, and A.S. van Jaarsveld, 2004: *Nature Supporting People. The Southern African Millennium Ecosystem Assessment. Integrated Report.* SIBN-0–7988–5528–2, Council for Scientific and Industrial Research, Pretoria.

Binswanger, H.P., 1980: Income distribution effects of technical change: Some analytical issues. *South East Asian Economic Review,* **1(3),** 179–218.

Bodansky, D., 2003: Illegal Regulation of the Effects of Military Activity on the Environment. In: *BERICHTE des Umweltbundesamtes.* 5/03, Erich Schmidt Verlag, Berlin.

Bongaarts, J., 1994: Population Policy Options in the Developing World. *Science,* **263,** 771–76.

Bongaarts, J., 1997: The Role of Family Planning Programmes in Contemporary Fertility Transitions. In: *The Continuing Demographic Transition,* G.W. Jones, R.M. Douglas, J.C. Caldwell, and R.M. D'Souza (eds.), Clarendon Press., Oxford, England, 422–43.

Bongaarts, J., 1999: Fertility Decline in the Developed World: Where Will It End? *American Economic Review,* **89(2),** 256–260.

Bongaarts, J. and S.C. Watkins, 1996: Social Interactions and Contemporary Fertility Transitions. *Population and Development Review,* **22(4),** 639–82.

Bongaarts, J. and G. Feeney, 1998: On the Quantum and Tempo of Fertility. *Population and Development Review,* **24(2),** 271–291.

Bouwman, A.F., L.J.M. Boumans, and N.H. Batjes, 2002: Emissions of N2O and NO from fertilized fields: Summary of available measurement data. *Global Biogeochemical Cycles,* **16(4),** art-1058.

Brechin, S., 1999: Objective problems, subjective values, and global environmentalism: Evaluating the postmaterialist argument and challenging a new explanation. *Social Science Quarterly,* **80(4),** 783–809.

Brechin, S.R. and W. Kempton, 1994: Global Environmentalism: A Challenge to the Postmaterialism Thesis? *Social Science Quarterly,* **75,** 245–269.

Brechin, S.R. and W. Kempton, 1997: Beyond Postmaterialist Values: National versus Individual Explanations of Global Environmentalism. *Social Science Quarterly,* **78,** 16–20.

Bruinsma, J.E., 2003: *World Agriculture: Towards 2015/2030.* Earthscan, London, UK.

Buckley, R., 2001: Environmental Impacts. In: *The Encyclopedia of Ecotourism,* D.B. Weaver (ed.), CABI Publishing, Oxon, 193–204.

Buckley, R.C., 1998: Tourism in wilderness: M&M Toolkit. In: *Personal, Societal, and Ecological Values of Wilderness: Sixth World Wilderness Congress Proceedings on Research, Management, and Allocation,* A.E. Watson, G.H. Aplet, and J.C. Hendee (eds.). 1, Rocky Mountain Research Station, USFS, 115–116.

Bumpass, L., 1990: What's Happening to the Family? Interactions Between Demographic and Institutional Change. *Demography,* **27(4),** 483–98.

Burdon, J.J., 1991: Fungal pathogens as selective forces in plant-populations and communities. *Australian Journal of Ecology,* **16,** 423–432.

Buttel, F.H., 2003: Environmental Sociology and the Explanation of Environmental Reform. *Organization and Environment,* **16,** 306–344.

Byerlee, D. and E. Hesse de Polanco, 1986: Farmers' stepwise adoption of technological packages: Evidence from the Mexican Altiplano. *American Journal of Agricultural Economics,* **68(3),** 519–527.

Byerlee, D. and P. Moya, 1993: *Impacts of International Wheat Breeding Research in the Developing World, 1966–1990.* International Maize and Wheat Improvement Center (CIMMYT), Mexico, D.F.

Caldwell, J.C., 1982: *Theory of Fertility Decline.* Academic Press, London, England.

Carpenter, S.R., N.F. Caraco, D.L. Correll, R.W. Howarth, A.N. Sharpley, and V.H. Smith, 1998: Nonpoint pollution of surface waters with nitrogen and phosphorus. *Issues in Ecology.*

Cassman, K.G. and P.L. Pingali, 1995: Intensification of irrigated rice systems: Learning from the past to meet future challenges. *GeoJournal,* **35,** 299–305.

Cassman, K.G., A. Dobermann, and D.T. Walters, 2002: Agroecosystems, nitrogen-use efficiency, and nitrogen management. *Ambio,* **31(2),** 132–140.

Cassman, K.G., A. Dobermann, D.T. Walters, and H.S. Yang, 2003: Meeting cereal demand while protecting natural resources and improving environmental quality. *Annual Review of Environment and Resources,* **28,** 315–358.

Castells, M., 1996: *The Rise of the Network Society.* Vol. 1, *The Information Age,* Blackwell, Oxford.

Cater, C. and E. Cater, 2001: Marine Environments. In: *The Encyclopedia of Ecotourism,* D.B. Weaver (ed.), CABI Publishing, Oxon, 193–204.

Chenery, H., S. Robinson, and M. Syrquin (eds.), 1986: *Industrialization and Growth: A Comparative Study.* Oxford University Press, Oxford.

Christianson, V., 1998: Fishery-induced changes in a marine ecosystem: insight from models of the Gulf of Thailand. *Journal of Fish Biology,* **53,** 128–142.

Christianson, V., V. Guénette, S., J.J. Heymans, C.J. Walters, R. Watson, D. Zeller, and D. Pauly, 2003: Hundred-year decline of North Atlantic predatory fishes. *Fish and Fisheries,* **4,** 1–24.

Cleland, J. and C. Wilson, 1987: Demand Theories of the Fertility Transition: An Iconoclastic View. *Population Studies,* **41(1),** 5–30.

Cohen, B., 1998a: The Emerging Fertility Transition in Sub-Saharan Africa. *World Development,* **26(8),** 1431–61.

Cohen, J.E., 1995: How Many People Can the Earth Support. W.W. Norton & Co., New York, 532.

Cohen, J.E., 1998b: Should Population Projections Consider 'Limiting Factors'—and If So, How? In: *Frontiers of Population Forecasting, Supplement to Population and Development Review,* W. Lutz, J.W. Vaupel, and D.A. Ahlburg (eds.). 24, 118–38.

Cohen, J.E., 2003: Human population: The next half century. *Science,* **302,** 1172–1175.

Collings, A., 1999: Tourism Development and Natural Capital. *Annals of Tourism Research,* **26(1),** 98–109.

Congleton, R.D. (ed.), 1996: *The Political Economy of Environmental Protection: Analysis and Evidence.* University of Michigan Press, Ann Arbor.

Convention on Biological Diversity, 2003: Biological Diversity and Tourism Introduction. Cited December 8 2003. Available at http://www.biodiv.org/programmes/socio-eco/tourism/.

Cramer, J.C., 1997: A Demographic Perspective on Air Quality: Conceptual Issues Surrounding Environmental Impacts of Population Growth. *Human Ecology Review,* **3,** 191–196.

Cramer, J.C., 1998: Population Growth and Air Quality in California. *Demography,* **35,** 45–56.

Crawley, M.J., 1986: The population biology of invaders. *Philosophical Transactions of the Royal Society of London* **B, 314,** 711–731.

Crawley, M.J., 1989: Invaders. *Plants Today,* **1,** 152–158.

Crawley, M.J. and R.M. May, 1987: Population dynamics and plant community structure: competition between annuals and perennial. *Journal of Theoretical Biology,* **125,** 475–489.

Dahl, R.A. and E.R. Tufte, 1973: *Size and Democracy.* Stanford University Press, Stanford, California.

Daily, G.C. and P.R. Ehrlich, 1996: Global change and human susceptibility to disease. *Annual Review of Energy and the Environment,* **21,** 125–144.

Daily, G.C. and P.R. Ehrlich, 1997: Population diversity: its extent and extinction. *Science,* **278,** 689–692.

Daly, H.E., 1991: *Steady-State Economics.* 2 ed. Island Press, Washington.

Dasgupta, P., K.-G. Mäler, and T, 1997: The Resource Basis of Production and Consumption: An Economic Analysis. In: *The Environment And Emerging Development Issues,* P. Dasgupta and K.-G. Mäler (eds.). 1.

Daszak, P., A.A. Cunningham, and A.D. Hyatt, 2003: Infectious disease and amphibian population declines. *Diversity and Distributions,* **9,** 141–150.

David, C. and K. Otsuka (eds.), 1994: *Modern Rice Technology and Income Distribution in Asia.* Lynne Rienner, Boulder.

David, P.A., 1990: The dynamo and the computer: A historical perspective on the modern productivity paradox. *American Economic Review,* **80(2),** 355- 361.

De Bruyn, S.M. and J.B. Opschoor, 1994: *Is the Economy Ecologizing? De-or Relinking Economic Development with Environmental Pressure.* TRACE Discussion Paper TI-94–65, Tinbergen Institute, Amsterdam.

De Bruyn, S.M., J. van den Bergh, and J.B. Opschoor, 1995: *Empirical Investigations in Environmental-Economic Relationships: Reconsidering the Empirical Basis of Environmental Kuznets Curves and the De-linking of Pollution from Economic Growth.* TRACE Discussion Paper TI-95–140, Tinbergen Institute, Amsterdam.

de Moor, A., 2002: The perversity of government subsidies for energy and water. In: *Greening The Budget: Budgetary Policies for Environmental Improvement,* J.P. Clinch, K. Schlegelmilch, R.-U. Sprenger, and U. Triebswetter (eds.), Edward Elgar, 368.

de Vries, B. and J. Goudsblom (eds.), 2002: *Mappae Mundi: Humans and their Habitats in a Long-term Socio-ecological Perspective.* Amsterdam University Press, Amsterdam.

de Vries, H.J.M., M.A. Janssen, and A. Beusen, 1999: Perspectives on Global Energy Futures—Simulations with the TIME Model. *Energy Policy,* **27,** 477–494.

de Vries, H.J.M., J.G.J. Olivier, R.A. van den Wijngaart, G.J.J. Kreileman, and A.M.C. Toet, 1994: Model for calculating regional energy use, industrial production and greenhouse gas emissions for evaluating global climate scenarios. *Water, Air and Soil Pollution,* **76,** 79–131.

DeFries, R., R.A. Houghton, and H. M., 2002: Carbon emissions from tropical deforestation and regrowth based on satellite observations from 1980s and 90s. *Proceedings of the National Academy of Sciences, USA,* **99,** 14256–14261.

Delgado, C.L., J. Hopkins, and V.A. Kelly, 1998: *Agricultural Growth Linkages in Sub-Saharan Africa.* IFPRI Research Report 107, International Food Policy Research Institute (IFPRI), Washington, D.C.

Demeny, P., 1997: Replacement-Level Fertility: The Implausible Endpoint of the Demographic Transition. In: *The Continuing Demographic Transition,* G.W. Jones, R.M. Douglas, J.C. Caldwell, and R.M. D'Souza (eds.), Clarendon Press, Oxford, 94–110.

Dietz, T., 1996/1997: The Human Ecology of Population and Environment: From Utopia to Topia. *Human Ecology Review,* **3,** 168–171.

Dietz, T. and P.C. Stern, 1998: Science, values and biodiversity. *BioScience,* **48,** 441–444.

Dietz, T. and P.C. Stern (eds.), 2002: *New Tools for Environmental Protection: Education, Information and Voluntary Measures.* National Academy Press, Washington, DC, 356 pp.

Dietz, T., E. Ostrom, and P.C. Stern, 2003: The Struggle to Govern the Commons. *Science,* **301,** 1907–1912.

Dillon, P.J. and F.H. Rigler, 1974: The phosphorus-chlorophyll relationship in lakes. *Limnology and Oceanography,* **19(5),** 767–772.

Dobermann, A. and K.G. Cassman, 2002: Plant nutrient management for enhanced productivity in intensive grain production systems of the United States and Asia. *Plant and Soil,* **247,** 153–175.

Dobermann, A., K.G. Cassman, C.P. Mamaril, and J.E. Sheehy, 1998: Management of phosphorus, potassium and sulfur in intensive, irrigated lowland rice. *Field Crops Research,* **56,** 113–138.

Dobermann, A., C. Witt, D. Dawe, G.C. Gines, R. Nagarajan, S. Satawathananont, T.T. Son, P.S. Tan, G.H. Wang, N.V. Chien, V.T.K. Thoa, C.V. Phung, P. Stalin, P. Muthukrishnan, V. Ravi, M. Babu, S. Chatuporn, M. Kongchum, Q. Sun, R.S. Fu, G.C., and M.A.A. Adviento, 2002: Site-specific nutrient management for intensive rice cropping systems in Asia. *Field Crops Research,* **74,** 37–66.

Dobson, A.P. and P.J. Hudson, 1986: Parasites, disease and the structure of ecological communities. *Trends in Ecology and Evolution,* **1,** 11–15.

Dobson, A.P. and M.J. Crawley, 1994: Pathogens and the structure of plant communities. *Trends in Ecology and Evolution,* **9,** 393–398.

Dobson, A.P., P.J. Hudson, and A.M. Lyles, 1992: Macroparasites: worms and others. In: *Natural Enemies. The Population Biology of Predators, Parasites and Diseases,* M.J. Crawley (ed.), Blackwell Scientific Publications., Oxford, 329–348.

Dollar, D. and P. Collier, 2001: *Globalization, Growth, and Poverty: Building an Inclusive World Economy.* Oxford University Press, Oxford.

Dosi, G., K. Pavitt, and L. Soete, 1990: *The Economics of Technical Change and International Trade.* Harvester Wheatsheaf, London.

Dregne, H., 2002: Land degradation in the Drylands. *Arid land research and management,* **13,** 99–132.

Dregne, H.E. and N. Chou, 1992: Global desertification and costs. In: *Degradation and restoration of arid lands,* H.E. Dregne (ed.), Texas Tech University, Lubbock, 249–282.

Dunlap, R.E. and A.G. Mertig, 1995: Global Concern for the Environment: Is Affluence a Prerequisite? *Journal of Social Issues,* **51(4),** 121–137.

Dunlap, R.E. and A.G. Mertig, 1997: Global Environmental Concern: An Anomoly for Postmaterialism. *Social Science Quarterly,* **78,** 24–29.

Dunlap, R.E., J.K. Grieneeks, and M. Rokeach, 1983: Human Values and Pro-Environmental Behavior. In: *Energy and Mineral Resources: Attitudes, Values and Public Policy,* D.W. Conn (ed.), American Association for the Advancement of Science, Washington, D.C., 145–168.

Dunlap, R.E., G.H.J. Gallup, and A.M. Gallup, 1993: Global Environmental Concern: Results from an International Public Opinion Survey. *Environment,* **35(November),** 7–15, 33–39.

Dunlap, R.E., K.D. Van Liere, A.D. Mertig, and R.E. Jones, 2002: Measuring Endorsement of the New Ecological Paradigm: A Revised NEP Scale. *Social Science Quarterly,* **56(3),** 425–442.

Dupre, A.H., 1986: *Science in the Federal Government: A History of Policies and Activities.* Johns Hopkins University Press, Baltimore.

Duvick, D.N. and K.G. Cassman, 1999: Post-green revolution trends in yield potential of temperate maize in the North-Central United States. *Crop Science,* **39,** 1622–1630.

Easterlin, R.A., 1969: Towards a Socio-Economic Theory of Fertility. In: *Fertility and Family Planning: A World View,* S.J. Behrman, L. Corsa, and R. Freedman (eds.), University of Michigan Press, Ann Arbor, MI, 127–56.

Easterlin, R.A., 1975: An Economic Framework for Fertility Analysis. *Studies in Family Planning,* **6,** 54–63.

Eckberg, D.L. and T.J. Blocker, 1989: Varieties of Religious Involvement and Environmental Concerns: Testing the Lynn White Thesis. *Journal for the Scientific Study of Religion,* **28,** 509–517.

Eckberg, D.L. and T.J. Blocker, 1996: Christianity, Environmentalism, and the Theoretical Problem of Fundamentalism. *Journal for the Scientific Study of Religion,* **35,** 343–355.

Edington, J.M. and M.A. Edington, 1986: *Ecology and Environmental Planning.* Chapman & Hall, London.

Ehrlich, P.R. and J. Holden, 1971: Impact of Population Growth. *Science,* **171,** 1212–1217.

Ehrlich, P.R. and G.C. Daily, 1993: Population extinction and saving biodiversity. *Ambio,* **22,** 64–68.

Ehrlich, P.R. and A. Ehrlich, 2004: *One with Nineveh: Politics, Consumption and the Human Future.* Island Press, Washington, D.C.

Ehrlich, P.R., J. Harte, M.A. Harwell, P.H. Raven, C. Sagan, and G.M. Woodwell, 1983: Long-term biological consequences of nuclear war. *Science,* **222,** 1293–1300.

El-Hinnawi, E., 1985: Environmental Refugees. United Nations Environment Programme, Nairobi.

Evenson, R.E. and D. Gollin, 2003: Assessing the impact of the green revolution: 1960–2000. *Science,* **300,** 758–762.

Falcon, W. and C. Fowler, 2002: Carving up the commons—emergence of a new international regime for germplasm development and transfer. *Food Policy,* **27,** 197–222.

Fan, S., P. Hazell, and S. Thorat, 1998: *Government Spending, Growth, and Poverty: An Analysis of Interlinkages in Rural India.* EPTD Discussion Paper 33, International Food Policy Research Institute (IFPRI). Washington, D.C.

FAO (Food and Agriculture Organization of the United Nations), 2001a: *The Global Forest Resources Assessment 2000 (FRA 2000): Main Report.* FAO forestry paper 140, FAO, Rome, 479 pp.

FAO, 2001b: *Global Forest Fire Assessment 1990–2000.* Working Paper 55, Forest Resources Assessment Programme, FAO, Rome, 495 pp.

FAO, 2002: *World agriculture: towards 2015/2030.* Food and Agriculture Organization of the United Nations, Rome, 97 pp.

FAO, 2004: FAOSTAT. Available at www.fao.org.

Fischer, F., 2000: *Citizens, Experts, and the Environment: The Politics of Local Knowledge.* Duke University Press, Durham.

Freeman, C. and C. Perez, 1988: Structural crises of adjustment: business cycles and investment behaviour. In: *Technical Change and Economic Theory,* G.e.a. Dosi (ed.), Pinter, London, 38–66.

Frejka, T. and G. Calot, 2001: Cohort Reproductive Patterns in Low-Fertility Countries. *Population and Development Review,* **27(1),** 103–32.

Frey, R.S. and I. Al-Mansour, 1995: The Effects of Development, Dependence and Population Pressure on Democracy: The Cross-National Evidence. *Sociological Spectrum,* **15,** 181–208.

Frost, W., 2001: Rainforests. In: *The Encyclopedia of Ecotourism,* D.B. Weaver (ed.), CABI Publishing, Oxon, 193–204.

Galloway, J.N. and E.B. Cowling, 2002: Reactive nitrogen and the world: 200 years of change. *Ambio,* **31,** 64–71.

Galloway, J.N., E.B. Cowling, S.P. Seitzinger, and R.H. Socolow, 2002: Reactive nitrogen: Too much of a good thing? *Ambio,* **31,** 60–63.

Galloway, J.N., W.H. Schlesinger, H. Levy, A. Michaels, and J.L. Schnoor, 1995: Nitrogen fixation: atmospheric enhancement—environmental response. *Global Biochemical Cycles,* **9,** 235–252.

Gardner, G.T. and P.C. Stern, 1995: *Environmental Problems and Human Behavior.* Allyn and Bacon, Needham Heights, MA.

Garenne, M. and V. Joseph, 2002: The Timing of the Fertility Transition in Sub-Saharan Africa. *World Development,* **30(10),** 1835–1843.

Geist, H.J. and E.F. Lambin, 2002: Proximate causes and underlying driving forces of tropical deforestation. *BioScience,* **52(2),** 143–150.

Geist, H.J. and E.F. Lambin, 2004: Dynamic Causal Patterns of Desertification. *BioScience,* **54(9),** 817–829.

Giddens, A., 1990: *The Consequences of Modernity.* Stanford University Press, Stanford.

Gössling, S., 2002a: Global environmental consequences of tourism. *Global Environmental Change,* **12,** 283–302.

Gössling, S., 2002b: Human-Environmental relations with tourism. *Annals of Tourism Research,* **29(2),** 539–556.

Gray, B., 1997: Landmines: The Most Toxic and Widespread Pollution Facing Mankind. In: *A Colloquium, Towards Ottawa and Beyond- Demining the Region, The environmental impacts. International Campaign to Ban Landmines (Australia).* http://members.xoom.com/dassur/icblcol1.doc or http://wwwfn2.freenet.edmonton.ab.ca/~puppydog/bgray.htm.

Grenfell, B.T. and A.P. Dobson, 1995: *Ecology of Infectious Diseases in Natural Populations.* Cambridge University Press, Cambridge.

Grossman, G., 1995: Pollution And Growth: What Do We Know? In: *The Economics Of Sustainable Development,* I. Goldin and L.A. Winters (eds.), Cambridge University Press, Cambridge.

Grossman, G. and A. Krueger, 1992: *Environmental Impacts of the North American Free Trade Agreement.* CEPR Working Paper 644, Centre for Economic Policy Research, London, April.

Grübler, A., 1994: Technology. In: *Changes in Land Use and Land Cover: A Global Perspective,* W.B. Meyer and B.L. Turner II (eds.), Cambridge University Press, Cambridge, 287–328.

Grübler, A., 1998: *Technology and Global Change.* Cambridge University Press, Cambridge.

Guagnano, G.A., P.C. Stern, and T. Dietz, 1995: Influences on attitude-behavior relationships: A natural experiment with curbside recycling. *Environment and Behavior,* **27,** 699–718.

Halpenny, E.E., 2001: Islands and Coasts. In: *The Encyclopedia of Ecotourism,* D.B. Weaver (ed.), CABI Publishing, Oxon, 193–204.

Hanada, A., 2003: *"Culture and Environmental Values: A Comparison of Japan and Germany."* Environmental Science and Policy, George Mason University, Fairfax, Virginia.

Harbaugh, W.T., A. Levinson, and D.M. Wilson, 2000: *Re-Examining The Empirical Evidence For An Environmental Kuznets Curve.* 7711, National Bureau of Economic Research, Cambridge, May.

Hardin, G., 1968: The Tragedy of the Commons. *Science,* **162,** 1243–1248.

Harvell, C.D., K. Kim, J.M. Burkholder, R.R. Colwell, P.R. Epstein, D.J. Grimes, E.E. Hofmann, E.K. Lipp, A.D.M.E. Osterhaus, R.M. Overstreet, J.W. Porter, G.W. Smith, and G.R. Vasta, 1999: Emerging Marine Diseases—Climate Links and Anthropogenic Factors. *Science,* **285(3 September),** 1505–1510.

Hawkins, D.E. and K. Lamoureux, 2001: Global Growth and Magnitude of Ecotourism. In: *The Encyclopedia of Ecotourism,* D.B. Weaver (ed.), CABI Publishing, Oxon, 63–83.

Hayami, Y. and R.W. Herdt, 1977: Market price effects of technological change on income distribution in semi-subsistence agriculture. *American Journal of Agricultural Economics,* **59(2),** 245–256.

Hayami, Y. and V.W. Ruttan, 1985: *Agricultural development: an international perspective.* Johns Hopkins University Press, Baltimore, 506 pp.

Hayami, Y., M. Kikuchi, P.F. Moya, L.M. Bambo, and E.B. Marciano, 1978: *Anatomy of a Peasant Economy: A Rice Village in the Philippines.* International Rice Research Institute (IRRI), Los Baños.

Hazell, P. and S. Haggblade, 1993: Farm-nonfarm growth linkages and the welfare of the poor. In: *Including the Poor,* M.L.a.J.v.d. Gaag (ed.), The World Bank, Washington, D.C.

Herdt, R.W., 1987: A retrospective view of technological and other changes in Philippine rice farming, 1965–1982. *Economic Development and Cultural Change,* **35(2),** 329–49.

Hirschman, C., 1994: Why Fertility Changes. *Annual Review of Sociology,* **20,** 203–33.

Hoelscher, D., R.F. Mueller, M. Denich, and H. Fuelster, 1996: Nutrient input-output budget of shifting agriculture in Eastern Amazonia. *Nutrient Cycling in Agroecosystems,* **47,** 49–57.

Houghton, J.T., Y. Ding, D.J. Griggs, M. Noguer, P.J. van der Linden, X. Dai, K. Maskall, and C.A. Johnson (eds.), 2001: *Climate Change 2001: The Scientific Basis. Contribution of Working Group I to the Third Assessment Report of the Intergovernmental Panel on Climate Change.* Cambridge University Press, Cambridge.

Huffman, W.E. and R.E. Evenson, 1993: *Science for Agriculture—A Long-term Perspective.* Iowa State University Press, Ames.

Hughes, J.B., G.C. Daily, and P.R. Ehrlich, 1998: The loss of population diversity and why it matters. In: *Nature and Human Society,* P.H. Raven (ed.), National Academy Press, Washington, DC.

Hugo, G., 1996: Environmental Concerns and International Migration. *International Migration Review,* **30(1),** 105–131.

IFA (International Fertilizer Industry Association), 2002: *Fertilizer use by crop.* 5 ed. IFA, IFDC, IPI, PPI, FAO, Rome.

IFA, 2003: *IFADATA statistics.* http://www.fertilizer.org/ifa/statistics.asp. Paris.

IFA, 2004: IFADATA statistics. Available at http://www.fertilizer.org/ifa/statistics.asp.

Inglehart, R., 1995: Public Support for Environmental Protection: Objective Problems and Subjective Values in 43 Societies. *PS: Political Science and Politics,* **15,** 57–71.

International Campaign to Ban Landmines, 2002: *Landmine Monitor Report 2002: Toward a Mine-Free World.* Human Rights Watch, New York.

International Civil Aviation Organization, 1995: *Civil Aviation Statistics of the World, 1994.* Doc. 9180/20, International Civil Aviation Organization, Montreal, Canada.

IPCC (Intergovernmental Panel on Climate Change), 2002: *Climate Change 2001: Synthesis Report.* Cambridge University Press, Cambridge.

Ironmonger, D.S., C.K. Aitken, and B. Erbas, 1995: Economies of scale in energy use in adult-only households. *Energy Economics,* **17,** 301–310.

Irwin, A., 1995: *Citizen Science: A Study of People, Expertise and Sustainable Development.* Routledge, London.

Jacobson, J.L., 1989: Environmental Refugees: Nature's Warning System. *Populi,* **16(1),** 29–41.

Jasanoff, S., 1990: *The Fifth Branch: Science Advisors as Policy Makers.* Harvard University Press, Cambridge.

Jeltsch, F., S.J. Milton, W.R.J. Dean, and N. van Rooyen, 1997: Analysing shrub encroachment in the southern Kalahari: a grid-based modelling approach. *Journal of Applied Ecology,* **34,** 1497–1508.

Jones, S., 2002: A Framework for Understanding On-farm Environmental Degradation and Constraints to the Adoption of Soil Conservation Measures: Case Studies from Highland Tanzania and Thailand. *World Development,* **30(9),** 1607–1620.

Kanagy, C.L. and F.K. Willits, 1993: A 'Greening' of Religion? Some Evidence from a Pennsylvania Sample. *Social Science Quarterly,* **74,** 674–683.

Karp, D.G., 1996: Values and Their Effects on Pro-Environmental Behavior. *Environment and Behavior,* **28,** 111–133.

Kasemir, B., J. Jager, C. Jaeger, and M.T. Gardner (eds.), 2003: *Public Participation in Sustainability Science.* Cambridge University Press, Cambridge.

Kasperson, J.X., R.E. Kasperson, and B.L. Turner II, 1995: *Regions at Risk: Comparisons of Threatened Environments.* United Nations University Press, Tokyo.

Kates, R.W. and V. Haarmann, 1992: Where people live: Are the assumptions correct? *Environment,* **34,** 4–18.

Kato, N., 1996: Analysis of structure of energy consumption and dynamics of emission of atmospheric species related to global environmental change (SO_x, NO_x, CO_2) in Asia. *Atmospheric Environment,* **30(5),** 757–785.

Kaufmann, D., A. Kraay, and M. Mastruzzi, 2003: *Governance Matters III: Governance Indicators for 1996–2002.* World Bank Policy Research Working Paper 3106, World Bank, Washington, DC.

Keeling, C.D. and T.P. Whorf, 2005: Atmospheric CO_2 records from sites in the SIO sampling network. In: *A Compendium of Data on Global Change.* Carbon Dioxide Information Analysis Center Oak Ridge National Laboratory, US Department of Energy, Oak Ridge, TN. (cdiac.esd.ornl.gov/trends/coz/sio-mlo.htm).

Kempton, W., J.S. Boster, and J.A. Hartley, 1995: *Environmental Values in American Culture.* The MIT Press, Cambridge, Massachusetts.

Keyfitz, N., 1982: Can Knowledge Improve Forecasts? *Population and Development Review,* **8(4),** 729–51.

Kluckholm, C., 1952: Values and Value-Orientation in the Theory of Action: An Exploration in Definition and Classification. In: *Toward a General Theory of Action,* T. Parsons and E. Shils (eds.), Harvard University Press, Cambridge, Massachusetts, 395–418.

Knodel, J. and E.V.D. Walle, 1979: Lessons from the Past: Policy Implications of Historical Fertility Studies. *Population and Development Review,* **5(2),** 217–45.

Kohler, H.-P., F.C. Billari, and J.A. Ortega, 2002: The Emergence of Lowest-Low Fertility in Europe During the 1990s. *Population and Development Review,* **28(4),** 641–680.

Kropp, J., K.B.L. Matthias, and F. Reusswig, 2001: Global Analysis and Distribution of Unbalanced Urbanization Processes: The Favela Syndrome. *GAIA,* **10(2),** 109–120.

Krug, E.G. (ed.), 2002: *World Report on Violence and Health.* World Health Organization, Geneva, 346 pp.

Kuznets, S., 1955: Economic Growth and Income Inequality. *American Economic Review,* **65,** 1–28.

Lambin, E.F., H. Geist, and E. Lepers, 2003: Dynamics of land-use and land-cover change in tropical regions. *Annual review of environment and resources,* **28.**

Lambin, E.F., B.L. Turner II, H.J. Geist, S.B. Agbola, A. Angelsen, J.W. Bruce, O. Coomes, R. Dirzo, G. Fischer, C. Folke, P.S. George, K. Homewood, J. Imbernon, R. Leemans, X. Li, E.F. Moran, M. Mortimore, P.S. Ramakrishnan, M.B. Richards, H. Skånes, W.L. Steffen, G.D. Stone, U. Svedin, T.A. Veldkamp, C. Vogel, and J. Xu, 2001: The causes of land-use and land-cover change: Moving beyond the myths. *Global Environmental Change,* **11(4),** 261–269.

Landes, D.S., 1969: *The Unbound Prometheus: Technological Change and Industrial Development in Western Europe from 1750 to the Present.* Cambridge University Press, Cambridge.

Lantican, M.A., P.L. Pingali, and S. Rajaram, 2003: Is research on marginal lands catching up? The case of unfavourable wheat growing environments. *Agricultural Economics,* **29(3),** 353–361.

Lanz, T.J., 1996: Environmental Degradation and Social Conflict in the Northern Highlands of Ethiopia: The Case of Tigray and Wollo Provinces. *Africa Today,* **43,** 157–182.

Leaning, J., 2000: Environment and health: 5. Impact of war. *CMAJ,* **163(9),** 1157–1161.

Lee, R., 2003: The demographic transition: Three centuries of fundamental change. *Journal of Economic Perspectives,* **17,** 167–190.

Leichenko, R.M. and K. O'Brien, 2002: The Dynamics of Rural Vulnerability to Global Change: The Case of Southern Africa. *Mitigation and Adaptation Strategies for Global Change,* **7(1),** 1–18.

Lepers, E., E.F. Lambin, A.C. Janetos, R. DeFries, F. Achard, N. Ramankutty, and R.J. Scholes, in press: A synthesis of rapid land-cover change information for the 1981–2000 period. *BioScience.*

Lesthaeghe, R., 1983: A Century of Demographic and Cultural Change in Western Europe: An Exploration of Underlying Dimensions. *Population and Development Review,* **9(3),** 411–35.

Lesthaeghe, R. and P. Willems, 1999: Is Low Fertility a Temporary Phenomenon in the European Union? *Population and Development Review,* **25(2),** 211–28.

Leung, B., D.M. Lodge, D. Finnoff, J.F. Shogren, M.A. Lewis, and G. Lamberti, 2002: An ounce of prevention or a pound of cure: bioeconomic risk analysis of invasive species. *Proc R Soc Lond B Biol Sci,* **269,** 2407–13.

Lindberg, K., 2001: Economic Impacts. In: *The Encyclopedia of Ecotourism,* D.B. Weaver (ed.), CABI Publishing, Oxon, 193–204.

Liu, J., G.C. Daily, P.R. Ehrlich, and G.W. Luck, 2003: Effects of household dynamics on resource consumption and biodiversity. *Nature,* **421,** 530–533.

Lodge, D.M., C.A. Taylor, D.M. Holdich, and J. Skurdal, 2000: Nonindigenous crayfishes threaten North American freshwater biodiversity: lessons from Europe. *Fisheries,* **25,** 7–20.

Lonsdale, W.M., 1999: Global patterns of plant invasions and the concept of invasibility. *Ecology,* **80,** 1522–1536.

Lubell, M., 2000: Cognitive conflict and consensus building in the National Estuary Program. *American Behavioral Scientist,* **44,** 629–648.

Lutz, W. and A. Goujon, 2001: The World's Changing Human Capital Stock: Multi-State Population Projections by Educational Attainment. *Population and Development Review,* **27(2),** 323–339.

Lutz, W., W. Sanderson, and S. Scherbov, 2001: The End of World Population Growth. *Nature,* **412,** 54545.

Lutz, W., B.C. O'Neill, and S. Scherbov, 2003: Europe's population at a turning point. *Science,* **299,** 1991–1992.

Mack, R.N., D. Simberloff, W.M. Lonsdale, H. Evans, M. Clout, and F.A. Bazzaz, 2000: Biotic invasions: Causes, epidemiology, global consequences, and control. *Ecological Applications,* **10,** 689–710.

MacKellar, F.L., W. Lutz, A.J. McMichael, and A. Suhrke, 1998: Population and Climate Change. In: *Human Choice and Climate Change,* S. Rayner and E.L. Malone (eds.). 1, Battelle Press, Columbus, Ohio, 89–193.

Maddison, A., 1991: *Dynamic Forces in Capitalist Development.* Oxford University Press, Oxford.

Maddison, A., 1995: *Monitoring the World Economy, 1820–1992.* OECD Development Centre Studies, Organisation for Economic Co-operation and Development, Paris.

Maddison, A., 2003: *The World Economy: Historical Statistics.* OECD, Paris, 273 pp.

Marco, D.E., S. Páez, and S. Cannas, 2002: Modelling species invasiveness in biological invasions. *Biological Invasions,* **4,** 193–201.

Marshall, M.G., T.R. Gurr, J. Wilkenfeld, M.I. Lichbach, and D. Quinn, 2003: *Peace and Conflict 2003- A Global Survey of Armed Conflicts, Self-Determination Movements, and Democracy.* Center for International Development and Conflict Management at the University of Maryland, College Park.

Martin, P. and J. Widgren, 1996: International Migration: A Global Challenge Population. Population Reference Bureau, Washington, DC.

Massey, D.S., J. Arango, G. Hugo, A. Kouaouci, A. Pellegrino, and J.E. Taylor, 1998: Worlds in Motion: Understanding International Migration at the End of the Millennium. Oxford University Press, Oxford, England.

Matson, P.A., R.L. Naylor, and I. Ortiz-Monasterio, 1998: Integration of environmental, agronomic, and economic aspects of fertilizer management. *Science,* **280,** 112–115.

McIsaac, G.F., M.B. David, G.Z. Gertner, and D.A. Goolsby, 2001: Nitrate flux in the Mississippi river. *Nature,* **414,** 166–167.

McNaughton, S.J., 1992: The Propagation of Disturbance in Savannas Through Food Webs. *J. Veg. Sci.,* **3,** 301–314.

Meadows, D.H., D.L. Meadows, J. Randers, and W.W. Behrens, 1972: *The Limits To Growth.* Earth Island Ltd., London.

Meisnesz, A., 1999: *Killer algae.* University of Chicago Press, Chicago.

Merchant, C., 1992: *Radical Ecology: The Search for a Livable World.* Routledge, New York.

Michigan Department of Environmental Quality, 2004: Aquatic Invasive Species. Available at http://www.michigan.gov/deq/0,1607,7–135–3313_3677_8314–83004—CI,00.html.

Middleton, N. and D. Thomas, 1997: *World Atlas of Desertification.* 2 ed. E. Arnold, London, 182 pp.

Millward, R., 1990: Productivity in the UK services sector: Historical trends 1865–1985 and comparisons with the USA 1950–85. *Oxford Bulletin Economic Stat.,* **52(4),** 423–436.

Mishima, S., 2001: Recent trend of nitrogen flow associated with agricultural production in Japan. *Soil Science and Plant Nutrition,* **47,** 157–166.

Mitchell, C.E. and A.G. Power, 2003: Release of invasive plants from fungal and viral pathogens. *Nature,* **421,** 625–627.

Mol, A.P. and D.A. Sonnenfeld (eds.), 2000: *Ecological Modernisation Around the World.* Frank Cass Publishers, London.

Moll, H.C., 1989: *Aanbod van en vraag naar metalen; ontwikkelingen, implicaties en relaties; een methodische assesment omtrent subsitutie van materialen.* 9, Institute for Energy and Environment (IVEM), University of Groningen, Groningen, the Netherlands.

Mosier, A.R., M.A. Bleken, P. Chaiwanakupt, E.C. Ellis, J.R. Freney, R.B. Howarth, P.A. Matson, K. Minami, R. Naylor, K.N. Weeks, and Z.L. Zhu, 2001: Policy implications of human-accelerated nitrogen cycling. *Biogeochemistry,* **52(3),** 281–320.

Munroe, D.K., J. Southworth, and C.M. Tucker, 2002: The Dynamics of Land-Cover Change in Western Honduras: Exploring Spatial and Temporal Complexity. *Agricultural Economics,* **27(3).**

Myers, N., 2002: Environmental Refugees: A Growing Phenomenon of the 21st Century. *Philosophical Transactions Of The Royal Society Of London Series B-Biological Sciences.,* **357(1420),** 609–613.

Myers, N. and J. Kent, 1995: Environmental Exodus: An Emergent Crisis in the Global Arena. Climate Institute, Washington, DC.

Nachón, C.T., 2000: Environmental Aspects of the International Crisis of Antipersonnel Landmines and the Implementation of the 1997 Mine Ban Treaty: Thematic Report. In: *Landmine Monitor Report 2000.*

Nakićenović, N., A. Grübler, and A. McDonald (eds.), 1998: *Global Energy Perspectives.* Cambridge University Press, Cambridge.

Nakićenović, N., J. Alcamo, G. Davis, B.d. Vries, J. Fenhann, S. Gaffin, K. Gregory, A. Grübler, T.Y. Jung, T. Kram, E.L.L. Rovere, L. Michaelis, S. Mori, T. Morita, W. Pepper, H. Pitcher, L. Price, K. Riahi, A. Roehrl, H.-H. Rogner, A. Sankovski, M. Schlesinger, P. Shukla, S. Smith, R. Swart, S.v. Rooijen, N. Victor, and Z. Dadi, 2000: *Special report on emissions scenarios: a special report of Working Group III of the Intergovernmental Panel on Climate Change.* Cambridge University Press, Cambridge, 599 pp.

National Academy of Sciences, 2001: *Growing Populations, Changing Landscapes: Studies from India, China, and the United States.* National Academy Press, Washington, DC.

National Research Council, 1992: *Restoration of aquatic ecosystems: science, technology, and public policy.* National Academy Press, Washington, D. C., USA.

National Research Council, 2000: Beyond Six Billion: Forecasting the World's Population. Panel on Projections. J. Bongaarts and R.A. Bulatao (eds.), National Academy Press, Committee on Population, Commission on Behavioral and Social Sciences and Education, Washington, DC.

Nelson, G.C. (ed.), 2001: *Genetically modified organisms in agriculture: economics and politics.* Academic Press, San Diego, 344 pp.

Nelson, G.C. and J. Geoghegan, 2002: Modeling deforestation and land use change: Sparse data environments. *Agricultural Economics,* **27,** 201–216.

Norgaard, R.B., 2004: Learning and Knowing Collectively. *Ecological Economics,* **49(2),** 231–241.

Nriagu, J.O., 1996: A history of global metal pollution. *Science,* **272,** 223–224.

O'Neill, B.C., 2005: Population scenarios based on probablistic projections: An application for the Millennium Ecosystem Assessment. *Population & Environment,* **26(3),** 229–254.

O'Neill, B.C. and B. Chen., 2002: Demographic determinants of household energy use in the United States. *Methods of Population-Environment Analysis, A Supplement to Population and Development Review,* **28,** 53–88.

O'Neill, B.C., F.L. MacKellar, W. Lutz, and L. Wexler, 2001: *Population and Climate Change.* Cambridge University Press, Cambridge.

OECD (Organisation for Economic Co-operation and Development), 1997: *Reforming Energy and Transport Subsidies.* OECD, Paris.

OECD, 1998: *Eco-Efficiency.* OECD, Paris.

OECD, 2004: *OECD Agricultural Policies 2004 At A Glance.* OECD.

Oenema, O., H. Kros, and W. De Vries, 2003: Approaches and uncertainties in nutrient budgets: implications for nutrient management and environmental politics. *European Journal of Agronomy,* **20,** 3–16.

Oeppen, J. and J.W. Vaupel, 2002: Broken Limits to Life Expectancy. *Science,* **296,** 1029–1030.

Oldeman, L.R. and G.W.J. van Lynden, 1997: Revisiting the GLASOD methodology. In: *Methods for Assessment of Soil Degradation,* R. Lal, W.H. Blum, C. Valentine, and B.A. Steward (eds.), CRC Press., New York, 423–439.

Oldeman, L.R., R.T.A. Hakkeling, and W.G. Sombroek, 1991: *World map of the status of human-induced soil degradation: an explanatory note.* second revised edition, International Soil Reference and Information Centre/United Nations Environment Programme, Wageningen/Nairobi, October 1991, 34 pp.

Olshansky, S.J., B.A. Carnes, and C. Cassel, 1990: In Search of Methuselah: Estimating the Upper Limits to Human Longevity. *Science,* **250,** 634–40.

Olshansky, S.J., B.A. Carnes, and A. Desesquelles, 2001: Prospects for Human Longevity. *Science,* **291,** 1491–1492.

Oppenheim-Mason, K., 1997: Explaining Fertility Transitions. *Demography,* **34(4),** 443–54.

Ostrom, E., J. Burger, C.B. Field, R.B. Norgaard, and D. Policansky, 1999: Revisiting the commons: Local lessons, global challenges. *Science,* **284,** 278–282.

Ostrom, E., T. Dietz, N. Dolsak, P.C. Stern, S. Stonich, and E. Weber (eds.), 2002: *The Drama of the Commons.* National Academy Press, Washington, D.C.

Panayotou, T., 1993: *Empirical Tests and Policy Analysis of Environmental Degradation at Different Stages of Development.* Working Paper 238, Technology And Employment Programme. International Labour Office, Geneva.

Panayotou, T., 2000: Economic Growth and the Environment. Center For International Development at Harvard University. Available at http://www.cid.harvard.edu/cidwp/056.htm.

Patterson, K.L., J.W. Porter, K.B. Ritchie, S.W. Polson, E. Mueller, E.C. Peters, D.L. Santavy, and G.W. Smith, 2002: The etiology of white pox, a lethal disease of the Caribbean elkhorn coral, *Acropora palmata. PNAS,* **99,** 8725–8730.

Pendersen, D., 2002: Political violence, ethnic conflicts, and contemporary wars: broad implications for health and social well-being. *Social Science & Medicine,* **55,** 175–190.

Peng, S. and K.G. Cassman, 1998: Upper thresholds of nitrogen uptake rates and associated N fertilizer efficiencies in irrigated rice. *Agronomy Journal,* **90,** 178–185.

Petschel-Held, G. and K.B.L. Matthias, 2001: Integrating Case Studies on Global Change by Means of Qualitative Differential Equations. *Integrated Assessment,* **2(3),** 123–138.

Petschel-Held, G., A. Block, M. Cassel-Gintz, J. Kropp, M. Lüdeke, O. Moldehauer, F. Reusswig, and H.J. Schellnhuber, 1999: Syndromes of global change: A qualitative modeling approach to assist global environmental management. *Environmental Modelling and Assessment,* **4,** 295–314.

Pimentel, D., 2002: *Economic costs of biological invasions: Biological Invasions: Economic and Environmental Costs of Alien Plant, Animal, and Microbe Species.* CRC Press.

Pimentel, D., L. Lach, R. Zuniga, and D. Morrison, 1999: Environmental and economic costs of nonindigenous species in the United States. *Bioscience,* **50,** 53–65.

Pingali, P. and G. Traxler, 2002: Changing locus of agricultural research: will the poor benefit from biotechnology and privatization trends. *Food Policy,* **27,** 223–238.

Pingali, P. and Y. Khwaja, forthcoming 2004: Globalization of Indian Diets and the Transformation of Food Supply Systems. *Indian Journal of Agricultural Marketing.*

Pingali, P.L. and S. Rajaram, 1999: Global wheat research in a changing world: Options for sustaining growth in wheat productivity. In: *CIMMYT 1998–99 World Wheat Facts and Trends. Global Wheat Research in a Changing World: Challenges and Achievements,* P.L. Pingali (ed.), CIMMYT, Mexico, D.F.

Pingali, P.L. and P.W. Heisey, 2001: Cereal-Crop Productivity in Developing Countries: Past Trends and Future Prospects. In: *Agricultural Science Policy,* J.M. Alston, P.G. Pardey, and M. Taylor (eds.), Johns Hopkins University Press for IFPRI, Baltimore.

Pinstrup-Andersen, P. and P.B.R. Hazell, 1985: The impact of the Green revolution and prospects for the future. *Food Review International,* **1(1),** 1–25.

Pinstrup-Andersen, P., N. Ruiz de Londoño, and E. Hoover, 1976: The impact of increasing food supply on human nutrition: Implications for commodity priorities in agricultural research. *American Journal of Agricultural Economics,* **58(2),** 131–42.

Pionke, H.B., W. J. Gburek, A.N. Sharpley, and J.A. Zollweg., 1997: Hydrologic and chemical controls on phosphorus loss from catchments. In: *Phosphorus loss from soil to water,* Tunney (ed.), CAB International, New York.

Plowright, W., 1982: The effects of rinderpest and rinderpest control on wildlife in Africa. *Symposia of the Zoological Society of London.,* **50,** 1–28.

Postel, S., 1997: Freshwater ecosystem services. In: *Nature's Services,* G. Daily (ed.), Island Press, Washington, D. C., 195–213.

Power, M.E., D. Tilman, J.A. Estes, B.A. Menge, W.J. Bond, L.S. Mills, G.C. Daily, J.C. Castilla, J. Lubchenco, and R.T. Paine, 1996: Challenges in the quest for keystones. *Bioscience,* **46(8),** 609–620.

Pretty, J., C. Brett, D. Gee, R.E. Hine, C.F. Mason, J.I.L. Morison, H. Raven, M.D. Rayment, and G. van der Bijl, 2000: An assessment of the total external costs of UK agriculture. *Agricultural Systems,* **65,** 113–136.

Princen, T. and M. Finger, 1994: *Environmental NGOs in World Politics: Linking the Local and the Global.* Routledge, London.

Prins, H.H.T. and F.J. Weyerhaeuser, 1987: Epidemics in populations of wild ruminants: anthrax and impala, rinderpest and buffalo in Lake Manyara National Park, Tanzania. *Oikos,* **49,** 28–38.

Putnam, R.D., R. Leonardi, and R.Y. Nanetti, 1993: *Making Traditions in Modern Italy.* Princeton University Press, Princeton.

Rabalais, N.N., 2002: Nitrogen in aquatic ecosystems. *Ambio,* **31,** 102–112.

Ramankutty, N. and J.A. Foley, 1998: Characterizing patterns of global land use: An analysis of global cropland data. *Global Biogeochemical Cycles,* **12,** 667–685.

Read, D.J., D.H. Lewis, A.H. Fitter, and I.J. Alexander (eds.), 1992: *Mycorrhizas in ecosystems.* CAB International, Wallingford, Oxon.

Redman, C.L., 1999: *Human Impact on Ancient Environments.* The University of Arizona Press, Tucson, AZ.

Renkow, M., 1993: Differential technology adoption and income distribution in Pakistan: Implications for research resource allocation. *American Journal of Agricultural Economics,* **75(1),** 33–43.

Renn, O., T. Webler, and P. Wiedemann (eds.), 1995: *Fairness and Competence in Citizen Participation: Evaluating Models for Environmental Discourse.* Kluwer Academic Publishers., Dordrecht.

Richardson, D.M., P. Pysek, M. Rejmánek, M.G. Barbour, F.D. Panetta, and C.J. West, 2000: Naturalization and invasion of alien plants: concepts and definitions. *Diversity and Distribution,* **6,** 93–107.

Richerson, P.J., R. Boyd, and B. Paciotti, 2002: An evolutionary theory of commons management. In: *The Drama of the Commons,* E. Ostrom, T. Dietz, N. Dolsak, P.C. Stern, S. Stonich, and E. Weber (eds.), National Academy Press, Washington, D.C., 403–442.

Riley, W.J., I. Ortiz-Monasterio, and P.A. Matson, 2003: Nitrogen leaching and soil nitrate, nitrite, and ammonium levels under irrigated wheat in Northern Mexico. *Nutrient Cycling in Agroecosystems,* **61,** 223–236.

Robinson, W.C., 1997: The Economic Theory of Fertility Over Three Decades. *Population Studies,* **51(1),** 63–74.

Rokeach, M., 1968: *Beliefs, Attitudes and Values: A Theory of Organization and Change.* Jossey-Bass, San Francisco.

Rokeach, M., 1973: *The Nature of Human Values.* Free Press, New York.

Rosa, E.A., N. Matsuda, and R.R. Kleinhesselink, 2000: The Cognitive Architecture of Risk: Pancultural Unity or Cultural Shaping? In: *Comparative Risk Perception,* O. Renn and B. Rohrmann (eds.), Kluwer, Dordrecth, The Netherlands, 185–210.

Rotmans, J. and H.J.M. de Vries (eds.), 1997: *Perspectives on Global Futures: The TARGETS approach.* Cambridge University Press, Cambridge.

Roy, R.N., R.V. Misra, and A. Montanez, 2002: Decreasing reliance on mineral nitrogen—yet more food. *Ambio,* **31,** 177–183.

Ruiz, G.M., P.W. Fofonoff, J.T. Carlton, J. Marjorie, and A.H. Hines, 2000: Invasion of coastal marine communities in North America: Apparent Patterns, Processes, and Biases. *Annual Review of Ecology and Systematics,* **31,** 481–531.

Ruttan, V.R., 2001: *Technology, Growth and Development.* Oxford University Press, New York.

Ryder, N.B., 1983: Fertility and Family Structure. *Population Bulletin of the United Nations,* **15,** 15–33.

Sakai, A., F.W. Allendorf, J.S. Holt, D.M. Lodge, J. Molofsky, K.A. With, S. Baughman, R.J. Cabin, J.E. Cohen, N.C. Ellstrand, D.E. McCauley, P. O'Neill, I.M. Parker, J.N. Thompson, and S.G. Weller, 2001: The population biology of invasive species. *Ann. Rev. Ecol. Syst.,* **32,** 305–32.

Sala-i-Martin, X., 2003: *The World Distribution of Income, 1970–2000.* Unpublished, Columbia University, NY.

Salzman, J. and D. Rejeski, 2002: Changes in pollution and implications for policy. In: *New Tools for Environmental Protection: Education, Information and Voluntary Measures,* P.C. Stern (ed.), National Academy Press, Washington, D.C., 17–42.

Scepan, J., 1999: Thematic validation of high resolution global land-cover data sets. *Photgrammetric engineering & remote sensing,* **65(9),** 1051–1060.

Scherr, S.J., 1997: People and environment: what is the relationship between exploitation of natural resources and population growth in the South? *Forum for Development Studies,* **1,** 33–58.

Schindler, D.W., 1977: Evolution of phosphorus limitation in lakes. *Science,* **195,** 260–262.

Schindler, D.W., F.A.J. Armstrong, S.K. Holmgren, and G.J. Brunskill, 1971: Eutrophication of Lake 227, Experimental Lakes Area, northwestern Ontario, by addition of phosphate and nitrate. *Journal of Fisheries Research Board of Canada,* **28,** 2009–2036.

Schipper, L., 1996a: Lifestyles and the environment: The case of energy. *Daedalus,* **125,** 113–138.

Schipper, L.J., 1996b: Excel spreadsheets containing transport energy data, PASSUM.XLS and FRTSUM.XLS. Versions of 9 February 1996. N. Nakićenović (ed.), International Energy Agency, Paris/ France/Lawrence Berkeley National Laboratory, Berkeley, CA.

Schmalensee, R., T. Stoker, and R. Judson, 1998: World carbon dioxide emissions: 1950–2050. *The Review of Economics and Statistics,* **LXXX(1),** 15–27.

Schnaiberg, A., 1980: *The Environment: From Surplus to Scarcity.* Oxford University Press, New York.

Schultz, P.W. and L.C. Zelezny, 1998: Values and Proenvironmental Behavior: A Five-Country Study. *Journal of Cross-Cultural Psychology,* **29(4),** 540–558.

Schultz, P.W. and L.C. Zelezny, 1999: Values as predictors of environmental attitudes. *Journal of Environmental Psychology,* **19,** 255–265.

Schwartz, S., 1996: Value Priorities and Behavior: Applying a Theory of Integrated Value Systems. In: *The Psychology of Values,* C. Seligman, J.M. Olson, and M.P. Zanna (eds.), Lawrence Erlbaum Associates, Mahwah, New Jersey, 1–24.

Schwartz, S.H., 1987: Toward a Universal Psychological Structure of Human Values. *Journal of Personality and Social Psychology,* **53,** 550–562.

Schwartz, S.H., 1992: Universals in the Content and Structure of Values: Theoretical Advances and Empirical Tests in 20 Countries. *Advances in Experimental Social Psychology,* **25,** 1–65.

Schwartz, S.H. and W. Bilsky, 1990: Toward a Theory of the Universal Content and Structure of Values: Extensions and Cross-Cultural Replications. *Journal of Personality and Social Psychology,* **58,** 878–891.

Schwartz, S.H. and S. Huismans, 1995: Value Priorities and Religiosity in Four Western Religions. *Social Psychology Quarterly,* **58,** 88–107.

Schweigert, P. and R.R. van der Ploeg, 2000: Nitrogen use efficiency in German agriculture since 1950: facts and evaluation. *Berichte ueber Landwirtschaft,* **80,** 185–212.

Scobie, G.M. and R.T. Posada, 1978: The impact of technical change on income distribution: The case of rice in Colombia. *American Journal of Agricultural Economics,* **60(1),** 85–92.

Secretariat of the Convention on Biological Diversity, 2003: *Interlinkages between biological diversity and climate change. Advice on the integration of biodiversity considerations into the implementation of the United Nations Framework Convention on Climate Change and its Kyoto protocol.* Technical Series 10, Secretariat of the Convention on Biological Diversity, Montreal, 154 pp.

Sekercioglu, C.H., 2002: Impacts of bird watching on human and avian communities. *Environmental Conservation,* **29(3),** 282–289.

Semken, H.A., 1983: Holocene mammalian biogeography and climatic change in the Eastern and Central United States. In: *Late Quaternary environments of the United States. Volume 2. The Holocene,* H.E. Wright (ed.), University of Minnesota Press, Minneapolis.

Senauer, B. and M. Sur, 2001: Ending global hunger in the 21st century: projections of the number of food insecure people. *Review of Agricultural Economics,* **23(1),** 68–81.

Shafik, N., and S. Bandyopadhay, 1992: *Economic Growth and Environmental Quality: Time Series and Cross-Country Evidence.* World Bank, Washington DC.

Sharpley, A.N., 1995: Identifying sites vulnerable to phosphorus loss in agricultural runoff. *Journal of Environmental Quality,* **24,** 947–951.

Sharpley, A.N., T.C. Daniel, and D.R. Edwards, 1993: Phosphorus movement in the landscape. *Journal of Production Agriculture,* **6(4),** 492–500.

Sharpley, A.N., S. C. Chapra, R. Wedepohl, J.T. Sims, T. C. Daniel, and K.R. Reddy, 1994: Managing agricultural phosphorus for protection of surface waters: issues and option. *Journal of Environmental Quality,* **23,** 437–451.

Sheldrick, W.F., J.K. Syers, and J. Lingard, 2003a: Soil nutrient audits for China to estimate nutrient balances and output/input relationships. *Agriculture, Ecosystems and Environment,* **94,** 341–354.

Sheldrick, W.F., J.K. Syers, and J. Lingard, 2003b: Contribution of livestock excreta to nutrient balances. *Nutrient Cycling in Agroecosystems,* **66,** 119–131.

Simon, N., 1962: *Between the Sunlight and the Thunder. The Wildlife of Kenya.* Collins., London.

Slimak, M.W., 2003: *Values and Ecological Risk Perception: A Comparison of Experts and the Public.* Environmental Science and Policy, George Mason University, Fairfax, Virginia.

Small, C., 2001: *Global Analysis of Urban Population Distributions and the Physical Environment.* Open Meeting of the Human Dimensions of Global Environmental Change Research, Rio de Janeiro, October, 11 pp.

Smil, V., 1997: Some unorthodox perspectives on agricultural biodiversity. The case of legume cultivation. *Agriculture, Ecosystems and Environment,* **62,** 135–144.

Smil, V., 1999: Nitrogen in crop production: An account of global flows. *Global Biochemical Cycles,* **13,** 647–662.

Smil, V., 2001: *Enriching the earth: Fritz Haber, Carl Bosch, and the transformation of world food production.* The MIT Press, Cambridge, Mass., USA, 338 pp.

Smith, K.A., I.P. Mctaggart, and H. Tsuruta, 1997: Emissions of N2O and NO associated with nitrogen fertilization in intensive agriculture, and the potential for mitigation. *Soil Use and Management,* **13,** 296–304.

Smith, K.R. and S. Mehta, 2003: The burden of disease from indoor air pollution in developing countries: comparison of estimates. *International Journal of Hygiene and Environmental Health,* **206,** 279—289.

Smith, V.H., 1998: Cultural eutrophication of inland, estuarine, and coastal waters. In: *Successes, Limitations, and Frontiers in Ecosystem Science,* M.L. Pace and P.M. Groffman (eds.), Springer-Verlag, New York.

Solow, R., 1957: Technical change and the aggregate production function. *Review of Economics and Statistics,* **39,** 312–320.

Spalding, R.F. and M.E. Exner, 1993: Occurrence of nitrate in groundwater—a review. *Journal of Environmental Quality,* **22,** 392–402.

Spinage, C.A., 1962: Rinderpest and faunal distribution patterns. *African Wildlife,* **16,** 55–60.

Stark, O., 1991: *The Migration of Labor.* Basil Blackwell, Cambridge, England.

Stern, D.I., 2004: The Rise and Fall of the Environmental Kuznets Curve. *World Development,* **32(8),** 1419–1439.

Stern, P.C., 2000: Toward a Coherent Theory of Environmentally Significant Behavior. *Journal of Social Issues,* **56(3),** 407–242.

Stern, P.C. and H. Fineberg, 1996: *Understanding Risk: Informing Decisions in a Democratic Society.* National Academy Press, Washington, D.C.

Stern, P.C., T. Dietz, and J.S. Black, 1986: Support for Environmental Protection: The Role of Moral Norms. *Population and Environment,* **8,** 204–222.

Stern, P.C., T. Dietz, and L. Kalof, 1993: Value Orientations, Gender and Environmental Concern. *Environment and Behavior,* **25,** 322–348.

Stern, P.C., T. Dietz, and G.A. Guagnano, 1995: The New Environmental Paradigm in Social Psychological Perspective. *Environment and Behavior,* **27,** 723–745.

Stern, P.C., T. Dietz, V.W. Ruttan, R.H. Socolow, and J. Sweeney (eds.), 1997: *Environmentally Significant Consumption: Research Directions.* National Academy Press, Washington, D.C.

Stern, P.C., T. Dietz, T. Abel, G.A. Guagnano, and L. Kalof, 1999: A Social Psychological Theory of Support for Social Movements: The Case of Environmentalism. *Human Ecology Review,* **6,** 81–97.

Stigliani, W.M., Peter Doelman, Wim Salomons, Rainer Schulin, Gera R. B. Smidt, and Sjoerd EATM Van der Zee, 1991: Chemical time bombs. *Environment,* **33(4),** 4–30.

Stoorvogel, J.J., E.M.A. Smaling, and B.H. Janssen, 1993: Calculating soil nutrient balances in Africa at different scales. 1.Supra-national scale. *Fertilizer Research,* **35,** 227–235.

Suhrke, A., 1994: Environmental Degradation and Population Flows. *Journal of International Affairs,* **47,** 473–96.

Suri, V. and D. Chapman, 1996: *Economic Growth, Trade and Environment: An Econometric Evaluation of the Environmental Kuznets Curve.* Working Paper WP-96–05, Cornell University, Ithaca.

Suzuki, A., 1998: *Fertilization of rice in Japan.* Japan FAO Association, Tokyo, 111 pp.

Thogersen, J., 2002: Promoting "Green" Consumer Behavior with Eco-Labels. In: *New Tools for Environmental Protection,* T. Dietz and P.C. Stern (eds.), National Academy Press, Washington, D.C., 83–104.

Thomas, D.S.G. and N.J. Middleton, 1997: *World atlas of desertification.* Second edition ed. Arnold, London, 180 pp.

Thucydides, 1989: *The Peloponnesian War / Thucydides; the complete Hobbes translation, with notes and a new introduction by David Grene.* University of Chicago Press, Chicago.

Tiessen, H., ed., 1995: *Phosphorus in the Global Environment: Transfers, Cycles, and Management.* Vol. 54, *SCOPE,* John Wiley & Sons, Chichester.

Tilman, D., K.G. Cassman, P.A. Matson, R.L. Naylor, and S. Polasky, 2002: Agricultural sustainability and intensive production practices. *Nature,* **418,** 671–677.

Tilton, J.E., 1990: *World Metal Demand: Trends and Prospects.* Resources for the Future, Washington, DC.

Todaro, M., 1976: International Migration in Developing Countries. International Labour Office, Geneva.

Torchin, M.E., K.D. Lafferty, A.P. Dobson, A.P. Mckenzie, and A.M. Kuris, 2003: Introduced species and their missing parasites. *Nature,* **421,** 628–630.

Townsend, A.R., R.B. Howarth, F.A. Bazzaz, M.S. Booth, C.C. Cleveland, S.K. Collinge, A.P. Dobson, P.R. Epstein, E.A. Holland, D.R. Keeney, M.A. Mallin, C.A. Rogers, P. Wayne, and A.H. Wolfe, 2003: Human health effects of a changing global nitrogen cycle, **1,** 240–246.

Traxler, G. and D. Byerlee, 1992: Economic returns to crop management research in post-green revolution setting. *American Journal of Agricultural Economics,* **74(3),** 573–582.

Tuler, S. and T. Webler, 1999: Voices from the Forest: What Participants Expect of a Public Participation Process. *Society and Natural Resources,* **12,** 437–453.

Turner II, B.L., Y. Geoghegen, and D.R. Foster (eds.) *Integrated Land Change Science and Tropical Deforestation in Southern Yucatan: Final Frontiers.* Clarendon Press of Oxford University Press, Oxford, UK.

Tyrell, T., 1999: The relative influences of nitrogen and phosphorus on oceanic primary production. *Nature,* **400,** 525–531.

Uexkuell, H.R.V. and E. Mutert, 1995: Global extent, development and economic impact of acid soils. *Plant and Soil,* **171,** 1–15.

UNCED (United Nations Conference on Environment and Development), 1992: *Agenda 21.,* United Nations, New York.

U.N. Environment Programme, 1999: UNEP Global Environment Outlook 2000. Cited May 30, 2004. Available at http://www1.unep.org/geo-text/0138.htm.

UNICEF, 1996: *The State of the World's Children.* Oxford University Press, Oxford.

United Nations, 1998: *World Population Monitoring 1997: International Migration and Development.* ST/ESA/SER.A/169, United Nations, New York.

United Nations, 2003a: *World Population in 2300.* ESA/P/WP.187, United Nations.

United Nations, 2003b: *International Migration Report 2002.* United Nations, New York.

United Nations, 2003c: *Population, Education and Development: The Concise Report.* ST/ESA/SER.A/226, United Nations, New York, 56 pp.

United Nations, 2003d: World Population Prospects: The 2002 Revision Population Database. 2003. Available at http://esa.un.org/unpp/.

United Nations, 2003e: *World Population Prospects: The 2002 Revision Highlights.* ESA/P/WP.180, United Nations, New York, 22 pp.

United States Census Bureau, 2003: International Data Base, updated 17 July 2003. Cited 16 December 2003. Available at http://www.census.gov/ipc/www/idbnew.html.

United States Geological Survey, 2003: *Mineral commodity summaries 2003.* U.S. Department of the Interior, U.S. Geological Survey, Washington, D.C.

USDA (U.S. Department of Agriculture), 2001: *Oil Crops Situation and Outlook Yearbook.* Economic Research Service, U.S. Department of Agriculture, Washington DC.

U.S. Environmental Protection Agency, 1996: *Environmental indicators of water quality in the United States. EPA 841-R-96-002,* USEPA, Office of Water Criteria and Standards Division (4503F), Washington, D.C., USA.

U.S. National Research Council: Panel on Public Participation in Environmental Assessment and Decision Making. Available at http://www7.nationalacademies.org/hdgc/Public_Participation.html.

van Auken, O.W., 2000: Shrub invasion of North American Semiarid Grasslands. *Ann. Rev. Ecol. Syst.,* **31,** 197–215.

van de Kaa, D.J., 1996: Anchored Narratives: The Story and Findings of Half a Century of Research Into the Determinants of Fertility. *Population Studies,* **50(3),** 389–432.

van der Duim, R. and J. Caalders, 2002: Biodiversity and Tourism. *Annals of Tourism Research,* **29(3),** 743–761.

Van der Pol, R.A. and B. Traore, 1993: Soil nutrient depletion by agricultural production in southern Mali. *Fertilizer Research,* **36,** 79–90.

Van Liere, K.D. and R.E. Dunlap, 1978: Moral Norms and Environmental Behavior: An Application of Schwartz's Norm-Activation Model to Yard Burning. *Journal of Applied Social Psychology,* **8,** 174–188.

Van Vuuren, D.P., B. Strengers, and H.J.M. de Vries, 2000: Long-term perspectives on world metal use—a systems dynamics model. *Resources Policy,* accepted for publication.

Vaupel, J.W., 1997: The Remarkable Improvements in Survival at Older Ages. *Philosophical Transactions Of The Royal Society Of London Series B-Biological Sciences.,* **B352(1363),** 1799–804.

Viguier, L., 1999: Emissions of SO2,NOx, and CO2 in transition economies: Emission inventories and Divisa index analysis. *Energy Economics,* **20(2),** 59–75.

Vitousek, P.M., J.D. Aber, R.W. Howarth, G.E. Likens, P.A. Matson, D.W. Schindler, W.H. Schlesinger, and D.G. Tilman, 1997: Human alteration of the global nitrogen cycle: Sources and consequences. *Ecological Applications,* **7,** 737–750.

Vollenweider, R.A., 1968: The scientific basis of lake and stream eutrophication with particular reference to phosphorus and nitrogen as eutrophication factors. *Technical report OECD, DAS/C81/68,* Paris, France, 1–182.

Vringer, K. and K. Blok, 1995: The direct and indirect energy-requirements of households in the Netherlands. *Energy Policy,* **23,** 893–902.

Wackernagel, M. and W. Rees, 1996: *Our Ecological Footprint.* New Society Publishers, Gabriola Island, B.C.

Waggoner, P.E. and J.H. Asubel, 2002: A framework for sustainability science: A renovated IPAT identity. *Proceedings of the National Academy of Sciences,* **99(12),** 7860–7865.

Walker, L. and P. Vitousek, 1991: An invader alters germination and growth of a native dominant tree in Hawaii. *Ecology,* **72,** 1449–1455.

Watson, R., M.C. Zinyowera, and R. Moss (eds.), 1996: *Climate Change 1995. Impacts, Adaptations and Mitigation of Climate Change: Scientific Analyses. Contribution of Working Group II to the Second Assessment Report of the Intergovernmental Panel on Climate Change.* Cambridge University Press, Cambridge, 861 pp.

Wearing, S., 2001: Exploring socio-cultural impacts on local communities. In: *The Encyclopedia of Ecotourism,* D.B. Weaver (ed.), CABI Publishing, Oxon, 193–204.

Weaver, D.B. (ed.), 2001: *The Encyclopedia of Ecotourism.* CABI Publishing, Oxon, 193–204 pp.

Welch, R.M. and R.D. Graham, 1999: A new paradigm for world agriculture: meeting human needs. Productive, sustainable and nutritious. *Field Crops Research,* **60,** 1–10.

Westing, A.H., 1988: Constraints on Military Disruption of the Biosphere: An Overview. In: *Cultural Norms, War and the Environment,* A.H. Westing (ed.), Oxford University Press.

W. Ghiorghis, A., 1993: The Human and Ecological Consequences of War in Eritrea. In: *Conflicts in the Horn of Africa: Human and Ecological Consequences of Warfare,* T. Tvedt (ed.), Reprocentralen HSC, Uppsala. Research Programs on Environmental Policy and Society, Department of Social and Economic Geography, Uppsala University.

White, L.J., 1967: The Historical Roots of Our Ecological Crisis. *Science,* **155,** 1203–1207.

Whitfield, S., T. Dietz, and E.A. Rosa, 1999: *Environmental Values, Risk Perception and Support for Nuclear Technology.* Human Ecology Research Group, George Mason University, Fairfax, Virginia.

Williamson, M. and A. Fitter, 1996: The varying success of invaders. *Ecology,* **77,** 1661–1666.

Wolfe, A.H. and J.A. Patz, 2002: Reactive nitrogen and human health: acute and long-term implications. *Ambio,* **31,** 120–125.

Wong, J., 2001: China's sharply declining fertility: Implications for its population policy. *Issues & Studies,* **37(3),** 68–86.

World Bank, 1992: *World Development Report 1992: Development and the Environment.* Oxford University Press, Oxford.

World Bank, 2002: *World Development Indicators 2002.* World Bank, Washington, DC, 432 pp.

World Bank, 2003: *World Development Indicators, 2003.* World Bank, Washington, DC, April.

World Bank, 2004: *World Development Report 2004: Making Services Work for Poor People.* World Bank, Washington, DC.

World Commission on Environment and Development, 1987: *Our Common Future.* Oxford University Press, Oxford.

World Conservation Monitoring Center, 1991: *Gulf War Impacts on Marine Environment and Species.* Gulf War Environmental Information Service, Cambridge, England.

World Resources Institute, 1997: *World Resources 1996–97.* Oxford University Press, New York.

World Resources Institute, United Nations Development Programme, United Nations Environment Programme, and World Bank, 2000: *World Resources 2000–2001: People and Ecosystems, The Fraying Web of Life,* Washington, D.C.

World Resources Institute, United Nations Development Programme, United Nations Environment Programme, and World Bank, 2002: *World Resources 2002–2004: Decisions for the Earth: Balance, Voice, and Power.* Washington, D.C.

World Tourism Organization, 1998: Ecotourism. *WTO News,* **I.**

World Trade Organization, 2003: *Understanding the WTO.* World Trade Organization, Geneva, September.

World Trade Organization, 2004: Why is the liberalization of services important? Cited May 30, 2004. Available at http://www.wto.org/english/tratop_e/serv_e/gats_factfiction2_e.htm.

York, R. and E.A. Rosa, 2003: Key challenges to ecological modernization theory. *Organization and Environment,* **16,** 273–288.

York, R., E.A. Rosa, and T. Dietz, 2003a: Footprints on the Earth: The Environmental Consequences of Modernity. *American Sociological Review,* **68(2),** 279–300.

York, R., E. Rosa, and T. Dietz, 2003b: STIRPAT, IPAT and IMPACT: analytic tools for unpacking the driving forces of environmental impact. *Ecological Economics,* **46,** 351–365.

Young, O.R., 2002: *The Institutional Dimensions of Environmental Change: Fit, Interplay and Scale.* The MIT Press, Cambridge.

Zambon, M.C., 2001: The pathogenesis of influenza in humans. *Reviews in Medical Virology,* **11,** 227–241.

Zlotnik, H., 1998: International Migration 1965–96: An Overview. *Population and Development Review,* **24(3),** 429–6.

Chapter 8

Four Scenarios

Coordinating Lead Authors: Steven Cork, Garry Peterson, Gerhard Petschel-Held
Lead Authors: Joseph Alcamo, Jacqueline Alder, Elena Bennett, Edward R. Carr, Danielle Deane, Gerald C. Nelson, Teresa Ribeiro
Contributing Authors: Colin Butler, Eduardo Mario Mendiondo, Willis Oluoch-Kosura, Monika Zurek
Review Editors: Antonio Alonso Concheiro, Yuzuru Matsuoka, Allen Hammond

---

BOXES

FIGURES

TABLES

*This appears in Appendix A at the end of this volume.

Main Messages

This chapter presents four internally consistent scenarios that explore aspects of plausible global futures and their implications for ecosystem services. Scenario development is a way to explore possibilities for the future that cannot be predicted by extrapolation of past and current trends. The four scenarios in this chapter are structured around the assumptions and rationale described in Chapter 5, the methods described in Chapter 6, and the drivers described in Chapter 7. The probability of any one of these scenarios being the real future is very small. Each scenario might resemble some people's ideal world, but one lesson to us as we developed them was that all four scenarios have both strengths and potentially serious weaknesses. An ideal future would probably involve a mix of all four, with different elements dominating at different times and in different places. The future could be far better or far worse than any of the scenarios, depending on choices made by key decision-makers and other people in society who bring about change. Our purpose in developing the stories is to encourage decision-makers to consider some positive and negative implications of the different development trajectories.

The Global Orchestration scenario depicts a worldwide connected society in which global markets are well developed. Supra-national institutions are well placed to deal with global environmental problems, such as climate change and fisheries. However, their reactive approach to ecosystem management makes them vulnerable to surprises arising from delayed action or unexpected regional changes. The scenario is about global cooperation not only to improve the social and economic well-being of all people but also to protect and enhance global public goods and services (such as public education, health, and infrastructure). There is a focus on the individual rather than the state, inclusion of all impacts of development in markets (internalization of externalities), and use of regulation only where appropriate. Environmental problems that threaten human well-being (such as pollution, erosion, and climate change) are dealt with only after they become apparent. Problems that have little apparent or direct impact on human well-being are given a low priority in favor of policies that directly improve well-being. People are generally confident that the necessary knowledge and technology to address environmental challenges will emerge or can be developed as needed, just as it has in the past. The scenario highlights the risks from ecological surprises under such an approach. Examples are emerging infectious diseases and other slowly emerging problems that are hard to control once they are established. Other benefits and risks also emerge from the inevitable and increasing connections among people and nations at social, economic, and environmental scales.

The Order from Strength scenario represents a regionalized and fragmented world concerned with security and protection, emphasizing primarily regional markets, and paying little attention to the common goods, and with an individualistic attitude toward ecosystem management. Nations see looking after their own interests as the best defense against economic insecurity. They reluctantly accept the argument that a militarily and economically strong liberal democratic nation could maintain global order and protect the lifestyles of the richer world and provide some benefits for any poorer countries that elect to become allies. Just as the focus of nations turns to protecting their borders and their people, so too their environmental policies focus on securing natural resources seen as critical for human well-being. But, as in Global Orchestration, people in this scenario see the environment as secondary to their other challenges. They believe in the ability of humans to bring technological innovations to bear as solutions to environmental challenges after these challenges emerge.

The Adapting Mosaic scenario depicts a fragmented world resulting from discredited global institutions. It sees the rise of local ecosystem management strategies and the strengthening of local institutions. Investments in human and social capital are geared toward improving knowledge about ecosystem functioning and management, resulting in a better understanding of the importance of resilience, fragility, and local flexibility of ecosystems. There is optimism that we can learn, but humility about preparing for surprises and about our ability to know all there is to know about managing socioecological systems. Initially, trade barriers for goods and products are increased, but barriers for information (for those who are motivated to use it) nearly disappear due to improving communication technologies and rapidly decreasing costs of access to information. There is great regional variation in management techniques. Some local areas explore adaptive management, using experimentation, while others manage with command and control or focus on economic measures. Eventually, the focus on local governance leads to failures in managing the global commons. Problems like climate change, marine fisheries, and pollution grow worse, and global environmental surprises become common. Communities slowly realize that they cannot manage their local areas because global problems are infringing, and they begin to develop networks among communities, regions, and even nations to better manage the global commons. The rebuilding is more focused on ecological units, as opposed to the earlier type of management based on political borders that did not necessarily align with ecosystem boundaries.

The TechnoGarden scenario depicts a globally connected world relying strongly on technology and on highly managed and often-engineered ecosystems to deliver needed goods and services. Overall, eco-efficiency improves, but it is shadowed by the risks inherent in large-scale human-made solutions. Technology and market-oriented institutional reform are used to achieve solutions to environmental problems. In many cases, reforms and new policy initiatives benefit from the strong feel for international cooperation that is part of this scenario. As a result, conditions are good for finding solutions for global environmental problems such as climate change. These solutions are designed to benefit both the economy and the environment. Technological improvements that reduce the environmental impact of goods and services are combined with improvements in ecological engineering that optimize the production of ecosystem services. These changes co-develop with the expansion and development of property rights to ecosystem services, such as requiring people to pay for pollution they create or paying people for providing key ecosystem services through actions such as preservation of key watersheds. These rights are generally created and allocated following the identification of ecological problems. Because understanding of ecosystem function is high, property rights regimes are usually assigned long before the problem becomes serious. These property rights are assigned to a diversity of individuals, corporations, communal groups, and states that act to optimize the value of their property. We assume that ecological management and engineering can be successful, although it does produce some ecological surprises that affect many people due to an over-reliance on highly engineered systems.

Some additional insights emerge from the scenarios, as follows:

A path of accelerated global cooperation and a focus on global public good is likely to improve overall human well-being, but it may have costly consequences for ecosystem services and some aspects of human well-being. Local problems can become unmanageable due to complacency and delayed action; ecological crises can accentuate inequalities as they tend to affect poor regions and countries; reactive solutions can carry unbearable social costs for less favorable areas.

A global development that emphasizes environmental technology and engineered ecosystems will contribute to sustainable development by allowing for greater efficiency and optimal control of ecosystems, but possibly at the cost of loss in local, rural, and indigenous knowledge and

cultural values. Increasing confidence in human ability to manage, tame, and improve nature may lead people to overlook factors that sometimes cause breakdowns of ecosystem services. Large-scale technological solutions carry the internal risks of failure and can engender technology-related ecological surprises.

Emphasis on adaptive management and learning at local scales may be achieved at the cost of overlooking global problems that may result in global environmental surprises with serious local repercussions. While local problems become more tractable, and can be addressed by citizens, attention to global problems such as climate change and marine fisheries may decrease, leading to increasing magnitude of their impacts.

Strategies that focus on local and regional safety and protection may disregard cross-border and global issues, restrict trade and movement of people, and increase inequalities, but protection of key natural resources in richer regions could see an improvement. Such a world encourages boundaries at all levels. It might offer security in the face of aggression, environmental pests, and diseases, but increases risks of longer-term internal and international conflict, ecosystem degradation, and declining human well-being. Alleviation of problems in poorer countries through strategic intervention by richer countries to reduce inequalities and environmental degradation would be likely to occur in cycles, leading to a fluctuating and unstable world for many people.

Institutional development, feedbacks between local and global processes, and the risks entailed by the substitution of ecosystem services by human, social, or manufactured capital determine society's ability to cope with ecological surprises. Globally controlled institutions can be too large and rigid to respond effectively to ecological surprises, yet local institutions may neglect important linkages for anticipating and managing such surprises. Concentrating ecosystem management on a single level of control (local, regional, national, continental, or global) is highly likely to fail to manage cross-scale ecological feedbacks in appropriate way. Solutions that substitute for ecosystem services carry risks, due to limited rates of learning, barriers to development, or the inherent brittleness of human-made solutions, including human or technical failures.

Local management of ecosystems provides better opportunities for more effective and fairer access to ecosystem services on local scales, but local strategies are more likely to be effective when accompanied by measures to ensure regional and global coordination. Local management can be particularly relevant when learning about ecosystems is adaptive and management is more proactive. If risks or surprises are to be reduced, however, local strategies should be accompanied by regional and global coordination measures (such as international treaties like the Convention on Biological Diversity or the Convention on Combating Desertification) that focus on the management of common pool resources and public goods.

A globally compartmentalized, environmentally reactive world could mask developing ecological and social disasters for several decades. The Scenarios Working Group agreed that Order from Strength is ultimately unsustainable in terms of ecosystems and the societies they support, but the group was surprised at the diversity of viewpoints on how such a scenario could unfold and over what time scale. It is plausible that if current trends toward increasing compartmentalization continue, the world could develop a false sense of security in coming decades unless efforts are made to understand and monitor ecosystems and their services.

Current understanding of ecological and social systems is inadequate to predict when and how ecological-social systems will produce adverse feedbacks or unpleasant surprises. There is considerable evidence of adverse cross-scale interactions between ecological processes and human activities (including interventions intended to fix problems). There also is emerging understanding of how these adverse outcomes have been caused. However, we found it surprisingly difficult to locate when within the next five decades the impacts would emerge or precisely what combinations of policies and interventions are most likely to produce them.

8.1. Introduction

The four major components of this chapter are the four scenarios: Global Orchestration, Order from Strength, Adapting Mosaic, and TechnoGarden. Interspersed among the scenarios are boxes drawing out specific issues relating to individual scenarios or comparing issues across scenarios.

The scenarios are fictional stories written from the point of view of someone looking back from 2050 at what has happened in the world since 2000. The stories are designed to draw out key aspects of the questions raised in Chapter 5; these are recapped later in this section. The storylines are based on the logic—or guidelines—developed in Chapter 5, but they do not try to include all elements of these guidelines. The stories would be too complex and the messages would be lost if they did.

The scenarios have been developed from input from all members of the Scenarios Working Group, but they have been woven into storylines by a smaller number of writers. The writers have researched elements of the stories and placed their own interpretation on this research. While our primary aim was to draw out the consequences of several plausible future worlds for ecosystem services, we needed to provide plausible explanations that considered social and economic drivers of change.

The explanations given in the scenarios are only some of the ways in which the worlds could develop. Each member of the Scenarios Working Group would have written each scenario differently if it had been his or her task. The purpose of the scenarios is to get the reader thinking about how the world might develop rather than to provide predictions. The writers of later chapters of this report have drawn their own conclusions based partly on the scenario storylines but also on their own imagination of how the logic behind the scenarios could have played out.

The logic of the scenarios was developed in Chapter 5. The key question addressed in the MA scenarios is:

- What are the consequences of plausible changes in development paths for ecosystems and their services over the next 50 years and what will be the consequences of those changes for human well-being?

  Four more specific questions were also considered:

- What are the consequences for ecosystem services and human well-being of strategies that emphasize economic policy reform (reducing subsidies and internalizing externalities) as the primary means of management?
- What are the consequences for ecosystem services and human well-being of strategies that emphasize local and regional safety and protection and that give far less emphasis to cross-border and global issues?

- What are the consequences for ecosystem services and human well-being of strategies that emphasize the development and use of technologies allowing greater eco-efficiency and adaptive control?
- What are the consequences for ecosystem services and human well-being of strategies that emphasize adaptive management and local learning about the consequences of management interventions for ecosystem services?

Additional questions for comparing the scenarios include:

- What are the most robust findings concerning changes in ecosystem services and human well-being across all four scenarios?
- What are critical uncertainties that we are confident will have a big impact on ecosystem services and human well-being?
- What are gaps in our understanding that we can identify right now that will affect our ability to model ecosystem services and human well-being?
- What opportunities exist for managing ecosystem services and human well-being?
- What is surprising in these results?

The four scenarios differed with respect to most of the direct and indirect drivers that are part of the Millennium Ecosystem Assessment framework. (See Chapter 5.) Two critical uncertainties emerged as primary issues in the discussions of the Scenarios Working Group. (See Figure 8.1 in Appendix A and Table 8.1.)

One of the working group's main foci was on the potential impact of ecological feedbacks on human well-being in the future. Key issues with respect to the emergence and management of these feedbacks include:

- whether governments and other decision-makers consider ecosystems to be fragile or robust to human impacts,
- whether they see ecosystems as primary underpinners of value to humans or of secondary importance after economic and social issues, and
- whether they therefore manage ecosystems proactively to avoid undesirable ecological feedbacks or reactively in the belief that problems can be fixed after they become apparent.

Thus, two of the scenarios were developed around proactive environmental management policies and two around reactive policies.

In TechnoGarden, proactive policies arise due to recognition of the economic value of ecosystem services. In Adapting Mosaic, they emerge from a strong recognition of the broader value of ecosystem services in underpinning human life and human well-being and the need to work with rather than against nature. In Global Orchestration, belief in the ability of humans to find technological approaches to repair or replace lost ecosystem functions is high, and ecosystems are considered to be robust to the impacts of humans. In Order from Strength, an inward focus on national security and economic growth by individual wealthy countries, together with a belief that ecosystems are robust and that reservation of representative ecosystems is enough to keep future options open, means that ecosystems are considered only after more pressing economic and social issues. In poorer countries, the struggle of people to survive economically and physically, combined with poor understanding of the relationships between human well-being and ecosystem health, makes conservation of ecosystems a low priority and exploitation an apparent necessity.

Our combined experience led us to identify the degree and scale of connectedness among and within institutions as the other major driver of how ecosystems will be managed in the future and what ecosystem services outcomes are possible. Two broad future trends seemed to be plausible from past and emerging present-day trends. A continuation and escalation of the present trend toward globalization and connectedness across country borders, with associated reductions in trade barriers and barriers to movement of people, were seen as plausible in the future. In Global Orchestration there is confidence that the right global policies will achieve economic equity among all countries and that environmental impacts will take care of themselves as prosperity grows. In TechnoGarden there is enthusiastic development of environmental technologies that are adopted

Table 8.1. Defining Characteristics of the Four Scenarios

Scenario Name	Dominant Approach for Sustainability	Economic Approach	Social Policy Foci	Dominant Social Organizations
Global Orchestration	sustainable development; economic growth; public goods	fair trade (reduction of tariff boundaries), with enhancement of global public goods	improve world; global public health; global education	transnational companies; global NGO and multilateral organizations
Order from Strength	reserves; parks; national-level policies; conservation	regional trade blocs; mercantilism	security and protection	multinational companies
Adapting Mosaic	local-regional co-management; common-property institutions	integration of local rules regulate trade; local nonmarket rights	local communities linked to global communities; local equity important	cooperatives, global organizations
TechnoGarden	green technology; eco-efficiency; tradable ecological property rights	global reduction of tariff boundaries; fairly free movement of goods, capital, and people; global markets in ecological property	technical expertise valued; follow opportunity; competition; openness	transnational professional associations; NGOs

by global companies and shared across international boundaries.

Alternatively, it was considered that a less connected, regional focus could emerge in two ways. One way could be by a continuation of the emerging trend toward decentralized decision-making, especially with respect to the environment. A failure or partial failure of global systems could see regional processes emerge as the major routes for environmental decision-making (Adapting Mosaic). The other way could be the escalation of concern about national security, filtering down to actions by wealthier parts of society within countries to protect their security and defend their access to scarce resources (Order from Strength).

Subsequently, similar uncertainties about global and local institutions emerged from some sub-global assessments. (See Box 8.1.)

8.2. Linkages with Chapter 9 on Quantitative Modeling

Readers may notice some inconsistencies between this chapter and Chapter 9. The quantitative models in the other chapter are based primarily on assumptions for which there is good published evidence. Three of the assumptions that have strong influence on both the models and our thinking in the storylines are about trends in population and income and the sizes of regional economies. Figures 8.2 and 8.3 illustrate just how large the changes from today are likely to be in all four scenarios, primarily due to the inevitable effects of population growth. Other important elements in the quantified scenarios include relationships between income growth and demand for goods and provisioning ecosystem services such as food, energy, and timber. The consequences of increases in the consumption of such goods and services in terms of global environmental change is relatively well explored in these models. In contrast, the models are less developed in dealing with ecological processes that occur at the local scale and in dealing with the impacts of ecological changes on human development and human behavior. Regulating services are generally less well covered by the global models, and supporting and cultural services are not able to be addressed at all in the models.

The storylines differ from the models in that the storylines can explore some plausible changes for which no models or little data currently exist. These largely have to do with people's attitudes and the ways in which they adapt to challenges. The biggest differences between the storylines and the models occur for Adapting Mosaic because the models inadequately address ecosystem feedbacks, which are a key factor in people's approach to management in this scenario. We think people in this scenario would be successful in developing cooperative approaches to overcome many of the challenges that the models suggest they would be faced with (such as declining production yields, falling food availability, and high food prices). Taking the models together with the storylines, we can see both the optimistic future in which humans adapt appropriately and the less

BOX 8.1

Issues Arising from the Southern African Millennium Ecosystem Assessment

In terms of standard indicators of human well-being, southern Africa has some of the poorest conditions to be found on the globe. Many of these conditions are related to the services provided by ecosystems. Life expectancy in sub-Saharan Africa has declined from 50 to 47 years of age since 1990, mainly due to the high infant mortality rate. HIV/AIDS is the leading cause of death; malaria and tuberculosis also present serious health problems. Lack of clean, affordable energy sources increases susceptibility to illness and malnutrition, and contributes to the poor domestic air quality experienced in many African cities. Cyclical droughts, political instability, and ongoing conflict in parts of the region disrupt food production systems and have displaced large numbers of people, increasing pressure on resources in asylum areas.

The Southern African Millennium Ecosystem Assessment consisted of five component studies at three nested spatial scales (a regional, two mega-basin, and two sets of local community assessments) in mainland Africa south of the equator. (See the MA *Multiscale Assessments* volume.) Each component of SAfMA, in consultation with its specific stakeholders, developed its own set of scenarios as a framework for considering the future of ecosystem services in the particular study area.

Governance emerged as a key uncertainty in scenarios developed at all scales of SAfMA. Uncertainties regarding economic growth, the type of policies pursued, and the effectiveness of policy implementation were all linked to governance. One important uncertainty of policy implementation was whether policies enhanced social equality and distribution of wealth, or whether wealth remained concentrated among the elite and powerful. Also related to this was the uncertainty regarding the degree of decentralization in governance and the strength of civil society.

Most scenarios indicate that the conditions in southern Africa remain stable or worsen over the next three decades; only under a limited set of circumstances is a significant improvement anticipated. It is expected that biodiversity, freshwater quality and quantity, biomass fuel, and air quality will decline under both scenarios of strong and poor central governance, although the degree of decline as well as the underlying processes will differ between the two. Food production is expected to remain stable or decline slightly under conditions of poor governance, whereas improved governance is anticipated to result in a strong improvement in food security. Mapped onto the MA global scenarios, most global-scale scenarios translate into similar outcomes for southern Africa, probably due to its marginal status and relative disconnect from many global socioeconomic processes.

The SAfMA local-scale scenarios highlight that general trends in ecosystem services at the regional and basin-scale may be reversed in particular local situations. The multiscale structure of the SAfMA also showed that certain responses or developments at larger scales are experienced as surprises or shocks at local scales—such as mega-parks and large irrigation schemes implemented without adequate local stakeholder participation and consideration of impacts. The Gorongosa-Marromeu scenarios, in particular, emphasized the need for good operational and transparent governance structures at all scales to ensure the equitable distribution of wealth and use of natural resources, and thus to also ensure longer-term sustainability.

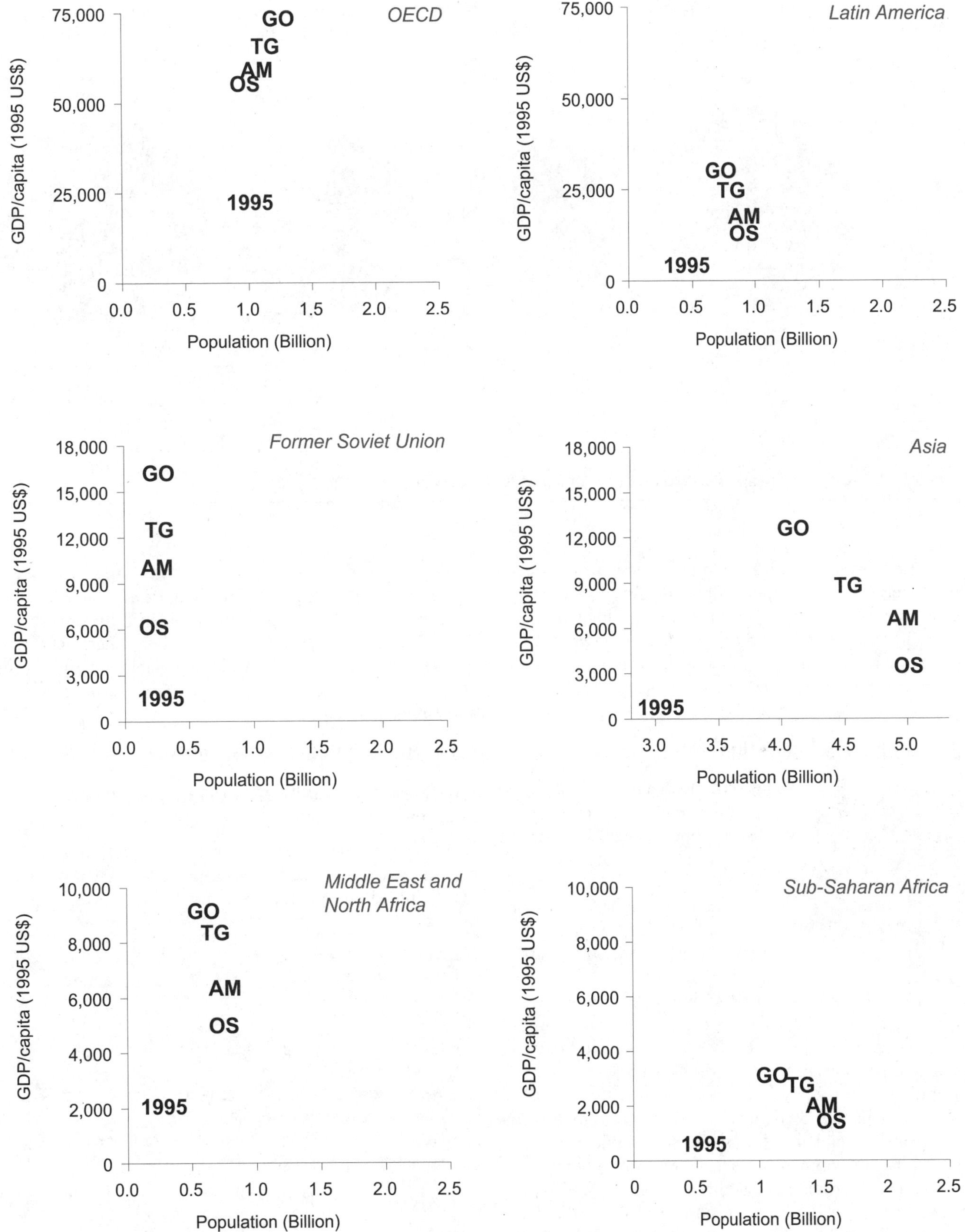

Figure 8.2. Projected Changes in Population and Income in MA Scenarios in 1995 and 2050. The number 1995 indicates the 1995 data while the projected data for 2050 are indicated by the code for each scenario. GO: Global Orchestration, TG: TechnoGarden, AM: Adapting Mosaic, and OS: Order from Strength. (Data from Chapter 7.)

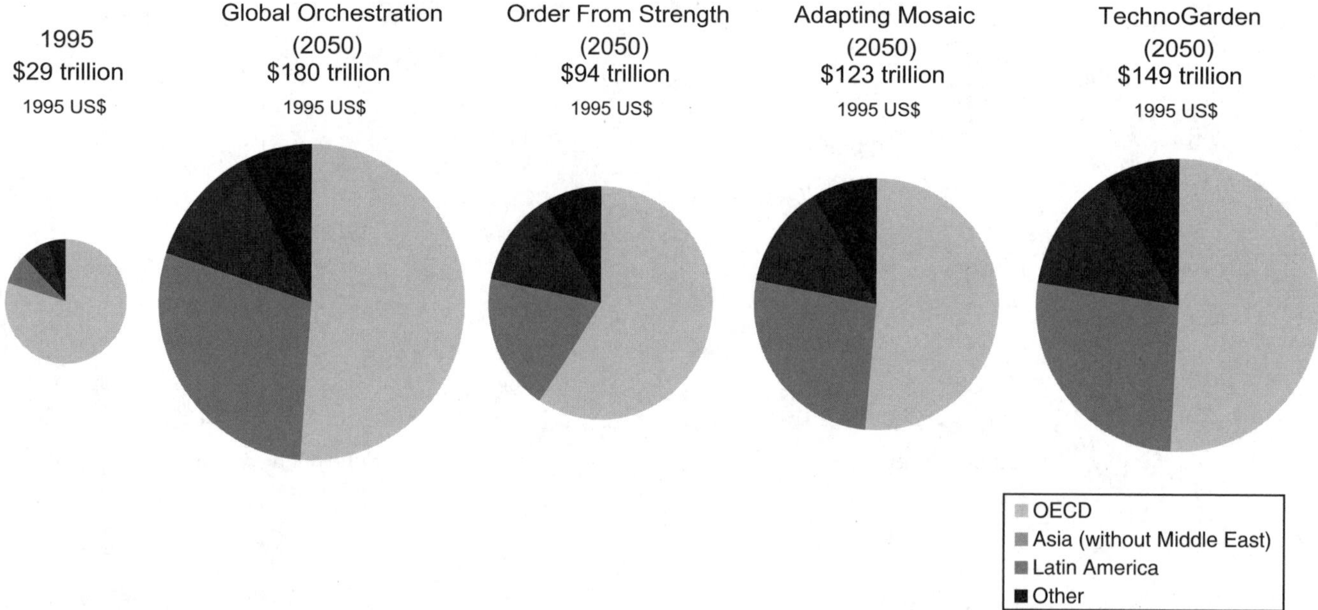

Figure 8.3. Projected Changes in Relative Size of Economies in MA Scenarios (Data from Chapter 7.)

optimistic view in which responses are less pronounced. The models do, however, contain some assumptions about human responses (see Chapter 5), some of which are optimistic and others not so optimistic. We have inserted footnotes in the relevant places in each scenario to highlight and explain the major differences between chapters.

8.3. Global Orchestration

This scenario is about global cooperation not only to improve the social and economic well-being of all people but also to protect and enhance global public goods and services (such as public education, health, and infrastructure). There is a focus on the individual rather than the state, inclusion of all impacts of development in markets (internalization of externalities), and use of regulation only where appropriate. Trade liberalization and free markets are key ingredients, but decision-makers in this scenario have gone beyond philosophies such as the Washington Consensus.[1]

People generally recognize that the environment provides a suite of global public goods and services, but the approach to environmental management is largely reactive. Environmental problems that threaten human well-being are dealt with after they become apparent. Those that have little apparent or direct impact on human-well being are given a low priority in favor of policies that directly improve well-being. People are generally confident that the necessary knowledge and technology to address environmental challenges will emerge or can be developed as needed, just as it always has in the past. The scenario highlights the risk that this lack of proactivity in addressing environmental concerns can lead to increased risks from ecological surprises, particularly emerging infectious diseases and other slowly emerging problems that are hard to control once they are established. Other benefits and risks (see Box 8.2) emerge from the inevitable and increasing connections among people and nations at social, economic, and environmental scales.

8.3.1 Global Orchestration 2000–15

8.3.1.1 The Legacy of the Twentieth Century: Tensions, Inequalities, and Concern for the Environment

At the beginning of the twenty-first century, poverty and inequality, together with environmental degradation, were being singled out by more and more concerned individuals

> **BOX 8.2**
> **Potential Benefits and Risks of Global Orchestration**
>
> Potential benefits:
>
> - Economic prosperity and increased equality due to more efficient global markets
> - Wealth increases the demand for a better environment and the capacity to create such an environment
> - Increased global coordination (such as markets, transport, fisheries, movement of pests and weeds, and health)
>
> Risks:
>
> - Progress on global environmental problems may be insufficient to sustain local and regional ecosystem services
> - Breakdowns of ecosystem services create inequality (disproportionate impacts on the poor)
> - Reactive management may be more costly than preventative or proactive management

and groups as some of the world's most pressing problems, although they were not always at the top of the agendas of global and national decision-makers. Social tensions were arising from inequalities in wealth, civil liberties, and access to food, water, other basic human needs, and global markets. Trust in vested interests was declining as governments and big business in all parts of the world were exposed for withholding and manipulating information about finances, people, and security issues.

These tensions were seen as the underlying causes of growing civil unrest, uncontrolled migration, and conflicts. Terrorism was on the minds of people in richer countries after an unprecedented attack on New York in 2001, although deaths of far more people from conflict, disease, poor nutrition, and inadequate sanitation were regular events in many poorer countries. While some saw terrorism as an evil that needed to be stamped out with force, others called for greater equality of access to the fundamentals of human well-being as the way to remove terrorism by removing the underlying causes.

There were increasingly vocal calls for the world's leaders to accept the responsibilities as well as the benefits of globalization.

8.3.1.2 Global Connectedness

Attempts by global aid and development agencies to reduce the number of people with low well-being led to a more connected world, as open markets joined countries and peoples together through trade. Borders—real and virtual—were under threat. Electronic communication allowed ideas, information, and misinformation to be rapidly propagated and shared. A global generation of teenagers shared thoughts and beliefs electronically. They had heightened concerns for the environment but also a yearning for the latest in consumer products and fashions. No force could hold back the movement of ideas and merchandise across borders. The power of global brands equaled that of governments when it came to manipulating the values and viewpoints of this generation. Music was bought and sold illegally with seeming impunity via worldwide computer networks, and independent television channels broadcast in competition with major international networks during conflicts.

While movies, music, tourism, and printed media promoted elements of the culture of the most industrialized countries globally, the influence of poorer countries was increasing through growing investments in technological innovation and their increasing impact on demand as consumers.

Airline travel was affordable to middle-income people in many parts of the world. People from all walks of life traveled frequently between east and west, north and south, around the globe. This not only made global business possible but also allowed regular face-to-face meetings between politicians and public servants of different countries. This high level of connectedness meant that, more than ever, most issues were global, as solutions applied to social, economic, or environmental problems in one country or region had major social and economic impacts on other countries and regions.

There were some counter-trends to this early globalization,[2] but some early successes in dealing reactively with these challenges through cooperation and the application of social or technological solutions increased the global majority's faith in globalization. These successes increased confidence, at least among the dominant countries, that money, technology, and human ingenuity could fix any problem that arose and that the benefits of a global community far outweighed the costs.

8.3.1.3 Focus on Social and Economic Policy

At the beginning of the twenty-first century, there was a wide variety of views about how to solve the world's problems. Since the mid-twentieth century there had been forecasts by some ecologists that natural resources were declining to critical limits.[3] The concept of resilience (see Box 8.3) had emerged to explain the way in which the complex interactions among species in ecological systems allow those systems to absorb perturbations and keep functioning. At the turn of the century, a major multi-partner multidisciplinary appraisal of the world's ecosystems, the Millennium Ecosystem Assessment, made use of the popular concept of ecosystem services as a focus for its assessments of how trends in ecosystem function might affect human welfare into the twenty-first century.

This assessment gave examples of emerging ecological disasters,[4] but many believed that smart policies and technological solutions could fix these problems as they arose, and pointed to examples of how this had been done in the past. This was a time of strong confidence in people's ability to take charge of their own destiny. The predominant attitude among wealthy and powerful countries, the development community, and international organizations was that a rational application of knowledge about economic and social

BOX 8.3

Ideas about the Resilience of Ecosystems and Society
(Adapted from the Resilience Alliance at www.resalliance.org)

- The systems that include people interacting with the environment rarely remain in a single, unchanging state. Shifting among alternative, distinctly different states (over various time periods, depending on the system) is the norm, not the exception.
- Surprises in these systems are created in cycles of long phases of increasing growth, efficiency, and predictability followed suddenly by brief phases of reorganization.
- Resilience is the capacity of a system to absorb disturbance, undergo change, and still retain essentially the same function and structure.
- Variability and flexibility are needed to maintain the resilience of nature and people. Attempting to stabilize such systems in some perceived optimal state (the command-and-control approach to management), whether for conservation or production, reduces resilience and often achieves the opposite result of driving a system toward rapid change.

processes would not only achieve economic growth but would also make sure the benefits flowed throughout societies. The emphasis of these reforms was on creating markets that would allow equitable participation and provide open access to goods and services. Transparent governance systems, like participatory democracies, were seen as the necessary foundations to economic growth.[5]

Occasional failures, though, brought up questions about how resilient ecosystems really are and how to balance this uncertainty with the lost economic and employment opportunities that could arise if the environment were protected more than it needed to be. At the time, these questions could not be resolved because the information was not available for most ecosystems. As time went on, investment in obtaining this sort of information waned because no major ecological disasters were apparent and it appeared that ecosystems could withstand anything that people did to them (subsequently it was found that this was often due more to luck than good management).[6]

It was becoming clear that the dream of a better world built on a foundation of reformed social and economic policy would live or die depending on how several major challenges were addressed. (See Box 8.4.) Wealthy countries, and especially the large corporations based in them, were optimistic that improving the flow of financial capital to poorer countries would lead to beneficial social reform and would encourage both demand for sustainable management of the environment and the development of appropriate institutions to achieve sustainable economic growth. But many people in poorer countries were concerned about an overly rapid inflow of financial capital before formal and informal institutions were ready to cope. This, they feared, would undermine maintenance or enforcement of environmental protection, encourage inflow of materials and practices harmful to the environment and people, and lead to overexploitation of natural resources to capitalize on new opportunities and to pay for debts incurred when foreign investments go wrong.

The hopes for resolving these issues rested on various multinational negotiations to seek agreement on rules of trade and environmental management. A measure of global agreement about the need to work together to address issues of environmental and human welfare had emerged from the Rio and Johannesburg Summits in 1992 and 2002.

BOX 8.4

Branch Points for a Globally Orchestrated World

This scenario is moderately optimistic about a globally orchestrated world. This reflects the belief of several members of the Working Group in what reform to global social and economic policy can achieve. Others find it easier to imagine disastrous outcomes from this scenario. The nature and quality of outcomes depends on a number of critical challenges being overcome. These can be likened to the world taking the right paths or branches in a journey by road or river.

Critical branch points include:

- Whether globalization and trade liberalization are accompanied by strategies for developing appropriate institutional arrangements in poorer countries and for avoiding undesirable activities by richer countries and corporations
- Whether global cooperation, including globally consistent standards for health, business, and intergovernmental relations, is developed without losing or threatening cultural diversity, including culturally appropriate approaches to democracy
- Whether ecosystems reach critical thresholds resulting in adverse changes that happen too fast or too extensively for reactive remediation to be effective
- Whether dominance of intellectual property by a few countries is substantially reduced—if not, global cooperation is much less likely to be achieved or maintained
- Whether ways are found to reduce the heavy debts carried by poorer countries

8.3.1.4 Promising Developments

A number of small but hopeful examples of successful international cooperation in Asia, Africa, Europe, and the Middle East in the years either side of 2010 built confidence in the connected-cooperative model. This confidence tipped the balance of world opinion toward global cooperation. The United Nations, with renewed confidence, reinvented itself as an organization primarily focused on promoting social and economic equity and relinquished its role as an international peacemaker and policing body.

Quietly, China addressed its long-standing environmental problems, including dismantling of unsustainable and polluting industries.[7] It took the first major steps to address desertification and began to develop agricultural and other industries in its western provinces based on ecological principles. Its economy continued to grow, accompanied by social reforms that developed a peculiarly Chinese form of participatory democracy,[8] which took until late into the 2020s to really become entrenched.

India continued its economic growth, which started toward the end of the previous century and would not plateau until the 2020s. This growth was largely due to innovation in communication and information technologies and services.

In Africa, the widespread unrest and instability of the early 2000s continued until around 2010. By this time, the growing prosperity of a few nations allowed them to make virtuous investments to assist their neighbors. These initiatives spread slowly through Africa, as national leaders united to develop cooperative policies for dealing with disease and poverty and to strengthen the continent's trading position globally. Despotic leaders were encouraged to stand down, and participatory democracy began to develop in many countries throughout the 2010s. Some African cities became centers for innovation in digital technologies.

Changes in these three places encouraged governments and private investors in rich countries to increase their investments in improving the wealth and well-being of poorer countries.

In Europe, Russia became more accepted as part of the European community, and the threat of conflict in the Balkans decreased as all nations worked hard at maintaining cooperative and friendly relations.

8.3.1.5 Environmental Issues Overlooked

With the strong focus on economic, social, and political issues throughout the 2000s and 2010s, thinking and research into the dynamics of socioecological systems proceeded only slowly. Things were going well for many people. Ecologists and environmental activists were still saying that society existed at a dangerous edge, but they had been saying this for a long time and, so far, most things had gotten better. It gradually became easier and easier to ignore their calls for stricter controls and more caution with respect to the environment.

Although the mantra of the 2000s was "triple bottom line," the environment was considered to be secondary to, and reliant on, economic and social issues. By 2015 it was becoming clear that information on the meaning and state of sustainable development was lacking and that public understanding of the underlying issues had been allowed to remain at a very low level. It had been assumed that increased income and access to the consumer goods of the day would improve the confidence and happiness of the public, but people were beginning to suspect that this was only true up to a point. As people's material well-being improved, cultural issues like identity, spiritual connection with the land, and harmony with other life-forms began to dwell more and more on their minds. Additionally, while people were able to reverse some environmental problems when they became wealthier—like local air pollution caused by factories[9]—there were other environmental problems, such as toxic waste, that did not appear to be easily fixable. The best they could hope for was to become wealthy enough to move away from the waste.

Thus the decade 2010–20 saw renewed efforts to address the environment directly, but progress was slow because information about how ecosystems worked was scarce.

8.3.2 Global Orchestration 2015–30

8.3.2.1 Improving Globalization by Including More Cultures

Starting in the late 2010s and continuing into the 2020s, a slow but vital process of change in world thinking happened. Up until then, increasing connectedness of the world had seen western culture, language, and thinking permeating other cultures, with little return of information in the other direction. The belief that western values and styles of governance were universally appropriate and beneficial had grown stronger.

There were, however, growing numbers of people voicing concern about global homogenization of culture and values. The improving economies of Asian, African, and Central European countries brought increased confidence in questioning the desirability of universal styles of governance and beliefs.

By 2012, these trends and concerns led to a flurry of international meetings to contemplate better ways of global governance. Models were sought from times in the past when economic prosperity was achieved while diverse cultures coexisted.[10] Using these, people recognized that everyone requires cultural identity and security as well as wealth and that cultural identity can be supplied in diverse ways throughout a country's development.

This was no sudden change in world values. The lessons were learned slowly and sometimes painfully. It was eventually realized, however, that global markets and the continued economic growth of both rich and poor nations could only be achieved by allowing the return to diverse regimes and values but within globally agreed rules of engagement between countries.

From then into the 2020s, there were changes in many cultures. For example, the western business community adopted a more Asian approach to business, with relationships, respect, cooperation, and trust becoming more prominent than they were in the highly competitive first decade of the century. Countries with large proportions of populations living in high-density urban or rural settings saw big changes too, as global institutions demanded uniformity of community health standards, business practices, and governance for reasons of health, security, and distribution and management of foods, building materials, and fuels.

A major unstoppable force that encouraged nations to work together was technology. Since the first transatlantic telegraphic cables were laid in the second half of the nineteenth century, nations had seen the benefits of sharing technologies. One country doing something in a unique way could find itself isolated from the rest of the world and might lose out on trade or other benefits of globalization. Technological developments, especially with respect to agriculture,[11] food production, and energy generation, became more and more rapid as the twenty-first century unfolded.[12] Many of these arose from Asian, South American, and African countries, which had been investing in innovation for two decades or more and were less constrained by existing technologies than countries that had been developed for longer. As these technologies became used widely, it was critical for countries to work together to form accepted ways of using them. No one wanted to be left behind.

8.3.2.2 Global Health Concerns

Increases in wealth, and in the availability of technology, had led to improvement in the health of people in many poorer countries. However, many problems still remained. Obesity-related diseases remained a threat, particularly in rapidly developing areas as new food choices became available and societies shifted their eating habits to less healthy styles. Emerging infectious diseases were also a risk. Several times, new pathogens arose in parts of the world where ecosystems were suddenly exposed to massive human impacts. Decline in the natural processes for regulating animal populations, together with greater visitation and exploitation of wildlands by humans, caused humans to be a frequent host for these new pathogens. In some cases, deaths from these new diseases were limited to a few hundred people, and the spread was contained around the source by international cooperative action. The mechanisms for this action were progressively refined from the impressive responses to SARS in 2003. But in each case, industries related to international tourism suffered economic downturns as the scare of spread via travelers and transportation systems swept the world. Increasingly, there were concerns that social progress was being delayed by environmental decline feeding back on human well-being.

The international health community was concerned about the emergence of new pathogens due to close human contact with animals, as occurred with SARS in China in 2003. Impressive international procedures were put in place to prevent, detect, and cope with outbreaks of this sort, led by the United Nations in its redefined role as a human welfare agency. Risk of such outbreaks remained high due to high global mobility of people and goods and varying standards of health monitoring and sanitation. Some attempts to promote (or impose) uniform approaches and standards met with resistance and resentment when they threatened the rural-, village-, and family-focused cultures in poorer countries, especially Asia and Africa. Countries had long been exposed to progressive acceptance of trade and migration policies emphasizing global cooperation, however, and they often agreed to these procedures to minimize disease risks, especially after seeing the problems caused by outbreaks of disease.

High global mobility was also a concern because it threatened to move exotic species around the world. The possibility of spreading crop and livestock pathogens was of major concern because enthusiasm for opening borders and removing trade barriers outstripped research and monitoring.[13] Increasingly, the role of maintaining functional ecosystems that regulate pest populations was being recognized, but the costs of returning ecosystems to functionality were now very high.

8.3.2.3 Removal of Some Lingering Barriers to Global Equity

Several other barriers to increased equity among nations were addressed around this time. There had been hope that high debt loads carried by developing countries in the 2000s would be reduced by 2015 with the new trade policies and loan restructuring. But setbacks due to the costs of implementing new health security policies gave impetus to calls for writing off a proportion of these debts. These calls had previously remained subdued due to the hope that improved access to international markets would help poor countries pay off their debts. The ever more apparent impacts of global climate change increased pressure on countries as well as corporations to take remedial action. Between 2015 and 2020, around 50% of the debts of the developing world were written off in exchange for carbon credits.

8.3.2.4 Pluses and Minuses of the New Trade Order

The benefits of open trade became apparent early, as countries providing temperate-zone agricultural products, such as certain cereals, meat, and milk, realized the economic benefit of better market access and less distorted market prices. As new market opportunities emerged, economic growth improved in some countries of the Americas and in a number of Southern and East African countries. Agriculture in these countries became increasingly intensified and simplified. The economies of scale were initially very beneficial to economic growth in these countries. With a mixture of international competitiveness and cooperation, the areas under agriculture expanded in some countries and receded in others.

Developments with respect to trade were not all rosy, however. Countries that relied mainly on export of tropical crops such as coffee and cocoa did not necessarily get any benefit from improved trade opportunities. Economic growth in this group of countries depended on additional opportunities for diversification of the economic base. In many cases, these opportunities were held back by slow development of policies aimed at improving human capital and infrastructure.

8.3.2.5 A Moderate Greening of Attitudes

As prosperity among the global middle class increased, so did demand for cleaner cities, less pollution, and a more beautiful urban environment. In most countries, this led to efforts to maintain green spaces and protect representative areas of natural environment for people to visit. Pollution in and around cities was addressed with vigor. Before the need for communication towers disappeared, it became global best practice to disguise these towers as artificial trees. Some children grew up to think pine trees really are 200 meters tall. Amusing as it sounds, this was symptomatic of a general lack of appreciation of the need to understand the processes that maintain natural systems. We now realize, however, that longer-term ecological challenges that were harder to address were more or less ignored by the general population because so many other things were going well. Examples included the slow rundown in fertility of agricultural soils, decline in natural controls on pests and diseases, and the slow loss of wildlands and their biodiversity.

The nongovernmental organizations that played such a strong role in raising human consciousness about the environment during last century had waned in their influence by this time, as public confidence in the ability of governments to solve the world's problems increased. The groups still existed, but they focused more on human well-being issues and addressing immediately obvious environmental challenges.

8.3.2.6 Agriculture: Expanded Area But Narrower Diversity and Ecological Basis

Driven by policies aimed at increasing GDP and human well-being in poorer countries, human impacts on terrestrial ecosystems increased as the total area under agriculture expanded.[14] The world was focused on generating employment and feeding a growing population. Since the turn of the century, many had argued that hunger was a problem related to equitable distribution rather than an absolute shortage of food. Distribution and equity issues were steadily addressed, but increasing the area under agriculture was the faster, and therefore the preferred, option among most communities in the meantime.

Since late in the twentieth century, large transnational companies had been steadily increasing their control over agricultural crops and livestock through the development of new genetic strains with improved performance. Economies of scale caused fewer and fewer varieties to be distributed over larger and larger areas, leading to agricultural specialization and simplification. In both the richer and the poorer worlds, very little of the agricultural land was not consolidated into large, highly mechanized industrial farms. There were still people practicing low-intensity farming, either as a lifestyle choice or in areas of marginal land quality and economic hardship, but they contributed little to food production or the countries' economies.

While these trends initially brought economic efficiency and growth to many countries, others suffered because their local conditions did not suit the mass-distributed varieties and their market was not large enough to encourage the industries to develop locally appropriate varieties. Many commentators at the time were concerned about loss of local ecological knowledge. The prevailing policies for dealing with natural genetic diversity and other natural assets like vegetation and soil systems and their associated fauna was to preserve or reserve representative examples in parks and museums. A positive legacy of these policies was the establishment of major gene banks containing the vast majority of wild varieties of crops previously used by humans.

8.3.3 Global Orchestration 2030–50

8.3.3.1 Broken Stranglehold on Intellectual Property

By 2025, a major international tension that had been growing since the early 2000s came to a head. Had this not been resolved, the world could not have continued down a path of increasing cooperation.

In the last decades of the twentieth century and into the 2000s, a complex web of laws and regulations relating to intellectual property had developed worldwide. This had many impacts on attitudes toward and management of the environment. For example, it led to the dominance of the market for seeds and stock varieties by large multinational companies that could impose rules and regulations that often discouraged the use of locally sustainable varieties and practices. Genetic diversity declined, and wild varieties largely existed in museums. Dominance of intellectual property was maintained by formidable networks for commercial intelligence, and wealthy nations were more able to attract the best minds from poorer countries. Furthermore, the culturally diverse nations of Asia, Africa, the Middle East, and even, to a degree, Europe could not speak at this time with a strong and uniform voice.

Things were changing, however. The growing unity within Europe, Asia, South America, and Africa by the mid-2020s, together with the increasing impacts of developing nations as innovators and participants as consumers in global markets, forced a review of patent and intellectual property institutions.[15] (See Box 8.5.) While the changes were not wholesale, they were enough to allow local pharmaceutical, genetic, and nanotechnological industries to establish with a focus on regionally differentiated markets. In small but significant ways, the brain drain started to reverse. These trends were gradual but consistent starting in the second quarter of the twenty-first century.

8.3.3.2 Benefits and Risks of Global Orchestration

By 2030, most elements of the strategies for global economic and social reform were in place. Many countries and individuals had realized enormous benefits from globalization. Impacts on ecosystem services were variable but mostly either neutral or improved in poorer countries.

But big challenges have also emerged by 2050. During the last two decades the balance of benefits and costs of the globally orchestrated policies of this half-century has become clearer. Most countries have prospered under the economic and policy plans put in place in the mid- to late-2000s. Overall, advocates of the principle that economic growth produces improvement in human well-being feel themselves vindicated by the state of the world in 2050. Their opponents suggest that economic growth was only achieved without major collapses in Earth's life-support systems because massive efforts were put into reducing environmental impacts as problems arose. The costs have been major in all nations.

The unprecedented enthusiasm for global cooperation among governments, NGOs, and companies was spurred on by early successes in dealing with conflicts, diseases, and global equity issues in the early part of the century. This helped the world cope with a number of important social and environmental issues that economic growth alone could not have dealt with. Limiting population growth to a maximum of 8 billion by 2050, which was largely a consequence of economic development in the previously poor world, also helped to limit environmental impacts.

In our globally orchestrated world, we have seen great technological progress within the energy sector, which has provided low-cost energy for all people with a high level of reliability. But early global complacency about climate policy forced us to adapt to many problems caused by climate change, even though the institutions for global cooperative action could have addressed the problem earlier.[16] In hindsight, we realize that the slow response to climate change was a major missed opportunity.

Increasing prosperity brought increased demand for meat in people's diets, which in turn led to growing demand for food and feed and a rapid expansion of crop area in all regions. Particularly hard hit was sub-Saharan Africa (see Box 8.6), which saw 50% of its forests disappear between 2000 and 2050. In the early 2000s doctors in the industrial world were concerned about an obesity epidemic. Steps were taken to encourage greater levels of activity and low levels of fat consumption. This was outweighed glob-

BOX 8.5

A Story of India (Global Orchestration)

As I wait for the Australian to arrive, my thoughts turn to the changes I have seen in my lifetime. India is one of the leading economies in the world of 2050. Americans, Brits, Germans, Russians, South Americans, Africans, and more are looking to build partnerships with us and learn from our successes.

In my mother's youth, the world's teenagers yearned for American clothes, music, appliances, movies, and the like. India, of course, always resisted the global culture. Our attitudes to life and death, wealth and poverty, were never well understood by westerners. Whereas we look at those worse off and feel thankful, Westerners look at those better off and feel envious. True, we have been touched a little by the global culture of envy—at least according to my aunt. The huge popularity of Indian literature and movies early this century was just one sign of our ascendancy. We invested in business and technological innovation, starting with call centers last century, which gave us the confidence to innovate in many other fields. Now India, China, Brazil, and a handful of other growing economies are influencing global consumerism and global culture. In Paris, every second young girl I saw was wearing *salwar kamis* (Indian clothing).

As a woman, I am privileged to have received a good education. When I was born in 2010 the girls in poor rural towns like ours didn't go to school—it was too expensive and they had women's work to do. But by the time I was old enough to attend there were places in schools for girls whose families wanted them to go. Education of girls was starting to be seen as an investment in a "more prosperous future"—the slogan of the government change-programs when I was growing up. In many places the new philosophy has worked. Of course the young people, like me, usually had to move to the cities to get jobs, but at least the jobs were there.

"Rural life" has changed so much. When I go home to visit my brother, all I see is field after field of rice, sometimes a bit of maize, but that is all. Boring. The harvests are bigger and better some years, but they still fail regularly. The changing weather has brought bigger extremes of hot and cold, wet and dry, and storms that wash away our topsoil. Locally, pests and diseases break out from time to time, and no one seems to be expecting them. Our government played a major role in international efforts to address existing global issues like climate change and international fisheries. But at home it expected economic prosperity and good education to be enough to fix environmental problems. The government was right sometimes, but when it was wrong it hurt many people. My cousin drowned in the big floods of 2015, along with 200 others from our district.

The company I work for now was started in 2012 by two young Indians who trained in the United States and returned home as new opportunities arose here. Southern India was booming, and at one time it even seemed like it might break away from the north. But the two Indias stayed together, in the spirit of cooperation that was spreading the world. The company survived that first decade as trade arrangements and intellectual property regulations were sorted out in sometimes strained negotiations between established and emerging powers. I joined them first as a trainee in 2028 and then as a graduate after 2032. I didn't stay long, changing jobs 10 times and retraining with new skills four times before returning as a senior manager in 2043.

So what do I tell the Australian about Indian attitudes and tastes today? In the cities we want cleaner streets and more green areas, but the environmental researchers tell us we still don't understand the essential role of nature in our lives and economy. My company is seen as a leader in socially responsible production. We've minimized our environmental damage, too. But now the government is looking for industry leaders to help develop an environmental repair program as the costs of complacency over the past decades are biting back. The Australian's company is in the same situation in his country. Environmental repair looks like it will be the next big growth area.

It is good to be on a new wave again. Still, I can't help wondering why we didn't see this one coming sooner.

ally by the spread of obesity-related illness through the now rapidly developing countries of Asia and Africa.

Increased irrigation and intensive inputs of fertilizers and chemicals for control of agricultural pests, combined with low levels of environmental protection, led to compounding cross-scale interactions among outbreaks of resistant pests, groundwater contamination, soil degradation, and accumulation of nutrients to toxic levels for crops and water sources for humans.[17]

Between 2030 and 2050, many of the world's fisheries collapsed. While there was at least superficial global cooperation in managing species, open borders and reduced trade barriers led to insurmountable obstacles to effective monitoring of stocks. In turn, there was uncontrollable exploitation and overfishing of many stocks. Because the dynamics of fisheries are complex, problems were often not noticed until after they were so severe that they could not be fixed.

The reactive nature of environmental management resulted in a number of cases of regional ecological degradation that were difficult, costly, or even impossible to recover from. Marine ecosystems and coastal wetlands were affected most strongly, as urban growth in virtually all countries with coastlines and moderate climates was concentrated around river mouths and along a 100-kilometer band off the coastline. Tourism and fishing in the Caribbean, for example, both declined after the loss of many coral reefs. In Australia and Africa, the tourism industry took heavy hits due to almost complete loss of coral reef ecosystems. During the 2010s, aquaculture industries worldwide were threatened by pollution and sediment run-off from increasingly urbanized coastlines in virtually all countries. Technological solutions were developed and implemented in the 2020s, but the industries did not recover until the early 2030s. And the global impacts of invasive species increased the costs of timber and agriculture and decreased people's well-being.

8.3.4 Challenges for 2050–2100

In reviewing the years 2000–50, we are struck by how far the world has come, how much has changed, and yet how much stays the same. In 2000, debate was intense around the relative merits of economic growth, global open trade, and social equity policies versus the alternatives. Today, the debate goes on, although many believe we did the right thing entering a path of increased globalization. We have seen many benefits of these policies but have also felt the distress of lost opportunities. We must ask whether similar

> **BOX 8.6**
> **Sub-Saharan Africa under Global Orchestration**
>
> As the world moved into the twenty-first century, the growing focus of decision-makers on poverty and inequality as sources of conflict and civil unrest drove greater attention and openness toward the economies of sub-Saharan Africa. Countries like Ghana, with a growing pool of educated labor, became prized sites for the outsourcing of data entry jobs, such as the cataloguing of parking tickets from New York City. Yet Ghana, along with Botswana and a handful of others, was attractive not only for its educated population, but for its successful transition to democracy after an extended postcolonial history of coups and dictators. The success of participatory democratic institutions in places like Ghana reassured businesses seeking to invest, and therefore worked in concert with initiatives sponsored by governments of the industrial world, such as the U.S. African Growth and Opportunity Act, to drive investment in these economies.
>
> Such investment spurred economic growth in several sub-Saharan African nations and, with it, the development of a middle and an upper class with sufficient education and financial resources to spend time worrying about the degradation of their national environments. These countries funded environmental assessments and studies, gave priority to issues like deforestation, and promoted the sustainable use of their natural resources. At the national scale, these efforts gained momentum as several countries funded environmental preservation and management programs with the money saved when at least half of their external debt burden was forgiven in debt-for-carbon-credits trades.
>
> The remarkable progress of these nations, however, must be considered in light of the important divisions that arose across the period 2000–50, both within countries and between them. Within the most successful countries in sub-Saharan Africa, the rural-urban divide was heightened by the increased educational and economic opportunities afforded to those living in the cities and by the concentration of decision-makers in urban areas. In many cases, environmental management projects were designed at the national scale by urban residents, with little or no consideration of the local, rural livelihoods likely to be affected by such projects.
>
> Thus, the establishment of forest reserves in Ghana's Upper Guinea Forest effectively displaced tens of thousands of Akan farmers reliant on swidden agriculture for their subsistence from "preserved areas" into unprotected parts of the forest. This displacement had two important results. Environmentally, the result was a patchwork effect, where protected parts of the forest flourished and unprotected parts were heavily degraded by increased agricultural pressure. Socially, the rural poor felt less and less connected to their national decision-makers, and therefore less and less a part of the national project. In countries like Ghana and Botswana, the development of a national identity was a key part of their political and economic success at the turn of the twenty-first century. The internal divisions driven by this success, though, threatened this identity, and therefore the very foundation of that success.
>
> If intranational tensions are important to discuss, so too is it critical to note that economies and political systems such as that of Ghana and Botswana were not the norm in sub-Saharan Africa. The incentives of the U.S. legislation were not extended to many nations that failed to develop political transparency and pluralism, in effect reinforcing the concerns of business seeking to invest in this region. Thus while some countries saw unprecedented growth and opportunity in the period 2000–50, others were caught in a cycle of decline and closure in which authoritarian governments were shut off from the globalized world, and the people under them were closed off as well. This process then created a patchwork effect at the regional scale, where successful countries often directly abutted one or more "rogues" or "failed states." The improved conditions in the stable states encouraged immigration, which no sub-Saharan African state had the resources to control.
>
> These surges of migration, driven by conflict and by dramatic environmental shifts in Southern Africa, heavily stressed successful countries, causing some to experience political failures as migrants swelled the ranks of the disaffected rural people, sparking intranational conflicts. Even in the best cases, such migration stressed the capacity of successful states to manage their economies and environments, limiting their success and suggesting trouble on the horizon for even the most stable and successful nations.

benefits could have been achieved with fewer losses of life, property, and human well-being if greater attention had been paid to the ecological underpinnings of our economic and social systems. There are those who argue that we have been too ready to accept economic signals as signs of progress and too quick to assume that success in tackling fast-moving ecological variables is evidence that we are making the right choices.

8.3.5 Insights into Global Orchestration from a Southern African Perspective

In the Southern Africa Focal Region Assessment, several scenarios were developed that dealt with aspects of Global Orchestration. African Partnership is an economic development scenario and Policy Reform explores policies for intensification of agriculture. Global Orchestration includes, but is more than, these scenarios. Its drivers are not just economic development and agricultural reform but also broader social reform, including protection of global public goods like education, health care, and safety. Nevertheless, the issues explored in the Southern Africa scenarios are relevant to Global Orchestration and give insights into how such a scenario could play out at a sub-global scale. (See Box 8.7.)

8.4 Order from Strength

In this scenario, the world becomes progressively compartmentalized as governments and then businesses and citizens turn their focus inward in response to threats from global terrorism and the breakdown of several processes involving global cooperation. People see looking after their own interests as the best defense against economic insecurity. Citizens reluctantly accept the argument that a militarily and economically strong liberal democratic nation can maintain global order and protect the lifestyles of the wealthy world and provide some benefits for any poorer countries that elect to become allies.

In a fundamental departure from the early twenty-first century, even rhetoric about the importance of trade liber-

BOX 8.7

Insights in Elements of Global Orchestration from a Southern African Perspective

Under the African Partnership scenario in the SAfMA regional assessment (see MA, *Multiscale Assessments*), high economic growth is underpinned by the intensification of agriculture, using highly selected seeds (including genetically modified organisms), irrigation, pesticides, and fertilizers. This boosts productivity, relieving pressure to cultivate new lands. Regional-scale food security is greatly improved, but water pollution and pressure on water supplies increases.

A dominant focus on commercially grown cash crops and a strong linkage to the global economy marginalizes small growers and impoverishes agricultural diversity. Consequently, vulnerability to pest outbreaks grows, and together with an increased frequency of droughts and floods resulting from climate change leads to large swings in cereal production and intermittent food shortages. Rising wealth accelerates a higher dietary demand for meat products, largely satisfied by expanded cattle ranching north of the Zambezi. Reduced pressure for land facilitates the development of an extensive system of state, private, and community-protected areas, which forms the cornerstone of a growing tourism sector, serving both foreign visitors and a growing urban middle class. Good land management practices outside of protected areas contribute to the maintenance of biodiversity in the region.

Under the Policy Reform scenario in the Gariep basin, consequences of the recent policies on agricultural intensification (GMOs, irrigation, pesticides, and fertilizers) for ecosystem services in the basin are mixed. This intensification is met with some resistance from health and environmental advocates and from small farmers unable to invest in these inputs, yet organic farming practices are also on the rise. Overall, productivity is boosted, and the expansion of agriculture onto marginal land is prevented. Food security across the basin improves. Pressures on water supplies increase, but an effective system of water tariffs, together with the establishment of catchment management agencies, now ensures that irrigators are accountable for their abstractions.

Cash crops are widely produced by commercial farmers who trade in a global economy but, due to past biodiversity losses, these are based on an impoverished genetic stock. This marginalizes small growers except for those who are linked to designer markets, such as organic farmers. This makes the crops more vulnerable to pests and diseases and the more frequent occurrence of droughts and floods precipitated by climate change. Intermittent food shortages occur but do not threaten food security in the basin. Intensive livestock production—batteries and feedlots, for example—becomes more common across the Gariep. A drive toward more-intensive meat production leads to an expansion of game farming operations in the basin, thereby creating a link between protected areas. Reduced pressure for land means a more positive outlook for conservation in general.

alization disappears in a backlash against globalization, which is seen as a source of instability and threats. (See Box 8.8.)

Just as the focus of nations turns to protecting their borders and their people, so too their environmental policies focus on securing natural resources seen as critical for human well-being. But they see the environment as secondary to their other challenges. They believe in the ability of humans to bring technological innovations to bear as solutions to environmental challenges after they emerge.

8.4.1 Order from Strength 2000–15

8.4.1.1 Teetering between Fragmentation and Connection

The beginning of the twenty-first century saw the world teetering between getting more connected and becoming fragmented and compartmentalized. There was cause to be optimistic about world peace and prosperity, but leaders were unsure whether a focus on open interactions with other countries or an inward focus on national security would achieve the best outcome for their people.

With the collapse of the Soviet Union and destruction of the Iron Curtain, a major source of potential global conflict was removed.[18] Globalization of technology, travel, and economic markets, which had been progressing apace in the second half of the twentieth century, was reaching its peak. There were high expectations from multi-nation negotiations of the early 2000s that progressive liberalization of trade and the movement of people would decrease the gap between rich and poor countries and universally improve human well-being.

The world was getting superficially more and more connected. Electronic communications were spreading to more and more homes. Air travel was becoming faster and more affordable.

But conflict and terrorism, together with stagnant economies, were pressing problems on the agendas of global and national decision-makers. Many believed that social tensions arising from inequalities in wealth, civil liberties, and access to food, water, and other basic human needs were the root cause of these problems.

Nations were also wrestling with issues of equity internally. Everyone agreed in principle that all citizens should have access to the fundamentals of a healthy and fulfilling life, but there were many difficulties with achieving this. It was so easy to just leave a few people behind and so difficult to bring everyone up to the same economic standards. Would political will be strong enough to overcome vested interests and public indifference? Furthermore, would the emerging trend toward globalization of markets work for or against equity for the poor and disempowered within and between countries? Would the removal of barriers to trade and the international operations of business allow developing nations to improve their economies and social processes? Or would it allow domestic processes to be dictated by corporations who wanted low inflation and high returns on investment?

8.4.1.2 Fear of a Fragmented World

Around the turn of the century, futurists wrote scenarios that imagined a world in which inequity is unchecked, where the rich get richer and protect themselves and their assets with walls and razor wire while the poor get poorer and seek to break down the walls.[19] The stories portrayed a scary world in which civil society eventually breaks down and conflict is widespread.

BOX 8.8

Potential Benefits and Risks of Order from Strength

Potential benefits:

- Increased security for nations and individuals from investment in separation from potential aggressors
- Increased world peace if a benevolent regime has power to act as global police
- Less expansion of invasive pests, weeds, and diseases as borders and trade are controlled
- In wealthy countries, ecosystems can be protected while degrading impacts are exported to other regions or countries
- Ability to apply locally appropriate limits to trade and land management practice, as trade is not driven by open and liberal global policies
- Protection of local industries from competition

Risks:

- High inequality/social tension, both within blocs and within countries, leading to malnutrition, loss of liberty, and other declines in human well-being
- Risk of security breaches (from poor to well-off countries and sectors of society)
- Global environmental degradation as poorer countries are forced to overexploit natural resources and wealthier countries eventually face global off-site impacts like climate change, marine pollution, air pollution, and the spread of diseases that become too difficult to quarantine
- Lower economic growth for all countries as poor countries face resource limitations and rich countries have smaller markets for their products
- Malnutrition

Nobody wanted such a world. In fact, many refused to believe it was even possible. Yet many of the ruling elite in governments and businesses of the time practiced a form of controlled inequity, believing it to be both inevitable and desirable for economic growth. They had faith that they could control inequity within limits that would encourage the poor to consume in an effort to improve their economic position and would prevent widespread disruptive action.

And yet, as the century turned, the wealthy of most major cities around the world had bars on their windows and were protected by fences topped with razor or electrified wire.

8.4.1.3 The Fragmentation Escalates

The optimism of growing economies in the industrial world at the beginning of the century was deflated by a series of international events, including the attacks on the World Trade Center in 2001 and continued conflicts in the Middle East, Africa, and South Asia. These conflicts drove headlines and contributed to pessimism about future international relationships. Trust in other governments declined universally, as countries turned their focus inward to maintain their national security, access to resources, and the well-being of their own people.

Despite reforms in Russia and other former Soviet countries, there was ongoing friction with the United States and a number of European countries. The Balkans remained unsettled, and the role of Turkey in Europe was not getting any clearer.

At the same time, governments of the industrial world reluctantly accepted the argument that a militarily and economically strong liberal democratic nation could maintain global order and protect the lifestyles of the wealthy world and provide some benefits for any poorer countries that elected to become allies.

This mix of distrust in outsiders but belief in order from strength became clear in the debates in the global forum for addressing international issues of the time, the United Nations. Former allies undermined one another, and it was clear that domestic issues rather than international ones drove the agendas of the participating countries. Despite the efforts of many who believed that the world needed global cooperation more than ever, the trend toward inward focus on national security issues continued and intensified throughout the first decade of the twenty-first century.

Economic issues contributed to the fragmentation that had been started by national security issues. After the financial excesses of investing in Internet startup companies at the end of the previous century, and a mini global recession stretching into the first decade of the twenty-first century, the industrial-nation economies did not rebound. Many East Asian economies continued to suffer from the after effects of the Asian financial crisis of the 1990s.

But with the growth of digital technologies a new source of fragmentation emerged—the so-called digital divide. Some Asian, South American, and African nations took a lead and established digital networks across their whole countries. This did not guarantee economic success, but it did give them an edge in many aspects of business. Increasingly, countries that had dropped behind in this digital revolution found it hard to maintain cohesive and adaptive networks among industry groups, resulting in lack of competitiveness in markets.

The digital divide not only applied to the gap between poorer countries. Even some wealthy countries failed to get connected fast enough and found themselves on the wrong side of the divide. Within countries, too, people varied greatly in their access to computers and networks and this affected their life choices, financial prospects, and even their freedom and ability to make informed decisions. In poorer countries, some elites managed to get connected to the wealthy in the rich world through technology.

A key turning point was the trade negotiations set up by the meetings of trade ministers from around the world in various cities between 1986 and 2001 leading to the creation of a World Trade Organization. These negotiations addressed a wide range of issues, including setting up a fair and market-oriented global trading system for agricultural goods, market access to nonagricultural goods and services, and fair dealing with respect to intellectual property, especially with respect to indigenous knowledge and implications of intellectual property for public health. During 2003–10, these negotiations ran into deadlock as a number of rich countries could not find a mutually satisfactory mechanism to reduce trade barriers and domestic subsidies, and several poorer countries began to stand their ground on these issues. In 2010, the World Trade Organization collapsed because the most influential nations no longer believed in trade liberalization.

The collapse of the negotiations soured political relationships among OECD nations and between those nations and the developing world. Further negotiations on access to markets, intellectual property, and the responsibilities of poorer and richer countries did not materialize. Rules and standards for intellectual property and digital communications, which could have been used as a tool for maintaining international equity and peace, rapidly became tools for drawing the best minds into the richest countries and for restricting access to markets. Large and innovative poorer countries flouted the rules and established their own standards. Corporations based in rich and powerful countries minimized investment in those nations, except where they could export profits to the country where they were headquartered.

The tensions already apparent among members and prospective members of the European Union early in the century were exacerbated during 2010–15 by the increasingly inward focus of nations. Negotiations about enlargement of the EU took longer than planned and took most of the attention of senior EU politicians, especially since global negotiations seemed to be going nowhere. The economies of the EU countries suffered from a prolonged downturn that further reduced the flexibility of its leadership in global negotiations in a variety of forums.

8.4.1.4 *A Fragmented Approach to the Environment*

Just as the focus of nations turned to protecting their borders and their people, their environmental policies also focused on securing natural resources seen as critical for their people. Among rich nations there remained belief in the ability of humans to bring technological innovations to bear as solutions to environmental challenges as well as confidence that a strong economy could afford the costs of that innovation. Natural environments were seen as providers of some critical services, like regulation of climate and the oxygen level in the air, but most other benefits from ecosystem services were believed to be either substitutable or repairable by technology. For example, everyone knew that water ultimately came from nature; during this time, however, the widespread belief was that water cleansing could be done better by human-powered technology than by use of preserved ecosystems.

In affluent nations, maintenance of biodiversity for its recreational and existence values was seen as an affordable luxury, one that people believed could be maintained by reservation of comprehensive, adequate, and representative samples of remaining ecological systems—so long as those reservations did not impinge on economic development. While governments claimed that these were proactive environmental policies, they were in reality reactive ad hoc solutions to growing public concern about declining species numbers. As the first decade of the century rolled on, the ad hoc measures merely slowed the decline in biodiversity and ecosystem function and delayed the inevitable environmental pathologies until the next decade.[20]

8.4.2 Order from Strength 2015–30

8.4.2.1 *Economic Fragmentation Escalates*

Lack of trust between countries, together with the impossibility of totally controlling movements of people, goods, and ideas across borders, caused increasing nervousness about industrial espionage and the spread of agricultural pests and diseases. Competition was uneven due to wealth and availability of resources between nations, so attention turned to the use of subsidies and tariffs, ecolabeling, and quotas as weapons for protecting countries' interests. Sanctions were also imposed by dominant countries and alliances on those countries that were thought to be threats to world political and economic order.

Only the wealthy nations could afford to use subsidies and indirect forms of support for within-country agricultural production. Though local residents in poorer countries got access to cheap products, their own production was put at risk by these practices. In some places, this caused a temporary environmental benefit, as exploitation of natural resources was reduced by low production.

The high level of concern about protecting farmers in wealthy countries with export subsidies had widespread negative impacts on the world's ecosystems in rich and poor countries. In poorer countries, economic activity was maintained through the export of alternative commodities not produced by wealthier nations. Typically these included nonagricultural products such as forest products, minerals, fish, wildlife, and the like. This introduced exploitation and degradation of ecosystems that affected human well-being by lowering the supply of raw materials, polluting air and water,[21] degrading soils, and eroding biodiversity and its associated services and roles. In rich countries, subsidized cropping systems expanded into former grasslands. This changed the habitat and carbon sequestering capacity and increased soil erosion and nutrient pollution.

Import tariffs and taxes were used in rich countries to raise money for government and to limit the amount of imports entering the country. These policies encouraged local industries and other economic activities. In poorer countries these same policies had the counterproductive effect of making agricultural technologies expensive so that farmers in poorer areas could not afford best-practice in relation to maintaining soil fertility and other ecosystem functions. As a result, food production there lagged behind other areas, and risks from pests and diseases were increased by continued use of old technologies. In response, people expanded agriculture over greater areas to maintain earnings, leading to increasing clearing of native ecosystems and reductions in their biodiversity.[22]

Many poorer countries used export bans and quotas on natural resources such as timber, minerals, and fish in order to foster domestic processing industries. While there were positive economic outcomes for the countries that employed quotas, such policies made it increasingly difficult to address global environmental issues, such as deforestation and exploitation of endangered species, because they made market pressures ineffective.

8.4.2.2 Health and Safety

As countries became more and more isolationist and as global environmental problems worsened, nations turned their attention to the use of health and safety regulations to reduce the chance that diseases and pests would be introduced to their countries through trade. Wealthier countries increasingly used ecolabeling and certification schemes to guide consumer behavior toward preferred producers.

These and other safety measures were effective in reducing the spread of pests and diseases. Regulations in many cases reduced pollution by discouraging use of fertilizers and chemicals and encouraging better soil management. However, they also frequently had a negative impact on the livelihoods of local people, especially in poor places, who depended on exporting environmental products. Increasingly, these people turned to alternative uses of the land and to illegal activities that sometimes increased ecosystem degradation.

Throughout the early part of the century, new diseases broke out in Africa and Asia, many arising from greater contact between humans and either wild or domestic animals. Although thousands of people died, many of these diseases stayed confined to local areas because the people had limited mobility. The rest of the world was generally concerned only to the extent that it was not under threat. While global cooperation was possible when diseases threatened wealthy nations and the freedom of movement of their people, diseases like AIDS that primarily affected poorer people and nations were not effectively addressed. (See Box 8.9.) Wealthy nations did not care, and poor nations could not afford to do much without the help of rich countries.

In general, foreign aid was provided on an ad-hoc basis, to tackle particular problems or crises and emergency situations, not to deal with longer-term development needs unless there was some obvious benefit to the wealthier country. Where development aid was offered, it was on terms that suited the donor country. Many people in poor countries argued against taking this aid on grounds that the terms often made things worse rather than better.

8.4.2.3 Environmental Impacts of Sanctions on Poor Countries

In this world, sanctions were an easier way than international cooperation to deal with failure to comply with trade or political measures. During 2010–20, such sanctions severely limited the amount of trade between poorer and richer countries, and many of the former were forced to rely on what their own ecosystems could offer for survival. Reduced imports of basic products, such as food, medicines, and industrial raw materials, caused local environmental strain as ecosystems were pushed to produce beyond capacity to provide these necessities.

Forests, grasslands, fisheries, farmlands, and even resources normally under state intervention were opened to continuous and heavy exploitation, particularly in areas of high population.[23] Many ecosystems showed dramatic decline in function and were not able to keep up with demand for goods. Sometimes, these resources deteriorated to the point of affecting human well-being through disease, malnutrition, and other impacts from flooding, drought, erosion, and loss of cultural identity.

8.4.2.4 Impact of Security Measures on Tourism

As human well-being was reduced in poor countries, rich countries began to increase security measures against immigration and terrorism. Nations often barred or tightly restricted certain classes of people or particular nationalities from visiting. This was very hard on the tourism industry, which not only affected livelihoods of many people, especially in poorer countries, but reduced funds that previously had been used to maintain natural places that attracted tourists. With tourism and the income it used to bring to poor places reduced, exploitation of natural ecosystems for survival increased even more.

In poor countries, the need to strive for economic growth in competition with rich countries drove resource depletion. Forests were cleared, and agriculture expanded and intensified in and around the places where people lived. Soil erosion increased, and the incidence of loss of property and deaths from land slippage and flooding also increased steadily.

8.4.2.5 Development of New Trading Blocs

Of course, a total inward focus would have been suicidal for poorer nations, and the breakdown of the previous global negotiations opened the door for new trade alliances. The new trading blocs were somewhat different from those that sprung up in the previous century. These started to form in the middle of the first decade and by 2015 were the new world order.

After a series of rapprochements with India, an Asian economic union was formed. The remainder of the Southeast Asian countries formed various arrangements with this new union. Other tentative blocs formed, such as a bloc of European and African nations and a bloc of countries in the Americas. Each bloc was generally made up of a dominant economic and military power that was joined by several wealthy economies and many countries with low per capita incomes.

The westernmost members of the former Soviet Union continued and accelerated the economic growth begun in the late 1990s. After years of debate, discussion, and deliberation, these countries signed preferential trade agreements with the EU similar in many ways to the arrangements of EU-affiliated nations in the late 1990s. African and Middle Eastern countries pushed hard for membership in the EU.

BOX 8.9

A Story of India (Order from Strength)

The harvest was poor again, despite the headman's prayers and the revival of some of the old rituals. Babies died of vomiting, and the nurse hadn't come for months. Some children still went to school, but rarely after the age of seven—they were of more use in the fields. The people from the development NGO also had stopped coming, even though a couple of the older people had listened to them and learned the alphabet, developed way back in the 1800s by Welsh missionaries.

In the city a new health minister had been appointed. At least she knew how to read and write, and she sometimes came to work, but otherwise there wasn't much to say. No one dared criticize the politicians, and the vote banks meant they answered only to the illiterates, who were easy to fool by promises of food and a fresh shirt. The AIDS awareness workers had at first hoped that the health minister might take more interest in their subject than her predecessor, who had continued to deny that the disease had reached their area, even when her own son became infected. It had been brought here quietly, spread by drivers trucking coal to the border and by infected girls too frightened to either complain or to ask their clients to use a condom—that could only mean one thing. At least dying quietly seemed better than being beaten to death, as some of the drug addicts had been.

The old custom of caring for and preserving the sacred groves was almost forgotten, just as was the reason for and the importance of a fallow period. Well, perhaps it wasn't entirely forgotten—the reality was that in order to feed a family, even for a few months a year, not only did all the fields need to be used, but day laboring jobs were needed, too. That's why extra children were handy—with big brother shoveling coal, little sister could at least herd goats and scare the few birds away.

Following the collapse of the reformed United Nations, after the default of the New Alliance, foreign officials had stopped lecturing the central government about the new disease. The law preventing people from selling their blood more than once a month still existed, but a small "commission" cleared that obstacle. It seemed an easy way to earn some extra cash. For a few extra pennies the certificate of hygiene could be forged as well—anyway, the officials knew that was irrelevant since the new Health Minister had declared the area to be infection free.

But something else seemed wrong this year. True, even the police had become accustomed to paying "safety" money—once 11%, but now 23%—to the insurgents who had crossed the border, but there was something strange about my new job offer. Five months pay for one month's work—even if it was far away; it just seemed odd. But how could one have the power to ask? And how could one resist?

When I started at the camp it was, I have to admit, rather exciting. For the first time in two years I could eat twice a day. The sore on my knee even started to fill in. The instructors were different from anyone I had ever met, and that was exciting, too. They taught me how to light fuses, and how to use a gun. By the time the armed motorcade came to cross the bridge, I was ready.

Regional trade agreements between the EU and African countries that had been in existence for many years were eventually merged into an umbrella agreement due to pressure from activists and those who thought that some terrorism could be avoided by improving the well-being of the poorest people. After years of negotiation, African countries were granted affiliate status that provided lower trade barriers for some important commodities. In the Americas, regional trade agreements further increased the integration of the North and South American economies and reduced trade and investments to the rest of the world.

The United Nations, never very efficient in its deliberations, became essentially irrelevant in a world where regional arrangements predominated and change happened quickly. The planned World Summit on Sustainable Development meeting of 2012 was called off after it became obvious that participation from the major blocs would be low. Regions viewed the Millennium Development Goals as a list to choose from selectively, with an emphasis on issues important to the dominant members of the blocs. OECD meetings become more acrimonious, and consensus was even more difficult to reach. As a result, there were fewer attempts to reach agreements, even within the wealthy countries.

8.4.2.6 Considerable Environmental Decline

As the attention of governments turned toward maintaining economic and military security, attention to the environment was highly variable between countries and depended mostly on the type of environmental issue at hand. Global issues like climate change, air pollution,[24] marine fisheries, and the emergence of new strains of infectious diseases from wild or domestic animals were almost impossible to address —there was always at least one key nation unwilling to cooperate due to national interests, and the international institutions that might have been able to address international issues were unstable if they existed at all. Global climate change increased less than had been expected at the turn of the century due to a larger than expected proportion of the world's population being forced to live a simpler and less materialistic existence, but few among those people took consolation from the figures.

Local environmental issues were dealt with very differently from country to country. Increasingly, local governments felt powerless to solve environmental problems and were overwhelmed by global issues no one seemed able to fix.

In wealthier countries, the obvious problems arising from degrading ecosystems, such as soil erosion, nutrient pollution of water, dust pollution of air, and damage from floods and other extreme weather, were minimized by using state-of-the-art technologies or by siting ecosystem-degrading industries in other countries. Cultural values from ecosystems were treated as luxuries and maintained in a similar way. Early in the century, various research and assessments, including the Millennium Ecosystem Assessment of 2005, alerted the world to the potentially serious impacts of declining soil fertility, loss of pollinators, loss of genetic diversity, and loss of water filtration capacity in watersheds if steps were not taken to better understand the processes providing these services from ecosystems. The rich nations generally allowed agriculture and urbanization to proceed as before, but at increasing intensities and over greater areas as population increased, assuming that ecosystems could absorb the impacts or if problems arose they could be addressed with technology and human ingenuity. As a consequence, urban sprawl continued to compete with agriculture for prime farmland. Agricultural land use was intensified where it could be and was extensified elsewhere. Problems were solved only when they could be easily solved with technology. When problems could not be solved or when areas became highly polluted, they were left for the poorest of the poor to inhabit.

In poorer countries, the situation was different. As aid from wealthier countries was progressively reduced and populations grew, communities were forced to extend the area under agriculture simply for survival. Where export markets were accessible, demands on the environment were increased to take advantage of opportunities. Where access to markets was lost or could not be maintained, the consequence was usually exploitation of natural resources. Two thirds of the central African forest present in 1995 disappeared by 2050, while about 40% of Asia's forests and 25% of Latin America's forests were depleted. By 2020, loss of soil, declining air quality, damage from flooding, and decline in coastal fisheries were severe in many poor countries. (See Box 8.10.)

These types of linked environment-economic problems in poorer countries caused increasing movement of people to wealthier countries searching for new opportunities. Most rich countries developed ever more sophisticated ways to keep unwanted refugees out but as the trickle became a flood, maintenance of border security was less and less possible. The problem became acute around 2020, and many of the wealthier countries sent funds and personnel into the main centers of unrest. For a while, this slowed down the pressure for immigration to wealthy countries, but the problems inevitably returned. Meanwhile, massive accumulations of people along borders started to generate their own ecological problems such as vegetation clearing, erosion, water pollution, and disease outbreaks that threatened adjacent wealthier countries. Wealthier countries also were threatened with particulate air pollution as forests burned in poorer countries and as soil mobilized from vegetation clearing was blown across continents.

8.4.3 Order from Strength 2030–50

8.4.3.1 Variable Economic Performances

By the mid 2020s, the differences between prospering and other regions were stark. The philosophy of protecting a personal patch permeated all aspects of life. The general sense of insecurity and the need to protect national interests and assets made the trading blocs tentative and always vulnerable to threats that countries would retreat if they were not getting enough benefits. Income disparities within blocs, both across the countries of the bloc and within individual countries, grew larger as rich nations were more easily able to take advantage of the opportunities of the bloc. There was great tension between the rich and poor countries within the blocs as they tried to maintain the delicate new balance among nations.

8.4.3.2 Declining Ecosystem Services

Until late in the first half of the twenty-first century, most residents of affluent nations were only partially aware of the decline in global ecosystems because the pollution, soil degradation, and habitat destruction associated with their consumption occurred primarily in other parts of the world. Reduced international travel only heightened the general lack of awareness among the public.

Demand for water increased beyond the realistic capacity to divert and collect rainfall for human use. Since countries were unable to develop cross-border agreements about water sharing, countries that happened to have enough water were extremely lucky and those that did not have enough were very unlucky. Availability of water had many follow-on effects in terms of the location of businesses that required water and other economic development factors. This often had the effect of keeping wealthy nations, which were also often rich in natural resources, rich and keeping poor nations poor. It also led to increased conflict over water, especially in water-scarce regions.

As investments in environmental protection and attention to ecological feedbacks diminished, marginal environments became increasingly vulnerable to extreme and surprising events. Since these environments were generally home to populations with low income and little economic resilience, there was pressure for migration both within poor countries and from poor countries to the more wealthy economies of a country's bloc.

8.4.3.3 Downward Cycles

Cycles of escalating poverty, environmental pressures, and potential conflict in less industrialized countries were interspersed with periods of ameliorative policies and investments by richer nations. This investment eased short-term problems like starvation and some illnesses. Longer-term problems, however, were made progressively worse because the underlying problems were not addressed. These included desertification from land clearing and overgrazing, the buildup of nutrients through excessive intensification of inputs to agriculture and poor infrastructure for dealing with urban wastes, chronic diseases caused through depleted ecosystem services for controlling disease-carrying species, and damage to crops and the environment from invasive species.

The lack of strategic and continuous investments by affluent nations in poorer ones led to increasing resentment by the less affluent countries and to conflict as they took violent action. Attempts to control terrorism after the high-profile outbreaks of the first decade of the new century kept a lid on this form of protest, but as time wore on resentment grew and it became harder and harder for governments to appeal to the morality of underprivileged, starving, and unwell populations.

Increasing expenditures on security, both by well-off nations and by well-off segments of populations within na-

BOX 8.10

What Happens to Marine and Coastal Fisheries in Order from Strength?

In this scenario, coastal habitats as well as some significant upwelling systems are significantly more susceptible to climate change, which ultimately has an impact on fisheries globally, with local and regional declines. Major events such as El Niño, extreme storm events, severe flooding, and significant changes in oceanographic process (such as in the Gulf Stream) have increasingly severe consequences, especially in the poor countries of each bloc and the poor within each country. Efforts to establish global fishing and environmental management agreements break down, but regional agreements expand to include some of the deep-water fisheries. With some exceptions that are important to selected species (such as squid harvested off the east coast of Argentina and sold to Europe), fleets from outside a bloc are discouraged from fishing in coastal waters, and some efforts to partition the deep sea take place. For some stocks, the regional regulation provides a degree of actual sustainability, but it is unclear whether this can last through 2050.

A focus on sustainable fisheries does not necessarily improve the ecosystem structure. Some nations focus on export-driven high-value fisheries that are often short-lived invertebrate species (shrimp, for instance). This results in larger, long-lived species that rely on the short-lived species as food being eliminated from the systems. These changed systems are vulnerable to severe events, and therefore food and fishmeal supplies are highly variable, which results in fluctuating profits. In areas without enforcement, destructive fishing practices continue, overexploitation is not reigned in, and stocks eventually decline along with inshore ecosystems. This has a cascading effect on coastal communities, which lose their source of food and income.

The wealthier nations of each bloc reduce their net outflows of fish products in order to secure food supplies and social benefits. There is a significant reduction in effort, starting with distant water fleets that are seen as threats to national food security. Fisheries that supply significant amounts of palatable fish biomass or contribute to the production of animal protein are well managed, along with their associated habitats. Areas are closed to fishing, where appropriate fisheries with low biomass production and destructive impacts (such as long-lining) are phased out. Marine protected areas, trade restrictions, and habitat restoration policies to sustain stocks and charismatic marine fauna are strengthened where selected fisheries are not threatened. In some areas, conflicts emerge between aquaculture and capture fisheries for space and marine water quality.

Exports of small pelagic species are eliminated in economically secure countries, as these fish are diverted to secure fishmeal and livestock feed. Aquaculture of low- to medium-valued species with high turnover rates expands, again to secure food supplies. Coastal environments are converted to accommodate the increase in aquaculture, but the profitability of the operations becomes questionable in areas where ecosystems have been degraded or areas are subject to increased major storm events.

Although some efforts are made to capture control of the deep oceans by the blocs, it proves impossible. Oceanic fishing continues to expand, with fleets exploiting stock farther offshore and in deeper waters. Soon long-lived stocks collapse, and fishing down the food web increases to the point where small planktonic invertebrates such as krill or species technically impossible to exploit in abysses are all that is left in the ocean.

tions, left little room for needed investments in human capital and local natural capital. In many ways, the moral character of many wealthy nations deteriorated, as altruism and ethical behavior were seen as inconsistent with the dominant paradigm of looking after personal interests.

8.4.4 Order from Strength beyond 2050

As we look back from 2050, we despair at the short-sightedness of our leaders over the past 50 years. Over short decision-making time frames, protection of borders and resources made sense to both politicians and their constituents. As the world became more compartmentalized, it became even harder for anyone to see clearly what was happening. Somehow we avoided the bleak forecasts of the early part of the century, but the seeds of those disasters are still with us. (See Box 8.11.)

The world's environmental condition has deteriorated substantially since 2000. In the first decade there was debate about when the consumption of the world's people would exceed the capacity of its ecosystems to maintain a sustainable supply of goods and services. (Some argued that this capacity had been exceeded at least two decades earlier.) It is now clear that in 2050 we are well and truly past the sustainable limit and the debate is around how far we can repair the damage done. Virtually all governments recognize they have major environmental problems, and they blame everyone except themselves.

Clearly, the lack of global cooperation played a major role in making environmental problems worse and allowing them to spread much farther than they might have with more cooperation in addressing them or a greater ability to address them earlier. Unless some form of global cooperation emerges in the next decade, associated with concerted efforts to improve equity among and within nations, the world seems destined to slip into deepening conflict and cultural decline. While the political will to make these changes is emerging, many of us fear that the resource base is too depleted to allow measurable progress within three decades. Our best hope is to minimize suffering during that time and aim for a stabilized world by the turn of the next century.

8.5 Adapting Mosaic

Underlying this scenario is society's strong emphasis on learning about socioecological systems through adaptive management. This focus on learning is linked with some local emphasis on balancing human, manufactured, and natural capital. There is optimism that we can learn about ecosystem management, but humility about our ability to prepare for surprises and know all there is to know about managing socioecological systems.

Initially, trade barriers for goods and products are increased, but barriers for information nearly disappear due to improving communication technologies and rapidly decreasing costs of access to information.

BOX 8.11

Sub-Saharan Africa under Order from Strength

As increasing fragmentation marked the world economy across the first decades of the twenty-first century, different areas in sub-Saharan Africa saw different results. West Africa remained in a state of uneasy stability after the removal of Liberia's Charles Taylor. This uneasy stability hinged on Nigeria, itself unsteady but propped up by increasing European and American interest in the oil reserves partially located beneath Nigeria's territorial waters and lands.

Southern Africa initially saw an overall decline in its economic fortunes, as it was forced to turn to its poorer African neighbors to replace industrial-country trading partners hiding behind ever-higher tariffs and other barriers. As the environmental impacts of intensive industrial farming began to make agriculture untenable in the industrialized world (for example, with the depletion of the Ogallala aquifer under the Midwestern United States), southern Africa saw something of an economic renaissance as it regained its role as the breadbasket of Africa, supplying both neighboring African regions as well as wealthier industrial markets.

For sub-Saharan Africa, the real problems began between 2020 and 2030, as changes in the global climate resulted in greatly decreased rainfall in southern Africa. The environmental impact of this change could not be offset by irrigation or GM technology, and farm output in this region went into sharp decline. As much of the income from this output was coming from international as opposed to domestic markets, the food that was produced continued to be exported, much as happened during the Ethiopian famine of the 1980s. Millions of people living in southern Africa found themselves without adequate food and without local resources with which to make up for this deficit. Unable to migrate to richer countries, the result of this widespread food insecurity was massive migration within sub-Saharan Africa, as millions moved from southern Africa north into East and West Africa.

The arrival of hundreds of thousands of hungry, unemployed poor in countries like Nigeria overtaxed the already limited social, political, and economic resources that held society together. In Nigeria, the results were catastrophic as large portions of the southern half of the country, overwhelmed by migrants seeking work and food in its cities, attempted to secede from the north in an effort to concentrate the petroleum resources found in this part of the country. This triggered internal fighting that was less coherent even than the Biafra conflict of the 1960s.

The Nigerian civil war was exacerbated by Cameroon's efforts to take advantage of this confusion to extend its own claims to the oilfields in the Gulf of Guinea, which spread the fighting to this nation as well. Migrants from these countries fled the conflict, joining those displaced from southern Africa, further swelling the ranks of the displaced in nearby countries. Opportunists throughout the region used this instability to lay claim to local territory, to stage coups (successful and unsuccessful), and to mobilize ethnic groups against one another for political gain. The wealthy countries did little to intervene, save for the dispatch of an American carrier group to the Gulf of Guinea to protect ongoing petroleum exploitation, especially by securing the oilfields of southern Nigeria.

By 2050, most of sub-Saharan Africa found itself in far worse shape than in 2000. Civil society and nation-building had failed in many places, and the redistribution of population throughout this region was wreaking untold havoc on local ecosystems as new migrants contributed to agricultural extensification at the costs of forests and other ecosystems. The industrial markets that once served as a key hope for sub-Saharan Africa's future turned their backs on this region as a "lost cause," ignoring the role that the industrial world itself had played in the sad state of this region, both in economic terms and in terms of its global environmental impact.

Power devolves to regions partly due to national government–led decentralization and partly because of disillusionment in the abilities of national governments to govern. There is a great regional variation in management techniques. Some local areas explore adaptive management, trying alternatives through experimentation, while others manage with command and control or focus on economic measures. The key idea behind the local management, though, is learning while managing.

Eventually, the focus on local governance leads to failures in managing the global commons. Problems like climate change, marine fisheries, and pollution grow worse, and global environmental surprises become common. Communities slowly realize that they cannot manage their local areas because global problems are infringing, and they begin to develop networks among communities, regions, and even nations to better manage the global commons. The rebuilding is more focused on ecological units, as opposed to the earlier type of management based on political borders that did not necessarily align with ecosystem boundaries.

People in this scenario have beliefs about the way the world works that drive ecosystem management. They believe that ecosystem services are important and that functioning ecosystems are an important part of providing ecosystem services; that cross-scale feedbacks happen, are important, and can be strong enough to change management policy and governance; and that focus on natural capital is enough to maintain adequate provision of ecosystem services. This changes later in the scenario, when there is increased focus on human and social capital. (See Box 8.12.)

8.5.1 Adapting Mosaic 2000–15

At the beginning of the century, several global trends were unfolding that were seen as a major threat to human well-

BOX 8.12

Potential Benefits and Risks of Adapting Mosaic

Potential benefits:

- High coping capacity with local changes (proactive)
- Win-win management of ecosystem services
- Strong national and international cooperative networks eventually built from necessity and bottom-up processes

Risks:

- Neglect of global commons
- Inattention to inequality
- Less economic growth than maximum because of less trading
- Less economic growth than the maximum possible because of diversion of resources to management

being on the global scale. Global trade expanded rapidly, institutionally regulated by the World Trade Organization rules of free trade. Conflicts in many regions were still present and often led to violent confrontations, both between as well as within countries. At the time, multinational companies played an important role in the process of globalization. Still, there were some major attempts to cope with arising environmental and social problems through international efforts, starting with the World Conference on Environment and Development in Rio de Janeiro in 1992 and the World Summit on Sustainable Development in Johannesburg in 2002. Global conventions, including ones on biodiversity or desertification, were signed to achieve better management of global environmental problems.

8.5.1.1 Growth of Civil Society: Investments in Social Capital

Though global efforts were still dominated by national governments, they were only part of the broader, self-organized system of global governance. In most of these efforts, civil society started to play a major role by the involvement of NGOs, the predecessors of the modern global civil society networks of 2050. The strengthening of local government and nongovernmental organizations was in large part due to an increasing perception that globalization was having negative impacts on the environment and social structures. Though this perception was only partly supported by scientific evidence, it was fostered by broad media coverage of examples like increasing unemployment rates in many rich countries or some major environmental disasters due to neglecting environmental standards in the developing world. Science played an increasingly important role, such as in support of the international conventions, and business was beginning to realize that some of these issues would endanger its long-term prospects.

At the same time, new partnerships emerged, both within and outside the global conventions and agreements. Systems for certification of renewable resources were developed or extended to new products, such as from forest goods to agricultural products.[25] These systems consisted of diverse groups of representatives from environmental and social groups, the trade and managing professions, indigenous people's organizations, community groups, and certification organizations from around the world. Although at the turn of the century only about 2 million hectares of the world's terrestrial ecosystems were certified under these systems, there was a growing demand for certified products due to increasing environmental concern among consumers in the rich world. As a result, the total portion of certified products rose to 18% of the global trade volume by 2010. The increased demand for environmentally sound products led to some of the first success stories of new, locally bounded partnerships for ecosystem management.

Though these initiatives brought some success to certain regions, they also exhibited some of the key problems of the century. On a local scale, many attempts to get a certificate brought about significant conflicts with powerful actors. In many instances, national decision-makers gained personal benefit from the traditional ways of ecosystem management. Corruption was abundant in many regions, and the people benefiting from it were afraid that the new system would put an end to their personal advantage. In other regions, professional lobbies strongly opposed the new systems, as they were reluctant to give control to local communities and actors. In some regions these conflicts were resolved by the proliferation of community-based land management authorities, and many people suspected that these opposition groups prevented an early boom of falsely certified forest sites.

The rise of new cooperation between NGOs and businesses was not the only sign of the emergence of new types of governance. The "organizational explosion" of civil society was evident in the first decade of the century. In the Philippines, for example, the number of nonprofit organizations grew from 18,000 in 1989 to 58,000 in 1996 and 134,000 in 2008. Many of these locally initiated groups became part of transnational advocacy coalitions on a broad range of issues, like human rights or the environment. Though access to the Internet still was limited in the developing world, the possibility of quickly exchanging information was an important condition for the proliferation of globally linked groups. The increasing number of broadband cell phones further helped to improve quick and fast information exchange. The number of mobile phones in China, India, and Brazil by far outnumbered those in North America and Europe already in the early years of the century.

These changes of civil society structures also had a qualitative dimension. An increasing number of organizations aimed at a nonprofit transfer of knowledge and skills throughout the world. Prototypical examples of this were Médicins sans Frontières (Doctors Without Borders) and the Academic Training Association. The first started off by bringing fast and nonbureaucratic medical help to civilians in areas of violent conflict or natural catastrophes. During the first decade of the century, it extended its aims by training and capacity-building activities. Similarly, the second group aimed to help restore academic training capacities in these areas, such as the region of the former Yugoslavia, in the early years of the century. These kinds of NGOs not only relied on the participation of politically interested people but were expert networks. Later in the decade, similar organizations developed for water management,[26] fisheries, labor safety, and pharmacy. These organizations turned out to be highly flexible and were thus able to incorporate the local contexts and peculiarities much better than the bureaucratic international or governmental aid organizations.

8.5.1.2 Education: Investments in Human Capital

The years of the first decade were also characterized by a strong increase in public and private education expenditures throughout the world. Whereas in the 1990s only a few low- or middle-income countries increased their spending for education, this was the standard trend in the first years of this century. Particularly in Latin America, but also in many other parts of the world, most countries spent more than 13% of their GDP on education in 2010. These countries reduced military expenditures significantly after democracies had stabilized. Also, in some pioneering countries the success of educational intensification in terms of poverty reduction was significant. The examples of those countries strengthened the view that highly skilled and professional workers had a significant positive impact on human well-being. In addition, when it came to development aid or financial credits, some international organizations and industrial nations favored countries that had increased public spending for education.

But it was not only public spending that helped to improve educational standards. Educational initiatives by private companies were starting to play a major role, and though often used for reasons of improving their images, these initiatives were soon producing benefits as graduating students sought jobs within the philanthropic companies. These highly educated and culturally aware employees thus helped with the "glocalization" that began to take place—the development and marketing of products adapted to and produced in a region and its specific cultural background.

8.5.1.3 *Managing Ecosystems: Successes and Failures*

The increase in investment in social and human capital can be seen as an important first step toward the world as we see it now. Many of these changes were directly related to changes in the ecosystems in the regions. In general, the trends of degradation and loss of ecosystems that were visible at the end of the last century were continuing, and in many regions the livelihood of an increasing number of persons became endangered.

There were, however, some isolated successes in better management of marginalized socioecological systems due to strengthened local institutions for learning. The Alternatives to Slash-and-Burn initiative, for example, a science-based organization with a network of some 50 sites with this traditional form of cultivation, was able to improve the livelihoods of most people within its network. As progress was not widespread, however, the effect at the time was more symbolic. It revealed that alternatives are possible when knowledge is locally relevant and appropriate institutional settings are met. (See Box 8.13.)

The significance of global conventions on environmental issues, though officially still valid, started to decline. In 2008, the starting year of the commitment period of the Kyoto protocol, it became obvious that its goal to reduce carbon emissions by a total of 5.2% in the signing countries would not be reached. This was to a large extent due to the limited cooperation to effectively implement the mechanisms designed in the protocol to better achieve the targets, such as joint implementation, green development projects, or emission trading schemes. In the years before 2008, the negotiations on agreements for the period beyond 2012 made only slow progress, as some countries still were reluctant to discuss any binding reduction targets. Though discussions had continued, it was clear by 2010 that there would not be any binding reduction targets within the next decade. In parallel, initiatives were intensified to agree on an "adaptation protocol," finally endorsed in 2019. These efforts were supported by the insight that climate change by this time had already brought about some major disruptions of ecosystems.

Nevertheless, there was some success with regard to climate protection. Western Europe continued its own efforts to reduce emission. It was still convinced that at longer terms the technological advantage gained by early implementation of new technologies would offer high economic benefits. In North America, the state initiatives for mitigation showed some success, though estimates of the overall reduction ranged from 1% to 3% below the baseline only. Though scientists continued to claim that climate change might have major impacts, especially in the developing world, mitigation of climate change as a global effort began to disappear from the global agenda.

8.5.1.4 *Economic Struggling*

In the beginning of the century, economic development was mainly seen as quantitative growth, and global cooperation and open multilateral trading were seen as the basic vehicles for growth. In view of a general slowdown of the global economy, the so-called Doha round of negotiations for better cooperation was started to maintain the process of reform and liberalization of trade policies. The negotiations were also intended to improve the effective participation of developing countries in the world trading system. It turned out, however, that the targets set were far too ambitious, and in 2005 negotiations finally failed. The primary consequence was a continuation of low growth rates on a global scale, although in some countries in transition, high growth rates prevailed throughout 2015.

Second, due to the failure of the negotiations, existing trade barriers and market distortions for agricultural products continued. These were seen as a major obstacle to agricultural development in poorer countries, as numerous subsidies in wealthy countries brought a substantial price advantage to their farmers.

Third, the growing limitations of free trade had some impact on the business strategies of multinational companies. Due to the raising of trade barriers, the multinationals started to look for new strategies to extend their business activities. At the beginning of the 2010s, it turned out that they could find these strategies by strengthening their local profiles in many regions of the world.

Due to the failure of the Doha talks, but also due to a trend of increased demand for environmentally sound products in richer countries, the process of globalization acquired a strong local component. People sought out local peculiarities and properties in order to maintain their cultural identity. This trend now increased, and global trade got a stronger local component. That is why we call the years between 2015 and 2030 the Era of Glocalization—a time when there was both a high degree of integration in the world economy and devolution to local and regional institutions in most countries.

8.5.2 Adapting Mosaic 2015–30

The Era of Glocalization started with a highly integrated world economy. Though the Doha round of negotiations for further liberalization of trade had failed, global trade was

BOX 8.13

A Story of India (Adapting Mosaic)

I was waiting for the Malaysian delegation when my thoughts started to wander back through time. It all started 10 years ago. No Indian will ever forget that summer, as it was like nature had become unnatural. Scientists told us that it was manmade, but how could it have been? How could man ever change the laws of nature? Even our grandfathers couldn't remember any stories of the past like this. Stories of a summer without any monsoon!

I can well remember the trouble we had in our village. We had been able to produce our own rice for centuries, but not that summer. Thanks to help from the government and from abroad we had only three casualties, but everyone was starving. And worst of all, one of our holy cows died because of the water shortage. It was simply too late when the water cars arrived.

There was intense debate on what we should do. The elders trusted nature and were convinced something like this would never happen again. Others said we should call for help from the government. Finally, four of us, who appealed to our own strength, were asked to create a better plan. My friend Yogesh had heard of an international water management organization that might help us. In the end, this help came in the form of Internet access for our village.

During the following years we collected information on integrated water management from all over the world. More important, however, we asked our elders about their knowledge of our local ecosystems. They, as well as many others in the village, remained highly skeptical and cautious. The challenge was to bring a range of knowledge—western and eastern, traditional science and local experience—together. Three years later we got permission to install three experimental fields to tests our ideas about ways to deal with water stress. Well, two years later the story repeated—no monsoon.

Two of the experimental fields completely failed and one of them was performing even worse than the traditional ones. Yet the third field actually was a big success, and its rice yield and water storage were better than five years before. The skeptics started to become curious about our system, and in the following months there was a lot of debate about how to further improve the approach. Many people from the neighboring villages also participated, and we were amazed about what they contributed. We experimented a lot and finally we got a system that not only helped us cope when the monsoon failed to appear, but also worked better in regular years. We now even sell water from this scheme—something that has become possible due to the new law on water trade. Thanks to this income, everyone is online in our village, the system is less labor-intensive, and more and more kids can now go to school.

Finally, the flight gate opened and Balan, the head of the delegation from Malaysia, was approaching me. I had had a regular e-mail exchange with him over the last weeks, and we were both looking forward to finally meet in person. In his first message he had told me a story—a story about a year without any monsoon.

still higher than in the early 1990s and than it is today in 2050.

8.5.2.1 The Rise of Regions

In 2015 the World Trade Organization had recovered from its struggle for a new strategy after the failure of the Doha talks. Within a new round of international trade negotiations, instruments to promote global economic growth were discussed. Perhaps the most important, in particular with regard to its long-term effect, was the Initiative for Regionalizing Global Trade. In this initiative, it was realized and acknowledged that, despite the prevailing barriers in the trade of goods, information could flow freely around the world. It was also conceived that the networks of civil society organizations might serve as a powerful vehicle for promoting the exchange of information and skills, and governments and others recognized that there was a still-growing demand for products that reflected the local roots of peoples' culture, attitudes, and lifestyles.

Within the Initiative for Regionalizing Global Trade, these trends were reflected by a system that gave leeway for transnational cooperation between business companies if these "cooperations" were based on local production of goods and the value-adding was documented so it was apparent for all participating partners. Though many NGOs criticized the agreement because it lacked human rights considerations, many large companies started to participate in these networks. Today there is no doubt that this initiative, though of limited success through 2020, brought remarkably positive stimuli to the world economy in the 2020s and 2030s, particularly after the initiative was extended to include civil society organizations in 2032.

The United Nations underwent a significant metamorphosis toward a stronger recognition of the emerging civil societies at smaller scales than the nation-state. In 2015, the United Nations Regional Organization (UNRO) was founded to seek to promote the "exchange, proliferation, and sustainability of regional development." UNRO got wide-ranging responsibilities, including educational, environmental, and human rights issues.

At the same time, the decade saw a further increase in the number and, more important, the effectiveness of civil society organizations. The success of the certification systems of the 2010s was still most impressive, though in the face of increasing trade barriers, global trade in ecosystem products was declining. Nevertheless, according to some estimates by the World Resources Institute in 2020, 34% of global ecosystem production came from certified sites. Due to these increases, the peak certification bodies were no longer able to ensure the quality of all certified operations. To improve coordination and quality control, they reorganized themselves in 2022 to become the Global Sustainable Ecosystem Business Organization.

8.5.2.2 Ecosystem Services

The most visible change in the Era of Glocalization was taking place in the agricultural sector, particularly with respect to property rights. At the beginning of the period, agriculture in the wealthy world was still highly intensified, and fertilizers and herbicides were widely used. Though the share of environmentally friendly and healthy production had risen to 12% in Europe and 7% in North America, the ecological problems induced by traditional means of cultivation prevailed. The 2020s, however, saw a rapid increase in the demand for high-quality, healthy food—a demand that could no longer be met. Major struggles between the various interest groups set in, which in some cases had to be resolved in courts. It turned out that most national legal systems were not capable of settling the disputes. As a consequence, major reforms were put in place, and in many countries the right for healthy food was included in the constitution. Many of the poorer countries followed suit, and by the end of the decade organic and naturally produced food had a market share of 34% in Europe and 21% in North America.

At the same time, the economic refocus on local and regional cultural values strengthened people's pride in their regions, and they started to consider "their" ecosystems as an integral part of their culture and local identity. This was also the reason why people in many dryland regions started to improve their well-being by adapting new agricultural techniques, which though based on traditional knowledge also incorporated innovations and experiences from other regions. In many of these regions, people were able to cope rather well with increased water shortages due to climate change.

In 2019 an adaptation protocol was signed within the Framework Convention on Climate Change. The pressure to agree on such a protocol was partly due to an increasing visibility of climate change effects. A significant increase in the frequency of extreme events, such as droughts in Europe or hurricanes in Central America and Southeast Asia, highlighted the need for adaptation. Hong Kong Metropolitan Area, at the time with a total of 25 million inhabitants and one of the largest metropolitan areas in the region, lost about 25% of its manufactured capital due to a single cyclone event in 2017. The Miami Protocol for adaptation to climate change established the Global Adaptation Facility, a fund that was set up to finance adaptation measures throughout the world. It was remarkable that this fund was administered by a partnership of the World Bank, the World Wide Fund for Nature, and the World Business Council for Sustainable Development.

A major challenge was the increased withdrawals of water as population increased in many parts of the world.[27] One of the most impressive success stories of this decade was the Euphrates-Tigris water scheme (see Box 8.14), which not only had massive impact with respect to water availability but also increased human well-being in the region. There were many more projects of this type, including experiments to combine local and network knowledge with regional specifics to gain better access to ecosystem services. Unfortunately, between 2015 and 2030 many of these experiments failed. There was, for example, the Midwest Organic Agriculture Project in the United States (see Box 8.15), which sought to meet the increasing demand for organic food in that country. Two of the main faults of the project were disregard of local traditions and an insufficient participation of local farmers in the design of the project.

BOX 8.14

The Euphrates–Tigris Integrated Watershed Management Project

One of the most impressive success stories of the 2010s was the Euphrates–Tigris water scheme, which not only had massive impact with respect to water availability but which also brought a new level of human well-being to the region. The project came from local initiatives, supported by the UNRO.

At the turn of the century, Turkey was trying to bring prosperity to its southeast region Anatolia through a large-scale water regulation scheme and the cultivation of irrigated cotton fields in the Anatolian highlands. Soon after, however, it turned out that many marginal groups in the region were excluded from the benefits of the project. Also, people working in the fields had major health problems, and salinization of soils progressed more rapidly than projected. Since Turkey finally had become a member of the European Union in 2012, there was moderate progress in the power these people actually had, and in 2015 the problems were so abundant that Turkey no longer could ignore the failure of the project. In its initialization phase, the project was heavily debated also because of its downstream effects—potential water shortages in Syria and Iraq. After the Iraq war in 2003, these conflicts were sharpened as the United Nations urged Turkey to release more water from the dams than it intended.

Altogether, in 2015 national and international pressure to revise and rebuild the project was immense. At the time, the international network of Water Engineers International, a nonprofit transnational organization similar to the Médicins sans Frontières initiative of the early years of the century, had ready-made plans for an alternative scheme, which would make use of the dams but which also sought to include ecosystem management in all three countries. When the project finally was realized with the help of UNRO, there were many win-win situations where the goal of "water by ecosystem management" was not only bringing a high degree of water safety to the region, but also increasing human well-being with respect to food security, health, and poverty alleviation. The trick was an adaptive mosaic of conserved areas, similar to the traditional three-field crop rotation system of central Europe in the fourteenth and fifteenth centuries.

At the time the world saw a rich portfolio of governance schemes for ecosystems and ecosystem services. In some regions, centralized management strategies were still prevailing by and large under national control. Other regions saw a profound switch to much more localized governance structures, often embedded into transnational networks of civil society organizations providing fruitful platforms for skill and knowledge exchange. The conditions and trends of ecosystems also showed a rich pattern of improvements and restoration, but also degradation.

The rate of biodiversity loss, though still a subject of global political and scientific debate, had been stabilized at low levels compared with 2000 in the regions where local management had been successfully implemented. In other areas, the trends of loss prevailing since the middle of the last century continued to threaten ecosystem functioning and human well-being. Yet attitudes toward biodiversity had changed quite substantially. Whereas at the beginning of the century people responded most to the loss of single, highly symbolic species, awareness now had switched to

BOX 8.15

The Midwest Organic Agriculture Project

In October 2014, the Midwest Organic Agriculture Project was launched in the United States with the goal of meeting the increasing domestic demand for organic food. The project was initiated by a national network of scientists and food co-ops that were actually able to gain support from major corporations in the food sector.

Based on the prevailing knowledge paradigm, the project developed a detailed strategy of land use throughout many areas in the region. They also designed training courses for the farmers and gave marketing guarantees for a range of products, including grain, maize, and beef. Unfortunately, the traditional networks of farming and marketing were overlooked, and the farmer's traditional decision-making with respect to what to cultivate and where to sell turned out to be much stronger than anticipated. Though 46% of the farmers in Nebraska participated, most of them applied the new techniques only for a limited time or on a small portion of their land.

The project was ended in 2018, and its impact on the national food market was negligible. There was, however, a thorough analysis of the project and its failures—an experience that helped make the next 15 years a period of rapid diffusion of success. One of the key faults of the project was the disregard of local traditions and an insufficient participation of local farmers in the design of the project.

BOX 8.16

Managing Malaria in Africa

In the 2010s, malaria was a highly limiting factor for economic development in Eastern Africa. In its 2020 Global Sustainability Outlook, the United Nations Regional Organization assessed that about 55 million people had died of malaria in the 2010s and that the overall costs of malaria in the region amounted to about 20% of its GDP and thus limited growth rates by another 50%. This was due in part to the fact that malaria took hold in new regions due to a significant warming of the region in the first decade of the century.

Facing these facts, UNRO decided to set up a competition in local ecosystem management for better prevention for malaria. Instead of earlier attempts of transferring knowledge into a region, the initiative sought to foster local initiatives and to build on existing networks of transnational NGOs. Within the first five years, most of the initiatives failed, but in 2028 a network of communities in Kenya and Tanzania came up with an idea of cooperative management of their local ecosystems, including far-reaching drainage into underground basins so that the breeding grounds of the vectors would be destroyed without the loss of water for agriculture and households. These small-scale constructions were accompanied by a medical treatment program with the aid of Médicins sans Frontières, which helped reduce the number of newly infected people significantly. After great success in the first five years, this combination of local aquatic ecosystem management and cross-scale knowledge transfer quickly spread to other localities. By 2035, the number of newly infected people per year in the region at large was down to 50,000 people, compared with 900,000 some 10 years earlier.

perceiving biodiversity not only as an essential supporting ecosystem service, but as a goal in and of itself. This understanding, though part of the scientific discussions for a long time, now took hold in the public debate in many regions. Demand for a cooperative management of provisioning and supporting ecosystem services grew steadily within the mosaic of success. The successful experiments of local and networked management made the message "we, the people, can do something about it" rather popular throughout the world. This attitude was largely responsible for the success of the initiative for malaria control in East Africa in the late 2020s. (See Box 8.16.)

The same holds true for other provisioning ecosystem services. The example of the Alternatives to Slash-and-Burn program and its achievements for improving human well-being was spreading further, and by the early 2020s it had reached 50 million people throughout the poorer world. Ecosystems in many of these regions had started to show signs of recovery, and people's livelihoods were safe and balanced. Nevertheless, due to a still-growing population and a general increase in the demand for ecosystem services, global trends showed a significant deterioration of the global environment. Global climate was changing rapidly and had been finally proved to be related to anthropogenic burning of fossil fuels and land use change.

Due to these global trends and the visible success of local strategies, the 2032 Global Conference on Sustainable Development, Rio + 40, further strengthened the proliferation of local and cross-scale learning initiatives. At the same time, the benefits of the educational imperatives of the first and early second decade showed up in many countries. This not only strengthened the efforts to manage ecosystems in a locally adapted manner, it also helped to broaden the expertise, skills, and knowledge base of the transnational civil society organizations. NGOs had gradually changed into highly professional companies, which in many instances served as lobby groups at international conferences and the UNRO meetings and also on the national level.

8.5.3 Adapting Mosaic 2030–50

This era was characterized by a rapid diffusion of civil societal management throughout the world. Based on the example of the Euphrates-Tigris integrated basin management scheme, there was increasing demand for new, small-scale, ecologically sound water basin management programs. Supported by UNRO or with participation of business coalitions within the World Trade Organization's Global Business Network Initiative, these schemes were put in place in as many as 23% of the main global basins. Though these programs were successful in many regions, there were quite a number of failures where the goals to integrate ecological and socioeconomic interests were not met. Some companies used these attempts as vehicles for their own interests. In other cases, local decision-makers used the schemes for their own enrichment. In many other cases, knowledge simply was not sufficient to realize ambitious targets.

8.5.3.1 A New Attempt at Free Global Trade

Though the signs of civil society were abundant at this time, economic development was still lacking in many regions.

The hope that glocalization would bring prosperity and wealth was fading, and the attempts to stimulate economic growth by a new round of negotiations showed only partial success. It was realized that the limits to growth induced by the prevailing high trade barriers could not be compensated for by the free flow of information and skills. The Global Business Network Initiative, though showing some success, was perceived as being too strict and not flexible enough to react to the rapid diffusion of the civil society and the increasing needs of a still-growing population.

Nevertheless, it was not possible to turn the clock back to the kind of world economy that was envisioned before the failure of the Doha talks at the beginning of the century. Though local and regional trade brought about some progress, the chances to invest in other countries were still not satisfying. As a consequence, in 2034 a World Conference on Economic Development took place at which new elements of a glocalized economy were discussed and put in place. The vision was to stimulate growth by bringing business interests together with the interests and needs of the emerging global civil society. (See Box 8.17.)

8.5.3.2 Tragedy of the Global Commons

Despite the widespread progress in managing local ecosystems, the inability to take care of the global commons—places and resources owned by all of the world's people and not just by one nation—brought forth major drawbacks for development. In 2022, world fish catch had declined to only 30 million tons, about 70 million tons below the peak catch at the turn of the century. This brought about a significant decline in protein supply for the world, in particular for those people heavily dependent on marine resources, such as those in the coastal zones of southern Africa, Latin America, and Southeast Asia. Consequently, the demand for meat increased significantly, which put additional pressure on land resources. Thus the hope to reduce land use change in the near future was fading.

Climate change also entered a new phase, bringing about major water crises in some regions of the world, including Togo, Mexico, and Turkey,[28] putting pressure on land resources in other areas, and finally having direct impacts on human well-being. Though the Miami Protocol for adaptation did show some success and probably helped many regions to cope with a changing climate, the funds in the Global Adaptation Facility were by no means sufficient. Also, the institution itself was heavily criticized for its ineffectiveness due to the major coordination problems of its leading organizations. It was not until 2038 that the fund was reorganized to address climate change directly by adding value to local programs.

Experts agreed that the danger of major disruptions of the climate system was still growing. Besides "traditional" disruptions, such as a significant weakening of the North Atlantic thermohaline circulation, new discontinuities were discussed, including a runaway climate change due to large-scale deforestation of the Congo basin.

As already described, the late 2030s and early 2040s were molded by a period of further promotion of civil society. Although there were still failing experiments, including the Mweru National Park tragedy (see Box 8.18), these efforts finally showed up in global trends by the early 2040s. The civil society–business partnership endorsed in the New Agenda for Development brought major progress with respect to economic development, and growth rates slowly but steadily increased in the 2040s. The broadening of skills and expertise continued due to an increased involvement of people from poorer nations, and thus the rates of technological progress in terms of sensible effectiveness of resource use and environmental pollution increased. Nevertheless, population was still growing quickly, and the need for ecosystem services was increasing in most regions and globally still outpacing any major progress.

8.5.3.3 Civil Societies Breakthrough

With respect to civil society–business partnerships, the example of the Vivanto-INESI partnership is most impressive. In 2038 many NGOs, including professional networks, formed the International Network of Sustainability Initiatives (INESI) in order to better coordinate their activities. Attempts made by UNRO to make INESI part of its own structure were rejected as the networks wanted to remain independent. INESI was a global player that not only brought together knowledge and skills of unprecedented depth and broadness, but also had economic power. By 2043, many of its professional network members were collectively self-organized and economically self-sufficient.

One of the major achievements of INESI was the founding of the Cooperative University for Sustainable Development in 2046. This completely new university system

BOX 8.17

The New Agenda for Development

In 2034, the World Conference on Economic Development took place in the Middle East. The conference agreed on a new framework for free trade. The most central tenets were:

- Free trade of end products certified to comply with the ecological and social standards of the region of origin.
- Free investments in regions under the condition of a sufficient and wide-ranging participation of local people and civil society organizations as well as professional networks (like Water Engineers International).
- Free flow of labor migration as long as both the country of origin as well as the country of destination complied with the minimal social standards set by UNRO.

The decisive point of the agreement was that the conditions were no longer bound to the nation-state, but much more to the regions. Nation-states agreed to these directives with the hope that within their countries those regions benefiting from the agreements would serve as a locomotive of growth for the whole country. Yet history has shown that although growth was stimulated in many regions, there were rising conflicts within states where not everyone had equal opportunities to take part in free trade. The late 2030s were therefore molded by the attempts of the nation-states to further extend the civil society within their countries.

BOX 8.18

The Mweru National Park Tragedy

After the success story of the tourism industry in the Okavango Delta late last century and in the first years of the twenty-first century, other places in sub-Saharan Africa sought to follow the same track and benefit from the growing revenues of tourism. The Okavango story also demonstrated that ecotourism can be designed in a fruitful and appropriate manner to sustain incomes, ecosystem integrity, and ecosystem services, in particular the aesthetic values of the deltas. Though the first national park in the Lake Mweru region in the border region between Congo and Zambia was established in the 1990s, initial efforts to increase the incomes by tourism were made in the 2010s. Based on the experiences in the Okavango, an ecotourism concept was developed by local authorities within a participatory process involving local stakeholders and indigenous people. This process guaranteed a high degree of ownership by the local people, both in terms of "mental" as well as legal ownership. Within its first five years, tourism brought a significant increase of income into the regions and people's well-being increased, in particular in its material dimension.

The high degree of partnerships and cooperation on the local level as it was established, however, was not sufficiently supported by regional measures. Thus a strong income gradient developed, in particular on the Congo side of the lake region. This induced a strong domestic migration, and in 2025 a total of 1.2 million people migrated to the Lake Mweru region. Though some of them were able to participate in the revenues by taking on minor jobs within the tourism facilities, many of the new settlers weren't successful. The informal settlements developing within the region undermined the positive image of the region, and the number of tourists started to decline—but not the number of migrants.

The new settlers also caused significant deforestation in the upstream region of Lake Mweru. By 2028, already 37% of the upstream area was deforested, and in 2029 a heavy rainfall event induced major landslides in the region, finally leading to a flash flood event in December. The flood not only destroyed almost all the tourism infrastructure, including the regional airport, but also directly caused a total of 1.2 million deaths in the region. The region did not recover well, as only 20% of the value of property was insured by international companies, and the local networks designed for taking over possible losses from small events as envisaged in the original concept were not able to cover the damage costs of the flash flood. This example again showed that the concentration on a single scale might induce major backlashes if cross-scale interactions are neglected.

was meant to promote mutual learning from local experiments, including learning from failures.

Finally, the decades between 2030 and 2050 witnessed the emergence of new types of global business players. Though many companies could still be considered global players, their internal structures had changed due to the need to adapt to the variety of local and regional standards and regulations. The most successful companies developed as a network of loosely linked subdivisions that have far-reaching competencies. They have internal trade systems that have taken up ideas developed for emission trading systems during the climate change debate at the turn of the century (such as ecosystem disturbance rights and proactive ecosystem management skills).

The most successful example in this regard is Vivanto, which makes 85% of its total revenues by ecosystem service trade. It had made its way through the turbulent decades of the 2010s through the 2030s by innovatively seeking new partnerships and reorganizing itself into a loosely coupled multinational company with rather independent subdivisions. After the New Agenda for Development was in place, Vivanto was able to increase its business with ecosystem service trading, and in 2045 the time was ripe for a closer cooperation with the civil-society sector—with INESI.

The 2040s also saw an increase in the efforts to extend the principles of civil society to the global scale and to seek new ways to cope with the problem of the global commons. In 2046 a partnership of the United Nations, INESI, and the World Business Council initiated a new round of negotiations for a globally orchestrated reduction of carbon emissions. Negotiations are still going on, though in 2048 an agreement was reached to extend the criteria of compliance of the Baghdad Treaty to the issue of carbon emissions. It is unclear what the effect of this agreement will be.

8.5.4 Adapting Mosaic: Where We Are in 2050

Over the last 50 years, a rich mosaic of local strategies to manage ecosystems and ecosystem services has emerged. There have been many successes in how individuals and communities have learned to manage the ecosystems they live in more proactively. They have achieved a high level of resilience of the coupled human-ecological system they are part of. This is, for example, reflected in the quick recovery of large areas in Central America after the most intensive hurricane ever, which scientists say was caused by climate change, which over the last 60 years has increased the global mean temperature by almost 2 K.

Though the direct damage was enormous, people's capacity to restore ecosystem services has been sufficient to recover from the impacts within two or three years in most cases. For the restoration of freshwater supply, for example, people used strategies based on the experiences gained in Viet Nam and Southern China after a typhoon hit that area in 2013. As key upstream ecosystems were identified and conserved, and as their water retention capacity and storm vulnerability were well managed, these ecosystems were hardly damaged by the hurricane.

The world has also seen the emergence of new networks and cooperations both across regions, like the one between Central America and Southeast Asia, and across various actors.[29] The Vivanto-INESI example described above is an example of cross-actor cooperation. The extent and content of such cooperation was barely conceivable at the turn of the last century. At the time, only a very few selective and small-scale cooperations of this kind were present. INESI itself constitutes a network that in its connectivity is by no means comparable to the so-called nongovernmental organizations of the turn of the century.[30] The effectiveness of the information exchange across the network is unprecedented and its Cooperative University for Sustainable Development, founded in 2046, is a success story in itself. The university has a strong focus on transferring and generalizing knowledge on ecosystem functioning and managing and does not, for example, have an engineering department, as a focus on one discipline is seen as being not helpful for its overall objective.

On the other hand, there are a large number of substantial failures where attempts did not bring about the necessary improvements of ecosystem functioning and human well-being. These failures led to unexpected breakdowns of ecosystems, either on local scales due to failed experiments or on global scales, as for fisheries, due to the general neglect of common property problems, especially during the first two decades of the century. Many ecosystems in sub-Saharan Africa, for example, are still under high pressure. (See Box 8.19.) The trends of losing ecosystem services are still increasing because in these places local knowledge for better management has been lost a long time ago or because insufficient institutional changes hindered the proliferation and implementation of local knowledge and learning. The strategies to manage the few successful areas in a much better way are specific to these regions, and their transfer to other regions were more or less a complete washout.

More recently, progress in understanding the conditions for success and failure helped to promote a better diffusion of good and effective management strategies. Accordingly, global trends of deterioration of ecosystems and human well-being have slowed down, in some instances even turned around. Nevertheless, in many areas major challenges remain in order to improve management strategies. In most of these cases, institutional settings are not in a shape that favors the development or adoption of new strategies. It is therefore highly important to reflect on the economic, political, and institutional changes of the last 50 years that enabled the changes we are witnessing. There are, however, some major indicators that give significant hope that "dooming" effects of large-scale disruptions of Earth's system can indeed be avoided. Most prominently, the improvement of local knowledge led to a decline in fertilizer input, and consequently nitrogen and phosphorus inputs in freshwater ecosystems have been reduced. This is seen as a good sign of hope for avoiding large-scale eutrophication.

The geography of success with respect to proactive ecosystem management based on local learning varied widely over the last 50 years. Some larger regions did quite well as a whole, whereas other regions with some very small exceptions by and large failed. In most regions of the world,

BOX 8.19

Sub-Saharan Africa under Adapting Mosaic

Sub-Saharan Africa emerged into the twenty-first century as a region under great stress from environmental degradation, poverty, and conflict. While the industrial world recognized the potential problems that might spring from such stresses, such as terrorism and global economic slowdowns, the efforts made by the wealthier countries to improve these conditions without sacrificing their economic superiority were largely ineffective. For example, in many countries the efforts to foster education as a means to manage population growth—a common effort in sub-Saharan Africa—found little governmental support, as these countries focused their financial resources on issues of debt and the costs of war (both financial and in terms of human capital). Further, these efforts tended to be focused in urban areas, bringing benefits to those already privileged by contact with money and education. Thus the NGO-organized education efforts that did not peter out after an initial push due to lack of funds or interest served to improve the situation of many urban dwellers, but in the process they heightened the urban-rural divide seen in most sub-Saharan African countries.

As governance devolved to regions, and as international accords on subjects ranging from economics to human rights fell by the wayside, the barriers to trade experienced here drove an increasing focus on regionally focused economic development. This created a situation in which the urban-rural divide was perpetuated. The urbanizing populations of this region needed food, which was available through the farms of rural dwellers. The continuing urbanization of the population became a threat to the social order in many states, as the migration drained the agricultural sector of their labor pool and threatened the urban food supply. Highly self-interested efforts to promote intensification in rural sites by urban elites were aimed at improving food supplies in the cities without concern for rural conditions. The result was, for example, the reintroduction of DDT into farming systems in several countries that, while improving crop yields and felicitously reducing mosquito populations and malaria rates, also induced enormous water resource degradation. Transnational organizations and groups that tried to connect with sub-national civil society groups often found themselves drawn into the growing tension between urban and rural, a tension expressed more often than not through violence.

While their economies were growing, many countries in sub-Saharan Africa were dealing with heavy debt loads that did not permit extensive expenditure on the adaptation measures called for in the 2019 adaptation protocol. Efforts on the part of organizations from wealthy countries to work with local civil society in various parts of the region succeeded in identifying key local knowledge resources that proved effective in managing local ecosystem change, but in the face of broad global shifts in precipitation and climate, as well as national pressures to increase agricultural production, much of this local knowledge was quickly obsolete. In southern Africa, there was simply no means for local farming systems to adapt to the ongoing loss of precipitation seen from 2000 to 2050, and no resources through which to construct the extensive irrigation systems necessary to preserve the agricultural systems of this subregion. Adaptive management, while productive in many other parts of the world, could not keep up with the changes in this region.

It was not until the period 2040–50 that the breakthrough of international civil society created an environment in which adaptive management could take effect in sub-Saharan Africa. The resources marshaled by these broad coalitions, both intellectual and financial, have served to supplement local knowledge and supplant local funding for such management. Struggles over the direction of such management continues, however, as international networks attempt to negotiate the regional and national politics of the urban-rural divide in order to ensure an adaptive management that is both sustainable and just. Whether such efforts will succeed in managing the urban-rural split or in overcoming the tremendous environmental problems experienced in sub-Saharan Africa over the previous 50 years remains to be seen.

however, a mixture of successes and failures can be observed. In a recent study by UNRO, major trends over the last 50 years were analyzed. The study revealed that the world divides into four major types of regions.

Some countries at the turn of the century started with a high potential for local, proactive ecosystem management but over time lost this capability by missing major opportunities for sustaining and further improving their social and human capital. Prominent examples are parts of southern North America, some countries within the European Union, and the former Pacific OECD. This group is sometimes called the *Northern Sleepers.*

Some countries managed to sustain the high level of social and human capital that they had at the turn of the century by strong investments in education and in improving networking capabilities, both in terms of infrastructure as well as institutions. The UNRO study called these countries *Sustainers.* Some countries within the European Union, major parts of North America, Australia, and some regions of what was known as the developing world at the turn of the century, such as India or South Africa, belong to this group.

Another group of countries that started off with a rather low potential for proactive, local ecosystem management was able to improve local institutions significantly to increase these capacities over the last 50 years. Educational programs further helped build human and social capital, and these countries have now experienced a significant improvement of human well-being. In analogy to the booming economies of the late twentieth century in South and Southeast Asia, the UNRO study refers to these countries as the *Pumas.* This includes some countries in Latin America, particularly Central America, plus a few large countries in Africa, like Nigeria or Mozambique, and some countries in Asia like Mongolia or the Philippines.

Finally, a last group of countries, in the UNRO study referred to as the *Trapped,* did not manage to sufficiently raise their human and social capital. This includes major regions in sub-Saharan Africa and a few countries in Latin America and Central Asia. These countries need further support to improve their social capital in order to catch up.

Due to the dynamics of investment in social capital in the different country groups, some previously developing countries have caught up with wealthier countries that were inattentive in taking care of the capabilities for learning and in facilitating networks for ecosystem management. (See Figure 8.4.)

Because of differences in social capital over time and because of natural conditions for resilience, ecosystem services developed rather differently in the different regions. Whereas the *Northern Sleepers* for some time profited from existing knowledge for proactive management, ecosystem services in these countries started to level off and even declined later in the century. In contrast, the further rise of services in the *Sustainer* countries, and later in the *Pumas,* was largely due to the increased capabilities for learning about ecosystem functioning. The end of the spectrum is constituted by the *Trapped* countries, where a significant decline in ecosystem services was observed, in particular after a number of failed experiments in the 2030s.

The trends in human well-being resulted from the interplay between economic, social, and political development with the trends and levels of ecosystem services. Note that the increase in richer countries, even for the *Sustainers,* was not as strong as for the previously poorer countries. This shows the important role of growth in social capital as a prerequisite for human development. This is also consistent with the decline in human well-being for the *Northern Sleepers.*

8.5.5 Insights into Adapting Mosaic from a Southern African Perspective

In the Southern Africa Focal Region Assessment, the Adapting Mosaic scenario was interpreted at regional (African Patchwork), basin (Local Learning), and local (Stagnation) scales. These scenarios are summarized in Box 8.20.

8.6 TechnoGarden

In the TechnoGarden scenario, technology and market-oriented institutional reform are used to improve the reliability and supply of ecosystem services. In this scenario, society's focus is on investing in human, manufactured, and natural capital. (See Box 8.21.) In this case, however, biotechnology and ecological engineering blur the distinction between natural and manufactured capital.

There is a strong belief that "natural capitalism," which focuses on looking for profits in working with nature, is advantageous for both individuals and society (Hawken et al. 1999). Technological improvements that reduce the amount of material and energy required to produce goods and services are combined with improvements in ecological engineering. Ecological understanding and technology allow people to alter ecological functioning to reduce trade-offs and increase synergies among ecological services. These technical advances are stimulated by, and in turn encourage, the development and expansion of markets in ecosystem services, such as requiring payment for carbon emissions and paying for ecological management that improves water quality.

Generally, ecological markets are established and property rights are assigned to ecosystem services following the identification of ecological problems. Because of investment in ecological understanding and natural capital, problems often are identified before they become severe. Ecological markets are established at local, national, regional, and global scales. Depending on the social and ecological context, property rights are granted to different actors.

This scenario explores the belief that ecological engineering will be fairly successful[31] and produce tolerably few major unexpected breakdowns of ecosystem services. Many

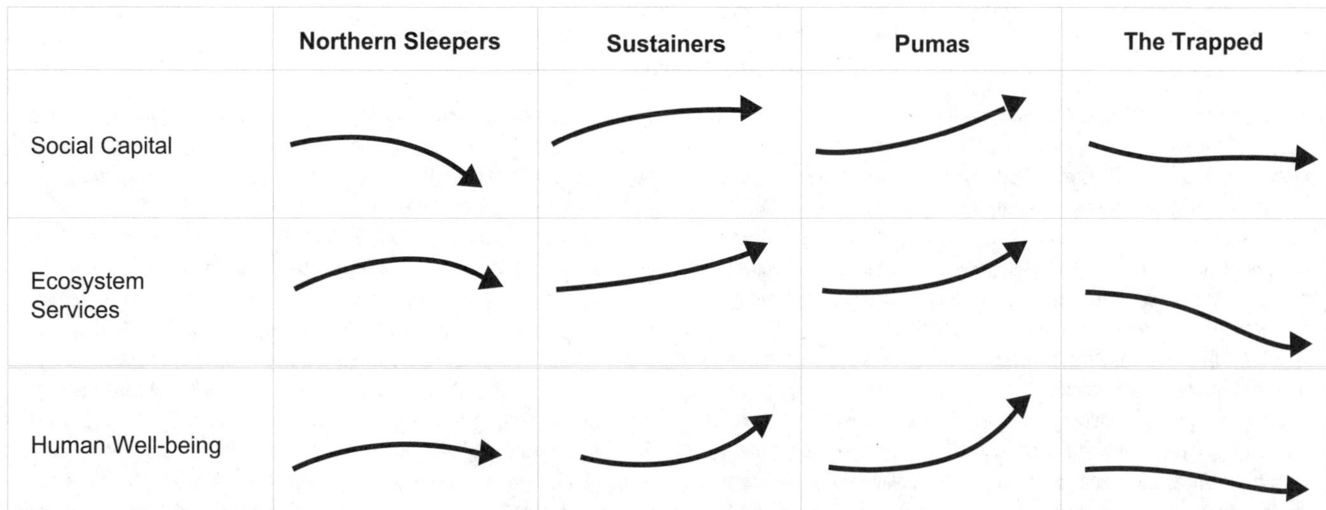

Figure 8.4. Trends in Social Capital, Ecosystem Services, and Human Well-being for Different Country Groups in Adapting Mosaic Scenario as Assessed by UNRO Study. For more details on country groups and the UNRO study, see Section 8.5.4. Note that the comparisons have been standardized for all countries for the year 2000.

ecologists agree that this assumption is plausible, but suspect that it is overly optimistic.

8.6.1 TechnoGarden 2000–15

At the beginning of the twenty-first century, poverty, inequality, and unfair global markets, together with environmental degradation, were pressing problems on the agendas of global and national decision-makers.

8.6.1.1 A Doubly Green Revolution

A key activity in which these issues intersected was agriculture. Agriculture was, and remains, the most extensive human modification of Earth's surface. At the start of the twenty-first century, world markets for agriculture were unequal. Trade barriers and perverse subsidies encouraged pollution in the rich world, impoverished rural communities, and undercut development in poor countries. A broad coalition of neo-liberals, development advocates, and environmentalists organized against agribusiness to stimulate a global transformation of agriculture in the early twenty-first century.[32]

These changes began to take hold in the 1990s, when, following a series of agricultural disease and food safety crises in Europe, governments in several European countries began to remove perverse subsidies from agriculture.[33] New EU policies encouraged farmers to manage their land to produce a bundle of ecosystem services rather than focusing on crop production alone. These policies required the development of property rights for ecosystem services. That is, people were paid for improving water quality by preserving key watersheds, while others had to pay to release pollution.[34] Soon farmers began providing additional ecosystem services, such as breeding habitat for birds, trout fishing, carbon sequestration, and improved water quality. These policies were reinforced during a decade of floods and droughts that raised the political importance of water and water quality across Europe. The initial success of policies to reduce agricultural runoff, which was lowering water quality and contributing to toxic algal blooms, combined with the realization that farmers could increase their income by receiving money to provide additional services, stimulated the expansion of multifunctional landscapes.[35]

Agriculture changed rapidly.[36] By 2015, roughly 50% of European agriculture, and roughly 10% of North American agriculture, was aimed at balancing the production of food with production of other ecosystem services. In other words, the land was dedicated to the goal of providing multiple services at once—food and some other service, such as recreational opportunities or biodiversity. This resulted in the diversification of agricultural production and lower yields, but increased profits for local farmers.

Buoyed by the success of these changes, the agricultural reform coalition convinced governments to remove export subsidies and trade barriers from global agricultural trade. Agricultural producers in poorer countries used the World Trade Organization and other international organizations to remove subsidies and other agricultural trade barriers. The liberalization of agricultural markets lead to a huge growth in food imports into richer countries, and further stimulated alternative forms of production of agricultural land. Increased imports attracted investment from agribusiness and supermarket chains in agriculture in Eastern Europe, Latin America, and Africa, which resulted in agricultural intensification in these regions. However, this intensification did not follow the same path as in Europe or North America. Agricultural entrepreneurs in these regions bred new varieties of existing crops and created locally adapted genetically modified crops and farming systems. Despite initial opposition in the EU, quick response to several minor ecological problems led to the increasing use, spread, and development of genetically modified crops, especially after the success of locally engineered varieties increased farm production and profitability in Asia, Africa, and Latin America. (See Boxes 8.22, 8.23, and 8.24.)

BOX 8.20

Adapting Mosaic in the Southern Africa Focal Region Assessment

The Adapting Mosaic scenario was interpreted in Southern Africa Focal Region Assessment at the regional (African Patchwork), basin (Local Learning), and local (Stagnation) scales. These storylines are based on an extrapolation of current trends in the region (see the MA *Multiscale Assessments* volume). While democracy and good governance take hold in some countries in southern Africa, severely limited state effectiveness, economic mismanagement, and conflict in other countries prevent the region from improving the well-being of its citizens.

Under the regional African Patchwork scenario, development trends apparent in the South African Development Community region over the past few decades generally persist until 2030. Low economic growth rates and declining foreign investment lead to the increased economic marginalization of Africa. Localized military conflicts continue to drain resources, damage infrastructure, and impede the provision of services.

Improvements in agricultural productivity per hectare are not sufficient to meet the needs of the growing population, resulting in large-scale conversion of woodlands to crops and the expansion of agriculture into marginal lands. While the maintenance of agricultural diversity affords some protection against pest outbreaks, climate change brings more frequent droughts and consequently crop failures, especially in marginal areas. The rural population relies heavily on a declining natural resource base for their subsistence, and many people migrate to cities, where they remain impoverished. Large quantities of food aid are needed to support the urban poor in particular; delivery of food aid in rural areas is impeded by poor infrastructure and conflict. Those rural people with access to land and resources are highly self-reliant, and locally organized. Protected areas are encroached, and wildlife and high-value plants virtually disappear from many areas.

Most governments are unable to ensure the provision of reliable, safe water or modern energy sources, resulting in continued high mortality from waterborne diseases and indoor air pollution, and large-scale deforestation for charcoal production. Poor enforcement of environmental standards, where they exist, make the region a dumping ground for dirty industries and waste from wealthier regions, resulting in deteriorating water and air quality. Water quality is further degraded by increased soil erosion and untreated sewage. A water supply crisis in the shared river basins in the southern part of the region is a major source of regional tension.

In the Gariep basin, the Local Learning scenario illustrates that most local authorities, facing severe budget constraints, are unable to make good on the promises of the free basic water and electricity programs. Lack of access to water, land, and mining rights increasingly causes local tensions and conflict across the basin. The remnants of commercial agricultural are sufficient to feed the urban markets until 2030 but are expanded onto more marginal lands with devastating environmental consequences. The conditions for the urban poor deteriorate rapidly due to the absence of a resource base and a lack of service delivery. The rural poor are isolated by a declining and impoverished infrastructure.

In the absence of effective central governance and market mechanisms, strong civil society networks encourage local infrastructure development, with service provision dependent on community initiative. The rural population, growing steadily and faced with a declining resource base for subsistence farming, becomes increasingly locally organized. Local tourism initiatives that emphasize conservation do spring up in places, and catch the eye of international NGOs, which lend them support.

Throughout the basin, regions are increasingly self-sufficient in obtaining the services they need. Their well-being varies: while there is less ambitious development of resources, there are also lower levels of pollution, producing only moderate declines in water quantity and quality region-wide. The sparsely populated arid west manages to maintain its energy, food, and biodiversity at constant levels. Small declines in energy occur in all other regions. Food production drops drastically in the Great Fish River, where the effects of climate change, land degradation, and a reduced labor force as a result of HIV/AIDS curb the capacity of the remaining arable land to feed its growing population. In the urban centers, reduced economic activity means a slight, though not severe, deterioration in water, energy, food, biodiversity, and air quality, while the minerals industry's output slowly increases, still fueled by foreign investment. For its affluent inhabitants, life is slightly worse; it is much worse for the poor. The arid west elites maintain their well-being at constant rates, while the poor in this region are slightly worse off. In the regional "grain basket," many who rely on a now-reduced agricultural income are slightly worse off regardless of affluence. Lesotho, recognizing the need for economic independence from South Africa, embarks on a program to reform its agricultural productivity, but needs international assistance. Foreign interest in Lesotho is piqued when a local discovery is made that a plant endemic to the Lesotho grasslands has high pharmaceutical value, calling attention to the need for formal conservation of this biome, as well as stronger legislation to protect intellectual property rights.

In the local communities of the Gariep basin, a Stagnation scenario takes hold. With the contraction of the macroeconomy, policies are weakly implemented, and projects are put in place in a heavy-handed way. There are no prospects of employment; human lives and values in the cities are cheap, and people migrate back to rural areas. People wait in vain for relief projects. Pensions and state grants stagnate and are delivered intermittently. The youth remain in rural areas. Basic services and infrastructure are not supplied, and health services are limited to mobile clinics and schools. Schools deteriorate. There are too many children, and classes are held outdoors. People realize that their survival and future is in their own hands, and human capital becomes the main asset. Community identity and bonds are strong. However, the few projects that are implemented from outside create a few jobs, and this tends to undermine community structures. Cohesion suffers because of strong competition for meager resources, and corruption increases.

Land use becomes destructive and land is limited. Grazing lands are taken over by protected areas, and village committees allocate the rest. Less water is available for drinking, irrigation, mining, and livestock. Fuelwood becomes scarcer. Most agricultural production is for home consumption. There are land invasions and uncontrolled extraction, with people making extensive illegal use of plant and animal resources on state land.

The combination of market-based environmental regulation of agriculture with the creation of air pollution emissions trading schemes stimulated a broader set of ecological property rights regimes that encouraged businesses, states, and individuals to adjust their practices and consumption. Decreasing costs and increasing quality operated in tandem with environmental concern to catalyze a broad set of technological changes.

8.6.1.2 Transportation Innovation from Latin America

Small and large high-technology companies in some poorer countries experienced rapid growth during the first decade

BOX 8.21

Potential Benefits and Risks of TechnoGarden

Potential benefits:

- Win-win solutions to conflicts between economy and environment
- Optimization of ecosystem services
- Societies that work with rather than against nature

Risks:

- Technological failures have far-reaching effects with big impacts
- Wilderness eliminated as "gardening" of nature increases
- People have little experience of non-human nature; leads to simple views of nature

BOX 8.22

Genetically Modified Organisms in TechnoGarden

Genetic modification serves as an evolving source of conflict in TechnoGarden. A succession of conflicts arises from continual advances in molecular biology that encourage the expansion, diversification, and regulation of genetic manipulation and modification.

Well-designed agricultural biotechnology, encouraged by multiphase physiological, medical, and ecological regulation, lead to great expansion of the use of GM crops in the early part of the twenty-first century. This expansion leads to conflicts between farmers growing GM crops and those growing organic and other non-GM crops. Farmers not using GM had their crops contaminated by plants and genes from GM crops, preventing the sale of their produce. While these cases lead to the additional expansion of ecological property rights and further encouraged the expansion of multifunctional agriculture, these changes are cold comfort to the small organic farmers who had their livelihoods eliminated.

Conflict over GM crops paled in comparison to the use of GM to produce crops to create pharmaceuticals. "Pharming," while lowering the cost of many drugs, rather predictably lead to cases where drugs contaminated the food supply, causing a range of serious and subtle health problems. The food scares and lawsuits lead to massive losses for farmers and some biopharmaceutical firms, as well as the reorganization, formulation, and regulation of the public use of GM. Pharmers abandoned the use of agricultural crops as producers of drugs and concentrated their farming in areas that contained few relatives of the pharming hosts. This gave a huge boost to the nascent biotech industries in isolated species-poor areas. Iceland, New Zealand, Japan, and Canada all used their industrial infrastructure and remoteness to a huge advantage.

Despite these problems, GM biofuels, such as rapeseed, and fast-growing trees were planted and developed with little opposition. Concern over the use of GM was limited because it did not involve the food supply, and the demand for renewable energy fuels was stimulated by carbon taxes. For decades biofuel plantations were successes for GM technology; however, in the late 2020s a previously unknown fungal disease caused catastrophic forest fires in a large number of biofuel plantations. These agro-industrial accidents led to a further refinement in biosafety protocols and stimulated the growth of multifunctional agriculture, based on the general principle that more diverse biofuel plantations would be less vulnerable to catastrophic surprises.

Many specialized uses of GMs, such as detecting mines, were quite successful. However, the spread of GM technology resulted in the casual use of GM technology that produced substantial environmental damage. For example, the accidental release of GM organisms produced by artists resulted in a series of catastrophic incidents. A GM algae escaped from an art installation at the Bio-(Diverse) City Art Bienalle in Tokyo, for instance. The algal infestation of Tokyo's water system resulted in over 189 deaths, $18 billion in renovation and decontamination costs, 49 criminal convictions, and the bankruptcy of the University of Melbourne (which was operating the exhibit).

Despite these setbacks, the continual development of professional, legal, and technological control methods allowed GM technology to be continually improved and used. GM organisms have produced decreases in soil erosion, salinization, biodiversity loss, and global carbon emissions, but these benefits must be evaluated against the death, disease, genetic contamination, and ecological disruption that GM organisms have also produced. In 2050, many molecular biologists compare GM to other technologies and assert its incredible environmental scorecard, while other scientists object to the loss of non-GM nature.

of the twenty-first century. Advances in computation and communication enabled substantial improvements in logistics, which lowered both the price and environmental impact of a wide variety of businesses and helped ensure that development in poorer countries during the twenty-first century did not replicate that of the countries that became rich during the twentieth. This new kind of development, driven in part by the costs of ecological services, took advantage of profitable opportunities for growth that did not impose huge environmental costs. These differences were perhaps best exemplified by the development of flexible low-cost transportation systems in rapidly growing cities.

Beginning at the end of the twentieth century, the combination of the economic costs of congestion, social concerns over fair access to transportation, and the impact of air pollution on urban health stimulated the development of advanced bus systems in South American cities such as Curitiba and Bogota. These systems used scheduling software, clean-fuel buses, and safe, enclosed subway-style stations. They demonstrated that even in poor, rapidly growing urban centers, mass transit could provide a faster commute, in a fraction of the implementation time of subways or light rail, with a lower economic and social cost. In the early twenty-first century, the replication of these advanced bus systems in other developing cities was aided by improvements in rapid construction techniques and logistics and by the availability of low-pollution vehicles. (See Box 8.25.) These systems required minimal new infrastructure and so were set up rapidly by removing a few highway lanes from the use of the minority of the population who owned automobiles.

The establishment of a variety of national, regional, and global tradable emission permits helped stimulate the spread of these transit systems from Latin America to other rapidly urbanizing cities such as Mumbai, Dhaka, and Lagos. This spread was facilitated by Latin American transit and logistics companies, as well as rival Indian and Chinese transnational bus corporations. By the second decade of the twenty-first century, intense competition between these transit compa-

BOX 8.23

A Story of India (TechnoGarden)

The market was crowded as always. While the bus filled, Raja looked out over the crowd of people. He was tired. Last night he had returned late from his weekend pilgrimage. The travel and excitement had left him exhausted. It had been great to see his cousins again. It had been years since they had been together. His father always talked about the old days, about all the time his brothers and cousins spent together. But now that all his uncles had moved to the city, he and his father were the only ones left, and they didn't farm anymore. Because Raja had never been interested in farming, they had sold most of the land to one of the local farm companies.

His father sold the farm after his mother died from a deadly bacterial infection. She had been healthy, but a week later she was dead. His dad changed after his mother died, he just played on his computer all day. Raja didn't know what to do for his father—other than to play video games with him. It was too bad that his father hadn't come on the pilgrimage. Raja knew that he would have loved to have seen the family, but he was old and weak. The doctors didn't know exactly what the problem was, but he had an unidentified immunological condition that left him exhausted.

After seeing his cousins, and hearing their tales of Bombay, Toronto, and Bangkok, the market seemed dull and provincial to him. His uncles had tried to convince him to use his genomics database administration skills to get a contract overseas, but that would mean leaving his father, who was sick, and he still hadn't been able to get his second-class genomics license.

As the bus pulled away from the market station onto the highway, it accelerated through the large irrigated fields of GM cotton that surrounded the town. The bus briefly stopped at the big temple complex outside of town, where a few French-speaking European monks got off. They were all part of the international project to rebuild a "famous" twelfth century temple garden that Raja, at least, had never even heard of before the monks had arrived. But they were fairly friendly, polite, and good gardeners.

After 10 minutes speeding through multihued agricultural fields they finally stopped at the grove of trees that held the offices of Raja's employer, Naidu Biosystems. Raja left the bus and saw his colleagues Sam, Kiran, and the new engineering intern from Bangalore, but he was too tired to chat right now. He shuffled along the gravel path, through the trees, and into his group's air-conditioned office. He was the last of his team to arrive and his coordinator, Ms. Patel, wished him good morning while she looked at the clock. She then asked him to check the field's aphid distribution data and compute the intervention options. It was a long, dull job. He slumped into his chair and then began poking and tapping his way though the data that streamed from the surrounding fields into the building—he sighed and got down to work.

nies and car manufacturers led to substantial improvements in vehicle technology, systems management, and emission reduction technology. New urban centers operated transit systems that provided better service at less cost and had fewer health and environmental impacts than their equivalents in wealthier countries.[37]

Some cities in the rich nations also adopted these systems, but not all. In most cities in these countries, transit services were increasingly diversified and flexible. Especially in dense cities, the costs of car ownership, particularly parking fees and congestion, stimulated the rapid growth of car-sharing companies. Urban residents increasingly avoided the costs associated with car ownership and chose instead to use a combination of taxis, mass transit, local car-sharing networks, and longer-term car rental agencies, which provided them with a range of affordable, flexible transportation services.[38]

8.6.1.3 Islands in the Net

In the early years of the twenty-first century, national governments became more deeply enmeshed within some global institutions while withdrawing from others. The success of EU federalism served as a model for the creation of regional partnerships among nations. The lessons of the EU were used to argue for its expansion and to design regional confederations, such as those that loosely joined more prosperous countries in South America, southern Africa, Southeast Asia, the Caribbean, and Polynesia. As a result, economic integration within Asia and Latin America increased.

Due to security concerns and other stresses of globalization, rich countries attempted to increase restrictions on trade and immigration, but this was prevented by pressure from a coalition of transnational companies, retailers, and cosmopolitan communities, particularly the Chinese and Indian diasporas. This cosmopolitan alliance pushed for globalization, the relaxation of trade barriers, and efficient systems of immigration (especially for the well educated and the rich). The alliance was able to overcome isolationist interest in a number of rich countries in Asia, Europe, and North America, which led to an even stronger set of global organizations and treaties.

Within this dense global net, some islands remained isolated from the expansion of global civic society and benefited little from increases in wealth and health in the world at large. Regions in conflict failed to attract international investment or much help from global civil society. In particular, drylands in Central Asia and the Sahel regions in Africa were subject to droughts and suffered from food shortages and chronic malnutrition. Weak or failed states remained havens for criminal networks, drug manufacturing, financial fraud networks, pirates, and guerillas. (See Box 8.26.) These groups benefited from lack of control in these areas, and often maintained conflict. Along with these local troubles, these areas often exported violence, disease, and pollution. While some war zones became involuntary parks—areas in which political and technological collapse allowed natural processes to reassert themselves (see Box 8.27)—the majority of the war zones suffered ecological degradation from unsustainable hunting and fishing, water pollution, mining, and deforestation.[39]

8.6.2 TechnoGarden 2015–30

8.6.2.1 Consolidation of Globalization

Transnational corporations, international NGOs, action groups, and global associations continued to flourish and

BOX 8.24

Sub-Saharan Africa under TechnoGarden

While sub-Saharan Africa entered the twenty-first century facing the challenges of poverty, disease, environmental degradation, and conflict, the increasing ecological awareness of the global population had rapid and beneficial effects on these challenges. The gradual elimination of agricultural subsidies in richer countries, especially in Europe, opened these markets to agricultural products from sub-Saharan Africa. Tens of billions of dollars flowed into African economies, allowing for the management of debt and the development of key infrastructure in transportation, communications, health, and education. Many major cities in this region became cleaner and more livable from these developments, and urban life expectancies rose in response.

Though market forces were keys to the success of agriculture in sub-Saharan Africa between 2000 and 2015, the introduction of GM crops to this region was also a critical development. Sub-Saharan Africa became something of a testing ground for these and other environmental technologies, with the result that many early feedbacks and problems were concentrated in this region. While these issues were eventually managed, the ecological damage they caused is still visible in many parts of this region through biodiversity loss and landscape change. Further, the widespread focus on agriculture in this world region has also contributed to massive intentional landscape transformations, such that today entire ecosystems, such as the Upper Guinea Forest, no longer exist in any functional manner. Instead, small remnants remain as protected areas—tourist attractions for those who seek to visit "the forests of Africa."

While much of sub-Saharan Africa saw its economic fortunes improve across this early period, several states lacked either the political will or ability to muster the resources necessary to take advantage of these changes. Further, true-cost pricing of fossil fuels struck a major blow to the economies of several nations around the Gulf of Guinea, as oil became a far less lucrative export. Smaller petrostates experienced coups and political instability, while larger states with more diversified economies were able to weather these economic changes through agricultural transformation. Thus sub-Saharan Africa found itself dealing with nations of strong growth next to nations of little or no growth and great conflict.

The nations experiencing strong growth recognized the threat that regional conflict presented to their long-term stability and growth, especially in the form of cross-border migration and international image. Rather than allow their neighbors to continue unchecked, as they had in the days of the Organization for African Unity, under the African Union stronger nations like South Africa, Ghana, and Nigeria began to exert political, economic, and, where necessary, military influence on their neighbors to ensure regional stability. The long-term result of these early efforts was the development of two strong regional organizations, one growing out of the Southern Africa development community and the other from the Economic Community of West African States. These organizations created shared currencies, promoted economic integration, and greatly enhanced the stability of their regions.

The increasing political stability and economic growth in sub-Saharan Africa enabled the use of highly developed technologies for ecological management. Further, the development of pop-up infrastructure allowed for the provisioning of economic, social, and ecological services to a wider constituency than ever before. In 2050, sub-Saharan Africa is hard to recognize from the perspective of the late twentieth century. Much of the ecology that shaped the industrial world's imagination about this region no longer exists or does so only in small reserves. The violence that so characterized this region in the late twentieth century has largely abated thanks to regional and subregional organizations. In the place of this mysterious, violent territory stands one of the globe's "breadbaskets," with some of the cleanest cities and most rational land use in the world.

BOX 8.25

Bogota's Success Story: The Bus Rapid Transit—A Solution for Megacities in Developing Countries

By 2030, 3 billion more people will be on Earth. Developing-country governments have noted with alarm that most of this growth will be in their already congested and polluted megacities and could be crippling. Novel transportation systems will be essential. One such system, Bus Rapid Transit, or BRT, is a high-speed, low-cost public transportation innovation first developed in Curitiba, Brazil, in the early 1980s.

The BRT model has helped to transform developing-country cities like Bogota and could help to avert severe crises in the transportation, environment, and health sectors (Ardila and Menckhoff 2002; Fouracre et al. 2003). Major cities in China and Mexico are looking to follow suit. Bogota's Transmilenio, as its BRT system is called, transports subway-level capacities of 1 million people a day at high speeds, but has two critical advantages over a subway system. It costs only 5% of what a subway with similar reach would, and it can be built in just three years—a fraction of the time needed for a subway.

What distinguishes BRT from conventional bus systems so that even the middle-class residents with cars prefer to use Transmilenio in Bogota? A successful Bus Rapid Transit system like Transmilenio is an integrated one where buses move at very high speeds, since they are physically separated from car lanes and can signal traffic lights to turn green; where people can get on and off in seconds, since riders prepay before boarding and buses have wide, low-level doors like a subway; where there are several express busses that do not stop at every station, but connect to local buses, similar to a subway model; and where stations are well maintained and have parking lots and taxi access plus excellent bicycle, pedestrian, and disabled facilities.

In 2004, the success of BRT systems led the Chinese government to begin developing policies to bring Bus Rapid Transit to its major cities. BRT in Bogota not only averted a crisis, it transformed the public space by lowering air pollution (since significantly fewer cars come into the city) and improving the quality of life in the urban core. Well executed, it promises to do the same for more of the developing world's overcrowded and polluted cities.

BOX 8.26

Failed States, Pirate Zones, and Ecological Disruption across the Scenarios

Many areas with the lowest state of human well-being and the greatest poverty are areas in which states have failed to maintain order. Most of these areas are within the remnants of failed authoritarian states or in the interior of states whose internal control has failed.[a] Following the end of the cold war, powerful countries stopped investing in some of the governments they had been supporting because the governments had become expensive and unimportant. In these areas, government sometimes collapsed, leaving a void where central government has not been able to enforce its rule. Frequently, militias supported by smuggling or extraction are based in regions like this that are not governed by states, where the central government has failed to provide any type of security, prosperity, or freedom to its inhabitants.

In these areas, various armed groups engage in violence to disrupt state intervention and to promote their own power and profits. These groups engage in resource extraction and smuggling—growing, processing, and distribution of narcotics. Often, groups in these areas have links with international criminal organizations, armed militias, and guerillas. For example, in many parts of the world drug production and smuggling is used to fund militias, while in other places militias are paid a tax to protect drug production from state intervention. Often the presence of valuable resources in a region can sustain conflict among various armed groups, because the presence of valuable, easily transportable commodities such as gems, gold, or narcotics provides funds to attract and arm militias. During the early twenty-first century, there were a number of such areas in which conflict combined with resource extraction in areas as diverse as the Congo River basin, the Thai/Burma border, and Colombia.

Improved transportation, communications, and logistics facilitate the activities of these groups by providing better opportunities for the movement of illicit drugs, weapons, and money and illegal immigrants. These groups innovatively fuse cutting-edge technology, local knowledge, and family ties to coordinate their networks, broadcast propaganda, and move people, guns, drugs, gemstones, and other valuable commodities across national borders.

These areas are often bad places for people to live. In particular, in areas where contesting militias exist, local people are often the targets of theft and violence. Disease is common due to a lack of public health infrastructure. These conflict zones have mixed ecological consequences. Small areas of heavy exploitation are used for mining or drug production, while bushmeat hunting and timber cutting occur over more substantial areas. Drug manufacturing can result in forest clearing and chemical pollution of streams and groundwater, while attempts at narcotic control by the larger world can also pollute local ecosystems. For example, anti-narcotic programs have resulted in the haphazard use of pesticides; stimulated land clearing, deforestation, and erosion; and increased the mortality of wildlife.

Impoverished, unsafe people are unlikely to be able to organize themselves to steward local resources effectively; indeed, there is every incentive for them to extract resources as quickly as possible. On the other hand, if conflicts prevent the clearing of large areas of land for agriculture, they may inadvertently conserve some aspects of local fauna. In other cases, however, there are more negative consequences for ecosystem services.

The world outside these areas both takes advantage of their chaos and suffers from it. The lack of any government to protect the public interest allows foreign companies to dump toxic waste and extract unprotected natural resources. For example, fishers from all over the world have been fishing off the unregulated coast of Somalia, while other boats have been dumping toxic waste there. At the same time, international shipping is affected by the disorder there by frequent pirate attacks (Christian Science Monitor 1997; The International Chamber of Commerce Commercial Crime Service).

While the piracy, trafficking, and conflict in these areas often spills out into the outside world, these spillover effects are frequently tolerated due to their relatively minor impact and the difficulty of intervening in the lawless areas. The persistence of these areas depends on the inability of local governments to control the area, the disinterest of global institutions or neighboring countries to intervene, and the continued profitability of conflict, smuggling, or piracy. For marginal places, with minimal resources of interest to the outside world, it takes extreme events, such as the September 11, 2001, attacks, to stimulate world forces to intervene in these areas.

The likelihood of lawless regions is higher in some scenarios than others. Pirate zones arise due to local issues. But the response to local problems depends on the role of the region within the world and the attitude of the larger world toward such regions. The likelihood of such regions both arising and persisting varies across the scenarios.

These sites are mostly likely to arise in a more fragmented, less globalized world like that of Order from Strength. The presence of extensive global trade, multilateral treaties, and international NGOs is likely to result in intervention from outside a region, either humanitarian or military, to reincorporate such regions within the global system. These regions are also more likely to arise when conflict and injustice are allowed to thrive within countries, and there is little effective regulation of global trade.

[a] As part of the MA, the Colombia Sub-global Assessment focuses on Colombia's main coffee region, which is located in the central part of the country and includes the mountainous regions of Antioquia, Caldas, Risaralda, Quindio, and Valle. In this region, traditional agriculture has transformed mountain ecosystems into rural landscapes. The assessment attempts to understand the factors that cause changes in ecosystems and how those drivers generate serious impacts that can in the long term cause environmental, economic, and social imbalances.

thrive. Easy travel, cheap communication, social software, and "intelligent" databases aided the management of these groups.[40] International migration increased, producing increasingly intertwined diasporas and an increasingly interconnected, powerful, and diverse global civil society. The impact of new civil organizations was diverse and multifaceted. Decreased costs of communication and travel stimulated the formation of many NGOs, but the main beneficiaries of these advances were the many technical and professional associations, who globalized and intensified their activities. International technical standards groups, formed by technical associations and endorsed by transnational corporations and national governments, developed increasing numbers of global technical standards for telecommunications, transport, manufacturing, and environmental management. One of the major symbolic steps in the process was the adoption of the metric system by the United States in 2029.

The proliferation of international standards in turn increased the power of professional associations; facilitated

BOX 8.27

Involuntary Parks across the Scenarios

Buffer regions between conflicted states as well as contaminated areas can become perverse nature preserves. The health and security risks of people entering these regions are so great that human activity is excluded. People are unable to use these areas to produce local ecological services, but they can benefit from the refuge that the areas provide to wildlife. The migration of people out of these regions, the presence of minefields, and continuation of conflict can in some ways act as a perverse form of conservation by creating involuntary parks.

The Demilitarized Zone between North and South Korea provides an example produced by conflicts between states. The Korean DMZ, which is roughly 250 kilometers across and 4 kilometers wide, contains the largest area of forest in the Korean peninsula and has been only minimally affected by human activity since the end of the Korean War in 1954. It contains a variety of species, including white-naped crane and the red-crowned crane as well as the Asiatic black bear, that are rare or endangered in the rest of the Korean peninsula (Government of Korea (website); Kim 1997).

The Chernobyl accident in the 1980s produced a large radioactively contaminated area in the Ukraine. While the radiation in this area increases mortality of people and wildlife, the absence of people has resulted in forest growth and a great increase in wildlife population. The slow decline in radioactivity has led to an increase in disaster tourism. The soil remains contaminated with radioactive fallout, however, preventing human use.

These circumstances could arise in all four scenarios, but the likelihood is greater in some than others. Conflicts and industrial disasters are both more frequent in Order from Strength, and there is less interest in and capacity for restoring such areas. In Global Orchestration there may be industrial disasters, but there are fewer unresolved conflicts, and there is global interest in repairing contaminated areas, leading to fewer involuntary parks. In TechnoGarden there are more involuntary parks, since more ecological engineering provides more potential for disaster, and pursuit of profit rather than equity leads to less repair of involuntary parks. However, increasing capacities in ecological engineering and potential profits available from repairing such areas leads to a gradual increase in their repair. Adapting Mosaic results in fewer large involuntary parks, as local organizations frequently occupy and attempt to make the best of small contaminated or conflicted areas.

more transnational collaborations of engineering, medical, and legal associations; and increasingly influenced government and corporate policies. While these groups encouraged the free flow of information within their professional organizations, they raised barriers to entry into these professions.

The process of globalization also focused on regulating many global and regional ecological commons. The success of a revised Kyoto treaty, and the new commitments taken up for the post-Kyoto period, combined with the growth of global institutions, resulted in a number of frameworks to govern the high seas, the atmosphere, and transboundary seas, rivers, and wildlife. One high-profile case was the increased regulation of high seas fishing in an attempt to deal with declines in catch from pelagic fisheries. Tuna, and later other large fish, were "ranched" using implanted transponder chips, radiotelemetry, and satellite monitoring. While these global regulations were often successful, formalized use tended to displace many local resource users, such as small-scale fishers.

8.6.2.2 Urban Eco-Development

The rapid urbanization of Asia and Africa during the twenty-first century resulted in both health disasters and unexpected successes. Asian cities experienced dynamic periods of urban reinvention. The regional dialogue between cities about their approaches to urban reinvention stimulated a new Asian urbanism and an approach to urban living that was emulated and adapted worldwide through its promotion and implementation by a competitive group of Asian-based transnational construction and maintenance corporations.

New-Asian urbanism combined existing technologies in novel ways and stimulated new ways of thinking about city planning. People found new uses for composite materials, and produced flexible, green building materials, which lowered building energy and water use while improving quality.

Asian urban areas, with their new construction, transportation, and manufacturing, produced rapidly increasing air pollution and carbon emissions. Global mean temperature had increased by over 1 degree Celsius, leading to further restrictions in emission permits. Increasing emission costs and the demand for healthier, cleaner cities enabled the rapid growth of innovative Indonesian design and architecture companies. The successful franchising of bus systems by Latin American transit corporations in combination with locally developed "green" housing techniques helped achieve this goal and produced surprisingly attractive cities that increasingly became tourist centers.

Unfortunately, much of the innovation of New-Asian urbanism was driven by the need to help people cope with environmental problems caused by rapid urbanization. Many urban ecological engineering efforts were the result of catastrophic disease outbreaks, with especially severe problems in rapidly growing tropical cities, such as Dhaka, Bangkok, and Manila. Poor water quality was a major chronic cause of disease in many rapidly urbanizing cities. Poverty and poor sewage spread cholera and typhoid. Irrigation projects spread schistosomiasis. Poor drainage in slums allowed populations of disease-spreading mosquitoes to thrive.

However, the emergence of several families of new respiratory diseases with high social and economic costs resulted in a global movement to enact health-oriented reforms that used ecological engineering to improve water quality in many cities (World Resources Institute 1998). In many regions, people began to modify and convert existing irrigation systems, often using multiple-use, managed canals to better connect cities and countryside. These urban technologies were quickly transplanted and adapted by the cities of Asia, especially new cities in China, and then gradually spread to some parts of Africa and Europe.

In richer countries, there was a different set of urban issues: cities were aging and shrinking. "New-Asian" urbanism emerged in countries like Japan as a way to make cities more beautiful, livable, and healthy. The Japanese approaches sought to combine Asian traditions with large-scale ecological restoration, which resulted in a substantial return of species formerly displaced from urban areas. New-Asian urbanism became a model for revitalizing the stagnating major cities in all richer countries.

Urban restoration, using techniques from agriculture, ecological engineering, and simple nano-machines, led to some enormously beautiful and popular restoration projects on Japan's inland sea. But it also produced some of the first biomechanical disasters, such as vast algal blooms due to unexpected interactions between water purification nano-machines and invasive waterborne bacteria. Fortunately, these blooms were quickly controlled and eliminated, but not before the catastrophic loss of most large marine life off the coast of Japan.

8.6.2.3 Green Design and Ecological Agriculture

The gradual removal of subsidies for fossil fuels, nuclear energy, and large hydro projects was motivated by the combination of global trade liberalization, local health concerns, and the rising costs of oil and natural gas as a result of the development of a variety of smaller-scale energy projects in the developing world. Large energy companies funded some projects, but there were many successful start-up energy companies founded by Chinese and Indian entrepreneurs. These developments were stimulated by the Global Environmental Facility's New Energy Fund, which was set up to mitigate climate change by stimulating the development of low-carbon emission energy systems. These efforts led to substantial decreases in the cost of power produced by wind, solar, and biofuels, as well as great increases in effectiveness of fuel cells and low-pollution fossil fuel power plants.

Following successes in cities and agriculture, people increasingly began to apply ecological engineering to optimize the supply of desired ecosystem services. Consequently, ecosystems were increasingly shaped to provide different bundles of ecosystem services. Ecological engineering was done privately at local, small, or regional scales by a variety of private, public, and community and individual actors.

Over time, the economic benefits of ecological engineering increased due to advances in technology and the decreasing costs of engineering techniques. Additionally, the costs of not engineering skyrocketed due to increasing local ecological problems that were most easily and quickly solved through ecological engineering techniques. Ecological engineers used advances in computer, communication, and materials science to build ecological infrastructure that was increasingly flexible, dynamic, and adaptive. Distributed monitoring networks operated in conjunction with advances in precision agriculture, allowing ecological dynamics to be cheaply steered. Innovations such as pop-up infrastructure, which only existed when needed, allowed people to intervene in ecological dynamics with far fewer direct, inadvertent side effects. One persistent issue, however, was the discovery that desired populations of wildlife depended on some previously unknown aspect of local ecosystems that was eliminated once ecological engineering improved the system.[41] It frequently turned out that ecological engineering solved the problem it was aimed at, but created some decline or problem in provision of another ecosystem service.

Unintended consequences of the increasing use of genetically modified crops, including wildlife die-offs and allergic reactions, eventually led to strict programs of testing and certification of genetically modified organisms. These regulations stimulated the further development of large-scale agriculture that used micro-doses of pesticides and introduced insect predators to control insect damage to crops. The fusion of some of these techniques with contemporary low-input farming methods produced large-scale ecological precision agriculture that included mapping, tracking, and biomonitoring of trace elements and disease.

8.6.3 TechnoGarden 2030–50

8.6.3.1 Technocrats Ascendant

The complex interlocking nature of the global economy and increased use of technology made it difficult for companies to adopt anything other than global standards. These technical standards, which played such a powerful role in determining policy and action, were set by scientific societies, professional groups, and their affiliated corporations and NGOs. The unelected technocrats in these organizations became increasingly important legislators of the world. The increased enforcement of global environmental and civil regulations by technocrats lead to an intense series of "police and nation building" wars in the 2030s as regions once controlled by militias, guerillas, or local groups were forcibly incorporated into the global economy. Ecological restoration projects, backed by military force, were implemented in conjunction with large-scale health and education networks.

Continued gradual improvements in energy efficiency, alternative energy, and biofuels left the Middle East with much lower than expected oil wealth. Social problems were further complicated by migrants from degraded dryland areas in the Sahel and Central Asia. Social tension led a globally connected diaspora of young people to violently contest the future of countries within this region until, aided by their foreign allies, a technocratic civil society of economists and engineers prevailed over religious leaders and nationalists and began to open and reform the economy. Innovative experiments, which adapted and combined technology and institutions from Asia, Africa, North America, and Europe, allowed the region to develop solar-driven technology that transformed its coasts and river valleys.

The openness of science and the technocrats' great faith in scientific and technological progress to stimulate economic and social growth moved intellectual property away from closed proprietary systems and toward more-open systems.[42] One side effect of this was the general availability of sophisticated open source educational resources, which greatly improved the educational opportunities available to most of the world's population. While information was freely available, however, access to professions was strongly regulated. Universities increasingly served as places for social networking and were differentiated by the opportunities for practice, experimentation, and research.

8.6.3.2 Eco-technology

By the 2040s, cheap, reliable eco-technologies were available worldwide. While a few of these were developed in richer countries, a major growth of eco-technology occurred across the poorer section of the world. Small companies and cooperatives developed eco-technologies to address local needs and local markets. These technologies gave many local communities the power to improve their situation by providing valuable ecosystem services, such as fresh water to nearby cities and exotic organic produce to the rich world. In the poor nations, these technologies substantially controlled AIDS, malaria, and many other diseases. The reliable provision of ecosystem services, along with economic growth, lifted many of the world's poor into a global middle class.

The continued development of new energy technologies resulted in substantial increases in the cost-effectiveness of some alternative energy sources, such as wind and solar, and great increases in effectiveness of fuel cells and low-pollution fossil fuel power plants. In many poorer countries, local companies developed efficient biofuels to replace inefficient charcoal and wood-burning stoves, providing people with more time, more forests, and better health. In conjunction with the decline in coal use, Asian engineers developed low-pollution methods of using coal to produce natural gas and hydrogen. Thus conditions for an effective response to climate change were very good under TechnoGarden. Due to the combination of greener energy production and more-efficient energy use, global greenhouse gas emissions peaked in the 2020s and declined to levels below those of the start of the twenty-first century well before 2050. However, because declines in emissions of particulates from fossil fuel burning temporally offset decreases in carbon emissions, there was still a 1.5-degrees Celsius increase in global average temperature by 2050.

8.6.3.3 Eco-urbanism

By 2035, most of the world's large cities had become polyglot, cosmopolitan places due to global connections and migration. Though some aspects of local culture were eliminated, cheap global communications created a huge diversity of global subcultures. While many of these subcultures were technical and global, intersecting groups were often locally based. Most of the world's urban population lived in new cities that were strikingly different from cites 50 years earlier. Ecotechnology and ecological markets allowed cities to vary their zoning, development, and architecture to local climate and ecology, even as global technology and culture drove city policies to be more similar. Unique aspects of locality led to new cultural traditions, such as the salmon festivals of the American Pacific Northwest and the Gojiro festival that symbolically destroyed large areas of Tokyo every five years.

Many people owned cars, but rates of car ownership converged worldwide such that the countries that were industrialized at the beginning of the twentieth century had fewer cars per capita than they did at the turn of this century, and many other countries had more. Urbanization and the creation of diverse, flexible, and adaptive transit systems reduced the relative value of car ownership in big cities. Most city dwellers chose to rent cars when they needed one. As air pollution and other car-related pollution problems got worse in high-density cities, consumers demanded cars that produced less pollution. Chinese and South Asian automobile manufactures led the way in providing affordable, fuel-efficient, low-pollution vehicles. New urban transportation systems substantially reduced emissions, increasing urban air quality and decreasing health costs of transport and greenhouse gas emissions.

Ecological restoration adopted and adapted many technologies from green precision agriculture that allowed a more fine-tuned approach toward steering an ecosystem through succession and response to external shocks, such as flooding and drought. These technologies were used for ecological restoration, gardening, and occasionally the creation of entirely novel designer ecosystems. These led to fierce conflicts between restorationists, who wanted to "re-wild" landscapes and bring back the Pleistocene era, and those who wanted to create ecosystems that functioned in completely novel ways.

8.6.4 The TechnoGarden World in 2050

The world in 2050 is cosmopolitan and wealthy, but unequal. Nine billion people occupy the planet, half of whom live in Asia. Poorer countries have almost the same share of the world economy as richer ones, but a much lower per capita income. Migration, urbanization, and global interconnectedness have transformed or eliminated a huge number of local, rural, and indigenous cultures. While the number of people living in absolute poverty has declined, huge differences in income exist between richer and poorer countries as well as within nations. This inequality is blamed for social tension, crime, and sometimes terrorism.

There has been considerable progress in addressing environmental problems related to pollution[43] and human health. Global agreements led to new management approaches to global and regional commons, such as the atmosphere, the oceans, and large rivers. The global energy system remains dominated by fossil fuels, but the dominance is gradually declining. Emissions of greenhouse gases are lower now than they were at the start of the century, despite a much larger global population and economy. The establishment of property rights has provided further incentive to the production of clean energy sources, efficient transit options, multifunctional agriculture, and various forms of eco-technology.

Societies have engaged in many ecological engineering activities to provide desired ecological services. Increasing wealth, expansion of education, and growth of the middle class led to demand for cleaner cities, less pollution, and a more beautiful environment. For example, in many rich countries new housing developments include rain gardens and wetlands to clarify runoff and provide wildlife habitat. People engage in different activities depending on the ecosystem services they desire and the difficulty of providing those services. For instance, planting trees for local climate moderation is much easier than designing an ecosystem to sustain a tiger population. These differences in abilities and potential produce regional differences in types of ecological modification. In general, richer countries focus on providing water regulation services, amenity services, and cultural services while poorer countries focus more on the production and regulation of water and the production of ecological goods and services. Within the rich and the poor worlds there are regional differences due to culture and the way that property rights are organized and the density, wealth, and activities of people within a region.

Ecosystem services have changed since 2000. In some areas they have improved beyond what experts expected at the end of the twentieth century. Despite successes of ecological restoration, in some cases ecological simplification has been irreversible. For example, despite substantial efforts at restoration, coral reefs have not improved greatly. In general, provisioning services have increased, but biodiversity has declined.

Many regulating ecosystem services have become less resilient and dependent on continual human management. Access to basic ecosystem services has improved for most of the world's poor, but at the cost of wilderness and the loss of populations of large mammals.

8.6.5 Challenges for TechnoGarden 2050–2100

Earth's ecosystems have been transformed by human intervention. While management has often been successful, it has become increasingly intense. Despite significant advances in rates of recycling and reuse in all sectors, increases in the consumption and use of materials and increases in human population mean that humanity's impact on global and local ecologies continues to intensify. Due to ecological engineering, however, some of these changes produce positive impacts for nature and humanity.

While the provision of basic ecosystem services has improved the well-being of the world's poor, the reliability of the services, especially in urban areas, is increasingly critical. While management has been able to cope with this complexity in the past, today people question whether depending on so much management is wise.

The privatization, professionalization, and ubiquity of ecosystem management have led to increasing risks of management failure. Ecosystem management tends to simplify ecosystems, particularly as the intensity of management increases, because the more obscure, apparently unimportant, or simply unknown processes are not supported or maintained.

The highly engineered ecosystems and ecosystem services that are now found in the world make these systems vulnerable to disruptions. Even subtle, successful ecosystem management poses a number of risks that arise from the loss of process diversity, loss of local knowledge, and increasing reliance on ever-decreasing variance in the supply of ecosystem services.

Increasing social reliance on the reliability of the provision of ecosystem services has led to a gradual decline in the maintenance of alternative mechanisms of supplying these services within regions, leaving the systems and people who depend on them increasingly vulnerable to fluctuations in supply. In cases where the increased risk of variability has led to more interventions and attempts at greater control, the need for extremely reliable provision of services has created a spiral of increasing vulnerability.

Furthermore, ecological engineering tends to focus on particular processes and services, which leads to problems at the boundaries between ecosystems and emergent problems where subtle local effects of engineered ecosystems interact at large scales to produce surprising fluctuations in the functioning of ecosystems. For example, agricultural areas rely on the existence of pollinators, which depend for their survival on surrounding ecosystems. Changes in these surrounding ecosystems that reduce the population of pollinators can increase the vulnerability of the agricultural areas. Sometimes these conflicts can be solved with changes that are mutually beneficial, but in other locations conflicting goals result in ecological instability. Difficulty in managing the flow of material and species among ecosystems and difficulties in managing subtle, slow changes have caused society to overlook a set of emerging cross-scale feedbacks that threaten to bite back beyond 2050. These problems are exacerbated by confident application of technological fixes without understanding various cross-scale system dynamics.

Although there is a belief that nature is important and should be monitored, increasing confidence in technology has allowed the monitoring effort to wane somewhat. Furthermore, a disproportionate amount of effort is put into technology for landscape manipulation rather than monitoring technology, with the result that by 2050 there is a dangerous imbalance between our ability to create unintended ecological feedbacks and our ability to detect them in time to deal with them and then respond to them effectively.

While the world has grown wealthier overall, persistent hunger remains in many regions due to poverty and conflict. Coping with ecological variation and environmental surprises are the main challenges facing the world at the start of the second half of the twenty-first century.

8.7 Cross-cutting Comparisons

The scenario storylines have not attempted to include discussion of all ecosystem services. This would have made the stories too cumbersome. Detailed quantitative consideration of impacts on some ecosystem services across scenarios is given in Chapter 9.

The reader might be surprised that all scenarios have both positive and negative aspects with respect to ecosystem services and human well-being. This was intentional. Some of the extreme risks in each scenario are explored in Boxes in each section.

The scenarios were developed around a set of logics (assumptions), and we tried to keep those assumptions constant throughout the stories. This enabled us to explore the implications of those assumptions. As our storylines developed we realized that no world based on fixed assumptions can hope to achieve sustainability of ecosystem services and human well-being.

This might seem self-evident, but commentators often give "recipes" for the perfect world that are essentially lists of fixed strategies assumptions. One of the strongest lessons for us from this project has been that success in achieving a sustainable world will require the strategic application of the full range of approaches to social, economic, and environmental management—including at times, perhaps, strategies that tend toward compartmentalization (for example, policies aimed at limiting the spread of invasive species). Our intent in developing four extreme stories was to draw out the situations in which any of the strategies could have unintended negative impacts.

Later chapters will draw out comparisons across scenarios in more detail. This section draws comparisons across scenarios with respect to issues of importance to:

- the major Conventions supporting the Millennium Ecosystem Assessment (Convention on Biological Diversity, U.N. Convention on Combating Desertification, the Convention on Wetlands, and the Convention on Migratory Species);
- a set of five key issues of concern identified by the MA (emerging diseases, decline of fisheries, climate change, eutrophication, and desertification); and
- a few other issues that have arisen as important in the development of the storylines (including invasive species, urbanization, and an example of how a present-day ecological challenge in the United States might be addressed in the future under the four scenarios).

8.7.1 Biodiversity, Wetlands, and Drylands

Figure 8.5 depicts a scorecard for the four scenarios with respect to biodiversity, wetlands, and drylands—indicators important to the conventions on biological diversity, wetlands, and desertification. For each variable, the two arrows indicate how we expect the indicator to develop within this scenario over the next 50 years. The full lines indicate the "best" possible case, whereas the dashed lines show the "worst" case. Note that these results have been derived by combining quantitative results from modeling (see Chapter 9) with qualitative considerations based on the individual storylines. In order to depict differences between scenarios more clearly, we have included statements on the major drivers of changes in the respective indicator.

8.7.1.1 Biodiversity

In Chapter 10 the outlook for biodiversity in the four scenarios is considered with respect to published trends and predictions from quantitative models. Here we explore ecological feedbacks that cannot presently be included in models, due to insufficient data, so the curves described are sometimes more complex than those in Chapter 10.

Chapter 10 argues that over the near future, TechnoGarden and Adapting Mosaic are likely to see a decrease in the rate of biodiversity loss. In TechnoGarden, this is due to a significant reduction in land use change. In Adapting Mosaic, it is due to a reduction in invasions of non-native species. Movements of invasive species between countries also are reduced in Order from Strength, but the number of outbreaks within poorer countries is higher and the overall chance of invasives reaching richer countries is still high. Also, land use changes, particularly in poorer countries, put high pressure on habitats. Land conversion pressures are much more significant for Global Orchestration due to the scale of conversion.

In the longer term, two different types of risk might counteract the positive developments within TechnoGarden and Adapting Mosaic. TechnoGarden bears the risk of significant, major technological failures that can lead, for example, to outbreaks of new pests and diseases threatening biodiversity. Thus, the worst-case curve for biodiversity in Figure 8.5 shows a rather abrupt increase in biodiversity loss midway through the period, which flattens out after appropriate countermeasures are developed. Toward the end of the scenario period, the rate of loss starts to decline as new knowledge is developed and brought to bear.

In Adapting Mosaic, failed experiments and climate change might also increase the rate of biodiversity loss in the longer term, but this is not seen as happening as abruptly as for TechnoGarden. The other two scenarios see some chances for reducing the rate of loss on longer terms, though this is more pronounced in Global Orchestration than in Order from Strength (and mainly in richer countries, where economic prosperity is higher). Within Global Orchestration, a decrease in the rate of biodiversity loss would occur if increased economic prosperity brings reduced pressure on the environment (a central assumption of this scenario). However, climate change might counteract the optimistic outlook for both Order from Strength and Global Orchestration.

8.7.1.2 Wetlands

Although population pressures assumed in the quantitative modeling are high for both Order from Strength and Adapting Mosaic, we could see a relatively slow loss of wetlands in the short term in these two scenarios, for different reasons. The "disconnected" character of these scenarios and the reduction in international trade and direct investments, particularly in poorer countries, could curtail the construction of new large-scale water schemes within the

Key:
++ strongly increasing pressure by this driver
\+ increasing pressure
0 no change when compared with today
– decreasing pressure
– – strongly decreasing pressure
→ a change in the pressure of the driver during the scenario

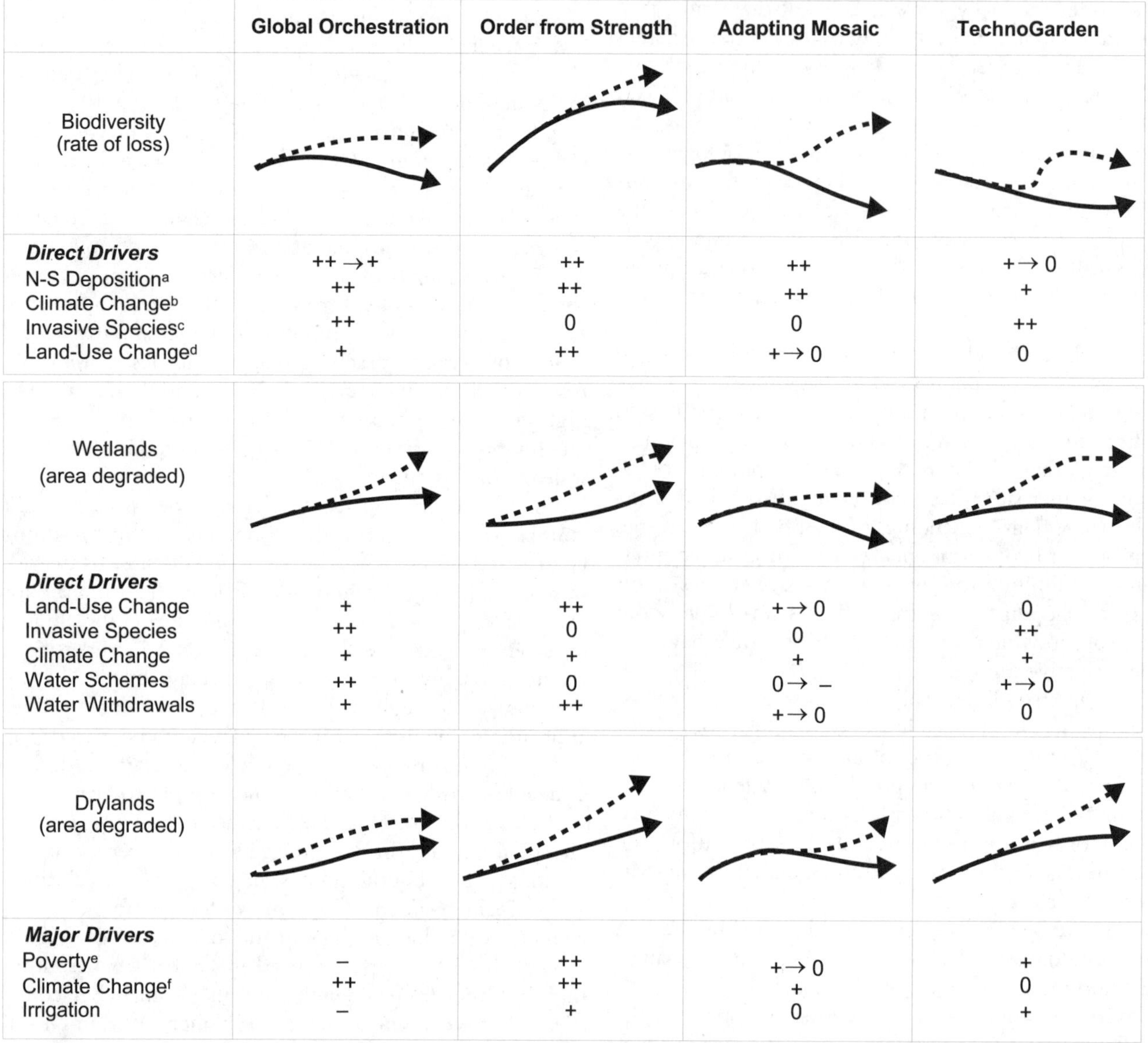

Figure 8.5. The Convention Scorecard. Arrows indicate the development over time of the issues named in the left-most column, which are important to the Convention on Biological Diversity, the Convention on Combating Desertification, and the Wetlands Convention. Solid lines indicate the best case, dashed lines indicate the worst case envisioned for each scenario. The row below the arrows for each issue contains a qualitative indication of changes in the relevant drivers.

[a] These two categories are merged into a single one, though one might observe rather different trends in the future. General trends throughout all scenarios suggest a reduction of sulfur emissions over the next 10–20 years more or less equally across all four scenarios, maybe to a lesser extent in Order from Strength. (See also Chapter 9) In contrast, emissions of NO_x will differ across scenarios according to technological progress which is most pronounced in TechnoGarden (See, for example, Box 8.23 on new transportation systems). Technological innovations do not play an important role in Order from Strength and Adapting Mosaic.

[b] Here, climate change is not restricted to particular regions or ecosystems. (See Chapter 9 for an extensive discussion)

[c] See Section 8.7.7

[d] See Chapter 9

[e] See Section 8.7.2 and Chapter 10

[f] See Table 14.7 this volume and its discussion

very near future. Within Adapting Mosaic, small-scale and integrated water management projects are developed. This releases some pressure on wetlands, which is not the case in Global Orchestration and is true only to a much lesser extent in TechnoGarden. In the latter scenario we can expect some improvements by technical solutions, although some wetlands are allowed to decline as their ecosystem services are provided elsewhere by technological alternatives.

In the most optimistic version of Order from Strength and Adapting Mosaic, there would be relatively low risks of invasive species due to the low levels of global trade and movements of people. In the worst-case versions of these scenarios (the dotted lines in Figure 8.5), population and invasive species have larger impacts. In addition, Order from Strength bears the risk that conversion of wetlands to other uses, particularly for inefficient production of food and to fuel economic growth, counteracts the positive effects mentioned and further exacerbates loss and degradation of inland wetlands.[44] A more detailed consideration of implications for coastal wetlands is given in Appendix 8.1.

On longer time scales, however, two other major drivers lead to further bifurcations in the trends of wetland loss. In both Order from Strength and Global Orchestration, there is a long-term increase of conversion to agricultural land use. For TechnoGarden and Adapting Mosaic, however, the technologies or skills for ecosystem management in place for the second half of the period until 2050 can even induce a restoration of wetlands, in the optimistic case. In addition, climate change, which only becomes significant in the second half of the period, might put further pressure on wetlands. Though various scenarios of climate change, including those presented in Chapter 9, do not show a significant change in effective precipitation (the difference between precipitation and evapotranspiration), we expect that sea level rise leads to loss of coastal wetlands like estuaries or tidal flats and deltas. This effect is most pronounced in Global Orchestration, Order from Strength, and Adapting Mosaic, where it might even overcompensate for the effects of learning. The effect is not so strong in TechnoGarden, thus further allowing for the leveling or turnaround of the trends of loss of the first 20 years of the century.

8.7.1.3 Drylands

Drivers for dryland degradation can be considered in two groups: the "disposition" of a region—that is, its climate and water availability—and the pressure that is put on the environment by land managers. Chapter 14 indicates that, globally, changes in arid areas as a result of climate change are relatively small up to 2020. Thus, in the short term it is the pressures of land management that play a more significant role. This leads to different risks within the four scenarios.

If the assumptions of Global Orchestration are borne out, the scenario sees a significant decrease in material poverty, which could induce a decrease in dryland degradation. If the poverty reduction is not sufficient enough, however, degradation can be expected to occur at a similar pace to today. In the short term, the other three scenarios do not see this kind of relieved pressure, and degradation can be expected to carry on.

In the longer term, the chances for reduced degradation appear in TechnoGarden and Adapting Mosaic. The reduction in TechnoGarden comes from technological progress bringing about new methods for production in dryland areas. It is not certain, however, that these methods will be available for the marginalized people who actually need them. Boundaries for technological diffusion will still exist, and if these persist no positive effect on drylands in marginalized regions may be possible. Adapting Mosaic sees improvement of local knowledge and property rights for better managing agriculture and ecosystem services. Yet some risk exists, either due to failed experiments or to major changes in climate, which makes the present skills inappropriate.

Global Orchestration might at the end of the scenario period see a halt of degradation due to the further reduction of poverty. Yet the risk remains that this reduction is not significant enough. Order from Strength sees the strongest degradation throughout the whole period, and the present trends might even increase due to climate change and further institutional failures.

8.7.2 Human Well-being

Human well-being is considered to have five main components: the basic materials needed for a good life, health, good social relations, security, and freedom and choice. (See Chapter 11.) As such, well-being differs across the scenarios, not only in its overall level but also in its composition. Furthermore, the scenarios differ with respect to the underlying direct causes of changes in human well-being—for example, different trends in development of human, social, manufactured, or natural capital. Figure 8.6 depicts the scorecards of the different scenarios for all five dimensions of human well-being.

8.7.2.1 Basic Material Needs

The four scenarios differ with respect to the form of capital that is the focus of development. Global Orchestration has a strong emphasis on manufactured capital (technological innovation for production and for ecosystem repair) on a global scale and a weaker focus on human capital (education about most things is a high priority, but learning about the environment is not) and natural capital (it is assumed that the environment will take care of itself if human capital is high). In contrast, Order from Strength focuses almost entirely on manufactured capital (industry) on a local to regional scale. TechnoGarden concentrates on manufactured capital in a way that is built on accumulated human capital (learning, tightly targeted as ecosystem management) and on building natural capital to provide financial and other benefits to humans. Finally, Adapting Mosaic puts major emphasis on the development of social and human capital through learning and development of cooperative networks.

Due to these considerations, the scenarios differ in terms of the dynamics of material income and its composition.

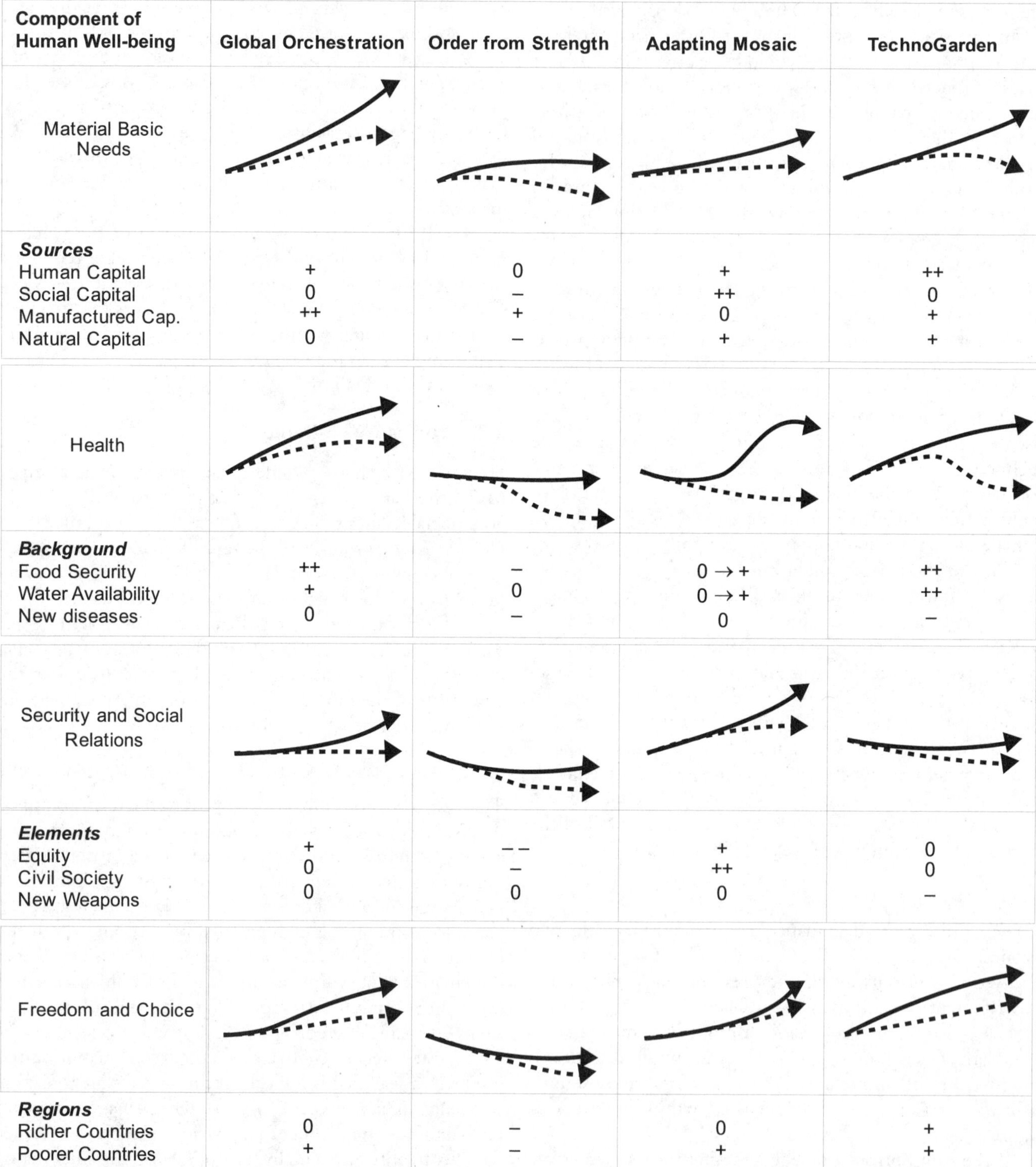

Key:

++	strongly increasing pressure by this driver
+	increasing pressure
0	no change when compared with today
–	decreasing pressure
– –	strongly decreasing pressure
→	a change in the pressure of the driver during the scenario

Component of Human Well-being	Global Orchestration	Order from Strength	Adapting Mosaic	TechnoGarden
Material Basic Needs				
Sources				
Human Capital	+	0	+	++
Social Capital	0	–	++	0
Manufactured Cap.	++	+	0	+
Natural Capital	0	–	+	+
Health				
Background				
Food Security	++	–	0 → +	++
Water Availability	+	0	0 → +	++
New diseases	0	–	0	–
Security and Social Relations				
Elements				
Equity	+	– –	+	0
Civil Society	0	–	++	0
New Weapons	0	0	0	–
Freedom and Choice				
Regions				
Richer Countries	0	–	0	+
Poorer Countries	+	–	+	+

Figure 8.6. Human Well-being Scorecard. Arrows indicate the development over time of the component of human well-being named in the left-most column. Solid lines indicate the best case, dashed lines indicate the worst case envisioned for each scenario. The row below the arrows for each issue contains a qualitative indication of changes in the relevant drivers.

Assessing the overall tendencies of material income for the different scenarios is complicated by uncertainty about if and how substitution among the different forms of capital will happen. It appears, however, that Global Orchestration and TechnoGarden offer the best opportunities for ensuring basic material needs. Nevertheless, the success of the development paradigms behind the policy reform scenarios in Global Orchestration is still debatable. On the other hand, TechnoGarden bears major risks of technological failures and ecological feedbacks, which would directly bring about major disruptions for material income. Order from Strength can be expected to show the weakest income growth, though some leeway does exist depending on the resilience of ecosystems. Finally, Adapting Mosaic offers opportunity for growth later in the century, depending on the degree of learning and success.

8.7.2.2 Health

The scenarios differ with respect to the general development of human health due to income change, changes in food consumption, or changes in the health system, and also with regard to the risk of new diseases. Emergence of new diseases is a direct outcome of ecological changes, whereas the other factors are only indirectly affected. Due to high income growth and increased equity in Global Orchestration and TechnoGarden, we expect improvements in human health globally. The introduction of new biotechnologies in TechnoGarden, however, introduces a high risk for new diseases. The high degree of global interconnectedness in both of these scenarios creates increased chances for the spread of diseases, but also the opportunity for global cooperation in fighting them.

The introduction of new diseases also constitutes a major risk within Order from Strength. The impact, however, is restricted to the poor regions of the world. In these regions, this risk comes from a general decline in human health due to low overall socioeconomic development, which also prevents effective measures to counteract diseases like malaria or AIDS. In Adapting Mosaic, we can expect an initial decline in health due to the limited growth and international cooperation. In the course of time, as new networks and skills develop, we expect improved health.

8.7.2.3 Social Relations and Security

Investment in social capital and the strengthening of civil societies constitute the core of Adapting Mosaic, thus the scenario sees the strongest improvement of social relations. Some risk exists if the conflicts that accompany the redistribution of political power persist over time and hinder the development of social networks. In Global Orchestration, economic equity improves over the short term, which makes the world a safer place. This also contributes to improved social relations in poorer countries, though economic competition might intensify over time and might even counteract the benefits of improved equity. Also, ecological feedbacks can introduce new sources of inequality.

TechnoGarden sees a world in which the new means and methods for managing ecosystems and ecosystem services are owned by an elite group of scientists, engineers, and business people. This leads to an increase in inequality and a general decline of social relations. If, however, the options for open-source ownership start to play a stronger role, relations might improve again, also within the emerging groups of professionals. The scenario also bears the danger of developing new biological weapons, which can make use of the technological skills and knowledge gained. Finally, Order from Strength sees a decline in both social relations and (ironically) security, especially in the poorer countries. Although the scenario is about nations protecting their own interests, leading, especially in richer countries, to stronger physical, economic, and political barriers at borders, this is likely to cause people to feel less secure personally. It might also induce some internal conflicts that further reduce people's feelings of security.

8.7.2.4 Freedom and Choice

The development of freedom and choice within the scenarios strongly depends on the regions. As we expect that the policy reforms in Global Orchestration come with an increase in strong and stable governance in many poorer countries, this scenario sees an increase in freedom and choice in the developing world. This outcome is also expected for Adapting Mosaic but for other reasons—the growth of the civil society and the devolution of power to lower scales also increases freedom and choice in poorer countries but might also change the character of participatory forms of governance in richer nations. In contrast, Order from Strength by its very character restricts the rights of people to choose and move and thus sees a strong decrease of freedom and choice in all parts of the world. Finally, technological development in TechnoGarden is connected with new forms of property rights and new ways to enter careers that increase freedom and choice in all regions.

8.7.3 Emerging Diseases

The complex issues surrounding emerging diseases are discussed in more detail in Chapter 11.

8.7.3.1 Global Orchestration

In Global Orchestration, the lowest population growth of the four scenarios is itself likely to minimize the outbreak and spread of infectious diseases. (See Table 8.2.) For example, diseases such as SARS and new forms of influenza arising from cohabitation of humans and domestic animals might be expected to be less likely with lower population growth. On the other hand, current population densities are sufficient for these diseases to arise, so the risk may remain high. Similarly, increasing wealth in developing nations and international cooperation to address global health threats should allow more effective control of disease outbreaks both locally and globally.

While the chance of new chronic diseases crossing from a nondomesticated animal species into humans, as HIV is thought to have done, is small, the environmental stresses encouraging such an occurrence may not be detected early in a Global Orchestration scenario. Furthermore, if the flow of wealth to poorer countries is slower than contemplated

Table 8.2. Emerging Diseases across the Scenarios

Factors	Global Orchestration	Order from Strength	Adapting Mosaic	TechnoGarden
Population growth	low	high	moderate	low
Nutrition	good	low in poorer countries	moderate	good
Sanitation	good	low in poorer countries	moderate	good
Exposure to non-developed ecosystems	low	high	moderate (but with strong learning)	low (but new ecosystems are created)
Management of known environmental risks	high	high in richer countries; low in poorer countries	high	high
Monitoring for unexpected environmental risks	low	low	high	high
Ability to detect and control outbreaks locally	high for expected outbreaks; low for unexpected	low	high	high
Global cooperation to control outbreaks	high	very low	low	high
Overall assessment	good outlook if optimistic assumptions are met—high risks if assumptions not met	poor outlook for poor countries; substantial risks to wealthy countries if inequalities are not kept within manageable bounds	good outlook for dealing with local problems; substantial risks from broad-scale outbreaks until cooperation among local entities is established	good outlook if technology is successful; low risks of manufactured diseases but the consequences could be devastating

in the scenario, if states fail through escalating corruption, if the expected decrease in pressure on the environment does not occur, or if other assumptions of the currently optimistic scenario fail, then diseases associated with poor sanitation or the emergence and spread of infectious diseases could become a major concern.

Even in the optimistic version of Global Orchestration, diseases of affluence (such as diabetes and kidney and heart diseases) and environmental contamination with heavy metals and other persistent pollutants (such as intellectual impairment and cancers) will be significant risks that could become a more expensive burden on health services than predicted.

8.7.3.2 Order from Strength

The hope for Order from Strength is that one or a few powerful nations can act in the interests of all nations to maintain global peace and economic stability. Even the most optimistic expectations, however, would see technology, knowledge, and wealth gaps among countries either maintained or worsening, due to the strong focus on national economic, social, and physical security. We expect to see high population growth in the less wealthy countries; low investment in human capital, including basic literacy and numeracy; increasing degradation of ecosystems; poor governance; and high corruption brought on by adversity. People in poorer countries will be driven to greater exploitation of undeveloped ecosystems, thus increasing exposure to known and unknown diseases (current examples include hemorrhagic fever and sleeping sickness).

Whereas health probably would increase in richer countries, due to their wealth and strong protection of borders, epidemic diseases are likely to be encouraged in poorer countries by these conditions. Known diseases like AIDS and TB would be hard to control, and those affected would include many knowledge-rich young and middle-aged adults. Declining nutrition would likely exacerbate poor health, fueling additional epidemics. The global impact of emerging disease would depend on how effectively wealthy countries could control emerging diseases that threaten their people. With poor international cooperation, this is likely to be a major challenge that from time to time will be impossible to meet, resulting in diseases affecting wealthy populations. What seems inevitable, however, is that local disease outbreaks in poorer countries would be frequent and would kill many people.

Although there is only a low probability of emergence of totally new, chronic diseases that cross from a non-domesticated animal species into humans, as HIV is thought to have done , this scenario provides the highest likelihood of such an event.

8.7.3.3 Adapting Mosaic

Adapting Mosaic brings the hope of better functioning social institutions leading to improved mental health and lowered incidences of alcoholism, domestic violence, diabetes, and depression in all cultures. There might be greater reliance on traditional health systems, and closer attention to managing environmental factors that could encourage the emergence and spread of human, animal, and plant diseases.

The risks in this scenario relate to how well ideas, technology, and capital circulate internationally. This could be a problem for development and distribution of vaccines, for example. The local focus of environmental management could improve detection of outbreaks, but coordinated management across regions and nationally could be a challenge at times. It is unclear how disease outbreaks like SARS or the periodic epidemics of meningitis that sweep the Sahel would be managed—probably poorly in the early stages of the scenario.

There are major ironies in this scenario with respect to health and emergent diseases. The strong focus on the environment and local learning come about because national and global governance structures have become ineffective and have lost credibility. So we see the second highest population growth among the scenarios, slower rates of technological and agricultural breakthroughs, and probably high impacts of climate change. This could put strains on ecosystems services, sanitation, and food supplies, producing the ingredients for outbreaks of old diseases and emergence of new ones in some places. Capacity to deal with broad-scale outbreaks would be relatively low.

In summary, this scenario brings both hope of effective control of disease emergence locally, but fears that management of outbreaks across broad scales will be hard to deal with effectively.

8.7.3.4 TechnoGarden

The focus on environmental and other technologies in TechnoGarden could see improved human, animal, and plant health through greater disease resistance in crops, improvements in human nutrition, extended life, reduced need for surgery, cheaper and more effective vaccines that confer lifetime immunity to multiple diseases, and reduced pollution of water and air. These factors, together with the second-lowest population increase among the four scenarios and a proactive approach to environmental monitoring and management, could cause this scenario to see low levels of heart and kidney disease, cancers, and mental impairment from heavy metal pollution. The conditions for emergence and spread of new diseases should be minimized, and global action to deal with emergence should be coordinated and effective.

Health issues relating to new pressures on a highly technological society are potential problems, and the easy and increased availability of calorie-dense food could exacerbate the nascent global epidemic of obesity and diabetes.

With respect to emerging diseases, there is also the (small) risk of the escape or deliberate release of devastating diseases engineered in environmental, health, or military laboratories. The genetic homogenization of food and other crops, for example by genetic engineering, also creates vulnerability to new agricultural diseases, which could have flow-on effects to human health.

8.7.4 Fisheries

The consequences of all four scenarios for fisheries are quite different in richer and poorer countries. (See Table 8.3.)

8.7.4.1 Global Orchestration

Although Global Orchestration is about cooperation among nations, we expect to see this cooperation emerge more slowly for marine resources than for other aspects of global governance. This is because of the already established power imbalances and the promise of rapid economic returns from exploiting these resources.

In some areas of richer countries, sustained catches are achieved (especially for high-value fisheries like tuna) through economic incentives, regulation, and creation of marine protected areas to eliminate destructive fishing practices. The need to deal with global climate change is a driver, as is economic gain from biodiversity and tourism. In other areas it is too late. Fisheries are abandoned, resulting in localized social impacts. The richer countries that continue to focus on optimizing profits from fishing continue to experience ecosystem degradation and eventual reduction in fisheries and, ultimately, jobs and other social conditions.

In poorer countries with stable governance, removal of trade barriers and support for institutional reform allows economic benefits to be gained from fisheries while ecosystem management is reformed, based on lessons and experience passed on by other countries. There is a race against climate change, as coral reefs are lost, coastlines are degraded, and river flows change in the tropics.

Poorer countries with poor governance experience short-term economic returns but see overexploitation of marine ecosystems through illegal activities, corruption, and lack of enforcement. Those with cheaper labor become fish-processing centers. Countries with few environmental controls have a significantly higher concentration of facilities but they also have to deal with degraded coastal waters.

Coastal aquaculture continues to expand in wealthy countries, although limited by the rate of development of technologies for feeding the fish. Often, economically depressed coastal communities are targeted. Conflict between the industry and conservation advocates continues, as biodiversity decline is noted in some areas and as social impacts become apparent. Offshore aquaculture expansion is slow at first but accelerates, especially for high-value species like tuna, once technology is developed to reduce the costs of operating far from land in heavy seas.

Coastal aquaculture (land- and water-based) expands even more rapidly in poorer countries, especially where environmental controls are minimal, as richer countries begin to be constrained by environmental and social policies as well as by rising labor costs. Coastal environments are severely affected, coastal fisheries degrade, small-scale fishers are displaced, and food security becomes an issue for many areas. Poorer countries with good governance and appropriate environmental controls use some of the economic benefits gained from improved fisheries management to develop appropriate aquaculture programs that provide cheap sources of protein for the domestic market with minimal environmental impacts but with significant social and economic benefits.

Wealthy countries continue to expand fisheries into the high seas. There is increasing exploitation of pelagic resources, but deep-sea fisheries such as those on seamounts and deep-sea corals cease. A system of high-seas marine protected areas is initiated after much negotiation between fishing nations. Some stocks become threatened but are re-

Table 8.3. Fisheries across the Scenarios. The final row indicates relative impacts on ecosystems and ecosystem services, from 4 being best and 1 being worst.

Factors	Global Orchestration	Order from Strength	Adapting Mosaic	TechnoGarden
Fishing practices	Decline in fisheries and ecosystems addressed once their economic importance becomes apparent.	Decline in fisheries and ecosystems addressed by rich nations expelling foreign fleets from exclusive economic zones, expanding EEZs, and pressuring poorer countries for access to additional fish resources.	Fisheries policy focuses on maintaining and repairing marine ecosystems.	Decline in fisheries and ecosystems addressed through repair using environmental technologies.
	In rich countries: • Ongoing ecosystem decline in many places due to optimization of fisheries for economic return. • Protection for valued species.	In rich countries: • Policies focus on maintaining production. • Protected areas supported only where not in conflict with fisheries or where there is tourism potential.	In rich countries: • Various interventions tried, with considerable regional variation in success and learning. • Protection and adaptive management given high priority.	In rich countries: • Engineering and simplification of many fisheries for economic return. • Attention to a broader range of ecosystem services in some systems.
	In poorer countries: • Fisheries in tropical countries at risk from climate change as coral reefs are lost, coast lines are degraded, and river flows change. • Free trade improves economic returns from fisheries. • Benefits from access fees and processing industries that are established, but coastal environments are severely affected.	In poorer countries: • Already exploited systems bear many of the impacts of climate change. • Some countries exploit high-value fisheries and minimize ecosystem impacts, while others maximize biomass through short-lived species and have high ecosystem impacts. • Rebuilding stocks depends on investment from richer countries wanting to further secure food supplies.	In poorer countries: • Over time, lessons are shared and many ecosystems are stabilized (temperate areas taking longer than tropical). • Developing nations struggle to find the right balance between maintaining ecosystems and economic development.	In poorer countries: • Technologies made available by investment from big corporations based in industrial countries. • Free trade improves economic returns from fisheries. • Benefits from recreational values of technologically enhanced coastal ecosystems, but danger that the value will flow out of the country.
Technology	Technology is developed to replace wild-caught fish meal for aquaculture.	Technology is slow to develop in rich countries because they are able to appropriate additional resources and in poor countries as they cannot afford it.	Low investment and uncoordinated effort in technology to replace fish meal slows the expansion of aquaculture, especially in poorer countries, where the high cost of environmentally appropriate technologies is another issue.	Rapid development of artificial food to replace wild-caught fish meal for aquaculture. New, simplified ecosystems engineered for one or a few valuable species.
Institutions	Effective development of, and compliance with, international conventions and treaties to reduce and ultimately eliminate illegal fishing. Developing countries with stable governance achieve economic returns and more sustainable fisheries, while those with weak governance get short-term economic gain but eventual collapse of fisheries.	International conventions ignored and power used to appropriate marine resources. Developing countries with sufficient marine resources to host distant water fleets negotiate as a regional block so that financial returns improve, and enforcement and management are more effective and efficient.	As inshore coastal areas and stocks are stabilized, fishers look outward to larger areas and regional fishing bodies become more relevant. Developing countries with strong governance attract financial aid to implement interventions, but those with weak governance and corruption struggle to implement appropriate interventions.	Global institutions strong—allows development and enforcement of international conventions.

High seas	Elimination of deep-sea fisheries for environmental reasons but increased pressure on pelagic resources on the high seas. The developing world benefits less than rich nations from the high seas. Benefits that do accrue are from processing facilities that are established by big corporations based in industrial countries.	Distant water fleets expand, international agreements are ignored, and a "slash and burn" attitude prevails with high seas resources. Countries with significant aquaculture industries heavily exploit small pelagic fisheries. Many long-lived species collapse. Some countries defend their high-seas aquaculture sites with force.	Initially, distant water fleets continue to expand into the high seas with few restrictions, and illegal, unregulated, and unreported fishing in these areas is ignored as countries struggle to find the right management intervention and as regional fish bodies are ignored.	Technology supports high-seas aquaculture and a grab for ocean estates in which poorer countries are left out.
Aquaculture	Coastal aquaculture expands, especially in economically depressed coastal communities. High-seas aquaculture develops slowly due to the economics of the technology and the cost of operation in the high seas. Richer countries try to secure disproportionate access rights to the high seas for aquaculture, bioprospecting, and carbon sequestration—open international dialogue required to ensure equity.	Inshore aquaculture expands rapidly for food security in poorer countries. Significant areas are converted, losing a number of ecosystem services. Pelagic stocks are overexploited to supply fishmeal. Climate change worsens the problem. Offshore aquaculture expands slowly as the technology is slow to develop and the costs of protection are high.	In wealthy countries, the expansion of aquaculture is based on current technologies, which have widespread and expensive impacts on coastal ecosystems. Aquaculture practices are reexamined and new practices are tested. Further offshore, aquaculture is only experimental due to the limited understanding of its impacts. Water-based coastal aquaculture is limited in poorer countries due to the cost of using environmentally appropriate technologies as well as the high price of fish food.	Technology-driven rapid increase in aquaculture in rich countries. Slower growth in poorer countries until technology is made available through investment by large corporations. Poorer countries focus on high-turnover, herbivorous species to meet food demands, but this has high environmental impact.
Overall outcome	Balance between unintentional overexploitation, due to a primary focus on food and dollars, and reinvestment of some profits in ecosystem repair. Both risks of ecosystem collapse and the costs of preventing it increase due to lack of a proactive approach. Strength is high degree of international cooperation. (3)	Risks of fisheries collapse are high worldwide due to unchecked exploitation at all scales and lack of international cooperation to address global processes. (1)	Increasingly informed local management achieves alot, but global processes are not dealt with well until late in the scenario. This is the classic problem of institutional responses being at the wrong scale for many of the ecological processes. (2)	High economic and social returns due to investment in technologies for environmental repair and enhancement. But technologies are initially expensive and some engineered ecosystems are vulnerable to unexpected perturbations. (4)

built through various economic instruments and trade negotiations.

Poorer countries do not benefit as much from the high seas as richer nations; benefits that do accrue are from processing facilities that are established in poor countries.

High-seas aquaculture is at first limited by excessive operating and technology costs and risks. It takes three to four decades to overcome these problems. Richer countries try to secure access rights to the high seas for aquaculture, bioprospecting, and carbon sequestration, but poorer countries are reluctant to agree. Ultimately, a global oceans commission is formed to manage the use of the oceans.

8.7.4.2 Order from Strength

Rich nations expel foreign fleets to ensure their food and economic security. They also try to extend their exclusion zones and to pressure poorer countries for access to fish resources. Fisheries policies of wealthy countries are focused on production and not necessarily ecosystem maintenance. In some ecosystems, fish or landing biomass is maintained or enhanced, but biodiversity or ecosystem services are not necessarily. These systems are vulnerable to disturbances such as disease and climate change. In countries where there is a significant aquaculture sector, small pelagic stocks are heavily exploited for fish feed. The highly variable nature of these stocks makes them difficult to manage, and some stocks that are affected by climate change collapse while others are sustained. Marine protected areas are supported only in areas not in conflict with fisheries or where there is considerable tourism potential. Protected areas, along with trade restrictions and habitat restoration, are used also to sustain stocks or protect economically important fauna such as whales and dolphins.

This scenario has the most significant negative impact on poorer countries since they bear many of the impacts of climate change—El Niño, storm events, flooding, and erosion. For many, the marine ecosystems are heavily exploited, and as the scenario progresses, exploitation increases further as wealthy nations try to secure food supplies. Some countries maximize high-value fisheries and thus are able to minimize ecosystem impacts. Others are forced to maximize biomass through short-lived low-trophic species so that food security improves but ecosystem impacts are major. Countries with effective enforcement eventually bring fisheries management under control and some stocks are sustained; however, some stocks collapse because intervention is too late. Efforts to rebuild stocks depend on financial and technical assistance from richer countries wanting to secure more food supplies. Countries with poor enforcement continue to have their stocks heavily exploited except where richer countries see this lack of management as a threat to their food security. Eventually, enforcement improves and the remaining stocks are sustained. Poorer countries with sufficient marine resources to host distant water fleets also begin to negotiate as a regional block so that financial returns improve, and enforcement and management is more effective and efficient.

Aquaculture expands rapidly inshore and offshore in richer countries for food security reasons—but at a high cost to the environment. Significant areas are converted, losing a number of ecosystem services. Primarily low-value, short-lived species are farmed, since they are economically efficient at producing protein, although some high-valued species are also farmed. Offshore aquaculture expands slowly as the technology is slow to develop and most fishmeal is used for either the aquaculture sector or the livestock sector. Also, the cost to protect offshore farms is expensive compared with inshore systems.

Poorer countries develop water- and land-based aquaculture to provide food security and to generate foreign exchange. Consequently there is a mix of high-value and low-value species farmed. The expansion of aquaculture is based on current technologies, which result in widespread impacts on coastal ecosystems. Some of these impacts are long-term, and rehabilitation is prohibitively expensive.

On the high seas, distant water fleets from wealthy nations continue to expand. International agreements are ignored and a "slash-and-burn" attitude prevails with high seas resources. Countries with significant aquaculture industries heavily exploit small pelagic fisheries. Many of the long-lived species in pelagic as well as in the deep-sea systems collapse, and soon the system is fished down to the point where small invertebrates dominate the landings. Initially, high-seas aquaculture develops slowly due to the high costs of technology and of operating in the high seas.

Some poorer countries attempt to address food security by securing sites for aquaculture through extending their exclusive economic zones. This creates significant problems between countries that have distant water fleets and those that do not. The latter countries see expansion of exclusive economic zones as a "sea grab." They are also concerned about the potential impacts from high-seas aquaculture (such as disease and genetic dilution). Conflicts arise, with some countries defending their high-seas aquaculture sites with force. There are also localized problems as disease and genetic dilution arise in countries that ignore international standards (such as the FAO Code of Conduct).

8.7.4.3 Adapting Mosaic

Fisheries policy in many richer countries focuses on maintaining marine ecosystems and, where it is economically feasible, rebuilding them as adaptive experiments. Destructive fishing practices are phased out. Initially, various interventions are tried, including modified individual transferable quotas, community quotas, marine protected areas, and construction of artificial reefs. There is considerable regional variation in the measures that are tested. Most international and regional fishing agreements are ignored, however, as are regional fishing bodies. Over time, as lessons are learned and shared, many ecosystems are stabilized, enabling them to buffer extreme events and climate change. Temperate areas take considerable time, compared with sub-tropical and tropical zones, to show the benefits of the interventions. Consequently there is considerable variation in areas and fisheries that recover or stabilize. As inshore coastal areas and stocks are stabilized, fishers look outward to larger areas, and regional fish bodies become more relevant. The interventions that succeeded previously are re-examined and modified for larger-scale management.

Poorer nations struggle to find the right balance between rebuilding ecosystems and economic development, since they lack the technical expertise and financial resources to reduce domestic and foreign fishing effort. Regional fishing bodies are important sources of technical assistance. Countries with strong and stable governance attract financial aid to implement interventions, especially once the lessons learned in richer countries are evaluated. Some ecosystems, especially in the tropics, respond quickly to reduced effort and the elimination of destructive fishing practices. Countries where governance is weak or corruption prevalent continue to struggle to implement appropriate interventions and stocks collapse.

In richer countries, aquaculture practices are re-examined and new practices are tested. Expansion of aquaculture is slowed due to reliance on wild-caught fishmeal. Alternative feed technologies take 20–30 years to develop. Further offshore, aquaculture is only experimental due to the limited understanding of its impacts.

Water-based coastal aquaculture is limited in poorer countries due to the cost of using environmentally appropriate technologies as well as the high price of fish feed. Countries with strong and stable governance are provided with technical and financial assistance to develop coastal low-tech aquaculture for both high-value species (to generate foreign exchange) and low-value species (to contribute to food security in the area). Countries with weak governance continue to expand and establish aquaculture that affects the coastal ecosystems and wild stocks. Such developments are economically viable for approximately 10 years, and then they are abandoned with irreversible loss of a range of ecosystem services.

Initially, distant water fleets continue to expand into the high seas with few restrictions, and illegal, unregulated, and unreported fishing is ignored as countries struggle to find the right management intervention after regional fish bodies are ignored. Only a few countries develop high-seas aquaculture, and much of it is experimental due to the ecological risks and high technology and operating costs. The rebuilding of inshore ecosystems and the development of appropriate technologies for coastal or inshore and freshwater aquaculture eliminates the need to expand aquaculture into the high seas.

8.7.4.4 TechnoGarden

In richer countries, destructive fishing practices are eventually eliminated, and some ecosystems are reconstructed (such as artificial reefs), although the costs are high and sometimes prohibitive. Many countries engineer marine ecosystems to provide high-value food—large shrimp, salmon, or cod systems, for example. But these systems suffer, due to their simplification, from surprises such as pests and diseases. Wild species are maintained for genetic purposes (similar to breeding zoos). As new technologies remove the need for wild-caught fish as food for aquaculture, some wild capture fisheries in wealthy countries service the gourmet and luxury food market, while aquaculture feeds the masses.

Economic imperatives drive most poorer countries to convert their waters to high-value fisheries, mostly financed by large corporations based in the wealthy world. Much of the production is exported with few economic and even fewer social benefits to the host countries. Initially, the low-value fisheries in these countries are used to service the fishmeal market, due to the rising price of fishmeal, but eventually new technology for producing artificial food for farmed fish puts these fisheries out of business, requiring international aid for industry and social restructuring.

New technologies reduce the environmental impacts of aquaculture and remove the need for wild-caught fishmeal in richer countries. Aquaculture expands rapidly in coastal areas to ensure a consistent source of fish for domestic consumption and trade. Much of the production is of high-value fish. The big risk here is maintenance of good water quality and prevention of surprises from disease. The highly managed, but simplified, ecosystems in shallow coastal areas are found to be unstable and vulnerable to even small perturbations. Offshore, deeper aquaculture is developed within national boundaries for high-value fisheries. These operations are technologically advanced with a high initial capital cost, but have low running costs since much of it is automated. Only high-value species such as tuna are farmed this way.

The cost of water-based aquaculture technology slows the growth of aquaculture in poorer countries. There is some investment by large corporations, since labor and other operating costs are cheaper. However, environmental quality is not as secure as in richer countries and therefore the coastal aquaculture sector in many poorer nations focuses on lower-value, high-turnover herbivorous species. These species become important for food security in many poorer countries, but the intensified aquaculture also results in conversion of large areas of coastal land and loss of ecosystem services, such as erosion control and maintenance of habitat for fish breeding.

Fleets (primarily from wealthy countries) continue to expand into the high seas, and landings of deep-water species maintain catch levels early in the century. However, soon these long-lived stocks are depleted, and catches decline to pre-1990 levels as large pelagic species such as tunas and sharks are all that remain. Pressure on pelagic fisheries as well as the krill fisheries in the Southern Oceans decreases as new technology produces artificial food for aquaculture. The richer countries, which have the technology for high-seas aquaculture, negotiate regional agreements to provide some form of security for aquaculture operations (fixed and floating) to develop outside of national exclusive economic zones. These operations are highly risky economically since large marine mammals as well as collisions with other vessels can destroy the crop. Poorer countries are left behind in this race for ocean real estate since, compared with richer countries, they lack the technology to stake a claim.

8.7.5 Climate Change

The complex issues surrounding climate change are discussed in more detail in Chapter 9. In the storylines in this chapter, climate change is discussed in different contexts in different scenarios. This is because the drivers of climate change, especially the amount and type of technology used, differ considerably between scenarios, as do the approaches to addressing impacts versus causes. As indicated in Chapter 9, the biggest differences between the scenarios with respect to the rate of climate change are seen in the rate of temperature change. (See Table 8.4.)

8.7.5.1 Global Orchestration

In this scenario, the mechanisms for global cooperative action to address climate change are in place, but the importance of the problem and its ecological causes are not recognized, and action is not taken until things get really bad. The crucial question under Global Orchestration is when interest will grow strong enough among all major partners to start cooperating on building an effective response to climate change.

The fact that causes and impacts of climate change are decoupled both in time and place complicates an effective response. It complicates the evidence that will be required to make climate change a priority under Global Orchestration, but it also makes it more difficult to negotiate a treaty among parties that can be both losers and winners from global climate change. In the quantification (see Chapter 9), no explicit climate policy is assumed under this scenario. Based on the storyline, however, we find it just as likely that after some time a final agreement can be made on a delayed response. Such a response would need to include major adaptation action and also mitigation action based on market-based incentives. The financial flows that are connected to trade in carbon credits could in fact be a great stimulus for poorer economies.

Table 8.4. Climate Change across the Scenarios

Factors	Global Orchestration	Order from Strength	Adapting Mosaic	TechnoGarden
Temperature change—degrees Celsius per decade (see Chapter 9) Current rate assumed to be 0.20° Celsius per decade	gradual rise to around 0.35° Celsius by 2050 (declining to 0.20° by 2100)	fluctuating but reaching around 0.26° Celsius by 2050 (increase to 0.30° by 2100)	gradual increase to around 0.28° Celsius by 2050 (declining to 0.16° by 2100)	initial rise to around 0.23° Celsius by 2020, then decline to 0.15° by 2050 (and to 0.06° by 2100)
Drivers	strong emphasis on economic growth; slow emergence of low-impact technologies	depressed industry and slow technological development in developing world keeps global emissions lower than other scenarios	initially poor international cooperation allows climate change to continue unchecked; eventually, focus on ecologically friendly industries lowers emissions and also economic growth	focus on economic growth but rapid emergence of low-impact technologies
Mechanisms	Market-based (e.g., carbon credits); international trade in carbon	displacement of impacts by wealthy countries; global cooperation almost impossible	local adaptation to impacts by better ecosystem management at first; later, global environmental management through cooperating networks	Market-based incentives; environmental technologies to reduce consumption and repair and enhance ecosystem services like carbon sequestration
Geographic differences	Impacts greater in developing countries until late in the period	Much greater impacts in poorer countries but feedback impacts on wealthy ones	Local adaptation to impacts of climate change until later in the scenario where global networks develop to address causes	Different environmental technologies in different regions depending on culture, opportunities, and resources

Despite policies and commitment to reducing inequity, the impacts of climate change are greater in poorer than in richer countries.

8.7.5.2 Order from Strength

Natural environments are seen as providers of some critical services, like regulation of climate and the oxygen level in the air, but most other benefits from ecosystem services are seen as either substitutable or repairable by technology.

Global issues like climate change are almost impossible to address, because there is always at least one key nation unwilling to cooperate due to national interests and because the international institutions that might have been able to address international issues are unstable (if they exist at all). Consequently, formulation of an effective response to climate change will be very complex in this world. The best hope might be partial deals among like-minded nations.

Ironically, up to 2050 global climate change increases less in this scenario than in the other three because a large proportion of the world's population is forced to live a simpler and less materialistic existence. (But climate change is still increasing at 2050 and is likely to be worse than in all the other scenarios by 2100; see Chapter 9.)

8.7.5.3 Adapting Mosaic

Problems like climate change grow worse because of disenchantment with global and national governments and an increasing focus on local issues. This leads to a focus on adapting to climate change for much of the period between 2000 and 2050, using sophisticated learning and intervention in ecosystem management. Eventually, however, it is recognized that problems like climate change require global action, so networks of cooperating local groups and businesses develop and take action toward the end of the period. The rate of climate change is relatively slow because the focus on environmental rather than technological solutions leads to lower environmental impacts and slower economic growth.

8.7.5.4 TechnoGarden

Among the four scenarios, TechnoGarden is clearly best equipped to deal with the issue of climate change, based on its international cooperation and proactive attitude toward solving ecological problems. Removal of subsidies for fossil fuels, nuclear energy, and large hydro projects stimulates the development of a variety of alternative energy sources. Large energy companies and entrepreneurs from a range of countries, especially poorer ones, support these developments. Big business invests in environmental technologies to ameliorate and address the causes of climate change, and global agencies are established to implement strategies.

The type of environmental technology and action employed within the richer and poorer worlds varies due to culture, the way that property rights are organized, and the density, wealth, and activities of people within a region. As under Global Orchestration, international trade in carbon credits could become a major financial flow under TechnoGarden.

8.7.6 Eutrophication

Nutrient pollution (eutrophication) of fresh and coastal waters expanded greatly over the second half of the twentieth century. These nutrients came mainly from three major sources: agricultural runoff, sewage, and the burning of fossil fuels. Nutrient pollution stimulates the growth of algae that consume oxygen and produces areas of low oxygen in waterways. These low-oxygen areas decrease the ability of fish, shellfish, and other organisms to persist. Furthermore, algal blooms can make fresh water toxic to people. These consequences of nutrient pollution decrease fishing, the supply of clean water, the aesthetic value of waterways, and habitat available for many species.

The extent and impact of eutrophication in the future will be determined by a number of factors:

- Algal blooms are worse when populations of top predators are reduced or when wetlands, which can remove excess nutrients before they reach waterways, are absent.
- Agriculture's impacts will be determined by the area under agriculture and the amount of nutrient runoff from land (nutrient runoff in turn will be affected by the type of agricultural practice, rainfall intensity, slope of the land, fertilizer application, and the amount of nutrients accumulated in the soil).
- The impacts of fossil fuels (which release NO_x) will depend on the total amount used and the technology that is used to burn them.
- The impacts of sewage will be determined by the volume produced and the ability to treat it for nutrient removal.

The technologies developed in TechnoGarden should reduce the risks of eutrophication, despite a rise in area under agriculture, by reducing the impacts of agriculture, sewage, and energy use.

In Adapting Mosaic, better local management of agricultural runoff, fossil fuel burning, and sewage will reduce the chances of eutrophication. But these could be offset by increasing use of fossil fuels and expansion in area under agricultural areas. Consequently, water quality in inland waterways should improve, but coastal eutrophication—especially where populations are dense, such as in the tropics—will increase.

Global Orchestration can expect increases in eutrophication. Minimal population growth combined with technological advances will increase energy efficiency and result in more-efficient agriculture. However, these moderate improvements will not purposefully address the nutrient runoff from agriculture and fossil fuel burning, resulting in an increase in nutrient pollution, especially in poorer countries. In addition, the reactive management style favored in this scenario is likely to be too slow to address changes in soil and sediment phosphorus, which may set the course for eutrophication that is very problematic to reverse if it is addressed after it is a problem.

Order from Strength is very vulnerable to increases in nutrient pollution, and areas of eutrophication should greatly expand. In this scenario, there is little investment in environmental technology or management to offset substantial increases in agricultural area and fossil fuel use.

In all scenarios, people will have to cope with eutrophication caused by storage of phosphorus in soils and recycling of phosphorus from lake sediments—two factors that make eutrophication extremely difficult to reverse once it is a problem.

8.7.7 Invasive Species

A recent meeting of experts on invasive species (Various Authors 2003) identified several major issues:

- *Species spread:* Species invasions are not natural and have been greatly accelerated by people. The world's biota are being homogenized by moving species accidentally and purposefully. Species are being mixed between places that have not been in contact for millions of years.
- *Regulating valuable invasive species:* It is difficult to get rid of invasive species that have high economic value to some people.
- *Anticipating invasions:* Only a small proportion of introduced species become invasive, but it is very hard to predict which ones will.
- *Invasion control:* Control of species invasions is complicated and difficult. Chemical and biological control frequently backfire. Physical removal is very expensive.
- *Monitoring and detection:* Due to the difficulty of control, the best approach to invasives is to reduce the rate of arrivals and to monitor for introductions. The best approach from a conservation point of view is a combination of no tolerance and immediate eradication.

This section looks at how people might deal with these challenges in the four scenarios. (See Table 8.5.)

8.7.7.1 Global Orchestration

The openness of borders, growth in economies of countries all around the world, and the strong emphasis on trade creates increasing opportunities for intended and unintended movement of species across borders. The confidence in technology to solve environmental problems as they arise is likely to discourage research to anticipate invasions or the scaling back of species that have economic value unless they became serious threats to economic growth in other sectors. Response to invasions should be facilitated by well developed global cooperation, but with invasives such approaches are likely to be too late to be cost-effective.

8.7.7.2 Order from Strength

This scenario has the highest risks with respect to invasive species. Global trade is expected to be lower than in other scenarios but still high enough to transmit invasive species. Global cooperation for controlling the spread of invasive species is likely to be poor. Outbreaks of invasive species, especially in poorer countries, are much more likely than in other scenarios due to poverty and poorly resourced management. Research into understanding, detecting, and controlling invasive species is likely to be low, consistent with the low priority given to proactive environmental management. Thus the only defense against invasions is tight border controls, which are unlikely to be effective since only small numbers of individuals need to get through to start an invasion. Control is likely to be unaffordable in poorer countries, and richer countries are likely to have many other demands on their treasuries, such as maintaining national security in the face of terrorism and other conflicts.

Table 8.5. Invasive Species across the Scenarios (Various authors 2003). Within each issue, the scenarios are ranked from best (4) to worst (1). See text for further explanation.

Factors	Global Orchestration	Order from Strength	Adapting Mosaic	TechnoGarden
Species spread	The large scale of global trade combined with the fivefold expansion of the global economy will increase the risks of biological invasions. (1)	Global trade will roughly double under this scenario. Strong border control is unlikely to be effective. The lack of international guidelines and attention to environmental processes means that invasions are likely to increase. (2)	Reduced long-distance transport of materials should reduce pressure on ecosystems, and a focus on local management should help respond to local invasions. Overall, risk is likely to decline. (4)	The large scale of global trade combined with the fourfold expansion of the global economy will increase risks; however, ecological technology should help mitigate this. Overall, the risk of species spread is likely to increase. (3)
Regulating introduced species that are valuable	The focus on economic growth in this scenario is unlikely to reduce this problem. (2)	There will be little attempt to control such species until after they have spread and have negative impacts. Ecological management is not a priority in rich countries, and poor countries cannot afford it. (1)	Effective local action but lack of national or global cooperation means that local groups will have to deal with frequent re-invasions. (3)	Use of tracking technologies (e.g., bio, nano, IT) will allow for the partial control of the spread of valuable potentially invasive species while allowing their continued economic exploitation. (4)
Anticipating invasions	Global cooperation helps but is usually too late, due to reactive environmental policies. (2)	All countries caught off guard. (1)	Early detection and action locally, and sharing of information globally, but lack of coordinated action to reduce risks. (3)	Attempts at early detection and effective control using environmental and other technologies partially effective. (4)
Invasion control	Use of combination of processes, but not in any organized or integrated way. Frequent mistakes result in species spread and negative ecological consequences. (2)	Little control. When applied, done haphazardly with little consideration of risks. Failures discourage further control. (1)	Probably beyond local management to develop. Focus would be more on managing landscapes to prevent invasion or to increase competitiveness of desirable species. (4)	Sophisticated regimes of biotechnologies, possibly including biological control. Danger that confidence in ability to manage nature leads to rediscovering the mistakes of previous generations. (3)
Monitoring and detection	Detection would not be early, but national and international action could be rapid once the problem is evident. (3)	Late detection and poorly coordinated response. (1)	Early detection with rapid response locally. However, lack of coordinated regulations on global trade leads to frequent surprises. (2)	Detection early, rapid, and effective due to extensive and internationally coordinated ecological monitoring by public, private, and citizen groups. (4)
Overall damage from invasive species	Major expansion in global trade combined with lackluster monitoring and control increase spread of invasive species. Monitoring and control are partially effective and mitigating damage. (2)	Despite lower species spread, the failures in monitoring, control, and management produce the largest risks. (1)	Local regulation prevents arrival of invasive species, and local monitoring, management, and control are effective at reducing species damage. However, a lack of coordination at national and global scales results in increased invasions. (3)	Despite an expansion of global trade, the effective international regulation, monitoring, control, and management work to reduce damage produced by invasive species. (4)

8.7.7.3 Adapting Mosaic

The local focus in this scenario will reduce long distance transport of materials. This should reduce the chances of species introductions, and close monitoring and learning by local managers should effectively deal with most invasions before they get established.

However, poorly developed national or global cooperation means that local groups could have to deal with frequent reinvasions. Similarly, until mechanisms develop for sharing the results and lessons from local monitoring and for developing coordinated response strategies, broad-scale outbreaks could be difficult to control.

8.7.7.4 TechnoGarden

This scenario also sees open and increasing trade, but with a proactive attitude toward environmental management supported by investment in environmental technologies. Depending on whether detection and control technologies improve fast enough to counteract the increasing opportunities for species movements, this scenario could see an increase or decrease in species introductions. On balance, the opportunities for species movement should grow faster than control technologies and the risk of species spread will increase.

Technologies (such as tracking technologies using biotechnology or nanotechnology) could help follow and control the spread of invasive species and could be used to keep the economically valuable ones contained. Investment in research for understanding, detecting, and controlling invasions—combined with international cooperation in amelioration—would give this scenario a distinct edge over the others. There is the danger that overconfidence and complacency could lead to surprises, however, and that technology could create new pests that are hard to control.

8.7.8 Urbanization

Table 8.6 summarizes some of the roles of ecosystem services in supporting the major urban center of São Paulo, Brazil. Table 8.7 summarizes the different challenges for urban areas in the four scenarios, which are discussed in more detail in the storylines.

8.7.9 Example: Agriculture and Hypoxia in the Gulf of Mexico

The Mississippi River drainage basin extends over roughly 3.2 million square kilometers, covering almost half of the continental United States. This basin is the third largest in the world and is home to about 70 million people. The Mississippi River basin's agricultural economy is worth $100 billion per year and produces about 40% of the world's corn and 40% of the world's soybeans (Donner et al. 2002).

However, the focus of the regional economy on increasing agricultural production has reduced the ability of the region's ecosystems to provide other ecosystem services, such as clean drinking water, fisheries, and wildlife. One of the major impacts of agriculture has been the decline in

Table 8.6. Ecosystems Services Supporting Urban Areas. The São Paulo Green Belt Sub-global Assessment explores the importance of ecosystem services to the 17.8 million inhabitants of the Metropolitan Region of São Paulo, Brazil, which is the world's fourth largest city. The Table summarizes some of the interactions between urban processes and the surrounding Green Belt.

Ecosystem Service	Environmental Good/Service	Description/Importance of Environmental Good / Service
Supporting service	ecological processes and biodiversity	The Atlantic Forest is one of the planet's richest biomes in terms of biodiversity; maintaining its integrity is crucial to the population of Brazil and important to the global population in general (for potential new medicines and other products). Locally, the Green Belt woods are important ecological corridors, acting as links connecting different forested regions of Brazil.
Provisioning services	underground and surface water supply	Water resources within the GB supply water to over 20 million people. Their endangerment can lead to a collapse in public water supply: water shortage is already common during the dry season. There is also a strong correlation between forest intactness and water quality; the trade-offs between these two services (or synergies) have serious economic implications.
	food safety	Today, 15% of the world's food is produced in backyards and small land tracts (Ian Douglas, Univ. of Manchester, 2002, personal communication). The GB has this tradition and today is one of Brazil's top organic produce regions (sustainability). Besides, the choice for agriculture in areas surrounding cities is regarded as an alternative to the outspread of big cities.
	forest resources	The forest products originating in the Green Belt are the mainstay of São Paulo State's economy, mainly coming from reforestation. The natural forests also provide important resources to several communities, including fuelwood and building material.
Regulating services	climatic regulation	The SPGB acts to counter the urban cask that causes temperature to rise (heat islands). This phenomenon has been linked to thermal imbalances that influence rainfall patterns and can lead to heavy floods in urban areas.
	soil protection and run-off regulation	The forests in the GB prevent soil erosion and minimize floods.
	carbon sequestration and pollutant reduction	Carbon sequestration: The GB has 311,407 hectares of undergrowth and 84,620 hectares of reforestation, which add up to help to balance a large fraction of the CO_2 generated by the urban population.
		Pollutant reduction: Forests act as a physical barrier against the movement of pollutants from metropolises. This has both global and local health implications.
Cultural services	social use	Metropolises like São Paulo and Santos lack green areas. The GB is often the only alternative for the population to be in contact with the natural environment. This is crucial for humanizing as well as for the physical and psychological health of the population.
	sustainable tourism	Brazil's forests generate income for urban-based tourism operators and other supporting businesses.

Table 8.7. Urbanization and Its Issues across the Scenarios

Factor	Global Orchestration	Order from Strength	Adapting Mosaic	TechnoGarden
Rates of urbanization	high as wealth and technological lifestyles grow in all countries	high in rich countries as wealth and technological lifestyles grow high in poorer countries due to poverty and rural decline	moderate as people reconnect with nature and many decide to live in rural areas	high as ecosystems are managed remotely
Infrastructure	global uniformity of community health standards, business practices, and governance for reasons of health, security, and distribution and management of foods, building materials, and fuels	urban sprawl competes with agriculture for the best land water pollution, poor infrastructure for dealing with urban wastes, chronic diseases caused through depleted ecosystem services for controlling disease-carrying species, and damage to crops and the environment from invasive species, are all made progressively worse because underlying problems were not addressed	damage to urban centers due to climate change is sometimes large, but people's capacity to restore ecosystem services has been sufficient to recover from the impacts within two to three years in most cases using strategies based on lessons learned elsewhere	improved transport and pollution management systems arising from innovation in developing countries increased use of public transport cities become multicultural and designed around local climate and ecology
Challenges	major impacts on coastal systems as coastal cities grow poorer cities face challenges as money is injected into their economies and urban populations grow rapidly due to rural people moving to cities and people moving from other countries	cities in wealthy countries face challenges managing their poor and preventing terrorism inflamed by poor living conditions in less wealthy countries cities in less wealthy countries face challenges building and maintaining basic infrastructure due to poverty and resource degradation that affects water quality, food production, waster disposal, and amenities	there is growing interest in life outside cities, increasing the challenge for urban planners to keep cities interesting for the residents city planners struggle to get high-quality scientific input to establish ecologically sustainable systems, and lessons learned by other cities are not freely shared	rapid urbanization, especially in Asia and Africa, creates pollution and disease challenges in richer countries, environmental technologies improve the livability of cities but create some new environmental challenges tensions exist between landscape designers and restorationists some cultures, especially indigenous, threatened by global homogenization
Demand for greener cities	high demand in all parts of the world as prosperity grows	low as ecosystems not seen as fundamentally important in richer countries and survival takes precedence in poorer countries	very high as the connection of humans with nature is seen as fundamentally important	high demand due to appreciation of ecosystem services rapid growth of Asian cities leads to innovation and periods of dynamic urban reinvention
Technology	technology to make life in cities more comfortable and rewarding is a high priority increased sharing of technologies across the world as global markets continue to grow environmental impacts are dealt with as modifications to technology once they are detected	moderate development of technologies to improve urban infrastructure in richer countries poorer cities struggle with outdated technologies	a strong emphasis on using ecosystem services instead of technology wherever possible, but this requires cooperation among several watersheds for big cities, and this is initially a challenge in this fragmented world	new-Asian urbanism combines existing technologies in novel ways and stimulates new ways of thinking about city planning
Sharing of solutions for livable cities	high	very low	low initially, then higher through networks of individuals	high

Gulf fisheries caused by increased flow nutrients from agricultural fertilizer into the region's surface waters. These nutrients flow through the Mississippi basin into the Gulf of Mexico, where they are the cause of seasonal low oxygen levels and hypoxia, which many aquatic species cannot survive. In 2002, the area of hypoxia extended over 20,000 square kilometers (Rabalais et al. 2002a). This region of hypoxia, know as the "dead zone," forms in the center of the most important commercial and recreational fishery in the continental United States (Goolsby et al. 1999).

During the second half of the twentieth century, land use and land management practices changed greatly in the Mississippi river basin. Following the invention of cheap industrial fertilizers in the middle of the twentieth century, the use of fertilizer has increased massively, from almost non-existent levels to yearly applications of 150 kilograms of nitrogen per hectare (Foley et al. in review). The expansion of industrial agriculture has increased nutrient runoff. In the second half of the twentieth century, the amount of nitrate entering the Gulf of Mexico tripled, due to increased runoff and a sixfold increase in fertilizer use (Donner et al. 2002).

The increase of nutrients in the Gulf increases aquatic net primary production, which consumes oxygen. In the warm summer, the water in the Gulf of Mexico is stratified into layers that do not mix much. The diffusion of oxygen into the water from the air is also reduced. This stratification is controlled by river outflow, wind mixing, the circulation of the Gulf, and summer temperatures. The combination of stratification and fertilization produces the area of low oxygen called the hypoxic zone. This area began to appear around the start of the twentieth century and became more severe after the 1950s. The nitrate flux leaving the Mississippi increased by a factor of three between the middle and the end of the twentieth century (Rabalais et al. 2002b).

Hypoxia simplifies the ocean bottom communities, reducing populations of shrimp and other commercially valuable species. Important fisheries are variably affected by increased or decreased food supplies, mortality, forced migration, reduction in suitable habitat, increased susceptibility to predation, and disruption of life cycles (Rabalais et al. 2002b). Since the 1960s, many other regions of coastal anoxia have emerged worldwide. Between 1990 and 2000 the area affected by anoxia has doubled (UNEP 2003).

8.7.9.1 How Will Hypoxia in the Gulf of Mexico Change across the Scenarios?

At the start of the twenty-first century, U.S. federal and state officials have developed a set of plans to reduce nitrogen export from the Mississippi. Will these programs be successful? Will they be sufficient to reduce the dead zone? We compared the four global scenarios using the issue of hypoxia in the Gulf of Mexico. First, we identify the factors that influence the dead zone (see Figure 8.7) and then we assess how these factors will vary across the scenarios.

8.7.9.2 Factors Influencing the Dead Zone

We identified five factors influencing the dead zone: climate, agricultural practices in the Midwest, the management of the main stem of the Mississippi, the management of the delta and New Orleans, and fishing.

Climate will influence hypoxia in the Gulf by influencing runoff from the Midwest, stratification of the water, and the agricultural practices of farmers in the Midwest. The timing and amount of rainfall influences how much fertilizer runs off from agricultural fields. Intense rainfall on frozen ground will transport larger amounts of nutrients into the Mississippi, while the same amount of precipitation spread out over a longer period will produce less erosion. It is uncertain whether climate change will or will not increase the flow of the Mississippi River (Justic et al. 2003). The temperature of the Gulf of Mexico is also expected to increase. This would increase productivity and increase stratification, both of which would increase the vulnerability of the Gulf to hypoxia (Justic et al. 2003). While past climate changes appear to be indirectly linked to increases in yield, it appears that these increases have approached a ceiling and that further improvements in climate will not substantially increase yields or area planted (Foley et al. in review).

The extent and type of agriculture practiced will influence the vulnerability of fields to runoff. The use of artificial drainage over the past 200 years in the Mississippi has eliminated once-common wetlands and riparian zones, decreasing the ability of the landscape to retain and recycle nutrients. Intensive agriculture, which uses large amounts of fertilizer in large fields, will produce more runoff than would agriculture that is adapted to local topography, disturbs the soil less, and uses less fertilizer. Nutrient runoff can be decreased by lowering the amount of fertilizer applied and reducing the amount that runs off the landscape. Less fertilizer can be used on the land by changing from row crops, such as corn and soybeans, to perennial crops such as alfalfa. Another approach is to apply nitrogen in the spring rather than in the fall, when there is more runoff. A third approach is to fertilize crops only when needed. This means using it only in areas of a field where it is required, and not adding nitrogen that is already being supplied through nitrogen fixation by soybeans, manure application, and atmospheric deposition. Runoff can be reduced on the farm by lowering runoff from livestock farming operations and by the construction, maintenance, and restoration of wetlands and riparian buffers (Mitsch et al. 2001).

The management of the river and its floodplain alters the transport of nitrogen to the Gulf. Wetlands have been destroyed and the Mississippi River's main channel has been shortened, dredged, and bordered by flood protection levees. These changes reduce the amount of nitrogen that is taken up by plants and animals living in the river, increasing the amount that flows into the Gulf (Turner and Rabalais 2003). Levees prevent seasonal flooding, protecting buildings near the river and preventing the movement of sediments and nutrients into nearby wetlands. Riparian buffers and floodplain wetlands can prevent nutrient-rich runoff from reaching the river. However, care has to be taken that water does not spread over the buffer, because if runoff cuts channels through a buffer it is not useful (Mitsch et al. 2001).

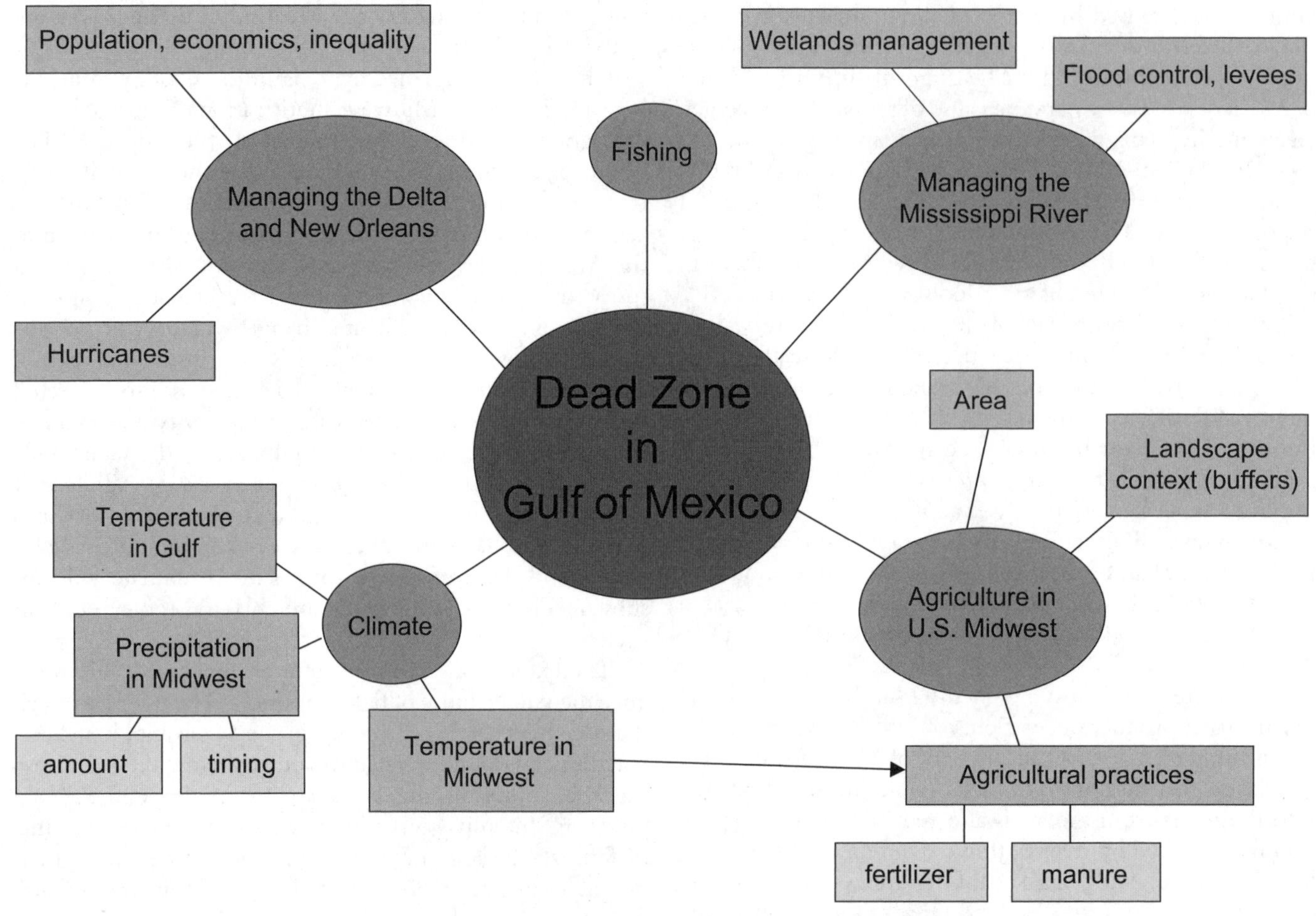

Figure 8.7. Conceptual Map of Direct and Indirect Drivers of the Dead Zone in the Gulf of Mexico

Hypoxia in the Gulf of Mexico is influenced by the separation of the Mississipi River from its delta. The channelization of the river prevents it from spreading over the floodplain. This prevents nitrogen from being removed in delta wetlands and is contributing to land erosion in the delta. Attempts to restore the delta could also decrease the movement of nutrients into the Gulf. Efforts to protect New Orleans from flooding and hurricanes could make the export of nutrients either worse or better, depending on how they modify the river and the delta (Mitsch et al. 2001).

Fishing is influenced by hypoxia, but the presence of substantial populations of algae-eating fish would likely decrease the vulnerability of the Gulf to hypoxia by reducing the production of phytoplankton. There has been increasing evidence that overfishing can increase the vulnerability of coastal areas to eutrophication, which contributes to hypoxia (Jackson et al. 2001).

8.7.9.3 Comparison of Factors across the Scenarios

The global scenarios differ in changes to global trade, investments in ecological technology, and approaches to regional ecological management. These differences shape how these scenarios could influence the five key factors that influence the extent of hypoxia. Climate change will not be significantly different across the scenarios—warming in the Gulf of Mexico and increased rainfall in the Midwest will increase the vulnerability of the Gulf to hypoxia in all scenarios. Differences among the factors influencing vulnerability are summarized in Table 8.8.

Overall, we expect that the dead zone should be reduced the most in the TechnoGarden scenario due to changes in agricultural practices, more sophisticated management of the delta and main stem, and better management coordination among upstream and downstream parts of the river. The coordination and ecological engineering across large scales required by this problem are well addressed by the society and technologies developed in TechnoGarden. Furthermore, ecological engineering of water and biogeochemical cycles are more predictable and better understood aspects of ecosystem dynamics.

Adapting Mosaic would decrease the area of hypoxia due to the cumulative effects of many local attempts to solve local ecological problems in the Mississippi basin. However, the decrease in global trade could lead to a higher amount of agriculture to meet local needs. Furthermore, problems of coordination among the Midwest and the Mississippi delta limit the effectiveness of local action. In this scenario, the dead zone would initially increase in area due to the existing nitrogen in the system, but this worsening will gradually slow and then be reversed.

Global Orchestration sets up a number of forces that decrease and increase nitrogen movement into the Gulf of

Table 8.8. Consequences of Each Scenario for the Factors Affecting Hypoxia in the Gulf of Mexico

Factor	Global Orchestration	Order from Strength	Adapting Mosaic	TechnoGarden
Farming	decrease in area; no change in nutrients; some improvement in land management; constant or minor decrease in nutrient runoff	increase in area; increase in fertilizer use; limited improvement in land management; increased nutrient runoff	increase in area; less fertilizer use; better land management practices; less nutrient runoff	decrease in area; less fertilizer use; better land management practices; less nutrient runoff
Managing the river	management of river for barges eliminates wetlands and increased channelization; some increase in wetlands and buffers; no change in proportion of nutrients entering the Mississippi	some local addition of riparian buffers and wetlands combined with decrease in wetlands and building levees; increased proportion of nutrients entering the Mississippi	some levee removal driven by farming and flood protection; restored wetlands and riparian buffers; decreased proportion of nutrients entering the Mississippi	levee removal and re-engineering of floodplains with ecologically sophisticated levees and engineered wetlands; decreased proportion of nutrients entering the Mississippi
Managing the river delta	investment in human well-being in delta results in many local improvements; however, river channelization leads to only small increases in flow through delta	some area abandoned; regulation of river; further decrease in delta despite some local increases in wetland	local projects, but disagreements about what to do about the river; slightly increased flow through the delta	federal ecological re-engineering of the delta leads to greatly increased area of wetlands
Changes in hypoxia	slow growth in area	substantial growth in area	initial increase in area, then gradual decline	reduction in area
Changes in fishery	sport fishery persists, commercial fishery closed due to low profitability	fishery eliminated	local management and improvement of fishery	fishery increased and combined with delta aquaculture maintained

Mexico. On the decreasing side, a rise in global agricultural trade and the cutbacks of agricultural subsidies reduce the extent of agriculture in the Mississippi basin. Furthermore, the decrease in the importance of agriculture in the economy leads to greater demand for clean water. These concerns lead to local mitigation measures. However, the management of the river for trade, including keeping it deep enough for boats, and the lack of basinwide integrated nutrient management work against the decrease in nutrient movement into the river. Local development in New Orleans and education and social development programs ease some of the problems of the delta and reduce its vulnerability to hurricanes. These positive and negative effects roughly cancel one another out, and ongoing transport of nutrients at present levels leads to an increase in the dead zone.

Order from Strength has largely negative consequences for hypoxia in the Gulf of Mexico. The decrease in global trade leads to continued high levels of agriculture in the United States. The non-local impacts of nutrient runoff are not addressed, and there is substantial conflict among upstream and downstream states over river management. Lack of integrated policy and political struggles among various jurisdictions over who is responsible for the nutrient pollution, together with social inequality and a lack of effective responses to rising sea level in the delta, combine to ensure that higher levels of nutrients enter the river and are transported into an increasingly vulnerable Gulf of Mexico. There would be substantial increases in the size and severity of area of hypoxia in the Gulf in this scenario.

8.7.9.4 Discussion

The comparison of Gulf of Mexico hypoxia across the scenarios provides lessons on the legacies of past decisions, the relationship between society and nature, and the scale of ecological management. Land use decisions over past centuries, in particular the use of large amounts of fertilizer following World War II, have put huge amounts of nutrients into the Mississippi River. Even with radical changes in land use and river management, these nutrients will remain at high levels within the Mississippi for decades.

Poor and rural people are most vulnerable to the loss of ecosystem services. Social policies, such as providing alternative livelihoods for fishers and investing in human capital in Mississippi delta communities, can mitigate the impact of the dead zone on human well-being.

The dead zone in the Gulf of Mexico has been produced by land use decisions decades ago and hundreds of kilometers away. The impact of these decisions on the Gulf depends on river management, delta management, and fishing, however. In such a situation, local ecological management can help address local problems but is insufficient to deal with the broader problem. Management responses that are integrated at a variety of scales are required to deal with large-scale emergent ecological problems, such as coastal eutrophication.

8.8. Insights into the Key Questions Posed in the Scenario Project

The following section draws insights from the scenario storylines with respect to the key questions posed in the scenario project, which are outlined in Chapter 6. Further insights are drawn in later chapters and brought together in Chapter 14.

The key questions are:

- What are the consequences of plausible changes in development paths for ecosystems and their services over the next 50 years, and what will be the consequences of those changes for human well-being?
- What are the most robust findings concerning changes in ecosystem services and human well-being across all four scenarios? What are critical uncertainties that we are confident will have a big impact on ecosystem services and human well-being?
- What are gaps in our understanding that we can identify right now that will affect our ability to model ecosystem services and human well-being?
- What opportunities exist for managing ecosystem services and human well-being?
- What is surprising in these results?

8.8.1 Consequences of Plausible Changes in Development Paths

Strategies for human development that emphasize economic policy reform (that is, reducing subsides and internalizing externalities) as the primary means of management potentially improve human well-being most rapidly of all, but they also carry risks of major social collapses and adverse environmental backlashes due to unforeseen feedbacks. Such strategies (our Global Orchestration scenario, for example) are likely to be based on the assumptions that increasing economic well-being decreases environment impact (Kuznets greening) and that ecosystems are sufficiently robust and human (technological) ingenuity is sufficiently great that environmental problems can be dealt with once they become apparent.

Because they address economic inequality directly, such strategies promise the most rapid improvements in human well-being. However, they carry several risks. If they are wrong about the robustness of ecosystems, then ecological crises could accelerate inequality because they disproportionately affect the poor. Although humans have a long history of technological solutions to environmental problems, post hoc technological solutions usually are much more costly than early detection and action. Such strategies are most likely to succeed in a world of centralized decision-making and global cooperation. If this is accompanied by a de-emphasis on local responsibility and authority for environmental management, then there is a risk that potentially unmanageable national and global issues could arise from aggregations of local problems that are not detected and addressed early enough. Global climate change, the salinization of landscapes, and overexploitation of fisheries are examples.

Strategies that emphasize local and regional safety and protection and that give far less emphasis to cross-border and global issues offer security in a world threatened by acts of aggression and the spread of environmental pests and diseases, but they lead to heightened risks of longer-term internal and between-country conflict, ecosystem degradation, and declining human well-being. In the Order from Strength scenario, countries turn inwards in response to terrorism, war, and loss of faith in global institutions. Trade and movement of people are restricted, leading to temporary security from human aggressors and some environmental benefits from control of the movements of exotic plants and animals.

Such a world encourages boundaries at all scales, and inevitably the gap between rich and poor grows within and between countries and the need for multiscale proactive management of ecosystems is neglected. Inequity leads to increasing demands on the environment for survival by the poor, and the threat of both aggression and environmental degradation increase. The only benefits for ecosystem services imaginable relate to protection of key natural resources in rich countries, although if (as is likely) this is combined with reactive rather than proactive environmental policies, then ecological problems are likely to emerge at both local scales and across scales from local to global.

Virtually all environmental problems currently imaginable, including outbreaks of pests and diseases, famine, erosion and flood damage, soil and water degradation, air pollution, and decline in fisheries, are likely under extremes of this type of strategy. In moderate forms, however, we could see periods of increasing inequity and environmental degradation interspersed with periods of alleviation of the problems by strategic intervention by the wealthy to keep cross-scale problems in check. Such cycles might continue for some decades but are unlikely to be sustainable in the long term.

Strategies that emphasize the development and use of technologies allowing greater eco-efficiency and optimal control offer hope for a sustainable future but still carry risks that need to be addressed. In our TechnoGarden scenario, such a strategy develops due to recognition of the importance of ecosystem services for human well-being and the emergence of markets that encourage trade in these services. Managing ecosystem services becomes a focus for business, and rights of ownership and access to ecosystems become more formal. As population continues to grow and demand for resources intensifies, people increasingly push ecosystems to their limits of producing the maximum amount of ecosystem services.

The risk under this strategy is destruction of local, rural, and indigenous cultures. The increasing confidence in human ability to manage, or even improve on, nature leads to overlooking slowly changing variables that are not easily reversible, like the loss of local knowledge about ecosystem services and eutrophication of fresh waters and coastal oceans. The new technologies themselves can even engender their own ecological surprises. But in general, people around the world have better access to resources, and managers think more about multifunctionality and systems approaches rather than single goals.

Strategies emphasizing adaptive management and local learning about the consequences of management interventions for ecosystem services are the hope of many ecologists and environmental groups, but creating a world like this might bring some challenges that cannot be overlooked. Our Adapting Mosaic scenario arises from some of the same inward-looking tendencies that produced Order from Strength. In this case, there is dissatisfaction with the results of global environmental summits and other global approaches, and trust in centralized approaches to environmental management declines. The focus is on learning about socioecological systems from multiple sources of knowledge, experimentation, and monitoring. The compartmentalization implied by these trends fosters competition and limits the sharing of learning for a time.

In Adapting Mosaic, there are failures as well as successes and there is the risk that inequality can increase rather than decrease, both within and between countries. Economic growth is likely to be less than in Global Orchestration or TechnoGarden, and this could threaten the establishment and persistence of this strategy unless gains in other aspects of human well-being are obvious in a majority of communities. While local problems become more tractable and are addressed by citizens, global problems like climate change and marine fisheries sometimes grow worse, and global environmental surprises become more common. The hope of this scenario is that there is scope for communities to reconnect at larger scales and to turn their attention back to addressing environmental challenges at multiple scales from local to global.

8.8.2 Robust Findings across the Scenarios

Adverse and unforeseen ecological feedbacks are risks in all trajectories of human development. In all four scenarios there was significant potential for adverse cross-scale ecological interactions. The key message is that whatever trajectory of human development is taken, we need to be alert to the nature of the cross-scale feedbacks that could be overlooked due to the characteristics of the trajectory.

In a globally connected but environmentally reactive world (Global Orchestration), global environmental issues like global climate change or large-scale national issues like salinization of landscapes could be dealt with at national and global scales because institutions at these scales would be well developed and strong. However, such problems often could be well advanced and difficult to reverse in time to avert disaster for two reasons: First, institutions involving local communities who are in touch in real time with ecosystem change could be relatively poorly developed. Second, the assumption that economic prosperity will equip society to deal with environmental challenges could entail lower investment in understanding and monitoring the ways in which small-scale processes aggregate to become larger-scale problems.

In a globally connected but environmentally proactive world (TechnoGarden), we imagine the risk of overconfidence in environmental technologies to the point where humans manipulate landscapes to get more and better ecosystem services. Presumably in this world, fast (short-term) processes will be well understood, but the risk is that slower (longer-term) processes that determine the long-term persistence of ecosystems could be overlooked and that ecological collapses could sneak up on future TechnoGardeners.

In a globally compartmentalized but environmentally proactive world (Adapting Mosaic), there is strong focus on and a dedication to learning about local-scale processes. But we explore how the social forces that allow local institutions to gather such strength and authority could also cause barriers to communication between localities that are competing with one another for better environmental outcomes (such as entry to green markets) and how the ability to spot the aggregation of processes across landscapes and regions might be diminished. Thus, new challenges akin to salinization or global climate change might grow to dangerous proportions before they are identified and action is taken. We see dangers in this scenario for world fisheries, for example.

And in a globally compartmentalized, environmentally reactive world (Order from Strength), we see potential for undesirable environmental impacts at all scales. For poorer individuals, regions, or nations, environmental management becomes a luxury that ranks after the essentials of economic and physical survival. The wealthy not only face the risks of a reactive strategy with respect to their own ecosystems, they also face risks from problems arising in poorer regions and nations that can have an impact across scales. An example is increased risks from major flooding, land movements, dust and other forms of air pollution, and the spread of pests and diseases, including human pathogens.

All trajectories of human development have potential benefits for ecosystem services and human well-being, although these benefits are distributed differently and are rarely uniform. Each of the scenarios lists both benefits and risks. (See Table 8.9 and Figure 8.8.)

8.8.3 Critical Uncertainties

Two fundamental ecological uncertainties underpin the ecosystem services outcomes from future human development. Two of the main ways in which our scenarios differ are around whether the world is more or less connected and whether policy-makers and decision-makers are more or less proactive about environmental management. This differentiation was chosen because we wanted to explore the implications of at least two critical uncertainties about ecosystems that arise from lack of information on ecosystem function and from the uncertainty associated with predicting outcomes from the interactions between human social systems and ecosystems.

The first uncertainty is about the ways and circumstances in which impacts on the environment at different scales interact, sometimes just among themselves but often with human interventions, to produce unexpected and undesirable outcomes.

The second uncertainty is about when and where different ecosystems are robust (able to withstand impacts without fundamentally changing in terms of the services they provide) versus fragile (likely to change rapidly and dramatically in terms of the services they provide).

Because of the tight linkages between ecosystem services and human well-being, virtually all policies for human development involve assumptions about these two uncertainties. Usually these assumptions are implicit rather than explicit. Policies that are reactive to environmental issues assume robustness, while proactive policies at least anticipate the possibility of fragility in some places and at some times. Policies seeking more or less connectedness do not

Table 8.9. Comparison of Potential Benefits and Risks across the Scenarios

Scenario	Potential Benefits	Potential Risks
Global Orchestration	economic prosperity and increased equality due to more-efficient global markets	progress on global environmental problems may be insufficient to sustain local and regional ecosystem services
	wealth increases demand for a better environment and the capacity to create a better environment	breakdowns of ecosystem services create inequality (disproportionate impacts on the poor)
	increased global coordination (e.g., markets, transport, fisheries, movement of pests and weeds, and health)	reactive management may be more costly than preventative or proactive management
Order from Strength	increased security for nations and individuals from investment in separation from potential aggressors	high inequality/social tension, both within blocs and within countries, leading to malnutrition, loss of liberty, and other declines in human well-being
	increased world peace if a benevolent regime has power to act as global police	risk of security breaches (from poor to middle well-off countries and sectors of society)
	less expansion of invasive pests, weeds, and diseases as borders and trade are controlled	global environmental degradation as poorer countries are forced to overexploit natural resources and wealthier countries eventually face global off-site impacts like climate change, marine pollution, air pollution, and spread of diseases that become too difficult to quarantine
	in wealthy countries, ecosystems can be protected while degrading impacts are exported to other regions or countries	lower economic growth for all countries as poor countries face resource limitations and rich countries have smaller markets for their products
	ability to apply locally appropriate limits to trade and land management practice as trade is not driven by open and liberal global policies	malnutrition
	protection of local industries from competition	
Adapting Mosaic	high coping capacity with local changes (proactive)	neglect of global commons
	"win-win" management of ecosystem services	inattention to inequality
	strong national and international cooperative networks eventually built from necessity and bottom-up processes	less economic growth than maximum because of less trading
		less economic growth than the maximum possible because of diversion of resources to management
TechnoGarden	creating "win-win" solutions to conflicts between economy and environment	technological failures have far-reaching effects with big impacts
	optimization of ecosystem services	wilderness eliminated as "gardening" of nature increases
	developing societies that work with rather than against nature	people have little experience of non-human nature; leads to simple views of nature

usually arise from environmental concerns, but they often involve assumptions about how environmental issues will be dealt with. These assumptions can make different ecological surprises more or less likely. Our scenarios draw out the assumptions and their potential consequences.

8.8.4 Gaps in Our Understanding

Critical gaps exist in our understanding of the robustness and resilience of ecosystems generally, the qualitative and quantitative nature of their response to human impacts and repair efforts, and the ways in which ecological processes can interact across scales of space and time. Our scenarios suggest that these are critical elements in determining when and where reactive environmental policies will be adequate and whether scales of institutional arrangements are appropriate for detecting and dealing with threats to ecosystem services and human well-being.

8.8.5 Opportunities for Managing Ecosystem Services and Human Well-being

Three of the four trajectories for human development explored in the scenarios offer both benefits and drawbacks with respect to ecosystem services and human well-being. A globally compartmentalized, environmentally reactive world offers few benefits and many drawbacks.

The major likely sources of adverse impacts on ecosystem services and human well-being are failure to address inequity within and between countries, failure to monitor and learn adaptively from changes in ecosystems across scales of time and space, failure to establish and maintain cross-border cooperation in monitoring and addressing cross-scale ecological processes, and failure to ensure that institutional capacity is appropriate before applying political and economic solutions to perceived social challenges. The diversity of viewpoints and expertise brought to the dialogue about the scenarios consistently found this set of issues arising as problems as the storylines were developed and tested for plausibility.

8.8.6 Surprises

A globally compartmentalized, environmentally reactive world could mask developing ecological and social disasters for several decades. Several other global scenario analyses include a fortress world scenario leading to breakdown of

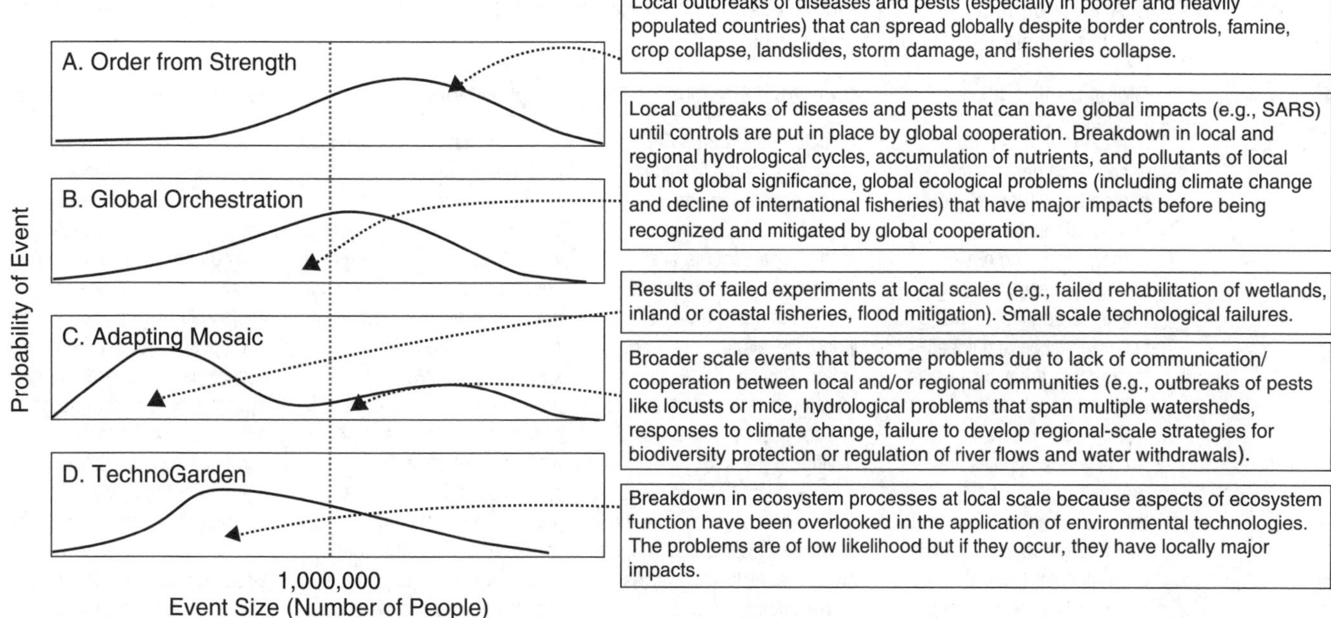

Figure 8.8. Types of Plausible Extreme Events across Scenarios. The distribution curves come from Figure 5.3, this volume. The x-axis is the magnitude of the disturbance of ecosystem services, measure by the number of people affected. The y-axis is the likelihood of an extreme ecosystem event of a given magnitude.

society. The Scenarios Working Group was unanimous that the Order from Strength scenario is unsustainable and ultimately disastrous in terms of ecosystems and the societies they support. But we were surprised at the diversity of viewpoints on how such a scenario could unfold and over what time scale. Despite the heightened risk of massive disasters from conflict and ecosystem breakdown in this scenario, some of us could imagine an ongoing cycle of escalations of social and ecological problems among poorer countries and sectors of society interspersed with strategic mitigation efforts by the wealthy to limit problems before they become global. This raises the possibility that, if current trends toward increasing compartmentalization continue, the world could develop a false sense of security in coming decades unless efforts are made to understand and monitor ecosystems and their services.

Current understanding of ecological and social systems is inadequate to predict when and how ecological-social systems will produce adverse cross-scale ecological interactions. There is considerable evidence of adverse cross-scale interactions between ecological processes and human activities (including interventions intended to fix problems). There also is emerging understanding of how these adverse outcomes have been caused. This understanding allowed us to agree on the types and relative risks of adverse cross-scale interactions in the four scenarios. However, we found it surprisingly difficult to locate when within the next five decades the impacts would emerge or precisely what combinations of policies and interventions are most likely to produce them. This uncertainty and lack of control is precisely the reason for employing a scenario approach (see Chapter 6), but a very large number of scenarios would be necessary to address all possible combinations of drivers that could produce desirable or undesirable outcomes.

Appendix 8.1. Potential Impacts of the Four Scenarios on Major Coastal Wetlands

Land management and climate change effects throughout catchments combine to have an impact on coastal and near-coastal systems. Addressing these impacts requires coordinated actions nationally and internationally, which are big challenges in the compartmentalized scenarios. The time lags between taking action within catchments and seeing responses in coastal systems can be many decades, due to the slow change in ecological processes and the accumulation of nutrients and other chemicals in the catchment soils. Even proactive approaches will face these time lags, but the problem will be much greater for reactive approaches.

Appendix Table 8.1 gives one view of the impacts on various types of coastal wetlands under the four scenarios.

Appendix Table 8.1. A View on Impacts of the Four Scenarios on Major Coastal Wetlands. This view is based on an interpretation of the scenario storylines (this chapter) and the modeling (Chapter 9).

Area	Global Orchestration	Order from Strength	Adapting Mosaic	TechnoGarden
Coral reefs	Although there is considerable coordinated national and international effort to address combined catchment-management and climate change impacts, the late response and long lag time between action and impact means that reefs continue to decline throughout much of the world under this scenario. Areas where impacts are greatest (e.g., many areas of the Caribbean and Southeast Asia) do not recover to their original state. Coastal communities relying on reefs for food and shoreline protection are at the greatest risk, especially with increased storm frequencies. Tourism in many areas declines or is lost altogether. Costs of amelioration are high, but the ability to mount internationally cooperative strategies prevents out-of-control escalation of environmental problems in many areas.	Climate change causes major bleaching events to occur. There also are major outbreaks of diseases, and food chains become unstable so that predators have major impacts in many areas of the world. Developing countries suffer substantially as the loss of coral reefs affects the lives of poor coastal communities relying on reefs for tourism as well as for food. Developing countries attempt to maintain their reef systems and in some areas the decline slows. As sea temperatures rise, however, there is little that can be done. In these countries the major impact is on the tourism industry, which shifts its focus from reefs to island/sandy beach destinations. The lateness of responses combined with inability to mount internationally cooperative approaches causes many of these problems to spiral out of control.	Various initiatives are instituted locally to reduce the impact of climate change. Community-based management becomes widespread and there are local successes first at maintaining the coral reefs and later at addressing external issues such as predators and pollution. Initially, there is lack of clarity in some areas about who has responsibility for reefs, but as communities realize the benefits of cooperative networks, reefs get more attention. Ultimately the rate of loss is slowed down in much of the world, but many areas remain vulnerable to disease and require careful management by local users. Restoration of lost areas is usually not successful; there is net loss of coral reefs by 2050, and reefs are no longer a source of food for many communities in developing countries. Where reefs are still in good condition, tourism is a strong incentive for good local management.	Coral reefs continue to decline. Services such as shoreline protection are provided through engineering works in rich countries, while in poorer countries communities are vulnerable to increased storms and coastal erosion. Food security is addressed through aquaculture, and the greatest benefits to local communities come if they have property rights over the sites. Some communities lose their tourism industry, but others develop stronger industries based on selling tourism and other ecosystem services. Some coral reef areas that are not as affected by climate change are maintained as reference sites and to safeguard genetic resources for future uses. Often these areas are owned and managed by private interests or NGOs because of their recognized ecosystem services values.
Seagrasses	Areas outside major urban/industrial centers and agriculture catchments survive in the long term. Impacts on coastal systems decline later in the scenario period as action is taken to increase the efficiency of agriculture. Beds in proximity to large urban areas are lost and cannot be recovered due to permanent changes in coastal features (e.g., port developments, break walls, etc.). In areas where coastal modifications are absent, reactive policies and actions to reduce pollution, freshwater loading, sedimentation, and destructive fishing produce some recovery. For many coastal communities in developing countries, food security is threatened.	Climate change and invasive species reduce biodiversity in some areas, especially where pollution from urban or industrial centers and agricultural catchments is not controlled. In other areas the development of coastal aquaculture to ensure food security destroys vast areas of seagrass and contributes to coastal pollution. While some wealthy nations are able to institute development and pollution controls, and therefore buffer the impact of climate change and reduce habitat conversion, other countries cannot. Direct impacts on seagrasses and indirect impacts on fisheries threaten food security in many poor countries.	Local communities soon appreciate the importance of seagrass to maintaining fisheries and other ecosystem services and various approaches to managing these habitats are tried. Trade-offs between protection of seagrasses and other social and economic benefits of development are major challenges, especially for coastal communities experiencing rapid urban growth. These are addressed through integrated regional initiatives to address broader coastal issues, but these initiatives take two decades to develop. In some cases, seagrass beds are eventually restored, but in others continued pollution means that seagrasses do not recover. The greatest progress in protecting and managing seagrass is accomplished outside of urban/industrial	There is a general loss of seagrass beds because many of the ecosystem functions can be provided through aquaculture, pollution controls, and engineering works. The loss of seagrasses is more rapid in wealthy countries, where the technology is readily available and can be funded by private enterprise. In poorer countries the rate of loss is less since they cannot afford the technology and there are incentives to protect seagrass beds that can be sold to corporations as offset credits to allow development elsewhere.

			areas, where changes to coastlines are limited.	
Mangroves	Countries continue to lose areas where sea level rises too fast for mangroves to adapt. Losses are highest in poorer countries, where conversion to other uses, such as aquaculture, housing, and agriculture, continue to expand to accommodate economic and population growth. Communities relying on mangroves are at risk in terms of food security and shoreline protection early in the scenario, although social reforms may alleviate this situation later. Eventually the value of mangroves in providing ecosystem services is recognized and measures are initiated that focus on sustainable use (e.g., crab farming, sustainable harvesting regimes, etc.) Rich countries fair much better, but storm frequencies increase so that remaining stands are threatened in some areas. Fisheries decline until measures for protecting mangroves are instituted, but even then recovery is slow and often fails because it has been left too late. Where there is sea temperature rise, some mangrove beds expand but are susceptible to episodic events such as disease and predation from other organisms.	The increasing storm frequencies, rising sea levels, and high freshwater inundation resulting from climate change results in many mangrove areas being environmentally stressed. Pollution from upstream sources as well as in the coast adds further stress to the mangroves. This has impacts on local fisheries (food security) and other services, such as flood control and carbon sequestration. Some stands are lost, exposing coastal communities to the direct impacts of storms and flooding and threatening their food security. To ensure food security, some countries convert many areas of mangroves to aquaculture ponds. However, the increased areas of standing water result in a rise in mosquito-borne diseases.	In some areas the conversion of mangroves to other uses is halted through targeted community-based management projects, and there are attempts to restore some of the services of mangroves. A few work, and the lessons learned are shared with other communities and other management agencies. As examples emerge to show that the benefits of converting mangroves are short-term, many communities cease conversion and initiate restoration works as part of larger coastal management programs. Once the loss of mangrove is slowed down, and especially where there is rehabilitation, fish stocks improve along with other socioeconomic benefits (e.g., improved coastal protection) and appropriate fisheries management measures are implemented so that the benefits are distributed equitably throughout the community, furthering the incentive to protect and manage mangroves within a larger coastal system. Much of the protection and management of mangroves takes place outside of large urban/industrial areas since fewer people are displaced.	Although technological developments can provide many of the provisioning services of mangroves through aquaculture, substitutes for other important services such as nutrient cycling and bioregulation are not well developed by 2050, and therefore the conversion of mangroves to other uses is reduced significantly once the limitations of technology are realized. Technology, however, does enable local communities to use mangrove resources sustainably through environmentally appropriate aquaculture in constructed ponds/cages. This allows less industrialized countries to take advantage of expanding markets for a range of marine products while avoiding overfishing. However, the concentration of aquaculture developments results in systems being more vulnerable to disease and other surprises.
Kelp beds	Beds close to the coast are stressed by a loss of predator fish species when pollution, storms, sedimentation, and in many areas elevated sea temperature cause beds to shift to being dominated by sea urchins. This results in a further cascading effect as other species that rely on kelp are stressed. Once measures are in place to improve water quality and reduce sedimentation and to improve the management of fisheries, some areas recover, but the return of some marine mammal species takes a very long time and some never recover. Many inshore areas are highly susceptible to stresses, and therefore	In coastal areas, kelp beds in many areas are exposed to increased storm action, increased freshwater inputs, sedimentation, and overfishing, resulting in increased stress and susceptibility to disease, predation, and competition from urchins and other invasive species. In wealthier countries action is taken to manage pollution and coastal development, but the long lag times mean that benefits take decades to emerge. Poorer countries cannot afford the trade-offs with production and development, and ecosystems continue to decline due to overfishing, pollution, and coastal development.	Kelp beds decline for at least a decade as they are overlooked by the focus on local community action. However, as communities recognize the benefits of cooperating to manage common goods and services, a range of initiatives emerge that protect coastal kelp beds as part of integrated coastal and fisheries management initiatives. The loss of kelp beds is halted or reduced, and in rare cases recovery takes place depending on the levels of disturbance. In those kelp beds where overfishing of top predators is addressed, the state of kelp beds and the services they provide improves. This requires the	A focus on better fisheries management and recognition of the financial benefits of ecosystem services encourages the protection of kelp beds as fish spawning areas. But climate change and lack of management success limit major benefits to rich countries. In regions with a history of overfishing, kelp beds shift to urchin-dominated areas in which kelp infrequently recovers. In this scenario there is a risk of overlooking the slow processes that maintain kelp beds in the long term in favor of shorter-term processes that are amenable to ecological engineering.

(continues)

Appendix Table 8.1. Continued.

Area	Global Orchestration	Order from Strength	Adapting Mosaic	TechnoGarden
	restoration efforts wax and wane depending on how well climate change is brought under control. Farther offshore, beds are less stressed, and in some areas beds expand as temperatures rise.		development of international cooperation, which is a big challenge in this scenario. In those areas where there has been a long history of overfishing and beds have shifted to urchin-dominated areas, the recovery of kelp systems is sporadic, and some areas never recover.	
Estuaries and embayments	Estuaries undergo major structural changes as flow regimes change with increased storm events and flood events, resulting from land conversion and climate change in both wealthy and poorer countries. This results in major changes in species and productivity of estuaries. Estuaries close to urban/industrial areas lose many ecological functions as pollution levels, freshwater flows, and coastal infrastructure increase and as flooding becomes more common. Hypoxia continues to be problematic in more and more areas, especially in the developing world. In areas, especially in wealthier countries, where estuaries have undergone minimal modifications in the past (e.g., military installations) or where pollution levels have been managed through foresighted or fortuitous management of catchments and domestic and industrial discharges, hypoxic events are reduced. Non-urban areas are also stressed and susceptible to surprises but not to the same extent (relatively more resilience in these systems). A major challenge for this scenario is that although the declining quality of estuaries is recognized, the actions take decades to produce improvements, and remedial programs are difficult to maintain over several terms of government.	Estuaries and lagoon systems are exposed to increased flooding, increased freshwater inputs, and sedimentation due to climate change and land conversion. This results in changes in the diversity of flora and fauna in these areas and alters the scope and nature of ecosystem services as well as increasing the risk of episodic events such as hypoxia and disease (e.g., cholera). These problems are made worse in areas where nutrient pollution is not controlled. Similarly, where coastal developments including aquaculture (pond and net) increase to ensure food security and economic growth, then there is a further decline in ecosystem services such as erosion control. In regions where estuary quality is determined by the actions of several countries, cooperative action is needed but hard to achieve under this scenario.	Communities try various approaches to managing estuaries. A problem early in the scenario is that the quality of estuaries is determined by the decisions of many communities, and cooperation among them is difficult to achieve while they focus on their own constituencies. As cooperative thinking emerges and lessons are shared, it becomes clear that estuaries are best managed as part of an integrated system and that the loss of ecosystem services cannot necessarily be replaced by technology or other parts of the ecosystem. As further lessons are learned, these management programs are then placed in regional strategies and plans that address external issues (e.g., hypoxia) that local communities cannot address as well. The focus on the importance of maintaining and where possible restoring ecosystem services increases. This results in a slowing of the rate of loss of estuaries and corresponding lowering of the extent and size of hypoxic zones. An important element of this scenario is that local learning allows communities to appreciate the slow ecological processes that cause lags between action and results.	Some of the impacts of climate change (storms and flooding) on estuaries are managed through engineering works, and provisioning services such as food are provided through aquaculture. The issues of pollution and freshwater flows are also managed through appropriate upstream practices. Ecological simplification contributes to the loss of genetic diversity, which reduces the long-term sustainability of aquaculture. Communities that maintain estuaries prosper by benefiting from improved water quality, reduction of disease, and maintenance of fisheries.
Intertidal areas	Flooding and storm events degrade low-lying shallow intertidal areas. Coastal erosion continues to be a problem, while conservationists and engineers	In poorer countries erosion continues to be a problem since there is limited assistance from the rich countries to implement engineering solutions. In	Intertidal areas are often managed within a broader integrated coastal management program that considers intertidal areas as well as others. In areas where	Soft engineering solutions are the focus of managing erosion as part of integrated management of coasts in both rich and poorer countries. This becomes the business of a

	debate hard and soft engineering solutions to manage the coast. In the wealthier world, hard engineering is used as an interim measure to halt erosion. Poorer countries are helped with aid from wealthier countries but, even so, engineering approaches are too expensive economically and socially.	wealthy countries action is taken to address coastal environmental problems only when they outweigh or impinge on economic imperatives. In poorer countries, there are constant conflicts between the ecological and social values of areas and basic needs for food, shelter, and economic development. In both rich and poor countries, erosion becomes an increasing issue because of failure to address the role of well-maintained ecosystems in stabilizing landforms. Both public and private property are threatened, as well as recreation, tourism, and food provisioning.	shorelines have been severely altered through engineering works (channeling, break walls, etc.), many ecosystem functions have been lost, while in other areas further degradation or restoration is possible, especially in the rich countries, where such expensive works can be funded. In the developing world, once better environmental management programs are in place and corruption is managed, the loss rate for intertidal areas slows, but not as slow as in the wealthy world.	number of specialist environmental engineering companies, who make money by providing not just stabilization services but a range of other ecosystem services. Despite the many successes of these companies, there are several spectacular failures due to underestimation of surprises from coastal systems and the power generated by rare combinations of natural processes.
Coastal lagoons (open and closed)	Lagoons are highly disturbed due to increasing storm frequencies resulting from climate change. Poorer countries are hardest hit, since many coastal communities depend on lagoons for food and income. Rich countries cope better, especially where urban areas are sparse and protection works are funded for large urban areas. The hydrology for many lagoons changes, resulting in decreased fisheries production and increasing the threat of food security in coastal communities in developing countries. With decreasing water quality, the risk of diseases and invasive species increases. Overall, these lagoon systems are not lost, but there is significant degradation. However, where environmentally appropriate protection works (hard and soft engineering) as well as where integrated coastal management is implemented, fewer ecosystem functions are lost.	Climate change impacts are not addressed in many countries until late in the scenario, resulting in a decrease in the number of functioning lagoons and substantial degradation of coastal lagoons, especially in poorer countries. Lagoons with sufficient water quality are used for aquaculture production and therefore continue their food provisioning services, but other services such as flood control, biodiversity, and bioregulation are lost. These systems are more vulnerable to major disease outbreaks and invasive species. Lagoon systems that are exposed to pollutants or subjected to shoreline changes are less resilient than lagoons that have some of their structural integrity intact and are not threatened with pollution.	Sea level rise and storms affect coastal lagoons through changes in water quality and hydrological processes. Communities develop a range of management systems to adapt and buffer these impacts. Over time, communities develop integrated management systems that are more effective and easier to implement. Systems managed outside an integrated system are more vulnerable to eutrophication, lose ecosystem services, and are at risk of diseases and invasive species.	Technological solutions (soft engineering) are developed to reduce the impact of climate change; to reduce pollution, including nutrient inputs; and to increase food production in lagoons. Some of this technology, especially relating to intensive aquaculture, is transferred to poorer countries, increasing food production, food security, and economic development. However, ecosystem services such as biodiversity and bioregulation are reduced, and lagoon systems become less resilient to disease and invasive species (ecological surprises). Where lagoons are part of an integrated management system, many of the stresses are addressed, and the delivery of ecosystem services improves.

Notes

1. The phrase "Washington Consensus" was coined by John Williamson in 1990 (Williamson 2000) to "refer to the lowest common denominator of policy advice being addressed by the Washington-based institutions to Latin American countries as of 1989." These policies were fiscal discipline; a redirection of public expenditure priorities toward fields offering both high economic returns and the potential to improve income distribution, such as primary health care, primary education, and infrastructure; tax reform (to lower marginal rates and broaden the tax base); interest rate liberalization; a competitive exchange rate; trade liberalization; liberalization of inflows of foreign direct investment; privatization; deregulation (to abolish barriers to entry and exit); and secure property rights. The term has been used since in a variety of ways. Williamson says: "Audiences the world over seem to believe that this signifies a set of neo-liberal policies that have been imposed on hapless countries by the Washington-based international financial institutions and have led them to crisis and misery. There are people who cannot utter the term without foaming at the mouth" (www.cid.harvard.edu/cidtrade/issues/washington.html).

2. For example, after the worldwide outbreak of SARS, some people were concerned that closer global linkages among nations could bring more problems than it was worth. Difficulty getting some major nations to ratify the Kyoto Protocol, along with a number of major flooding and fire events related to climate change, caused some people to think that achieving global cooperation was hopeless idealism. This feeling was reinforced by problems achieving consensus in free trade discussions and some major multinational conflicts that for awhile had nations talking more about national security than international cooperation. Many activists in rich countries worried about the potential for negative impacts on the poor if globalization was not carefully controlled and balanced to benefit the poor.

3. For example: Carson (1962), Ehrlich (1968, 1974), Hardin (1993), Meadows et al. (1972).

4. Including decline in many ocean fisheries, uncontrollable pollution impacts on rivers and estuaries arising from rapidly increasing industrial activity in both rich and poorer countries, massive damage from floods arising from clearing of vegetation on floodplains, and impacts from the movement of refugees due to civil and military conflicts.

5. The rise of the Brazilian Worker's Party to government in Brazil, for example, illustrates a strong desire of the electorate to have more say in government (see discussion at www.pt.org.br and www.npr.org/programs/npc/2002/021210.lula.html).

6. Nobel Prize winner Paul Crutzen (1995) claims that the decision to use chlorine in chlorinated hydrocarbons was a lucky chance based on the interests and expertise of the chemists involved. Chlorine reacts with ozone only under certain conditions. An alternative would have been bromine. Bromine is 100 times more reactive. If by chance bromine had been used, there could have been devastating effects on humans before the problem was detected and action could be taken.

7. It did this to minimize input costs (e.g., coal) to industry and it did it because the political system allowed strong central control and rapid decision-making.

8. "Peculiarly" alludes to the observation (Gray 2003) that there is more than one potential approach to democracy, depending on such variables as a country's culture, history, and economic circumstances, and that China, in particular, has resisted copying an American model.

9. The possible trends in air pollution are complex and are explored in more detail in Chapter 9. Under Global Orchestration we expect lower levels of sulfur-related air pollution in all regions except sub-Saharan Africa. For NO_x, however, the trends are favorable only for OECD and Latin America. Elsewhere (*with low to medium certainty*) NO_x-related pollution increases substantially (Asia and the former Soviet Union) or moderately (sub-Saharan Africa and Middle East/North Africa).

10. John Gray (2003) gives examples of Buddhist India (centuries ago), the Ottoman Empire, the Moorish Empire of Medieval Spain, and China up to the nineteenth century as societies in which cultural and political diversity were tolerated and even celebrated.

11. The quantitative models (see Chapter 9) illustrate the delicate balance of trends that are required for this scenario to succeed. The trends include:

- a rapid rate of technology development and investments in agricultural research that lead to substantial yield increases rendering large expansion in new crop areas unnecessary (only in sub-Saharan Africa is a large expansion of cropland projected);
- irrigation of much more agricultural land (the projected increase in area of irrigated land is the highest among the four scenarios spurred by high economic growth and large investments in irrigation systems—this will require appropriate increases in efficiency of water use to avoid demand exceeding supply);
- total agricultural land is projected to grow because of the demand for pasture land and biofuels;
- trade liberalization and economic opening will create new markets and more trade among increasingly prosperous poorer countries; and
- large investments in agricultural research and infrastructure, particularly in poorer countries, help bring down international food prices for livestock products and rice.

12. Various authors discuss the exponentially increasing rate of advancement of some technologies, leading to a potential singularity (rate of change beyond the ability of society to adapt without fundamental and unpredictable change). See, for example, Vinge (1993), Brand (1998), Moravec (1998), Kurzweil (1999), Calvin and Loftus (2000), Broderick (2001), and Whole Earth Magazine (2003).

13. Issues associated with invasive species are discussed later in this chapter and in Various Authors (2003).

14. In the quantitative models (Chapter 9), all four MA scenarios result in increased global food production, both total and per capita, by 2050 compared with the base year. Demand for and prices of food increase with population and for other reasons that differ between scenarios. As would be expected, variation between regions of the world is very different from scenario to scenario. Global Orchestration achieves the largest global production increases, fuelled by trade liberalization and economic opening. The richer countries help to meet demand from poorer countries, but as agriculture expands regions like sub-Saharan Africa and Latin America become net exporters of certain foods, while the OECD and Asia are projected to increase net imports. Large investments in agricultural research and infrastructure, particularly in poorer countries, help bring down international food prices for livestock products and rice.

15. For example, China recently stood its ground in standards for electronic communication and won a compromise from the United States (Anon 2004a; Markus 2004).

16. Climate change represents a typically difficult problem to deal with under the Global Orchestration storyline, as the impacts of climate change are disconnected from its causes in both time and place. This means that it is difficult to pass a test of "sufficient evidence" for people to decide to start mitigating this problem. It is for this reason that development of climate change policy is very slow under Global Orchestration, even though the international focus of the scenario could create the required condition for formulating climate responses.

17. The projections from the modeling in Chapter 9, which were unable to take future social and technological responses fully into account, indicated that unless there are large improvements in the efficiency of water use and treatment of return water, both water withdrawals and wastewater discharges will have increased by 40% worldwide by 2050. Return flows were predicted to decrease on average in OECD and former Soviet countries because of a levelling off of population, decreasing irrigated area, and improving efficiency of water use. Even though low priority is given to environmental protection in Global Orchestration, the richer societies maintain their current efforts at environmental management. Hence it is reasonable to assume that the level of wastewater treatment in OECD countries will remain at least at its current level. Because of the booming water withdrawals, however, return flows increase by a factor of 3.6 in sub-Saharan Africa, a factor of 2 in Latin America, and more moderately in the Middle East and North Africa (22%) and Asia (48%). Therefore, it is likely that wastewater will remain untreated in many areas and that the level of water contamination and degradation of freshwater ecosystems may increase (*low to medium certainty*). Chapter 9 estimates that nearly 60% of the world will live in these areas in 2050. Thus a major requirement for Global Orchestration to have an optimistic outcome is significant improvement in treatment of wastewater, which will be expensive and technically challenging if the roles of ecosystems in minimizing these challenges are addressed reactively rather then proactively.

18. We acknowledge than many commentators are currently arguing that a multipolar world is in fact more stable than a unipolar world

19. For example: "Fortress World," "Barbarization," "Breakdown," and "Mad Max" (see Chapter 2 and Costanza 2000).

20. Lack of international cooperation is also likely to have direct consequences for the type of response options that are available to humanity under this scenario. It seems very unlikely that people could negotiate global solutions and treaties to address problems of global environmental change under the disaggregated Order from Strength. Moreover, the reduction of international trade also meant that countries tended to focus on using resources within their own regions. This can imply significant ecological consequences for regions like Asia that are likely to rely on using domestic coal resources (with relatively large environmental impact), or land-scarce regions that will need to consider using even more marginal land resources for food supply.

21. In the quantitative modeling (Chapter 9) this scenario has the largest withdrawals of water because of its slower improvement of the efficiency of water use and faster population growth. Accordingly, it also has the largest return flows, with a doubling of worldwide total flows by 2050. The smallest increase is in the former Soviet Union (9%), followed by OECD (nearly 40%). All other regions experience much larger increases—Asia and the Middle East/North African countries with approximately a doubling, Latin America more than a factor of three, and sub-Saharan Africa a factor of 4.7. The combination of exploding wastewater discharges and negligence of the environment could lead to large risks to freshwater ecosystems and water contamination. An additional dimension of this scenario is that return flows continue to increase rapidly after 2050.

22. In the quantitative models (Chapter 9), all four MA scenarios result in increased global food production, both total and per capita, by 2050 compared with the base year. Demand for and prices of food increase with population and for other reasons that differ between scenarios. As would be expected, variation between regions of the world is very different from scenario to scenario. Under Order from Strength, food production in poorer countries stalls because of low investments in technology and infrastructure, and this puts pressure on many to import food. This is only partly offset by low income growth, which dampens food demand in poorer countries. As a result, prices for most foods are projected to increase over the coming decades, and food shortages are a major risk in this scenario.

23. The quantitative models in Chapter 9 project that in Order from Strength there is the fastest depletion of forest area worldwide among the scenarios (at a rate near the historic average, only to slow down after 2050 because of slowing population growth). The loss of forest cover is in particular large in sub-Saharan Africa.

24. The possible trends in air pollution are complex and are explored in more detail in Chapter 9. The level of sulfur-related air pollution declines only slightly worldwide in Order from Strength, because of declines in OECD, the former Soviet Union, and Latin America. Asia and Middle East/North Africa have the largest emission increases of all scenarios. There is a significant decline in NO_x-related pollution in OECD countries, and a major increase elsewhere.

25. For present examples of certification, see the Forest Stewardship Council (www.fscoax.org) or the Marine Stewardship Council (www.msc.org).

26. In the quantitative modeling (Chapter 9), large water withdrawals and large returns of wastewater are projected. This could cause major health challenges, especially in regions where the largest increases might occur (sub-Saharan Africa, Latin America, Asia, and the Middle East/North African countries). However, these projections are based on the larger populations and lower investments in technology projected in this scenario. The models could not take account of adaptations in land management and possible application of inexpensive approaches to waste generation and management that would take advantage of ecosystem services like waste assimilation services. In our storylines, we are optimistic about the emergence of new approaches like these supported by local learning and the sharing of lessons learned between regions. Nevertheless, the emergence of these approaches is vital for the success of this scenario.

27. Chapter 9 projects large increases in water withdrawals in this scenario, probably beyond what would be available, which could dramatically affect food production. However, the modeling could not take account of people's reaction to increasing water withdrawal. In this scenario, we expect that people will recognize this problem early and learn to deal with it through more efficient use of water, including reduction of water losses in irrigation, greater return of water to the environment, and better choice of crops and cropping areas. The contrast between the storylines and the modeling illustrates that dealing with water withdrawals by increasing human populations will be one of the major challenges of the future, even for a scenario based on adaptive learning at regional scales.

28. These examples are based on projections by McCarthy et al. (2001), Chapter 4.5.2.

29. The reader might notice some apparent inconsistencies between these storylines and the quantitative models for this scenario in Chapter 9, especially with respect to expansion of agriculture and production and availability of food. The differences are because the storylines have been more optimistic about the development of better environmental management and the ability to minimize losses in productivity through local learning and the emergence of cooperation among local groups at regional to global scales. When the models were developed, we expected this scenario to feature a disconnected world throughout. However, when developing the storylines we realized that despite disaggregated global institutions, people would still take advantage of the increasing opportunities for electronic communication and that "bottom-up" connections would inevitably develop. The models could not take account of these developments, but they provide a stark statement of the increasing challenges that a world based on local learning would have to address to avoid failure. The quantitative models assumed low investments in food production technologies and no breakthroughs in yield-enhancing technologies. Globally, irrigated area is expected to grow very slowly up to 2050, but to increase particularly in sub-Saharan Africa and Latin America. Crop harvested area is projected to increase, particularly in sub-Saharan Africa, Latin America, and the Middle East/North Africa. Food prices are projected to rise, and increases in calorie availability are very small. The number of malnourished children is projected to increase initially but then decline globally by 2050, but there could be major increases in some poorer countries. (See Chapter 9 for details.)

30. See also Jaeger (1994), in particular Section 5.4.

31. Some people are skeptical that large improvements in technological efficiency are possible in only a few decades (Jacobs 1991). Many other authors, exemplified by the Rocky Mountain Institute and Wuppertal Institute, have made convincing cases that massive improvements in efficiency are possible even with current technology. As TechnoGarden is based upon technological improvement, we have maintained this assumption.

32. The recent collapse of the WTO talks in Cancun provides an example of the coalitions and issues.

33. For example, England Rural Development Programme (www.defra.gov.uk/erdp/default.htm) and FOE's campaign (www.choosefoodchoosefarming.org/).

34. The quantitative modeling (Chapter 9) projects lower water withdrawals and returns of wastewater than in other scenarios due to a stronger emphasis on improving water efficiency and the development of environmental technologies to minimize pollution and maximize ecosystem services for maintaining water quality. The development of these technologies during the first decade of this scenario is vital for its success. Otherwise, it could become more like Global Orchestration, with reactive approaches to emerging environmental problems, except that it could also create new problems through ill-equipped attempts at manipulating ecosystems.

35. See Web sites and reports on multifunctional landscapes in the Netherlands and the EU, www.urban.nl/index.htm, and in Denmark, www.naturraadet.dk/start.htm#english/default.htm.

36. In the quantitative models (Chapter 9), all four MA scenarios result in increased global food production, both total and per capita, by 2050 compared with the base year. Demand for and prices of food increase with population and for other reasons that differ between scenarios. As would be expected, variation between regions of the world is very different from scenario to scenario. The TechnoGarden scenario operates somewhat similarly to the Global Orchestration scenario, with substantial improvements in crop yields combined with a lower preference for meaty diets reducing pressure on crop area expansion. Increased food demand is also met through exchange of goods and technologies. Both calorie consumption levels and the reduction in the number of malnourished children are similar, albeit somewhat lower than, the Global Orchestration scenario.

37. Some of these possibilities are discussed by United Nations Population Division (2001). China's energy changes are discussed by China Energy Group at Berkley Labs, at eetd.lbl.gov/EA/partnership/China.

38. Car-sharing is discussed at www.carsharing.net.

39. The Korean Demilitarized Zone provides an example of an involuntary park. Hunting of bushmeat, mining, and logging during the war in the Congo is an example of ecological degradation. (See Price 2002.)

40. Cheap communication and organization is allowing vast online groups to organize and function. Linux provides an example, but so does the anti-globalization movement (Arquilla and Ronfeldt 2001; Ronfeldt et al. 1998).

41. The UK's Royal Society's report on recent large-scale field trials discusses the impacts of GM agriculture on wildlife populations. This is available at www.defra.gov.uk/environment/gm/fse/index.htm.

42. See Creative Commons at creativecommons.org for a current nongovernmental organization advocating more open copyright laws. They have recently moved to create international licensing agreements.

43. The possible trends in air pollution are complex and are explored in more detail in Chapter 9. Under TechnoGarden, major reductions in sulfur-related pollution are achieved globally (except for sub-Saharan Africa). Meanwhile, NO_x pollution declines substantially only in OECD and Latin America, while elsewhere it continues to increase about the same as in Global Orchestration.

44. Note that the quantitative models (Chapter 9) were unable to take social factors like these into account, and their outputs more strongly resemble the worst than the best case in Figure 8.5.

References

Anon, 2004a: U.S. Chip Makers Raising Stakes in Wi-Fi China Row. *Mobile Tech News,* Wednesday 26-May-2004. Available at www.mobiletechnews.com/info/2004/01/30/014317.html.

Ardila, A., and G. Menckhoff, 2002: Transportation policies in Bogota, Colombia: Building a transportation system for the people. *Transportation Research Record,* **1817,** 130–136. www.dcdata.com/dbtw-wpd/exec/dbtwpub.dll

Arquilla, J., and D. Ronfeldt, editors, 2001: *Networks and Netwars: The Future of Terror, Crime, and Militancy.* RAND Corporation, Santa Monica.

Brand, S., 1998: Freeman Dyson's Brain. *Wired Magazine,* **February.**

Broderick, D., 2001: *The Spike: How Our Lives Are Being Transformed by Rapidly Advancing Technologies.* Tom Doherty Associates, New York.

Calvin, W.H., and E. F. Loftus, 2000. The poet as brain mechanic: A 2050 version of Physics for Poets. An occasional paper for the Global Business Network. Available at williamcalvin.com/2000/2050.htm.

Carson, R. 1962: *Silent Spring.* Houghton Mifflin, Boston, MA.

Christian Science Monitor International Edition Monday October 20, 1997: Post-US, Somalia Finds Many Cash In on Chaos. Available at search .csmonitor.com/durable/1997/10/20/intl/intl.4.html.

Costanza, R, 2000: Visions of alternative (unpredictable) futures and their use in policy analysis. *Conservation Ecology,* **4(1),** 5. [online] URL: www .consecol.org/vol4/iss1/art5.

Crutzen, P., 1995: Nobel acceptance speech. www.nobel.se/chemistry/laureates/1995/crutzen-lecture.html

Donner, S.D., M.T. Coe, J.D. Lenters, T.E. Twine, and J.A. Foley, 2002: Modeling the impact of hydrological changes on nitrate transport in the Mississippi River Basin from 1955 to 1994. *Global Biogeochemical Cycles* **16(3),** Article No. 1043.

Ehrlich, P., 1968: *The Population Bomb.* Ballantine, New York.

Ehrlich, P., 1974: *The End of Affluence.* Ballantine, New York.

Ekins, P., and M. Jacobs, 1995: Environmental sustainability and the growth of GDP conditions for compatibility. In: *The North, the South and the Environment—Ecological Constraints and the Global Economy,* V. Bhasker and A. Glynn (eds.). Earthscan, pp. 9–46.

Foley, J. A., C. J. Kucharik, S. D. Donner, T. E. Twine, and M. T. Coe, in review: Land use, land cover, and climate change across the Mississippi Basin: impacts on land and water resources. In: *Ecosystem Interactions with Land Use Change.* R. DeFries, G. Asner, and R. Houghton (eds.). American Geophysical Union.

Fouracre, P., C. Dunkerley, and G. Gardner, 2003: Mass rapid transit systems for cities in the developing world. *Transport Reviews* **23,** 299–310.

Goolsby, D.A., W.A. Battaglin, G.B. Lawrence, R.S. Artz, B.T. Aulenbach, R.P. Hooper, D.R. Keeney, and G.J. Stensland, 1999: *Flux and sources of nutrients in the Mississippi-Atchafalaya River basin.* Topic 3 Report for the Integrated Assessment on Hypoxia in the Gulf of Mexico. NOAA Coastal Ocean Program Decision Analysis Series 17, NOAA Coastal Ocean Office, Silver Spring, Md.

Government of Korea website: www.korea.net/kwnews/pub_focus/content .asp?cate = 01&serial_no = 305

Gray, J., 2003: *Al Qaeda and What it Means to be Modern.* Faber & Faber, London

Hardin, G., 1993: *Living Within Limits.* Oxford University Press, New York.

Hawken, P., A. Lovins, and L.H. Lovins, 1999: *Natural Capitalism: Creating the Next Industrial Revolution.* Little, Brown and Co., Boston.

Jackson, J. B. C., M. X. Kirby, W. H. Berger, K. A. Bjorndal, L. W. Botsford, B. J. Bourque, R. H. Bradbury, R. Cooke, J. Erlandson, J. A. Estes, T. P. Hughes, S. Kidwell, C. B. Lange, H. S. Lenihan, J. M. Pandolfi, C. H. Peterson, R. S. Steneck, M. J. Tegner, and R. R. Warner, 2001: Historical overfishing and the recent collapse of coastal ecosystems. *Science* **293,** 629–638.

Jacobs, M., 1991: *The Green Economy: Environment, Sustainable Development and the Politics of the Future.* Pluto Press, London.

Jaeger, C. C., 1994: *Taming the Dragon: Transforming Economic Institutions in the Face of Global Change.* Gordon and Breach Science Publishers, Yverdon.

Justic, D., N.N. Rabalais, and R.E. Turner, 2003: Simulated responses of the Gulf of Mexico hypoxia to variations in climate and anthropogenic nutrient loading. *Journal of Marine Systems* **42,** 115–126.

Kim, K.C., 1997: Preserving Biodiversity in Korea's Demilitarized Zone. *Science,* **278,** 242–243.

Kurzweil, R., 1999: *The Age of Spiritual Machines: When Computers Exceed Human Intelligence.* Viking, New York.

Markus, F., 2004: China-US talks reach wi-fi deal. *BBC News* Thursday, 22 April, 2004. Available at news.bbc.co.uk/1/hi/business/3648653.stm.

McCarthy, J. J., O. F. Canziani, N. A. Leary, D. J. Dokken, and K. S. White, editors, 2001: *Climate Change 2001: Impacts, Adaptation, and Vulnerability.* Contribution of Working Group II to the Third Assessment Report of the Intergovernmental Panel on Climate Change. Cambridge University Press, Cambridge.

Meadows, D. H., D. L. Meadows, J. Randers, and W. W. Behrens III, 1972: *The Limits to Growth.* Universe Books, New York.

Mitsch, W. J., J. W. Day, J. W. Gilliam, P. M. Groffman, D. L. Hey, G. W. Randall, and N. M. Wang, 2001: Reducing nitrogen loading to the Gulf of Mexico from the Mississippi River Basin: Strategies to counter a persistent ecological problem. *BioScience* **51,** 373–388.

Moravec, H., 1998: When will computer hardware match the human brain? *Journal of Evolution and Technology,* **1**-March 1998 Available at www.trans. humanist.com/volume1/moravec.htm.

Price, S.V. (ed), 2002: *War and Tropical Forests: Conservation in Areas of Armed Conflict.* The Haworth Press, Binghamton, New York.

Rabalais, N.N., R.E. Turner, and D. Scavia, 2002a: Beyond science into policy: Gulf of Mexico hypoxia and the Mississippi River. *Bioscience* **52,** 129–142.

Rabalais, N.N., R.E. Turner, and W.J. Wiseman, 2002b: Gulf of Mexico hypoxia, aka "The dead zone". *Annual Review of Ecology and Systematics* **33,** 235–263.

Ronfeldt, D., J. Arquilla, G. Fuller, and M. Fuller, 1998: *The Zapatista Social Netwar in Mexico.* RAND, Santa Monica, CA.

The International Chamber of Commerce Commercial Crime Service: Available at www.iccwbo.org/ccs/imb_piracy/weekly_piracy_report.asp

Turner, R. E., and N. N. Rabalais, 2003: Linking landscape and water quality in the Mississippi river basin for 200 years. *Bioscience* **53,** 563–572.

United Nations Environment Programme, 2003: *Emerging challenges—New findings.* GEO Yearbook 2003. UNEP. www.unep.org/GEO/yearbook/index.htm.

United Nations Population Division, 2001: *World Urbanization Prospects: The 2001 Revision.* United Nations Population Division.

Various authors, 2003: Special Section: Population Biology of Invasive Species. *Conservation Biology* **17,** 24–92. Summarized by Fred Allendorf at www .eurekalert.org/pub_releases/2003-01/sfcb-ssi012803.php.

Vinge, V., 1993: The Singularity. *Whole Earth Magazine* at www.whole earthmag.com [reprinted in Spring 2003].

Whole Earth Magazine, 2003: What Keeps Jaron Lanier Awake at Night—an interview in *Whole Earth,* Spring 2003. The entire issue is about singularity, see it at www.wholeearthmag.com.

Williamson, J., 2000: What should the World Bank think about the Washington Consensus? *World Bank Research Observer* **15(2),** 251–264. www.world bank.org/research/journals/wbro/obsaug00/pdf/(6)Williamson.pdf.

World Resources Institute, 1998: *World Resources 1998–1999: Linking Environment and Human Health.* WRI, Washington DC.

Zemlo, T., 2002: Medical and life science professionals are squeamish about biotech art. The Science Advisory Board, www.scienceboard.net/community/news/news.108.html.

Part III

Implications of the Millennium Ecosystem Assessment Scenarios

Chapter 9

Changes in Ecosystem Services and Their Drivers across the Scenarios

Coordinating Lead Authors: Joseph Alcamo, Detlef van Vuuren, Wolfgang Cramer

Lead Authors: Jacqueline Alder, Elena Bennett, Stephen Carpenter, Villy Christensen, Jonathan Foley, Michael Maerker, Toshihiko Masui, Tsuneyuki Morita,* Brian O'Neill, Garry Peterson, Claudia Ringler, Mark Rosegrant, Kerstin Schulze

Contributing Authors: Lex Bouwman, Bas Eickhout, Martina Floerke, Rattan Lal, Kiyoshi Takahashi

Review Editors: Bach Tan Sinh, Allen Hammond, Christopher Field

---

---

*Deceased.

APPENDIXES

BOXES

FIGURES

*This appears in Appendix A at the end of this volume.

TABLES

*This appears in Appendix A at the end of this volume.

Main Messages

The demand for provisioning services, such as food, fiber, and water, strongly increases in all four scenarios (*with medium to high certainty*). This is due to expected increases in population, economic growth, and changing consumption patterns. Increasing demand for provisioning services leads (*with high certainty*) to further stress on the ecosystems that provide these services. By 2050, global population increases (*with medium to high certainty*) to 8.1–9.6 billion, depending on the scenario. At the same time, per capita GDP expands by a factor of 1.9–4.4, again depending on the scenario (*low to medium certainty*). Increasing income fuels increasing per capita use of most resources in most parts of the world. The combination of increasing population and per capita consumption increases the demand (*with high certainty*) for ecosystem services, including water and food. Demand is dampened somewhat by increasing efficiency in the use of resources.

Trade-offs between ecosystem services continue and perhaps intensify. The gains in provisioning services, such as food supply and water, will come partly at the expense of losses of other services. Providing additional food to match increased demand will lead (*with low to medium certainty*) to further expansion of agricultural land, and this in turn will lead to the loss of natural forest and grassland, as well as the loss of ecosystem services associated with this land (genetic resources, wood production, habitat for fauna and flora). Water use will increase in poorer countries (*with high certainty*), and this is likely to be accompanied by a deterioration of water quality and the loss of the ecosystem services provided by clean freshwater systems (genetic resources, fish production, habitat for aquatic and riparian flora and fauna).

Overall, the largest decrease in the quality of ecosystems and the provision of ecosystem services (*with medium certainty*) occurs under the Order from Strength scenario. This is driven by a relatively large increase in population, a reactive attitude toward ecological management, the low level of technological development, and restrictions on trade.

The scenarios indicate (*with medium certainty*) certain "hot spot regions" of particularly rapid changes in ecosystem services, including sub-Saharan Africa, the Middle East and Northern Africa, and South Asia. To meet its needs for development, sub-Saharan Africa is likely to rapidly expand its withdrawal of water, which will require an unprecedented investment in new water infrastructure. Under some scenarios (*medium certainty*), this rapid increase in withdrawals will cause a similarly rapid increase in untreated return flows to the freshwater systems, which could endanger public health and aquatic ecosystems. Sub-Saharan Africa could experience not only accelerating intensification of agriculture but also extensification through expansion of agricultural land into natural areas. Further intensification could lead to a higher level of contamination of surface and groundwaters. Extensification will come at the expense of a large fraction of sub-Saharan Africa's natural forest and grasslands (*medium certainty*), as well as the ecosystem services they provide.

In all scenarios, rising income in the Middle East and Northern African countries leads to greater meat demand, which could lead to a still higher level of dependence on food imports (*low to medium certainty*). There is a *medium certainty* that rising incomes put further pressures on limited water resources in the hot-spot regions, which will either stimulate innovative approaches to water conservation or could limit development. In South Asia, deforestation continues in all scenarios, despite increasingly intensive industrial-type agriculture. Here, rapidly increasing water withdrawals and return flows further intensify water stress. There may be regions (*with low certainty*) where the pressure on ecosystems causes breakdowns in these ecosystems, and these breakdowns could interfere with the well-being of the population and its further economic development.

The four scenarios describe contrasting pathways for the development of human society and ecosystems. At the same time, similar outcomes for ecosystem services can be achieved through multiple pathways. For example, food demand across scenarios can be fulfilled either through expansion in cropping area or through an increase in crop yields. Similarly, comparable rates of land use change can result from different combinations of growth rates of population, economic activity, and technology developments.

There are several conclusions regarding specific drivers and ecosystem services:

- **Vast changes are expected in world freshwater resources and hence in the ecosystem services provided by freshwater systems.** A deterioration of the services provided by freshwater resources (aquatic habitat; fish production; water supply for households, industry, and agriculture) is expected under the two scenarios that are reactive to environmental problems (Global Orchestration and Order from Strength). A less severe decline is expected under the two scenarios that are proactive about environmental issues (TechnoGarden and Adapting Mosaic) (*medium certainty*). Water withdrawals are expected to increase greatly outside wealthy countries (as a result of economic and population development) but will continue to decline in other regions (as a result of saturation of per capita demands, efficiency improvements, and stabilizing population) (*medium certainty*).

  The extent of the increases outside the rich countries is scenario-dependent. In sub-Saharan Africa, domestic water use greatly increases in all scenarios, and this implies (*with low to medium certainty*) an increased access to fresh water. However, these estimates do not factor in the technical and economic feasibility of increasing domestic water withdrawals. Under the Global Orchestration and Order from Strength scenarios, massive increases in water withdrawals are expected to lead to an increase in untreated wastewater discharges (in poorer countries), causing a deterioration of freshwater quality. Climate change leads to both increasing and declining river runoff, depending on the region. The combination of huge increases in water withdrawals, decreasing water quality, and decreasing runoff in some areas leads to an intensification of water stress over wide areas.

- **Land use change is a major driver of changes in the provision of ecosystem services up to 2050 (*medium to high certainty*).** The scenarios indicate (*low to medium certainty*) that 10–20% of current grassland and forestland may be lost between now and 2050, mainly due to the further expansion of agriculture (and secondarily, because of the expansion of cities and infrastructure). This expansion mainly occurs in low-income and arid regions, while in the high-income regions, agricultural area declines. The provisioning services associated with affected biomes (genetic resources, wood production, habitat for terrestrial biota and fauna) will also be reduced. The degree to which natural land is lost differs among the scenarios. The Order from Strength scenario has the greatest implications from land use changes, with large increases in both crop and grazing areas. TechnoGarden and Adapting Mosaic, in contrast, are the most land-conserving scenarios because of increasingly efficient agricultural production, lower meat consumption, and lower population increases. Existing wetlands and the services they provide (such as water purification) are faced with increasing risk in some areas due to reduced runoff or intensified land use in all scenarios.

- **After 2050, climate change and its impacts (such as sea level rise) have an increasing effect on the provision of ecosystem services (*medium certainty*).** Under the four MA scenarios, global temperature is expected to increase significantly: 1.5–2.0° Celsius above pre-industrial in 2050, and 2.0–3.5° Celsius in 2100, depending on the scenario and using median estimates for climate change variables (*medium certainty*). This is in the low to middle range of the scenarios developed for the IPCC Third Assessment Report (2.0–6.4° Celsius). The main reasons for this are that the MA range does not include the effect of the uncertainty in climate sensitivity and the MA set includes one scenario that assumes climate policy, in contrast to the climate policy–free IPCC scenarios. There is an increase in global average precipitation (*medium certainty*), but some areas will become more arid while others will become moister. Climate change will directly alter ecosystem services, for example, by causing changes in the productivity and growing zones of cultivated and noncultivated vegetation. Climate change also alters the frequency of extreme events, with associated risks to ecosystem services. Finally, it will indirectly affect ecosystem services in many ways, such as by causing sea level to rise, which threatens mangroves and other vegetation that now protect shorelines.

- **Food security remains out of reach for many people, and child malnutrition cannot be eradicated by 2050 (*with low to medium certainty*), even though the supply of food increases under all four scenarios (*medium to high certainty*) and diets in poorer countries become more diversified (*low to medium certainty*).** On a global basis, food supply increases significantly in all four scenarios. On a per capita basis, however, basic staple production stagnates or declines for all scenarios in the Middle East and North Africa and increases very little in sub-Saharan Africa (*low to medium certainty*). Resulting shortfalls in these regions are expected to be covered through increased net food imports (*medium certainty*). Even though cereal production in 2050 will be 50% larger and the per capita availability of food increases, child malnutrition is not eradicated (*low to medium certainty*). Moreover, higher grain prices under the Order from Strength and Adapting Mosaic scenarios indicate a tightening of world food supplies. Order from Strength leads to the highest estimated number of malnourished children in 2050—181 million, compared with 166 million children today. Also in Adapting Mosaic, we estimate (*with low certainty*) that by 2050 some 116 million children might still be malnourished. The large number of malnourished children arises because of inadequate investments in food production and its supporting infrastructure and high population growth. Larger investments in health and education and enhanced community development could reduce the number of malnourished children (*with high certainty*).

- **Demand for fish as food will expand, and the result will be an increasing risk of the major long-lasting decline of regional marine fisheries (*low to medium certainty*).** The demand for fish from both freshwater and marine sources as well as from aquaculture will increase across all scenarios because of increasing human population, income growth, and increasing preferences for fish. Increasing demand will raise the pressure on marine fisheries, which may already be near their maximum sustainable yield, and could cause a long-term decline in their productivity (*low to medium certainty*). The production of fish via aquaculture will add to the risk of decline of marine fisheries if aquaculture continues to depend on marine fish as a feed source.

- **The future contribution of terrestrial ecosystems to the regulation of climate is uncertain.** Deforestation is expected to reduce the carbon sink, most strongly under the Order from Strength scenario (*with medium certainty*). Carbon release or uptake by ecosystems affects the CO_2 and CH_4 content of the atmosphere at the global scale and thereby affects global climate. Currently, the biosphere is a net sink of carbon, absorbing about 1 to 2 gigatons of carbon per year, or approximately 20% of fossil fuel emissions. It is very likely that the future of this service will be greatly affected by expected land use change. In addition, a higher atmospheric CO_2 concentration is expected to enhance net productivity, but this does not necessarily lead to an increase in the carbon sink. The limited understanding of soil respiration processes generates uncertainty about the future of the carbon sink. There is a medium certainty that climate change will increase terrestrial fluxes of CO_2 and CH_4 in some regions (in Arctic tundras, for example). Among the four scenarios, the greatest reduction of the terrestrial biosphere's carbon sink will occur (*with low certainty*) under the Order from Strength scenario because of its high level of deforestation.

9.1 Introduction

The capacity of ecosystems to provide services is determined by many different direct and indirect driving forces operating at the local to global level. (See Chapters 1 and 7.) Changes in driving forces will catalyze changes in the provision of ecosystem goods and services. In this chapter, we estimate the future changes in ecosystem services according to changes in driving forces described in the MA scenarios. The expectations about the future of ecosystem services are consistent with the storylines presented in Chapters 5 and 8.

We present estimates of changing ecosystem services in the form of both qualitative and quantitative information. The qualitative information is based on our interpretation of the storylines presented in Chapters 5 and 8, while the quantitative information is based on a modeling analysis, also related to the storylines, as explained below.

Qualitative expectations for future ecosystem services are summarized in Table 9.1 for provisional and regulating services. Since it is not feasible to present these expectations in natural units, such as tons of grain or cubic meters of potable water, we use a simple, three-level indicator system: zero if the ecosystem service changes little between 2000 and 2050, +1 if it is in better condition in 2050 than in 2000, and −1 if it is in worse condition. These qualitative expectations were not calculated from computer models but are based on our assessment and judgment of the storylines presented in Chapters 5 and 8. But these judgments are analogous to the model results in some ways. Model results and qualitative expectations were both constructed to be consistent with the logic and rationale of the storylines. They both have substantial uncertainties. By explicitly comparing outcomes for ecosystem services using both qualitative and quantitative approaches, we gain some perspective on the uncertainties.

The storylines also imply certain conclusions about the vulnerability of ecosystems. An ecosystem is vulnerable if it is sensitive to anthropogenic or non-anthropogenic disturbances. If society is highly dependent on a service provided by a threatened or sensitive ecosystem, then society too is vulnerable.

The quantitative conjectures come from a modeling exercise described in Chapter 6 and are reported in the text and the figures of this chapter. The exercise had the following basic steps. First, from the storylines we derive a set of

Table 9.1. Qualitative Expectations for Provisioning and Regulating Ecosystem Services in MA Scenarios. Ecosystem services are defined in the MA conceptual framework volume (MA 2003: 56–59). "Industrial Countries" stands for nations that are relatively developed and wealthy in 2000; "Developing Countries" stands for nations that are relatively undeveloped and poor in 2000. Note that any particular nation could switch categories between 2000 and 2050. Scores pertain to the endpoint, 2050. A score of +1 means that the ecosystem service is in better condition than in 2000. A score of zero means that the ecosystem service is in about the same condition as in 2000. A score of −1 means that the ecosystem service is in worse condition than in 2000.

Vulnerability of an ecosystem service is defined as the sensitivity of the service to external disturbances multiplied by the sensitivity of the socioecological system to changes in the ecosystem service. If society is highly dependent on an ecosystem service that is sensitive to disturbance, then the ecosystem service has high vulnerability. If society is not dependent on an ecosystem service, or if the ecosystem service is not sensitive to disturbance, then the ecosystem service has low vulnerability.

	Global Orchestration		Order from Strength		Adapting Mosaic		TechnoGarden	
Ecosystem Service	**Industrial Countries**	**Developing Countries**	**Industrial Countries**	**Developing Countries**	**Industrial Countries**	**Developing Countries**	**Industrial Countries**	**Developing Countries**
Provisioning services								
Food (extent to which demand is met)	+1	+1	0	−1	0	−1	+1	+1
Fuel	+1	+1	+1	+1	+1	+1	+1	+1
Genetic resources	0	0	−1	−1	+1	+1	0	+1
Biochemical discoveries	−1	+1	−1	−1	0	0	+1	+1
Ornamental resources	0	0	0	−1	+1	+1	0	0
Fresh water	+1	+1	0	−1	+1	−1	+1	0
Comments on vulnerability of provisioning services	increased vulnerability of ecosystem services to stochastic shocks; breakdowns have greater impact on the poor		in industrial countries, ecosystem services are vulnerable due to small patch effects and climate change; in developing countries, ecosystem services are vulnerable due to these factors as well as overexploitation, degradation of ecosystems, and lower trade		vulnerability of ecosystem services is reduced, especially in developing countries; smoother adaptation of ecosystem services to changing environmental drivers		vulnerability of ecosystem services is increased by maximization of efficiencies; generally high performance of ecosystem services but frequent massive interruptions due to environmental changes and shocks; tendency to increase provisioning services due to investment in technology	

Regulating services								
Air quality maintenance	0	0	0	−1	0	0	+1	+1
Climate regulation	0	0	−1	−1	0	0	+1	+1
Water regulation	0	−1	−1	−1	+1	+1	0	+1
Erosion control	0	−1	−1	−1	+1	+1	0	+1
Water purification	0	−1	−1	−1	+1	+1	0	+1
Regulation of human disease	0	+1	0	−1	0	+1	+1	+1
Biological control	0	−1	−1	−1	+1	+1	0	0
Pollination	−1	−1	−1	−1	0	0	−1	−1
Storm protection	0	−1	0	−1	+1	+1	+1	0
Comments on vulnerability of regulating services	improvements in targeted problem areas, but increased vulnerability of ecosystem services to stochastic shocks, especially in developing countries; indirect effects and cross-sector interactions are the root of increasing problems; low attention to underlying processes, so regulating services should generally diminish; some improvements for ecosystem services that have clear direct effects on people		increased vulnerability of ecosystem services to stochastic shocks leads to severe breakdowns of ecosystem support to people, especially in developing countries; wealth allows some adaptation in industrial countries		vulnerability is reduced in water management and some aspects of ecosystem management, but shows little change for large-scale commons (atmosphere, global fisheries)		improvements are achieved through increased efficiency, but the engineered ecosystems are increasingly vulnerable to shocks and require frequent technological fixes; more focus on underlying processes, but emphasis on efficiency creates new problems that require new fixes	

quantitative assumptions for the indirect drivers of ecosystem changes (such as population and economic growth). Second, we use information about indirect drivers to derive assumptions about the direct drivers of ecosystem change (such as energy use and irrigated area). In some cases (land cover change, for instance), models are used to derive these direct drivers. Next, the direct drivers are input to a suite of numerical simulation models. These models generate first estimates of temporal and spatial changes in a wide range of ecosystem services. As noted in Chapter 6, an important point is that these models can only cover a small part of the attributes and processes having to do with ecosystem services. For example, while the quantitative models address many provisioning and regulating ecosystem services, we do not have quantitative models for estimating future provision of supporting and cultural services. We try to fill in some of the missing information with the qualitative expectations.

This chapter starts with a discussion of the assumed changes in indirect and direct drivers and then describes estimates for each of the provisioning and regulating ecosystem services in turn. This is followed by brief sections describing qualitative estimates of supporting and cultural services. We focus on results for 2050, which is a compromise between the shorter time horizon of a typical agricultural or urban prospective study and the longer time horizon of climate impact studies. The year 2050 also gives us a long-term perspective on the ecological consequences of current actions and policies. Nevertheless, where appropriate we also provide information about the year 2100 and temporal trends throughout the twenty-first century.

9.2 Indirect Drivers of Ecosystem Services

Drivers of ecosystem services, as the term is used in the MA, are human-induced factors that directly or indirectly cause a change in an ecosystem. The difference between indirect and direct drivers is that the latter unequivocally influence ecosystem processes, while the former operate more diffusely, often by altering one of the more direct drivers.

Chapter 7 discusses the role of the different drivers of change in ecosystem services, their historic changes, and the range of possible changes in the future. In this chapter, we estimate their changes under each of the storylines. Models were used to provide quantitative estimates for most of the relevant drivers. (See Table 9.2.) For other drivers, qualitative judgment was used.

The key indirect driving forces of the MA scenarios are population, income, technological development, and changes in human behavior. The future trends of these driving forces are quite different, as implied by the story-lines of the four scenarios.

9.2.1 Population

9.2.1.1 Methodology and Assumptions

Change in population is important because it will influence the number and kind of consumers of ecosystem services. Furthermore, it will directly affect the amount of energy used, the magnitude of air and water pollutant emissions, the amount of land required, and the other direct drivers of ecosystem change. Population scenarios are developed on a regular basis by demographers at the United Nations and the International Institute for Applied Systems Analysis (IIASA; Lutz et al. 2001). Both groups also try to express the uncertainties of the population projections, by giving either more than one scenario (the United Nations) or probabilistic projections (IIASA).

By 2050, most projections are in the range of 7–11 billion people. After 2050, this range widens significantly, with some scenarios showing increasing population levels while others show decreasing levels. In recent years there have been several downward revisions of population projections. Thus the population projections of the MA scenarios, as discussed here, have a lower range than those used in earlier global environmental assessment studies (such as the Intergovernmental Panel on Climate Change and the Global Environmental Outlook of the U.N. Environment Programme). The four population projections used here have been based on the IIASA 2001 probabilistic projections for the world (Lutz et al. 2001), but they are designed to be consistent with the four MA storylines. (See Table 9.3.) The IIASA projections are generally consistent with those from the other major institutions that produce global population scenarios (United Nations, World Bank, and U.S. Census Bureau).

The first step in deriving the projections was to make qualitative judgments about trends in the components of population change (fertility, mortality, and migration) in 13 world regions for each of the MA storylines. Next, the qualitative judgments were converted into quantitative assumptions based on conditional probabilistic projections. Using this approach, the high, medium, and low categories in Table 9.3 were mapped to three evenly divided quantiles of the unconditional probability distributions, as defined in the IIASA projections, for each component of population change. Single, deterministic scenarios for fertility, mortality, and migration in each of 13 regions were derived for

Table 9.2. Driving Forces and Their Degree of Quantification

Quantified Drivers	Unquantified Drivers
Indirect	**Indirect**
Population growth	Sociopolitical
Economic activities	Culture and religion
Technology change	
Direct	**Direct**
Energy use	Species introduction/removal
Emissions of air pollutants (sulfur, nitrogen)	
Emissions of GHG and climate change	
Land use/cover change	
Harvest and resource consumption	
External inputs (irrigation, fertilizer use)	

Table 9.3. Assumptions about Fertility, Mortality, and Migration for Population Projections in MA Scenarios. In the IIASA projections, migration is assumed to be zero beyond 2070, so all scenarios have zero migration in the long run.

Variable	Global Orchestration	Order from Strength	Adapting Mosaic	TechnoGarden
Fertility	HF: low LF: low VLF: medium	HF: high LF: high VLF: low	HF: high/ medium LF: high/ medium VLF: low	HF: medium LF: medium VLF: medium
Mortality	D: low I: low	D: high I: high	D: high/ medium I: high/ medium	D: medium I: medium
Migration	high	low	low	medium

Key: I = industrial country regions; D = developing-country regions; HF = high fertility regions (TFR>2.1 in year 2000); LF = low fertility regions (1.7<TFR<2.1); VLF = very low fertility regions (TFR<1.7). (Total fertility rate is the number of children that a woman would have at the end of her fertile period if current age-specific fertility rates prevailed.)

each storyline, defined by the medians of the conditional distributions for these variables. Population projections for each MA scenario were then produced based on the deterministic scenarios for each component of population change. Regional population projections were then downscaled to the country level to facilitate impact assessments and to allow modeling groups to reaggregate the country level results to their own regional definitions. (More information on the methodology for deriving population projections is given in Chapter 6.)

Table 9.3 lists the qualitative assumptions about fertility, mortality, and migration for each storyline. These assumptions are expressed qualitatively as high, medium, or low and in relative rather than absolute terms. That is, a high fertility assumption for a given region means that fertility is assumed to be high relative to the median of the probability distribution for future fertility in the IIASA projections. Since the storylines describe events unfolding through 2050, the demographic assumptions specified here apply through 2050 as well. For the period 2050–2100, assumptions were presumed to remain the same in order to gauge the consequences of trends through 2050 for the longer term. This is not intended to reflect any judgment regarding the plausibility of trends beyond 2050.

Trends in fertility and mortality in currently high-fertility countries were based on demographic transition reasoning. In Global Orchestration, higher investments in human capital (especially education and health) and greater economic growth rates are assumed to be associated with a relatively fast transition, implying lower fertility and mortality than in a central estimate. In Order from Strength, lower investments in human capital and slower economic growth lead to a slower transition (that is, higher fertility and mortality). TechnoGarden, with more moderate investments and economic growth assumptions, is assumed to undergo a moderate pace of change in both fertility and mortality. The Adapting Mosaic storyline begins similarly to Order from Strength but diverges later because large investments in education pay off in an acceleration of economic growth and technological development in all regions. Demographic trends in Adapting Mosaic are therefore specified to follow Order from Strength for 10 years and then to diverge to "medium assumptions" of mortality and fertility by mid-century.

The determinants of long-term fertility change are poorly known in countries that have completed the demographic transition to low fertility, and therefore there is little basis for preferring one set of assumptions over another for a given storyline. In the face of this uncertainty, the overarching rationale for specifying trends for given storylines was chosen to be the scope of convergence in fertility across low fertility countries. Since Order from Strength describes a regionalized, divergent world, and Global Orchestration a globalizing, convergent world, these characteristics were applied to future fertility. Thus the low fertility countries were divided into two groups (one with "very low fertility" and one with "low fertility," see note to Table 9.3), and fertility assumptions were adopted such that fertility in these two groups would tend to converge in the Global Orchestration scenario to around 1.6 and diverge in the Order from Strength scenario to span a range from 1.3 to 2.2. In Adapting Mosaic, fertility initially follows the Order from Strength assumptions, then diverges toward medium levels. In TechnoGarden, medium fertility is assumed.

Mortality in wealthy country regions is assumed to be lowest in the Global Orchestration scenario, consistent with its high economic growth rates, relatively rapid technological progress (assumed to occur in the health sector as well), and reductions in inequality within the region. In contrast, Order from Strength, which assumes growing inequality within wealthy countries and even the potential for reemergence of some diseases, is assumed to have the highest mortality. TechnoGarden assumes a medium pace of mortality change, and Adapting Mosaic follows the Order from Strength assumptions for 10 years before diverging to medium levels in 2050.

Net migration rates are assumed to be low in the regionally oriented scenarios (Adapting Mosaic and Order from Strength), consistent with higher barriers between regions. In Global Orchestration, permeable borders and high rates of exchange of capital, technology, and ideas are assumed to be associated with high migration. TechnoGarden assumes a more moderate migration level.

9.2.1.2 Comparison of Population Size among Scenarios

Table 9.4 shows the results for global population size through 2100 for each of the four scenarios. The range between the lowest and the highest scenario is 8.1–9.6 billion in 2050 and 6.8–10.5 billion in 2100. These ranges cover 50–60% of the full uncertainty distribution for population size in the IIASA projections. The primary reason that these scenarios do not fall closer to the extremes of the full uncertainty distribution is that they correlate fertility and mortality: Order from Strength generally assumes high fertility and

Table 9.4. Population by Region in 1995 and Assumptions in MA Scenarios (IIASA)

Region	Population in 1995	Global Orchestration			Order from Strength			Adapting Mosaic			TechnoGarden		
		2020	2050	2100	2020	2050	2100	2020	2050	2100	2020	2050	2100
						(million)							
Former Soviet Union	285	290	282	245	287	257	216	288	273	246	292	281	252
Latin America	477	637	742	681	710	944	1,309	708	933	1,155	672	831	950
Middle East and North Africa	312	478	603	597	539	774	972	537	765	924	509	692	788
OECD	1,020	1,136	1,255	1,153	1,076	998	856	1,079	1,068	978	1,117	1,154	1,077
Asia	3,049	3,861	4,104	3,006	4,210	5,023	5,173	4,201	4,992	4,753	4,039	4,535	3,992
Sub-Saharan Africa	558	858	1,109	1,132	956	1,570	1,988	951	1,492	1,775	907	1,329	1,516
World	**5,701**	**7,260**	**8,095**	**6,814**	**7,777**	**9,567**	**10,514**	**7,764**	**9,522**	**9,830**	**7,537**	**8,821**	**8,575**

high mortality, and Global Orchestration generally assumes low fertility and low mortality. Both of these pairs of assumptions lead to more moderate population size outcomes.

Adapting Mosaic is nearly identical to Order from Strength at the global level over most of the century, even though it is designed to follow Order from Strength only for 10 years and then diverge from it. This is because the effects of deviations in fertility in the Adapting Mosaic scenario do not become apparent in population size for many decades due to population momentum and because both fertility and mortality trends diverge. Thus, although fertility declines in Adapting Mosaic relative to Order from Strength after 2010, tending (eventually) toward a smaller population size, mortality declines relative to Order from Strength as well, tending toward a larger population size. The net result is little difference, especially in the short to medium term.

The relationship across scenarios differs by region. While in poorer-country regions the ranking is the same as in the global results (that is, Global Orchestration produces the lowest population size, and Order from Strength the highest), this ranking is reversed in many of the wealthy-country regions (Western Europe, Eastern Europe, Soviet Europe, and Pacific OECD). The main reason is that Order from Strength is assumed to have divergent fertility trends coupled with low migration among the wealthy regions.

Thus the regions with currently very low fertility rates (less than 1.5 births per woman) are projected to see little change in fertility levels in the future, maintaining the fertility difference between these regions and North America and China, where fertility remains around replacement level of about 2 births per woman. These assumptions, in the absence of countervailing increases in net migration into the region, produce substantial population declines in the very low fertility regions. For example, in Western Europe population declines by nearly 20% by 2050 and by more than 50% by 2100. Declines are even greater in other European regions. By contrast, in the Global Orchestration scenario, fertility rates are assumed to converge across wealthy countries, leading to increases in the regions where fertility is currently very low. In addition, migration into the region is assumed to be high in this scenario. The combined effect is to make Global Orchestration the highest population scenario for the richer-country regions.

The range of outcomes for one region, North America, is particularly small over all four scenarios, despite widely differing sets of assumptions about input variables. The reason is that assumptions about the different components of population change, as dictated by the storylines, tend to offset each other. When fertility is assumed to be low, mortality is low as well, and migration (which has a substantial influence on population growth in this region) is high. A similar situation holds, in reverse, when fertility is high. Thus the range of population size outcomes is only 426–439 million in 2050 and 420–540 million in 2100.

In sub-Saharan Africa, the HIV/AIDS epidemic takes a heavy toll in all scenarios. Life expectancy for the region as a whole is assumed to decrease and not to return to current levels for 15–25 years, depending on the scenario. In individual countries where HIV prevalence rates are highest, population is projected to decline. Yet the population of the region as a whole is projected to grow in all scenarios, driven by the large countries of the region whose HIV/AIDS prevalence rates are estimated to be relatively low and either past or near their peaks (UN 2003), by the momentum inherent in the young age structure of the region, and by relatively high fertility.

9.2.1.3 Comparison of Aging among Scenarios

The age distribution of the population will have an important influence on future consumption patterns as well as on the vulnerability and adaptive capacity of society. This is reflected, for example, in a computation of the number of malnourished children later in this chapter.

In all scenarios, substantial aging of the population occurs. The least amount of aging occurs in Order from Strength, due to its high fertility and mortality assumptions in poorer countries, but even in this case the proportion of the population above age 65 more than doubles from about 7 to 17% by 2100. In Global Orchestration, the proportion above age 65 triples by 2050 (to 22%) and increases by a factor of six (to 42%) by 2100. This result is driven by low fertility assumptions in poorer-country regions, along with low mortality assumptions for all regions.

Within these general trends at the global level, results vary by region. In all richer-country regions, the proportion over 65 doubles to at least 30% by 2100 in nearly all scenarios (the only exception is the Order from Strength scenario in North America). In contrast, while aging is extraordinarily fast in poorer regions in most scenarios—the proportion over 65 increases, for example, from 5% currently to over 40% by the end of the century in Global Orchestration—in Order from Strength the older age group never accounts for more than 20% of the population in any of these regions. In fact in sub-Saharan Africa, where fertility and mortality are the highest, little aging occurs over the first half of the century in any scenario. And even by the end of the century, the proportion of the population there over 65 years of age reaches only 22% in the most extreme outcome (the Global Orchestration scenario).

9.2.2 Economic Development

9.2.2.1 Methodology and Assumptions

Economic development as a driver of the use of ecosystem services comprises many dimensions—including income levels, economic structure, consumption, and income distribution. Often, however, levels of per capita income (GDP or GNP) are used as a measure of the degree of economic development. In fact, per capita income is typically the only development indicator used in the literature for long-term scenarios.

Assumptions about economic development influence the future of ecosystem services by affecting the direct drivers of ecosystem changes such as energy use and food consumption and the indirect drivers such as technological progress. The relationship between income development and direct drivers differs greatly among ecosystem services. For several services, model calculations assume that the higher the income, the greater the per capita consumption of commodities, up to some saturation level (for example, energy consumption per sector or domestic water use). For other services, high income may lead to a decrease in consumption because of a change in consumption patterns (fuelwood consumption, say).

Income levels are best measured in local currencies for many analyses with a national focus. However, for international comparison they need to be converted into a common unit. Historically, most scenario analyses have used conversion into U.S. dollars based on market exchange rates. An alternative measure is based on "purchasing power parity". PPP values show the ratio of the prices in national currencies of the same good or service in different countries and reflect the fact that many products have lower prices in low-income countries. Although PPP comparisons are considered to be a better indicator of relative wealth, the measurement is somewhat more problematic. PPP values can be determined by measuring price levels of a representative set of goods and services in different countries; it is not, however, straightforward to define such a set across a range of very different economies, also taking into account differences in quality. Hence the advantages and disadvantages of both approaches are being intensively debated.

An important aspect of this debate has been the recent discussion about MER-based income projections underlying the scenarios of IPCC's Special Report on Emission Scenarios (Nakićenović et al. 2000). While most scenarios in SRES indeed use MER numbers, some of them have reported PPP values too, assuming real exchange rates to change dynamically with increasing degree of development. In the view of the SRES researchers, changing the metric of monetary income levels does not change the underlying real activity levels that are relevant for ecological impacts. They argue that the use of different income measures implies a different relationship between income and these physical indicators, implying that it does not matter whether PPP or MER values are used (that is, all effects are cancelled out).

Castles and Henderson (2003), however, questioned the use of MER-based income projections in the SRES. They argued that underestimation of real income in developing countries (by using MER numbers) led SRES modeling teams to overestimate activity growth rates in the next 100 years (and therefore the growth of greenhouse gas emissions) (see also Economist 2003; Maddison 2004). In response, the SRES researchers indicated that the IPCC growth projections are consistent with historic growth trajectories and that using alternative metrics for growth will not fundamentally change the scenarios (Nakićenović et al. 2003).

Several researchers explored the issue more quantitatively. Manne and Richels (2003) found some differences between using PPP and MER estimates as a result of counteracting influences in their model. Differences found by McKibbin et al. (2004) were larger, but they too concluded that possible impacts are within the range of other uncertainties impacting emissions. Finally, Holtsmark and Alfsen (2004) showed that, in their model, consistent replacement of the metric of monetary proxies (PPP for MER) throughout (for income levels but also for underlying technology relationships) led to a full cancellation of the impact. Using PPP values might give rise to lower growth rates for developing countries, but also to a different relationship between income and demand for energy. On the basis of these studies, it seems that although impacts on economic growth projections are uncertain, using PPP-based values instead of MER-based ones would at most only mildly change future estimates of resource consumption.

In the MA, we use income levels mostly as a proxy to derive activity levels measured in physical units in different models. The final results in terms of demand for ecological services have been checked against historic trends and among different regions and were found to be consistent and convincing. It should be noted that the income numbers themselves (expressed in MER-based values) should be used with some reservation in light of the debate just described—and should certainly not be directly interpreted as to express real differences in economic welfare among different regions. In a more qualitative way, however, they do express the storylines of the different scenarios.

Historically, global GDP has increased by a factor of 20 over the last 110 years, or at a rate of about 2.7% per year. Per capita GDP growth was 1.5% per year (Maddison 1995). There has been, however, a substantial variation in the rates of economic growth over time and across countries. For the OECD region, economic growth has accelerated to over 1% per year since about 1870. For most developing countries, comparable conditions for economic growth existed only in the second half of the twentieth century. It is important not to conceptualize economic development as a quasi-autonomous, linear development path. Numerous socioinstitutional preconditions have to be met before any "takeoff" into accelerated rates of productivity and economic growth can materialize.

Different strategies to create such conditions have been successful (Freeman 1990; Chenery et al. 1986). Once these preconditions are met, it is not uncommon for countries to experience an "acceleration phase" in which they catch up relatively quickly to wealthier countries. The most obvious examples have been Japan, South Korea, and China (all experienced economic growth rates over 6% over a period of at least 20 years). At the same time, income gaps in both absolute and relative terms have not disappeared from the world. For instance, per capita GDP growth in Africa has been below OECD levels since 1950, and even negative in several periods since 1980. Since 1990, other important economic trends have been the serious economic setbacks in Eastern Europe and the former Soviet Union after the transition to market economies and the more recent slow recovery of economies in Latin America.

Most economic growth scenarios found in literature only encompass periods of 10–20 years (e.g., World Bank 2002). An important exception have been the economic scenarios developed as a basis for building energy and environmental scenarios, such as those reviewed in Alcamo et al. (1995) and Nakićenović et al. (2000). Typically, such scenarios show annual economic growth rates (GDP per capita measured at MER) between 0.8 and 2.8% over the 1990–2100 period. In most cases, economic growth slows down in the second half of the century as a result of (assumed) demographic trends (aging of the population), saturation of consumption, and slower reduction in technological change. Moreover, most scenarios assume that incomes in different regions will converge in relative terms (that is, higher growth rates are assumed in poorer countries than in OECD ones).

The MA scenarios for income cover a range of economic growth rates consistent with the scenario storylines described earlier in this volume. Table 9.5 shows the qualitative asassumptions for economic variables fitting to these storylines. Using these assumptions together with the World Bank's economic prospects to 2015 (World Bank 2002) and IPCC's SRES scenarios (Nakićenović et al. 2000) as starting points, we have selected economic growth rates for each scenario. Compared with the SRES scenarios, this means that growth rates in developing regions (in particular, Africa and West Asia) have been slowed down somewhat and now bracket the World Bank prospect. As a result, the degree of convergence in the scenarios is also somewhat lower. For the period after 2015, the more detailed IMAGE implementation of the SRES scenarios were used. The SRES scenarios were scaled down earlier to the level of 17 regions (see IMAGE-team 2001) using the macroeconomic model "WorldScan," following a procedure described by Bollen (2004).[1] Assumptions range from high economic growth for Global Orchestration and low economic growth for Order from Strength, with TechnoGarden and Adapting Mosaic falling between (and partly branching off of these).

9.2.2.2 Comparison of Economic Development among Scenarios

In Global Orchestration, economic growth is assumed to be above historic averages for several regions, due to a combination of trade liberalization, economic cooperation, and rapid spread of new technologies. Among the scenarios, Global Orchestration also assumes the highest rates of investment in education and health care. The wealthier countries have a per capita growth rate of about 2.4% per year in the 2000–25 period, slowing down to around 1.8% per year afterwards. (See Table 9.6.) The Asian economies return to rapid growth rates during most of this period (with growth rates of 5–6% per year). The Latin American region overcomes its debt and balance-of-trade problems and finds itself back on track with strong economic growth. Africa carries out institutional reforms that enable strong economic growth after 2025, when it finally exploits its rich natural and human resources. After 2025, Africa achieves growth rates that are only slightly below the Asian economies in the 1980s and 1990s. As poorer countries grow much faster than others, the income gap between richer and poorer regions closes in relative terms—but hardly in absolute terms. (See Table 9.7.) In all scenarios, growth rates for the countries of the former Soviet Union are relatively high because the region uses its highly skilled labor force to recover from the economic downturn of the 1990s.

Economic development in TechnoGarden follows a similar pattern to Global Orchestration, but with lower growth rates from 2000 to 2050. By the end of the period, however, earlier investments in technology pay off with higher economic growth rates similar to Global Orchestration. Investments in human resources are likely to be lower than under Global Orchestration, partly as a result of the emphasis of TechnoGarden on technology investments.

Under the Order from Strength scenario, global economic growth is sluggish (staying below historic rates) because of the low level of international trade (except for food staples) and limited exchange of technology. The high-income countries manage to maintain a growth rate of per capita GDP of 1.9% per year during the first half of the century, but this drops to 1.2% during the second half. The income gap between rich and poor regions widens between 2000 and 2025. Despite the sluggish economy, average GDP per person increases by a factor of two between now and 2050. Investments in education and health care outside of current high-income regions will be low because of the lack of financial capital.

Table 9.5. Qualitative Assumptions on Economic Growth in MA Scenarios. The terms low, medium, and high are relative to the normal development pathways that are assumed for these regions.

Variable	Global Orchestration	Order from Strength	Adapting Mosaic	TechnoGarden
Average income growth	high	industrial countries: medium developing countries: low	begins like Order from Strength, then increases in tempo	somewhat lower than Global Orchestration, but catching up
Income distribution	income distribution becomes more equal	income distribution remains similar to today	begins like Order from Strength, then becomes more equal	similar to Global Orchestration

Table 9.6. Annual Growth Rates of GDP per Capita, 1971–2000, and Assumptions in MA Scenarios

	Historic	Global Orchestration			Order from Strength			Adapting Mosaic			TechnoGarden		
Region	**1971–2000**	**1995–2020**	**2020–50**	**2050–2100**	**1995–2020**	**2020–50**	**2050–2100**	**1995–2020**	**2020–50**	**2050–2100**	**1995–2020**	**2020–50**	**2050–2100**
					(percent per year)								
Former Soviet Union	0.4	3.5	4.9	3.1	2.2	2.6	2.7	2.6	4.0	3.1	2.9	4.5	3.1
Latin America	1.2	2.8	4.3	2.2	1.8	2.3	1.8	2.0	3.0	2.2	2.4	3.9	2.2
Middle East and North Africa	0.7	2.0	3.4	2.5	1.5	1.8	1.9	1.6	2.4	2.4	1.7	3.3	2.5
OECD	2.1	2.45	1.9	1.3	2.1	1.3	0.9	2.0	1.6	1.2	2.2	1.7	1.4
Asia	5.0	5.06	5.3	3.1	3.2	2.4	2.1	3.8	4.1	2.5	4.2	4.7	3.1
Sub-Saharan Africa	−0.4	1.69	4.0	4.1	1.0	2.1	2.1	1.2	2.9	3.3	1.4	3.8	4.1
World	**1.4**	**2.38**	**3.0**	**2.3**	**1.4**	**1.0**	**1.3**	**1.5**	**1.9**	**1.9**	**1.9**	**2.5**	**2.3**

Table 9.7. Annual GDP per Capita by Region in 1995 and Assumptions in MA Scenarios

	GDP, 1995	Global Orchestration			Order from Strength			Adapting Mosaic			TechnoGarden		
Region	**1995**	**2020**	**2050**	**2100**	**2020**	**2050**	**2100**	**2020**	**2050**	**2100**	**2020**	**2050**	**2100**
					(dollars per capita)								
Former Soviet Union	1,630	3,853	16,223	76,107	2,837	6,198	23,708	3,093	10,109	46,010	3,365	12,560	58,898
Latin America	4,337	8,660	30,427	92,226	6,747	13,293	31,952	7,229	17,489	52,575	7,769	24,682	74,738
Middle East and North Africa	2,068	3,363	9,223	31,630	3,010	5,070	13,214	3,085	6,337	20,711	3,186	8,353	28,757
OECD	22,657	41,496	73,607	143,151	37,752	55,734	85,678	37,188	59,114	106,588	39,235	65,876	128,822
Asia	784	2,694	12,600	57,296	1,733	3,564	9,913	1,972	6,612	22,961	2,212	8,781	40,947
Sub-Saharan Africa	637	969	3,117	23,035	820	1,540	4,492	860	1,997	10,169	910	2,787	20,629
World	**5,102**	**9,190**	**22,282**	**68,081**	**7,204**	**9,838**	**18,377**	**7,338**	**12,932**	**32,808**	**8,162**	**16,941**	**51,546**

The Adapting Mosaic scenario initially follows the pattern of the Order from Strength scenario, but because of large investments in education and health care, economic growth rates increase over time and approach those of the TechnoGarden scenario in the last half of the century.

9.2.3 Technological Change

9.2.3.1 Methodology and Assumptions

The rate of technological change is an indirect driver of changes in ecosystem services because it affects the efficiency by which ecosystem services are produced or used. Most relevant in this context are the factors related to energy, water, and agriculture. A higher rate of improvement of crop yields, for instance, could lead to a lower demand for cropland (to produce the same amount of food), reducing the need to convert forest or grassland. Technological change, however, can also lead to increased pressure on ecosystem services because technological advancements often require large amounts of goods and materials themselves and can cause new ecological risks. (For example, the application of chemical fertilizers for increasing crop yield can also lead to nitrogen contamination of surface water and groundwater.)

Technological change is a complex and dynamic process. It is linked to the economic and cultural environment beyond individual "innovating firms," as described by Rostow (1990) and Grübler (1998). Innovations are highly context-specific in that they emerge from local capabilities and needs and evolve from existing designs. Numerous examples illustrate the messiness and complexity of the innovation process (e.g., Grübler 1998; Rosenberg 1994). Nevertheless, some generalizations can be applied to the concepts of innovation and technological change (see Nakićenović et al. 2000): Innovation draws on underlying scientific or other knowledge. Many innovations depend on knowledge obtained through experience. The social and economic environment should encourage a situation in which innovators are willing and able to take some risks. Technology change may be both supply- and demand-driven. And technological diffusion is an integral part of technology change (and can thus be slowed down by protectionist measures).

It is notable that technology development is typically driven by factors unrelated to ecosystem services. For example, efforts to improve crop yield depend on factors such as the profitability of farmland and general investments in education and research. The same holds for the development of new energy technologies such as solar and wind power, which are influenced by trends in fossil fuel prices and environmental policies. Michaelis (1997), for instance, showed the strong relationship between fuel prices and the rate of energy efficiency improvement.

9.2.3.2 Comparison of Technological Change among Scenarios

In order to maintain consistency across the MA scenarios, it is necessary to assume some general trends in technological change over the scenario period and to apply these general trends to all scenario variables that are strongly influenced by technological change. This section presents the general assumptions made about technological development used to select future trends in improvement in irrigation efficiencies, crop yield improvements, improvements of water use efficiency, improvement of energy use efficiency, costs reductions of new energy technologies, and the rate of emission control technologies.

The assumed overall trend in "technological efficiency" for the four MA scenarios is given in Figure 9.1. We assume the highest rates of technological development under Global Orchestration because this scenario has several features that are favorable to technology development (see Table 9.8): high economic growth rates, which in principle are consistent with new capital investments; large investments in education; low trade barriers, leading to relatively rapid dispersal of knowledge and technologies; and an accent on entrepreneurship, possibly providing a stimulus to human ingenuity. It should be noted, however, that it is also assumed that the technology development under Global Orchestration will not necessarily be environmentally friendly. Fossil fuel–based technologies could develop at the same rate as, for instance, solar or wind power. In the example for irrigation efficiency, we assume that careful market-oriented reform under Global Orchestration in the water sector (with coordinated government action) could lead to greater water management investments in efficiency-enhancing water and agricultural technology, particularly in Asia and sub-Saharan Africa.

Under TechnoGarden, a somewhat lower rate of technology development is expected than under Global Orchestration, given the fact that several of the factors just mentioned are less dominant. A central tenet of this scenario is that technology development is directed to reduce (or at least mitigate) existing ecological problems. That implies relatively high technological growth rates for environmental technologies, but lower rates of development for technologies in general. Later in the century, the technology rate improvement could start to catch up with Global Orchestration (as it is less close to the frontier). For the example of irrigation efficiency, under the TechnoGarden scenario technological innovations could help boost irrigation efficiency levels across the world to previously unseen levels. Gradual introduction of water price increases in some agricultural areas induce farmers in these regions to use water more efficiently. As a result, high efficiency levels are reached, particularly in regions where little or no further improvement had been expected, like the OECD and the Middle East and Northern Africa.

Under Adapting Mosaic, regionalization and higher barriers could be expected to slow economic growth and the dispersion of technologies and to slow down overall technological development (2000–25). Increased (decentralized) learning could at the same time build up a new basis from which technologies can be developed. Therefore, technologies under this scenario develop slowly at first but speed up later in the century compared with the other scenarios. For the example of irrigation efficiency, local adaptations—including expansion of water harvesting and other water conservation technologies as well as the increased application of agro-ecological approaches—could help boost efficiency levels in some regions and countries. Efficiency increases are achieved but remain scattered in areas and regions within countries, and the global and regional impacts are smaller than under the TechnoGarden and Global Orchestration scenarios.

Finally, under the Order from Strength scenario, technology development will be relatively slow throughout the whole period, especially in low-income countries. The main reasons include the lack of international cooperation and the low potential for investment. This is reflected in the assumptions made for irrigation efficiency. Government budgetary problems are assumed to worsen, resulting in dramatic government cuts in irrigation system expenditures. Water users strongly oppose price increases, and a high degree of conflicts hinder local agreements among water users for cost-sharing arrangements. Rapidly deteriorating infrastructure and poor management reduce system- and basin-level water use efficiency under this scenario. As a result, efficiency levels are assumed to decline in both wealthy countries and in poorer countries where efficiencies are already quite low.

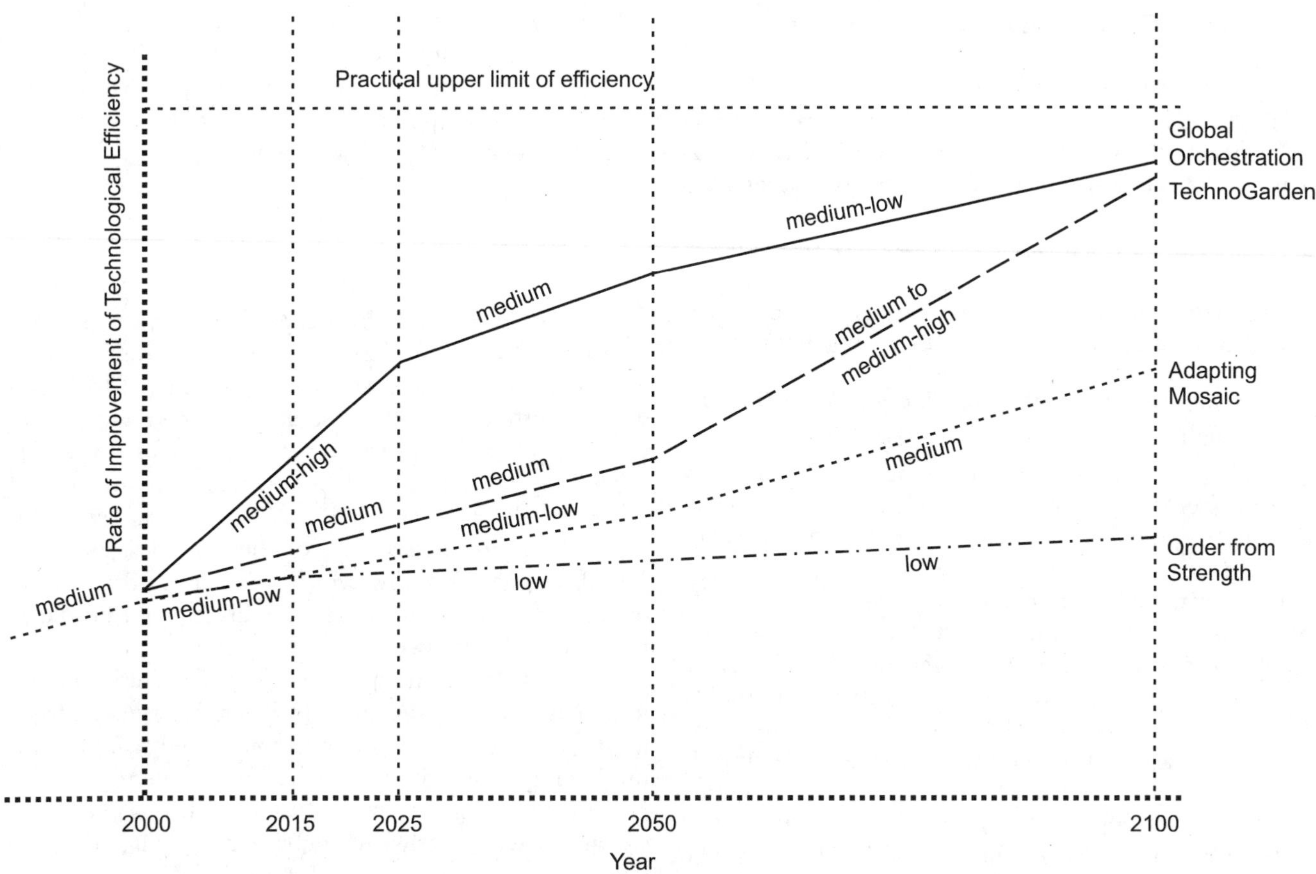

Figure 9.1. Global Trends of Technological Efficiencies in MA Scenarios. Depicted are the qualitative assumptions made for changes in technological efficiency under the four MA scenarios. Technological Efficiency refers, for example, to the conversion efficiency of power plants, or the yield of all crops per hectare. As a reference point for the scenarios, we designate the current rate of improvement of all technologies as "medium." Therefore, a "high" scenario implies an acceleration, and a "low" scenario implies a slowing of the current rate of improvement. These qualitative assumptions are used for setting technology-related parameters in the models used for quantifying the scenarios (e.g., the rate of increase of crop yield due to technological improvements in crops). For the TechnoGarden scenario, a faster rate of improvement than shown in the curve was assigned to the technologies directly related to pollution control such as air pollution filtering devices. This is consistent with the storyline of the scenario which specifies that the environmental orientation of TechnoGarden leads to a faster improvement in pollution control technologies than under the Global Orchestration scenario, but a slower improvement in all other technologies.

Table 9.8. Qualitative Assumptions for Technology Development in MA Scenarios

Variable	Global Orchestration	Order from Strength	Adapting Mosaic	TechnoGarden
Investments into new produced assets	high	industrial countries: medium; developing countries: low	begins like Order from Strength, then increases in tempo	high
Investments into human capital	high	industrial countries: medium; developing countries: low	begins like Order from Strength, then increases in tempo	medium
International relationships (stimulating technology transfer)	high	low (medium among cultural groups)	low-medium	high
Overall trend	high	low	medium-low	medium for technology in general; high for environmental technology

9.2.4 Social, Cultural, and Political Drivers

9.2.4.1 Methodology and Assumptions

Social, cultural, and political drivers are important indirect drivers of ecosystem services. Assumptions about these drivers influence the trends of the direct drivers of ecosystem services, such as the trends in producing energy or consuming food. Here we briefly review the trends of these important indirect drivers, as they are implied by the storylines in Chapter 8, and give some examples of their impact on the direct drivers of ecosystem services. (See Table 9.9.)

9.2.4.2 Comparison of Social, Cultural, and Political Drivers among Scenarios

Although it is difficult to represent social and cultural factors in global scenarios, we have two preliminary examples of including the influence of these factors on resource consumption.

The scenarios differ in people's attitude toward international cooperation, and this leads to other assumptions for drivers of ecosystem change in the scenarios. A positive attitude toward international cooperation is assumed to lead to a higher level of international trade in Global Orchestration and TechnoGarden. Conversely, a more negative attitude in Adapting Mosaic and Order from Strength is assumed to inhibit the formulation of international environmental policies, and, in particular, climate policies. Hence, these two scenarios assume a low level of controls of greenhouse gas emissions.

The scenarios also differ in people's attitudes toward environmental policies, and this leads to other assumptions for drivers of ecosystem change. The projected attitudes toward environmental policies also lead to assumptions about other variables. For instance, a generally reactive attitude with regard to environmental policies is assumed in Global Orchestration. This is consistent with the scenario's optimistic view on the robustness of ecosystems and the abilities of humans to deal with environmental problems when they are observed, combined with a strong focus on improving human well-being by means of social policies and economic development. This was interpreted to mean that there will be no incentive in the future to reduce the amount of meat consumed per person (despite the connection between meat consumption, livestock grazing, deforestation, and soil degradation). Similarly, under Global Orchestration it was assumed that society is not likely to subsidize use of renewable energy for environmental protection reasons. Figure 9.2 shows that Global Orchestration gets a much lower share of energy use from renewable energy than the two scenarios that emphasize proactive ecological policies, TechnoGarden and Adapting Mosaic.

Table 9.9. Assumed Changes for Selected Indirect Drivers in MA Scenarios

Variable	Global Orchestration	Order from Strength	Adapting Mosaic	Techno-Garden
International cooperation	strong	weak—international competition	Weak—focus on local environment	strong
Attitude toward environmental policies	reactive	reactive	proactive—learning	proactive

9.2.5 Energy Use and Production

9.2.5.1 Methodology and Assumptions

Energy use has many indirect effects on ecosystem services. The use of fossil fuel determines the rate of air pollutant emissions and therefore the load-on quality of the atmosphere. The level of biofuel use affects the type and distribution of land cover and the services provided by forest and other land cover types, while the magnitude of thermal-generated electricity will influence water withdrawals. Energy production is also one of the principal sources of greenhouse gas emissions, which are the main determinants of climate change, which itself is a direct driver of changes in ecosystem services.

The amount of energy used in the different scenarios is influenced by the demand for energy services (driven mostly by economic and population growth) and by continuing improvements in the efficiency of energy use.

Wide-ranging "reference" and storyline-based energy scenarios have recently been published, including the regularly updated scenarios of the International Energy Agency (IEA 2002), the U.S. Department of Energy (DOE 2004), the Shell Oil Company, the IPCC SRES scenarios (Nakićenović et al. 2000), and the World Energy Assessment (WEA 2000). Nakićenović et al. (2000) provide an extensive overview of the energy scenarios found in the literature. Almost all scenarios show substantial increases in energy use in the period from 2000 to 2050. While new energy carriers (such as renewables) increase their market share, energy use continues to be dominated by fossil fuel use in nearly all scenarios. After 2050, some scenarios indicate stabilizing or even decreasing energy use, while others show continuous growth. In the compilation of the MA scenarios, we have combined assumptions of the IPCC SRES scenarios with the drivers discussed earlier. (See Table 9.10.)

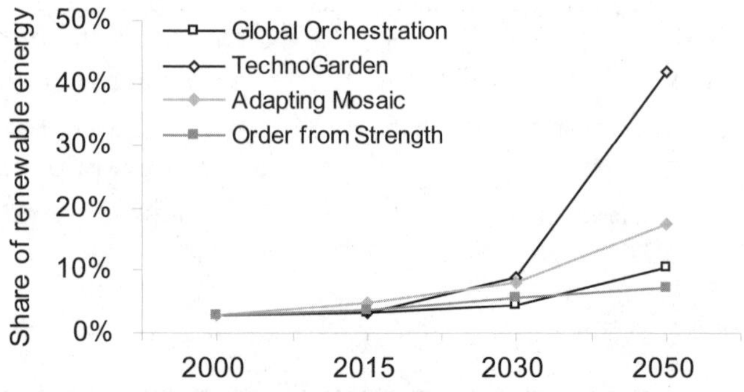

Figure 9.2. Share of Renewable Energy in Total Primary Energy Consumption in MA Scenarios. Renewable Energy is defined here as solar, wind hydropower, and the use of modern biofuels. (IMAGE 2.2)

Table 9.10. Main Assumptions about Energy in MA Scenarios

Variable	Global Orchestration	Order from Strength	Adapting Mosaic	TechnoGarden
Energy demand	lifestyle assumptions and energy efficiency investments based on current North American values	regionalized assumptions	regionalized assumptions	lifestyle assumptions and energy efficiency investments based on current Japanese and West European values
Energy supply	market liberalization; selects least-cost options; rapid technology change	focus on domestic energy resources	some preference for clean energy resources	preference for renewable energy resources and rapid technology change
Climate policy	no	no	no	yes, aims at stabilization of CO_2-equivalent concentration at 550 ppmv

9.2.5.2 Comparison of Energy Use and Production among Scenarios

The dominant themes in the Global Orchestration scenario are a rapid increase in energy demand (driven by strong economic growth), a minimization of energy costs, and provision of a reliable energy production system. Environmental considerations receive little attention, as society believes the environmental impacts of energy production to be either small or manageable by future technological change (if signals of severe environmental deterioration become apparent). Based on these considerations, we have assumed that there is no attempt to control greenhouse gas emissions in the first decades of the scenario period. At the same time, technology development in the energy sector is relatively fast, which leads to indirect reductions in emissions of greenhouse and other pollutant gases.

Since the effects of climate change are apparent later in Global Orchestration, we assume that society responds by adapting to impacts rather than reducing emissions (since by that time a certain degree of climate change will be unavoidable). As a result of these assumptions, fossil fuel use expands rapidly, in particular the use of gaseous fuels (for households and electricity production) and liquid fuels (in the transport sector, possibly replaced by hydrogen). Total energy use increases up to 1,200 exajoules by 2050 (compared with a current level of 400 exajoules) and levels off toward the end of the century. (See Figure 9.3 in Appendix A.) Trends are very similar to IPCC's A1b scenario (Nakićenović et al. 2000), while consumption levels are somewhat lower due to lower population and economic growth. In the second half of the century, new (non-fossil) fuel options rapidly penetrate the market.

The TechnoGarden scenario assumes that society will be convinced that environmental degradation decreases human well-being and therefore supports long-term reductions of greenhouse and other air pollutant emissions. To mitigate climate change, the international community adopts a goal of limiting global mean temperature increase to 2° Celsius by 2100 over preindustrial levels (similar to the current target for climate policy in the EU and several European countries). Assuming medium values for the relevant parameters in a simple climate model, the attainment of this temperature goal implies that global emissions must fall below half the current emissions before 2100. Since emissions stem mostly from energy use, this requires a reduction in the use of fossil fuels, which is brought about by energy efficiency, increasing use of "zero-carbon" energy sources (modern biofuels and solar and wind energy, as examples), and more low-carbon fuels (principally natural gas). As a result, total energy use reaches a level of 510 exajoules in 2050 and slowly increases thereafter, despite relatively high economic growth rates. This energy scenario is similar to others that aim to achieve comparable climate goals, such as those that aim to stabilize carbon dioxide at 450 parts per million by volume or total greenhouse gas concentration at 550 parts per million CO_2-equivalent (Morita et al. 2001; van Vuuren and de Vries 2001).

A central theme of Order from Strength is securing reliable energy supplies, and this leads to a focus on developing domestic energy sources. Slow diffusion of new technologies and increased barriers for global energy trade (particularly important for natural gas and oil) also contribute to a continued intensive use of domestic fossil fuels. For China and India, this implies a continued reliance on coal. Total energy use increases almost linearly throughout the century, reaching about 800 exajoules in 2050. This is much lower than Global Orchestration because the Order from Strength scenario has lower economic growth, particularly in poorer countries. Energy use is higher than in TechnoGarden because Order from Strength assumes slower improvements in the efficiency of energy use. This scenario is similar in character to IPCC's A2 scenario.

Adapting Mosaic is similar to Order from Strength in that it has lower economic growth rates than Global Orchestration and lacks global climate policies. Global energy use in 2050 (880 exajoules) is between Global Orchestration and TechnoGarden. However, it differs from the Order from Strength scenario in that there is great concern about environmental degradation. Thus, local approaches are adopted for improving efficiency of energy use and for exploiting environmentally friendly fuels. As a result, total energy use stabilizes soon after mid-century, and non-fossil fuels play an increasing role in the energy economy.

9.2.6 Summing Up Trends in Indirect Drivers

Figure 9.4 provides a graphical overview of the global trends for several crucial indirect drivers. As concluded by

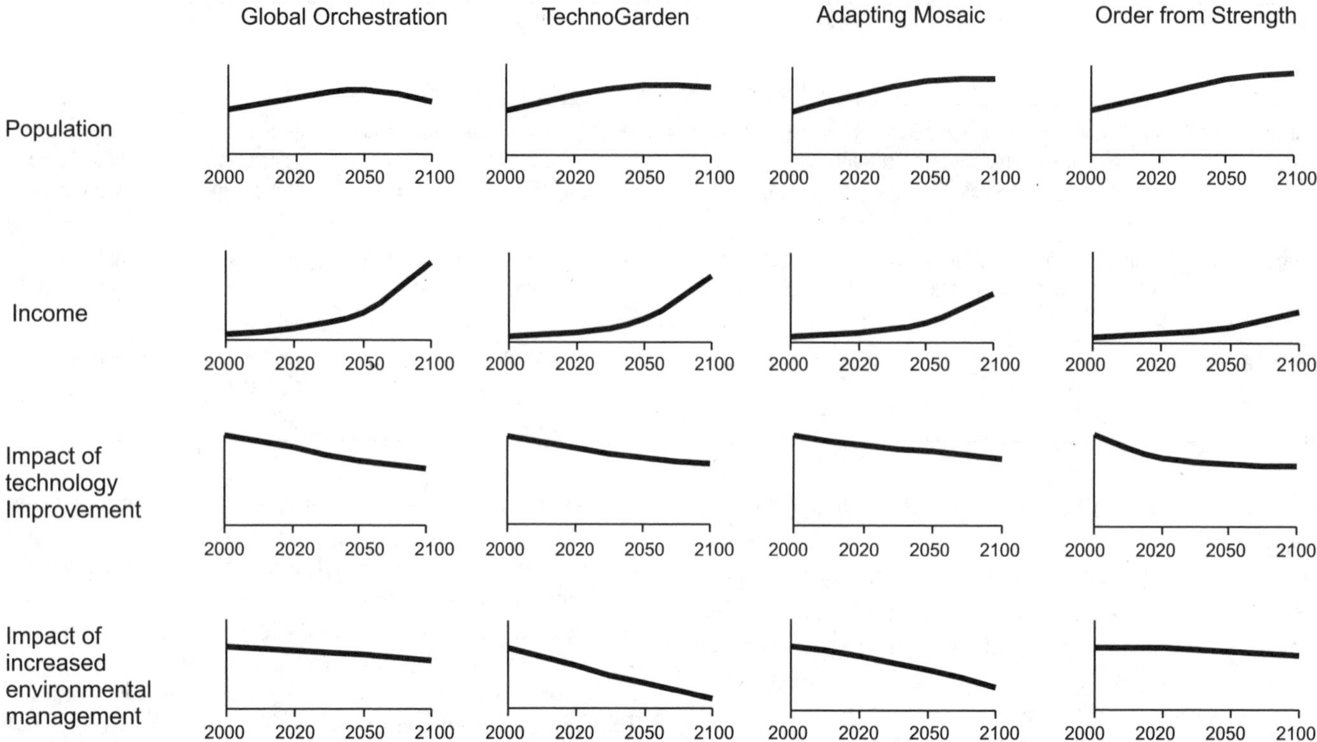

Figure 9.4. Impact of Trend in Crucial Indirect Drivers on Pressures on Ecosystems in MA Scenarios. Population and activity growth lead to increased pressures; technology improvement and increased impact of increased environmental management to fewer pressures.

Nakićenović et al. (2000), it is not advisable to assume that future indirect drivers, such as population and economic growth, will be independent of one another. Their coupling can be taken into account in scenario storylines and, where possible, in the models used to produce quantitative scenarios. In the MA scenario we have, for example, assumed that there is a higher probability of high population growth in poorer countries under low economic growth scenarios (due to a slowdown of the demographic transition). We have also assumed that there is a higher likelihood of faster technological development under higher economic growth (because of higher investments in research and education).

We note here that some assumed relationships between indirect drivers tend to create compensating effects in the scenarios. For example, since we combine the assumption of highest population growth with the lowest economic growth, and the lowest population growth with the highest economic growth, we compute a narrower range of demands for goods and services among the scenarios than the range of assumptions of population growth and economic growth.

9.3 Direct Drivers of Ecosystem Services

Direct drivers are mainly physical, biological, or chemical processes that tend to directly influence changes in ecosystem goods and services. In some cases it is difficult to distinguish between drivers and ecosystem services. An example is the case of the ecosystem service "food provisioning," which itself is a prime determinant of the direct driver "land use change" (discussed here as a direct driver). The direct drivers discussed here are:

- greenhouse gas emissions,
- air pollution emissions,
- risk of acidification and excess nitrogen emissions,
- climate change,
- sea level rise,
- land use and land cover change,
- use of nitrogen fertilizers and nitrogen loading to rivers and coastal marine systems, and
- disruption of landscape by mining and fossil fuel extraction.

9.3.1 Greenhouse Gas Emissions

9.3.1.1 Methodology and Assumptions

Greenhouse gas emissions determine to a large degree both the rate and intensity of future climate change. The main sources of these emissions are energy use, agricultural activity, industrial processes, and deforestation. The most important greenhouse gas (in terms of the contribution to increased forcing) is CO_2. Emissions of methane and nitrous oxide, stemming mainly from agricultural sources, account for about one fifth of total greenhouse gas emissions (in units of equivalent carbon dioxide).

In recent years, several long-term greenhouse gas scenarios have been published. Alcamo and Nakićenović (1998) and Nakićenović et al. (2000) published extensive overviews of available scenarios in the literature, showing that the range covered by the IPCC SRES scenarios reasonably coincides with the range drawn up by "non-intervention" scenarios in the literature (that is, scenarios that do not assume specific policies to reduce greenhouse gas emissions or stimulate additional uptake of CO_2 by the atmosphere). Using the SRES scenarios or other reference scenarios as departure points, researchers have developed scenarios that incorporate climate policies (examples include Alcamo and Kreileman 1996; Hyman et al. 2003; Manne and Richels 2001; Morita et al. 2001; Reilly et al. 1999; van Vuuren et al. 2003).

The greenhouse gas trends for the MA scenarios (see Table 9.11) can be derived almost directly from the energy and land use trends discussed elsewhere in this chapter. Their range coincides well with those found in the literature, in particular the IPCC scenarios and, in the case of the TechnoGarden scenario, the derived climate policy scenarios. Obviously, the range presented is not exhaustive—for instance, lower greenhouse emission pathways are possible, but at relatively high costs.

9.3.1.2 Comparison of Greenhouse Gas Emissions among Scenarios

Under Global Orchestration, greenhouse gas emissions peak at mid- century just above 25 Gt C-eq, compared with around 10 Gt C-eq in 2000.[2] Emissions decline afterwards because total energy use stabilizes and a greater percentage of low carbon fuels are used. (See Figure 9.5 in Appendix A.) CO_2 emissions are projected to grow somewhat faster than those of other important greenhouse gases such as CH_4 and N_2O, since the drivers of CO_2 (energy production) grow somewhat faster than the drivers of the other gases (agricultural variables).

The share of emissions coming from the OECD and former Soviet Union regions declines from 48% to 30% as a result of larger economic and population growth in the other regions. In terms of emissions, Global Orchestration is comparable to IPCC-SRES A1b scenario or other scenarios with relatively high emissions.

The strong climate policies in TechnoGarden limit the increase in fossil fuel consumption in that scenario. Hence emissions grow much more slowly; they peak around 2020 at 12 GtC-eq and decline by 2050 to 30% below their level in 2000. Several studies indicate that technical options exist for such emission reductions (IPCC 2001). The economic costs of these emission controls are much more uncertain and generally range from 1% to 4% of world GDP (IPCC 2001). Here, we use an implementation of a multigas reduction strategy calculated by IMAGE that aims to limit global temperature increase to 2° Celsius above preindustrial levels. In the OECD and former Soviet Union regions, emissions decline by 2050 to about 30% of emissions in 2000. In Asia and Latin America, emissions return to their 2000 values around 2050. For Africa and the MENA regions, emissions growth (coming from low levels in 2000) is reduced. The difference between the regions is caused by the much faster increase in population and economic activities in developing regions. The emissions under this scenario are representative of low emissions scenarios found in the literature.

The trend in emissions under Order from Strength follows the linear increase in total global energy use of this scenario. Emissions almost double between 2000 and 2050, and again between 2050 and 2100. Up to 2050 the emissions of all greenhouse gases increase, whereas afterwards the only significant increase is from CO_2 emissions. Order from Strength is the only scenario where continuing deforestation implies that land use change will remain an important source of CO_2 emissions. Quantitatively, the emissions in this scenario are similar to IPCC's A2 scenario (which implies that it compares well to relatively high emission scenarios in the second half of the century).

Emissions under Adapting Mosaic grow steadily to a level of 18 Gt C-eq around mid-century. After 2050, emissions gradually decline to a level slightly above 16 Gt C-eq as energy growth slows and more low-carbon fuels are used. This emission path is comparable to that of the IPCC-B2 scenario. The scenario is representative of the medium range of scenarios found in the literature.

9.3.2 Air Pollution Emissions

9.3.2.1 Methodology and Assumptions

A large number of activities contribute to air pollution. Burning of fossil fuels and biomass contribute to air pollutants such as sulfur dioxide, carbon monoxide, nitrogen oxides, particulate matter, volatile organic compounds, and some heavy metals. In addition, industrial activities and agriculture also contribute to air pollution. In our assessment, we concentrate on sulfur dioxide and nitrogen oxides. Emissions

Table 9.11. Kyoto Greenhouse Gas Emissions in 1995 and Assumptions in MA Scenarios (IMAGE 2.2)

Greenhouse Gas	Emissions in 1995	Global Orchestration	Order from Strength	Adapting Mosaic	Techno-Garden
		(emissions in GtC-equivalent[a])			
CO_2	7.3	20.1	15.4	13.3	4.7
CH_4	1.8	3.7	3.3	3.2	1.6
N_2O	0.7	1.1	1.1	0.9	0.6
Other GHG	0.0	0.7	0.5	0.6	0.2
		(percent)			
OECD and former Soviet Union as share of total emissions	48	30	34	29	22

[a] GtC-equivalent emissions are the contribution of different greenhouse gases expressed in tons of carbon based on 100-year global warming potentials.

of these compounds lead to problems both near and far from their source. In the vicinity of pollution sources, high emissions (in particular, when combined with unfavorable meteorological conditions) can lead to the buildup of high concentrations of SO_2 , ozone, and other gases and pose a threat to human health. The local level of SO_2 and ozone can be high enough to cause long-term damage to vegetation. SO_2 and NO_x emissions are also transported hundreds of kilometers from their source and are then deposited via precipitation and diffusion to vegetation and soils, where they cause acidification of soils and freshwater systems (as well as direct impacts on vegetation). Because of their important role in many key air pollution-related problems, SO_2 and NO_x are good indicators of air pollution.

Trends of SO_2 and NO_x emissions are somewhat different. SO_2 emissions are relatively easy to control, either by filtering them from smokestacks or reducing the sulfur content of fuels. As a result, SO_2 emissions trends tend to follow a pattern that is sometimes referred to as the Environmental Kuznets Curve. First, emissions increase with growing energy use, but they eventually decrease as impacts of emissions increase and society demands control of air pollution. SO_2 emissions are currently decreasing in most OECD countries, but some researchers claim that emissions may again increase with economic growth once the cheaper measures for abating SO_2 emissions are exploited. Measures for reducing NO_x emissions are usually more expensive. As a result, NO_x emissions have been less controlled than SO_2 emissions and only in high-income countries.

9.3.2.2 Comparison of Sulfur Dioxide Emissions among Scenarios

Most published scenarios of global SO_2 emissions follow the historical trends, showing declines in emissions in most high-income countries and initially increasing emissions followed by a decline in low-income countries (see, e.g., Mayerhofer et al. 2002; Bouwman et al. 2002). For NO_x, in general a similar pattern is noted, but later in time and with less stringent reduction in emissions.

Based on the storylines of the MA scenarios (see Table 9.12), both the AIM and IMAGE modeling groups independently made assumptions on the development of the major drivers of emissions and emission control policies. Both results are discussed in order to capture some of the uncertainty of estimates. (If only one result is quoted, then it refers to AIM model results.) While the scenarios have been worked out at the regional scale, we concentrate here on the global results.

Under Global Orchestration, the elaboration of both models shows an initial increase followed by declining emissions, which is a result of decreasing emissions in high-income countries and initially increasing emissions in low-income regions. The rate of decline after 2020, however, is uncertain. As a result, estimates for 2050 emissions differ between near-current emission levels (IMAGE) to a 45% drop worldwide in AIM (compared with 2000). (See Figure 9.6.) The regional results (shown for AIM in Figure 9.7), show that the reductions are much stronger in the OECD and former Soviet regions, assuming a continuation of current controls of SO_2 emissions and a major shift to lower sulfur fuels. Sulfur dioxide emissions in most other regions initially grow, but by 2050 drop considerably. (See Figure 9.7.) In sub-Saharan Africa, in contrast, emissions more than double because the economic level is still not high enough to support sulfur emission controls.

Under TechnoGarden, both stricter environmental policies and the benefits of climate policies contribute to reducing sulphur emissions (high carbon fuels often also contain high sulphur levels). The two models agree on very substantial drops in global sulphur emissions. Emission reductions in this scenario outside the OECD and former Soviet region can also be impressive, such as for the MENA and for Latin America. This is partly caused by the fact that the assumed climate policies in TechnoGarden are effective in all regions (possibly financed through emission trading schemes). In sub-Saharan Africa, emissions still grow significantly, resulting mainly from the low 2000 values.

For Adapting Mosaic, there is quite some difference between the AIM and IMAGE 2020 values, but in 2050 reductions are in both cases 30–40%. Differences between the models are mainly caused by different expectations of when air pollution control policies will become important. In this scenario, trends can differ widely in different regions. Reductions are strong in the OECD, Latin America, and former Soviet region, but other regions show stable emissions or even an increase.

Finally, elaboration of Order from Strength indicates that this scenario has the highest emissions of the four MA scenarios—in fact, showing a net increase of emissions in 2050. In this scenario, emissions reductions in OECD, Latin America, and former Soviet regions are offset with strong emission increases in sub-Saharan Africa and Asia.

Table 9.12. Main Assumptions about Air Pollution Emissions in MA Scenarios (IMAGE 2.2)

Variable	Global Orchestration	Order from Strength	Adapting Mosaic	TechnoGarden
SO_2 policies and NO_x policies	environmental Kuznets type, thus decreasing after sufficient income	environmental Kuznets type; low income growth slows down policies	proactive going beyond environmental Kuznets type; reduced income growth slows policies somewhat	proactive going beyond environmental Kuznets type
Characteristic driving force	strong increases in energy use and transport	coal dominant energy carrier in Asia	most drivers have medium values	climate policies have large co-benefits

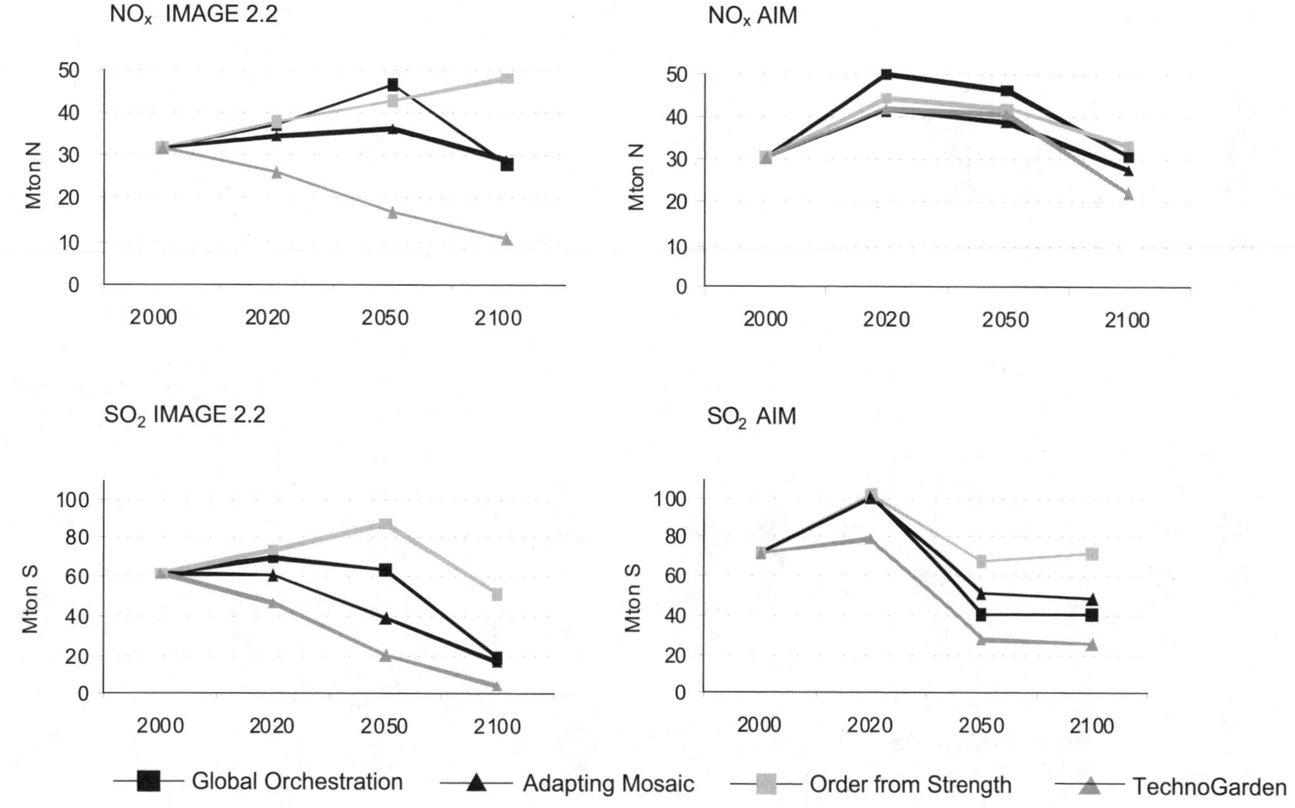

Figure 9.6. Trends in SO_2 and NO_x Emissions in MA Scenarios (IMAGE 2.2 and AIM)

9.3.2.3 Comparison of Nitrogen Oxides Emissions among Scenarios

Globally, the trend of nitrogen oxides emissions differs from that of sulfur dioxide. Global emissions of NO_x increase under every scenario except TechnoGarden.

Under Global Orchestration, worldwide emissions between now and 2050 increase by over 50%, following similar trends in IMAGE and AIM. At the same time, emissions decrease by nearly 60% in OECD countries. (See Figure 9.8.) Elsewhere, emissions are driven upwards by the expansion of energy use for transportation and power generation. The biggest increase is in Asia and the former Soviet region (by a factor 2 to 3), owing to their high economic growth rates. The increase in emissions is lower in sub-Saharan Africa and the MENA because of their lower economic growth, which leads to lower energy use.

In TechnoGarden, emissions tend to increase because of rapidly expanding transportation energy use and to decrease because of tighter controls and the co-benefits of climate policies. The final balance can result in increasing emissions in AIM and decreasing emissions in IMAGE (also depending on the type of climate action taken). In general, emissions decrease in currently high-income countries and increase in currently low-income countries. Increases in sub-Saharan Africa and the MENA are about the same as in Global Orchestration.

For Adapting Mosaic, the worldwide increase in emissions up to 2050 in both models is about 20–30%. NO_x emissions drop by two thirds in OECD countries and by over 40% in Latin America because of pollution controls. Increases in other regions are substantial because of expanded transportation energy use (a factor of 2.8 increase in Asia and the former Soviet Union).

In the Order from Strength scenario, emissions increase worldwide by 38% between now and 2050. Slow economic growth and other priorities for policy-making (poverty and security) lead in IMAGE to a continuous increase in global emissions (driven by low-income regions), while in AIM, emissions peak. The lower economic growth (compared with Global Orchestration) leads to a lower rate of emissions, but the lack of pollution controls in most regions leads to a higher rate. The growth of emissions is substantial in other regions: about 50% in MENA and Latin America, and about a factor of 2.6 in Asia and the former Soviet region.

9.3.2.4 Summing Up Air Pollution Emissions

The following more general observations regarding emission trends can also be made:

- Under Global Orchestration, emission trends are balanced between increasing sources of emissions and increasing commitments to emission controls as a result of increasing demand for clean air. Global SO_2 emissions are expected to stabilize while NO_x emissions increase between 2000 and 2050. Most of this increase occurs in Asia, the former Soviet Union, Africa, and MENA.
- Under TechnoGarden, we expect strong reductions in SO_2 and NO_x emissions as a result of strong investments in emission controls and the co-benefits of climate change policies.

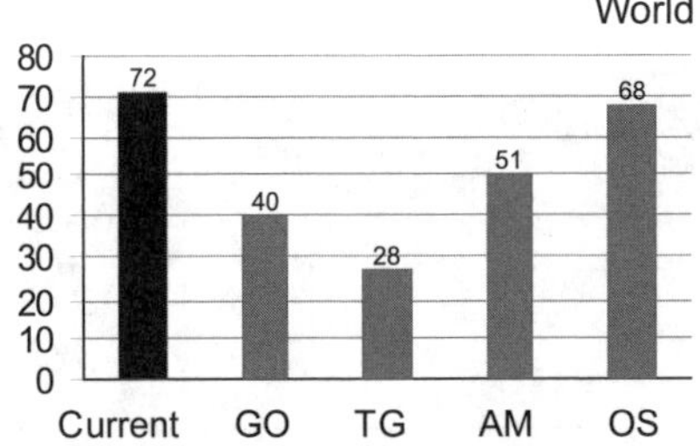

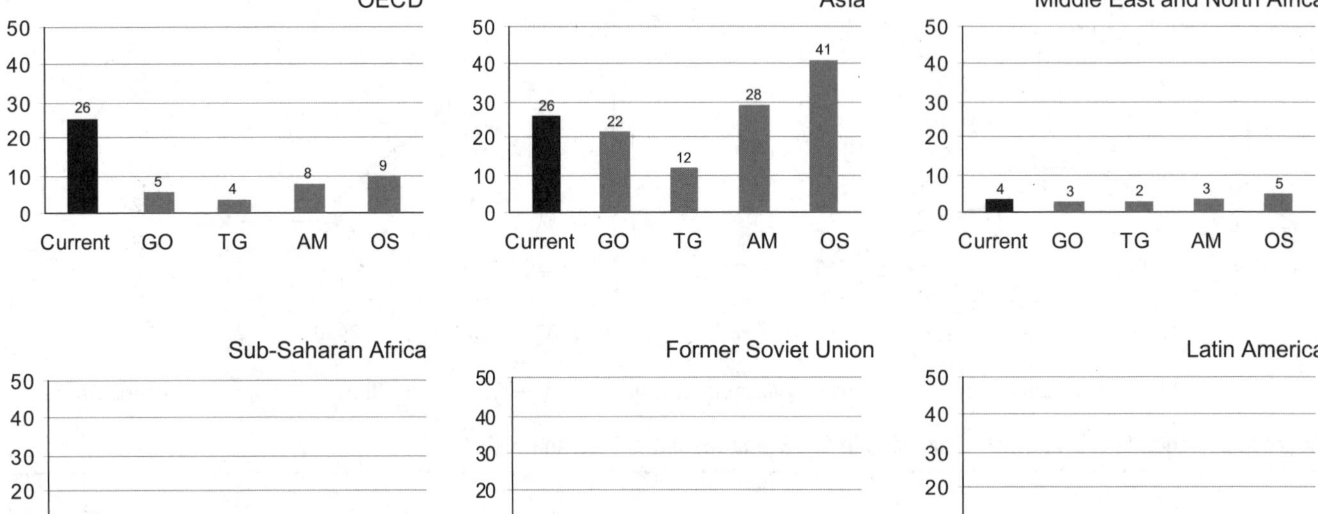

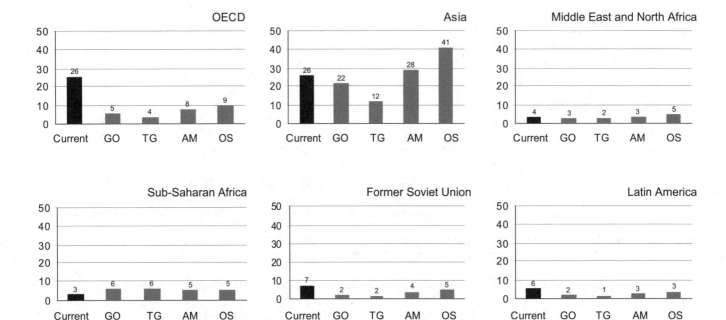

Figure 9.7. Emissions of Sulfur Dioxide in Different World Regions for MA Scenarios in 2050. Scenario names: GO: Global Orchestration; TG: TechnoGarden; AM: Adapting Mosaic; OS: Order from Strength. (AIM)

- Under Adapting Mosaic, environmental awareness is higher than under Global Orchestration, but lower economic growth in developing regions implies less energy use (and thus lower emissions) but also less investment in emission control technology. The result is that sulfur-related pollution declines in all regions except Asia, where it has a slight net increase. Trends for NO_x are similar to those in the Global Orchestration scenario.
- The level of sulfur-related air pollution declines only slightly worldwide under the Order from Strength scenario. Emissions decline in OECD, the former Soviet region, and Latin America. Asia and MENA have the largest emission increases of all scenarios. There is a significant decline in NO_x-related pollution in OECD countries, and a major increase elsewhere.

9.3.3 Risks of Acidification and Excess Nitrogen Loading from Air Pollution

9.3.3.1 Methodology and Assumptions

Atmospheric deposition of nitrogen and sulphur can lead to degradation of ecosystems as a result of the accumulation of excess nitrogen (also called terrestrial eutrophication) and acidification. These have been prominent environmental problems in North America and Europe for about 30 years. Recently, they also have been recognized as potential threats to ecosystems in other parts of the world. Excess quantities of nitrogen can alter ecosystems by causing shifts in species composition, increased productivity, decreased species diversity, and altered tolerance to stress conditions (Pitcairn 1994). Increases in sulfur and nitrogen input to ecosystems can also cause acidification of soils and thereby interfere with the growth processes of vegetation.

The risk to terrestrial ecosystems of the accumulation of nitrogen has been mapped at both the regional and global scale. For these estimates, the concept of "critical loads" has been used. A critical load is defined as "a quantitative estimate of an exposure to one or more pollutants below which significant harmful effects on specified sensitive elements of the environment do not occur according to present knowledge" (Nilsson and Grennfelt 1988). Two types of critical loads are evaluated here: the critical load for acidification, which is an estimate of the threshold of impacts for acid deposition (sulfur and nitrogen), and the critical load for terrestrial eutrophication, which is a measure of the threshold for the impacts of excess nitrogen deposition.

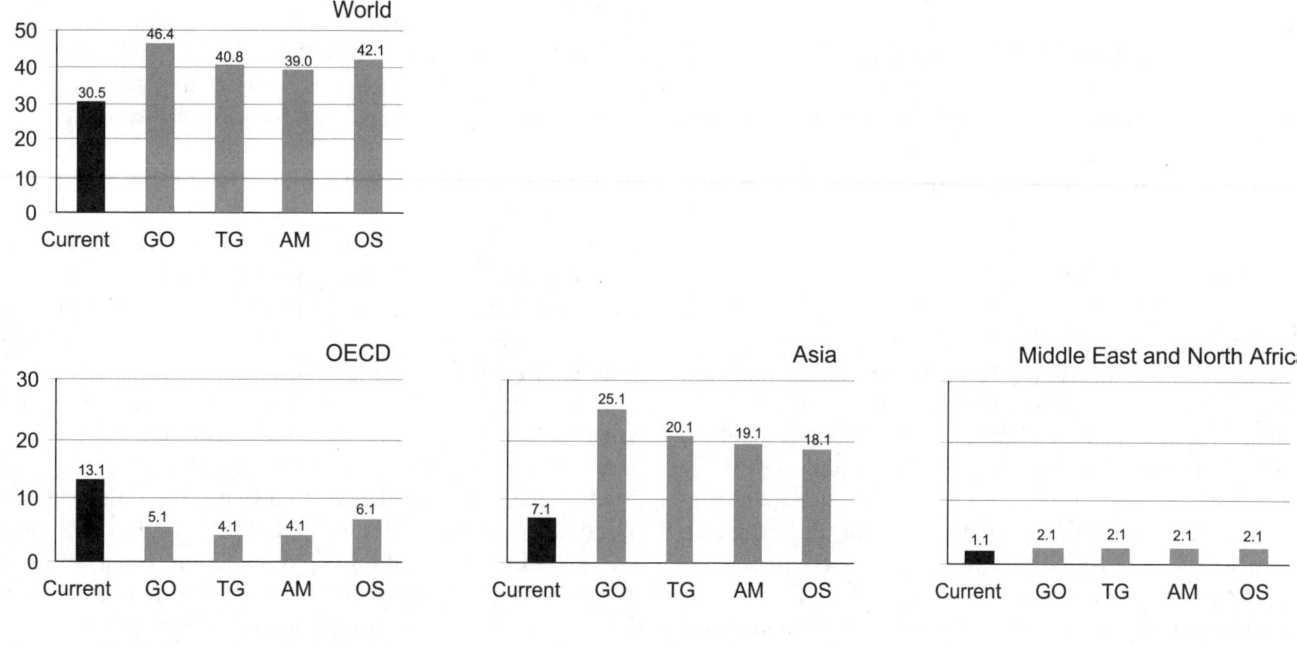

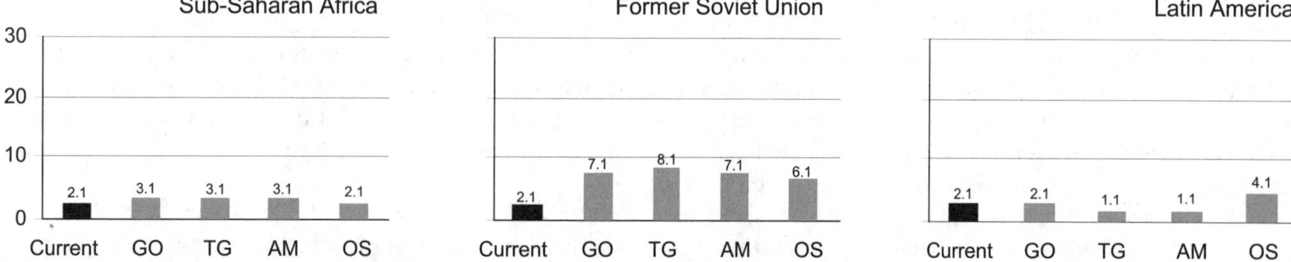

Figure 9.8. Emissions of Nitrogen Oxides in Different World Regions for MA Scenarios in 2050. Scenario names: GO: Global Orchestration; TG: TechnoGarden; AM: Adapting Mosaic; OS: Order from Strength (AIM)

In Europe, calculations with the RAINS model show considerable areas to be exposed to deposition levels above critical loads. The RAINS model has also been used in Asia and found high risks of acidification in Eastern China that were projected to increase in the future. Kuylenstierna et al. (1998), Rhode et al. (2002), and Bouwman et al. (2002) assessed acidification risks and nitrogen deposition risks at the global scale by overlaying deposition maps of S and N with critical loads maps. These studies indicate that current acidification risks are, relatively speaking, most severe in Europe and North America. Bouwman et al. (2002) evaluated scenarios of sulfur and nitrogen deposition and concluded that risks of acidification and nitrogen would increase in parts of China, Latin America, Africa, and Siberia.

It is possible to obtain a first crude estimate of air pollution–related risks under the MA by scaling the map of Bouwman et al. (2002) by the emission scenarios from the MA scenarios. This assumes as a first rough approximation that the deposition of SO_2 and NO_x in each region will linearly change along with the change in emissions in each region. The ratio between deposition and critical load is an indication of risks of acidification and nitrogen deposition, with values above one indicating that the "local" critical load for either acidification or excess nitrogen is exceeded and indicates a high risk to ecosystems.

9.3.3.2 Comparison of Risks of Acidification and Excess Nitrogen from Air Pollution among Scenarios

Figure 9.9 (see Appendix A) shows the results for acidification risks for the Order from Strength and TechnoGarden scenarios in 2050 (those with, respectively, the highest and lowest global emissions). Under Order from Strength, acidification risks decrease in OECD but increase in East Asia, Africa, and Latin America. Under TechnoGarden, the risks decrease markedly in North America, Europe, and East Asia and remain at current levels in Africa and Latin America. The low risk levels in this scenario are a consequence of both stringent emission control policies and the co-benefits of climate change policies (which reduce fossil fuel combustion).

The figure shows that similar trends occur for excess nitrogen deposition. In the Order from Strength scenario, risks of excess nitrogen deposition increase, especially in East and South Asia, while under TechnoGarden they decrease in OECD countries and stabilize in the rest of the world. Compared with acidification, the risks of excess nitrogen deposition occur farther away from industrial centers or densely populated regions. An important reason is that nitrogen emissions result from not only industrial activities and transport but also agricultural emissions. Moreover, several ecosystems are rather sensitive to excess nitrogen deposition. As discussed earlier, in contrast to sulfur emissions (which together with emissions of nitrogen oxides are the main cause of acidification), nitrogen emissions are expected to increase in most scenarios.

9.3.4 Climate Change

9.3.4.1 Methodology and Assumptions

The Intergovernmental Panel on Climate Change concluded in its latest assessment that there is new and stronger evidence that most of the climate change observed over the twentieth century is attributable to human activities (IPCC 2001). The report also indicates that future climate change is to be expected, as a function of continuing and increasing emissions of fossil fuel combustion products, changes in land use (deforestation, change in agricultural practices), and other factors (for example, variations in solar radiation).

Assessments of the potential influence of these factors indicate that increased greenhouse gas concentrations (caused by fossil fuel emissions and land use change) are the dominant factor in both historic and future changes of global mean temperature (IPCC 2001). The contribution of land use change to the increase in global mean temperature increase is assessed to be small compared with the fossil fuel emissions. At the local scale, however, changes in biophysical factors (surface roughness, albedo) related to land use change can be as important as changes in greenhouse gas concentrations. Moreover, under particular circumstances (for instance, in the case of a large-scale dieback of the Amazon), changes in land cover could also have a large contribution globally (Cox et al. 2000; Cramer et al. 2004).

The emissions of the MA scenarios cover the range of emission scenarios of the IPCC. The IPCC scenarios have been assessed in terms of their possible climate change, using both simple models (e.g., MAGICC; Wigley and Raper 2001) as well as state-of-the art climate models. IPCC (2001) concluded that the increase of greenhouse gas concentrations under the IPCC scenarios could cause a 1.4–5.8° Celsius increase in global mean temperature (in the absence of climate policies) between 1990 and 2100 (compared with preindustrial level, approximately 0.5° Celsius needs to be added).

Here, the influence of the MA scenarios is assessed using methods consistent with IPCC assessments and guidelines. The results are based on estimates of regional change in temperature and rainfall, made through an adapted version of the IPCC pattern-scaling approach (Carter et al. 2001; Schlessinger et al. 2000). This method combines global mean temperature trends estimated from the global energy balance model MAGICC (Wigley and Raper 2001) with a normalized pattern of climate change from the general circulation model HadCM3 (IPCC 1999).

Although trends in emissions vary considerably between the MA scenarios, the differences in calculated global temperatures in 2050 are not very large. By 2050, the results of the four scenarios ranges from a 1.6° Celsius (TechnoGarden) to 2.0° Celsius (Global Orchestration) increase (relative to pre-industrial levels) for a medium value for climate sensitivity (2.5° Celsius). This relatively small difference between scenarios is because of the lag time between the buildup of emissions in the atmosphere and the response of the climate system to this buildup. Moreover, low greenhouse gas emissions scenarios usually also have low sulfur dioxide emissions (as emissions stem from the same activity). While low greenhouse gas emissions lead to a slower increase of radiative forcing (and thus global mean temperature increase), lower sulfur emissions lead to a reduced cooling effect from sulfur aerosols.

Some recent studies have estimated climate policy scenarios that focus strongly on reduction of non-CO_2 greenhouse gases (e.g., Manne and Richels 2001; Hyman et al. 2003). Such studies generally find that costs savings can be obtained from also reducing non-CO_2 greenhouse gases. The implementation of the TechnoGarden scenario is consistent with the latest insights in emissions reduction of these gases, optimizing the reduction of the different gases on the basis of marginal costs (Delhotal et al. 2005; Schaefer et al. 2005). More extreme scenarios have been published (Hansen et al. 2000) in which even stronger reductions of non-CO_2 gases are achieved. Such scenarios can further reduce short- to medium-term climate change (to 2050), producing important ecological benefits but also at (probably) significant costs.

9.3.4.2 Comparison of Climate Change among the Scenarios

The calculated temperature increase in the 2000–50 period in all scenarios (1.0–1.5° Celsius) exceeds the increase in global mean temperature since 1850 (about 0.6° Celsius). Differences between the scenarios are much sharper by the end of the century. Under TechnoGarden, the increase in global average surface temperature in 2100 is slightly over 2° Celsius (above preindustrial). The increase is nearly 3.5° Celsius under the higher emissions growth of Global Orchestration. (See Figure 9.10.) Acknowledging the uncertainty in climate sensitivity in accordance with the range indicated by IPCC (1.5–4.5° Celsius) would lead to a wider range of temperature increase. Both the upper and lower end of this range would be shifted downward somewhat compared with the range for the IPCC SRES scenarios described earlier. This is because the TechnoGarden scenario includes climate policies (while the IPCC scenarios did not cover climate policies), and because the highest emissions scenarios (Global Orchestration and Order from Strength) show somewhat lower emissions than the highest of the IPCC scenarios, as explained earlier.

Among the scenarios, there are sharp differences in their decadal rate of temperature change. (See Figure 9.11.) This is of particular importance from the point of view of climate impacts because it is presumed that the faster the rate of climate change, the more difficult the adaptation of society and nature to the changes. Ecosystems differ greatly in their ability to adapt to this expected temperature change. The rate of temperature change during the 1990s was in the

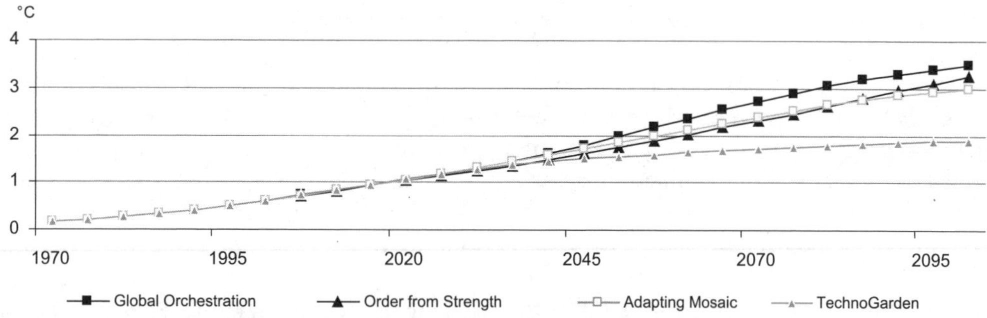

Figure 9.10. Change in Global Average Surface Temperature in MA Scenarios 1970–2100 (IMAGE 2.2)

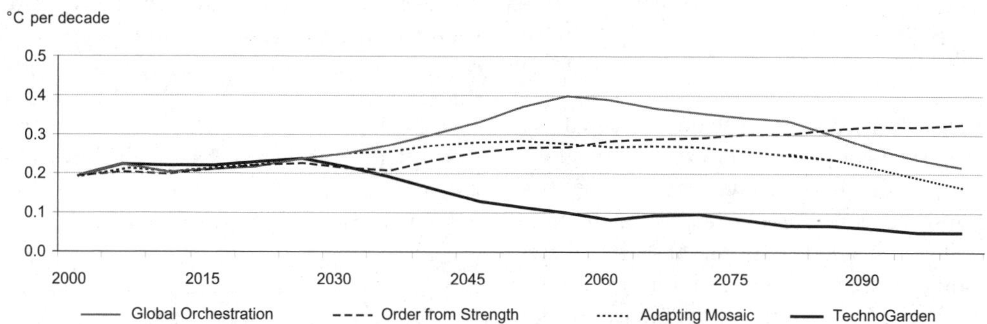

Figure 9.11. Decadal Rate of Change of Global Temperature in MA Scenarios (IMAGE 2.2)

order of 0.2° Celsius. Projections of future climate change are generally in the range of 0.1–0.4° Celsius per decade, assuming no major regime shifts in the global climate systems (such as breakdown of the North Atlantic Oscillation).

The rate of temperature change under the TechnoGarden scenario becomes slower and slower, reaching about 0.1° Celsius per decade in the middle of the scenario period. Meanwhile, the rate sharply increases under the Global Orchestration scenario until mid-century, when it reaches more than 0.4° Celsius per decade and then declines. The rate in the Adapting Mosaic scenario lies between these two scenarios, leveling off at mid-century at around 0.3° Celsius per decade and then declining. Meanwhile, at mid-century the rate in the Order from Strength scenario is lower (around 0.25° Celsius per decade) than that of Adapting Mosaic but is still increasing, so that it has the highest value of all scenarios (0.3° Celsius) at the end of the century.

Although these values may be uncertain, in each of the MA scenarios climate change is expected to be very likely. The benefits of assumed climate policies under TechnoGarden will help to slow down the rate of climate change during the 2000–50 period and will lead to much lower temperature increases compared with the other scenarios by the end of the century. The rate of climate change is likely to increase to at least mid-century in three of the four scenarios (all except for TechnoGarden), as a result of the reduced sulfur cooling effect and increases in greenhouse gas emissions. Likewise, it is likely that three out of four will have a declining rate of temperature increase after mid-century (all except Order from Strength).

While the computation of global mean temperature is uncertain, the patterns of local temperature change are even more uncertain. In its comparison of temperature calculations from different climate models, IPCC (2001) noted some areas of agreement (such as temperature increase likely being higher at higher latitudes than near the equator) but also many areas of disagreement. Disagreements, for example, typically occur in areas with complex weather patterns.

9.3.4.2.1 The influence of biophysical factors on climate change

As noted, land use changes can affect various biophysical factors that have a major impact on climate (and that form a direct linkage between ecosystems and climate change). With the MA scenarios, the impact of biophysical factors will be most pronounced for the scenarios with the largest land use changes. These include in particular Order from Strength and TechnoGarden. In Order from Strength, a continuously increasing population leads to a major expansion of agricultural lands, causing further deforestation in tropical areas. While impacts in tropical zones via albedo changes are relatively small, other influences of large-scale deforestation of tropical rain forests on (local) climate are highly uncertain but may be significant. In contrast, TechnoGarden is the scenario with the most reforestation in temperate zones. An even higher rate of reforestation might be expected under this scenario if reforestation is used as a climate policy for sequestrating CO_2. As indicated by Betts (2000), this actually could lead to increased warming as a result of reduced albedo. Again, these effects are still highly uncertain.

9.3.4.2.2 Precipitation changes

While future regional temperature is uncertain, still more uncertain are the computations of precipitation patterns within regions. Climate models can provide insight into overall global and regional trends but cannot provide accurate estimates of future precipitation patterns when the landscape plays an important role (as in the case of mountainous or hilly areas). Recognizing this uncertainty, we use a standard integrated assessment approach to estimate uncertain but plausible future changes in precipitation. Figure 9.12 shows a typical spatial pattern of changes in precipitation up to 2050 in Global Orchestration. According to this scenario, approximately three quarters of the land surface has increasing precipitation. This is a typical but not universal result from climate models. Some arid areas become even drier according to Figure 9.12 (see Appendix A), including the Middle East, parts of China, southern Europe, the northeast of Brazil, and west of the Andes in Latin America. This will increase water stress in these areas, as described later.

Although climate models do not agree on the spatial patterns of changes in precipitation, they do agree that global average precipitation will increase over the twenty-first century. This is consistent with the expectation that a warmer atmosphere will stimulate evaporation of surface water, increase the humidity of the atmosphere and lead to higher overall rates of precipitation. In general, climate models give a more consistent picture for temperature change than for precipitation.

9.3.4.2.3 Climate change impacts

Figure 9.13 (see Appendix A), from the IPCC assessment, summarizes the findings from a large number of climate impact studies. The main result is that risks of different types increase with increasing temperature, but at different tempos. Comparing the temperature increases from 2000 to 2100 with the risks indicated by the IPCC, the lowest temperature increase scenario, TechnoGarden, will still have high risks for unique and threatened systems and extreme climate events. For aggregate impacts, the 2° Celsius temperature increase experienced in this scenario falls in the middle category; while the risks of large-scale discontinuities (breakpoints in natural systems) are assessed to be low. The higher temperature increase scenarios (Global Orchestration, Adapting Mosaic, and Order from Strength) reach the range in which there are higher risks of large-scale breakdowns in natural systems.

9.3.5 Sea Level Rise

9.3.5.1 Methodology and Assumptions

One of the major impacts of climate change will be a rise in average global sea level as warmer temperatures melt currently permanent ice and snow and cause a thermal expansion of ocean water. Furthermore, climate change may cause stronger and more persistent winds in the landward direction along some parts of the coastline, and this will also contribute to rising sea level at these locations.

9.3.5.2 Comparison of Sea Level Rise among Scenarios

We have made a first-order estimate of the expected (global average) sea level based on the climate change scenarios corresponding to the four MA scenarios. The average rise up to 2100 ranges from 50 centimeters (in TechnoGarden) to 70 centimeters (in Global Orchestration). (See Figure 9.14.) The actual increase in different regions might be higher or lower, depending on changes in ocean currents, prevailing winds, and land subsidence rates.

Note in Figure 9.14 that sea level still has a rising tendency at the end of the century, even though Figure 9.10 indicated that air temperatures stabilize under three of the four scenarios. The increase in sea level lags decades behind the increase in temperature because there is a long delay in heating the enormous volume of the world's oceans. This means that the temperature could stabilize over the course of the scenario period while sea level continues to rise. For example, the trends shown imply that sea level could further rise by at least an additional 1m during the course of the twenty-second century.

9.3.6 Change in Land Use or Land Cover

9.3.6.1 Methodology and Assumptions

Land use change and its consequences for the land cover form an important component of global change (Turner et al. 1995). The type of land use and land cover has direct consequences for most ecosystem services, including provisioning services for food, fiber, and water; regulating services of carbon storage and erosion control; most cultural services; and biodiversity. Historically, large areas of natural ecosystems have been converted into agricultural areas; since 1700, for instance, more than 41 million square kilometers of ecosystems have come into production as either cropland or pasture (30% of the non-ice-covered land area) (Klein-Goldewijk 2004; Ramankutty and Foley 1999).

Land use changes, however, are not easy to capture in large-scale environmental models. They often evolve from diverse human activities that are heterogeneous in spatial and temporal dimensions. They also strongly depend on local environmental conditions and ecological processes. As a result, global models tend to focus on a selected number of major processes. The discussion here focuses on changes in forestland and agricultural land (pasture plus cropland for food, feed, and biofuel crops).[3]

It should be noted that comparisons with other land use change scenarios are difficult, since many published scenarios focus only on local and regional issues or on certain aspects of land use such as the environmental consequences of different agrosystems (e.g., Koruba et al. 1996), agricultural policies (e.g., Moxey et al. 1995), and food security (e.g., Penning de Vries et al. 1997).

Nevertheless, some typical trends can be observed in published scenarios. First of all, most of them show in the near-future a continuation of recent trends: that is, a steady increase of agricultural land (cropland and pastureland) in developing countries and constant or declining coverage of agricultural land in industrial countries (e.g., FAO 2003; IMAGE-team 2001). Crucial factors in existing scenarios involve population change, changes in agricultural output (mostly through intensification, but sometimes also extensification), changes in dietary practices, and agricultural trade.

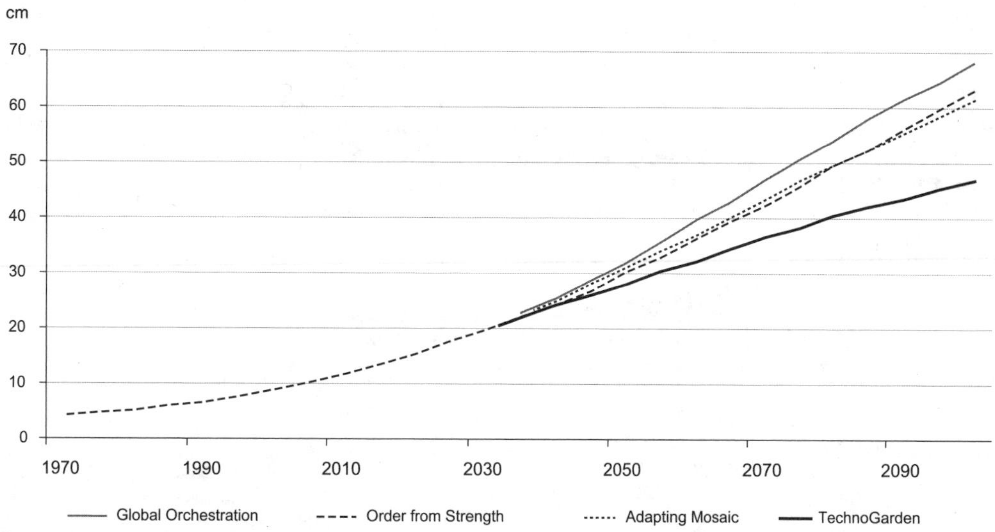

Figure 9.14. Sea Level Rise in MA Scenarios 1970–2100 (IMAGE 2.2)

In all published scenarios, increases in agricultural production in low-income countries are mostly achieved through increasing yields, but at the same time there is also a further expansion of agricultural land. This increase in desired production comes mainly from steep increases in food demand, especially the demand for animal products. For example, meat consumption in China increased yearly by 2.6 kilograms in the 1990s (FAO 2003). Such an increase is also expected in other developing countries in the next decades. Existing scenarios also show a further loss of forest cover in developing countries and a net gain in forest cover in high-income countries. Old-growth forest in industrial countries can be further reduced for timber production, however, and the net gain is achieved by an increase in new forest. In terms of ecological functions, there are important differences between primary and secondary forests.

9.3.6.2 Comparison of Land Use or Land Cover among Scenarios

In the twentieth century, major transformations in land use and land cover have created a large downward pressure on the potential of ecosystems to provide ecological services.[4] Over the last decades, however, this trend has become rather diverse, with increases in forest area in some regions (industrial regions) and further decline in forest area in others (developing regions). At the global level, this trend continues in the four MA scenarios.

In the first decades of the scenario period, all scenarios show an ongoing expansion of agricultural land replacing current forest and grassland. This expansion occurs mainly in poorer countries, while agricultural land in the OECD and former Soviet regions actually declines. (See Figure 9.15.) Despite the considerable differences in individual driving forces among the scenarios, differences in land use among them remain somewhat small. This is partly a result of counteracting trends in the driving forces (low population growth and high economic growth—so high caloric diets for fewer people—versus higher population growth but lower economic growth, which means more people eating less per capita). In addition, it is also a result of increases in different kinds of land use (for instance, a strong increase in land for fodder and grass under Global Orchestration to feed the animals versus a stronger increase in land used for biofuels to meet the climate targets in TechnoGarden).

Compared with the other three scenarios, Order from Strength exhibits by far the fastest rate of deforestation at the beginning of the scenario period. (See Figure 9.16.) The rate of loss of "original" forests actually increases from the historic rate (of about 0.4% annually between 1970 and 2000) to 0.6%. (See caption in Figure 9.16 for definition of "original" forest.) The estimation of annual historic loss of "original" forests is consistent with upper estimates in Chapter 5 in the MA *Current State and Trends* volume but is not strictly comparable because of different averaging periods and definitions of forests. This increase in the deforestation rate comes from the faster expansion of agricultural land, resulting mainly from rapidly growing population combined with slow improvements in crop yield in low-income regions. Since crop yield remains low compared with increasing demand for food products, more agricultural land is needed (although many increases in crop production are also achieved through intensification of existing agricultural land). In the other scenarios, the rate of loss of undisturbed forests is at the historic rate (Global Orchestration and Adapting Mosaic) or slightly below (TechnoGarden).

In 2020, the Order from Strength scenario shows an increase of arable land in poorer countries of almost 13% over the 2000 figure. This is almost twice the figure in the FAO prognosis for 2015 (a 6% increase) (FAO 2003). The increase of arable land in the other three scenarios is close to the FAO projection (5–6% increase). For pastureland, the TechnoGarden scenario shows a decrease, which can be explained by the assumed decrease in meat consumption and a shift toward high-efficient feed instead of grass for animals. This trend, however, is offset by a larger demand for cropland. The other scenarios all show increases in the amount

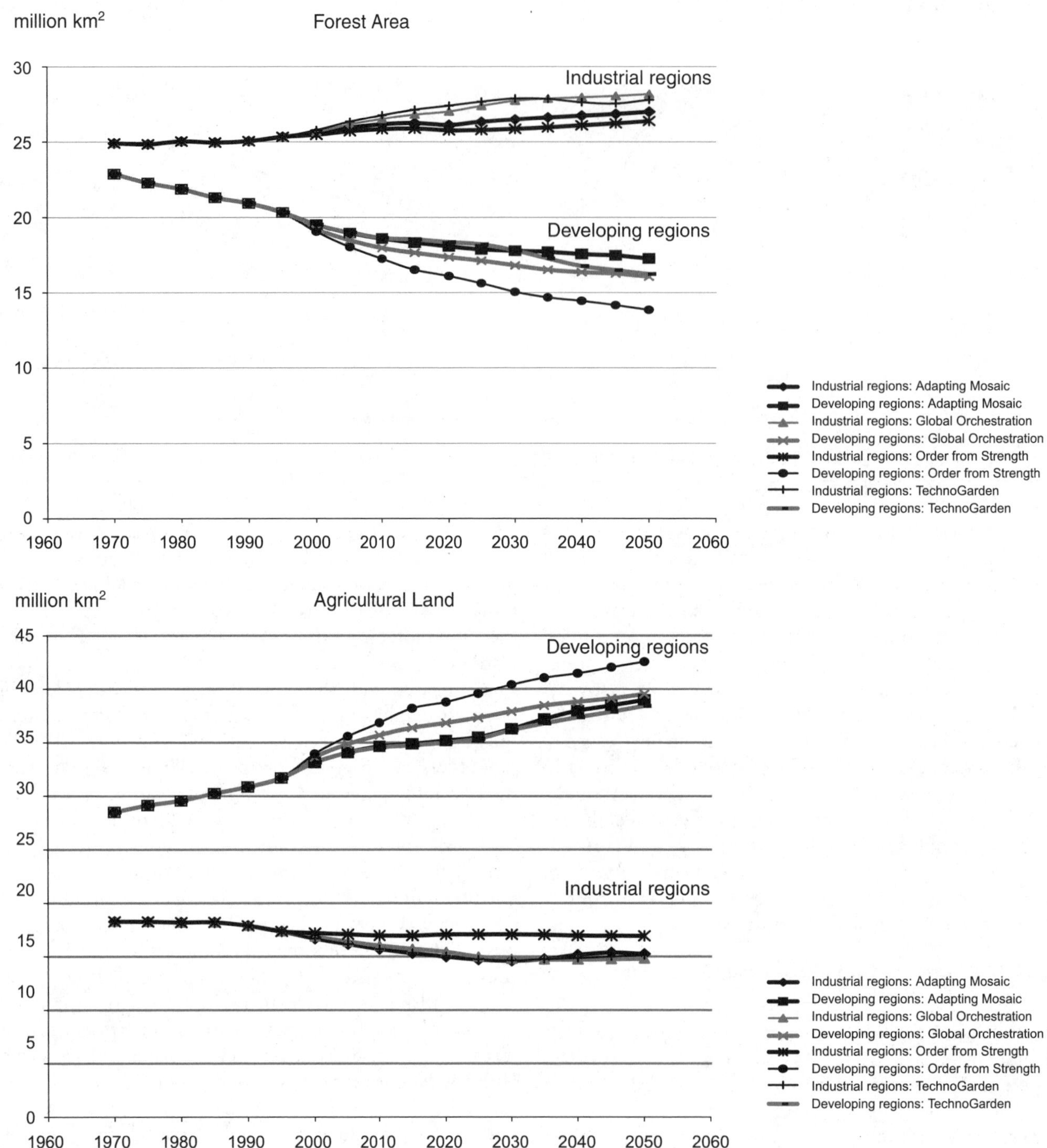

Figure 9.15. Changes in Global Forest Area and Agriculture Land for MA Scenarios. Agriculture land is defined as pasture and crop land. (IMAGE 2.2)

of pastureland. These trends are well in line with the constant prognosis of FAO (FAO 2003).

Figure 9.17 indicates the total land use in 2050 per region and scenario. While rapid depletion of forest area continues under the Order from Strength scenario, under TechnoGarden we expect an increase in net forest cover. Production of biofuels, particularly under the TechnoGarden scenario, is an important category of land use, especially in former Soviet countries, OECD countries, and Latin America. Although the coverage of energy crops remains relatively small, it actually has a large influence on the trends in land use.

Under the Order from Strength scenario (see Figure 9.18 in Appendix A), there is a continuous increase of agricultural area in poorer countries, particularly in sub-Saharan Africa and Latin America. Important factors include a relatively fast population growth and a limited potential to import food (particularly relevant for Africa). As a result, the depletion of forest area continues worldwide at a rate near the historic average, only to slow down after 2050 be-

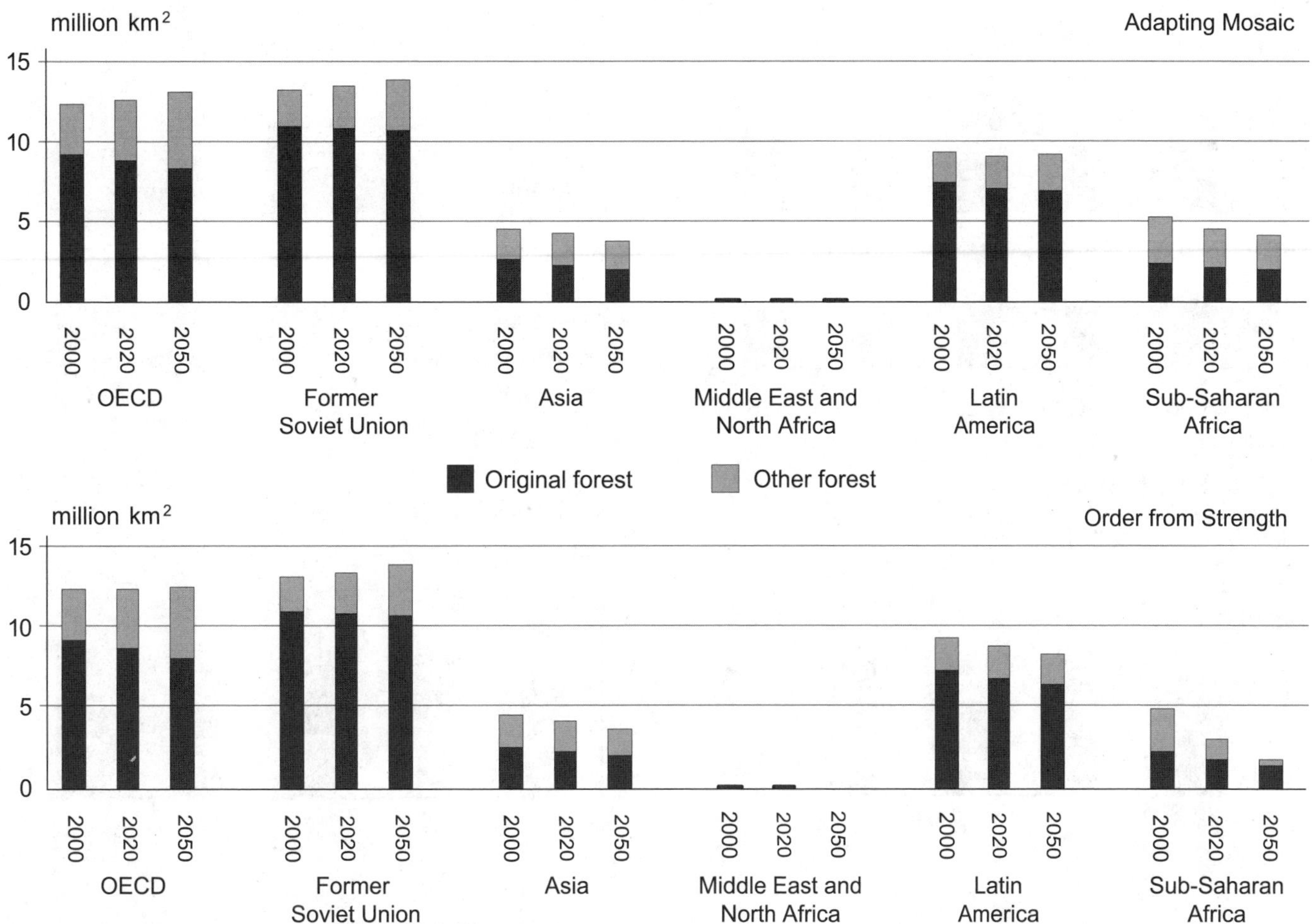

Figure 9.16. Trends in Forest Area by World Region in Two Scenarios. This figure distinguishes between "original" and "other" forests. "Original" forests are defined here as forests that were present in 1970 and have not changed their attributes through either expansion of agricultural land, timber production or climate change since then. "Other" forests are those forests which have been grown from abandoned agricultural or other land, or have been established from other types of land because of climate change. (IMAGE 2.2)

cause of slowing population growth. As a result, two thirds of the Central African forest present in 1995 will have disappeared by 2050. For Asia and Latin America, these numbers are 40% and 25%, respectively. In other regions the rate of forest loss slows down.

The land use conversion in this scenario clearly exceeds that of the FAO reference scenario in 2030 (FAO 2003). This difference mainly comes from a considerably lower improvement in agricultural yields that is expected under the Order from Strength scenario than under the FAO scenario. Fischer et al. (2002) also show a major increase of total agricultural land in a scenario that assumes a regionalized world (increase of nearly 20% in the second half of the twenty-first century for the IPCC A2 scenario compared with 15% in Order from Strength). Similarly, Strengers et al. (2004) report a similar result for the IPCC A2 scenario (increase of 22% in 2050).

Agricultural area under the Global Orchestration scenario also expands at a fast rate, but for other reasons than in Order from Strength. Here, rapid income growth and stronger preferences for meat result in growing demand for food and feed, leading to a rapid expansion of crop area in all regions. There is no net increase of pastureland, as low-input extensive grazing systems are replaced by more intensive, crop-intake forms of grazing, a result that is comparable to the FAO analysis for 2030 (FAO 2003). Undisturbed forests disappear at a slower rate than in Order from Strength, but still at near-current global rates. About 50% of the forests in sub-Saharan Africa disappear between 2000 and 2050.

The TechnoGarden scenario results in the lowest conversion of natural land to agricultural land. One important factor is the assumed decrease in meat consumption, which leads to lower land demands for feed crops and grazing. This is partly offset, however, by a strong increase in food demand in poorer countries. The improvement of yields (as a result of a widely available technology) ultimately leads to a slower expansion of agricultural land. In terms of total land, the results for TechnoGarden are comparable with those of FAO projections for 2030 (FAO 2003). Under TechnoGarden there is a small decrease in pastureland and a small increase in arable land or food production, mainly in poorer regions. However, there is a large increase in land for growing energy crops as part of climate policies for reducing greenhouse gas emissions. Although this scenario has the lowest rates of land conversion, the depletion of forestland is still significant in Africa and Southeast Asia. Global deforestation rates, however, are far lower than in the other scenarios.

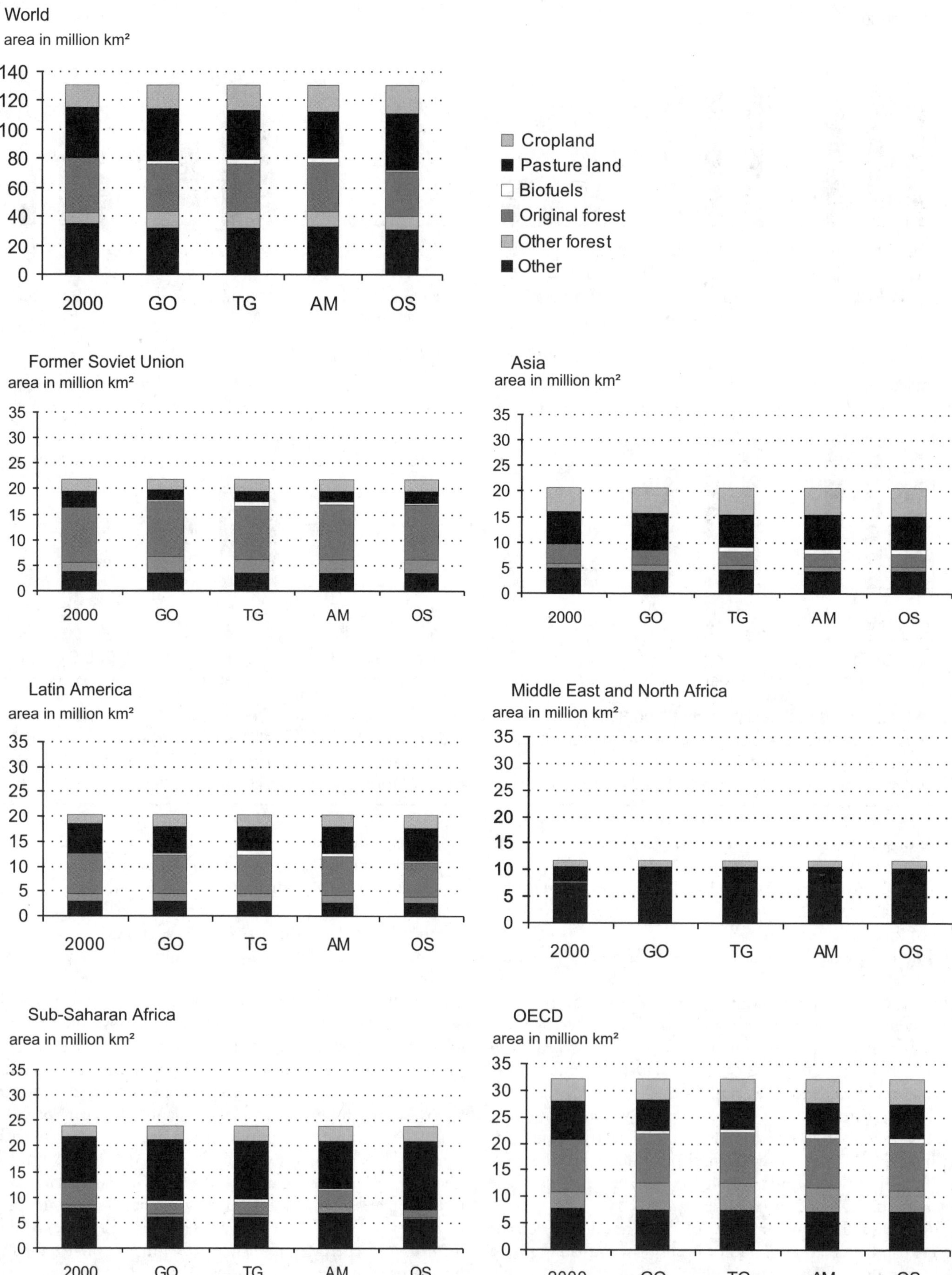

Figure 9.17. Land Use Patterns by Region in MA Scenarios in 2050. Scenario names: GO: Global Orchestration; TG: TechnoGarden; AM: Adapting Mosaic; OS: Order from Strength. (IMAGE 2.2)

The Adapting Mosaic scenario, like Order from Strength, also assumes relatively slow yield improvement in the first decades. However, a lower increase in population and locally successful experiments in innovative agricultural systems (translated into an increasing rate of improvement of crop yields) mitigate a further expansion of agricultural land in other regions after 2040. This is particularly important for trends in Africa; in fact, Adapting Mosaic shows the lowest deforestation rates for this region of all four scenarios. In contrast, however, the relatively low yield improvement causes a virtual depletion of forest areas in South Asia up to 2100. Globally, the long-term deforestation rates in this scenario are slightly above those of TechnoGarden.

The changes in land use just described will also have a tremendous impact on the vulnerability of different regions. Figure 9.19 shows the land use of each region in 2050 compared with the total potential area of productive arable land (that is, areas with potential productivity—based on soil and climatic condition—that is more than 20% of the maximum achievable yield of the best-growing crop). By 2050, under Order from Strength, Africa and Asia have put virtually all productive land under cultivation to fulfill the demand for crops and animal products. This clearly indicates a high vulnerability to abrupt changes in the natural system. A similar but less extreme situation occurs for Africa under the Global Orchestration scenario, and for Asia and Africa under both Global Orchestration and TechnoGarden. In these cases, however, large-scale global trade could help overcome problems of suddenly declining production levels as a result of abrupt ecological changes. The above-mentioned processes result in a less vulnerable situation for Africa in Adapting Mosaic.

Based on these results, we can conclude land use change will continue to form a major pressure on ecosystem services in the four MA scenarios. At the same time, all four scenarios find the loss of natural forests to slow down compared with historic rates. This mainly results from increases in natural areas in industrial regions (consistent with trends of the past few decades). In developing regions, the conversion rates slow down in three out of four scenarios. In Order from Strength, however, the rate of conversion continues at nearly the historic rate of the past three decades.

9.3.7 Use of Nitrogen Fertilizer and Nitrogen Loads to Rivers and Coastal Marine Systems

The presence of excess nutrients in water can lead to eutrophication. This nutrient enrichment of waters can lead to

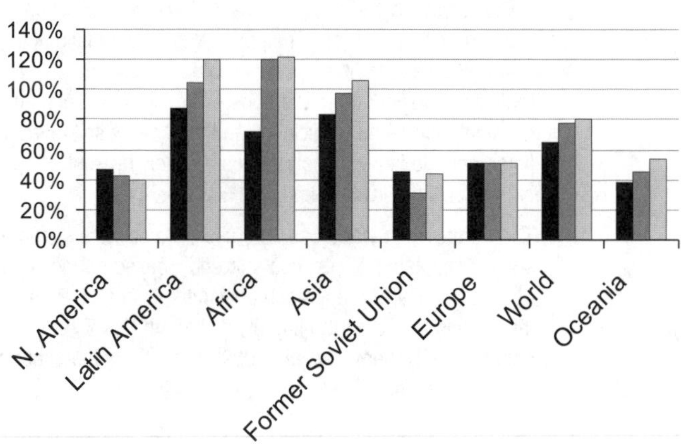

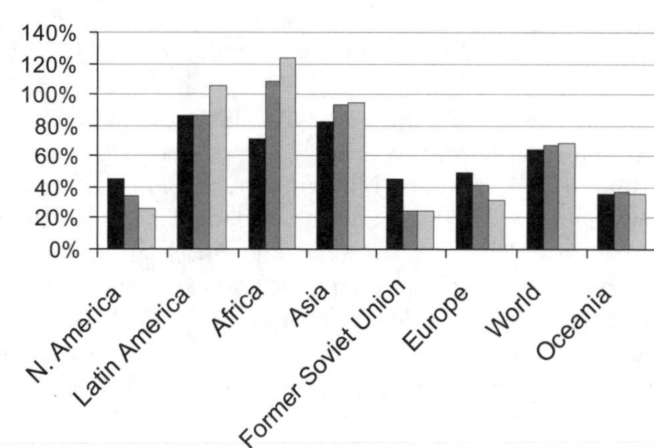

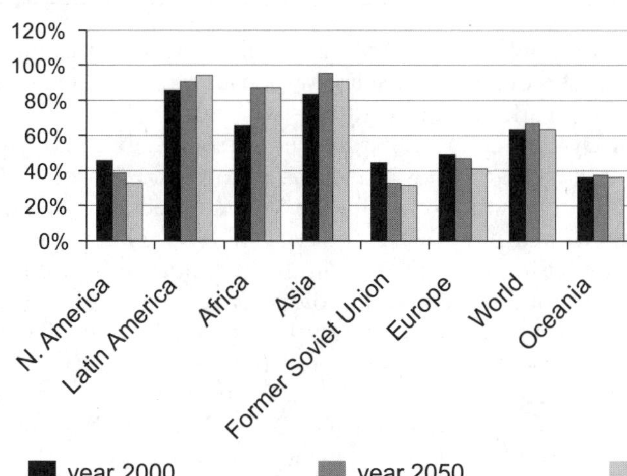

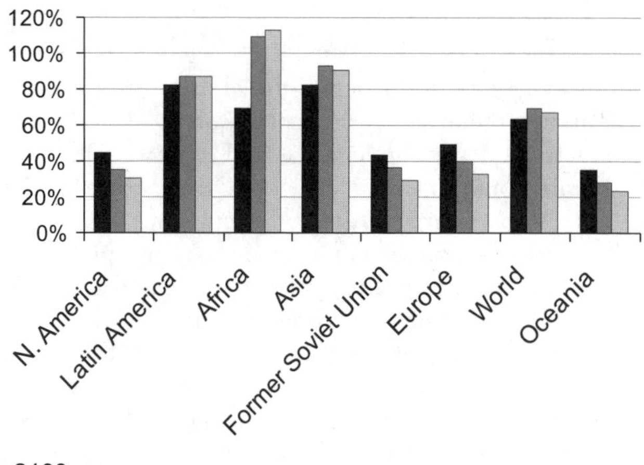

year 2000 year 2050 year 2100

Figure 9.19. Ratio of Agricultural Land to Total Productive Arable Land in MA Scenarios (IMAGE 2.2)

algal blooms, changes in the organisms present, low oxygen levels in the water, and generally lower water quality. Nitrogen and phosphorus are commonly the nutrients contributing to eutrophication. In the context of the MA scenarios, we concentrate on changes in nitrogen loading given the presence of information that could be coupled to the scenarios. There is ongoing concern about nitrate leaching to waters because of eutrophication, about other environmental effects associated with high nitrate levels, and about the risk that high nitrate levels in drinking water may present to human health.

9.3.7.1 Methodology and Assumptions

9.3.7.1.1 Trends in the use of nitrogen fertilizer

Projections of global nitrogen fertilizer use cover a range of time horizons, scenarios, and underlying assumptions. Table 9.13 and Figure 9.20 (see Appendix A) summarize a set of recent fertilizer use scenarios. The Figure shows that all scenarios expect an increase of fertilizer use. The range among the different scenarios is considerable, with the highest scenarios indicating increases in N-fertilizer use of 80% or more until 2020, while the lowest show an increase of less than 10%.

Based on current insights in changes in nitrogen efficiency and agricultural scenarios, we expect the outcomes for the Global Orchestration scenario to be near the Constant Nitrogen Efficiency scenario of Wood et al. (2004) or the A1b scenario of IMAGE-team (2001)—that is, around 110 million tons in 2020 and 120–140 million tons in 2050. The TechnoGarden scenario is likely to correspond to the outcomes of the Improved Nutrient Use efficiency scenario of Wood et al. (2004), around 100 million tons in 2020 and 110–120 million tons in 2050. The Adapting Mosaic scenario is likely to fall between these two extremes, while fertilizer use under Order from Strength could be near the outcomes for Global Orchestration. Clearly, there are important uncertainties in projections, including the effective potential for improving efficiency, the paucity of data on crop-specific nutrient application rates, the area fertilized and corresponding yield responses, and the lack of explicit incorporation of market prices of fertilizers (Wood et al. 2004).

Table 9.13. Overview of Scenario Studies on Nitrogen Fertilizer Use

Reference	Method
Bumb and Baanante 1996	Projections of N fertilizer to 2000 and 2020 using three approaches—the Nutrient Removal Approach and the Cereal Production Method to assess N requirements to meet projected cereal needs in 2020 (Rosegrant et al. 1995) and the Effective Fertilizer Demand Method projecting N use on the basis of a range of economic, demographic, and other factors
Tilman et al. 2001	Projection based on linear regressions of N fertilizer usage and time, population, and GDP for the period 1960 to 1999, extrapolated mean values of N fertilizer use
Daberkow et al. 1999	Projection built on crop area and yield projections developed by FAO in support of the *Agriculture Towards 2015/30* study (Bruinsma 2003) to assess corresponding fertilizer needs. The authors used the Fertilizer Use By Crop database (FUBCD IFA, IFDC, FAO 1999) to derive crop-specific nutrient application and response rates for three scenarios: Baseline, Improved Nutrient Use Efficiency, and Nitrogen Use on Cereals
Galloway et al. 2004	Projection based on the Daberkow et al. (1999) "baseline" scenario for 2030, and extrapolated an N fertilizer use of 135 million tons in 2050, assuming a constant N fertilizer growth rate to 2050
Wood et al. 2004	Used the newest FUBCD data—Trend Analysis was based on an update of the Bumb and Baanante (1996) Effective Demand Approach and assumed that N fertilizer applications would be higher in areas of significant soil degradation between 2020 and 2050 as part of a broader strategy of soil fertility restoration; given the conservative assumptions about constant nitrogen use efficiency and the goal of soil rehabilitation embedded in this analysis, results likely present an upper bound on N fertilizer needs
	The Future Food Need Scenario used two scenarios, one assuming constant nitrogen use efficiency, based on the Nitrogen Use on Cereals approach of Daberkow et al. (1999), and the second based on the Improved Nutrient Use Efficiency approach of Daberkow et al. (1999), but with region-specific nitrogen use efficiencies
IMAGE-team 2001	Scenarios based on expansion of crop area and assumed changes in fertilizer used per hectare for the four IPCC scenarios

9.3.7.1.2 Nitrogen loads to rivers and coastal marine systems

Anthropogenic disturbance of the global nitrogen cycle is an important global environmental problem. On one hand, production on some agricultural land is not as high as it can be because of nitrogen deficiencies. On the other hand, the runoff of excess nitrogen from agricultural land and from other anthropogenic sources causes the eutrophication of rivers and other freshwater systems. Nitrogen loads in rivers eventually find their way to the coastal zone, where they also cause eutrophication. Here, we focus on the nitrogen loading to rivers and its routing to the coastal zone.

Several studies have estimated the past and current river nitrogen transport to oceans (Green et al. 2004; Meybeck 1982; Seitzinger and Kroeze 1998; Seitzinger et al. 2002; Turner et al. 2003; Van Drecht et al. 2003). Despite the fact that scenarios of nitrogen fluxes are still under development, there is a growing interest in the potential threat of further increases of nitrogen loading on aquatic systems. More qualitative work on nitrogen fluxes was published earlier as part of UNEP's Global Environment Outlook (UNEP 2002).

In order to assess changes in nitrogen fluxes from rivers to oceans in the context of the MA, we used a global model developed by Van Drecht et al. (2003). This model describes the fate of nitrogen in the hydrological system up to river mouths, at a spatial resolution of 0.5 by 0.5 degree and an annual temporal resolution. This model was used earlier to describe the development of river nitrogen fluxes based on the Agriculture Towards 2030 projection of the FAO (hereinafter referred to as AT 2030) (Bruinsma 2003), and a projection for sewage effluents (Bouwman et al. 2005b). We used these results as a reference to estimate the change in river nitrogen export on the basis of the four MA scenarios.

9.3.7.2 Comparison of Nitrogen Fertilizer and Nitrogen Loads to Rivers among Scenarios

On the basis of projections for food production and wastewater effluents, the global river nitrogen flux to coastal marine systems may increase by 10–20% in the coming three decades. While the river nitrogen flux will not change in most wealthy countries, a 20–30% increase is projected for poorer countries, which is a continuation of the trend observed in the past decades. This is a consequence of increasing nitrogen inputs to surface water associated with urbanization, sanitation, development of sewerage systems, and lagging wastewater treatment, as well as increasing food production and associated inputs of nitrogen fertilizer, animal manure, atmospheric nitrogen deposition, and biological nitrogen fixation in agricultural systems. Growing river nitrogen loads may lead to increased incidence of problems associated with eutrophication in coastal seas.

Regarding the oceans receiving nitrogen inputs from river systems, our results indicate that strong increases in the 1970–95 period occurred in the Pacific (42%), Indian (35%), and Atlantic Oceans (18%) and in the Mediterranean and Black Seas (35%), with a global increase of 29%. (See Table 9.14.) For the coming three decades, the increase will be even faster in the Indian Ocean (50%), while the increase for the Pacific (31%) and Atlantic Oceans (8%) is slower than in the 1970–95 period. For the Mediterranean and Black Seas (at −5%), a slow decrease of river nitrogen export is estimated.

It is possible to estimate the nitrogen fluxes of each MA scenario by assessing their relative differences for the various nitrogen-emission sources with respect to the AT 2030 projection.[5] This is possible because there are firm relationships between the total inputs of nitrogen in terrestrial systems (deposition, biological fixation, fertilizers, and animal manure) and the river transport of nitrogen.

Changes in the inputs from natural ecosystems to total river transport were assessed on the basis of estimates for nitrogen deposition for each scenario. The river transport from agricultural systems was assessed for each scenario on the basis of total nitrogen fertilizer use and animal manure production. Fertilizer use was assumed to be correlated with total crop production in dry matter, while animal manure production was assumed to be related to total livestock production in dry matter. The river nitrogen load from sewage effluents was assumed to be related to total population, whereby the human emissions and wastewater treatment were assumed to be related to GDP. Finally, the number of people connected to sewerage systems was held the same for each scenario, as we assumed development of sewerage systems has a high priority for human health reasons in all scenarios (although this could be seen as relatively optimistic in the case of Order from Strength).

The results of this comparison show considerable differences between the scenarios. (See Figure 9.21.) In Global Orchestration, fast economic development causes a shift toward more protein-rich food consumption and higher human-waste production. At the same time, the nitrogen removal in wastewater treatment will be higher than in the AT 2030 scenario. Agricultural production is not much different from the AT 2030 scenario, so that the river loads stemming from fertilizers and animal manure are similar. However, atmospheric nitrogen deposition rates in Global Orchestration are much higher than in any of the other scenarios, causing higher river nitrogen loads.

In TechnoGarden, a proactive attitude with regard to ecological management is assumed to lead to lower per capita meat consumption, while wastewater treatment has a

Table 9.14. River Nitrogen Export to Atlantic, Indian, and Pacific Oceans and to Mediterranean and Black Seas and Contributions from Natural Ecosystems, Agriculture, and Sewage for 1970, 1995, and 2030. Columns may not add up due to separate rounding. (IMAGE 2.2)

River Export by Source	Atlantic Ocean	Indian Ocean	Pacific Ocean	Arctic	Mediterranean and Black Seas	World
			(million tons per year)			
In 1970						
Natural	15	3	5	1	1	25
Agriculture	4	1	2	0	1	7
Sewage	1	0	1	0	0	2
Total	19	4	7	1	2	34
In 1995						
Natural	16	4	4	2	1	28
Agriculture	5	2	4	0	1	13
Sewage	1	0	1	0	1	3
Total	23	6	10	2	3	44
In 2030[a]						
Natural	16	5	5	1	1	28
Agriculture	6	3	6	0	1	17
Sewage	1	1	1	0	1	4
Total	24	9	13	2	3	50

[a] Results for 2030 are based on the AT 2030 projection (Bruinsma 2003) and presented by Bouwman et al. (2005b).

high policy priority for the prevention of eutrophication of surface water. Atmospheric deposition is much less than in the Global Orchestration scenario, causing a reduction of the nitrogen load in the coming three decades.

The river nitrogen loads in 2030 for the other scenarios will be lower than for Global Orchestration but higher than for TechnoGarden. Although the river nitrogen load stemming from agricultural sources and deposition levels are comparable between the scenarios, there is a difference in the inputs from sewage effluents. The population for 2030 is not different between Order from Strength and Adapting Mosaic, but in Order from Strength economic growth is slower, leading to a slower growth in removal during wastewater treatment than in Adapting Mosaic.

The results indicate that in three of the four scenarios, there is a further increase in nitrogen transport in rivers. The increase is in particular large under the Global Orchestration and Adapting Mosaic scenarios. Only TechnoGarden shows a decrease in nitrogen transport by rivers. The major drivers of increased nitrogen loading are agriculture and sewerage systems. Nitrogen deposition (from atmospheric emissions) increases less—or is even reduced.

Assuming similar regional patterns of increase among the different scenarios, it can be concluded on the basis of these global differences that the increase in nitrogen inputs to the Indian Ocean and Pacific Ocean will be faster in Global Orchestration than in the AT 2030 projection and slower in the TechnoGarden scenario. The development for the other two scenarios is comparable to that in the AT 2030 scenario. Although there are large uncertainties in such scenarios and there may be important regional differences between them, the global trends and expected changes for oceans and seas as a whole may be more robust.

9.3.8 Disruption of Landscape by Mining and Fossil Fuel Extraction

One factor affecting the degree of disruption of landscape will be the intensity and type of mineral exploitation. The MA scenarios only focused on energy production, but it can be assumed that extraction of other key resources will follow a similar trend. From the scenarios we can deduce that the biggest disruption by far will be caused by Order from Strength, where total fossil fuel use increases by more than a factor of 2.5 by 2100 compared with 2000. Not only is the magnitude of fossil fuel use large, but in this scenario society gives environmental protection low priority. This combination of factors suggests that mineral exploitation will have the largest impacts on the landscape under this scenario.

The Global Orchestration scenario will have the next largest impact, with fossil fuel use increases of about a factor of two over the same period and environmental management also largely neglected. The impact is likely to be the smallest under the TechnoGarden scenario, because fossil fuel use substantially declines up to 2100 and because environmental management is given high priority. An intermediate case is Adapting Mosaic, which also gives priority to environmental protection, but fossil fuel use nearly doubles up to 2100.

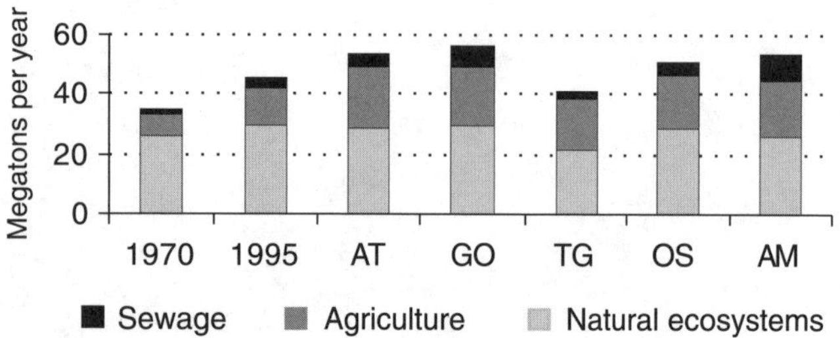

Figure 9.21. Global River Nitrogen Export Stemming from Natural Ecosystems, Agricultural Systems, and Sewage Effluence for 1970 and 1995 with Projections for 2030 and Model Results for MA Scenarios (FAO; IMAGE 2.2)

9.4 Provisioning Ecosystem Services

Provisioning ecosystem services include services that directly produce goods that are consumed by humans. The conceptual framework of the MA lists the following provisioning services:

- food (including a vast range of food products derived from plants, animals, and microbes);
- fiber (including materials such as wood, jute, hemp, silk, and several other products);
- fuel or biofuel (including wood, dung, and other biological material that serves as a source of energy);
- fresh water;
- genetic resources (including the different aspects of genetic information used for animal and plant breeding and biotechnology);
- biochemicals, natural medicines, and pharmaceuticals; and
- ornamental resources.

This section describes some of the possible changes in these services under the four MA scenarios. It focuses on the services where adequate differentiation between the scenarios can be achieved, based on model calculations, qualitative interpretation of the scenario storylines (see Chapter 8), assessment of recent literature, and interpretation of changes in possible drivers of these services. The services for which a sufficient assessment can be made include food, fiber, fuel, and fresh water. A short concluding section discusses possible changes for other provisioning services.

Overall, considerable differences in the pressure on ecosystems to produce provisioning ecosystem services can be found across the scenarios. An important factor here is that the (strongly) increasing demand for provisioning services is driven by population growth, economic growth, and consumption changes. Increases in demand are on a global scale particularly large in Global Orchestration (with increased welfare being an important driver), but also in Order from Strength (with lower welfare but higher population growth). Increases in demand for services are partly offset by increases in the efficiency at which these services are provided (for example, agricultural yields). However, they will also lead to increased pressures on the ecosystems that are providing these services or on service quality.

Another important factor in the relationship between provision of ecosystem services and pressure on ecosystems is the human attitude toward ecosystem management. Under Global Orchestration, ecosystem management is reactive, driven primarily by response to environmental crises. Consequently, vulnerability of provisioning ecosystem services grows as demands on ecosystems grow due to population growth, economic expansion, and other factors. In Order from Strength, vulnerabilities of provisioning ecosystem services also increase. In wealthy countries, ecosystem services are vulnerable because of the vulnerability of small patches to disturbance and climate change. In poorer countries, services are vulnerable due to these same factors, exacerbated by overexploitation, degradation of ecosystems, and expanding poverty.

In Adapting Mosaic, ecosystem management is often directed at reducing vulnerability; in many regions, decentralization and a focus on adaptive change allow ecosystem services to adjust smoothly to changes in climate and other environmental drivers. In TechnoGarden, the emphasis is on efficiency, which often increases the supply of provisioning ecosystem services at the cost of also increasing their vulnerability. The drive for efficiency leads to dependence on a narrow range of production systems that successfully produce high levels of ecosystem services but are vulnerable to unexpected change. Thus there is a generally high performance of provisioning ecosystem services, but with some surprising, dangerous disruptions that are difficult to repair.

9.4.1 Food

9.4.1.1 Methodology and Assumptions

The ecosystem services provided by agriculture are assessed in two ways: the services delivered by agriculture, using total food production as a measure of the services, and the services "delivered" to each person or the outcomes of these services, using per capita food availability and the number of malnourished children as a measure. Both are equally important: total food production is related to the amount of agricultural land, water, and other ecosystem resources required to deliver food services, while per capita food consumption and kilocalorie availability establishes a connection between ecosystem services and human well-being.

9.4.1.2 Comparison of Food Production among Scenarios

Various factors determine the global and regional food production in the MA scenarios. The most general drivers have been discussed in sections 9.2 and 9.3. Some more-specific drivers are discussed in Appendix 9.1. In addition, Appendix 9.2 indicates some of the assumptions of modeling dietary preferences and yield increases. All four MA scenarios result in increased global food production, both total and per capita, by 2050 compared with the base year. (See Figures 9.22 to 9.25.) Yet different means are used to achieve production increases, and—most important—outcomes vary for the food-insecure.

Under Global Orchestration, rapid income growth in all countries, increasing trade liberalization, and urbanization fuel growth in food demand. Global cereal and meat demand grow fastest among the four scenarios, with cereals being used increasingly as livestock feed. Grain production growth is driven by growth in yield as a result of large investments in the areas of agricultural research and supporting infrastructure, making large crop area expansion unnecessary; rapid growth in food demand is also met through increased trade. By 2050, international food prices are lower for livestock products and rice, whereas pressure on maize from demands for animal feed and wheat as a direct food item leads to increased prices for these commodities. Per capita calorie availability under this scenario in 2050 is highest among the four scenarios, and the number of malnourished children drops to just under 40% of current levels.

Under Order from Strength, economic growth in wealthy countries is somewhat lower and in poorer countries is much reduced, protectionist trade policies prevail, and total population in 2050 is highest among the four scenarios. Per capita food availability in 2050 is also higher, on average, but reaches only 83% of Global Orchestration levels. Moreover, production growth is achieved through significant expansion in crop-harvested area, as reduced investments in yield improvement are insufficient to keep up with demand levels. A second reason for crop area expansion lies in remaining trade protection levels , such as import tariffs and quotas, or trade-distorting subsidies, implemented by trading partners, which increase the cost of procuring food, particularly for poor people in low-income countries, at the same time that elites in wealthy as well as poorer countries continue to expand and diversify their diets.

As food production levels cannot keep pace with (albeit somewhat depressed) food demand, international food prices for major crops increase significantly. (Depressed livestock demand from slow income growth results in reduced livestock prices, on the other hand.) As high levels of crop prices surpass the cost of protection, food-deficit countries resort to food imports. As a result, trade levels are not much reduced under the Order from Strength scenario, compared with Global Orchestration, but the cost of procuring food is much higher. Calorie consumption levels improve only very slowly up to 2050, and the number of malnourished children by 2050 under Order from Strength is the highest among the four scenarios.

Under the TechnoGarden scenario, growing incomes in all countries are combined with medium-level population growth, increasing trade liberalization, and a drive for innovations in all sectors, including food production. TechnoGarden operates somewhat similarly to Global Orchestration, with substantial improvements in crop yields but here combined with a lower preference for meaty diets, both of which reduce pressure on crop area expansion. Increased food demand is also met through exchange of goods and technologies. Both calorie consumption levels and the reduction in the number of malnourished children are similar

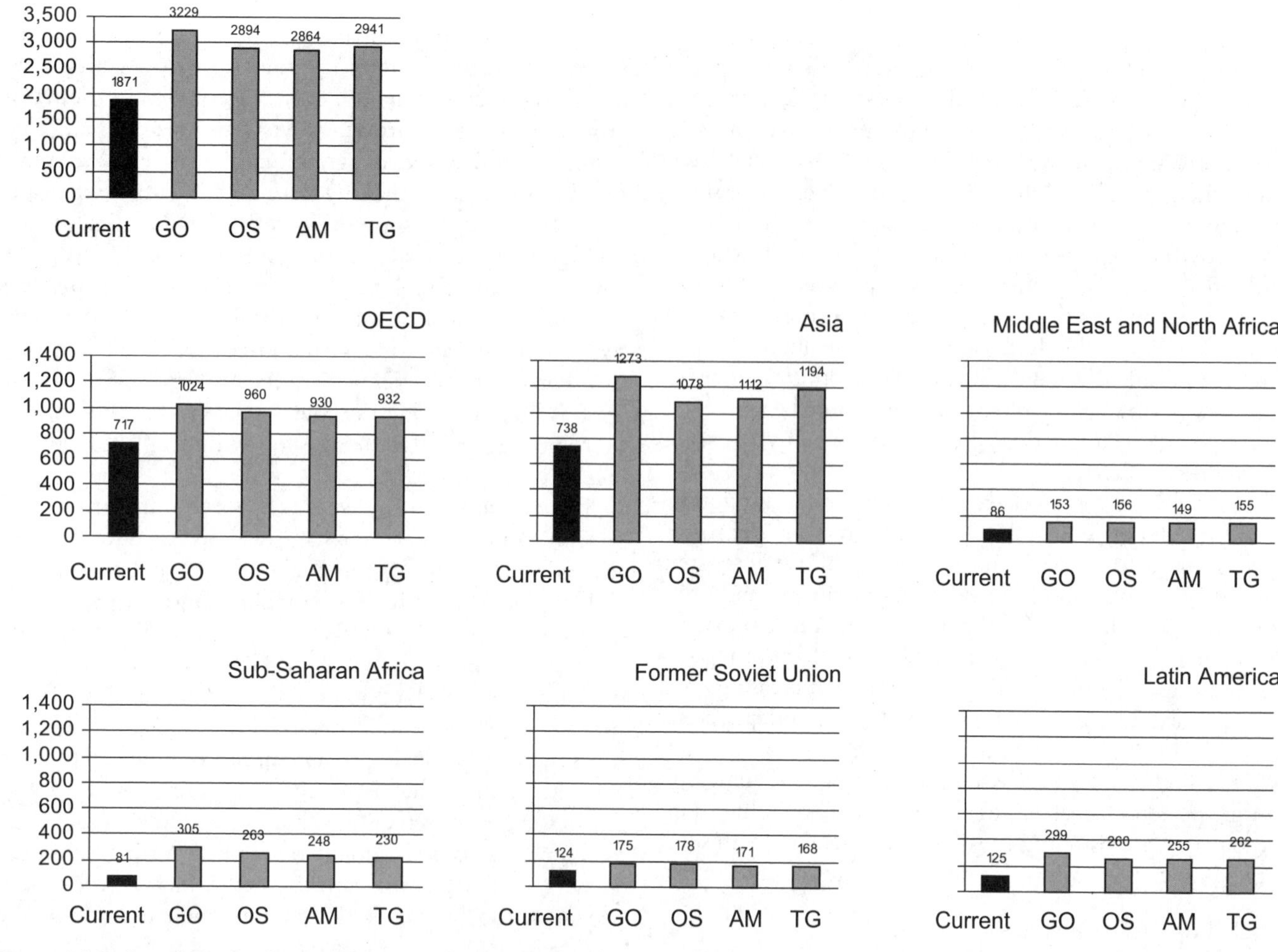

Figure 9.22. Cereal Production by World Region in MA Scenarios in 2050. Scenario Names: GO: Global Orchestration; OS: Order from Strength: AM: Adapting Mosaic; TG: TechnoGarden. (IMPACT)

to, albeit somewhat lower than, the Global Orchestration scenario.

Finally, under the Adapting Mosaic scenario, the focus is on the adaptation of local approaches to the improvement of ecosystem services. Incomes grow slowly, while populations continue to grow steadily up to 2050. Food production outcomes are achieved in ways similar to the Order from Strength scenario. Food is produced locally, on expanded crop areas, with little attention to yield growth; but expansion is insufficient to meet effective demand at current price levels in many areas of poorer countries; as a result, pressure on both food prices and demand for net imports increases. While calorie availability improves only very slowly, the number of malnourished children is reduced slightly more than under Order from Strength due to a focus on social investments under this scenario.

9.4.1.2.1 Food supply and demand to 2050

Under the Global Orchestration scenario, demand for food crops (including cereals, roots and tubers, soybean, sugar crops, vegetables, and fruit crops) is projected to increase by 3,321 megatons to 7,227 megatons in 2050; cereal production alone is expected to increase by 73% by 2050, the largest increase across the four scenarios, while global demand for livestock products is expected to grow by 357 megatons or 63%. Globally, average per capita demand for cereals as food is projected to increase slightly by 10 kilograms to reach 172 kilograms in 2050. While the relatively low OECD cereal consumption levels indicate highly diversified diets, Asia's and MENA's much higher cereal consumption levels are characteristic of far less diversified diets. In sub-Saharan Africa, low cereal consumption levels indicate gaps in food availability rather than diets diversified away from staple cereals, but also reflect the more root- and tuber-oriented diets of the region.

Per capita demand for livestock products is likely (*with high certainty*) to increase much more rapidly worldwide, driven by strong income growth and increasing preference for livestock products. Globally, annual per capita consumption is expected to increase from 36 kilograms in 1997 to 70 kilograms by 2050; with large increases in Asia, the former Soviet Union, and OECD. However, sub-Saharan Africa and MENA are unlikely to experience significant increases in per capita meat consumption, reaching levels of only 27 and 34 kilograms, respectively, by 2050.

Under the Order from Strength scenario, assumptions about how the world will respond to growing food produc-

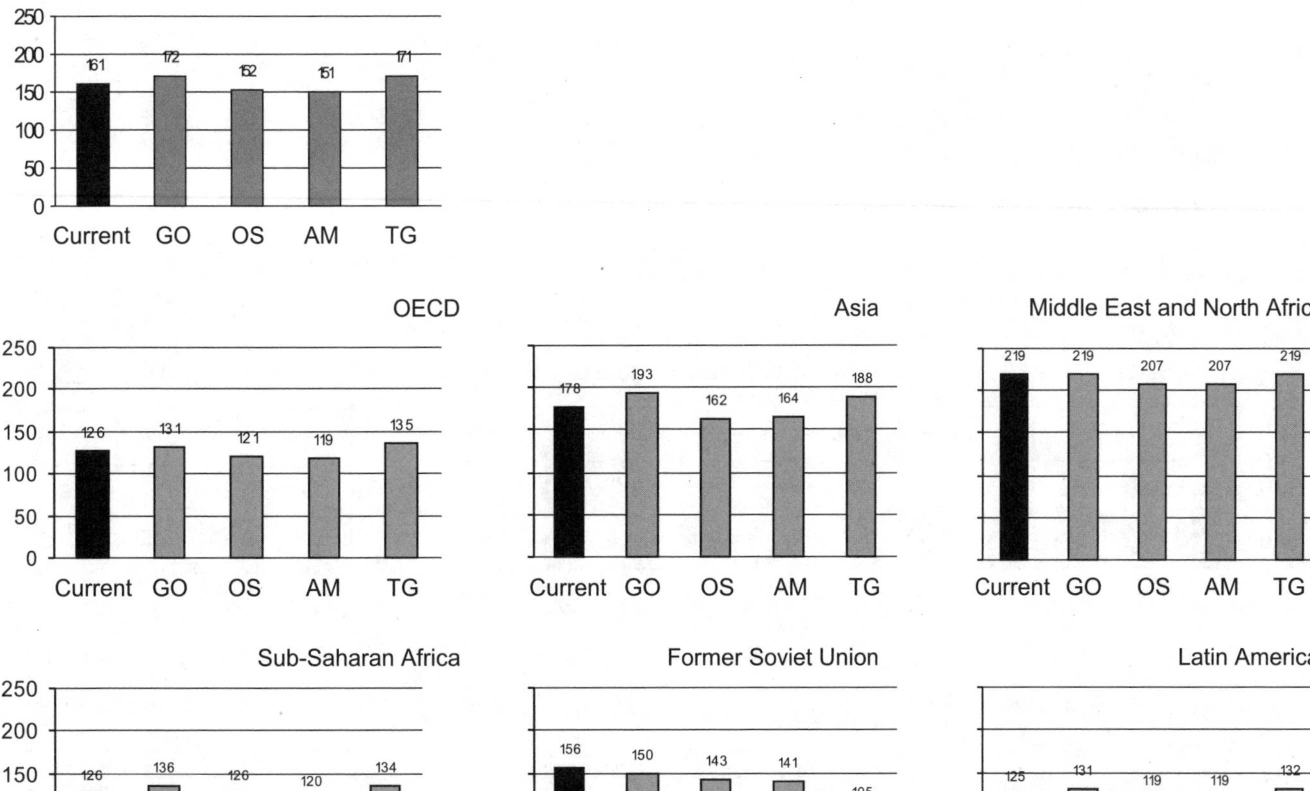

Figure 9.23. Cereal Consumption (as Food) by World Region in MA Scenarios in 2050. Scenario Names: GO: Global Orchestration; OS: Order from Strength; AM: Adapting Mosaic; TG: TechnoGarden. (IMPACT)

tion challenges play out in depressed demand for meat and grains in poorer countries. Food production in all categories increases substantially compared with today, but grain and particularly meat production levels are far below those achieved under Global Orchestration. By 2050, sub-Saharan Africa only achieves 20% of OECD's 1997 meat consumption level of 88 kilograms, and even Latin America, a region well known for meat consumption with 54 kilograms per capita in 1997, only increases consumption levels to 65 kilograms by 2050. Moreover, global average per capita cereal consumption as food declines by 10 kilograms by 2050.

Under the TechnoGarden scenario, total food crop demand increases by 3,017 megatons up to 2050; cereal demand goes up by 1,070 megatons; and meat demand, by 166 megatons. Per capita cereal as food demand is expected to increase by 9 kilograms overall, with the largest increases in South Asia (23 kilograms), the OECD, and the former Soviet regions (10 kilograms). Preference for meat products is lower under TechnoGarden than under Global Orchestration. As a result, per capita demand for livestock products grows by only 6 kilograms globally during 1997–2050. The increase is largest in Asia, at 12 kilograms, followed by Latin America, with 11 kilograms.

Under the Adapting Mosaic scenario, by 2050 demand for all food products is somewhat depressed as people cannot afford higher-value foods and focus on locally adapted production methods and consumption. Total food crop demand grows by 2,797 megatons to reach 6,704 megatons by 2050. Cereal demand increases by 994 megatons, and demand for meat products grows by 179 megatons. Average per capita cereal food demand decreases by 10 kilograms to 151 kilograms in 2050. The former Soviet region and Asia experience the sharpest declines, at 15 kilograms and 14 kilograms. Per capita meat consumption levels under Adapting Mosaic only increase significantly for the OECD region, by 24 kilograms.

9.4.1.2.2 Extensification versus intensification of agriculture

Under the Global Orchestration scenario, the rapid rate of technology development and investments in agricultural research will lead to substantial yield increases, rendering large expansion in new crop areas unnecessary. (See Figure 9.26.) Globally, harvested area for grains is projected to expand at 0.01% annually from 1997 to 2050 and then to contract at 0.28% annually up to 2100. Only in sub-Saharan Africa will a large expansion of cropland be necessary for increasing production.

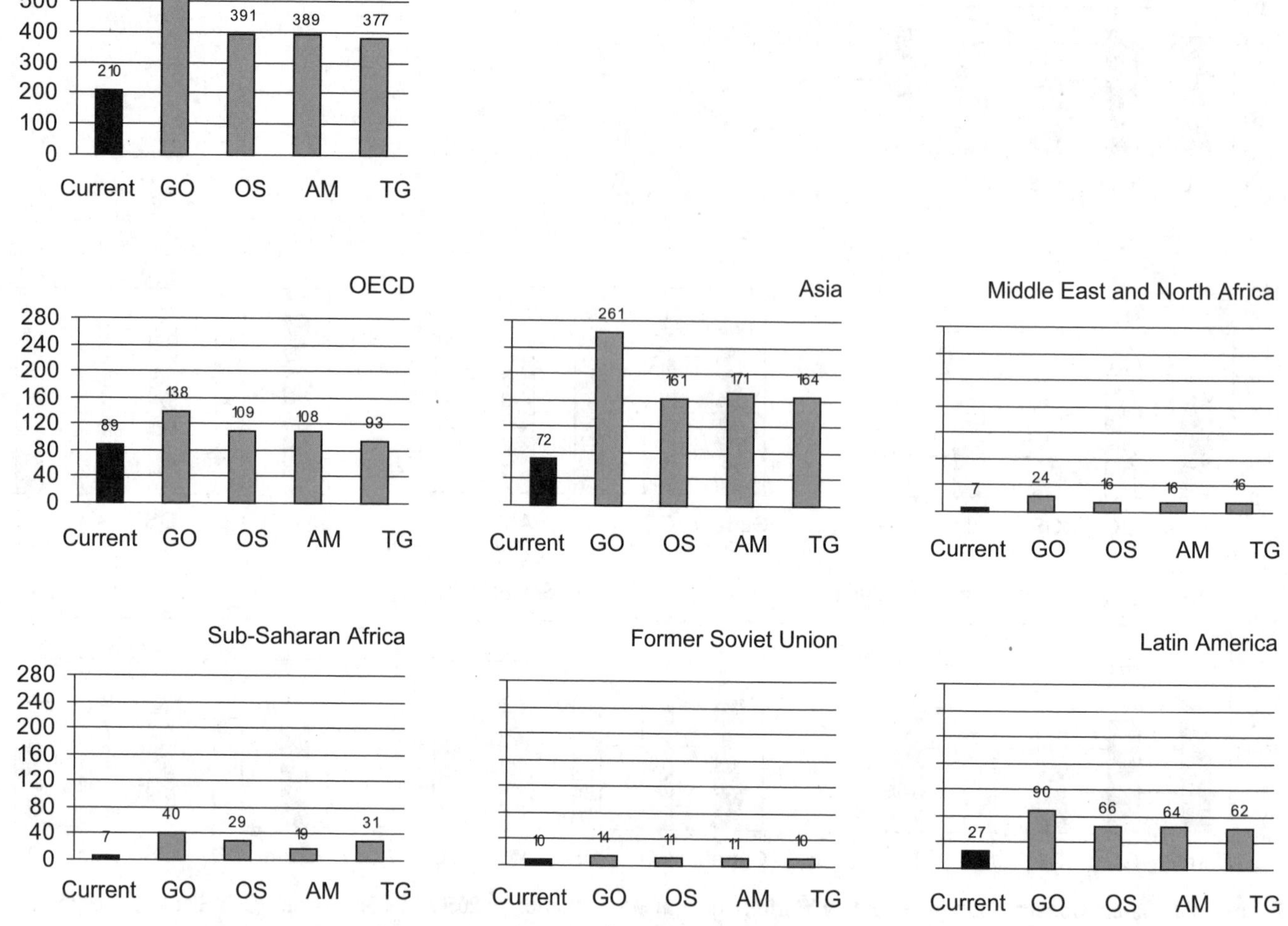

Figure 9.24. Meat Production by World Region in MA Scenarios in 2050. Scenario Names: GO: Global Orchestration; OS: Order from Strength; AM: Adapting Mosaic; TG: TechnoGarden. (IMPACT)

Although total cropland will not greatly expand, much more of it will be irrigated in 2050 then is today. Irrigated area will grow under Global Orchestration from 239 million to 262 million hectares (the largest increase among all four MA scenarios) spurred by large investments in irrigation systems. (See Figure 9.27.) The growth of irrigation is one of the main factors explaining productivity increases. The growth rate differs among the four MA scenarios based on the quantification of the storylines. (See Box 9.1.) Furthermore, total agricultural land will grow because of the demand for pastureland and biofuels, as described earlier.

Under Order from Strength, society invests relatively little in crop technology and supporting infrastructure. As a result, expansion in area will need to carry the brunt of food supply increases. Globally, crop area is projected to increase by 137 million hectares to reach 823 million hectares to supply future food needs, equivalent to an annual rate of 0.34%, before slowing to 0.25% per year from 2050 to 2100. Area expansion for cereals will be spread out among the poorer regions, with Latin America, sub-Saharan Africa, and MENA all experiencing harvested area expansion in the order of more than 40%. Expansion will be slightly lower in the former soviet and OECD regions and lowest in Asia. At the same time, irrigated area is expected to contract by 1 million hectares from 1997 to 2050 and a further 7 million hectares from 2050 to 2100, with area declines in Asia and the former Soviet Union more than offsetting net increases in the other regions.

The TechnoGarden scenario, characterized by innovations in agricultural technology and crop productivity but also less meat-based diets, requires even less area expansion than Global Orchestration. Up to 2050, irrigated area grows substantially, but less so than in Global Orchestration. In later periods, growth in irrigated area slows considerably due to slowing pressure on ecosystem services and food production. Globally, cereal harvested area contracts by 0.01% annually from 1997 to 2050 and a further 0.14% annually from 2050 to 2100, to 637 million hectares. However, total food crop area is expected to increase by 0.11% annually during 1997–2050 before contracting by a similar rate from 2050 to 2100. Although most regions will achieve production growth by intensification of existing cropland, expansion of cultivated land will still be important in sub-Saharan Africa, (accounting for 30% of total production growth up to 2050) and Latin America and MENA (accounting for about 11% of total production growth).

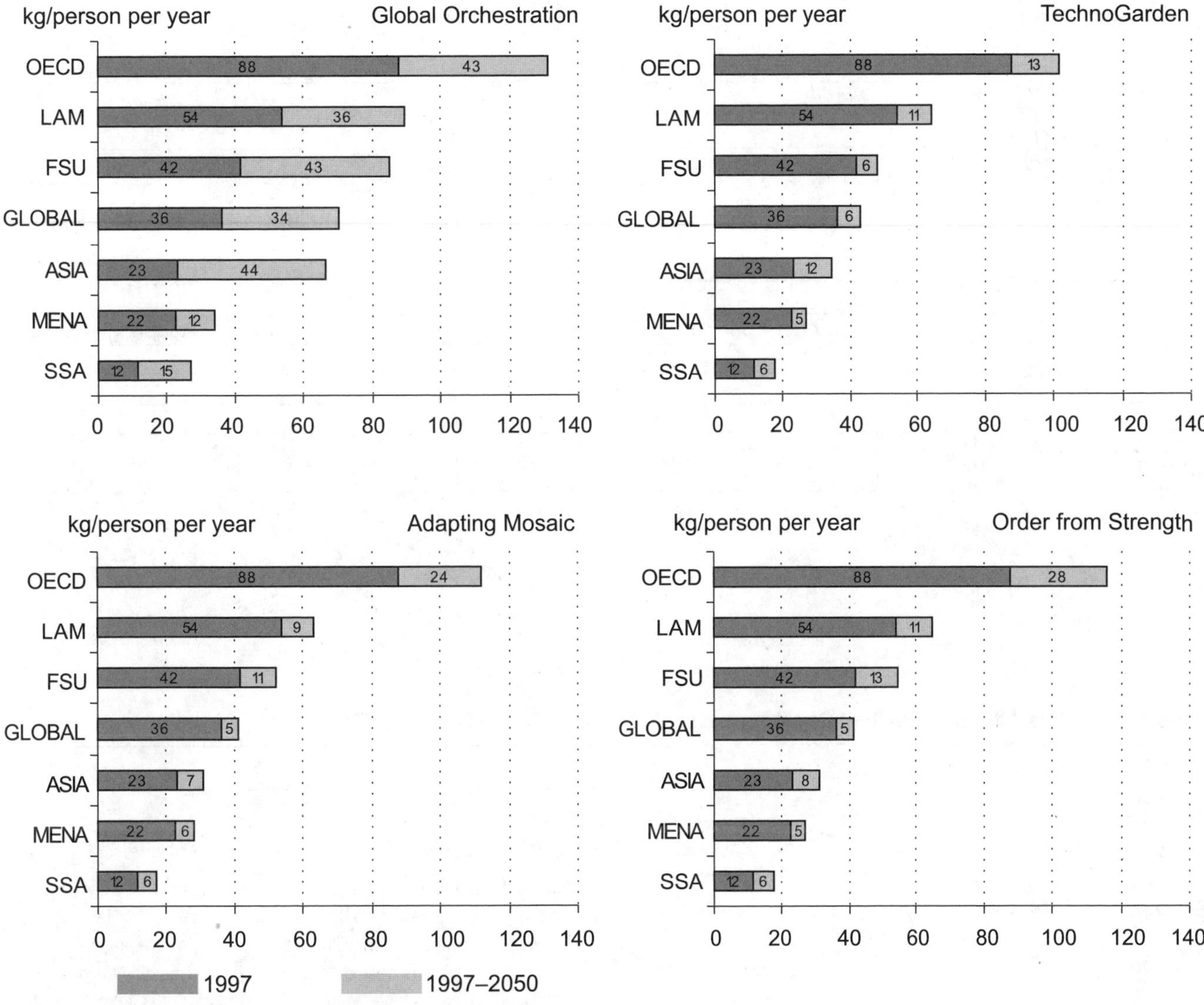

Figure 9.25. Meat Consumption by World Region in MA Scenarios in 2050. World regions: LAM: Latin America; FSU: Former Soviet Union; MENA: Middle East and Northern Africa; SSA: Sub-Saharan Africa. (IMPACT)

The Adapting Mosaic scenario postulates a combination of slow growth in food demand, low investments in food production technologies, and no breakthroughs in yield-enhancing technologies. Globally, irrigated area is expected to grow very slowly up to 2050, and then decline slightly. It will increase however in sub-Saharan Africa and Latin America. Depressed food demand under the Adapting Mosaic scenario will not be able to compensate for stagnant crop yields. As a result, crop harvested area is expected to increase at 0.16% per year for cereals and at 0.23% annually for all food crops, from 1997 to 2050, before contracting at −0.06% and −0.04% annually, respectively. Similar to the other MA scenarios, most cereal harvested area will be added in sub-Saharan Africa, at 39 million hectares, followed by Latin America (10 million hectares) and MENA (7 million hectares).

9.4.1.2.3 The potential impact of climate change on future agricultural yields

The impacts of climate change on crop yields have been assessed by IPCC (2001) in its Third Assessment Report. In fact, two combined effects have to be accounted for: the impacts of climate change and those of a rising atmospheric CO_2 concentration. The latter (also referred to as carbon fertilization) can increase yields and make plants more stress-resistant against warmer temperatures and drought. Climate change can lead to both increases and decreases in yields, depending on the location of changes of temperature and precipitation (climate patterns) and the crop type. IPCC concluded, with medium confidence, that a few degrees of projected warming will lead to general increases in temperate crop yields, with some regional variation. At high amounts of projected warming, however, most temperate crop yield responses could become negative. In the tropics, where some crops are already near their maximum temperature tolerance, yields could, depending on the region and the exact pattern of climate change, become adversely affected. Adaptation could mitigate these impacts.

Studies indicate that taking into account the carbon fertilization impact could be very important for final outcomes. For example, from studies in Montana in the United States it was concluded that climate change only decreases

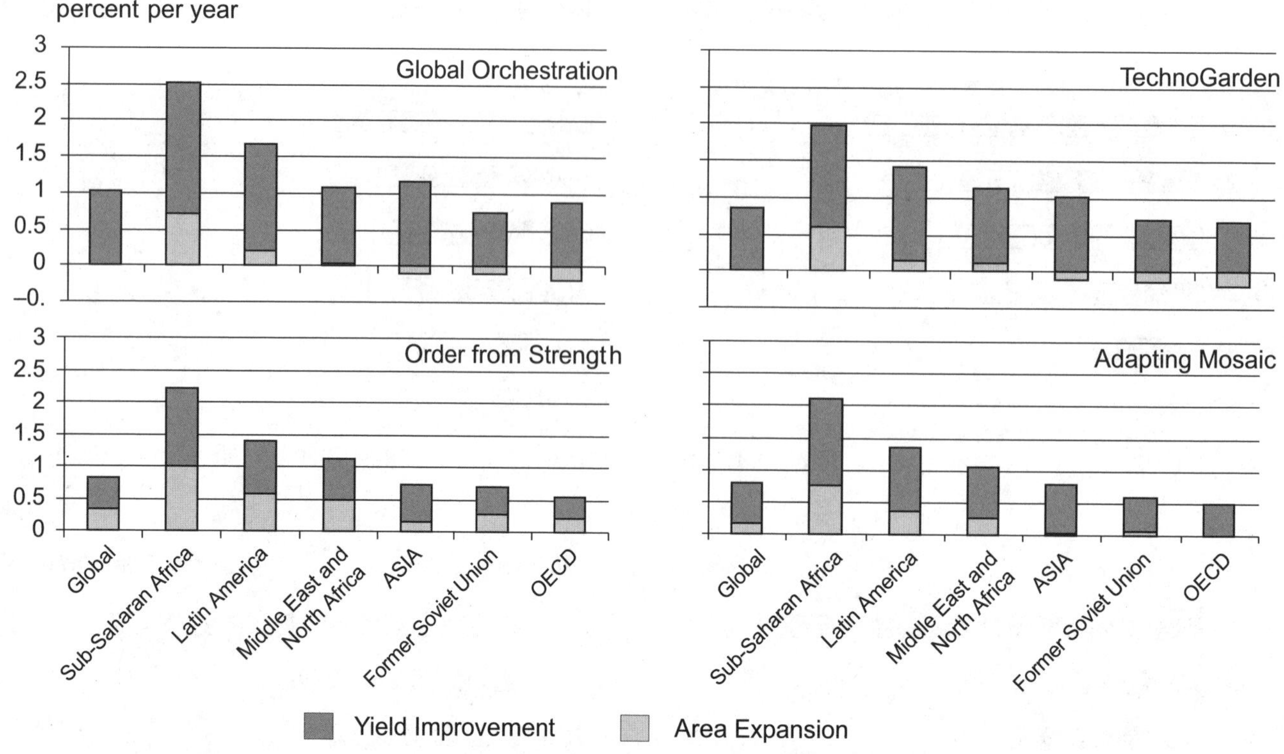

Figure 9.26. Factors Affecting Growth of Cereal Production in MA Scenarios, 1997–2050 (IMPACT)

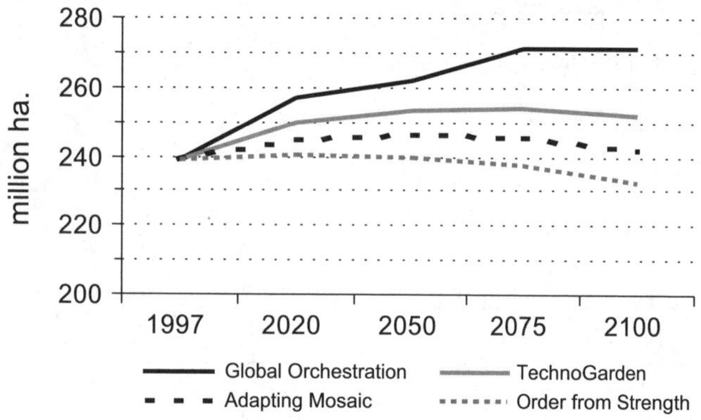

Figure 9.27. Global Irrigated Area in MA Scenarios, 1997–2100 (IMPACT)

BOX 9.1

Rate of Irrigated Area Growth among Scenarios

Among the four scenarios, effective growth in irrigated area is largest for Global Orchestration, at 0.18 percent per year during 1997–2050, followed by TechnoGarden, with growth of 0.11 percent annually.

Annual growth in irrigated area is much lower under Adapting Mosaic, at 0.06 percent per year, and under the Order from Strength scenario (0.01 percent annually). During 2050–2100, irrigated area declines under all but the Global Orchestration scenario.

Up to 2050, under Global Orchestration area expands most rapidly in Latin America, at 0.5 percent per year, followed by sub-Saharan Africa, at 0.3 percent annually. In the TechnoGarden scenario, growth in Latin America, sub-Saharan Africa, West Asia, and North Africa is similar to Global Orchestration, but growth slows in Asia, the former Soviet Union, and the OECD region.

Under the Adapting Mosaic scenario, area actually contracts in the former Soviet Union at 0.1 percent per year (by 0.8 million hectares) and in Asia at 0.03 percent per year (by 2.3 million hectares). Growth in sub-Saharan Africa and Latin America remains strong, however. Finally, under the Order from Strength scenario, irrigated area reductions in Asia and the former Soviet Union are even larger, at 3.3 million and 1.7 million hectares, respectively, whereas irrigated area growth remains strong in Latin America and sub-Saharan Africa.

the wheat yield from around −50 to −70%, while CO_2 fertilization leads to yield increases of +17 to +55%. The combination of both factors returns changes in wheat yield from −30% to +30% (Antle et al. 1999; Paustian et al. 1999). Parry et al. (1999) also conclude that the changes in crop yield range from negative impacts (–10% in North America, Latin America, Asia, and Africa) to positive impacts (+10% in Latin America) when both climate change and CO_2 fertilization are taken into account.

Because temperature increase enhances photo-respiration in C3 species,[6] such as wheat, rice, and soybean, the positive effects of CO_3 enrichment on photosynthetic productivity usually are greater when temperature rises (Bowes et al. 1996; Casella et al. 1996). However, the grain yield of CO_2-enriched rice shows about a 10% decline for each 1° Celsius rise above 26° Celsius. This decline is caused by a shortening of growth duration. Similar scenarios have been reported for soybean and wheat (Mitchell et al. 1993; Bowes et al. 1996). With rice, the effects of elevated CO_2 on yield may even become negative at extremely high temperatures (above 36.5° Celsius) during flowering (Horie et al. 2000).

For the MA scenarios, we have used the calculations of the IMAGE model to assess the impacts of yields. These impacts are fully included in the land use and food results shown in this chapter. It should be noted that the regional impacts are very uncertain, as the patterns of climate change are uncertain. Some signals, however, are visible across most models. In Figure 9.28 (see Appendix A), this is shown for selected regions in the Order from Strength scenario (the other scenarios show, in general, a smaller climate change impact). Regions that are positively affected in terms of yield changes include the United States and the former Soviet Union. In other regions, however, the impacts will clearly be negative—including, in particular, South Asia (that is, India), which in turn will have a strong negative impact on two very important crop types, rice and temperate cereals. On top of already existing difficulties feeding growing populations, this type of stress could have significant consequences. Other negatively affected regions under this climate change pattern include OECD Europe and Japan. These impacts are taken into account in the production levels discussed in this chapter.

9.4.1.2.4 The role of trade and international food prices

Under the Global Orchestration scenario, trade liberalization and economic opening helps fuel rapid increases in food trade. Total trade in grain and livestock products increases from 196 megatons to 670 megatons by 2050, the largest increase among the MA scenarios. (See Figure 9.29.) Net grain trade increases more than 200% from 1997 to 2050. The OECD region, in particular, responds to the increasing cereal demands in Asia and MENA with an increase in net cereal exports of 89 megatons. Moreover, the very rapid yield and area increases projected for the sub-Saharan Africa region turn the region from net cereal importer at present to net grain exporter by 2050. Net trade in meat products increases 674%, albeit from presently low levels. Net exports will increase particularly in Latin America, by 23 megatons, while the OECD region and Asia are projected to increase net imports by 15 megatons and 10 megatons, respectively.

Large investments in agricultural research and infrastructure, particularly in poorer countries, help bring down international food prices for livestock products and rice. Over the 1997–2050 period, livestock prices decline by 9–13% and rice prices drop by 31%, whereas maize and wheat prices increase by 14% and 39%, respectively, because of demand for animal feed. (See Figure 9.30.)

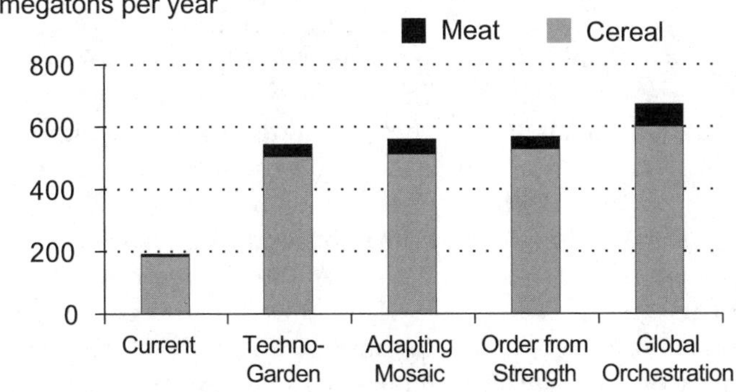

Figure 9.29. International Trade in Cereals and Meat Production in MA Scenarios in 2050 (IMPACT)

Under Order from Strength, countries maintain current protection levels. At the same time, food production stalls because of low investments in technology and infrastructure, and this puts pressure on countries to import food. Finally, low income growth dampens food demand somewhat in poorer countries. Hence, even though this is a scenario in which trade is not encouraged, total trade in food commodities more than doubles relative to 1997. The combination of population growth and lagging food production leads Asia to import from the OECD, despite existing barriers to trade. Meanwhile, sub-Saharan Africa is a net importer, albeit at reduced levels, due to higher costs of trading and depressed demand. Net trade in meat products is much lower than in the Global Orchestration scenario, reaching 41 megatons by 2050, but most trade is carried out intra-regionally.

Depressed demand from lower income levels cannot compensate for even lower investments in food production and supporting infrastructure and for high population growth. As a result, prices for all cereals are projected to increase over the coming decades: with price increases ranging from 19% (maize) to 46% (rice). Meat prices, on the other hand, continue to decline by 3–12%.

Under the TechnoGarden scenario, trade liberalization continues apace. Pressure on trade is somewhat reduced due to the preference for a diet with less meat, relatively good production conditions in the various countries and regions, and somewhat lower income growth than in Global Orchestration. Total trade for grains and meat products grows to 543 megatons by 2050. Net cereal trade is dominated by Asian net imports of 124 megatons and OECD net exports of 159 megatons, followed by net imports in MENA of 70 megatons. Net meat trade is dominated by net imports in the OECD region (17 megatons in 2050), supplied through net exports from Latin America, sub-Saharan Africa, and Asia. Growth in production and trade will more than compensate for increased demand, resulting in declines for international food prices across the board. By 2050, prices for wheat, rice, and maize are projected to decline by 11–26% and prices for beef, pork, and poultry by 6–23%.

Under Adapting Mosaic, the focus is on local food production and conservation strategies, with limited exchange of goods and services. Low income growth depresses food demand, but the large increase in population puts upward pressure on food, which is being produced without technological breakthroughs or enhancements due to lack of investment in this area. Total grain and meat trade increases to 560 megatons by 2050. Cereal trade increases by 175% over 1997 levels, most of which is accounted for by increased net imports in Asia and MENA and increased net exports of the OECD region. Similarly to the other MA scenarios, the former Soviet Union can improve its net export position. Appropriate technologies and conservation strategies help sub-Saharan Africa become a small net cereal exporter by 2050. Total net meat trade increases by 31

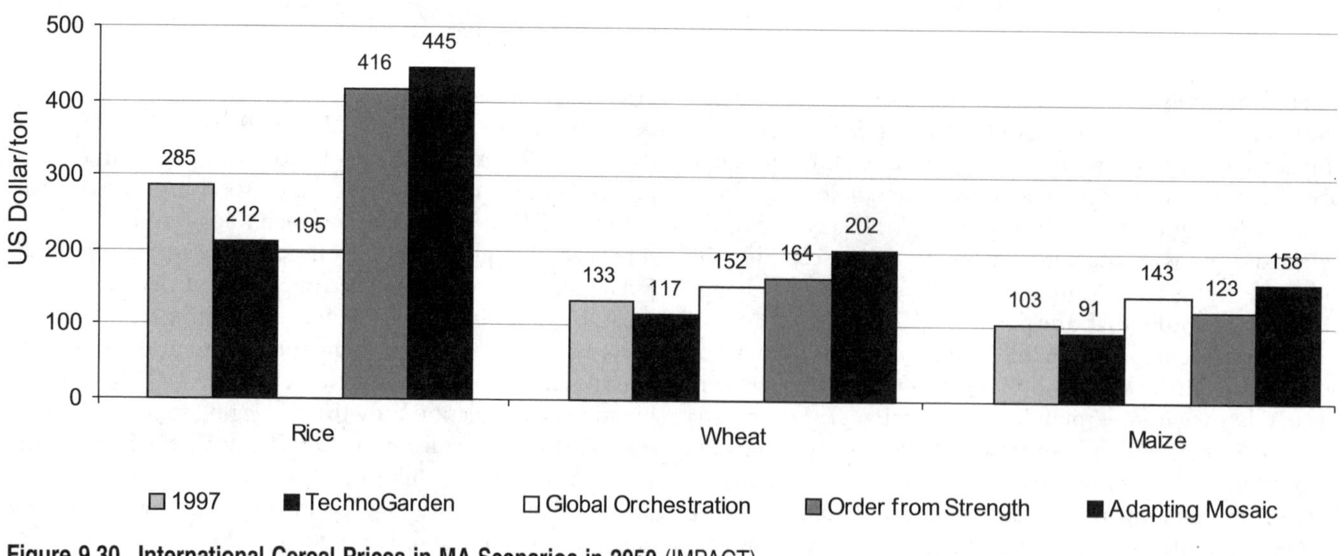

Figure 9.30. International Cereal Prices in MA Scenarios in 2050 (IMPACT)

megatons, the smallest increase among the MA scenarios. By 2050, Asia is projected to supply about 20 megatons of livestock products to all other regions except Latin America.

Insufficient food production causes international cereal prices to increase by 52–56% for wheat, maize, and rice, whereas livestock prices decline by 2% (beef and pork) and 15% (poultry).

9.4.1.2.5 Outcomes for calorie availability and child malnutrition

Although total food production levels by 2050 are similar across scenarios, outcomes for calorie availability and child malnutrition levels in poorer countries vary considerably. The increase in global average caloric availability is largest under Global Orchestration, at 818 kilocalories per capita per day between 1997 and 2050, followed by Techno-Garden at 507 kcal/cap/day, whereas increases are only 207 kcal/cap/day and 250 kcal/cap/day, respectively, under the Adapting Mosaic and Order from Strength scenarios. Under Global Orchestration, all regions experience large increases in calorie availability, led by Asia with an increase in 1,035 kcal/cap/day.

Under the Order from Strength scenario, on the other hand, the increase in per capita calories is largest in the OECD region, at 616 kcal/cap/day. Under TechnoGarden, caloric availability in all regions but sub-Saharan Africa surpasses 3,000 kilocalories. Under Adapting Mosaic, increases in calorie availability are very low. Similar to the Order from Strength scenario, by 2050 improvements in caloric availability in sub-Saharan Africa and Asia, particularly South Asia, are slow. The kilocalorie availability remains particularly low in sub-Saharan Africa, at less than 2,500 kcal/cap/day, and only reaches 3,000 kcal/cap/day in two regions, Latin America and MENA.

Food consumption together with the quality of maternal and child care and of health and sanitation are important determinants for child malnutrition outcomes. Three out of the four MA scenarios result in reduced child malnutrition by 2050. Under the Order from Strength scenario, there are 18 million more malnourished children in 2050 than in 1997. (See Figure 9.31.)

Today, South Asia accounts for slightly more than half of all malnourished children in developing countries, followed by sub-Saharan Africa, home to 20% of all malnourished children. Under the Global Orchestration scenario, the number of malnourished children is projected to decline by 50 million children in South Asia and by 15 million in sub-Saharan Africa. Under Order from Strength, on the other hand, the number of malnourished children is projected to increase by 18 million in sub-Saharan Africa and by 6 million children in South Asia as a result of depressed food supplies, higher food prices, and low investments in social services. Under the Adapting Mosaic scenario, the number of malnourished children would still increase by 6 million children in sub-Saharan Africa, but it would decline by 14 million in South Asia. Finally, under TechnoGarden, the number of malnourished children declines by 5 million in sub-Saharan Africa and by 32 million in South Asia.

9.4.2 Fish for Food Consumption

9.4.2.1 Methodology and Assumptions on Fish Consumption

Aquatic ecosystems of the world provide an important provisioning service in the form of fish and seafood. Fish are an important source of micronutrients, minerals, essential fatty acids, and proteins, making a significant contribution to the diets of many communities. Globally, 1 billion people rely on fish as their main source of animal proteins, and some small island nations depend on fish almost exclusively. Currently, 79% of fish products are harvested from marine sources (FAO 2000).

To assess future production of world fisheries, it is important to understand the different interpretations of trends in the last decades. Since 1970, total fisheries production has more than doubled, to more than 90 million tons, with most of the increase in the last 20 years from aquaculture. Global capture food fisheries, however, have been stagnant at around 60 million tons since the mid-1980s, as most of the world's capture fisheries stocks are fully or overex-

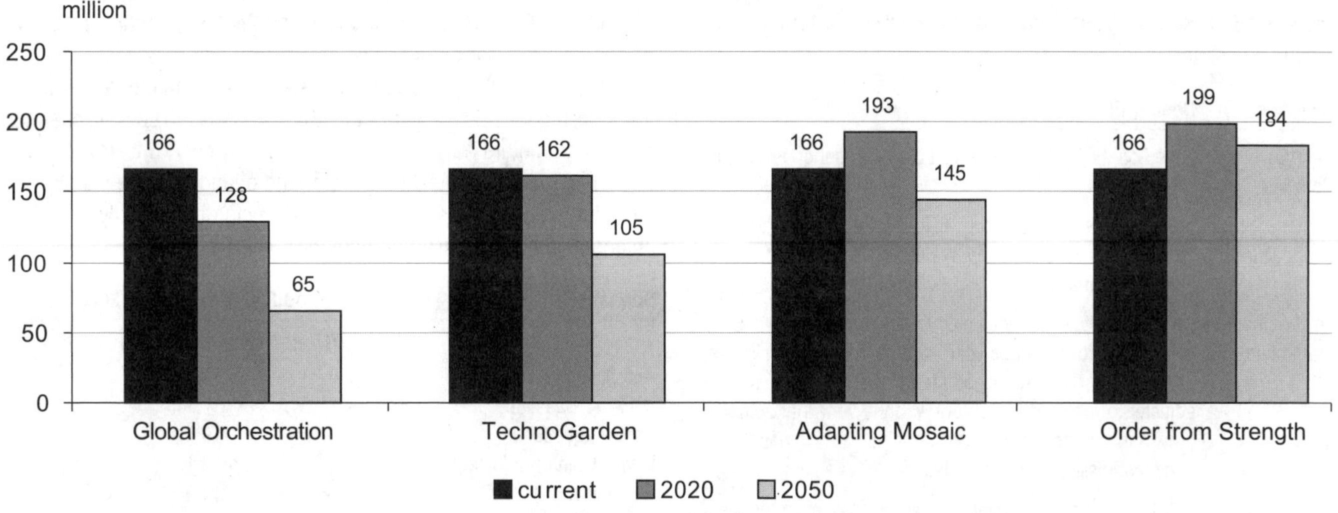

Figure 9.31. Number of Malnourished Children in Developing Countries in MA Scenarios in 2020 and 2050 (IMPACT)

ploited (Delgado et al. 2003; FAO 2000). In fact, some experts indicate that marine capture fisheries might actually have been declining for more than a decade if statistics are corrected for overreporting (Watson and Pauly 2001; Lu 1998). At the same time, developing countries have become major exporters of marine fish products. Developing-country aquaculture production rose from less than 2 million tons in 1973 to more than 25 million tons by 1997.

Yet aquaculture production relies partly on the supply of fish meal and fish oil. While some experts indicate that growth in aquaculture production could lead to greater pressure on stocks of fish used for feed (Naylor et al. 2000), others report instead that efficiencies in the use of fish feeds are improving and that substitute products based on plant matter are being developed (Delgado et al. 2003; Wada, N., personal communication, May 2004). This controversy plays a crucial role in the future of global fisheries. Other adverse impacts of rapid increases in aquaculture production could include the destruction of coastal ecosystems, like mangroves, and increased pollution levels in the form of effluent, chemicals, and escaped farm fish (Goldburg and Triplett 1997). Socioeconomic impacts include loss of property rights and declining incomes for local fishers who rely on capture fisheries (Alder and Watson submitted).

Based on the above, most experts agree that most unmanaged fisheries are near maximum sustainable exploitation levels and that their production will only grow slowly until 2020 (see, for example, FAO 2003 or Delgado et al. 2003) or will even decline (Watson and Pauly 2001). Moreover, there might be important trade-offs between production levels and other ecological services provided by marine ecosystems (such as biodiversity). Large fisheries collapses cannot be ruled out, as observed historically for specific coastal systems (Jackson et al. 2001). As a substitute for marine fishery production, aquaculture (including marine culture) has the greatest potential for satisfying future production increases (FAO 2003). Emerging land, water, and input constraints, however, will place additional pressure on technology to find alternative ways to increase productivity levels.

The forecast of fisheries outcomes needs to be based on stock assessment, on fish population dynamics, on biophysical modeling, on market interactions among producers, consumers, and traders, and on interactions with outcomes for other foods and feeds. Future trends in fish demand have been analyzed from an economic point of view by the International Food Policy Research Institute using the IMPACT model (Delgado et al. 2003, see Chapter 6 for discussion of methodology), as well as from an ecological point of view for various marine fisheries by Pauly et al. (2003) and others. Both perspectives are crucial for an understanding of the prospects of future world fisheries, but as of yet they have not been combined in a single consistent quantitative framework. Hence here we draw on results from both perspectives and combine them in a preliminary way. We first present computed trends of changing fish demand, and then draw on ecological modeling results to assess whether these future demands can be met from the ecological point of view. The assumptions used in the ecological modeling exercise are discussed in Appendix 9.2.

9.4.2.2 Comparison of Fish Consumption among Scenarios

For our estimations of future fish supply and demand we draw on a series of scenarios already available or especially prepared for the MA scenario analysis. (See Table 9.15.) Under the IMPACT baseline scenario, global fish production will increase slightly faster than global population through 2020, to 130 million tons (40% increase) with an increasingly tight supply situation indicated by jumps in real fish prices of 4–16% (livestock product prices, in contrast, are expected to decline). Aquaculture is projected to account for 41% of total production by 2020.

A scenario projecting even faster aquaculture expansion (which might fit well with Global Orchestration and TechnoGarden) suggests that despite short-term tendencies in the opposite direction, over the long run lower food fish prices resulting from more rapid aquaculture expansion could possibly reduce pressure on capture fishing efforts and generally benefit the health of fish stocks (see also Bene et

Table 9.15. Scenario Description for Alternative Outcomes in Future Fisheries

Scenario	Description
IMPACT Baseline	based on IFPRI/WorldFish most plausible set of assumptions
Faster aquaculture	production growth trends for 4 aquaculture output aggregate commodities are increased by 50 percent relative to the baseline scenario
Lower China production	Chinese capture fisheries production is reduced by 4.6 million tons for the base year; income demand elasticities, production growth trends, and feed conversion ratios are adjusted downward, consistent with the view that actual growth in production and consumption over the past two decades is slower than reported
Ecological collapse	a contraction by 1 percent annually in production for all capture fisheries commodities
MA scenarios	based on MA macroeconomic drivers, but without specific changes to fish parameters

Table 9.16. Projected Per Capita Food Fish Production in 2020, Alternative Scenarios. The first four scenarios are based on Delgado et al. 2003. Outcomes for fish supply, demand, and trade are reported for 2020 only. For the MA scenarios, no specific changes to fish parameters have been introduced. However, drivers, such as economic growth, population growth, and changes in the various substitutes and complements of fish products do, indirectly, affect outcomes for fish supply and demand.

Scenario	Food Fish Production, 2020
	(kilograms per person per year)
Actual in 1997 (MA)	15.7
IMPACT Baseline	17.1
Faster aquaculture	19.0
Lower China production	16.1
Ecological collapse	14.2
Global Orchestration	17.3
Order from Strength	14.8
Adapting Mosaic	15.1
TechnoGarden	16.0

al. 2000; Clayton and Gordon 1999; Anderson 1985; Ye and Beddington 1996; Pascoe et al. 1999; and similar supply elasticity assumptions made by Chan et al. 2002). Total food fish production under this scenario is projected to increase to 145 million tons.

Experts do have different visions on whether such production levels are attainable, as it requires considerable technology advances in aquaculture. If the Watson and Pauly (2001) values of overestimated fish catch (mainly in China) are incorporated into IMPACT, then fish production in 2020 would be 7 million tons lower than the IMPACT baseline, and annual per capita consumption would decline by 1–16 kilograms. An additional scenario exploring the outcomes of potential large fisheries collapses results in production declines of 17%, with shortfalls mitigated by production responses to major output price increases of 26–70% in both capture food fisheries and aquaculture. Under this scenario, per capita food fish consumption would drop to 14 kilograms by 2020.

For the MA scenarios, rapid increases in urbanization and income growth result in the highest per capita demand levels for the Global Orchestration scenario (17.3 kilograms), and production of 128 million tons by 2020. Under Order from Strength, on the other hand, rapid population growth combined with slower economic progress result in the lowest production increases, at 117 millions tons in 2020, corresponding with depressed per capita demand levels of 14.8 kilograms. (See Table 9.16.). The values of the other MA scenarios fall between these two extremes.

9.4.2.3 Methodology and Assumptions on Fish Landings

The future of wild capture fisheries depends on several factors, such as changes in average climate and climate variability causing shifts in species distributions and abundance (increase of species at some locations, decline at others), fishing subsidies that will affect the catch at the fisheries level, the danger of overfishing due to the absence or failure of fisheries management, and factors such as population growth and food preferences affecting the demand for marine products. In this section and in Figure 9.32, we describe changes in the fisheries of three marine ecosystems—the Gulf of Thailand, Central North Pacific, and North Benguela—for the four MA scenarios.[7] These case studies were selected because they represent a variety of fishery conditions. The qualitative assumptions for the three case studies under the MA scenarios are summarized in Table 9.17.

9.4.2.4 Comparison of Fish Landings Among Scenarios

All four scenarios maintain the weight and value of current landings in the Gulf of Thailand. However, the consequence of this is a severe decline in the diversity of landings (see Chapter 10), which could increase the vulnerability of the fishery to disease, climate change, and other stresses. In Global Orchestration, the weight of landings (primarily high-valued invertebrates) remains stable until 2010, while profits are increased. The policy focus changes in 2010 to a balance between increasing profits, jobs, and ecosystem structure (where "increasing ecosystem structure" means rebuilding the trophic structure of the fishery). This policy results in a temporary and slight decline in the weight of landings until the system responds and stabilizes to a level similar to the year 2000. In 2030, the policy focus changes to a balance between increasing profits and ecosystem structure resulting in a slight increase in the weight of landings and a substantial increase in their value.

Order from Strength has high weight and value of landings until 2010, when policies are reoriented to optimizing profits and jobs. Under this policy there is a slow and steady increase in the weight of landings and their value. After 2030, the policy is to rebuild demersal species as well as optimizing jobs rather than profits, with the system responding and then stabilizing at a slightly higher level of weight and substantially higher value.

In the TechnoGarden scenario, landings in the Gulf of Thailand are initially stable. In 2010 the weight and value

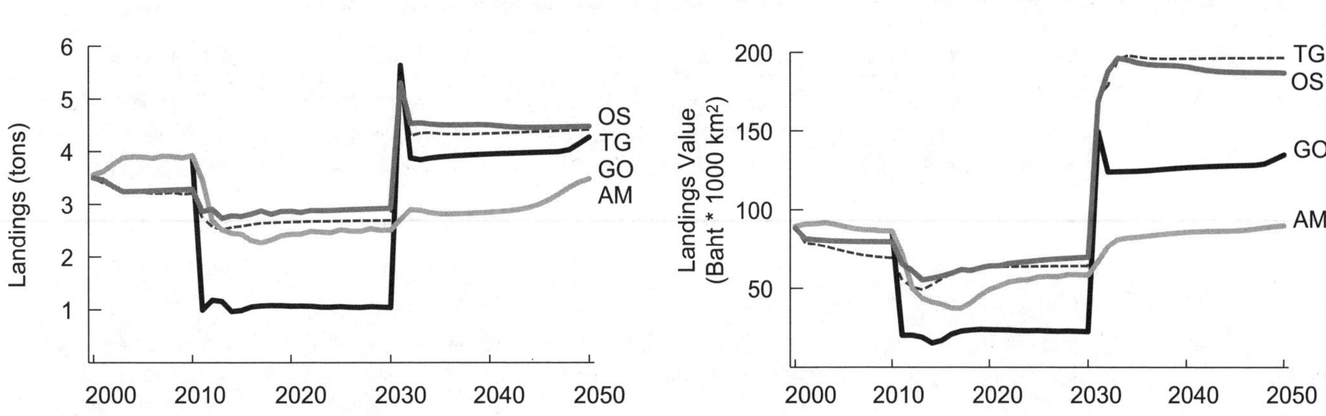

Central North Pacific

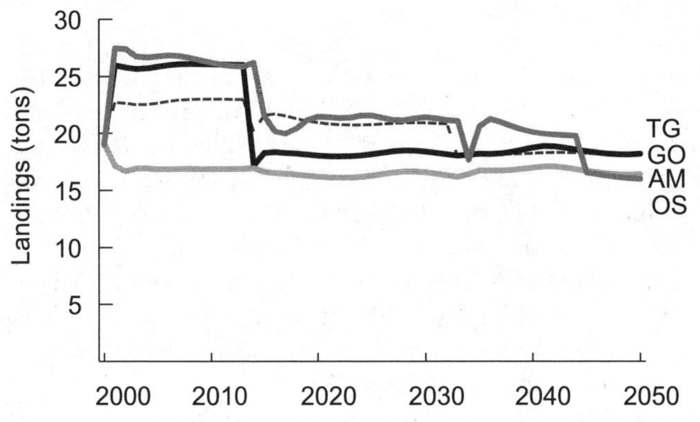

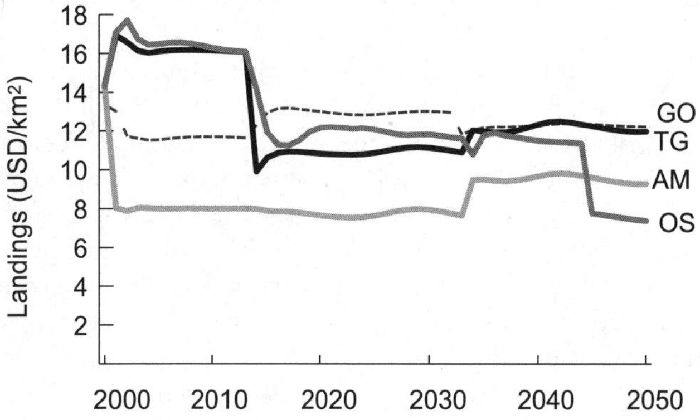

North Benguela

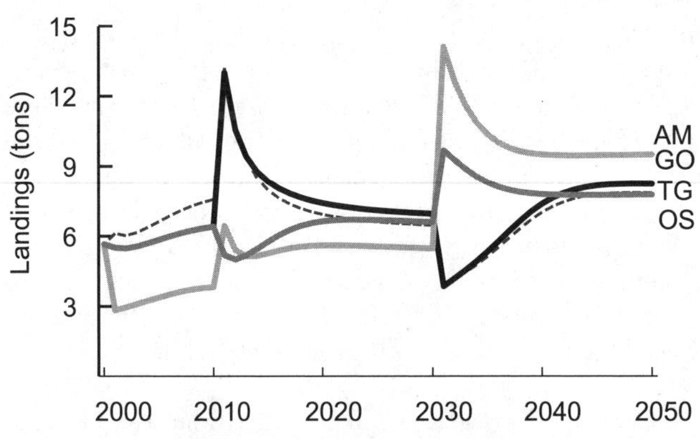

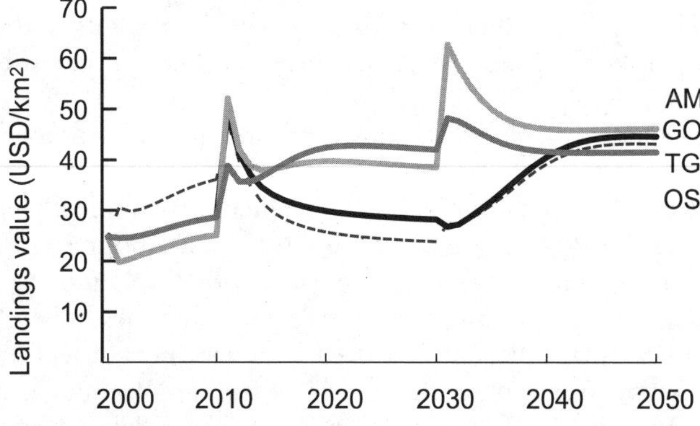

Figure 9.32. Comparison of Fish Landings in Three Specific Regions for MA Scenarios. Scenario names: AM: Adapting Mosaic; GO: Global Orchestration; OS: Order from Strength; TG: TechnoGarden. (Ecopath/Ecosim)

of landings decline until the fish industry is reoriented from optimizing the catch for small pelagic species to support of the growing aquaculture industry. The system responds quickly, and the weight of landings soon stabilizes until the next policy change, which aims to further optimize small pelagics directly or as bycatch from the invertebrate fisheries. The value of the landings has a similar trend.

There is an overall decline in the weight of landings to 25% of the 2000 level in the Adapting Mosaic scenario, but this trend is reversed in 2030 when levels begin to increase. In this scenario, profits are maximized first and jobs have a minor focus. However, by 2010 the policy focus changes to a rebuilding of the ecosystem, including demersal species, and therefore there is a detectable decline in landings and substantial decline in profits. The ecosystem continues to rebuild and demersal stocks and their value increase. However, landings do not recover, and by 2050 they are the lowest level of any scenario.

Table 9.17. Qualitative Assumptions for Case Studies of Regional Marine Fisheries in MA Scenarios (EcoSim/EcoPath)

Assumption	Global Orchestration	Order from Strength	Adapting Mosaic	TechnoGarden
Objective for fisheries production	optimize profits (mostly) and jobs—later in the scenario, also attention to preserving ecosystems	optimize profits; also some attention to ecosystem preservation	mixed focus among profits, jobs, and ecosystems	optimize profits and jobs (more mixed)
Climate change	GoT: medium—high NB: Medium—Low CNP: Low	GoT: med NB: low-med CNP: low	GoT: high NB: low to high CNP: low	GoT: high NB: med CNP: low
Specific additional assumptions	jobs less important in the CNP because much of the fishing is done by distant water fleets	concentration on fishing for fishmeal for aquaculture	assumes that CNP continues to have a significant distant water fleet while GoT and NB fleets are primarily domestic	increased fishing efforts from more distant (high-income countries) that aim for food security

Key: GoT = Gulf of Thailand, NB = North Benguela, and CNP = Central North Pacific.

None of the four scenarios are able to increase or even maintain current landing levels in the Central North Pacific. In Global Orchestration, the weight of landings declines well below the 2000 level and remains low. The value of landings follows a similar trend because fishery policy focuses primarily on profits, followed by jobs, until 2030. After 2030 the focus changes to a balance between maximizing profits and ecosystem structure, which is reflected in the decline in landings in 2030.

Under Order from Strength, the weight and value of landings significantly decline over the scenario period with brief intervals of recovery. The weight of landings declines under TechnoGarden and Adapting Mosaic, but the value of landings increases slightly. The value of landings is maintained under the TechnoGarden scenario because of the development of a highly profitable aquaculture industry that uses fish feed from sources not based on small pelagic fisheries.

The North Benguela ecosystem landings and profits can be maintained under the four scenarios. In the Global Orchestration scenario the North Benguela system initially increases slightly from the 2000 level. When policies shift after 2010 from a balance between profits and jobs to a focus on jobs followed by profits, some fisheries begin to harvest small pelagic fish as feed for a growing aquaculture industry. In response, the weight of landings declines until 2030, when landings again increase to a level slightly higher than in 2000. The value of landings follows a similar trend, with the weight of landings peaking in 2010, declining until 2030, and then increasing again to a higher level in 2050.

Order from Strength results in an overall increase in the weight and value of landings as profits and jobs are maximized. Rebuilding of ecosystem structure starts in 2025 but jobs remain the priority until 2030, when profits are first maximized, followed by jobs.

Under TechnoGarden, landings initially increase as profits and jobs are optimized. When the focus of policy changes to maximizing profits followed by jobs, landings decline to levels slightly higher than in 2000. The trend continues until 2030, when policies are again changed, this time to the simultaneous optimization of profits, jobs, and ecosystem structure. As part of this policy shift, fisheries focus on harvesting small pelagic fish to supply the aquaculture industry. The net effect of this policy shift is a substantial decline in the weight and value of landings. There is an overall increase in landings and value in Adapting Mosaic from 2000, peaking in 2030 but declining to a stable level within about eight years. Initially, the policy focus is on jobs, followed by profits, until 2010.

Very dynamic changes are found for the weight and value of landings for the different scenarios under all case studies. Not only total fish landings change but also the type of species. (See Chapter 10.) Overall, no single scenario was superior in its performance across the three modeled ecosystems for landings or landed value. The pattern that does emerge is that to maintain or improve the provisioning services or the economic value of these three ecosystems, there is a trade-off between the magnitude of production and the diversity of the landings, especially in the Gulf of Thailand. (See Chapter 10.) In some ecosystems, there is a trade-off between increasing the number of landings (food provisioning) and the economic value of the landings (profits), as seen in the Central North Pacific model, where landings declined in the TechnoGarden scenario while profits improved.

The conclusions of these three case studies can be summarized as follows: Policies that focus on maximizing profits do not necessarily maintain diversity or support employment. Similarly, policies that focus on employment do not necessarily maximize profits or maintain ecosystem structures. The diversity of the stocks exploited can be enhanced if policy favors maximizing the ecosystem or rebuilding stocks. Diversity, however, is lost if the sole objective of management is to maintain or increase profits.

9.4.2.5 Comparing Two Approaches to Model Fish Consumption and Production

Two different approaches have been used to explore the possible development of the provisioning service of fish for food consumption under the four MA scenarios—an economic modeling approach and an ecological modeling approach. The two highlight different results. The IMPACT modeling framework shows that total demand for food fish will continue to increase under the MA scenarios. The important question, however, is whether it is feasible to increase production to meet this demand. As shown in our assessment, aquaculture may play an important role here, but it is constrained by its current dependence on marine fish as a major feed source. This dependence must be reduced by advances in, for example, feed efficiency and alternative, plant-based sources of feed.

The ecological modeling of three regional fisheries show that maintaining or increasing current levels of landings will lead to the depletion of predators at the top of the food web, and ecosystems could become dominated by short-lived species at lower trophic levels. This development can compromise the diversity of the ecosystem and make it more vulnerable to external perturbations (such as those stemming from the variability of climate, nutrient availability, or demand). Diversity is also lost if the sole objective of management is to maintain or increase profits. The diversity of the exploited stocks can only be enhanced if policy favors maximizing the ecosystem structure or rebuilding stocks.

9.4.3 Uncertainty of Agricultural Estimates and Ecological Feedbacks to Agriculture

As a whole, the quantified scenarios show a confident picture of the future—both food supply and demand increase into the future along with economic development, while global food trade smoothes out the differences in food-growing ability among nations, assuming that importing countries find the financial resources to do so. But the Order from Strength scenario shows that unfavorable developments in the food sector could threaten this relatively positive global picture. Moreover, the global picture masks significant regional problems. Although average food availability continues to increase, access to sufficient food will continue to remain out of reach for many people in poorer countries, particularly in sub-Saharan Africa and South Asia, leading to a continuing substantial level of child malnutrition in these regions. Moreover, it is uncertain whether the global growth projected in these scenarios is feasible from the standpoint of ecological sustainability.

On the one hand, this confident view of the future is not unlike our experience over the last 100 years, in which food production and consumption have steadily increased as countries have gotten richer, despite temporary setbacks due to political crises, poor planning, or the occasional drought. On the other hand, we should not assume that the global agricultural system will remain as robust as it apparently is now. Several in particular could pose increasing risks to the agricultural production computed in these scenarios.

First, scarcity of water is a concern. Many of the areas where crop and fish production will intensify or expand are also areas currently in the "severe water stress category," as described later, and are expected to have an increasing level of water stress across all scenarios (such as the Middle East, sub-Saharan Africa, parts of China, and India). This is particularly important because irrigation will continue to play an important role in the agriculture of these regions. It is also shown that wastewater discharges are likely to double over much of this area, also endangering the source of freshwater fish not coming from aquaculture. Unfortunately, the model results for agricultural production do not take water scarcity into account in their calculations. While solutions for water scarcity may be found, this should not be taken as a given. Therefore, the role of water as a limiting resource should be kept in mind when interpreting the food production scenarios.

Intensification of agricultural inputs is a second factor to consider. Nearly all scenarios assume improvements in efficiency of agricultural input use, including increased efficiency of land use through yield increases and multiple cropping, increased efficiency of irrigation water use, and increased uses of agricultural machinery, fertilizers, and pest control. However, whereas some of these drivers were explicitly varied across scenarios, like yield growth or efficiency of irrigation, others, like fertilizer applications or changes in nitrogen-use efficiency, were not quantified or changed. We also have not evaluated the long-term risks of intensive agricultural inputs on pest outbreaks, groundwater contamination, soil degradation, and other ecological impacts. We expect these risks to be of greatest concern in Global Orchestration, which has the highest level of agricultural inputs and a low level of environmental protection. Next in line could be either Order from Strength, because of its low environmental consciousness, or TechnoGarden, because of the possibility of technological failure. Perhaps the Adapting Mosaic scenario would have the lowest level of risk because of its lower level of agricultural inputs and higher level of actions to protect the environment.

Sustainability of marine fisheries must be considered as well. The scenarios show a medium to large increase in fish production and consumption in all regions of the world. But we have not yet analyzed in detail the ability of the world's marine fisheries to sustain the computed fish production.

The fourth factor of concern is food insecurity and the affordability of food. In the model calculations, rising food demand will be met through increased production and food trade. If production levels are below food demand, then prices adjust upwards until a lower or depressed effective demand can be met. Moreover, higher food prices induce additional small supply expansion. As of yet we have not analyzed in detail the affordability of increased food prices for income groups. We have noted, however, that lack of technological development under the Order from Strength and Adapting Mosaic scenarios will force lower-income countries to import cereals at prices substantially above those of today. Cereals at these prices may not be within easy reach of lower-income groups, as indicated by the large number of malnourished children under these two scenarios.

The outcomes for food production do not vary significantly across scenarios at the global level, with cereal production in 2050 projected to be 50% larger under all four scenarios and with basic staple production projected to stagnate or decline in MENA and increase very little in sub-Saharan Africa by 2050. However, scenario outcomes do vary by region and within regions, particularly for the poor. Moreover, the means by which food production levels are increased vary significantly by scenario, with some focusing on area expansion and local production, whereas others rely on yield improvements and enhanced trade.

Differing means of increasing production could have an impact on pressure on ecosystems due to agricultural production. In the Order from Strength and Adapting Mosaic scenarios, protectionist policies, together with lack of investments in agricultural research and agriculture-related infrastructure, result in increased food prices, depressed food demands, and slow improvements in food consumption on a caloric basis. Moreover, under Order from Strength, the number of malnourished children will be higher by 2050 than today. The outcome is different for the Global Orchestration and TechnoGarden scenarios, where more food is produced by boosting crop yield and increasing the international exchange of goods, services, and knowledge. In these scenarios, crop area can be conserved, food prices increase much less, per capita food consumption increases faster, and the number of malnourished children declines.

9.4.4 Fuel

9.4.4.1 Methodology and Assumptions

The biosphere provides humanity with both traditional fuelwood and so-called modern biofuels, a category that includes alcohol derived from fermenting maize and sugar cane, fuel oil coming from rape seed, fast-growing tree species that provide fuel for power-generating turbines, and agricultural wastes, also burned to generate power. While fuelwood has been steadily replaced by other energy carriers, it still accounts for a large percentage of total energy use in some places. At the same time, the current use of modern biofuels is quite modest, although it could greatly expand, according to some energy scenarios (as described earlier). An important advantage of these is that in terms of greenhouse gas emissions they are neutral (the CO_2 emitted by burning biofuels has been absorbed first by plant growth). For this reason they play a significant role in the TechnoGarden energy scenario, where climate policy is given high priority. In the MA scenarios, biofuels play a role both for electricity production and as transport fuel. While many existing scenarios agree on an increase in modern biofuel use, major uncertainties exist regarding where the biofuels are produced and consumed and when the major penetration of biofuels into the energy mix will occur.

9.4.4.2 Comparison of Fuels among Scenarios

Under Global Orchestration, the global production of biofuels increases from its current level by a factor of six—mainly driven by cost increases for fossil fuels. (See Figure 9.33.) The regions making the biggest contribution to this increase are Asia (factor of eight), followed by the MENA countries (nearly a factor of six), and sub-Saharan Africa (about a factor of 4.5). There are two main factors leading to this large expansion in biofuel use. First, good land is available for biofuel production because competition from food production is low—food crops are grown very efficiently on existing crop areas in most regions (because of the high crop yield achieved from investments in agricultural research and fertilizer and other inputs). Second, the demand for electricity is high because of strong economic growth. Hence there is a large demand in general for energy and for biofuel electricity in particular because biofuels can be grown on relatively cheap and productive land. However, one unwelcome consequence of this intense use of biofuels is a high rate of deforestation in the regions committed to biofuel production.

Global production of biofuels under TechnoGarden increases by about a factor of four, mainly driven by climate policy. Production lags behind that scenario, however, because income (and therefore energy demand) is lower in the TechnoGarden scenario.

The level of investment in agricultural technology is low in the Order from Strength scenario, and as a result crop productivity is also relatively low. At the same time, population growth is larger than the other scenarios, and food demand is proportionately large. Since productivity is low on existing cropland, the increased demand for food has to come at least partly from new croplands. Energy crops must compete with food crops for land, and this makes land and biofuels more expensive. In addition, slower economic growth leads to lower growth in energy demands. These factors result in the slowest growth of biofuel production among all the MA scenarios. Nevertheless, global biofuel production still grows by more than a factor of two to fulfill the needs of the growing population.

The Adapting Mosaic scenario is an intermediate case compared to the others. Economic growth and crop productivity are higher than in Order from Strength but lower than in the others. As a result, energy demand is somewhat higher and competition with food production somewhat lower than in Order from Strength, and biofuel production is also somewhat higher. Globally, biofuel production increases by a factor of 2.8 over today, led by the MENA countries (factor of six), sub-Saharan Africa (factor of four), and Asia (factor of three).

9.4.4.3 Major Uncertainties of Fuel Estimates

Although calculation of the land requirement of energy crops takes into account current productivity of soils, it does not factor in the degradation that will result from these crops. Biofuel crops tend to degrade soils faster than many other crop varieties because they have high productivities and require large amounts of fertilizer and other inputs. Therefore, it is important to keep in mind that because of soil degradation, energy cropping is ecologically damaging over the long term and may thus be less economical.

9.4.5 Freshwater Resources

9.4.5.1 Methodology and Assumptions

The ecosystem services provided by freshwater systems have many dimensions. This section looks especially at water supply for households, industry, and agriculture and

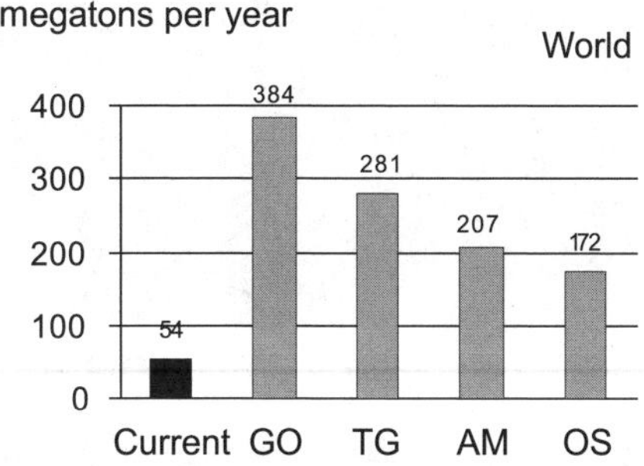

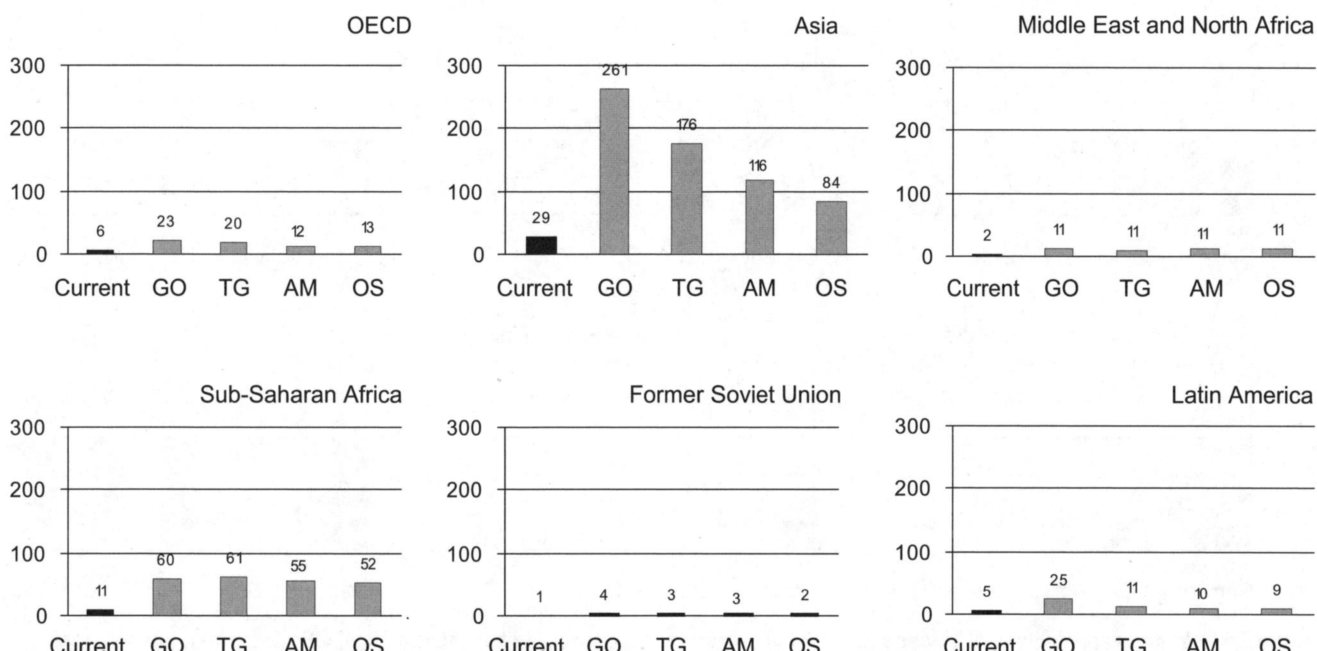

Figure 9.33. Total Biofuel Production by World Region in MA Scenarios in 2050. Scenario names: GO: Global Orchestration; TG: TechnoGarden; AM: Adapting Mosaic; OS: Order from Strength. (AIM)

at habitat for freshwater ecosystems, including fisheries. As indicators of these services, we describe the changing state of water availability, water withdrawals, water stress, and return flows. Each of these topics is useful for describing a different aspect of the ecosystem services delivered by freshwater. The end of the section summarize their consequences for the different MA scenarios.

9.4.5.2 Comparison of Water Availability among Scenarios

''Water availability'' is used here to mean the sum of average annual surface runoff and groundwater recharge. This is the total volume of water that is annually renewed by precipitation and theoretically available to support society's water uses and the needs of freshwater ecosystems. As used here, the term does not refer to availability in a technical or economic sense. In reality, society can exploit only a small fraction of this volume because water-rich areas are not necessarily near high population areas, because water is ''unusable'' as it rushes past cities in the form of floods, or because society cannot afford adequate water storage facilities. One estimate is that only about 30–60% of typical river basin water resources are ''modifiable'' (Falkenmark and Lindh 1993). On the other hand, water availability may be underestimated in the sense that we do not take into account the possible availability of water from desalination or waste recycling in the future. Despite its drawbacks, we believe that the concept of water availability used here gives a useful estimate of the total quantity of water available to meet the freshwater needs of society and ecosystems.

Since estimates of current water availability vary greatly, two independent estimates (from the WaterGAP and AIM models described in Chapter 6) are presented in Figure 9.34. Current global availability is estimated to be from 42,600–55,300 cubic kilometers per year.

The differences between scenarios are not as large as the differences between regions. By 2050, global water availability increases by 5–7% (depending on the scenario), with Latin America having the smallest increase (around 2%, depending on the scenario), and countries from the former Soviet Union the largest (16–22%). The changes in availability are small up to 2050 because of two compensating effects: increasing precipitation tends to increase runoff while warmer temperatures intensify evaporation and transpiration, which tends to decrease runoff. Hence, the direction of change of runoff does not correspond exactly to the

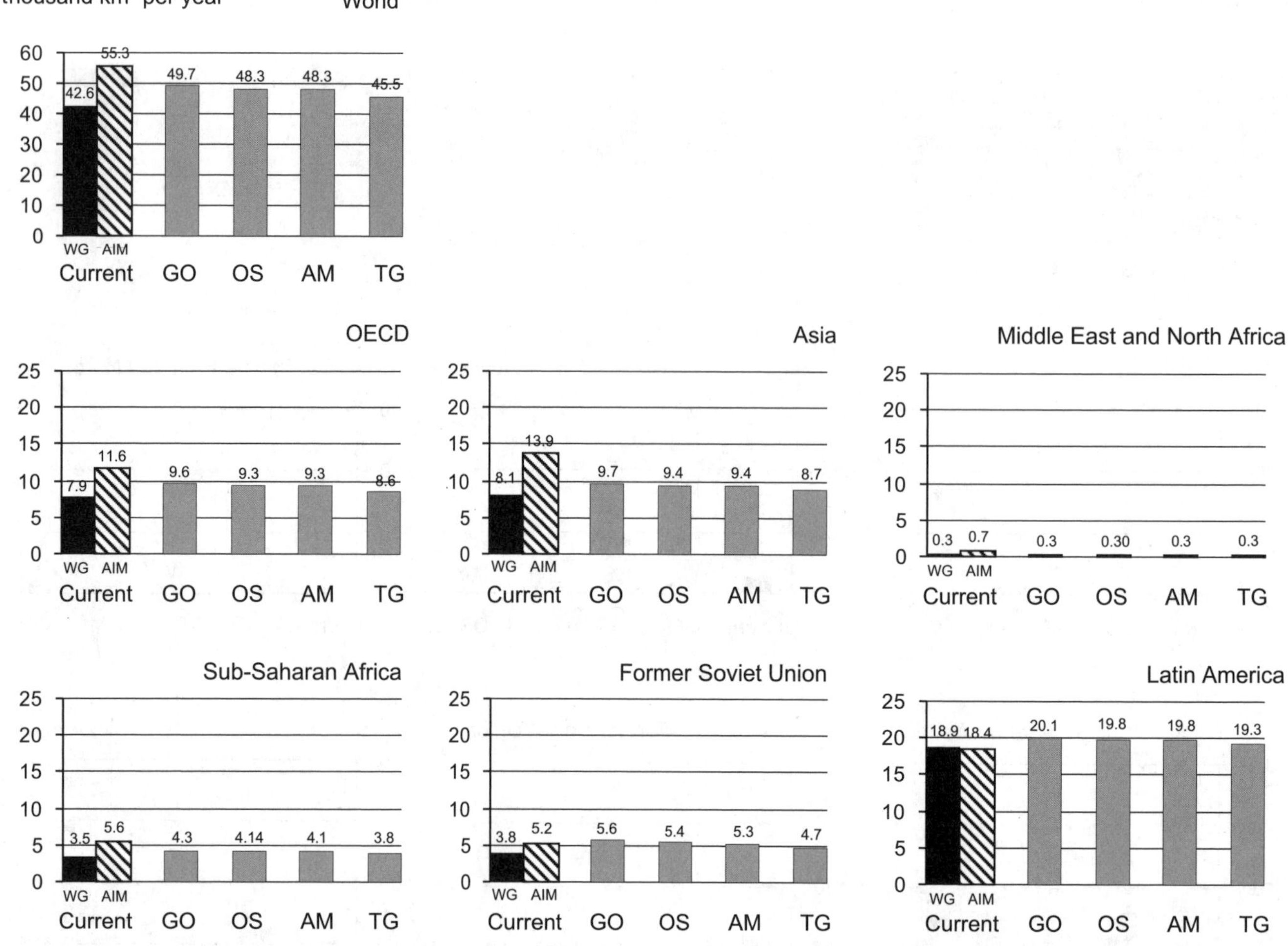

Figure 9.34. Water Availability in MA Scenarios in 2100. Scenario names: GO: Global Orchestration; OS: Order from Strength; AM: Adapting Mosaic; TG: TechnoGarden. (WaterGAP; AIM)

direction of change of precipitation shown earlier in Figure 9.12.

By 2100, the effect of increasing precipitation becomes more important, and runoff increases over most land areas. Still, the differences between scenarios are not as large as the differences between regions. Large areas on each continent have 25% or more runoff by 2100 (relative to the current climate period). Although availability increases in most areas, there are important arid regions where availability drops 50% or more under all scenarios, including Southern Europe, parts of the Middle East, and Southern Africa.

The largest increase in water availability occurs under the Global Orchestration scenario (17%) because it has the fastest rate of climate change through most of the scenario period. The smallest change occurs under TechnoGarden (7%) because it has the lowest rate of climate change. For these scenarios, the water availability in the already-arid MENA countries sinks by 1.5% under Global Orchestration and 3.5% under TechnoGarden scenario. Results for the Order from Strength and Adapting Mosaic scenarios fall between these figures (except for the MENA countries, in which the decrease is 4% in both cases).

While the increase in availability makes more water available for water supply, an increase in runoff can also correspond to more frequent flooding. We estimate that regions with the largest increases in water availability will also have more frequent high runoff events. We did not analyze this effect because no validated model is currently available in the literature for making worldwide calculations of flooding.

9.4.5.3 Comparison of Water Withdrawals and Use among Scenarios

While water availability indicates the amount of water theoretically exploitable, water withdrawals give an estimate of the water abstracted by society to meet its domestic, industrial, and agricultural needs. Hence it is a useful indicator of ecosystem services. (The water requirements for supporting a freshwater fishery are discussed later.) Compared with availability, water withdrawals show large changes over time and between scenarios up to 2050. Worldwide withdrawals in 1995 are estimated to have been about 3,600–3,700 cubic kilometers per year, or approximately 7–8% of estimated water availability, depending on the model used for calculations. While this does not seem like much, the intensity of withdrawals is high relative to water availability in several regions of the world.

Under Global Orchestration, strong economic growth coupled with an increase of population leads to a worldwide increase in withdrawals of around 40%. (See Figure 9.35.) But the changes are only slight in OECD, MENA, and former Soviet countries because of compensating effects—continuing improvements in water efficiency and stabilization or decrease of irrigated land tend to lower water use, while economic and population growth tend to increase water use. Although the efficiency of water use also improves over time in other regions, the effect of increasing population and economic growth leads to fulfillment of pent-up demands in the domestic and industrial sectors and to very large increases between 1995 and 2050 in sub-Saharan Africa, Latin America, and Asia. According to this scenario, many more people gain access to a water supply, as domestic water use substantially increases in nearly all regions. (See Figure 9.36.) The only exception is OECD, where domestic water use declines because nearly the entire population already has access to an adequate water supply and because the efficiency of water use continues to improve.

In TechnoGarden, strong structural changes in the domestic and industrial sectors and improvements in the efficiency of water use in all sectors lead to decreases in water withdrawals in OECD (10%) and the former Soviet Union countries (23%). The same factors lead to a slowdown in the growth of withdrawals in the rest of the world. Nevertheless, water withdrawals grow by a factor of 2.4 in sub-Saharan Africa because of pent-up demand for household water use and growing industrial water requirements.

Although Adapting Mosaic and Order from Strength do not have the largest economic growth, they have the largest water withdrawals because of slower improvement of the efficiency of water use and faster population growth. Withdrawals increase substantially worldwide (52–82%) and moderately in the OECD (7–34%) (under Adapting Mosaic and Order from Strength, respectively). In the former Soviet countries, withdrawals decrease under Adapting Mosaic (9%) and level off under the Order from Strength scenario. Increases in withdrawals are very substantial in sub-Saharan Africa (a factor of 3 under both scenarios), in Latin America (factor of 2.5–3), and Asia (60–100%), while they are more moderate in the arid climate of the MENA countries (28–46%).

9.4.5.4 Comparison of Water Scarcity and Water Stress among Scenarios

The changes in water availability and withdrawals just described have consequences on water stress in freshwater sys-

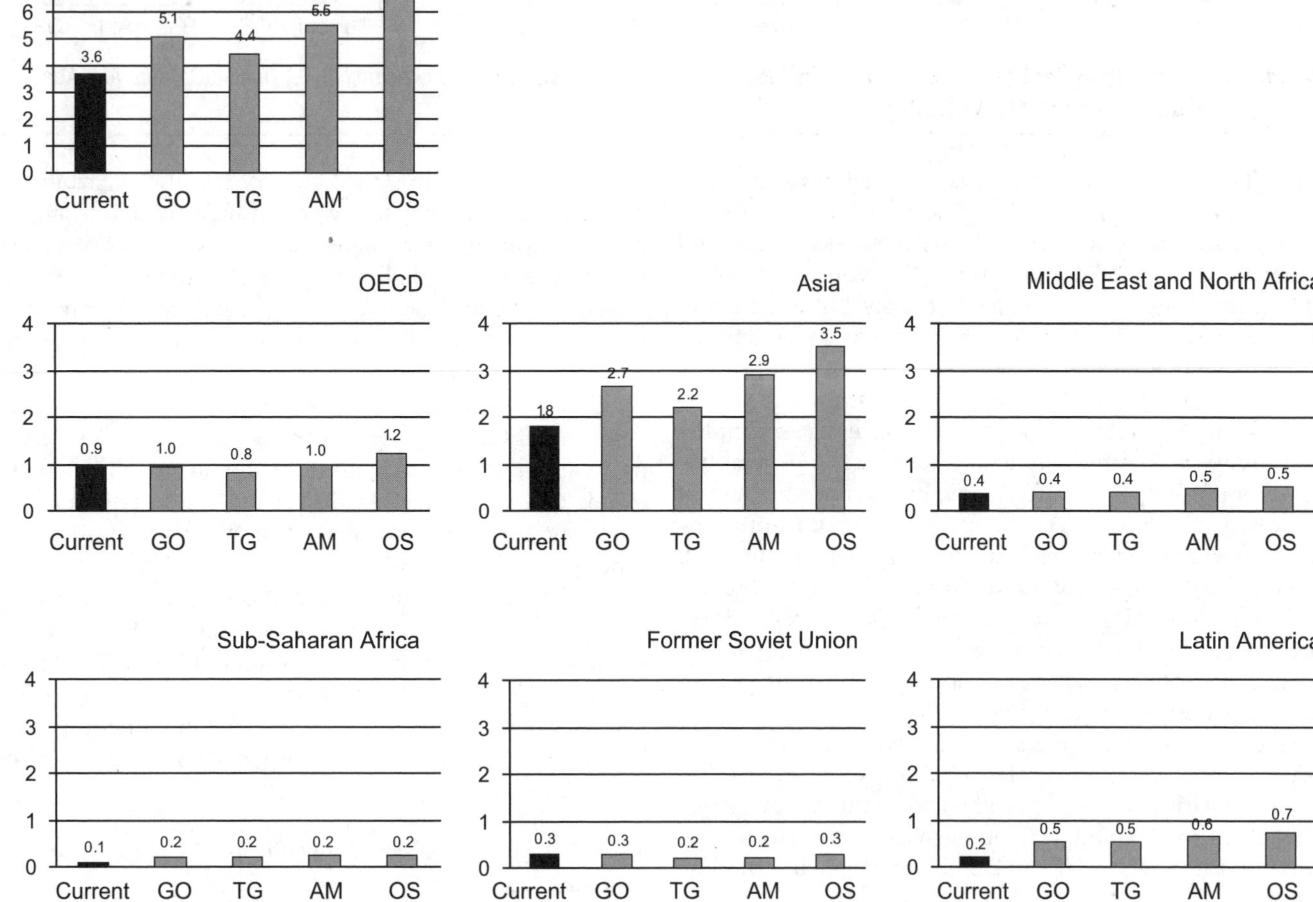

Figure 9.35. Water Withdrawals in MA Scenarios in 2050. Scenario names: GO: Global Orchestration; TG: TechnoGarden; AM: Adapting Mosaic; OS: Order from Strength. (WaterGAP)

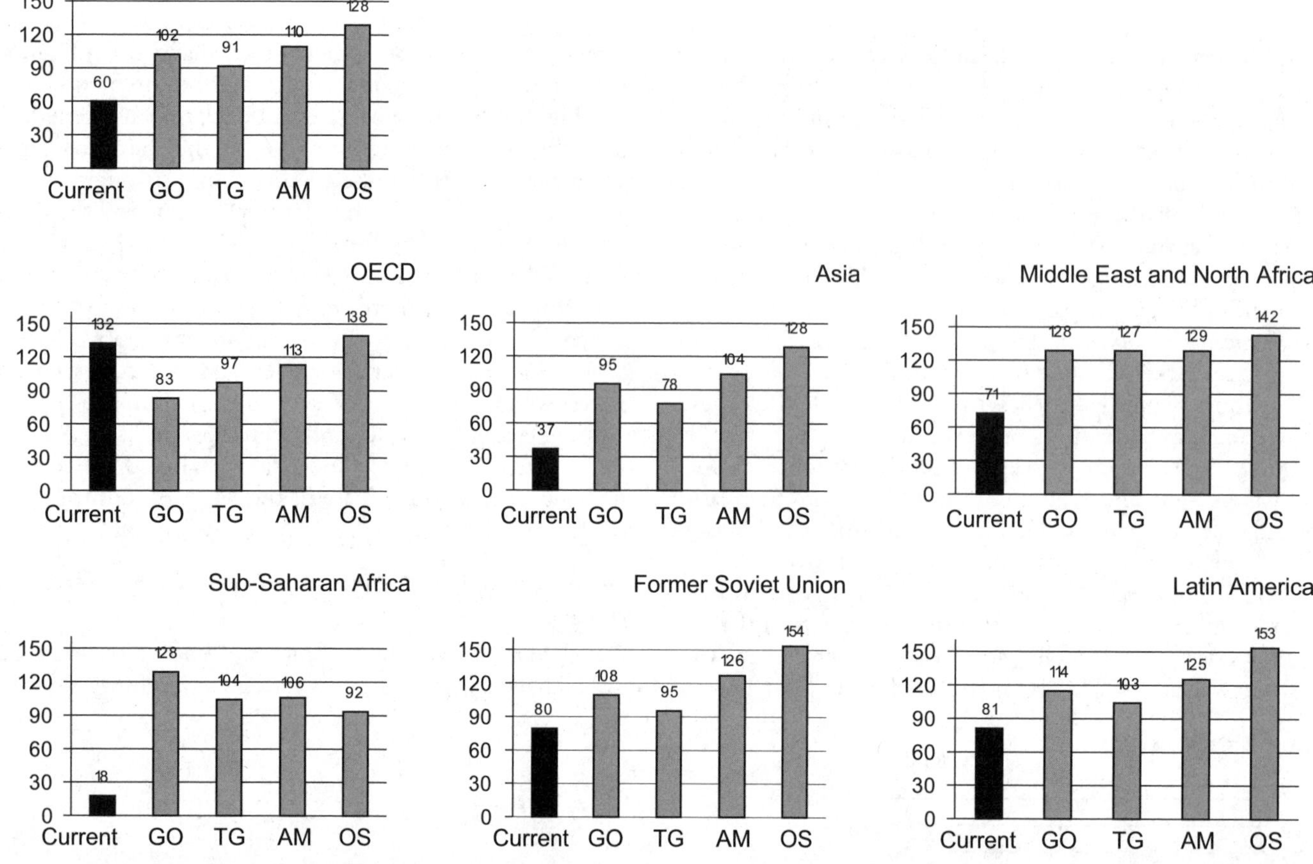

Figure 9.36. Domestic Water Use in MA Scenarios in 2050. Scenario names: GO: Global Orchestration; TG: TechnoGarden; AM: Adapting Mosaic; OS: Order from Strength. (WaterGAP)

tems. The concept of "water stress" is used in many water assessments to obtain a first estimate of the extent of society's pressure on water resources (Alcamo et al. 2000 and 2003; Cosgrove and Rijsberman 2000; Vörösmarty et al. 2000). It is assumed that the higher the level of water stress, the greater the limitations to freshwater ecosystems, and the more likely that chronic or acute shortages of water supply will occur. A common indicator of water stress is the withdrawals-to-availability ratio, or wta. This indicator implies that future water stress will tend to decrease in general because of growing water availability, but increase because of increased withdrawals. An often used approximate threshold of "severe water stress" is a wta of 0.4 (Alcamo et al. 2000 and 2003; Cosgrove and Rijsberman 2000; Vörösmarty et al. 2000). River basins exceeding this threshold, especially in developing countries, are presumed to have a higher risk of chronic water shortages and thus greater threats to freshwater ecosystems.

Figure 9.37 depicts the area of the world in the "severe water stress" category in 1995. Much of northern and southern Africa, as well as central and southern Asia, is included. In total, about 18% of the world's river basin area falls into this category. About 2.3 billion people live in these areas.

The area in the severe water stress category under the Global Orchestration scenario in 2050 is shown in the bottom half of Figure 9.37. Some areas, especially in OECD, fall out of the severe stress category because of stabilizing withdrawals and increasing water availability due to higher precipitation under climate change. The areas of severe water stress expand slightly in the rest of the world. A total of about 4.9 billion people live in these areas. Over most of these areas, increasing withdrawals tend to increase the level of water stress over today's level.

Under TechnoGarden, water withdrawals up to 2050 drop in OECD and the former Soviet Union and grow more slowly in other regions. Water stress follows these trends and declines in many parts of these two regions, while increasing more slowly than the other scenarios in other parts of the world.

Under the Adapting Mosaic and Order from Strength scenarios, water withdrawals increase sharply as just described, and the area under severe water stress in 2050 covers about 17–18% of the total watershed area. Water stress increases over all these areas. About 5.3–5.5 billion people live in river basins with severe water stress—some 60% of the world's population.

9.4.5.5 Comparison of Water Quality and Return Flow among Scenarios

The concept of "return flows" is used here to assess changes in water quality. Return flows are the difference between withdrawals and consumption and therefore provide a

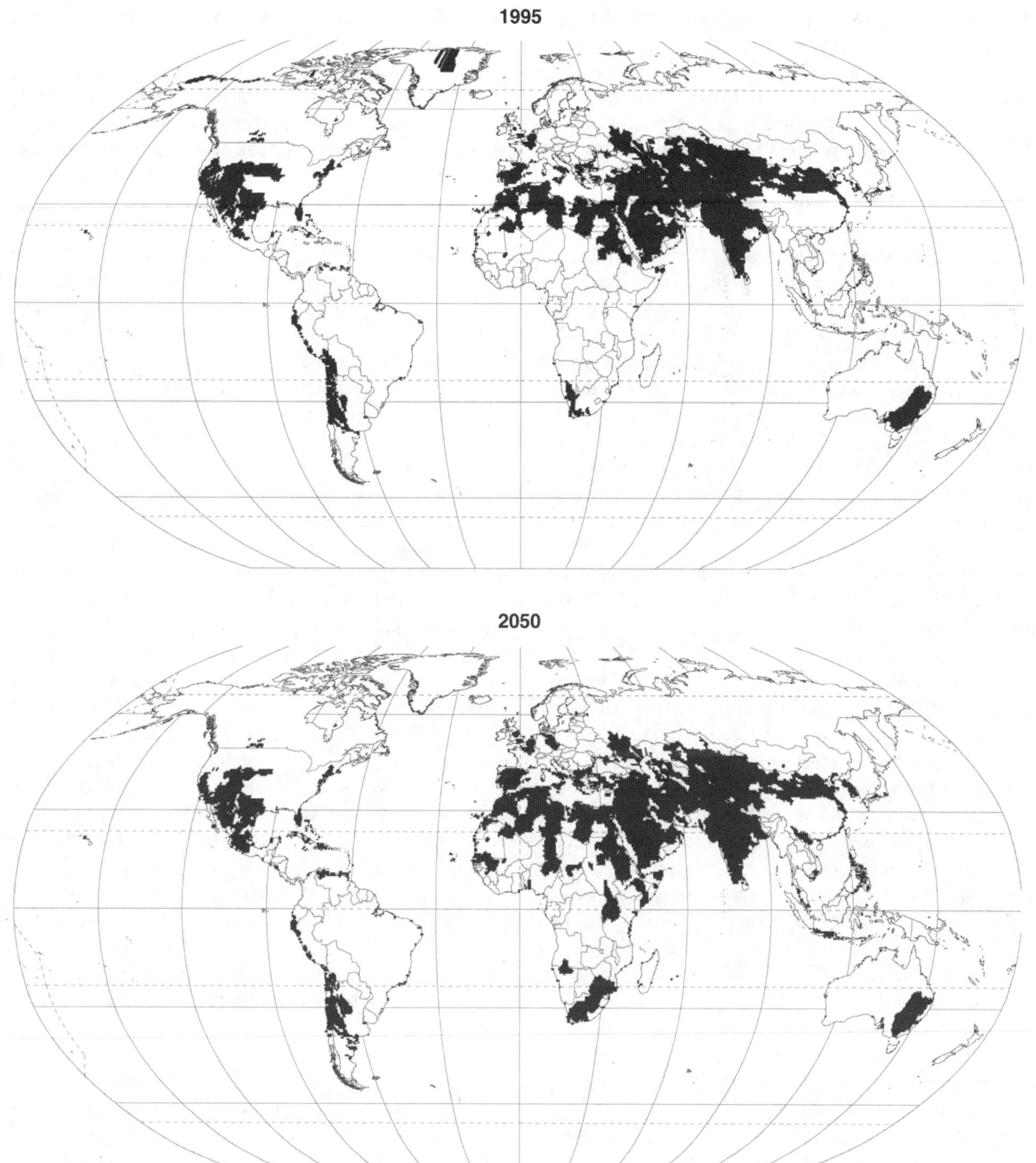

Figure 9.37. Areas under Severe Water Stress in the Global Orchestration Scenario in 1995 and 2050 (WaterGAP)

rough estimate of the magnitude of wastewater discharged into the receiving water in a watershed. Depending on the type of return flows, high rates of these flows could correspond (*with medium certainty*) to low water quality and high levels of water contamination and pressure on freshwater ecosystems. We use return flows as a surrogate variable for water quality because it is not possible at this time to compute worldwide changes in water quality for the different scenarios.

Since irrigation usually consumes more water than domestic or industrial uses, the return flows for irrigation are also usually a smaller fraction of its withdrawals. Another important point is that the type and concentration of water pollutants is quite different from different sources. The wastewater discharged by a power plant after it is used for turbine cooling is hot but relatively clean compared with wastewater discharged by a typical municipality. Untreated municipal wastewater contains pathogens, organic wastes, and toxic materials that typically contaminate a receiving water; only the simplest of aquatic ecosystems can survive and the contaminated water cannot be used for human contact or water supply.

Irrigation return flows are normally not returned to rivers in a single large discharge pipe as a typical municipal or industrial source, but they enter the river in a diffuse way along many kilometers. These return flows are an important source of herbicides and pesticides that are bio-concentrated in aquatic ecosystems and can interfere with or poison various organisms in the ecosystem, as well as of nutrients that promote eutrophication of natural waters. Hence, the impact of irrigation return flows will not be the same as the impact of return flows from a city or industry.

Perhaps the most important factor to take into account in assessing the impact of return flows is whether they will be treated or not. In OECD, the partial treatment of wastewater (removal of organic wastes and pathogens) is very common, and the trend is toward at least partially treating nearly all municipal and industrial wastewater discharges. It is much more uncertain whether agricultural return flows will be collected and treated (because of the high costs of collecting and treating high volumes of diffuse wastewater sources). Wastewater treatment elsewhere, however, is now quite uncommon except in some cities, and it is difficult to anticipate the future coverage of wastewater treatment (except perhaps that there is a trend toward disinfecting wastewater before discharge.)

Under Global Orchestration, by 2050 worldwide return flows increase by 42%. The magnitude of these flows follows that of withdrawals, meaning the larger the volume of withdrawn water, the larger the size of wastewater discharges. Return flows decrease on the average in OECD and the former Soviet countries because of leveling off population and improving efficiency of water use. These factors tend to decrease withdrawals and hence return flows. Furthermore, even though low priority is given to environmental protection, the richer societies in this scenario maintain their current efforts at environmental management. Hence it is reasonable to assume that the level of wastewater treatment in OECD countries will remain at least at its current level.

Because of the rapidly increasing water withdrawals under Global Orchestration, return flows also substantially increase between 2000 and 2050 in sub-Saharan Africa (factor of 3.7), Latin America (factor of 2), and Asia (49%), and more moderately in the MENA countries (24%). Figure 9.38 illustrates the large area where return flows are estimated to at least double under this scenario between now and 2050. Over 78% of the watershed area of sub-Saharan Africa is in this category, as is substantial parts of MENA (37%), Asia (26%), and Latin America (38%). (See Table 9.18.) Consistent with the storylines of this scenario, low priority will be given to environmental management in the world's poorer regions. Therefore, it is likely that wastewater will remain untreated in many areas and that the level of water contamination and degradation of freshwater ecosystems may increase. Since much of this return flow will come from agricultural areas, under this scenario we expect a large increase in nitrogen loading to rivers and subsequently to coastal areas, as described earlier.

We estimate that 4.4 billion people or nearly 55% of the world will live in these areas in 2050. (See Table 9.19.) We emphasize, however, that return flows will cause major problems only if they remain untreated.

Under TechnoGarden, the trends up to 2050 are in the same direction as Global Orchestration, but the stronger emphasis on improving water efficiency and somewhat lower economic growth rates lead to a stronger decrease in return flows between now and 2050 in OECD (18%) and the former Soviet countries (43%). The same factors lead to slower growth of return flows in sub-Saharan Africa (factor of 3.5), MENA (17%), and Asia (nearly 20%). The change in Latin America is the same as in Global Orchestration (increase by a factor of two). Similarly, large areas will have increases of 100% or more return flows, and a total of 3.9 billion people will live in these areas. Since the emphasis in this scenario is on environmental management, and since return flows do not increase too much in MENA or Asia, it may be that most of the wastewater flows in these regions will be treated. It is less likely that the enormously increasing return flows of sub-Saharan Africa and Latin America will be fully treated.

In Adapting Mosaic, return flows decrease in the former Soviet Union between now and 2050 because withdrawals decrease. In all other regions, return flows increase much more than in Global Orchestration or TechnoGarden because of the lower level of water use efficiency and the larger population, which leads to higher withdrawals and more return flows. Return flows increase very substantially in sub-Saharan Africa (factor of 5.5), Latin America (factor of 2.6), Asia (76%), and the MENA countries (56%) and increase slightly in OECD (4%). The area of watersheds with at least a 100% increase in return flows between now and 2050 is considerably larger than in Global Orchestration or TechnoGarden, and 6.4 billion people—67% of the world's population in 2050—live in these areas. Since Adapting Mosaic puts a strong emphasis on local environmental protection, and since wastewater treatment technology is simple and can be applied easily on the local level, we expect (*with medium certainty*) a high level of wastewater treatment.

As noted, Order from Strength has the largest withdrawals because of its slower improvement of the efficiency of water use and faster population growth. Accordingly, it also has the largest return flows, with a doubling of worldwide total flows between now and 2050. The smallest increase is in former Soviet countries (9%), followed by OECD, with a nearly 40% increase. All other regions experience much larger increases—Asia and MENA countries (approximately a doubling), Latin America (more than a factor of 3), and sub-Saharan Africa (a factor of 4.7). The area with a doubling of return flows is somewhat larger than in Adapting Mosaic, and 6.7 billion people live in these areas (70% of global population). The level of environmental concern here is much lower than in Adapting Mosaic, and therefore the expected level of wastewater treatment is also much lower. The combination of exploding wastewater discharges and negligence of the environment could lead to large risks to freshwater ecosystems and water contamination. An additional dimension of this scenario is that return flows continue to increase rapidly after 2050. For example,

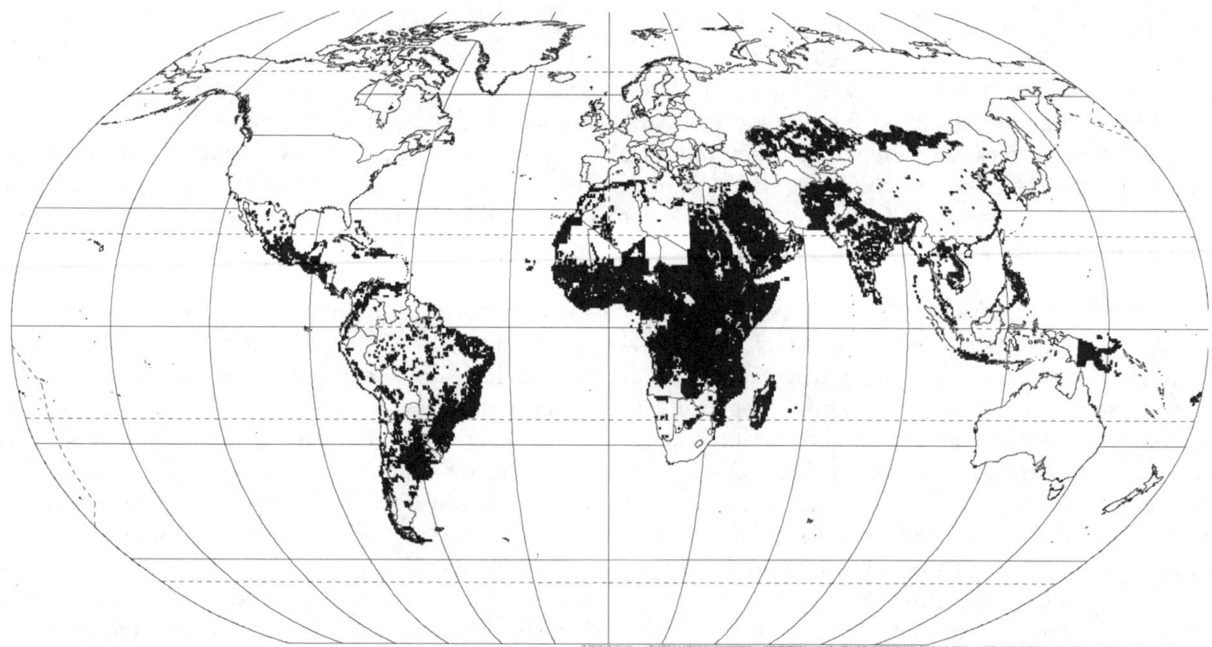

Figure 9.38. Areas Where Return Flows Increase at Least 100% in the Global Orchestration Scenario, Present–2050 (WaterGAP)

Table 9.18. Total Area of River Basins, by Region, Where Return Flows Increase at Least 100% between Now and 2050 in MA Scenarios (WaterGAP)

Region	Global Orchestration	Order from Strength	Adapting Mosaic	TechnoGarden
		(thousand square kimometers)		
Former Soviet Union	4,810	6,103	7,009	5,092
Latin America	7,798	18,441	17,286	10,814
Middle East and North Africa	4,306	9,148	8,798	4,856
OECD	7,271	8,837	13,699	7,230
Asia	5,375	10,220	7,244	4,587
Sub-Saharan Africa	18,724	22,299	22,215	17,661
World	**48,284**	**75,047**	**76,251**	**50,239**

Table 9.19. Total Number of People, by Region, Living in Areas Where Return Flows Increase at Least 100% between Now and 2050 in MA Scenarios (WaterGAP)

Region	Global Orchestration	Order from Strength	Adapting Mosaic	TechnoGarden
		(million)		
Former Soviet Union	1.5	32.8	70.4	5.8
Latin America	719.1	926.1	908.2	701.9
Middle East and North Africa	327.8	633.9	684.2	333.2
OECD	13.9	31.5	66.3	12.4
Asia	2,156.2	3,582.0	3,238.3	1,689.5
Sub-Saharan Africa	1,222.5	1,518.8	1,445.7	1,196.6
World	**4,440.9**	**6,725.0**	**6,413.1**	**3,939.4**

return flows increase in sub-Saharan Africa by a factor of 4.6 between 1995 and 2050, and double again between 2050 and 2100.

9.4.5.6 Uncertainty of Freshwater Estimates

The concept of "water availability" used in this section does not refer to the freshwater resource available to water users in an economic or technical sense, but only to the total theoretical volume of water annually available in each watershed due to precipitation. It is currently not possible to estimate water availability more precisely on a global basis.

Return flows are used as an indicator of water quality, but it would be more desirable to have a direct indicator of future water quality so there is a more certain connection with the state of freshwater ecosystems and risk of water contamination. This is not yet possible globally.

An important source of uncertainty in estimating return flows is the uncertainty of the ratio of consumption to withdrawals in different water use sectors. In addition, the tools used to estimate the indicators of ecosystem services of fresh water are too aggregated to include the role of local water policies and management.

9.4.5.7 Summary: Freshwater Services

In the Global Orchestration scenario, water availability increases in most countries because of climate change, but it decreases in some important arid regions, especially in poorer countries. Water withdrawals and domestic water use level off in wealthy countries but increase substantially in poorer ones that are providing access to adequate water supply for many needy people. Water stress and the volume of return flows go down in the rich countries because of climate change and stabilization of water withdrawals. Water-related problems (eutrophication, for instance, and competition between human and environmental water requirements) continue in these nations but do not intensify greatly.

Water stress goes up in poorer countries under Global Orchestration because of a massive increase in withdrawals and return flows. Wastewater discharges are left mostly untreated because the priority is given to expanding economic and industrial capacity without incurring extra costs for environmental protection. This leads to an intensification and expansion of water resource problems in poorer countries, including more frequent water shortages during low flow periods, as well as deterioration of aquatic ecosystems. The increased wealth in poorer countries allows society to deal with some of these problems on a case-by-case basis, and luckily there are no major ecological collapses, such as water-related disease outbreaks, or hunger shortages because of the disappearance of freshwater fisheries.

Under the TechnoGarden scenario, the water availability situation is similar to Global Orchestration except the changes are not as great because the rate of climate change is not as great. Water withdrawals and domestic water use decrease in wealthy countries and slowly grow in poorer ones except in Africa, where they increase very rapidly. Although per capita domestic water use does not increase as rapidly as in Global Orchestration, a greater emphasis is put in TechnoGarden on providing minimum adequate water supply to those needing it.

Water stress goes down in wealthy countries in TechnoGarden because precipitation increases and withdrawals decrease. Return flows decrease there too and therefore society can afford advanced treatment of municipal and industrial wastes, as well as the collection and control of agricultural runoff. These actions greatly reduce the load of nutrients and toxic substances to freshwater systems. Furthermore, wealthy countries intervene to physically restore natural habitat in freshwater systems. As a result, there is a significant restoration of aquatic ecosystems. At the same time, the overconfidence of society that it can engineer solutions to water resource problems leads it to overlook continuing problems. For example, heavy storm runoff overloads wastewater treatment plants, and contaminated sediments of riverbeds continue to leach toxic materials accumulated during the twentieth century.

The increase in withdrawals and return flows is slower than in Global Orchestration but still very fast, especially in Africa and Latin America. Society puts a heavy emphasis on bringing wastewater treatment up to current OECD standards (secondary treatment of municipal wastes, control of toxic discharges from industry). But agricultural sources are not controlled, and a resulting rapid increase in nutrient and pesticide discharges causes eutrophication, toxicity, and other water quality problems. Furthermore, Africa and Latin America cannot keep up with the increase in return flows and are not able to achieve OECD standards.

In Adapting Mosaic, changes in climate change and hence water availability are similar to Global Orchestration and TechnoGarden scenarios, but intermediate in intensity. Water withdrawals and domestic water use stabilize in wealthy countries and are moderate to large in poorer nations. While per capita domestic water use is lower than in Global Orchestration and TechnoGarden, more local effort is invested in providing people with a minimum amount of household water supply. Water stress goes down in rich nations because of increased precipitation and stabilization of withdrawals, and the amount of return flows stabilizes. Local authorities and communities in these countries do not have to deal with increasing wastewater loadings and therefore have time to find local solutions to the competition for water resources between water use sectors and human and ecosystem requirements.

The level of water stress and volume of return flows explodes in poorer countries under Adapting Mosaic, but in many cases local solutions are found for allocating water supply to different sectors (integrated watershed management) and for preserving the viability of many aquatic ecosystems. But it is difficult to find local solutions fast enough in Latin America and Africa, where return flows increase by a factor of 3.6 and 5.5, respectively. Hence water problems in these regions increase over the first half of the twenty-first century. One advantage of the local or watershed approach to water management everywhere is that it is usually well tailored to local ecosystems, and thus failures in water management can be easily corrected. Hence, under this scenario there is slow but steady progress in protecting or restoring aquatic ecosystems and providing freshwater ecosystem services.

The water availability situation under Order from Strength is similar to that of Adapting Mosaic. Water withdrawals and domestic water use moderately increase or level off in wealthy countries and are moderate to large in poorer ones. There is much lower access to adequate water supply than in the other scenarios, and it is likely that the Millennium Development Goal for access to adequate water supply would not be met under Order from Strength. The level of water stress stabilizes in wealthy countries but increases substantially in poorer ones. The volume of return flows also has a massive increase in the first half of the twenty-first century (in Latin America a factor of 4, and in

Africa a factor of 5.6). Three quarters of the world lives where return flows double in the first half of century. Environmental management is not given a priority, and the technology and capacity for ecological management are not built up. Ignorance of the implications of the large increase in water stress and return flows in poorer countries leads to severe regional water quality crises, with widespread destruction of aquatic ecosystems, the contamination of water supply, and widespread water shortages. Poorer countries fall behind in development.

9.4.6 Other Provisioning Ecosystem Services

9.4.6.1 Methodology and Assumptions

Although the future states of genetic resources, biochemicals, and ornamental resources were not directly evaluated by model calculations, we examine here the trends of some related indicators. These include the extent of natural versus agricultural land, since these resources usually require undisturbed habitat; the rate of change of this habitat as indicated by the rate of deforestation and the rate of climate change, since the faster the change, the more doubtful that plants and animals can adapt to these changes; and the level of water stress in freshwater resources, which indicates the pressure on aquatic and riparian species. Moreover, the storylines in Chapter 8 also indicate certain trends in human behavior and policies that will affect these services. By examining the trends of these variables we make some preliminary judgments about the future trends of genetic resources, biochemicals, and ornamental resources. (Note that these are only a few of the many important factors that will influence the state of genetic resources, biochemicals, and ornamental resources in the future; for example, these do not include an indicator for the marine environment.)

9.4.6.2 Comparison of Genetic Resources among Scenarios

UUnder the Global Orchestration scenario, pressures grow on remaining undisturbed terrestrial and aquatic ecosystems. Throughout the twenty-first century, existing forests disappear at rates comparable to the last few decades. The decadal rate of temperature change is much higher than at present, and ranks as the highest among the four MA scenarios throughout most of the period. As a result of changing temperature (and precipitation), the type and viability of current vegetation also changes over extensive areas, especially in wealthy countries. To meet growing food demand due to higher incomes and population, the level of agricultural production on existing cultivated land is intensified in poorer countries by increasing the application of fertilizer and other inputs, and these chemicals also contaminate nearby protected natural areas. In freshwater ecosystems, the level of water stress increases over wide areas, especially in poorer countries, because of rapidly increasing withdrawals. In addition to these pressures, society is also not particularly mindful of the connection between its activities and the state of ecosystem services. In sum, it is possible that genetic resources may severely decline under this scenario.

Under the TechnoGarden scenario, global climate protection policies lead to lower rates of temperature change (compared with the first decades of the twenty-first century), but vegetation areas still change extensively, especially in richer countries. Stabilizing water demands and efficiency improvements lead to decreases in water withdrawals and reduced stress on freshwater ecosystems. In wealthy countries, the drive for efficiency narrows the range of genetic resources used by people. This offsets the effects of reduced water stress, conservation, and genetic technology, leading to little net change in genetic resources.

Under TechnoGarden, the rate of deforestation in poorer countries is high, but it eventually drops below current rates. Efficient water use leads to lower growth in water withdrawals and slower increases in stress on freshwater ecosystems. However, high levels of fertilizer and pesticides are used on agricultural land to boost crop yields, which leads to contamination of natural areas. Counter to these trends, genetic diversity is enhanced in poorer countries by intensified efforts to preserve landraces. Under this scenario, we also expect that ecological engineering of plants and animals will have an influence on overall genetic resources. But at this time we cannot estimate what this influence will be. Finally, we also expect (*with low certainty*) only a small change in genetic diversity in poorer countries.

Under the Order from Strength scenario, the rate of forest disappearance in poorer countries is even greater than under Global Orchestration (because of more inefficient agricultural production). Also, growing population and inefficient water use in these countries leads to rapid growth in water withdrawals and stress on freshwater ecosystems. A side effect of the lower level of wealth in this scenario is that farmers in poorer countries cannot afford to apply as many pesticides and fertilizer to cropland, meaning that the loading of these chemicals onto nearby natural areas is somewhat lower than under Global Orchestration. On the other hand, climate change is not as great overall; therefore, while climate-related changes in vegetation still occur in wealthy countries, they are not as extensive as in Global Orchestration. Society in this scenario also gives low priority to environmental protection. Summing up the different factors, we expect (*with low certainty*) that genetic resources could decline at around the same rate as in the Global Orchestration scenario.

In Adapting Mosaic, the rate of climate change is not as high as in Global Orchestration, nor as low as in TechnoGarden. Therefore, the extent of area with changed vegetation in wealthy countries due to climate change is also between these two scenarios. The rate of forest disappearance in poorer countries under Adapting Mosaic drops below current rates but is still high. Water withdrawals significantly increase in these countries, but not as much as under Order from Strength because water is used more efficiently. Under this scenario, society is mindful of the connection between its activities and ecosystem services. Therefore, the use of fertilizer and other inputs on agricultural land is somewhat lower than in Order from Strength. Moreover, genetic diversity used by people is increased by the greater spatial heterogeneity of ecosystem management in all countries. Considering the different factors, we expect (*with low certainty*) that genetic diversity could either remain about the same or slightly increase under this scenario.

9.4.6.3 Comparison of Biochemical Discoveries and Ornamental Resources among Scenarios

The trends just described for Global Orchestration—high deforestation rates, steadily increasing temperature and climate-related changes in vegetation, intensification of agricultural land, increasing water withdrawals and water stress—tend to threaten ecosystems in poorer countries and eventually to decrease biodiversity. This is somewhat compensated for by increasing investments in biochemical exploration, so that the net rate of biochemical discoveries is roughly constant in poorer countries up to 2050. At the same time, the sum of ornamental resources declines (*with low certainty*) along with biodiversity.

While these trends pertain especially to poorer countries, pressures on biodiversity also increase elsewhere because of intensification of agriculture and a failure to devise policies to deal with current threats to biodiversity. We expect (*with low certainty*) that the decline in biodiversity will be accompanied by a decline in biochemical discoveries and ornamental resources.

As noted, the pressures on ecosystems under TechnoGarden—climate change, rate of increase of water withdrawals, deforestation—will be somewhat lower than under the other scenarios. Moreover, biochemical innovation is also a high priority for society. Hence, we expect (*with low certainty*) that biochemical discoveries will increase in all countries up to 2050. At the same time, the TechnoGarden scenario emphasizes the utilitarian uses of ecosystems. Therefore, we estimate (*with low certainty*) that ornamental resources receive no special attention and remain about the same as today.

Pressures on ecosystems, as noted above, are relatively high in Order from Strength in all countries, with a resulting decrease in biodiversity. In addition, conflict and a poor security situation will hamper biochemical exploration in some parts of the world. In sum, we expect (*with low certainty*) that biochemical discoveries and the availability of ornamental resources will decline up to 2050 under the Order from Strength scenario.

In Adapting Mosaic, there are lower pressures on ecosystems as compared with Global Orchestration and TechnoGarden, and biodiversity is conserved. Because of the scenario's focus on local and regional development, however, there is relatively low impetus for the international development and trade in biochemicals. Hence we estimate (*with low certainty*) that both the level of biodiversity and the rate of biochemical discovery are maintained at roughly today's levels. Since this scenario emphasizes the local individuality of ecosystem management, we estimate (*with low certainty*) that the availability of ornamental resources will increase.

9.5 Regulating Ecosystem Services

Regulating ecosystem services are defined as the benefits obtained from regulation of environmental conditions through ecosystem processes. The conceptual framework of the MA lists the following clusters of regulating services:

- air quality maintenance (through contribution to or extraction of chemicals from the atmosphere, as a result of ecosystem function);
- climate regulation (through the influence of ecosystems on the energy, water, and carbon balance of the atmosphere);
- water regulation, erosion control, and water purification (through the effect of ecosystems on runoff, flooding, aquifer recharge, and water quality);
- human disease control (through the effect of ecosystems on human pathogens, such as disease vectors);
- biological pest and disease control (through the influence of ecosystems on the abundance of animal and plant pathogens);
- pollination (through influences of ecosystems on the abundance and distribution of pollinators); and
- coastal protection (through the protecting effect of ecosystems such as coral reefs and mangroves on coastal structures).

On the basis of the analyses in the MA *Current State and Trends* volume, this section describes the impact of MA scenarios on some of these services. The presentation focuses on services where differentiation between scenarios can be achieved, based on either calculations with numerical models or on an assessment of recent scientific literature or both. The impacts of ecosystems on human disease control are treated in Chapter 11. Additional information on changes in regulating services can be found in Chapter 8.

Overall, the vulnerability of most regulating services contrasts clearly across the scenarios. In Global Orchestration, a predominantly reactive approach to ecosystem management rarely addresses regulating ecosystem services. The net result is greater vulnerability of regulating ecosystem services, especially in poorer countries. The exceptions are a few cases in which the connection between ecosystem services and human welfare is direct and clearly understood. In Order from Strength, the vulnerability of regulating ecosystem services generally increases as the availability of regulating ecosystem services declines. The wealth of richer countries sometimes allows adaptations that conserve regulating ecosystem services, but in poorer countries the regulating ecosystem services become much more vulnerable due to the effects of population growth, conflict, slow economic growth, and expanding poverty.

In Adapting Mosaic, society emphasizes local or regional ecosystem management. Maintenance or expansion of regulating ecosystem services will often be the goal of this ecosystem management, leading to declines in the vulnerability of these services. However, the primary focus is local or regional ecosystem issues. Global regulating services, such as those related to climate or marine fisheries, could become more vulnerable during Adapting Mosaic. In TechnoGarden, society emphasizes engineering of ecosystems to provide regulating ecosystem services. While this approach is successful for some ecosystem services in some places, in other cases oversimplification of ecosystems increases the system's vulnerability to change and disturbance. Impacts of unforeseen disturbance create the need for new technological innovations. In some cases, this leads to a spiral of increasing vulnerability.

9.5.1 Climate Regulation/Carbon Storage

9.5.1.1 Methodology and Assumptions

The biosphere, and the ecosystems it consists of, plays a key role in the climate system, for example, by respiring and taking up CO_2, by emitting other trace gases such as CH_4, and by reflecting and absorbing solar energy. On the global scale, the biosphere currently helps "regulate" climate by capturing carbon due to increased growth, thereby reducing the concentration of CO_2 in the atmosphere and slowing down climate change (Schimel et al. 2001).

For the theoretical case of a biosphere/atmosphere equilibrium, the biosphere takes up as much CO_2 for plant growth as it emits by plant and soil respiration. But under most circumstances, one or the other of these fluxes dominates, with frequent oscillations due to temporal and spatial environmental variability. Overall, the land biosphere currently takes up 2.3 gigatons per year ($\pm$ 1.3 gigatons) more carbon than it emits (Bolin et al. 2000). Contributing factors to this important global service are increasing forest area in some regions and the stimulation of plant productivity through increasing temperature or atmospheric CO_2. The result is that warming, and other climate change, occurs at a slower rate than would be expected in the absence of the carbon sink.

9.5.1.2 Comparison of Climate Regulation among Scenarios

As described earlier, climate policy is not assumed to be a priority under Global Orchestration. Nevertheless, being a relatively low-cost measure, climate regulation by ecosystems could be a focus of global policy during Global Orchestration, being implemented through the protection of old-growth forests for their soil carbon stocks and through other measures that avoid unnecessary release of carbon from the biosphere. Consequently, the role of ecosystems in climate regulation becomes more important in all countries. It is, however, not clear how much the carbon sequestration capacity of ecosystems could increase in wealthy nations during Global Orchestration, nor how long this effect might last.

During Order from Strength, the capacity to regulate climate is expected to decline in both rich and poorer countries due to lack of international coordination. Global issues are not a primary focus of ecosystem management in Adapting Mosaic. However, climate regulation would be a secondary consequence of improving ecosystem management in many regions. On balance, we expect little change in climate regulation by ecosystems during Adapting Mosaic. During TechnoGarden, great strides are made in all countries in engineering ecosystems to regulate climate. It is unclear, however, whether biospheric carbon storage can be much enhanced beyond what is already achieved by protective measures in Global Orchestration.

It is outside the scope of this analysis to assess the effect of vegetation on local climate, but we use model simulations to estimate here the effectiveness of the land biosphere in taking up CO_2 from the atmosphere. Net primary productivity of the land biosphere has been used as an indicator for a change in this service—however, since NPP is mostly balanced by soil respiration processes and the results of natural disturbance (fire, windstorms, and so on), it does not in itself allow for the direct estimation of climate regulation. At the local scale, for example, high NPP occurring after a clear-cut during regrowth of quickly growing trees and their understory can be more than balanced by the respiration flux from decaying organic material in the soil, thereby turning the ecosystem into a carbon source for some time (WBGU 1998). At the broader scale, however, NPP changes, such as those estimated by the IMAGE model, may give at least a hint at the changing regulating capacity of the land biosphere, because there is, at any point in time, only a small percentage of the land in early successional stages.

NPP was estimated in 2000 by IMAGE to be about 61.4 gigatons of carbon per year (within the typical range of other estimates; cf. Cramer et al. 1999). Based on IMAGE calculations, NPP increases across all scenarios and regions because of increasing temperature and atmospheric CO_2. Global estimates for 2050 range from 70.4–74.6 gigatons. Global Orchestration has the largest increase because it has the fastest pace of increasing temperature and atmospheric CO_2. Conversely, the TechnoGarden scenario has the smallest carbon uptake because it has the lowest temperature and CO_2 levels. The largest uptakes of CO_2 occur in regions with extensive forests, such as Russia and Canada.

On one hand, these estimates give a realistic representation of the future climate regulation function of the biosphere because they take into account the effect of deforestation in reducing the area of the biosphere, as well as shifts in vegetation zones caused by climate change. On the other hand, they may be overly optimistic because they do not factor in possible changes in soil processes that may lead to a net release rather than uptake of CO_2 by the biosphere (cf. the conflicting findings of Cox et al. 2000 and Friedlingstein et al. 2001). Moreover, the processes by which higher CO_2 stimulates greater carbon uptake by plants are not yet sufficiently understood and may be incorrectly represented in current models. Finally, the estimates of CO_2 uptake presented here do not take into account the future establishment of large-scale forest plantations for storing CO_2 from the atmosphere.

In conclusion, therefore, no scenario can count on a great effectiveness of the land biosphere as a climate-regulating factor independent of management. If an additional mitigation effect is achieved, then this will be due to favorable circumstances and probably not last much longer than for the twenty-first century (Cramer et al. 2001).

9.5.2 Risk of Soil Degradation

9.5.2.1 Methodology and Assumptions

The world's land resources play an important role in the production of food. The capacity of soils to perform this function can be seriously impaired by soil degradation (such as wind or water erosion), chemical degradation (such as salinization), and physical deterioration (such as soil compaction). Water erosion, as one of the degradation processes occurring most extensively at global level, has been singled out for this study. It is influenced by natural conditions, but also in the way that soil is used. Important factors that influence the rate of soil erosion (in the future) include agricultural practices, land·use change (in particular loss of vegetative cover), and changes in precipitation as a result of climate change.

9.5.2.2 Comparison of Erosion Risk among Scenarios

Historical trends in cropland degradation rates have been reported in UNEP's Global Assessment of Soil Degradation study (Oldeman et al. 1991; GRID/UNEP 1991) and in other studies (e.g., Kendall and Pimentel 1994). At present, the global hot spots of soil erosion by water are China, the Himalayan Tibetan ecosystem, the Andean region, the Caribbean (Haiti), the highlands of East Africa and Central America, southeastern Nigeria, and the Maghreb region. Similarly, the global hot spots of wind erosion are West and Central Asia, North Africa, China, sub-Saharan Africa, Australia, and the southwestern United States. Asia and Africa are the worse off regions in terms of land areas affected by at least moderate erosion. Anthropogenic causes responsible for erosion in these regions are deforestation, overexploitation of natural vegetation, overgrazing, and extension of agricultural activities to marginal land (such as steep land).

Changes in the risk of water-induced erosion from land use and climate change can be assessed with a methodology used for UNEP's Global Environmental Outlook (Hootsman et al. 2001; Potting and Bakkes 2004).[8] Here, water erosion risks are calculated by combining three indices: terrain erodibility, rainfall erosivity, and land cover:

- The terrain erodibility index is based on soil (bulk density, texture, soil depth) and terrain properties (slope angle), both of which are assumed to be constant in time.
- The rainfall erosivity index is determined by changes in monthly precipitation.
- The land cover pressure expresses the type of land cover. It is large for most agricultural crops, and small for natural land cover types such as forests.

The resulting index is a measure of the potential risk of water erosion, but it does not capture management practices. Such practices can make an enormous difference in actual erosion. Susceptibility to erosion is exacerbated by soil tillage and other mechanical disturbance. However, mechanical conservation measures (such as contour plowing, deviation ditches, and terracing and agronomic soil conservation practices) will prevent much water erosion in the real world.

Current climate models expect global precipitation to increase as a result of climate change (as described in the section on climate change). As a result, rainfall erosivity will also increase. Precipitation increase is likely to be strongest under the Global Orchestration scenario (see Table 9.20), but as noted before, in the comparatively short period until 2050 differences among the scenarios are still relatively small. The risk of water erosion is largest in agricultural areas, independent of the soil and climatic conditions. In the section on land use change, it was found that the largest increase in agricultural land will occur for the Order from Strength scenario. In the other scenarios, however, agricultural land also increases, particularly in poorer countries.

Table 9.20. Overview of Trends for Water-induced Erosion in MA Scenarios (IMAGE 2.2)

Variable	Global Orchestration	Order from Strength	Adapting Mosaic	Techno-Garden
Precipitation increase	+ +	+	+	+
Land use change	+	+ +	+	+
Agricultural practices	0	0/+	–	–

Key: + = Increased pressure on erosion control; O = neutral impact; – = decreasing pressure on erosion control.

Combining trends in climate and land use change and the erosibility index allows a calculation of the water erosion risk index. Compared with the present situation, the soil area with a high water erosion risk more than doubles by 2050 in all scenarios. (See Figure 9.39.) Differences among the scenarios up to 2050 are relatively small, with risks under TechnoGarden and Global Orchestration being somewhat less than under the other scenarios. Increases in risk areas occur in nearly all regions, with the exception of parts of the OECD region (Central Europe, Australia, and New Zealand)). Here, the area with a high erosion risk decreases, mainly as a result of gradually decreasing grazing areas. Areas with the most apparent increases in risk include North America (OECD region), Latin America, sub-Saharan Africa, and parts of Asia. (See Figure 9.40.) Increases are largest under Order from Strength, mainly due to higher larger food demand (due to larger population growth) combined with slower technological improvements. These two trends lead to the most rapid expansion of agricultural land.

In terms of potential agricultural practices that could mitigate the changes in the risk factors just calculated, we expect under Global Orchestration mainly a continuation of today's practices. Reforming socioeconomic policies in poorer countries could, however, lead to a much higher awareness of soil degradation. In Order from Strength, degradation rates in non-OECD regions for land owned by the poor could be more rapid, as they work low-quality land with insufficient resources, while the most productive agricultural land is managed by the elite. Here, changes in agricultural practices are not likely to reduce erosion risks. In Adapting Mosaic, local objectives on prevention of soil erosion could somewhat reduce erosion rates, slowing degradation on active agricultural land and significantly restoring currently degraded land. Under TechnoGarden, finally, the relatively low population levels and more ecologically proactive agricultural practices could in fact lead to a decline in net cropland degradation rates over the course of the scenario.

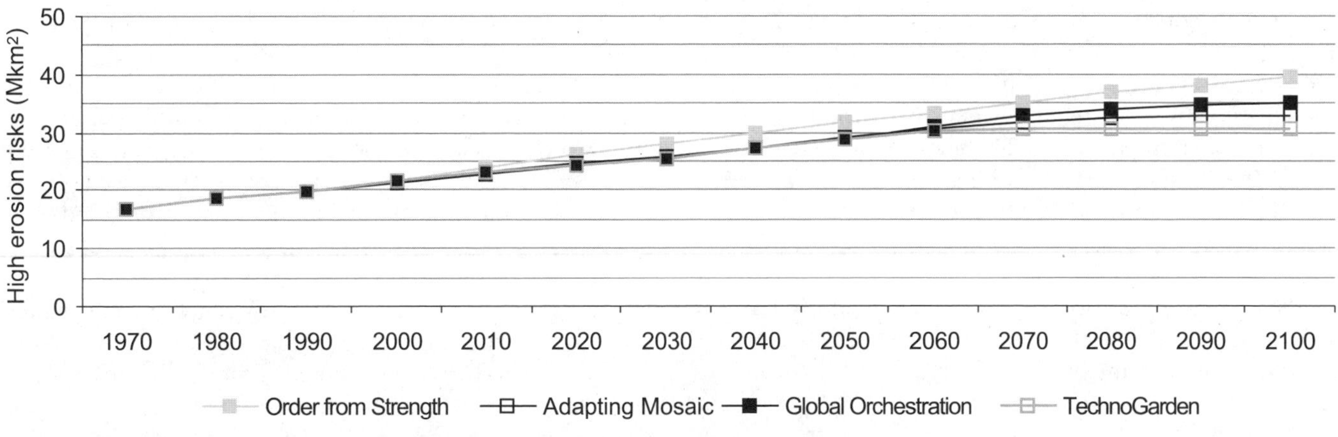

Figure 9.39. Global Area of Soils with High Water Erosion Risk in MA Scenarios (IMAGE 2.2)

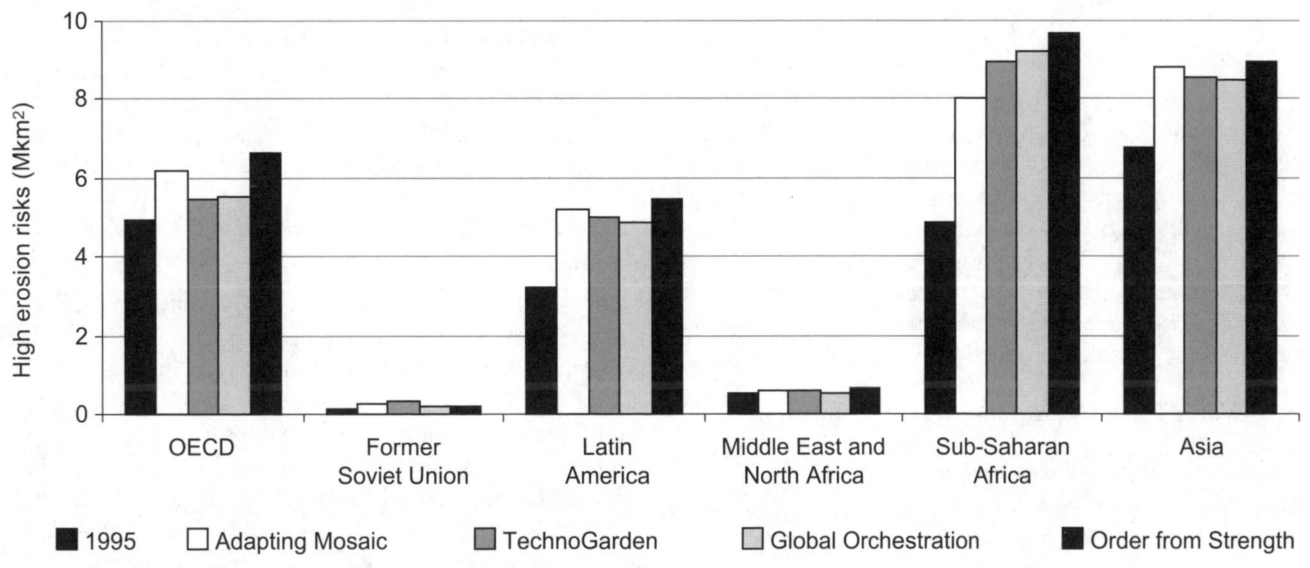

Figure 9.40. Global Area of Soils with High Water Erosion Risk in MA Scenarios in 2050 (IMAGE 2.2)

9.5.2.2.1 Regional trends

As indicated, soil erosion risks will be exacerbated in densely populated countries of the tropics and sub-tropics, where natural resources are already under great stress. The projected increase in soil erodibility is attributed to the decrease in soil organic matter content, reduction in the magnitude and stability of aggregates, and increase in the proportion of rainfall lost as surface runoff. The problem of soil erosion by water will be exacerbated in China, South Asia, Central Asia, the midwestern United States, East African highlands, the Andean region, the Caribbean, northern Africa, and the Maghreb. It should be noted that, in contrast to water erosion, the wind erosion hazard may not increase with the projected climate change. It may either stay the same or decrease slightly, because the projected increase in rainfall may improve the vegetative cover and decrease wind erosion.

9.5.2.2.2 Soil erosion and climate change

Soil erosion is not only influenced by climate change, it also contributes to greenhouse gas emissions. Soil organic matter (that is, carbon) is preferentially removed by both water and wind erosion. The fate of this matter is determined by a series of complex processes. Of the 4.0–6.0 gigatons of carbon per year translocated by water erosion, 2.8–4.2 gigatons are redistributed over the landscape, 0.4–0.6 gigatons are transported into the ocean and may be buried with sediments, and 0.8–1.2 gigatons are emitted into the atmosphere (Lal 2003). As this is a relatively large flux (cf. the 6 gigatons of carbon per year emitted from fossil fuel burning), changes in erosion can be relevant. Increases in soil erosion risks under each of the scenarios could lead to an increasing contribution of soil erosion to climate change.

9.5.2.2.3 Soil erosion and world food security

Productivity loss by soil erosion is attributed to the decline in effective rooting depth, reduction in available water-holding capacity, decline in nutrient reserves, and other short-term and long-term adverse effects on soil quality. Although no estimates of future yield losses are available, we can get some indication of potential losses by looking at review studies on current impacts on agricultural yields. Estimates on global productivity losses, in different periods, range from about 0.5% to 12.7% (Crosson 1994; Oldeman 1998; Den Biggelaar et al. 2001, 2004a, 2004b).

There are several regions where the productivity impacts are much higher, up to 20% per year or more (Dregne 1990, 1992, 1995; Lal 1995, 1998). For instance, Oldeman (1998) indicates impacts of 25% for Africa, 36.8% for Central America, 12.8% for Asia, and 13.9% for South America. Lal (1995) estimated that the reduction in crop yield due to past erosion was 8.2% for the African continent. In terms of future change, Lal (1995) estimated that if the accelerated erosion continues unabated, yield reductions in Africa by 2020 may double (to 16.5%). In our scenarios, this situation could especially occur under Order from Strength and Global Orchestration. In the other two scenarios, the impacts might increase for some time, although improvements might be possible later on.

9.5.2.2.4 Summary

All four scenarios are likely to experience an increased risk of water-induced erosion due to increased precipitation and further conversion of forest areas into cropland or pastureland. Changing agricultural practices, and other types of measures, will determine whether this will also result in increased erosion levels. Such measures could include adapting to climate change, taking soil conservation practices, and preventing further expansion of agricultural land, for instance by intensifying livestock production where possible.

9.5.3 Water Purification and Waste Treatment

9.5.3.1 Methodology and Assumptions

"Water purification" and "water regulation" refer to services provided by freshwater ecosystems, including wetlands, that help break down and remove substances harmful to humans and ecosystems. "Waste processing" is a more general term applied to all wastes and ecosystems. Here we focus on the ecosystem service of "water purification" although we believe (*with medium certainty*) that outcomes for "waste processing" are similar.

Although changes in water purification and waste processing depend on many factors, we can only quantify (and in an approximate manner) a few of these factors:

- *Dilution capacity of receiving waters.* Wastewater discharged into receiving waters is diluted and dispersed, although not necessarily below harmful concentrations in the vicinity of the wastewater discharge. Nor does dilution necessarily protect society or ecosystems from downstream impacts of these substances or the bio-concentration of harmful substances. As a surrogate of this dilution capacity we use runoff (see earlier description). In principle, an increase in runoff (outside of flooding periods) also increases dilution capacity.
- *State and areal extent of wetlands.* Wetland processes remove undesirable substances and treat and detoxify a variety of waste products (see MA *Current State and Trends,* Chapter 15). Denitrification processes convert nitrogen from the form that promotes eutrophication (nitrate) to nitrogen gas. Concentrations of easily degraded chemicals are reduced by the long residence time of water in wetlands. More persistent metals and organic chemicals in water are adsorbed to wetland sediments and therefore removed from the water column, but this can create hot spots of contamination in sediments. While we do not compute the state or extent of wetlands, we use two surrogate variables for this information: runoff and land encroachment. First, a large enough reduction in runoff can reduce the area and effectiveness of wetlands for processing wastes; the larger the reduction, the higher the risk to the waste processing ability of wetlands. Second, wetlands are drained and occupied because of the expansion of agricultural or urban land; the larger the expansion of agricultural land and population, the greater the risk of disappearing wetlands.
- *Magnitude of wastewater load.* The ability of wetlands and other aquatic ecosystems to detoxify wastewater can be overwhelmed by high waste loading rates (see MA *Current State and Trends,* Chapter 15). For freshwater ecosystems, this means the higher the loads of wastewater, the higher the risk that the ecosystem's waste processing ability will be overloaded.

9.5.3.2 Comparison of Water Purification Capacity among Scenarios

Under Global Orchestration, geographically we expect little net change in water purification capacity in wealthy countries. Dilution capacity of most rivers increases because higher precipitation leads to increases in runoff in most river basins. However, some smaller regions have decreasing precipitation and hence their rivers have decreasing runoff and dilution capacity. Wetland areas decrease because of the expansion of population and agricultural land, but this is a small change compared with in the other scenarios.

Under this scenario, wastewater flows increase by 40% (and hence increase the risk of overloading the detoxification ability of freshwater systems), but this is the second lowest increase among the scenarios. These factors may lead to a reduction in the ability of freshwater systems to handle wastewater loadings, but the reduction may be lower than in Order from Strength and Adapting Mosaic. Moreover, under this scenario the wealth of rich countries is used to repair breakdowns in water purification as they occur. In poorer countries, however, there are net losses in water purification by ecosystems. The pace of ecosystem degradation, the overtaxing of ecosystems by high waste loads, the decline of wetland area because of the expansion of population and agricultural land all tend to drive a deterioration of water purification.

Water purification declines in all countries under Order from Strength. In this scenario, the expansion of agricultural land and population is the largest of all the scenarios and poses the greatest risk to the state and extent of wetlands (and hence their capacity to process wastes). Likewise, the magnitude of wastewater discharges is the largest. In wealthy countries, lack of international coordination complicates the management of transnational watersheds, leading to further deterioration of water purification. In poorer countries, losses of water regulation capacity of ecosystems are more severe than during Global Orchestration.

The expansion of agricultural land and population (and risk to wetlands) is large under Adapting Mosaic, but not as large as in Order from Strength. The magnitude of wastewater discharges is second largest among the scenarios. Although these factors tend to reduce the ability of freshwater ecosystems to purify water, society gives local water management special priority and therefore ensures that wetlands are protected and wastewater discharges are treated. Hence in all countries we expect an improvement in the water purification capacity of ecosystems.

Under TechnoGarden, re-engineering advances in wealthy countries are slow because of existing ecological problems, such as the high levels of nutrients in soils and lags in ecosystem regrowth and turnover of infrastructure. On the other hand, this scenario has the smallest increase in pressure on the environment (smallest expansion of population and agricultural land, and smallest increase in volume of wastewater discharges). The net result is little change in water regulation by 2050. In poorer countries, there are improvements by 2050 because the time lags for ecosystem engineering are shorter, and in some cases the countries learn from, and avoid, errors made earlier rich countries.

9.5.4 Coastal Protection

9.5.4.1 Methodology and Assumptions

"Storm protection" describes the role of ecosystems in protecting society from storm damage. Here we focus on the ecosystem service of coastal protection. Although many different factors influence the level of coastal protection in a particular scenario, we take into account the adaptive capacity of nature, the adaptive capacity of society, and the extent of sea level rise.

The adaptive capacity of nature depends largely on the existence of natural buffers against storms such as coral reefs, mangrove forests, and sand bars (see MA *Current State and Trends,* Chapter 16). Meanwhile, the adaptive capacity of society (in the sense of coastal protection) is a function of many economic, social, and political factors, including the priority society gives to preserving or restoring natural buffers. The extent of sea level rise will depend on various tectonic processes over geologic time, but more so on climate change over the time horizon of the MA scenarios.

9.5.4.2 Comparison of Coastal Protection among Scenarios

Coastal protection remains about the same in wealthy countries during Global Orchestration. Under the reactive ecosystem management that prevails in this scenario, it is thought to be more cost-effective to address storm damage after it occurs than to maintain ecosystem configurations that mitigate storm damage. In poorer countries, coastal protection declines due to degradation of ecosystems. A similar viewpoint leads to little change in coastal protection by ecosystems in wealthy countries during Order from Strength. In poorer countries, ecosystem degradation leads to extensive losses of coastal protection during Order from Strength.

Under Adapting Mosaic, society emphasizes a configuration of ecosystems to meet regional goals. Storm protection is likely to be one of those goals, leading to improvements under this scenario. In TechnoGarden, ecosystems are deliberately engineered to provide ecosystem services such as coastal protection. This leads to improved coastal protection in wealthy countries. In poorer countries, improvements are sometimes offset by unforeseen responses of ecosystems. The net result is little overall change in coastal protection from ecosystems.

Earlier in this chapter we described the sea level rise that is expected (*with high certainty*) to accompany climate change in the MA scenarios. IPCC assessments indicate that sea level will rise under climate change because warmer air temperatures will cause ocean water to expand, and warmer air temperatures will melt the ice and snow that now persist on the ice caps and glaciers from year to year. Furthermore, climate change may cause stronger and more persistent winds in the landward direction along some parts of the coastline, and this will also contribute to rising sea level at these locations. In the four scenarios (given a medium climate sensitivity), the global-mean sea level is expected to increase in the range of 50 centimeters (in TechnoGarden) to 70 centimeters (in Global Orchestration) between 1995 and 2100 (but there is a considerable uncertainty attached to these numbers).

While the precise impact of sea level rise on reducing coastal protection is difficult to assess, we estimate (*with medium to high certainty*) that populated coastal areas under all scenarios will require new coastal protection measures such as stronger and higher dikes or flood gates in estuaries. These are all expensive undertakings and they might be affordable only in the world's richer countries. Hence, for all scenarios we expect (*with medium certainty*) a higher storm risk to coastal populations because of sea level rise and a relatively higher risk in poorer than in wealthy countries.

9.5.5 Other Regulating Ecosystem Services

Here we briefly address the trends of two other processes that fit in the category Regulating Ecosystem Services.

9.5.5.1 Comparison of Pollination among Scenarios

The MA *Current State and Trends* volume describes the deterioration of pollination due to species losses, use of biocides, climate change, and diseases of pollinators. This trend will continue during the Global Orchestration, Order from Strength, and TechnoGarden scenarios. In addition, the continuing deforestation and urbanization in these scenarios is likely to be accompanied by landscape fragmentation (that is, the degree to which natural landscapes are broken up by different land uses of society), which will further reduce the effectiveness of pollinators. In TechnoGarden, there are some successful efforts to engineer pollination and produce crops that do not need pollinators—for example, development of self-pollinated strains.

During Adapting Mosaic, maintenance of pollinators is a goal of some regional ecosystem management programs. Some of these succeed in maintaining populations of pollinators or in adapting to shifting ranges of pollinators as the climate changes. Thus in some regions the pollination capacity is maintained or even improves. On balance, pollination is maintained during Adapting Mosaic.

9.5.5.2 Comparison of Biological Pest and Disease Control among Scenarios

Biological control is expected to change little in wealthy countries during Global Orchestration, since increased wealth should improve biological control research and practices, but the spread of invasive species will present new challenges. In poorer countries, losses of biodiversity during Global Orchestration will compromise the capacity for biological control. In Order from Strength, biological control is expected to deteriorate in all countries due to the decline in local ecosystems and biodiversity. In Adapting Mosaic, there is emphasis on adjusting ecological feedbacks to meet local goals for ecosystem management. This is likely to lead to improvements in biological control in at least some regions of both rich and poorer countries. In TechnoGarden, society invests in engineering of biological controls, but as ecosystems are simplified the biological controls become more difficult to implement. On balance, there is little net change in the capacity of ecosystems to provide biological control in this scenario.

9.6 Supporting Ecosystem Services

Supporting ecosystem services are those that are necessary for the production of all other ecosystem services. Their impacts on people are indirect or occur over a long time frame. Examples of supporting ecosystem services are soil formation, primary production, nutrient cycling, and provisioning of habitat. Since the impacts of these services occur over such a long time period, management does affect many of them in a time period relevant for 50-year scenarios. However, even small changes in the provision of these services will eventually affect all other types of ecosystem services.

In general, the scenarios in which people handle environmental problems in a reactive manner more often than not—Global Orchestration and Order from Strength—do not focus on maintaining supporting services. The short-term approach to fixing the most immediate problems does not allow for full consideration of long-term services like the ones in this category. Thus supporting services undergo a slight, gradual decline in these two scenarios. This decline is likely to go unnoticed until it causes major surprises. On the other hand, the two scenarios in which some environmental actions are proactive, Adapting Mosaic and TechnoGarden, may give some consideration to the management of certain supporting services, causing them to remain steady throughout these scenarios.

9.7 Cultural Ecosystem Services

Cultural ecosystem services are nonmaterial benefits obtained from ecosystems. The conceptual framework of the MA lists the following clusters of cultural services:

- cultural heritage and diversity,
- spiritual and religious,
- knowledge systems (diversity and memory),
- educational and aesthetic values,
- inspiration,
- sense of place, and
- recreation and ecotourism.

This section describes some of the possible changes in these services under the four MA scenarios on the basis of the qualitative assessment. (See Table 9.21.) The presentation focuses on the services where differentiation between scenarios can be achieved, based on either calculations with numerical models or on our best qualitative assessment derived from the scenarios and an assessment of recent scientific literature. In general, global models have been less successful at quantitatively estimating changes in cultural ecosystem services (see Chapters 4 and 12); therefore, most of the discussion here will focus on a qualitative assessment of changes in cultural ecosystem services across the four scenarios.

Overall, cultural services decline slightly in Global Orchestration. People have less contact with nature and therefore have less personal familiarity with it. This lack of personal experience generally reduces the benefits of cultural services. The world in this scenario experiences some loss of indigenous knowledge systems and other cultural diversity, but recreation possibilities do increase, particularly in poorer countries. On the other hand, cultural services generally decline in Order from Strength, especially in poorer countries. People in wealthy countries have far less contact with nature and less familiarity with it. Adapting Mosaic shows a different pattern: an increase in basically all cultural services. This scenario, with its focus on preservation of local knowledge and innovation, prizes cultural ecosystem services and emphasizes retaining or improving them. TechnoGarden is focused on education and knowledge but ignores local or traditional knowledge in favor of global technologies. Thus, the results for cultural ecosystem services under this scenario are mixed. Knowledge shifts away from traditional knowledge to technological information, leading to a loss of cultural diversity.

When we consider the many particular types of cultural ecosystem services, it seems that each scenario offers a different mix, so that there are no really strong consistent patterns of decline or improvement in cultural services across all scenarios. Overall, it appears that the details of each particular path into the future have a considerable, but path-specific, impact on the provision of cultural ecosystem services.

The cultural services associated with a sense of place—inspiration, aesthetic values, cultural heritage, social relations, knowledge systems, and sense of place itself—follow approximately the same pattern across the scenarios. These cultural services stay the same as they were in the year 2000 in Global Orchestration, mostly stay the same in wealthy countries in Order from Strength, improve in Adapting Mosaic, and stay the same or decline somewhat in TechnoGarden. The declines in poorer countries in Order from Strength are largely due to the difficulty of simply meeting

Table 9.21. Qualitative Expectations for Cultural Ecosystem Services in MA Scenarios. Ecosystem services are defined in the MA conceptual framework volume (MA 2003: 56–59). "Industrialized Countries" stands for nations that are relatively developed and wealthy in 2000; "Developing Countries" stands for nations that are relatively underdeveloped and poor in 2000. Note that any particular nation could switch categories between 2000 and 2050. Scores pertain to the endpoint of the scenarios, 2050. A score of +1 means that the ecosystem service is in better condition than in 2000. A score of 0 means that the ecosystem service is in about the same condition as in 2000. A score of −1 means that the ecosystem service is in worse condition than in 2000.

	Global Orchestration		Order from Strength		Adapting Mosaic		TechnoGarden	
Ecosystem Service	Industrial Countries	Developing Countries	Industrial Countries	Developing Countries	Industrial Countries	Developing Countries	Industrial Countries	Developing Countries
Cultural diversity	−1	−1	−1	−1	+1	+1	−1	−1
Spiritual and religious values	0	0	0	−1	+1	+1	−1	−1
Knowledge systems (diversity and memory)	0	−1	−1	−1	+1	+1	0	0
Educational values	0	0	−1	−1	+1	+1	+1	+1
Inspiration	0	0	0	−1	+1	+1	0	0
Aesthetic values	0	0	0	−1	+1	+1	0	0
Social relations	0	0	0	−1	+1	+1	−1	−1
Sense of place	−1	0	0	−1	+1	+1	−1	−1
Cultural heritage values	0	0	−1	−1	+1	+1	0	0
Recreation and ecotourism	−1	+1	−1	+1	−1	−1	+1	+1
Comment on cultural services	loss of some indigenous knowledge systems; people have less contact with nature and therefore less personal familiarity with it		loss of many indigenous knowledge systems; emphasis on security inhibits innovation in ecosystem management; ecotourism is less safe in the developing countries; especially in industrial countries, people have less contact with nature and therefore less personal familiarity with it; there is an increase in fundamentalism with respect to spiritual and religious values		emphasis on preservation of local knowledge for ecosystem management, and innovation of new ways of managing ecosystems		shift of knowledge from traditional forms of technological information; changing knowledge through spiral of technical innovation; tourism in engineered ecosystems is fundamentally different from that in wilderness; designer ecosystems should increase perceived value of nature; people have less contact with nature and therefore less personal familiarity with it	

basic needs in these areas. The declines in TechnoGarden, on the other hand, are due to the tendency in this scenario to engineer ecosystems that are similar despite differences in location.

Other types of services follow different patterns across the scenarios. For example, cultural diversity declines under all scenarios except Adapting Mosaic, where it improves.

9.8 Cross-cutting Synthesis

The MA scenarios about the future of ecosystem services, as presented in this chapter, have been developed using multiple quantitative and qualitative methods, each of which comes with its own assumptions and uncertainties. In most cases, a rigorous assessment of this uncertainty is not yet possible. Here, we provide a synthetic analysis of ecosystem change based on a comparison of results across the four scenarios.

9.8.1 What Drives the MA Scenarios?

The key driving forces of the MA scenarios include population, income, technological development, changes in consumption patterns, land use change, and climate change. (Other driving forces are described in the early part of this chapter.) The future trends of these driving forces are quite different under the storylines of the four scenarios.

Change in population is important because population size directly influences the demand for future ecosystem services. Global population estimates for 2050 range from approximately 8 billion (Global Orchestration) to 9 billion (Order from Strength). Assumptions about future income affect the amount of ecosystem services required or consumed per person. The low estimate (Order from Strength) implies that global annual average income levels per person (as measured on market exchange basis) increase by a factor of two between now and 2050. Under the high estimate

(Global Orchestration), the increase is more than a factor of four during the same period. The income gap between the richest and the poorest world regions remains about the same under Order from Strength but narrows under the other three scenarios.

The rate of technological change affects the efficiency by which ecosystem services are produced or used. For example, a higher rate of technological change leads to more-rapid increase in the yield of crops per hectare or in the efficiency of water use by power plants. The rate of technological change in general is highest under the Global Orchestration scenario and lowest under the Order from Strength scenario. With respect to environmental technology (such as control of emissions, efficiency of water use) developments under TechnoGarden are as rapid as under Global Orchestration.

Consumption patterns also form an important driver determining the provision of ecosystem services. Consumption patterns are strongly affected by development pathways (and therefore, all other factors being equal, by economic growth). Relative to the overall trends, however, consumption patterns can be assumed to be more oriented toward low ecological impacts in the two environmentally proactive scenarios, TechnoGarden and Adapting Mosaic, and to be more material-intensive under the other two scenarios. One example of this is the consumption of meat products.

For land use change, results critically depend on the scenario results for food production and trade. The amount of land used by humans has been increasing significantly at the expense of natural biomes (such as forest and grasslands). In the MA scenarios, expansion of agricultural land slows down and even stabilizes in TechnoGarden, but continues to grow under Order from Strength. Critical factors include population size, diets (in particular consumption of meat products), and development of agricultural yields. Climate change, finally, takes place in all four scenarios. The increase of global mean temperature ranges from 1.6° to 2.0° Celsius in 2050, and from 2.0° to 3.5° Celsius in 2100.

9.8.2 Patterns in Provisioning and Regulating Ecosystem Services across the Scenarios

In the world of Global Orchestration, outcomes for ecosystem services are mixed. There is a very rapid increase in demand for provisioning services, such as food and water. As the means for investments exists, it can be expected that there is a simultaneous improvement in the provisioning services for food, fiber, and fuel. However, there is also degradation of ecosystem services related to biodiversity, such as biochemical discoveries and biological control. The ad hoc approach to the management of ecological issues that comes from addressing each issue as it becomes important leads to neglect of the underlying causes of watershed degradation. Degradation of watersheds leads to breakdowns in freshwater availability, water regulation, erosion control, water purification, and storm protection. Particularly in poorer countries, there is degradation of regulating ecosystem services. In general, the tendency to neglect the underlying processes that provide ecosystem services creates vulnerability during Global Orchestration. Probably one of the most important threats to sustainability under this scenario is climate change. The increase of global mean temperature is the largest of the four scenarios.

Under Order from Strength, there is a major risk of collapsing ecosystem services. These are maintained only for a few cases in wealthy countries. In poorer countries, all provisioning and regulating ecosystem services are in worse condition in 2050 than they were in 2000. Nationalism and lack of international cooperation make it difficult to address transnational ecosystem problems. Overarching concerns about security push ecosystem issues into the background for policy-makers, except for times when ecosystem services such as food or freshwater supply fail catastrophically. These failures have greatest impact in poorer countries. In rich countries, disaster-relief efforts address the short-term consequences of ecosystem breakdowns but rarely tackle underlying causes or reduce vulnerabilities. In all countries, the vulnerability of ecosystem services is greater in 2050 than it was in 2000.

There will be a strong focus under the Adapting Mosaic and TechnoGarden scenarios to maintain (or even improve) provisioning and regulating ecosystem services.

In Adapting Mosaic, ecosystem management focuses on comanagement of local or regional resources. People attempt to reduce vulnerability of ecosystem services using approaches that "design with nature." These attempt to achieve an optimal balance among local social, ecologic, and economic needs. They involve ongoing adjustments of management practices to changing conditions and opportunities. The adjustments sometimes require experiments that are inconsistent with management for maximizing the production of desired ecosystem services. The approaches of Adapting Mosaic tend to maintain or increase both genetic resources and the diversity of landscapes. They decrease vulnerability of ecosystem services, particularly those related to food and fresh water. Moreover, the lack of focus on global ecological change means a great risk of leaving these unaddressed. For instance, under this scenario a considerable increase in global mean temperature is expected to occur.

In TechnoGarden, the prevailing approach to ecosystem management seeks innovative environmental technologies that allow the supply of desired ecosystem services to be increased. This approach enables the production of more food and fiber per unit land area, thereby reducing the footprint on wild lands of agriculture and forestry. The technological emphasis also provides advances in manipulating genetic resources, utilizing natural biochemicals, and introducing ecological engineering of air quality, fresh water, and climate. These approaches are efficient, but they create vulnerabilities because they neglect the ecological infrastructure and diversity that ensure production of ecosystem services. Some of the vulnerabilities are discovered only when they trigger breakdowns. Some of the breakdowns are catastrophic and require expensive re-engineering of ecosystem service systems. Only under this scenario have we assumed climate policies to be implemented. As a result, global mean temperature increase is limited to 2° Celsius above preindustrial levels.

9.8.3 Hotspot Regions with Particularly Rapid Changes in Ecosystem Services

In three regions, several different pressures on ecosystems and human well-being seem to be relatively high under all the scenarios.

- *Central Africa*—The African region sees a rapidly increasing population in all four MA scenarios. As a result, the demand for provisioning services such as food and water also increases rapidly, in fact in some cases even beyond the potential of this region to supply these services. Increased food imports will be an important strategy to deal with these problems. Nevertheless, in most of the MA scenarios the area for natural biomes will be strongly reduced. Moreover, to meet its needs for development, the central part of Africa will need to rapidly expand its withdrawal of water, and this will require an unprecedented investment in new water infrastructure. Under some scenarios, this rapid increase in withdrawals could also cause a similarly rapid increase in untreated return flows to the freshwater systems, which could endanger public health and aquatic ecosystems. The further expansion of agriculture would lead to losses of ecosystem services provided by forests in this region and to undesirable side effects of agricultural intensification, such as contamination of surface and groundwaters.
- *Middle East*—The MA scenarios tend to indicate that rapid increases in population and (secondary) rising incomes in MENA countries will lead to greater demand for food (including meat), which could lead to a still higher level of dependence on food imports. Rising incomes will also put further pressures on limited water resources, which will either stimulate innovative approaches to water conservation or possibly limit development.
- *South Asia*—The MA scenarios point toward continuing deforestation in these areas, increasingly intensive industrial-type agriculture, rapidly increasing water withdrawals and return flows, and further intensive water stress. This may be a region where the pressure on ecosystems causes breakdowns in these ecosystems, and these breakdowns could interfere with the well-being of the population and its further economic development.

9.8.4 Trade-offs between Ecosystem Services

A robust preliminary conclusion is that all scenarios in general depict an intensification of the trade-offs already observed between different ecosystem services. In particular, the demand for services increases rapidly under each of the four scenarios (but with a wide range across them). This increase in demand could place significant pressure on ecosystems.

9.8.4.1 Possible Gains in Provisioning Services

- World agricultural production increases. For example, world total production of grains increases around 50% for all scenarios, with larger differences between scenarios for poorer regions. However, per capita consumption of grain (for food and feed) remains near its current level of around 300 kilograms per year. Consumption in the sub-Saharan region does not substantially increase under any scenario.
- Domestic water use per person per year grows in the sub-Saharan and other poorer regions by a factor of five or more (depending on the scenario and region), and this implies increased access of the population in these regions to fresh water. In OECD countries, there is a decline in domestic water use per person because of more-efficient usage. Because of stabilization of food consumption and gains in access to water supply and other factors, the percentage of malnourished children falls by 40% in sub-Saharan Africa under the Global Orchestration scenario. The decline is much smaller under the Order from Strength scenario.
- The amount of wood extracted from remaining forests for fuelwood and fuel products is likely to increase greatly up to 2050 (despite a loss of land) under all scenarios. The sustainability of this wood extraction has not been analyzed.
- The demand for the provisioning service of fish production increases under all scenarios. Whether this demand can be met depends critically on assumptions on the sustainability of aquaculture.

9.8.4.2 Possible Losses in Provisioning Services

- Some of the gains in agriculture will be achieved through expansion of agricultural land and at the expense of uncultivated natural land. This applies to all scenarios. A rough estimate is that, by 2050, 10–20% of current grassland and forestland will be lost, mainly due to the expansion of agriculture (and secondarily because of the expansion of cities and infrastructure). The provisioning services associated with this land (genetic resources, wood production, habitat for terrestrial biota and fauna) will also be lost. The loss of wood production on this land might be compensated for by more-intensive production elsewhere.
- Although gains are made in access to fresh water, all scenarios also indicate a likely increase in the volume of polluted fresh water (especially in poorer countries if the capacity of wastewater treatment is not greatly expanded). Moreover, the expansion of irrigated land (which contributes to the increased production of grains) leads to substantial increases in the volume of water consumed in arid regions of Africa and Asia. These and other changes in the freshwater system are likely to cause a reduction in the provisioning services now provided by freshwater systems in developing countries (such as genetic resources, fish production, and habitat for other aquatic and riparian animals).

9.8.4.3 Uncertain Changes in Regulating Services

- It is not clear whether climate regulation will be increased or decreased under the scenarios. On one hand, the warmer, moister climate will, on average, increase primary productivity and the uptake of CO_2 in the atmosphere. On the other hand, the depletion of natural forest and grassland may lead to a decrease in standing biomass on Earth. Uncertain factors are the longer-term direct influence of CO_2 on plant growth and soil carbon pools, as well as the uncertainty of rainfall changes in water-limited regions. This question must be further analyzed.

9.8.4.4 Possible Changes in Regulating Services

- All scenarios assume an increase in per capita income and imply an increase in material well-being. This is likely to lead to higher consumption of electricity and fuel for transport, as well as to a higher production of industrial products. The result will be a decline in air quality maintenance, as indicated by a substantial rise in the emissions of SO_2 and NO_x, especially in poorer countries. While wealthy countries are expected to maintain or expand their control of local and regional air pollution, the same is not expected for poorer regions. The loss of natural land will also affect the regulating services it provided (erosion control, regulation of human diseases, and water regulation).
- Reduction in the size of natural ecosystems might have strong repercussions for regulation services associated with these ecosystems (such as erosion control, regulation of human diseases, and water regulation).

9.8.5 Uncertainty

Uncertainty in the assessment of this chapter arises from two principal sources: the methods being applied (primarily numerical models, expressing currently available understanding in mathematical formalisms), and the scenario storylines themselves.

9.8.5.1 How Certain Are the Model Results Used to Quantify the Scenarios?

As discussed in Chapter 6, all model results transport uncertainty since they make assertions about complex ecosystems over several decades into the future. A typical problem arises when output from one model is used as input for another model, and the uncertainty of the models propagates and multiplies. However, this problem can be lessened by interpreting modeling results in a conservative way.[9] Another drawback with current modeling approaches for ecosystem services is the lack of connections and feedbacks between human and environmental systems.

Despite these uncertainties, numerical models are a useful tool that allows us to combine complex ideas and data from the social and natural sciences in a consistent way. It is evident that this combination provides useful information to supplement the storylines presented in Chapters 5 and 8. Indeed, the modeling results show certain tendencies that can help us anticipate coming risks to ecosystem services. Moreover, they can provide information that is useful for developing policies to lessen these risks. In many areas, qualitative assessment methods can supply further valuable information.

9.8.5.2 Do the Scenarios Cover the Entire Range of Possible Futures?

The quantitative scenarios do not cover the entire range of possible futures because they do not include major surprises that we know from history will have a profound effect on ecosystem services (such as breakthroughs in technology, unexpected migration movements, and major industrial accidents). The models used to quantify the scenarios also rarely generate plausible "breaking points" in which ecological thresholds are exceeded (such as rapid changes in water quality or pest outbreaks over large agricultural areas). Exceeding these thresholds could have important consequences on the future of the world's ecosystems. Models do not generate breaking points because they poorly represent global feedbacks and linkages. This reflects the state of the art of global modeling, which needs to be improved to address urgent MA-relevant questions. On the other hand, the scenario storylines do include many examples of surprises and breaking points.

9.8.5.3 Likelihood of Surprises in Different Scenarios

Each scenario carries a certain risk of surprising disturbances. The level of risk is based on the pressures on ecosystems and on society's concern for learning about how ecosystems work and understanding threshold behaviors in ecosystems. Societies also have different vulnerabilities to these surprises based on aspects of well-being, including social networks, education, flexibility, and economic well-being. (The risk of extreme events is discussed in more detail in Chapter 5.)

For most services, the pressure on ecosystems is greatest from the Order from Strength scenario because of the combination of slow, unrelenting growth of population, slower development of new technologies, and a lack of interest in environmental management. Consequences include increases in global energy use throughout the century and an acceleration of current deforestation rates, with near depletion of forests in Africa and parts of Asia and the Amazon by the end of the century. High demand for ecosystem services, combined with a lack of concern for the ecosystems providing these services, leads to a high likelihood of situations in which society is surprised by sudden changes in ecosystems.

In Global Orchestration, the world may be confronted with major unexpected consequences of climate change, since here the average rate of change of temperature and precipitation is likely to be the fastest in the first half of the century. Of all the scenarios, this world could see the greatest consequences of climate change on water resources, changing natural vegetation, and crop yield. Intensification of agriculture is also extreme. Fast economic growth and a focus on reducing poverty lead to high levels of consumption of ecosystem services. This, combined with a lack of concern for understanding the ecosystems that provide these services, again leads to a high likelihood of surprise. This is offset by a high potential to respond to surprise due to higher incomes and economic well-being.

TechnoGarden also has a high risk of surprise. TechnoGarden's focus on new technologies leads to a high risk of technological failure. The risk is that each new technology can lead to a new and unexpected problem, which is then solved by another new technology. People in the TechnoGarden scenario are able to reduce their risk through a focus on understanding ecosystems and maintaining supporting and regulating ecosystem services. This scenario features high levels of agricultural intensification, which may yield unexpected outcomes for ecosystem services. Moreover, the assumptions on climate change policy in this scenario rely on strong technological development and continuous support to bear the costs that are associated with these policies. There is a risk that one of these conditions might not be met.

The Adapting Mosaic scenario has the lowest risk of surprising events. Moreover, expansion of agricultural land in poorer countries is lowest in this scenario. Nevertheless, the major risk factor in this scenario is represented by major ecological changes that occur at the global scale, as environmental management is oriented at the local scale. The most obvious example is climate change (we have not assumed climate policies under this scenario), but other examples could include disturbance of the global nitrogen cycle.

9.8.6 Outlook

Using four major storylines and a set of numerical models to investigate the potential risks for the future provision of ecosystem services may appear an impossible task under conditions of scientific rigor and quality standards. However, it must be noted that uncertainty in itself is not a factor disqualifying systematic analysis. Advances in global ecological research can only be achieved on the basis of searching for the limits of our understanding. Areas for improvement, in fact, could include the following: better coverage of provisioning, regulating, and supporting ecological services; better coverage of the feedbacks for ecosystem change on human development; better insights into the sustainability of some provisioning services, such as water supply and wood production; and further disaggregation of the analysis, for instance by using local-scale models within the global context laid out by the tools used in this chapter.

Appendix 9.1. Selected Drivers of the Ecosystem

Crop Area and Livestock Numbers Growth

Exogenous assumptions were made for crop area and livestock number development by scenario. Crop area and livestock numbers are then further adjusted endogenously based on other parameters (yield growth, population, income growth, and so on) to meet effective food demands. The ranking resulting from effective crop area growth over the projections period is (1) Order from Strength, (2) Adapting Mosaic, (3) TechnoGarden, and (4) Global Orchestration, where crop area expansion is largest under Order from Strength and lowest under Global Orchestration. Total livestock population is a function of the livestock's own price and the price of competing commodities, the prices of intermediate (feed) inputs, and a trend variable reflecting growth in the number of livestock slaughtered. Numbers and weight growth for the group of livestock products is heavily influenced by the increasing share of chicken (low weight, large numbers) in total livestock production and the correspondingly declining share of beef and pig (larger weight and relatively lower numbers) in total production. Whereas for crop production, yield growth is the major contributor to future production increases, for livestock products, growth in numbers will remain dominant for production increases into the future.

Crop area expansion under the Order from Strength scenario is driven by a combination of high population growth and low yield improvements. Increased food demand coupled with low output per unit area induces farmers to expand cultivation on marginal lands. In terms of regional implications, between the base year and 2050 sub-Saharan Africa sees the largest expansion, with an overall increase in harvested cereal area of 70%, followed by Latin America with an increase of 36%, and MENA at 30%. Cereal area is projected to increase in Asia by 8%, in the former Soviet Union by 17%, and in the OECD region by 11%. The evolution is similar for livestock numbers. Over the 1997–2050 period, the number of slaughtered cattle is expected to increase by 75%, that of pigs by 34%, and that chicken by 100%, with numbers growth accounting for 78%, 74%, and 83% of total production growth respectively.

Similarly, under the Adapting Mosaic scenario, high population and relatively low crop yield growth also lead to substantial crop area expansion, led by sub-Saharan Africa, with 50%, followed by Latin America, 21%; MENA, 16%; former Soviet countries, 2.8%; and Asia, 1.5%. Area is projected to slightly decline in the OECD region, however (−0.1%). Livestock numbers growth under Adapting Mosaic is much slower compared with the Order from Strength scenario, with the slaughtered cattle numbers increasing by 47%, pigs by 29%, and chicken by 104%, with numbers growth accounting for 66%, 67%, and 80% of production growth, respectively.

Rather than expanding in all MA regions, as in the case of Order from Strength, crop areas in TechnoGarden and Global Orchestration are projected to contract considerably in certain countries and regions. The decline under TechnoGarden is mainly due to an increase in conservation programs to retire land for biodiversity, improved land use through improved technology applications on existing areas, and sufficient yield enhancements making expansion into marginal areas unnecessary. By 2050, cereal crop areas are expected to decline in the OECD region by 10%, in the former Soviet Union by 7%, and in Asia by 6%. On the other hand, harvested areas are set to rise 37% in sub-Saharan Africa, 9% in Latin America, and 7% in MENA. Under the TechnoGarden scenario, the global number of buffaloes and other cattle increases by 48%, the number of pigs rises by 26%, and the number of chickens goes up by 73% from 1997–2050. Numbers growth is projected to account for 65%, 61%, and 73% of livestock production growth, respectively.

Under the Global Orchestration scenario, cereal areas also decline in some regions, but less so than under TechnoGarden, due to higher income growth in poorer regions and more demand for meatier diets. Globally, cereal-harvested area still expands slightly. By 2050, cereal-harvested area in the OECD and former Soviet regions is 10% and 6% lower than in the base year, and in Asia, it is 6% lower. On the other hand, expansion in cereal-harvested area in sub-Saharan Africa will still be considerable, at 45%. The area in Latin America is projected to expand by 12% and in MENA by 2%. Driven by rapid increase in demand for livestock products, the number of buffaloes and other cattle is projected to expand by 101%, the number of pigs by 71%, and the number of chickens by 172%, with growth in numbers accounting for 72% (beef), 74% (pigs), and 80% (poultry) of production growth.

Crop Yield Improvement

In IMPACT, crop yield is a function of the commodity price, the prices of labor and capital, and a projected non-price exogenous trend factor reflecting technology improvements. The non-price trend reflecting technological change is affected by a number of indirect determinants. These include public and private R&D, agricultural extension and farmers' schooling, development of infrastructure and markets, and irrigation capacity. The MA storylines provide different descriptions for the development of individual components of these technological trends depending on the scenario, with the two major determinants of the trend being changes in (agricultural) investment levels and water and energy use efficiency.

In Global Orchestration, crop yield improvement over time is assumed to range from medium to high for both rich and poorer countries. Improvements in water use efficiency and energy use efficiency, as well as large investments in agricultural research and supporting infrastructure, particularly in poorer countries, are major drivers behind the crop yield improvement for Global Orchestration. The greatest yield increases are seen in sub-Saharan Africa, with increases of 159% over the base levels, followed by Latin America at 114%, Asia at 84%, and MENA at 74%, compared with 49% for the former Soviet states and 47% for the OECD region.

Under TechnoGarden, crop yield improvement over time is assumed to be lower in the wealthy world due to a greater focus on organic farming. However, investments in biotechnology and other crop innovations are sufficient enough to bring about significant crop yield growth. Under TechnoGarden, the OECD and former Soviet Union are expected to experience cereal yield growth up to 2050 of about 45%; Asia, 72%; Latin America, 93%; sub-Saharan Africa, 106%; and MENA, 70%.

For Order from Strength, crop yield improvement is assumed low in all countries, as are improvements in water use and energy use efficiency, resulting from low investments in these sectors. In the OECD and former Soviet countries, yield improvements are only 20% and 24% over base levels, respectively. In sub-Saharan Africa, by 2050 cereal yield levels are still below 2 tons per hectare, after yield increases of 90% over base levels. In Latin America, yield levels are 54% higher; in MENA, 40%; and in Asia, 36%.

For Adapting Mosaic, crop yield improvement is assumed to start out at a medium level and then decrease over time in the rich world due to the adoption of organic farming. In the OECD and former Soviet countries, cereal crop yields increase by 38% and 34%, respectively. In poorer countries, crop yield improvements are somewhat larger, due to successful local adaptation mechanisms. Regionally, sub-Saharan Africa will lead with improvement in cereal yields of 103% by 2050. Latin America, Asia, and MENA follow with 69%, 49%, and 50% improvements respectively over base year cereal yields.

Average final cereal crop yields by 2050 are highest for Global Orchestration (4.7 tons per hectare), followed by TechnoGarden (4.3 tons), Adapting Mosaic (3.8 tons), and Order from Strength (3.5 tons).

Changes in Livestock Slaughtered Weight

Livestock slaughtered weight in the model is affected mainly by expected changes in technological development, without additional price effects. The assumptions made in terms of technological change in this case are along the same lines as those made with respect to crop yield changes, and the same ranking of scenarios is observed as a result.

Livestock slaughtered weight improves most under Global Orchestration and least under the Order from Strength scenario. For example, by 2050 slaughtered weight of cattle is projected to reach 260 kilograms per head under Global Orchestration compared with 229 kilograms per head under Order from Strength, 242 kilograms under Adapting Mosaic, and 245 kilograms under TechnoGarden. Among regions, growth in slaughtered weight for cattle under Global Orchestration is projected to be particularly high in Asia, followed by sub-Saharan Africa and MENA.

Under TechnoGarden, livestock weight improvement over time is assumed to be low in wealthy countries due to the already high level of weights achieved and relatively lower demand for livestock products, while it will be medium to high in poorer nations due to innovations in livestock breeding.

In Order from Strength, yield improvement for livestock is assumed low in all countries. Finally, for Adapting Mosaic livestock yield improvement is assumed to start at a medium level and to decline in the wealthy world, and to start out from medium-low to reach a medium level in poorer nations, driven by locally adapted breeding and a lack of investments in modern techniques in both rich and poor countries.

Basin-level Irrigation Efficiency

Under the Order from Strength scenario, government budgetary problems are assumed to worsen, resulting in dramatic government cuts on irrigation system expenditures. Water users fight price increases, and a high degree of conflicts results in lack of local water-user cooperation for cost-sharing arrangements. The turnover of irrigation systems to farmers and farmer groups is accelerated but not accompanied by the necessary reform of water rights and necessary funding. Rapidly deteriorating infrastructure and poor management reduce system- and basin-level water use efficiency under this scenario. As a result, efficiency levels, which are already quite low in most of Asia, drop by 23–28% to reach levels of only 0.25–0.30 by 2050. In East and South Africa, levels decline slightly less, by about 20%, to reach 0.44 by 2050. Declines are similar in Latin America, to reach 0.32–0.34 by 2050. In MENA, where efficiency levels were very high in 2000, at 0.6–0.7, levels are expected to decline to 0.56 by 2050. Efficiencies are more resilient in wealthy countries as elites concentrate resources on some systems to maintain minimum food self-sufficiency levels. As a result, levels decline by about 8% in the OECD and slightly more, about 15%, in the former Soviet nations.

Under the Global Orchestration scenario, careful market-oriented reform in the water sector and more comprehensive and coordinated government action will lead to greater water management investments in efficiency-enhancing water and agricultural technology. The effective price of water for the agricultural sector is assumed to increase more rapidly to induce water conservation as well as to free up agricultural water for other environmental, domestic, and industrial uses. Large investments in poorer countries lead to rapid increases in efficiency levels in Asia and sub-Saharan Africa, where levels increase to as high as 0.4–0.5 and 0.56–0.74, respectively. Selected—economically viable—investments also enhance efficiency levels in those countries and regions, where relatively high efficiency had been achieved by 2000, including the OECD and MENA, although increases are small. The highest irrigation efficiency level is achieved in the region facing the greatest water scarcity, North Africa, at 0.8.

Under TechnoGarden, technological innovations for on-farm efficiency increases help boost irrigation efficiency levels across the world to previously unseen levels. Moreover, river basins make progress toward more integrated basin management through real-time measuring and management of water resources. Gradual introduction of water price increases in some agricultural areas induce farmers in these regions to use water more efficiently. As a result, high efficiency levels are reached, particularly in regions where little or no further improvement had been expected, like the OECD and MENA. There, levels increase to 0.75–0.9 by 2050. Advances are also significant in Asia, reaching 0.5 by 2050, and in sub-Saharan Africa, where levels of 0.75 can be reached.

Under the Adapting Mosaic scenario, local adaptations, including expansion of water harvesting and other water conservation technologies, as well as the increased application of agro-ecological approaches, help boost efficiency levels in some regions and countries. Successful efficiency increases—although important—remain scattered in areas and regions within countries, and the global and regional impacts are smaller than under TechnoGarden and Global Orchestration. In Asia, the results are mixed, with increases in efficiency in East Asia balanced by declines in South Asia. In sub-Saharan Africa, the outcomes are more favorable, with conservation strategies boosting efficiency levels by 2–11%. The former Soviet Union is less successful with efficiency-enhancing methods, experiencing a slight decline in efficiency levels. The OECD region, as a whole, does successfully apply locally developed methods to enhance irrigation efficiency, with increased efficiency levels in some countries more than balancing declines in other ones.

Appendix 9.2. Additional Description of the Modeling Done for Chapter 9

Productivity Increase

Agricultural productivity growth can be due to area expansion or yield growth for crops and to an increasing number of animals slaughtered or improvement in slaughtered weight per head for livestock. In IMPACT, the factors influencing productivity growth include management research, conventional (plant) breeding, wide-crossing and hybridization breeding, and biotechnology and transgenic breeding. Other sources of growth considered include private-sector agricultural research and development, agricultural extension and education, markets, and infrastructure. In short, productivity drivers include greater public/private investment, better management practices, and improved technologies. Drivers were not further subdivided among technology, management, and infrastructure because the outcomes on the drivers are a function of all three of these factors. Area/numbers and yield/slaughtered weight growth were differentiated by scenario as deviations from our best estimates of future productivity and area increases for the 45 IMPACT countries and regions. Results from IMAGE on yield reduction factors from climate change were incorporated into IMPACT production growth assumptions for the four MA scenarios.

Dietary Preferences

Food demand is a function of the price of the commodity and the prices of other competing commodities, per capita income, and total population. Per capita income and population increase annually according to country-specific population and income growth rates by scenario.

Price elasticities of demand, which govern sensitivity of food consumption to a change in prices, are one important factor relating to price effects. The impact of changes in income on food demand is captured by the income elasticities of demand.

Price elasticities of demand are assumed to stay the same across the four MA scenarios. Income elasticities of demand, on the other hand, do vary by scenario. In general, income elasticities are considerably higher for high-valued commodities such as meat, milk, fruits, and vegetables compared with the elasticities for basic staple crops such as rice and wheat. With increasing incomes, the elasticity of demand for rice actually turns negative in some countries. In IMPACT, income elasticities of demand for meat and fish for wealthy countries is assumed to be lower than for poorer ones; as incomes increase, the elasticity of demand with respect to income declines.

The scenarios vary in their assumed income elasticities of demand for meat. (Income elasticities of demand for fish are not varied across scenario.) Among wealthy countries, demand for meat is assumed to be less sensitive to income changes for non-Global Orchestration scenarios. Between TechnoGarden and Order from Strength, income elasticities of demand are more inelastic for TechnoGarden in rich countries whereas they are similar for poorer countries. Adapting Mosaic is assumed to have the most inelastic income elasticities of demand for all countries.

Drivers Affecting Rates of Malnutrition in Addition to Caloric Consumption

IMPACT generates projections of the percentage and number of malnourished preschool children (0 to 5 years old) in poorer countries. Projections for the proportion and number of malnourished children are derived from an estimate of the functional relationship between the percentage of malnourished children, the projected average per capita kilocalorie availability of food, and non-food determinants of child malnutrition, including the quality of maternal and child care (proxied by the status of women relative to men as captured by the ratio of female to male life expectancy at birth), education (proxied by the share of females undertaking secondary schooling), and health and sanitation (proxied by the percentage of the population with access to safe drinking water). The equations used to project the percentage and numbers of malnourished children are as follows:

$$\%MAL_t = -25.24 * \ln(KCAL_t) - 71.76\ LFEXPRAT_t - 0.22\ SCH_t - 0.08\ WATER_t \quad (1)$$

and

$$NMAL_t = \%MAL_t \times POP5_t \quad (2)$$

where *%MAL* is the percentage of malnourished children, *KCAL* is per capita kilocalorie availability estimated in IMPACT, *LFEXPRAT* is the ratio of female to male life expectancy at birth, *SCH* is the percentage of females with secondary education, *WATER* is the percentage of the population with access to safe water, *NMAL* is the number of malnourished children, and *POP5* is number of children below five years of age.

Average per capita consumption per day is determined for the four different MA scenarios from IMPACT runs up to 2050 incorporating quantified parameters from the four storylines, including assumptions on area and yield growth, population and income growth, food preferences, investment levels, and assumptions regarding openness to trade. The non-food determinants of child malnutrition are synthesized from the storylines and assumed to improve the least under the Order from Strength scenario and the most under the TechnoGarden scenario.

Detailed Assumptions of the EwE Models for Fisheries Calculations

Appendix Table 9.1 shows the detailed assumptions that were made in the driving forces in each case study under the four MA scenarios.

Appendix Table A9.1. Summary of Harmonizing Storylines and Case Studies of Regional Marine Fisheries. Numbers in parentheses represent ratio of optimizing two or more policy options (EcoSim/EcoPath)

Time	Gulf of Thailand	North Benguela	Central North Pacific
Global Orchestration			
2000–10	optimize profits from shrimp and jobs (70/30); climate change medium-high	optimize profits and jobs (50/50); climate change medium	optimize profits from tuna and jobs (80/20); climate change low
2010–30	optimize profits, jobs, and then ecosystems (50/30/20); climate change impact reducing	optimize profits and jobs (30/70); climate change medium-low; increase catch of fish for fish food	optimize profits from tuna and jobs (70/30); climate change stable
2030–50	optimize profits and ecosystem (biomass) (50/50)	optimize profits, jobs, and ecosystems (50/20/30); increase the catch of small pelagics	optimize profits and ecosystems (50/50) through building of bigeye tuna
Order from Strength			
2000–10	optimize profits of the invertebrate fishery and jobs (50/50)	optimize profits and jobs (50/50) of high value fisheries; DWF increases effort (mod–high of current species as EU pushes for food security and African debts mount)	optimize profits from the tuna fishery as well as jobs (75/25); DWF effort remains stable since countries focused on national issues
2010–30	optimization mix continues (50/50) but effort increasing since Thailand feels the effects of national EEZs and despite agreements it has no room to expand DWF which is now concentrated in the Gulf	climate change starts low with build up over this decade to medium impact; rebuilding of biomass starts late in this period but there is still concern with maintaining jobs (30 profits/50 jobs/20 ecosystem)	optimize profit and jobs (85/15); Japan returns to drift netting; DWF has moderate increase as United States secures food and increases presence in Pacific for security
2030–50	climate change has significant impact (high impact) and ecosystem severely destabilized; rebuilding stocks of demersal species continues with objective of optimizing jobs rather than profits	mix of profit and job optimization (60/40) increased fishing effort with switch through time to fish meal species for domestic and international aquaculture operations and also internal food security	profit optimization not as important but jobs are (60/40); Japan stops drift netting by 2040, DWF effort remains stable
Adapting Mosaic			
2000–10	optimize profits of the invertebrate fishery and jobs (70/30)	optimize profits and jobs (40/60) and maintain food and fish meal fisheries	optimize profits from the tuna fishery; turtle exploitation ceases
2010–30	climate change starts in earnest (medium-high impact); optimize for profits; shift to rebuilding stocks of demersal species starts	climate change starts low with build up over this decade to medium impact; rebuilding of biomass starts late in this period but there is still concern with maintaining jobs (30/50/20)	climate change minimal if any impact; severe exploitation of bigeye tuna until close to 2030 when stock rebuilding commences at the same time as shift to optimizing for jobs with profit (70 profit/30 jobs)
2030–50	climate change has significant impact (high impact) and ecosystem severely destabilized; rebuilding stocks of demersal species continues with objective of optimizing jobs rather than profits	climate change continues to have high impact with some destabilization of the system; food security becomes an issue and therefore focus is on maximizing biomass for fish feed since it goes to aquaculture that ensures a stable supply of food (0 profits/100 jobs /0 ecosystems)	climate change has a low impact, bigeye tuna rebuilding continues; optimize for ecosystem, especially for top predators; international MPA to rebuild stocks (50 profits/50 jobs)
TechnoGarden			
2000–10	optimize profit	optimize profit	optimize profit
2010–30	optimize pelagic catch (cost of fishing lower) followed by ecosystem optimization (since impacts can be engineered)	optimize profits while increasing pelagics (50/50)for fish food since technology makes aquaculture widespread and demand for fish meal up despite artificial feed improvements	optimize profit, but with costs lowered since technology improves; possible to have more tuna caught younger for ranching (2015–2030)
2030–50	optimize pelagic catch—by 2040 ecosystem irrelevant due to technology advances—profits maximized by using Gulf to produce quality fishmeal for prawn aquaculture	optimize profits from fish used in fishmeal; basically supplies European demand for aquaculture	optimize profits, but fish changes to species for fishmeal since technology cracks tuna hatchery technology

Key: DWF = Distant Water Fleet; EEZ = Exclusive Economic Zone; MPA = Marine Protected Area

Notes

1. The WorldScan economic model has been set up to reproduce the GDP per capita numbers of the IPCC SRES scenarios at the level of the four aggregated IPCC regions, thus providing detailed information on a consistent macroeconomic trajectory for 12 WorldScan regions (Bollen 2004). In a next step, this information was further disaggregated into 17 IMAGE regions using simple desaggregation rules (See IMAGE-team 2001).

2. Gt C-eq is gigatons (thousand million tons) of greenhouse gas emissions, expressed in equivalent carbon dioxide emissions (in units of carbon). The conversion of different gases is done on the basis of "global warming potential" reported for 100 years, which measures the contribution of the different greenhouse gases over a 100-year-time period relative to carbon dioxide. The other greenhouse gases included in the numbers above are methane, nitrous oxides, HFCs, PFCs, and SF_6.

3. In assessing trends in land use and land cover change under the MA scenarios, we have used the IMAGE 2.2 model, with scenarios starting in 1970. For the historic 1970–95 period, land use (size of agricultural area) trends of IMAGE have been calibrated against FAO data. The size of 14 natural biomes (including ice, a large number of forest types, grasslands, desert, etc.) have been determined on the basis of the BIOME model that is included in IMAGE using climate and soil data. While the BIOME model represents overall patterns in existing land cover maps very well, the area of each biome type does not match exactly to available databases on land cover (typically, differences can be the order of 10–20% on the level of continents).

4. Here, we use the land use-change scenarios as calculated by IMAGE 2.2 under the MA storylines to assess the possible changes in land use. The changes in agricultural demand and agricultural management are derived from IMPACT as described in the section on provisioning of food. In addition, timber demand and demand for biofuels are taken into account. Climatic changes have been taken into account, as they drive changes in natural vegetation but also influence crop growth and thus yields.

5. Data on agricultural production for 1970–95 (FAO 2001) and for 2030 according to the AT 2030 projection (Bruinsma 2003) were implemented by Bouwman et al. (2005a) in the Integrated Model to Assess the Global Environment model (IMAGE-team 2001) to generate 0.5 by 0.5 degree global land cover maps. These were used to allocate fertilizer and animal manure inputs, ammonia volatilization, and crop nitrogen export. Country data on sanitation coverage, connection to sewerage systems, and wastewater treatment were taken from several sources (EEA 1998; EEA 2003; WHO/UNICEF 2000, 2001a, 2001b). For countries where data were lacking, the percentage of the population with connection to sewerage systems was estimated on the basis of the fraction of the urban population with improved sanitation and the degree of urbanization. In combination with the AT 2030 projection, target values for the year 2030 for the connection to sewerage systems and wastewater treatment were modified from WHO/UNICEF (2000) with adjustments for many countries on the basis of past developments or trends observed in other countries.

6. During the first steps in CO_2 assimilation, C3 plants form a pair of three carbon-atom molecules. C4 plants, on the other hand, initially form four carbon-atom molecules. An important difference between C3 and C4 species for rising CO_2 levels is that C3 species continue to increase photosynthesis with rising CO_2, while C4 species do not. So C3 plants can respond readily to higher CO_2 levels, and C4 plants can make only limited responses. C3 plants include more than 95 percent of the plant species on Earth. (Trees, for example, are C3 plants.) C4 plants include such crop plants as sugarcane and corn. They are the second most prevalent photosynthetic type.

7. The Gulf of Thailand is a shallow, tropical coastal shelf system that has been heavily exploited since the 1960s. This has caused the system to change from a highly diverse ecosystem with a number of large long-lived species (such as sharks and rays) to one that is now dominated by small, short-lived species that support a high-valued invertebrate fishery (Pauly and Chuenpagdee 2003). In the Central North Pacific, tuna fishing is one of the major economic activities. Recent assessments of the tuna fisheries indicate that top predators such as blue marlin and swordfish declined since the 1950s while small tunas, their prey, have increased (Cox et al. 2000). The North Benguella Current is an upwelling system off the west coast of Southern Africa. This system is highly productive, resulting in a rich living marine resource system that supports small, medium, and large pelagic fisheries (Heymans et al. 2004).

8. The water erosion index of Hootsman et al. (2001) can be compared with the erosion severity classes of GLASOD (Oldeman et al. 1991). The modeled estimates for the global land area for 1990 corresponded for approximately 85% with the GLASOD inventory.

9. For example, results from a simple climate model are input to a global water model to compute changes in runoff due to climate change. In this case, the uncertainties of the climate model are propagated to the water model. This problem can be reduced by recognizing that the uncertainty of the climate model is relatively high for computed spatial patterns of temperature and precipitation, but much less for the magnitude and direction of these changes. Therefore, statements about the changes in runoff at particular locations will be highly uncertain and should be avoided, whereas statements about the size of the area in a large region affected by increasing or decreasing runoff have a lower level of uncertainty and are appropriate for the MA scenario analysis. The key is to aggregate results either spatially or temporally because uncertainties that are important on the fine scale partly cancel out when data are aggregated.

References

Alcamo, J., P. Döll, T. Henrichs, F. Kaspar, B. Lehner, T. Rösch, and S. Siebert, 2003: Global estimation of water withdrawals and availability under current and "business as usual" conditions. *Hydrological Sciences,* **48(3),** 339–348.

Alcamo, J., T. Henrichs, T. Rösch, 2000: *World Water in 2025:Global Modeling Scenarios for the World Commission on Water for the Twenty-first Century.* World Water Series Report 2, Center for Environmental Systems Research, University of Kassel, Germany.

Alcamo, J. and N. Nakićenović (eds.), 1998: Long-term greenhouse gas emission scenarios and their driving forces. *Mitigation and Adaptation Strategies for Global Change,* **3(2–4)**

Alcamo, J. and G.J.J. Kreileman, 1996: Emission scenarios and global climate protection. *Global Environmental Change,* **6(4),** 305–334.

Alcamo, J., A. Bouwman, J. Edmonds, A. Grübler, T. Morita, and A. Sugandhy, 1995: An evaluation of the IPCC IS92 emission scenarios. In *Climate Change 1994, Radiative Forcing of Climate Change* and *An Evaluation of the IPCC IS92 Emission Scenarios,* J.T. Houghton, L.G. Meira Filho, J. Bruce, Hoesung Lee, B.A. Callander, E. Haites, N. Harris and K. Maskell (eds.), Cambridge University Press, Cambridge, pp. 233–304.

Alder, J. and R. Watson: Fisheries Globalization: Fair Trade or Piracy in Globalization. In: *Effects on Fisheries Resources,* W.W. Taylor, M.G. Schechter, and L.G. Wolfson (eds). Cambridge University Press, New York. *submitted.*

Anderson, J.L. 1985. Private aquaculture and commercial fisheries: bioeconomics of salmon ranching. *Journal of Environmental Economics and Management,* **12(4),** 353–370.

Antle, J.M., S.M. Capalbo, and J. Hewitt, 1999: *Testing hypotheses in integrated impact assessments: climate variability and economic adaptation in great plains agriculture.* National Institute for Global Environmental Change, Nebraska Earth Science Education Network, University of Nebraska-Lincoln, Lincoln, NE, USA, pp. T5-4 and 21.

Bene, C., M. Cadren, and F. Lantz, 2000: Impact of cultured shrimp industry on wild shrimp fisheries: Analysis of price determination mechanisms and market dynamics. *Agricultural Economics,* **23(1),** 55–68.

Betts, R.A., 2000: Offset of the potential carbon sink from boreal forestation by decreases in surface albedo. *Nature.* **408,** 187–190.

Bolin, B., R. Sukumar, P. Ciais, W. Cramer, P. Jarvis, H. Kheshgi, C.A. Nobre, S. Semenov, and W. Steffen, 2000: Global Perspective. In: *Land Use, Land-Use Change, and Forestry: A Special Report of the IPCC,* R.T. Watson, I.R. Noble, B. Bolin, N.H. Ravindranath, D.J. Verardo, and D.J. Dokken (eds.), Cambridge University Press, Cambridge, UK, pp. 23–52.

Bollen, J.C., 2004: *A Trade View on Climate Change Policies. A Multi-Region Multi-sector Approach.* PhD-thesis. Amsterdam University.

Bouwman, A.F., K.W. Van der Hoek, B. Eickhout, and I. Soenario, 2005a, in press: Exploring changes in world ruminant production systems. *Agricultural Systems,* **89,** 121–153.

Bouwman, A.F., G. v. Drecht, J.M. Knoop, A.H.W. Beusen, and C.R. Meinardi, 2005b, in press: Exploring changes in river nitrogen export to the world's oceans. *Global Biogeochemical Cycles,* in press.

Bouwman, A.F., D.P. Van Vuuren, R.G. Derwent, and M. Posch, 2002: A global analysis of acidification and eutrophication of terrestrial ecosystems. *Water, Air, and Soil Pollution,* **141,** 349–382.

Bowes, G., J.C.V. Vu, M.W. Hussain, A.H. Pennanen, and L.H. Allen Jr., 1996: An overview of how rubisco and carbohydrate metabolism may be regulated at elevated atmospheric (CO_2) and temperature. *Agricultural and Food Science in Finland,* **5,** 261–270.

Bruinsma, J.E., 2003: *World Agriculture: Towards 2015/2030.* An FAO perspective. Earthscan Publications Ltd., London.

Carter, T.R., E.L. La Rovere, R.N. Jones, R. Leemans, L.O. Mearns, N. Nakícenovíc, B.A. Pittock, S.M. Semenov and J.F. Skea, 2001: Developing and applying scenarios. In: *Climate Change 2001. Impacts, adaptation and vulnerability.* J.J. McCarthy, O.F. Canziani, N. Leary, D.J. Dokken and K.S. White(eds.), Cambridge University Press, Cambridge. pp. 145–190.

Casella, E., J.F. Soussana, and P. Loiseau, 1996: Long term effects of CO_2 enrichment and temperature increase on a temperate grass sward, I: productivity and water use. *Plant and Soil,* **182,** 83–99.

Castles, I. and D. Henderson, 2003: The IPCC Emission Scenarios: An Economic-Statistical Critique. *Energy and Environment,* **14(2&3),** 159–185.

Chan, H.L., M.C. Garcia, and P. Leung, 2002: *Long-Term World Projections of Fish Production and Consumption.* Report to the Food and Agriculture Organization of the United Nations.

Chenery, H., S. Robinson, and M. Syrquin (eds.), 1986: *Industrialization and Growth: A Comparative Study.* Oxford University Press, Oxford.

Clayton, P.L., and D.V. Gordon, 1999: From Atlantic to Pacific: price links in the U.S. wild and farmed salmon market. *Aquaculture Economics & Management,* **3(2),** 93–104.

Cosgrove, W. and F. Rijsberman, 2000: *World Water Vision: Making Water Everybody's Business.* World Water Council. Earthscan Publications, London.

Cox, P.M., R.A. Betts, C.D. Jones, S.A. Spall, and I.J. Totterdell, 2000: Acceleration of global warming due to carbon-cycle feedbacks in a coupled climate model. *Nature,* **408,** 184–187.

Cramer, W., A. Bondeau, S. Schaphoff, W. Lucht, B. Smith, and S. Sitch, 2004: Tropical forests and the global carbon cycle: impacts of atmospheric carbon dioxide, climate change and rate of deforestation. *Philosophical Transactions of the Royal Society, Ser. B,* **359,** 331–343.

Cramer, W., A. Bondeau, F.I. Woodward, I.C. Prentice, R.A. Betts, V. Brovkin, P.M. Cox, V. Fisher, J. Foley, A.D. Friend, C. Kucharik, M.R. Lomas, N. Ramankutty, S. Sitch, B. Smith, A. White, and C. Young-Molling, 2001: Global response of terrestrial ecosystem structure and function to CO_2 and climate change: results from six dynamic global vegetation models. *Global Change Biology,* **7(4),** 357–373.

Cramer, W., D.W. Kicklighter, A. Bondeau, B. Moore III, G. Churkina, B. Nemry, A. Ruimy, A.L. Schloss, and Participants of "Potsdam'95", 1999: Comparing global models of terrestrial net primary productivity (NPP): Overview and key results. *Global Change Biology,* **5(1),** 1–15.

Crosson, P., 1994: Degradation of resources as a threat to sustainable agriculture. First World Congress of Professionals in Agronomy. Santiago, CA, 35pp.

Delgado, C.L., N. Wada, M.W. Rosegrant, S. Meijer, and M. Ahmed, 2003: *Fish to 2020: Supply and Demand in Changing Global Markets.* International Food Policy Research Institute, Washington, D.C., 226 pp.

Delhotal, K. C., F.C. DelaChesnaye, A. Gardiner, J. Bates, and A. Sankovski, 2005, in press: Mitigation of methane and nitrous oxide emissions from waste, energy and industry. *The Energy Journal.*

den Biggelaar, C., R. Lal, K. Wiebe, and V. Breneman, 2004a: The global impact of soil erosion on productivity. I. Absolute and relative erosion-induced yield losses. *Advances in Agronomy,* **81,** 1–48.

den Biggelaar, C., R. Lal, K. Wiebe, H. Eswaran, V. Breneman, and P. Reich, 2004b: The global impact of soil erosion on productivity. II. Effects on crop yields and production over time. *Advances in Agronomy,* **81,** 49–95.

den Biggelaar, C., R. Lal, K. Wiebe, and V. Breneman, 2001: Impact of soil erosion on crop yields in North America. *Advances in Agronomy,* **72,** 1–52.

DOE (U.S. Department of Energy), 2004: *International Energy Outlook 2004.* Energy Information Agency. Washington D.C.

Dregne, H.E, 1995: Erosion and soil productivity in Asia and New Zealand. *Land Degradation and Rehabilitation,* **6(2),** 71–78.

Dregne, H.E, 1992: Erosion and soil productivity in Asia. *Journal of Soil and Water Conservation,* **47,** 8–13.

Dregne, H.E, 1990: Erosion and soil productivity in Africa. *Journal of Soil and Water Conservation,* **45,** 431–437.

Economist, 2003: Hot potato. The Intergovernmental Panel on Climate Change had better check its calculations. *Economist,* February 13, 2003.

EEA (European Environment Agency), 2003: *Europe's Environment.* The third assessment. Environmental Assessment report 10, European Environmental Agency, Copenhagen.

EEA, 1998: *Annual topic update 1997.* Topic report 02/98, European Environment Agency, European Topic Centre on Coastal and Marine Environment, Copenhagen.

Falkenmark, M. and G. Lindh, 1993: Water and economic development. In: *Water in Crisis.* P. Gleick (ed.), Oxford University Press, New York, pp. 80–91.

FAO (Food and Agriculture Organization of the United Nations), 2003: *World Agriculture: Towards 2015/2030,* J. Bruinsma ed. Earthscan Publications, London, 432 pp.

FAO, 2001: FAOSTAT database collections Food and Agriculture Organization of the United Nations, Rome. Available at http://www.apps.fao.org.

FAO, 2000: The state of world fisheries and aquaculture. FAO, Rome.

Fischer, G., M. Shah, and H. Van Velthuizen, 2002: *Climate Change and Agricultural Vulnerability.* Contribution to the World Summit on Sustainability Development, Johannesburg 2002. International Institute for Applied Systems Analysis (IIASA), Vienna, Austria.

Freeman, C., 1990: Schumpeter's business cycles revisited. In *Evolving Technology and Market Structure—Studies in Schumpeterian Economics.* A. Heertje, M. Perlman (eds.), University of Michigan, Ann Arbor, MI, pp. 17–38.

Friedlingstein, P., L. Bopp, P. Ciais, J.-L. Dufresne, L. Fairhead, H. LeTreut, P. Monfray, and J. Orr, 2001: Positive feedback between future climate change and the carbon cycle. *Geophysical Research Letters,* **28,** 1543–1546.

Goldburg, R. and T. Triplett, 1997: *Murky Waters: Environmental Effects of Aquaculture in the U.S.* Environmental Defense Fund, New York.

Green, P., C.J. Vörösmarty, M. Meybeck, J.N. Galloway, B.J. Petersen, and E.W. Boyer, 2004: Pre-industrial and contemporary fluxes of nitrogen through rivers: a global assessment based on typology. *Biogeochemistry,* **68,** 71–105.

GRID/UNEP (United Nations Environment Programme), 1991: Global Assessment of Human Induced Soil Degradation (GLASOD). Global Digital Database (GRID). Nairobi.

Grübler, A., 1998: *Technology and Global Change.* Cambridge University Press, Cambridge.

Hansen, J., M. Sato, R. Ruedy, A. Lacis, and V. Oinas, 2000: Global warming in the twenty-first century: an alternative scenario. PNAS. Proc Natl Acad Sci U S A. 2000 Aug 29;97(18):9875–80.

Heymans, J.J., L.T. Shannon, and A. Jarre, 2004, in press: Changes in the Northern Benguela ecosystem over three decades: 1970s, 1980s and 1990s. *Ecological Modelling.*

Holtsmark, B.J. and K.H. Alfsen 2004, in press: PPP-correction of the IPCC scenarios: Does it matter? *Climatic Change.*

Hootsman, R.M., A.F. Bouwman, R. Leemans, G.J. Kreileman, 2001: *Modelling land degradation in IMAGE 2.* RIVM report 481508009. National Institute for Public Health and the Environment. The Netherlands.

Horie, T., J.T. Baker, H. Nakagawa, and T. Matsui, 2000: Crop ecosystem responses to climatic change: rice. In: *Climate Change and Global Crop Productivity.* K.R. Reddy and H.F. Hodges (eds). CAB International, Wallingford, United Kingdom, pp. 81–106.

Hyman, R.C., J.M. Reilly, M.H. Babiker, A. De Masin, and H.D. Jacoby, 2003: Modeling non-CO_2 Greenhouse Gas Abatement. *Environmental Modeling and Assessment,* **8(3),** 175–186.

IEA (International Energy Agency), 2002: *World Energy Outlook.* Paris.

IMAGE-Team, 2001: The IMAGE 2.2 implementation of the SRES scenarios. A comprehensive analysis of emissions, climate change and impacts in the twenty-first century. Main disc. *RIVM CD-ROM publication 481508018,* National Institute for Public Health and the Environment, Bilthoven, the Netherlands, see www.mnp.nl/image.

IPCC (Intergovernmental Panel on Climate Change), 1999: Data Distribution Center for Climate Change and Related Scenarios for Impacts Assessment.

IPCC, 2001: *Climate Change 2001: The scientific basis,* Cambridge University Press, 881 pp.

Jackson, J.B.C., M.X. Kirby, W.H. Berger, K.A. Bjorndal, L.W. Botsford, B.J. Bourque, R.H. Bradbury, R. Cooke, J. Erlandson, J.A. Estes, T.P. Hughes, S. Kidwell, C.B. Lange, H.S. Lenihan, J.M. Pandolfi, C.H. Peterson, R.S. Steneck, M.J. Tegner, and R.R. Warner, 2001: Historical overfishing and the recent collapse of coastal ecosystems. *Science,* **293,** 629–638.

Kendall, H. and D. Pimentel, 1994: Constraints on the expansion of the global food supply. *Ambio,* **23(3),** 198–205.

Klein-Goldewijk, K., 2004: Changes in pasture and crop land as reported by the HYDE database. Available at: www.mnp.nl/hyde.

Koruba, V., M.A. Jabbar, and J.A. Akinwumi, 1996: Crop-livestock competition in the West African derived savannah: application of a multi-objective programming model. *Agricultural Systems,* **52,** 439–453.

Kuylenstierna, J.C., S. Cinderby, and H. Cambridge, 1998: Risks from future air pollution. In: *Regional Air Pollution in Developing Countries.* J. Kuylenstierna and K. Hicks (eds.). Stockholm Environment Institute. York. UK.

Lal, R., 2003: Soil erosion and the global carbon budget. *Environment International,* **29,** 437–450.

Lal, R., 1998: Soil erosion impact on agronomic productivity and environment quality. *Critical Reviews in Plant Sciences,* **17(4),** 319–464.

Lal, R, 1995: Erosion crop productivity relationships for soils of Africa. *Soil Science Society of America Journal,* **59,** 661–667.

Lu, F, 1998: *Output data on animal products in China. How much they are overstated.* China Center for Economic Research, Bejing, China.

Lutz, W., W. Sanderson, S. Scherbov, 2001: The end of world population growth. *Nature,* **412(6846),** 543–545.

MA (Millennium Ecosystem Assessment), 2003: *Ecosystems and Human Well-being. A Framework for Assessment.* Island Press. Washington DC.

Maddison, A., 2004: *The ppp-price is right.* Letter to The Economist. 8 July 2004.

Maddison, A., 1995: *Monitoring the World Economy 1820–1992.* OECD Development Centre Studies. Organisation for Economic Cooperation and Development, Paris.

Manne, A. and R. Richels, 2003: *Market exchange rates or purchasing power parity: Does the choice make a difference in the climate debate.* AEI-Brookings Joint Centre for Regulatory Studies. Working paper 03-11.

Manne, A. and R. Richels, 2001: An Alternative approach to establishing trade-offs among greenhouse gases. *Nature,* **410,** 675–677.

Mayerhofer, P., B. de Vries, M. den Elzen, D. van Vuuren, J. Onigkeit, M. Posch, and R. Guardans, 2002: Long-term consistent scenarios of emissions, deposition and climate change in Europe. *Environmental Science and Policy.* **236,** 273–305.

McKibbin, W.J., D. Peace, and A. Stegman, 2004: *Long-run projections for climate change scenarios.* Lowy Institute. Sydney. Working papers in international economics. No. 1. 04.

Meybeck, M., 1982: Carbon, nitrogen and phosphorous transport by world rivers. *American Journal of Science,* **282,** 401–450.

Michaelis, L., 1997: *Special Issues in Carbon/Energy Taxation: Carbon Charges on Aviation Fuels.* Working Paper 12 in the Series on Policies and Measures for Common Action Under the UNFCCC, OECD, Paris.

Mitchell, R.A.C., V.J. Mitchell, S.P. Driscoll J. Franklin, and D.W. Lawlor, 1993: Effects of increased CO2 concentration and temperature on growth and yield of winter wheat at 2 levels of nitrogen application. *Plant Cell Environment,* **16,** 521–529.

Morita, T., and J. Robinson (eds.), 2001: Greenhouse Gas Emission Mitigation Scenarios and Implications in IPCC (2001). *Climate Change 2001: Mitigation.* Cambridge University Press. Cambridge. UK.

Moxey, A.P., B. White, and J.R. O'Callaghan, 1995: CAP reform: an application of the NELUP economic model. *Journal of Environmental Planning and Management,* **38,** 117–123.

Nakićenović, N., A. Grübler, S. Gaffin, T. Tong Jung, T. Kram, T. Morita, H. Pitcher, K. Riahi, M. Schlesinger, P. R. Shukla, D. van Vuuren, G. Davis, L. Michaelis, R. Swart and N. Victor, 2003: IPCC SRES Revisited: A Response. *Energy and Environment.* **14(2&3),** 187–214.

Nakićenović, N., and R. Swart (eds). 2000: *Special Report on Emissions Scenarios. A Special Report of Working Group III of the Intergovernmental Panel on Climate Change.* Cambridge University Press, Cambridge (United Kingdom).

Naylor, R.L., R.J. Goldburg, J.H. Primavera, N. Kautsky, M.C.M. Beveridge, J. Clay, C. Folke, J. Lubchenco, H. Mooney, and M. Troell, 2000: Effect of aquaculture on world fish supplies, *Nature,* **405,** 1017–1024.

Nilsson, J. and P. Grennfelt, 1988: *Critical loads for sulfur and for nitrogen.* Nord 1988: 97. Nordic Council of Ministers. Copenhagen.

Oldeman, L.R., 1998: *Soil Degradation: A Threat to Food Security.* ISRIC Report 98/01, Wageningen, Holland, 14pp.

Oldeman, L., R. Hakkeling, and W. Sombroek, 1991: *World Map of the Status of Human-Induced Soil Degradation: An Explanatory Note.* 2nd rev. ed. Nairobi. UNEP.

Parry, M., C. Fischer, M. Livermore, C. Rosenzweig, and A. Iglesias, 1999: Climate change and world food security: a new assessment. *Global Environmental Change,* **9,** 51–67.

Pascoe, S., S. Mardle, F. Steen, and F. Asche, 1999: *Interactions Between Farmed Salmon and the North Sea Demersal Fisheries: A Bioeconomic Analysis.* Centre for the Economics and Management of Aquatic Resources (CEMARE) Research Paper No. 144, University of Portsmouth, U.K.

Pauly, D. and R. Chuenpagdee, 2003: Development of fisheries in the Gulf of Thailand large marine ecosystem: Analysis of an unplanned experiment. In: *Large Marine Ecosystems of the World: Change and Sustainability,* G. Hempel and K. Sherman (eds.). Elsevier Science.

Pauly, D., J. Alder, E. Bennett, V. Christensen, P. Tydemers, and R. Watson, 2003: The future for fisheries. *Science,* **302 (5649),** 1359–1361.

Paustian, K., E.T. Elliott, and L. Hahn, 1999: Agroecosystem boundaries and C dynamics with global change in the central United States. In: *Great Plains Region Annual Progress Reports,* Reporting Period: 1 July 1998 to 30 June 1999. National Institute for Global Environmental Change, U.S. Department of Energy, University of California, Davis, CA, USA.

Penning de Vries, F.W.T., R. Rabbinge, and J.J.R. Groot, 1997: Potential and attainable food production and food security in different regions. *Philosophical Transactions of the Royal Society of London, Series B—Biological Sciences,* **352,** 917–928.

Pitcairn, C.E.R. (ed.), 1994: *Impacts of Nitrogen Deposition on Terrestrial Ecosystems.* Deparment of Environment, London.

Potting, J. and J. Bakkes, 2004: The GEO-3 Scenarios 2002–2032. UNEP/RIVM. United Nations Environment Programme. Nairobi.

Ramankutty, N., and J. A. Foley, 1999: Estimating historical changes in global land cover: Croplands from 1700 to 1992. *Global Biogeochemical Cycles,* **13,** 997–1027.

Reilly, J., R.G. Prinn, J. Harnisch, J. Fitzmaurice, H.D. Jacoby, D. Kickligther, P.H. Stone, A.P. Sokolov, and C. Wang, 1999: Multi-gas assessment of the Kyoto Protocol. *Nature,* **401,** 549–555.

Rhode, H., F. Dentener, M. Schulz, 2002: The global distribution of acidifying wet deposition. *Environmental Science and Technology,* **36,** 4382–4388.

Rosenberg, N., 1994: *Exploring the Black Box: Technology, Economics and History.* University Press, Cambridge.

Rostow, W.W., 1990: *The Stages of Economic Growth, Third Edition.* Cambridge University Press, Cambridge.

Schaefer, D. O., D. Godwin, and J. Harnisch, 2005, in press: Estimating future emissions and potential reductions of HFCs, PFCs and SF6. *The Energy Journal.*

Schimel, D.S., J.I. House, K.A. Hibbard, P. Bousquet, P. Ciais, P. Peylin, B.H. Braswell, M.J. Apps, D. Baker, A. Bondeau, J. Canadell, G. Churkina, W. Cramer, A.S. Denning, C.B. Field, P. Friedlingstein, C. Goodale, M. Heimann, R.A. Houghton, J.M. Melillo, B. Moore III, D. Murdiyarso, I. Noble, S.W. Pacala, I.C. Prentice, M.R. Raupach, P.J. Rayner, R.J. Scholes, W.L. Steffen, and C. Wirth, 2001: Recent patterns and mechanisms of carbon exchange by terrestrial ecosystems. *Nature,* **414,** 169–172.

Schlesinger, M.E., S. Malyshev, E.V. Rozanov, F. Yang, N.G. Andronova, B. de Vries, A. Grübler, K. Jiang, T. Masui, T. Morita, N. Nakićenović, J. Penner, W. Pepper, A. Sankovski, and Y. Zhang, 2000: Geographical distributions of temperature change for scenarios of greenhouse gas and sulphur dioxide emissions. *Technological Forecasting and Social Change,* **65,** 167–193.

Seitzinger, S.P., C. Kroeze, A.F. Bouwman, N. Caraco, F. Dentener, and R.V. Styles, 2002: Global patterns of dissolved inorganic and particulate nitrogen inputs to coastal systems: Recent conditions and future projections. *Estuaries,* **25,** 640–655.

Seitzinger, S.P. and C. Kroeze, 1998: Global distribution of nitrous oxide production and N inputs in freshwater and coastal marine ecosystems. *Global Biogeochemical Cycles,* **12,** 93–113.

Strengers, B., R. Leemans, B. Eickhout, B. De Vries and L. Bouwman, 2004: The land use projections and resulting emissions in the IPCC SRES scenarios as simulated by the IMAGE 2.2 model. *Geojournal,* **61,** 381–393.

Turner, R.E., N.N. Rabelais, D. Justic, and Q. Dortch, 2003: Global patterns of dissolved N, P and Si in large rivers. *Biogeochemistry,* **64,** 297–317.

Turner, B.L., D.L. Skole, S. Sanderson, G. Fischer, L. Fresco, and R. Leemans, 1995*: Land-use and Land-cover change: Science/Research plan.* International Geosphere-Biosphere Programme, Human Dimensions of Global Environmental Change Programme, Stockholm, 132 pp.

UN (United Nations), 2003: World Population Prospects: The 2002 Revision Population Database. [online] United Nations Population Division, New York, Cited October 14, 2004. Available at http://esa.un.org/unpp.

UNEP (United Nations Environment Programme), 2002: *Global Environment Outlook 3.* United Nations Environment Programme. EarthScan. London.

Van Drecht, G., A.F. Bouwman, J.M. Knoop, A.H.W. Beusen, C.R. and Meinardi, 2003: Global modeling of the fate of nitrogen from point and nonpoint sources in soils, groundwater and surface water. *Global Biogeochemical Cycles,* **17(4),** 26-1 to 26-20.

Van Vuuren, D.P., M.G.J. Den Elzen. M.M. Berk, M.M., P. Lucas, B. Eickhout, H. Eerens, and R. Oostenrijk, 2003: Regional costs and benefits of alternative post-Kyoto climate regimes. Comparison of variants of the Multi-stage and Per Capita Convergence regimes. *RIVM Report no. 728001025,* National Institute of Public Health and the Environment, Bilthoven, the Netherlands.

Van Vuuren, D.P. and H.J.M. De Vries, 2001: Mitigation scenarios in a world oriented at sustainable development: the role of technology efficiency and timing. *Climate Policy,* **1,** 189–210.

Vörösmarty, C.J., P. Green, J. Salisbury, and R.B. Lammers, 2000: Global water resources: vulnerability from climate change and population growth. *Science,* **289,** 284–288.

Watson, R., and D. Pauly, 2001: Systematic distortions in world fisheries catch trends. *Nature,* **414,** 534–536.

WBGU (German Advisory Council on Global Change), 1998: *The Accounting of Biological Sinks and Sources Under the Kyoto Protocol—A Step Forwards or Backwards for Global Environmental Protection?* Special Report 1998. WBGU, Bremerhaven, Germany.

WEA (World Energy Assessment), 2000: *World Energy Assessment. Energy and the challenge of sustainability.* United Nations Development Programme. New York.

WHO/UNICEF (World Health Organization/United Nations Children's Fund), 2001a: Joint monitoring programme for water supply and sanitation. Coverage estimates 1980-2000, World Health Organization / UNICEF, Geneva.

WHO/UNICEF, 2001b: Water supply and sanitation assessment 2000. Africa regional assessment, World Health Organization / UNICEF. Available at www.whoafr.org.

WHO/UNICEF, 2000: Water supply and sanitation assessment 2000, World Health Organization / UNICEF. Available at http://www.childinfo.org/eddb/sani and: http://www.who.int/docstore/water_sanitation_health/Globassessment/.

Wigley, T.M.L. and S.C.B. Raper, 2001: Interpretation of high projections of global-mean warming. *Science.* **293,** 451–545.

Wood, S., J. Henao, and M. Rosegrant, 2004, in press: The role of nitrogen in sustaining food production and estimating future nitrogen fertilizer needs to meet food demand. In: *Agriculture and the Nitrogen Cycle: Assessing the Impacts of Fertilizer Use on Food Production and the Environment,* Arvin R. Mosier, J. Keith Syers and John R. Freney (eds.). Scientific Committee on Problems of the Environment (SCOPE), Paris.

World Bank, 2002: *Global Economic Prospects 2002.* Washington, DC.

Ye, Y. and J.R. Beddington, 1996: Bioeconomic interactions between the capture fishery and aquaculture. *Marine Resource Economics,* **11(2),** 105–123.

Chapter 10

Biodiversity across Scenarios

Coordinating Lead Author: Osvaldo E. Sala

Lead Authors: Detlef van Vuuren, Henrique Miguel Pereira, David Lodge, Jacqueline Alder, Graeme Cumming, Andrew Dobson, Volkmar Wolters, Marguerite A. Xenopoulos

Contributing Authors: Andrei S. Zaitsev, Marina Gonzalez Polo, Inês Gomes, Cibele Queiroz, James A. Rusak

Review Editors: Gerardo Ceballos, Sandra Lavorel, Stephen Pacala, Jatna Supriatna, Gordon Orians

---

FIGURES

TABLES

*This appears in Appendix A at the end of this volume.

Main Messages

This chapter discusses the consequences of the four scenarios developed by the Millennium Ecosystem Assessment for biodiversity and focuses on two different aspects of biodiversity—losses of local populations and global species extinctions. In this assessment, local extinctions occur from a reduction in habitat availability. On a longer time scale, global extinctions may occur when species reach equilibrium with the altered habitat. Global and local extinctions occur on a time scale that we cannot accurately anticipate.

Habitat loss in terrestrial environments will lead, *with high certainty,* to a sharp decline in local diversity and the ecosystem services it provides (*very certain*) in all four MA scenarios during the 2000–50 time period. Scenario analysis demonstrated a decline of habitat availability by 2050 that ranged from 20% in Order from Strength to 13% in TechnoGarden (*medium certainty*) relative to habitat availability in the year 1970.

Habitat loss in terrestrial environments will lead *(with high certainty)* in all four MA scenarios to global species extinctions and associated losses of ecosystem services (such as the development of new drugs). Analyses using the well-established species-area relationship and a state-of-the-art model of land use change indicate that 12–16% (*low certainty*) of species will potentially be lost at ecological equilibrium with the altered habitat. Also, significant loss of ecosystem services will occur long before a species becomes globally extinct.

Order from Strength is the scenario (*with high certainty*) with the largest losses of habitat and local plant populations, whereas TechnoGarden and Adapting Mosaic had the smallest losses. The Order from Strength scenario showed the highest expansion of cropland resulting from slow rates of yield improvement and higher population growth. The Adapting Mosaic scenario showed a relatively low rate of habitat losses in part because of the slow development rate in developing countries, which reduced the demand for food and the change in land use. Scenarios that showed the largest losses of habitat also put the largest number of species in trajectories that may lead to global extinctions (*medium certainty*), although the time lags from habitat reduction to extinction are unknown.

Scenario results showed that the different terrestrial biomes of Earth will lose habitat and local plant species populations at different rates (*high certainty*) during the 2000–50 period. The biomes with the higher rates of habitat and local species diversity losses are warm mixed forests, savannas, scrub, tropical forests, and tropical woodlands. Biomes that lose species at the lowest rate include those with low human impact as well as those where land use changes and human intervention have already occurred.

It is unlikely that the Convention on Biological Diversity target for reducing the rate of biodiversity loss by 2010 will be met for terrestrial ecosystems under the explored scenarios. The two scenarios that take a more proactive approach to the environment (TechnoGarden and Adapting Mosaic) have more success in reducing loss rates of terrestrial biodiversity than the two that take a reactive approach (Global Orchestration and Order from Strength) (*medium certainty*).

For the three drivers tested globally across scenarios, land use change was the dominant driver of biodiversity change in terrestrial ecosystems, followed by changes in climate and nitrogen deposition (*medium certainty*). Some individual biomes showed different patterns. For example, climate change was the dominant driver of biodiversity change in tundra, boreal and cool conifer forest, savanna, and deserts. Nitrogen deposition was found to be a particularly important driver in warm mixed forests and temperate deciduous forest (ecosystems that are sensitive to deposition and relatively close to densely populated areas). In addition, the impact of other drivers, such as invasive species, could not be assessed as fully and may therefore be underestimated.

Under all scenarios, 70% of the world's rivers, especially those at higher latitudes, will increase in water availability, raising the potential for production of fishes adapted to higher flow habitats, which would likely be nonindigenous species (*low certainty*). No quantitative models exist that allow estimation of any additional consequences of increased discharge on biodiversity.

Under all scenarios, 30% of the modeled river basins will decrease in water availability from the combined effects of climate change and water withdrawal. Based on established but incomplete scientific understanding, this will result in eventual losses (at equilibrium) of 1–55% (by 2050; 1–65% by 2100) of fish species from these basins (*low certainty*). Climate change rather than water withdrawal was the major driver for the species losses from most (~80%) basins, with projected losses from climate change alone of about 1–30% (by 2050; 1–65% by 2100). Differences among scenarios were minor relative to the average magnitude of projected losses of freshwater biodiversity.

Losses of biodiversity of fishes predicted only on the basis of drying are underestimates. Drivers other than loss of water availability will cause additional losses that are likely to be greater than losses from declining water. Many of the rivers and lakes in drying regions will also experience increased temperatures, eutrophication, acidification, and increased invasions by nonindigenous species. These factors all increased losses of native biodiversity in rivers and lakes that are drying and caused losses of fishes and other freshwater taxa in other rivers and lakes. No algorithms exist for estimating the numbers of riverine and lake species lost from these drivers, but recent experience suggests that they cause losses greater than those caused by climate change and water withdrawal.

Rivers that are forecast to lose fish species are concentrated in poor tropical and sub-tropical countries, where the needs for human adaptation are most likely to exceed governmental and societal capacity to cope. The current average GDP in drying countries is about 20% lower than that in countries whose rivers are not drying.

Diversity of marine biomass was quite sensitive to changes in regional policy. Scenarios with policies that focused on maintaining or increasing the value of fisheries resulted in declining biomass diversity, while the scenarios with policy that focused on maintaining the ecosystem responded with increasing biomass diversity. However, rebuilding selected stocks did not necessarily increase biomass diversity as effectively as an ecosystem-focused policy.

Diversity of commercial fisheries showed large differences among scenarios until 2030, but all scenarios converge into a common value by 2050. Policy changes after 2030 generally included increasing the value of the fisheries by lowering costs, focusing on high-value species, substituting technology for ecosystem services, or a combination of the three approaches. However, no approach was optimal, since the approaches used in the scenarios reduced biomass diversity to a common level in each ecosystem.

As global trade increases, the numbers of intentional and unintentional introductions will increase in terrestrial, freshwater, and marine biomes. Unless greater management steps are taken to prevent harmful introductions that accompany increased trade, invasive species will cause increased ecological changes and losses of ecosystem services in all scenarios. Because of differences among scenarios in economic growth and openness to foreign trade, invasive species increase most in Global Orchestration, followed in order by TechnoGarden, Adapting Mosaic, and Order from Strength.

Lag times in species extinctions provide a window of opportunity for humans to deploy aggressive restoration practices that may rescue species that otherwise may be lost. Many actions that can be taken by policymakers (such as habitat restoration and establishment of protected areas) may change the fate of a species that would otherwise become extinct in a few generations.

Ecosystem services provided mostly by species in the upper trophic levels, such as biological control, tend to be lost first with increasing habitat loss (*low certainty*). Ecosystem services provided by species in the lower trophic levels, such as provisioning of food, fiber, and clean water, tend to be lost only after severe habitat loss has occurred. The relationship between habitat loss, biodiversity loss, and the provisioning ecosystem services depends on the notion that all species in an ecosystem do not have the same probability of extinction and all the ecosystem services are not provided by the same type of species. Increasing habitat loss leads first to the loss of species in the higher trophic levels (top predators), while only extreme losses of habitat result in the extinction of species in the lower trophic levels (plants and microorganisms).

10.1 Introduction

The four scenarios developed by the Millennium Ecosystem Assessment explore a broad set of possible socioeconomic trajectories for human society. Each scenario will have different consequences for biodiversity and the provisioning of ecosystem services. In this chapter, we consider the future of biodiversity under each scenario.

Since biodiversity forms the basis for ecosystem services, the current decline of global biodiversity is of great concern. Despite the ongoing conservation efforts of the international community, biodiversity loss continues to occur at an unprecedented rate of up to 100–10,000 times the background rate in the fossil record of the Cenozoic (Reid 1992; Barbault and Sastrapradja 1995; May et al. 1995; Pimm et al. 1995; Foote and Raup 1996). Changes in land use are expected to be the major driver of biodiversity change in this century, followed in importance by changes in climate, nitrogen deposition, biotic exchange (accidental or deliberate introduction of a species into an ecosystem), and atmospheric CO_2 levels (Sala et al. 2000). Depending on the assumptions that are made, the precise ranking of some drivers may vary; for instance, Thomas et al. (2004) have suggested that climate change may be as important as land use change in driving biodiversity loss over the next 50 years. In any case, it is clear that all these drivers will have major impacts on biodiversity.

Biodiversity is a composite measure of the number of species (species richness) and the number of individuals of different species (relative abundance). Most ecosystem services, such as the provisioning of food or clean water, depend on the presence of sufficient numbers of individuals of each species. These services will decline locally with the local extirpation or reduction of populations, long before global extinctions take place. For other ecosystem services, and in particular those that rely on genetic diversity, the central issue is species richness. For example, the provisioning of new pharmaceutical drugs to cure current and future diseases and the maintenance of genetic resources to improve current crop varieties are not directly related to the abundance of individuals within a species. In these instances, the provision of services only ceases after global extinction.

For terrestrial ecosystems, we considered changes in both local and global biodiversity. Local losses of biodiversity are important because they may anticipate global losses and because they affect local people who benefited from the services provided by the species that became extinct. In addition, local extinctions affect the global provisioning of ecosystem services that depend on the abundance of individuals, as noted. Global biodiversity changes are important because they are irreversible; species that go extinct globally will never reappear. Losses of global biodiversity affect the provisioning of both types of ecosystem services—those that depend on abundance and those that depend on the maintenance of unique genetic combinations.

The freshwater biodiversity exercise focused on local extinctions because freshwater communities are organized around watersheds, which means that extinctions are watershed-specific. The marine biodiversity assessment focused on the diversity of commercial fish species, both because these species are directly relevant to humans and because more comprehensive data sets were unavailable.

In all three cases, the assessment focused on species diversity because of the availability of published information. We note that diversity within species (genetic diversity) could be equally affected by human activity, with potentially large consequences for the provisioning of ecosystem services.

We used different approaches to assess changes in biodiversity in terrestrial, freshwater, and marine environments because the drivers of biodiversity change and our level of scientific understanding, as reflected in the available models, are different in the three different environments. We used the species-area relationship to assess the global impact of land use change on terrestrial ecosystems. The species-area relationship has been documented in more than 150 articles for many taxa and many systems, ranging from oceanic islands to isolated habitat patches in terrestrial landscapes. We estimated the area of habitat lost (and the ensuing local loss of species diversity) as a function of local changes in land use, as assessed by the IMAGE model. (See Chapters 6 and 9.)

The temporal aspect of the species-area relationship must be considered carefully, because extinctions do not occur immediately after a reduction in the area of available habitat (Tilman et al. 1994; Magsalay et al. 1995; Brooks and Balmford 1996). The scenarios for future biodiversity based on the species-area relationship in this chapter refer to the number of species that would be expected to go extinct when populations relax to an equilibrium in a reduced area of habitat. In our models, habitat reductions result from either land use or climate change. Determining the relaxation times for entire communities is particularly difficult because it requires tracking the species composition of habitat remnants through time. Furthermore, the time lag will depend on the life history of the species concerned; relaxation to an equilibrium may occur faster in species with shorter generation times (Brook et al. 2003).

Recent studies have placed some bounds on relaxation times. Brooks et al. (1999) fitted exponential decay curves to estimated bird species losses in forest fragments in Kenya (100–10,000 hectares) and found half-lives (the time to lose 50% of species predicted to go extinct at equilibrium) to be in a range of 23–80 years. Ferraz et al. (2003) used bird occurrence data taken during 14 years in Amazon forest fragments to show that half-lives where shorter for smaller fragments, with fragments in the size range 10–100 hectares having half-lives of about a decade. Finally, Leach and Givnish (1996) studied Wisconsin prairie remnants (0.2–6 hectares) and found that 8–60% of the original plant species had gone locally extinct over 32–52 years.

Overall, these results suggest that about half of the species losses predicted in this chapter may occur over a period of decades to 100 years. Our assessment yields estimates of the number of vascular plant species that are expected to go extinct when populations reach equilibrium with the reduced habitat. From a policy perspective, time lags between habitat reduction and species extinction provide a precious opportunity for humans to deploy aggressive restoration practices to rescue those species that would otherwise become extinct, although habitat restoration measures will not save the most sensitive species that go extinct soon after habitat loss. The time lags between habitat reduction and extinction can also mask serious problems; for example, long-lived tree species that have lost their pollinators may linger for hundreds of years before extinction finally occurs.

In addition to land use change, we considered climate change and nitrogen deposition as major influences on terrestrial biodiversity. Several other factors, including elevated CO_2, species invasions, and patterns of habitat fragmentation, are potentially important but were not included in this assessment because of a lack of appropriate data and models at the global scale. We used three complementary, published approaches to explore the effects of climate change on terrestrial biodiversity. These included analysis of changes in the locations of the boundaries of entire biomes, in potential biodiversity as a response to climate, and in tick diversity in Africa based on the summed predictions of models for individual species. We assessed the potential impacts of nitrogen deposition by estimating nitrogen loads in different regions and applying the concept of a critical load below which no damage occurs.

Freshwater ecosystems are among the most threatened on Earth. Consequently, understanding the relationships between aquatic species diversity and environmental drivers is of critical importance. Compared with terrestrial ecosystems, however, the patterns and determinants of biodiversity in freshwater ecosystems are poorly known. Quantitative information on species richness patterns and responses to anthropogenic environmental changes is largely lacking for freshwater taxa. This lack of information is particularly acute at large spatial and temporal scales. Freshwater taxa occupy the first four places in the IUCN list of the proportion of U.S. species at risk of extinction: freshwater mussels, crayfishes, amphibians, and freshwater fish. Globally, the best evidence suggests that freshwater biodiversity is more threatened than terrestrial taxa are by global changes (Ricciardi and Rasmussen 1999). This is partly because humans are drawn to riparian habitats, leading to a concentration of anthropogenic impacts near coastal and freshwater habitats. Furthermore, human consumption of water is reducing available habitat for freshwater organisms (see MA *Current State and Trends,* Chapter 8; Lodge 2001; Poff et al. 2001).

It is difficult to make quantitative predictions for how freshwater and marine biodiversity will be affected by future global changes. In addition to the rarity of freshwater biodiversity data (see MA *Current State and Trends,* Chapter 5), many of the statistical and conceptual tools available for use in conservation planning for terrestrial biodiversity are not readily transferable to analyses of freshwater biodiversity. Species-area curves, for example, cannot realistically be used to predict species loss in lakes. While dramatic examples of loss of lake habitat area exist (such as the drying of the Aral Sea from irrigation withdrawals), the biodiversity of lakes is in general more affected by a reduction in the quality than the quantity of water.

We examine four of the five most globally important, proximate drivers of biodiversity loss in lakes and rivers (Sala et al. 2000): eutrophication/land use change, acidification, climate change, and water withdrawal. The impact of climate change and water withdrawal on riverine fauna is addressed quantitatively using previously published species-river discharge relationships (Oberdorff et al. 1995). We focus on fishes and river discharge because those are the only previously published data and models that exist with suitably global coverage. (See MA *Current State and Trends,* Chapter 5.) The results for fish are of general importance because fish are an important controller of aquatic food webs and are often the taxa providing the most direct ecosystem goods to humans.

The biodiversity of marine systems is not as well described as that of terrestrial systems for a number of reasons. (See MA *Current State and Trends,* Chapter 18.) While there is a solid understanding of biodiversity changes in commercial fisheries, other areas such as the deep sea, the mid-water column, seamounts, and thermal vents are poorly described. We used a quantitative modeling approach to investigate how the diversity of fisheries and the biomass of different species might change under different MA scenarios in the three regions of the world for which we had good modeling tools.

This chapter describes the outcomes of the four MA scenarios for biodiversity in each of the three environments—terrestrial, freshwater, and marine. In the case of terrestrial ecosystems, we also assess the feasibility of achieving the Convention on Biodiversity target of significantly reducing the rate of biodiversity loss by the year 2010. Finally, we explore the uncertainties associated with our analysis and the regional differences in biodiversity changes across scenarios.

10.2 Terrestrial Biodiversity

10.2.1 The Approach

Our assessment of global changes in terrestrial biodiversity is based on the implementation of the MA scenarios in the IMAGE model, describing changes in native habitat cover, climate change, and nitrogen deposition over time. This section explains the calculation of native habitat cover using the IMAGE model, how the species-area relationship links changes in habitat area to global species extinctions, the different approaches used to estimate the effect of climate change, and how we estimated the effect of nitrogen deposition.

10.2.1.1 The IMAGE 2.2 Modeling Framework

The IMAGE 2.2 integrated assessment modeling framework consists of a set of linked and integrated models that together describe important elements of the cause-response chain of global environmental change. The framework and its submodels have been described in detail in several publications (Alcamo et al. 1998; IMAGE-team, 2001). Important elements include the description of emissions of greenhouse gases and regional air pollutants, climate change, and land use change. In the model, socioeconomic processes are mostly modeled at the level of 17 world regions, while climate, land use, and several environmental parameters are modeled at a 0.5x0.5 degree resolution.

The land cover model of IMAGE simulates the change in land use and land cover in each region driven by demands for food, forage, grass, timber, and biofuels and by changes in climate. It also includes a modified version of the BIOME model of Prentice et al. (1992) that is used to compute (changes in) potential vegetation. The potential vegetation is the equilibrium vegetation that should eventually develop under a given climate. The shifts in vegetation zones, however, do not occur instantaneously. In IMAGE 2.2, such dynamic adaptation is modeled explicitly according to the algorithms developed by Van Minnen and Ihle (2000).

10.2.1.2 The Species-Area Relationship Approach and Limitations

The relationship between species numbers and area is ubiquitous in nature (Rosenzweig 1995; Lomolino and Weiser 2001): the larger the area sampled, the larger the number of species found. The SAR is well described by the power law $S = cA^z$, where c is species local density and depends on the taxon and region being studied and z is the slope of the relationship and depends primarily on the type of SAR (oceanic islands, nested areas in a region, or different biological provinces). The value of z is also influenced by other factors, such as the scale of sampling (Crawley and Harral 2001). The strengths and weaknesses of using the SAR to forecast biodiversity loss are discussed in Chapter 4. Here, we briefly review those strengths and weaknesses and our theoretical understanding of the SAR.

Several factors contribute to the increase in species with area (Rosenzweig 1995). First, larger areas have a larger number of habitats, and therefore they will contain more specialized species. Second, when comparing among isolated units such as islands, habitat fragments, or biogeographic provinces, larger units will have lower extinction rates and, to a lesser extent, higher immigration rates. Third, at a geological time scale, larger units will have higher speciation rates. Fourth, there is a sampling issue; larger areas have a larger number of individuals and a higher probability of including rare species than smaller units. The second type of explanation, extinction versus immigration, is the basis for the theory of biogeography (Mac Arthur and Wilson 1967), which explains the variation of species diversity among islands of an oceanic archipelago.

Decrease in area of a habitat will lead to biodiversity loss through all four mechanisms just mentioned, but the loss of specialized species and the increase in extinction rates will be the first impacts felt (Rosenzweig 2001). The precise shape of the relationship describing the loss of species from an original habitat as a function of the remaining habitat area after conversion to agriculture is still an open question. There are three associated issues. First, many species are not restricted to their native habitat and can live in the agricultural landscape (Pereira et al. 2004). Second, the slope of the species-area relationship used for the loss of total area of a biome is still uncertain. Third, it is not clear how the SAR can account for the effect of habitat fragmentation. Although there is high certainty of the overall shape of the species-area relationship, there is uncertainty in the z parameter determining the exact slope of the relationship. We established bounds for the slope of the species-area relationship (z-value) based on an extensive literature search. The distribution of the values reported in the literature for the slope of the species-area relationship was the basis for a statistical analysis that provided confidence intervals for some of the estimates.

It is also uncertain whether the biodiversity loss associated with the decrease in biome area caused by climate change is well described by the SAR approach. Thomas et al. (2004) used this approach to forecast the impact of climate change for animal and plant species in six regions of the world. Here, we applied the SAR approach to assess the effect on biodiversity of biome-area decreases caused both by land use change and by climate change.

As discussed in the introduction, it is important to note that the extinctions predicted by SAR do not occur instantaneously because there is a time lag between habitat loss and species extinctions (Brooks et al. 1999; Tilman et al. 2002). We do not know precisely how long this is, and it will vary according to the life history of individual species. A few studies have suggested that many species extinctions would occur during the first 100 years after habitat reduction (Wilson 1992). In this chapter we estimate, for each scenario, the number of species that would go extinct when populations reach equilibrium with a reduced habitat resulting from either land use or climate change.

The number of endemic species is another source of uncertainty associated with the use of SARs to calculate species losses based on changes in habitat area. Species extinctions due to reductions in area represent global extinctions only for species that are endemic and do not exist outside this area. The proportion of endemic species is related to the level of disaggregation of vegetation units. If the vegetation of the world were lumped in a small number of large units, these would be quite different from each other in environmental conditions and species composition, and they would have very few species in common. In this case, the abundance of endemic species in each unit would be large. But if the vegetation of Earth were partitioned in a large number of small units, these would be similar to each other in environmental space and they may contain larger numbers of species in common and fewer endemic species. A large degree of disaggregation, consequently, would result in an overestimation of species losses estimated with the species-area approach. Here, we minimized this error by using large vegetation units.

Our units of analysis for the terrestrial biodiversity loss are the intersection of the 17 IMAGE biomes and the six biogeographic realms of Olson et al. (2001). These units were chosen to ensure that units were not too large as to miss regional patterns but at the same time not too small as to have a low percentage of endemics in each unit.

10.2.1.3 Estimating Original Biodiversity of Vascular Plants

Ideally, we would like to have species counts of vascular plants for each realm-biome unit, but with the exception of North America (Kartesz and Meacham 1999) these data are not available at the regional scale. Therefore, we used an indirect estimate of the diversity of vascular plants by scaling up the species-area relationship from local diversity of vascular plants to the realm-biome units. Using the SAR to estimate regional species counts based on local data is a common practice (e.g., Groombridge and Jenkins 2002), but it has limitations (Crawley and Harral 2001). In order to perform this scaling, we did an assessment of the species local density and z-values to be used for each realm-biome unit.

Data on the local species density, the c value of the SAR (which represents the intrinsic diversity of each system for a unity size, in our case 10,000 square kilometers), were obtained for each set of realm-biome combination by comparing the 1995 IMAGE land cover map to a map of the local diversity of vascular plants (Barthlott et al. 1999). Because each realm-biome combination spanned a range of classes of local diversities of vascular plants (the c-values), mean values of the local diversity in each realm-biome were calculated using a Geographic Information System. While in general each realm-biome unit is only moderately heterogeneous in diversity; this averaging may underestimate the diversity of species in cases where both species-rich and species-poor areas are found within a unit.

The slope of the SAR, the z-value, depends on the type (Rosenzweig 1995) and scale (Crawley and Harral 2001) of the sampling, and it has been hypothesized to depend on other variables as well, such as latitude/biome and taxon (Preston 1962; Connor and McCoy 1979). We compiled 82 values of z reported in the literature for species-area relationship of vascular plants. For each study, we recorded the author and the original data source, the location, the biome, the z-value, the c-value, the scale of the study (minimum and maximum area sampled), and the type of sampling. We considered three types of sampling units: continental, islands, and provinces. The continental SAR was obtained by sampling nested areas within a region. The island SAR was obtained by sampling the number of species of each island of an archipelago, where islands and archipelago included not only oceanic islands but also mountaintops. The provincial SARs were obtained from sampling the number of species in different biogeographic regions.

In our database, the most important variable in determining the z-value was the type of sampling. (See Figure 10.1, ANOVA $p<0.001$, $r^2=0.52$.) Continental SARs produced the lowest z-values, intermediate z-values were obtained for islands SARs, and the highest z-values came from provincial SARs. After accounting for the effect of the SAR type, the minimum area and the maximum area sampled did not have significant effects on the z-value (ANCOVA, $p>0.05$; see also Table 10.1).

To calculate the number of actual species in each biome/region, we needed to scale up from the unit area of 10,000 square kilometers of Barthlott's map (1999) to the total area of the biome-region. To compute future loss of biodiversity, we scaled down from the original biome area to the smaller biome area remaining after habitat conversion. It could be argued that the continental SAR is the most appropriate to scale up, whereas the island SAR is the most

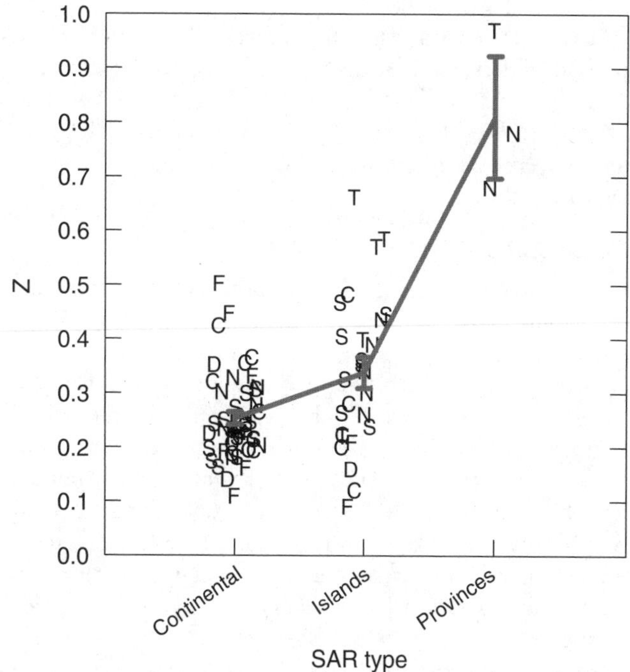

Figure 10.1. Z-Values Reported in Studies of the Species-Area Relationship in Vascular Plants. Each point corresponds to a study and is labeled with the biome category of the studied area: T-tropical forest and tropical woodland; F-temperate deciduous forest; C-boreal forest, coniferous forest, wooded tundra, and tundra; S-warm mixed forest, scrubland, and savanna; D-grassland and desert; N-no specific biome. The line joins the means in each type, and the error bars are standard errors.

Table 10.1. Mean and Standard Deviation of Z-Values and of the Natural Logarithm of Z-Values in Studies of Species-Area Relationships in Vascular Plants. Values for a dataset restricted to studies where the minimum area sampled was greater than 1 square kilometer are in parentheses.

Area		Mean	Standard Deviation	Natural Logarithm of z-Values	
	N	(μ_z)	(σ_z)	$(\mu_{Ln(z)})$	$(\sigma_{Ln(z)})$
Continental	47 (5)	0.253 (0.265)	0.079	−1.417	0.299
Islands	26 (7)	0.338 (0.308)	0.144	−1.186	0.480
Provinces	3	0.810	0.148	−0.222	0.180

appropriate to scale down. Several studies have used the island SAR to predict biodiversity loss (May et al. 1995; Pimm et al. 1995; Brooks et al. 1999), based on the argument that habitat conversion results in islands of native habitat in a sea of human-modified habitat. However, Rosenzweig (2001) has suggested the use of the provincial SAR, arguing that in the long term each native habitat fragment will behave as an isolated province. At the other extreme, it could be argued that in the short term the species that go extinct are the ones endemic to the area of lost habitat, and that is best described by the continental SAR (Kinzig et al. 2001; Rosenzweig 2001). In order to give the full range of possibilities, we made our calculations using the three types of SAR.

Finally, we found that tropical forest and tropical woodland had higher z-values than other biomes. Therefore, for those biomes and for each type of SAR we used z-values 20% higher than the means reported in Table 10.1. At the other extreme, for the tundra biome, we used z-values 20% lower than the mean.

10.2.1.4 Estimating Global Species Losses

We estimated changes in global biodiversity by calculating the change in area as a result of both habitat loss (due to agricultural expansion) and climate change. We assumed that the SAR could be applied independently for each realm-biome combination—thus assuming that the overlap in numbers of species was minimal (relative to the number of species that are endemic to each realm-biome combination). Furthermore, we assumed that diversity loss would occur as a result of the transformation of natural vegetation into a human-dominated land cover unit and that human-dominated vegetation had a diversity of zero endemic native species.

In our calculations, we did not assume any extinction rate with time, but simply assumed that at some point in time, the number of species will reach the level as indicated by the SAR. This means that our results should not be interpreted in terms of an immediate loss of number of species but in terms of species that may go extinct when populations eventually reach equilibrium with the reduced habitat.

We also applied SAR in only one direction: habitat loss leads to extinctions of species, but subsequent increase in area of a habitat would not lead to a similar increase. We did so because the processes of extinction and speciation occur in different time scales, with losses occurring much faster than the evolution of new species.

Finally, in IMAGE 2.2 fast climate change can lead to a difference between actual and potential vegetation. Areas where the colonization of the potential vegetation has not yet occurred can be assumed to have a significantly reduced biodiversity (Leemans and Eickhout 2004). Here, we assumed a loss of 50% of the species in the grid cells where adaptation of the vegetation lags. Our estimates of species losses at equilibrium have low certainty because they are based on a series of models that are linked sequentially, with the output of one being the input to the next. Moreover, each of them has its own assumptions and uncertainties.

10.2.1.5 Estimating Local Species Losses

We estimated local losses of biodiversity as a direct function of habitat loss. Species inhabiting a patch of native vegetation go locally extinct as the patch is converted into habitats such as agricultural land or urban patches in which they cannot survive. Losses of species are directly proportional to losses in native habitat. Moreover, in contrast to the SAR calculations for global biodiversity, this indicator is assumed to be fully reversible under the time scales considered. This difference from the global losses calculated in the previous section is based on the fact that species that go locally extinct in one patch may survive in other patches.

Local and global losses of biodiversity are important for different reasons. Global extinctions are particularly important for humans because they are irreversible and they eliminate some ecosystem services, such as the maintenance of the genetic library. This type of ecosystem service depends on the existence of unique genetic combinations that can be used to develop new pharmaceutical drugs as well as new varieties of plants and animals that may cope with new diseases or climate change. Local extinctions are important because they affect local human populations and global provisioning of ecosystem services that depend on the abundance of individual species.

Local species losses were estimated as a function of the transformation of native habitat into another category such as several agricultural categories and urban patches. We report changes in native habitat availability for 2050 relative to habitat availability for 1970 for the four MA scenarios and disaggregated by biome and biological realm.

10.2.2 Uncertainties in Extinction Predictions

As explained, one important source of uncertainty in our predictions is the slope of the species-area relationship. This results both from not knowing which type of SAR is more appropriate to describe biodiversity loss and also from the wide range of the z-values for a given type of SAR. To quantify the uncertainty associated with the type of SAR used, we present results for the three types. To quantify the uncertainty associated with the wide range of z-values, we use Monte Carlo simulations based on the distribution of z-values in the literature.

The z-values for the continental SAR follow a lognormal distribution. (See Figure 10.2; Kolmogorov-Smirnov, p = 0.764.) A similar pattern is observed for the island SAR (Kolmogorov-Smirnov, p = 0.994). We used Monte Carlo simulations to convert this probability distribution of the z-values into confidence intervals for the extinctions. In each Monte Carlo simulation, a random z-value was drawn from the lognormal distribution and used for all region/biome combinations (except tropical forest/woodland, with a value 20% higher than the drawn value, and tundra, with a value 20% lower). Then, using IMAGE, we calculated the number of extinctions per biome and region and also the total number of extinctions. Five hundred Monte Carlo simulations were performed. The mean, the standard deviations, and the range of the predicted extinctions are reported.

Our biodiversity loss estimates assume a low overlap in the species composition of the different biomes. In order to examine this assumption, we studied the distribution of North American plants, using the database of Kartesz and Meacham (1999), which lists for each state in the United States and for each Canadian province the composition of vascular plants. We selected for each biome a set of states that were covered only by that biome, based on the 1995 IMAGE land cover map. (See Table 10.2.)

The matrix of vascular plant species overlap between the different biomes is shown in Table 10.3. The matrix is asymmetric around the diagonal. For instance, although 73% of the plant species in tundra and ice are also present in the boreal forest and cool conifer biomes, only about 12% of the latter biomes' species are present in tundra and ice. This asymmetry is caused by forest and cool conifer biomes being more spacious than tundra and ice.

Overlaps vary widely, but the general pattern is that the bigger the distance between the biomes, the smaller the species overlap. On average, the maximum overlap with a neighbor biome is about 60%. Thus, assuming the extreme and unlikely case of one biome disappearing and the other remaining intact, we would predict a little more than twice the extinctions that would in fact occur. One possible approach to avoid this problem would be to consider only the species endemic to each biome. However, it has been found (Borges personal communication) that restricted range species may follow a SAR with a higher z-value than broadly distributed species. Therefore the problem of not having a very large proportion of endemics to each biome may be smaller than what Table 10.3 could suggest.

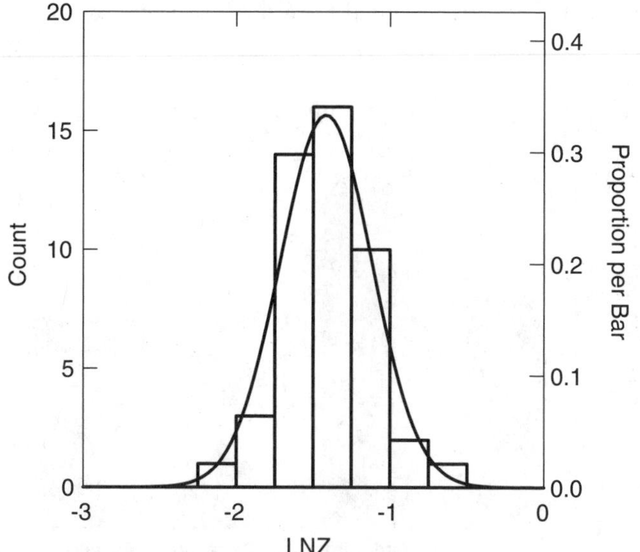

Figure 10.2. Histogram of Z-Values of the Species-Area Relationship in Continental Studies of Vascular Plants, Fitted with a Log-normal Distribution

Table 10.2. States or Provinces Selected as Representatives of Each Biome in Order to Analyze Species Overlap

Biome	State or Province
Tundra and ice	Franklin
Warm mixed forest	South Carolina, Mississippi, Louisiana, Alabama
Temperate deciduous forest	Virginia, Kentucky, West Virginia
Temperate mixed forest	Nova Scotia, Maine, Vermont, New York, Michigan, Iowa, Wisconsin
Scrubland, grassland, and hot desert	California, Arizona, New Mexico
Tropical forest	Puerto Rico, Virgin Islands
Boreal forest, cool conifer	Newfoundland, Manitoba

Our approach focused primarily on vascular plants because of the numerous studies of SARs and the availability of global data sets of species density. How representative are scenarios based on vascular plants of the patterns expected in other taxa? Can we extrapolate results obtained with vascular plants to other taxa? The answers to these questions are related to another one. How related are richness patterns among taxa? In order to address these questions, we undertook an assessment of the literature within the MA framework.

The review on richness correlations among taxa was largely based on the literature published over the last decade, but some earlier works have also been included. Of the more than 100 publications reviewed, only 48 were appropriate for extracting data. The attention paid to richness correlations is strongly different for different groups, with mammals (10 publications), vascular plants (13), beetles (12), butterflies (19), and birds (26) being the best-represented taxa.

Most of the authors of the literature focused on a certain continent or country or even on smaller regions. Some claimed that certain groups such as tiger beetles and large moths can serve as reliable predictors of the richness of other taxa at the global level, but these conclusions seem to be driven by the wish to make generalizations from a few charismatic groups rather than by detailed analyses of complex assemblies. The diversity of trees and shrubs as well as that of butterflies, birds, and mammals showed a comparatively high share of correlations with other taxa that were significantly positive. More than 50% of the total amount

Table 10.3. Overlap in Vascular Plant Composition between Different Biomes in North America and the Caribbean

Biome	Tundra and Ice	Warm Mixed Forest	Temperate Deciduous Forest	Temperate Mixed Forest	Scrubland, Grassland, and Hot Desert	Boreal Forest and Cool Conifer	Tropical Forest
			(percent)				
Tundra and ice		4.6	11.8	40.2	30.0	73.2	0.8
Warm mixed forest	0.3		63.6	50.1	28.9	17.3	14.1
Temperate deciduous forest	1.1	76.6		73.8	32.9	29.6	9.1
Temperate mixed forest	3.2	52.5	64.1		37.8	40.1	6.0
Scrubland, grassland, and hot desert	1.2	15.2	14.3	19.0		12.4	4.1
Boreal forest, cool conifer	12.1	37.9	53.6	83.7	51.5		3.9
Tropical forest	0.1	21.2	11.3	8.5	11.5	2.7	

of the significant positive correlations reported focused on birds, mammals, and vascular plants. Significant negative correlations, in contrast, have been recorded with species richness of lichens, ants, and beetles (only one case available for each group). Unfortunately, most authors do not provide the complete results of regression analyses but only mention correlation coefficients. Thus, the list of regression equations that could be used for modeling exercises only includes four studies covering four pairs of taxa.

Correlation between identical taxa can be different in grids of different size. Significant positive correlations for butterflies, birds, and mammals prevail at the low-resolution scales, while correlation results for vascular plants and beetles are more randomly distributed across scales. For vascular plants, resolution scale distribution is much more scattered than for all other groups. In the size class used in the MA model (100–999 square kilometers), six correlations were discovered but no promising results on correlations of vascular plants with other groups could be found. Similar checks could not be made for large moths and tiger beetles (the groups assumed to be good indicators for global biodiversity) (Pearsson and Cassola 1992).

The current state of knowledge generally indicates that we cannot generalize spatially explicit estimates of changes in global biodiversity based on diversity changes of particular focal taxa. Prendergast et al. (1993) showed that coincidence of biodiversity hotspots for several taxa is rare, suggesting the difficulties of extrapolating data from one taxa to others.

10.2.3 Terrestrial Biodiversity Change across Scenarios

10.2.3.1 Local Loss of Species through Loss of Habitat

Conversion of a patch of native vegetation into a cropland or part of the urban landscape results in the immediate extirpation of populations. In the case of the shift to agriculture, a simple community made up of one or a few cultivated species and a reduced number of cosmopolitan weeds replaces a diverse ecosystem. Conversion may involve logging of the forest, burning remnants with no commercial value, plowing the soil, and planting of a monospecific crop. The losses of the local populations with the conversion of a unit of native vegetation into a human-dominated patch are rapid and directly proportional to habitat loss.

Local losses are very important because without adequate conservation efforts they can lead to global losses of species. Furthermore, local losses affect the provisioning of ecosystem services derived from biodiversity to local people and after some time affect the global services provided by ecosystems.

Habitat losses by 2050 relative to habitat availability in the year 1970, and the corresponding extirpation of population and local losses of species, increase dramatically in all scenarios, ranging between 13 and 20%. (See Figure 10.3.) Order from Strength had the largest losses (20%) as a result of the relatively large increase in food demand due to the high population growth rate and the relatively slow increase in yield. The latter resulted from the relatively low transfer of technology to regions far from the centers of technology development, where the highest population growth is also expected. This means that under this scenario, increasing

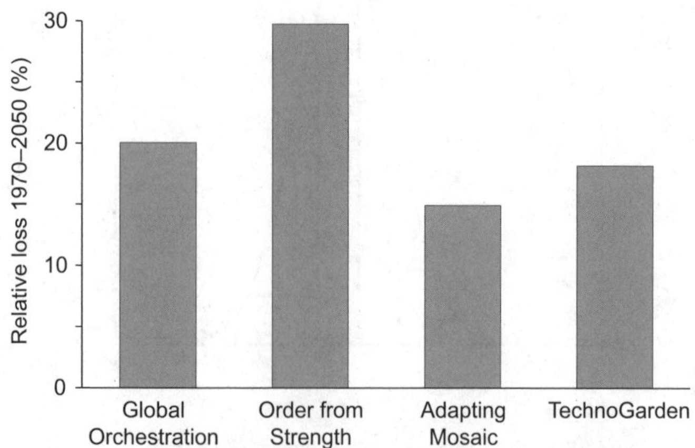

Figure 10.3. Losses of Habitat in 2050 Relative to Habitat Availability in 1970 for the MA Scenarios

the crop area was the necessary response to meet the increasing food demand. Finally, the increase in agricultural area occurred at the expense of a reduction of native habitat, with the consequent extirpation of local population and the local losses of species.

TechnoGarden was the scenario with the lowest loss of habitat (13%), mostly as a result of optimistic estimates in the increase in crop yield resulting from fast technological progress. Adapting Mosaic and Global Orchestration yielded intermediate results resulting from the relatively slow increase in food demand in the former and relatively low human population growth in the latter. It should be noted that in these calculations, losses have been assumed to be fully reversible.

Habitat loss was not uniformly distributed across the different biomes of the world. (See Table 10.4 and Figure 10.4.) Warm mixed forests and savannas in the Order from Strength scenario were the biomes with the highest losses. These biomes are located in areas such as Africa with the highest growth in human population and corresponding growth in food demand. Order from Strength constrained the trade of food and consequently most of the food demand had to be met locally, resulting in a large-scale transformation of native habitat into agricultural land and large losses of local populations. Adapting Mosaic showed the largest losses of habitat for warm mixed forests in addition to large losses in temperate deciduous forests due to similar mechanisms as those described for Order from Strength.

Habitat availability increased slightly in some biomes, such as the temperate mixed forest, mostly as a result of abandonment of pastures. A large fraction of the temperate mixed forest is located in the industrial world, where most of the land use changes have already occurred. In addition, in the Order from Strength and Adapting Mosaic scenarios, which emphasize regionalization (as opposed to globalization), demands for food and agricultural land in the temperate mixed forest region are stable.

Local and global losses of biodiversity differ in reversibility. Local losses could be reversed as a result of abandonment of agricultural land or active conservation practices. Populations can invade from adjacent patches naturally or with human intervention. Ecosystem services derived from local diversity can therefore increase or decrease as a result of gains and losses of habitat. Habitat and population losses can occur very rapidly, but gains in habitat from abandonment take longer periods of time, depending on the ecosystem of interest.

10.2.3.2 Global Loss of Vascular Plant Species through Loss of Habitat

The loss of vascular plant species that would occur when they reach equilibrium in 2050 differs among the four MA scenarios because of differences in expansion of agricultural area. (See Figure 10.5.) Worldwide, the changes in habitat availability experienced during the 1970–2050 period may result in a decrease of 12–16% in biodiversity at equilibrium. By far the strongest decrease occurred for the Order from Strength scenario (which had the largest expansion of cropland due to slow yield improvement and high population growth). In fact, by 2050 the changes in habitat and consequent species losses of this scenario had hardly slowed down compared with historical rates.

The TechnoGarden and Adapting Mosaic scenarios, in contrast, showed the slowest losses—on the order of 12% of species lost at equilibrium compared with current biodiversity. In TechnoGarden, the lower species decline rate was mainly due to much higher yield improvements in developing countries and a stabilizing population. Consistent with the storyline of the scenario, the assumptions of this scenario are relatively optimistic. In Adapting Mosaic, slower development rates in developing countries slowed down increases in food demand. Global Orchestration fell in between the other three scenarios, with a 13% loss of species at equilibrium with 2050 land use changes.

Table 10.4. Change in Land Cover in 2050 in Four Scenarios

Category of Land	2000	GO	OS	AM	TG	GO	OS	AM	TG
	(million hectares)					(percent change)			
Agricultural land	3,357	3,646	4,162	3,580	3,660	109	124	107	109
Extensive grassland	1,711	1,700	1,704	1,704	1,707	99	100	100	100
Regrowth forests	446	630	523	550	462	141	117	123	103
Ice	231	224	225	222	221	97	97	96	96
Tundra	768	727	727	726	724	95	95	95	94
Wooded tundra	106	84	83	86	89	79	78	81	83
Boreal forest	1,509	1,554	1,551	1,556	1,553	103	103	103	103
Cool conifer	168	196	188	192	194	117	112	114	116
Temperate mixed forest	201	262	236	250	287	130	117	124	143
Temperate deciduous forest	145	133	110	119	155	91	76	82	107
Warm mixed forest	95	79	62	76	109	83	65	80	115
Steppe	804	750	692	749	730	93	86	93	91
Desert	1,678	1,643	1,637	1,660	1,665	98	98	99	99
Scrubland	207	170	122	183	182	82	59	88	88
Savanna	705	404	316	511	450	57	45	73	64
Tropical woodland	483	517	426	524	503	107	88	109	104
Tropical forest	670	568	520	597	594	85	78	89	89

Key: GO = Global Orchestration; OS = Order from Strength; AM = Adapting Mosaic; TG = TechnoGarden

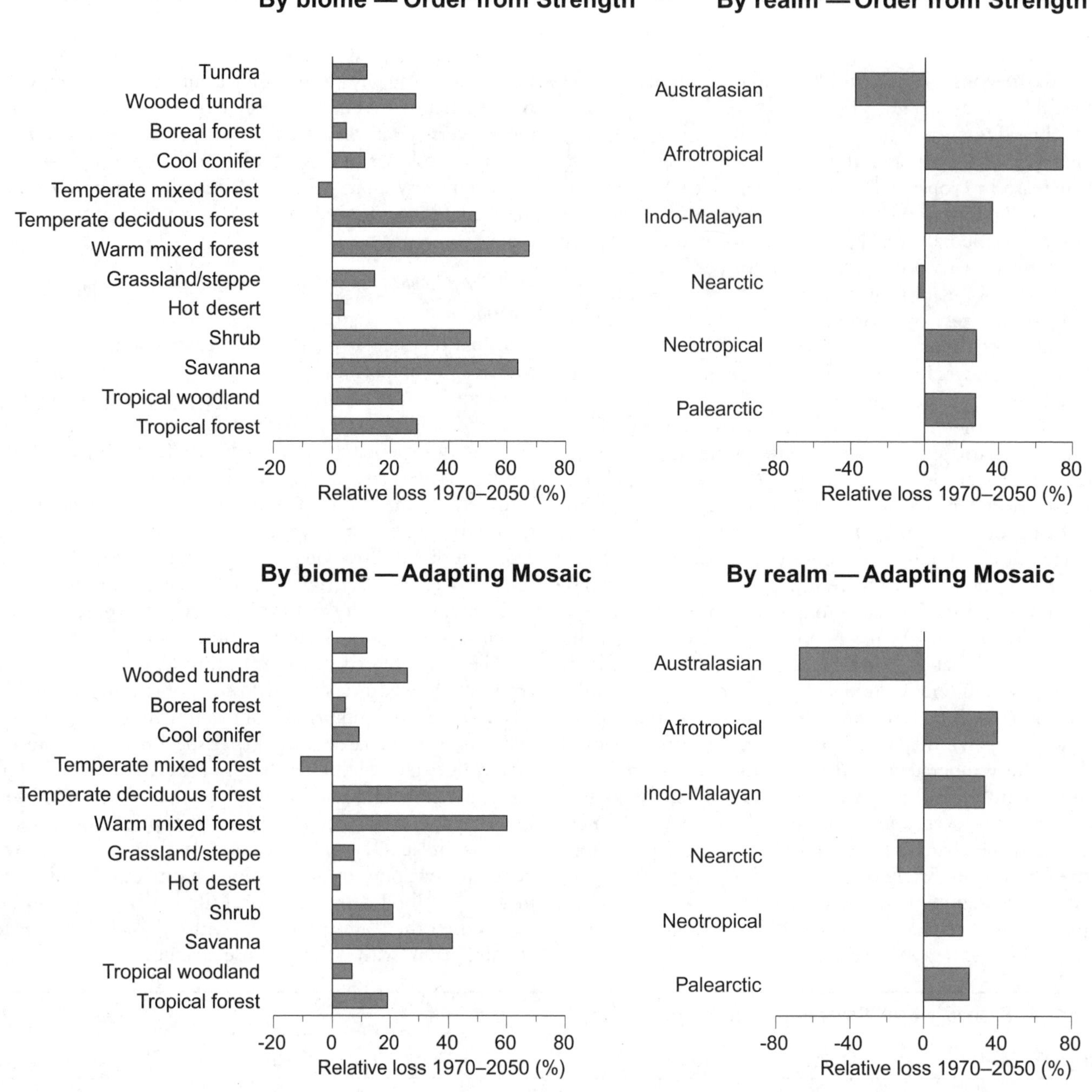

Figure 10.4. Losses of Habitat in 2050 Relative to 1970 for Different Biomes and Realms for Two Scenarios (Negative values indicate net increase.)

There are major differences in the species loss trends among the different biomes. Figure 10.6 shows the relative losses by major biome type and ecological realm for Order from Strength and Adapting Mosaic. The results indicate that warm mixed forests, savanna, scrub, tropical forests, woodlands, and temperate deciduous forests seem to suffer most from biodiversity losses through loss of habitat. In particular, tropical forest, tropical woodland, savanna, and warm mixed forest account for 80% of all species lost (in total, nearly 30,000 species).

While all biomes show lower habitat and species losses under Adapting Mosaic than under Order from Strength, this is particularly so for the tropical biomes. When comparing both scenarios, there are also differences in the time frames in which the different biomes suffer habitat losses. In Order from Strength, it can be seen that for the temperate biomes and warm mixed forests almost all habitat losses occur before 2020, while for the tropical habitats the 2020–50 period sees almost similar habitat losses as the 2000–20 period. In contrast, under Adapting Mosaic a considerably different time dynamic is seen, with most of the habitat losses in warm mixed forest occurring in the second period.

The largest relative habitat and species losses occur in the Afrotropic region, which has the largest expansion of agricultural land (driven jointly by a rapidly increasing population and strong increases in per capita food consumption) under all scenarios. The second most important region in terms of relative losses is the Indo-Malayan region. The Paleartic region, in contrast, sees the lowest losses in biodiversity through loss of habitat.

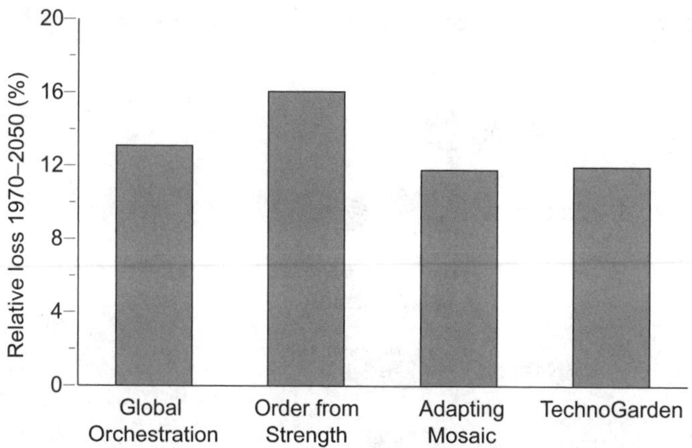

Figure 10.5. Relative Losses of Global Vascular Plant Biodiversity when Populations Reach Equilibrium with Reduced Habitat for MA Scenarios

A major uncertainty in the analysis is represented by the z value used in the SAR equation. To analyze the uncertainty of z, two experiments were conducted: a Monte Carlo analysis sampling z's from the distribution indicated earlier and replacement of the z's from the studies on the island scale by the provincial scale. The analysis indicates that a 16% loss of vascular plant species for the Order from Strength scenario falls within a range of plus and minus the standard deviation of 10–20%. (See Figure 10.7.) The highest and lowest runs (within the total set of 500 runs) show a 6% and 29% loss, respectively.

Replacing the z's from the island scale to those established for provinces resulted in an increase of forecast losses. For Order from Strength, losses increased from 16% to 26%. On the other hand, using the z values as established for the continental scale decreased losses to 13%. The relative results and the changes over time, however, remain the same. The left-hand side of Figure 10.7 shows the results of a Monte Carlo analysis sampling from the log-normal distribution of island scale z-values for vascular plants as shown in Figure 10.5 (from a set of 500 runs). The right-hand side of Figure 10.7 compares the results of using z values from studies at the continent, island, and provincial scale for Adapting Mosaic and Order from Strength.

The level of aggregation (in terms of the total number of biomes in which the terrestrial vegetation was divided) represents an uncertainty within the analysis. In our analysis, we assumed that regions have a high proportion of endemism. Increasing the number of regions under this assumption will increase the global number of species lost. Having too few regions would underestimate species loss, since it would not take into account different ecological regions. Alternatively, having too many regions would overestimate species loss, as it would double count too many species when aggregating from local to global extinction.

The influence of the regional definitions was analyzed by varying the number of regions in the analysis from 4 to 75 (versus 65 in standard run). (See Figure 10.8.) The highest aggregated regional definition used in this analysis corresponded to the highest aggregation level in Bailey's map of global ecoregions (four domains: arctic, humid temperate, dry tropic, and humid tropical). The next level corresponds to the 14 biomes recognized in IMAGE at the global level (loosely corresponding to the level of divisions in the Bailey set). The third level combines the four domains of the Bailey set with the realms of the WWF ecoregion map, creating a total set of 24 regions. The fourth level corresponded to the standard regional definitions used in this analysis, while the fifth level adds an additional 10 regions by assuming that East Asia and Japan can be identified as separate realms (which can be concluded from the province level map of the Bailey ecoregion definitions). The results indicate that indeed our results do vary for these different definitions, with a larger species loss with increasing disaggregation of vegetation units. The number of species lost increased with the number of units in which Earth vegetation was partitioned.

10.2.3.3 Convention on Biological Diversity Target

The target of the Convention on Biological Diversity is to bring the rate of loss of biodiversity by 2010 significantly lower than the current rate. To test this, we assumed that the current rate should be interpreted as the historic average of the last two decades as calculated using the SAR approach on the basis of the IMAGE global change data. First, we calculated the habitat loss that occurred during the last two decades and estimated its effect on species diversity when populations reach equilibrium with the reduced habitat. Second, we calculated the average rate of change in habitat loss and the consequent change in the number of equilibrium species for each of the four scenarios for the period 2000–20, centered in the year 2010. In order to estimate the probability that each scenario meets the CBD target, we plotted the loss of equilibrium species relative to historic rates. (See Figure 10.9.)

It is unlikely that the CBD target will be met for terrestrial ecosystems under the scenarios explored by the MA. Order from Strength and Global Orchestration would probably not meet the target because the estimated rates of habitat loss and the consequent losses of species at equilibrium exceeded those of the previous 20 years. Order from Strength presented a rate of loss that was considerably higher than the historic rate, mostly as a result of the relatively slow improvement of agricultural efficiency in combination with a sharp increase in food demand. Global Orchestration also showed relative rates of loss that were somewhat higher than the historic rates, resulting from an improvement of the historic rate of food consumption (and therefore agricultural expansion). TechnoGarden and Adapting Mosaic have more success at reducing the loss rates of terrestrial biodiversity relative to the historical rates. However, we expect with *high certainty* that our analysis is underestimating losses of the different scenarios because it does not take into account other pressures on biodiversity such as climate change and nitrogen loading, which are expected to increase.

10.2.3.4 Loss of Terrestrial Biodiversity through Climate Change

Climate change will certainly influence several aspects of ecosystems. Grabherr et al. (1994) were among the first to

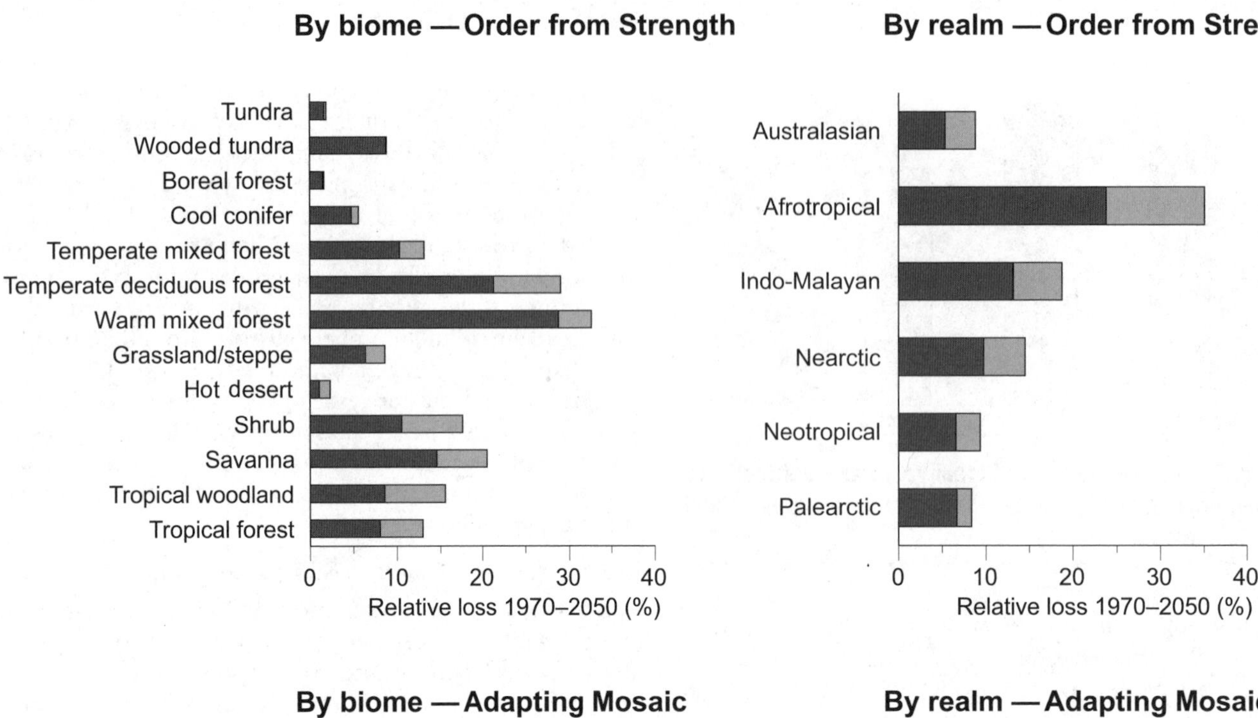

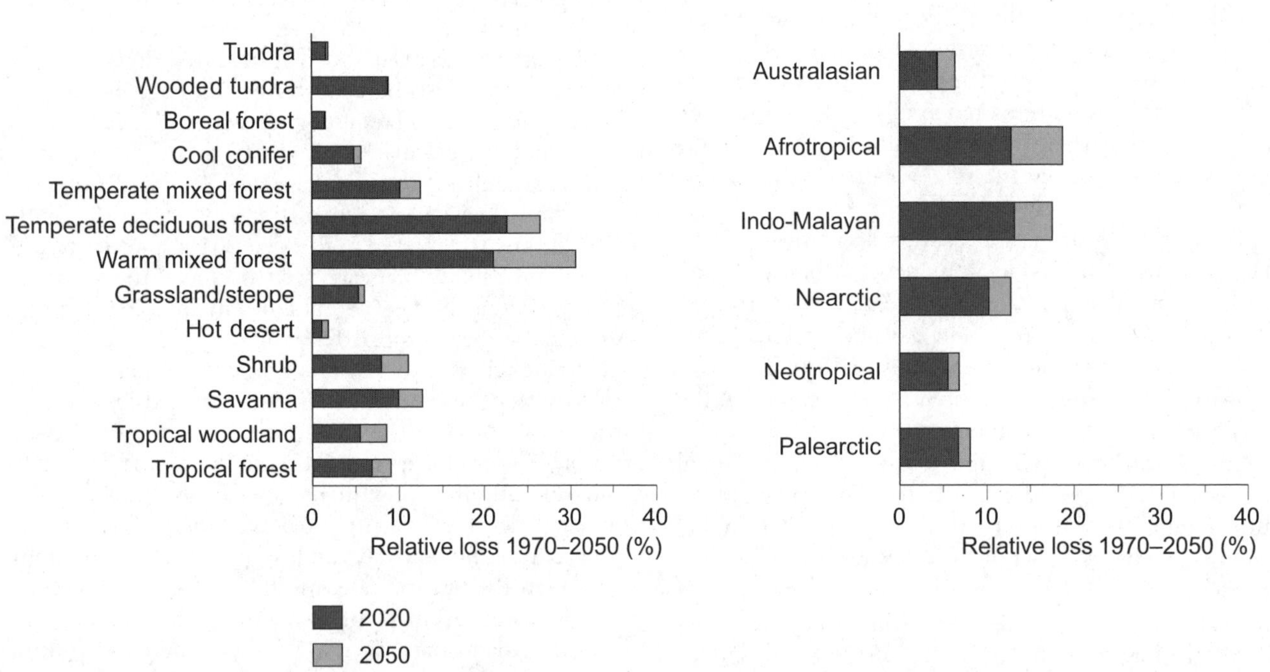

Figure 10.6. Relative Losses of Biodiversity of Vascular Plants through Habitat Loss for Different Biomes and Realms for Two Scenarios. Losses would occur when populations reach equilibrium with habitat available in 2050 and are relative to 1970 values.

report that ecosystems are already changing as a result of climate change. They used long-term observations from alpine vegetation and demonstrated that the distributions of many species had increased in altitude. Parmesan and Yohe (2003) analyzed the response of more than 1,560 plant and animal species in both marine and terrestrial environments and reported a clear effect of climate change on their distribution. Their analysis documented an average range shift of 6 kilometers per decade toward the poles or meters per decade upward. Other similar observations have been made of the impacts of climate change on the distribution of several plant and animal species (Both and Visser 2001; Root et al. 2003).

One of the most common ways to study the impact of climate change on the distribution of ecosystems is to describe their climatic envelope and compare them against climate-change scenarios provided by global circulation models (e.g., Prentice et al. 1992; Cramer and Leemans 1993; Malcolm and Markham 2000). Van Minnen and Ihle (2000) have attempted to add some form of transient impacts to the "climate envelope" approach by modeling the migration process of total biomes as a function of distance,

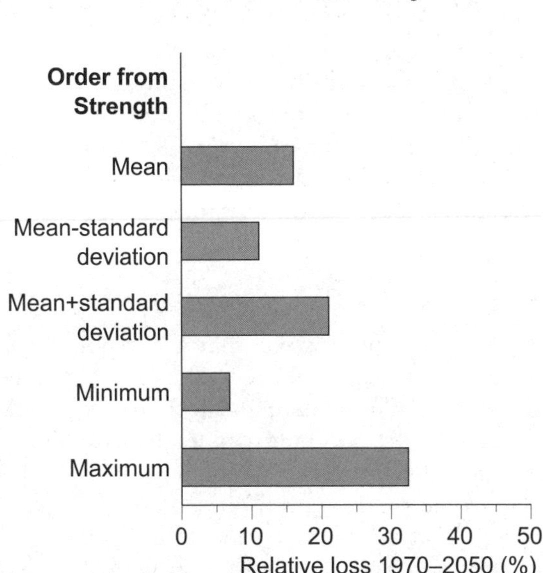

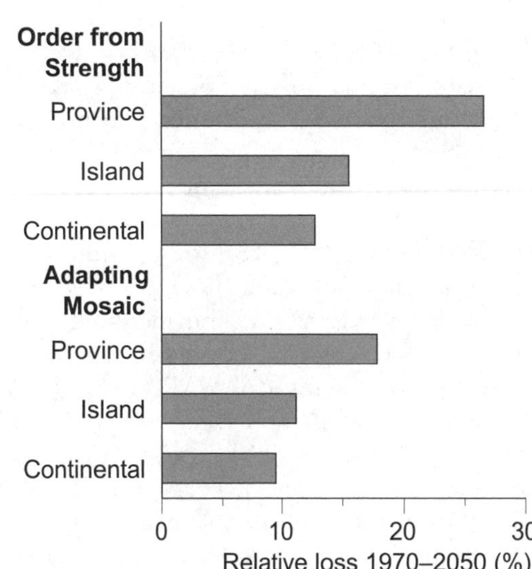

Figure 10.7. Uncertainty Analysis for Different Z Estimates. Influence of using different estimates of z calculated using a Monte Carlo technique on relative species loss for the Order from Strength Scenario shown left. Influence of using z-values for continents, islands, and provinces on the estimate of biodiversity change for two Scenarios shown right.

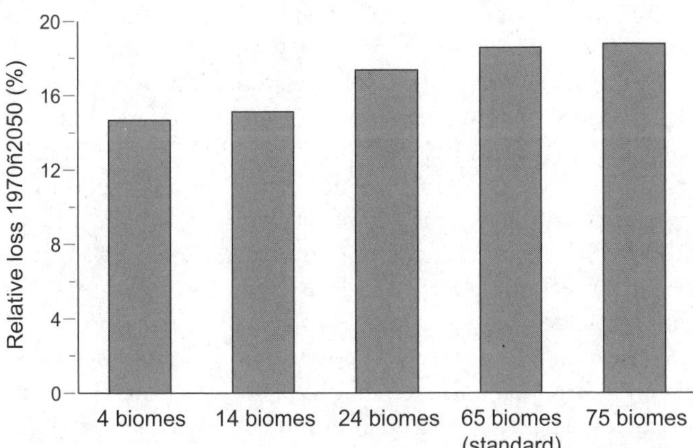

Figure 10.8. Influence of Using Different Regional Definition on Loss of Vascular Plant Estimates at Equilibrium for Order from Strength Scenario

migration rates, and original and new vegetation types. Real changes could be much more complex because individual species are the units that will respond to climate change. Solomon and Leemans (1990) concluded that future climate change could lead to large-scale synchronization of disturbance regimes, leading to the emergence of early-phase succession vegetation, with opportunistic generalist species dominating over large areas.

The decline of individual plant species due to climate change results from either competitive exclusion or the direct effect of climate change through increased drought frequency. Species changes resulting from competitive exclusion occur much more slowly than the response due to increased drought occurrence. Therefore, the fastest impacts might be expected in areas that show increased drought as a result of climate change. A comparison of global circulation models identified Central America, Southern Africa, Southern Europe, and Northern Australia as areas where all models agree that water availability will decline.

To test the impacts of climate change on biodiversity under the MA scenarios, we used three different approaches. First, we used a process-based model developed by Kleidon and Mooney (2000) that simulates the response of randomly chosen parameter combinations ("species") to climate processes. The model mimics the current distribution of biodiversity under current climate, and modeled "species" can be grouped into categories that closely match currently recognized biomes. Second, we used a model of African tick species diversity to show possible changes in species ranges as result of climate change. Finally, we con-

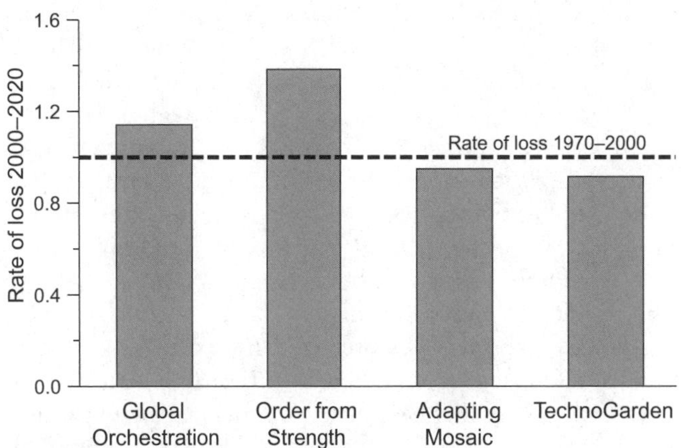

Figure 10.9. Potential Species Loss for MA Scenarios Compared with Historic Rates

sidered the possible impact of climate change at the level of biomes as calculated by IMAGE.

10.2.3.4.1 Climate change and potential biodiversity change

A warmer climate with altered hydrological regimes will affect plant functioning and the distribution of plant species. The impact of these changes on the potential distribution of plant diversity within the context of the MA scenarios was estimated by using the simulation modeling approach of Kleidon and Mooney (2000). Using the monthly mean temperature and precipitation anomalies of the Global Orchestration scenario, we altered the climatic forcing of the model. Light use efficiency was increased by 23%, consistent with the simulated increase of NPP by the IMAGE model. Modeling results indicated (*with low certainty*) that global environmental changes would lead to an increased environmental capacity for plant diversity in most regions of the world. (See Figure 10.10 in Appendix A.) These higher levels are mainly attributable to the increase in light use efficiency. Since the evolution of new species is unlikely to happen in this time frame, the increased capacity might increase the vulnerability of ecosystems to invasions.

Note that the magnitude of change is especially sensitive to the assumed increase in light use efficiency. If, for instance, the increase of productivity with elevated CO_2 were less, the increase would be less or counteracted by the decrease in precipitation. In this sense, Figure 10.10 mainly illustrates the point that biodiversity—the c in the species-area curves—is likely to be affected by global climatic change.

10.2.3.4.2 Climate change and tick diversity in Africa

We used a well-documented set of tick distribution data (Cumming 1998, 1999, 2000, 2002) to develop multivariate logistic regression models of the pan-African occurrences of individual tick species from climate data at 0.5-degree resolution. We used ticks as a model organism for invertebrates as their distributions are tightly linked to climate. Ticks are also significant to the public because they can be vectors of diseases. Maps for 73 species (from a fauna of about 240 species) were stacked to produce species-richness estimates under current (see Figure 10.11 in Appendix A) and projected conditions. The most severe impacts of climate change are likely to be on species that are highly specialized, including those with small species ranges and limited physiological tolerances. Although some African ticks probably fall into this category, insufficient data were available to model the distributions of rare ticks or those with limited species ranges. Consequently, the analysis should be interpreted as giving insight into how hardy, long-lived invertebrates with relatively high dispersal ability may respond to climate change.

Although the analysis predicted no extinctions among the 73 tick species considered in this analysis, changes in local tick species richness as a consequence of expansion and contraction of species ranges are likely to occur in all scenarios. (See Figure 10.12 in Appendix A.) The differences among scenarios are subtle and reflect the spatial nature of changes in weather patterns. Local biodiversity increased sharply in some areas and decreased sharply in others. TechnoGarden is the scenario in which the least expansion of tick species ranges occurs; the most expansion occurs in Order from Strength, although tick species richness in this scenario is reduced in areas of Angola and Tanzania. The dominant trend across all scenarios is an increase in local tick species richness (probably accompanied by increased tick burdens on livestock), with reductions in a smaller subset of locations.

10.2.3.4.3 Climate change and biome shifts

In many cases, biomes will shift geographically along with changes in climate. In general, the edges of current ecological zones are affected the most. It should be noted that each species would respond independently because each of them has different environmental requirements and a different capacity to adapt. However, our current understanding and models do not allow us to model shifts at the species level.

This exercise has modeled shifts with climate of vegetation types, which may have some structural similarity with current vegetation types but likely will not have the same species composition. The loss of species belonging to vegetation types that lost area as a result of climate change was calculated using the SAR approach described earlier. Increases in area of vegetation types did not result in increases in diversity. Vegetation types that would not have enough time to shift along the rate of climate change will result in degradation of remaining systems. In this assessment, we assumed that areas that shifted to new potential vegetation have lost all the endemic species of the original vegetation type but that those areas where species did not have time to adjust to the new climatic conditions lost only 50% of the original species. Figure 10.13 in Appendix A indicates these different categories according to IMAGE calculations.

The number of affected ecosystems increases with time, given increasing climate change. Only in the TechnoGarden scenario, after 2050, does the number of ecosystems without adaptation decrease because of the stabilization of GHG concentrations and the lower rate of temperature change. In the four scenarios, about 5–20% of the ecosystems will be seriously affected by climate change—without possible adaptation, the worst would be Global Orchestration. Focusing on protected areas only reveals similar numbers. Under Global Orchestration, in 20% of the protected areas the originally protected ecosystem will have been either replaced or severely damaged as a result of climate change. In the case of protected areas, any change (with and without adaptation) can be assumed to be negative, as in most cases they were selected because of the uniqueness of the ecosystems.

Figure 10.14 summarizes the results by biome for the two extreme climate change scenarios, Order from Strength and TechnoGarden. It shows that climate change is going to have an impact on biodiversity under both scenarios. The impacts, however, are more severe in Order from Strength. The Figure shows that most heavily affected biomes (in terms of percent change) include cool conifer forests, tundra, shrubland, savanna, and boreal forest.

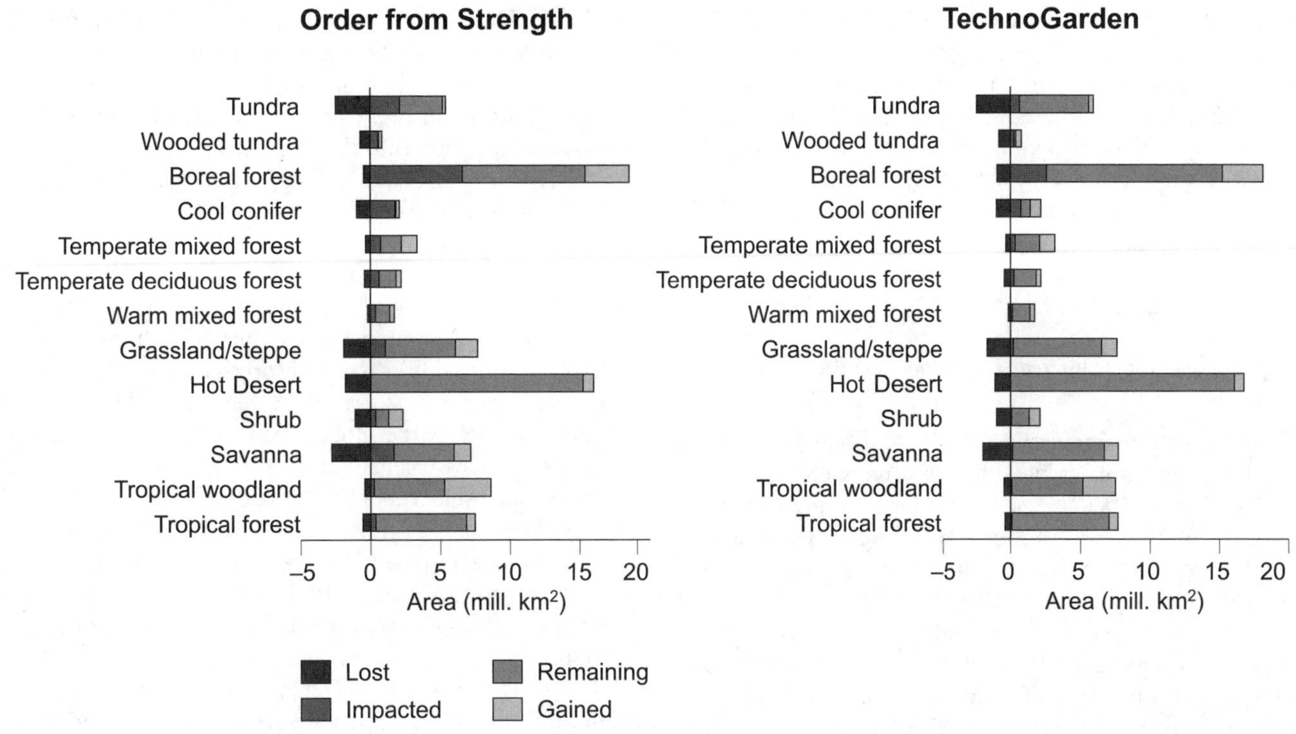

Figure 10.14. Impact of Climate Change in 2100 on Area of Different Biomes in Order from Strength and TechnoGarden Scenarios

Thomas et al. (2004) performed a similar analysis of the potential effect of climate change by 2050 on global extinctions for selected regions of the world, which accounted for 20% of the area of terrestrial ecosystems. The authors evaluated three rates of climate change and two hypothetical cases with or without dispersal limitations. Results of this analysis encompass the results of the four MA scenarios reported in Figure 10.14. Maximum extinction due to climate change, as reported by Thomas et al. (2004), was 100% of the plant species in Amazonia, assuming no dispersal capabilities and maximum rate of climate change; there was a minimum value of 3% for Europe, assuming no dispersal constraints and minimum climate change. On average, across regions and taxa, the Thomas et al. (2004) exercise yielded losses of species at equilibrium with reduced habitat that ranged between 15% and 37%.

10.2.3.5 Loss of Terrestrial Biodiversity through Changes in Atmospheric Deposition

Atmospheric deposition of nitrogen can lead to change of ecosystems as a result of nitrogen excess (also called terrestrial eutrophication). Nitrogen excess can be an important cause of ecosystem degradation as it is the primary nutrient limiting plant production in many terrestrial environments. Increases in nitrogen input can therefore alter these ecosystems and lead to shifts in species composition, increased productivity, decreased species diversity, and altered tolerance to stress conditions (Pitcairn 1994). The most important anthropogenic sources of nitrogen emissions are fossil fuel burning and industrial and agriculture activities. Excess nitrogen deposition has been a prominent environmental problem in North America and Europe since 1970.

Bouwman et al. (2002) made a global assessment of acidification and excess nitrogen deposition effects on natural ecosystems by overlaying current and future deposition maps of sulfur and nitrogen with sensitivity maps for both acidification and nitrogen deposition (at 0.5x0.5 degrees). These sensitivity maps are expressed in so-called critical load values below which no damage is assumed to be negligible. They are calculated on the basis of soil, ecosystem, and climate data and on soil dust deposition. For future emissions, they used deposition maps calculated by the STOCHEM environmental chemistry and transport model based on the IPCC IS92a scenario.

To estimate excess nitrogen deposition risks under the MA scenarios, we followed the approach of Bouwman et al. (2002) and scaled their deposition map on the basis of changes in nitrogen emissions in each of the 17 global regions of the IMAGE 2.2 model (on their turn, based on changes in energy use, agriculture, and environmental policy). At a 0.5x0.5 degree map, we calculated the deposition of nitrogen and compared it to critical loads of these grid cells. The ratio between deposition and critical load was used as an indication of risks of nitrogen deposition. A ratio below 1 implies limited risks (at least based on the average grid cell; obviously within the grid cell many ecosystems will occur that are more sensitive than the average, and this approach will therefore result in an underestimation of the actual risk); a high ratio indicates a very high-risk level of disturbance.

Bobbink (2004) reported biodiversity losses for different levels of nitrogen loading for a large number of different ecosystems, and these results were summarized in a large number of ecosystem-specific relationships. Wedin (1996) (see also Tilman et al. 1996; Haddad et al. 2000) showed changes in plant biodiversity for grasslands for different levels of nitrogen loading. (See Figure 10.15.) These studies indicated a 25% reduction of diversity for a ratio of three times critical load, a 50% reduction for eight times the critical load, and a 60% reduction for 25 times the critical load. We used those numbers to calculate our overall threat indicators—and aggregated our results to the level of the IMAGE 2.2 biomes (by WWF region). The numbers show the average reduction in diversity by biome. (See Figure 10.16 in Appendix A.)

10.2.3.6 Integrating Different Environmental Pressures on Biodiversity

On the basis of the indicators calculated above, it is possible to compare the impact of different drivers of biodiversity loss. (See Figure 10.17.) For habitat loss, we used the SAR approach described earlier, but accounting only for changes due to agricultural expansion and timber production. For climate change and excess nitrogen, we also used the SAR approach to aggregate grid-level effects at the level of complete biomes. Figure 10.18 shows the impacts at the global level of the combination of habitat loss, climate change, and nitrogen deposition under the four MA scenarios in 2050.

Earlier, Sala et al. (2000) performed an exercise in which they developed the same kind of graphs on the basis of expert judgment, supported by selected modeling results. In general, the current calculations confirmed the findings of Sala et al. (2000), with some exceptions. As in the earlier study, habitat loss was found to be the most important driver of future biodiversity loss. However, in some biomes climate change was identified as the major cause of biodiversity loss, including tundra and deserts and to some degree boreal forests. The overall impacts of climate change (and other drivers) on boreal forests was assessed to be higher by Sala et al. (2000), which might be due to the limitations of the present method (focusing mainly on the total size of the different biomes) but also to the assessment year (2050 versus 2100). Deposition of nitrogen has been identified as a major driver of species loss in temperate forests, warm mixed forest (particularly in Asia), and, to a lesser degree, savanna. This is consistent with the earlier assessment. Habitat loss, finally, was found to be particularly important for species loss in temperate forests, warm mixed forests, savanna, and tropical forest.

The differences among the scenarios are relatively small due to delays within the system and counteracting assumptions. The 50-year modeling window chosen for this exercise may not be enough for the different climate change scenarios to unfold fully. In addition to delays in the drivers, there are important delays in the response of biodiversity. Losses of species at the global level do not occur immediately after the loss of habitat or alteration of the environment. For example, reduction of habitat lowers the number of individuals in a population and puts this species on an extinction trajectory. However, the species extinction would not effectively occur for quite some time, depending on the life cycle of the species and the characteristics of the ecosystem.

The four MA scenarios have assumptions about effects on biodiversity that compensate each other. The compensatory nature of the different assumptions reduces the differences in biodiversity effects among scenarios. For example, one scenario assumes lower food demand and other assumes higher demand but also higher yield. The end result is that the differences in land use change and biodiversity loss between these two scenarios were relatively small. Similarly, in different scenarios the increase in pastureland is compensated by the decrease in cropland and vice versa. In general, TechnoGarden results in the least amount of pressure on biodiversity—although the difference with Adapting Mosaic is small, and mainly due to lower pressure from climate change. In contrast, the highest pressures were found for Order from Strength—in particular, for land use change and deposition.

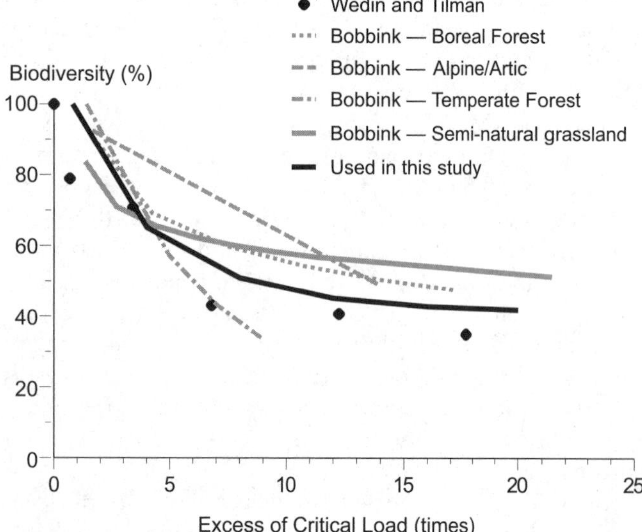

Figure 10.15. Effect of Ratio of Nitrogen Deposition to the Critical Load on Plant-Species Diversity

10.3 Freshwater Biodiversity

10.3.1 The Approach

Species-discharge curves are similar to species-area curves for terrestrial biota in the sense that richness numbers increase logarithmically with discharge (Oberdorff et al. 1995). They are subject to some of the same hypotheses to explain their occurrence, including the theory of island biogeography and the dependence of species immigration and extinction on river size and the theory on the increase types of resources and habitats, certainly including more open-water, floodplains, backwater, and high-flow habitats that only high discharges provide.

Although a positive relationship exists between riverine species richness and catchment size for fish, mussels, and aquatic invertebrates (Sepkoski and Rex 1974; Welcomme 1979; Livingstone et al. 1982; Strayer 1983; Brönmark et al. 1984; Eadie et al. 1986; Angermeier and Schlosser 1989; Hugueny 1989; Oberdorff et al. 1993), discharge and other indices of habitat volume are better predictors of species

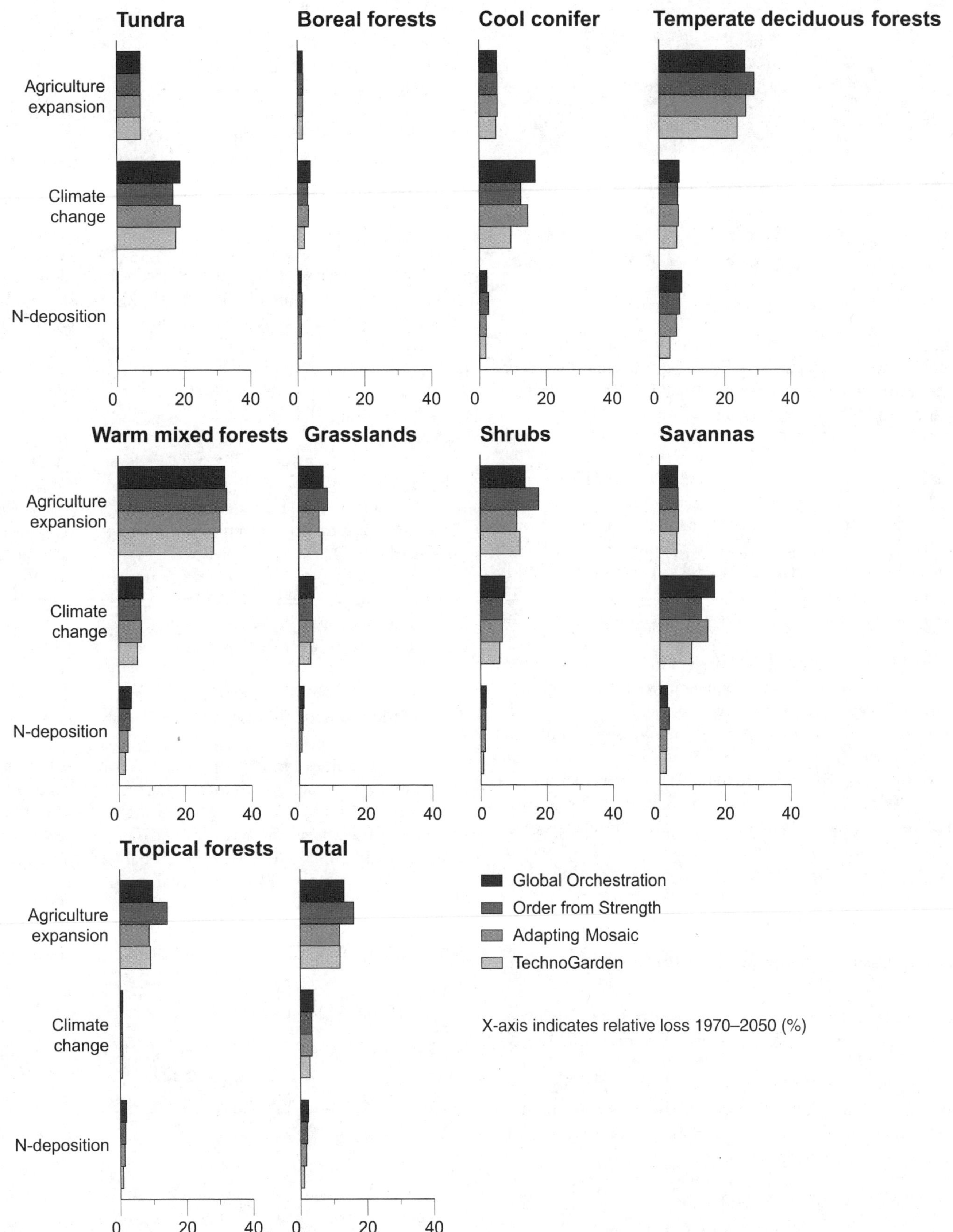

Figure 10.17. Comparison of Effects of Agriculture Expansion, Climate Change, and Nitrogen Deposition on Relative Biodiversity Loss in Four Scenarios for Different Biomes and the World

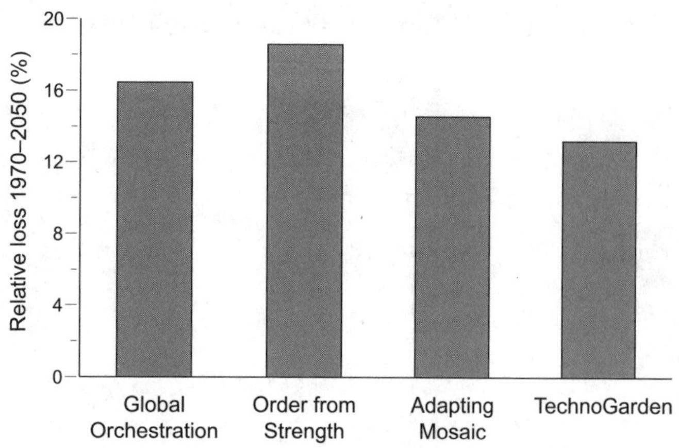

Figure 10.18. Relative Losses of Global Vascular Plant Biodiversity when Populations Reach Equilibrium with Reduced Habitat, Taking into Account Agricultural Expansion, Climate Change, and Nitrogen Deposition for Four Scenarios

richness than area (Livingstone et al. 1982; Angermeier and Schlosser 1989). Thus we address the effects of reductions in water discharge (an index of habitat type and amount) on fish species number. We address only fishes because no regional or global data sets exist for other taxa. Reductions in discharge result not only from climate change, but also from consumption of water for agriculture and other human uses.

To forecast the impact of drivers in addition to climate and water withdrawal (eutrophication/land use, acidification) on rivers and for all drivers on lakes, we rely exclusively on semi-quantitative or qualitative algorithms, such as previously published experimental or regional studies summarized for lakes by Lodge (2001) and for rivers by Poff et al. (2001). The only top driver (Sala et al. 2000; Brönmark and Hansson 2002) that we are largely unable to address is nonindigenous species; we address these only in the most qualitative way because no rigorous algorithms exist to forecast changes in the occurrence or impact of such species.

10.3.1.1 Quantitative Algorithms for Forecasting Biodiversity Loss

We used a species-discharge regression published in Oberdorff et al. (1995) to forecast loss of riverine fish. We obtained fish species numbers for 344 global rivers from Oberdorff et al. (1995) and from FishBase (www.fishbase.org). This quantitative approach should be regarded as speculative because this is the first application to forecasting of existing fish species-discharge relationships. For many rivers, fish species numbers include native and established nonindigenous species because most data sets (e.g., Oberdorff et al. 1995) did not distinguish between them. For rivers where it was possible to distinguish native from nonindigenous species, the percentage of nonindigenous species was low (< 5%) and simply added a minor amount of uncertainty to the species-discharge model. Although some human influence may be incorporated in the data (such as dams and nonindigenous species), we assume that such effects are minimal and that our species-discharge model reflects evolutionary and ecological outcomes roughly in equilibrium with natural discharge.

We used the Water Global Assessment and Prognosis model (Alcamo et al. 2003) to compute current and future discharge. The effects of climate change and water consumption are calculated separately in WaterGAP (global hydrology model and global water use model), thus allowing us to assess their independent impacts on discharge and hence on fish biodiversity. Briefly, in the global hydrology, model river discharge is computed by performing a grid-based water balance of the vegetation canopy and soil, driven by precipitation and other climate data. Water withdrawal and consumptive water use are computed in the global water use model using national estimates of domestic and industrial use in addition to estimates of irrigated areas and livestock. Additional details are available in Alcamo et al. (2003) and in Chapter 9.

We calibrated the discharge values used in the published model (Oberdorff et al. 1995) against the baseline discharge values used by WaterGAP in order to obtain a relationship that we could use to forecast future biodiversity with the WaterGAP output for the scenarios. Discharges from Oberdorff et al.(1995) and WaterGAP were highly correlated (r^2 = 0.84). The final species discharge regression we used was constructed for rivers located between 42° N and 42° S, the latitudinal band within which reduced discharge is predicted to occur under the scenarios (see Figure 10.19):

$$\text{log fish species number} = 0.4{\star}(\text{log WaterGAP discharge, } m^3\ s^{-1}) + 0.6242$$

Discharge values (annual means) were obtained at the river mouth and represent discharge during the baseline climate period. This regression explained slightly more of the variance in fish species number ($r^2 = 0.57$) than the regression originally published by Oberdorff et al. (1995) (log fish species number = 0.3311★(log discharge, $m^3\ s^{-1}$) + 0.83, $r^2 = 0.52$). We assessed the relative contribution to fish biodiversity of loss of climate and water withdrawal by comparing fish species numbers based first on discharge with

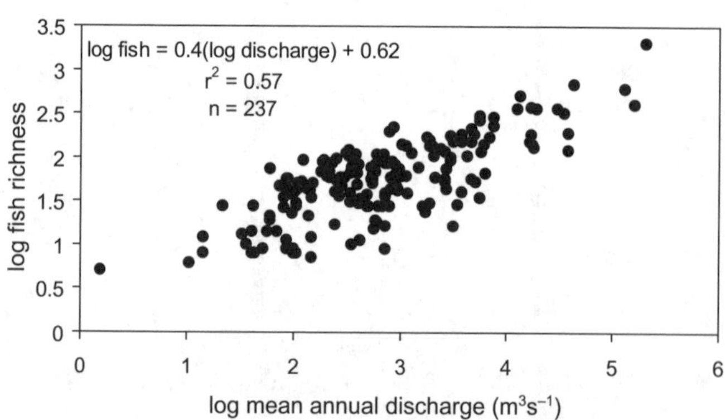

Figure 10.19 Fish Species Discharge Curve Used to Build Scenarios of Fish Loss. The regression was modeled with rivers found between 42N and 42S, where reduced discharge is predicted to occur.

only climate change in the scenario and then on discharge affected by both climate and human withdrawals of water. We calculated fish loss in this order because humans are more likely to manage water than climate; the incentives for a country or region to manage water withdrawal are stronger than the ones to reduce greenhouse gases. In other words, calculating fish loss in this way treats climate change as a given (because it is beyond the control of individual countries or regions) and water withdrawal as a driver that is more likely to be managed effectively.

As a consequence of this order of calculations, the forecast impact of water withdrawal on fish loss is maximized because of the nonlinear relationship of species with discharge (that is, the loss of species per unit discharge reduction increases as discharge is reduced). Forecasts of fish loss from water withdrawal include the effects of current and past water use in addition to any additional consumption that occurs in the scenarios. Confidence intervals (95%) were calculated from the critical values of the *t* distribution using the standard deviation of the predicted fish species number to which we added the uncertainty (slope * standard deviation of predicted discharge) generated from WaterGAP for the future discharge. (See Chapter 9.)

We selected for analysis the two scenarios for 2100 that produced the most fish losses (Global Orchestration) and the least (TechnoGarden), plus one intermediate scenario (Order from Strength); we also conducted fish loss analyses for 2050 on Global Orchestration and Order from Strength.

Because change in the magnitude of extreme discharge events could have strong biological consequences (Poff et al. 2001), we also tested whether the WaterGAP index of the magnitude of low flow (Q90 = the discharge exceeded by 90% of monthly averages) changed from the 1995 baseline. Q90 correlated strongly and positively with 2100 annual discharge by river basin for all scenarios ($r^2 = 0.94$). Furthermore, the slope of this relationship for each scenario did not differ from the slope for 1995. Thus, this index of flow did not change from baseline conditions in any scenario. However, other features of the hydrograph important to fishes and other aquatic biota (such as the timing and duration of low or high flows) may change under future climate and other drivers. We could not, however, assess those potential impacts in this analysis.

Even in the absence of any mitigation or conservation measures, the forecast loss of species would not, of course, occur instantaneously; rather, the expectation would be that these species would be likely to become extinct on a schedule that we cannot accurately anticipate (Minckley et al. 2003; see also MA *Current State and Trends* for a discussion about the difficulty of accurately calculating extinctions). The slow pace of many extinctions would provide time to plan and implement measures to prevent some losses in biodiversity. In North America, spacious river basins are also rich in endemics, and this pattern likely holds for other continents (Oberdorff et al. 1999). However, endemic species lists are unavailable for most rivers, especially those in the parts of the world that dry in the scenarios (see description of results that follows). Therefore, our fish richness data include both species endemic to each basin and those that occur in multiple basins.

Potential fish losses that we calculate here are thus a combination of extinctions at equilibrium for that river basin (local biodiversity losses) and global losses. In terms of ecosystem services provided by fisheries, local extinctions are a more relevant metric than global extinctions.

We believe that all plausible biases in this species-discharge approach are likely to underestimate long-term species losses at equilibrium, because this method does not account for interactions between the effects of discharge and the effects of other habitat features that will no doubt be affected simultaneously by decreases in discharge and by other drivers (eutrophication, acidification, dams, and so on), which we can only address qualitatively (see following description).

Increases in discharge (which occur for 60–70% of the world's river basins in all four scenarios) would not necessarily lead to increases in fish richness on the time scale of the scenarios, even on a local scale, because species migration from other river basins might happen only slowly or at least at rates that we cannot model in this context. At the global scale, species richness would not increase appreciably in 100 years because evolution typically happens more slowly. Thus we assumed no discharge-related change in fish species richness for river basins that experience increases in discharge, although many of these river basins will be strongly affected by other drivers (anthropogenic introductions of nonindigenous fish species). While increases in discharge may increase the production of fishes, nonindigenous species would probably be favored as a result of the habitat changes that come with increased discharge.

Overall, then, we consider these regression analyses to be a speculative guide to plausible outcomes for biodiversity of fishes. They provide a conservative index of river system-specific extirpation of fish species as a function of the drivers that affect discharge in the WaterGAP model (climate and water withdrawals), assuming steps to prevent such extinctions are not taken.

10.3.1.2 Qualitative Approaches to Forecasting Biodiversity Change

To supplement the quantitative fish species-discharge approach to riverine fish species number, we also provide qualitative analyses of the potential impact of other important drivers. We rely on qualitative scenarios of these other drivers because no quantitative algorithms exist to relate them to biodiversity. Our scenarios of impact on biodiversity, community structure, and ecosystem productivity are thus qualitative.

10.3.1.2.1 Eutrophication

We used WaterGAP's return flow (see Chapter 9) as an index of poor water quality derived from human water use. We interpreted this as an index of eutrophication in rivers, lakes, and wetlands, because human use is likely to result in increases in the nitrogen and phosphorus content of waters, especially in countries where water treatment capacities are poor or lacking. In more industrialized countries, where water treatment exists, return flow may be a poor indicator of eutrophication, and we put less emphasis on this for those areas. In some regions, return flow is likely to be laden with industrial pollutants, but this will depend on what types of local industry exist, which is beyond the scope of the present analysis. As an index of aerial deposition of N, we use the same approach as described earlier for terrestrial ecosystems: we combine estimates of N deposition from IMAGE with spatially explicit estimates of sensitivity to N based on soils (Bouwman et al. 2002). Abundant literature (at least for selected regions) on the biological impacts of eutrophication provides a basis for assessing potential biological impacts. Because the productivity of many freshwater ecosystems is limited more by P than by N (Elser et al. 1990; see also MA *Current State and Trends*), our conclusions about eutrophication of freshwaters remain highly speculative.

10.3.1.2.2 Acidification

As an index of SO_x aerial deposition, we use the same approach as for aerial deposition of N: we combine estimates of S deposition from IMAGE with spatially explicit estimates of sensitivity to acidification based on soils (Bouwman et al. 2002; see also Chapter 9). Because acidification sensitivity as estimated globally by Bouwman et al. (2002) is consistent with more local studies on the biological impacts of acidification (e.g., Schindler et al. 1985; Brezonik et al. 1993b; Frost et al. 1995; Vinebrooke et al. 2003), these data provide a basis for assessing potential biological impacts in different scenarios.

10.3.1.2.3 Temperature

The direct effects of temperature per se on biodiversity are difficult to assess. Many previous studies, especially of fishes in lakes and streams, have illustrated that relative distributions and abundance of fish species are likely to change within a basin or within a lake as a result of temperature increases; likewise, the edges of geographic ranges of fishes will move toward higher latitudes (e.g., Lodge (2001); Rahel (2002)). More specific scenarios of the impact of temperature increases at a global scale are currently impossible and we do not attempt to construct any here.

10.3.2 Quantitative Results for Fish Biodiversity Based on River Discharge

The major patterns of discharge changes from baseline conditions are very similar across scenarios. Under all scenarios, approximately 70% of the world's rivers have increased discharge. Fish production may increase and could benefit humans. There is little basis on which to forecast consequences of increasing discharge for freshwater biodiversity. Some native species would no doubt decline as conditions change. If nonindigenous species are introduced, they would have an increased probability of success as new habitats appear to which native species are not adapted. Because of the highly uncertain consequences of increased discharge on fish and other aquatic biota, we do not consider these river basins further.

In contrast, under all scenarios, approximately 30% of the world's rivers have decreased discharge and decreased fish species diversity, resulting largely from climate change and, to a lesser extent, from increasing water withdrawal by humans. Basin-specific reductions in fish species numbers differ much more widely between basins than between scenarios. By 2050, for the 110 modeled river basins that are drying, the basin-specific percentage of fish species likely to face extinction ranges from about 1% to about 60%; for 2100, analogous values range from about 1% to about 65%. (See Figure 10.20 here and Figure 10.21 in Appendix A for 23 representative rivers.) Water withdrawal contributed little (generally an additional 1–5%) to potential fish species loss in most (~80%) rivers. In some regions, however, including India, Australia, and parts of Eastern Europe, water withdrawal was a substantial driver. In the Middle East and India especially, water withdrawal caused most of the extinctions. Considering both climate and water withdrawal, Global Orchestration resulted in the highest fish species losses overall, Order from Strength marginally lower losses, and TechnoGarden produced the fewest species losses. (Adapting Mosaic was not modeled.)

Losses of fish biodiversity were concentrated in southern Africa, northern Africa, eastern Europe and the Middle East, India, Australia, south-central South America, northern South America, and southern Central America. In many of these countries, fishes are an important indigenous source of protein, and governments and society have less capacity to cope with losses of such ecosystem services than in countries experiencing lower losses of fish biodiversity (see MA *Current State and Trends*). As documented by the IPCC, areas predicted to experience drying differ under different global climate models. However, most of the rivers predicted by WaterGAP to lose discharge are in areas predicted to dry by most general circulation models (see Figure 3.3 in the Third IPCC Assessment). Thus these patterns are robust for most areas of the world.

10.3.3 Qualitative Results for Fish Biodiversity

10.3.3.1 Eutrophication

Increases in return flows (estimated by WaterGAP) differ across scenarios (see Chapter 9), with the greatest increases over baseline conditions in Global Orchestration, followed by Order from Strength and TechnoGarden. Almost all areas with large increases in return flows are also areas with decreased discharge. This is especially true for central and southern Africa, the Middle East, India and neighboring states, Central America, and eastern and southern South America. For freshwater taxa in these areas, habitat quality will be declining simultaneously with habitat volume, as pollution by nutrients and other chemicals increases. Under Global Orchestration and Order from Strength (but less for TechnoGarden), most of these same areas experience increased atmospheric deposition of nitrogen, further increasing the potential for eutrophication in water bodies that are N-limited. Many other regions also experience increased atmospheric deposition of N. Thus even in areas of steady or increasing freshwater habitat, nitrogen enrichment is

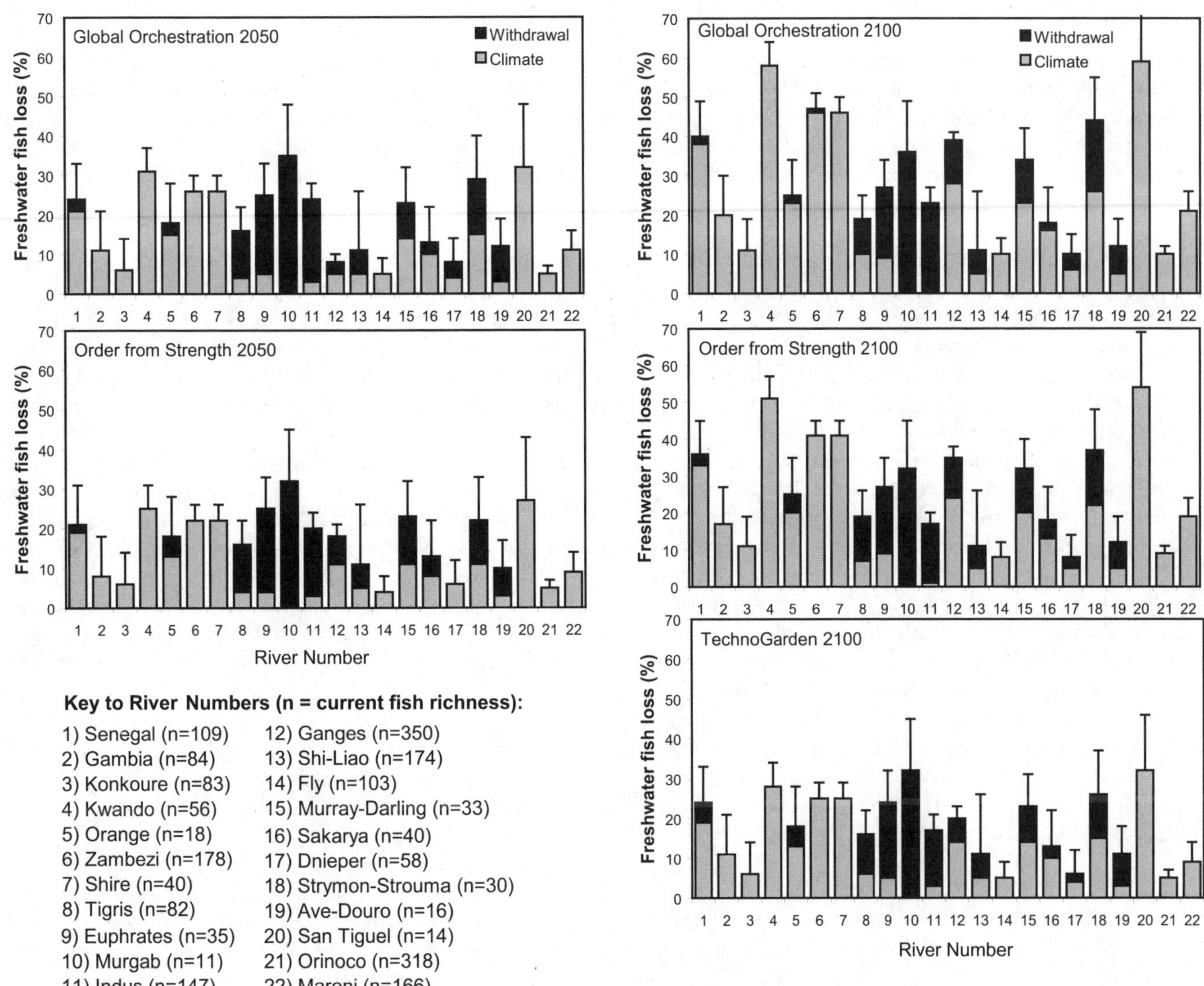

Figure 10.20. Percentage Losses of Fish Species Predicted from Decreases of River Discharge Resulting from Climate Change and Water Withdrawal by Humans for Two Scenarios in 2050 and Three Scenarios in 2100. The 22 rivers depicted are representative of the 110 modeled rivers that experience losses of discharge under the scenarios. Percentage loss is ± 95% confidence intervals. Gray indicates discharge resulting from climate change; black indicates discharge resulting from water withdrawals by humans.

likely to reduce freshwater biodiversity and change species composition of freshwater taxa. Of the three scenarios we examined closely, only under TechnoGarden were there regions of steady or declining N deposition.

In general, for the majority of the world under both Global Orchestration and Order from Strength, the symptoms of eutrophication (including both P and N enrichment) can potentially be strong. These include increased concentration of noxious algal blooms (while decreasing total species richness of all taxa) (Schindler 1977), decreases in oxygen, water quality, and aesthetic value, and severe reduction of fish populations and species (see MA *Current States and Trends*).

10.3.3.2 Acidification

Acidification increased in some parts of the world in all three scenarios considered and was especially severe in Global Orchestration and Order from Strength. Some regions affected by water loss and eutrophication also experienced increasing acid deposition—for instance, the Middle East in all three scenarios. Likely consequences of acidification are well established from many previous observations and experiments. We know from the acidification of many waterways in North America and Europe that substantial ecological and biological changes have occurred. Entire food webs have been affected, with most species disappearing while others increased (Vinebrooke et al. 2003). Fish are particularly vulnerable to decreases in pH and can disappear completely from acidified systems (Schindler et al. 1985; Brezonik et al. 1993a; Frost et al. 1995).

10.3.3.3 Hotspots of Freshwater Biodiversity and Ramsar Sites

According to Groombridge and Jenkins (2002), the major global hotspots for fishes are the Amazon basin and neighboring parts of South America and the basins of central Africa. Although the cores of these areas do not experience drying, some of the edges of these large hotspots do, especially in northern South America. In these scenarios, the global fish hotspots did not experience large increases in eutrophication and acidification. Other freshwater taxa suffered more from one or more drivers. For example, one of the two major hotspots for crayfish, southeastern Australia, will suffer extreme drying. In contrast, two of the major hotspots of freshwater crab diversity occur in Central America/northern South America and India, where drying and water pollution increase greatly in Global Orchestration (both regions) and Order from Strength (Central America/northern South America). The major global hotspot for fairy shrimp, which inhabit wetlands, is southern Africa (Groombridge and Jenkins 2002), where drying, eutrophication, and acidification are all increasing.

Because wetland ecosystems are by definition low-volume aquatic ecosystems and are often seasonally absent under current conditions, they are particularly vulnerable to changing conditions of climate, human water use, and pollution (Revenga and Kura 2003). These scenarios present severe threats to wetlands. For example, in many drying river basins of the world, a large proportion of the basin area is currently wetland: Orinoco 15%, Parana 11%, Ganges 18%, Fly 42%, Sepik 33%, Krishna 16%, and Brahmaputra 21%. Multiple Ramsar wetlands occur in the following river basins that dry and suffer strongly from other drivers in the scenarios: Senegal (4 sites), Parana (7), Indus (10), Ganges (4), and Murray-Darling (10) (Revenga et al. 1998).

10.3.3.4 Multiple Drivers and Interactions

Data on the global distribution of freshwater biodiversity are fragmentary at best, but it is clear from existing information that some taxonomic groups are likely to continue to experience very high extinction rates as a result of combinations of drivers (Jenkins 2003). Other anthropogenic influences are likely to increase freshwater species loss above what is reflected in the species-discharge model. Increased water temperatures would further exacerbate the stress experienced by fish and other taxa (Matthews and Zimmerman 1990; Casselman 2002). Secondary infections of fishes may increase in areas of low water flow (Steedman 1991; Chappel 1995; Janovy et al. 1997).

Overfishing, particularly in poor developing countries, will continue to reduce fish populations (see MA *Current State and Trends*; Bradford and Irvine 2000; Odada et al. 2004). Xenobiotics (human-made organic chemicals) that may alter survival, reproduction, and growth for aquatic biota are forecast to increase (Brönmark and Hansson 2002). Salinity can increase in highly irrigated rivers and lakes, with subsequent negative effects on many freshwater taxa (Williams 2001). Dams may significantly reduce populations of migratory fish (see MA *Current State and Trends*) and negatively affect species richness in general (Cumming in press). Additional dams are planned for some rivers that are forecast to lose discharge and thus potentially lose fish, which may further reduce fishes. For example, there are six new dams planned in the Tigris and Euphrates basins, seven in the Ganges Basin, and two in the Orange River (Revenga et al. 1998). These river basins all have high numbers of endemic fish (see MA *Current State and Trends*).

All these changes and their interactions are likely to decrease habitat for native species and favor the survival of any nonindigenous species that are introduced (Kolar and Lodge 2000). And the likelihood of introduction of nonindigenous species will increase where human population and trade increases (Levine and D'Antonio 2003). Trade differs among scenarios (see later section), but history suggests in general that we should expect increases in freshwater nonindigenous species in all scenarios in the absence of new prevention and control efforts. The fish fauna of the United States, for example, has been largely homogenized over the last decades (Rahel 2000), while dams and other anthropogenic drivers increase the occurrence and impact of nonindigenous species (Marchetti and Moyle 2001). Thus without major efforts to prevent the introduction of nonindigenous species, and without additional conservation efforts to reduce the impact of other drivers, much greater declines in freshwater biodiversity than those implied by our quantitative and qualitative models are likely to result from multiple, interactive, cumulative, and long-term effects in rivers that experience decreasing or increasing discharge.

10.3.3.5 Aggregated Effects of Drivers and Human Well-being

In all the scenarios, but especially in Global Orchestration and Order from Strength, freshwater ecosystems in some parts of the world changed in major ways. In many parts of the world, declining water quantity and quality occurred simultaneously in all scenarios. While this caused human hardship directly, it also caused large losses of ecosystem services in the form of harvest of freshwater fishes, fiber from freshwater wetlands, and other freshwater taxa. The most negative combinations of drivers coincide with geopolitical regions where the capacity of governments and society to cope with the loss of biodiversity and ecosystem services is low. For example, GDP per capita (CIA 2003) in drying countries is about 20% lower than in countries that do not get dryer in the scenarios.

10.4 Marine Biodiversity

10.4.1 The Approach

A global ecological model of marine systems does not exist yet, but there are more than 100 ecosystem models of various ecosystems throughout the world based on the Ecopath with Ecosim software, which is described in Chapter 6. Ecopath with Ecosim uses a combination of trophic levels (functional groups) and species to describe the ecosystems rather than complex webs of individual species, which limits how well it can describe "biodiversity" changes. However, changes in the diversity of the system based on the various functional groups can be described using Kempton's Q for the biomass of groups with a trophic level of 3 or more. Three models—the Gulf of Thailand (shallow coastal shelf system), Benguela Current (coastal upwelling system), and the Central North Pacific (pelagic system)—were modeled using the four MA scenarios. These three systems are described in detail in Chapter 6, while changes in their biomass diversity are described here.

Kempton's Q is a relative index of biomass diversity and is based on a modified version of Kempton's Q75 index originally developed for expressing species diversity (Kempton 2002). The index is estimated as Q75 = S/(2 log(N0.25•S/N0.75•S)), where S is the number of species (here functional groups) and N times i•S is the number of individuals (here biomass) in the sample of the (i•S) most common species (or of a weighted average of the species closest if i•S is not an integer). The Q75 index thus describes the slope of the cumulative species-abundance curve between the lowest and highest quartiles. A sample with high diversity will have a low slope, so an increase in diversity will manifest itself through a lower Q75 index. To reverse this relationship, and to make the Q75 index relative to the baseline run in the ECOPATH with ECOSIM simulations, we expressed the biodiversity index as (2 – Qrun/Qbaserun), truncating the index at zero in the unlikely case that the Q75 index should more than double.

The Q75 index and the inverse diversity index are sensitive to the number of species (functional groups), and have merit mainly for expressing relative changes for a given model or for models with the same group structure. To focus the index on the exploited part of the ecosystem—that is, the part for which there is the most information and where human impact is most likely to be seen—the analysis is limited to groups with a trophic level of 3 or more. This excluded from the index the primary producers and groups that are primarily herbivores or detritivores (such as zooplankton and most benthos groups). In the analysis here, a high biomass diversity index refers to an ecosystem of greater evenness (that is, even distribution of biomass among a number of species), and a low index value refers to an ecosystem where one or two species are much more abundant among a small number of species.

The number of studies focused on the future of marine biodiversity is limited compared with ones on terrestrial systems. Field et al. (2002) examined the future of the world's oceans to 2020, but biodiversity was not a significant focus of their study. Culotta (1994) noted that studies of marine biodiversity were in their infancy. Progress has been made since then, albeit slowly, especially in less accessible environments and in assessing future changes in ecosystems such as the deep sea (Glover and Smith 2003), polar seas (Clark and Harris 2003), and vents and seamounts (Koslow et al. 2001). Potential climate change impacts on coastal marine biodiversity have also been the focus of recent studies (e.g., Kennedy et al. 2002).

However, none of these studies has quantitatively examined changes in biodiversity; they have all been qualitative, based on projections of current trends. The lack of quantitative methods for examining changes in marine biodiversity at the ecosystem scale is limited by methods and robust, broad-scale information. The species-area curves and other species-specific methods are of limited use in marine ecosystems because species extinctions are rarely observed and because information at the species level is not available for many systems.

10.4.2 Marine Biodiversity Change across Scenarios

10.4.2.1 The Gulf of Thailand Model

The results of the modeling of the Gulf of Thailand using the four scenarios are illustrated in Figure 10.22. By 2050, all four scenarios approached a similar Kempton Q index (biomass diversity index) that was less than the value for 2000. Initially, in the TechnoGarden scenario the biomass diversity index declined slightly as the Gulf of Thailand ecosystem was managed so that the profits from the high-value fisheries were increased. This resulted in the number of species and the biomass of some species declining more

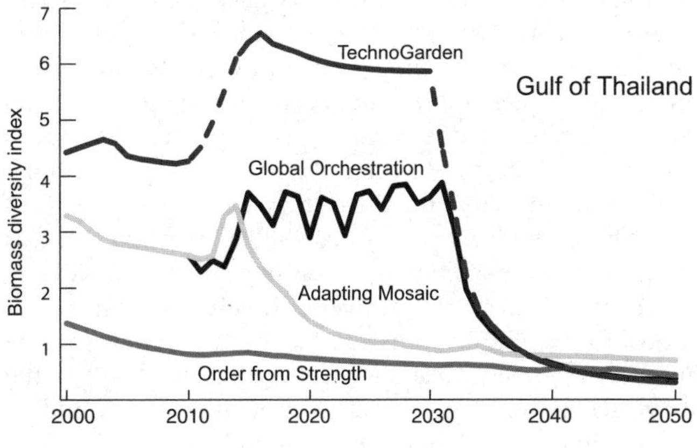

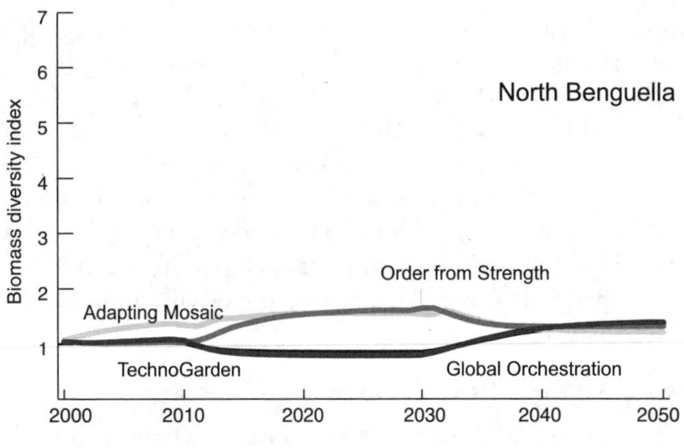

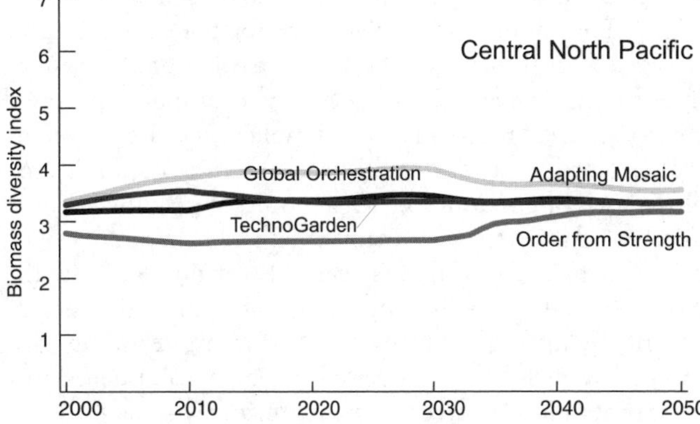

Figure 10.22. Changes in Biomass Diversity by 2050 in Four MA Scenarios in Three Specific Regions. Diversity is for groups with a trophic level equal to or higher than three.

relative to other species. In 2010, the policy shifted to rebuilding the ecosystem, which accounted for the increasing biomass diversity index as the number of species increased and the biomass was more evenly distributed among these species. This resulted in a more stable ecosystem structure. As technology developed and was able to provide some ecosystem services, the policy focus turned to providing fish that were used to produce fishmeal for the aquaculture sector, which had taken over the role of food provisioning. This resulted in the biomass diversity index declining rapidly when the ecosystem changed from an even distribution of species and associated biomass to an ecosystem dominated by a few functional groups.

The Gulf of Thailand responded to the Global Orchestration scenario in a similar manner to the TechnoGarden, but the changes in the biomass diversity index (number of species and abundance) were relatively lower. The Adapting Mosaic scenario also commenced with a focus on increasing the profits of high-value fisheries and as a consequence the biomass diversity index declined slowly, with some species and species abundance within functional groups declining. In 2010, attempts to rebuild the ecosystem experiencing climate change within a policy that still had increasing profits as a priority were of limited success since the biomass diversity increase was much less than in the TechnoGarden and Global Orchestration scenarios. In this case, changes in the number of species and abundance were small relative to the above scenarios. Despite policies that included rebuilding selected high-value stocks, the biomass diversity index declined rapidly to a system with a few functional groups dominating the ecosystems. The Order from Strength scenario, which had a focus on increasing the value of the fisheries throughout the 50 years of the scenario, resulted in a steady decline in the biomass diversity index.

The results of the four scenarios suggest that it is (moderately) likely that the Gulf of Thailand would lose biomass diversity—that is, the number of species will decline. For those species that remain, a few of them will dominate (in abundance) the ecosystem.

10.4.2.2 Benguela Current Model

The results of the modeling of the North Benguela Current system using the four scenarios are also presented in Figure 10.22. The biomass diversity index for the four scenarios changed very little over the 50 years of modeled results. In Adapting Mosaic, the biomass diversity index increased consistently, as the management policy was a mix of increasing profits from food fisheries and maintaining employment opportunities. This could be achieved if a number of fisheries were maintained and fisheries that employed a number of fishers were also maintained. In the North Benguela ecosystem, this resulted in a diversity of fisheries, with the abundance of species more evenly distributed as well as the number of species, but not necessarily increasing (biomass diversity index relatively stable).

In the Order from Strength scenario, the focus was also on maintaining profits and employment opportunities; however, in this scenario high-value fisheries were targeted initially and therefore there was little change in the biomass diversity index in the first 10 years. In 2010, ecosystem rebuilding commenced, and in response the biomass diversity index also increased to a level similar to the Adapting Mosaic scenario. This was not surprising, since this phase of the scenario also tried to maintain employment opportunities and value. In this case, it was likely the species that were maintained were not identical to those in the Adapting Mosaic scenario, but biomass abundance was similar.

The TechnoGarden and Global Orchestration scenarios behaved similarly in this ecosystem: both scenarios had food security as a component as well as maintaining or increasing the value of the fisheries. Neither scenario had a focus on ecosystems rebuilding except Global Orchestration after 2030, which accounted for the increasing biomass diversity index. The increasing biomass diversity index in the TechnoGarden scenario was due to a policy shift to maintain fisheries for fishmeal, which encompassed many of the species found in the Benguela system.

The differences between scenarios for the North Benguela ecosystem were not as dramatic as in the Gulf of Thailand, possibly due to different ecosystems and different states. The Benguela and Central North Pacific ecosystems were not as disturbed as the Gulf of Thailand.

10.4.2.3 The Central North Pacific Model

In the North Pacific ecosystem, current diversity was maintained (*low to moderate certainty*) under the four scenarios (see Figure 10.22). However, if increasing value or employment opportunities were the policy imperatives, biomass diversity could not be sustained. In the Central North Pacific, the Order from Strength scenario did not maintain biomass diversity compared with the other three scenarios. The biomass diversity index in the Order through Strength scenario declined for the first 20 years as the value of the fisheries increased. The slight decline in evenness reflected the slight lowering of biomass abundance for some species as distant water fleets continued to fish in the area. After 2010, drift net fishing resumed as companies continued to improve the value of the fisheries. This resulted in a lowered biomass diversity index until around 2040, when drift net fishing was banned again. This was reflected in an increasing biomass diversity index as more fisheries are sustained.

The Adapting Mosaic scenario, in the Central North Pacific, governments closed the turtle fisheries and focused on increasing the value of the tuna fisheries, which resulted in a slight increase in the biomass diversity index as species abundance increased in evenness until 2030. After 2030, the focus was on rebuilding bigeye tuna stocks, one of the most valuable species, and therefore increasing the value of the fisheries. This resulted in declining biomass diversity index as species abundance for some species decreased. The biomass diversity index was maintained in the TechnoGarden and Global Orchestration scenarios despite the initial divergence between them.

The biomass diversity index in TechnoGarden increased initially as the status quo was maintained and the biomass abundance was distributed more evenly among the species. After 2010, fishing costs declined as technology overcame several cost issues associated with catching the fish. Consequently, the biomass diversity index declined, with some species declining in abundance. Ocean ranching continued to develop in the open sea, resulting in younger tuna being caught for this sector. This resulted in a minimal change in the biodiversity diversity index until around 2030, when the technology for hatchery production of valuable species such as tuna was developed. However, this change in target species did not affect the biomass diversity index since the species abundance did not change overall but instead was redistributed in different functional groups.

The Adapting Mosaic scenario also maintained the Central North Pacific's diversity with little change over the 50 years of modeled results. The focus throughout this scenario was the continued optimizing of the value of the fisheries, which was dominated by large pelagic species. This modeling suggested that waiting until 2040 to rebuild bigeye stocks would have had no impact on the biomass diversity.

It was likely that biomass diversity could have been maintained in the Central North Pacific across all four scenarios; however, diversity was not improved unless the management imperatives for increasing the value of fisheries were substantially reduced.

10.5 Invasive Species across Biomes and Scenarios

Human-caused invasions by non-native species are a major cause of reductions in native biodiversity and consequent changes in ecosystem structure, ecosystem services, and human well-being (Mack et al. 2000; see also MA *Current State and Trends,* Chapters 4, 5, and 23). Although quantitative evidence for invasion-caused extinctions is poor for some ecosystems (Gurevitch and Padilla 2004), the dramatic impact of invasions on ecosystem functioning and ecosystem goods and services is well established (Mack et al. 2000). This section briefly explains the scale of the current problem, the major impacts of anthropogenic species introductions, and the worrying outlook for the impacts of non-native species on ecosystems.

While most non-native species do not cause harm, a small proportion do (harmful non-native species are termed "invasive") (Mack et al. 2000). Nonnative species are pervasive; the number of species introductions being discovered is accelerating over time (Cohen and Newman 1991; Ricciardi 2001). Anthropogenic species introductions with severe effects on ecosystems have occurred in terrestrial, freshwater, and marine systems (Mack 2000). Approximately half of the currently threatened or endangered species in the United States are affected by invasive species (Wilcove and Chen 1998), and some of the largest vertebrate extinctions in recent times have been driven in large part by invasives (Witte et al. 2000). A conservative estimate of the annual cost of invasive species to the U.S. economy is $137 billion (Pimentel et al. 2000). While intended introductions in agriculture, aquaculture, and other sectors can have great net benefits to society (for example, most crop plants in most parts of the world are non-native), the number of unintended and harmful introductions is increasing as trade increases (Levine and D'Antonio 2003; Drake and Lodge 2004).

Species are transported intentionally and unintentionally into every country by every conceivable conveyance. Most nations have few safeguards to prevent the escape of non-native species into natural environments (Invasive Species Advisory Committee 2003). Virtually no economic incentive structure exists for limiting species invasions because the unintended costs of harmful invasions are usually borne by all citizens while the benefits of importation of species are concentrated in commercial interests (Perrings et al. 2002).

Unless steps are taken to reduce the unintended transport of species, the numbers of non-native species established in most countries are expected to continue to increase with increasing trade (Levine and D'Antonio 2003; Drake and Lodge 2004). For example, because of extensive trade between northern Europe and the North American Great Lakes region over the last two to three centuries, a high proportion of the roughly 100 non-native species in the Baltic Sea are native to the Great Lakes, and 75% of the recent arrivals of the about 170 non-native species in the Great Lakes derive from the Baltic Sea (some by way of earlier trade between the Ponto-Caspian region and the Baltic region) (Ricciardi and MacIsaac 2000). In recent decades, this species exchange has been driven largely by the unintentional release of organisms from the ballast tanks of ships.

Nonnative species are also introduced deliberately into many ecosystems to enhance food production, provide aesthetic services, or reduce disease (Mack and Lonsdale 2001). Such introductions may involve either the direct release of a non-native species into the wild or its secondary release from a captive environment. For example, the Nile perch was introduced into Lake Victoria as a food species; African grasses have been introduced to many parts of the United States and Latin America as forage for cattle; pigs and goats introduced for food and milk have a history of causing erosion and ecological degradation on islands; rabbits and cats introduced for aesthetic purposes and pest control, respectively, have reduced biodiversity in Australia; black wattle trees (*Acacia mearnsii*) were originally introduced to southern Africa for use in the leather industry and have now become a target for biocontrol using gall-forming wasps introduced from Australia; and releases of unwanted aquarium specimens have been the source of numerous harmful introductions of invasive species into freshwater systems, including water hyacinth (Padilla and Williams 2004).

Introductions of species that are perceived as economically valuable and ecologically benign are likely to increase with increasing globalization—for example, as farmers become more aware of food production methods in other countries and consumer demand for a greater variety of produce increases. Such introductions carry a substantial risk. Although many non-native species introductions are not harmful, the impacts of the small percentage of non-native species that become invasive may be severe in both environmental and financial terms (Leung et al. 2002; Pimentel 2002).

With even the most conservative forecast of the relationship between trade and species introductions, it is likely that between 2000 and 2020 some 115 new insect species and five new plant pathogen species will become established in the United States (Levine and D'Antonio 2003). Experiences in North America with the introduction and spread of chestnut blight, Dutch elm disease, and, more recently, sudden oak death (*Phytopthora ramorum*) illustrate how ecologically and economically damaging single-plant pathogen species can be. Early in the twentieth century, for example, the American chestnut was the dominant overstory tree in the deciduous forests of the North American Appalachian Mountains. Not only was it ecologically important, but it provided many large ecosystem services, especially nuts and lumber. Since the mid-twentieth century, those ecosystem services completely disappeared and were replaced only in part, if at all. (See MA *Current State and Trends,* Chapter 4.) It is now possible that sudden oak death could have a similar effect on many oak species in North America.

As commerce develops or grows between countries not previously linked strongly by trade, especially those with similar terrestrial or aquatic climates, whole new sets of species will become established. For example, imports from China into the United States have increased about sixfold over the last decade; over this period a subset of the Chinese species discovered in the United States have become very damaging, including snakehead fishes (*Channa spp.*), the Asian longhorned beetle (*Anoplophora glabripennis*), and the emerald ash borer beetle (*Agrilus planipennis*). Thus, scenarios about the impact of changes in trade need to consider not just changing volumes of trade, but also which countries are linked by new trading patterns.

While the MA scenarios do not specify trading partners or quantify levels of trade, they do allow the assessment of differences in overall trade volume based on calculations of GDP and openness to trade in the scenario storylines. (See Table 10.5.) Global Orchestration and TechnoGarden both assume relatively rapid income growth and further globalization leading to a strong increase in global trade. This increase will be greatest under Global Orchestration. The TechnoGarden scenario also emphasizes increased use of GMOs and other forms of human intervention in ecosystems, which could increase the risks of escape of intentionally introduced species into natural ecosystems. On the other hand, the higher awareness of ecosystem functioning under TechnoGarden might be reflected in tighter management of non-native species.

Table 10.5. Scenario Characteristics Relevant to Estimating Relative Magnitude of International Trade

Characteristic	Global Orchestration	Order from Strength	Adapting Mosaic	Techno-Garden
GDP (annual growth rate of real total GDP)	3.5%	2.0%	2.6%	3.0%
Economic openness	+ +	– –	–	+ +
Rank of magnitude of species transport	1	4	3	2

According to the storylines, global trade will be much smaller under Order from Strength because of lower growth rates and protective trade policies. In Adapting Mosaic, trade levels will be lower than in Global Orchestration and TechnoGarden, but probably higher than in Order from Strength. Nevertheless, global trade levels in both cases could still be similar to those of today as a consequence of net expansion of the global economy. Assuming that the positive relationship between imports and non-native species that exists for the United States (Levine and D'Antonio 2003) applies globally, invasive species would cause the severest ecological changes and losses of ecosystem services in Global Orchestration, followed in order by TechnoGarden, Adapting Mosaic, and Order from Strength. Effects of invasive species could, in fact, play a much larger role in environmental changes than other drivers (such as climate, land use, or water consumption) that we have been able to assess more quantitatively. They will certainly interact strongly with other drivers (Mooney and Hobbs 2000).

10.6 Opportunities for Intervention

This assessment focused on losses of biodiversity at local and global scales. These two scales represent different opportunities for intervention. Local species losses can be reverted by a series of active management actions that range from abandonment and natural colonization to artificially increasing immigration rates. Ecosystem services provided by species in the original ecosystem would be restored with different delays, depending on the ecosystem service and the ecosystem.

Global extinctions are irreversible, and no human action can reverse this loss for future generations. However, major time lags occur between a reduction in habitat availability (as described in the Introduction) and the global extinction of species. Changes in habitat availability beyond a certain threshold may reduce the size of populations to a point that in a number of generations they would not be able to sustain themselves. These lags provide a wonderful opportunity for policy-makers to react and deploy actions that may reverse the trend that could have led to the global extinction of species. Many actions may change the trajectory of a species that was bound to become extinct in a few generations, including the location of protected areas, establishing corridors connecting small patches, and other steps that fall within the realm of the discipline of habitat restoration.

10.7 Ecosystem Services Derived from Biodiversity

The study of the effects of biodiversity on the functioning of ecosystems and their ability to provide goods and services has recently attracted a lot of attention from theoreticians and experimentalists. Ecologists predict a negative relationship between decreasing biodiversity and functioning of ecosystems and provisioning of services (Chapin et al. 2000). The exact shape of this relationship depends on the ecosystem process and service as well as the order in which species are lost or added (Mikkelson 1993; Sala et al. 1996; Petchey et al. 1999; Petchey and Gaston 2002).

Two kinds of ecosystem services can be identified, depending on their dependence or not on the abundance of individuals. This classification of ecosystem services has important consequences for the relationship with biodiversity and has broad policy implications. Type-I ecosystem services depend on the abundance of individuals and include provisioning services such as food and fiber production, regulating services such air quality maintenance and erosion control, and cultural services such as aesthetic values. Biodiversity affects the rate of these ecosystem services at a local scale, and the provisioning depends on the abundance of each species.

Type-I ecosystem services are related to the disappearance of species at the local scale and the extirpation of populations. A decline in habitat availability and in the presence of a species results in a proportional decline in the service. For example, a 50% decline in the abundance of a fruit-tree species determines a proportional decline in the provisioning of that food type. Biodiversity declines affect the provisioning of this kind of ecosystem service before global extinctions occur. Habitat loss and local extinctions provide good estimates of the loss of this type of ecosystem services. Another important characteristic of Type-I ecosystem services is that changes in their availability are reversible. A reduction in the abundance of one species and the services that it provided could be reversed as a result of reduced pressure or active conservation practices.

Type-II ecosystem services are independent of the abundance of individuals of a given species. This service type includes the provisioning of genetic resources, which are the basis for animal and plant breeding and biotechnology. Another example of Type-II ecosystem service is the provisioning pharmaceuticals that modern medicine depends on heavily. The service is provided by the unique genetic combination resident in native populations and not by the number of copies of this combination. The availability of Type-II services is affected by global extinctions because local extinctions do not affect the availability of the genetic code. Consequently, changes in the provisioning of type-II ecosystem services are completely irreversible. Species that become globally extinct are lost forever. Global losses of species also result in the irreversible loss of the ecosystem service that the species was providing or was going to provide. Global extinctions are the best way of estimating losses in the provisioning of Type-II ecosystem services.

In this assessment, we discussed provisioning of Type-I ecosystem services earlier in the chapter when we analyzed patterns of habitat loss and local extinctions in the different MA scenarios. The loss of this type of ecosystem services was estimated as directly proportional to the loss of habitat. Type-II ecosystem services were covered when we discussed global extinctions in the different scenarios. Global extinctions are important beyond the Type-II ecosystem services that they affect.

Empirical support of the relationship between biodiversity and ecosystem functioning and provisioning of ecosystem services is currently lagging behind the development of models. The first group of experiments on this topic focused on aboveground primary production and plant-species diversity. Large-scale manipulative experiments using grassland species in different regions of the world showed a similar pattern, with the first species losses resulting in small decreases in primary production while further reductions in species diversity resulted in an accelerated decrease in production (Tilman et al. 1996, 1997; Hector et al. 1999). Empirical evidence of the effects of biodiversity on other services and for other ecosystem types is still not available.

A number of possible functional forms have been suggested for the relationships that couple biological diversity to the rate with which different types of ecosystem processes are undertaken (Sala et al. 1996; Tilman et al. 1996; Kinzig et al. 2001). Central to all of these is the argument that there is some asymptotic maximum rate at which the activity is undertaken that declines to zero as species diversity and abundance are reduced (Mikkelson 1993; Tilman et al. 1997; Loreau 1998; Crawley et al. 1999; Loreau et al. 2001). (See Figure 10.23.) The shape of this relationship depends on the service under consideration.

The allocation of ecosystem services to these different shapes of the relationship depends on two basic assump-

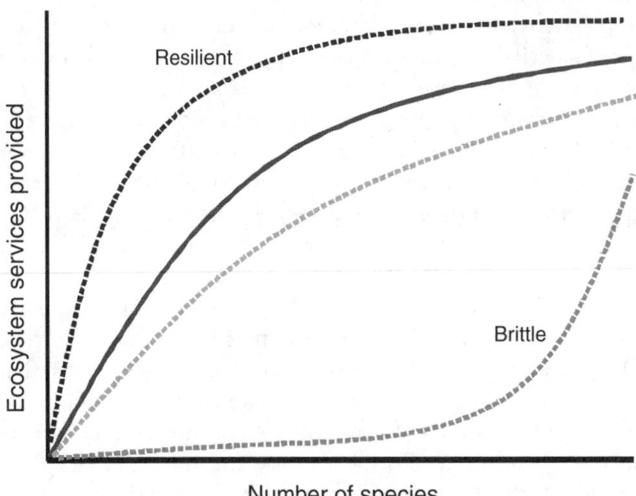

Figure 10.23. Relationship between Species Abundance and Ecosystem Function for Resilient and Brittle Ecosystem Services. In some cases decline may be rapid as the abundance of the species undertaking the activity declines (for example, population regulation of herbivores by top carnivores). In others, there may be considerable redundancy, and the relative efficiency with which any function is undertaken declines only slowly with loss of species diversity and abundance declines. Arguably, this is the case for nutrient cycling and water cleansing, though it is worth noting that the net amount of nutrients and water processed will remain dependent upon the net area (and quality) of land available.

tions: species at different trophic levels perform different ecosystem services, and species at higher trophic levels will be lost more rapidly than those at lower trophic levels. A number of examples of faunal collapse support our contention that species at higher trophic levels are lost more rapidly than those at lower trophic levels. (See Figure 10.24.) The classic studies of John Terborgh and colleagues at Lago Guri illustrate that the loss of top predators first and then mesopredators in fragmented natural systems lead to cascading effects that sequentially disrupt regulatory processes at lower trophic levels (Terborgh et al. 1997). Recent studies of a variety of organisms also suggest that species on higher trophic levels are more sensitive to climate-induced perturbations (Voight et al. 2003).

Each of these cases provides an example of faunal collapse following either a change in habitat quality or in response to exploitation that has removed species from the higher trophic levels. Although there are no studies that explicitly explore the change in trophic structure as a habitat is fragmented and reduced in size, a number of studies suggest that food-chain length is a function of habitat size (Cohen and Newman 1991; Post et al. 2000; Post 2002) and that species at higher trophic levels have steeper slopes in their species-area curves (Holt et al. 1999). All these empirical studies suggest that declines in habitat quality or quantity will lead to decreases in the length of food chains and hence a more rapid loss of services provided by species at higher trophic levels.

Using expert opinion, we assigned the different MA ecosystem services as belonging to one of the many different shapes of the biodiversity-ecosystem service relationship. Higher numbers reflect services that are brittle and are mostly performed by species in upper trophic levels. Lower numbers indicate ecosystem services that are quite resilient and performed by species in lower trophic levels. The central criterion to allocate services to a particular shape was that functions undertaken by species at higher trophic levels are more brittle than those at lower ones. (See Table 10.6.)

Different ecosystem services tend to be undertaken by species at different trophic levels. While top carnivores such as tigers and wolves provide a heightened spiritual quality to ecosystems, species such as nematodes, mites, beetles, fungi, and bacteria undertake many of the processes that cleanse air and water. At intermediate trophic levels, autotrophs (plants) provide not only structure and buffering against erosion but also most of the nutrients that are then passed up the food chain by primary and secondary consumers. Because species at the top of the food chain tend to be lost from declining habitats before those lower in the food web, it is likely that ecosystem services supplied by these species will be lost before those supplied by species at the base of the food chain. We would thus expect to see an initial sequential reduction in economic goods and services as natural systems are degraded, followed by a more rapid sequential collapse of goods and services. This implies that the sequence of ecosystem service loss is likely to be predictable; the relative position of the thresholds at which services breakdown requires further attention.

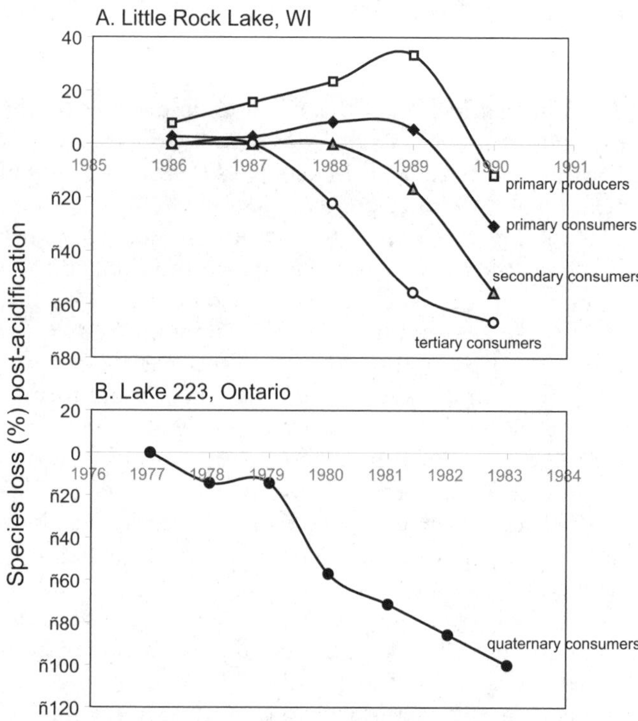

Figure 10.24. Annual Species Loss in Response to Gradual Experimental Acidification in Two North Temperate Lakes. Loss is measured as percent of pre-acidification species number. A) Four lower trophic levels in Little Rock Lake, Wisconsin, United States: primary producers (initial N = 51phytoplankton species); primary consumers (initial N = 36 primarily herbivorous zooplankton species); secondary consumers (initial N = 9 omnivorous zooplankton species); and tertiary consumers (initial N = 9 primary carnivorous zooplankton Species). B) Quaternary consumers in Lake 223, Ontario, Canada: (initial N = 7 fish species). For A, initial pH = 5.59, final pH = 4.75; unavailable for B. For B. the cessation of recruitment (absence of young-of-the-year) was treated as species extirpation. Additional experimental details are available. Additional experimental details are available for A) in Brezonik et al. (1993) and B) in Schindler et al. (1985).

This chapter has distinguished between two types of biodiversity losses: extirpations of local populations (local extinctions) and global extinctions. We concluded with *high certainty* that under all scenarios in the near future there will be important losses of habitat and consequently losses of local populations. Global extinctions will occur at uncertain times because of lags between environmental change and the occurrence of global extinctions. This last section of the chapter links the two types of biodiversity losses with two types of ecosystem services. Furthermore, we suggested a relationship between the rate of decrease of an ecosystem service with species losses and the location in the trophic hierarchy of the species associated with that particular service.

Table 10.6. Response of Different Ecosystem Services in Different Ecosystem Types to Changes in Biodiversity. Responsiveness is described in an arbitrary scale of 1–5, with higher values describing services and ecosystems that are performed by species in upper trophic levels and therefore are brittle in comparison with services and ecosystems with lower values that are performed by species in lower trophic levels and are resilient.

Ecosystem Service	Urban Systems	Cultivated	Drylands	Forest and Woodlands	Coastal	Inland Water Systems	Island	Mountain	Polar	Marine
Provisioning										
Food	1	1	1	5	1	5	1	1	5	5
Biochemicals and pharmaceuticals	0	1	3	3	5	3	3	3	5	5
Genetic resources	0	5	3	3	5	3	3	3	3	3
Fuelwood	1	5	1	1	5	0	1	1	5	5
Fiber	1	1	1	1	1	5	1	1	5	1
Ornamental resources	0	1	5	5	5	5	5	5	3	5
Fresh water	1	5	1	1	0	3	1	1	1	0
Regulating										
Air quality	2	1	1	1	1	1	1	1	1	1
Climate regulation	3	1	1	1	1	1	1	1	1	1
Erosion control	3	1	1	1	5	5	1	1	1	0
Water purification and waste treatment	3	1	2	2	5	1	3	3	1	1
Regulation of human diseases	5	5	2	3	0	4	3	3	1	1
Biological control	4	5	4	4	5	5	3	3	1	5
Detoxification	3	1	3	3	5	1	3	3	5	1
Storm protection	1	1	1	3	5	3	1	1	1	0
Cultural										
Cultural diversity and identity	3	1	4	4	3	5	5	5	3	3
Recreation and ecotourism	5	1	4	4	3	5	5	5	3	3
Supporting										
Primary production	1	1	1	1	1	1	1	1	1	1
O_2 production	1	1	1	1	1	1	1	1	1	1
Pollination	3	5	3	3	1	0	3	3	5	0
Soil formation and retention	1	1	1	1	5	1	1	1	1	0
Nutrient cycling	3	5	3	3	3	1	3	3	5	1
Provision of habitat	4	5	3	3	5	4	3	3	1	5

References

Alcamo, J., P. Doll, T. Henrichs, F. Kaspar, B. Lehner, et al., 2003: Development and testing of the WaterGAP 2 global model of water use and availability, *Hydrological Sciences Journal–Journal Des Sciences Hydrologiques,* **48,** pp. 317–37.

Alcamo, J., E. , M. Kreileman, R. Krol, J. Leemans, J. Bollen, et al., 1998: Global modelling of environmental change: An overview of IMAGE 2.1. In: *Global Change Scenarios of the 21st Century, Results from the IMAGE 2.1 Model,* J. Alcamo, R. Leemans, E. Kreileman,. a. E.K.J. Alcamo (eds.), Elsevier Science, London, UK, pp. 3–94.

Angermeier, P.L. and I.J. Schlosser, 1989: Species-Area relationships for stream fishes, *Ecology,* **70,** 1450–62.

Barbault, R. and S. Sastrapradja, 1995: Generation, maintenance and loss of biodiversity. In: V.H. Heywood and R.T. Watson (eds.), *Global Biodiversity Assessment,* Cambridge University Press, Cambridge, UK, pp. 192–274.

Barthlott, W., N. Biedinger, G. Braun, F. Feig, G. Kier, et al., 1999: Terminological and methodological aspects of the mapping and analysis of global biodiversity, *Acta Botanica Fennica,* **162,** pp. 103–10.

Bobbink, R., 2004: Plant species richness and the exceedance of empirical nitrogen critical loads: An inventory, *Landscape Ecology,* Utrecht University, The Netherlands.

Both, C. and M.E. Visser, 2001: Adjustment to climate change is constrained by arrival date in a long-distance migrant bird, *Nature,* **411(6835),** pp. 296–8.

Bouwman, A.F., D.P. Van Vuuren, R.G. Derwent, and M. Posch, 2002: A global analysis of acidification and eutrophication of terrestrial ecosystems, *Water, Air and Soil Pollution,* **141,** pp. 349–82.

Bradford, M.J. and J.R. Irvine, 2000: Land use, fishing, climate change, and the decline of Thompson River, British Columbia, coho salmon, *Canadian Journal of Fisheries and Acquatic Sciences,* **57,** pp. 13–6.

Brezonik, P.L., J.G. Eaton, T.M. Frost, P.J. Garrison, T.K. Kratz, et al., 1993a: Experimental acidification of Little Rock Lake, Wisconsin: Chemical and biologcal changes over the pH range 6.1 to 4.7, *Canadian Journal of Fisheries and Acquatic Sciences,* **50,** pp. 1101–21.

Brezonik, P.L., J.G. Eaton, T.M. Frost, P.J. Garrison, T.K. Kratz, et al., 1993b: Experimental acidification of Little Rock Lake, Wisconsin: Chemical and biologcal changes over the pH range 6.1 to 4.7, *Canadian Journal of Fisheries and Aquatic Sciences,* **50,** pp. 1101–21.

Brönmark, C. and L.A. Hansson, 2002: Environmental issues in lakes and ponds: Current state and perspectives, *Environmental Conservation,* **29,** pp. 290–306.

Brönmark, C., J. Herrman, B. Malmqvist, C. Otto, and P. Sjöström, 1984: Animal community structure as a function of stream size, *Hydrobiologia,* 112, pp. 73–9.

Brook, B.W., N.S. Sodhi, and P.K.L. Ng, 2003: Catastrophic extinctions follow deforestation in Singapore, *Nature,* **424,** pp. 420–3.

Brooks, T.M. and A. Balmford, 1996: Atlantic forest extinctions, *Nature,* **115,** p. 115.

Brooks, T.M., S.L. Pimm, and J.O. Oyugi, 1999: Time lag between deforestation and bird extinction in tropical forest fragments, *Conservation Biology,* **13,** pp. 1140–50.

Casselman, J.M., 2002: Effects of temperture, global extremes, and climate change on year-class production of warmwater, coolwater, and coldwateer fishes in the Great Lakes Basin. In: *Fisheries in a Changing Climate,* N.A. McGinn, (ed.), American Fisheries Society, Bethesda, MD, pp. 39–60.

Chapin, F.S., E.S. Zavaleta, V.T. Eviner, R.L. Naylor, P.M. Vitousek, et al., 2000: Consequences of changing biodiversity, *Nature,* **405,** pp. 234–42.

Chappel, L.H., 1995: The biology of Diplostomatid eyeflukes of fishes, *Journal of Helminthology,* **69,** 97–111.

CIA (Central Intelligence Agency), 2003: *World Factbook 2003,* CIA, Washington, DC.

Clark, H. and C.M. Harris, 2003: Polar marine ecosystems: Major threats and future change, *Environmental Conservation,* **30,** pp. 1–25.

Cohen, J.E. and C.M. Newman, 1991: Community area and food-chain length: Theoretical predictions, *The American Naturalist,* 138, 1542–54.

Connor, E.F. and E. McCoy, 1979: The statistics and biology of the species-area relationship, *The American Naturalist,* **113,** 791–833.

Cramer, W.P. and R. Leemans, 1993: Assessing impacts of climate change on vegetation using climate classification systems. In: *Vegetation Dynamics Modelling and Global Change,* A.M. Solomon and H.H. Shugart (eds.), Chapman-Hall, New York, NY, pp. 190–217.

Crawley, M.J., S.L. Brown, M.S. Heard, and G.R. Edwards, 1999: Invasion-resistance in experimental grassland communities: Species richness or species identity? *Ecology Letters,* **2,** pp. 140–8.

Crawley, M.J. and J.E. Harral, 2001: Scale dependence in plant biodiversity, *Science,* **291,** pp. 864–8.

Culotta, E., 1994: Is marine biodiversity at risk? *Science,* **263,** pp. 918–20.

Cumming, G.S., 1998: Host preference in African ticks: A quantitative data set, *Bulletin of Entomological Research,* **88,** pp. 379–406.

Cumming, G.S., 1999: Host distributions do not limit the species ranges of most African ticks (Acari: Ixodida), *Bulletin of Entomological Research,* **89,** pp. 303–27.

Cumming, G.S., 2000: Using habitat models to map diversity: Pan-African species richness of ticks (Acari: Ixodida), *Journal of Biogeography,* **27,** 425–40.

Cumming, G.S., 2002: Comparing climate and vegetation as limiting factors for species ranges of African ticks, *Ecology,* **83,** pp. 255–68.

Drake, J.M. and D.M. Lodge, 2004: Global hot spots of biological invasions: Evaluating options for ballast-water management, *Proceedings of the Royal Society of London Series B–Biological Sciences,* **271,** pp. 575–80.

Eadie, J.M., T.A. Hurly, R.D. Montgomerie, and K.L. Teather, 1986: Lakes and rivers as islands: Species area relationships in the fish faunas of Ontario, *Environmental Biology of Fishes,* **15,** pp. 81–9.

Elser, J.J., E.R. Marzolf, and C.R. Goldman, 1990: Phosphorus and nitrogen limitation of phytoplankton growth in the freshwaters of North america: A review and critique of experimental enrichments, *Canadian Journal of Fisheries and Aquatic Science,* **47,** pp. 1468–77.

Ferraz, G., G.J. Russell, P.C. Stouffer, R.O.J. Bierregaard, P.S. L., et al., 2003: Rates of species loss from Amazonian forest fragments, *Proceedings of the National Academy of Sciences,* **100,** pp. 14069–73.

Field, J.G., G. Hempel, and C.P. Summerhayes, 2002: *Oceans 2020: Science, Trends, and the Challenge of Sustainability,* Island Press, Washington, DC.

Foote, M. and D.M. Raup, 1996: Fossil preservation and the stratigraphic ranges of taxa, *Paleobiology,* **22,** pp. 121–40.

Frost, T.M., S.R. Carpenter, A.R. Ives, and T.K. Kratz, 1995: Species compensation and complementarity in ecosystem function. In: *Linking Species and Ecosystems,* C.G. Jones and J.H. Lawton (eds.), Chapman and Hall, New York, NY, pp. 224–39.

Glover, A.G. and C.R. Smith, 2003: The deep-sea floor ecosystem: Current status and prospects of anthropogenic change by the year 2025, *Environmental Conservation,* **30,** pp. 219–41.

Grabherr, G., M. Gottfried and H. Pauli, 1994: Climate effects on mountain plants, *Nature,* **369,** p. 448.

Groombridge, B. and M.D. Jenkins, 2002: *World Atlas of Biodiversity,* University of California Press, Berkeley, CA.

Gurevitch, J. and D.K. Padilla, 2004: Are invasive species a major cause of extinctions? *Trends in Ecology and Evolution,* **19,** pp. 470–4.

Haddad, N.M., J. Haarstad, and D. Tilman, 2000: The effects of long-term nitrogen loading on grassland insect communities, *Oecologia,* **124,** pp. 73–84.

Hector, A., B. Schmid, C. Beierkuhnlein, M.C. Caldeira, M. Diemer, et al., 1999: Plant diversity and productivity experiments in European grasslands, *Science,* **286,** pp. 1123–27.

Holt, R.D., J.H. Lawton, G.A. Polis, and N.D. Martinez, 1999: Trophic rank and the species-area relationship, *Ecology,* **80,** pp. 1495–1504.

Hugueny, B., 1989: West African rivers as biogeographic islands: Species richness of fish communities, *Oecologia,* **79,** pp. 236–43.

IMAGE-team, 2001: *The IMAGE 2.2 Implementation of the SRES Scenarios: A Comprehensive Analysis of Emissions, Climate Change and Impacts in the 21st Century,* RIVM (Rijksinstituut voor Volksgezondheid en Milieu/National Institute of Public Health and the Environment) CD-ROM publication 481508018, Bilthoven, The Netherlands.

Janovy, J., Jr., S.D. Snyder, and R.E. Clopton, 1997: Evolutionary constraints on population structure: The parasites of Fundulus zebrinus (Pisces: Cyprinodontidae) in the South Platte River of Nebraska, *Journal of Parasitology,* **83,** pp. 584–92.

Jenkins, M., 2003: Prospects for biodiversity, *Science,* **302,** pp. 1175–77.

Kartesz, J.T. and C.A. Meacham, 1999: *Synthesis of the North American Flora,* Chapel Hill, NC.

Kempton, R.A., 2002: Species diversity. In: *Encyclopedia of Environmetrics,* A.H. El-Shaarawi and W.W. Piegorsch, (eds.), John Wiley and Sons, Chichester, UK, pp. 2086–9.

Kennedy, V.S., R.R. Twilley, J.A. Kleypas, J.H. Cowan, and S.R. Hare, 2002: *Coastal and Marine Ecoystems and Global Climate Change: Potential Effonts on US Resouces,* Pew Centre on Global Climate Change, Arlington, VA.

Kinzig, A.P., S.W. Pacala, and D. Tilman, 2001: *The Functional Consequences of Biodiversity: Empirical Progress and Theoretical Extensions,* Princeton University Press, Princeton, NJ.

Kleidon, A. and H.A. Mooney, 2000: A global distribution of diversity inferred from climatic constraints: Results from a process-based modelling study, *Global Change Biology,* **6,** pp. 507–23.

Kolar, C. and D.M. Lodge, 2000: Freshwater nonindigenous species: Interactions with other global changes. In: *Invasive Species in a Changing World,* H.A. Mooney and R.J. Hobbs (eds.), Island Press, Washington, DC, pp. 3–30.

Koslow, J.A., K. Gowlett-Holmes, J.K. Lowry, T. OHara, G.C.B. Poore, 2001: Seamount benthic macrofauna off southern Tasmania: Community structure and impacts of trawling, *Marine Ecology Progress Series,* **213,** pp. 111–25.

Leach, M.K. and T.J. Givnish, 1996: Ecological determinants of species loss in remnant prairies, *Science,* **273,** pp. 1555–8.

Leemans, R. and B. Eickhout, 2003: *Analysing Changes in Ecosystems for Different Levels of Climate Change,* OECD (Organisation for Economic Co-operation and Development) Project on the benefits of climate policy, Paris, France.

Leung, B., D.M. Lodge, D. Finnoff, J.F. Shogren, M.A. Lewis, and G. Lamberti, 2002: An ounce of prevention or a pound of cure: Bioeconomic risk analysis of invasive species, *Proceedings of the Royal Society of London Series B–Biological Sciences,* **269,** pp. 2407–13.

Levine, J.M. and C.M. D'Antonio, 2003: Forecasting biological invasions with increasing international trade, *Conservation Biology,* **17,** pp. 322–6.

Livingstone, D.A., M. Rowland, and P.E. Bailey, 1982: On the size of African rivererine fish faunas, *American Zoologist,* **22,** pp. 361–9.

Lodge, D.M., 2001: Lakes. In: *Global Biodiversity in a Changing Environment,* F.S. Chapin III, O.E. Sala, and E. Huber-Sannwald (eds.), Springer, New York, NY, pp. 277–313.

Lomolino, M.V. and M.D. Weiser, 2001: Toward a more general species-area relationship: Diversity on all islands, great and small, *Journal of Biogeography,* **28,** pp. 431–5.

Loreau, M., 1998: Biodiversity and ecosystem functioning: A mechanistic model, *Proceedings of the National Academy of Sciences of the United States of America,* **95,** pp. 5632–6.

Loreau, M., S. Naeem, P. Inchausti, J. Bengtsson, J.P. Grime, et al., 2001: Biodiversity and ecosystem functioning: Current knowledge and future challenges, *Science,* **294,** pp. 804–8.

Mac Arthur, R.H. and E.O. Wilson, 1967: *The Theory of Island Biogeography,* First ed., Princeton University Press, Princeton, NJ.

Mack, R.H., D. Simberloff, W.M. Lonsdale, H. Evans, M.Clout, et al., 2000: Biotic invasions: Causes, epidemiology, global consequences, and control, *Issues in Ecology,* **5.**

Mack, R.N. and W.M. Lonsdale, 2001: Humans as global plant dispersers: Getting more than we bargained for, *Bioscience,* **51,** pp. 95–102.

Magsalay, P., T.M. Brooks, G. Dutson, and R. Timmins, 1995: Extinction and conservation on Cebu, *Nature,* **373,** p. 294.

Malcolm, J.R. and A. Markham, 2000: *Global Warming and Terrestrial Biodiversity Decline,* WWF report, WWF (World Wide Fund for Nature), Gland, Switzerland.

Marchetti, M.P. and P.B. Moyle, 2001: Effects of flow regime on fish assemblages in a regulated California stream, *Ecological Applications,* **11,** pp. 530–9.

Matthews, W.J. and E.G. Zimmerman, 1990: Potential effects of global warming on native fishes of the Southern Great Plains and the Southwest, *Fisheries,* **15,** pp. 26–32.

May, R.M., J.H. Lawton, and N.E. Stork, 1995: Assessing extinction rates. In: *Extinction Rates,* J.H. Lawton and R.M. May, (eds.), Oxford University Press, Oxford, UK, pp. 1–24.

Mikkelson, G.M., 1993, How do food webs fall apart? A study of changes in trophic structure during relaxation on habitat fragments, *Oikos,* **67,** pp. 593–47.

Minckley, W.L., P.C. Marsh, J.E. Deacon, T.E. Dowling, P.W. Hedrick, et al., 2003, A conservation plan for native fishes of the Lower Colorado River, *Bioscience,* **53,** pp. 219–34.

Mooney, H.A. and R.J. Hobbs, 2000: *Invasive Species in a Changing World,* Island Press, Washington, DC.

Oberdorff, T., J.-F. Guégan and B. Hugueny, 1995: Global scale patterns of fish species richness in rivers, *Ecography,* **18,** pp. 345–52.

Oberdorff, T., E. Guilbert and J.-C. Lucchetta, 1993: Patterns of fish species richness in the Seine River basin, France, *Hydrobiologia,* **259,** pp. 157–67.

Oberdorff, T., S. Lek and J.-F. Guéguan, 1999, Patterns of endemism in riverine fish of the Northern Hemisphere, *Ecology Letters,* **2,** pp. 75–81.

Odada, E.O., D.O. Olago, K.A.A. Kulindwa, M. Ntiba, and S. Wandiga, 2004: Mitigation of environmental problems in Lake Victoria, East Africa: Causal chain and policy options analyses, *Ambio,* **33,** pp. 13–23.

Olson, D.M., E. Dinerstein, E.D. Wikramanayake, N.D. Burgess, G.V.N. Powell, J.A. Underwood, et al., 2001: Terrestrial ecoregions of the world: A new map of life on Earth, *Bioscience,* **51,** pp. 933–8.

Padilla, D.K. and S.L. Williams, 2004: Beyond ballast water: Aquarium and ornamental trades as sources of invasive species in aquatic ecosystems, *Frontiers in Ecology and the Environment,* **2,** pp. 131–8.

Parmesan, C. and G. Yohe, 2003: A globally coherent fingerprint of climate change impacts across natural systems, *Nature,* **421,** pp. 37–42.

Pearson, D.L. and F. Cassola, 1992: World-wide species richness patterns of tiger beetles (Coleoptera: Cicindelidae): Indicator taxon for biodiversity and conservation studies, *Conservation Biology,* **6,** pp. 376–91.

Pereira, H., G.C. Daily, and J. Roughgarden, 2004: A framework for assessing the relative vulnerability of species to land-use change, *Ecological Applications.* In press.

Perrings, C., M. Williamson, E.B. Barbier, D. Delfino, S. Dalmazzone, et al., 2002: Biological invasion risks and the public good: An economic perspective, *Conservation Ecology,* **6(1),** p. 1.

Petchey, O., P. McPhearson, T. Casey, and P. Morin, 1999: Environmental warming alters food-web structure and ecosystem function, *Nature,* **402,** pp. 69–72.

Petchey, O.L. and K.J. Gaston, 2002: Extinction and the loss of functional diversity, *Proceedings: Biological Sciences,* **269,** pp. 1721–7.

Pimentel, D., 2002: *Biological Invasions: Economic and Environmental Co of Alien Plant, Animal, and Microbe Species,* CRC Press, Boca Raton, FL.

Pimentel, D., L. Lach, R. Zuniga, and D. Morrison, 2000: Environmental and economic costs of nonindigenous species in the United States, *Bioscience,* **50,** pp. 53–65.

Pimm, S.I., G.J. Russell, J.L. Gittelman, and T.M. Brooks, 1995: The future of biodiversity, *Science,* **269,** pp. 347–50.

Pitcairn, C.E.R., 1994: *Impacts of nitrogen deposition in terrestrial ecosystems,* The United Kingdom Review Group in Impacts of Atmospheric Nitrogen.

Poff, N.L., P.L. Angermeier, S.D. Cooper, P.S. Lake, K.D. Fausch, et al., 2001: Fish diversity in streams and rivers. In: *Global Biodiversity in a Changing Environment: Scenarios for the 21st Century,* F.S. Chapin III, O.E. Sala, and E. Huber-Sannwald (eds.), Springer, New York, NY.

Post, D.M., 2002: The long and short of food-chain length, *Trends in Ecology and Evolution,* **17,** pp. 269–77.

Post, D.M., M.L. Pace, and N.G.J. Hairston, 2000: Ecosystem size determines food-chain length in lakes, *Nature,* **405,** pp. 1047–9.

Prendergast, J.R., R.M. Quinn, J.H. Lawton, B.C. Eversham, and D.W. Gibbons, 1993: Rare species, the coincidence of diversity hotspots and conservation strategies, *Nature,* **365,** pp. 335–7.

Prentice, I.C., W. Cramer, S.P. Harrison, R. Leemans, R.A. Monserud, et al., 1992, A global biome model based on plant physiology and dominance, soil properties and climate, *Journal of Biogeography,* **19,** pp. 117–34.

Preston, F.W., 1962: The canonical distribution of commonness and rarity: Part I, *Ecology,* **43,** pp. 185–215.

Rahel, F.J., 2000, Homogenization of fish faunas across the United States, *Science,* **288,** pp. 854–6.

Rahel, F.J., 2002: Using current biogeographic limits to predict fish distributions following climate change. In: *Fisheries in a Changing Climate, American Fisheries Society Symposium,* N.A. McGinn (ed.), Bethesda, MD, pp. 99–112.

Reid, W.V., 1992: How many species will there be? In: *Tropical Deforestation and Species Extinction,* T.C. Whitmore and J.A. Sayer, (eds.), Chapman and Hall, London, UK, pp. 53–73.

Revenga, C. and Y. Kura, 2003: *Status and Trends of Biodiversity of Inland Water Ecosystems,* Secretariat of the Convention on Biological Diversity, Montreal, Canada.

Revenga, C., S. Murray, J. Abramovitz, and A. Hammond, 1998: *Watersheds of the World: Ecological Value and Vulnerability,* World Resources Institute, Washington, DC.

Ricciardi, A., 2001: Facilitative interactions among aquatic invaders: Is an "invasional meltdown" occurring in the Great Lakes? *Canadian Journal of Fisheries and Aquatic Sciences,* **58,** pp. 2513–25.

Ricciardi, A. and J.B. Rasmussen, 1999: Extinction rates of North American freshwater fauna, *Conservation Biology,* **13,** pp. 1220–2.

Ricciardi, A. and H.J. MacIsaac, 2000: Recent mass invasion of the North American Great Lakes by Ponto-Caspian species, *Trends in Ecology and Evolution,* **15,** pp. 62–5.

Root, T.L., J.T. Price, K.R. Hall, S.H. Schneider, C. Rosenzweigh, et al., 2003: Fingerprints of global warming on wild animals and plants, *Nature,* **421,** pp. 57–60.

Rosenzweig, M.L., 1995: *Species Diversity in Space and Time,* Cambridge University Press, Cambridge, UK.

Rosenzweig, M.L., 2001: Loss of speciation rate will impoverish future diversity, *Proceedings of the National Academy of Sciences of the United States of America,* **98,** pp. 5404–10.

Sala, O.E., F.S. Chapin, J.J. Armesto, E. Berlow, J. Bloomfield, et al., 2000: Global biodiversity scenarios for the year 2100, *Science,* **287,** pp. 1770–4.

Sala, O.E., W.K. Lauenroth, S.J. McNaughton, G. Rusch, and X. Zhang, 1996: Biodiversity and Ecosystem Function in Grasslands. In: *Functional Role of Biodiversity: A Global Perspective,* H.A. Mooney, J.H. Cushman, E. Medina, O.E. Sala, and E.D. Schulze (eds.), J. Wiley and Sons, Hoboken, NJ, pp. 129–49.

Schindler, D.W., 1977: Evolution of phosphorus limitation in lakes, *Science,* **195,** pp. 260–2.

Schindler, D.W., K.H. Mills, D.F. Malley, D.L. Findlay, J.A. Shearer, et al., 1985: Long-term ecosystem stress: The effects of years of experimental acidification on a small lake, *Science,* **228,** pp. 1395–1401.

Sepkoski, J.J. and M.A. Rex, 1974: Distributions of freshwater mussels: Coastal rivers as biogeographic islands, *Systematic Zoology,* **23,** pp. 165–88.

Solomon, A.M. and R. Leemans, 1990: Climatic change and landscape-ecological response: Issues and analyses. In: *Landscape Ecological Impact of Climatic Change,* M.M. Boer and R.S. de Groot, (eds.), IOS Press, Amsterdam, The Netherlands, pp. 293–317.

Steedman, R.J., 1991: Occurrence and environmetal correlates of black Spot disease in stream fishes near Toronto, Ontario, *Transactions of American Fisheries Society,* **120,** pp. 494–9.

Strayer, D., 1983: The effects of surface geology and stream size on freshwater mussel (Bivalvia, Unionidae) ditribution in southeastern Michigan, U.S.A., *Freshwater Biology,* **13,** pp. 253–64.

Terborgh, J.T., L. Lopez, J. Tello, D. Yu, and A.R. Bruni, 1997: Transitory states in relaxing ecosystems of land bridge islands, In: *Tropical Forest Remnants: Ecology, Management, and Conservation of Fragmented Communities,* J.W.F. Laurance and R.O. Bierregaard (ed.), University of Chicago Press, Chicago, IL, pp. 256–74.

Thomas, C.D., A. Cameron, R.E. Green, M. Bakkenes, L.J. Beaumont, et al., 2004: Extinction risk from climate change, *Nature,* **427,** pp. 145–8.

Tilman, D., J. Knops, D. Wedin, P. Reich, M. Ritchie, et al., 1997: The influence of functional diversity and composition on ecosystem processes, *Science,* **277,** pp. 1300–2.

Tilman, D., R. May, C.L. Lehman, and M.A. Nowak, 1994: Habitat destruction and the extinction debt, *Nature,* **371,** pp. 65–6.

Tilman, D., R. May, C.L. Lehman, and M.A. Nowak, 2002: Habitat destruction and the extinction debt, *Nature,* **371,** pp. 65–6.

Tilman, D., D. Wedin, and J. Knops, 1996: Productivity and sustainability influenced by biodiversity in grassland ecosystems, *Nature,* **379,** pp. 718–20.

Van Minnen, J.R. , R. Leemans, and F. Ihle, 2000: Defining the importance of including transient ecosystem responses to simulate C-cycle dynamics in a global change model, *Global Change Biology,* **6,** pp. 595–611.

Vinebrooke, R.D., D.W. Schindler, D.L. Findlay, M.A. Turner, M. Paterson, et al., 2003: Trophic dependence of ecosystem resistance and species compensation in experimentally acidified Lake 302S (Canada), *Ecosystems,* **6,** pp. 101–13.

Voight, W., J. Perner, A.J. Davis, T. Eggers, J. Schumacher, et al., 2003: Trophic levels are differentially sensitive to climate, *Ecology,* **84,** pp. 2444–53.

Wedin, D.A. and D. Tilman, 1996: Influence of nitrogen loading and species composition on the carbon balance of grasslands, *Science,* **274,** pp. 1720–3.

Welcomme, R.L., 1979: *Fisheries Ecology of Floodplain Rivers,* Longman, London, UK.

Wilcove, D.S. and L.Y. Chen, 1998: Management costs for endangered species, *Conservation Biology,* **12,** pp. 1405–7.

Williams, W.D., 2001: Anthropogenic salinisation of inland waters, *Hydrobiologia,* **466,** pp. 329–37.

Wilson, E.O., 1992: *The Diversity of Life,* Harvard University Press, Cambridge, MA.

Witte, F., B.S. Msuku, J.H. Wanink, O. Seehausen, E.F.B. Katunzi, et al., 2000: Recovery of cichlid species in Lake Victoria: An examination of factors leading to differential extinction, *Reviews in Fish Biology and Fisheries,* **10,** pp. 233–41.

Chapter 11

Human Well-being across Scenarios

Coordinating Lead Authors: Colin Butler, Willis Oluoch-Kosura
Lead Authors: Carlos Corvalan, Julius Fobil, Hillel Koren, Prabhu Pingali, Elda Tancredi, Monika Zurek
Contributing Author: Simon Hales
Review Editors: Faye Duchin,* Richard Norgaard, David Rapport

---

---

*Member of Board for first round of review only.

BOXES

FIGURES

TABLES

*This appears in Appendix A at the end of this volume.

Main Messages

Human well-being is considered to have five main components: the basic materials needed for a good life, health, good social relations, security, and freedom of choice and action. Human well-being is a context-dependent, multidimensional continuum—from extreme deprivation, or poverty, to a high attainment or experience of well-being. Ecosystems underpin human well-being through supporting, provisioning, regulating, and cultural services. Well-being also depends on the supply and quality of human services, technology, and institutions. Technological and social changes cannot fully substitute for ecosystem services.

Adverse effects to human well-being occur when ecosystem service loss exceeds certain thresholds. The scenarios have different implications for ecosystem services, institutional evolution, and other determinants of human well-being. Together, the storylines and the model results provide a foundation for estimating future human well-being. However, models cannot yet predict critical turning points in the relationship between supply and human demand for ecosystem services. Beyond such points, the scenarios themselves may transform in ways that are difficult to predict.

Social and ecological systems are characterized by threshold points and alternative states, which are qualitatively and quantitatively different from their pre-threshold condition, often in unpredictable ways. Such systems produce surprises when system states change in ways that are counterintuitive, unexpected, difficult to reverse, or disproportionate to the magnitude of external forcing. Surprises may be beneficial or adverse. The magnitude, sign, and significance of future surprises for the socioecological system are intensely contested.

A bi-directional causal relationship exists between some adverse ecological and social surprises. Such interactions may occur, for example, between runaway climate change, desertification, fisheries collapse, eutrophication, new diseases, major violent conflict, governance failure, and increased fundamentalism and nationalism. Feedbacks between ecological and social surprise may generate new, comparatively durable social conditions that are harmful to human well-being.

The vulnerability of human well-being to adverse ecological and social surprises varies among the scenarios. High levels of human and other forms of capital do not always guarantee preparedness for adverse surprise and in some cases may even generate complacency. The vulnerability of human well-being to adverse surprise in each scenario is determined by interactions among the likelihood of, social preparedness for, resilience to, and magnitude of adverse surprise. These qualities vary among the scenarios.

Scenarios vary in their assumed rate and direction of economic, social, and institutional change. Scenarios characterized by socially beneficial institutional change may inadvertently undermine human well-being by increasing the vulnerability to adverse ecological feedbacks. However, scenarios that stress high ecological protection may prove unexpectedly vulnerable to adverse social change. The degree of improvement for human well-being, though uncertain, is probably positive in three of the scenarios but negative in Order from Strength, which has the highest vulnerability to social and ecological change.

Scenarios exchange impenetrable complexity for simpler but less realistic conceptualizations. The future is likely to include elements from all four scenarios, as well as others both undescribed and unimagined. Given sufficient cooperation, information flow, preparedness, and adaptiveness, the human future may respond to a dynamic interchange between various scenarios, at varying temporal and geographic scales, in ways that lead toward sustainability.

11.1 Introduction

This chapter describes the implications for human well-being across the four Millennium Ecosystem Assessment scenarios. These implications are derived from a qualitative analysis of the storylines, quantitative model results, and an assessment of the assumptions that underpin the scenarios. Case studies and examples are used to illustrate principles, but the chapter does not attempt to be exhaustive. The primary focus is through 2050, though some categories of major adverse surprise are more likely to occur in the following 50 years.

11.1.1 Ecosystems and Human Well-being

Human well-being lies along a multidimensional continuum—from extreme deprivation, or poverty, to a high attainment or experience of well-being. Human well-being is complex and value-laden; context- and situation-dependent; and reflects social and personal factors such as geography, ecology, age, gender, and culture. Well-being is experiential—that which people value being and doing. Despite this diversity and subjectivity, there is wide agreement that human well-being has five key, reinforcing components: the basic material needs for a good life, health, good social relations, security, and freedom and choice (MA 2003). It follows that poverty is more than material lack and that material sufficiency alone does not guarantee human well-being.

Ecosystems underpin human well-being through their supporting, provisioning, regulating, and culturally enriching services. Shortages of food, fiber, and other material goods adversely affect human well-being via direct and indirect pathways, as does scarcity and failure of ecosystem-regulating services. Ecosystems also affect well-being through many cultural services, which influence the aesthetic, recreational, cultural, and spiritual aspects of human experience. Many causal routes link ecosystems and ecosystem service change and human well-being, forming a complex weave with social, economic, and political threads. Often, though by no means always, the quality, frequency, and availability of ecosystem services is a present or historical contributor to these social, cultural, and economic factors. That is, these factors are not always independent of ecosystem services. Turning points, or thresholds, further obscure the causal connections between ecosystem services, the social milieu, and human well-being. Sometimes, a minor increment of change in ecosystem services can trigger a substantial change to human well-being, operating through causal routes that are often obscure or in some cases widely denied.

Human well-being is also intimately related to the supply and quality of human-generated services, institutions, and technology. Institutions—formal and informal—are the body of laws, customs, economic relationships, property rights, norms, and traditions that operate within society (Dietz et al. 2003; Arrow et al. 2004). Among other functions, institutions regulate many of the links between ecosystem services and the constituents and determinants of human well-being. Inequitable access to ecosystem services is also influenced by institutions, which in turn are vulnerable to the influence of powerful individuals or groups.

There is growing concern over the degree to which the necessary suite of ecosystem services can be maintained at both geographical and temporal scales. Global population and average living standards continue to increase, but at the expense of many ecosystem services, particularly those that are non-provisioning. While ingenuity, technology, and human services have partially substituted for many ecosystem services, other services have been damaged, lost, or appropriated, often for the benefit of some human populations at the expense of others.

Future human well-being is a function not only of what happens to its five constituents, but also of their interactions. In turn, these constituents and interactions depend on economic circumstances, ecosystem services, global and local institutions, and feedbacks and interactions between these determinants. Over the long run, many decisions that have an impact on ecosystem services also affect human well-being.

11.1.2 Ecological and Social Surprises

Social and ecological systems are characterized by alternative states and thresholds (Kay et al 1999; Scheffer et al. 2001), which we call "surprises." Surprises occur when system states change in ways that are disproportionate to the magnitude of external forcing. These changes are often counterintuitive, unexpected by most people, and difficult to reverse. Human well-being, in aggregate, is vulnerable to many adverse ecological surprises, and to most if not all adverse social surprises (Levin 1999; Folke et al. 2005). The distribution of adverse changes to human well-being as a result of adverse surprise is rarely uniform, however, so that subgroups of humans can often increase their well-being even when aggregate well-being declines. The potential magnitude, sign, and significance of future surprise for the socioecological system are intensely contested, not least because of this unevenness of distribution (Arrow et al. 2004; Johnson 2001; Lomborg 2001).

The trends described by the MA scenarios reflect this wider debate. While the model results described in Chapter 9 suggest a general improvement in most aspects of human well-being in most scenarios, this should not be interpreted as meaning that the risk of a major decline in human well-being on a regional or even global scale is trivial. The results may more accurately reflect the inability of scientific knowledge to as yet adequately conceptualize, quantify, and model surprise. While the comparatively modern science of computer model-based prediction is likely to be improved by better theory, more accurate monitoring and data, and more powerful computers, uncertainty and debate will be perennial (Gunderson and Holling 2002). Nevertheless, it may be possible to model, at least approximately, how critical scarcities of human, social, natural, physical, and financial capital and flows could interact to cause a major discontinuity at a local, regional, or even global scale (Butler et al. 2005).

Surprise, by its nature, cannot be precisely forecast by time, place, nature, or scale. Even so, at any time the likelihood of some form of surprise at some place and scale is high. In the next century, a non-trivial possibility exists in all scenarios that one or several adverse surprises will occur that are of sufficient magnitude to substantially reduce human well-being in ways that are not captured by the model results described in Chapter 9.

We distinguish between ecological and social surprises. While some positive surprises are likely, we judge that the frequency and magnitude of positive surprise is unlikely to fully balance that of adverse social and ecological surprise. Plausible adverse ecological surprises include runaway climate change, major emerging diseases (including to plants, crops, and animals as well as humans), wide-scale eutrophication, desertification, and multiple fisheries collapse. Plausible adverse social surprises include the collapse or erosion of beneficial institutions, a global stock market collapse, a major energy shock, and many types of violent conflict. A combination of these surprises is also plausible.

Surprises of one category often have an impact on other systems. For example, a major adverse social surprise is likely to have ecological effects, and vice versa. Causality between ecological and social surprises is frequently bidirectional and sometimes operates through incompletely understood, often controversial, psychosocial factors. In some important cases, major adverse social surprises are likely to be precipitated by reductions in ecosystem services. Beyond certain thresholds, feedbacks can generate downward spirals, establishing new forms of social conditions harmful to aggregate well-being. Examples of such social phenomena include major violent conflict and may also include intensified fundamentalism, nationalism, or a failure of markets and governance. Other forms of major adverse surprise include an increased frequency of earthquakes or volcanic eruptions. Such events would interact with ecological and social systems, as well as have many direct effects upon human well-being.

Decisions taken to enhance human well-being over short time scales can sometimes impair well-being over a longer period, such as the prolonged collapse of the North Atlantic cod fishery that was precipitated by years of overfishing, once sanctioned in the pursuit of shorter-term well-being. In the future, several trajectories approaching thresholds could interact, forcing at least one trajectory to exceed a threshold.

Understanding the linkages between the components of human well-being and future changes to ecosystem services, though important, is insufficient to describe future well-being. As mentioned, well-being also depends on human-generated services, institutions, technology, and energy.

11.1.3 Risk, Vulnerability, and Social Coping Ability

The vulnerability of human well-being to adverse ecological and social surprises varies among the scenarios. Vulnerability is conceptualized as an interaction between the likelihood of an adverse surprise, social alertness to such a possibility, and the social coping ability to deal with an adverse surprise. Opportunity costs restrict levels of alertness to only those events perceived by society as both likely and partly remediable. Although high levels of human, social, physical, and other forms of capital are correlated with social resilience and alertness to some risks, this does not guarantee adequate preparation to adverse ecological and social surprise. High levels of capital induce complacency and thus may perversely reduce alertness and preparedness to some forms of adverse surprise.

Complacency concerning the possibility of adverse ecological surprises is likely to be higher in societies where the connection between ecosystem services and human well-being is poorly recognized, seen as highly indirect, or both. Societies may also be vulnerable to adverse ecological surprises if their quality of preparation is low, even if they are acutely aware of its possibility. However, even where there is inadequate preparation for adverse surprise, the quality and richness of human and other forms of capital can still provide a kind of safety net. In summary, vulnerability is a function of the likelihood of adverse surprise, the adequacy of alertness and other forms of preparation, and the presence or absence of concurrent stresses, whether ecological, social, or another form.

11.2 Material Needs

11.2.1 Indicators and Determinants

Basic needs include access to a secure and adequate livelihood, income, and assets. Livelihood implies the activities, claims, and access that help individuals meet material needs (food, water, shelter, fiber, clothing, medicinal, and other materials) from ecosystems in a sustainable way.

An indication of the ability of populations to meet basic material needs in the scenarios is provided by the average consumption of materials, by purchasing power, and by nutritional indicators such as the number of malnourished children. Averages ignore the equity aspects of material access, however. In some cases, parts of the population may consume an excessive quantity of a material such as food, while another group may be undersupplied with food. Data concerning the distribution and the trend of distribution of basic needs are desirable to make inferences about human well-being, including for many nonmaterial aspects of well-being. At the same time, quality differences of some material needs are difficult to disentangle.

11.2.2 Global Orchestration

Global Orchestration is broadly characterized by increased income for rich and poor, by many forms of social and economic convergence between wealthy and poorer countries, and by investments in human capital, especially education. Institutions that promote fairness, equity, and property rights are strengthened, both globally and locally. Technology registers advances, and even though ecosystem services are stressed in many regions, important indicators of basic needs—including average income and average per capita food production—improve substantially during the period under examination. Many inequalities within and between nations decline. Food insecurity in poorer countries is greatly alleviated.

While policies that favor market forces are encouraged, it is recognized that these policies must be harnessed with strategies geared at improving the quality and structure of economic growth, including generating better governance mechanisms and reducing inequality. However, there is little explicit focus in this scenario on the connection between natural systems and human well-being. Instead, it is argued that a materially wealthier society will in the long run demand and be able to afford improved ecological conditions (UNEP 1999). Losses in ecosystem services are thus viewed as a temporary and correctable trade-off for greater material wealth. Improved technologies are used to increase the effectiveness of restoration and to substitute for lost or damaged ecosystems and their services. People also assume that other components of well-being, such as freedom and choice, will be enhanced as long as material levels of consumption increase, no matter what happens to ecosystems and ecosystem services.

In this scenario, globalization processes continue through the dynamic and intensive interaction of economic, social, political, communicative, ethical, humanistic, and environmental spheres (UNEP 1999). Specific reform policies adopted by governments, corporations, and others include trade and other forms of deregulation, reduced public expenditure for social services, and privatization of state enterprises, goods, and services (Kydd et al. 2000).

The cross-border flow of ideas, capital, people, goods, and services, including of transformed and disguised ecosystem services, expands. Migration is uncontrolled. Most agricultural subsidies and other barriers to trade are removed in order to allow more equal participation and access to global markets for all trade partners. The opposition of trade unions, farmers, and others who predicted a "race to the bottom" as a result of free trade gradually fades as it becomes clear that these fears were exaggerated. New technologies and economic niches increase employment in both rich and poorer countries, especially in the service industries that tend to the increasingly aged populations in Europe and parts of Asia. The strong focus on education and effective governance systems in poorer countries aims to enhance the ability of these populations to take advantage of the new opportunities.

Most ecosystem services in wealthy countries continue to improve because of both protection and deliberate restoration. In poorer countries, increased demand for eco-tourism slows tropical deforestation and strengthens protection for endangered ecosystems such as coral reefs and trophy species such as the white rhinoceros. Higher levels of education, urbanization, information access, and press freedom increase the demand for family planning services in poorer countries. Freer migration compensates for the low fertility rate in many rich countries, including in European nations and Japan (Vallin 2002; Demeny 2003). Declines in fertility rates further contribute to increased economic takeoff (Coale and Hoover 1958; Kelley 2001). In turn, positive feedbacks develop that build on deliberate policies to improve transparency and governance.

In this scenario, growing incomes due to economic and policy reforms, urbanization, and increased global trade lead to a rapid growth in effective food demand (Sen 1981). Populations that are mostly vegetarian shift their food consumption patterns to higher-protein diets, increasing the demand for fish and meat, as described in Chapter 9.

Among all four scenarios, the increase in food supply is predicted to be the greatest in Global Orchestration. (See Chapter 9.) This is achieved mainly by raising yields on land currently in production through techniques such as genetically modified crops, more-intensive cropping practices, and increased use of fertilizers. Aquaculture continues to expand, and although this contributes to the depletion of oceanic fish stocks (Pauly et al. 2003; Powell 2003), both fish and other food prices remain affordable. Therefore, the basic supply of food, a foundation for alleviating poverty, improves, including in poorer countries.

The risk of adverse ecological surprise in Global Orchestration is high. If these adverse changes are of sufficient magnitude and effectively irreversible, then people in this scenario are likely to find that their efforts to reduce poverty and increase wealth will be undermined (McMichael and Butler 2004). Coastal dead zones may increase in size, reducing the provisioning and cultural services of coastal services, such as fishing, swimming, and water sports. Global temperature increase and sea level rise are highest in this scenario until 2050, hence there is a risk of declining regional or even global agricultural productivity that is not fully captured by the model results. Beyond 2050, however, the effects of climate change in several other scenarios approach that of this one, as described later. This scenario also projects large increases in water withdrawal and water pollution, particularly in poorer countries.

In summary, the basic material needs component of human well-being is predicted to improve in Global Orchestration. (See Figure 11.1 in Appendix A.) But this result is predicated on the proper functioning of local and global institutions; the ability of markets, governments, and technological systems to respond rapidly and appropriately to price signals; and the absence of major adverse ecological or social surprises of sufficient power to radically alter the social and institutional trends described.

11.2.3 Order from Strength

Order from Strength is characterized by increased emphasis on security and national protection, at the expense of global issues. International and domestic inequality are largely ignored while people become more concerned with securing boundaries between rich and poor. As with Global Orchestration, there is a widespread assumption that ecosystems are either robust or unimportant, and this is used to justify the weak protection of a few representative ecosystems within richer countries. In poorer countries, the fate of ecosystems is left largely to unrestrained market forces, magnified by the growing population.

In both rich and poorer countries, priority is given to short-term interest, disregarding the effects of policies on ecosystem services and on less powerful peoples. This has far-reaching implications for the material supply of goods, especially to the poor. Although some wealthy populations are able to increase consumption of material goods, for most people material well-being decreases. (See Figure 11.1b.) Market forces, unencumbered by mechanisms designed to protect the interests of the poor, exacerbate many kinds of inequality, both within and between nations. Declining international cooperation and trust (Wright 2000) causes the collapse of global negotiations, including for climate protection and trade. Reduced trade reinforces poverty traps in wealthy and poorer countries. Expensive, inefficient industries and farming practices are sheltered in this scenario, which is also marked by increased smuggling, including of drugs, weapons, and scarce and valuable ecosystem goods, such as endangered fish stocks and charismatic species.

Restricted human movement limits trade and employment as well as the flow of ideas, information, trust, and understanding between the poor and rich world. This worsens poverty wherever it exists. The demographic transition in poorer countries falters, as the spread of education slows, leading to larger populations that damage local ecosystems because of their need to obtain an increased proportion of their material goods locally.

Decreased international cooperation harms many global environmental public goods. In wealthy countries, reliance is placed on adaptation rather than mitigation to deal with climate change. The low rate of technology transfer to poorer countries increases the rate of greenhouse gas accumulation, however, as those economies remain fossil-fuel dependent for longer, especially by burning poor-quality coal. Toxic wastes are produced in large quantities in those countries, with much of it dumped and burned locally. Some pollutants, including persistent organic compounds and mercury, are disseminated widely in the atmosphere and food chain.

11.2.4 Adapting Mosaic

In Adapting Mosaic, decision-making is decentralized to local authorities and communities. As well, there is a shift in thinking toward policies that proactively try to manage ecosystems. Civil society is strengthened, including among indigenous populations. The proximity of communities and their ecosystems, combined with devolved political power, facilitates intense engagement between local actors and local ecosystems. Rather than helplessly witnessing the erosion of valued ecosystems by distant, often unknown actors, local populations are better able to protect (or, in some cases, elect to transform) their local ecosystem. This facilitates the sustainable use of many of these local ecosystems for material provision (Ros-Tonen 2003).

The connectedness between resource users and ecosystems implies the existence of strong, locally powerful institutions, able to protect and regulate common resources. (See Box 11.1.) Alliances may also be forged on a wider

BOX 11.1

Management of Common Forest Resources in Tanzania (Kayambazinthu et al. 2003)

In the pre-colonial period, the common forest and pasture resources of the Miombo woodlands in Tanzania were managed through institutions that ranged from local to national and formal to informal. These operated independently and sometimes in combination. While some still function, many have been weakened.

Traditionally, chiefs were the holders of powerful, stable, and valued user-group rules, which encapsulated spiritual, cultural, economical, and ecological beliefs and practices that in turn facilitated the durable use of valued natural forest resources. These included rules and taboos to manage and protect sacred woodlands, specific tree species, and sometimes even individual trees. Cutting of living trees was prohibited.

These institutions operated at complementary temporal and hierarchical levels. They included *Dgasinga,* instrumental in the communication and articulation of indigenous knowledge, attitudes, and practices in regulating access to natural resources. The bylaws of *Ngitili* arose from the need to reduce overgrazing at a regional scale. At the local scale, *Lyabujije* provided clear rules to regulate pasture and woodland resources within and between villages.

The strength of these institutions rested with the village elders, who acted as custodians and enforcers of the forest-preserving bylaws. Their authority was reinforced by their subjects' acceptance and faith in their political, religious, and spiritual power. This resulted in a stable and durable social and resource cohesion for extended periods. Yet elders were not beyond the law; those who committed misconduct could be dismissed.

Following colonization, however, these traditional institutions were weakened. Colonial powers tended to view the traditional institutions with disdain, and they also promoted new religious beliefs that undermined traditional authority, including taboos. Though intended to facilitate development, many new institutions had little local legitimacy, and undermined previously successful resource management mechanisms. The establishment of forest reserves, which restricted access to local communities (except for some sacred groves and graveyards), was often unsuccessful, in part because of reduced local autonomy. Centralized management raised tension between newly established committees and traditional leaders.

This situation was exacerbated by many enforced institutional changes from structural adjustment programs during the 1980s and 1990s, characterized by commercialization of forest products and reduced subsidies. Declining productivity and increased poverty, in part because of increased input prices, also forced many farmers to sell forest products and to neglect the traditional values of the woodlands. In summary, exploitation of Miombo woodlands increased.

Belatedly, government institutions are rediscovering the value and relevance of reviving and harnessing local institutions for effective management. There is a growing realization that ancient institutions that devolve power and that use traditional belief are not only more enduring and culturally resilient, they are better suited to resource management than fiats from centralized governments and committees.

scale. In some cases these alliances provide complementary knowledge and goods, building a beneficial, large-scale symbiosis. Less nationalism leads to reduced military spending and the freeing of resources for the improvement of human and social capital. Greater autonomy reduces local and regional disputes and, consequently, civil war and terrorism.

Material well-being in this scenario stays, on average, at about the same levels as in 2000 (see Figure 11.1c), but inequality is reduced, increasing material goods available to poor people while decreasing material consumption of the wealthy. While economic growth, as conventionally measured, is slower in wealthy countries in Adapting Mosaic than in Global Orchestration and TechnoGarden, this is due to more saturated demand for material goods, with more people voluntarily reducing consumption (Hamilton 2003). There is also slower growth in the demand for meat, as people adapt to the health and ecological concerns arising from high meat consumption. This reduces the pressure to convert forests to pasture for export purposes. There is less global trade and most food is produced locally. This improves food security, including through greater self-sufficiency.

However, protection of the global commons—including the climate, deep sea fisheries, Antarctica, and some aspects of biodiversity—is impaired in this scenario due to the shift of focus from global and regional to local (Buck 1998; Kaisiti 2003; McMichael et al. 2003). A trade-off occurs between improved local conditions and the continuous strain on and poor management of the global commons, which suffers in the absence of effective global protective strategies.

11.2.5 TechnoGarden

TechnoGarden shows an increase of material human well-being. (See Figure 11.1d.) This is achieved by spectacular improvements in many kinds of technologies that are geared at improving access to goods and services and at the same time protecting ecosystems for the long term. Environmental engineering advances include improving ecosystems such as wetlands and polluted rivers, lakes, and coastal zones (Odum and Odum 2003; Palmer et al. 2004). Agricultural landscapes become more diverse and less dependent on fertilizers and pesticides as farmers increasingly focus on producing multiple ecosystem services in addition to food. Farmers in both rich and poorer countries are paid for providing non-provisioning ecosystem services, such as reducing erosion, sequestering carbon, increasing pollination, and providing recreation.

Nutrients and water are better metered and used more efficiently, reducing waterlogging and salinization, eutrophication, and coastal dead zones. Crop yields increase because of many ingenious improvements in plants, such as modified flowering times to improve heat tolerance (Sheehy 2001). Other advances, including indoor cropping, hydroponics, and genetic engineering and conventional breeding, may improve photosynthesis and tolerance of current limits, such as drought, frost, heat, flooding, pests, diseases, and soil deficiencies (Johnson 1990; Botkin 2001; BIO 2003).

The development of synthetic timbers helps preserve native forests, sparing habitat for endangered species and simultaneously increasing ecotourism. Implanted computer chips, radio telemetry, and satellite surveillance are used to better regulate common resources, such as the open ocean and the Amazonian rain forest (Nepstad et al. 1999; Laurance et al. 2001). Technological improvements reduce wastage by facilitating "just in time" manufacturing and delivery systems. Improved recycling and design greatly reduce material and embedded energy wastage (Prabhakar 2001; WBCSD and UNDP 2003).

Increased trade and technological exchange between rich and poorer countries spurs agricultural and other forms of development. Cheap greenhouse gas–neutral techniques for large-scale desalinization of seawater are developed, facilitating the transformation of vast areas of coastal desert into productive farmland, fostering millions of new jobs and thousands of new industries. Improved energy efficiency along with renewable, decentralized energy technologies and a greater use of biofuels reduce the income and influence of Middle East oil exporters, accelerating their economic diversification.

The modeling results show a similar increase in global food production as in Global Orchestration, with substantial variation in demand between the regions of the world. The growth in demand for livestock products is slower than in Global Orchestration because of health and ecological concerns. Food prices decline, especially in poorer countries. Increased food supply is achieved mainly through agricultural intensification and more-intensive livestock production. There is also a substantial expansion of irrigated area until 2050, when there is a decline. The scenario results also assume an emphasis on technical water use efficiency for agricultural, domestic, and industrial use. Therefore, water stress is foreseen to grow only slowly on a global scale, though large regional differences are observed.

Reliance on technological approaches increases the risk of technology failure. More important, unless accompanied by a parallel development of institutions, this scenario is at a high risk of adverse social surprises. The analysis of the scenarios (see Chapters 8 and 9) identifies two possible problems in this regard. First, most innovations are likely to be aimed at improving provisioning ecosystem services, which have an immediate, direct impact on human well-being, without similar attention to supporting or regulating services. Second, increasingly complex technologies are vulnerable to breakdown. New technological solutions are likely to cause new problems, which in turn will require new fixes, in a cycle that may never end and that, in fact, may lead to solutions that are increasingly complex and therefore increasingly prone to breakdown.

The unfolding of this scenario depends not only on the successful development of new technologies but also on their worldwide dissemination and adoption by ecosystem users and managers. As is evident from the last 30 years, this process is slow and difficult, particularly in low-income and disadvantaged populations. Techniques for better natural resource management are not automatically adopted. Location-specific needs, poor education, and the cost and difficulty of finding appropriate local information complicate the process.

11.3 Health

11.3.1 Indicators and Determinants

The World Health Organization defines health as a state of "complete physical, mental and social well-being and not merely the absence of disease or infirmity," a concept similar to that for well-being. But more commonly, health is viewed as a desirable physical and mental state, characterized by strength, stamina, equanimity, and a lack of pain.

Life expectancy, the best-known measure of health, is insensitive to quality of life issues. Instead, disability-adjusted life years are often used to establish priorities between different health problems (James and Foster 1999). However, DALY-based burdens of disease assessments do not fully account for complex causal pathways, long time scales, potential irreversibility, and individual versus community responsibility properties, which ideally would be included in assessing the health burden of ecological change.

Individual health depends on interacting genetic, environmental, social, and medical factors. Adverse change in any single factor is rarely, if ever fully, compensated for by increase in other factors. Civil society is an important determinant of health through means such as education, leadership, and the distribution of limited resources. Technologies also depend on social factors, especially cooperation in their design and implementation. Health is also related to perceptions of individual and collective freedom and hope. And it depends on community factors. For example, even the most prudent person living in an air-polluted city will sustain lung damage, while only the most informed fish-eater will know the likely mercury content of the next fish dish.

Human health depends on all four forms of ecosystem services. Most obviously, it depends on the provisioning services that generate food and fresh water. Regulating services enhance health through means such as reducing floods and drought. Psychological health is influenced by many culturally enriching ecosystem services (Frumkin 2002). Supporting ecosystem services, such as the recycling of nutrients underpin the other services and are thus indirectly essential to health. At the same time, the health of many people relies on income obtained from the extraction and transformation of ecosystems, often for consumption by wealthier populations.

Predicting global population health over the next decades is difficult (McMichael 2001; Butler et al. 2005). But future health is still likely to be mainly determined by these social, political, and environmental factors. The evolution of many health determinants is implicit in the storylines of the different scenarios, while others are explicit in the model results. These determinants include the degree and trend of regional and global inequality, the quality of institutions, the degree of technological innovation, population growth, and the productivity, distribution, and accessibility of ecosystem services. It is increasingly feasible to predict the future range and severity of several important infectious and noninfectious diseases (Hales and Woodward 2003; Murray and Lopez 1997; Tanser et al. 2003; Webby and Webster 2003). Predicting the regional and global disease burden of emerging diseases is a bigger challenge, however.

11.3.2 Global Orchestration

Global Orchestration is characterized, broadly, by increased income in all countries, by strong institutions conducive to human development, and by investments in human capital, including education. Food production per person improves and the percentage of energy- or protein-undernourished children is reduced by 2050 to 20% from its current level of over 30%. The absolute number of undernourished children also declines. Total population growth is lowest in this scenario, minimizing the chance of regional population decline through catastrophic conflict, disease, or famine.

These projections suggest that health will improve substantially, particularly in poorer countries. (See Figure 11.1a.) The burden of epidemic diseases such as HIV/AIDS, malaria, and tuberculosis are reduced compared with today and also with the other scenarios. Mental depression, currently predicted to constitute an important burden of disease in 2020, is comparatively reduced in poorer countries in this scenario, as poor populations gradually improve their living standards, benefit from better, more inclusive governance, see that their children have greater opportunity, and yet retain significant social connectiveness, cultural pride, and a sense of meaning. In wealthy countries, depression may also be reduced if recent trends toward increased atomization and loss of meaning can be reversed (Eckersley 2004).

Improved vaccine development and distribution allow people in this scenario to cope comparatively well with the next influenza pandemic (Webby and Webster 2003). The impacts of other new diseases, such as SARS, are also limited by well coordinated public health measures, including vaccines (Gao et al. 2003). On the other hand, the lack of a precautionary approach to public health, combined with a shallow understanding of ecological risk, generates more zoonoses, such as through poorly regulated "wet markets" (Webster 2004). Global health organizations are better funded, as is regional health capacity, including for primary health care, laboratories, and hospitals. Regional shortfalls in food harvests are adequately managed by effective food relief programs.

There are, however, some important caveats. The most important is that the increased resources in this scenario are distributed in ways that benefit the public good. The scenario is highly vulnerable to adverse ecological shocks. Many environmental conditions needed for good public health could worsen. For example, microbiological water pollution in poorer countries could become an even more important source of ill health than it is now. Environmental contamination with persistent pollutants (Webster 2003), including heavy metals, could also become more pervasive, with many adverse health effects. These include a further reduction in cognitive potential for affected populations (Kaiser 2000; Tong et al. 2000), in addition to that from macro- and micronutrient deficiency (Couper and Simmer 2001; Berkman et al. 2002; Grantham-McGregor 2002) and increased endocrine diseases and cancer (Butler and McMichael in press). For wealthy populations in poorer countries, the complications of diabetes such as renal and cardiac disease could entail a large and expensive burden on health services (Zimmet 2000). Although total calories per capita increase in this scenario, dietary diversity may fall, narrowing micronutrient intake.

11.3.3 Order from Strength

Population growth is highest in this scenario, and there is the lowest investment in human capital, including basic literacy and numeracy. There is decreased commerce and scientific and cultural exchange between richer and poorer countries. Inequality increases between and within these nations. In both rich and poor countries, there are more enclosed, gated communities, whose inhabitants are tended for by less wealthy and generally less healthy service populations. This scenario has the least convergence between health and social conditions for rich and poorer countries, and it is likely that these will actually diverge, thus widening the existing "health gap" between these groups (WHO 2002). (See Figure 11.2 in Appendix A.)

In many regions of poorer countries, the supply of critical ecosystem services reaches critical levels of scarcity, leading to new forms of poverty traps, conflict, and impaired governance, such as was recently seen in Rwanda (André and Platteau 1998) and Haiti. Institutions conducive to development and good governance in poorer countries are weak, including those designed to improve global governance. They are overwhelmed by powerful lobby groups with narrow interests, including calls for greater security, and by schemes that promote corruption.

Stocks of human capital in many poor regions weaken because of the death and migration of knowledge-rich adults (Piot 2000; de Waal and Whiteside 2003). Infant and maternal mortality rates remain high in poorer countries, as do the health consequences of difficult births and obstructed labor, such as infections, epilepsy, and fistulas. Nutrition in these countries deteriorates further, exacerbating ill health, including by further reductions in cognitive development and immunity.

"Orphan" diseases—poorly researched and little-understood conditions that primarily harm poor populations—remain neglected, as do orphan drugs and orphan crops. Efforts such as the Roll Back Malaria campaign (Teklehaimanot and Snow 2002) remain underfunded. The limited research into health problems of poorer countries focuses primarily on ways to reduce short-term rather than long-term risk, such as vaccines for tourists rather than insecticide impregnated bednets (Sachs and Malaney 2002). Increased population pressure in these countries forces more contact between humans and nonagricultural ecosystems, especially to obtain and trade bushmeat and other forest goods. This exposes non-immune populations to new viruses, leading to more outbreaks of hemorrhagic fever and zoonoses. Sleeping sickness increases as poverty forces humans to penetrate tsetse fly–infested regions.

In many parts of poorer countries populations decline, at least for short periods, because of epidemics such as AIDS and TB, as well as episodes of violent conflict. The modeling results predict substantial population increase in these countries in this scenario over the next 50 years, but this is questionable, perhaps best illustrating a case where the modeling constraints and assumptions lead to implausible results.

New and resurgent diseases increase in poorer countries, but few, if any, become major disease burdens in richer ones (Glass 2004). However, it is possible, though with low probability, that a more chronic disease could cross from a nondomesticated animal species into humans, slowly and then more rapidly colonizing human populations, as HIV is thought to have done (Wolfe et al. 2004).

Some aspects of health improve in wealthy countries, but at a lower rate of improvement than in Global Orchestration. In part this is because the higher emphasis on security causes an opportunity cost to health research. The higher risk of terrorism in this scenario causes increased anxiety for people in rich nations. The increased emphasis on competition and the free market also reduces aspects of population health in comparison to scenarios with more cooperation.

In summary, even if a modest improvement to health in rich countries occurs it is unlikely to compensate for the deterioration of health in poorer ones that is likely in this scenario. Also, an ever-growing fraction of the world's population live in these poorer nations and experience chronic ill health.

11.3.4 Adapting Mosaic

In Adapting Mosaic, local solutions are developed for ecosystem and political management. This scenario is characterized by greater regional pride and more cultural and social diversity. This improves mental health, including of minority populations, and leads to reduced alcoholism, domestic violence, depression, and intravenous drug use. Knowledge and practice of traditional health systems is better preserved, but this could also mean the persistence of practices that some find offensive, such as child marriage and female circumcision.

Population growth is second highest in this scenario, and technological and agricultural breakthroughs are less marked than in TechnoGarden. The number of energy-undernourished children in 2020 is predicted to increase by about 16% before declining. At the same time, however, the percentage of malnourished children in every region declines.

The greater emphasis on small-scale cooperation leads to a more even distribution of ecosystem services. Greater social connectivity improves some aspects of health. However, the global capacity to provide disaster relief weakens. A lack of global leadership also undermines effective global environmental treaties. Climate change and other large-scale environmental problems are thus comparatively severe in this scenario, exacerbating their long-term adverse health effects.

Without an explicit focus on promoting development, many regions are unable to develop a sufficient critical mass of expertise to foster the new technologies needed to maintain high living standards and at the same time cope with the adverse effects of climatic and other harmful environmental change. While a few communities voluntarily adapt an energy-sparing lifestyle, their impact on global greenhouse gas emission reduction is small. However, this scenario is not very different from the others with regard to climate change until the second half of this century because of the many forms of inertia involved.

Crucial to health improvement is the degree to which ideas, technology, and capital circulate internationally. In other words, health standards are likely to fall behind in communities disconnected from the broader research community. For example, while education is well recognized as integral to the improvement of health in the materially poor Indian state of Kerala, this improvement is also dependent on the existence and availability of modern health services and technologies; education in this case may be necessary, but it is not sufficient.

In summary, health does not improve as much in poorer countries in this scenario as it does in Global Orchestration. (See Figures 11.1c and 11.2c.) In wealthy countries, on the other hand, greater local cooperation and connectivity improve psychological and mental health. Communities in poorer countries become more culturally distinct and more resistant to the forces that operate through the mass media to promote phenomena such as "coca-colonization" and tobacco consumption. As a result, conditions such as obesity, diabetes, and cancer may not become as common in those countries as now seems likely. Still, the prevalence of these conditions increases in other communities that voluntarily adopt health-damaging behavior.

11.3.5 TechnoGarden

TechnoGarden sees spectacular improvements in many kinds of technology and agriculture, improving nutrition globally. Cheaper communications technology facilitates improved literacy and access to useful information. Medical breakthroughs extend life expectancy and improve the quality of life in old age. Heat-stable, single-dose oral vaccines that confer lifetime immunity to multiple diseases are developed. Water and indoor air pollution, currently responsible for the sixth and tenth highest component of the global health burden (Ezzati and Kammen 2002; Ezzati et al. 2002), are virtually eliminated.

Technological and scientific advances greatly increase the global human carrying capacity. For example, solar energy breakthroughs allied with desalinization enable the irrigated agricultural development of deserts, allowing the migration of millions of people from areas that are currently densely populated.

The rate of institutional evolution is particularly important in this scenario, since rapid technological development could undermine many institutions, either inadvertently or through its deliberate use by powerful actors. Optimistically, new health technologies and better nutrition could themselves trigger social and economic improvements, especially among poor tropical populations, by reducing the development-stalling impact of disease, undernutrition, and high birth rates (Birdsall et al 2001; Sachs and Malaney 2002; de Waal and Whiteside 2003).

Though cheap robots reduce danger, drudgery, and servitude, they are also used to increase unemployment and the exploitation of people. Virtual reality is misused to pacify and condition people in ways that reduce their freedom. Family and social ties loosen when children bond to "virtual" nurses rather than to flesh-and-blood playmates. Audiences desensitized by an excessive diet of virtual violence and pornography challenge civil society norms when whetted appetites demand ever-increasing doses. Alternatively, societies could use the new technologies for greater expression, strengthening social, family, and human capital, while unemployment could be reduced by managers insisting that overworkers take more leisure time, freeing job opportunities for others.

The easy and increased availability of calorie-dense food exacerbates the nascent global epidemic of obesity and diabetes, in both rich and poorer countries. Obesity also increases the rate of some forms of cancer. Lack of large muscle use in childhood and poor gross motor coordination fostered by sedentary lifestyles reduces mobility and bone density in later life, canceling surgical improvements. Designer drugs prove more dangerous and addictive than promised. Discrimination based on genetic profiles, for employment and insurance, for example, becomes commonplace.

On the positive side, the development of genetic engineering leads to cheap and widespread "nutraceuticals," providing both micronutrients or individually tailored medications. New surgical techniques greatly extend life expectancy, for those with sufficient means.

TechnoGarden could also—though with low probability—see the development of truly devastating diseases. These could escape or be deliberately released from biowarfare laboratories (Anonymous 2001a). Diseases targeting specific genetic characteristics could be engineered for "ethnic cleansing" or other forms of genocide. New diseases could also arise or be more widely disseminated by new technologies, as occurred with several infectious diseases in the twentieth century. The genetic homogenization of food and other crops creates vulnerability to new agricultural diseases, with adverse knock-on effects to human health. Although the scenario assumes remarkable technological ingenuity, a struggle could develop between increasingly sophisticated technologies and increasingly complex problems. Some problems are likely to be deliberately caused by human techno- and eco-vandals. On balance, health improvements in all countries are not as marked as in Global Orchestration.

11.4 Social Relations and Security

11.4.1 Forms, Expressions, and Determinants

Social relations refer to the degree of influence, respect, cooperation, and conflict that exists between individuals and groups. These relations underlie security and, in some cases, violent conflict. Social relations, expressed through manners, customs, traditions, diplomacy, and other means, operate at many scales—including among and within families, neighborhoods, genders and within religious, cultural, economic, political, and ethnically linked communities. Social relations influence and are influenced by the distribution and management of limited resources, including of ecosystem and human services. Social relations are also influenced by current and previous institutions and by past and present social and economic relationships. Differing perceptions over complex issues are inevitable (Adams et al. 2003), but good social relations can reduce tension and, in many cases, prevent violent conflict.

Many societies and groups have developed institutions that have maintained a durable and adequate stock of common resources by mediating access and use of ecosystems (Ramakrishnan et al. 1998). Sometimes, these institutions have evolved to cope with periodic ecosystem service shortages, such as by reciprocal trade between climatically distinct regions (Cordell 1997). In recent centuries, many indigenous populations have experienced a profound loss of local ecosystem services, especially of their intertwined provisioning and culturally enriching aspects. In many cases, these ecosystems have been appropriated by more powerful populations, who have then transformed them. While this has often multiplied the provisioning aspect of the original ecosystem, there has been little regard for the culturally enriching aspect. In addition, the material benefits of such transformations have rarely been distributed equitably. In some cases, the effects of such loss have been transmitted through generations.

Position and rank, influenced by such factors as age, gender, birth order, family, wealth, income, class, caste, education, ethnic group, and ability, are universal within societies (Price and Feinman 1995). But changes in either the rank or entitlement of individuals and groups can both alter or reflect changed social relations.

Important shifts in social relations may occur if the supply of desired goods (including, in extreme cases, the loss of an entire ecosystem) reaches a critical threshold (sometimes far above zero) and may lead to intensified grievance and resentment. Losses that reflect and alter social relations often have a nonmaterial aspect, such as eroded rights and freedoms, including of cultural expressions and physical movement. While social relations can deteriorate at any scale, violent conflict is often most intense between cohesive groups, such as tribes, states, and multinational alliances. In some cases, grievance and deprivation (including perceived deprivation) can spawn rigid positions, based on core cultural beliefs that are often religious in character. Such positions, frequently characterized by opponents as "fundamentalist" or "extremist," often provoke similarly strong responses.

Security includes the ability to gain access to natural and other resources and to safely retain personal safety and physical property. It also refers to a person's sense of the future, especially to periods characterized by increased vulnerability, such as old age, sickness, and economic downturn.

Causality between social relations, ecosystem services, and other valued ends is often multidirectional. While poor economic circumstances are usually associated with reduced physical and economic security, violent conflict can also often reduce the stock and flow of material goods, not only by destroying physical and ecological capital, but also by repelling the flow of financial and human capital, which can worsen poverty traps and insecurity (Bloom and Canning 2001).

11.4.2 Global Orchestration

On balance, social relations improve in Global Orchestration as wealth increases, democracy spreads, and inequality declines. (See Figure 11.1a.) Reduced international and domestic inequality is a major step toward solving hostilities and widening the scope of cooperative society. The power and authority of global organizations such as the United Nations increases, fostering improved international relations. In parallel, strengthened participatory democracy increases decentralized decision-making, especially within poorer countries. This motivates governments and other leaders to work genuinely to reduce disease and poverty, including in sub-Saharan Africa. However, the reduction of agricultural and other forms of subsidy is bitterly resisted by some farming and other groups, who resent the loss of income and privilege that this entails for them.

In combination, these factors that enhance cooperation strengthen institutions that potentially improve management and conservation of global commons. Because of the low value given to ecosystems, however, this scenario is likely to see the continued transformation—and in some cases the destruction—of ecosystems that are of high value to indigenous but relatively powerless populations. This may lead to a deepened sense of resentment, which cannot be completely assuaged, even by financial compensation.

While the aesthetic, material, and spiritual loss of a comparatively small number of such people may be viewed by the majority as acceptable, such loss and grievance could still fuel guerilla wars, insurgencies, protest movements, and legal action. Coalitions between concerned populations in wealthy countries and indigenous populations are likely in an attempt to protect particularly charismatic ecosystems, such as mountain gorilla habitat. But many less famous ecosystems are likely to be forcefully transformed in this scenario.

Beyond these foreseeable but small-scale conflicts, social relations and security could deteriorate if the scenario unravels because of institutional failure or unexpectedly severe ecological surprise or simply because the scale of problems in poorer countries proves intractable to the improvements that have been posited. For example, accelerated climate change could interact with tropical deforestation and poor coastal ecosystem service management to cause repeated and severe landslides, storm damage, and coastal flooding, eroding other forms of progress in populous poorer countries. This could damage social relations, both within the affected area and between affected populations and others who are less affected yet at least equally responsible. At the worst, such disaffected populations could contribute to locally and regionally active terrorist networks, for whom recruitment increases with infrastructural damage, material shortage, unemployment, and non-conciliatory governance. Deepening inequalities, if allowed to occur, could then weaken the capacity to solve social, economic, and cultural problems. At the worst, civil unrest, terrorism, and resource-based conflict could further undermine social cohesion, leading to emergence of an Order from Strength scenario. It is important to note that some forms of adverse ecological surprise, of which the most stark and foreseeable example is runaway climate change, could relentlessly erode good governance, year after year, because of their intractable, inexorable, and essentially irreversible nature.

11.4.3 Order from Strength

Out of the four scenarios, social relations are the most impaired in Order from Strength, especially in poor regions where investment in natural and human capital is particularly limited. (See Figure 11.1b.) This is likely even in the absence of severe adverse ecological surprise. The collapse of global trade talks and the abandonment of poorer countries lead to a compartmentalized world, where economically powerless populations are left to their fate, which can include local tyrants. There is very little interest in the public good. Consequently, social relations deteriorate on many scales, from local to international. Civil society deteriorates, especially in poorer countries. In the worst case, "barbarization" could develop, characterized by widespread lawlessness, corruption, prejudice, and terrorism (Raskin et al 2002).

Among countries and regions where order prevails, security is very strict, particularly along borders and at official entry points. Many borders are physically and electronically fortified, with constant surveillance. The high and increasing transaction costs of security reduce both trade and travel, lowering the material and social quality of life for many people. This is especially true when extended families straddle both sides of these barriers. Legal migration from poor to wealthy regions is strictly controlled, and the few migrants who are admitted face substantial discrimination, low wages, demeaning jobs, and limited opportunities. Racism and other forms of appearance-based and culturally derived prejudice increase, while direct experience of other regions and cultures falls, increasing the likelihood of stereotypical descriptions of alien peoples and cultures. Thus, social relations at the global scale deteriorate, locking in even more discrimination, apathy, and disrespect.

Inequality is increased within as well as between countries. Civil wars and rebellions become even more frequent within poorer countries. Domestic and other forms of hidden violence also rise in richer countries, including through the promotion of a culture that is more militarized, fearful, and discriminatory. Refugee numbers outside the fortified borders increase, and many refugees are confined indefinitely in large squalid camps. International organizations lose their autonomy, or even cease to exist. For a time, a few poorer countries struggle to maintain educational standards, but the repeated successful luring of their brightest and best graduates by wealthy countries gradually lowers the morale and expertise in poor areas. Consequent feedbacks in poorer nations reduce investment, increasing debt defaults, poverty, economic failure, mistrust, crime, corruption, epidemics, and violent conflict.

11.4.4 Adapting Mosaic

In Adapting Mosaic many locally cooperative networks and actors emerge. Civil society strengthens local government and organizations. Science and businesses play an increasing role in the support of conventions to address persisting environmental problems. Decentralized management helps conserve local ecosystems that are of importance for religious, spiritual, recreational, educational, and aesthetic reasons (Kaisiti 2003). New partnerships emerge, incorporating professionals, indigenous peoples' organizations, community groups, and product certification organizations, to bring change at the local level. While the devolution of power to the local level initially triggers conflicts with the former holders of power, the scenario envisages that these struggles are successfully resolved.

Military expenditure and conflict are reduced, freeing enormous resources to improve human well-being. For example, investment currently channeled to improve weapons is used to build roads and develop improved crops. On the other hand, the power of the United Nations and other global organizations to tackle any global crises, including disputes that might still arise over the management of common resources, is weaker in this decentralized scenario. New actors may have conflicting interests, objectives, and values not reflected in previous management regimes. Thus, tensions and conflicts may still arise, even within the new power structures. This is most likely where there are changes in customary rights and practices.

11.4.5 TechnoGarden

Several unique characteristics influence social relations in TechnoGarden. The most important is the extent to which technology changes human relationships, shaped by evolution over millions of pre-TechnoGarden years. Humans have always had some technologies, and for generations many humans have experienced an increasing rate of technological change. Humans remain social animals, however, requiring physical affection and contact for proper development. This scenario sees the development of technologies that lead to cheaper and easier forms of multilingual communication. This enhances social development, including between NGOs, social activists, and special interest groups. But new technologies also distort social development by, for example, creating people who feel more bonded with synthetic forms of reality and stimulation rather than other people and life-forms. In times of stress, such relationships are unlikely to be fulfilling. Such people, even if relatively few, not only feel alienated from society but also behave destructively.

Cloning, "designer children," and other forms of social manipulation also affect social relations. Too late, people realize that genetic tinkering releases undesirable traits, including behavioral. On the positive side, violent conflicts lessen and social relations improve as new technologies facilitate the expansion of agriculture to areas that are currently sparsely inhabited. But without sufficient institutional evolution, new technologies—including of weapons—generate new arms races at regional, international, and even local scales. In summary, as previously indicated with regard to this scenario, the rate of institutional evolution is vital in steering and controlling humanity's growing technological prowess in ways that improve human well-being.

On the negative side, cheaper and better communication could also be used by groups such as drug smugglers and terrorists, whose aim or effect is to disrupt social relations. Disaffected individuals and groups could also make use of new weapons. The diffusion of eco-technology may reduce indigenous and other local knowledge of ecological processes and management, creating vulnerability and dependence.

11.5 Freedom and Choice

11.5.1 Determinants and Expressions

Freedom and choice includes the ability to acquire, to experience, and to select what someone likes, including from ecosystems. It also includes the capability of fulfilling personal choices. Freedom is much more than material. It relates also to the ability to participate in debate, to travel, and to hold, study, and express personal beliefs, including views that differ from the majority. There can be an asymmetric relationship between freedom and choice and the other components of well-being. For example, it is possible to have enough material goods to survive comfortably and yet to feel far from free. It is also possible to feel secure, to enjoy good social relations and health, and yet not be free (Sen 1999). And reduced freedom means reduced well-being.

Freedom implies the ability to pursue the other components of well-being, though without guaranteeing their attainment. Yet freedom must also be limited, both for individuals and groups, because an excess of freedom for one party results in the dearth of another's. The distribution of access to resources, ecosystems services, and the quality of institutions are vital determinants of the degree of freedom and choice. Excessive material inequality can not only create but also result from a sense of grievance. This can contribute to the generation of religious and other forms of fundamentalism, reducing freedom of expression. It is also possible to have societies where most people have fairly equal access to material goods yet there is a highly asymmetrical distribution of influence and a consequent lack of freedom.

11.5.2 Global Orchestration

In Global Orchestration, many forms of material freedom and choice increase because of the greater purchasing power predicted in this scenario, the greater supply of many goods and services, and the lessening of material and other forms of inequality. (See Figure 11.1a.) Democracy increases greatly, causing a range of virtuous feedbacks, such as better education and a greater, less skewed flow of information. This liberation of human potential is particularly striking in societies that have previously been repressed. The convergence between wealthy and poorer countries leads to more freedom of movement both within and across borders. Reduced inequality sees less envy, crime, and discrimination, generating new freedoms of expression, especially for minorities.

Some freedoms are reduced, however. Although the total supply of many goods and services increases, many consumers have less choice. Goods, both manufactured and grown, are relatively homogenous culturally, technologically, and genetically. Many services are delivered anonymously, supplied by providers based in physically distant economies with cheaper price structures.

Scientists, ecotourists, and deep ecologists have less opportunity to visit, experience, study, and honor ecosystems destroyed or irrevocably altered to further material progress. Others who feel deeply connected to natural systems, such as many indigenous people, also experience a sense of loss, including lost freedom. It is also plausible that micromanagement, applied repeatedly with the best of intentions, could reduce freedom through a combination of market and benevolent government tyranny.

11.5.3 Order from Strength

Order from Strength clearly results in a marked restriction of freedom and choice in many regions. (See Figure 11.1b.) Goods, trade, and travel are tightly controlled, including by the manipulation of information, the suppression of protest, and the control of the mass media using public relations techniques, particularly to benefit and protect powerful individuals and groups (Stauber and Rampton 1995; Beder 1998). Communication technologies, including the Internet, are censored and manipulated by governments and powerful corporations, with limited freedom for NGOs and other groups attempting to provide countering views.

In this scenario, many individuals and groups receive arbitrary treatment according to qualities such as their name, nationality, dress, ethnicity, and appearance. Freedom of speech and self-expression suffer. Religious, ethnic, and other forms of fundamentalism strengthen, further reducing freedom of expression, including, paradoxically, of religion.

Global initiatives and conventions for ecosystem management decline, in part because of the opportunity cost of constantly responding to the numerous security problems likely to occur in this scenario. The freedom of individuals to share information, even to ameliorate environmental problems, is restricted. The access by many individuals to ecosystem services in this scenario is also likely to fall as regional inequalities increase. This reduces human freedom.

11.5.4 Adapting Mosaic

In Adapting Mosaic, freedom of choice and action are improved on average over conditions in 2000. (See Figure 11.1c.) Resource users acquire many local forms of freedom and choice. These include the ability to manage ecosystems through local institutions and the freedom to experiment with different forms of management and to modify and manipulate ecosystems to produce services consistent with their needs and wants. Poor people in poorer countries gain the freedom to restrict foreign investment by imposing taxes and tariffs, including those that could weaken environmental protection.

Yet the limited international reach of global society is likely to extend beyond that of an impaired capacity to form global environmental agreements to also include impaired international policing and a weakening of international cooperation in spheres such as global human rights protection. Although the scenario assumes a high degree of social harmony, it is in fact likely that some groups will try to take advantage of the comparative legal and peacekeeping vacuum likely to form. This could lead to gross human rights violations that go unnoticed by the wider community. Freedom and security will be reduced for vulnerable groups.

11.5.5 TechnoGarden

Freedom and choice are, on average, improved in TechnoGarden compared to 2000. (See Figure 11.1d.) The rate of increase of freedom in TechnoGarden depends crucially on the accompanying rate of institutional evolution. Technologies themselves are benign, having no capacity to either constrain or liberate freedom. If they can be used successfully to increase the per capita supply of well-managed ecosystems, however, and at the same time allow a greater preservation and restoration of wild ecosystems, then many forms of freedom and choice are likely to increase. For example, higher food production, cleaner water, and, possibly, expanded areas suitable for human habitat should increase freedom.

Consumers with sufficient means have new choices of foods, production systems, technologies, and entertainment. Nanotechnology, robotics, and other technologies enable niche markets to become economic. Books never go out of print, as microprinting technologies become inexpensive and widely available. Other technologies preserve languages and customs, broadening cultural and consumer choice.

Freedom and choice for some populations, such as those living in areas flooded by new dams or whose land has been appropriated for other uses, are reduced, however. The scenario could also unravel, with new technologies being adapted to reduce freedoms, including by new forms of surveillance, policing, and military techniques. Some of these technologies could also be used illicitly or accidentally in ways that reduce freedom.

11.6 Ecosystem Services and Human Well-being across the Scenarios

11.6.1 Provisioning Services

Without adequate provisioning services, human well-being will clearly decline. Material needs and health are obviously vulnerable to reduced provisioning services (below a threshold), but such declines are likely to lead to reductions in the other aspects of human well-being as well. This relationship is not linear. Once adequate provisioning services are available to deprived populations, additional increases in provisioning ecosystem services are unlikely to lead to commensurate improvements in human well-being.

In all four scenarios and in all regions analyzed, the model results show that per capita cereal consumption (as food, rather than animal feed) remains little changed in 2050 compared with the present, although in Adapting Mosaic it falls slightly in sub-Saharan Africa, the region with the lowest current consumption. (See Chapter 9, Figure 9.23.) However, per capita meat consumption is found to increase by 2050 in sub-Saharan Africa in all four scenarios. In Adapting Mosaic and Order from Strength, the absolute number of energy-undernourished children in poorer countries is predicted to increase by 2020, but this improves by 2050, particularly in Adapting Mosaic. (See Chapter 9, Figure 9.31.)

These model results raise deep questions about the assumptions used to generate them. The finding of a reduced percentage and absolute number of energy-malnourished children in sub-Saharan Africa in 2050 in Order from Strength seems optimistic, particularly in the context of the tripling of population projected for this region in this scenario. Instead, several factors appear likely to reduce the capacity of sub-Saharan Africa to lower the number and percentage of energy-undernourished children. These plausible factors include a substantial increase in the price of energy, harmful effects to domestic agricultural productivity because of HIV/AIDS and probably other diseases, and, perhaps most important, the damage to infrastructure and human capital from repeated violent conflict. The models conclude that the demand for food increases strongly in all four scenarios, but this conclusion is in doubt with regard to a large increase in effective food demand in Order from Strength (Sen 1981).

Biofuel includes trees, agricultural wastes, and crops grown specifically as energy carriers, such as maize fermented to produce alcohol. Most biofuel is burned to produce electricity rather than used for heating and cooking, as traditional firewood would be. Biofuel production increases substantially in the Global Orchestration and TechnoGarden scenarios. (See Chapter 9, Table 9.33.) If there is a co-existent oil shortage, together with an immature or costly substitute fuel for transport, large areas of agriculturally productive land may be harnessed to produce biofuel, with adverse effects for total food production.

11.6.2 Regulating Services

Regulating services are also essential for human well-being. Most obviously, these include ecosystems sufficiently intact to reduce flooding, landslides, and storm surges and to maintain river flow in dry areas. Less obviously, there is a strong interaction between land cover and climate at the regional and global scales. Many ecosystems can be transformed to supply markets (such as forests to lumber, or mangroves to fish farms) but this is at the loss of regulating services, for which markets are weak. Beyond thresholds of loss, diminished regulating services lead to disproportionate harm to human well-being, especially for vulnerable communities.

Globally, regulating services are best preserved in TechnoGarden and Adapting Mosaic. For example, tropical deforestation is lowered in TechnoGarden by a combination of reduced tropical hardwood consumption by rich populations, technological developments leading to substitution, and slower population growth in poorer countries. In Adapting Mosaic there is in general greater protection of local ecosystems, including through greater resistance to the organized criminal systems that have contributed to much illegal logging in Southeast Asia (Dauvergne 1997; Jepson et al. 2001).

In contrast, in both Order from Strength and Global Orchestration, a combination of market forces, undervaluation, and feedbacks lead to substantial deforestation, not only in the mostly tropical poorer countries but also in large swathes of Siberia. In Central America and the Caribbean this leads to more local flooding (Hellin et al. 1999) and development reversals, as occurred following Hurricane Mitch in 1998. As the century passes, deforestation increasingly interacts with climate change and fires, reducing the terrestrial carbon sink and enhancing the chance of runaway climate change (Cox et al. 2000). This risk exists in all scenarios, but is highest in Order from Strength and Global Orchestration.

11.6.3 Cultural Services

Cultural ecosystem services are also essential for human well-being, particularly by supplying aesthetic, recreational, psychological, and spiritual benefits. In turn, the sense of loss from or connection with such services is a factor in broader issues of security, cultural relations, and freedom. The protection of sites, species, or ecosystems of special significance can improve the morale of large populations, just as the loss of such icons can exacerbate despair, violence, and loss.

The cultural services of ecosystems are best protected in Adapting Mosaic. In this scenario, the greater autonomy of local groups facilitates stronger protection of valued landscapes, streams, and species, though some species remain vulnerable to circumstances well beyond local control, such as the lack of protection of distant landing sites for migratory birds or the dissemination of persistent chemicals that may reduce fertility.

In Global Orchestration, the cultural services of many ecosystems are undervalued unless they already generate substantial income, such as through tourism. Economic growth is viewed primarily in monetary terms rather than as an incomplete indicator of what is really important (Arrow et al. 2004). Consequently, old-growth forests are converted to palm oil plantations and wetlands to fish farms.

In TechnoGarden, there is also an undervaluation of many cultural ecosystem services. There is a pronounced belief, merging with hubris, in the ability of technology to provide adequate substitutes, such as through virtual life forms and ecological engineering. Ultimately, these are likely to prove less compelling to humans than living species and genuine ecosystems. However, the greater efficiency of transformed ecosystems, such as those used to grow food, also means that there is less demand to transform ecosystems that are currently little changed. This will result, de facto, in greater protection for some cultural services, especially in poorer countries.

In Order from Strength, many cultural services of ecosystems decline, especially in poorer countries. Numerous species, including birds, mammals, and fish, are forced into extinction, both directly by hunting and indirectly by habitat loss and pollution. The more privileged populations in the wealthy countries also lose access to many cultural services, not only through the probable extinction of charismatic species, such as some of the great apes, but also because firsthand experience of species that do survive in poorer countries is increasingly rare. The genetic health of species protected in richer countries in zoos and wildlife parks also suffers, as they become more inbred.

The decline of cultural ecosystem services in Global Orchestration, TechnoGarden, and Order from Strength is likely to have a generally negative impact on human well-being, especially for people with a high degree of biophilia (Wilson 1984; Frumkin 2002).

11.7 Vulnerability

11.7.1 Ecological Surprises

Major adverse ecological surprises are more likely to occur in Order from Strength and Global Orchestration than in TechnoGarden. Adapting Mosaic has the lowest rate of adverse ecological surprise. Social resilience to such surprises is seen to be strongest in Adapting Mosaic and TechnoGarden but is particularly low in Order from Strength.

The vulnerability of Global Orchestration to a major ecological surprise is an important weakness of this approach to development. The reduced emphasis on ecological issues in this scenario means that society and individuals will be poorly equipped to detect early indicators of major ecological change. Symptoms such as localized eutrophication, strange but limited animal and plant diseases, or fisheries collapse are likely to be dismissed rather than recognized as significant. This is likely to further delay and weaken action, risking ineffectiveness. By the time a major ecological change is occurring, such as runaway climate change, it may be impossible to stop. Order from Strength has a similarly reactive approach to ecosystem management, which, coupled with societal ignorance about ecology, gives this scenario a very high potential for adverse ecological surprise.

Any major ecological surprise is likely to have important social consequences, even in a society as well-educated, harmonious, and technologically advanced as postulated in Global Orchestration. Runaway climate change is a particularly serious threat in this scenario. Its likely manifestations would include increased extreme weather events such as droughts, floods, fires, and, perhaps, more violent storms and fires. It could also precipitate many adverse agricultural effects in tropical poorer countries, including in sub-Saharan Africa and South Asia, both of which could experience increased food shortages and famine. In the subcontinent, climate change could exacerbate pre-existing security tensions between nuclear-armed rivals.

Some technologies developed in TechnoGarden are likely to have unexpected effects. While some will be trivial, others could be serious. The twentieth century provides many examples of how technologies and new techniques conceived as benign or beneficial in fact had harmful consequences (European Environmental Agency 2001). Some—such as the spread of bovine spongiform encephalopathy (Butler 1998) and of HIV among Chinese blood donors (Anonymous 2001b)—have cost billions of dollars and affected thousands of lives. An apparently more benign case was the accidental exposure between 1955 and 1963 of over 100 million people to a polio vaccine potentially contaminated with a simian virus (McCarthy 2002). While this did not cause a major public health problem (although there are concerns that this virus may be associated with lymphoma) (Vilchez et al. 2002), it could have. The natural but unforeseen arsenic contamination of tubewells developed to extract microbiologically safe water in Bangladesh and parts of India (Smith et al. 2000) is another example of adverse technological surprise. On the positive side, TechnoGarden's proactive and learning-focused approach may help people in this scenario avoid some surprises.

New agricultural technologies are also likely to have unexpected results, not all of which will be benign. For example, genetically modified plants may spread new genes into wild species. Uncertainty about the scale of these changes exceeds our current ability to monitor them. Groundwater extraction may promote locally higher crop production but reduce production elsewhere by disturbing the underwater hydrology (Alley et al. 2002). Aquifers may also become excessively contaminated by pesticides and other substances with uncertain but potentially long-lasting effects. Clearing of water catchments and wetlands for farming and aquaculture may change micro-climates, reduce fish breeding, and make densely populated areas more vulnerable to disasters such as storm surges and flooding. New dams may reduce the productivity of downstream fisheries and agriculture.

In the absence of institutional changes that protect poorer populations, new technologies—including ones that facilitate the conversion of large areas of low value land for more productive uses—tend to have a disproportionately negative effect on women and marginal populations, including pastoralists and indigenous peoples.

Adapting Mosaic tends to avoid surprises by being proactive and by taking small steps and carefully monitoring the results. Some small-scale surprises do happen, particularly when experiments fail, but these generally do not affect large numbers of people negatively.

11.7.2 Social Surprises

Major adverse social surprises include large-scale violent conflict, governance failure, and increased fundamentalism and nationalism. Global Orchestration is likely to have the greatest capacity to cope with social surprises and—in the absence of major adverse ecological surprises—the least chance of having them occur. This social resilience is fostered by the improvement of human, social, and physical capital in this scenario, but at the expense of natural capital. There is a belief in this scenario that as these nonenvironmental forms of capital improve, the self-organizational capacity of society to respond appropriately to various forms of stress—even ecological—will also improve (Johnson 2001). Reduced inequality (including of opportunity and respect) should also mean greater social cohesion and cooperation in the face of stress. On the other hand, the increasingly homogenous global culture in this scenario is likely to promote many forms of cultural reaction, especially among people who lag or resist identification with this global average. This could stimulate cults and religious and other forms of fundamentalism.

The amount of social surprise is highest in Order from Strength. Fundamentalism and other forms of extremism are likely to attract considerable support in this scenario, especially in poorer countries depleted of knowledge and human capital. Corruption and violent conflict are also likely to increase in these countries in Order from Strength, repeatedly undermining human well-being. In addition, grievance-motivated criminal and terrorist groups based in these countries are likely to attempt to wage a guerilla war against the heavily fortified richer countries. This could take the form of attacks on enclaves of rich populations in poorer nations or tourists who venture into such countries, cyberattacks, and the establishment of terrorist sleeper cells within rich countries.

11.7.3 Other Surprises

Other categories of adverse surprise, such as an energy shock, are also possible. While technological optimists and many economists (Johnson 2001) argue that higher energy prices are likely to drive a successful and rapid energy transition, this proposition is unproven. Even if methods to develop alternative portable fuels, including from renewable sources of energy, are technically feasible, there is no guarantee that the rate of this development and its diffusion will be sufficiently rapid to prevent widespread hardship, especially in poorer countries. Multiple interactions between adverse social, ecological, and other surprises are possible. These could have ripple effects, both temporally and geographically, that impair the response to crisis, causing a downward spiral.

11.8 Summary and Conclusion

Each scenario has different consequences and implications for the components of human well-being. (See Table 11.1 and Figure 11.1.)

The Global Orchestration scenario ostensibly has the greatest improvement in human well-being at the global level, especially for populations who are currently disadvantaged. (See Figure 11.2a.) Since the main focus of decision-makers in this scenario is enhancing the functioning of human systems via increased economic growth and improved social policies, improvements in the poor countries are expected to be quite substantial across all human well-being components. In rich countries especially, material well-being and health improve, but other components do not change significantly compared with today.

We believe many systematic obstacles will limit fulfillment of this scenario's promise, however. In particular, we think that the scenario underestimates the obstacles that will impede realization of its assumptions (such as the surprises that might be involved with the removal of agricultural subsidies; see Box 11.2). There is an inherent contradiction between its assumptions of both decreased regulation and increased privatization and the claim that this is consistent with policies that will improve the quality and structure of economic growth in ways that promote equity and good governance (Stiglitz 2003). In parallel, the benefits of "trickle down" promised by advocates for free trade are likely to be exaggerated in this scenario, even if subsidies are successfully lowered. Consequently, the institutional reform promised by this scenario is unlikely to lead to substantial increases in the ability of the poor to attain secure access to resources, including of ecosystem services.

We also find that this scenario has a high vulnerability to major adverse ecological surprise, such as runaway climate change. (See the Socioecological Indicators in Figure 11.1a.) Human well-being is predicted to improve because of a strong focus on increased provisioning ecosystem services, such as food and water availability, and the priority given to short-term goals over a long-term vision of how to deal with change. However, the maintenance and restoration of supporting and regulating services is comparatively ignored. This creates a high risk of eventual breakdown of key ecosystem functions, including provisioning services, with a direct impact on human well-being. In addition, this scenario is characterized by a lack of capacity to monitor and react quickly to adverse ecosystem changes, creating vulnerability to the passing of ecosystem thresholds. This vulnerability exists despite the improved education in this scenario, in part because there is little educational focus on the interaction of human systems with nature.

In the Order from Strength scenario, human well-being is clearly decreased. (See the HWB Indicators in Figure 11.1b.) The global distribution of resources and services that underpin human well-being is more skewed than at present. Wealthy populations meet material needs but experience increased psychological insecurity, while the resources of poor populations are depleted by more powerful groups. As Figure 11.2b demonstrates, there may be slight improvements in material well-being and health in rich countries, but social relations, freedom and choice, and security are likely to deteriorate due to huge inequalities between rich and poor populations both between and within countries. For poor countries, the picture looks even gloomier. Inequality, underinvestment in education, continuing population increase, and little regard for the maintenance of key ecosystems on which mainly poor people depend leads to an overall decrease in all components of human well-being.

Table 11.1. Benefits Compared with Risks and Costs for the Five Components of Human Well-being across the Four Scenarios

Component	Benefits	Risks and Costs
	Global Orchestration	
Material needs	increased incomes, employment, and food; convergence between wealthy and poorer countries; better education	adverse ecological surprises; privatization may act against some groups; food diseases if checks are not maintained in flow chain; more invasive species
Health	better nutrition; strengthened global health services; enhanced vaccine development and distribution; better mental health	increased obesity-related diseases, especially Type II diabetes; more cancer and osteoporosis
Security	proliferation of peace initiatives and civil societies; reduced crime as inequality narrows	increased violent conflicts over diminishing resources, at worst case transformation to Order from Strength
Social relations	convergence of cultures and aspirations between wealthy and poorer countries; greater democracy, friendship, and cultural exchange	lost traditions, cultures, knowledge, and valued ecosystems could trigger resentment or even terrorism against global culture; smaller families increase social isolation
Freedom of choice and action	increased political freedom, civil liberties, information flow, movement, expression, and association	more homogenous culture and ecosystems reduce choice
	Order from Strength	
Material needs	representative ecosystems will be maintained in rich countries to provide ecosystem services	increased material scarcity for poor populations; degraded ecosystems for poor populations; increased material inequalities
Health	improved health in rich countries that can afford health services and have expertise for technological breakthroughs	malnutrition; increased poverty-related diseases, drug resistance, violence, post- traumatic stress, and depression; increased anxiety disorders for rich and poor
Security	advanced technologies and services for those who can afford them	increased crime; hostility among and within nations; terrorism, civil wars, rebellions, fear, stigmatization
Social relations	reasonable among high-income populations, but overlay of anxiety	gangsterism, corruption, intimidation within poor populations; widespread fear, mistrust, intolerance, and hostility
Freedom of choice and action	limited freedom for protected high-income populations	increased surveillance and control; restricted freedom of speech, association, movement, travel, self-expression; censorship and propaganda
	Adapting Mosaic	
Material needs	engagement between ecosystem users and owners facilitates sustainable use; strong local resource institutions	reduced productivity from global commons due to misuse
Health	reduced stress; traditional health systems; better mental health, especially for indigenous populations	loss of global health capacity; uneven distribution of high technology medical services, including surgery and new vaccines and drugs
Security	lowered incidences of conflicts as spending on military action is reduced to cater to local development	increased risk of international criminal activity and hidden human rights abuses, including genocide
Social relations	strong local networks, regional pride; local alliances, local solutions for disputes	lack of global police force or strong influence of global social norms could lead to persistence of local forms of oppression for women and minorities
Freedom of choice and action	freedom to form partnerships, develop local solutions, maintain customs, and relate to and access local ecosystems	curtailed freedom and choice through continuation of existing traditions and customs that restrict freedoms for women and minorities
	TechnoGarden	
Material needs	increased income; ubiquitous computers and communication; robotics and intelligent buildings	increased unemployment due to robots; inadequate inclusion of displaced workers
Health	improved nutrition; cleaner air, water, and food: health informatics; better surgical techniques and new drugs; "nutraceuticals"	addiction to designer drugs and virtual reality; poor gross motor coordination; genetically engineered pathogens; adverse health effects of new foods and chemicals
Security	better surveillance technology and security systems; robotic guards	cheap, powerful weapons could allow small groups to threaten security; electronic fraud; arms races of many kinds and many scales
Social relations	better communication; more literacy; less conflict; technology links new groups; more understanding and trust	genetic-based discrimination; reliance on synthetic reality may create emotional vulnerability and antisocial behavior; weakened family links
Freedom of choice and action	diversified ecosystem services; choices for production systems; selection of desired genetic characteristics from ecosystems	new surveillance technologies (visual and electronic) could reduce freedom

BOX 11.2

How Global Orchestration Could Become Order from Strength

The benefits to well-being promised by Global Orchestration are likely to be greatly overstated if its key assumptions prove exaggerated or false. Rather then declining, multidimensional inequality could increase, leading to an increased "lock in" between rich and poor. Different rules, conditions, and feedbacks could be applied to and come to characterize wealthy and poor populations, exacerbating inequality. Developing-country debt burdens could increase, contributing to additional burdens on ecosystems—for example, through deforestation, increased erosion, ongoing biodiversity loss, and more pollution.

Mechanisms introduced to reduce inequality and to improve human capital could prove unexpectedly fragile and rapidly be abandoned. For example, attempts to increase education in poorer countries could be set back by an exacerbated "brain drain" or by a disease that particularly affects young adults. Funds pledged for development could be repeatedly diverted to address crises involving wealthy countries, such as wars, terrorism, or intractable difficulties in converting to post–fossil fuel energy.

Increased violent conflict or new diseases in poorer countries could also undermine development. Tensions and mistrust between rich and poorer countries could re-emerge if wealthy countries fail, despite repeated promises, to reduce their agricultural and other barriers to free trade. Tension and inequality could also worsen if it becomes clear that free trade, even though sincerely attempted by many parties, fails to surmount its zero-sum-gain obstacles, stubbornly creating new generations and populations of winners and losers (Mehmet 1995). As a result of such pathways, convergence between rich and poorer countries could weaken or even reverse. If this happened, then ecosystem services, especially in the latter nations, are also likely to continue to decline (Welch 2001).

Unanticipated adverse ecological surprise could interact with flawed human actors, greatly magnifying the destruction of built and other forms of capital. Nationalism, fundamentalism, protectionism, and terrorism could resurface, eroding the capacity of global institutions even if they are stronger than today. Because of these setbacks in development and human capital formation, population pressure may be even higher than currently anticipated, creating additional pressure on surviving ecosystems and also inhibiting the economically beneficial effects of the "demographic dividend" that an increasing number of observers have recognized, including as a crucial factor in China's growing prosperity (Williamson 2001; Vallin 2002). As the worst case, feedbacks could develop that lead to more corruption, more inequality, more material shortages, and additional famines and disease. In short, this could transform Global Orchestration to a scenario akin to Order from Strength even if in its early stages the trajectory follows that predicted in Global Orchestration.

(See the Socioecological Indicators in Figure 11.1b.) Poor people are repeatedly forced to give priority to the short-term survival of their families at the expense of ecosystem services and, ultimately, their long-term well-being. Little is done to foster the adaptation potential of societies to social and natural changes. As well, the resilience of these societies to adverse environmental change is likely to be low.

Like the others, this scenario is also probably unrealistic. While pessimists see considerable evidence of a nascent Order from Strength scenario in the present (Kaplan 1994), many countering forces exist that are likely to alleviate the most negative aspects of this scenario. These include the efforts by many governments, NGOs, and individuals to protect and improve governance and qualities such as human rights. The scenario also probably underestimates the "bootstrapping" capacity flowing from the self-organization of individuals and communities once they are given access to some resources, particularly education. That is, even in the generally dismal circumstances foreseen in poorer countries in this scenario, pockets of resilience and resistance are likely to remain, creating persistent hope for a reversal of the broader trends.

In the Adapting Mosaic scenario, most aspects of human well-being improve. Provisioning services in poorer countries are significantly increased through investment in social, natural, and, to a lesser extent, human capital at local and regional levels. (See Figure 11.2c.) In richer countries, provisioning services change little because a threshold of consumption has been achieved. Because of the partial devolution of decision-making power to smaller scales and increased cooperation and exchange at the local and subnational levels, social relations, freedom and choice, and security are also likely to improve. Furthermore, priority is given in this scenario to developing flexible management regimes that monitor, mitigate, or adapt to environmental change and that hence build social adaptive capacity. Decision-makers see maintaining ecosystems and the services they provide as key to human survival. This understanding is built on an overall long-term understanding of ecosystem changes and their impact on human systems and the recognition that these changes matter. Seen at the global level, human well-being improves, but not quite as quickly as in the Global Orchestration scenario. (See the HWB Indicators in Figure 11.1c.)

Yet this scenario makes the unrealistic assumption that populations of malcontented humans will either disappear or become of trivial importance within a very short period. This ignores history. In reality, substantial populations of raiders, pirates, criminals, and free-riders are likely to persist, and their influence is likely to impair successful realization of this scenario. Many such populations will have access to powerful weapons and other forms of coercion. In the absence of global policing implied by this scenario, it is likely that at least some of these "human predators" will forge complex, even if unstable, alliances. In response, this is likely to generate large-scale cooperation among the many human groups who would otherwise form vulnerable "prey."

The TechnoGarden scenario appears to offer many ways to improve human well-being. However, this depends crucially on the speed of institutional development. Turning again to history, it seems likely that institutional development will lag behind that of technology. Thus, social relations and freedom appear to be at particular risk everywhere in this scenario. (See Figure 11.2d.) Unless the distribution of the new environmentally friendly technologies is reasonably equitable, much of the potential for better human well-being in this scenario will remain elusive. Nevertheless, this environmentally pro-active scenario means that decision-makers focus on gaining a better understanding of the management of ecosystems. Consequently, this scenario shows improvements in social adaptive capacity, though not as great as in Adapting Mosaic. (See the Socioecological Indicators in Figure 11.1d.)

TechnoGarden also has a risk of a breakdown of many ecosystem and technological functions, potentially precipitating a cascading decline for human well-being if a limit is passed in the human capacity to control and successfully manage the complex hybrid of social, technological, and ecological systems that are foreseen.

In summary, each scenario exhibits a different package of benefits, risks, and adverse impacts for human well-being. They portray a range of trade-offs between development and ecological strategies currently discussed by policy-makers at different levels. Figure 11.1 depicts and compares some of the main trade-offs. While no single scenario is best, the Order from Strength scenario is clearly the least desirable. People who are currently poor and vulnerable experience the highest risk of future poverty and vulnerability in this scenario.

While all scenarios are claimed to be equally plausible, this is true only in the sense that none are very likely. On the other hand, the future is likely to contain recognizable elements of all four scenarios, just as the present does. Given sufficient cooperation, information flow, preparedness, adaptivity, and technological breakthrough, the human future may respond to a dynamic interchange between various scenarios, at varying temporal and geographic scales, in ways that lead to sustainability.

References

Adams, W.M., D. Brockington, J. Dyson, and B. Vira, 2003: Managing tragedies: understanding conflict over common pool resources. *Science,* **302,** 1915–1916.

Alley, W.M., R.W. Healy, J.W. LaBaugh, and T.E. Reilly, 2002: Flow and storage in groundwater systems. *Science,* **296,** 1985–1990.

André, C. and J.P. Platteau, 1998: Land relations under unbearable stress: Rwanda caught in the Malthusian trap. *Journal of Economic Behaviour and Organization,* **34,** 1–47.

Anonymous, 2001a: Don't underestimate the enemy. *Nature,* **409,** 269.

Anonymous, 2001b: China faces AIDS [editorial]. *The Lancet,* **358,** 773.

Arrow, K., P. Dasgupta, L. Goulder, G. Daily, P. Ehrlich, G. Heal, S. Levin, K.-G. Mäler, S. Schneider, D. Starrett, and B. Walker, 2004: Are we consuming too much? *Journal of Economic Perspectives,* **18,** 147–172.

Beder, S., 1998: *Global Spin: The Corporate Assault on Environmentalism.* Scribe Publications, Melbourne, Australia, 288 pp.

Berkman, D.S., A.G. Lescano, R.H. Gilman, S.L. Lopez, and M.M. Black, 2002: Effects of stunting, diarrheal disease, and parasitic infection during infancy on cognition in late childhood: a follow-up study. *The Lancet,* **359,** 564–571.

BIO (Biotechnology Industry Organization), 2003: Biotechnology information, advocacy and business support: agricultural production applications. Washington, D. C. Cited 18 October 2004. Available at www.bio.org/speeches/pubs/er/agriculture.asp.

Birdsall, N., A.C. Kelley, and S.W. Sinding (eds.), 2001: *Population Matters Demographic Change, Economic Growth, and Poverty in the Developing World.* Oxford University Press, New York, xvi, 440 pp.

Bloom, D. and D. Canning, 2001: Cumulative causality, economic growth, and the demographic transition. In: *Population Matters: Demographic Change, Economic Growth, and Poverty in the Developing World,* N. Birdsall, A.C. Kelley, and S.W. Sinding (eds.), Oxford University Press, Oxford ; New York, 164–197.

Botkin, D., 2001: Ecological risks of biotechnology. Cited 18 October 2004. Available at http://pewagbiotech.org/events/1204/presentations/Botkin.ppt.

Buck, S.J., 1998: *The Global Commons: An Introduction.* Island Press, Washington, DC 225 pp.

Butler, D., 1998: British BSE reckoning tells a dismal tale. *Nature,* **392,** 532–533.

Butler, C.D., C.F. Corvalan, and H.S. Koren, in press: Human health, well-being and global ecological scenarios.

Butler, C.D. and A.J. McMichael, 2005: Environmental Health. In: *Social Injustice and Public Health,* V. Sidel and B. Levy (eds.), Oxford University Press, Oxford, UK, (in press).

Coale, A.J. and E.M. Hoover, 1958: *Population Growth and Economic Development in Low Income Countries A Case Study of India's Prospects.* Princeton University Press, Princeton NJ. 389 pp.

Cordell, L., 1997: *Archaeology of the Southwest.* 2nd ed. Academic Press, San Diego, CA. 522 pp.

Couper, R.T.L. and K.N. Simmer, 2001: Iron deficiency in children: food for thought [editorial]. *Medical Journal of Australia,* **174,** 162–163.

Cox, P.M., R. Betts, C.D. Jones, S.A. Spall, and I.J. Totterdell, 2000: Acceleration of global warming due to carbon-cycle feedbacks in a coupled climate model. *Nature,* **408,** 184–187.

Dauvergne, P., 1997: *Shadows in the Forest: Japan and the Politics of Timber in Southeast Asia. Politics, Science and the Environment,* MIT Press, Cambridge MA, 336 pp.

de Waal, A. and A. Whiteside, 2003: New variant famine: AIDS and food crisis in southern Africa. *Lancet,* **362,** 1234–1237.

Demeny, P., 2003: Population policy dilemmas in Europe at the dawn of the twenty first century. *Population and Development Review,* **29(1),** 1–28.

Dietz, T., E. Ostrom, and P.C. Stern, 2003: The struggle to govern the commons. *Science,* **302,** 1907–1912.

Eckersley, R., 2004: *Well and Good: How We Feel and Why It Matters.* Text Publishing, Melbourne, Australia, 311 pp.

European Environmental Agency, 2001: *Late Lessons from Early Warnings: The Precautionary Principle 1896–2000.* Environmental Issue Report No 22. Luxembourg Office for Official Publications of the European Communities, Luxembourg, 210 pp.

Ezzati, M. and D. Kammen, 2002: The health impacts of exposure to indoor air pollution from solid fuels in developing countries: knowledge, gaps, and data needs. *Environmental Health Perspectives,* **110,** 1057–1068.

Ezzati, M., A.D. Lopez, A. Rodgers, S.V. Hoorn, C.J.L. Murray, and C.R.A.C. Group, 2002: Selected major risk factors and global and regional burden of disease. *The Lancet,* **360,** 1347–1360.

Folke, C., S. Carpenter, B. Walker, M. Scheffer, T. Elmqvist, L. Gunderson and C.S. Holling, 2005: Regime shifts, resilience and biodiversity in ecosystem management. *Annual Review of Ecology Evolution and Systematics,* **35,** 557–581.

Frumkin, H., 2002: Beyond toxicity: Human health and the natural environment. *American Journal of Preventive Medicine,* **20,** 234–240.

Gao, W., A. Tamin, A. Soloff, L. D'Aiuto, E. Nwanegbo, P.D. Robbins, W.J. Bellini, S. Barratt-Boyes, and A. Gambotto, 2003: Effects of a SARS-associated coronavirus vaccine in monkeys. *The Lancet,* **362,** 1895–1896.

Glass, R.I., 2004: Perceived threats and real killers [editorial]. *Science,* **304,** 927.

Grantham-McGregor, S., 2002: Linear growth retardation and cognition [commentary]. *The Lancet,* **359,** 111–114.

Gunderson, L. and C. Holling (eds.), 2002: *Panarchy: Understanding Transformations in Human and Natural Systems.* Island Press, Washington DC, 507 pp.

Hales, S. and A. Woodward, 2003: Climate change will increase demands on malaria control in Africa. *The Lancet,* **362,** 1775.

Hamilton, C., 2003: *Growth Fetish.* Allen & Unwin, Sydney, Australia, 262 pp.

Hellin, J., M. Haigh, and F. Marks, 1999: Rainfall characteristics of Hurricane Mitch. *Nature,* **399,** 316.

James, K.C. and S.D. Foster, 1999: Weighing up disability (Commentary). *Lancet,* **354,** 37 87–88.

Jepson, P., J.K. Jarvie, K. MacKinnon, and K.A. Monk, 2001: The end for Indonesia's lowland forests? *Science,* **292,** 859–861.

Johnson, B., 1999: Genetically modified crops and other organisms: Implications for agricultural sustainability and biodiversity. Cited 18 October 2004. Available at www.cgiar.org/biotech/rep0100/johnson.pdf.

Johnson, D.G., 2001: On population and resources: a comment. *Population and Development Review,* **27(4),** 739–747.

Kaiser, J., 2000: Mercury report backs strict rules. *Science,* **289,** 371–372.

Kaisiti, H., 2003: Redistribution of Indonesian forests: impacts of decentralization of power in forest management. Paper presented at the Globalisation, Localization and Tropical forest management in 21st century. Amsterdam, Netherlands.

Kaplan, R.D., 1994: The coming anarchy. *Atlantic Monthly,* **273(2),** 44–76.

Kay, J.J., H.A. Regier, M. Boyle, and G. Francis, 1999: An ecosystem approach for sustainability: addressing the challenge of complexity. *Futures,* **31,** 721–742.

Kayambazinthu, D., F. Matose, G.C. Kajembe and N. Nemarundwe, 2003: Institutional arrangements governing natural resource management of the Miombo woodland. In: *Policies & Governance Structures in Miombo Woodlands in Southern Africa,* G. Kowero, B.M. Campbell and U.R. Sumaila (eds). CIFOR, Bogor, Indonesia, pp. 45–79.

Kelley, A.C., 2001: The population debate in historical perspective : revisionism revised. In: *Population Matters: Demographic Change, Economic Growth, and Poverty in the Developing World,* N. Birdsall, A.C. Kelley, and S.W. Sinding (eds.), Oxford University Press, Oxford ; New York, pp. 24–54.

Kydd, J., A. Dorward and C. Poulton, 2000. Globalization and its implications for the natural resources sector: a closer look at the role of agriculture in the global economy. An issues paper for the DFID Natural Resources Advisers Conference, Winchester, 10th July 2000.

Laurance, W.F., M.A. Cochrane, S. Bergen, P.M. Fearnside, P. Delamonica, C. Barber, S. D'Angelo, and T. Fernandes, 2001: The future of the Brazilian Amazon. *Science,* **291,** 438–439.

Levin, S.A., 1999: *Fragile Dominion: Complexity and the Commons.* Helix, Reading MA, 264 pp.

Lomborg, B., 2001: *The Skeptical Environmentalist.* Cambridge University Press, Cambridge, UK. 515 pp.

MA (Millennium Ecosystem Assessment), 2003: *Ecosystems and Human Well-being.* Island Press, Washington, D.C., pp. 71–84.

McCarthy, M., 2002: Unclear whether monkey virus in old polio vaccines caused cancer, says IOM. *The Lancet,* **360,** 1305.

McMichael, A.J., 2001: *Human Frontiers, Environments and Disease: Past Patterns, Uncertain Futures.* Cambridge University Press, Cambridge, UK, 413 pp.

McMichael, A.J., C.D. Butler, and M.J. Ahern, 2003: Global environment. In: *Global Public Goods for Health,* R. Smith, R. Beaglehole, D. Woodward, and N. Drager (eds.), Oxford University Press, Oxford, pp. 94–116.

McMichael, A.J., and C.D. Butler, 2004: Climate change, health, and development goals: needs and dilemmas *Lancet,* **364,** 2004–2006.

Mehmet, O., 1995: *Westernizing the Third World: The Eurocentricity of Economic Development Theories.* Routledge, London UK, 186 pp.

Murray, C.J.L. and A.D. Lopez, 1997: Alternative projections of mortality and disability by cause 1990–2020: Global Burden of Disease study. *Lancet,* **349,** 1498–1504.

Nepstad, D.C., A. Verissimo, A. Alencar, C. Nobres, E. Lima, P. Lefebvre, P. Schlesinger, C. Potter, P. Moutinho, E. Mendoza, M. Cochrane, and V. Brooks, 1999: Large-scale impoverishment of Amazonian forests by logging and fire. *Nature,* **398,** 505-–508.

Odum, H.T and B. Odum, 2003: Concepts and methods of ecological engineering *Ecological Engineering* 20, 339–361.

Palmer, M., E. Bernhardt, E. Chornesky, S. Collins, A. Dobson, C. Duke, B. Gold, R. Jacobson, S. Kingsland, R. Kranz, M. Mappin, M.L. Martinez, F. Micheli, J. Morse, M. Pace, M. Pascual, S. Palumbi, O.J. Reichman, A. Simons, A. Townsend, and M. Turner, 2004: Ecology for a crowded planet. *Science,* **304,** 1251–1252.

Pauly, D., J. Alder, E. Bennett, V. Christensen, P. Tyedmers, and R. Watson, 2003: The future for fisheries. *Science,* **302,** 1359–1361.

Piot, P., 2000: Global AIDS epidemic: time to turn the tide. *Science,* **288,** 2176–2178.

Powell, K., 2003: Fish farming: Eat your veg. *Nature,* **426,** 378–379.

Prabhakar, V.K. (ed), 2001: *Biotechnology and Pollution Control.* Anmol, New Delhi, India. 266 pp.

Price, T.D. and G.M. Feinman (eds.), 1995: *Foundations of Social Inequality.* Plenum Publishing Corporation, New York, NY, 280 pp.

Ramakrishnan, P.S., K.G. Saxena, U.M. Chandrashekara, 1998. *Conserving the Sacred: For Biodiversity Management,* UNESCO, New Delhi, 480 pp.

Raskin, P., G. Gallopin, P. Gutman, A. Hammond, R. Kates, and R. Swart, 2002: *Great Transition: The Promise and Lure of the Times Ahead.* Stockholm Environment Institute, Boston, 99 pp.

Ros-Tonen, M., 2003: Background document. Paper presented at the Globalisation, Localization and tropical forest management in 21st century, Amsterdam, Netherlands.

Sachs, J. and P. Malaney, 2002: The economic and social burden of malaria. *Nature,* **415,** 680–685.

Scheffer, M., S. Carpenter, J.A. Foley, C. Folke, and B. Walker, 2001: Catastrophic shifts in ecosystems. *Nature,* **413,** 591–596.

Sen, A.K., 1981: *Poverty and Famines: An Essay on Entitlement and Deprivation.* Clarendon Press, Oxford, New Delhi, 257 pp.

Sen, A.K., 1999: *Development as Freedom.* Oxford University Press, Oxford, UK, 336 pp.

Sheehy, J., 2001: A new plant for a changed climate. International Rice Research Institute. Cited 18 October 2004. Available at http://www.irri.org/publications/annual/pdfs/ar2001/sheehy2.pdf.

Smith, A.H., E.O. Lingas, and M. Rahman, 2000: Contamination of water supplies by arsenic in Bangladesh. *Bulletin of the World Health Organization,* **78(9),** 1093–1103.

Stauber, J.C. and S. Rampton, 1995: *Toxic Sludge is Good for You. Lies, Damn Lies and the Public Relations Industry.* Common Courage Press, Monroe, Maine, 236 pp.

Stiglitz J. 2003. *The Roaring Nineties.* Penguin Books, London, UK, 389 pp.

Tanser, F.C., B. Sharp, and D. le Sueur, 2003: Potential effect of climate change on malaria transmission in Africa. *The Lancet,* **362,** 1792–1798.

Teklehaimanot, A. and R.W. Snow, 2002: Will the Global Fund help roll back malaria in Africa? [commentary]. *The Lancet,* **360,** 888–889.

Tong, S., T. Prapamontol, and Y. von Schirnding, 2000: Environmental lead exposure: a public health problem of global dimensions. *Bulletin of the World Health Organisation,* **78(9),** 1068–1077.

UNEP (United Nations Environment Programme), 1999: Environmental impacts of trade liberalization and policies for sustainable management of natural resources: a case study of India's automobile sector. UNEP, Geneva, Switzerland.

Vallin, J., 2002: The end of the demographic transition: relief or concern? *Population and Development Review,* **28(1),** 105–120.

Vilchez, R.A., C.R. Madden, C.A. Kozinetz, S.J. Halvorson, Z.S. White, J.L. Jorgensen, C.J. Finch, and J.S. Butel, 2002: Association between simian virus 40 and non-Hodgkin lymphoma. *The Lancet,* **359,** 817–823.

WBCSD and UNDP (World Business Council for Sustainable Development and United Nations Development Programme), 2003: Eco-efficiency and cleaner production: Charting the course to sustainability., UNDP online library. Cited 18 October 2004. Available at http://www.iisd.ca/consume/unep.html, last accessed 10.18.04

Webby, R.J. and R.G. Webster, 2003: Are we ready for pandemic influenza? *Science,* **302,** 1519–1522.

Webster, P., 2003: For precarious populations, pollutants present new perils. *Science,* **299,** 1642.

Webster, R.G., 2004: Wet markets—a continuing source of severe acute respiratory syndrome and influenza? *The Lancet,* **363,** 234–236.

Welch, C., 2001: Structural adjustment programs and poverty reduction strategy. The progressive response, *Foreign Policy in Focus,* **4(15).** Cited 18 October 2004. Available at http://www.fpif.org/progresp/volume4/v4n15_body.html.

WHO (World Health Organization), 2002: The World Health Report 2002. World Health Organisation, Geneva, Switzerland.

Williamson, J.G., 2001: Demographic change, economic growth, and inequality. In: *Population Matters: Demographic Change, Economic Growth, and Poverty in the Developing World,* N. Birdsall, A.C. Kelley, and S.W. Sinding (eds.), Oxford University Press, Oxford, UK, 106–136.

Wilson, E.O., 1984: *Biophilia.* Harvard University Press, Cambridge, MA, 157 pp.

Wolfe, N.D., W.M. Switzer, J.K. Carr, V.B. Bhullar, V. Shanmugam, U. Tamoufe, A.T. Prosser, J.N. Torimiro, A. Wright, E. Mpoudi-Ngole, F.E. McCutchan, D.L. Birx, T.M. Folks, D.S. Burke, and W. Heneine, 2004: Naturally acquired simian retrovirus infections in central African hunters. *The Lancet,* **363,** 932–937.

Wright, R., 2000: *Non Zero. The Logic of Human Destiny.* Pantheon Books, New York, NY, 435 pp.

Zimmet, P., 2000: Globalization, coca-colonization and the chronic disease epidemic: can the Doomsday scenario be averted? *Journal of Internal Medicine,* **247(3),** 301–310.

Chapter 12

Interactions among Ecosystem Services

Coordinating Lead Authors: Jon Paul Rodríguez, T. Douglas Beard, Jr.
Lead Authors: John R.B. Agard, Elena Bennett, Steve Cork, Graeme Cumming, Danielle Deane, Andrew P. Dobson, David M. Lodge, Michael Mutale, Gerald C. Nelson, Garry D. Peterson, Teresa Ribeiro
Review Editors: Bach Tan Sinh, Christopher Field

FIGURES

TABLES

Main Messages

Ecosystem service trade-offs arise from management choices made by humans, who intentionally or otherwise change the type, magnitude, and relative mix of services provided by ecosystems. Such trade-offs will be critical considerations for policy-makers over the next 50 years. Trade-offs can be classified in terms of their temporal and spatial scales, and their degree of reversibility. They can also be classified in terms of the type of service targeted and the type of service "traded-off." Identifying trade-offs allows policy-makers to understand the long-term effects of preferring one ecosystem service over another and the consequences of focusing only on the present provision of a service rather than its future.

Major decisions in the next 50–100 years will have to be made on the current use of nonrenewable resources and their future use. Important specific trade-offs are those between agricultural production and water quality, land use and biodiversity, water use and aquatic biodiversity, and current water use for irrigation and future agricultural production. These overarching trade-offs appear consistently throughout all four Millennium Ecosystem Assessment scenarios. Technological or institutional advances that mitigate such trade-offs will improve ecosystem services and simplify the factors that must be considered in making decisions.

Synergistic interactions allow for the simultaneous enhancement of more than one ecosystem service. Since increasing the supply of one ecosystem service can enhance the supply of others (for example, forest restoration may lead to improvements in several cultural, provisioning, and regulating ecosystem services), successful management of synergisms is a key component of any strategy aimed at increasing the supply of ecosystem services for human well-being.

Numerous trade-offs exist that are unknown and unanticipated by people acting within all four MA scenarios. These trade-offs may not manifest until long after the initial decisions are made, even though they are already affecting the mix of ecosystem services provided. Synergisms and trade-offs also often have unanticipated effects on secondary services, not just the primary ecosystem services that we intend to affect with a decision.

Because trade-offs exist and because policy-makers must make decisions about ecosystem services, they are sometimes forced to make decisions that prefer some ecosystem services over others. In general, across all four MA scenarios and case study examples, trade-off decisions showed a preference for provisioning, regulating, or cultural services (in that order). Supporting services are more likely to be "taken for granted."

Slowly changing variables, which tend to underlie supporting services, are often ignored by policy-makers in ways that seriously undermine the long-term existence of provisioning ecosystem services. Slowly changing variables are difficult to understand and rarely quantified within ecosystem models, and their change is difficult to detect. Examples of slowly changing variables or processes include geologic weathering, soil formation and condition, populations of long-lived organisms, and genetic diversity of organisms that directly affect people. Monitoring programs that focus on slowly changing variables may help decision-makers value supporting services appropriately.

Each of the MA scenarios takes a different approach to trade-offs. In Global Orchestration, society gives preference to provisioning ecosystem services. In Order from Strength, present use of ecosystem services is favored over potential future uses. Under Adapting Mosaic, there is no dominant type of trade-off because most decisions are made locally. However, the approach to trade-offs becomes more ecologically sound, as previously unidentified trade-offs and synergisms are revealed through learning and incorporated into decision-making. There is greater opportunity for institutional solutions to trade-off problems in Adapting Mosaic. In TechnoGarden, cultural services are undervalued and often traded-off in management decisions. There is greater opportunity for technological solutions to trade-off problems in TechnoGarden.

Current models are unable to capture all the interactions and secondary effects of trade-offs and synergisms; thus the quantitative model results are a crude lower boundary of the impact of potential ecosystem service trade-offs. Cultural ecosystem services are almost entirely unquantified in scenario modeling; therefore, the calculated model results do not fully capture losses of these services that occur in the scenarios. The quantitative scenario models primarily capture the services that are perceived by society as more important—provisioning and regulating ecosystem services—and thus do not fully capture trade-offs of cultural and supporting services.

12.1 Introduction

Ecosystem services do not operate in isolation. They interact with one another in complex, often unpredictable ways. Many services are provided by ecosystems in interdependent "bundles." (See Chapter 3.) By choosing one bundle, other services may be reduced or foregone. For example, impounding streams for hydroelectric power may have negative consequences for downstream food provisioning by fisheries. Knowledge of the interactions among ecosystem services is necessary for making sound decisions about how society manages the services provided by nature.

The models that we use to understand and make decisions about ecosystems are often inadequate for addressing interactions of multiple ecosystem services (Sterman and Sweeney 2002). In contrast, because of their nature as complex, logical stories, scenarios consider as many interactions as possible. Therefore, the Millennium Ecosystem Assessment scenarios, which focus on the future of ecosystem services and human well-being, provide an ideal opportunity to examine the interactions among ecosystem services. (For a short description of the four scenarios, please see the Summary for Decision-makers.)

This chapter explores two specific policy-relevant interactions among ecosystem services: synergisms and trade-offs. By highlighting these two types of interactions, we are recognizing that although some properties of ecosystems may be susceptible to human intervention and control, others are not; understanding this distinction is essential if we are to manage ecosystem services to maximize human well-being.

In the context of the provision of ecosystem services, a synergism is defined as a situation in which the combined effect of several forces operating on ecosystem services is greater than the sum of their separate effects (adapted from Begon et al. 1996). In other words, a synergism occurs when ecosystem services interact with one another in a multiplicative or exponential fashion. Synergisms can have positive and negative effects. Synergistic interactions pose a major challenge to the management of ecosystem services because the strength and direction of such interactions remains virtually unknown (Sala et al. 2000). But synergisms also offer opportunities for enhanced management of such services. For example, if society chooses to improve the delivery of an ecosystem service, and this service interacts in a positive and synergistic way with another ecosystem service, the resulting overall benefit could be much larger than the benefit provided by one ecosystem service alone.

Trade-offs, in contrast, occur when the provision of one ecosystem service is reduced as a consequence of increased use of another ecosystem service. Trade-offs seem inevitable in many circumstances and will be critical for determining the outcome of environmental decisions. In some cases, a trade-off may be the consequence of an explicit choice; but in others, trade-offs arise without premeditation or even awareness that they are taking place. These unintentional trade-offs happen when we are ignorant of the interactions among ecosystem services or when we are familiar with the interactions but our knowledge about how they work is incorrect or incomplete. As human societies transform ecosystems to obtain greater provision of specific services, we will undoubtedly diminish some to increase others.

Often, interactions among ecosystem services simply exist, and policy-makers cannot choose whether to allow a trade-off or not. For example, if we devote a particular piece of land to timber harvesting, its value for nature recreation will probably decrease. Although this will happen regardless of whether we acknowledge that a choice was made, timber harvesting techniques are susceptible to improvements that may improve recreation opportunities. Many trade-offs can be modified by technology or by human or institutional services that regulate access to and distribution of ecosystem services. For instance, a trade-off may exist between agricultural production and species richness, yet we can use technological advances to increase agricultural production and make our farms more diverse at the same time.

Decisions relating to natural resource management often revolve around ecosystem service trade-offs and involve services that interact synergistically. Robust decisions take careful account of their impacts on a range of ecosystem services and do not focus only on a single service of greatest apparent interest. A better knowledge of trade-offs and synergisms would simplify environmental decision-making. To illustrate ecosystem service trade-offs and their consequences for society, this chapter draws on the results of the scenario analyses and a variety of published case studies. We focus on synergisms when opportunities for the improved delivery of multiple ecosystem services simultaneously exist.

This chapter considers the interactions among ecosystem services in five major sections. First, we examine the results of both the quantitative and qualitative MA scenario models to derive an understanding of the major trade-offs common across all scenarios and the different trade-offs and synergisms illustrated by the scenarios. We also explore the links between ecosystem service trade-offs, synergisms, and the Millennium Development Goals. Second, we present a series of case studies from the literature and use the results of these case studies to develop two different approaches for understanding the nature of trade-offs. Third, we combine the results from the scenarios and the case studies to propose some characteristics that are common to all trade-off decisions. Finally, we illustrate some of the common dilemmas faced when making ecosystem service management decisions and discuss some of the problems of using modeling results when examining ecosystem service trade-offs.

12.2 Interactions among Ecosystem Services in the Scenarios

To help understand ecosystem service interactions, we propose a system with three axes: spatial scale, temporal scale, and irreversibility. Each interaction can then be classified in one of two categories for each one of the axes. (See Figure 12.1.) *Spatial scale* refers to whether the effects of the synergism or trade-off are felt locally or at a distant location. *Temporal scale* refers to whether the effects take place relatively rapidly or slowly. *Irreversibility* expresses the likelihood that the perturbed ecosystem service may return to its original state if the perturbation ceased.

Because many management actions affect more than one ecosystem service at a time and may operate at different scales simultaneously, it can be difficult to classify ecosystem service interactions in a single category. At the same time, knowledge of the different scales at which policies should be targeted is a key component of managing ecosystem services. Therefore, creating classifications is an important first step toward improving our understanding of the interactions among ecosystem services.

Classification schemes allow a manager to think strategically about the use of ecosystem services, understand the nature of the ecosystem services being considered, be aware of the spatial and temporal scale at which the ecosystem services operates, and determine how far-reaching the effects of particular decisions can be. The policy-maker can tailor management decision to the appropriate scale to mitigate any negative effects and thereby produce "win-win" solutions.

At many points throughout the scenario analysis, quantitative and qualitative results reflect the different underlying decision-making paradigms within a particular scenario.

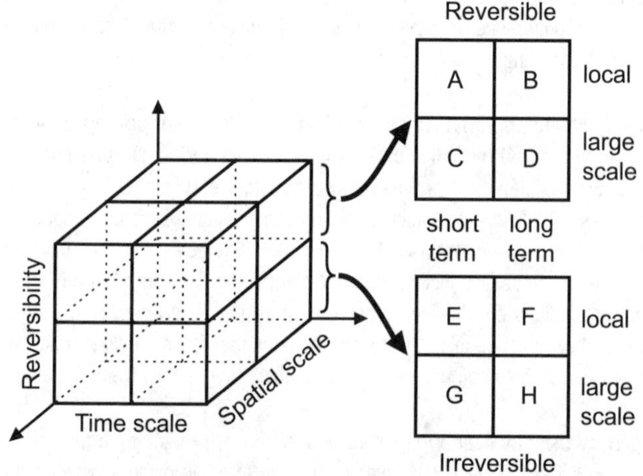

Figure 12.1. Eight Categories of Ecosystem Service Trade-offs, Classified According to Spatial and Temporal Scales and Degree of Reversibility

Despite the difference of worldviews represented in the scenarios, some major trade-offs are common to all of them, with major implications for the continuing delivery of supporting and regulating ecosystem services. The ecosystem service trade-offs that are present across the scenarios may be a result of the underlying assumptions of either the scenarios or the models used. However, cross-scenario commonalities also suggest that these trade-offs are likely to occur regardless of the path that society takes, largely because such trade-offs are driven by the short-term provisioning services that are necessary to assure human well-being. In each case, the scenarios show that our management and decisions about future trade-offs will have a significant effect on the provision of ecosystem services (and hence, human well-being) by the year 2050.

12.2.1 Agricultural Production, Water Quality, and Aquatic Habitats and Species

Agricultural production shows an inverse relationship with water quality and quantity: as we increase agricultural production, the quality of water and the quantity available tend to decrease. (See Chapter 9.) In general, increased efficiency of agricultural production has been accomplished through technology and the increased use of water, nutrients, and pesticides. Because the supply of water is finite, water used for agriculture cannot be used for other purposes. Thus we trade off having water available for other uses in order to increase agricultural production. Nutrients and pesticides can run off from agricultural fields into nearby streams, rivers, lakes, and estuaries, leading to declines in water quality. Thus, use of nutrients and pesticides to increase agricultural production can lead to critical declines in water quality. The negative impacts on water quality often propagate downstream. In the scheme presented in Figure 12.1, such water quality trade-offs can be local or large-scale, short-term or long-term, and are probably not reversible in short time frames (categories A-D). The case of agriculture and hypoxia in the Gulf of Mexico provides a very compelling example of the complexities involved in managing the impacts of agrochemicals. (See Chapter 8.)

Greater use of the world's water supply for agricultural production may improve basic food production and human health in many places. However, increases in pollution and water shortages caused by more-intensive agriculture may make many of these regions more vulnerable to surprises, such as drought, eutrophication, or floods that overwhelm sewage treatment plants. One unexpected consequence of agricultural intensification and climate change is that many rivers will have higher discharge rates, becoming more prone to flooding and drying, with few big differences between scenarios. Many areas that are already water-limited will face further water availability stress and will be more susceptible to environmental perturbations such as drought. These regions may find themselves facing water shortages or water that is undrinkable. Evolution of technology is projected to help the current situation, but only slightly; water limitations will be a concern regardless of which scenario is considered.

In all scenarios, higher income and increasing investments in technology lead to intensification and expansion of agriculture. (See Chapter 9.) Further, the increases in total agricultural production lead to the expansion of irrigated farmland, increased water stress, and increases in the volume of polluted water. Provisioning services such as access to water are traded off for increases in food supply. The heavy emphasis on food production leads to a multitude of uncertainties in relation to the integrity of other ecosystem services.

Changes in water quality also have negative impacts on freshwater biodiversity. As in the trade-off between food production and terrestrial plant biodiversity (described in the next section), short-term gains in water access that initially increase human well-being will lead to reductions in aquatic habitat (and biodiversity) and ultimately to greater regional vulnerability to water shortages (see Chapter 11) and a decline in human well-being. The decrease in available fresh water also has implications for the future productivity of freshwater fisheries, waste removal, and human settlement patterns. (See Chapter 10.)

According to the scenarios, fresh water is a commodity that will require significant planning and conservation in the future to assure that demands do not outstrip the necessary supply. In almost all instances, the scenarios suggest that numerous trade-offs will have significant impact on the quantity and quality of fresh water available for all aspects of human well-being. When making choices about the short-term provisioning needs gained from agricultural production, managers who incorporate the realities of limited freshwater supply in their models for management planning will be more successful than those who do not. Technologies that promote or conserve fresh water, similar to those emphasized in TechnoGarden, can also be used to mitigate some of the freshwater pressures. Finally, fresh water is unevenly distributed over the planet, and subsequent water shortages will also develop unevenly. Therefore, there will be spatial trade-offs among water-rich and water-poor regions.

12.2.2 Land Use and Biodiversity

The expansion of agricultural production that takes place in all scenarios has potentially severe consequences for biodiversity. Expansion of the total agricultural area decreases the area of forests and grasslands. This reduction leads to a decrease in total vascular plant biodiversity and limits soil formation. (See Chapter 10.) Even though the rate of loss of vascular plant biodiversity in TechnoGarden is slower than in the other scenarios, it still results in approximately 300 vascular plant species being lost each year. Order from Strength provides the worst scenario for terrestrial vascular plant diversity because of the high rate of human population growth and the low agricultural yields (requiring extensive rather than intensive agriculture) resulting from the small transfer of technology from rich to poor countries.

Expansion of agriculture leads immediately to local losses of biodiversity through extirpation of local populations and loss of landscape diversity and, most important, loss of ecosystem services. These losses occur even if species extinctions do not or if extinctions are delayed due to the slow approach to equilibrium.

A number of cascading effects result from the trade-off between land use and biodiversity. Perhaps the most important effects involve the unintentional impairment of supporting services, such as future soil formation, water purification capacity, or the maintenance of species habitat. Conversion of natural forests into croplands will also reduce ecosystem services such as climate regulation and carbon sequestration. The loss of supporting services does not often have immediate consequences. However, the slow degradation of supporting services makes it very hard for future policy-makers to reverse the trend in biodiversity loss. Thus, the heavy emphasis on food production across all scenarios is associated with future reductions of other ecosystem services.

Land use trade-offs may be mitigated by zoning plans that allow multiple uses of land resources within regions and by land use practices that maintain ecosystem services in combination with food production. Policy-makers can also capitalize on the synergistic interactions between land use and the delivery of multiple ecosystem services (forest restoration, for instance, may "create" several provisioning, regulating, cultural, and supporting services). Management regimes such as those outlined in the broad-scale policies developed in Global Orchestration may help alleviate land use problems globally, but global policies must also be intertwined with smaller-scale policies, such as those found in Adapting Mosaic, to help avert small-scale land use problems. Development of more-productive crops under TechnoGarden will also alleviate some land use problems.

Nevertheless, land use problems still remain across all scenarios because of the large increase in population. A good approach to managing land to minimize ecosystem service trade-offs will combine the best global policies (including free trade of food resources) with development of smaller-scale policies, such as protected areas and the use of technology that increases food production per square meter of agricultural land. Approaches that integrate continued support of forest areas along with agricultural production (such as shade-grown coffee) minimize land use versus biodiversity trade-offs.

12.3 Trade-offs Illustrated by the Scenarios

In all scenarios, society modifies the supply of a variety of ecosystem services. (See Figure 12.2.) Broadly speaking, under the two "reactive" scenarios (Global Orchestration and Order from Strength) the losses are greater than the gains. Even in the "proactive" scenarios (Adapting Mosaic and TechnoGarden), however, there are reductions in the supply of ecosystem services in one of the dimensions considered.

In Global Orchestration, society focuses primarily on the provisioning ecosystem services that generate tangible products to improve human well-being. When environmental problems arise, they are dealt with according to the belief that economic growth can always provide resources to substitute for lost ecosystem functions. Proactive management of ecosystem services is not pursued. Under this scenario, society will tend to trade off regulating and supporting services while trying to maximize provisioning ecosystem services.

The trade-off approach for regulating and supporting ecosystem services is slightly different from the approach for cultural ecosystem services. Regulating and supporting services are routinely ignored in trade-off decisions, because in many instances in this scenario, human well-being is very good. For example, increased human and economic well-being leads to urban growth into wetlands and along coastlines, which ultimately causes the diminishment of nutrient cycling and water purification and the elimination of fish habitat within these areas. People in this scenario typically ignore these negative effects until they are a serious problem. In contrast, there is some recognition that cultural ecosystem services or cultural differences are essential to maintain.

At the same time, the emphasis on free trade and global policy causes many cultures to be subsumed into an overall "global culture." For example, even though some aspects of Asian culture are integrated into western business practices, many of the traditional practices, such as religious ceremonies, are eliminated as these cultures strive to become part of the global community. The best example of the emphasis on provisioning ecosystem services in this scenario may be the increased importance of meat in the diet, which results from a general increase in human well-being. The increased production of meat causes extensification of agriculture to provide animal feed. Extensification happens at the cost of land-based biodiversity. This and other similar trade-offs are largely ignored in this scenario, as this change in diet is viewed as a benefit of Global Orchestration policies.

Order from Strength places little value on ecosystem services, because rich and poor countries are both focused on increasing their wealth and power through economic growth. All ecosystem services, but especially those that occur over large spatial or temporal scales, are likely to be traded off, as there are no international mechanisms or incentives to protect them. In rich countries, ecosystems are believed to be robust and therefore are used without restrictions in order to improve human well-being. All that is required is that representative samples are preserved in order to have a "natural data base" for developing appropriate technologies to repair or replace them. Provisioning ecosystem services are likely to be favored without considering the impacts on other ecosystem services, as they directly improve human well-being. In poor countries, the conservation of ecosystem services is not considered a priority, thus substantial trade-offs occur among all services. It is assumed that concerns over the delivery of ecosystem services will spontaneously evolve once more-pressing social and economic issues are resolved and that any problems incurred through trade-off decisions will be repairable at a later date.

The lack of value placed on ecosystem services in Order from Strength can perhaps best be illustrated by the exam-

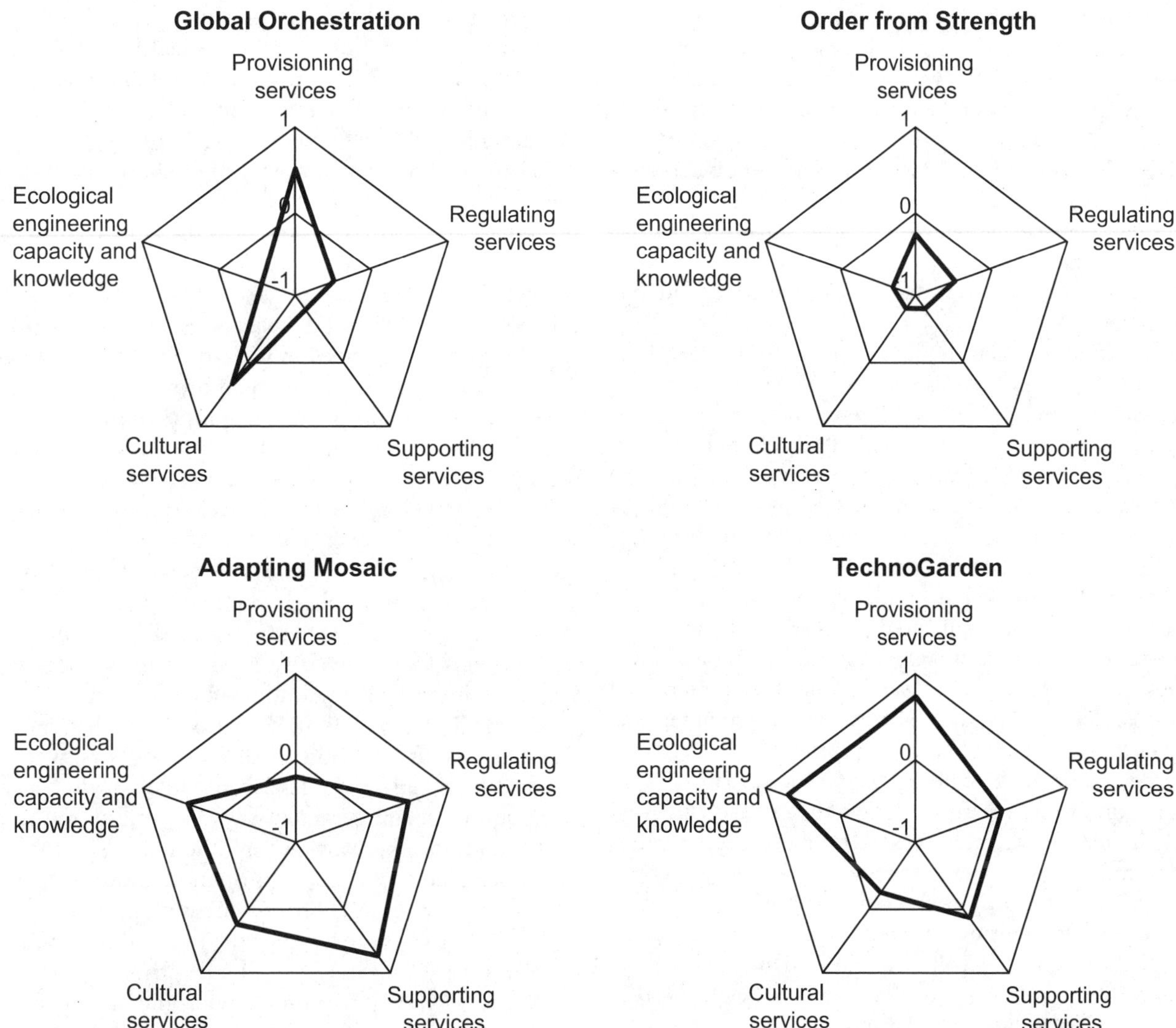

Figure 12.2. Relative Change in Provision of Ecosystem Services in MA Scenarios. Dark lines indicate the state of each Ecosystem Service (ES) at the end of the scenario storyline relative to a starting point of zero. A positive value (between 0 and 1) indicates an increase in the supply of a particular ES. A negative value (between 0 and −1) indicates a decrease in supply. Therefore, as the "stars" increase in size, the overall supply of ES increases, while as they decrease, the overall supply of ES decreases.

ples drawn from marine fisheries and the plight of sub-Saharan Africa. (See Chapter 8.) In Order from Strength, the rich countries use their wealth to control global fisheries while protecting their own stocks. Their emphasis is not on maintaining adequate provisioning resources for human well-being. Instead, they focus on controlling the global market for fisheries to maximize economic gain. Exports of small pelagic fishes are diverted for further production of meat (a luxury food resource in rich countries) instead of being exported as food products to poor countries. Trade-offs at the global scale are nonexistent, as the emphasis is on exploitation for economic gain. In contrast to the rich countries, most of sub-Saharan Africa no longer has food security in 2050, because of the effects of climate modification and population growth in this region. The decision for policy-makers is not about trading off provisioning services for other ecosystem services, but instead is solely focused on maintaining their own food security.

Under Adapting Mosaic, there is no dominant ecosystem service trade-off paradigm, although negative trade-offs tend to decline over time. In the short term, societies are likely to engage in a variety of ecosystem service trade-offs as they experiment with the supply of ecosystem services according to their local needs, especially provisioning services. No single trade-off dominates, since conditions vary globally and societies only focus on their local set of conditions and problems. Over time, local management improves throughout the world. Local institutions and innovations reduce the number and magnitude of negative trade-offs.

The Adapting Mosaic scenario leads to many local management examples that build on previous experiences and deal with each set of trade-offs independently. For example, in the Euphrates-Tigris river (see Chapter 8), the initial trade-off decisions provide more provisioning services (cotton production) at the expense of supporting and regulating services (soil formation, saline control on the land). However, working within the area, managers learn how to use the Adapting Mosaic of conserved areas to eventually craft solutions that provide for "win-win" interactions in provisioning, regulating, and supporting ecosystem services. Similarly, malaria control in Africa (see Chapter 8) involves the trade-off of a regulating ecosystem service (disease control) with a provisioning service (fresh water). Through the use of adaptive management on a fairly small scale, however, managers are able to craft solutions that produce "win-win" solutions that provide both fresh water and malaria control.

TechnoGarden assigns high value to ecosystem services, but mainly from a human-use perspective. This means that cultural ecosystem services are more likely to be traded off and lost than other types of services. Initially, there is great interest in the variety of provisioning, regulating, and supporting ecosystem services as models for possible technological developments, but as key societal ecosystem services are identified and replaced by technological equivalents, society becomes more likely to trade off any existing ecosystem services for their engineered alternatives. In the short term, society will predominantly trade off cultural ecosystem services for other types of services; in the long term, all types of services may be traded off as key ecosystem services are identified and technologically optimized.

The emphasis on technological fixes leads to the rapid urbanization of many parts of the globe, especially in Asia. As urban areas grow, traditional cultural resources such as temples and religious sanctuaries are traded off for urban areas. This is not a long-term solution, however, as there still is a need for cultural services, and many are "reinvented": the rebirth of Japanese urban gardens, for instance, or the creation of salmon festivals in the U.S. Pacific Northwest or the Gojiro festivals in Japan.

One of the most important conclusions from all scenarios is that the total pressure on ecosystem services worldwide will increase. Some of this is a consequence of the projected human population growth used in these scenarios. Even in cases such as TechnoGarden and Adapting Mosaic (which attempt to mitigate some of these environmental pressures), increases in provisioning ecosystem services will be traded off against supporting and regulating services. There is perhaps no more compelling example than the combined synergistic effect of greater use of greenhouse gases (through increased human population and a greater reliance on fossil fuels technology) and the decline in carbon sequestration that has resulted from the conversion of forested areas into agriculture. Thus, the ability of the biosphere to regulate climate change—even with the technological fixes expected in TechnoGarden or the localized controls of Adapting Mosaic—will not be easily restored, as the regulating and supporting services provided by forests are traded off by the additional expansion of agriculture, a provisioning service.

We also examined the trade-offs and synergisms among ecosystem services that might develop as governments work to achieve the Millennium Development Goals adopted at the UN General Assembly in September 2000. The eight goals are to eradicate extreme poverty and hunger; achieve universal primary education; promote gender equality and empower women; reduce child mortality; improve maternal health; combat HIV/AIDS, malaria, and other diseases; ensure environmental sustainability; and develop a Global Partnership for Development (UNDP 2003).

The scenarios offer a number of insightful illustrations of the ways in which ecosystem service trade-offs may affect the ability of governments to reach the MDGs. Let us consider the first goal: the eradication of extreme poverty and hunger. Each scenario indicates a different likelihood that this goal will be met. For example, Global Orchestration has the greatest reduction in poverty and hunger as a result of improvement in the delivery of provisioning services. In contrast, hunger and poverty regimes continue to exhibit strong rich-poor divides under Order from Strength, although this disparity is lessened among northern countries. Achieving poverty alleviation in the short term may also be accompanied by long-term costs, such as narrowing the genetic base of crops or increasing nutrient input to freshwater systems from fertilizers and pesticides.

The drive to eradicate extreme hunger and poverty has ramifications for biodiversity. It also demands actions that will carry important implications for the attainment of the other MDG. Analysis of the scenarios shows that one of the major trade-offs common to all scenarios is between land use and biodiversity (as described earlier in this chapter). Although there are certainly ways to mitigate the impacts of this trade-off (perhaps through the innovations found in TechnoGarden or the emphasis on more thorough environmental accounting in Global Orchestration), policy-makers will face choices that may favor the first Millennium Development Goal at the expense of biodiversity.

Another MDG is to ensure environmental sustainability. All scenarios indicate that the volume of polluted water will increase as a result of the projected increase in agricultural production. Further increases in the use of water for food production also indicate that there will be a decrease in freshwater biodiversity. Policy-makers will be forced to examine the trade-offs among the two goals (eradicating extreme poverty and hunger and ensuring environmental sustainability) and, where possible, to develop policies that produce "win-win" outcomes. This will be a complex process that draws heavily on the past experiences of natural resource managers, as illustrated in the case studies and the scenarios. Whether it is realistic to expect that the Millennium Development Goals can be reached without a significant loss of biodiversity remains to be seen.

12.4 Interactions among Ecosystem Services in Selected Case Studies

One way to understand the consequences of ecosystem services decisions is to examine the outcomes of past management activities. The following examples illustrate some of the dilemmas and trade-offs that society must face when deciding to enhance one ecosystem service without fully understanding the impacts on other services. The order of this presentation is arbitrary and does not reflect any attribute of the ecosystem services involved, the region, or anything else. Our goal is to provide a series of examples of the implications of human actions directed at modifying the outputs of one or more ecosystem services. We did not attempt to be comprehensive, just illustrative.

12.4.1 Vulture Declines in India

The recent sudden decline of Gyps vultures in eastern India provides a compelling example of how species declines can cause declines in provision of many ecosystem services, illuminating unexpected synergisms among species and socioecological processes. Vultures play an important role as natural garbage collectors in many parts of India. In particular, vultures help to dispose of cattle carcasses in areas where beef eating is forbidden. In Amritsar, center of the Parsi religion, they also help remove human corpses from traditional sites of "laying to rest."

In the last few years, vulture numbers suddenly declined, with consequences that cascaded throughout the region. Since there are too few vultures to clean the corpses, the Parsi are no longer able to lay their dead to rest without causing a health hazard. Instead, the dead are stored until a future time. But the less obvious consequences are leading to even more dramatic effects. Carcasses of cattle are transported to areas on the edge of towns and villages. These areas are now increasingly dangerous to visit because vultures do not rapidly remove the meat from carcasses, tempting other carnivores to the area. Feral dog populations have increased as a result of the lower competition with vultures for meat. Growing dog populations are likely to cause an increase in rabies risk, dramatically heightening the consequences of being attacked by a dog.

Vulture declines have recently been linked to the use of the veterinary drug diclofenac (Oaks et al. 2004). Thus, in this example, attempts to improve the health of domestic animals had a series of cascading, unanticipated, and unknown effects on many other services, even to the point of possibly having a negative effect on human disease in the area. Depending on whether the impact of diclofenac proves to be reversible or irreversible, this trade-off could be classified as Type A or E, as described in Figure 12.1. That is, this trade-off is local, rapid, and of unknown reversibility.

12.4.2 Lakeshore Development in the Northern United States

Property values surrounding lakes in northern Wisconsin in the United States are strongly linked to the development patterns around the lake. During the last 30 years, there has been a substantial increase in the development and building on lake shores (Peterson et al. 2003) that has resulted in the creation of a "lake community" on many lakes. The initial conversion of these lakes from undeveloped to developed shorelines resulted in an increase in property values around these waters. Although development was accompanied by an initial increase in cultural ecosystem services, changes in shoreline vegetation resulted in increased sedimentation (soil loss; soil provides a supporting ecosystem service), a reduction of the amount of habitat (a supporting ecosystem service) available for fishes (Christensen et al. 1996), and a decrease in fish growth rates (Schindler et al. 2000).

Although zoning regulations can help to control shoreline development, lake communities are often resistant to zoning and control, even though there is evidence that zoning results in even higher increases in property value (Spaltro and Provencher 2001). In addition, shoreline developments often lead to increases in primary production due to increased fertilizer use and sedimentation from runoff. The consequence is a decrease in water quality (regulating ecosystem service) and subsequent reduction in the aesthetic quality of the lake (cultural ecosystem service).

Resistance to zoning and government regulation by property owners in this area led to overdevelopment and the environmental impacts just discussed. It remains to be seen whether the long-term cumulative environmental impact will negatively affect property values. Several types of trade-offs are involved here. For example, the reduction of fish habitat is probably irreversible, local, and rapid (Type E), while decreases in water quality and aesthetic value of lakes may be reversible (with successful enforcement of regulations on fertilizer use), large-scale, and long-term (Type D).

12.4.3 Fisheries and Tourism in the Caribbean: Jamaica and Bonaire

Many ecosystem services are provided by the Caribbean Sea. Two of the most prized are fisheries and recreation. The Caribbean attracts about 57% of scuba diving tours worldwide. In the 1950s, 1960s, and 1970s, Jamaica was the prime dive location, and hard corals covered as much as 90% of shallow coastal areas (Goreau 1959). By the late 1960s, chronic overfishing had reduced fish biomass by about 80% compared with the previous decade (Munro 1969). Then, in the early 1980s, two extreme events hit Jamaican coral reefs, causing their collapse. In 1980, Hurricane Allen broke many large elkhorn and staghorn corals into pieces (Woodley et al. 1981). In 1983, an unidentified disease spread throughout the Caribbean and killed 99% of black-spined sea urchins (*Diadema antillarum*), the primary grazer of algae on the reefs (Lessios 1988). Without the ecosystem services provided by grazing fish or sea urchins, fleshy macro-algae came to dominate coral reefs (more than 90% cover) in just two years (Hughes 1994). The lucrative dive tourism industry in Jamaica declined.

When the sea urchin mass mortality occurred throughout the region, most sites suffered algal overgrowth, but a few sites did not. Sites like Bonaire, with abundant grazing fish, had no reported algal overgrowth. In Bonaire, the Reef Environmental Educational Foundation has recently generated statistics from about 60,000 coral reef fish surveys, which rate seven dive sites in Bonaire among the top 10 worldwide for fish species richness, with over 300 species (REEF 2003). Bonaire banned spear fishing from its reefs in 1971. In 1979, the Bonaire Marine Park was created to preserve for scuba divers the entire area surrounding the island, from the shoreline to 60m depth. In 1992, active management of the park started with the introduction of mandatory permits for divers, bringing in about $170,000 a year to support protected area management. Economic activities (dive operators, hotels, etc.) connected with the park attract about 10,000 people annually, valued at over $23 million per year. In contrast, the cost of park management is under $1 million per annum.

Thus, protecting the fish for the regulating ecosystem service they provide as algal grazers and for their aesthetic attraction to tourists yielded a positive financial return in the long term. In this case, regulating provision of one service (the fishery) maintained resilience in the system and led to a long-term gain in provision of recreation as well as a stable, long-term fishery. These synergistic interactions among ecosystem services allow for the simultaneous enhancement of the supply of more than one ecosystem service.

12.4.4 Fertilizer Use in the United States

Intensive agriculture within the United States has resulted in massive soil loss (a decrease in a supporting service) throughout the Mississippi drainage region (Malakoff 1998). The initial conversion of land in this area from prairie and grassland to agriculture was motivated by an interest in increasing food production (a provisioning service). To maintain high levels of crop output in spite of topsoil erosion, farmers have maintained soil fertility through the addition of either natural (manure) or chemical fertilizers.

The effects of the high level of artificial fertilization have also resulted in massive changes in downstream areas: many small-scale changes by individual farmers on their own fields have resulted in the creation of a hypoxic zone (a "dead zone") in the Gulf of Mexico. (See Chapter 8.) This zone of low oxygen has resulted in declines in the shrimp fishery as well as in other local fisheries in the Gulf region (Malakoff 1998). In this case, attempts to maintain and increase the provision of one service, food, have caused substantial declines in many ecosystem services in another location. The effects of this trade-off are felt over a large spatial scale and are likely to last for a long time. Whether they are reversible or not remains to be seen. The trade-off can therefore be classified as Type D or H.

12.4.5 Mine Effluent Remediation by Natural Wetlands on the Kafue River, Zambia

An example from Zambia demonstrates a trade-off in which protection of an extensive, unique ecosystem is achieved through the degradation of smaller, upstream wetland systems (von der Heyden and New in press-a). The Kafue River originates along the watershed between Zambia and the Democratic Republic of the Congo 100 kilometers northeast of the industrialized Copperbelt mining region. It is the dominant source of water and food for various urban and rural settlements and enterprises. Although the river only drains 20% of Zambia's surface area, it is the principal water source for all of the country's major towns (Mutale and Mondoka 1996). Wetlands, locally called *dambos,* are apparent throughout the Copperbelt Province, where they occur primarily as headwater features forming the source of the Kafue River.

Commercial mining on the Copperbelt began in the 1920s, and since then the region has been characterized by one of the highest densities of large-scale mines in the world. Mining-related contamination of the Copperbelt's water resources has been a matter of great concern over the past decades (Pettersson and Ingri 2001). Since many Copperbelt mines and their related infrastructure are located on or near the catchment's rim, effluent originating from these operations follows the natural drainage path, first entering the dambo wetlands before discharging into larger waterways and ultimately the Kafue River (Limpitlaw 2002). Although wetlands throughout the Copperbelt have been affected and degraded as a result of the discharge of mine effluent, these systems have given a considerable level of protection to the downstream ecosystem through the filtration, retention, and remediation of effluent contaminants within the wetland sediment and flora (von der Heyden and New in press-b, in press-c).

While the wetland systems demonstrate great efficiency in protecting downstream environments from mine-related pollutants, natural wetlands are known to be fragile ecosystems that are extremely valuable to local resource users and are a key component of the regional ecosystem. It is uncertain if the wetlands are able to provide their regulating ecosystem services indefinitely or at a constant level. Further understanding of the complexity of factors affecting the impact, capacity, and alternatives to the use of natural wetlands in mine effluent remediation is necessary to assess comprehensively the role of the wetlands in the management of the Copperbelt environment. Perhaps irreversibly, ecosystem services provided by wetlands have been traded off for the long term, over large spatial scales. Thus this trade-off can be classified as Type H.

12.4.6 No-take Zones in St. Lucia

As fisheries worldwide continue to decline (FAO 1996; Jackson et al. 2001; Myers and Worm 2003; Roberts 2002), there has been an increasing interest in fishery exclusion zones, both to allow for the recovery of targeted species and as a mechanism to increase the catch outside of protected areas. Recent research suggests that these objectives can be successfully achieved by the designation of no-take marine reserves (Gell and Roberts 2003a).

For example, the Soufrière Marine Management Area, created in 1995 along 11 kilometers of the coast of St. Lucia in the Caribbean, includes five small marine reserves alternating with areas where fishing is allowed. Roughly 35% of the fishing grounds in this area has been set aside and protected. The initial cost of restricting access to fishers in about a third of the available area (a decline in a provisioning ecosystem service) has been easily compensated for by the benefits. As may be expected, fish biomass inside the reserves tripled in just four years, but, more important, biomass in the fished areas doubled during the same period and remained stable thereafter (Roberts et al. 2001). In less than the typical term of an elected governmental official, the fishery recovered and landings increased. There is growing evidence from around the world supporting marine reserves and fishery closures as an effective tool for managing fish, one of the most important provisioning ecosystem services (Gell and Roberts 2003b). Wise local management of fisheries averted a negative impact, possibly for the long term. Therefore this is an example of a Type B interaction.

12.4.7 Lobster Fishing in Maine

Lobster fishing in the northeastern United States has important social and economic consequences for many of the coastal communities in this region. Perhaps nowhere is it more important than in the state of Maine, which harvests the majority of lobster produced in the country (Acheson and Steneck 1997). Since 1870, this fishery has experienced a period of bust followed by a fairly extended period of boom in the numbers of lobster produced. A combination of formal state-based and informal social regulations set and enforced by territorial harbor cooperatives has contributed to the expansion and continued success of the lobster fishery, even as other fisheries in the same area have failed (Acheson et al. 1998; Jackson et al. 2001).

The lobster fishery provides important provisioning services such as food and economic well-being for communities. The development of harbor cooperatives for social enforcement of regulations also provides members and communities with a sense of identity, which is important for social reinforcement of informal regulations on the fishery (Acheson et al. 1998). The strong bonds created within the harbor cooperatives help limit the total effort within the fishery (resulting in a short-term economic cost), assuring that harvest is limited and the lobster fishery is preserved for the long term (Acheson et al. 1998).

The cultural services provided by the lobster cooperatives may have also had synergistic effects, because one of the contributing factors to the current lobster boom is an increased conservation attitude among lobster fishers (Acheson and Steneck 1997; Acheson et al. 1998). Formation of lobster cooperatives provided the social fabric and peer pressure necessary for lobster fisheries within to adhere to a conservation ethic. This "win-win" outcome in a fairly small-scale system was a product of synergistic interactions among ecosystem services and it helped play a part in the lobster boom and maintain the cultural identity of the lobster communities.

12.4.8 Water Quality and Biological Invaders in the U.S. Laurentian Great Lakes

Beginning about 1870, a set of connected canals was opened in Chicago, Illinois, that reversed the flow of the Chicago River. The purpose of the engineering project was to flush waste from the burgeoning number of human households and slaughterhouses away from Lake Michigan, the drinking water supply for the growing city. The Chicago River, which had naturally flowed into Lake Michigan, was thereby linked via the Chicago Ship and Sanitary Canal to the Mississippi River drainage; over time this became an important conduit for commercial and recreational navigation, as well as a huge open sewer. Because the canal was filled largely with untreated sewage and animal waste, dissolved oxygen concentrations were too low for most organisms to survive for many miles downstream in the Des Plaines and Illinois rivers. This caused a complete loss of riverine fisheries until the 1970s, when Clean Water Act regulations made the waterway habitable again for fish and other organisms.

Paradoxically, the consequence of improved water quality in the last three decades has been a surge in invasive species moving in both directions in the canal. The best documented example is the rapid spread of zebra mussels (*Dreissena polymorpha*). From its initial site of introduction in the Great Lakes about 1986, zebra mussel larvae were transported down the canal into the Illinois and Mississippi rivers, all the way to New Orleans (just north of the Gulf of Mexico) in about four years (Stoeckel et al. 1997). The consequence of zebra mussel spread within the Great Lakes has been $100 million in annual costs to the power industry and other users, extirpation of native clams in Lake St. Clair, and large changes in energy flow and ecosystem function (Lodge 2001). Other nonindigenous species in the Great Lakes—two fish, for example, the round goby and the Eurasian river ruffe (*Gymnocephalus cernuus*)—are also nearing the canals and could join the zebra mussel in its southward migration. Other species that have had large impacts elsewhere—two Asian carp species are of special concern—are migrating northward and nearing Lake Michigan (Stokstad 2003).

12.4.9 Flood Control by the Three Gorges Dam in China

The construction of the Three Gorges Dam in China is an effort to provide a technological substitution for the ecosystem services of flood control while also producing electricity through hydropower. Flood control is important for the well-being of the millions of people, mostly rice farmers, who live on the floodplain of the Yangtze. Sedimentation from the Tibetan plateau has raised the height of the Yangtze channel to the point where it now sits several meters above its floodplain. Once the Three Gorges Dam is constructed, it is anticipated that large floods on the Yangtze will be controllable.

Construction of the dam will have other effects as well, however: Once the dam is full, levels of schistosomiasis near Chongqing, at the north end of the impoundment, are predicted to rise dramatically as a consequence of the decreased water speed. The capacity of the Yangtze to remove wastes, including industrial effluent and sewage, will also be significantly reduced. Water quality within the long, narrow impounded area is likely to decline. The reservoir that resulted from the construction of the Three Gorges Dam has necessitated the relocation of around 2 million people and caused flooding of numerous villages and historical monuments.

The decision to build the dam is in part a consequence of earlier decisions that encouraged people to settle in the wetland areas that would formerly have provided flood control services. Some of the ecosystem services provided by the Yangtze that will be lost, such as disease regulation, food production, and waste removal, have been assigned a relatively low priority compared with energy and flood control, which will be gained. Interestingly, the communities negatively affected by schistosomiasis (upstream) will be different from those that benefit from flood control (downstream).

As shown in this case, it is not uncommon that management of ecosystem services may result in an inequitable distribution of the benefits and costs of management actions. This example also shows that when a management decision is focused on a small subset of ecosystem services (flood control and electricity production, in this case), the impact of the decision on interrelated secondary services may be largely ignored. This is an example of a Type H trade-off: irreversible, large-scale, and long-term.

12.4.10 Dryland Salinization in Australia

Dryland salinization has been a major issue facing farmers in Australia since the 1930s. It was not until the late 1980s and early 1990s, however, that the problem moved from being individual to collective (Anderies et al. 2001; Greiner and Cacho 2001; Briggs and Taws 2003). To increase agricultural production (a provisioning service), many farmers cleared the original woody vegetation and replaced it with pastures and crops (Schofield 1992; Farrington and Salama 1996). The natural tree landscape of Australia had provided an important but undervalued regulating service by maintaining the groundwater at low enough levels that salts were not carried upwards through the soil. Once the woody vegetation was removed, the groundwater table moved toward the surface, bringing salt into the surface soils. As the salt content in soils increases, lands become unusable for traditional agriculture (Anderies et al. 2001; Greiner and Cacho 2001; Briggs and Taws 2003).

Dryland salinization motivated the development of the Hunter river salinity trading scheme (www.epa.nsw.gov.au/licensing/hrsts/index.html) and a political push to move toward salt-trading schemes that start with the development of salinity targets (www.mdbc.gov.au/naturalresources/salinity/factsheets/fsa1002_101.html). Ecological restoration efforts include planting trees in plots contiguous to fields to recover the ecosystem services provided by native vegetation (Schofield 1992; Farrington and Salama 1996). The total amount of land available for grazing decreases since trees take up some of the space, but tree plots help maintain the water table low enough to avoid salinization (Anderies et al. 2001; Briggs and Taws 2003). In areas of the Murray River catchment, the establishment of salt quota allocation systems is also necessary to assure that salt levels in the drinking water supply for Adelaide remain low (Anderies et al. 2001). Dryland salinization therefore has both local and distant effects, illustrating the spatial segregation of trade-offs among ecosystem services.

12.5 Characteristics of Trade-offs in the Scenarios and Case Studies

One way to look at the implications of policy-makers' actions on the delivery of ecosystem services is to ask which ecosystem service is traded off (explicitly or implicitly) when another service is selected as a target of a policy prescription. Though trade-offs may lead to the "sacrifice" of one service for another, this is not always so. In some cases, non-target ecosystem services may be enhanced, leading to a synergistic increase in the services provided. Analysis of results from the case studies allow for the identification of policies that have led to "win-win," "win-lose," and "lose-lose" situations, according to whether the policy recommendation resulted in a positive response in both the targeted and other ecosystem services, a negative and a positive response, or two negative responses. (See Table 12.1.)

We find examples of all three types of trade-off. Two cases stand out as clear "win-win" situations. Lobster fishing in the northeastern United States and no-take areas and fishery production in Saint Lucia show how short-term losses in catch due to the policies implemented led to long-term increases in production. Human well-being and fishery production both increased by enlightened management. In contrast, the remediation by natural wetlands on the Kafue River in Zambia is a candidate for a "win-lose" case: the quality of highland wetlands was "sacrificed" by mining effluents, though the wetlands still continue to provide this regulating ecosystem services. (It remains unclear if they can maintain the service at the same level or do it in perpetuity.) In addition, society can benefit from the income generated from mining, and the quality of water is maintained.

Identification of common characteristics found among trade-off decisions will allow policy-makers to develop better-informed decisions about the choices that they face. Understanding typical trade-off patterns associated with

Table 12.1. Types of Ecosystem Service Trade-offs in Case Studies. The plus and minus signs next to the numbers for each case study indicate positive and negative impacts, respectively, over the ecosystem service or services traded off. Two plus or minus signs indicate more than one service traded off in that category. Two signs separated by a slash indicate short-term/long-term differences in the trade-off or spatially segregated costs and benefits. Key to the case studies (see section 12.4 for their full names): 1: vulture declines in India, 2: value of lakeside property in the United States, 3: fisheries and tourism in the Caribbean, 4: fertilizer use in the United States, 5: remediation by natural wetlands on the Kafue River in Zambia, 6: no-take areas and fishery production in Saint Lucia, 7: lobster fishing in the northeastern United States, 8: Great Lakes of the United States, 9:Three Gorges Dam in China, and 10: dryland salinization in Australia.

Ecosystem Service Traded off	Ecosystem Service Targeted			
	Provisioning	Regulating	Cultural	Supporting
Provisioning	6(−/+), 7(−/+)	8(− −), 9(+)	2(−)	
Regulating	1(− −), 3(−), 5(−/+), 10(−)	8(− −), 9(− −)	2(−)	
Cultural	7(+)	1(−)	2(−)	
Supporting	4(−)	8(−)	2(−)	4(− −)

ecosystem management decisions may help managers comprehend the implications of their choices, even when they cannot predict the secondary services that will be affected. Common characteristics arise from analyses of the trade-offs found across all four scenarios and can be illustrated by examples drawn from real-world decision-making. In this section we summarize some of the main issues that must be considered when making decisions about ecosystem service trade-offs.

12.5.1 Unknown and Unanticipated Trade-offs

In all four scenarios and in our real-world case study examples, numerous trade-offs exist that are unknown and unanticipated. These may not manifest themselves until long after the initial decisions are made, even though they are already affecting the mix of ecosystem services provided. Illustrating such examples from within the scenario results themselves is difficult, because if the unknown and unanticipated trade-offs were known, they could be planned for. Instead the scenarios present many surprises (based on known interactions) that, in the real world, could be a result of unknown and unanticipated events. For example, in TechnoGarden, allergies to the pollen of genetically modified organisms develop, and massive exotic algal blooms occur as a result of failed water-supply manipulations. These surprises are a result of management trade-off decisions that result in unpredictable changes, forcing managers to make additional, unanticipated trade-off decisions.

While the previous example comes from the scenarios, the case studies also clearly show that unanticipated trade-offs are common and indicate that we can expect more unexpected trade-offs and synergies in the future. For example, vulture declines in India are remarkable in demonstrating how a change in the abundance of one species can have unexpected consequences over something as seemingly unconnected as the presence of rabies in dogs. Similarly, in the Great Lakes ecosystem, the efforts to increase waste removal and, later, to improve water quality in the waste canal led to a subsequent increase in non-native species, which has contributed to the long-term decline of biodiversity within the Great Lakes ecosystem.

Even the best models, classification schemes, or processes used to understand the trade-offs inherent in management decisions will not be able to anticipate all the effects of these decisions. Ultimately, there will always be some unanticipated effects of management decisions. Yet there are management techniques that can be used to mitigate the impact of unanticipated trade-offs. Management designed to maintain or improve resilience may help mitigate the impact of unanticipated effects, as seen in the Bonaire example. Resilience can be incorporated into ecosystems, for example, by creating redundant approaches to providing similar ecosystem services within each ecosystem. Development of a protected areas network that has multiple protected areas within a broader ecosystem would be one example of incorporating redundancy into ecosystem management plans. The use of adaptive management, or learning by doing, allows lessons learned from unanticipated effects to be applied to future decisions.

12.5.2 Choice of Ecosystem Service Trade-offs

Policy-makers are often forced to choose some ecosystem services over others. Across all four scenarios, trade-off decisions show preference for provisioning, regulating, cultural and supporting services, in that order. In all instances the increase in population growth, a major assumption of the scenarios, forces trade-offs that tend to favor provisioning and, to some extent, regulating ecosystem services. This is not surprising, as management choices tend to increase the supply of services that are perceived by society as more important—provisioning and regulating services—and thus do not fully value trade-offs of cultural and supporting services. In addition, supporting services are more likely to be "taken for granted." Because supporting and regulating services contribute to the ability of ecosystems to provide provisioning in the future, these decisions may be seriously undermining the future of provisioning ecosystem services and human well-being.

Real-world examples support the contention that managers must make trade-offs that explicitly or implicitly lead to preferences among ecosystem services. For example, the Three Gorges Dam in China is expected mainly to prevent floods (regulating ecosystem service) and will also positively affect electricity and food production (provisioning ecosystem services), but will negatively affect disease regulation and waste removal (regulating ecosystem services) and biodiversity. Perhaps the most telling example is that of the value of lakeshore property in the United States: developments targeting the cultural ecosystem services provided by owning a home near the water create negative impacts on other provisioning, regulating, cultural, and supporting ecosystem services, which in turn undercut the cultural service that they initially sought to optimize.

The recognition that managers rank ecosystem services in specific sequences allows a better understanding of how trade-off choices are made. Managers can then acknowledge that their decisions have ramifications on the supply of other ecosystem services and provide support for examining all aspects of each trade-off decision.

12.5.3 Slowly Changing Factors

The slowly changing factors that underlie supporting and regulating services are often ignored by policy-makers and not actively pursued by policy processes. Because supporting services often depend on slowly changing factors such as soil fertility, groundwater levels, or soil formation, they may not generally be perceived to be responsive to policy intervention. Slowly changing factors are rarely quantified and may be difficult to monitor. However, as discussed in Chapter 3, it is often these slowly changing variables that lead to unanticipated changes in ecosystem services.

In many instances, society chooses to trade off supporting or regulating services in favor of short-term provisioning ecosystem services. The case study examples about fertilizer use in the United States and mine effluent in Zambia illustrate this type of trade-off. Inattentiveness to supporting and regulating services can lead to a loss of resilience, leaving socioecological systems more vulnerable to surprises in delivery of provisioning services. Surprise, which is often linked to the misunderstanding or non-identification of the slow variables that regulate ecosystem services, is a common part of ecosystem management (Gunderson and Holling 2002).

Across all scenarios, such surprises or unexpected consequences of ecosystem management lead to a litany of additional trade-offs that society must make to ensure maintenance of ecosystem services. That is, short-term choices for human well-being can be derailed by surprises, ultimately leading to negative impacts on long-term human well-being. Addressing the negative impacts after the fact may be more costly than effectively managing the slowly changing variables to avoid problems in the first place. People in the Global Orchestration scenario focus on short-term availability of provisioning services and generally ignore slowly changing variables, with the idea that they will be able to address the impacts of the trade-off on other services after people have enough provisioning services. In this sense, the scenarios indicate the importance of recognizing the existence of delays. Many ecosystem problems only become apparent after a long time period. The long-term implications of decisions means that in many cases management regimes are only put in place after meaningful change can happen.

The results of trade-off decisions in the scenarios and case studies can be used to help understand the implication of slowly changing factors. Recognizing the importance of slowly changing factors and their effects on the long-term delivery of ecosystem services will help us develop more successful management plans. For example, land use plans in agricultural areas that recognize that high fertilizer use will ultimately result in lower water quality will be more successful in the long-term provisioning of clean fresh water than plans that do not. Such management plans might limit the impacts of fertilizer through reduced use, development of buffers, or other technology to assure water quality in the future.

12.5.4 Temporal Trade-offs

Managers must clearly identify trade-offs to allow policy-makers to understand the long-term effects of preferring one ecosystem service over another. Many decisions are made to maintain provisioning services in the present, often at the expense of provisioning services in the future. The decision to provision now versus provision later is especially pervasive in the Order from Strength scenario. Long-term decision planning is very hard to do, because many managers are rewarded for short-term success. Achieving short-term success may mean forgoing opportunity for future rewards. However, long-term rewards are characteristic of some real-world examples, such as no-take zones in St. Lucia and lobster fishing in Maine. In these two fishery examples, a short-term loss caused by the implementation of fishing restrictions was compensated by a long-term increase in production as stocks recovered. Limitation of "free access" to these resources was also fundamental.

Formal acknowledgement that trade-off decisions operate across time will help managers and policy-makers understand the importance of thinking about ecosystem services beyond the immediate need. Development of management regimes for protection of ecosystem services will have to incorporate an understanding of the time scales at which each trade-off occurs (at least the known ones) and ways to assure there is balance between short- and long-term needs from ecosystem services. Recognizing and planning beyond the traditional short-term time frames common in traditional resource management will help build potential for success stories like the St. Lucia case study. Management schemes that do not recognize the long-term effects of trade-off decisions will not be as successful as those that do. Incentives that cause a decline in future discount rates and thus increase the willingness of people to invest for the long term will give managers tools that help mitigate the effect of short- versus long-term trade-off effects.

12.5.5 Spatial Trade-offs

Trade-offs are also often made spatially. Management decisions can have impacts in areas far removed from where the initial trade-off decision occurs. This is especially relevant for the trade-off decisions that are made within the Order from Strength scenario. Decisions made in that scenario rarely take into account the possible implications outside political borders. Lack of accounting for spatial considerations when making trade-off decisions within the Order from Strength scenario creates further pressure on resources in regions where resources are scarce. The Global Orchestration scenario, in contrast, has mechanisms for coping with trade-offs, which allow accounting for decisions outside traditional trade-off boundaries. In many instances this means that there can be more equitable resource distribution cross political borders. In contrast to decisions made about temporal resources, many policy-makers facing ecosystem service trade-off decisions do not account for the spatial effects of those decisions or for the kinds of landscape- and ecosystem-wide effects that are discussed in Chapter 3.

Case studies also portray the dilemmas associated with decision-making at multiple spatial scales. For example, consider the case of dryland salinization in Australia. Each farmer, caring only for his land, removed woody vegetation in order to have more space for crops and pasture. Unfortunately, the actions of many individual farmers added up to the serious ecological problem of dryland salinization. Ecological restoration efforts focused on planting trees affect the water table relatively quickly at the local level, but the establishment of a successful salt-allocation system for an entire watershed, as is needed to assure water quality for the city of Adelaide, is much harder to implement. Similarly, excessive nutrient use on farms in the Mississippi River watershed, which increases food production, is having a negative impact on ecosystem services far downstream in the Gulf of Mexico.

Many managers recognize the need to consider the effects of trade-off decisions outside of traditional geopolitical boundaries. However, there are few incentives for managers to make decisions for the greater good at the cost of local or small-scale well-being. The dilemma faced by policy-makers is that successful management of ecosystem service tends to occur at fairly small spatial scales, while trade-offs that occur at larger scales ultimately affect even the smallest-scale ecosystem. Incentives that encourage policy-makers to bring expert experience of small-scale "win-win" solutions to large-scale problems may help policy-makers think broadly about decisions. Further, development of models that allow small-scale systems to be applied to large-scale problems will ensure that these experiences can be used for the greater good.

12.6 Conclusions

Trade-offs are a matter of societal choice. The lessons gained from scenarios and the examination of case studies, including the explicit recognition of trade-offs and their importance for the long-term sustainability of ecosystem services, will help policy-makers to gain a better understanding of the choices that they face and their consequences. In this section, we summarize some of the major implications of the material in this chapter.

12.6.1 Cautions about Quantitative Models

We need to be cautious about using quantitative models, including the ones in the MA, because these rarely represent trade-offs with accuracy. As in any modeling exercise, the qualitative and quantitative results from the scenarios are built on a series of assumptions. For example, there are assumptions regarding fertility, mortality, and migration of humans and the qualitative and quantitative aspects of economic growth. (See Chapter 9.) These assumptions are designed to match the scenario storylines and drive the results of the modeling. The models, in turn, are able to project outcomes over a relatively small array of ecosystem services (see Chapter 9), allowing us to develop an expectation of the conditions of the world under the different scenarios, using a sample of the services provided by ecosystems to humanity. At this point, one could ask: Are trade-offs adequately described by the storylines or the models? Are there any important trade-offs left out? What does our collection of case studies tell us about the importance of trade-offs that are missed by the storylines or the models?

Though the answers to these questions might appear obvious, as nobody doubts that models are only a simplified version of reality, their consequences are very important. Let us consider initially the last two questions and return to the first one later. A quick glance at Table 9.1 in Chapter 9 highlights the biases in the quantitative scenario analysis: there is a strong dominance of provisioning and regulating ecosystem services. Interestingly, this lines up perfectly with our case studies: provisioning and regulating ecosystem services are targeted more often than other ecosystem services. Certainly not by accident, the models focus on the ecosystem services that appear to be perceived by society as more important (driving research agendas and funding) and give less attention to cultural and supporting services.

There are two consequences of this bias. First, cultural and supporting services are essentially left out of the quantitative modeling exercise altogether. We know from the scenario assumptions that Order from Strength and TechnoGarden are more likely to trade off cultural ecosystem services over others, but since we are unable to compare these services in the scenarios quantitatively, the relative gain or loss of cultural ecosystem services cannot be formally expressed. The challenge is probably greater for supporting ecosystem services, as "those are necessary for the production of all other ecosystem services" (MA 2003). The fact that existing models do not consider these services is a limitation. Not considering supporting services almost guarantees that we will face surprises and sudden shifts in provisioning services in the future. Since we know that addressing these after they are a problem is generally more costly and time-consuming, we are setting ourselves up for more expensive future management by ignoring supporting services.

Second, and perhaps more important, a clear message emerging from the case studies is that ecosystem services interact and that targeting one can affect many others. The fact that models are able to explore only a small subset of ecosystem services (even within provisioning and regulating services) means that a smaller set of potential trade-offs can be quantified. Thus even if the models were able to perfectly characterize all the trade-offs among the ecosystem services that they consider, they would still underestimate the consequences of any societal choice, as many other trade-offs would remain unquantified.

Given the complexity of interactions among ecosystem services and the limited set of services directly quantified (see Chapter 9), it is a great achievement of the scenario development and modeling teams that some trade-offs were successfully visualized. But again, if model results are a simplification of reality, the simplification of the trade-offs is certain to be greater. Model results, at best, represent a crude lower bound of the expected consequences of any specific scenario. Reality will certainly be characterized by many other unforeseen changes in ecosystem services. Models offer us a means for contrasting societal choices, but history shows us that these choices can lead to far more severe consequences than models could ever predict (Ehrlich and Mooney 1983).

12.6.2 Dilemmas in Ecosystem Service Decisions: Complex Interactions of Ecosystem Services and Human Societies

Making choices about the management of ecosystem services is a prevalent feature of all human societies. In many cases, these choices have directly affected the delivery of non-target services (either positively or negatively).

Synergisms occur when ecosystem services interact with each other in a multiplicative or exponential fashion. For example, invasions of exotic plants are promoted by human disturbance (Crawley 1987). In the case of the human modification of the U.S. Laurentian Great Lakes, disturbance of the aquatic landscape facilitated the invasion of zebra mussels both by changing the composition of the local biota and physically providing a route for their spread. But not all interactions between ecosystem services need to be negative. The conservation of fishes in Bonaire not only maintained the interest of tourists, it also protected the fishery for the future and made the reef resistant to the loss of the regulatory function provided by black-spined sea urchins. Achieving successful synergistic interactions remains a major challenge in the management of ecosystem services because the strength and direction of such interactions remains virtually unknown (Sala et al. 2000).

Trade-offs may arise without premeditation: regulation of lobster fishing in the northeastern United States was motivated by a need to increase the supply of lobsters, a provisioning service, and turned out to also enhance the cultural services related to strengthening the social fabric and community organization of fishing cooperatives. Mining along the Kafue River shows how Zambians have traded off the quality of upstream wetlands while retaining the properties of drinking water and food (provisioning services) provided by the lower portions of the watershed.

As the human domination of Earth increases in extent and intensity, three important dilemmas arise:

- To what degree can human-created services substitute for ecosystem services?
- What degree of ecological complexity is needed to provide reliable ecosystem services?
- Are there limits to successfully engineering ecosystems, and what are they?

Understanding these dilemmas may help improve our decisions about trade-offs and management of complex socioecological systems.

12.6.2.1 Ecosystem Services and Human Services

Many people believe that the products of ecosystems, ranging from clear water to the beauty of a tiger, cannot be substituted for by other services. Numerous studies, however, implicitly or explicitly assume that the products of human ingenuity can provide good or at least satisfactory replacements for most ecosystem services. The degree to which ecological services can be replaced by technologically generated alternatives is very uncertain. Replaceability depends upon what services people want to replace, what technologies are available, and what other ecosystem services are (intentionally or accidentally) traded off by the technological replacement. Future technologies may allow feats that are impossible or prohibitively expensive today. On the other hand, formerly unknown or unimportant ecosystem services may be discovered to be fundamental to people or the maintenance of other ecological services.

An example of this type of dilemma is provided by water management. Humans have always altered rivers to regulate water levels. While these interventions were often successful, changes in rivers and their floodplains decreased their ability to provide regulating and supporting services, resulting in water contamination and floods. People have begun to realize that it may be less costly to enhance flood control and water quality ecosystem services via ecosystem protection rather than construct artificial water control and purification systems. In the United States, for example, New York City manages watersheds in the Catskills in ways that improve the quality of New York's drinking water. This model of ecological management improved the quality of the drinking water at a far lower cost than building a water treatment plant (Chichilnisky and Heal 1998; Heal 2000). Both forest habitat and water quality are enhanced.

This dilemma is illustrated primarily in the TechnoGarden scenario, in which societies favor the use of technology to enhance direct provision of services by ecosystems. This type of trade-off is also common in Global Orchestration, in which societies believe that, when needed, human ingenuity will find acceptable replacements for ecosystem services.

12.6.2.2 How Much Ecological Complexity Is Enough?

Humans are simplifying Earth's ecosystems, and the consequences of this simplification on the continued production of ecosystem services is uncertain. Some ecological work suggests that relatively few species performing different ecological functions can provide many ecosystem services (Tilman et al. 1996; Ewel and Bigelow 1996). However, other research indicates that while this may be true over small areas and short periods, the loss of species increases the variability of ecosystem services and increases ecosystems' vulnerability to disturbance (Peterson et al. 1998; McCann 2000).

If ecosystems could be simplified with minimal loss of ecosystem services, ecological simplification would be an ethical issue, peripheral to sustainable development. If ecosystem services are vulnerable to ecological simplification, however, maintaining and creating complex ecosystems should lie at the center of sustainable development efforts. Evidence to date suggests that complexity and redundancy are indeed fundamental to maintaining the supply of ecosystem services (Hobbs and Cramer 2003). The management dilemma that will arise is whether or not to create policies to encourage maintenance of ecological complexity. This may require discovering ways to balance short-term, local loss against long-term, regional gain. This question is, of course, related to the question of how much biodiversity (that is, landscapes, ecosystems, species, populations, and genes) is needed to effectively and sustainably produce desired ecosystem services. This dilemma is illustrated in Global Orchestration, in which ecosystems are simplified to produce immediate benefits to human well-being without regard to the future provision of ecosystem services.

12.6.2.3 To What Extent Can Ecosystems Be Engineered?

Ecological engineering offers potential for people to increase the quality and amount of ecological and human-produced services they use, while maintaining the ability of ecosystems to continue to produce ecological services (McDonough and Braungart 2002). But the goal of producing a "Garden Earth" requires that people reliably engineer ecosystems to produce desired services sustainably. Unfortunately, past ecological engineering efforts have frequently produced surprising consequences (Cohen and Tilman 1996; Holling 1986; Holling and Meffe 1996; Gunderson and Holling 2002), which suggests that we still lack the sophistication or necessary understanding to engineer ecosystems. This dilemma is illustrated in TechnoGarden, in which societies value understanding ecosystems and use that understanding to control and improve provision of ecosystem services.

There are a number of examples of land management by local people in both the old and new world that suggest that people can improve the productivity of ecosystems in a relatively sustainable way and in a fashion that does not eliminate the ability of surrounding ecosystems to provide nonagricultural ecological services. For example, research over the past decades in the Amazon found that approximately 10% of its land area is anthropogenically produced fertile soil that is more resilient to disturbance than non-anthropogenic soil (Glaser et al. 2001). Similarly, it has been suggested that pre-Columbian societies generated productive mixed aquaculture/agriculture systems in relatively unproductive parts of Bolivia (Erickson 2000). Another example of an integrated approach is the water temple system in Bali. A system of water temples was used to balance rice production, which is increased by the staggered availability of water to different fields, with the need to control rice pest populations, which increase rapidly with synchronized production across large areas (Lansing 1991). The ability of engineered ecosystems to produce a broad variety of ecosystem services rather than optimizing a single service remains largely untested, however.

12.6.3 Complex and Cascading Effects of Trade-offs

The case studies and the results from the scenarios demonstrate that trade-offs are complex and often have ramifications far beyond the decision that led to the trade-off itself. Trade-offs can affect service provision in places that are far away, they can affect other services nearby, and they can affect the future provision of ecosystem services. Whether they affect nearby services, faraway services, or future services, trade-offs are usually involve unanticipated effects on secondary services. Decisions frequently cascade through multiple ecosystem services following both known and unknown pathways. Unanticipated effects on secondary services and the multiple pathways trade-offs may take add to the complexity of ecosystem service management choices. This complexity can have serious implications for making trade-off decisions.

Lessons from the examples presented in this chapter suggest that managers can benefit by classifying their trade-off decisions, identifying the characteristics common to their decisions, and understanding the potential dilemmas that their decisions must address. Although it will be impossible to mitigate all the unknown and unanticipated effects of each decision, management schemes that focus on "win-win" outcomes and include sufficient redundancy within each plan will be more successful than other management schemes. Structured approaches to making decisions about ecosystem services, which take advantage of existing models and include an adaptive management approach, will have a higher likelihood of mitigating unintended consequences.

References

Acheson, J.M. and R.S. Steneck, 1997: Bust and then boom in the Maine lobster industry: perspectives of fishers and biologists. *North American Journal of Fisheries Management,* **17,** 826–847.

Acheson, J.M., J.A. Wilson, and R.S. Steneck, 1998: Managing chaotic fisheries. In: *Linking Social and Ecological Systems, Management Practices and Social Mechanisms for Building Resilience,* F. Berkes and C. Folke (eds.), Cambridge University Press, Cambridge, U.K., 390–413.

Anderies, J.M., G. Cumming, M. Janssen, L. Lebel, J. Norberg, G. Peterson, and B. Walker, 2001: A resilience centered approach for engaging stakeholders about regional sustainability: An example from the Goulburn Broken catchment in Southeastern Australia. *Technical Report, CSIRO Sustainable Ecosystems.*

Begon, M., J.H. Harper, and C.R. Townsend, 1996: *Ecology. Individuals, Populations and Communities, Third Edition.* Second ed. Blackwell Science, Cambridge, Massachusetts, 1068 pp.

Briggs, S.V. and N. Taws, 2003: Impacts of salinity on biodiversity-clear understanding or muddy confusion? *Australian Journal of Botany,* **51(6),** 609–617.

Chichilnisky, G. and G. Heal, 1998: Economic returns from the biosphere. *Nature,* **391,** 629–630.

Christensen, D.L., B.R. Herwig, D.E. Schindler, and S.R. Carpenter, 1996: Impact of lakeshore residential development on coarse woody debris in North temperate lakes. *Ecological Applications,* **6(4),** 1143–1149.

Cohen, J. E., and D. Tilman, 1996: Biosphere 2 and biodiversity: The lessons so far. *Science* **274,** 1150–1151.

Crawley, M.J., 1987: What makes a community invasible? In: *Colonization, Succession and Stability,* A.J. Gray, M.J. Crawley, and P.J. Edwards (eds.), Blackwell Scientific Publications, 429–453.

Ehrlich, P. R., and H. M. Mooney, 1983: Extinction, substitution and ecosystem services. *BioScience* **33,** 248–254.

Erickson, C.L., 2000: An artificial landscape-scale fishery in the Bolivian Amazon. *Nature,* **408,** 190–193.

Ewel, J.J. and S.W. Bigelow, 1996: Plant life-forms and tropical ecosystem functioning. In: *Biodiversity and Ecosystem Processes in Tropical Forests,* J.H. Cushman (ed.), Springer-Verlag, Heidelberg, Germany, 101–126.

FAO (Food and Agriculture Organization of the United Nations), 1996: The State of World Fisheries and Aquaculture (SOFIA). Fishery Department. Available at: http://www.fao.org/docrep/003/w3265e/w3265e00.htm, Rome, Italy.

Farrington, P. and R.B. Salama, 1996: Controlling Dryland Salinity by Planting Trees in the Best Hydrogeological Setting. *Land Degradation & Development,* **7,** 183–204.

Gell, F.R. and C.M. Roberts, 2003a: *The Fishery Effects of Marine Reserves and Fishery Closures.* World Wildlife Fund, Washington, D.C., USA, 90 pp.

Gell, F.R. and C.M. Roberts, 2003b: Benefits beyond boundaries: the fishery effects of marine reserves. *Trends in Ecology and Evolution,* **18(9),** 448–455.

Glaser, B., L. Haumaier, G. Guggenberger, and W. Zech, 2001: The 'Terra Preta' phenomenon: a model for sustainable agriculture in the humid tropics. *Naturwissenschaften,* **88,** 37–41.

Goreau, T.F., 1959: The ecology of Jamaican coral reefs .1. Species composition and zonation. *Ecology,* **40(1),** 67–90.

Greiner, R. and O. Cacho, 2001: On the efficient use of a catchment's land and water resources: dryland salinization in Australia. *Ecological Economics,* **38,** 441–458.

Gunderson, L. and C. Holling (eds.), 2002: *Panarchy: Understanding Transformations in Human and Natural Systems.* Island Press, Washington, D.C.

Heal, G., 2000: Valuing ecosystem services. *Ecosystems,* **3,** 24–30.

Hobbs, R. J., and V. A. Cramer, 2003: Natural ecosystems: Pattern and process in relation to local and landscape diversity in southwestern Australian woodlands. *Plant and Soil* **257,** 371–378.

Holling, C.S., 1986: The resilience of terrestrial ecosystems: local surprise and global change. In: *Sustainable Development of the Biosphere,* R.E. Munn (ed.), Cambridge University Press, Cambridge, 292–317.

Holling, C.S. and G.K. Meffe, 1996: Command and control and the pathology of natural resource management. *Conservation Biology,* **10(2),** 328–337.

Hughes, T.P., 1994: Catastrophes, phase-shifts, and large-scale degradation of a Caribbean coral reef. *Science,* **265(5178),** 1547–1551.

Jackson, J.B.C., M.X. Kirby, W.H. Berger, K.A. Bjorndal, L.W. Botsford, B.J. Bourque, R.H. Bradbury, R. Cooke, J. Erlandson, J.A. Estes, T.P. Hughes, S. Kidwell, C.B. Lange, H.S. Lenihan, J.M. Pandolfi, C.H. Peterson, R.S. Steneck, M.J. Tegner, and R.R. Warner, 2001: Historical overfishing and the recent collapse of coastal ecosystems. *Science,* **293(5530),** 629–638.

Lansing, J.S., 1991: *Priests and programmers: technologies of power in the engineered landscape of Bali.* Princeton University Press, Princeton, N.J.

Lessios, H.A., 1988: Mass mortality of *Diadema antillarum* in the Caribbean: What have we learned? *Annual Review of Ecology and Systematics,* **19,** 371–393.

Limpitlaw, D., 2002: *An Assessment of Mining Impacts on the Environment in the Zambian Copperbelt.* University of the Witwatersrand, PhD dissertation, Johannesburg, South Africa.

Lodge, D. M., 2001: Lakes. In: *Global Biodiversity in a Changing Environment: Scenarios for the 21st Century,* F. S. Chapin, III, O. E. Sala, and E. Huber-Sannwald, (eds.) Springer-Verlag, New York, pp. 277–313.

MA (Millennium Ecosystem Assessment), 2003: *Ecosystems and Human Well-being,* Island Press, Washington, D.C., U.S.A.

Malakoff, D., 1998: Death by suffocation in the Gulf of Mexico. *Science,* **281,** 190–192.

McCann, K.S., 2000: The diversity-stability debate. *Nature,* **405(6783),** 228–233.

McDonough, W. and M. Braungart, 2002: *Cradle to Cradle: Remaking the Way We Make Things.* North Point Press.

Munro, J.L., 1969: The sea fisheries of Jamaica: past, present and future. *Jamaica Journal,* **3,** 16–22.

Mutale, M. and A. Mondoka, 1996: *Water Resources Availability, Allocation and Management, and Future Plans for the Kafue River Basin.* Paper presented 9 May 1996, at the Kafue River Basin Study Seminar, Ministry of Environment and Natural Resources, Lusaka, Zambia.

Myers, R.A. and B. Worm, 2003: Rapid worldwide depletion of predatory fish communities. *Nature,* **423(6937),** 280–283.

Oaks, J. L., M. Gilbert, M. Z. Virani, R. T. Watson, C. U. Meteyer, B. Rideout, H. L. Shivaprasad, S. Ahmed, M. J. I. Chaudhry, M. Arshad, S. Mahmood, A. Ali, and A. A. Khan, 2004: Diclofenac residues as the cause of population decline of vultures in Pakistan. *Nature,* **427,** 630–633.

Peterson, G.D., T.D. Beard, Jr., B.E. Beisner, E.M. Bennett, S.R. Carpenter, G.D. Cumming, C.L. Dent, T.D. Havlicek, 2003: Assessing future ecosystem services: a case study of the Northern Highlands Lake District, Wisconsin. *Conservation Ecology,* **7(3),** 1. [online] URL: http://www.consecol.org/vol7/iss3/art1.

Peterson, G.D., C.R. Allen, and C.S. Holling, 1998: Ecological resilience, biodiversity and scale. *Ecosystems,* **1,** 6–18.

Pettersson, U.T. and J. Ingri, 2001: The geochemistry of Co and Cu in the Kafue River as it drains the Copperbelt mining area, Zambia. *Chemical Geology,* **177,** 399–414.

REEF (Reef Environmental Education Foundation), 2003: Survey Project Statistics. Cited 1 September 2003. Available at http://www.reef.org/.

Roberts, C.M., 2002: Deep impact: the rising toll of fishing in the deep sea. *Trends in Ecology and Evolution,* **17(5),** 242–245.

Roberts, C.M., J.A. Bohnsack, F. Gell, J.P. Hawkins, and R. Goodridge, 2001: Effects of Marine Reserves on Adjacent Fisheries. *Science,* **294(5548),** 1920–1923.

Sala, O.E., F.S. Chapin III, J.J. Armesto, E. Berlow, J. Bloomfield, R. Dirzo, E. Huber-Sanwald, L.F. Huenneke, R.B. Jackson, A. Kinzig, R. Leemans, D.M. Lodge, H.A. Mooney, M. Oesterheld, N.L. Poff, M.T. Sykes, B.H. Walker, M. Walker, and D. Wall, 2000: Global Biodiversity Scenarios for the Year 2100. *Science,* **287(5459),** 1770–1774.

Schindler, D.E., S.I. Geib, and M.R. Williams, 2000: Patterns of fish growth along a residential development gradient in north temperate lakes. *Ecosystems,* **3,** 229–237.

Schofield, N.J., 1992: Tree planting for dryland salinity control in Australia. *Agroforestry Systems,* **20,** 1–23.

Spaltro, F. and B. Provencher. 2001. An analysis of minimum frontage zoning to preserve lakefront amenities. *Land Economics,* **77,** 469–481.

Sterman, J. D., and L. B. Sweeney, 2002: Cloudy skies: Assessing public understanding of global warming. *System Dynamics Review* **18,** 207–240.

Stoeckel, J. A., D. W. Schneider, L. A. Soeken, K. D. Blodgett, and R. E. Sparks, 1997: Larval dynamics of a riverine metapopulation: implications for zebra mussel recruitment, dispersal, and control in a large-river system. *Journal of the North American Benthological Society,* **16,**586–601.

Stokstad, E., 2003: Can well-timed jolts keep out unwanted exotic fish? *Science,* **301,** 157–159.

Tilman, D., D. Wedin, and J. Knops, 1996: Productivity and sustainability influenced by biodiversity in grasslands ecosystems. *Nature,* **379,** 718–720.

UNDP (United Nations Development Programme), 2003: *Human Development Report 2003: Millennium Development Goals: A Compact Among Nations to End Human Poverty.* Oxford University Press, New York, USA, and Oxford, U.K., 368 pp.

von der Heyden, C.J. and M.G. New, In press a: Mine effluent remediation within a natural Copperbelt wetland: mass balance and process-based investigations. *Applied Geochemistry.*

von der Heyden, C.J. and M.G. New, In press b: Wetland sediment chemistry: a history of mine contaminant remediation and an assessment of processes and pollution potential. *Journal of Geochemical Exploration.*

von der Heyden, C.J. and M.G. New, In press c: Natural wetlands for mine effluent remediation: the case of the Copperbelt. *Water Resources Research.*

Woodley, J.D., E.A. Chornesky, P.A. Clifford, J.B.C. Jackson, L.S. Kaufman, N. Knowlton, J.C. Lang, M.P. Pearson, J.W. Porter, M.C. Rooney, K.W. Rylaarsdam, V.J. Tunnicliffe, C.M. Wahle, J.L. Wulff, A.S.G. Curtis, M.D. Dallmeyer, B.P. Jupp, M.A.R. Koehl, J. Neigel, and S.E. M, 1981: Hurricane Allen's Impact on Jamaican Coral Reefs. *Science,* **214(4522),** 749–755.

Chapter 13

Lessons Learned for Scenario Analysis

Coordinating Lead Authors: Nebojsa Nakićenović, Jacqueline McGlade, Shiming Ma
Lead Authors: Joe Alcamo, Elena Bennett, Wolfgang Cramer, John Robinson, Ferenc L. Toth, Monika Zurek
Review Editors: Rusong Wang, Antonio La Viña, Mohan Munasinghe

---

FIGURES

Main Messages

The Millennium Ecosystem Assessment scenarios build on earlier scenarios and modeling efforts and also extend scenario analysis to include ecosystem services and their consequences for human well-being, which has not been done before. The main objectives of the MA scenarios are to assess future changes in world ecosystems and resulting ecosystem services over the next 50 years and beyond, to assess the consequences of these changes for human well-being, and to inform decisions-makers at various scales about these potential developments and possible response strategies and policies to adapt to or mitigate these changes. These objectives are reflected in the overall approach and are an integral part of the four narrative and quantitative scenarios about alternative futures.

MA scenarios further refine and extend a number of recent methodological improvements in the scenario formulation process. These include integration across social, economic, environmental, and ecosystems dimensions; disaggregation across multiple scales of global patterns down to regional and in some cases also place-specific developments; multiple futures across four alternative scenarios to reflect deep uncertainties of long-range outcomes; and quantification of key variables linked to ecosystem conditions and ecosystem services along alternative narrative storylines. In addition, the MA scenarios make four important new contributions: They extend the integrated assessment approaches to include ecosystem services and their consequences on human well-being. They model explicitly changes in biodiversity as an integral part of scenario development. They assess interactions and trade-offs among ecosystem services. And they assess possible replacement of some ecosystem services by other services and the emergence of new ways of providing these services, such as through technological change.

MA scenarios establish another important precedent in adopting the long time horizon of 50 years, and for some variables, a century. Many salient but slow trends in ecosystems will only become visible over this long time period, but decisions influencing these trends have already been taken or will be taken in the immediate future. Integrated scenarios, which portray human activities together with ecosystem dynamics, are the main tool available for the assessment of alternative futures and possible response strategies. Increasingly, long time horizons and global perspectives are required to understand complex interactions between human and ecological systems.

The MA scenarios provide rich and useful images of broad patterns of possible futures at the global scale and at the level of major world regions. However, the models are not able to perform detailed analyses of local processes and impacts. One possible remedy and a crucial improvement for similar future assessments in terms of usability for several stakeholders would be to "soft-link" sector- and region-specific models by using the global scenario framework and outputs of global models to drive them. A particularly useful feature of the present effort is that scenarios provide the information about the socioeconomic and technological development patterns that is necessary for the assessment of the viability and effectiveness of various instruments and response strategies currently available or that might become available in the future to different stakeholder groups.

A key goal of the MA scenarios is to help decision-makers gain a better understanding of the intended and unintended effects of various policy measures for maintaining ecosystem services and human well-being simultaneously. Human activities have become an important co-determinant of Earth systems, and decisions made now and in the immediate future will have consequences across both temporal and spatial scales. Alternative futures described in the MA scenarios are subject to human choices, both those already made and those to be made in the future. Multiple dynamics of change characterize human activities and ecological systems. Bringing these interactions to the foreground is one of the main goals of the MA scenarios.

While basic human conditions generally improve across three scenarios and decline for some people in one scenario, they all portray, to a varying extent, perilous paths of ecosystems change. This illustrates complex linkages and feedbacks of ecosystem service changes on human well-being. While material elements of human well-being generally increase, loss of ecosystem services leads to higher inequalities in some of the scenarios and even degradation of some aspects of human well-being in others. Degradation and loss of some ecosystem services also affects the trade-offs and relations between provisioning and regulating functions of ecosystems. This is one central and important development pattern shared by all four scenarios.

The paths of the four scenarios are fundamentally different. Each scenario includes inherent path-dependencies and irreversibilities that are the results of the long response times and cumulative nature of many changes that affect ecosystem services and human well-being. The scenarios do not converge, though the possibility exists that decisions could be made in one scenario to make it evolve (or branch out) into an alternative future resembling one of the other MA scenarios. They differ with respect to many of the drivers of global change identified by the MA. In particular, the scenarios differ with respect to whether the world becomes more or less interconnected and whether environmental management is reactive or proactive.

Land use changes are perhaps the most critical aspect of anthropogenic global change in influencing the future of ecosystems and their services. Nevertheless, indirect effects on future ecosystem services, which can potentially result from other global changes (climate change, biodiversity loss), will also be of importance and superimposed on effects of land use changes. In the four MA scenarios, land use changes will directly determine many of the provisioning and regulating functions of ecosystems. These will depend on the future changes in biodiversity, desertification, or wetlands—all a function of land use changes. Land use patterns are in turn directly dependent on some of the main scenario driving forces, such as demographic, economic, and technological changes.

The models used to quantify the MA scenarios were unable to explore or elaborate on the evolutionary path-dependencies among anthropogenic system and ecosystem development, possible emergence of thresholds, and the specific dynamics caused by bifurcations. The current quantitative methodological approaches are not well suited for assessment of cross-scale phenomena such as place-specific developments in relation to regional and global ones. Extending the scenario development to more than one set of integrated models might better encompass some of the deep uncertainties associated with alternative futures and the resulting ecosystem services. Since the qualitative development of the MA scenarios was able to address path-dependencies, thresholds, and bifurcation dynamics, the storylines should be consulted for additional depth and richness about path-dependencies, thresholds, and cross-scale feedback.

Important lessons learned in the development of the MA scenarios could help improve the development of global storylines in any future assessments:

- Development of regional and more place-specific scenarios would help inform and create better global scenarios. Regional scenarios can use more accurate local information to develop scenarios and might represent system dynamics more accurately. They can also pinpoint specific variables of interest.

- Better communication and interaction with policy-makers would help inform the development of the storylines by indicating the key variables that are of interest to decision-makers. This would be useful both for understanding which are the most pressing questions of policy-makers as well as for communicating results to policy-makers at the end of the process. Having more policy-makers and decision-makers within the working group may be one way to improve this communication.

- Improved communication and interaction across scientific disciplines would improve future scenarios. Differences among disciplines' core beliefs about how the world functions were also often the critical issues that policy-makers wanted to have addressed in the scenarios. Better interdisciplinary communication prior to initiation of future assessments might make an exceedingly complex process a bit easier.

13.1 Introduction

The main objectives of the MA scenarios are to assess changes in ecosystems and their services over the next 50 years and beyond, to assess the consequences of these changes for human well-being, and to inform diverse decisions-makers about these potential developments and how they can affect them through response strategies and policies. These objectives, as stated in Chapter 5, are reflected in the overall approach used to develop the new MA scenarios. The objectives are an integral part of the basic assumptions about the future of the main driving forces of the four scenarios.

The scenarios do not attempt to describe all possible futures that can be imagined. The MA scenario paths were developed to provide plausible answers to the major uncertainties and focal questions about the future of socioecological systems. (See Chapter 5.) In addition, the scenarios systematically follow through a number of assumptions and management approaches currently discussed by decision-makers around the world. The quantification of changes in ecosystem services across the four scenarios are based on an integrated modeling framework, which uses an existing set of models, but with new linkages designed explicitly to reflect the main objectives of the MA scenarios. A brief summary of the four scenarios and their main characteristics is given in the SDM. Chapters 8, 9, 10, and 11 synthesize a wide range of social, economic, environmental, and policy implications of changes in ecosystem services on human well-being, and include an assessment of possible policy responses to these changes.

The MA scenarios have been conceived and developed to provide insights into a broad range of potential future ecosystem changes. The objective was to portray plausible developments that are internally consistent rather than those that may be considered to be desirable or undesirable. The idea of what is "negative" or "positive" in any given scenario is inherently dependent on the eye of the beholder and thus highly subjective. Therefore, great attention was given in previous chapters to presenting both positive and negative aspects in the scenarios. Uniting only "positive" or "negative" features in a scenario would result in homogeneous and "unidimensional" futures that may not be plausible and consistent.

The overall objectives, scenario formulations, and development of the MA scenarios are new. However, our approach builds on previous scenarios and modeling efforts in the literature. The new and innovative element is the focus on ecosystem services and human well-being. The MA scenarios distinguish between provisioning, regulating, supporting, and cultural ecosystem services (MA 2003). Access to ecosystem services is one of several factors affecting human well-being, which is considered along a multidimensional continuum—from extreme deprivation or poverty to a high attainment of experience of well-being—and has five major components: material well-being, health, good social relations, security, and freedom and choice (MA 2003, Chapter 3). Human well-being is context-dependent, reflecting factors such as age, culture, geography, and ecology.

Many insights were gained in the process of developing scenarios that focus on ecosystem services and human well-being. In the models used to quantify the MA scenarios, land use changes were perhaps the most critical aspect of anthropogenic global change in influencing the future of ecosystems and their services. As a first approximation, the results of MA scenarios modeling exercises suggest that land use changes directly determine many of the provisioning and regulating functions of ecosystems. These will depend on future changes in biodiversity, desertification, or wetlands, all a function of land use changes. Land use patterns are in turn directly dependent on some of the main scenario driving forces such as demographic, economic, and technological change. Another important finding across all scenarios is that indirect effects on future ecosystem services that might result from other global changes, such as those of climate, will be of secondary importance compared with land use changes. In the scenarios, the global changes generally tend to amplify effects, especially the adverse consequences of changes in ecosystem services.

The MA scenarios also have important weaknesses that point to potential areas for future improvements. Many needed improvements relate to our ability to quantify the future of ecosystem services. These include improving capabilities for modeling ecosystem services and human well-being, especially for supporting and cultural ecosystem services; consideration of the consequences of a larger range of driving forces; and more-explicit treatment of the deep uncertainties associated with alternative futures, especially those related to quantifying ecosystem services.

Perhaps the most important deficiency in the new scenarios is that the current models are not well suited for the assessment of cross-scale phenomena such as place-specific developments in relation to regional and global ones. This means that the models are not able to sufficiently explore the evolutionary path-dependencies among anthropogenic systems and ecosystem developments or the possible emergence of resilience or irreversibilities. However, many of the cross-scale feedbacks and many other scenario characteristics that could not be captured by models were treated in greater detail in the narrative storylines. Extending scenario development to more than one set of integrated models might be another way to better encompass some of the deep uncertainties associated with alternative futures and the resulting ecosystem services.

13.2 New Contributions of the MA Scenarios

The MA scenarios make a number of new contributions to the method of scenario analysis for exploring (inherently unknown) alternative futures, to the process of how scenarios are developed, and to the inclusion of an integrated assessment of ecosystem services and human well-being.

13.2.1 An Analytical Typology of the MA Scenarios

The MA was developed in the context of major advances in the methodology of scenario analysis. Figure 13.1 shows a typology for assessment based on the distinction made by Rayner and Malone between descriptive social science research, based on an analysis of mostly quantitative energy and material flows, and interpretive social science, focused on the values, meaning, and motivations of human agents (Rayner and Malone 1988; see also Robinson and Timmerman 1993). The figure further distinguishes between more global and more local analysis and attempts to indicate typical forms of analysis that correspond to the four quadrants identified. The distinctions among the quadrants shown in Figure 13.1 underlie many of the problems of interdisciplinary communication and analysis in the sciences. It is well known that it is difficult to combine, for example, interpretive place-based analysis of human motivations with, say, a quantitative analysis of energy systems and emissions. Much of the early work in the climate field, whether global or local, was located on the descriptive side of the typology.

It is particularly noteworthy therefore, that recent developments in scenario analysis are beginning to bridge this difficult gap (Morita and Robinson 2001; Swart et al. 2004; see also Chapter 2). Over the past decade, the global scenario analysis community has begun to combine the primarily qualitative and narrative-based scenario analyses undertaken by Royal Dutch/Shell and other companies (Wack 1985a, 1985b; Schwartz 1992) with global modeling work in the form of analyses that bring together the development of detailed narrative storylines with their "quantification" in various global models (Raskin et al. 1998; Nakićenović et al. 2000). For example, the Special Report on Emissions Scenarios work (Nakićenović et al. 2000), undertaken for the Intergovernmental Panel on Climate Change, cut across the interpretive/descriptive divide, though it still focused mainly on the global and regional level.

As shown in Figure 13.2, the MA scenarios work also cuts across the divide between interpretive and descriptive research by combining narrative storylines and quantitative modeling. However, it also begins to reach across the global/local gap, with a stronger focus on local analysis of ecosystem effects. This was mainly accomplished by incorporating information from a few of the sub-global assessments in the global scenario effort and vice versa. Also, a few methodological steps were explored to link or nest the development of the local, regional, and global scenarios. Linking and nesting different scale scenario exercises is a field that needs further exploration in the future. In this way, the MA work contributes to the trend toward more integrated and more interdisciplinary work on the relationships among human and natural systems.

Figure 13.2 demonstrates the place of the MA analysis along two axes describing the geographical scale of work and the degree to which the scenarios are based on interpretive, qualitative storylines or grounded in model-based descriptions. The MA scenarios combine the storyline approach with a modeling exercise. (See Chapter 6.) The four scenarios presented in this volume are primarily global scenarios, but the MA did go one step in the direction of developing multiscale scenarios (MA *Multiscale Assessments*, Chapter 10).

13.2.2 Contributions to the Process of Scenario Development

The MA further refined and extended a number of recent methodological improvements in the scenario formulation process. They include integration across social, economic, environmental, and ecosystems dimensions; disaggregation across multiple scales of global patterns down to regional and in some cases also place-specific developments; multiple futures across four alternative scenarios to reflect deep uncertainties of long-range outcomes; and quantification of key variables linked to ecosystem conditions and ecosystem services along alternative narrative storylines (See Chapter 2).

	Local	Global
Interpretive	place-based case studies	global storylines
Descriptive	regional science	global modeling

Figure 13.1. Analytical Typology of Scenarios Analysis. This figure illustrates local and global scenarios exercises and exercises that are more based on interpretive, qualitative, or descriptive modeling-based approaches.

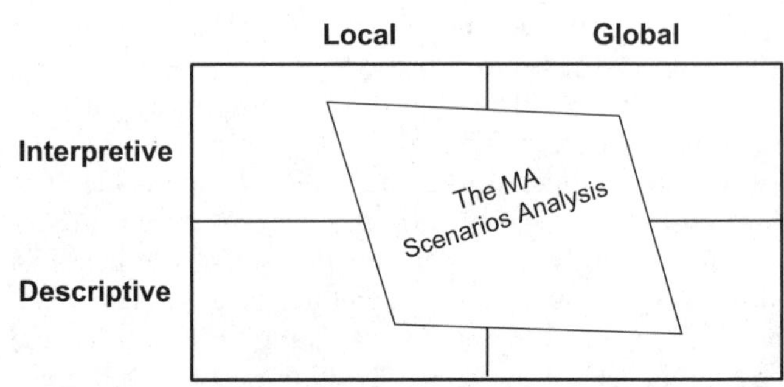

Figure 13.2. Placing of MA Scenarios Analysis in Analytical Typology. For a description of the typology see Figure 13.1.

Integration across dimensions is needed because multiple anthropogenic stresses have an impact on the environment via direct driving forces such as pollution, climate and hydrological change, resource extraction, and land use changes, and they play an important role in co-determining future evolution of ecosystems. These changes in direct drivers result from long causal chains of indirect drivers such as population, economic, and technological patterns, that are, in turn, conditioned by such ultimate drivers as human values, culture, interest, power, and institutions. To capture this nexus of interactions and emergent systems properties, a systemic framework is required that includes the key driving forces, possible consequences of their unfolding or collapse, and feedbacks. (See Chapter 1.) MA scenarios achieve this by introducing these integrative elements:

- formulation of extensive narrative stories, or storylines, of four alternative sets of main driving forces to 2050 and beyond to provide a context for unfolding ultimate drivers and impacts on ecosystem services and human well-being;
- quantification of the narrative storylines by extended integrated assessment models that include quantitative assessment of biodiversity, water, and fisheries; and
- qualitative assessment of possible trade-offs among ecosystem services and substitutions by other services, the implications for human well-being, and possible policy responses.

Many of these integrative elements are present in other scenarios in the literature, but the emphasis on and integration of ecosystem services and human well-being is new.

Disaggregation across multiple scales is essential because multiple interactions between anthropogenic and ecological systems occur within and across scales from global and regional to local and place-specific levels. These different spatial scales are often associated with characteristically different temporal scales and provide mutually enhancing perspectives into possible futures. Many of the relevant global processes operate over very long time periods ranging from decades to centuries, and their impacts on human well-being are usually indirect and complex but fundamental, in some cases even threatening human existence itself. In contrast, place-specific and local processes are usually much more direct in the ways that they affect human well-being, such as health, air and water quality, or nutrition.

Scientific capabilities to model links across temporal and spatial scales are very modest. Often such attempts are reduced to disaggregating global development into a (small) number of global regions. Narrative storylines are richer in the sense that they can provide apparently seamless connections across multitudes of scales, but compared with numerical and analytical models they are not quantitative and do not provide reproducibility under varying assumptions about main driving forces. In Chapter 2, the hope was expressed that perhaps future scenario-building techniques and models would evolve to allow seamless views across scales and levels of analysis, representing each spatial unit as an interacting component of an integrated global system. The MA scenarios provide a first step in this direction by providing, for example, quantitative links between different scales of land use patterns, climate change, and biodiversity. Another method of bridging across scales is more anecdotal and is provided in Chapter 10, where some place-specific, scenario-dependent developments of marine biodiversity are briefly described for places such as the Gulf of Thailand and the North Pacific.

Multiple futures are fundamental to any scenario enterprise, because prediction of complex and evolving systems is not possible. They are required for indicating the range of plausible futures and for encompassing some of the deep uncertainties associated with the evolution of complex systems. Examples of deep uncertainties are nonlinear responses of complex systems, emerging properties and path-dependencies, and generally unpredictable behavior that emerges due to branching points, bifurcations, and complex temporal and spatial dynamics. Complex systems are inherently unpredictable, especially when human response strategies that have yet to be defined are involved.

The overall time horizon of the four MA scenarios is to 2050 and beyond. Such long time horizons are required to encompass fundamental changes in anthropogenic and ecological systems and their interactions. It is likely that they will unfold in unexpected ways and will embody important surprises. Such surprises could include unexpected emergent properties, path-dependencies, and the crossing of critical thresholds, leading to irreversibilities. Given the modest modeling techniques available today, development of a rich set of alternative scenarios is the main method used to encompass these different possibilities and the associated uncertainties. This approach is also followed in the four MA scenarios. In addition to the quantitative formulation of many of the alternative scenario characteristics with an extended integrated assessment modeling framework, the MA scenarios also have elaborate narratives that extend across a multitude of levels and scales. They provide the background information about the main driving forces, the associated fundamental drivers, and their consequences. In this way, they link various analytical and numerical methods that are not (yet) an integral part of the IAM used to quantify the four scenarios.

The long time horizon for the scenarios of 2050 and beyond extend the state-of-the-art in scenario building. Such long time horizons were used first in the scenarios for assessing anthropogenic climate change, its impacts, and possible response strategies. The time horizon of a century or more was imposed by long-time constants in the climate system response to anthropogenic forcing, such as the emissions of greenhouse gases, aerosols, and particulate matter. In contrast, most of the economic scenarios, whether global, regional, or national, are associated with relatively short time horizons of a decade or two at most; this does not allow for fundamental changes in economic system, but it does capture the accumulation of more gradual and incremental changes based on current trends and tendencies. Demographic scenarios are usually somewhere in between due to the large inertia associated with population momentum and slow cumulative changes arising from migration patterns or the emergence of pandemics. Fundamental technological, institutional, and infrastructure changes can also take many decades or centuries, so that scenarios of the possible emergence and diffusion of new technologies usually have time horizons of 50 years or more. Integrated climate scenarios that included possible human response strategies were the first applications of integrated scenarios with time horizons of century or more.

The MA scenarios establish another important precedent in adopting a similarly long time horizon so as to encompass alternative future developments of ecosystem services and human well-being. This is important not merely because some of the ecosystem services such as biodiversity may significantly decrease over these longer time horizons, threatening at least some aspects of human well-being, but also because these futures are subject to human choices that have not yet been made. It is difficult to separate all complex interactions that co-determine changes in human well-being components related to biodiversity.

So far it is not possible to quantify all these interactions, but an innovative part of the MA is a first attempt in this direction. Decision-makers and various stakeholders need to understand how policies and other measures can influence and affect future provision of ecosystem services and human well-being. Human activities have become an important co-determinant of Earth systems, and decisions made now and in the immediate future will have consequences across both temporal and spatial scales. Integrated scenarios are the main tool available for the assessment of alternative future developments and possible response strategies. Increasingly, long time horizons and global perspectives are required to understand complex interactions between human and natural systems.

The combination of narrative storylines and their quantification in integrated scenarios of alternative futures is the main method for capturing complexity and uncertainty and transcending limits of conventional deterministic models of change. (See Chapter 2.) MA scenarios address a highly complex set of interactions between human and natural systems, a scientific challenge that is compounded by the cumulative and long-term character of the phenomena. While the world of many decades from now is indeterminate, scenarios offer a structured means of organizing information and gleaning insight into the possibilities. Scenarios can draw on both science and imagination to articulate a spectrum of plausible visions of the future and pathways of development. Some characteristics of the MA scenarios are assumed to evolve gradually and continuously from current social, economic, and environmental patterns and trends; others deviate in fundamental ways. A long-term view of a multiplicity of future possibilities is required in order to be able to consider the ultimate risks of maintaining adequate ecosystem services, assess critical interactions with other aspects of human and environmental systems, and guide policy responses.

The development of methods to effectively blend quantitative and qualitative insights is at the frontier of scenarios research today. The narrative storylines give voice to important qualitative factors shaping development such as values, behavior, and institutions, providing a broader perspective than is possible by analytical and numerical modeling alone. (See Chapter 2.) Storylines are rich in detail, texture, metaphors, and possible insights, while quantitative analysis offers structure, discipline, rigor, and reproducibility. The most relevant recent efforts are those that have sought to balance these attributes. They provide important insights into how current tendencies and trends might become amplified in different future worlds across the four storylines and provide a multitude of different details across scales and systems. They are embedded in extensive assessment of the main driving forces and their future developments across scenarios in the literature. (See Chapter 7.)

13.2.3 Ecosystem Services and Human Well-being Development Paths across Scenarios

The MA scenarios make four important new contributions to the contents of global environmental scenarios exercises compared with the literature:

- extending the integrated assessment approaches to include ecosystem services and their consequences on human well-being;
- modeling changes in biodiversity as an integral part of scenario development;
- assessing the interactions and trade-offs among ecosystem services; and
- assessing the possible replacement of some ecosystem services by others and the emergence of new ways of provisioning of these services, such as through technological change.

Perhaps the most important new contribution of MA scenarios to global scenario analysis is the extension of integrated assessment and modeling of alternative futures to explicitly include ecosystems and their services and human well-being. The vast body of environmental scenarios literature deals primarily with climate-related issues. The first integrated assessment models and frameworks were developed to link driving forces to possible climate change consequences, including impacts of climate changes, and various response strategies, including mitigation and adaptation.

Figures 13.3 and 13.4 compare the MA modeling system to that of the IPCC, which is an integrated assessment of

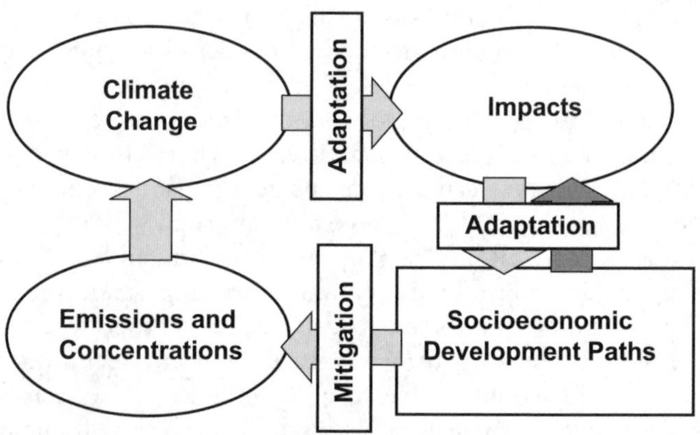

Figure 13.3. Integrated Assessment Framework of Intergovernmental Panel on Climate Change (IPCC 2001)

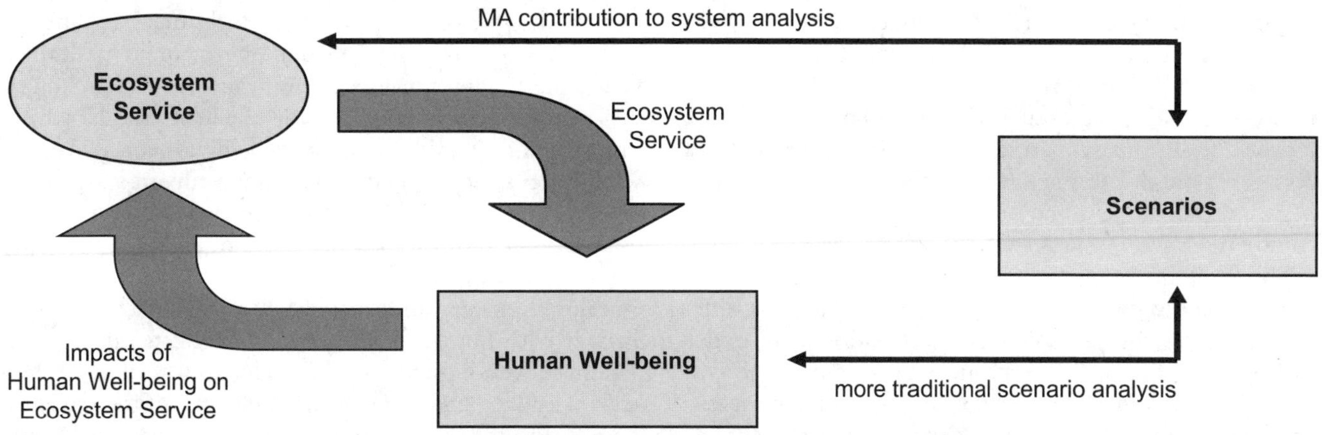

Figure 13.4. Approach and Contributions of MA to Focus of Scenarios Development

climate change that takes account of the human drivers of, and responses to, changes in the physical climate system. The MA scenarios add a new element by exploring feedbacks from the responses to environmental change to the drivers that affect this change. Prior to the MA, the majority of scenario exercises explored changes in socioeconomic systems and their impacts on the variable of interest (such as climate change or water supply and quality). This approach was enhanced in the MA scenarios by focusing on parallel changes in ecosystem services and human systems and their impacts on one another. Figure 13.4 demonstrates this new approach to integrating across the various driving forces and simultaneously describing the feedback loops within the socioecological system.

In some sense, there is a symmetry between integrated approaches to assess future climate change and the MA approach to assessing future ecosystem changes. MA scenarios extend the state-of-the-art by including both an integrated treatment of climate and ecosystems change as well as the resulting changes in ecosystem services and human well-being. However, they do not explicitly deal with the ethical dimension of the social developments suggested by some of the scenario features. For example, the scenarios do not provide details on possible implications for equity or lifestyle changes, but they directly address changes in human well-being that result from changes in ecosystem services. While the perspective of services is not new in itself, the explicit inclusion of a multitude of ecosystem services in MA scenarios is unique in scenario analysis. Assessing multiple ecosystem services is a formidable task because ecosystem services do not act in isolation; they interact with one another and with anthropogenic systems in complex and usually unpredictable ways.

Technology and institutions also play a role in the provision of ecosystem services; the nature of the interaction between ecosystems, technology, and institutions is not always entirely clear, however. The MA scenarios also address the issue of the extent to which alternative services, derived from new technologies, institutions, and other human activities, offer new solutions to improve the efficiency, accessibility, affordability, and quality of the provision of ecosystem services.

It may be easy to replace some services with others or to improve their provision through technological or institutional advances. More often, however, ecosystem services may not be easily substitutable by other services. Thus the degree to which the provision of ecosystem services can be improved and pressure on ecosystems reduced by technologically generated alternatives is highly uncertain. It will depend on what services are demanded in the future, what new technologies are available, and what other ecosystem services are (purposefully or accidentally) traded off by the substitution processes. (See Chapter 12.)

It is true that future technologies may offer feats that are impossible or prohibitively expensive today. The costs of many technologies are likely to decrease, making their widespread diffusion possible and affordable. However, some technologies may have unpleasant and unexpected negative effects and cause unanticipated interactions among ecosystem services. Similarly, currently unknown ecosystem services, or those considered to be less important, may be found to be fundamental to people or the future maintenance of ecosystems. These are some of the dimensions of ecosystem change that are addressed across the four MA scenarios.

While an important step forward, the MA scenarios work is incomplete. Particularly the quantitative analyses of ecosystem services and human well-being were driven by the underlying scenario analysis, with no explicit modeling feedbacks from those analyses back to the underlying anthropogenic driving forces. In other words, while in the quantitative models changes in ecosystems affect ecosystem services and human well-being, changes in ecosystem services and human well-being do not lead to further changes in the anthropogenic drivers of change. These feedbacks were, however, included in the narrative analysis of the interaction between human well-being and ecosystem services outcomes.

While incomplete, the inclusion of ecosystem effects in the MA represents a major step forward. It points the way toward a truly integrated assessment of possible future relationships among human and natural systems.

The MA scenarios develop four different futures, each plausible in its own right, and detail the resulting changes in ecosystem services and their consequences for human well-being. While basic human conditions generally improve across all four scenarios, the scenarios nevertheless all result in perilous paths of ecosystem changes. Global population increases in all four scenarios, as does the affluence of an average person, but this generally occurs at the cost of ecosystems degradation and consequently also lower per capita ecosystem services.

The individual development paths of the scenarios are fundamentally different and do not converge. They differ with respect to the paths of many of the drivers of global change. In particular, the scenarios differ with respect to whether the world becomes more or less interconnected and whether environmental management is reactive or proactive. Global Orchestration is connected and reactive, Order from Strength is disconnected and reactive, Adapting Mosaic is disconnected and proactive, and TechnoGarden is connected and proactive.

These basic differences are explored in more detail in the full storylines. The scenarios differ with regard to institutions, technology, social organization, and ecological change. None of these four development paths leads to a full transition to sustainability, even though the two proactive scenarios include response to ecosystem change. TechnoGarden follows a "technology strategy" to deal with these threats, while Adapting Mosaic takes a more institutional and behavioral path. A complementary mixture of these two different strategies is required for the sustainability transition and a future where the loss of ecosystem services is avoided both through substitutions by other services or through human change itself. In contrast, the two reactive scenarios put more emphasis on material human well-being but take different strategies toward this goal. Global Orchestration is a world where globalization of economic activities is fairly successful and results in the highest economic growth rates of all four scenarios. Order from Strength is a fragmented world where economic and material interests are focused on more local and regional solutions.

This rudimentary and "caricature-like" brief description of the scenarios illustrates the fundamentally different nature of their development paths. Chapters 8 and 9 provide a rich qualitative and quantitative image of these four futures. Despite this detail, each scenario represents a kind of "pure" strategy. In reality, elements of each described possible world will be presented in some aspects of the future development but not others. Therefore, it should be noted that the scenarios jointly give the full range of heterogeneity of future developments. Some combinations of them are a real possibility, especially at the more local and regional level. Even though they portray divergent development paths, they jointly hold the full richness of future ecosystem services and human well-being.

One common finding is that all these different possibilities lead to a general decrease of regulating and supporting ecosystem services per capita while ecosystem managers try to maintain provisioning services. However, the chosen development paths all lead to different outcomes for the five components of human well-being. Global Orchestration leads to the greatest improvement in human well-being. TechnoGarden improves all but one element of human well-being. Adapting Mosaic significantly improves human well-being in the South with little change in the North. In Order from Strength, human well-being is decreased.

While basic human conditions generally improve across three scenarios and decline in one scenario, they all result, to a varying extent, in perilous paths of ecosystem changes and lower per capita ecosystem services. This illustrates the complex linkages and feedbacks of ecosystem service changes with human well-being. While material elements of human well-being generally increase, loss of ecosystem services leads to higher inequalities in some of the scenarios and even degradation of some aspects of human well-being in others. Degradation and loss of some ecosystem services also affects the relationship between provisioning and regulating functions of ecosystem services.

The Global Orchestration scenario has the highest rates of economic development associated with lowest population increase. This leads to a high capacity to both invest in the future and respond to potential threats. However, this carries an environmental price as it leads to the highest rates of global change, including the highest rates of forest land disappearance and the highest levels of energy consumption. Nevertheless, the overall improvement of human well-being is the highest with this scenario, particularly in the currently developing countries. This may be because increases in wealth and equity lead to a high capacity to respond to changing circumstances and offset some loss of ecosystem services.

But the key uncertainty of the Global Orchestration path is whether or not the impressive rates of economic development and human well-being can be sustained if all externalities are accounted for. It remains unclear whether this development path is only achieved at the cost of endangering the biosphere and other planetary systems irrespective of apparent improvement of human well-being, thus undermining economic systems in the long term. Full consideration of the feedbacks from ecosystems to human systems in the quantitative models might reduce the level of economic development and the relative improvement of human well-being.

The Order from Strength development path comes close to disaster. It is the world with the highest population growth, some of the highest adverse environmental impacts, and the lowest rates of economic development. The income disparities in this possible future are similar to those that prevail today. Ecosystem services are seriously eroded. Society lacks the economic or technological capacities to adapt or respond to these threats to human well-being. For example, while high trade boundaries reduce the chance of invasive species, the risk of adverse impacts on human well-being when invasions do occur is high due to low social capacity for adaptation.

There is an utter disregard, under Order from Strength, for human interference in the biophysical processes and a general erosion of the main pillars of sustainability: environment, equity, and development. The consequences are the second highest demand for energy services (after Global

Orchestration) associated with the highest greenhouse gas emissions for all scenarios, the highest levels of water abstraction, and the lowest per capita incomes. It is not surprising that the incidence of disease, poverty, loss of security and freedom, and other adverse human conditions are most prevalent in this scenario.

The Adapting Mosaic development path is the most proactive about emerging challenges. It sets parallels between the adaptive development of humans and ecosystems. From this perspective it appears to be most attractive. It achieves high levels of human well-being with relatively low risks and intrusion on ecosystems and nature. Only TechnoGarden has lower impact on ecosystems. But TechnoGarden has high and unknown risks of collapse due to technology failure, while Adapting Mosaic is low risk. This development path highlights the benefits of evolutionary responses fostered by investments in social, natural, and human capital at local and global levels. There is recognition that natural capital underpins other forms of capital, such that property rights are attached to ecosystem services. In other words, they are internalized in decision processes.

Thus, Adapting Mosaic is the scenario with the highest degree of assumed feedback from nature to the human sphere and that includes explicit consideration of ecosystems. In the other three scenarios, ecosystems are reduced to the benefits provided by ecosystem services. In this sense, Global Orchestration and TechnoGarden come closest to achieving a sustainability transition, but somewhat ironically, Global Orchestration does so without vigorous diffusion of new technologies and TechnoGarden without high rates of economic development. This might indeed constitute a possible internal contradiction between these two storylines.

TechnoGarden also improves human well-being, but with higher risk and uncertainty. Technological approaches alone, without appropriate institutional and social embedding, cannot resolve the challenges of the future. In addition, technology itself may be the source of additional problems due to the potential for breakdowns in highly controlled systems. Also, the assumption in this scenario is that the described changes toward "greener technologies" might take some time to occur. Therefore, technology does not lead to a higher level of economic growth right away.

In the past, technological change has generally been the main driver of economic development in the long run. In TechnoGarden, technological change is used to improve provision of ecosystem services and does not provide the means for rapid rates of development and elimination of poverty, as in the Global Orchestration strategy. The difference is primarily in the focus in Global Orchestration on global institutions to improve equality and eliminate poverty. For example, technology leads to a rapid shift away from the current reliance on fossil energy in this scenario so that its consequences for climate change are lower than in other scenarios; however, there is no attempt to improve access to energy sources for all people. It is possible that the realized rates of development in this scenario would be higher if some of these positive environmental externalities were included as a resource. The cumulative effect of technological change and lower environmental impacts in this future world could indeed lead to the highest level of human well-being. However, the high risk of collapse would remain.

The Order from Strength development path is the least sustainable both because of its disregard for the impacts on ecosystems and because of the lack of development. It comes close to embodying many of the fears associated with a perilous development path or a failure in human development. It is indeed surprising that despite these monumental failures, the development path manages to improve some aspects of human well-being in the currently rich countries. (See Chapter 11.)

The scenarios also do not include a development path characterized by a "best case"—a success of sustainable development associated with a high degree of environmental protection at all scales. This would be a scenario that combines some of the positive characteristics of Global Orchestration, TechnoGarden, and Adapting Mosaic into a development path that is both evolutionary and has high social and technological innovative capacities without a disregard for natural or human capital, such as where there is a high dependency by the poor on biodiversity. Such a scenario might serve to illustrate the policies and strategies that would be required over the next decades to reduce human intrusion on nature while improving human well-being for all future generations. The MA nevertheless chose not to develop this "rosy" scenario, as for each country and in each location-specific case the mix of these different elements is likely to differ.

The MA scenarios show increasing pressures on biodiversity and many other adverse impacts on ecosystems up to 2050 and beyond. However, ecosystems play a crucial role in overall biophysical processes. For example, they are an important regulating component of the global carbon cycle. The potential impacts of ecosystem service changes on global and climate change are twofold. They include the interplay of ecosystem services and other service in co-determining water, air, and land use as well as the emissions of greenhouse gases. They also include the regulating function of ecosystems in global and climate change. MA scenarios describe the former impacts explicitly. The latter ones are more difficult to capture in the models. For example, climate models do not include changes in the carbon cycle that might emerge due to anthropogenic causes of ecosystems changes, nor do they include the effects of albedo change arising from land use change.

13.3 Robust Findings of the MA Scenarios

A striking feature of the four MA scenarios is that they all show increased demand for ecosystem services and place increased pressure on ecosystems to supply those services. People in each scenario pursue different methods for ameliorating the increased pressure on ecosystems. In some scenarios, such as TechnoGarden, the greater pressure on ecosystems is somewhat relieved through technological advances to improve efficiency of service delivery. In Global Orchestration, society focuses on equal access to ecosystem services in the hopes that this will eventually lead to greater care of ecosystems. In general, scenarios in which increased demand for ecosystem services is coupled with disregard for regulating and supporting services (Order from Strength and Global Orchestration) suffer from high risk of breakdowns in access to provisioning ecosystem services. Those scenarios in which learning about ecosystem function is a priority (TechnoGarden and Adapting Mosaic) tend to do better in maintaining supporting services.

Together, the scenarios illustrate that it is difficult to replace the provision of ecosystem services with alternatives without fundamentally changing human well-being. That is, replacing one service with another often creates demands on ecosystems in another place, in a future time, or for a different service. For example, we can replace biofuels with fossil fuels, which may improve local forests but which may also lead to increased demand for carbon sequestration globally. One of the most common types of these trade-offs is to favor provisioning services at the expense of supporting services. As noted earlier, however, ignoring supporting services leads to high risk of future breakdown in access to provisioning services. One of the implications of this may be that ecosystems may not be able to support large human populations at high levels of material human well-being—levels that are associated with increasing consumption and production patterns.

Highlighting this trend is an important message for policymakers. Describing irreversible trends, such as the loss of forest area, which occurs in all scenarios, can help to focus attention on developing strategies to ameliorate the consequences of these developments and devise adaptation mechanisms. The MA scenarios provide some hints about how this might be done. In addition, the scenarios alert policymakers to possible unexpected irreversibilities that they might foster through the decisions they take today.

A major driver for the portrayed irreversible trends in the MA scenarios is increasing land use for human purposes, which in return reduces the area available to unmanaged ecosystems. This trend is further exacerbated by climate change and other adverse environmental developments such as loss of water and air quality. For example, changes in land use are expected to be the major driver of biodiversity changes (based on quantitative assessment of species-area relationships in the next century across all four scenarios), followed in importance by changes in climate, nitrogen deposition, biotic exchanges, and atmospheric concentrations of carbon dioxide. (See Chapter 10.)

Storylines and qualitative assessment of ecosystem services and their impacts on human well-being provide more insights into possible future events and developments that models are not currently capable of quantifying. As noted earlier, human well-being has five key, reinforcing components: basic material needs for a good life, health, good social relations, security, and freedom and choice. Ecosystems underpin human well-being through their supporting, provisioning, regulating, and culturally enriching services. (See Chapter 11.) World populations increase in all four MA scenarios along with incomes and material consumption, which help determine human well-being. The tacit assumption is that there are biophysical limits to ecosystems' ability to produce services for human use. In scenarios, this decline cannot be completely compensated for by technological and social changes. This critical result is associated with high scientific confidence. (See Chapter 11.)

In some ways, this important result unearths one of the most interesting findings of the MA scenarios. All four scenarios support larger populations with higher levels of income and consumption, which puts increasing stress on ecosystem services. Ecosystems in the MA scenarios are generally able to support the increasing demand for services, but at the cost of supporting and regulating services and therefore at the cost of increased risk of breakdowns in provisioning services. This seeming contradiction raises a question: Is it possible to imagine sustainable futures with more people on the planet all enjoying higher human well-being compared with today but without further degradation of ecosystems and their services? Such a future would imply that the human "footprint" on the planetary processes would need to decrease through more efficient provision of services or that provisioning ecosystem services would need to be substantially decreased without affecting human well-being. The ultimate limits in the substitutability of ecosystem services is a critical research topic that clearly emerges from the findings of the MA scenarios.

The MA scenarios explore management and policy options that are currently being discussed by decision-makers. In this way, the scenarios provide a long-term perspective for different near-term decisions about ecosystem management and use. This can give useful concrete policy guidance to decision-makers along with the options discussed today. For example, new technologies could be developed that are able to improve the efficiency of ecosystem service provision. Bioengineering and tremendous progress in genetically modified organisms are one possibility for reducing the burden on natural ecosystems to meet human demands, but this is already controversial. The scenarios show that there are possible benefits, as well as enormous risks to use of these technologies. They hold the promise of limiting human demands for space and ecosystem services. At the same time, they may lead to a multitude of unintended adverse consequences.

The possible convergence of some new technologies discussed in Chapter 7 could lead to revolutionary changes and a new wave of economic development not based on current consumption patterns. There is a considerable research effort on the possible benefits, risks, and ethics of nano, genetic, information, and new cognitive science and technologies. Their convergence toward meeting new human needs and providing services could be a powerful economic drive that could in principle reduce the impact on ecosystems. However, they carry great risk of problems and unforeseen consequences on ecosystem services and human well-being. A lower risk possibility is a fundamental change in human behavior that leads to much better ecosystem management and more humble demands on ecosystem services despite more affluent and larger future populations.

13.4 Lessons Learned about Policy, Planning, and Development Frameworks

13.4.1 Global Policies in Context of Other Scale Policies

A complex and interacting set of processes that span environmental, economic, technological, social, cultural, and political dimensions drive the global system. Policies aimed at one aspect, such as poverty alleviation, may exacerbate other problems, such as environmental degradation. Contradictory sectoral policies can negate one another or backfire. Scenario analysis helps identify opportunities for mutual reinforcement between sectoral policies.

Policies need to be consistent across scales as well as across sectors. For instance, national policies should neither inhibit local initiatives nor undermine global policies. Moreover, since each human community and ecosystem is unique, policies should allow for and encourage adaptations to the local context and conditions. Again, the integrated scenario approach can illuminate the requirements for a multilevel policy framework. In the new century, environmental problems will continue to cross borders. Water pollution, air pollution, and depletion of the stratospheric ozone layer are problems that do not respect national borders. Neither does the buildup of greenhouse gases in the atmosphere that contributes to climate change. Nor do the birds, fish, and zooplankton that may be contaminated in one country but become part of the food web in another. Facing these challenges will require cooperative regional, continental, and global solutions.

Dealing with cross-border problems requires much the same kind of institutional apparatus at the local and national levels as described earlier: problems must be detected and diagnosed, interests must be balanced within and across borders, and agreements need to be implemented. There is, however, one big difference: at the global level, commitment is a more difficult problem, and there is no central authority to enforce agreements, although the emergence of a world environment organization could be envisaged under certain scenarios.

Social, political, economic, and ecological processes can be more readily observed at some scales than others, and these may vary widely in terms of duration and extent. Furthermore, social organization has more- or less-discrete levels, such as the household, community, and nation, which correspond broadly to particular scale domains in time and space.

A long time horizon is an obvious requirement for policies that aim to affect development over many decades. But long-term policy is often, by default, the cumulative result of a series of policies with a shorter-term outlook. Therefore, the long-range impacts of short-term policies should be designed and assessed in advance. Alternative policies may achieve similar short-term goals but have very different long-term impacts. Ideally, policies should create a platform for the next round of new and more advanced policies. The scenario approach is particularly appropriate for incorporating long-term considerations into today's policy discussions.

New social actors are becoming increasingly important, complementing traditional modes of decision-making and action. In particular, NGOs encompass a broad variety of interests, including educational organizations, trade unions, religious organizations, aid and development organizations, charities, and the media. The policy process needs to involve and mobilize all relevant institutions, tapping into their diverse capacity and potential for interaction, synergy, and complementarity. This means that participation, negotiation, and the articulation of multiple goals should substitute for antagonism and exclusion. Again, a scenario-building process that can cultivate contrasting visions and perspectives is a critical technique for fostering pluralistic dialogue.

13.4.2 The Usefulness of the MA Scenarios for Stakeholders

This section explores what we learned about the usefulness of scenarios (development, modeling, analysis) for generating policy-relevant information. It also assesses the usefulness of the information in the scenarios for preparing policy analyses for selected stakeholder groups discussed in greater detail in Chapter 14. Most of the selected indicators, however, refer to the quantitative results of the scenario modeling exercise. Understanding many of the assumptions that the qualitative storylines portray in more detail and their consequences for different stakeholder groups can add substantially to the analysis. Chapter 14 also addresses what different user groups are likely to find most useful or missing when they conduct their own policy analyses of the MA scenarios.

13.4.2.1 Background

Scenarios have become a popular tool in environmental assessment and management in the past two decades. They played a limited role in the 1960s, when local and largely short-term problems dominated the environmental agendas. Increasing concerns over multifaceted, continental- to global-scale, and especially long-term problems led to the growing use of scenarios. Although policy-making is only one of many possible uses of scenarios, an increasing number of scenario applications attempt to provide useful information for decision-makers in public policy or private entities. Global scenarios represent a special cluster within this realm.

The work of the MA Scenarios Working Group has built on other global environmental assessments to design a relatively "user-driven" process. This is an ambitious objective. The range of targeted stakeholders includes the main international environmental agreements explicitly concerned with specific ecosystems or their services (the U.N. conventions on biodiversity, desertification, and wetlands), national governments (both parties to the international agreements and regulators of domestic policies), the private sector (extending from local resource operators to multinational companies using ecosystem services), and civil society (communities crucially depending on the local ecosystems and their services as well as NGOs protecting specific community interests or broader environmental values). This user-driven objective has been pursued from the beginning of the overall MA process.

Representatives of this wide-ranging and diverse intended audience have been surveyed concerning the information they would most need from the MA. The user questions have been sorted, analyzed, and summarized in Chapter 5.

Yet this goal of stakeholder participation must confront the reality of the scenario development process (scenarists getting carried away by their own storylines and visions of the future) and, most important, the constraints of the tools currently available to obtain the information users require. In this respect, an important question is related to the distance between the variables in the models and scenarios on the one hand and the indicators related to the stakeholders interests, values, and mandates on the other.

The quickest way to obtain information from the scenarios for policy analysis is when the indicators of interest are directly available as input (assumptions about driving forces) or output (results of the model calculation or the qualitative assessment based on the input assumptions) variables in the storylines or in the assessment models. An almost equally simple way to acquire information is when scenario or model variables can serve as precursors that can be converted by generally accepted procedures into policy-relevant indicators. Post-processing is required when the relationship between scenario variables and policy indicators are indirect. This can take the form of statistical processing of model output or running additional models fed by the primary model output to generate the required policy-relevant indicator. Inferred indicators require expert judgments and special procedures to be derived from one or several scenario or model output variables. Finally, distant indicators are those on which some sparse information is available in the model results but obtaining them would entail special post-processing arrangements (such as an expert panel). Given the limited time and resources available to conduct the policy analyses in Chapter 14, most indicators were taken directly from scenario storyline and model results (Chapters 8 and 9) and their interpretation and analysis by fellow experts in Chapters 10, 11, and 12.

13.4.2.2 Usefulness of the MA Scenarios to Specific Stakeholder Groups

The United Nations Convention on Biological Diversity has evolved into a complex web of thematic issues. A detailed comprehensive assessment of the implications for all themes and subprograms is not possible here. However, a good deal of relevant information on biodiversity is available in the scenarios.

Among the major threats to biodiversity, habitat transformation is the most important one, and quantitative results are available on key drivers: population growth and urbanization, fossil fuel extraction, change in agricultural area, forest conversion, and land fragmentation. Agricultural intensification and water withdrawal are available as indicators for the threat of overexploitation and inappropriate management. The information about pollution as a threat to biodiversity is limited to SO_2 and NO_x emissions, critical load excess, and return flows as proxies for water pollution. Climate change as a threat is characterized by changes in temperature and precipitation and by biome shifts. In contrast, there is no quantitative scenario output that would indicate the evolution of invasive species that is an increasing concern among threats to biodiversity.

The Convention on Wetlands (the Ramsar Convention) is concerned with wetlands, which exist in all continents, are diverse, and provide important functions and services. The quantitative model output of the scenarios does not contain direct or precursor indicators of wetlands change. This is because global models are designed to capture broad patterns of global change and because modeling techniques have not advanced much for wetlands processes, which generally take place at much smaller scales. One good way to generate wetland-related indicators would be to use the output of the global models operated in the MA scenarios exercise as input to drive general or location-specific wetland models to explore the potential impacts of different global scenarios. This was not possible in the present assessment. Therefore, conceivable impacts for wetlands are assessed on the basis of coarse indicators like regional water withdrawal, return flows, and water quality. This leads to a rather rough assessment of the emerging risks to wetlands, but the characteristic differences among the scenarios are apparent even at this level.

The mandate of the United Nations Convention to Combat Desertification is to address and alleviate degradation of land in arid, semiarid, and dry subhumid areas. The causes and processes of desertification are multiple and diverse. The desertification process itself is not directly depicted in global models because the process involves many local biophysical and socioeconomic factors that the global models cannot address. One possibility could be to use such models as post-processors of the global model runs. This was not feasible in the present assessment. Instead, an attempt was made to define suitable proxy variables from which indicators of desertification could be inferred. One of the main socioeconomic drivers behind desertification is unsustainable land use in arid areas. Models provide output that can be used to generate indirect indicators by taking the extent of arid areas under agriculture in different regions as a proxy to assess the desertification risk. These indicators do not provide a rich information base, but they are suitable for conducting a broad assessment of desertification risk and response options under the MA scenarios.

The main insights from the MA scenarios for national governments are analyzed at two time horizons in Chapter 14. The medium-term assessment looks at the implications for the prospects of reaching the Millennium Development Goals. The long-term assessment seeks to estimate the prospects for implementing the long-term objectives of sustainable development as confirmed and proclaimed by the Johannesburg Declaration at the World Summit on Sustainable Development in 2002.

The MA scenarios and model results contain a lot of useful information about the evolution of ecosystem services that is related to the achievement of these near- and long-term objectives. In some cases, the information is available on the relevant MDG indicator or its close proxy or precursor version (such as the proportion of land area covered by forests or of population with sustainable access to an improved water source). For many MDG indicators, implications of the scenarios can be assessed from indirect indicators. Good examples are the indicators defined to measure progress on Target 2: "halve the proportion of people who suffer from hunger." The indicator of prevalence of underweight children is well approximated by the model output indicating the percent of malnourished children (a composite indicator, see Chapter 14 for details), while the indicator on the proportion of population below minimum level of dietary energy consumption can be inferred, albeit roughly, from the model output indicating the amount of calories available per capita per day.

A lot of qualitative information can also be used as inferred indicators relating to the MDGs, like income growth in regions for the prospects of achieving Target 1: halving the proportion of people whose income is less than $1 a day. In contrast, the scenarios are rather vague about general development issues like education (school enrollment, literacy rates), overall health status (under-five and infant mortality rates, immunization against measles). Finally, the MA scenarios are totally silent about the financial aspects of development, such as the prospects for official development assistance, for debt and debt relief, especially for heavily indebted developing countries, or for foreign direct investments.

The situation is similar in the case of the indicators that can be used to measure progress on the objectives of the Johannesburg Declaration up to 2050. General economic development indicators (GDP growth, for example, or the gap between rich and poorer regions), many directly or indirectly environment-related social indicators (food security, hunger, access to energy), and many environmental indicators are readily available and provide useful information for policy analysis. However, the more specific characterization of the economic and financial development (access to financial resources, sharing the benefits from opening markets, or access to health care) is omitted due to the primary focus on global ecosystems futures.

One of the most useful aspects of the scenarios for communities and NGOs is the effort to assess the relationship between ecosystem changes and human health and well-being. Given their global nature, however, the scenarios are not able to fully model all the trade-offs and interactions between ecosystem services and human well-being, especially in reaction to specific response options and adaptation. Given the scenarios' inability to fully model all ecosystem services (cultural and supporting, along with provisioning and regulating), as well as the complex interactions between ecosystem services, human well-being, and response options over time, it is difficult to address the thresholds at which further ecosystem degradation and reductions in human well-being might occur. The scenarios do not fully meet the needs of civil society stakeholders to have the MA address the impact of ecosystem change on the vulnerability and resilience of human communities and on their cultural concerns. These issues are more successfully addressed in the sub-global assessments. Communities are interested in learning about site-specific impacts in relation to global changes, but the scenario methodology is not sufficiently advanced yet to make such cross-scale assessments readily possible.

There is a wealth of information in the MA scenarios that is useful and relevant for the private sector, but it is not easy to extract and summarize it. First, virtually all issues that are important in these scenarios are likely to have some degree of private-sector implications, for the private sector that is closely related to or has indirect stakes in ecosystem services is widespread and diverse—from large multinational companies to small local resource operators. Second, the MA scenarios address a large number and diversity of ecosystems and human well-being issues. The intersection of these features implies that it is a major challenge to consolidate the complex ecological data, analytical information, and modeling results in a succinct assessment for the private sector. There are two principal ways this task could be accomplished. The first is to focus on the specific interests of a small and carefully selected set of stakeholders (logging companies interested in timber, pharmaceutical firms pursuing genetic resources, and so on) and to prepare targeted assessments of the implications of different scenarios for them. The second possibility is to consider the interests of the private sector as a whole in ecosystem services and then derive general insights about the risks and opportunities emerging under different assumptions about the future. The second option was taken in Chapter 14, which means that there remains a lot of unveiled information in Chapters 8–12 that might be useful for the private sector and worth exploring.

In summary, the MA scenarios provide rich and useful pictures of broad patterns of possible futures at the global scale and at the level of world regions. However, it is impossible to perform detailed quantitative analyses of local processes and impacts with the set of models adopted in the present endeavor. The qualitative scenarios and storylines address many issues that the models cannot, providing a more rich and detailed investigation. One crucial improvement for the quantitative models in future assessments would be to soft-link sector- and region-specific models by using the global scenario framework and outputs of global models to drive them. A particularly useful feature of the current effort is that scenarios provide information about socioeconomic and technological development patterns, which is necessary for the assessment of the viability and effectiveness of various instruments and response strategies. Such tools may be currently available or might become available in the future for different stakeholder groups to protect their interests or fulfill their mandates in the contexts of widely diverging but plausible futures.

13.4.3 Path-dependencies, Irreversibilities, and Their Implications

Most complex systems display non-linear behavior in which relevant phenomena drastically change after certain threshold values are exceeded or where initially small but cumulative effects become ever more important as the system evolves. They also display hysteresis (a history or path dependence of systems behavior, which in the case of complex systems means that they do not return to their original state even when the influence of the driving forces that changed them ceases), leading to important irreversibilities in their behavior. This is often referred to as path-dependency. Many systems are characterized by such irreversibilities; examples include technological change, global climate change, and biodiversity loss. Often very similar initial conditions can lead to fundamentally different outcomes, and these are usually very sensitive to the actual development path taken. They can be characterized by emergent properties that can evolve in fundamentally different ways along alternative future development paths.

The MA scenarios are themselves examples of such path-dependencies. By the end of the twenty-first century, they have evolved and branched out into fundamentally different futures that depend on a myriad of intervening changes and decisions taken along the way. Some of the path-dependent phenomena could undergo abrupt changes over the time frames considered in the MA scenarios. Climate-related abrupt changes could include loss of the Greenland ice sheet, shutdown of the North Atlantic Thermohaline Circulation, or the release of methane from permafrost or from deep sea clathrates deposits. In technology, abrupt and path-dependent changes could include rapid deployment and improvement of renewable energy technologies leading to low emissions futures, but also major breakthroughs in availability of fossil energy sources rendering them practically inexhaustible (such as methane clathrates as energy source). Similarly, overfishing could irreversibly deplete stocks, while new fish farming methods with low environmental impacts could allow for a recovery of now-endangered fish stocks. Finally, desertification could become irreversible beyond some critical levels of vegetation and soil loss.

Future anthropogenic climate change is characterized both by path-dependencies and numerous irreversibilities. Human activities have caused and will continue to cause climate change. The main direct causes are the global emissions of greenhouse gases primarily due to energy and land use changes. Emissions of particulate matter, aerosols, and many other substances also affect current and future climate change. For the wide range of IPCC emissions scenarios (Nakićenović et al. 2000), Earth's mean surface temperature change is projected to warm 1.4–5.8° Celsius by the end of the twenty-first century, with land areas warming more than the oceans and the high latitudes warming more than the tropics (Cubasch et al. 2001). The associated sea level rise is projected to be 9–88 centimeters.

The range of future mean surface temperature changes for the MA scenarios is narrower, with 1.6–2° Celsius by 2050 because of two important factors, namely that one single integrated assessment model and one climate model were used to estimate future climate changes. Six different IAMs and different climate models were used by the IPCC, adding significantly to the range of uncertainties and widening the range up to 5.8° Celsius. It is indeed possible that the full range of climate change for the MA scenarios would be comparable if the scenario approach were extended to a wider range of models and quantifications. An important indication that this may be the case is that cumulative carbon emissions across scenarios are comparable with the IPCC ranges. In other words, the climate implications are likely to be underestimated through the approach taken by the MA.

The seemingly small changes in temperature and sea level rise across the scenarios mask the more fundamental path-dependencies. For example, the scenario with the highest emissions by 2050, Global Orchestration, leads to almost five times higher emissions than the scenario with the lowest emissions, TechnoGarden. It is clear that reducing Global Orchestration emissions fivefold would dramatically change the nature of the scenario, requiring massive emissions mitigation measures and policies ranging from radical technology change to modified human behavior. In other words, greenhouse gas emissions (and many other scenario characteristics) differ in their emerging properties, associated irreversibilities, and resulting development paths.

Despite this possible underestimation of the future climate changes across MA scenarios (given the current uncertainties), the impacts on ecosystem services and human well-being will be significant and pose a major reason for concern. For example, anthropogenic climate change will have fundamental impacts on biodiversity. It affects individual organisms, populations, species distributions, and ecosystem composition and function both directly (through increases in temperature and change in precipitation, for instance, and in the case of marine and coastal ecosystems also through changes in sea level and storm surges) and indirectly (through changing the intensity and frequency of disturbances such as wildfires, for example). Processes such as habitat loss, modifications, and fragmentation and the introduction and spread of non-native species will affect the impacts of climate change.

The general effect of human-induced climate change is that the habitats of many species will move poleward (toward higher latitudes) or upward (toward higher altitudes) from their current locations (Gitay et al. 2002). It is clear that such significant impacts on biodiversity by anthropogenic climate change would also lead to significant additional loss of ecosystem services beyond those that occur due to other pressures of human development on ecosystems in MA scenarios.

Ecological systems can also be involved in abrupt changes both at small and large scales, usually acting in concert with physical and chemical components of the Earth system and frequently also due to the influence of human systems. One of the best-known examples from the past of such an abrupt change is the transition from a green to an arid Sahara in the mid-Holocene (Claussen et al. 1999; deMenocal et al. 2000). About 6,000 years ago, the climate in northern Africa was much more humid than today, supporting savanna vegetation throughout the region, with little or no desert. The change that occurred was both abrupt and severe, leading to a complete desertification of much of this area—the formation of the present Sahara Desert. The ultimate trigger for the shift was a small change in the distribution of incoming solar radiation in the region due to a subtle change in Earth's orbit (Steffen et al. 2004). This change by itself was not significant enough to drive the vegetation shift but rather nudged the Earth system across a threshold that triggered a number of biophysical feedbacks that led rapidly to a drying climate and then to an abrupt change in vegetation (Steffen et al. 2004). This episode demonstrates the complexity of the dynamics that lie behind threshold-abrupt change behavior.

The behavior of the terrestrial carbon cycle is an aspect of Earth-system functioning that may experience abrupt change, particularly in the second half of this century, which is over the time frame adopted in the MA scenarios. At present, terrestrial ecosystems absorb about 25–30% of the CO_2 emitted by human activities, thus providing a valuable free ecosystem service that slows the rate of climate change. Simulations of the evolution of terrestrial carbon sinks from 1850 to 2100 (Cramer et al. 2001) show the development of the current strong sink through the second half of the twentieth century. The sink will continue to grow in size through the first half of this century, according to these simulations, but is likely to saturate around 2050, with no further increase. One simulation shows a rapid collapse of the sink through the second half of the century, with the terrestrial biosphere as a whole (Cox et al. 2000) perhaps even becoming a net source of CO_2 to the atmosphere by 2100 because of a much drier climate in Amazonia and a subsequent loss of the remaining forests there (Steffen et al. 2004). This again indicates the possible irreversibilities associated with high carbon emissions paths such as those of Order from Strength and to a lesser extent Global Orchestration and Adapting Mosaic futures.

Marine ecosystems commonly show threshold-abrupt change behavior, sometimes called regime shifts. For example, there appear to have been dramatic and synchronous changes to marine ecosystems in the North Pacific Ocean in the late 1970s. Such changes cannot be ascribed to local ecological interactions only. They involve many different biological and environmental parameters (over 100 in the case of the North Pacific), show coherence over large spatial scales, and are correlated to very large-scale external forcings, often teleconnections in the climate system. The 1977 regime shift in the North Pacific, for example, is correlated to a sharp increase in mean global surface temperature (Steffen et al. 2004).

Human impacts can also trigger abrupt changes in marine ecosystems, particularly through overfishing and eutrophication. Recent reports (Myers and Worm 2003) claim that about 90% of the large predatory fish biomass has been removed from the world's oceans, with removal rates being highest with the onset of post–World War II industrial fisheries. Given the importance of top-down controls on the dynamics of marine ecosystems, there is the possibility that such overfishing could lead to regime shifts in marine ecosystems, with reverberations through to lower trophic levels such as zooplankton. On a smaller scale, overfishing is already known to cause sharp regime shifts in coastal ecosystems (Jackson et al. 2001).

Human-dominated waste loading on the coastal zone has also led to abrupt changes (from an Earth system perspective) in the functioning of marine ecosystems in the form of eutrophication. If the level of nutrient loading is high enough, significant changes can occur to the species composition of the ecosystem, often leading to a simplification of ecosystem structure (domination by one or a few species) (Gray et al. 2002). Severe eutrophication can lead to the formation of hypoxic (oxygen-depleted) zones, in which the dissolved oxygen concentration is below that necessary to sustain animal life (Rabalais 2002), usually resulting in drastic changes to ecosystem structure. The region of the Gulf of Mexico near the mouth of the Mississippi River and the Baltic Sea in northern Europe are regions where hypoxic zones commonly occur (Steffen et al. 2004). In certain cases, hypoxic zones such as those that seasonally occur on the west Indian shelf release nitrogen oxide, a greenhouse gas (Naqvi et al. 2000).

It remains to be seen how overfishing and eutrophication in concert will alter global biogeochemical cycles and the resulting global inventories of carbon, nitrogen, phosphorus, and silica. Despite the seemingly large capacity of marine ecosystems to assimilate the impacts of waste loading and overfishing, the imminent collapse of many coastal ecosystems is a warning that human and systemic global pressures may act synergistically to trigger large-scale regime shifts in global marine ecosystems (Steffen et al. 2004). The loss of such ecosystem services and their possible adverse impacts on human well-being cannot be treated adequately in the current IAM approaches. The MA has made a significant contribution by linking a fisheries model to its IAM. This is an important area and direction for future improvements in our capacity to model complex global systems and their future development paths.

Critical thresholds and irreversible changes are probably hidden in the largely unexplored domain of interactions among climate and environmental change, socioeconomic development, and human and animal health. Scenarios remain the main tool for gaining a better understanding of these critical interactions and numerous feedbacks. Human activities have already become a geophysical and biogeochemical force that rivals natural processes, and this is likely to increase, but to differing degrees, across all four MA scenarios. This implies that major discontinuities in the socioeconomic domain may lead to corresponding disruptions in the biogeochemical/physical and ecosystem domains—that is, that abrupt changes can be expected in the coupled human-environment system. Thus abrupt changes in socioeconomic systems could attenuate or amplify changes occurring in other aspects of the coupled system (Steffen et al. 2004).

Abrupt changes in coupled socioeconomic and natural systems have occurred in the past. The archeological and paleoecological records indicate that major shifts in societal conditions in the past often appear to have been linked with abrupt changes in the biophysical environment (Alverson et al. 2003), including, perhaps, the collapse of the Classic Period and lowland Maya civilization and various "pulses" of Mongul expansion from Mongolia, which had significant consequences for Imperial China and eastern Europe (Steffen et al. 2004).

One of the most important potential discontinuities is the spread of a new disease vector, resulting in a pandemic. High population densities in close contact with animal reservoirs of infectious disease facilitate rapid exchange of genetic material, and the resulting infectious agents can spread quickly, with few barriers to transmission through a worldwide contiguous, highly mobile human population. The almost instantaneous outbreak of SARS in different parts of the world is an example of such potential, although rapid and effective action contained its spread. Warmer and wetter conditions as a result of climate change may also facilitate the spread of diseases such as malaria. Malnutrition, poverty, and inadequate public health systems in many developing countries provide large immune-compromised populations with few immunological and institutional defenses against the spread of an aggressive infectious disease. An event similar to the 1918 Spanish Flu pandemic, which is thought to have killed 20–40 million people worldwide, could now result in over 100 million deaths within a single year. Such a catastrophic event, the possibility of which is being seriously considered by the epidemiological community, would probably lead to severe economic disruption and possibly even rapid collapse in a world economy dependent on fast global exchange of goods and services (Steffen et al. 2004).

Another important area of emergent properties and a possible source of abrupt and irreversible changes is the interaction of technological change and the natural environment, including ecosystems. An obvious case is the current advances in bioengineering that can affect and interact with natural ecosystems. The possible interactions include adverse and irreversible impacts on regional ecosystems. Historical examples are many, including the introduction of new species into foreign environments leading to dramatic ecosystem changes and shifts. The other possibility is production of resistant and better-adapted species to overcome some future challenges. MA scenarios explore many but not all possibilities that might emerge during the twenty-first century as the result of technological changes that may either directly affect ecosystem services or indirectly affect both ecosystem services and human well-being.

A century is ample time for pervasive diffusion of fundamentally new technologies and systems. In fact, whole new technoeconomic paradigms have emerged in the past over similar time scales—from the emergence of the coal, steam, steel, and telegraph eras to the ages of oil, gas, internal combustion, gas turbines, petrochemicals, pharmaceuticals, mass production, and so on. Also in the future, new technologies could lead to new combinations of technologies and human activities. Today, the possible convergence of nano-, cogno-, bioengineering, and information technologies is seen as a possible way of enhancing human performance, modifying organisms into components of larger technoeconomic systems, and directly interacting with many micro- and nano-scale systems of both inanimate and biological origin.

All four MA scenarios consider further economic development in the world despite increases in human population. Much of the increased human productivity would stem from ecosystem services, some of which might indeed be enhanced by convergence of advanced technologies into new paradigms during the twenty-first century. At the same time, many of these future technological possibilities may bring with them unanticipated effects—some that might threaten human well-being and ecosystems and some that might enhance them and reduce human interference with natural systems. MA scenario storylines tackle many of these complex issues. However, dramatic effects of such technological developments were not pursued in scenario quantifications, presumably partly because models used in the MA were not designed for the assessment of technological changes and technology diffusion. Technology might indeed prove to be one of the most fundamental drivers of future human development and ecosystem services.

The question of how robust an increasingly interlinked, globalized world economy is must be addressed urgently (Steffen et al. 2004). There will almost surely be significant increases in need in the future for the provisioning of natural resources and ecosystem services. And despite technological advances, meeting these needs will have impacts on the Earth system and especially on many already-threatened ecosystems. There is a high probability that droughts, floods, and severe storms will occur more frequently, and an increasing probability that the more drastic, abrupt changes of the type described earlier could also occur. Coping with such stresses would take an increasing share of economic activity away from the evolution and growth of the economy in general (Steffen et al. 2004). How many such stresses, occurring when and where, would it take for the global economic system to begin a downward, self-reinforcing spiral that would lead to a rapid collapse? Should such a collapse occur, it could lead to a significant and probably long-lasting change in the fundamental human-ecosystem relationship.

13.5 Information Gaps and Research Needs

This section describes the research needed to improve the development of global scenarios in the future. It includes both research needed in the formal sense as well as improvements that could be made in methodology and in the manner in which the scientific community operates.

13.5.1 Global Storyline Development

Determining the nature of the MA global storylines by choosing the key drivers that would vary across scenarios and those drivers that would follow the same trend in all scenarios was a long and sometimes difficult process. This was partly the result of divergent views on the use of scenarios, and partly the result of divergent disciplinary approaches to science. Here, we describe some research and changes in research methods that could improve the development of global storylines in the future.

- Development of regional and local scenarios linked to the global scenarios would help create better global scenarios. Regional and local scenarios can use more accurate local information and might represent some system dynamics more accurately. They can also pinpoint specific variables of interest. Expanding methodologies for linking scenarios developed at different geographical scales or nesting them within one another to foster the exchange of information across scales will be an important step to help improve scenario development methods.

- Better communication and interaction with policy-makers would help inform the development of the storylines by indicating the key variables that are of interest to decision-makers. Improved communication with policy-makers would be useful both for understanding policy-makers' most pressing questions and for communicating results to them at the end of the process. Having a greater presence of policy- and decision-makers within the working group may be one way to improve this communication.
- Better communication and interaction across scientific disciplines would help facilitate future global scenario development projects. During the MA scenario development process, it turned out that the differences among disciplines' core beliefs about how the world functions were also often the critical issues that policy-makers wanted to have addressed in the scenarios. (See a more detailed discussion in Chapter 5.) However, it took our working group several meetings to come to a full understanding of the differences of core beliefs among disciplines. Better interdisciplinary communication prior to initiation of this project might have made this process easier.

13.5.2 Modeling Complex Systems

The response of complex systems to environmental change is assessed using models that are based on multiple driving variables, along with their interactions. Models currently can provide part but not all of the information needed for scenarios of ecosystem services. For example, temperature trends at the broad scale can be assessed relatively well through current climate models, whereas rainfall patterns cannot. Nevertheless, rainfall is important for many ecosystem services in many regions, often more important than temperature. The assessment therefore is incomplete and can be improved once more-reliable rainfall simulations are made or the uncertainty range of rainfall simulations are considered explicitly (which was not possible within the MA models).

While the assessment models can provide useful information to scenarios, their coverage of the Earth system is incomplete, and their description of essential ecosystem functioning is better for some processes than others. Climate and carbon cycle impacts are relatively well understood while even the most basic nutrient cycles are less reliable in current models.

With regard to analyzing ecosystem services, the current models have a critical deficiency in that they are not able to simulate the important feedbacks—the changes, often small, in ecosystems, that feed back and affect social systems, sometimes in large ways—that were described in the MA storylines. Therefore, the qualitative assessments by the storylines of some of these feedbacks loops became an important means to describe likely changes under the four MA scenarios.

The models are also incomplete in the sense that we need to better understand the interactions among variables across models. For example, we urgently need better models that link likely changes in such things as land cover to likely changes in essential ecosystem processes, including nutrient cycles, primary production, energy flow, and key community dynamics, as well as the relationship of these changes to ecosystem services. One of the important weaknesses of our current understanding is the lack of data for broad-scale (long-term and large-scale) ecological dynamics. Improved models and approaches are required to better understand ecosystems and their interaction with human systems, but such more-advanced models will still not lead to a complete and full integration and may never lead to full understanding of the complex interactions. The better models based on broad-scale dynamics would enhance our understanding of the ecosystem patterns and ecosystem processes. The benefits and requirements for these improved models and approaches are as follows:

- Better models for the relationship between ecosystem change and provision of ecosystem services would greatly improve quantification of the scenarios. Even in cases where we can develop decent models of the ecosystem, it is extremely difficult to predict the end result for provision of ecosystem services.
- Much better models for the relationship between ecosystem services and human well-being are desirable. In cases where we were able to model changes in ecosystem services (only provisioning and regulating services), we were rarely able to estimate the impact on human well-being. Most of the quantifiable indicators of human well-being related to population and demographics. It proved much more difficult to estimate spiritual well-being, recreation opportunities, or even impacts on human health.
- Models currently estimate only provisioning and regulating services. As pointed out in Chapter 12, this aligns perfectly with our understanding that provisioning and regulating services are often given greater priority than cultural and supporting services in management decisions. Certainly not by accident, the models focus on the ecosystem services that are perceived by society as more important (driving research agendas and funding); they give less attention to cultural and supporting services. There are two consequences of this bias.

First, cultural and supporting services are left out of the quantitative modeling exercise altogether: changes in these services simply are not quantified. This has critical implications. Supporting services are necessary for the production of all other ecosystem services, yet we cannot quantify how they may change in the future. If supporting services are declining, we may face severe and possibly sudden loss of provisioning and regulating services in the future.

Second, the fact that models are able to only explore a small subset of ecosystem services (even within provisioning and supporting services) means that a smaller set of potential trade-offs can be quantified. Thus, even if the models were able to perfectly characterize all the trade-offs among the ecosystem services that they considered, this would plainly underestimate the consequences of any societal choice, as many other trade-offs would remain unquantified. The consequence is that model results, at best, represent a crude lower bound of the expected consequences of any specific scenario.

- Improvements in modeling interactions among drivers or services would improve the quantification of scenarios. For example, it was very difficult to model how changes in agricultural production interacted with changes in water quality.
- Developments are needed to improve comparison of results across different models. Because the models we used calculated different variables or used different region boundaries, many variables were not comparable. Even in cases where two models calculated, say, land cover change, we could not always easily compare the results across models to ensure that our models were giving similar results. Comparison across models was even more difficult in the cases where the models were calculating different variables.
- There is a great deal of research needed on focused scientific topics. This is covered in Chapter 4.

13.5.3 Harmonizing Models and Storylines for Understanding Complex Systems

A major challenge for future scenario exercises will be to improve the level of harmonization between storylines and quantifications of the scenario. This involves three main tasks.

First, adequate time has to be given between the development of storylines, deriving model inputs from the storylines, running the models, interpreting model output, and revising the storylines. Perhaps two or three full iterations of this cycle are required to achieve a high level of consistency between the storylines and model calculations. Iterations are also required for achieving convergence among various subcomponents of the scenarios in addition to facilitating scenario consistency.

Second, the models need to be able to incorporate some of the important factors in the storylines, such as cross-scale feedbacks, thresholds, and small-scale changes. We know that these types of changes, which the models cannot fully address, are important determinants of the future. If the models are not able to address these factors, they will never match the richness or plausibility of the storylines in developing pathways to the future.

Third, it is important to make the conversion of information between the storylines and models more transparent and less arbitrary. Currently, information from the storyline is used to prescribe model inputs in an ad hoc (although consistent) fashion. For example, general statements about technological progress in the storylines were used to specify important input parameters to the modeling exercise, such as the rate of change of crop yield and the rate of improvement of domestic water use efficiency. Likewise, results from the modeling exercise (such as estimates of land use or cover change) are used ad hoc to modify or enhance the qualitative statements of the storylines. Rather than perform this conversion ad hoc, systems analysis techniques should be used to make the conversion more transparent and scientifically rigorous. For example, future exercises should consider the usage of fuzzy sets or agent-based approaches to convert from the linguistic statements of the storylines to the numerical information needed for model inputs, and from numerical model outputs to linguistic statements. Various techniques of qualitative modeling may also be useful for this conversion of information.

13.5.4 Research on Vulnerability

Further research on ecosystems and human well-being is needed. At the moment, scientific models for assessing thresholds of vulnerability in ecosystems are very few and not sufficiently developed (Peterson et al. 2003). Similarly, possible ways these thresholds will affect human well-being could be better understood. Research on thresholds is needed to more fully understand socioecological resilience and human well-being.

While we can sometimes quantify the trajectory of provision of a given ecosystem service, we cannot always determine whether the trajectory will continue the same way or whether it will change radically upon crossing some unknown threshold. Yet these threshold changes are often the most important changes in ecosystem services to understand. Research about thresholds and socioecological resilience would greatly improve our understanding of how to quantify and anticipate thresholds in management. A key issue here though is whether we have enough information to assess the thresholds or the vulnerability of ecosystems and human well-being to extreme events (ecological and socioeconomic surprises). Integrated assessments can be one of the main tools used to understand the resilience (buffer) of ecosystems and human well-being.

13.6 Conclusions

The MA scenarios have broadened global scenario exercises in their scope and methodology. By including and focusing on the many services that ecosystems provide to sustain anthropogenic systems, the MA scenarios explore the manifold linkages that exist between ecological and human systems. Previous scenario exercises have focused on some of the links between specific driving forces of environmental change and their impacts. For example, the IPCC explored connections among energy and land use and climate change. However, previous scenarios have not included ecological dynamics in their storylines or analysis and have not attempted to understand the effects of change on a suite of ecosystem services and human well-being. The MA scenarios expand the reach of analysis by including multiple ecosystem services and by linking environmental changes to their impacts on human well-being.

In addition, the MA scenarios contribute to the methodology of scenario analysis in various ways. They advance the role of qualitative and quantitative information and highlight the advantages of combining the two in the scenario development process. The scenarios also demonstrate the importance of integration across various disciplines to derive internally consistent, detailed pictures of the future. They also stress the significance of including various stakeholder perspectives in the scenario development process, so that the scenarios focus on questions about the future that are relevant for their potential users. Furthermore, the scenarios reveal the imminent path-dependencies and irreversibilities of plausible development pathways, which helps highlight the implications of decisions taken today.

With this analysis, the MA scenarios provide important insights for various stakeholder groups, such as the U.N. conventions, national governments, NGOs, local communities, and the private sector. Each group can derive implications from the set of scenarios in order to develop robust strategies for their policy decisions.

The MA scenario analysis can be improved in future scenario projects for ecosystem services. Modeling complex socioecological systems with their interactions and feedback loops remains a key challenge. Providing information not just on services with the highest immediate priority for many people and organizations (provisioning and regulating services) but also on supporting and cultural ecosystem services will enhance the level of analysis for decision-making. Furthermore, linking global models to models that operate at smaller geographical scales can enhance the consistency and quality of the derived information.

The qualitative analysis of ecological feedback loops, thresholds, risks, and vulnerabilities as part of scenario development can provide important insights that existing global ecological change models have so far not been able to capture. Global modeling and integrated assessment efforts, though, can provide important consistency checks of assumptions on key driving forces and their interactions. Enhancing the methodology for combining quantitative and qualitative analysis in the future will greatly improve our ability to deal with the complexities that lie ahead of us.

References

Alverson, A., R. Bradley, and T. Pedersen, eds., 2003: Paleoclimate, global change, and the future, *IGBP,* Global Change Series, Springer-Verlag, Berlin and New York.

Claussen, M., C. Kubatzki, V. Brovkin, A. Ganopolski, P. Hoelzmann, and H.J. Pachur, 1999: Simulation of an abrupt change in Saharan vegetation at the endo fo the mid-Holocene. *Geophysical Research Letters,* **24,** 2037–2040.

Cox, P.M., R.A. Betts, C.D. Jones, S.A. Spall, and I.J. Totterdell, 2000: Acceleration of global warming due to carbon-cycle feedbacks in a coupled climate model. *Nature,* **408,** 184–187.

Cramer, W., A. Bondeau, F.I. Woodward, I.C.Prentice, R.A. Betts, V. Brovkin, P.M. Cox, V. Fisher, J.A. Foley, A.D. Friend, C. Kucharik, M.R. Lomas, N. Ramankutty, S. Sitch, B. Smith, A. White, C. Young-Molling, 2001: Global response of terrestrial ecosystem structure and function to CO_2 and climate change: Results of six dynamic global vegetation models. *Global Change and Biodiversity,* **7,** 357–373.

Cubasch, U., G.A. Meehl, G.J. Boer, R.J. Stouffer, M. Dix, A. Noda, C.A. Senior, S. Raper, and K.S. Yap, 2001: Projections of Future Climate Change. *Climate Change 2001: The Scientific Basis,* Third Assessment Report, Working Group I of the Intergovernmental Panel on Climate Change (IPCC), 525–582. Cambridge University Press, Cambridge. Available at http://www.ipcc.ch.

deMenocal, P.D., J. Ortiz, T. Guilderson, J. Adkins, M. Sarnthein, L. Baker, M. Yarusinsky, 2000: Abrupt onset and termination of the African humid period: Rapid climate response to gradual insolation forcing. *Quaternary Science Review,* **19,** 347–361.

Gitay, H., A. Suarez, R.T., and D.J. Dokken, 2002: Climate Change and Biodiversity, the Intergovernmental Panel on Climate Change, IPCC Technical Paper V, IPCC, Geneva.

Gray, J.S., R.S. Wu, and Y.Y. Or, 2002: Effects of hypoxia and organic enrichment on the coastal marine environment. *MEPS,* **238,** 249–279.

IPCC (Intergovernmental Panel on Climate Change), 2001: *Climate Change 2001: Synthesis Report,* IPCC R.T. (ed.), Cambridge University Press, Cambridge.

Jackson, J.B.C., M.X. Kirby, W.H. Berger, K.A. Bjorndal, L.W. Botsford, B.J. Bourque, R.H. Bradbury, R. Cooke, J. Erlandson, J.A. Estes, T.P. Hughes, S. Kidwell, C.B. Lange, H.S. Lenihan, J.M. Pandolfi, C.H. Peterson, R.S. Steneck, M.J. Tegner, and R.R Warner, 2001: Historical overfishing and the recent collapse of coastal ecosystems, *Science,* **293,** 629–637.

MA (Millennium Ecosystem Assessment), 2003: *Ecosystems and Human Well-being: A Framework for Assessment.* Island Press, Washington, DC.

Morita, T., and J. Robinson, 2001: Greenhouse Gas Emission Mitigation Scenarios and Implications. *Climate Change 2001—Mitigation.* Report of Working Group III of the Intergovernmental Panel on Climate Change. B. Metz, O. Davidson, R. Swart, and J. Pan. Cambridge University Press, Cambridge, 115–166.

Myers, R.A. and B. Worm, 2003: Rapid worldwide depletion of predatory fish communities. *Nature,* **423,** 280–283.

Nakićenović, N., J. Alcamo, G. Davis, B. de Vries, J. Fenhann, S. Gaffin, K. Gregory, A. Grübler, Tae Yong Jung, T. Kram, E. Lebre La Rovere, L. Michaelis, S. Mori, T. Morita, W. Pepper, H. Pitcher, L. Price, K. Riahi, A. Roehrl, H.-H. Rogner, A. Sankovski, M. Schlesinger, P. Shukla, S. Smith, R. Swart, S. van Rooijen, N. Victor, Zhou Dadi 2000: *Special Report on Emissions Scenarios* (SRES). Working Group III of the Intergovernmental Panel on Climate Change (IPCC), 595 pp. Cambridge: Cambridge University Press. Available at http://www.grida.no/climate/ipcc/emission/index.htm.

Naqvi, S.W.A., D.A. Jayakumar, P.V. Narvekar, H.Naik, V.V.S.S.Sarma, W.D. Souza, S. Joseph and M.D. George, 2000: Increased marine production of N_2O due to intensifying anoxia on the Indian continental shelf. *Nature,* 2000, Nov 16, **408 (6810),** 346–349.

Peterson, G.D., S.R. Carpenter, and W.A. Brock, 2003: Uncertainty and the management of multistate ecosystems: An apparently rational route to collapse. *Ecology,* **84(6),** 1403–1411.

Rabalais, N., 2002: Nitrogen in aquatic ecosystems. *Ambio,* **31,** 102–112.

Raskin, P., G. Gallopin, A. Hammond, and R. Swart, 1998: *Bending the Curve: Toward Global Sustainability, A Report of the Global Scenario Group.* Stockholm, Stockholm Environment Institute.

Rayner, S. and E. Malone, 1988: The challenge of climate change to the social sciences. In: *Human Choice and Climate Change, Volume 4—What Have We Learned.* S. Rayner and E. Malone. Battelle Press, Columbus, OH.

Robinson, J., and P. Timmerman, 1993: Myths, rules, artifacts, ecosystems: framing the human dimensions of global change. In: *Human Ecology: Crossing Boundaries,* T. D. S. Wright, R. Borden, G. Young and G. Guagnano. The Society for Human Ecology, Colorado, pp. 236–246.

Schwartz, P., 1992: *The Art of the Long View.* London, Century Business.

Steffen, W., M.O. Andreae, B. Bolin, P.J. Crutzen, P. Cox, U. Cubasch, H. Held, N. Nakićenović, R.J. Scholes, L. Talaue-McManus, B.L. Turner, 2004: Abrupt changes: the Achilles heels of the earth system. *Environment,* **46(3),** 8–20.

Swart, R., P. Raskin, J. Robinson, 2004: The problem of the future: sustainability science and scenario analysis. *Global Environmental Change,* **14,** 137–146.

Wack, P., 1985a: Scenarios: uncharted waters ahead. *Harvard Business Review,* **5** (Sept./Oct.), 72–89.

Wack, P., 1985b: Scenarios: shooting the rapids. *Harvard Business Review,* **6** (Nov./Dec.), 139–150.

Chapter 14

Policy Synthesis for Key Stakeholders

Coordinating Lead Author: Ferenc L. Toth
Lead Authors: Eva Hizsnyik, Jacob Park, Kathryn Saterson, Andrew Stott
Contributing Authors: Douglas Beard, Danielle Deane, Claudia Ringler, Detlef van Vuuren
Review Editors: Wang Rusong, Antonio La Viña, Mohan Munasinghe, Otton Solis

---

FIGURES

TABLES

Main Messages

The MA scenarios demonstrate the fundamental interdependence between climate change, energy, biodiversity, wetlands, desertification, food, health, trade, and the economy—and thus the need for relevant international agreements to work together to sustain life on Earth. This interdependence between environmental and development goals stresses the importance of partnerships and the potential for synergies among multilateral environmental agreements. As the basis for international cooperation, all global environmental agreements will operate under profoundly different circumstances in the four scenarios, and their current instruments (such as exchange of scientific information and knowledge, technology transfer, benefit sharing, and financial support) might need to be revised and complemented by new ones suited to changing sociopolitical conditions.

The interdependence between socioeconomic development and ecosystems services also requires national governments and intergovernmental organizations to provide the enabling conditions and to regulate the actions of the private sector, communities, and nongovernmental organizations. The responsibility of national governments to establish good governance at the national and subnational levels is complemented by their obligation to shape the international context and enabling conditions by negotiating, endorsing, and implementing international environmental agreements. Current and improving international instruments have better prospects to promote sustainable use of ecosystem services in the Global Orchestration and TechnoGarden futures, while national and local eco-management initiatives play a central role in the Adapting Mosaic scenario.

The MA scenarios show that the present focus of activity within the Convention on Biological Diversity on meeting the World Summit on Sustainable Development's target of significantly reducing rates of biodiversity loss will be difficult to achieve. The pressures on biodiversity will continue to grow during the twenty-first century, particularly through population and economic growth and the additional effects of climate change and pollution. All development pathways described by the MA scenarios have potentially significant negative impacts on biodiversity and its related ecosystem goods and services. The work programs of the CBD already include many of the actions needed to reduce these impacts, and these actions are implemented with varying degrees of success within the differing scenarios. For example, targets and associated actions within the CBD's Global Strategy for Plant Conservation emphasize the issues of habitat loss, conservation of protected areas, and sustainable management, but they may need to be expanded by actions to address the increasing threats of climate change and air pollution. The scenarios also anticipate the exacerbating regional disparity of impacts due to growing populations and economies of Asia, Latin America, and sub-Saharan Africa.

The nature and magnitude of future stress on wetlands and the prospects under the Ramsar Convention for helping to protect them are diverse across the scenarios: some stresses are stronger in the globalization scenarios, others are larger in the regional fragmentation scenarios. Existing international protection mechanisms have better prospects for success in the globally connected worlds and might need to be reformed in response to weakened global institutions of the locally oriented development paths. Greater pressure for agricultural land and massive increases in water withdrawals pose larger threats of wetland drainage and conversion in the regionally fragmented scenarios (Adapting Mosaic and Order from Strength) than the significant but smaller land and water stresses in the high-growth globalized worlds. In addition to more efficient technologies (TechnoGarden) or institutions (Global Orchestration), the latter scenarios imply stronger motivation to undertake and more effective instruments to implement wetland conservation under a global environmental agreement. An important feature of the Adapting Mosaic scenario is nonetheless that it pictures environmentally oriented societies that find practices and resources for cleaning return flows and for restoration of wetlands even in the absence of economic value, although the success of land and ecosystems co-management varies across regions.

The magnitude of future pressures causing desertification and the opportunities for the Desertification Convention to help mitigate the process vary across the scenarios: pressure is largest in Order from Strength, more modest in Adapting Mosaic and TechnoGarden, and lowest in Global Orchestration. Prospects for financial and technology transfers to help combat desertification are better in the globalization scenarios and more difficult in the fragmentation scenarios. All combinations of slow-to-fast population growth and improving economic conditions over the next decades will exert additional pressure on land resources and pose additional risk of desertification in dryland regions. Opportunities for the Desertification Convention will differ according to the diverse sociopolitical, economic, and technological conditions described in the specific scenarios. In a globalizing world, prospects for international environmental cooperation and resource transfers to support their implementation are likely to be better due to the institutional reforms (Global Orchestration) or because of the fast rate of technological development (TechnoGarden). It also requires political willingness in the affected countries to rank land degradation high on their political agenda and to commit national resources to fighting it. In a fragmented world, the role of a global agreement is more limited either because of the diminished interest in resource transfers (Adapting Mosaic) or because of the total lack of interest in what is going on beyond national or regional boundaries (Order from Strength). Yet in Adapting Mosaic, proactive local strategies might mitigate land degradation and reduce the need for global instruments.

Prospects for reaching the Millennium Development Goals by 2015 vary across the scenarios, geographical regions, and the goals themselves: halving poverty by 2015 is more likely to be achieved in a globalizing world in Latin America, South Asia, and India, while hunger will remain in most regions in all scenarios. Income growth is fastest in Global Orchestration and slowest in Order from Strength, on average. Halving the share of population (by 2015 relative to 1990) living on less than $1 a day has already been achieved in the East-Asia/Pacific region and in China. This target is likely to be achieved in Latin America, South Asia, and India under Global Orchestration but not under Order from Strength. Reaching this target in the Middle East and North Africa and in sub-Saharan Africa is unlikely under all four scenarios. The scenarios almost uniformly indicate that it will be difficult to halve undernourishment by 2015 in most regions except China under Global Orchestration and TechnoGarden and in Latin America under TechnoGarden despite sufficient, stable, or slightly increasing average availability of per capita dietary energy (except in sub-Saharan Africa). This implies that hunger remains an economic (income) and social (equity and distribution) issue rather than solely a natural resource/ecosystem problem.

Global environmental sustainability goals, which are part of the MDGs, largely fail, while local environmental quality is projected to improve in some scenarios. These general patterns, however, disguise considerable heterogeneity across regions and over the scenarios' time horizons. Total area covered by forests declines slightly globally, but a strong contrast exists between increasing forest areas in the OECD and the former Soviet Union and declining forest areas in all developing regions, especially sub-Saharan Africa and Middle-East/North Africa. Greenhouse gas emissions are projected to increase under all four scenarios in the OECD, to decline somewhat (except under Order from Strength) in the former Soviet Union, and to increase drastically in all developing regions. The prospects for improving local environmental quality are better. There is a good chance to reach the MDG of halving the proportion of people without sustainable access to safe drinking water in most regions except sub-Saharan Africa (despite fast progress) and Latin America (due to slow progress).

The MA scenario implications for ecosystem services and human well-being are of primary interest to local communities, NGOs, and other participants in civil society, as they often depend more directly on ecosystem services for daily well-being than other institutional actors (such as corporations) do. While human well-being and GDP per person improves on average in all the scenarios except Order from Strength, this masks increased inequity. The resulting ecosystem degradation causes a decline in per capita aggregated ecosystem services in all scenarios. Opportunities and priorities for community and NGO response differ across the scenarios. The "worst case" scenario for communities and NGOs is Order from Strength, in which community health and well-being are threatened by loss of biological diversity and associated ecosystem services, decreases in the availability and quality of fresh water, climate change, and decreases in air quality. The reactive, regionalized Order from Strength focus would offer little opportunity for success in community and NGO attempts at co-management of resources or at partnerships with other actors at multiple scales due to limited financial support for NGO activity and the challenge of finding ways for global policies to also be reactive to local problems.

While the Global Orchestration scenario might offer significant financial support for social and environmental NGOs, it also describes high risk of adverse impacts from climate change and little opportunity for NGOs and communities to foresee and prevent the thresholds at which further ecosystem degradation and reductions in human well-being might occur. The greater political commitment to address environmental issues in the TechnoGarden and Adapting Mosaic scenarios contributes to less severe implications for biodiversity loss, loss of ecosystems services, and impacts on human well-being. These two scenarios offer the greatest opportunity for communities to obtain land and resource tenure, maintain and use traditional knowledge, and partner with NGOs and other actors to respond successfully to emerging threats. The more institutional and behavioral strategy of Adapting Mosaic might encourage monitoring of indicators of ecosystem change at all levels in order to enhance the ability of communities to anticipate and adapt to change that threatens community livelihoods and health.

Local communities and NGOs can work together with government and the private sector to advocate policies and to execute on-the-ground practices that protect, mitigate, and restore some of the ecosystem services that are threatened by the development paths and assumptions in the four scenarios. NGOs and communities often know what needs to be done; they just need partnerships and financial resources to make it happen. In all scenarios, NGOs and communities can be more strategic in their efforts to integrate environmental imperatives with political realities. The synthesis of changes in ecosystem provisioning and regulating services indicates that in 2050 the trade-offs between ecosystem services will be more intense than at present, there will be greater inequities between rich and poor nations and regions, and there will be greater adverse impacts from unanticipated disasters. This implies that environmental justice and ethics should be of even greater concern to communities and NGOs than they are today.

A critical component of better understanding and managing the interrelationship between human well-being and ecosystem services is the identification of crucial links between ecosystems and the private sector. Climate change, water, and biodiversity loss are likely to pose the greatest policy concerns to the private sector in 2000–50. Climate change is likely to have a significant level of private-sector involvement in the near term (~2010), since it is an issue with sufficient media attention and institutional capacity. The emphasis on globalization and international technology cooperation in Global Orchestration and TechnoGarden highlights the important role that private-sector actors, particularly multinational corporations, will have to play in addressing global environmental policy concerns. At the same time, increased involvement of multinational corporations may lead to greater "privatization" of global environmental governance, diminished government/civil society oversight, and greater criticisms of western eco-business strategies from poor countries. The Order from Strength scenario is likely to lead to the greatest conflict between wealthier countries and the poorer nations as well as within rich countries (most notably, between the United States and Europe). Regionally structured business–civil society partnerships are likely to be an important feature in the Adapting Mosaic scenario. Yet a more geographically sensitive approach may also result in greater fragmentation and the duplication of policy approaches.

None of the scenarios can be singled out as the most desirable future. Each scenario has several positive and negative characteristics because each entails different combinations of relatively smaller or larger ecosystems stresses and more or less stakeholder capacity to cope with the emerging risks. Because of the need to make socioeconomic choices among mutually exclusive options and because of the biophysical trade-offs among ecosystems functions and services, it is not possible to handpick a combination of drivers and ecosystem management strategies to achieve what might appear to be the best selection of features across scenarios. Thus, not even the most brilliant and committed policy-makers operating in a highly cooperative international community could achieve such dreamworld futures. The cornerstone of masterly policy-making is finding the best compromises among conflicting objectives, making appropriate interventions to achieve them, and doing regular reassessments of policies against anticipated and unanticipated outcomes.

14.1 Introduction

The MA scenarios—four plausible pathways into the future—were conceived and developed to provide insights for a broad range of private stakeholders and public policy-makers into the risks and opportunities that might emerge for ecosystems and their provisions of various functions and services under four distinctively different but plausible futures. Preceding chapters in this volume present the social, economic, and political characteristics of the four development paths, their consequences for the demand for ecosystems services, the principal ways societies manage their relations with nature to fulfill those demands, the fundamental implications for ecosystems, and the consequences of how ecosystems respond to the combinations of driving forces. Chapters 9, 10, and 11 provide cross-scenario comparisons of provisioning/regulating functions, biodiversity, and human well-being.

This final chapter summarizes the implications of the scenarios for diverse groups of stakeholders, ranging from local communities to those managing international environmental agreements. Moreover, it seeks to assess the most promising response options that might be available to different actors under the four scenarios to manage emerging ecosystem conditions—both threats and opportunities—according to the stakeholders' objectives (government, communities, the private sector) or mandates (international agreements).

A general trend of accelerating globalization can be observed in recent decades: most national governments delegate smaller or larger parts of their sovereignty to supranational or multinational institutions (to the European Union, for example, or international economic, environmental, and other agreements), an increasing number of private enterprises operate across national boundaries, communities organize themselves into international networks (such as Klimabündnis), environmental movements establish global organizations (WWF, for instance, and Greenpeace), and even antiglobalization movements are globalizing themselves: witness the recent mega-gatherings at the World Social Forum in Porto Alegre, Brazil, and in Mumbai, India. Nonetheless, national governments are likely to maintain their central role in coordinating and regulating most aspects of socioeconomic development, including societies' relationships with ecosystems. The nature and the exact form of the regulation and the distribution of responsibilities among central governments, communities, and the private sector may vary depending on the broad sociopolitical features of the scenario, but the key role of national governments is likely to continue.

The first part of this chapter deals with the three main international conventions concerned with broad environmental issues or specific ecosystems: the biodiversity convention, the wetlands accord, and the desertification agreement. (Although the MA scenarios contain some information on climate change and its impacts, this chapter does not assess the implications for the climate change convention; this could be done in the IPCC's Fourth Assessment Report.) These and other international agreements regulating the many facets of international relations among nation-states are negotiated and signed by national governments. Governments also provide the institutional framework for domestic implementation. Moreover, the central role of the government in the domestic sphere involves delineating and negotiating the distribution of power and responsibilities between communities as well as demarcating guidelines for the private sector and NGOs. Key aspects of the relationships among all the stakeholders shaping the fate of ecosystems and human well-being are highlighted in the final section of the chapter.

Each stakeholder section starts with a brief overview of the main mandates related to or key interests of the group in ecosystems and their services. This is followed by brief assessments of the main threats and opportunities concerning those interests and mandates under the four MA scenarios. Finally, a set of response options are considered and analyzed that are available to the stakeholder group in order to identify those that might be potentially effective and successful under the social and political circumstances of a scenario. At the end of each section, a summary table presents qualitative assessments of the threats and opportunities and of the prospects for response options and interventions to manage them. The only exception is the section of national governments, as their main function, in addition to keeping policies and regulations sufficiently flexible to accommodate changes in external conditions, is to shape future trends and driving forces rather than just adapt to them.

The remainder of this section recaps the four MA scenarios, which are based on contrasting assumptions about the driving forces that are currently changing the world—demographic developments, the rate and structure of economic growth, sociopolitical developments like changing governance systems, cultural factors, and possible developments in science and technology. None of these factors work in isolation, and thus the scenarios contain a number of explicit and implicit assumptions about how the different driving forces interact and what their weight is and will be in the years to come. All these factors determine how natural systems are used to provide the services required for human survival and thus change direct factors of ecosystem changes, such as land use or pollution regimes.

Each of the trajectories that the scenarios portray begins with a number of choices made today or in the very near future. Many of these decisions are quite substantial and require wider changes in policies worldwide. Quite a few of these decisions are based on possibilities we currently see emerging and that are being discussed in various policy fora around the world. All these general policy directions nevertheless require concrete measures to make political choices a reality. The direction these choices go in the real world will determine how we and our children will live in the future, and the real future is likely to represent a mix of various strategies and options described in the scenarios.

The trajectory of the Global Orchestration scenario is based on the strong commitment of governments and other policy-makers to tackle the problems currently plaguing societies. Eradicating hunger and poverty worldwide and fostering the creation of more equitable, democratic societies that give citizens equal opportunities is seen by policy-makers in this scenarios as the foremost task in the years come. Therefore the main focus is developing human and social capital and restructuring economic and social systems. Measures to reach these goals include the creation of equitable access of all players to global markets by eliminating distorting subsidies and trade barriers (the Doha Round of WTO negotiations was to be a first step in this direction), overhauling social systems, investing in education, and ensuring the creation and maintenance of global public goods by rethinking and redefining the role of public and private-sector investments in science and technology. Environmental problems are not forgotten, but they only enter the policy-making arena if they are large-scale or affect a bigger number of people. Otherwise they are dealt with in a reactive manner, fixing what is possible to remedy in the short run but not putting particular attention to the development of long-term solutions that prevent mismanagement of ecosystems.

The Order from Strength scenario trajectory starts off with growing mistrust in global institutions, like the United Nations, and in their ability to find solutions to today's problems. Strong countries feel increasingly that they need to take matters into their own hands to ensure that their integrity and security is not threatened by outside forces they cannot control. These nations focus mainly on internal issues and are only concerned with developments outside their own borders if they directly affect their own country.

This attitude also leads to a retreat from a number of global agreements such as WTO or the Kyoto Protocol if they are seen as being out of harmony with country interests. These developments eventually result in a growing fragmentation between stronger and weaker countries. But this attitude also affects developments within nations. More powerful or wealthier groups try to make sure that things work for them, neglecting some of the costs this might have for others. This attitude then results in a growing fragmentation within society. Although the environment is not forgotten, growing environmental problems are only dealt with whenever they directly affect people or if the benefits from environmentally friendly management are perceived to substantially outweigh costs. Particularly in currently developing countries, scarce financial and deteriorating natural capital forces decision-makers to make tough choices between long-term solutions and short-term fixes to arising problems.

The Adapting Mosaic scenario starts in a similar way as Order from Strength, in that it sets off with the growing conviction of decision-makers around the globe that the solutions to many problems need concrete remedies at the local and national level. A second notion though makes this scenario very different: The focus on local solutions is not driven by overall security concerns but by the growing understanding of human-ecosystem connections and the importance of maintaining the functioning of the whole suite of local ecosystem services that underpin local economic systems. Increasingly the diversity of local systems is seen as an important asset that needs to be fostered, as it provides a variety of new solutions to old problems. Human and ecological systems are seen as evolving together. This nevertheless also requires changes in resource management and governance systems, leading to the devolution of power to local resource users, which is not always and in all nations a smooth process. This development, though, is thought to eventually result in the emergence of new governance systems and organizations not just at the global level, but also at the regional and global scale.

The TechnoGarden scenario trajectory also starts off with a change in the definition of the importance of ecosystem services and their relationship to economic systems. As in Adapting Mosaic, maintaining all categories of ecosystem services and taking a proactive approach to their management is increasingly felt to be necessary in order to guarantee the smooth functioning of human systems. In this scenario, however, technology is seen as the key to managing ecosystems; "natural capitalism," which focuses on obtaining profits by working with nature, is perceived to be advantageous for both individuals and society. Policymakers all over the world push for and invest in the development of environmentally friendly, "green" technologies that allow for a better management of the ecosystems for human purposes. Examples are new technologies for "cleaner" transportation systems or new urban planning and building schemes. One example of a measure that can set off this trajectory is the move of the European Union from production-based agricultural subsidies to payments for environmental services of farmers.

Each scenario trajectory together with decisions taken along the way will result in quite different outcomes by 2050, and each outcome will encompass different trade-offs. None of the future worlds described have only positive or negative outcomes. In Global Orchestration, the main trade-offs consist in managing ecosystems for their provisioning services at the expense of regulating, supporting, and cultural services. In addition, long-term maintenance of all services is traded off for current benefits to human societies. This trade-off is even stronger in the Order from Strength scenario. In Adapting Mosaic, trade-offs between ecosystem service categories and between services and human well-being components exist, but due to the varying nature of pursued management strategies around the globe (the "mosaic" of different experiments, approaches, and strategies), no overall trade-off paradigm exists. Rather a diversity of trade-off decisions emerges. The Techno-Garden world explores the double- edged sword of technology, which can have large beneficial effects but is also prone to failures. In addition, cultural ecosystem services are undervalued, and they are traded off for improvements in other services.

Improvements for human well-being can be found in all four scenarios but with very different rates of improvement and very different groups of society or countries winning or losing. And the environmental costs for human gains also differ widely between the scenarios. In three of them, human well-being overall improves but the costs and the risks of each development path on the environmental side vary. None of the scenarios portrays a complete breakdown of all ecosystem services, but many decisive steps and decisions have to be taken to change trajectories and avert some of the currently existing risks of ecosystem degradation and depletion. In reality, the future will be a mix of all the different approaches, strategies, and decisions that the scenarios portray, but many tough choices will have to be made along the way.

14.2 Implications for the Convention on Biological Diversity

The objectives of the Convention on Biological Diversity are the conservation of biological diversity, the sustainable use of its components, and the fair and equitable sharing of the benefits arising out of the utilization of genetic resources. Biological diversity means the variability among living organisms from all sources including, among other components, terrestrial, marine, and other aquatic ecosystems and ecological complexes of which they are part; this includes diversity within species, between species, and of ecosystems. Sustainable use means the use of components of biological diversity in a way and at a rate that does not lead to the long-term decline of biological diversity, thereby maintaining its potential to meet the needs and aspirations of present and future generations.

The objectives are translated into policies and concrete action through the agreement of international guidelines and the implementation of work programs of the Convention and of National Biodiversity Strategies and Action Plans. The Convention is developing seven thematic work programs—on forest diversity, dry and subhumid lands, biodiversity of inland waters, marine and coastal biodiversity, agricultural biodiversity, mountain biodiversity, and island biodiversity. Cross-cutting issues include, among others, biosafety; access to genetic resources; traditional knowledge, innovations, and practices; indicators; taxonomy; public education and awareness; incentives; and invasive alien species. Some cross-cutting initiatives directly support work under the thematic programs, such as the work on indicators. Others are developing discrete products that may be separate from the thematic programs. The convention has adopted the "ecosystem approach" as a strategy for the integrated management of land, water, and living resources that promotes conservation and sustainable use in an equitable way.

The sixth meeting of the Conference of the Parties in April 2002 adopted the Strategic Plan for the Convention, which commits Parties to "achieve by 2010 a significant reduction of the current rate of biodiversity loss at the global, regional, and national level as a contribution to poverty alleviation and to the benefit of all life on earth" (Decision VI/26). The Strategic Plan also commits Parties to a more effective and coherent implementation of the three objectives of the Convention.

At the World Summit on Sustainable Development in Johannesburg in August/September 2002, governments adopted a Plan of Implementation that reconfirmed the role of the CBD as the key instrument for the conservation and sustainable use of biological diversity and the fair and equitable sharing of benefits arising from its use. With respect to the 2010 target, the WSSD Plan of Implementation recognizes that "the achievement by 2010 of a significant reduction in the current rate of loss of biological diversity will require the provision of new and additional financial and technical resources" (paragraph 44).

While world political leaders have agreed that "biodiversity loss" constitutes a serious challenge at the global, regional, and national level, there is as yet no widely accepted definition of what biodiversity loss means or how it can be monitored or assessed. The following definition of biodiversity loss was proposed at the 2010–The Global Biodiversity Challenge Conference in London in 2003 (UNEP/CBD/SBSTTA/9/INF/9), and adopted by the seventh meeting of the CBD Conference of the Parties in Kuala Lumpur in 2004 (Decision VII/30): "the long term or permanent qualitative or quantitative reduction in components of biodiversity and their potential to provide goods and services, to be measured at global, regional and national levels."

COP7 also decided to establish a small number of global goals and sub-targets to clarify the 2010 global biodiversity target, covering six focal areas of the convention. Further work is required to integrate the goals and targets into the work programs of the convention. In order to assess progress at the global level toward the 2010 target, COP7 agreed that a balanced set of indicators should be identified or developed (Decision VII/30), as described later.

The outcomes of the MA scenarios are highly relevant to the immediate work of developing global goals, subtargets, and indicators for assessment of progress toward the 2010 target. The Subsidiary Body on Scientific Technical and Technological Advice has recommended that the targets should be challenging but realistic, recognizing the constraints of Parties, especially developing countries. The MA scenarios can help in the process of setting realistic and attainable outcome-oriented targets within the work programs of the convention, as these are reviewed over the next few years. COP7 invited other related assessment processes such as the MA to contribute reports and information that assist in monitoring progress toward the 2010 target.

At its sixth meeting, the Conference of the Parties adopted the Global Strategy for Plant Conservation as a pilot approach for the use of outcome targets for the convention and to consider the application of the approach to other areas (Decision VI/9). The GSPC includes 16 specific and measurable targets for 2010. It therefore offers a case study to evaluate the MA scenarios against specific CBD targets and to provide feedback to the convention on the general use of outcome targets.

No consideration has yet been given by international policy-makers to establishing targets over longer time scales (up to 2050). However, the CBD objectives imply that biological diversity, at ecosystem, species, and genetic levels, should be conserved indefinitely in order to maintain its potential to meet the needs and aspirations of present and future generations. For this longer time scale, the MA scenarios help inform future policy direction within the CBD by identifying the future risks to biological diversity and how these risks vary with different response options.

14.2.1 Threats to Biodiversity in the MA Scenarios

The main global-level threats to biodiversity identified within the work programs of the CBD are habitat transformation (such as conversion to agriculture, urbanization and infrastructure development, fragmentation, and mining and engineering works); overexploitation (such as overgrazing, overharvesting, overfishing, loss of plant and animal genetic resources, and water abstraction); inappropriate management (such as undergrazing, changes in fire regimes, and soil erosion); invasive alien species; pollution (such as sulfur and nitrogen emissions); and climate change (such as long-term changes in temperature and rainfall, extreme events, and sea level change).

The quantitative outputs of the MA scenarios are mapped onto the main threats to biodiversity in Table 14.1. The association between MA output variables and threats is not precise, and there are significant aspects of the threats that are not represented within the MA outputs. Gaps in coverage relate particularly to fisheries, inappropriate management, and invasive alien species. However, the association between MA scenario output variables and biodiversity threats is sufficient to identify the general, long-term risks to meeting the objectives of the CBD. (Chapter 10 provides quantitative information of expected impacts on global bio-

Table 14.1. MA Quantitative Scenario Outputs Related to Main Threats to Biodiversity

Threats to Biodiversity	Quantitative Scenario Outputs[a]
Habitat transformation	change in agriculture area conversion of forests fragmentation and biodiversity loss population growth (urbanization) fossil fuel extraction
Overexploitation and inappropriate management	agricultural intensification water abstraction
Invasive species	
Pollution	emissions of SO_2 and NO_X excess of critical loads return flows to rivers
Climate change	temperature rainfall biome shift

[a] This is not a comprehensive list of possible threats but a list of threats that have been quantified in the MA scenarios.

diversity, in particular loss of habitats, loss of plant species, and shifts in terrestrial biomes due to climate change.)

14.2.1.1 Habitat Transformation

In the Global Orchestration and Adapting Mosaic scenarios, global rates of forest loss due to agricultural expansion are similar to present rates, while in TechnoGarden they are slightly lower. Rates increase by 50% in Order from Strength up to 2020. In all scenarios there is a large increase of the rate of forest loss in sub-Saharan Africa and a lesser increase in OECD countries.

Biodiversity losses occur directly through loss of habitat and indirectly through fragmentation. The results show a decline of 12–16% in vascular plant species diversity as a consequence of global habitat loss between 1970 and 2050, assuming that species diversity eventually reaches an equilibrium with the area of habitat available. The highest losses occur in Order from Strength and the lowest in the TechnoGarden and Adapting Mosaic. Rates of loss in plant diversity increase between the two time periods 1980–2000 and 2000–20 in Order from Strength and Global Orchestration by 40% and 10%, respectively, but decline by 15–20% in the other two scenarios. There are major differences in plant diversity between the different biomes, and tropical forest, tropical woodland, savanna, and warm mixed forest account for 80% of all plant species lost. The severity of impact of habitat transformation on biodiversity depends largely on details of habitat conversion. If biodiversity hot spots and functioning ecological networks are maintained within protected areas or by other conservation mechanisms, then risks of massive biodiversity loss may be reduced. Nonlinear and lagged responses may occur as habitats become progressively isolated and reduced in size.

14.2.1.2 Overexploitation

Agricultural intensification occurs under all scenarios, but especially in Global Orchestration and TechnoGarden, where intensification enables increased food production with less land-take for agriculture. Intensification, including introduction of new crop/livestock varieties, management, fertilizer, and pesticide regimes, is likely to be detrimental to wildlife species and genetic varieties of crops and livestock that are associated with traditional/low intensity agricultural habitats. Risks to biodiversity may be reduced in TechnoGarden by adoption of appropriate management regimes or traditional practices (such as preservation of uncultivated areas and linear habitats) in agricultural ecosystems of high importance for biodiversity. Under the Adapting Mosaic scenario, genetic diversity used by people is increased by spatial heterogeneity of ecosystem management.

Water abstraction and water stress are critical threats to wetland ecosystems. Water abstractions increase to meet population growth and irrigation demands in all scenarios by between 20% and 80% globally, with two- to threefold increases in sub-Saharan Africa and Latin America. Increased abstractions exceed expected increases in precipitation (due to climate change) and create water stress under Global Orchestration and especially under Adapting Mosaic and Order from Strength. Geographical variations in future precipitation are highly uncertain. Wetland habitats in sub-Saharan Africa and Latin America, in catchments where increased demand coincides with lower precipitation, are most vulnerable to reduced water levels. However, under TechnoGarden, reduced abstractions may enable restoration of wetlands in the former Soviet Union.

GDP per person increases in all scenarios, especially in Global Orchestration and TechnoGarden, and especially in Asia and the OECD. Growing income levels, coupled with increased populations, are likely to intensify pressure from tourism, leading to habitat loss and overexploitation. However, there will also be more opportunity for tourism to provide self-funding opportunities for biodiversity conservation. Both positive and negative impacts of tourism are likely to be highly localized. Global tourism is most likely to increase under Global Orchestration and TechnoGarden.

14.2.1.3 Pollution

Sulfur dioxide emissions can cause acidification impacts, especially in freshwater ecosystems, where high levels of deposition occur on acidic soils with low buffering capacity (as in Scandinavia and North America). Global sulfur dioxide emissions fall in all scenarios, but especially in TechnoGarden and Global Orchestration. Regionally, however, increases occur from existing low levels in sub-Saharan Africa in all scenarios. Following large reductions in emissions in OECD, Asia becomes the dominant source of sulfur dioxide under all scenarios. The scenarios indicate a reduced acidification risk in OECD and the former Soviet Union. There is a high risk of acidification becoming a localized problem within vulnerable ecosystems in Asia under Adapting Mosaic and Order from Strength scenarios.

Nitrogen oxide emissions can cause eutrophication (artificially raised nutrient levels), especially where high levels of deposition occur in low nutrient status terrestrial ecosystems (such as in lowland heaths in northern Europe). Global nitrogen oxide emissions increase under all scenarios by 20–

50%, with the highest increase occurring in Global Orchestration. (Ammonia emissions have not been modeled.) Emissions are likely to be reduced in OECD but increase two- to fourfold in Asia and the former Soviet Union. There is a high risk of eutrophication becoming a significant problem within vulnerable ecosystems in Asia under all scenarios.

The combined impacts of acidification and eutrophication result in an overall estimated decline in plant species diversity of 2–5% across all terrestrial habitats by 2050. Temperate and warm mixed woodlands are most severely affected, with plant species diversity decline of 5–10% across the scenarios. Losses are highest in Global Orchestration and lowest in TechnoGarden.

Return flows, as an indication of freshwater and estuarine pollution, increase under all scenarios by 40–200%. Return flows are generally stable or reducing in OECD and the former Soviet Union, but there are large increases in sub-Saharan Africa and Latin America. There are therefore high risks of increased pollution of freshwater and recipient coastal habitats in those regions.

14.2.1.4 Climate Change

The impacts of climate change will be most severe where the rates of change in climatic variables exceed the rate of species dispersal and adaptation within biomes. In the four scenarios, about 5–20% of ecosystems will be seriously affected by climate change, the worst being Global Orchestration. In that case, in 20% of protected areas the originally protected ecosystem will have either been replaced or seriously damaged as a consequence of climate change alone. The most heavily affected biomes are boreal and cool conifer forests, tundra, shrubland, and savanna. In addition to shifts in zonal climates, coastal habitats are also affected by an increasing rate of sea level rise, reaching around 25 centimeters above 2000 levels by 2050 under all scenarios. Coral reefs, mangrove forests, and salt marshes are particularly vulnerable, but estimates of potential global losses are not available.

14.2.1.5 Combined Threats

The above threats to biodiversity do not act in isolation. Under most scenarios, and in most regions, there is a high risk that rapid climate change will occur concurrently with continuing loss and fragmentation of natural habitats and with increasing overexploitation of natural resources and pollution.

The combined impacts on biodiversity of land use change, climate change, emissions of greenhouse gases, and regional air pollutants have been modeled using the IMAGE integrated assessment framework. The outputs show that the area of agricultural land increases at the expense of natural habitats in all scenarios. The increase in area of agricultural land is as much as 24% in Order from Strength by 2050 but only 7–9% in the other scenarios. Tropical savanna is the most severely affected biome, with losses of between 27% in Adapting Mosaic and 55% in Order from Strength. Forested land as a whole shows a slight increase in all scenarios except Order from Strength. But within the forest biomes, gains in regrowth, boreal, and temperate mixed forests are offset by losses in the more species-rich tropical, temperate deciduous, and warm mixed forests under most scenarios. Order from Strength is the most extreme, with losses of 22% of tropical forest, 24% of temperate deciduous, and 35% of warm mixed forest by 2050. In contrast, under TechnoGarden there are gains in most forest biomes except tropical forests, which decrease by 11% by 2050.

Of the three main threats to terrestrial biodiversity, habitat loss emerges as the most significant pressure on biodiversity under all scenarios up to 2050. Habitat loss leads to 11–16% decline in biodiversity across all habitats. According to these models, climate change and air pollution are associated with lesser declines of 2–5%. However, there is strong differentiation between impacts on different biomes. The greatest pressure in tundra and desert biomes is climate change, whereas in warm mixed and tropical forests, habitat loss and air pollution are most significant. Savanna and temperate forest have high levels of pressure from all three factors. Boreal forest has low pressure from all three factors. There is less distinction between the four scenarios. Overall, Order from Strength creates the highest rates of habitat loss and Global Orchestration has the highest risk of climate change and air pollution impacts. In most biomes TechnoGarden has the lowest pressures for all three impacts. The highest threats to biodiversity in most scenarios are in sub-Saharan Africa and Latin America. In these regions pressures on biodiversity by 2050 are increased by factors of two to four above present levels. These regions also contain many of the world's existing hot spots of biodiversity. The lowest threats in most scenarios are found in the OECD and the former Soviet Union. TechnoGarden emerges consistently as the scenario with lowest pressure on biodiversity.

Losses in biodiversity—that is, loss of habitats, decline in species abundance, and loss of genetic diversity—have implications for ecosystem goods and services and human well-being. The qualitative assessment of the future vulnerability of ecosystem services shows strong differentiation between the scenarios. The highest vulnerability occurs in the Order from Strength scenario, with decreases in provisioning services (such as genetic resources and biochemical discoveries) and decreases in regulating services (such as water regulation and biological control). In Global Orchestration, ecosystem services are maintained in the North but show some losses in the South. In Adapting Mosaic and TechnoGarden, ecosystem services generally increase or are unchanged. Adapting Mosaic in particular shows increases in ecosystem services associated with biodiversity (such as genetic resources, ornamental resources, and biological control).

14.2.2 Prospects for the CBD

14.2.2.1 2010 Target

COP7 adopted a limited number of global indicators for assessing progress toward the 2010 target (Decision VII/30). These trial indicators are not yet specified in detail and they have not been evaluated directly by the MA scenarios.

However, there is some evidence from the quantitative scenario results to suggest the possible short-term trends in the aspects of biodiversity covered by these indicators. (See Table 14.2.) The evidence is inconclusive, but it suggests that the target is very challenging though achievable—at least, in some regions.

The pressures identified by the MA up to 2010 are mostly similar in character, scale, and intensity to those that the international community has experienced over the past 20 years and that are already the subject of the CBD work programs. However, emerging pressures from climate change and air pollution may not be adequately addressed. For example, targets and associated actions within the CBD Global Strategy for Plant Conservation (see Table 14.3) emphasize issues of habitat loss, conservation of protected areas, and sustainable management and pay less attention to the less tangible but increasing threats of climate change and air pollution. As all these pressures on biodiversity increase under the MA scenarios up to 2010, the policy responses need to extend and become more effective at global, regional, national, and local levels. This shows the need for full implementation and provision of adequate resources for existing CBD work programs. There is also evidence that the growing populations and economies of Asia, Latin America, and sub-Saharan Africa will exacerbate regional disparity of impacts. There is a real prospect that rates of biodiversity loss will slow or halt in rich nations while accelerating elsewhere.

14.2.2.2 Response Strategies beyond 2010

The CBD encompasses a comprehensive range of detailed response strategies within its work programs. Although space does not permit a full analysis of how these responses may develop within each program under the different scenarios, the Expanded Work Programme on Forest Biological Diversity (Decision VI/22) is used as an example. This was chosen because it contains a comprehensive set of policy responses that address the main threats to biodiversity assessed within the MA.

Table 14.4 summarizes the responses currently planned within the expanded work program and shows how these may develop under each scenario, based on a qualitative interpretation of the scenario storylines up to 2030. The results show that the wide range of current policy responses in the forest work program is generally robust to the different possible futures. The CBD appears to have anticipated the major dimensions of change captured in the MA scenarios. For example, when we looked at the threat of habitat loss we saw that response strategies regarding establishment of networks of protected areas would develop a different emphasis in each of the scenarios. In Global Orchestration, we anticipate that global networks of protected forest areas will be established with an emphasis on promoting the economic and social benefits of global tourism. In Order from Strength, we anticipate that regional or national networks of protected forest areas and private reserves will be the main policy tool for maintaining forest goods and services in wealthy countries, with ineffective networks and accelerated loss of forests elsewhere.

The goals and sub-targets agreed to by COP7 in Kuala Lumpur provide a framework for assessing longer-term implications of the MA scenarios for the CBD. Although these goals and sub-targets are primarily intended to clarify the 2010 biodiversity target, facilitate assessment of progress, and promote coherence among the programs of work, they are sufficiently general to be used as a guide to the longer-term objectives of the convention.

Table 14.5 compares the outcomes of the four MA scenarios for the period 2030–50 with respect to these CBD goals. TechnoGarden and Adapting Mosaic provide the most positive outcomes for the CBD. TechnoGarden combines multilateral regulation and management of global commons with an integrated, "ecosystem approach" to conservation of biodiversity within sustainable production systems. Adapting Mosaic also provides positive outcomes, but these are more regionally differentiated, as the best practices and resources for conservation of biodiversity are not universally applied. Traditional knowledge and rights of indigenous communities receive greater recognition, but global commons are not managed collectively. The Global Orchestration and, especially, Order from Strength scenarios have poor outcomes for the CBD goals. In Global Orchestration there is some success in conserving biodiversity in protected areas, at least within wealthy countries, and in benefit sharing and transfer of resources, but the CBD is marginalized in the drive for economic growth. In Order from Strength, the outcomes are overwhelmingly negative as the lack of global cooperation is compounded by increasing regional inequality and a failure to share benefits or transfer resources.

Table 14.6 provides a concise summary of key stresses for the CBD and the prospects for success of relevant response options under the four scenarios. The most favorable future scenario for conservation of biodiversity may combine elements of the TechnoGarden and Adapting Mosaic scenarios by developing strong international institutions for the sharing of information, guidance, and resources but still enabling regional and national diversity and recognizing the value of local knowledge and solutions. The work programs of the CBD and the national strategies and action plans already provide an appropriate response framework. In particular, the CBD provides a basis for international cooperation, exchange of scientific information and knowledge, access and benefit sharing, and transfer of financial resources and technology. The CBD has already developed guidance on sustainable use and the "ecosystem approach" and is working to establish synergies with the other Rio conventions and related multilateral environmental agreements.

The CBD recognizes the sovereign right of states to exploit their own resources pursuant to their own environmental policies and the responsibility to ensure that activities in their jurisdiction do not cause damage to the environment of other states. The CBD therefore relies primarily on the voluntary participation and cooperation of Parties in the implementation of its work programs. Efforts to introduce a stronger regulatory component, such as a protocol on protected areas, have been resisted, and instead the emphasis is

Table 14.2. Evidence from MA Scenarios for Provisional CBD Indicators for Assessing Progress toward the 2010 Biodiversity Target (CBD Decision VII/30)

Provisional Indicators	Evidence from Scenarios up to 2010
Components of biodiversity	
Trends in extent of selected biomes, ecosystems, and habitats	rate of natural forest loss continues at current rates, or accelerates; warm mixed forest and savanna most at risk from habitat loss; some restoration of forest and wetlands in OECD and former Soviet Union
Trends in abundance and distribution of selected species	increased pressures from habitat loss, overexploitation, and pollution; sub-Saharan Africa, Latin America, and Asia most at risk; temperate and warm mixed woodland most at risk from air pollution
Change in status of threatened species	threatened species not modeled directly but rate of extinction of vascular plant species due to habitat loss accelerates in OS and GO scenarios and slows in TG and AM scenarios; likely to be exacerbated by climate change; tropical forest, tropical woodland, savanna, and warm mixed forest account for 80% of all plant species lost
Trends in genetic diversity of domesticated animals, cultivated plants, and fish species of major socioeconomic importance	increased pressure from agricultural intensification; genetic resources decrease in OS
Coverage of protected areas	coverage of protected areas not modeled; protected areas at risk from longer-term climate change impacts, air pollution, and overexploitation
Sustainable use	
Area of forest, agricultural, and aquacultural ecosystems under sustainable management	not modeled; expected to vary in accordance with scenario storylines; increases in TG and AM scenarios
Proportion of products derived from sustainable sources	
Threats to biodiversity	
Nitrogen deposition	increases under all scenarios by 20–50% by 2050
Numbers and cost of alien invasions	not modeled; expected to increase as a result of climate change and increased global trade and mobility
Ecosystem integrity and ecosystem goods and services	
Marine trophic index	marine biodiversity modeling results uncertain
Fragmentation	not modeled
Human-induced ecosystem failure	not modeled; expected to vary in accordance with scenario storylines; most significant failures in OS and GO scenarios
Health and well-being of people living in biodiversity-based resource-dependent communities	not modeled; expected to vary in accordance with scenario storylines; most significant failures in OS and GO scenarios
Water quality	decreases under all scenarios by 40–200% by 2050
Biodiversity used in food and medicine	not modeled; expected to vary in accordance with scenario storylines; most significant uses in TG and AM scenarios
Traditional knowledge, innovations, and practices	
Linguistic diversity and numbers of speakers of indigenous languages	not modeled; expected to vary in accordance with scenario storylines; greatest diversity maintained in AM scenario
Access and benefit-sharing	
To be defined	not modeled; access likely to be greatest with GO and TG, least with AM; total benefits likely to be greatest with TG
Resource transfers	
Overseas development assistance Technology transfer	not modeled; expected to vary in accordance with scenario storylines; greatest resource and technology transfers in TG and GO scenarios

Key: GO = Global Orchestration; OS = Order from Strength; AM = Adapting Mosaic; TG = TechnoGarden

Table 14.3. Analysis of Future Trends Identified in MA Scenarios and Planned Actions up to 2010 within the CBD

Threats to Biodiversity	Current GSPC 2010 Targets	Planned Actions within CBD	Response to Future Trends Identified in MA
Habitat loss *Increasing pressure for agricultural and development land*	At least 10% of the world's ecological regions effectively conserved. Protection of 50% of the most important areas of plant diversity.	About 10% of the land surface is currently protected but some ecosystem types are poorly represented. Actions are needed to improve the representation of different ecosystems within protected areas and increase their effectiveness.	Strengthen protection, management, sustainable use, and funding of protected areas. Improve markets for ecosystem services and for common property and community-based management. Maintain and restore connectivity within fragmented ecosystems. Enhance yields from productive ecosystems to reduce pressure for agricultural expansion. Adopt flexible and forward-looking approach to PA networks that recognizes that the distributions of habitats and species will change as a consequence of climate change.
Overexploitation and inappropriate management *Increasing agricultural intensification, use of new technologies, and overharvesting of natural products*	At least 30% of productive lands managed consistent with the conservation of plant diversity. No species of wild flora endangered by international trade. 30% of plant-based products derived from sustainable sources. 70% of genetic diversity of crops conserved.	Conserve biodiversity within production systems (e.g., agriculture or forestry). Use management practices that avoid adverse impacts. Use integrated, sustainable management practices. Apply ecosystem approach to land use decisions and management. Extend certified products. Extend gene banks and acquisition of indigenous and local knowledge.	Promote sustainable use of productive lands. Promote more-effective education, incentives, regulation, and enforcement. Maintain traditional knowledge about plant varieties. Improve markets for ecosystem services and for common property and community-based management.
Invasive species *Increased risk of invasion due to climate change and world trade*	Management plans in place for at least 100 major alien invasive species.	Establish risk assessment procedures and management strategies at national levels.	Implement control strategies.
Pollution *Increased impacts of acidification and eutrophication, especially in temperate and warm mixed woodland*	No targets.	None within GSPC, but actions included in forest work program.	Establish monitoring protocols for impact assessment. Extend multilateral agreements on control of emissions. Improve efficiency of nitrogen use.
Climate change *Evidence of biodiversity impacts and first losses attributed to climate change*	No targets.	None.	Establish monitoring protocols and assessment tools. Review implications for in situ conservation objectives and policy instruments.

on promotion of voluntary guidelines. Overall, progress is largely determined by the commitment, effective voluntary participation, and cooperation of Parties, other nations, and relevant stakeholders from local to international levels, as well as the provision of adequate human and financial resources necessary for the conservation of biodiversity.

The present focus of activity within the CBD is toward meeting the WSSD target of significantly reducing the rate of biodiversity loss by 2010, recognizing the fundamental contribution that biodiversity makes to ecosystem goods and services and poverty reduction. The MA scenarios show that this target will be difficult to achieve by 2010 and that the pressures on biodiversity will continue to grow during the first half of the twenty-first century, particularly through population and economic growth and the additional effects of climate change and pollution. The immediate challenge for the CBD is to translate the growing evidence of rapid biodiversity loss and ecosystem failure, both observed and projected, and their implications for human well-being into willingness by governments to fully implement their commitments under the CBD. An important step toward addressing this challenge was made at COP7 by agreeing on a framework and a process to set outcome-oriented targets for the work programs of the convention and to assess progress using a limited number of global indicators. Clarity about the issues and the gravity of the situation is an essential stimulus to government action.

The MA scenarios make an important contribution to the evidence base and will be a useful tool in the ongoing process of formulating attainable targets for the convention. Inevitably there is not an exact match between the MA outputs and the goals, targets, and associated indicators that have subsequently been agreed on as priorities within the CBD. In the future, a better match should be achievable.

Table 14.4. Analysis of Future Trends Identified in MA Scenarios up to 2030 and Possible Responses within the CBD

Major Threats to Biodiversity and MA Trends	Planned Responses within CBD Programme on Forest Biological Diversity	Possible Responses in MA Scenarios up to 2030			
		Global Orchestration	*Order from Strength*	*Adapting Mosaic*	*TechnoGarden*
Habitat loss *Increasing pressure of conversion for agriculture, urbanization, and infrastructure*	Ensure adequate and effective protected area forest networks. Assess adequacy of existing PAs and establish effective networks. Prevent and mitigate losses due to fragmentation and conversion. Encourage creation of private reserves. Establish ecological corridors. Promote cost-benefit analysis of development projects, taking into account impacts on biodiversity.	Global networks of protected areas established. However, remaining areas of forest depleted and ineffective ecological corridors. Development projects do not take full a account of forest biodiversity in cost-benefit analysis; greater emphasis on economic and social benefits. Emphasis on economic and social values of forest biodiversity. Protected areas managed to provide economic and social benefits through tourism.	Strongly regulated networks of protected areas and private reserves established in some regions or countries. Elsewhere, areas of forest severely depleted and fragmented. Approach lacks global representation and flexibility in face of climate change. Emphasis on national and regional PA networks and private reserves as policy tool.	Effective networks of protected areas and ecological corridors in some regions or countries. Elsewhere, areas of forest depleted and fragmented. Development projects take account of forest ecosystem services and importance for well-being of indigenous and local communities. Emphasis on establishment of protected areas to maintain ecosystem goods and services and support indigenous and local communities within the ecosystem approach.	Effective and representative global networks of protected areas established. Remaining areas of forest reduced, but ecological corridors retained and established. Development projects take account of ecosystem services. Emphasis on protected areas, integrated with ecological networks to maintain ecosystem goods and services. Guidelines/ protocol adopted on protected areas.
Overexploitation and inappropriate management *Increasing demand for timber and overharvesting of natural products; increased fire risk due to human pressures and climate change*	Promote sustainable use of forest resources. Support activities of indigenous and local communities involving the use of traditional knowledge. Develop programs for sustainable use of timber and other forest products. Prevent losses caused by unsustainable harvesting. Prevent and mitigate adverse effects of forest fires. Develop guidance and adapt ecosystem approach to forests both inside and outside protected areas. Promote restoration of forest biodiversity to restore ecosystem services.	Consumer preferences drive sustainable use of timber and other forest products. Regulated global trade and certification schemes. Forest fire management driven by commercial timber considerations. Emphasis on promoting economic and social values of sustainable forest production. Ecosystem approach adapted to optimize economic and social benefits from sustainable use.	Sustainable use of forest resources promoted in wealthier countries, with establishment of effective regional certification schemes. Forests regarded as important recreational resource. Elsewhere, unsustainable harvesting and fire risk increases. Emphasis on national and regional protected area networks and private reserves as policy tool.	Ecosystem approach developed and adopted in some places both within and outside protected areas. Activities of indigenous and local communities supported. Unsustainable harvesting reduced in some regions and countries. Emphasis on developing the ecosystem approach and promoting local solutions to management problems.	Watershed management issues and carbon trading drive sustainable use of timber, substitution for forest products, and restoration and management of forests for biofuels. Development and sharing expertise in forest management. Emphasis on maintenance and restoration of ecosystem services.
Invasive species *Increased risk of invasion due to climate change and world trade*	Prevent the introduction of invasive alien species and mitigate negative impacts.	Emphasis on developing appropriate control methods where economic interests are at risk.	Emphasis on developing appropriate control methods for protected areas.	Emphasis on risk assessment and developing control methods.	Emphasis on risk assessment, monitoring, and prevention, including regulation of genetically modified organisms. Development of technology for excluding or eradicating invasive species.

(continues)

Table 14.4. Continued

Major Threats to Biodiversity and MA Trends	Planned Responses within CBD Programme on Forest Biological Diversity	Possible Responses in MA Scenarios up to 2030			
		Global Orchestration	*Order from Strength*	*Adapting Mosaic*	*TechnoGarden*
Pollution *Increased impacts of acidification and eutrophication, especially in temperate and warm mixed woodland*	Increase understanding of impact. Support monitoring programs. Promote reduction of pollution levels (sulfur dioxide and nitrogen oxides) and mitigate impacts.	Multilateral regional agreements on control of emissions relaxed. Emphasis on monitoring, assessment, and mitigation of impact of pollution on commercial forest products.	Failure of multilateral regional agreements on control of emissions. Monitoring protocols for impact assessment established in some regions and countries. Research undertaken to develop mitigation techniques for protected areas in wealthier regions and countries. Emphasis on monitoring, assessment. and mitigation of impact on protected areas.	Failure of multilateral regional agreements on control of emissions. Mitigation methods developed within ecosystem approach at a local level. Emphasis on monitoring, assessment, and mitigation of impact of pollution within ecosystem approach.	Multilateral regional agreements on control of emissions extended. Monitoring protocols for impact assessment established. Research undertaken to develop mitigation techniques. Emphasis on monitoring, assessment and mitigation of impact on ecosystem services and developing synergies with regional agreements on control of emissions.
Climate change *Evidence of biodiversity impacts and first losses attributed to climate change*	Promote monitoring and research on impacts of climate change. Promote maintenance and restoration of forest biodiversity to enhance capacity to resist or adapt to climate change. Promote forest biodiversity conservation and restoration in climate change mitigation and adaptation strategies.	Forest restoration an important component of adaptation strategies, including development assistance. Emphasis on promoting appropriate forest restoration and management strategies to maintain forest productivity.	Forest restoration an important component of adaptation strategies in wealthier regions and countries. Emphasis on monitoring impacts on protected areas and developing management guidelines for forest restoration and management in protected areas.	Forest restoration an important component of mitigation and adaptation strategies within ecosystem approach. Emphasis on providing guidelines for mitigation and adaptation to maintain ecosystem services and support indigenous and local communities.	Forest restoration an important component of mitigation and adaptation strategies, including development assistance. Management seeks to enhance capacity of forest ecosystems to adapt to change, including attempts to improve ecological connectivity. Emphasis on monitoring and research to anticipate climate change effects and develop guidelines for forest management and restoration strategies. Developing synergies with mitigation and adaptation strategies in the climate change convention.

14.3 Implications for the Ramsar Convention

Currently, wetlands cover about 6% of Earth's land surface. Besides their direct contribution to local economies through water supply, fisheries, forestry, agriculture, and tourism, they provide various ecosystems services, most notably biodiversity conservation. The Convention on Wetlands of International Importance especially as Waterfowl Habitat (the Ramsar Convention) is one of the oldest global environmental agreements and to date the only one dealing with a particular ecosystem. It defines wetlands in an all-encompassing manner: "Wetlands are areas of marsh, fen, peatland or water, whether natural or artificial, permanent or temporary, with water that is static or flowing, fresh, brackish or salt, including areas of marine water the depth of which at low tide does not exceed six meters" (Article 1.1 of the Convention).

The primary objective of the convention is to provide a framework for national action and international cooperation for the conservation and wise use of wetlands and their resources. The Convention defines wise use of wetlands as "their sustainable utilization for the benefit of human kind in a way compatible with the maintenance of the natural properties of the ecosystem." Sustainable utilization, in turn, is explained as "human use of a wetland so that it may yield the greatest continuous benefit to present generations

Table 14.5. Qualitative Comparison between Scenarios with Respect to Global Goals and Targets of the CBD up to 2050. Note that CBD targets are specified in relation to the WSSD 2010 global target.

	MA Scenarios			
CBD Goals and Targets	**GO**	**OS**	**AM**	**TG**
Protect the components of biodiversity				
Goal 1. Promote the conservation of the biological diversity of ecosystems, habitats, and biomes				
Target 1.1: At least 10% of each of the world's ecological regions effectively conserved	+/−	+/−	+	+
Target 1.2: Areas of particular importance to biodiversity protected	+/−	− −	+/−	+
Goal 2. Promote the conservation of species diversity				
Target 2.1: Restore, maintain, or reduce the decline of populations of species of selected taxonomic groups	− −	− −	−	−
Target 2.2: Status of threatened species improved	−	− −	+/−	+/−
Goal 3. Promote conservation of genetic diversity				
Target 3.1: Genetic diversity of crops, livestock, and harvested species conserved and associated indigenous knowledge maintained	− −	−	+	−
Promote sustainable use				
Goal 4. Promote sustainable use and consumption				
Target 4.1: Biodiversity-based products derived from sources that are sustainably managed	+/−	− −	+/−	+
Target 4.2: Unsustainable consumption of biological resources reduced	− −	− −	+/−	+
Target 4.3: No species of wild flora or fauna endangered by international trade[a]	+/−	− −	−	+/−
Address threats to biodiversity				
Goal 5. Pressures from habitat loss, land use change and degradation, and unsustainable water use reduced				
Target 5.1: Rate of loss and degradation of natural habitats decreased	−	− −	+/−	+
Goal 6. Control threats from invasive alien species				
Target 6.1: Pathways for major potential alien invasive species controlled	−	+/−	−	+
Target 6.2: Management plans in place for major alien invasive species that threaten ecosystems, habitats, or species	+	−	+	+
Goal 7. Address challenges to biodiversity from climate change and pollution				
Target 7.1: Maintain and enhance resilience of the components of biodiversity to adapt to climate change	− −	− −	+/−	+
Target 7.2: Reduce pollution and its impacts on biodiversity	− −	− −	+/−	+/−
Maintain goods and services from biodiversity				
Goal 8. Maintain capacity of ecosystems to deliver goods and services and support livelihoods				
Target 8.1: Capacity of ecosystems to deliver goods and services maintained	− −	− −	+/−	+
Target 8.2: Biological resources that support sustainable livelihoods, local food security, and health care, especially of poor people, maintained	− −	− −	+/−	−
Protect traditional knowledge, innovations, and practices				
Goal 9. Maintain sociocultural diversity of indigenous and local communities				
Target 9.1: Protect traditional knowledge, innovations, and practices	− −	− −	+	−
Target 9.2: Protect the rights of indigenous and local communities over their traditional knowledge	+/−	− −	+	+/−
Fair and equitable sharing of benefits				
Goal 10. Ensure the fair and equitable sharing of benefits arising out of the use of genetic resources				
Target 10.1: All transfers of genetic resources are in line with CBD and other applicable agreements	+	−	+/−	+
Target 10.2: Benefits arising from the commercial exploitation of genetic resources shared with countries providing such resources	+	−	+/−	+
Ensure provision of adequate resources				
Goal 11. Parties have improved financial, human, scientific, technical, and technological capacity to implement CBD				
Target 11.1: New and additional financial resources are transferred to developing countries to allow for effective implementation of CBD	+	− −	−	+
Target 11.2: Technology is transferred to developing countries to allow for effective implementation of CBD	+	− −	+	+

[a] The CBD target refers to trade in endangered species.

Key: + trend toward target; − trend away from target; − − marked trend away from target; +/− strong regional differentiation of trends

GO = Global Orchestration; OS = Order from Strength; AM = Adapting Mosaic; TG = TechnoGarden

Table 14.6. Summary of Key Stresses and the Prospects for Success of Relevant CBD Response Options in MA Scenarios. All values are estimates of relative comparison among scenarios and stresses. Many responses apply to more than one stressor.

Stresses and Responses	GO	OS	AM	TG
Ecosystem stress—habitat loss	●●●●	●●●●●	●●	●●
Establish effective global network of protected areas	✻✻✻	✻	✻✻	✻✻✻✻
Maintain and restore connectivity	✻	✻	✻✻	✻✻✻
Reduce pressure for agricultural expansion	✻	✻	✻✻	✻✻✻
Ecosystem stress—overexploitation	●●●	●●●●	●●	●●
Promote sustainable use of productive lands	✻✻	✻	✻✻✻	✻✻✻✻
Promote more-effective education, incentives, regulation	✻✻	✻✻	✻✻✻	✻✻✻✻
Maintain traditional knowledge	✻	✻	✻✻✻✻	✻✻
Ecosystem stress—invasive species	●●	●●	●	●
Implement control strategies	✻	✻	✻✻✻	✻✻✻
Ecosystem stress—pollution	●●●	●●	●●	●
Reduce emissions of NO_x	✻	✻	✻✻	✻✻✻
Establish monitoring protocols	✻✻	✻	✻✻✻	✻✻✻✻
Ecosystem stress—climate change	●●●●	●●●	●●●	●●
Promote synergy between carbon storage and habitat conservation	✻✻	✻	✻✻	✻✻✻

Key: GO = Global Orchestration; OS = Order from Strength; AM = Adapting Mosaic; TG = TechnoGarden
Stresses: 5 ● = severe stress, 0 ● = no worse than 2004
Responses: 5 ✻ = success likely, 0 ✻ = unfeasible/ineffective

while maintaining its potential to meet the needs and aspirations of future generations."

This section summarizes the most characteristic implications for wetlands of the four MA scenarios and provides a comparative assessment of the relative importance of direct and indirect drivers of wetland change. This is followed by an appraisal of the promising response options and the prospects for action for the convention and its parties under the four scenarios.

14.3.1 Threats to Wetlands in the MA Scenarios

None of the models used in the MA scenario exercise deals directly with wetlands. What makes this assessment even more difficult is that modeling results provide very few clues from which information could be derived concerning the fate of wetlands under the four scenarios. The combined outcomes of climate and land use change calculations in the IMAGE model can be used to get a rough estimate of the main natural driver, climate. The WaterGAP model performs detailed calculations of water availability, water demand, and water stress indicators. Modeling results presented in Chapter 9 suggest that, on balance, besides a gradually increasing climate change impact, socioeconomic driving forces are likely to remain the main source of threats to wetlands over the next half-century.

Table 14.7 summarizes the findings of the modeling activities concerning water-related issues on the basis of results in Chapter 9. As the modelers correctly point out, these results need to be handled with extreme care. The magnitude of uncertainties involved is clearly demonstrated by the case of modeling water availability. Estimates about the present values of water availability vary up to a factor of two in some regions. There is a much bigger diversion among present water availability values across models than there is for diversions among projected values for 2050 or 2100 across the scenarios on the basis of the same models. In terms of water availability, models indicate that regions are affected differently, but regional precipitation modeling is still among the most uncertain parts of general circulation models. Nonetheless, the modeling results appear to be plausible and they are certainly useful for comparing the projected values across scenarios.

It is interesting to observe that similar water-related indicators may emerge from rather different socioeconomic scenarios. The two globalization scenarios involve rather similar water availability, water withdrawal values, scarcity/stress features, and even return flows, although TechnoGarden has only 10% more people who are on average about 30% less well off compared with the Global Orchestration scenario. The key difference is in water quality, which is much worse in Global Orchestration and not declining in TechnoGarden relative to the present. The explanation is the strong environmental orientation and the fast rate of technological development in TechnoGarden. Similar relationships can be observed between the two isolation scenarios. In the bleak world of high population and very low economic growth of Order from Strength, the drastically increasing pollution of the doubling return flows is posing a major threat to wetlands, especially in developing regions. In contrast, under similar demographic and economic conditions in Adapting Mosaic, the quality of the almost doubling return flows can even improve in many

Table 14.7. Water-related Indicators in MA Scenarios in 2050 Relative to 2000. Note that, for example, 2.4* indicates a factor increase of 2.4 by 2050 relative to 2000.

Model Results	Global Orchestration	Order from Strength	Adapting Mosaic	Techno-Garden
Water availability	largest change: 5%	4–5%	4–5%	smallest change: 4%
Water withdrawals	+40% 2.5* SSA 1.7* LA 1.5* Asia MENA decrease OECD, FSU slight increase	+80% 4* SSA 3.5* LA +40% MENA +90% Asia +32% OECD no change FSU	+50% 4* SSA 3* LA +25% MENA +60 Asia +5% OECD −9% FSU	+20% 2.4* SSA −11% OECD −24% FSU
Area affected by water scarcity or stress (18% in 2000)	slight expansion	23%	22%	slowly increasing
Water return flow	+40% 3.6* SSA 2.0* LA +22% MENA +48% Asia OECD, FSU decrease	+100% 5.6* SSA 4* LA +100% MENA +100% Asia +40% OECD +10% FSU	+60% 5.5* SSA 3.6* LA +55% MENA +75% Asia +3% OECD FSU decrease	+20% 3.6* SSA 2* LA +16% MENA +20% Asia −18% OECD −42% FSU
Water quality	worse	much worse	same/improve	same +/−

Key: SSA = Sub-Saharan Africa; LA = Latin America; MENA = Middle-East and North Africa; FSU = Former Soviet Union

regions thanks to the environmental orientation and the reliance on local knowledge and eco-management experimentation.

Table 14.8 presents the qualitative assessment of the impacts of different indirect and direct drivers of wetland change based on the MA scenarios. A caveat should be mentioned, however: the relationships among the indirect and direct drivers and their impacts on wetlands are much more complex than can be presented in a simple table. It is obvious that a larger and more affluent population would demand much more food and, all other factors being equal, this would imply pressure for more agricultural land and would threaten wetlands to be drained and converted into cropland. Yet if the food demand is satisfied from modestly increasing areas by adopting fast-improving technologies and relying on more-efficient production (TechnoGarden) or on the basis of more-efficient allocation of production fostered by fairer trade and the elimination of subsidies (Global Orchestration), then the pressure for more agricultural land and wetland drainage is significantly less than if the basic needs of larger, less affluent populations need to be satisfied entirely on the basis of local knowledge (Adapting Mosaic) and local resources (Order from Strength).

The globalization of agricultural and fishery markets can take effect in two directions. If the process involves an ever-tougher competition of perverse subsidies, the threats to wetlands can be significant. If, however, the globalizing markets are not distorted by preferential interventions, the risks for wetlands are likely to be much smaller and opportunities for conservation may even arise. Depending on the cultural and sociopolitical circumstances, the empowerment of communities to manage their own resources or the privatization of open-access resources (always exposed to the risk of overexploitation), both accompanied by appropriate conservation incentives and regulation and by internalizing all external costs, are good opportunities for wetland conservation. The prospects for social transformations with favorable impacts for wetlands from the global level are best under Global Orchestration, while positive regional influences will be stronger under Adapting Mosaic. The chances of favorable social effects for wetlands are more limited in the other two scenarios.

The effects of the small magnitude of climate change expected up to 2050 are likely to be minor compared with

Table 14.8. Relative Importance of Direct and Indirect Drivers of Wetland Change in MA Scenarios

Indirect Drivers	Direct Drivers	Global Orchestration	Order from Strength	Adapting Mosaic	TechnoGarden
Population growth	Drainage/conversion	−	− − −	− −	−
Economic growth		− −	−	−	−
Globalization of agriculture, fishery markets	Introduction of alien species	− −/+	−	−	−/+
Increasing demand for water	Water diversion,	−	− − −	− −	−
	Water pollution	− −	− − −	−/+	−/+
Privatization and empowerment		+	−/+	+	+ + +
Financial transfers		+	...	+	+ +
Climate change	Mean temperature/precipitation,	−	−	−	−
	Extreme events	−	−	−	−
	Sea temperature, sea level rise	−	−	−	−

Key: − − − high, − − medium, − low level of risk of degradation
+ + + high, + + medium, + low level of opportunity for conservation

the changes that might be triggered by the social, economic, and technological drivers in most places. The slowly emerging patterns of climate change may play a more significant role for wetlands at some locations close to boundaries of climatic zones. This is expected to change in the long term by 2100 and beyond if uncontrolled emissions of greenhouse gases continue. The gradually changing temperature and precipitation patterns are likely to be less of a problem than changes in the frequency and magnitude of extreme events triggered by climate change. Unfortunately, there is hardly any reliable information available on the latter. Nonetheless, the next few decades present challenges but also the opportunity for wetland managers to devise ways to help wetlands adapt to possibly more significant climate change in the second half of this century.

14.3.2 Prospects for the Ramsar Convention

The Ramsar Strategic Plan 2003–2008 was adopted by the Eighth Meeting of the Conference of the Contracting Parties in 2002. The plan lists specific WSSD objectives to which Ramsar could contribute, but it does not delineate near-term targets. In fact, it does not distinguish near-term goals and long-term objectives at all. Rather, general objectives of the Strategic Plan are specified as progress toward the ultimate objective of the convention over the long term. The five general objectives are stimulating the wise use of all wetlands by developing, adopting, and using the appropriate instruments and measures; stimulating and supporting the implementation of the Strategic Framework by monitoring and managing their listed sites; promoting international cooperation, particularly by mobilizing additional financial and technical assistance for wetland conservation and wise use; ensuring the necessary implementation capacity, resources, and mechanisms for the convention; and proceeding toward the accession of all countries to the convention.

The actual response options and implementation mechanisms available to the Ramsar Convention appear to be rather weak at first sight. Yet they have proved remarkably effective in most cases in the past, and their effectiveness could certainly be improved by making more resources available to foster some of the implementation mechanisms. Table 14.9 presents an assessment of the prospects for the various response options to provide effective support to wetland conservation under the four scenarios.

There is a clear and obvious pattern emerging from Table 14.9. A global environmental agreement based on the voluntary commitments of its parties has much better prospects to be an effective mechanism of wetland protection under the globalization scenarios than in the fragmented worlds. The motivation for and the perceived benefits from including ecological treasures on the List of Wetlands of International Importance are much larger in a future in which countries have a rich web of economic, cultural, and environmental linkages. The relative importance of policy guidelines versus technical guidelines differs slightly as a function of how technologically oriented societies are (TechnoGarden) versus the extent to which they pursue

Table 14.9. Prospects for the Ramsar Convention's Policy Instruments in MA Scenarios

Response Options and Implementation Mechanisms	Global Orchestration	Order from Strength	Adapting Mosaic	Techno-Garden
Listing	***	*	**	***
Policy guidelines	***	*	*	**
Technical guidelines	**	*	*	***
Financial mechanisms	***	*	*	**
Technical assistance	**	*	**	***
Regional initiatives for implementation (core fund)	***	*	**	***
Communication/ education/public awareness	**	*	**	***

Key: *** good; ** modest; * poor

policy coordination (Global Orchestration). This is also the case for the prospects for financial mechanisms as opposed to technical assistance as implementation mechanisms. In a dynamic, innovation-oriented future, technical assistance projects under the Ramsar Convention appear to be more dominant, whereas a free-market- and trade-oriented world biased toward reactive environmental management is more likely to use financial mechanisms to compensate occasional losers of environmental change and to support rehabilitative measures.

Given the vulnerability of many small wetland areas to irreversible changes triggered by relatively modest perturbations, proactive protection is ecologically more sensible. Funding to support regional initiatives for implementation through the Ramsar Convention is obviously more likely in the futures in which countries are interconnected than among largely segregated, introverted countries. Communication, education, and public awareness are more likely to be able to contribute to wetland conservation in the environmentally oriented scenarios (TechnoGarden and Adapting Mosaic), although Global Orchestration also offers good chances. In the globalization scenarios, the high level of affluence and the increasing leisure time of people are likely to give an unprecedented rise to ecotourism, and this in itself could provide a very strong economic motivation to pursue the wise use of wetlands. Eco-tourism is also an important connection back to the idea of listing as an implementation mechanism, because the List of Wetlands of International Importance could be an obvious source for guidebooks and tourism operators in selecting destinations.

Table 14.10 summarizes the key stresses of concern for the Ramsar Convention and the prospects for success of relevant response options under the four MA scenarios. While the pressure on wetlands is relatively modest in the Adapting Mosaic scenario, the role of the Ramsar Conven-

Table 14.10. Summary of Key Stresses and the Prospects for Success of Relevant Ramsar Convention Response Options in MA Scenarios. All values are estimates of relative comparison among scenarios. Many responses apply to more than one stressor.

Stresses and Responses	Global Orchestration	Order from Strength	Adapting Mosaic	TechnoGarden
Ecosystem stress—drainage and conversion	●●●	●●●●	●●●	●●
Listing	✻✻✻✻	✻	✻✻✻	✻✻✻✻
Technical guidelines	✻✻✻	✻✻	✻✻	✻✻✻✻
Financial mechanisms	✻✻✻✻	✻	✻	✻✻✻
Ecosystem stress—water diversion and pollution	●●●	●●●●●	●●●	●●
Policy guidelines	✻✻✻✻		✻✻✻	✻✻✻
Technical assistance from higher-income to developing countries	✻✻✻	✻	✻✻	✻✻✻✻
Regional initiatives for implementation	✻✻✻✻	✻	✻✻✻✻	✻✻✻✻

Key: Stresses, 5 ● = severe stress, 0 ● = no worse than 2004
Responses, 5 ✻ = success likely, 0 ✻ = unfeasible/ineffective

tion to help protect or counterbalance the risks is much more limited than in the globalization scenarios.

The obvious worst case is the Order from Strength world, in which the severe threats to wetlands from multiple sources (high population growth, slow technological development, and negligence of the environment) are combined with a severely weakened Ramsar Convention due to the breakdown of global institutions at large. Another important difference between Adapting Mosaic and Order from Strength is that in the proactive, environmentally oriented Adapting Mosaic world, the focus of the Ramsar Convention might shift from the global to the regional level. Regions with similar wetland problems could get into tighter regional cooperation networks, while the global agreement might serve as an umbrella of lesser importance. Since many regions are likely to be economically homogeneous, the emphasis in the operation of the regional Ramsar mosaics might shift from financial transfers to knowledge sharing and know-how transfer.

14.4 Implications for the Desertification Convention

Desertification is defined as the degradation of land in arid, semiarid, and dry subhumid areas. It has been identified as a major socioeconomic and environmental problem for many countries around the world. Direct drivers of desertification include overcultivation, overgrazing, deforestation, and inappropriate irrigation management. These drivers can be traced back to a range of economic and social pressures, lack of knowledge, war, and natural climate fluctuations such as drought. (See also MA *Current State and Trends,* Chapter 22.)

The objective of the United Nations Convention to Combat Desertification in Countries Experiencing Serious Drought and/or Desertification, Particularly in Africa, as specified by Article 2, is "to combat desertification and mitigate the effects of drought in countries experiencing serious drought and/or desertification, particularly in Africa, through effective action at all levels, supported by international cooperation and partnership agreements, in the framework of an integrated approach which is consistent with Agenda 21, with a view to contributing to the achievement of sustainable development in affected areas." The implementation of this objective involves long-term integrated strategies to improve the productivity of land and to rehabilitate, conserve, and sustainably manage land and water resources.

14.4.1 Risk of Desertification in the MA Scenarios

Desertification results from natural causes (such as a change in precipitation) or human causes (such as land clearance and inappropriate land uses) or a combination of these. In general, desertification results in lower biodiversity levels, shifts in species composition and natural areas, and lower productivity in cultivated areas. The decrease in vegetation cover and the subsequent loss of soil material and soil organic matter reduces soil fertility. Low soil fertility, in turn, reduces vegetative cover, leading to a vicious circle. The CCD uses the ratio of mean annual precipitation to mean annual potential evapotranspiration to identify drylands. They include arid, semiarid, and dry subhumid areas (in other than polar and subpolar regions) in which this ratio ranges from 0.05 to 0.65.

The MA adopted this definition to identify the total amount of dryland areas and their changes over time under the four MA scenarios using the IMAGE 2.2 model. Obviously, as these are modeling results, the 2000 results from IMAGE are somewhat different from those based on current actual climate estimates, but in general they approximate the reality reasonably well. Table 14.11 indicates that globally, changes in arid areas (as a result of climate change) are relatively small. This follows from the fact that climate change is expected to result in increasing precipitation but also increasing evaporation (as a result of temperature increase). The changes differ clearly among the different regions. It should be noted, however, that the regional results should be regarded as uncertain: both temperature and precipitation patterns differ strongly among the different climate models. In Latin America and the former Soviet Union, a considerable decrease in arid areas is observed. In contrast, in the OECD, Asia, and sub-Saharan Africa, a

Table 14.11. Changes in Dryland Areas in MA Scenarios. Note that year 2000 values correspond to 100%. (IMAGE 2.2 Model runs)

Region	Dryland Area: Area	Dryland Area: Share of Total Area	Change in 2050: Global Orchestration	Change in 2050: Order from Strength	Change in 2050: TechnoGarden	Change in 2050: Adapting Mosaic
	(thousand sq. km.)	*(percent)*	*(percent)*			
OECD	10,670	47	106	101	107	106
Latin America	5,004	25	97	97	96	97
Sub-Saharan Africa	13,024	55	102	101	101	102
Middle East and North Africa	11,351	97	101	101	101	101
Asia	8,440	41	103	102	101	103
Former Soviet Union	4,406	20	98	99	99	98
World	**52,896**	**44**	**102**	**101**	**102**	**102**

clear increase of arid areas is noticeable. Finally, in the Middle East and North Africa, the arid areas are more or less constant.

For desertification, however, the increase in arid areas is less important than the pressure on these areas. Therefore, Figure 14.1 indicates the size of arid areas that are used for agricultural purposes—that is, for cropland and intensive pastures but also (and mostly) for extensive grazing. It is worth noting that there is a large interannual variation in the use of drylands, and their use for agricultural and nonagricultural purposes is also changing over time. The resolution of the global models used in the MA scenario development is too coarse to depict such variations at the local scale. Moreover, the analysis of the desertification risk is based on the predicted increase of arid areas devoted to agriculture, including the area for free-ranging livestock. In reality, the desertification risks are more complex and numerous, but they are difficult to depict in a global model. Nonetheless, the broad patterns emerging from these models provide useful insights into the emerging risks and opportunities for dryland management under the four scenarios.

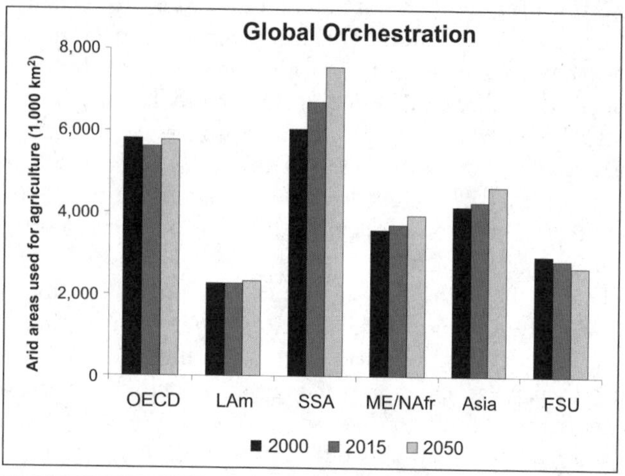

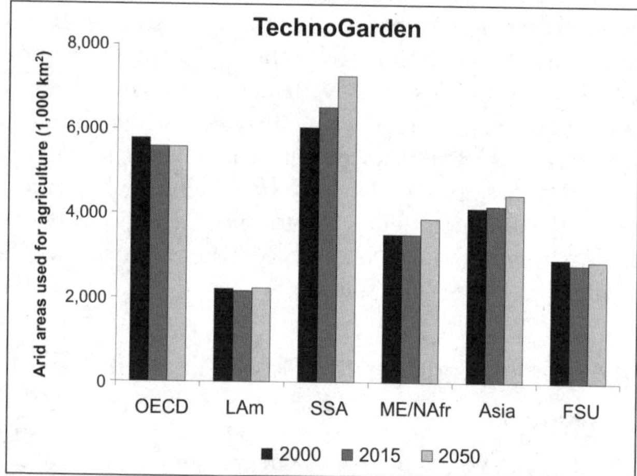

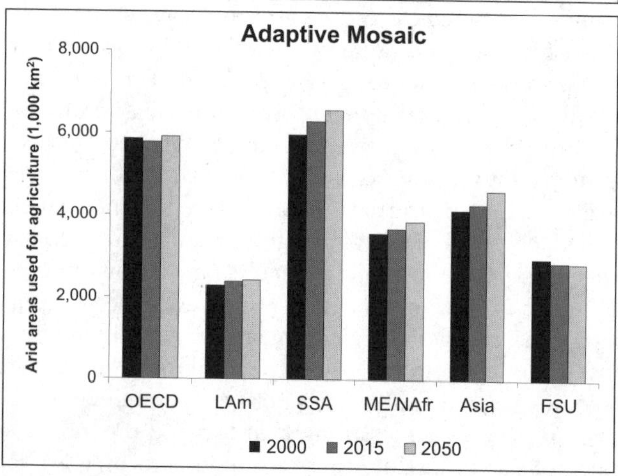

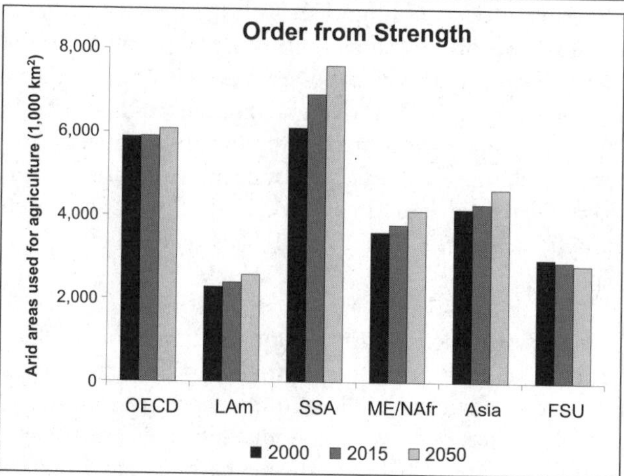

Figure 14.1. Arid Areas under Agriculture in MA Scenarios (IMAGE 2.2 Model)

Concerning the pressures on arid areas, some clear differences among the scenarios are noticeable. In three scenarios—TechnoGarden, Global Orchestration, and Order from Strength—there is a considerable expansion of agricultural land in Africa, driven by population growth and relatively rapid increases in food demand (TechnoGarden and Global Orchestration only). As shown in Figure 14.1, a considerable part of the expansion is likely to occur in arid areas—a trend that has been observed already over the last few decades. While the increase in food demand in Adapting Mosaic is comparable to Order from Strength but increases in agricultural efficiencies are assumed to be higher, this scenario turns out to be the most optimistic (although also here some expansion occurs). Other regions that are expected to see further expansion of agriculture into arid areas are Middle East and North Africa and Asia. For these regions, the differences across the scenarios are relatively small.

In Order from Strength there is a clear increase of the desertification risk in Latin America, while in the other scenarios the risks remain more or less constant. In OECD, under Order from Strength there is a small increase in the desertification risks; for Adapting Mosaic and Global Orchestration it is constant; and for TechnoGarden there is a small decrease. The latter is mainly caused by relatively low meat-intensive diets combined with rapid technological development. Finally, for the region of the former Soviet Union, most scenarios project a decrease in desertification risks, caused mainly by a decrease of the arid areas themselves as a result of climate change.

Table 14.12 summarizes the most plausible direct and indirect causes behind the desertification risk under the four scenarios. The reactive management scenarios involve the largest amount of cumulative risk of desertification. Under Order from Strength, the characteristics of socioeconomic development (high population growth, slow rates of technological development, and neglect of the environment) lead to severe stresses to land resources in dryland regions. Due to policy reforms (privatization and consolidation of property rights), relatively less pressure results under Global Orchestration, but market failures and policy failures can equally pose certain risks of desertification.

Table 14.12. Relative Importance of Direct and Indirect Drivers of Desertification in MA Scenarios

		Scenarios			
Indirect Drivers	**Direct Drivers**	**GO**	**OS**	**AM**	**TG**
Economic and social pressure / Lack of knowledge	overcultivation	*	***	*	*
	overgrazing	**	***	*	*
	deforestation	*	***	*	*
	poor irrigation	**	***	*	**
War		**	**	**	*
Drought		*	*	*	*

Key: GO = Global Orchestration; OS = Order from Strength; AM = Adapting Mosaic; TG = TechnoGarden

*** = major factor ** = medium factor * = minor factor

In the TechnoGarden world, technological development can make a dramatic contribution to reducing pressure in dryland areas. Improvements in crop varieties and agronomic techniques, including irrigation technologies, can contribute to the reduction of desertification and also to the reclamation of some already-degraded areas. The other environmentally proactive scenario, Adapting Mosaic, turns out to be relatively beneficial concerning desertification, but for different reasons. Here the basic mode of operation is to develop local combinations of technologies and organizations (formal institutions) that lead toward sustainable agriculture in dryland areas. Given the diversity of socioeconomic conditions across the regions in this scenario, it is difficult to detect comprehensive patterns. Nonetheless, abolishing open access in one way or another (through community management, local or regional government control, privatization, or combinations thereof) is the first crucial step to control overexploitation and reduce pressure on drylands in a proactive ecosystem management scenario.

14.4.2 Prospects for the Desertification Convention

What are the prospects and opportunities for action under the CCD in the contexts of the four scenarios? The primary form of implementation of the desertification convention is National Action Programs complemented by sub-regional and regional action programs where appropriate. The five regional implementation annexes of the convention specify the criteria for these programs.

Activities in the NAPs can be divided into general and specific categories. The general actions include addressing the underlying causes of desertification; promoting awareness about the risks, causes, and processes; and providing the enabling environment (institutional and legal framework) for managing the risk of desertification. A series of specific actions are included in the NAPs: establishing early warning systems, strengthening drought preparedness, establishing food security, establishing alternative livelihoods, and developing sustainable irrigation schemes.

The second main implementation vehicle of CCD is scientific and technical cooperation. This involves information collection, analysis, and exchange; technological research and development; and technology transfer. The third main category of implementation incorporates capacity building, education, and efforts to raise public awareness. The measures to support NAPs are based on various forms of financial cooperation. Such cooperation includes mobilizing financial resources directly; encouraging the mobilization of private finances; and promoting access to technology, knowledge, and know-how.

Table 14.13 provides an overview of the prospects of the various response options and implementation mechanisms of CCD under the four MA scenarios. The first strikingly bad news is that in the world of Order from Strength, in which the risk of desertification is the highest and the actual magnitude of desertification is likely to be the highest by far, there will be very little chance for the CCD to help countries halt or even slow desertification. The two main reasons for this are obvious. In a fragmented world with

Table 14.13. CCD Response Options and Implementation Relationships in MA Scenarios

Response Options	Global Orchestration	Order from Strength	Adapting Mosaic	TechnoGarden	Notes
NAP general					
Address underlying causes	***	*	**	**	
Promote awareness	***	*	***	**	
Provide enabling environment (legislation, institutions/legal)	***	*	**	**	
NAP specific					
Establish early warning system	*	*	***	***	drought
Strengthen drought preparedness	**	*	***	***	overcultivation
Establish food security	**	**	***	***	overgrazing,
Establish alternative livelihood	*	*	***	**	deforestation
Develop sustainable irrigation	**	*	***	***	poor irrigation
Scientific and technical cooperation					
Information collection, analysis, exchange	**	*	**	***	
Research and development Transfer of technology	**	*	***	***	
Research and development	**	*	***	***	
Transfer of technology	**	*	**	***	
Capacity building					
Education	**	*	***	***	
Public awareness	**	*	***	***	
Measures to support NAPs					
Mobilize financial resources	***	*	*	***	
Encourage private financing	***	*	*	***	
Promote access to technology, knowledge, know-how	***	*	**	***	

***Key: *** = good prospects ** = medium prospects * = poor prospects

inward-looking regions, the scope for global environmental agreements is rather poor in the first place. The outlook is bad even for "global commons" types of agreements, and there remains little motivation to arrange massive resource transfers from rich nations to poorer dryland regions in order to mitigate desertification. The second reason is the underlying management philosophy of this scenario. In an environmentally reactive ecosystem management mode, dryland degradation is likely to go further before its impacts (massive famines, environmental and hunger refugees) trigger a significant response.

As a global environmental agreement with resource transfer from North to South, the CCD has the best prospects in the scenarios assuming continuing globalization. The overall socioeconomic and political conditions under Global Orchestration provide better conditions to implement the general components of NAPs, like addressing the underlying causes and providing the necessary enabling environment to combat desertification. In a TechnoGarden world, CCD mechanisms involving direct and specific interventions by developing and transferring the appropriate technologies are more likely.

The most promising sources of funding are likely to differ as well. With the confidence in markets and secured property rights, it is likely to be much easier to mobilize private capital under Global Orchestration. TechnoGarden is more likely to mobilize public funds and publicly financed technological development and transfer. In Adapting Mosaic, the overall social and political conditions and the focus on environmentally sound management options are favorable for the CCD implementation mechanisms as well, but the disconnect among the regions of the world would probably allow only limited resource transfers. As the main focus in this scenario shifts away from global agreements to developing local solutions and experimenting and learning how to mange local systems better, it is likely that resources will have to be mobilized and used primarily within a region, which might not be easy to do in the currently poor regions of the world. Sharing and transfer of knowledge across regions, however, is not likely to be affected.

It is important to point out that NAP implementation requires not only resource transfers from donors but also political willingness and awareness by affected countries—for example, by ranking land degradation high in their political agenda and consequently also committing national resources to fight it. An equally important and closely related issue is that the mode of operation of CCD needs to change after the Sixth Conference of the Parties from issues of process to real implementation on the ground. Establishing the appropriate links between the CCD main instruments (National, Sub-regional, and Regional Action Programs) and development strategies of the affected countries (National Strategy for Sustainable Development, Poverty Reduction Strategy Programs, and so on) would be a first step to ensure that NAPs are not just purely theoretical exercises disconnected from reality but tools deeply anchored in the national context. Both issues appear to be major hurdles in many countries today, and the prospects for improvement will evolve differently in the four MA scenario worlds.

A summary of key stresses for the concerns of the CCD is presented in Table 14.14, together with the prospects for success of relevant response options under the four scenarios. In Global Orchestration and TechnoGarden, in which global agreements (including transparency and accountability of resource transfers) function well, an increasing flow of funds and technologies to poorer countries that establish the domestic frameworks of NAP implementation will help persuade other countries to get their domestic policies organized in order to secure their shares from those flows. Such a positive trend may also help in establishing appropriate relationships between national development frameworks and CCD implementing tools (the action programs), thus helping to overcome the experienced gap at country level between measures targeting land degradation and those aimed at eradicating poverty or achieving food security and sound water management, as well as between national agriculture sector priorities and the improvement of livelihoods for rural populations.

The incentive for NAP implementation in Adapting Mosaic may come from regional cooperation between local networks and groups of practitioners and ecosystem managers interested in NAP measures who also push to keep them on the agendas of national governments. Finally, neither social motivation (no interest in the environment) nor international economic motivation (resource or technology transfers) exists for caring much about desertification NAPs in the Order from Strength scenario.

In summary, continued population growth through the first half of this century and improving economic conditions are likely to exert a substantial amount of additional pressure on land resources worldwide. These trends enhance the risk of desertification in dryland regions. Since the scenarios involve diverse sociopolitical, economic, and technological features, the opportunities for CCD to fulfill its mission will differ as well. In a globalizing world, prospects for international environmental cooperation and resource transfers to support their implementation are likely to be better either due to the institutional reforms (Global Orchestration) or because of the fast rate of technological development and deployment (TechnoGarden). In a fragmented world, the role of a global agreement is more limited either because of the diminished interest in resource transfers (Adapting Mosaic, although the stress is also lower under this scenario) or because of the total lack of interest in what is going on beyond the national or regional boundaries (Order from Strength).

14.5 Implications for National Governments

National governments play a central role in regulating many activities affecting ecosystems and the use of their services. They represent sovereign nation-states at international negotiations and become parties to international environmental agreements that directly regulate international aspects of ecosystems management. Similarly, they decide whether to join international economic agreements (trade, finance, development) that often trigger indirect implications for the use and protection of ecosystems services. This section considers the domestic concerns of national governments and focuses on how the evolution of ecosystems under the four MA scenarios affects the chances of governments to accomplish their declared objectives of pursuing sustainable development.

The assessment of national-scale issues on the basis of global scenarios is no easy task. Countries differ widely in so many of the key attributes (geography, climate, economic development, social values, institutional arrangements) that make each of them rather unique and require country-specific analysis. This is clearly impossible in a global-scale study because neither the verbal scenarios (storylines) nor the adopted models provide information at the national level. Instead, we contemplate global and large-scale re-

Table 14.14. Summary of Key Stresses to Drylands and the Prospects for Success of Relevant CCD Response Options in MA Scenarios. All values are estimates of relative comparison among scenarios. Many responses apply to more than one stressor.

Stresses and Responses	GO	OS	AM	TG
Ecosystem stress—overcultivation	●●●	●●●●●	●●	●
Address underlying causes	✻✻✻✻	✻	✻✻✻	✻✻✻
Establish alternative livelihood	✻✻	✻✻	✻✻✻✻	✻✻✻
Develop sustainable irrigation	✻✻✻	✻	✻✻✻✻✻	✻✻✻✻✻
Transfer technology	✻✻✻✻	✻	✻✻✻	✻✻✻✻✻
Ecosystem stress—overgrazing	●●●	●●●●●	●●	●●
Address underlying causes	✻✻✻✻	✻	✻✻✻	✻✻✻
Establish alternative livelihood	✻✻✻	✻	✻✻✻	✻✻
Establish early warning and drought preparedness	✻✻✻	✻	✻✻✻✻	✻✻✻✻
Promote awareness	✻✻✻	✻	✻✻✻✻	✻✻✻

Key: GO = Global Orchestration; OS = Order from Strength; AM = Adapting Mosaic; TG = TechnoGarden

Stresses, 5 ● = severe stress, 0 ● = no worse than 2004 Responses, 5 ✻ = success likely, 0 ✻ = unfeasible/ineffective

gional (continental or subcontinental) patterns of the issues national governments are concerned about. The global scenario results and the assessment in this section might become useful starting points for national studies that seek to explore the country-specific prospects and challenges under the MA scenarios in more detail.

Notwithstanding the numerous specificities in their ecosystem-related interests and objectives, governments have repeatedly pronounced common principles and objectives concerning socioeconomic development and environmental management at various international fora over the past two decades. This section looks at two recent proclamations: the U.N. Millennium Development Goals and associated targets provide the framework for exploring the medium-term implications (to 2015) of the MA scenarios, and the Johannesburg Declaration on Sustainable Development serves as the basis to investigate the long-term (to 2030–50) outcomes. Since these were both approved at large intergovernmental conferences, they are the officially confirmed and documented concerns of national governments. These sections are followed by a more detailed assessment of the food-ecosystems-security relationships.

14.5.1 Medium-term Implications for the MDGs

The Millennium Summit in 2000 confirmed that progress toward sustainable development and poverty eradication has top priority. The Millennium Development Goals, derived from agreements and resolutions of relevant U.N. conferences in the post-Rio years, established rather ambitious goals. The most pressing challenges for humanity are organized into eight main goals and are specified in terms of 15 (+1) quantitative targets. Some goals are only very remotely related to the protection of ecosystems and the use of their services: Goals 2, 3, 4, and 5, for example, focus on crucial social (primary education, gender equality) and human health (child mortality, maternal health) concerns. Other goals have important indirect implications for ecosystems services and development: Goal 1 (halving the proportion of people who suffer from hunger), for instance, and Goal 7 (halving the proportion of people without sustainable access to safe drinking water).

At the macro policy level, Goal 7 calls for integrating the principles of sustainable development into country policies and mentions, among others, the land area covered by forests or under protection to maintain biological diversity, energy intensity, and per capita carbon emissions as indicators of measuring progress. Ample opportunities exist to make progress on this goal, and many economists suggest that eliminating perverse subsidies that distort the energy and agriculture sectors in many countries could make a good start. Ironically, some energy-related measures aimed at poverty alleviation would be likely to affect the sustainability indicators on energy intensity or emissions in the short run because they would increase energy use per unit of GDP (providing electricity to promote education, increased industrialization, and urbanization) and CO_2 emissions per capita (replacing unsustainable biomass, typically fuelwood, by commercial fossil energy in households, for example). However, once these investments in infrastructure and human capital (education, gender equality) start paying back, the energy and carbon intensity indicators should improve as well.

To ensure appropriate interpretation of the results in this section, it is important to note that the MDGs denote most quantitative targets as improvements relative to the 1990 situation. Most models that provide quantitative projections under the four MA scenarios use 2000 as their reference year. It is therefore difficult to assess the projected achievements until 2015 according to the MDG starting point. Another complication is that the MDGs specify most targets for 2015, whereas the models adopted in the MA scenario exercise have 50- or even 100-year time horizons and, in some cases, 5- or 10-year time steps. This means that these models make only two or three steps until 2015, and the scenario dynamics are hardly distinguishable at this time horizon.

The broad evolution patterns of the verbal scenario storylines are even more difficult to peg to specific years like 2015. Moreover, the early phase of any scenario exercise designed to explore long-term futures is dominated by the starting situation. The MA scenarios are no exemptions, and the marked diversions among the four storylines just begin to emerge by 2015.

Next it should be noted that the MDGs encompass key elements of the full span of social, economic, political, institutional, and environmental components of sustainable development. The MA scenarios are concerned with a specific subset: the main components of socioeconomic development that shape human impacts on ecosystems and the use of their services as driving forces of ecosystem changes, along with the repercussions on human well-being of the changes triggered in the quantities and qualities of ecosystems services. Therefore it is not possible even to infer information for some MDGs, and only remotely related information can be presented as proxy or "circumstantial evidence" for others.

The first MDG is to eradicate extreme poverty and hunger. The MA models do not break down populations into subcategories according to income levels. Hence it is impossible to obtain direct information about the proportion of population below $1 per day, the poverty gap ratio, or the share of poorest quintile in national consumption. The economic growth assumptions in the scenarios nevertheless can provide an indication. Per capita GDP growth is highest in Global Orchestration, followed by TechnoGarden and Adapting Mosaic, with Order from Strength lagging behind. Global Orchestration furthermore results in the greatest improvements for the poorest people, as the main focus of decision-makers in this scenario is placed on improving human systems. Despite slower increases in incomes in TechnoGarden and Adapting Mosaic, other aspects of human well-being improve in both scenarios and the number of hungry people also declines. In Order from Strength, the distribution of the modestly increasing material wealth deteriorates and all human well-being aspects decline compared with today.

Despite numerous international initiatives and national programs, hunger and malnutrition have been persistent problems in several world regions in the past few decades. All four MA scenarios project declining proportions of underweight children in the 0–5 age group, but these improvements are far from the ambitious target of halving the share of people suffering from hunger even if we consider the improvements between 1990 and 1997 (the year of the model's reference point). Moreover, the improvements are slowest in the regions with the biggest problems: South Asia and sub-Saharan Africa. Due to the lack of disaggregated population in the models, we cannot say much about the proportion of population below minimum level of dietary energy consumption. The per capita figures of available dietary energy improve in all developing regions (except West Asia and North Africa), more or less together with improving per capita incomes. This confirms that currently and in the near future hunger is more a social and economic problem than an environmental one. Thus the distribution of the available calories will remain a fundamental issue in determining the actual prevalence of hunger in 2015.

MDG 3 is on promoting gender equality and empowering women. The MA scenarios provide only one rather remote indicator on this topic. The percentage of females undertaking secondary schooling differed widely across developing regions in 1997. Improvements are projected in all regions under all scenarios, but the vast differences in female secondary education remain: one in five females getting secondary education in sub-Saharan Africa stand out against the 70% in China. (Nonetheless, gender disparity remains hidden in the absence of comparable indicators for males.)

The sixth MDG calls for combating HIV/AIDS, malaria, and other disease. Neither human health nor its linkages to ecosystems services are modeled in the MA scenarios. The storylines provide some indications, but they are more relevant for getting some ideas about the longer-term trends than as indicators of actual achievements up to 2015. The general patterns of change in human health mirror those of per capita incomes: substantial health improvements and considerable reductions in the burden of epidemic diseases (HIV/AIDS, malaria, tuberculosis), particularly in the South under Global Orchestration, and moderate progress in these areas, albeit elimination of diseases due to water and indoor air pollution in TechnoGarden. It is difficult to estimate how many of these improvements will take place by 2015. The scenarios speak of a number of obstacles to health improvements in the Adapting Mosaic future, as there is less technology transfer and cooperation across regions. This slower improvement of health gives little hope for any progress toward the MDG targets of halting the spread of HIV/AIDS and the incidence of malaria by 2015. The calamitous future of Order from Strength implies disastrous health trends for many low-income regions: the collapse of international malaria programs, the continued spread of HIV/AIDS, and the failure to manage tuberculosis might put the world on a trajectory that leads it away from and not toward this MDG target. The severity of this risk is illustrated by the fact that populations in some regions might actually decline as a result.

Target 9 of MDG 7—ensure environmental sustainability—embraces a policy-related principle (incorporate sustainable development into all relevant policies and programs) and an overall biophysical target (reverse the loss of environmental resources). The models adopted in the MA scenario work calculate several indicators relevant for the latter. It is a gloomy observation that, except for Latin America, none of the developing regions come even close to stabilizing their forested areas. In fact, deforestation continues in all scenarios in the Middle East and North Africa, sub-Saharan Africa, and Asia. The bleakest future awaits forests in the first two regions under Global Orchestration and Order from Strength, as about one third of their forests in 1995 are projected to disappear by 2015. The MA scenarios do not contain projections of the changes in land area protected to maintain biological diversity.

The MDG indicator list has GDP per unit of energy use as a proxy indicator of energy efficiency. The MA models project changes in the inverse of this indicator, energy intensity, which measures the amount of primary energy consumed per unit of GDP. This indicator shows impressive improvements in most regions in all scenarios. The two exceptions are the Middle East/North Africa, where energy intensity stagnates, and Latin America, where this indicator is projected to deteriorate through 2015 relative to 1995 in all four scenarios. The really bad news, however, is that the energy efficiency improvements are projected to be overwhelmed by fast-growing energy use and other activities emitting greenhouse gases, mainly CO_2. In the two decades between 1995 and 2015, GHG emissions increase around 50% in Asia under each of the four scenarios, more than triple in sub-Saharan Africa under Global Orchestration, and also double in the other three scenarios.

Access to safe (treated or uncontaminated) water appears to be a success story in the MA scenarios. Solid improvements are projected for all developing regions under all scenarios. Even in sub-Saharan Africa, where more than half the population had to use contaminated water in 1997, the share of population with access to safe water reaches 60% in all four scenarios.

The final MDG, on a global partnership for development, has three main components—official development assistance, market access, and debt sustainability—that are central elements of the "globalization-fragmentation" axis that splits global futures into these two main categories. Accordingly, it is not difficult to guess the prospects for the targets and indicators included in this category: in the future worlds in which global cooperation is a key element, it is also likely that development aid and fairer access to global markets will be a priority for decision-makers. The Global Orchestration scenario in particular focuses on these issues. Yet the scenario storylines do not give particular indicators that can be used to gauge progress on these important matters, and they are not included as variables in the MA models either.

As this short assessment demonstrates, the MA scenarios contain a lot of relevant information about the prospects for reaching the MDGs under four profoundly different scenarios. Yet 2015 is too near and the temporal resolution of the long-term MA scenarios and models is too coarse for spectacular diversions to emerge. "Fast variables" that respond to changes in their driving forces without delay—deforestation, energy efficiency improvements, deterioration of morbidity and mortality as a result of collapsing because of a lack of well-targeted,-organized, and -funded programs—can show large differences in their development paths across the scenarios even in one or two decades. In contrast, "slow variables" that have their own inertia and react to their determinants with delay—demographic factors, education achievements, infrastructure development like safe water and sanitation—show little variation between the scenarios over the short to medium term. The reason is that it takes years to decades until a change in, for example, demographic or educational policies has a discernible impact on the birth rates, age structure, human capital stock, and so on. Characteristic differences in the future of these variables take at least four to five decades to emerge.

14.5.2 Long-term Implications for the Johannesburg Declaration

The World Summit on Sustainable Development adopted the Johannesburg Declaration on Sustainable Development in 2002. The declaration recognizes that although some progress has been made, major challenges still must be overcome to implement the vision of sustainable development. The section on "the challenges we face" specifies poverty eradication, changing consumption and production patterns, and managing the natural resource base for economic and social development as overarching objectives of and essential requirements for sustainable development. The subsequent paragraphs list income gaps between the higher-income and developing worlds, environmental degradation (biodiversity loss, declining fish stocks, desertification, climate change, natural disasters, and pollution), and globalization (bringing both challenges and opportunities for the pursuit of sustainable development), whereupon the entrenchment of these global disparities may result in the poor losing confidence in the democratic systems. Paragraph 18 of the JDSD lists "essential needs" and suggests speedily increasing the "access to such basic requirements as clean water, sanitation, adequate shelter, energy, health care, food security and the protection of biodiversity" (UN 2002:3).

The MA scenarios resonate well with these concerns even though they do not address all of them explicitly or in full detail. Assuming that the issues listed in the JDSD are officially declared long-term concerns of national governments regarding sustainable development, the performance of the four MA scenarios can be assessed against these overall objectives over the long term. This section synthesizes relevant information by looking at persistent trends in the scenario storylines and presenting two snapshots of the modeling results for 2030 and 2050.

Table 14.15 presents the overall JDSD objectives and scenario results in three main groups: economic, social, and environmental. Many aspects of the scenarios are compared and analyzed in earlier chapters of this volume and in earlier sections of this chapter. Here we present a few emerging insights that are of particular importance for national governments.

Taking the economic objectives first, Table 14.15 does not contain entries about questions like access to financial resources and sharing the benefits of opening markets. However, the GDP growth figures imply the answers. In the Global Orchestration future, the sustained economic growth rate of approximately 8% per year that increases the volume of goods and services by a factor of 20 in 50 years in Asia, and the similarly impressive economic performance in all other currently developing and transitional economies are inconceivable without massive improvements in access to financial resources, both foreign direct investments and official development aid. Another implicit driver behind these remarkable trends in Global Orchestration is the more equitable sharing of the benefits of opening markets that channel a larger proportion of efficiency gains from foreign investments and international trade to developing regions, as the focus of this scenario is to combine more equitable access to markets with strong social policies. This is clearly not the "Washington Consensus."

The other three scenarios entail not only slower economic growth rates but also significantly slower convergence (TechnoGarden and Adapting Mosaic) or outright divergence of per capita incomes (Order from Strength) between OECD and the developing regions. In TechnoGarden and Adapting Mosaic, good governance structures, which are based on the elimination of corruption and political stability, are pursued in different ways and evolve slower than in Global Orchestration. Particularly in Adapting Mosaic, decision-makers experiment with a wide range of new, more localized governance structures, and not all experiments work equally well. Some of these might actually foster the proliferation of corruption and mismanagement within the local governance structures if transparency and oversight from either local groups or higher scale structure is missing.

Food security and the elimination of hunger are stated prominently in the JDSD as the most urgent social (but also economic) challenges. In order to explore food security effects, the IMPACT model projects the percentage and number of malnourished preschool children (those under age five) in developing countries. A malnourished child is a child whose weight-for-age is more than two standard deviations below the weight-for-age standard set by the U.S. National Center for Health Statistics/World Health Organization. This standard is adopted by many U.N. agencies in assessing nutritional status in developing countries. The projected numbers of malnourished children are derived from an estimate (for detailed information, see Smith and Haddad 2000) of the functional relationship between the percentage of malnourished children and several factors: average per capita calorie consumption, non-food determinants of child malnutrition such as the quality of maternal and child care (proxied for by percentage of females undertaking secondary schooling as well as by females' status relative to men as captured by the ratio of

Table 14.15. Prospects for Progress toward Long-term Sustainable Development in MA Scenarios, 2030–50

Sustainable Development Objectives for 2030 and 2050 from JDSD	Indicators			Global Orchestration		Order from Strength		Adapting Mosaic		TechnoGarden	
ECONOMIC											
GDP growth	million 1995 dollars,			***2030***	***2050***	***2030***	***2050***	***2030***	***2050***	***2030***	***2050***
(1995 = 100)	in 1995		OECD	259%	382%	203%	237%	203%	265%	232%	316%
	OECD	21,469,311	FSU	373%	920%	214%	315%	257%	538%	305%	691%
	FSU	854,712	LAC	394%	983%	299%	548%	334%	711%	362%	892%
	LAC	1,711,802	MENA	324%	807%	280%	497%	299%	634%	318%	793%
	MENA	875,642	Asia	794%	2118%	435%	720%	566%	1333%	632%	1614%
	Asia	2,945,748	SSA	319%	936%	290%	656%	305%	792%	321%	1001%
	SSA	283,642	World	330%	636%	236%	321%	254%	429%	287%	520%
	World	28,140,857									
Poverty eradication											
Gap between higher-income and developing countries	GDP per person (1995			***2030***	***2050***	***2030***	***2050***	***2030***	***2050***	***2030***	***2050***
	dollars), in 1995		OECD	221%	305%	189%	230%	187%	244%	204%	272%
	OECD	25,747	FSU	375%	954%	226%	379%	267%	602%	307%	735%
	FSU	2,061	LAC	276%	633%	180%	278%	202%	365%	235%	513%
	LAC	3,591	MENA	195%	427%	150%	222%	161%	283%	179%	380%
	MENA	2,502	Asia	596%	1564%	288%	435%	378%	810%	446%	1079%
	Asia	968	SSA	180%	455%	135%	226%	145%	286%	165%	406%
	SSA	482	World	243%	443%	157%	189%	170%	254%	201%	333%
	World	4,931									
Access to financial resources											
Benefits from opening markets											
Use of modern technology	assumption: overall trend		high			low		medium-low		medium for technology in general; high for environmental technology	
Technology transfer	assumption: international relationships (stimulating technology transfer)		high			low (medium among cultural groups)		low-medium		high	
SOCIAL											
Food security	percent of malnourished children (0–5 years old)										
		1997		***2025***	***2050***	***2020***	***2050***	***2025***	***2050***	***2025***	***2050***
	LatAm	9.1	LatAm	6.0	0.0	7.2	4.3	7.4	4.8	6.4	1.6
	SSA	32.8	SSA	29.0	18.6	30.6	26.3	30.8	23.7	29.3	20.0
	WANA	13.2	WANA	12.1	9.5	12.8	11.5	13.1	12.1	12.5	10.3
	S Asia	50.8	S Asia	44.5	37.3	47.3	45.6	47.2	42.7	45.5	38.7
	SE Asia	34.1	SE Asia	27.7	20.2	31.1	28.6	31.0	27.1	29.4	24.1
	China	17.4	China	11.0	7.2	14.4	14.4	14.4	14.1	12.8	10.9
	Developing	31.4	Developing	27.2	19.8	29.7	26.8	29.7	25.0	27.9	21.5
	number of malnourished children (0–5 years old), in thousands										
		1997		***2025***	***2050***	***2025***	***2050***	***2025***	***2050***	***2025***	***2050***
	LatAm	5,86	LatAm	2,661	0	4,977	3,396	4,937	3,101	3,709	833
	SSA	32,667	SSA	31,066	17,487	50,376	50,500	48,069	38,479		
	WANA	5,978	WANA	4,860	3,120	7,190	6,687	7,121	6,320	6,154	4,784
	S Asia	85,040	S Asia	59,542	34,832	99,693	91,046	94,591	70,913	75,892	53,054
	SE Asia	19,244	SE Asia	11,954	6,489	20,622	19,304	19,562	15,185	16,049	11,391
	China	18,364	China	6,660	2,855	13,307	12,958	12,767	10,842	9,795	7,171
	Developing	166,379	Developing	116,742	64,783	196,166	183,891	187,047	144,842	150,512	105,172

(continues)

Table 14.15. Continued

Sustainable Development Objectives for 2030 and 2050 from JDSD	Indicators			Global Orchestration		Order from Strength		Adapting Mosaic		TechnoGarden	
	kilocalories available per person per day										
				2025	***2050***	***2025***	***2050***	***2025***	***2050***	***2025***	***2050***
			LatAm	3,233	3,698	3,090	3,350	3,063	3,235	3,177	3,484
			SSA	2,539	2,972	2,432	2,617	2,378	2,495	2,500	2,801
			WANA	3,125	3,458	3,035	3,242	2,997	3,141	3,073	3,348
			Asia	3,181	3,702	2,823	2,938	2,829	2,955	3,019	3,291
			Industrial	3,645	3,967	3,522	3,770	3,461	3,612	3,552	3,780
			Developing	3,099	3,562	2,825	2,963	2,814	2,930	2,976	3,240
			World	3,201	3,636	2,939	3,068	2,921	3,025	3,078	3,325
Changing consumption and production patterns	meat consumption, in kilograms per person per year, in 1997		change in 2050 (1997 = 100)								
	OECD	88	OECD	149%		132%		127%		115%	
	FSU	42	FSU	207%		131%		126%		114%	
	LAM	54	LAM	167%		120%		117%		120%	
	MENA	22	MENA	155%		123%		127%		123%	
	Asia	23	Asia	291%		135%		130%		152%	
	SSA	12	SSA	225%		150%		150%		150%	
	World	36	World	194%		114%		114%		117%	
Access to adequate shelter											
Access to energy	primary energy use, in gigajoules per person		Change: 1995 = 100								
		1995		***2030***	***2050***	***2030***	***2050***	***2030***	***2050***	***2030***	***2050***
	OECD	204.7	OECD	114%	146%	140%	144%	131%	127%	101%	80%
	FSU	115.7	FSU	158%	213%	130%	159%	130%	165%	94%	89%
	LAC	45.6	LAC	291%	409%	191%	254%	202%	277%	174%	201%
	MENA	55.8	MENA	178%	276%	146%	177%	147%	195%	123%	136%
	Asia	31.9	Asia	243%	405%	150%	187%	164%	228%	134%	164%
	SSA	26.3	SSA	97%	156%	87%	102%	87%	113%	72%	91%
	World	65.3	World	169%	231%	122%	134%	125%	147%	103%	103%
	assumption: energy supply			market liberalization; selects least-cost options; rapid technology change		some preference for clean energy resources		focus on domestic energy resources		preference for renewable energy resources + rapid technology change	
	percent of renewable in world energy			***2030***	***2050***	***2030***	***2050***	***2030***	***2050***	***2030***	***2050***
				4%	11%	4%	11%	4%	11%	4%	11%
Access to health care											
Ensure capacity building											
Human resource development	assumption: investments into human capital			high assumes the highest rates of investment in education and health care		higher-income countries: medium developing countries: low investments in education and health care outside of current high-income regions will be low because of the lack of financial capital		initially follows the pattern of the Order from Strength scenario, because of large investments in education and health care; economic growth rates increase over time and approach those of the TechnoGarden scenario in the last half of the century		medium investments in human resources are likely to be lower than under Global Orchestration, partly as a result of the emphasis of Techno-technology investments	

Education	percentage of females undertaking secondary schooling										
		1997		***2025***	***2050***	***2025***	***2050***	***2025***	***2050***	***2025***	***2050***
	LatAm	56.6	LatAm	62.3	72.0	62.3	68.0	62.3	70.0	62.30	72.00
	SSA	15.8	SSA	21.8	47.3	21.8	32.1	21.8	45.3	21.83	47.25
	WANA	58.5	WANA	72.4	74.5	72.4	73.5	72.4	74.0	72.40	74.50
	S Asia	30.4	S Asia	45.1	60.4	45.1	53.4	45.1	68.0	45.10	68.90
	SE Asia	51.9	SE Asia	65.4	77.0	65.4	70.4	65.4	75.0	65.41	76.64
	China	63.5	China	74.4	75.3	74.4	74.8	74.4	78.1	74.40	75.30
ENVIRONMENTAL											
Protect natural resource base	disruption of landscape		the second-worse case because fossil use increases by about a factor of two over the same period, and environmental management is also largely neglected			the biggest disruption by far because total fossil fuel use increases by more than a factor of 2.5 by 2100 compared with 2000; society gives environmental protection low priority		an in-between case that also gives priority to environmental protection, but fossil fuel use nearly doubles up to 2100		the impact is likely to be the smallest because fossil fuel substantially declines up to 2100, but because environmental management is given high priority	
Biodiversity			high deforestation rates, steadily increasing temperature and climate-related changes in vegetation, intensification of agricultural land, increasing water withdrawals and water stress tend to threaten ecosystems in the South and eventually decrease biodiversity; decreasing biodiversity is compensated for somewhat by increasing investments in biochemical exploration so that the net rate of biochemical discoveries is roughly constant in the South up to 2050								
Fish stocks	sustainability of marine fishery—the scenarios show a medium to large increase in fish production and consumption in all regions of the world										
Desertification	see Section 14.4										
Climate change	GHG emissions, in gigatons of carbon equivalent		1995 = 100								
		1995		***2030***	***2050***	***2030***	***2050***	***2030***	***2050***	***2030***	***2050***
	OECD	3.854	OECD	136%	130%	142%	140%	115%	90%	71%	26%
	FSU	1.219	FSU	103%	104%	129%	151%	99%	101%	41%	11%
	LAC	0.922	LAC	215%	281%	232%	322%	189%	236%	139%	95%
	MENA	0.494	MENA	242%	349%	289%	477%	243%	380%	178%	154%
	Asia	2.729	Asia	203%	273%	270%	392%	206%	256%	134%	103%
	SSA	0.513	SSA	325%	337%	243%	298%	156%	273%	204%	241%
	World	9.731	World	173%	203%	198%	255%	154%	179%	104%	71%
	temperature increase, in degrees Celsius over preindustrial			***2030***	***2050***	***2030***	***2050***	***2030***	***2050***	***2030***	***2050***
				1.31	1.98	1.25	1.75	1.30	1.86	1.29	1.55
	CO_2-eq. concentration, in ppmv			***2030***	***2050***	***2030***	***2050***	***2030***	***2050***	***2030***	***2050***
	CO_2 equivalent 1995 = <411			561	719	550	666	534	629	503	516

(continues)

Table 14.15. Continued

Sustainable Development Objectives for 2030 and 2050 from JDSD	Indicators	Global Orchestration	Order from Strength	Adapting Mosaic	TechnoGarden
Air pollution	NO$_x$ emissions, in teragrams of nitrogen per year	1995 = 100			

1995		
OECD	15.916	
FSU	3.809	
LAC	4.693	
MENA	1.89	
Asia	9.5	
SSA	4.601	
World	40.409	

	Global Orchestration 2030	Global Orchestration 2050	Order from Strength 2030	Order from Strength 2050	Adapting Mosaic 2030	Adapting Mosaic 2050	TechnoGarden 2030	TechnoGarden 2050
OECD	88%	79%	102%	86%	80%	63%	48%	30%
FSU	110%	109%	104%	104%	92%	88%	62%	47%
LAC	166%	171%	148%	171%	132%	142%	106%	87%
MENA	181%	220%	169%	204%	160%	200%	113%	92%
Asia	236%	293%	188%	230%	174%	186%	96%	73%
SSA	102%	111%	114%	119%	107%	122%	89%	88%
World	140%	153%	132%	141%	116%	117%	75%	58%

Sustainable Development Objectives for 2030 and 2050 from JDSD	Indicators	Global Orchestration	Order from Strength	Adapting Mosaic	TechnoGarden
Water pollution	return flow 2050 water withdrawal—consumptive use-qualitative estimation	***2020:*** increasing except FSU ***2050:*** increasing except OECD and FSU wastewater flows increase by 40% (and hence increase the risk of overloading the detoxification ability of freshwater systems), but this is the second lowest increase among the scenarios; wealth of the North is used to repair breakdowns in water purification as they occur; in the South, there are net losses in water purification by ecosystems	***2020:*** increasing ***2050:*** increasing water purification declines in both the North and the South; large expansion of agricultural land and population poses the largest risk to the state and extent of wetlands (and hence their capacity to process wastes); likewise, the magnitude of wastewater discharges is the largest among the scenarios	***2020:*** increasing except FSU and MENA ***2050:*** increasing except FSU magnitude of wastewater discharges is second largest among the scenarios; although these factors tend to reduce the ability of freshwater ecosystems to purify water, society gives local water management special priority and therefore ensures that wetlands are protected and wastewater discharges are treated; hence in both North and South an improvement is expected in the water purification capacity of ecosystems.	***2020:*** increasing except FSU ***2050:*** increasing except OECD and FSU little change in water regulation by 2050; in the South, improvements by 2050 because the time lags for ecosystem engineering are shorter, and in some cases the South is able to learn from and avoid errors made earlier in the North
Marine pollution					
Access to clean water	percentage of population with access to treated surface water or untreated but uncontaminated water from another source				
Sanitation					

Access to clean water	Global Orchestration 2025	Global Orchestration 2050	Order from Strength 2025	Order from Strength 2050	Adapting Mosaic 2025	Adapting Mosaic 2050	TechnoGarden 2025	TechnoGarden 2050
LatAm	83.5	86.1	83.5	84.7	83.5	85.0	83.5	86.1
SSA	69.5	78.1	69.5	72.5	69.5	77.1	69.5	78.1
WANA	92.0	94.0	92.0	92.5	92.0	93.0	92.0	94.0
S Asia	83.8	92.8	83.8	86.6	83.8	92.5	83.8	93.0
SE Asia	82.8	91.7	82.8	86.9	82.8	91.3	82.8	91.7
China	80.0	84.3	80.0	83.0	80.0	81.9	80.0	84.3

Key: FSU = former Soviet Union; LAC = Latin America and the Caribbean; MENA = Middle East and North Africa; SSA = sub-Saharan Africa; WANA = West Asia and North Africa

female to male life expectancy at birth), and health and sanitation (proxied for by the percentage of the population with access to treated surface water or untreated but uncontaminated water from another source).

The parameters determining childhood malnutrition in addition to kilocalorie availability (access to water, female/male life expectancy at birth, and share of female secondary schooling) are collected from actual values for the baseline and are then estimated up to 2050 based on the qualitative scenario storylines. It is deeply disheartening to see that, even after half a century of unprecedented economic growth in the Global Orchestration scenario, almost 4 out of 10 children under five years of age remain malnourished in South Asia and about 20% on average in the poorer world, despite Latin America's successful elimination of the problem. Not surprisingly, malnourishment indicators are worse in all regions under all other scenarios. This reconfirms the importance of the economic development dimension of hunger, which needs to be dealt with urgently as opposed to the importance of the ecological/natural resource constraints as the main cause of hunger, even over the long term.

The MA models do not keep track of the intranational distribution of incomes and the access to food by different social groups. Yet scenario results in Table 14.15 suggest that, perhaps with the exception of sub-Saharan Africa in 2030, at least in the Global Orchestration future the minimum level of dietary energy consumption should be available to all on the basis of the average calories available per capita per day. At a global level, this indicator is less than 10% below the wealthy-country average in 2050 (except for sub-Saharan Africa). This, in turn, reconfirms the importance of the social equity and distribution dimensions of hunger and its mitigation.

These findings should not be taken to mean land, water, and other resources in food and agriculture are less important. Their degradation is likely to just exacerbate the problem. However, MA scenario results on these issues are perfectly plausible and congruent with the results of a large body of past empirical work: the principal underlying causes of persistent hunger are economic (poverty, lack of income to buy or grow enough food) and social (inequity, deprivation of the opportunity to earn incomes or obtain land) rather than environmental or natural resource–related. The three factors—land and natural resources, economic, and social—need to be managed in a coordinated way to foster progress toward eliminating hunger.

The key social factor behind the phenomenal income growth in the Global Orchestration scenario and the rather modest achievements under Order from Strength also becomes obvious from Table 14.15. Investments in human capital (education, health care) stagnate or even decline in Order from Strength. This forecloses the adoption of new technologies, retards the skills of the labor force, and undermines labor productivity because of the feeble health status of the workforce. Not even the most radical reshaping of the international order or the soaring abundance of financial resources would be of any help to accelerate sluggish economic development in these circumstances. In contrast, Global Orchestration exhibits the highest rates of investments in education and health care. The improving labor force facilitates the seizing of opportunities opened by expanding access to financial resources and markets.

There are two important messages emerging from the comparison of these two scenarios. First, international economic conditions (trade, markets, financial resources) and the domestic human capital situation are two equally important preconditions for rapid economic development. Second, these two factors are mutually reinforcing and involve positive feedback loops. Segregation from international markets obstructs efficiency gains and hampers income growth. The resulting shortage of funds impedes investments in human capital and further retards productivity improvements; the poverty trap of Order from Strength is closed. The causality is quite the opposite in Global Orchestration: under the premise that equitable market access is guaranteed, improving human capital can not only help generate faster productivity improvements domestically, it is also allows for better integration into the international economy and helps to reap additional benefits from trade. This makes raising funds for investments in human capital easy and self-enforcing. In this way, human capital formation, together with economic growth, can help establish the economic basis for other aspects (social, environmental) of sustainable development.

The balancing of economic and environmental concerns over time is a contentious issue among scientists and a controversial point in policy-making. Selected environmental features of the scenarios pertaining to international dimensions of biodiversity, wetland conservation, and desertification are discussed in preceding sections of this chapter. This section uses a global and a local environmental issue declared to be of high importance to national governments to illustrate how the MA scenarios differ.

The international environmental agenda has been dominated by the risk of anthropogenic climate change over the past two decades. The MA scenarios suggest that this problem will persist through the end of their time horizon (2050). Global GHG emissions (measured in terms of the Kyoto gases in gigatons of carbon equivalent) more than double by 2050 in Global Orchestration and almost double in Adapting Mosaic. While in both of these scenarios emissions decline after 2050, they continue to increase in Order from Strength. TechnoGarden shows the lowest overall increase in emissions and therefore exhibits the lowest overall temperature change or risk of crossing climate thresholds.

The reasons for these interesting patterns are complex. In the globalized worlds, well-functioning international agreements, such as the UNFCCC, are more likely to succeed in maintaining better control of global pollutants, such as GHGs. TechnoGarden, which describes a globalized and environmentally oriented world, shows the effects of international cooperation combined with strong global climate policies to curb a global commons problem. In Global Orchestration, in which the main focus lies on socially equitable economic growth but less on environmentally sound practices, environmental agreements manage to keep at least partial control on global problems, though progress will be slower. Emission levels are likely to decline after 2050, as more countries will be able to replace polluting industries due to higher revenues. However, the risks associated with the continued increase in GHG emissions over the first half of the twenty-first century in this scenario are substantial.

In the more fragmented worlds, global commons problems are more difficult to deal with, as the breakdown of global environmental regimes undermines action to reduce globally harmful pollutants. The fragmented global economy and the drying up of international financial flows result in markedly slower growth rates and, other things being equal, in lower GHG emissions under Adapting Mosaic. Nevertheless, the replacement of carbon-intensive industries is also likely to be slower. The preoccupation with local ecosystems and pollution issues is likely to lead at first to a disregard for global climate change problems. Later, however, many local improvements and renewed attention to the global commons will contribute to overall easing of the problem. This renewed attention arises as decision-makers realize that they need to deal with the impacts of global problems on local processes.

In the world of Order from Strength, agreements for the protection of global commons are not likely to be very effective. Despite the fact that economic growth will be relatively low, fragmentation and little consideration for environmental policies will exacerbate the climate problem. Measures to ameliorate the problem will likely be introduced too late (whenever severe impacts become visible) to have a real impact due to the inertia of the climate system.

The MA scenarios indicate that progress toward sustainable development is possible under very different circumstances and along different pathways. But they also demonstrate the potential threats to ecosystems and human well-being that might emerge along some paths. The choice of the actual direction and the implementation strategy rests mainly with national governments. The documentation of the relationships among driving forces, ecosystem change, and human well-being in the scenarios is intended to help governments and other actors make those choices.

14.5.3 Economic Growth, Food Security, and Stable Governance Structures in the MA Scenarios

14.5.3.1 Security Concerns of Governments

Security, defined as the continuation of stable governance structures and safety for citizens within state boundaries, is a key concern for many policy-makers in all countries. One of the key factors affecting global security is the distribution of wealth among individuals and the access to necessary ecosystem services. Within each scenario, many of the inequalities of wealth created among different regions result from the lack of movement of technology and free trade between regions. Many policies that could affect trade and movement could be the result of stringent security barriers. A synergistic effect occurs—the lack of free trade and movement of technologies that may result because of internal government security concerns further exasperate income disparities (especially in the Order from Strength scenario), which in turn could drive instability created by security concerns among regions. If we assume that wealth is one factor that affects access to various levels of ecosystem services, then we can assume that access to ecosystem services will vary in relationship to wealth distribution. The more skewed the distribution of wealth is within a country and between countries, the more likely it is that conflicts about the use of various ecosystem services will arise. Conflicts about provisioning services, such as access to food or water, can already be found today in many parts of the world, posing severe security risks to these regions.

The comparison between food consumption patterns in Order from Strength and Global Orchestration illustrates this point. The greatest disparity in total consumption of meat, fish, and grain among regions for all scenarios occurs between these two. Part of this difference is due to the unequal distribution of wealth between various regions and the ability to afford food production internally within each region. (Certainly other factors, such as the accessibility of fresh water, tillable land, and other ecosystem services, also affect food consumption.) Unequal access to ecosystem services, whether from disparities in wealth or other factors, can lead to development of regional ecosystem service hot spots. Within these hot spots there exist rapid changes in ecosystem services and an increasing need to gain access to these services. The hot spots that have developed across all scenarios are in the developing regions of the Central Part of Africa, the Middle East, and South Asia. Coincidentally, many of the security concerns of the higher-income world have recently focused on the same regions that the scenarios identify as facing rapid changes in delivery of ecosystem services.

What are the promising actions that could help alleviate security concerns for governments? Development of policies that favor high economic growth, equality in wealth distribution, and subsequent equal access to ecosystem services will likely lead to more stable governance within all regions and more security for all governments. More equitable distribution of wealth could be developed by policies that promote fair, free trade and encourage the transfer of technologies among all regions of the world. However, attention given to global hot spots will help alleviate economic growth and wealth disparities, which in turn could help create stable global security environments. If global or regional security is a major concern of governments, the attention should be paid to the regions where the potential exists for rapid change in ecosystem services. This is certainly an oversimplification of the issues driving global security, as a number of other factors, including cultural and human well-being, affect global security, but environmental degradation has become a pressing problem in many regions of the world and its impacts on security need to be considered.

14.5.3.2 Food Production and Ecosystem Services

Food production and food security have been long-standing concerns for national governments. Total food production will increase across all four scenarios. The measures used to achieve overall productivity increases and the importance of food imports differ across scenarios and depend on trade policies, investments in agricultural research and technology, technology adoption, and agricultural infrastructure implementation. Thus, governments face different dilemmas with respect to food production under each of the trajectories the MA scenarios describe.

In Order from Strength and Adapting Mosaic, food production increases are relatively slow, which results from a combination of protectionist policies and lack of investments in agricultural research. More food is being produced locally in each nation. Agricultural production extensifies and more land is taken into production. The results are increased food prices and slow improvements in caloric intake. In these scenarios, all nations face similar dilemmas over expansion of agricultural areas and subsequent negative impacts on ecosystem services. However, wealthier countries are largely able to keep up with food demand whereas poorer ones struggle to do the same. In contrast to these scenarios, the high investment in agricultural research and generally free trade strategies for food result in increased global food production, with less emphasis on agricultural expansion for all nations under Global Orchestration and TechnoGarden. In these scenarios, the open global trade policies result in more trade among nations for food and less reliance on internal production. All nations in these cases fare better with caloric intake, which is fairly even among all countries.

The threats and risks faced by governments concerning food production vary by scenario and between rich and poor countries, but in all cases increases in food production result in differences in irrigation and subsequent access to fresh water. Increases in irrigation will occur in particular in the Global Orchestration and TechnoGarden scenarios and could result in a trade-off between access to fresh water and access to food sources. Potentially devastating effects in countries where fresh water is limited will have to be offset by increased access to food from outside sources. Otherwise, water-poor countries are likely to lose most of their water supplies. Another option might be the development of global freshwater policies. Governments in these scenario worlds will be faced with the dilemma of having agricultural technology development that incorporates innovative approaches to save water match population growth and the subsequent increase in food demand.

Governments in Order from Strength and Adapting Mosaic face a different dilemma. The expansion of agricultural areas within each country needed to keep up with food production will result in a decline in other ecosystem services. For example, wetlands that are drained to establish fields will lose their water retention and purification capabilities, which are important regulating services. Access to technology and improved practices to overcome the problems created by increased cropland area will be different among governments. Particularly in the Order from Strength world, it is likely that rich countries will have a disproportionate ability to develop "technological fixes" because of the disparities in wealth. Poorer countries will be faced with the inevitable trade-offs between production of food and access to other ecosystem services, with limited resources to develop new solutions. This will further exasperate the differences among countries. In Adapting Mosaic, these tendencies are not likely to be quite as strong, as there will be a number of areas focusing on experimenting with specific solutions to food production problems. But as the exchange of technology and knowledge is more localized toward the beginning of this scenario, innovative solutions will not catch up with food demands in all areas of the globe. Therefore, poorer countries will have to rely on area expansion, at least for a while, to meet the demand.

What might be promising actions for food policy development? As with previous discussions, the type of policies that governments can implement will depend on the trajectory depicted by each scenario. Comparing strategies across scenarios can help governments adjust their policies and develop robust solutions that work under all the described worlds.

In the more globalized worlds of Global Orchestration and TechnoGarden, all countries continue policies that promote investment in agricultural research and the open and fair exchange of goods and technologies. This will help create novel solutions for food production and balance food supply and demand, particularly for areas in which environmental conditions, such as access to fresh water, constrain food production.

In Order from Strength, governments will have to pursue policies that address the trade-off between increased land use for agriculture and loss of other environmental services. Development of policies that promote increased investment in agricultural technology will inevitably be necessary. In wealthy countries, this will eventually mean a reordering of priorities for research and development funding, which could conflict with other demands on research funds. In poorer countries, this dilemma will be even more severe as resources are fewer. Some of them might have to have stringent protection for areas essential to provide a number of non-agriculture-related ecosystem services. However, it may be unrealistic to believe that countries facing persistent hunger will be able to take key natural resources out of production. Developing measures that couple short-term profitability aspects with conservation objectives may be an important step forward in protecting these natural resources.

In the Adapting Mosaic scenario, some decision-makers in poorer countries are likely to take this route by experimenting with local solutions to address the productivity-conservation problem.

14.6 Implications for Communities and NGOs

This section is addressed to the primary users of the Millennium Ecosystem Assessment, including NGOs working on environmental and social development issues and other civil society organizations involved in local development and environmental protection. This includes cooperatives, indigenous peoples' organizations, and indigenous communities. The section synthesizes the policy implications of all the scenarios for civil society, with a primary emphasis on local communities and local to international NGOs. Local communities are considered to include locally focused groups or movements of civil society and the managers of local common property/pool natural resources.

Nongovernmental organizations are nonprofit organizations independent of government that receive at least a share of their support from private sources. While most NGOs have a local focus, those that deal with regional or international issues can have significant local impacts. NGOs are considered to include academic institutions, foundations, and private voluntary organizations. They are defined here as focusing on protecting public goods rather than the well-being of their particular constituents (that is, not unions or cooperatives). This section will primarily address responses to the scenario implications requiring formal or informal policies and will give less attention to the other types of responses by communities and NGOs, such as altering management practices or use of technology, that are covered in greater detail in the *Policy Responses* volume.

While it is obvious that none of the scenarios have outcomes that only affect local communities or NGOs, some of the possible outcomes could have particular impacts on local community health and well-being and pose unique challenges and opportunities for NGOs. Table 14.16 summarizes some of the major stresses and response options for communities and NGOs under the four global scenarios.

14.6.1 Communities

14.6.1.1 Community Concerns and Specific Scenario Implications

Local communities are particularly concerned with direct impacts on their health and well- being. Loss of biological resources is often of greatest concern when it affects livelihood options; pollution of air and water is of concern when it has health impacts. Communities are also concerned with the indirect drivers of ecosystem change (such as economic and social justice and equity, population, and education), and they play critical roles in responding to both direct stresses and indirect drivers.

When considering community responses to the scenario outcomes, it is important to consider the impacts on different types of communities (indigenous peoples, fisherfolk, farmers, women's groups, and so on) and on the communities that are most vulnerable (such as poor communities in developing nations that are directly dependent on local biodiversity for survival). In order to determine which communities, in which locations, will be most vulnerable, it will be necessary to integrate the level of exposure to stresses with how sensitive and resilient people and ecosystems are to those stresses. Local communities are often able to determine what impacts will be, but they are challenged to obtain the necessary access to those with the political power or funding to pursue the necessary response options.

14.6.1.1.1 *Biodiversity: habitat loss and overexploitation*

Important links exist between biodiversity loss, loss of ecosystem services, and human livelihood impacts. All the scenarios indicate that habitat loss and fragmentation, climate change, and pollution will result in the loss of biological diversity. Habitat loss is the greatest threat in general, though climate change is the major threat to desert and tundra biomes. TechnoGarden has the lowest potential impacts on biodiversity. At the same time, this scenario could lead to the greatest use of new technologies to replace declining ecosystem services. Some of these technologies, however, have potentially serious impacts on biodiversity, such as the outbreak of new pests or genetically modified organisms. NGOs and communities therefore need to consider seriously the benefits and risks of new technologies.

Climate change could have severe impacts on biodiversity, particularly under the Global Orchestration scenario. Biomes at special risk are boreal forests, tundra, shrub, and savanna. Many local communities, such as the reindeer-herding nomadic Evensk in the far east of the former Soviet Union, could have their livelihoods affected by changes in the tundra ecosystem that affect food and habitat for reindeer.

The scenarios have different implications for different regions. Higher-income nations that are largely far less dependent on local biological resources for immediate human needs will be less affected by the possible changes. Developing-country communities (particularly in sub-Saharan Africa and Latin America) are expected to experience the greatest risk of biodiversity loss in the short and long term.

Agricultural land increases at the expense of natural habitat in all scenarios. Loss of tropical savanna is most severe. While the amount of forested land shows some increase due to regrowth in most scenarios, tropical forests decrease in all scenarios by 2050 (11% loss in TechnoGarden and 22% loss in Order from Strength).

14.6.1.1.2 *Provisioning and regulating ecosystem services*

One of the most critical concerns for communities will be changes in provisioning and regulating ecosystem services and the associated changes in environmental security. While in most scenarios provisioning services improve while regulating services decline, in Order from Strength there is high vulnerability and collapse of both regulating and provisioning ecosystem services. Adapting Mosaic shows an increase in ecosystem services resulting from co-management strategies aimed at managing a whole range of ecosystem functions, while TechnoGarden also shows an improvement but from its reliance on innovative technology. The increase in the spatial heterogeneity of ecosystems in Adapting Mosaic could lead to greater use of genetic diversity by local communities. Pollination is expected to be worse than present in both North and South in all scenarios except Adapting Mosaic, where it is unchanged. Since most of the models that the scenarios are based on were not good at predicting specific thresholds for changes in ecosystem services, local monitoring will be important for all scenarios to allow timely community responses.

All the scenarios indicate increases in return flows of water, which is an indicator of pollution of fresh water and estuaries, with the highest risk in sub-Saharan Africa and Latin America. The water stress identified as a particular risk in Order from Strength could lead to construction of dams for water storage, potentially resulting in greater habitat alteration and biodiversity loss. Similar water stress in the

Table 14.16. Community and NGO Primary Stresses and Selected Response Options in MA Scenarios. All values are estimates of relative comparison among scenarios. Many responses apply to more than one stressor.

Stresses and Responses	Scenario GO	OS	AM	TG
On communities				
Ecosystem stress—habitat loss and overexploitation of biodiversity	●●●●	●●●●●	●●●	●●●
Adapt local livelihood options: community forestry, ecotourism	✻✻	✻	✻✻✻✻	✻✻✻
Partnerships with NGOs, government, private sector to protect local habitats	✻✻	✻✻✻	✻✻✻	✻✻✻
Achieve more sustainable use of productive lands: IPM	✻✻	✻	✻✻✻	✻✻✻✻
Maintain and use traditional knowledge	✻	✻	✻✻✻✻	✻✻
Vulnerability of other ecosystem services (provisioning, regulating, cultural, and so on)	●●●● (D) ●●● (I)	●●●●●	●●	●●
Strengthen traditional community institutions	✻✻✻	✻	✻✻✻✻	✻✻✻
Seek resource tenure or use rights on state or private land and water	✻✻✻		✻✻✻✻	✻✻✻
Human health and well-being (livelihood, security, health, good social relations)		●●●● (D) ●●● (I)	●●	●
Participate in planning, implementation, and review of development projects	✻✻✻	✻✻	✻✻✻	✻✻
On nongovernmental organizations				
Ecosystems stress—habitat loss	●●●●	●●●●●	●●●●	●●●
Promote effective global network of protected areas for priority ecosystems	✻✻	✻	✻✻✻	✻✻✻✻
Publicize private-sector and government unsustainable resource use and impacts on communities (Home Depot store protests, certification, and so on)	✻✻✻✻	✻✻	✻✻✻	✻✻✻✻
Promote reduced pressure for agricultural expansion	✻	✻	✻✻✻	✻✻✻
Ecosystem stress—overexploitation and other ecosystem services	●●●●	●●●●	●●●	●●●
Promote sustainable use of productive lands	✻✻	✻	✻✻✻	✻✻✻✻
Promote more-effective education, incentives, regulation	✻✻	✻✻	✻✻✻	✻✻✻✻
Help maintain traditional knowledge	✻	✻	✻✻✻✻	✻✻
Obtain private-sector financial support and collaborate on voluntary environmental agreements	✻✻✻✻✻	✻	✻✻✻	✻✻✻
Human health and well-being (livelihood, security, health, good social relations		●●●● (D) ●●● (I)	●●	●
Monitor and report status of ecosystems and human well-being	✻✻✻✻	✻	✻✻	✻✻✻✻✻
Obtain public-sector financial support for advocacy to support public participation, and so on	✻✻✻	✻	✻✻	✻✻✻✻
Support capacity building	✻✻✻	✻	✻✻✻✻	✻✻✻✻

Key: GO = Global Orchestration; OS = Order from Strength; AM = Adapting Mosaic; TG = TechnoGarden

D = developing countries; I = higher-income countries

Stresses: 5 ● = severe stress, 0 ● = no worse than 2004

Responses: 5 ✻ = success likely, 0 ✻ = unfeasible/ineffective

Adapting Mosaic scenario might lead to innovative conservation approaches as well as greater dam building.

14.6.1.1.3 Human health and well-being

While per capita ecosystem services decline, per capita income increases in all scenarios, and the relative income differences across the globe narrow in all but the Order from Strength scenario. Global Orchestration's focus on the social and environmental policies that can improve well-being for the poorest communities will lead to the greatest improvement in human well-being for the communities currently the poorest and most vulnerable. Nevertheless, in this scenario the risk of unexpected environmental changes also increases due to a lack of attention to long-term environmental changes. The fact that the consequences of these changes will be first experienced by local communities underlines the importance of communities monitoring change.

Human well-being improves for the most part early in all scenarios, but this is coupled with significant inequalities in distribution, and human well-being then declines in the scenarios that experience greater ecosystem service loss and social instability (Order from Strength). All the scenarios indicate agricultural intensification and greater food production. TechnoGarden does so with the lowest increase in land under agriculture due to a strong focus on intensification efforts (Order from Strength shows a 24% increase in the area of agricultural land by 2050).

Ecosystems also supply many cultural services to local communities (such as recreation and sacred trees), which are directly connected to their well-being and often constitute an important part of their culture. The maintenance of local culture will be a challenge in the globalized worlds of Global Orchestration and TechnoGarden. In particular, TechnoGarden's reliance on spreading emerging technology globally could undermine the use of local techniques and practices for ecosystem management.

Also related to human health issues are land use change by humans, which will continue to have potentially adverse effects on community health by creating habitats (including dams and irrigation systems) in which mosquitoes can thrive. Mosquitoes are vectors of a wide variety of human and animal pathogens, including malaria, dengue, and filariasis. Current deforestation appears to be associated with the expansion of mosquito distributions and the increase in mosquito-borne disease transmission. The higher deforestation rates in Order from Strength could potentially lead to a greater incidence of mosquito-borne diseases in communities near tropical forests.

14.6.1.2 Priorities for Near-term and Long-term Community Responses

While the challenges of integrated responses are discussed in the *Policy Responses* volume, there are benefits to communities examining what the "worst-case scenario" might be for all stressors and exploring whether they could respond in an integrated way that involves various sectors and focuses on different stressors at a same time. An integrated response might involve multiple actors (community, government, NGOs, the private sector), multiple sectors and scales (water, biodiversity, human well-being), and multiple knowledge systems.

14.6.1.2.1 Biodiversity conservation

Communities can support policies that integrate biodiversity conservation with development policy at the local level by exercising their available opportunities for voice and vote (participating in opportunities to comment on programs, pressuring decision-makers, communicating with NGOs and the media, and so on). The local/regional and proactive focus of the Adapting Mosaic scenario indicates positive benefits for biodiversity and communities.

Many resource management practices of indigenous and local peoples are directly related to conserving biodiversity over much of the globe, yet many communities are socially marginalized and have lost rights to their resources and lands. Recently, indigenous and local communities have begun to obtain the recognition they deserve in global treaties (in the CBD, the Rio Declaration, the World Heritage Convention, and so on). Expanding community land and resource tenure (or acknowledging customary land tenure, such as the case for most of Papua New Guinea) is often seen as one of the best ways to conserve local biodiversity. However, just as is the case when the state retains those property rights, those rights could then be traded or sold or used in an unsustainable manner in the view of outsiders. Many studies have shown that local community resource management can effectively integrate local stakeholders into environmental governance and that integration of local environmental knowledge can lead to more effective resource management. The Adapting Mosaic scenario portrays one possible pathway for greater local control over resource management and depicts a variety of plausible outcomes.

14.6.1.2.2 Other ecosystem services

Land use change directly determines provisioning, supporting, and regulating ecosystem services. The scenarios differ in the role of institutions and property rights in local management of ecosystem services. Indigenous peoples' movements, with support from NGOs, can work with decision-makers to advocate and create polices that require governments to work in partnerships with indigenous peoples. The New Zealand Resource Management Act and Treaty of Waitangi requires local government to work in partnership with indigenous Maori peoples for local land and resource management, for example, and the Canadian government acknowledges property rights for indigenous peoples through individual tradable permits for fishing rights.

A response that could be used by communities in all four scenarios is to seek partnerships with their respective governments to integrate protection of ecosystem services with income generation. The Working for Water program initiated by the government of South Africa in 1995 is an excellent example of an innovative approach to maintaining water security, restoring the productive potential of land, and decreasing unemployment in marginalized communities. Over 300 projects help generate income for local people who remove alien species and maintain vegetative cover.

The knowledge of traditional and local managers, provided as a part of an informed public participation process, can be invaluable in defining ecological risks and ways of avoiding them. The Bedouin people in the deserts of Egypt, for instance, have thousands of years of experience managing limited water resources that would be of benefit to engineers designing irrigation schemes. Communities can use existing mechanisms for public participation in national and local environmental impact assessments and also seek to have such participation expanded. While public participation has greatly expanded in many nations in the last 15 years, there is still considerable room for improvement. Local people participate in reporting on the state of the environment in many countries in Africa, though the participation of women is still low.

14.6.1.2.3 Human health and well-being

The feedback links between ecosystem change and human health and well-being are frequently most obvious in impoverished communities that cannot afford the same "buffers" to a decline in ecosystem services as wealthy communities. Communities, through capacity-building initiatives of government, NGOs, and the private sector, can strengthen their understanding of the contribution of human landscape and habitat change to many human health outcomes and of options for disease management. This might result in changed community land use practices or community response to government or private development plans. For example, urban communities might respond to vector-borne transmission of disease by urging development planning to include consultation with entomologists, epidemiologists, and health care specialists to address issues such as storm water management and vector control.

Many of the options for the near term also apply to the long term, particularly developing partnerships between communities and NGOs in order to develop stronger voices with governments and international treaties. Capacity-building support will be essential in the long term. While it is an obvious point, the long-term scenarios continue to emphasize the importance of more proactive policies that integrate environmental issues with development issues. Communities and NGOs (both environment and development NGOs) can work together effectively to support such policies.

The global MA scenarios did not fully meet the expectations of civil society stakeholders to have the MA address the impact of ecosystem change on the vulnerability and resilience of human communities and on their cultural concerns. These issues were more successfully addressed in the MA sub-global assessments. Communities are interested in learning about site-specific impacts in relation to global changes, but integrating information across multiple geographical scales is a new challenge in the development of global scenarios. The global MA scenarios can be used to describe the main boundary conditions under which local communities are likely to operate in each of the described pathways, but they do not have the specific details that local or regional scenario exercises are able to portray.

14.6.1.3 Additional Community Response Strategies and Options by Scenario

14.6.1.3.1 Global Orchestration

This scenario attempts to improve the well-being of poorer countries by removing trade barriers and increasing investments in social and education policies, with the risk that local and regional environmental problems can become worse. One risk of Global Orchestration to biodiversity is that a narrow selection of high-yield commercial crops could spread around the globe, resulting in wild varieties existing only in gene banks that do not offer complete protection from extinction. A community priority for the near term might be to apply local ecological knowledge to helping conserve landraces and local varieties of agricultural species, possibly in partnership with NGOs, governments, and the private sector. Increased community focus on energy conservation and efficiency could help mitigate some of the adverse impacts expected to result from the development path described in this scenario.

14.6.1.3.2 Order from Strength

This scenario describes a world in which rich nations focus on protecting their borders, and environmental services are only protected (inadequately) in formal protected areas. In this regionally focused, reactive world, communities would benefit from more strategic understanding of how to influence government and private sector policy. This would be necessary to encourage policies that can create an enabling environment to stimulate and support changes in individual incentives for better resource management. This could include, for example, passing laws that mandate greater community management of protected areas, acknowledging the value of traditional approaches to conserving land such as sacred groves, or upholding community taboos and special resource harvest levels and/or seasons.

14.6.1.3.3 Adapting Mosaic

This scenario favors local management and control of ecosystems and offers communities the greatest opportunities to adapt to changing conditions and take control of managing their resources. There are many examples of local communities using traditional knowledge to alter their resource use patterns in response to the abundance or scarcity that they observe. For example, levels of family harvests from forest gardens in Southeast Asia are altered depending on actual fruit abundance to ensure sufficient fruits are left for wildlife support and for regeneration. The Adapting Mosaic scenario will increase success of this response for biodiversity conservation and other resource uses, as local success begins to be shared at the global level later in the scenario. However, the lack of early focus on global policy might cause effective local efforts to be less successful because of global problems. Success in conserving a local coral reef, for example, might be adversely affected by global changes in pollution, temperature, or exotic species.

Communities can encourage governments to expand policy options that acknowledge a role for traditional knowledge in management and conservation of local ecosystems. In South Africa, the value of both participation and traditional knowledge are encompassed in programs like CAMPFIRE (Communal Areas Management Program for Indigenous Resources). In India, joint forest management is an example of government devolving resource management to local communities. Empowering local communities and legitimizing their traditional knowledge can also be an effective way to control exotic species introduction and removal at local levels. Communities can also help demonstrate the importance of conserving cultural diversity along with biological diversity.

14.6.1.3.4 TechnoGarden

This scenario outlines how technology can maximize production of ecosystem services for humans. The implications for biodiversity indicate greater need for community policies for managing biodiversity sustainably for local consumption and other benefits (from ecosystems services such as water and nutrient cycling to generation of income from ecotourism). There is less that communities can do to affect climate change beyond working to influence national and global policies via election of government representatives and participating in opportunities to comment on government policies and plans.

Communities, in partnership with local NGOs and others, can draw attention to cases where the conservation priorities and policies of international NGOs and the foundations and governments who support them are not a reflection of local interests. For example, global-scale priorities for "hot spots" for conservation often do not acknowledge that most local biodiversity has local importance for livelihoods and other ecosystem services.

14.6.2 Nongovernmental Organizations

14.6.2.1 NGO Values and Concerns and Specific Scenario Implications

NGOs have a wide range of concerns and values, depending on the scale of their work (local, national, regional, or international) and their focus (environmental, social, or economic; policy or on-the-ground work). NGOs also play many roles, from protesting outside meetings like the World Economic Forum to having a "seat at the table" in crafting policy. Many NGOs work to support communities that might lack the capacity to protect ecosystems directly. International environmental NGOs often focus on global environmental resources regardless of national boundaries. Many international NGOs seek to conserve representative samples of biologically diverse habitats, with less focus on habitats that might be extremely important to local communities. NGOs and other civil society groups are actively involved in mobilizing public support for international environmental agreements. Environmental NGOs are increasingly advocating economic policies that provide incentives for improved conservation, such as taxation and subsidies, and that can help alter market-based incentives for overexploitation.

NGOs play an increasingly important role in protecting environmental resources and processes. While they do not have uniform approaches or views on environmental protection policies, they can play important roles in raising public awareness; organizing communities; pressuring governments, the private sector, foundations, and multilateral organizations in the design of international treaties; and providing legal assistance to local and indigenous communities. NGOs also directly own or help to co-manage protected reserves in many nations. The Nature Conservancy, for instance, manages reserves it owns in Latin America and elsewhere.

NGOs can help to monitor implementation and compliance with environmental policies from local-scale to global multilateral environmental agreements. They tend to support public participation (or at least transparency to increase the awareness of civil society) in government and private-sector decision-making, which can help enforce existing environmental policies and develop support for improved policies.

14.6.2.1.1 Biodiversity loss and overexploitation

The scenario implications for biodiversity loss as a result of land use change, exotic species introductions, and climate change create many imperatives for NGO action, since the impacts of such changes will be felt across political boundaries. Many impacts are particularly severe in developing nations, where financial resources are more limited. International NGOs can assist where national NGOs are weak, and they have a particular opportunity to focus attention on important conservation areas that cross multiple political boundaries, which no one state is likely to take responsibility for. While NGOs lack the ability to directly create and enforce biodiversity policies, they have an enormous opportunity to help shape international and national government and private-sector policy by raising awareness and conducting advocacy initiatives and then helping to implement and monitor those policies.

14.6.2.1.2 Ecosystem services: provisioning and regulating services

All the scenarios indicate decreases in provisioning and regulating ecosystem services, with more severe stress in Global Orchestration and Order from Strength than in TechnoGarden and Adapting Mosaic. Many creative approaches to conserving ecosystem services have been and can be stimulated by NGOs, which often can help to "broker" agreements between the private sector and government (as in the many market-based mechanisms for protecting environmental services provided by forests, including certification of sustainably harvested timber, or protection of water or biodiversity from forests). NGOs can play an important role in stimulating market-based responses to climate change, such as carbon investment funds. For example, in 2004 the NGO Rainforest Action Network helped Citigroup agree to social and environmental investment policies that help stem climate change and habitat loss in vulnerable ecosystems.

14.6.2.1.3 Human health and well-being

The scenario implications for human well-being indicate that there will be even greater need for NGOs to play a role translating and disseminating information on the state and sustainability of ecosystems and links to human well-being, including results from sub-global assessments and global scenarios, to communities and governments in the context of options for policy responses. For example, NGOs have played a major role in such communications at Conference of the Parties meetings for international treaties and at international fora such as the World Social Forum.

14.6.2.2 Priorities for Near-term and Long-term NGO Responses

14.6.2.2.1 Biodiversity loss and overexploitation

One of the most important policies for NGOs to support in the near term is the creation and maintenance of protected areas for biodiversity conservation. This can be done via policy advocacy, by directly purchasing or co-managing protected lands, by conservation easement management, and by partnerships with government, the private sector, and communities. International and national NGOs will increasingly need to develop and campaign for policies that address both the need for conservation in protected areas as well as the need for local and indigenous community access to common property or state- controlled biological resources. Policy incentives for conservation of the unprotected portions of the landscape will also be important and can involve public-private sector partnerships.

NGOs could ensure that the outcomes of their own projects are monitored. They could also advocate for government and private-sector policies to support appropriate monitoring, which can then provide data on specific conservation approaches that are seen as being more successful than others. If NGOs want to support adaptive management approaches, then it will be important, as threats to biodiversity increase, that policies to support monitoring are in place. NGOs have often been guilty of moving their support and advocacy from one conservation approach to another (from community-based conservation in the early 1990s to "direct payment" for conservation) without adequate information that any one particular approach actually works better than another.

NGOs have the opportunity in the near term to examine how geographic priorities for international, national, and local conservation are determined by international conventions, NGOs, and governments. The questions they can ask include: Where do various systems for setting global priorities for biodiversity conservation agree and disagree and why? Global "hot spot" declarations by NGOs do not necessarily overlap with local priorities, though they have been successful in attracting global financial resources. Also, how can global priorities among countries be reconciled with national priorities within countries? The scenarios indicate that sub-Saharan Africa, Latin America, and Asia are most at risk of species loss before 2010. Some NGOs will want to focus on policies in these most threatened areas. NGOs also have the opportunity in the near term to emphasize policies that will enhance the ability of institutions and ecosystems to adapt to the long-term changes highlighted in the scenarios.

In terms of policy, NGOs can advocate enhanced enforcement of the many international treaties that can help to conserve biodiversity (including the CBD, the International Tropical Timber Agreement, Convention on the Non-navigational Uses of International Watercourses, and the International Treaty on Plant Genetic Resources for Food and Agriculture, and CITES). The CBD, if enforced, could help prevent or mitigate the long-term threats to biodiversity by encouraging global accountability and local action.

14.6.2.2.2 Ecosystem services: provisioning and regulating services

NGOs can advocate incorporation of consideration of ecosystem services and biodiversity in integrated regional planning in order to demonstrate the important linkages to human health. Finding ways to conserve the ecosystem services of protected areas while also meeting the immense human needs that exist near most protected lands in developing nations will continue to be one of the major policy challenges.

NGOs can also develop and promote innovative approaches to ecosystem protection, many of which depend on market forces and partnerships with the private sector and government, such as tradable permits for species protection, debt-for-nature swaps, certification programs for sustainably produced timber and marine and food resources, and appropriate pricing of services and resources in order to create incentives for conservation.

14.6.2.2.3 Human health and well-being

One way to better integrate sectors and achieve more adaptive policies would be for partnerships to emerge between NGOs with a primarily environmental mission and those with a primarily social focus. The scenarios indicate the need for more interdisciplinary policies concerning both direct and indirect drivers of environmental changes. Missions of social and environmental NGOs might conflict in some cases and reinforce each other in others. The results in Chapter 9, for example, might lead social NGOs concerned with food availability and child health to support policies to raise agricultural production by increasing agricultural land area or policies to increase productivity on existing land already being used for agriculture. However, environmental NGOs might support land conservation to protect biodiversity along with increased productivity on already disturbed land. Yet both social and environmental NGOs might share common objectives in supporting polices for sustainable agricultural practices to decrease risks to biodiversity or in working to decrease deforestation in malaria areas, since deforestation has been linked to an increase in malaria. Support for technologies such as sustainable agriculture and pollution abatement systems can be included among the policy options for protecting food supplies and human health.

All the scenarios indicate that due to the threats emerging over the longer term (2030–50) it would be important for NGOs to support policies aimed at monitoring ecosystems in order to be better prepared for opportunities to act on the observed changes. Regularly updated indicators of ecosystem structure and function will allow communities and NGOs to engage in better prevention, mitigation, adaptation, or restoration of ecosystems. This monitoring would be enhanced if the institutions that emerge and evolve for global environmental management begin to focus more on biological/ecosystem units rather than political/national boundaries. Such a focus would also help in assessing ecosystem vulnerability, which is expected to grow and to have a disproportionately adverse impact on the poor.

The synthesis of changes in ecosystem services (Chapters 8 and 9) indicates that in 2050 the trade-offs between ecosystem services will be more intense than at present and that greater inequities between rich and poor nations and regions and greater adverse impacts from unanticipated disasters are very plausible future developments. This implies that environmental justice and ethics should be of even greater concern to communities and NGOs than they are at present.

One of the most useful aspects of the scenarios for NGOs and communities is the effort to assess the relationship between ecosystem changes and human health and well-being. Given their global nature, however, the scenarios are not able to fully describe all the trade-offs and interactions between ecosystems services and human well-being, especially in response to specific response options and adaptation possibilities.

14.6.2.3 Additional NGO Response Strategies and Options by Scenario

One of the important implications of all the scenarios is the opportunity that NGOs have now to help craft policy solutions that will support conservation before more irreversible loss occurs in species, habitat, and ecosystem function. Given that many of the areas of global conservation importance also have growing populations and poverty (such as South and Southeast Asia, the far eastern former Soviet Union, Equatorial Africa/the Congo Basin, and the Upper Amazon), there is opportunity to integrate conservation with the agendas to alleviate poverty, increase equity, sustain health, and so on.

14.6.2.3.1 Global Orchestration

While support for international treaties is expected to continue, the Global Orchestration scenario also points out the difficulty of adjusting global-scale policies to deal with local and regional issues as they arise.

This scenario also addresses the issue of global trade and its implications for different actors in society. It portrays a continuation of globalization, in which trade, information, and technology flows increase. But these are coupled with mechanisms to enhance the equal participation of all in global markets by, for example, decreasing market-distorting subsidies and focusing on global public goods creation and protection. NGOs and community groups might want to work with other actors to explore how to harness the best aspects of free-market capitalism while addressing those aspects that lead to negative environmental outcomes. An examination of the conditions under which free-market economies produce better environmental outcomes would be valuable. In addition, NGOs can take a lead in exploring mechanisms for protecting public goods and for developing social policies that enhance the chances of different groups in society for equal participation in global markets.

14.6.2.3.2 Order from Strength

The perennial challenge for NGOs to find funding support will be heightened under the regional, reactive focus in Order from Strength. International NGOs are likely to find less support from multilateral sources (the United Nations or the World Bank) and less interest in work on international treaties. That said, in this scenario decision-makers focus on use of protected areas to conserve biodiversity and ecosystem services. Thus NGOs will likely find support for parks creation, though funding for enforcement and monitoring will probably be limited. NGOs will also need to focus more attention on obtaining resources for conservation outside parks.

A theme across all the priorities for action by NGOs and communities is being able to adapt to emerging approaches and outcomes for environmental policy and to focus first on direct drivers of ecosystem change (with perhaps the exception of climate change). The regional, reactive focus in Order from Strength will require national NGOs that can change to focus from international treaties to more local issues. Social unrest could result from the large income inequities in this scenario, and NGOs will need to support vulnerable communities.

14.6.2.3.3 Adapting Mosaic

This scenario offers the greatest opportunity for NGOs to work with communities to develop ways to adapt and respond to changes in ecosystems and human health. Adapting Mosaic is likely to result in lower levels of global financial resources being available to NGOs, and less support for international environmental agreements. The scenario might result in more financial support from regional bodies for the work of national and local NGOs, and regional environmental agreements will require more NGO attention.

In addition, NGOs will be one of the key players to make the future world portrayed in this scenario work. They will have to play an important role in capacity building and monitoring activities for ecosystem change in this scenario. Furthermore, helping to build or rebuild community structures for ecosystem management will be one of the major tasks for NGOs in this scenario. Developing adaptive management systems requires a thorough understanding of ecosystem functions and possible management options. Communities will have to organize themselves to be active partners in management schemes and in order to negotiate with other actors. NGOs are likely to be very important in providing support for these activities.

As many NGOs also work in a variety of communities or areas, they will need to act as bridges for building networks with other communities, research agencies, and so on that can help in developing monitoring and management options. Particularly in poorer countries in which well-functioning governance structures will first have to be developed, this role of NGOs will be extremely important in an Adapting Mosaic world.

14.6.2.3.4 TechnoGarden

Monitoring of project and policy outcomes by NGOs will be important in this future world since there will be uncertainty about the degree to which ecosystem services can be replaced by technical alternatives.

In response to loss of biological diversity and associated ecosystem services, NGOs might have more success in this scenario advocating precautionary conservation policies that can justify action to regulate potentially irreversible environmental and social harms. NGOs might have success in encouraging increased attention to creation of a global network of protected areas to help stem the loss of biodiversity. Monitoring of outcomes, particularly those that are unexpected, will be important in this scenario.

14.6.3 Interactions between Communities, NGOs, and Other Response Actors

Local communities and NGOs can work together with government and the private sector to advocate policies and to execute on-the-ground practices that protect, mitigate, and restore some of the ecosystem services that are threatened by the development paths and assumptions in the four scenarios. Government and the private sector play important roles in creating the enabling environment for community and NGO action through policies and funding. NGOs and communities often know what needs to be done; they just need such partnerships to make it happen. In all scenarios, NGOs and communities can be more strategic in their efforts to integrate environmental imperatives with political reality.

Communities and NGOs bear twin burdens that must be addressed separately: determining what actions they can take to sustain ecosystems and human well-being and then obtaining the political and financial support for that course of action. Contributing to or developing the best course of action from a technical perspective, given the uncertainties and the array of possible options, is difficult enough. The bigger challenge, however, may be getting the sustained political will and financial resources needed for adoption and implementation. This becomes even more challenging in the face of uncertainty and the inevitable setbacks as different solutions are tried.

It is important therefore to look at what the scenarios indicate about the political sphere in which NGOs and communities will have to operate. Communities and NGOs work to address the incentives and structural problems in all sectors that produce harmful environmental and societal impacts. Whether communities and NGOs have any impact on powerful actors depends on a few key factors, some of which the scenarios infer and even address directly.

The particular strategies and degrees of success vary widely, of course, depending on a country's type of state, stage of economic development, and civil sector maturity and whether a particular NGO is local, national, regional, or international. Despite the best intentions, not all NGO and community efforts are successful. In general, though, NGOs' key instruments of success include using the power of community organizing, the media, science, government-mandated public participation forums, coordination between NGOs from economically developed countries and economically challenged ones, and, more recently, strategic collaboration with the private sector. Communities often do not have access to or knowledge of the powerful political elites, and thus they benefit from partnerships with NGOs and the private sector in obtaining this access.

Financial grant support from the philanthropy sector is critical to success for most NGO efforts, so the interests and assets of foundations in each scenario must also be considered. NGOs are sometimes considered to include foundations; there are over 60,000 registered foundations in the United States alone, and many hundreds of community foundations across the globe. The implications of the scenarios for foundation assets, priorities, and spheres of operation are important because NGOs depend on foundation support. It is important to note that the financial assets of corporate foundations may be affected differently than privately endowed foundations in each scenario. Consideration needs to be given to how each scenario might affect the enormous remittances sent home by overseas workers, and how those remittances help sustain community well-being.

Foundations play an important role in funding the efforts of communities, NGOs, and other actors in civil society. The issues, approaches, and groups that foundations support send strong signals to the rest of society about needs and confidence in expected outcomes. Across all scenarios, NGOs and community groups will need to devote resources to gaining a better understanding of political realities and tailoring their strategies accordingly. Developing a sound understanding of the forces that influence decision-making and finding a way to participate effectively in those arenas will be as important, if not more important, than developing tailored, feasible solutions that address policy-makers' concerns. For instance, some of the recent progressive solutions developed for air quality problems were successful in receiving political support because NGOs helped demonstrate economic benefits in addition to individual health benefits, thus successfully countering corporate claims that the costs of regulation were too high.

In Adapting Mosaic, where tailored solutions on more local and regional scales are described, avenues for communities and NGOs can be expected to be open and inclusive. This can lead to failures, though, for environmental solutions that require international coordination. Foundation support for NGO work might remain high, but multilateral and bilateral support could decrease.

In TechnoGarden, where advanced technological solutions are aggressively implemented and environmental problems are explicitly addressed, it is likely that NGOs that can contribute to this will be able to have influence. However, NGOs that advocate for changing some of the underlying economic and government incentives that cause the problems may find themselves sidelined or challenged for funding for the kind of solutions they would hope to advocate.

Most challenging for NGOs and communities is likely to be the Order from Strength world, where those with economic and political power will ignore the social and environmental problems outside their nations in order to defend their own wealth. The international collaboration and funding that has been critical for NGOs in countries that are not economically wealthy are likely to be severely compromised. The funds required to make any kind of significant change on the "outside" will become even more difficult, since societal chaos will be greater. But when complete catastrophic breakdown on the "outside" threatens to overwhelm those inside the barriers, it is likely they will provide multilateral and bilateral funds to help address the widespread hunger, disease, and environmental problems that threaten their security. If foundation endowments are not adversely effected under Order from Strength, foundations might increase NGO support when they see a decrease in bilateral and multilateral support.

The fortunes of communities and NGOs are most difficult to assess in the Global Orchestration scenario. The international coordination efforts are likely to be structured to provide for avenues of input for civil society. However, the degree to which environmental problems are seen as priorities is lower, so the results may be mixed. Global Orchestration could include greater international awareness of global environmental problems and result in more grants and other financial support for social and environmental NGOs and communities in poor nations. At the same time, it also describes the high risk of adverse impacts from climate change and little opportunity for NGOs and communities to foresee and prevent the thresholds at which further ecosystem degradation and reductions in human well-being might occur.

The many adverse outcomes envisioned for ecosystem services and human well-being under Global Orchestration and the other scenarios indicate that long-term sustainability will not result from policies that only address economic and social issues and that only focus on global scales. Current and future policies from local to global levels reflect complex relationships and interactions between governments, industry, NGOs, and, increasingly, civil society. By working together, all these actors might be able to create a world that integrates the best aspects of the various scenarios: a world that acknowledges the importance of natural and human capital and the fact that loss of some natural capital is completely irreversible, that uses adaptation and learning, and that fosters innovation in technology and institutions.

All the scenarios point to the importance of NGOs and communities obtaining an understanding of the policies that affect ecosystems and human health and then finding ways to interact with the decision-makers who bring about change. NGOs and communities can help to create policies and practices that:

- adapt to deal with uncertain outcomes by allowing for learning from experience and by using enough precaution to avoid irreversible outcomes;
- are appropriate to the scale, region, and sector of society affected while also attending to cumulative impacts across scales;
- integrate multiple disciplines and examine trade-offs among both direct and indirect values of ecosystems;
- incorporate both quantitative science and perhaps more qualitative traditional and practitioner knowledge in the proposed solutions; and
- are based on principles of transparency, participation, equity, and attention to vulnerable groups.

The policy options that NGOs and communities can advocate in negotiations with international, national, and local governments include a primary focus on ways to reduce direct drivers of ecosystem change (such as policies to encourage the sustainable harvest of forests or marine resources, to reduce greenhouse gas emissions, or to create protected areas) or on ways to protect human well-being. Some policy options will do both, and others will involve trade-offs between short-term human needs and longer-term ecosystem services.

International NGOs can play an important role in helping local communities directly with conservation initiatives, particularly when national and local government is weak or does not give priority to conservation. It will be important, however, for NGOs to simultaneously help increase the capacity of those national governments to address conservation in an integrated way with other development objectives, in order to ensure that conservation investments are sustainable after NGO involvement ends.

The priority that international NGOs give as to which ecosystems, habitats, and species to conserve from a global perspective might not coincide with local communities' perspective on conservation priorities. The scenarios indicate the importance of strong local and national NGOs that are not just sub-offices implementing the missions of international NGOs. Worldviews and values are different among NGOs at different scales, and the policy priorities of international NGOs are most often appropriate for global-scale concerns.

While bearing in mind the challenges, costs, and benefits of integrated responses outlined in Chapter 15 of the *Policy Responses* volume, vertical and horizontal integration of responses across sectors should be explored by partnerships among actors. Sustainable forest management is one example of an integrated response that attempts to conserve a number of different ecosystem services (biodiversity, hydrological processes, climate regulation, forest products, tourism, cultural values, and so on) while also improving human well-being. Sustainable forest management typically involves a wide range of actors, including local communities, NGOs, government, and the private sector. NGOs and communities can profit from the lessons of these management practices.

14.7 Implications for the Private Sector

14.7.1 Linkages and Stakes

This section documents the important yet often misunderstood nexus between ecosystem goods and services and the private sector. "Private sector" is used here in the broadest sense possible to include relevant private actors involved in commercial business activities in local, national or regional, and global settings, although the primary audience might be firms and industries involved in food and agriculture, biotechnology and pharmaceuticals, resource extraction (forestry, mining, fisheries, and so on), and energy (petroleum, natural gas, and so on), as well as sectors with high environmental impacts (such as basic industries like steel and chemicals) and high material or resource dependence (such as semiconductor and other high-technology products dependent on water and other resources). In many countries, publicly owned companies also operate in these sectors. They may be mandated to provide some public services. However, if they have commercial concerns, they are similar to those of private enterprises. Hence the key points summarized in this section are largely relevant for public companies as well.

Greater synergy between ecology and private-sector interests centers on two interrelated issues: first, how and under what circumstances can business interests "internalize" the "negative externalities" of resource extraction, production, and consumption even when there are no legal or short-term business interests to do so? Second, how can feedback mechanisms be created between ecological concerns and business interests through which signals of positive/negative trends can be channeled back between the two sectors without formal governmental interventions that are often too late and inadequate to prevent permanent environmental damage? Two other important issues or questions and implications from the MA process for the private sector are worth noting:

- What are the impacts of environmental change on firms and industries dependent on ecosystem services (such as private timber, fishing, and agriculture businesses) as well as firms and sectors affected by changes in ecosystem functions (such as private petroleum or other companies with high GHG impacts)?
- What key business opportunities and constraints are likely to arise under different scenarios based on local resource conditions (such as water availability and quality), public governmental rules (such as international environmental conventions and local resource conventions), and private governance regimes (such as international nongovernmental mechanisms like the Global Reporting Initiative and ISO 14001)?

14.7.2 Implications of Change in Ecosystem Services

14.7.2.1 Implications of Indirect and Direct Drivers of Ecosystem Services Change

The scenarios portray that an increase in per capita income and material well-being is likely to lead to higher consumption of electricity and production of industrial products. Whereas richer countries are expected to maintain or expand their control of local and regional air pollution, the same is not expected in poorer regions. For many multinational business enterprises, the rise in per capita income and material well-being in the poorer world are likely to translate into new business opportunities. These will be balanced everywhere by declines in the quality of the environment in which to conduct business. Due to the increased influence of local communities and NGO participation in global environmental governance, there is likely to be greater pressure, particularly on highly visible brand-driven multinational corporations, to go beyond prevailing rules and regulations and to play more of a role in improving local environmental conditions.

14.7.2.2 Implications of Change in Ecosystem Functions

World total production of grains increases around 50% for all scenarios, with larger differences between scenarios for the poorer world regions. Some of the gains in agriculture will be achieved through expansion of agricultural land and at the expense of uncultivated natural land. One robust finding across all scenarios is that, up to 2050, 10–20% of current grassland and forestland will be lost, mainly due to the expansion of agriculture (and, secondarily, because of the expansion of cities and infrastructure). While ecological degradation is often portrayed as a conflict between "public environmental interests" and "private business goals," different types of "business conflicts" are likely to emerge in the future. With tourism becoming the world's largest employer and an important economic factor in many poorer countries, native forestland and other natural resources will be increasingly perceived as "vital business assets" of many private companies.

Although gains are made in access to fresh water, the MA scenarios suggest a likely increase in the volume of polluted fresh water, particularly in poor countries. Moreover, the expansion of irrigated land (which contributes to the increased production of grains) leads to substantial increases in the volume of water consumed in arid regions of Africa and Asia. The availability of and access to clean water is likely to change the way private enterprises in the poor and rich worlds conduct business in the twenty-first century. For industries as different as agriculture and high technology (semiconductor plants require enormous amounts of water for chip production, for instance), water will increasingly be a factor in determining where, how, and with whom private enterprises conduct business.

14.7.2.3 Implications of Change in Biodiversity across Scenarios

Despite continued conservation efforts of the international community, biodiversity loss is occurring at an unprecedented rate. A number of important issues and questions arise for the private sector from the impact on biodiversity across the scenarios. First, what is the connection if any between biodiversity decline and economics or business intensity? If tropical Africa, which is not known as a rapidly growing economic corner of the globe, is the region that has lost the most vascular-plant species, what connection is there between biodiversity loss on one hand and business intensity on the other hand? Is it the case of just using the wrong type of economic and business model?

Second, if land use change is the dominant driver of biodiversity change (followed by changes in climate and nitrogen and sulfur deposition), how can private companies best facilitate the prevention of biodiversity decline? Even for private companies that are involved in the commercialization of biodiversity resources (such as pharmaceutical companies), it is not always clear what role they can play in these complex global ecological dilemmas.

Third, how can the private sector as a group better mobilize its efforts to prevent the continuing decline in the quality of freshwater species, which are estimated to be components of the most threatened ecosystems in the world? It is undoubtedly more difficult to mobilize business support for an issue like the protection of freshwater species, which is rarely regarded as a business priority of most firms and organizations.

14.7.3 Possible Response Strategies and Options for the Private Sector

Several near-term (~2010) and long-term (2030~50) private-sector response strategies and options can be identified for the four MA scenarios. (See also Table 14.17.)

14.7.3.1 Global Orchestration

Under the near-term Global Orchestration scenario, most firms and industrial sectors are likely to continue their business as usual strategies. Existing global policy frameworks like CBD and international organizations like the World Bank will remain the key institutional focal points for the private sector. The private sector will aim to manage its environment-related business risks through involvement in business and civil society forums like the Global Reporting Initiative and by giving nominal financial and organizational support to global ecological dilemmas like water and biodiversity loss. Certain business sectors like the financial and insurance industries may prove to be more proactive if the environmental management risk of lending to and insuring other businesses increases.

Table 14.17. Private-Sector Response Strategies and Options in MA Scenarios

Scenario	Near Term (2010)	Long Term (2030–50)
Global Orchestration	continuing reliance on existing global civil society-business-NGO forums to address environmental concerns	"ineffectively" managed environmental multistakeholder forums may lead to greater reliance on more-exclusive private sector–oriented policy forums
Order From Strength	with increasing number of boycotts and protests of western business interests, increasingly difficult to resort to multistakeholder forums to address environmental policy concerns	multistakeholder forums and international organizations likely to be abandoned by developing countries as instruments to govern the global ecological commons
Adapting Mosaic	trends toward regionally based models of public-private partnerships and collaborative activities to address environmental management and policy concerns	greater use of and reliance on a regionally based corporate environmental management strategic framework
TechnoGarden	global environmental frameworks are likely to be accepted as part of a "normal" global corporate governance	even more than the Global Orchestration scenario, clean energy business sectors are likely to represent more-promising business opportunities

Under the long-term Global Orchestration scenario, the scope of private-sector involvement in environmental matters is likely to depend on the "success" of existing environmental policy and management frameworks. If the U.N. and international policy framework turns out to be ineffective or businesses perceive it as an inefficient way to manage environmental risks, the private sector may create a private, business-directed policy alternative (such as the World Economic Forum) with little to no accountability to civil society. Key global ecological dilemmas will continue to get nominal support from the private sector, but business firms will still view them as under the primary stewardship responsibility of U.N. and other international institutions.

Business–civil society forums in the Global Orchestration scenario may develop a much stronger role in global governance, but the development of these forums is likely to come at the expense of national governmental capacity and sovereignty as well as the ability of NGOs and community groups to influence policy design and development.

Clean-energy business sectors under Global Orchestration, including hydrogen, solar, wind, and small-scale hydropower, are likely to become major revenue streams for the private sector and may usher in a radical transformation of the global energy infrastructure and system. The likelihood of this trend may depend a great deal on the relative scarcity of and ongoing prices of petroleum and other fossil fuel products.

14.7.3.2 Order from Strength

Under Order from Strength, there will be increased conflict in the near term on strategies between firms and industries in rich countries and those in poorer ones as well as within the wealthy western block of countries (particularly between private actors in the United States versus Europe) on a wide range of issues including genetically modified food, climate change, and bio-intellectual property concerns. Whereas private firms and industries in wealthy countries will stress "efficiency" and "continued access," their counterparts in poorer nations (led by large emerging market countries like China, India, and Brazil), with support from their respective government and civil society actors, are likely to demand "equity" and improved "terms of trade."

Business-civil society forums and global business networks are likely to become even more dominated by business and NGO interests from rich countries, with corresponding declining interest and support from private and nongovernmental actors in poorer nations. International government-business-civil society forums like the Global Reporting Initiative and the U.N. Global Compact will be increasingly perceived as institutional tools under the control of western private enterprises and governments. North American and European multinational enterprises, particularly those from the extractive and consumer products industries, will be increasingly subject to a wide array of boycotts, protests, and civil actions from NGO and community groups in poorer nations.

In the long term, the conflicts in Order from Strength between private business interests in wealthier countries and those from the poorer world are likely to intensify to the point that the work of the World Trade Organization and other global business and trade-setting bodies will be seriously impaired if not collapse altogether. Although the economies of poor countries are better integrated into the international political economy and more multinational corporations become "local businesses" in emerging markets, the recognition of mutual interests between multinational corporations and local businesses in those nations will be difficult to sustain in this scenario.

The serious deterioration and possible collapse of mutual interests between actors in the richer and poorer countries is likely to make it very difficult for different stakeholders to form much-needed partnerships and collaborative activities addressing sustainable ecosystem management. This situation will be further hampered by weakened global governance institutions that have traditionally helped ease tensions and broker collaborations between the public and private stakeholders in the two worlds. Highly visible multinational enterprises from North America and Europe are likely to be most negatively affected under this scenario.

14.7.3.3 Adapting Mosaic

In the near-term Adapting Mosaic scenario (as in TechnoGarden), there is likely to be great symmetry between the interests of the private sector, governments, and civil society actors in governing global ecological commons. Unlike the TechnoGarden scenario, however, there will be less emphasis on global environmental frameworks like the Convention to Combat Desertification and more emphasis on regional environmental policy mechanisms.

With the locus of governance moving toward individual regions around the world, regional organizations such as the Asian Development Bank, the U.N. Economic and Social Commission for Asia and the Pacific, Asia-Pacific Economic Cooperation, and so on as opposed to global organizations are more likely to serve as partners for the private sector in addressing sustainable development concerns. One impact of this trend is likely to be greater reliance on regional partnerships and collaborative activities in addressing ecosystem management. A wetland conservation project in Indonesia might involve such a diverse set of stakeholders as a Filipino NGO, a Japanese company, and the Asian Development Bank. In addition, the growing importance of local communities as co-managers of ecosystems and local resources in this scenario can open up new opportunities for partnerships with businesses, be they local, national, or international enterprises. Nevertheless, business will also have to seek new alliances with local communities in order to open up new opportunities, which might not always be an easy task.

Over the longer term, the threat of a global backlash to multinational corporations is lower under Adapting Mosaic due to the greater emphasis on a regional approach to environmental governance. At the same time, the likelihood of policy fragmentation, particularly in the way the private sector manages global environmental dilemmas like climate change, water, and biodiversity loss, is likely to increase. An important impact of Adapting Mosaic on the environmental strategy of the private sector might be greater reliance on and use of a regionally based corporate environmental strategic framework. A number of companies in Asia, North America, and Europe already use a regional environmental management framework, and under Adapting Mosaic this approach will probably become the environmental strategy of choice for many private enterprises, particularly for large multinational corporations.

14.7.3.4 TechnoGarden

In this scenario, there will be great symmetry in the near term between the interests of the private sector, governments, and civil society in governing the global ecological commons. Global environmental frameworks like the Convention on Biological Diversity and the Ramsar Convention are likely to be accepted as a normal part of doing business on a global scale and integrated into existing business regulatory frameworks like the ISO series. Even countries that are opposed to international environmental regimes will be pressured by companies in their own countries to accept prevailing global environmental norms.

Nongovernmental business policy forums and environmental business networks will grow in importance in terms of policy salience, especially those that have a strong technological component. Private-sector enterprises in the clean energy and technology sectors should benefit commercially from recognition of the economic value of ecosystem services and the mainstreaming of environmental technologies. Companies in the fossil fuel and carbon-intensive industries will be under growing pressure to reposition themselves and will possibly be forced to sell certain businesses (such as coal and certain carbon-intensive extractive sectors).

Over the long term, the proactive policies of TechnoGarden resulting from the recognition of the economic value of ecosystem services may be undermined if the bulk of the "perceived" benefits of these policies go to multinational corporations in wealthier countries at the expense of local businesses in poorer ones. There might also be increased global tensions if governments and private firms in North America and Europe block poor-country access to innovative green technologies due to intellectual property concerns. The type of criticism that is now leveled against western pharmaceutical companies for their role in posing barriers to AIDS medicine distribution in Africa may in the future extend to multinational companies involved in the research and development of green technologies.

A green technology fund similar to the model of establishing a global fund to finance the phaseout of ozone-depleting substances and to combat AIDS in the developing world may be established by private and public stakeholders to finance and disseminate cost-effective environmental technology systems to local communities in the poorer countries. Even more than in the Global Orchestration scenario, clean energy business sectors including hydrogen, solar, wind, and small-scale hydropower are likely to represent promising business opportunity in TechnoGarden, and the building blocks to establish a cleaner global energy infrastructure and system are likely to be established.

Table 14.18 summarizes the primary stresses and selected response options for the private sector under the four MA scenarios.

Table 14.18. Private-Sector Primary Stresses and Selected Response Options in MA Scenarios. All values are estimates of relative comparison among scenarios. Many responses apply to more than one stressor.

Stresses and Responses	GO	OS	AM	TG
Ecosystem stress—climate change	●●●●	●●●●●	●●●●	●●●
Reduce emissions of greenhouse gases	✻✻		✻✻	✻✻✻✻
Invest in clean energy and technologies	✻✻✻	✻	✻✻✻	✻✻✻✻
Ecosystem stress—food and land use	●●●	●●●●	●●●	●●
Reduce consumption of forestry and other ecological assets	✻✻	✻	✻✻✻	✻✻✻
Develop green labeling and purchasing policies	✻✻✻	✻	✻✻✻	✻✻✻✻
Ecosystem stress—water	●●●●	●●●●●	●●●	●●●
Reduce consumption of water	✻		✻✻	✻✻
Establish market-based pricing system	✻✻	✻	✻✻	✻✻✻
Ecosystem stress—biodiversity loss	●●	●●●●	●●	●●
Reduce economic activities around ecologically sensitive areas	✻✻		✻✻	✻✻✻
Invest in ecotourism, sustainable agriculture, and other forms of conservation enterprises	✻✻✻	✻✻	✻✻✻	✻✻✻✻✻

Key: GO = Global Orchestration; OS = Order from Strength; AM = Adapting Mosaic; TG = TechnoGarden
D = developing countries; 1 = higher-income countries
Stresses: 5 ● = severe stress, 0 ● = no worse than 2004
Responses: 5 ✻ = success likely, 0 ✻ = unfeasible/ineffective

14.8 Synthesis

The MA scenarios contain a huge amount of information about the possible directions of socioeconomic and environmental developments over the next few decades, including the direct and indirect driving forces of ecosystem changes. Model-based assessments and verbal scenario studies explore their implications for the condition and services of ecosystems in large world regions over time as well as the repercussions of ecosystem changes on human well-being. This chapter organizes this rich information base according to the explicitly declared or implicitly pursued interests, values, and mandates of selected key social groups and international organizations. The chapter's sections identify emerging ecosystems-related stresses, risks, and opportunities and provide assessments of conceivable response options to manage those threats in the contexts of the four MA scenarios for each stakeholder group.

Given the diversity of interests and mandates of the stakeholders on the one hand and the availability and prospective effectiveness of their response options on the other hand, the four futures hold rather different threats and reaction opportunities. Yet a meaningful attempt to synthesize the main insights for the global society as a whole needs to incorporate the scenario outcomes.

A comparative evaluation of the scenarios based purely on the nature and magnitude of ecosystem-related stresses would be inadequate. It is not only the threats and opportunities that count but also the capacity of the affected stakeholders to manage the risks and seize the opportunities. A seemingly minor ecosystem change might have grave implications if the affected stakeholders lack effective measures to cope with it, including the financial or other incapacity of implementing conceivable mitigation measures.

Although the implications of the MA scenarios for national governments constitute an important part of this chapter, governments are not included in the ranking of risks and response options of the four futures. There are two main reasons. The first is the multifaceted interests and responsibilities of governments in regulating how societies affect ecosystems and make use of their services. This encompasses an immense diversity of conceivable interventions, of which, however, the appropriateness of one or another depends on many societal and biophysical factors. But the main reason for not preparing a ranking for governments is that in the prevailing political structures governments have the largest potential to influence the driving forces that determine how the future will unfold. This means governments can best shape which of the four archetypes depicted in the MA scenarios will dominate the future of the global society. Accordingly, governments have the primary responsibility to foreclose unfavorable directions and to usher their nations toward sustainable development.

None of the scenarios can be singled out as the most desirable future. Each scenario has several positive and negative characteristics because each entails different combinations of relatively smaller or larger ecosystems stresses and more or less stakeholder capacity to cope with the emerging risks. Unfortunately, it is not possible to handpick a combination of drivers and ecosystems management strategies to achieve what might appear to be the best selection of features across scenarios because of the need to make socioeconomic choices among mutually exclusive options and the biophysical trade-offs among ecosystems functions and services. Chapter 12 provides ample evidence of the latter. Thus, not even the most brilliant and committed policymakers operating in a highly cooperative international community could achieve such dreamworld futures.

The cornerstone of masterly policy-making is finding the best compromises among conflicting objectives, making the appropriate interventions to achieve them, and doing regular reassessments of policies against anticipated and unanticipated outcomes. Our hope is that the MA scenarios in general and this chapter in particular provide useful insights for public policy-makers and private stakeholders to make informed choices and choose the appropriate measures now and as the future unfolds.

References

IPCC (Intergovernmental Panel on Climate Change), 2001: *Climate Change 2001. The Scientific Basis.* Cambridge University Press, Cambridge, UK.

Smith, L.C. and Haddad, L., 2000: *Explaining Child Malnutrition in Developing Countries: A Cross-country Analysis.* IFPRI Research Report No. 111, International Food Policy Research Institute, Washington, DC.

UN (United Nations), 2002: Report of the World Summit on Sustainable Development. A/CONF.199/20, United Nations, New York, NY.

Appendix A
Color Maps and Figures

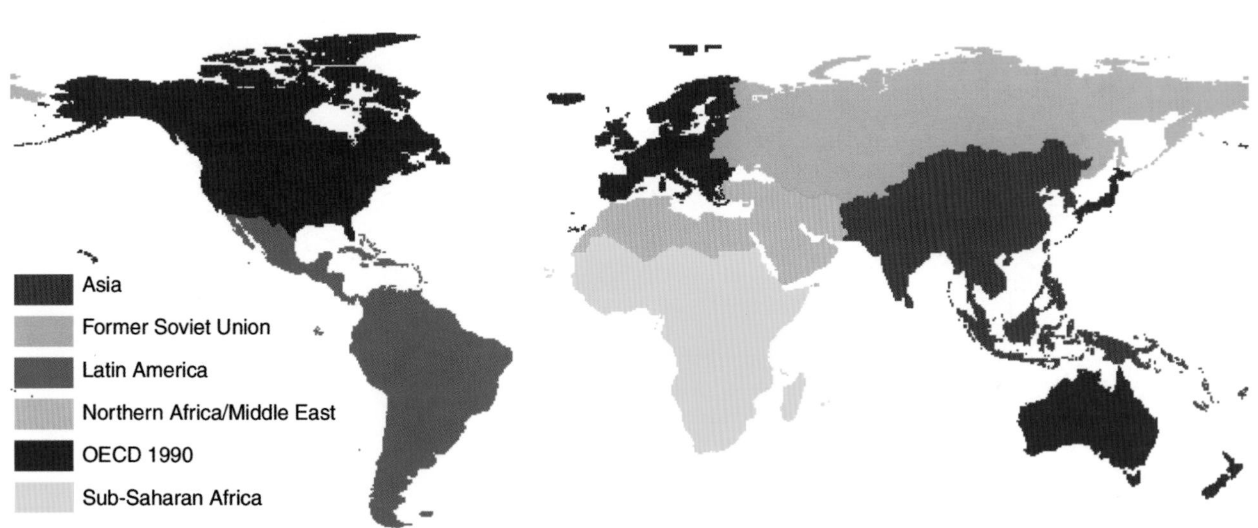

Figure 6.3. Reporting Regions for the Global Modeling Results of the MA. The region labeled OECD does not correspond exactly with the actual member states of the OECD. Turkey, Mexico, and South Korea, member states of OECD, are reported here as part of the regions Northern Africa/Middle East, Latin America, and Asia, respectively. All countries in Central Europe are reported here as part of hte OECD region. This reporting definition is used because regions have been aggregated from the regional definitions of the models used. IMAGE and WaterGAP models have used a slightly different definition. (Millennium Ecosystem Assessment)

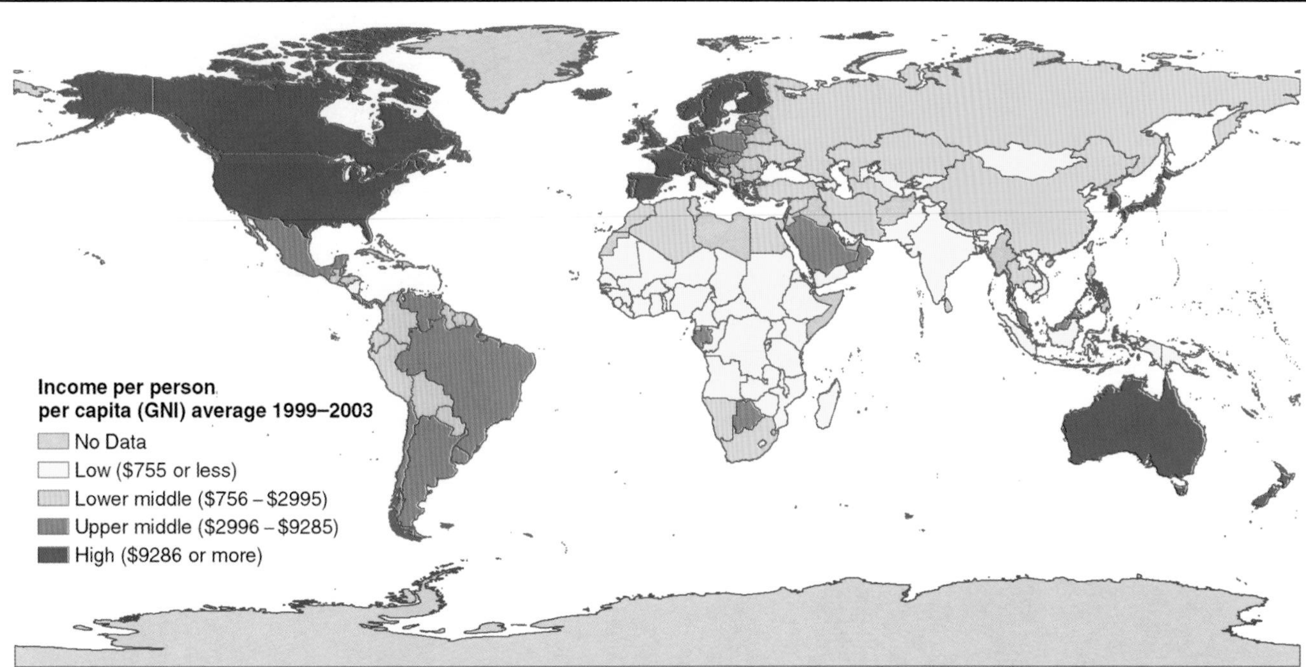

Figure 7.5. Income per Person, per Capita (GNI) Average, 1999–2003. National income is converted to U.S. dollars using the World Bank Atlas method. U.S. dollar values are obtained from domestic currencies using a three-year weighted average of the exchange rate. (World Bank 2003)

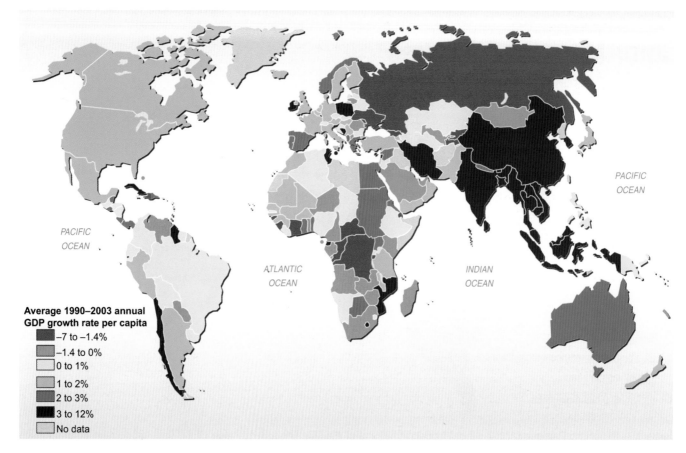

Figure 7.6a. Average GDP per Capita Annual Growth Rate, 1990–2003

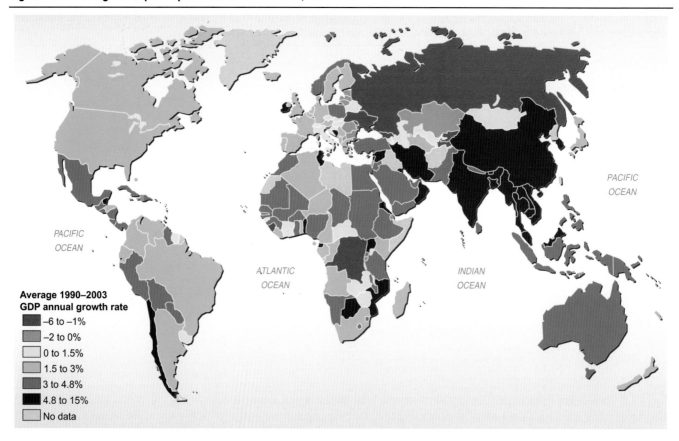

Figure 7.6b. Average GDP Annual Growth Rate, 1990–2003 (Based on data downloaded from the online World Bank database and reported in World Bank 2004.)

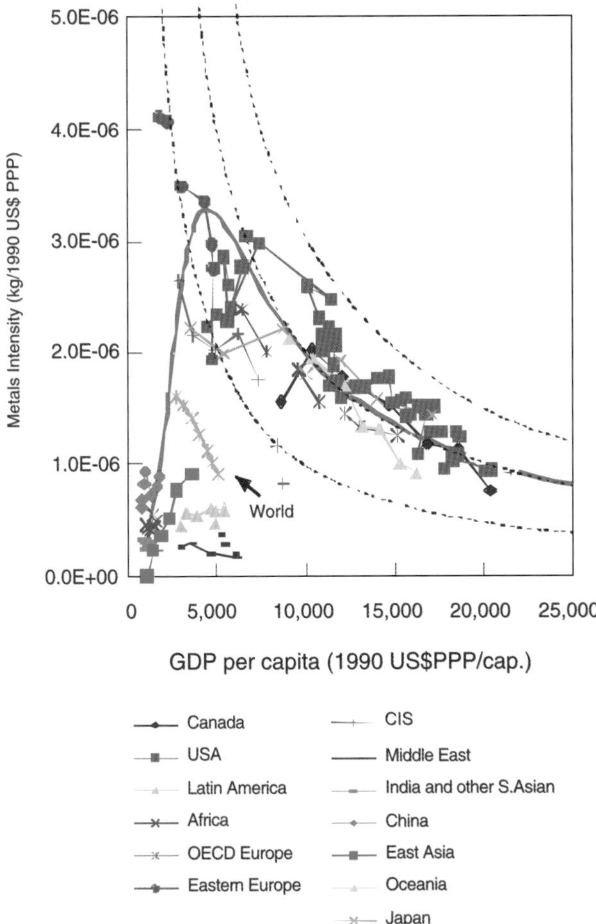

Figure 7.9. Metals Intensity of Use per Unit of GDP (PPP) as a Function of GDP (PPP) per Capita for 13 World Regions (Nakićenović et al. 2003). Metals include refined steel and MedAlloy (the sum of copper, lead, zinc, tin, and nickel). GDP here is measured in terms of purchasing power parities (PPP). The dashed curves are isolines that represent a constant per capita consumption of metals. The thick line indicates the inverse U-shaped curve that best describes the trends in the different regions as part of a global metal model. (Van Vuuren et al. 2000)

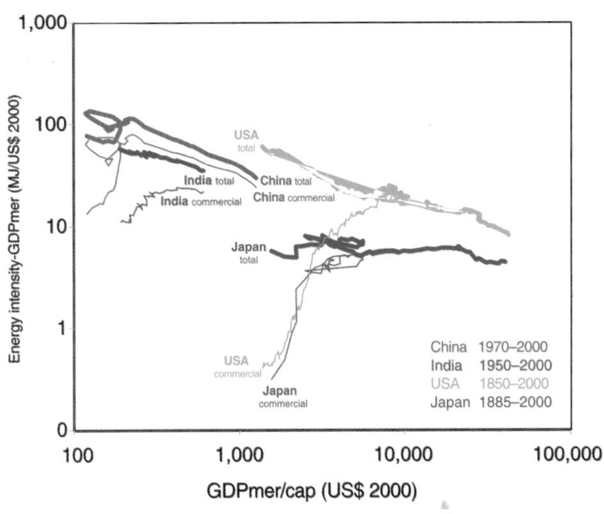

Figure 7.10. Energy Intensity Changes with Changes in per Capita Income for China, India, Japan, and United States. Historical data for the United States since 1800 are shown. Data are converted from domestic currencies using market exchange rates. (Nakicenovic et al. 1998)

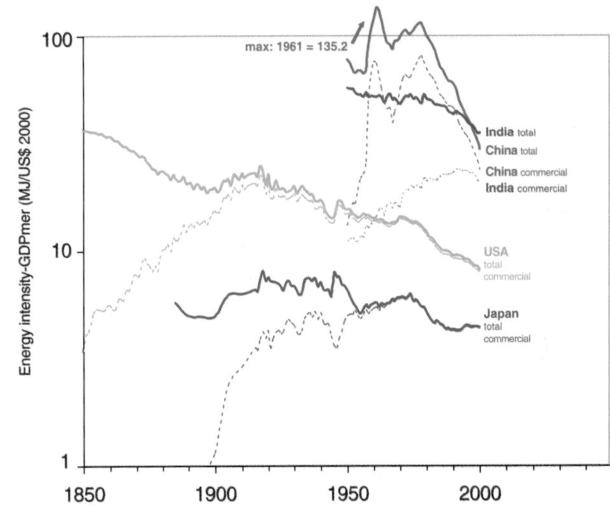

Figure 7.11. Energy Intensity Changes over Time for China, India, Japan, and United States. Data are converted from domestic currencies using market exchange rates. (Nakićenović et al. 1998)

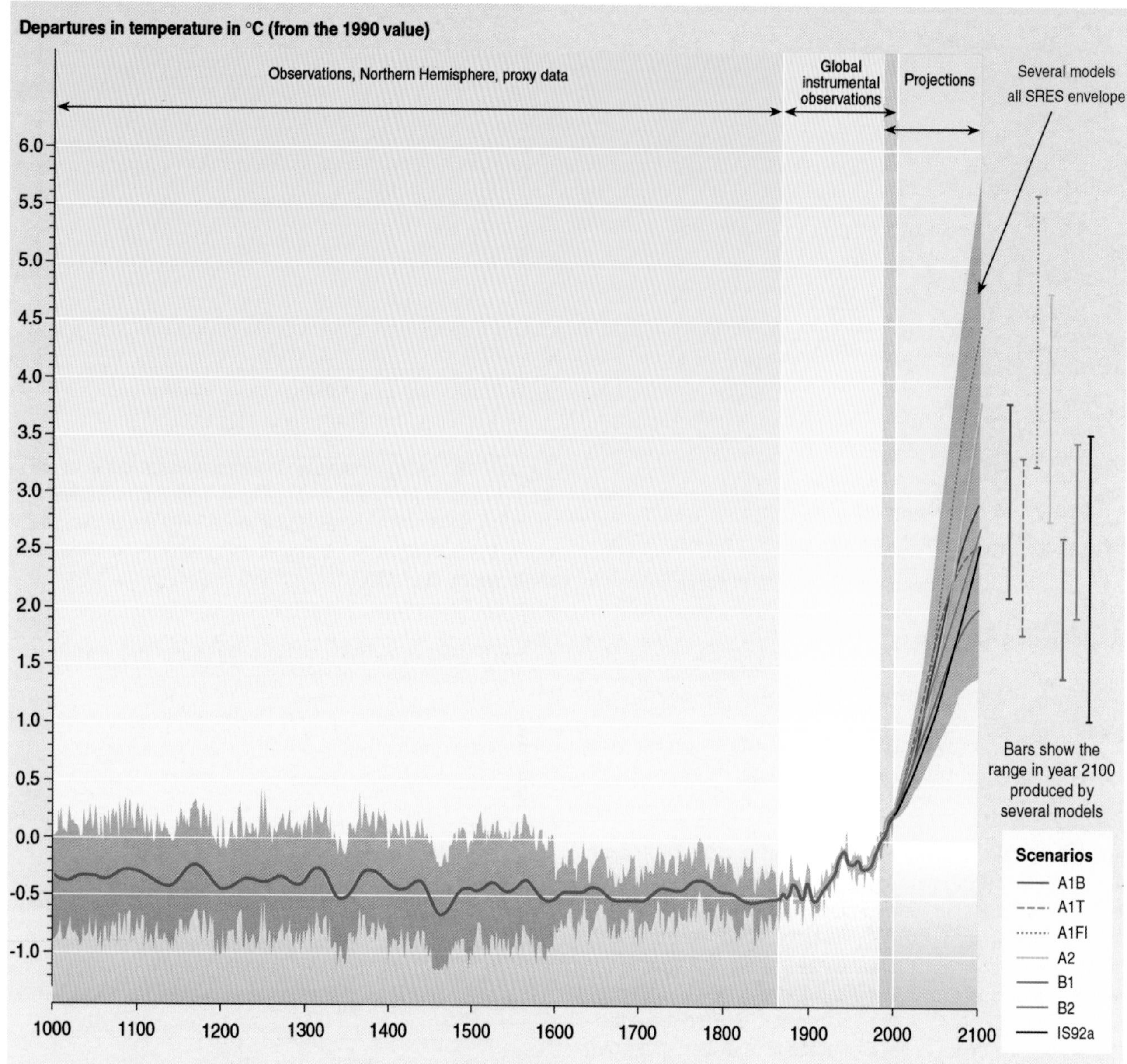

Figure 7.13. Variations of the Earth's Surface Temperature, 1000–2100. The temperature scale is a departure from the 1990 value. For 1000–1860: variations in average surface temperature of the Northern Hemisphere are shown (corresponding data from the Southern Hemisphere not available) reconstructed from proxy data (tree rings, corals, ice cores, and historical records). The line shows the 50-year average, the grey region the 96% confidence limit in the annual data. For 1860–2000: variations in observations of globally and annually averaged surface temperature from the instrumental record. The line shows the decadal average. For 2000–2100: scenarios and IS92a using a model with average climate sensitivity. The grey region marked "several models all IPCC SRES envelope" shows the range of results from the full range of 35 SRES scenarios in addition to those from a range of models with different climate sensitivities. (IPCC 2002)

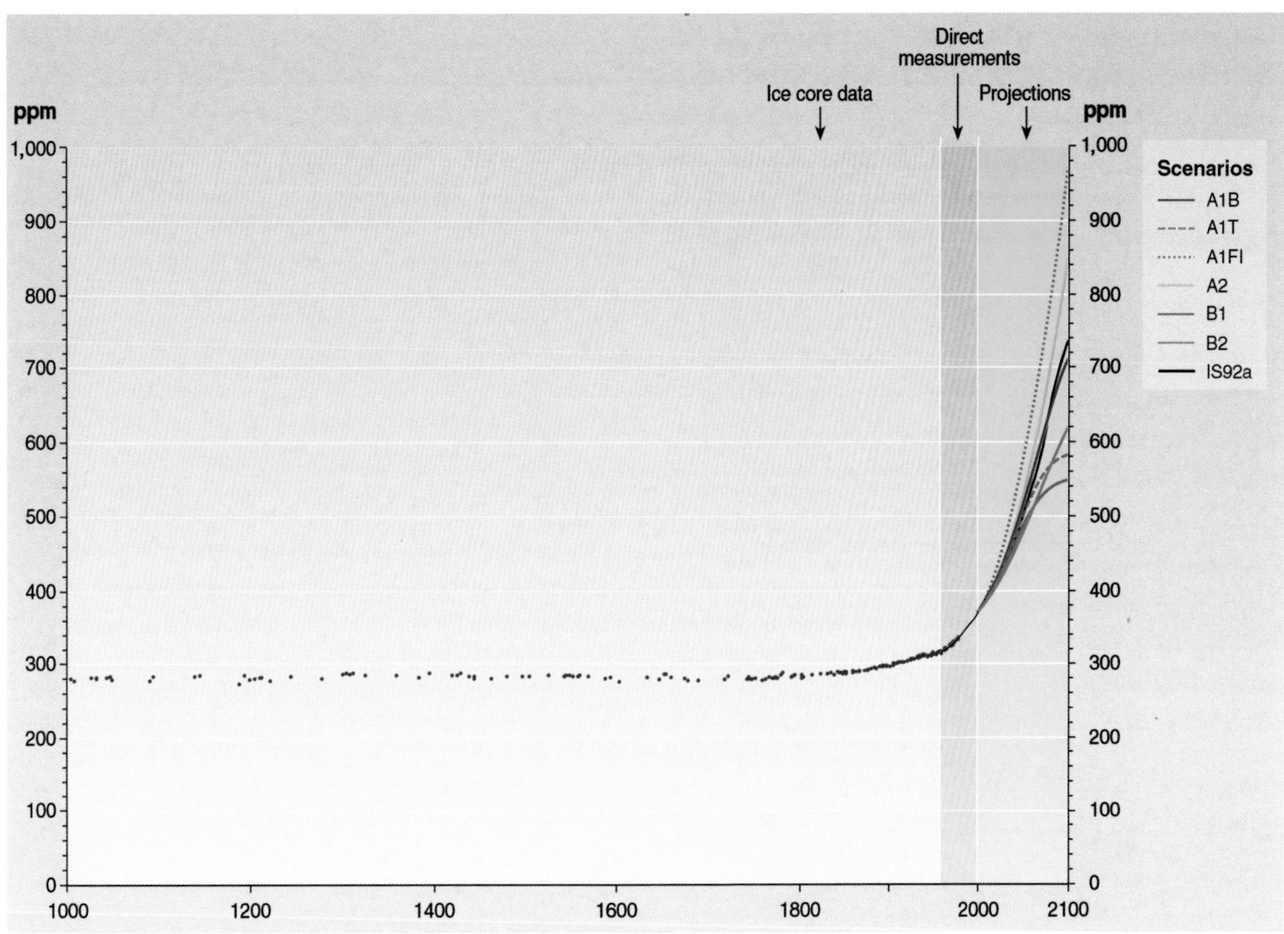

Figure 7.14. Past and Future Carbon Dioxide Concentrations. Atmospheric carbon dioxide concentrations from year 1000 to 2000 are from ice core data and from direct atmospheric measurements over the past few decades. Projections of carbon dioxide concentrations for 2000 to 2100 are based on six illustrative IPCC SRES scenarios and IS92a (for comparison with the IPCC Second Assessment Report). (IPCC 2002)

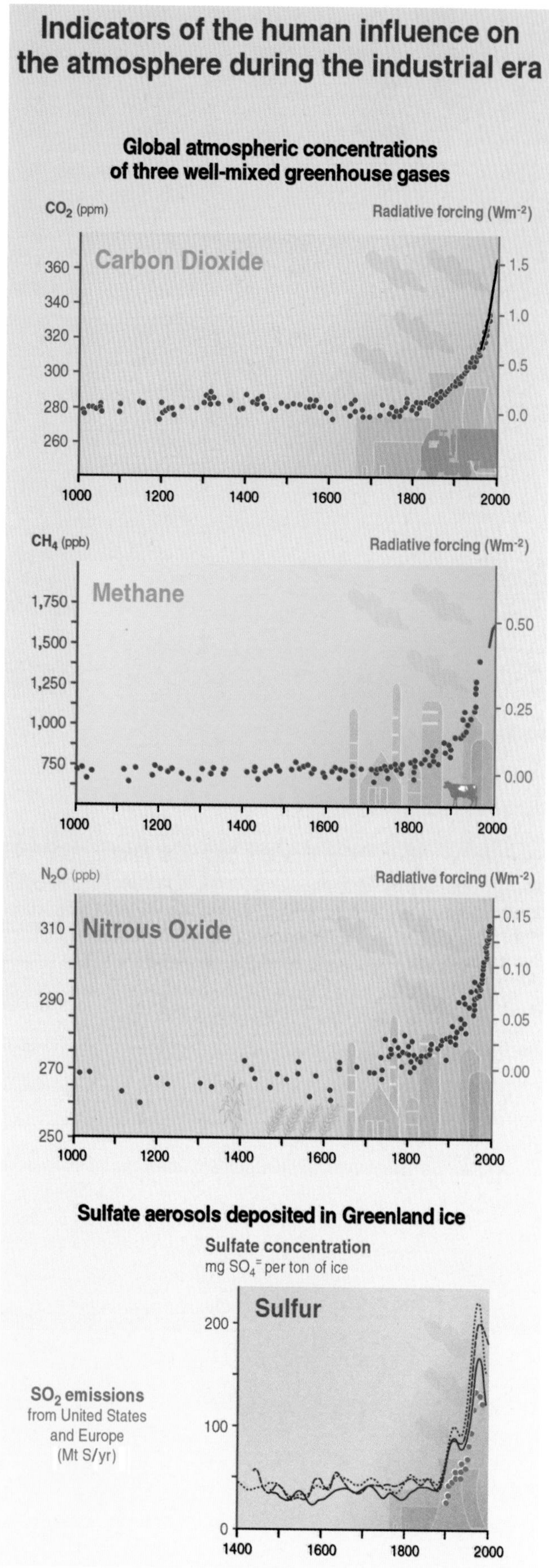

Figure 7.15. Concentration of Greenhouse Gases (IPCC 2002)

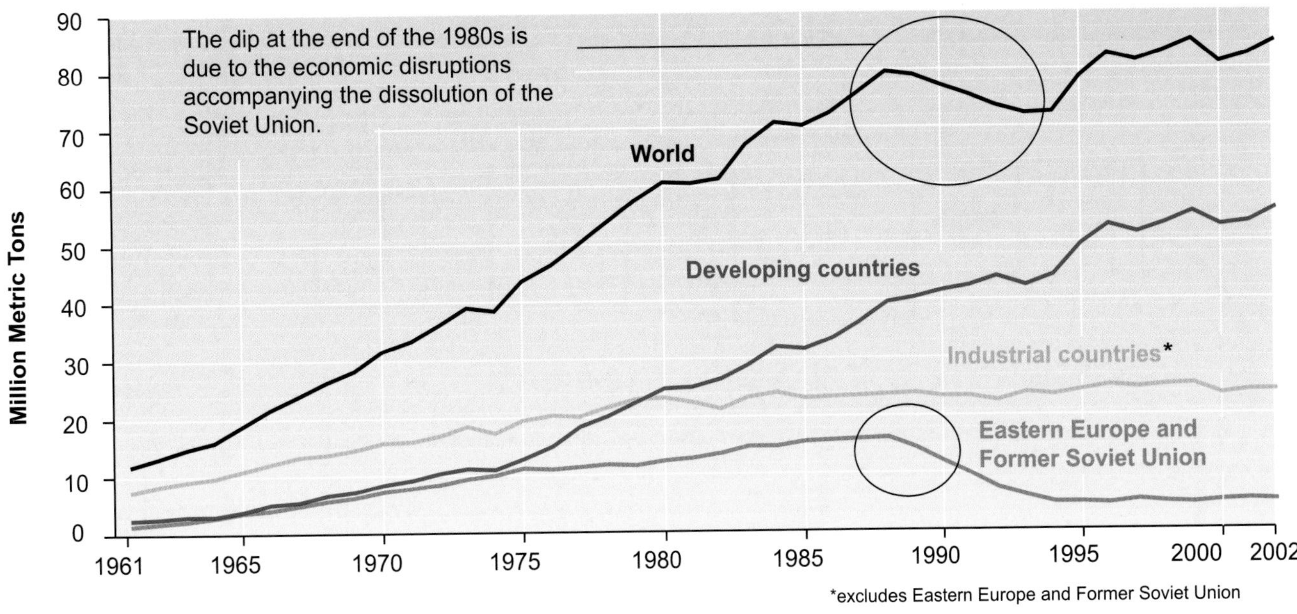

Figure 7.16. Trends in Global Consumption of Nitrogen Fertilizers, 1961–2001 (IFA 2004)

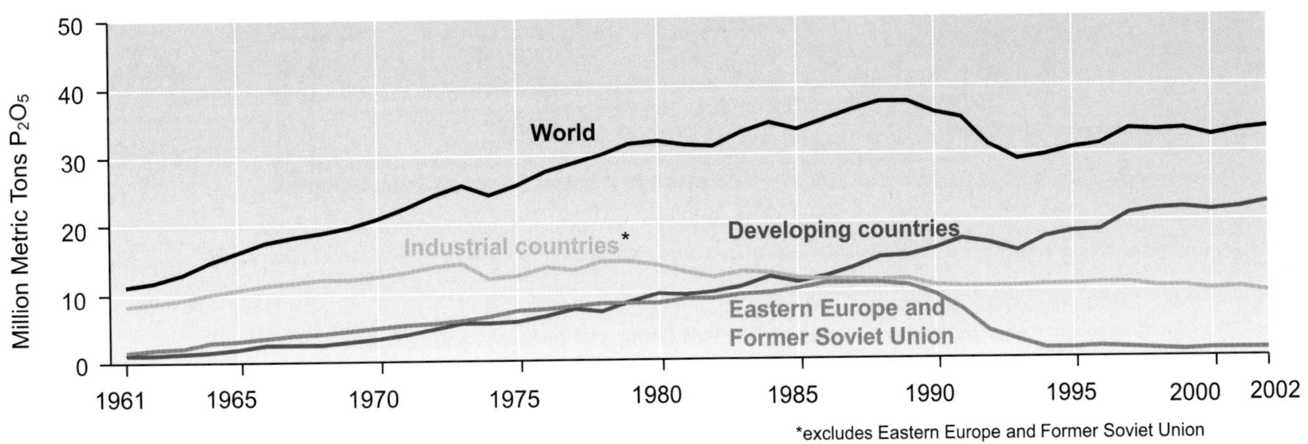

Figure 7.18. Trends in Global Consumption of Phosphate Fertilizer, 1961–2002 (IFA 2004)

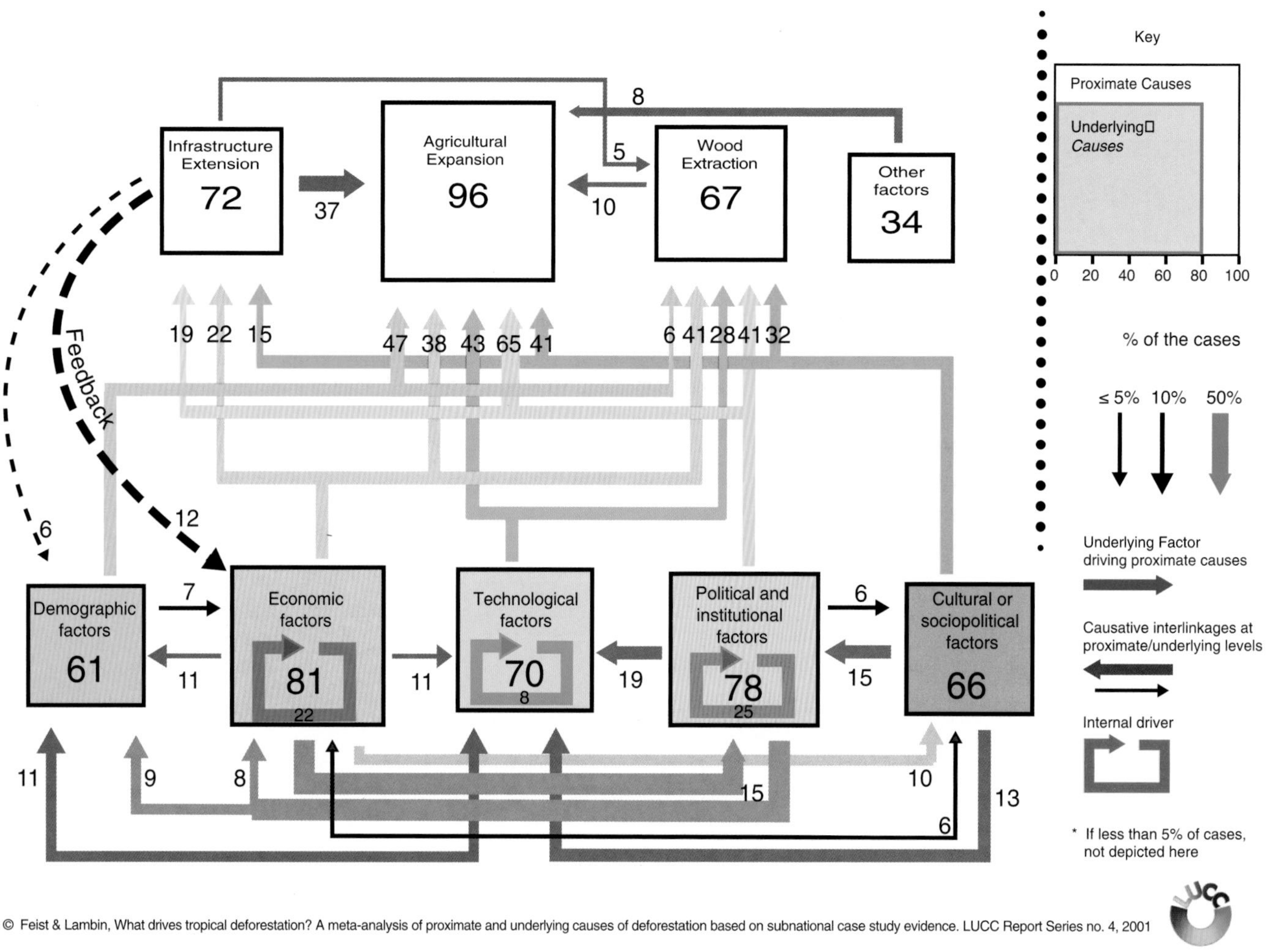

Key
Proximate Causes
Underlying Causes
0 20 40 60 80 100
% of the cases
≤ 5% 10% 50%
Underlying Factor driving proximate causes
Causative interlinkages at proximate/underlying levels
Internal driver
* If less than 5% of cases, not depicted here
Infrastructure Extension 72
Agricultural Expansion 96
Wood Extraction 67
Other factors 34
Demographic factors 61
Economic factors 81
Technological factors 70
Political and institutional factors 78
Cultural or sociopolitical factors 66
Feedback
© Feist & Lambin, What drives tropical deforestation? A meta-analysis of proximate and underlying causes of deforestation based on subnational case study evidence. LUCC Report Series no. 4, 2001

Figure 7.22. An Overview of the Causative Patterns of Tropical Deforestation (Geist and Lambin 2002)

Four scenarios
Global
Global Orchestration
Equity, economic growth, and public goods
TechnoGarden
Green technologies and ecological economics
Reactive
Proactive
Adapting Mosaic
Integrated management, local adaptation, and learning
Order from Strength
National security
Regional
PAST
2000
FUTURE
2050
Present conditions and trends

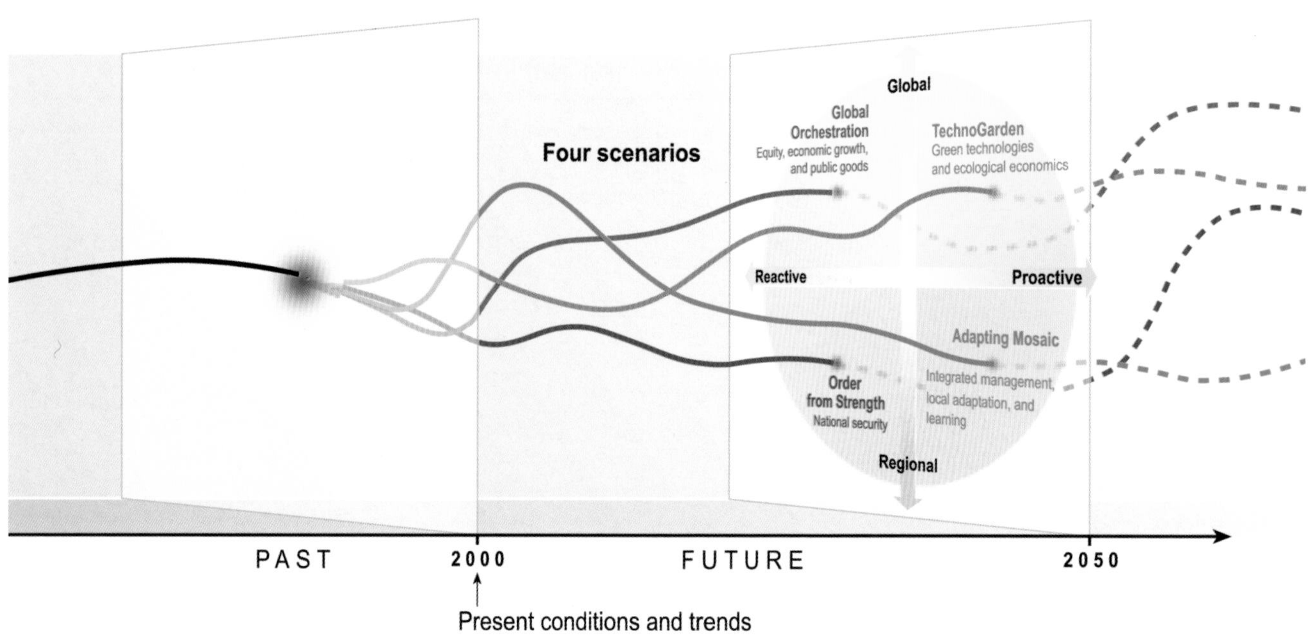

Figure 8.1. Prime Critical Uncertainties Distinguishing MA Scenarios

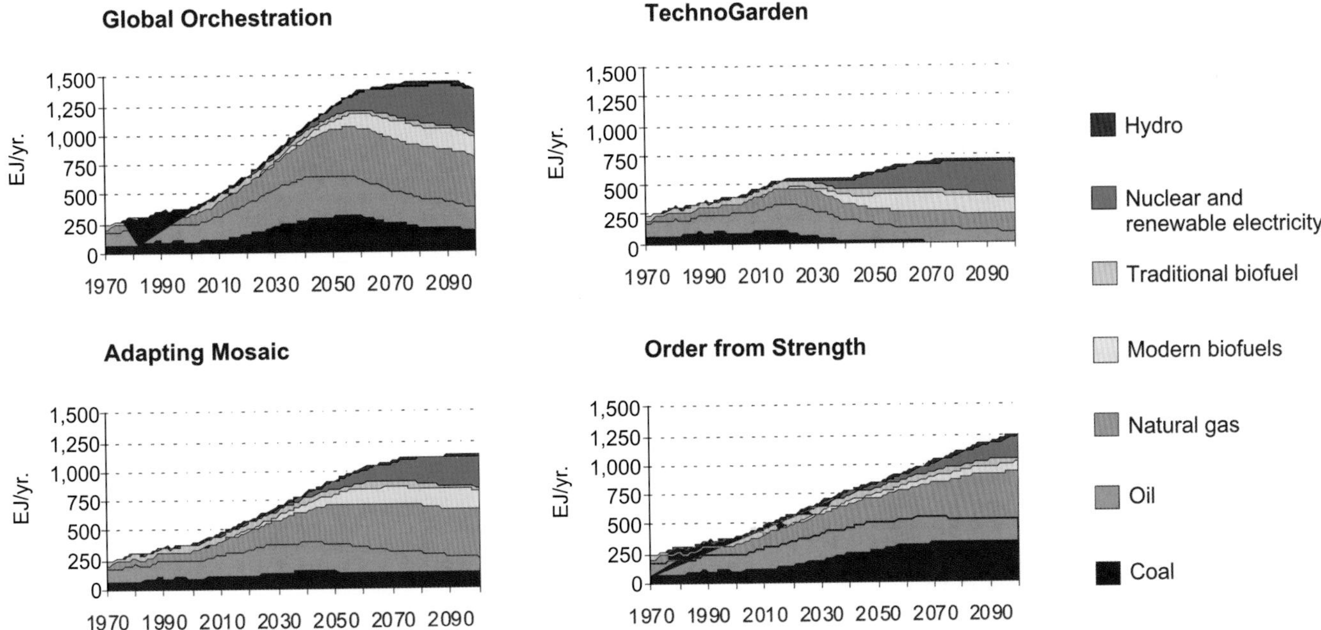

Figure 9.3. Global Energy Consumption in MA Scenarios (IMAGE 2.2)

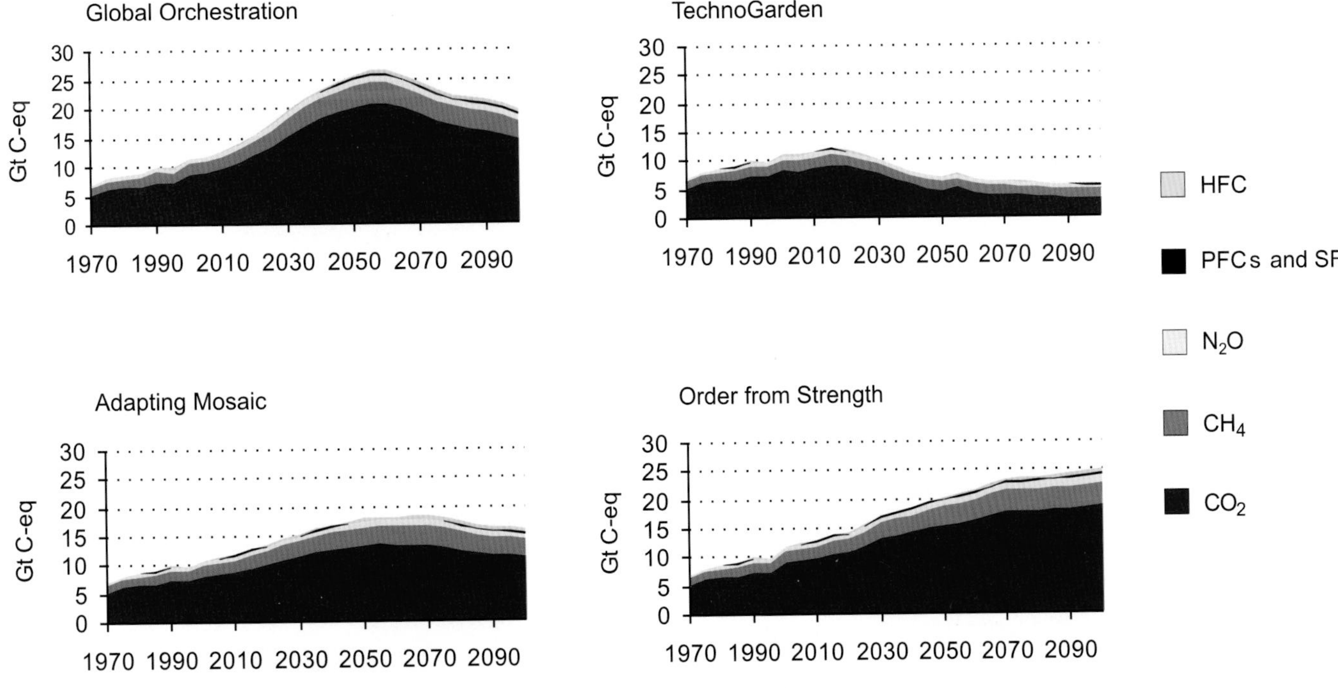

Figure 9.5. Global Greenhouse Gas Emissions in MA Scenarios (IMAGE 2.2)

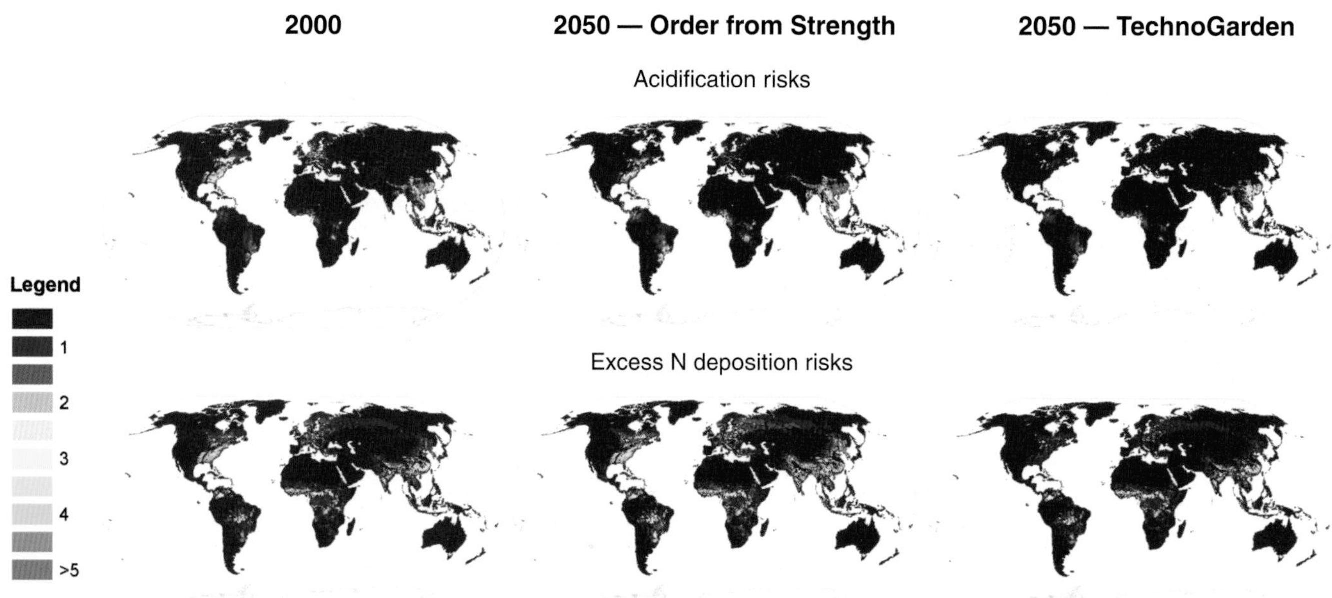

Figure 9.9. Exceeding of Acidification and Nitrogen Deposition Critical Loads in the Order from Strength and TechnoGarden Scenarios in 2050 (IMAGE 2.2)

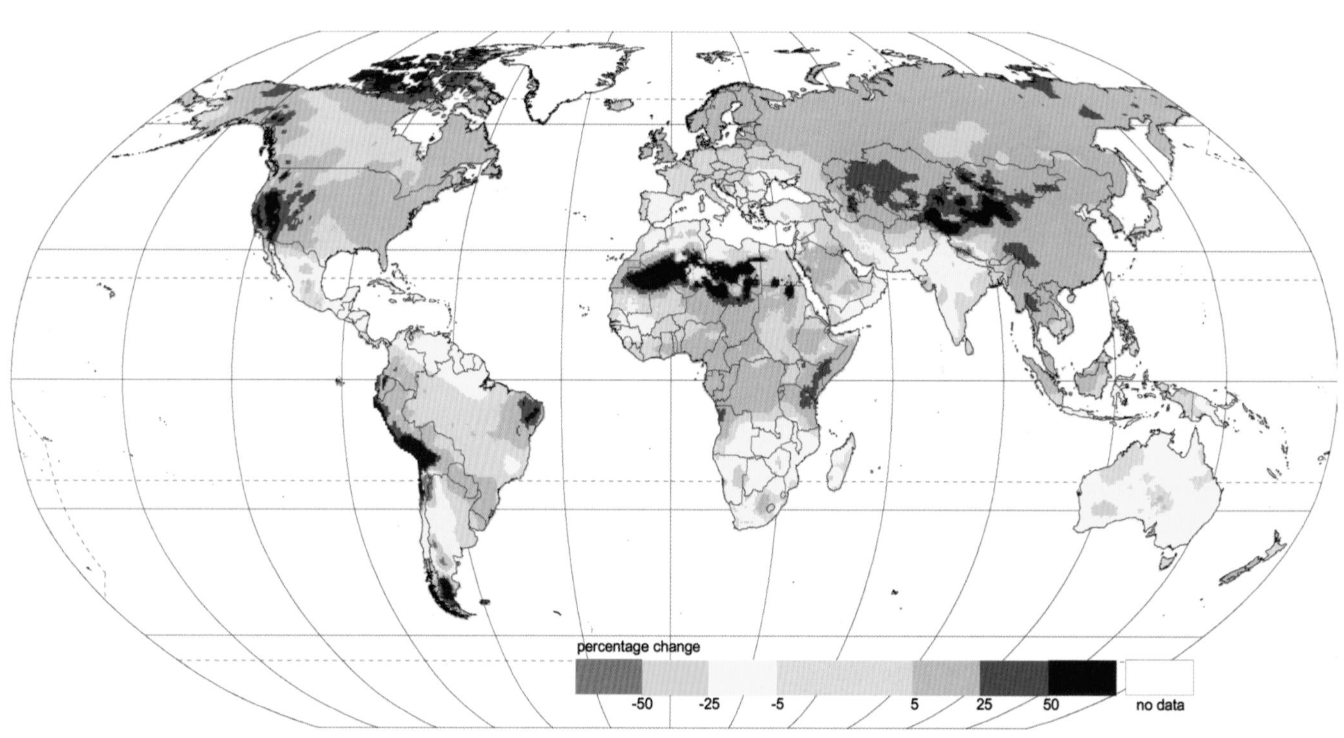

Figure 9.12. Change in Precipitation in 2050 Compared with Current Climate under the Global Orchestration Scenario (IMAGE 2.2)

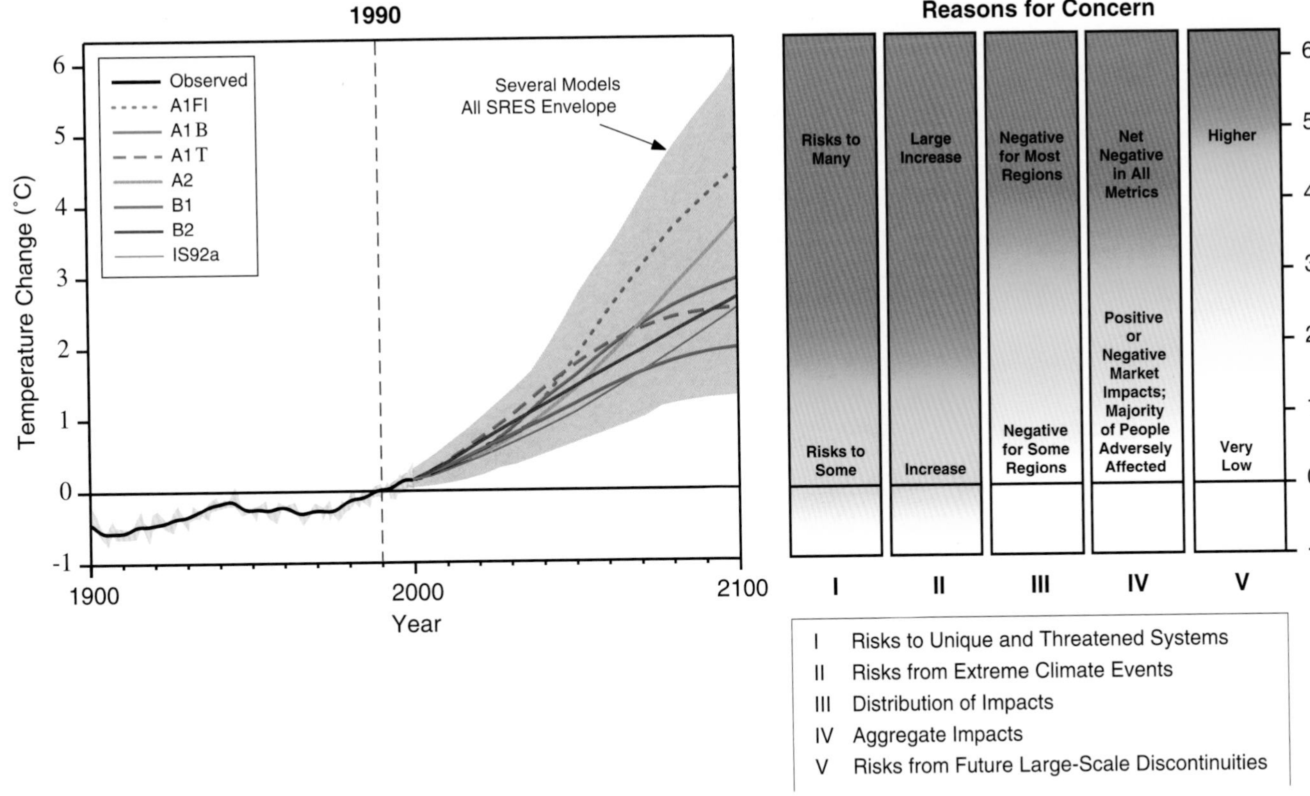

Figure 9.13. Causes of Concern in Third Assessment Report of the IPCC (IPCC 2001)

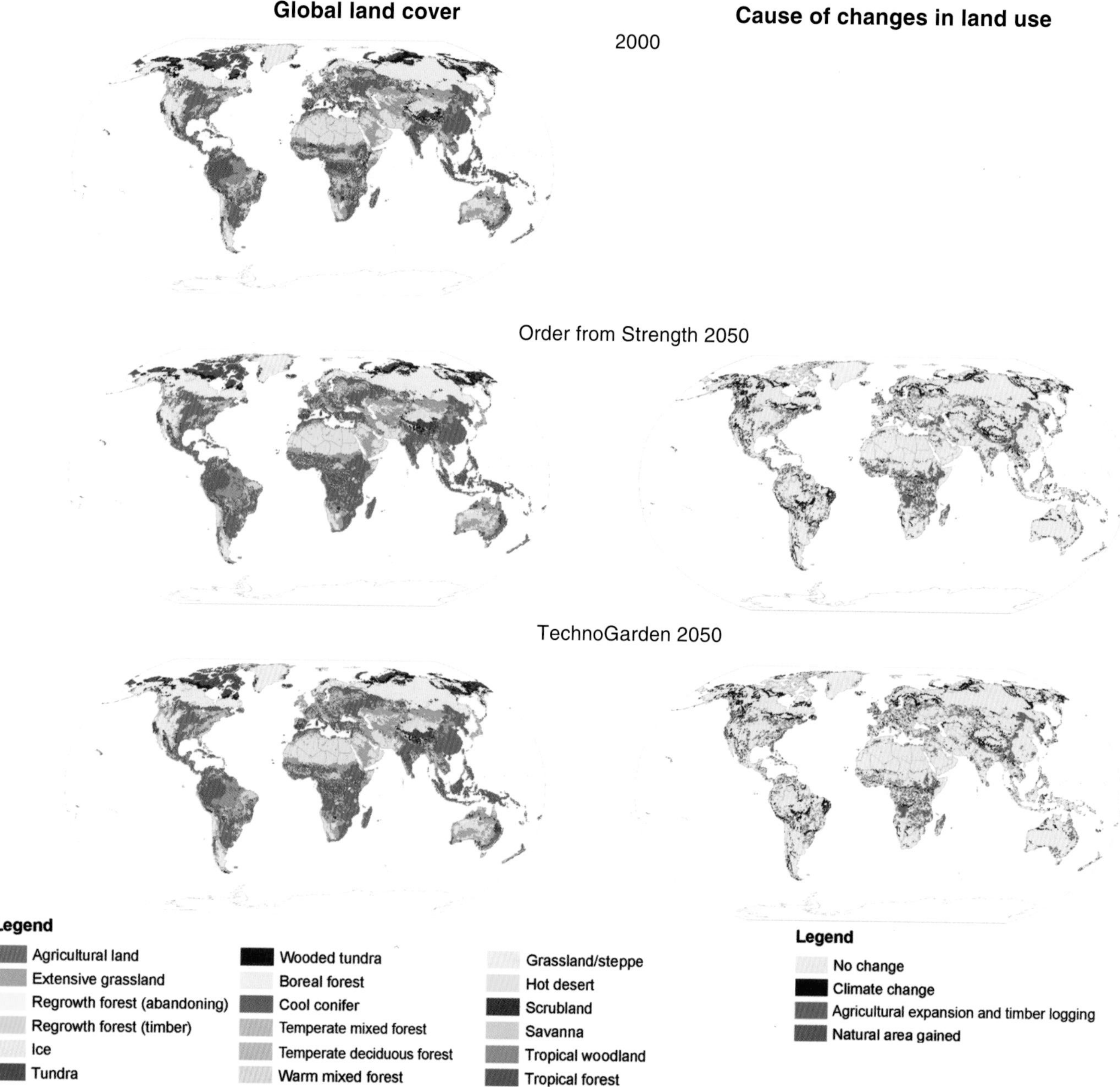

Figure 9.18. Land Use Patterns in Two Scenarios in 2050. The maps on the left indicate global cover in 2000 and 2050. The maps on the right indicate the cause of changes in land use between 2000 and 2050, including shifts in biome types as a result of climate change. (IMAGE 2.2)

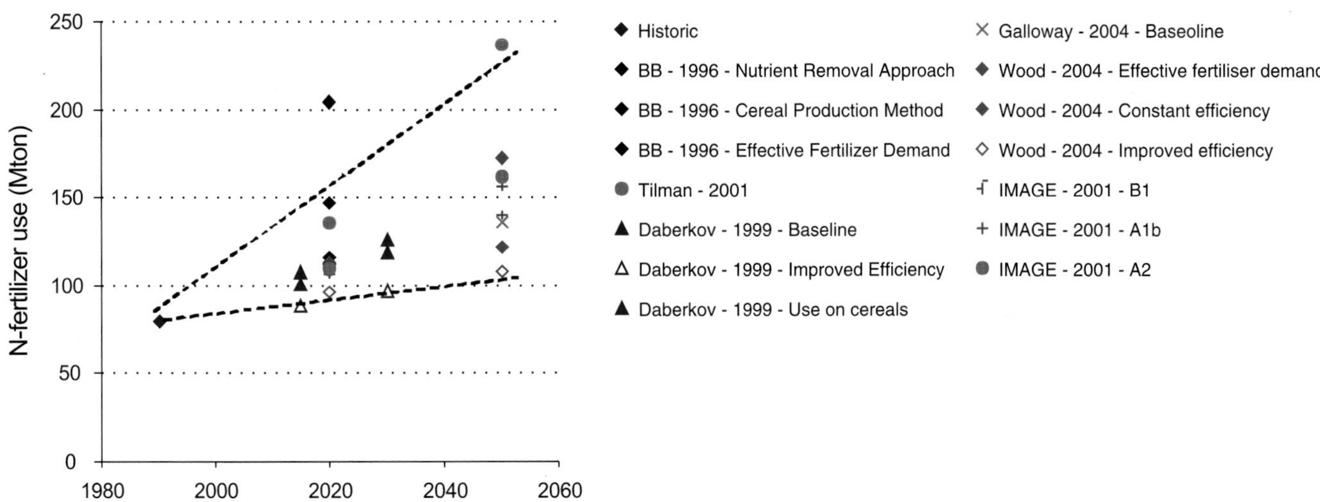

Figure 9.20. Nitrogen Fertilizer Use under Different Scenarios

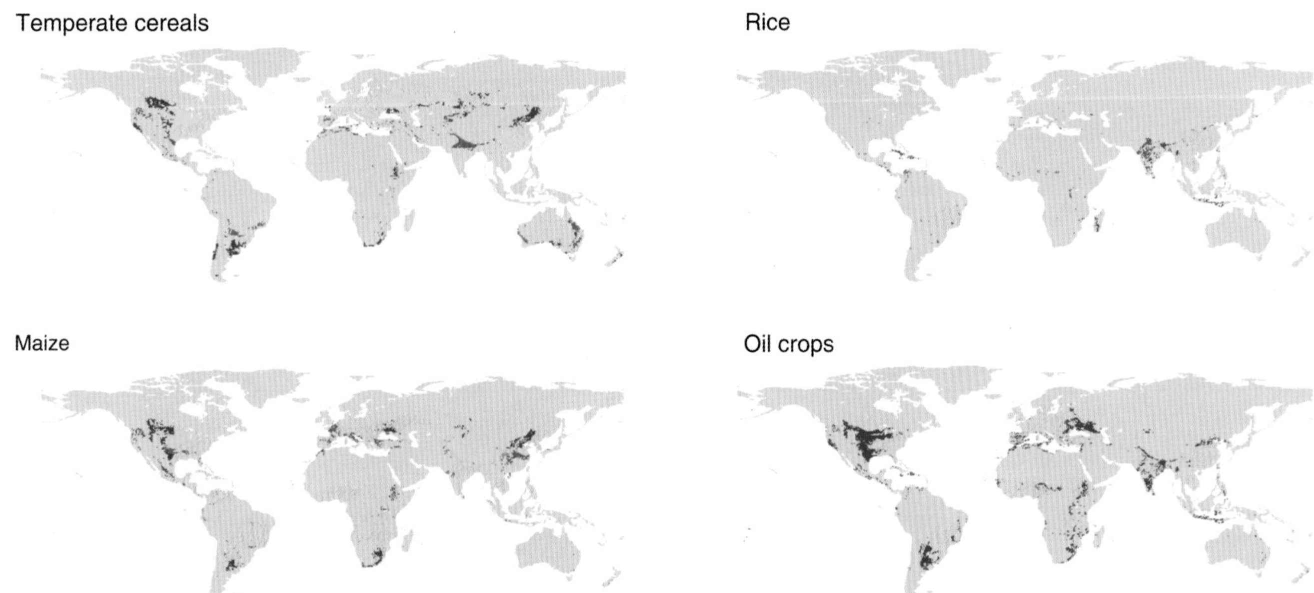

Figure 9.28. Crop Yield for the Order from Strength Scenario from 2000 to 2100. Red indicates a significant decrease; yellow for a stable yield; blue signifies a significant increase. (IMAGE 2.2)

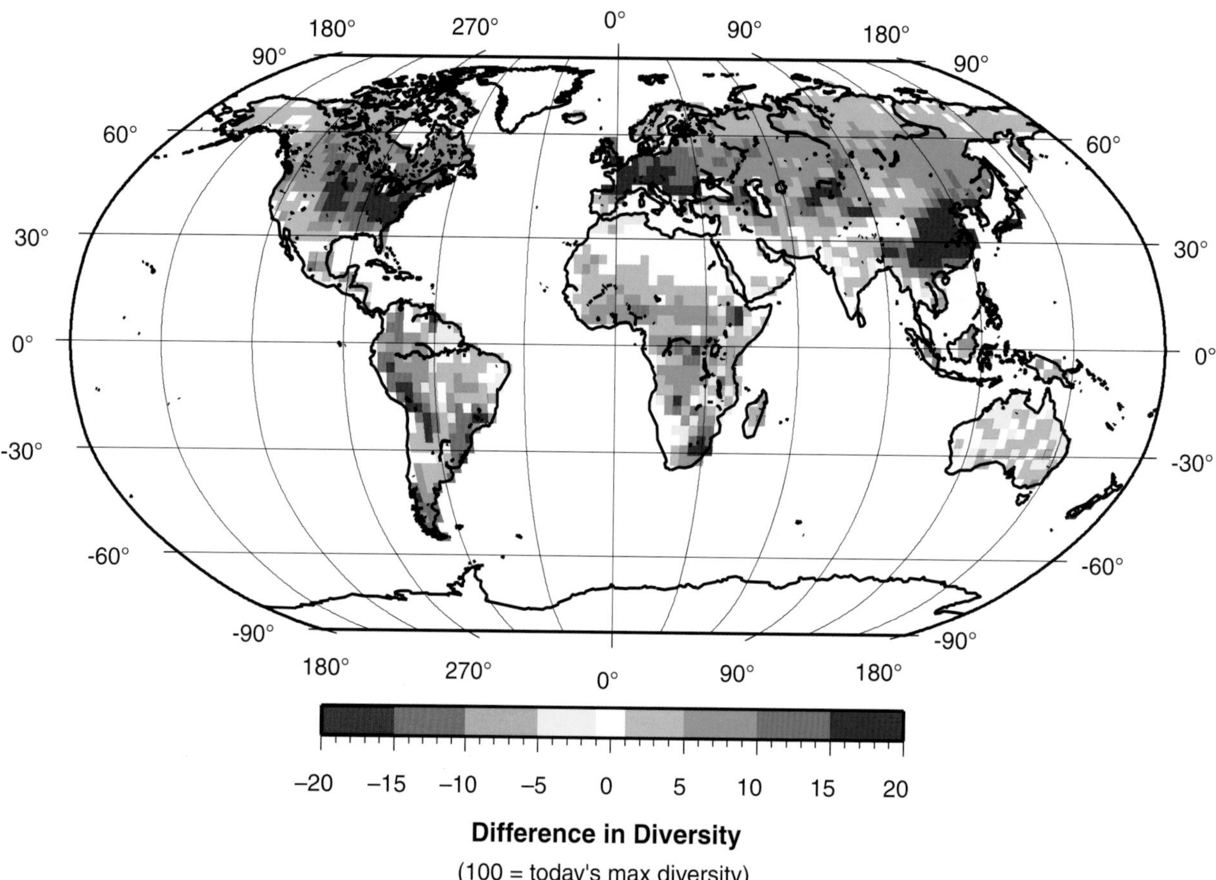

Figure 10.10 Potential Plant-Species Diversity as Determined by Climate Patterns. Blue tones represent increases in diversity relative to present, and reddish tones represent decreases in diversity. Potential plant-species diversity represents diversity when ecosystems reach equilibrium with climate. (Millennium Ecosystem Assessment)

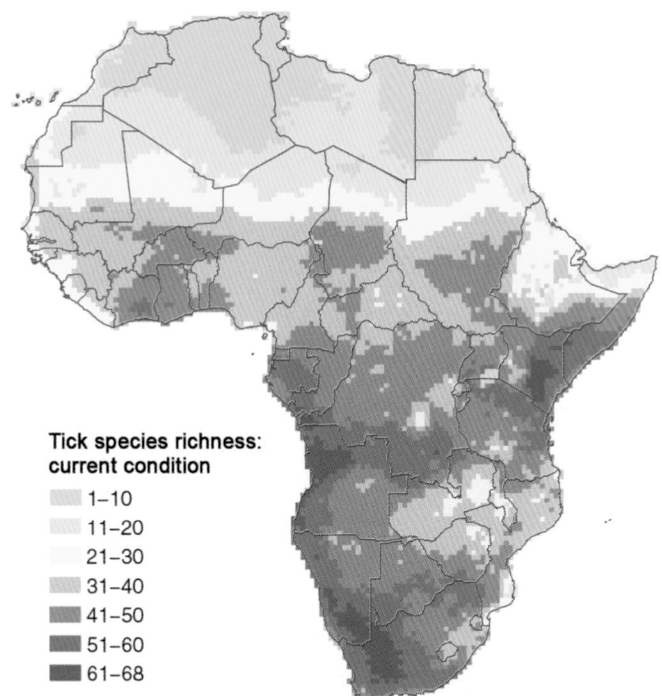

Figure 10.11. Species Richness of African Ticks in 2000, at a Resolution of 0.5 Degrees. This map is based on climate-driven estimates of species ranges for 73 of the approximately 240 African species. The numbers in the legend indicate the number of tick species by grid cell. Tick species richness is highest in East Africa, Kenya, and Tanzania. There are pockets of high diversity in the Eastern Highlands of Malawi and Zimbabwe, the Cape, and West Africa; the lowest species richness occurs in the desert areas.

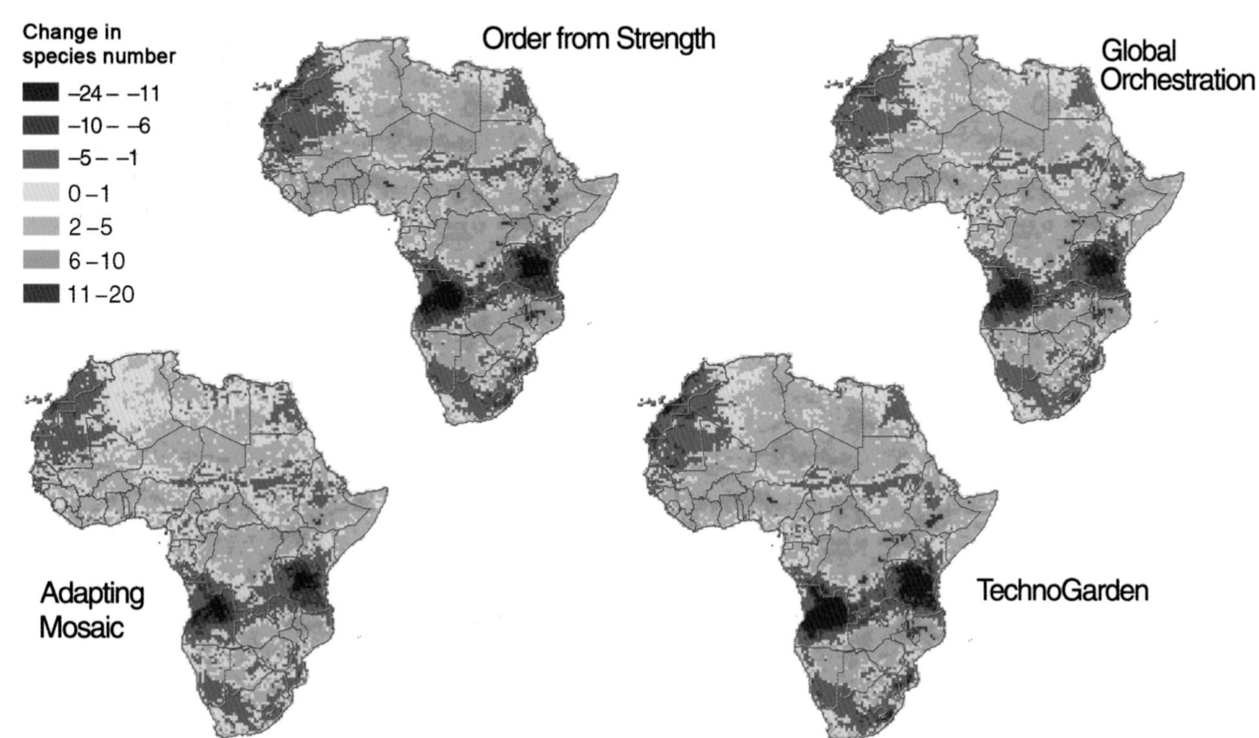

Figure 10.12. Predicted Changes in Tick Species Richness (per one-half degree cell) in Africa by 2100 in MA Scenarios. The number on the legend indicates the number of species that are gained or lost from each grid cell relative to a 2000 baseline estimate.

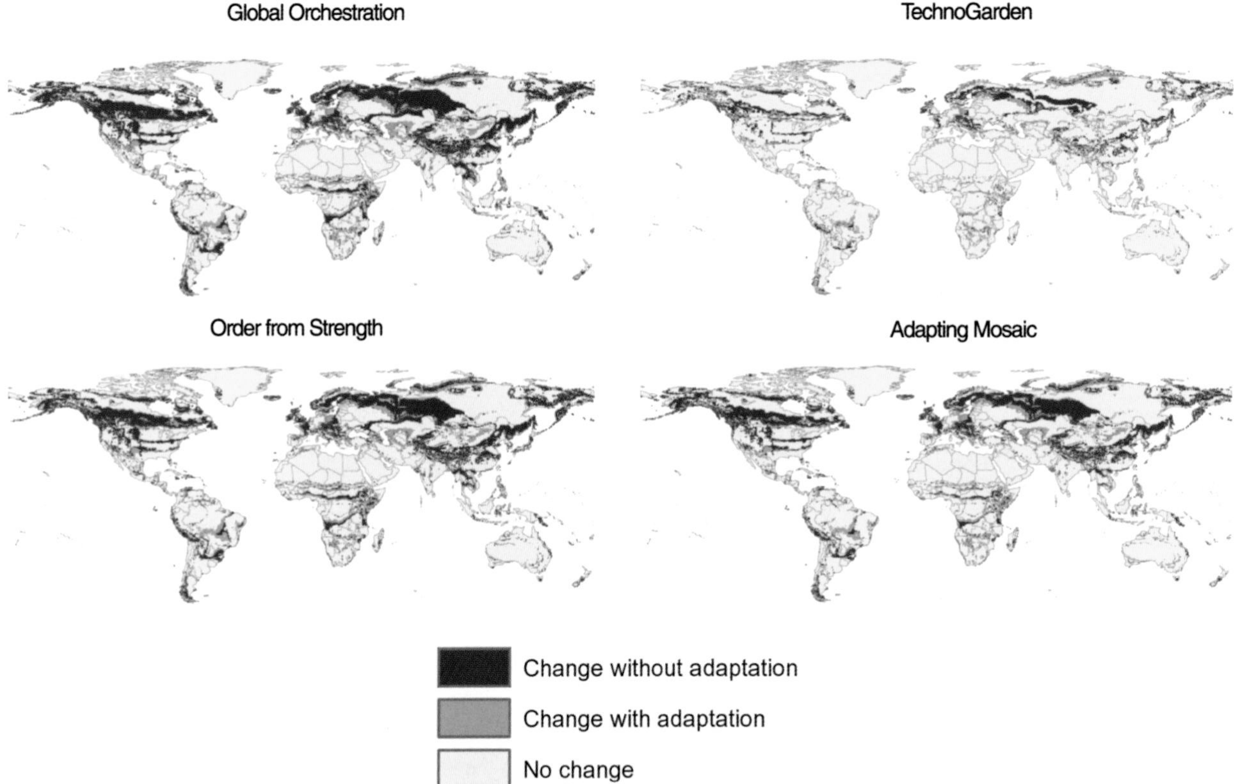

Figure 10.13. Threat to Natural Ecosystems from Climate Change Following the Biome Approach in the IMAGE 2.2 Model in MA Scenarios

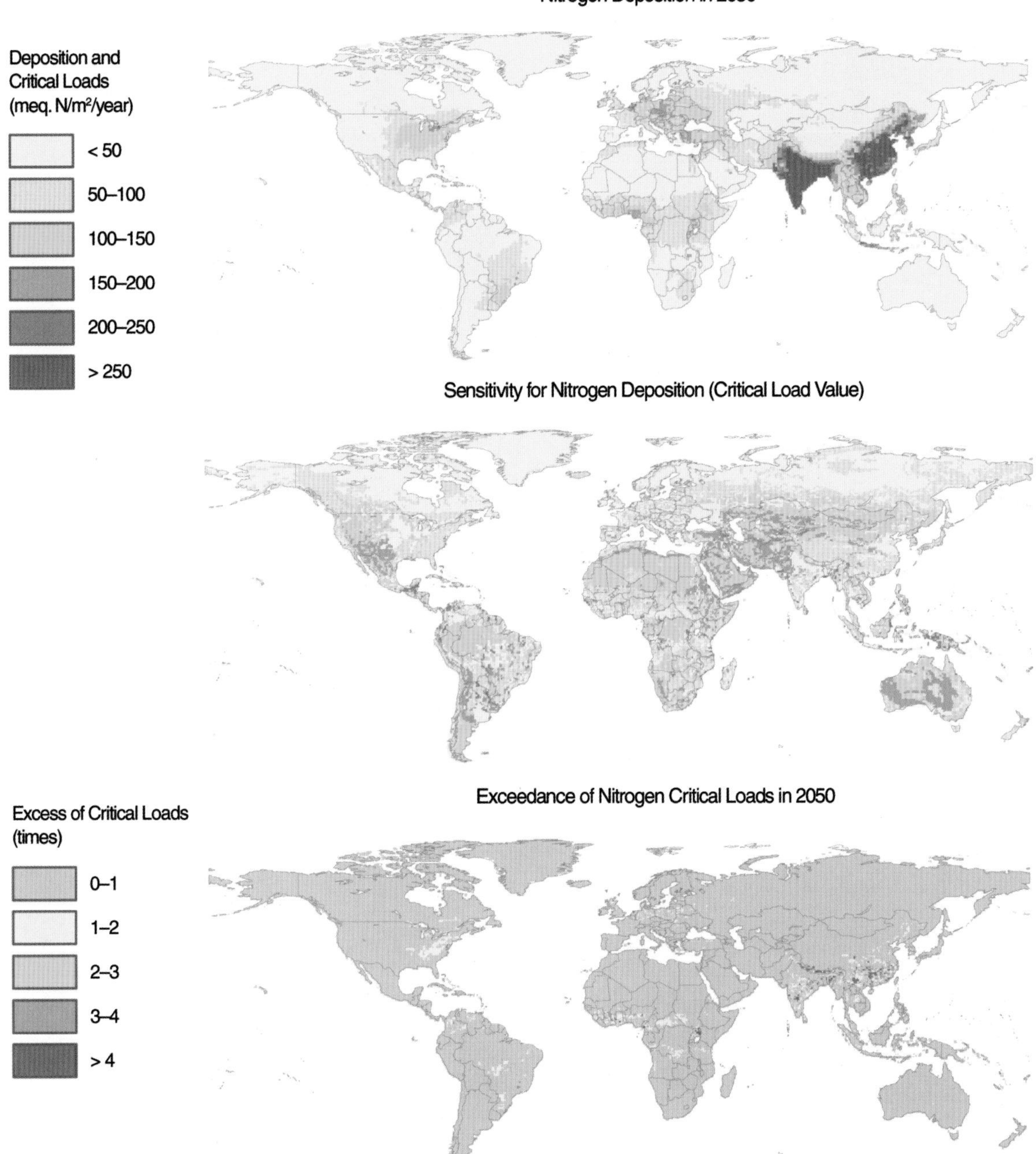

Figure 10.16. Nitrogen Deposition, Sensitivity, and Exceedance of Critical Loads for Order from Strength Scenario in 2050. In these maps for sensitivity, red tones indicate insensitive.

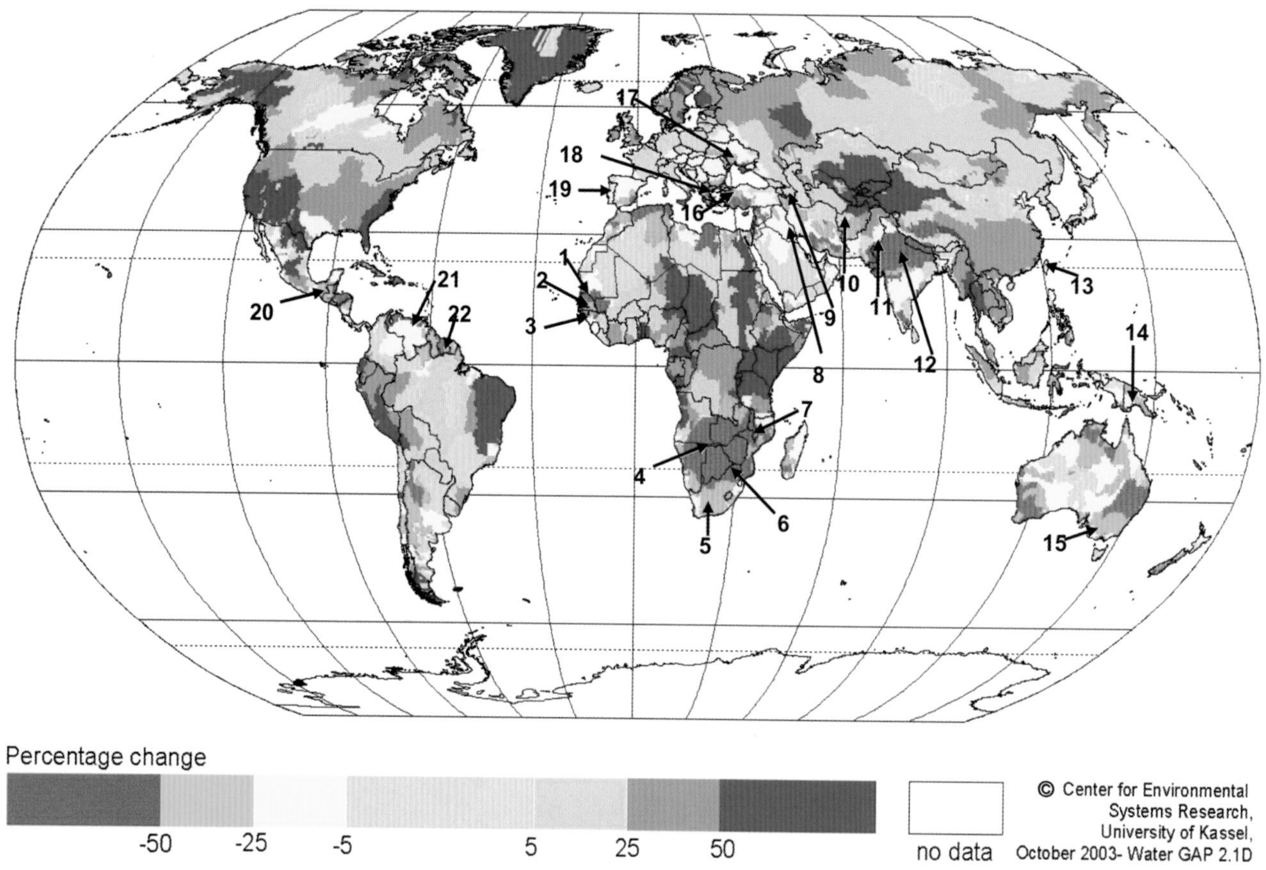

Figure 10.21. Change in Annual Water Availability in Global Orchestration Scenario in 2100. Numbers indicate the location of river basins in Figure 10.20. Shades from grey through red indicate regions that are drying.

a) Global Orchestration

HWB Indicators
Material Well-being
+1
0
-1
Health
Security
Freedom and choice
Social Relations

Socioecological Indicators
Long-term perspective over short-term goals
+1
0
-1
Social adaptive capacity to ecosystem change
Maintenance of ecosystem services

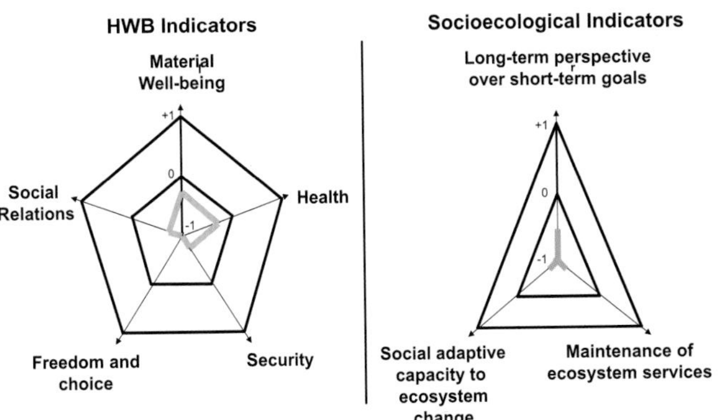

c) Adapting Mosaic

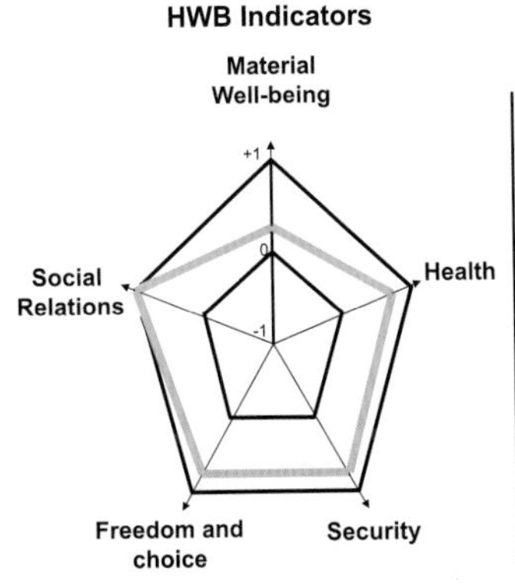

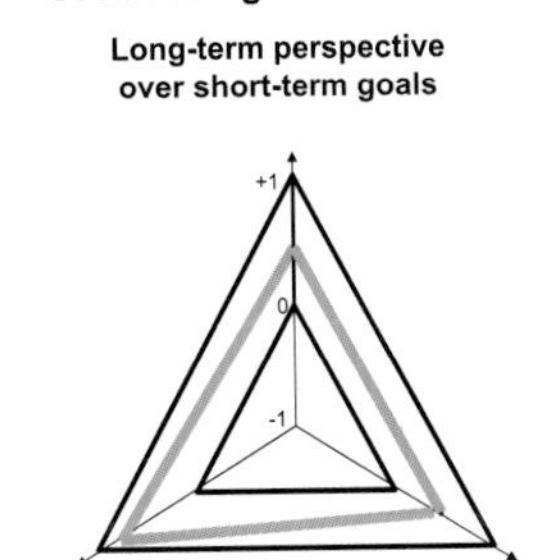
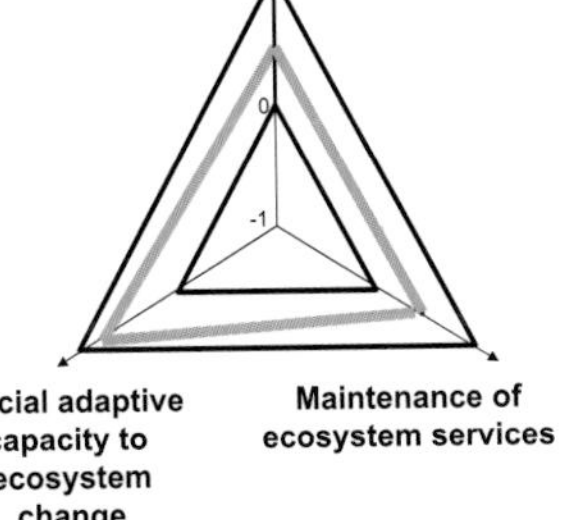

d) TechnoGarden

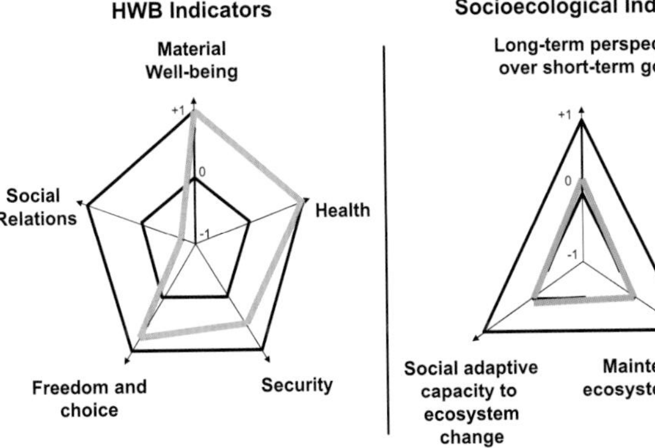

Figure 11.1. Changes in Human Well-being and Socioecological Indicators for MA Scenarios, 2000–50. Each axis in the star diagrams on the left represents one of the five human well-being (HWB) components as defined by the MA Conceptual Framework. The area inside the pentagon represents HWB as a whole. The '0' line represents the status of each of these components in 2000. If the yellow line moves more toward the center of the pentagon, this HWB component deteriorates in relative terms between today and 2050; if it moves toward the outer edges of the pentagon it improves. The diagrams on the right show the changes for three indicators representing socioecological variables. The '0' line represents the current status. If the green line moves toward the center of the triangle, the status of the indicator deteriorates in relative terms compared with today; if it moves more toward the outer edges of the triangle it improves.

a) Global Orchestration

Rich countries

Material Well-being

+1

0

-1

Social Relations

Health

Freedom and choice

Security

Poor countries

Material Well-being

+1

0

-1

Social Relations

Health

Freedom and choice

Security

b) Order from Strength

Rich countries

Material Well-being

+1

0

-1

Social Relations

Health

Freedom and choice

Security

Poor countries

Material Well-being

+1

0

-1

Social Relations

Health

Freedom and choice

Security

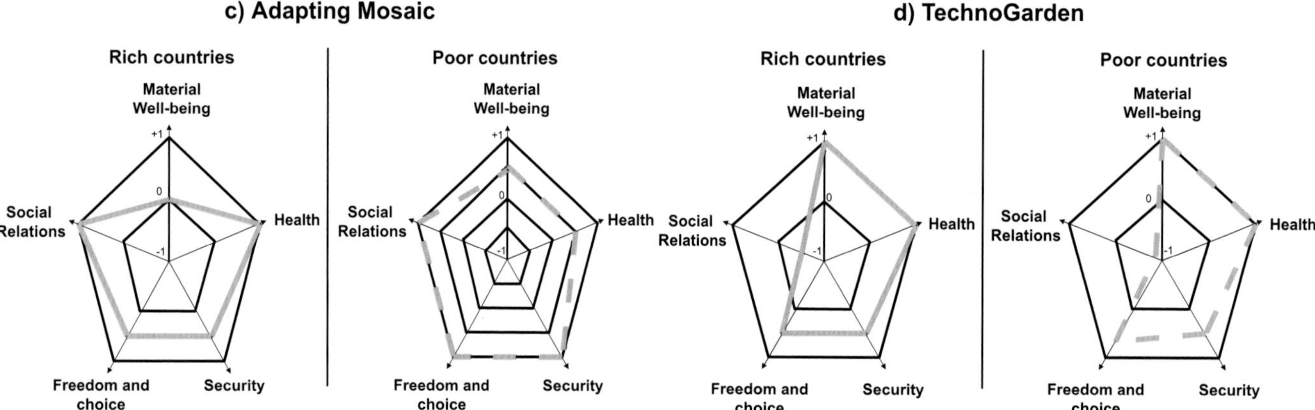

Figure 11.2. Changes in Currently Industrialized and Developing Countries for Human Well-being Indicators for MA Scenarios, Today–2050. Each axis in the star diagrams on the left represents one of the five human well-being (HWB) components as defined by the MA Conceptual Framework. The area marked by the lines between the arrows represents HWB as a whole. The '0' line represents the status of each of these components today. If the yellow line moves more toward the center of the pentagon, this HWB component deteriorates in relative terms between today and 2050; if it moves toward the outer edges of the pentagon HWB improves.

Appendix B
Authors

Argentina
Ana Parma, Centro Nacional Patagonico
Miguel Pascual, Centro Nacional Patagonico
Marina Gonzalez Polo, Universidad de Buenos Aires
Elda Tancredi, Lujan National University

Australia
Colin Butler, Australian National University
Steven Cork, CSIRO Australia and Land & Water Australia

Austria
Nebojsa Nakićenović, International Institute for Applied Systems Analysis and Vienna University of Technology
Brian O'Neill, International Institute for Applied Systems Analysis

Brazil
Eduardo Mario Mendiondo, Universidade de São Paolo

Canada
Jacqueline Alder, University of British Columbia
William Wai Lung Cheung, University of British Columbia
Villy Christensen, University of British Columbia
Garry Peterson, McGill University
John Robinson, University of British Columbia
Marguerite A. Xenopoulos, Trent University

China
Shiming Ma, Chinese Academy of Agricultural Sciences

Germany
Joseph Alcamo, University of Kassel
Wolfgang Cramer, Potsdam Institute for Climate Impact Research
Martina Floerke, University of Kassel
Michael Märker, University of Potsdam
Gerhard Petschel-Held, Potsdam Institute for Climate Impact Research
Kerstin Schulze, University of Kassel
Volkmar Wolters, Justus-Liebig-University Giessen

Ghana
Julius Fobil, University of Ghana

Hungary
Eva Hizsnyik, International Institute for Applied Systems Analysis
Ferenc L. Toth, International Atomic Energy Agency

Japan
Toshihiko Masui, National Institute for Environmental Studies
Tsuneyuki Morita,[1] National Institute for Environmental Studies
Kiyoshi Takahashi, National Institute for Environmental Studies

Kenya
Willis Oluoch-Kosura, University of Nairobi

The Netherlands
Lex Bouwman, Netherlands Environment Assessment Agency (MNP/RIVM)
Bert de Vries, Netherlands Environment Assessment Agency (MNP/RIVM)
Bas Eickhout, Netherlands Environment Assessment Agency (MNP/RIVM)
Pavel Kabat, ALTERRA Green World Research
Marja Spierenburg, Free University of Amsterdam
Detlef van Vuuren, Netherlands Environment Assessment Agency (MNP/RIVM)

New Zealand
Simon Hales, Wellington School of Medicine & Health Sciences

Portugal
Inês Gomes, Faculdade de Ciências da Universidade de Lisboa
Henrique Miguel Pereira, Faculdade de Ciências da Universidade de Lisboa
Cibele Queiroz, Faculdade de Ciências da Universidade de Lisboa

Russian Federation
Andrei Zaitsev, Institute for the Problems of Ecology and Evolution

Spain
Diana E. Marco, Estación Experimental del Zaidin, CSIC

Trinidad and Tobago
John R. B. Agard, University of the West Indies
Danielle Deane, The Hewlett Foundation

United Kingdom
Joanna House, University of Bristol
Andrew Stott, Department for Environment, Food and Rural Affairs
Paul Wilkinson, London School of Hygiene and Tropical Medicine

United States
T. Douglas Beard Jr., U.S. Geological Survey
Asmeret Asefaw Berhe, University of California-Berkeley
Elena M. Bennett, University of Wisconsin
Stephen R. Carpenter, University of Wisconsin
Edward R. Carr, University of South Carolina
Kenneth G. Cassman, University of Nebraska-Lincoln
Graeme S. Cumming, University of Florida
Ruth DeFries, University of Maryland
Robert Dickinson, Georgia Institute of Technology
Thomas Dietz, Michigan State University

[1] Deceased.

Achim Dobermann, University of Nebraska-Lincoln
Andrew Dobson, Princeton University
Jonathan Foley, University of Wisconsin
Jacqueline Geoghegan, Clark University
Elisabeth Holland, National Center for Atmospheric Research
Anthony Janetos, The H. John Heinz III Center for Science, Economics, and the Environment
Peter Kareiva, TNC-The Nature Conservancy
Axel Kleidon, University of Maryland
Hillel Koren, University of North Carolina, Chapel Hill, and U.S. Environmental Protection Agency
Rattan Lal, Ohio State University
Marc Levy, Columbia University
David Lodge, University of Notre Dame
Steven Manson, University of Minnesota
Fran Monks, Washington, DC
Harold Mooney, Stanford University
Gerald C. Nelson, University of Illinois
Richard B. Norgaard, University of California-Berkeley
Dennis Ojima, Colorado State University
Jacob Park, Green Mountain College
Paul Raskin, Tellus Institute/Stockholm Environment Institute
Claudia Ringler, International Food Policy Research Institute
Mark Rosegrant, International Food Policy Research Institute
James A. Rusak, University of Wisconsin
Osvaldo E. Sala, Brown University
Kathryn Saterson, Duke University
B.L. Turner II, Clark University
Diana Wall, Colorado State University
Robert Watson, The World Bank

Venezuela
Jon Paul Rodríguez, Instituto Venezolano de Investigaciones Científicas

Zambia
Michael Mutale, Department of Water Affairs

International Organizations
Carlos Corvalan, World Health Organization
Jacqueline McGlade, European Environment Agency
Prabhu L. Pingali, Food and Agriculture Organization of the UN
Teresa Ribeiro, European Environment Agency
Monika B. Zurek, Food and Agriculture Organization of the UN

Appendix C

Abbreviations and Acronyms

AI aridity index

AKRSP Aga Khan Rural Support Programme

AMF arbuscular mycorrhizal fungi

ASB alternatives to slash-and-burn

ASOMPH Asian Symposium on Medicinal Plants, Spices and Other Natural Products

AVHRR advanced very high resolution radiometer

BCA benefit-cost analysis

BGP Biogeochemical Province

BII Biodiversity Intactness Index

BMI body mass index

BNF biological nitrogen fixation

BOOT build-own-operate-transfer

BRT Bus Rapid Transit (Brazil)

BSE bovine spongiform encephalopathy

Bt *Bacillus thuringiensis*

C&I criteria and indicators

CAFO concentrated animal feeding operations

CAP Common Agricultural Policy (of the European Union)

CAREC Central Asia Regional Environment Centre

CBA cost-benefit analysis

CBD Convention on Biological Diversity

CBO community-based organization

CCAMLR Commission for the Conservation of Antarctic Marine Living Resources

CCN cloud condensation nuclei

CCS CO_2 capture and storage

CDM Clean Development Mechanism

CEA cost-effectiveness analysis

CENICAFE Centro Nacional de Investigaciones de Café (Colombia)

CFCs chlorofluorocarbons

CGIAR Consultative Group on International Agricultural Research

CIFOR Center for International Forestry Research

CITES Convention on International Trade in Endangered Species of Wild Fauna and Flora

CMS Convention on the Conservation of Migratory Species of Wild Animals (Bonn Convention)

CONICET Consejo de Investigaciones Científicas y Técnicas (Argentina)

COP Conference of the Parties (of treaties)

CPF Collaborative Partnership on Forests

CSIR Council for Scientific and Industrial Research (South Africa)

CV contingent valuation

CVM contingent valuation method

DAF decision analytical framework

DALY disability-adjusted life year

DDT dichloro diphenyl trichloroethane

DES dietary energy supply

DHF dengue hemorrhagic fever

DHS demographic and health surveys

DMS dimethyl sulfide

DPSEEA driving forces-pressure-state-exposure-effect-action

DPSIR driver-pressure-state-impact-response

DSF dust storm frequency

DU Dobson Units

EEA European Environment Agency

EEZ exclusive economic zone

EGS ecosystem global scenario

EHI environmental health indicator

EIA environmental impact assessment

EID emerging infectious disease

EKC Environmental Kuznets Curve

EMF ectomycorrhizal fungi

E/MSY	extinctions per million species per year
ENSO	El Niño/Southern Oscillation
EPA	Environmental Protection Agency (United States)
EPI	environmental policy integration
EU	European Union
EU ETS	European Union Emissions Trading System
FAO	Food and Agriculture Organization (United Nations)
FAPRI	Food and Agriculture Policy Research Institute
FLEGT	Forest Law Enforcement, Governance, and Trade
FRA	Forest Resources Assessment
FSC	Forest Stewardship Council
GATS	General Agreement on Trade and Services
GATT	General Agreement on Tariffs and Trade
GCM	general circulation model
GDI	Gender-related Development Index
GDP	gross domestic product
GEF	Global Environment Facility
GEO	*Global Environment Outlook*
GHG	greenhouse gases
GIS	geographic information system
GIWA	Global International Waters Assessment
GLASOD	Global Assessment of Soil Degradation
GLC	Global Land Cover
GLOF	Glacier Lake Outburst Flood
GM	genetic modification
GMO	genetically modified organism
GNI	gross national income
GNP	gross national product
GPS	Global Positioning System
GRoWI	*Global Review of Wetland Resources and Priorities for Wetland Inventory*
GSG	Global Scenarios Group
GSPC	Global Strategy for Plant Conservation
GtC-eq	gigatons of carbon equivalent
GWP	global warming potential
HDI	Human Development Index
HIA	health impact assessment
HIPC	heavily indebted poor countries
HPI	Human Poverty Index
HPS	hantavirus pulmonary syndrome
HWB	human well-being
IAA	integrated agriculture-aquaculture
IAM	integrated assessment model
IBI	Index of Biotic Integrity
ICBG	International Cooperative Biodiversity Groups
ICDP	integrated conservation and development project
ICJ	International Court of Justice
ICRAF	International Center for Research in Agroforestry
ICRW	International Convention for the Regulation of Whaling
ICSU	International Council for Science
ICZM	integrated coastal zone management
IDRC	International Development Research Centre (Canada)
IEA	International Energy Agency
IEG	international environmental governance
IEK	indigenous ecological knowledge
IFPRI	International Food Policy Research Institute
IGBP	International Geosphere-Biosphere Program
IIASA	International Institute for Applied Systems Analysis
IK	indigenous knowledge
ILO	International Labour Organization
IMF	International Monetary Fund
IMPACT	International Model for Policy Analysis of Agricultural Commodities and Trade
IMR	infant mortality rate
INESI	International Network of Sustainability Initiatives (hypothetical, in *Scenarios*)
INTA	Instituto Nacional de Tecnología Agropecuaria (Argentina)
IPAT	impact of population, affluence, technology
IPCC	Intergovernmental Panel on Climate Change
IPM	integrated pest management
IPR	intellectual property rights
IRBM	integrated river basin management
ISEH	International Society for Ecosystem Health
ISO	International Organization for Standardization
ITPGR	International Treaty on Plant Genetic Resources for Food and Agriculture
ITQs	individual transferable quotas
ITTO	International Tropical Timber Organization
IUCN	World Conservation Union
IUU	illegal, unregulated, and unreported (fishing)
IVM	integrated vector management

IWMI International Water Management Institute
IWRM integrated water resources management
JDSD Johannesburg Declaration on Sustainable Development
JI joint implementation
JMP Joint Monitoring Program
LAC Latin America and the Caribbean
LAI leaf area index
LARD livelihood approaches to rural development
LDC least developed country
LEK local ecological knowledge
LME large marine ecosystems
LPI Living Planet Index
LSMS Living Standards Measurement Study
LULUCF land use, land use change, and forestry
MA Millennium Ecosystem Assessment
MAI mean annual increments
MBI market-based instruments
MCA multicriteria analysis
MDG Millennium Development Goal
MEA multilateral environmental agreement
MENA Middle East and North Africa
MER market exchange rate
MHC major histocompatibility complex
MICS multiple indicator cluster surveys
MIT Massachusetts Institute of Technology
MPA marine protected area
MSVPA multispecies virtual population analysis
NAP National Action Program (of desertification convention)
NBP net biome productivity
NCD noncommunicable disease
NCS National Conservation Strategy
NCSD national council for sustainable development
NDVI normalized difference vegetation index
NE effective size of a population
NEAP national environmental action plan
NEP new ecological paradigm; also net ecosystem productivity
NEPAD New Partnership for Africa's Development
NFAP National Forestry Action Plan
NFP national forest programs
NGO nongovernmental organization
NIH National Institutes of Health (United States)
NMHC non-methane hydrocarbons
NOAA National Oceanographic and Atmospheric Administration (United States)
NPP net primary productivity
NSSD national strategies for sustainable development
NUE nitrogen use efficiency
NWFP non-wood forest product
ODA official development assistance
OECD Organisation for Economic Co-operation and Development
OSB oriented strand board
OWL other wooded land
PA protected area
PAH polycyclic aromatic hydrocarbons
PCBs polychlorinated biphenyls
PEM protein energy malnutrition
PES payment for environmental (or ecosystem) services
PFT plant functional type
PNG Papua New Guinea
POPs persistent organic pollutants
PPA participatory poverty assessment
ppb parts per billion
PPI potential Pareto improvement
ppm parts per million
ppmv parts per million by volume
PPP purchasing power parity; also public-private partnership
ppt parts per thousand
PQLI Physical Quality of Life Index
PRA participatory rural appraisal
PRSP Poverty Reduction Strategy Paper
PSE producer support estimate
PVA population viability analysis
RANWA Research and Action in Natural Wealth Administration
RBO river basin organization
RIDES Recursos e Investigación para el Desarrollo Sustentable (Chile)
RIL reduced impact logging
RLI Red List Index
RO reverse osmosis

RRA	rapid rural appraisal
RUE	rain use efficiency
SADC	Southern African Development Community
SADCC	Southern African Development Coordination Conference
SAfMA	Southern African Millennium Ecosystem Assessment
SAP	structural adjustment program
SAR	species-area relationship
SARS	severe acute respiratory syndrome
SBSTTA	Subsidiary Body on Scientific, Technical and Technological Advice (of CBD)
SEA	strategic environmental assessment
SEME	simple empirical models for eutrophication
SES	social-ecological system
SFM	sustainable forest management
SIDS	small island developing states
SMS	safe minimum standard
SOM	soil organic matter
SRES	Special Report on Emissions Scenarios (of the IPCC)
SSC	Species Survival Commission (of IUCN)
SWAP	sector-wide approach
TAC	total allowable catch
TBT	tributyltin
TC	travel cost
TCM	travel cost method
TDR	tradable development rights
TDS	total dissolved solids
TEIA	transboundary environmental impact assessment
TEK	traditional ecological knowledge
TEM	terrestrial ecosystem model
TESEO	Treaty Enforcement Services Using Earth Observation
TEV	total economic value
TFAP	Tropical Forests Action Plan
TFP	total factor productivity
TFR	total fertility rate
Tg	teragram (10^{12} grams)
TK	traditional knowledge
TMDL	total maximum daily load
TOF	trees outside of forests
TRIPS	Trade-Related Aspects of Intellectual Property Rights
TSU	Technical Support Unit
TW	terawatt
UMD	University of Maryland
UNCCD	United Nations Convention to Combat Desertification
UNCED	United Nations Conference on Environment and Development
UNCLOS	United Nations Convention on the Law of the Sea
UNDP	United Nations Development Programme
UNECE	United Nations Economic Commission for Europe
UNEP	United Nations Environment Programme
UNESCO	United Nations Educational, Scientific and Cultural Organization
UNFCCC	United Nations Framework Convention on Climate Change
UNIDO	United Nations Industrial Development Organization
UNRO	United Nations Regional Organization (hypothetical body, in *Scenarios*)
UNSO	UNDP's Office to Combat Desertification and Drought
USAID	U.S. Agency for International Development
USDA	U.S. Department of Agriculture
VOC	volatile organic compound
VW	virtual water
WBCSD	World Business Council for Sustainable Development
WCD	World Commission on Dams
WCED	World Commission on Environment and Development
WCMC	World Conservation Monitoring Centre (of UNEP)
WFP	World Food Programme
WHO	World Health Organization
WIPO	World Intellectual Property Organization
WISP	weighted index of social progress
WMO	World Meteorological Organization
WPI	Water Poverty Index
WRF	white rot fungi
WSSD	World Summit on Sustainable Development
wta	withdrawals-to-availability ratio (of water)
WTA	willingness to accept compensation
WTO	World Trade Organization
WTP	willingness to pay
WWAP	World Water Assessment Programme
WWF	World Wide Fund for Nature
WWV	World Water Vision

Appendix D

Glossary

Abatement cost: See *Marginal abatement cost.*

Abundance: The total number of individuals of a taxon or taxa in an area, population, or community. Relative abundance refers to the total number of individuals of one taxon compared with the total number of individuals of all other taxa in an area, volume, or community.

Active adaptive management: See *Adaptive management.*

Adaptation: Adjustment in natural or human systems to a new or changing environment. Various types of adaptation can be distinguished, including anticipatory and reactive adaptation, private and public adaptation, and autonomous and planned adaptation.

Adaptive capacity: The general ability of institutions, systems, and individuals to adjust to potential damage, to take advantage of opportunities, or to cope with the consequences.

Adaptive management: A systematic process for continually improving management policies and practices by learning from the outcomes of previously employed policies and practices. In active adaptive management, management is treated as a deliberate experiment for purposes of learning.

Afforestation: Planting of forests on land that has historically not contained forests. (Compare *Reforestation.*)

Agrobiodiversity: The diversity of plants, insects, and soil biota found in cultivated systems.

Agroforestry systems: Mixed systems of crops and trees providing wood, non-wood forest products, food, fuel, fodder, and shelter.

Albedo: A measure of the degree to which a surface or object reflects solar radiation.

Alien species: Species introduced outside its normal distribution.

Alien invasive species: See *Invasive alien species.*

Aquaculture: Breeding and rearing of fish, shellfish, or plants in ponds, enclosures, or other forms of confinement in fresh or marine waters for the direct harvest of the product.

Benefits transfer approach: Economic valuation approach in which estimates obtained (by whatever method) in one context are used to estimate values in a different context.

Binding constraints: Political, social, economic, institutional, or ecological factors that rule out a particular response.

Biodiversity (a contraction of biological diversity): The variability among living organisms from all sources, including terrestrial, marine, and other aquatic ecosystems and the ecological complexes of which they are part. Biodiversity includes diversity within species, between species, and between ecosystems.

Biodiversity regulation: The regulation of ecosystem processes and services by the different components of biodiversity.

Biogeographic realm: A large spatial region, within which ecosystems share a broadly similar biota. Eight terrestrial biogeographic realms are typically recognized, corresponding roughly to continents (e.g., Afrotropical realm).

Biological diversity: See *Biodiversity.*

Biomass: The mass of tissues in living organisms in a population, ecosystem, or spatial unit.

Biome: The largest unit of ecological classification that is convenient to recognize below the entire globe. Terrestrial biomes are typically based on dominant vegetation structure (e.g., forest, grassland). Ecosystems within a biome function in a broadly similar way, although they may have very different species composition. For example, all forests share certain properties regarding nutrient cycling, disturbance, and biomass that are different from the properties of grasslands. Marine biomes are typically based on biogeochemical properties. The WWF biome classification is used in the MA.

Bioprospecting: The exploration of biodiversity for genetic and biochemical resources of social or commercial value.

Biotechnology: Any technological application that uses biological systems, living organisms, or derivatives thereof to make or modify products or processes for specific use.

Biotic homogenization: Process by which the differences between biotic communities in different areas are on average reduced.

Blueprint approaches: Approaches that are designed to be applicable in a wider set of circumstances and that are not context-specific or sensitive to local conditions.

Boundary organizations: Public or private organizations that synthesize and translate scientific research and explore its policy implications to help bridge the gap between science and decision-making.

Bridging organizations: Organizations that facilitate, and offer an arena for, stakeholder collaboration, trust-building, and conflict resolution.

Capability: The combinations of doings and beings from which people can choose to lead the kind of life they value. Basic capability is the capability to meet a basic need.

Capacity building: A process of strengthening or developing human resources, institutions, organizations, or networks. Also referred to as capacity development or capacity enhancement.

Capital value (of an ecosystem): The present value of the stream of ecosystem services that an ecosystem will generate under a particular management or institutional regime.

Capture fisheries: See *Fishery.*

Carbon sequestration: The process of increasing the carbon content of a reservoir other than the atmosphere.

Cascading interaction: See *Trophic cascade.*

Catch: The number or weight of all fish caught by fishing operations, whether the fish are landed or not.

Coastal system: Systems containing terrestrial areas dominated by ocean influences of tides and marine aerosols, plus nearshore marine areas. The inland extent of coastal ecosystems is the line where land-based influences dominate, up to a maximum of 100 kilometers from the coastline or 100-meter elevation (whichever is closer to the sea), and the outward extent is the 50-meter-depth contour. See also *System.*

Collaborative (or joint) forest management: Community-based management of forests, where resource tenure by local communities is secured.

Common pool resource: A valued natural or human-made resource or facility in which one person's use subtracts from another's use and where it is often necessary but difficult to exclude potential users from the resource. (Compare *Common property resource.*)

Common property management system: The institutions (i.e., sets of rules) that define and regulate the use rights for common pool resources. Not the same as an open access system.

Common property resource: A good or service shared by a well-defined community. (Compare *Common pool resource.*)

Community (ecological): An assemblage of species occurring in the same space or time, often linked by biotic interactions such as competition or predation.

Community (human, local): A collection of human beings who have something in common. A local community is a fairly small group of people who share a common place of residence and a set of institutions based on this fact, but the word 'community' is also used to refer to larger collections of people who have something else in common (e.g., national community, donor community).

Condition of an ecosystem: The capacity of an ecosystem to yield services, relative to its potential capacity.

Condition of an ecosystem service: The capacity of an ecosystem service to yield benefits to people, relative to its potential capacity.

Constituents of well-being: The experiential aspects of well-being, such as health, happiness, and freedom to be and do, and, more broadly, basic liberties.

Consumptive use: The reduction in the quantity or quality of a good available for other users due to consumption.

Contingent valuation: Economic valuation technique based on a survey of how much respondents would be willing to pay for specified benefits.

Core dataset: Data sets designated to have wide potential application throughout the Millennium Ecosystem Assessment process. They include land use, land cover, climate, and population data sets.

Cost-benefit analysis: A technique designed to determine the feasibility of a project or plan by quantifying its costs and benefits.

Cost-effectiveness analysis: Analysis to identify the least cost option that meets a particular goal.

Critically endangered species: Species that face an extremely high risk of extinction in the wild. See also *Threatened species*.

Cross-scale feedback: A process in which effects of some action are transmitted from a smaller spatial extent to a larger one, or vice versa. For example, a global policy may constrain the flexibility of a local region to use certain response options to environmental change, or a local agricultural pest outbreak may affect regional food supply.

Cultivar (a contraction of cultivated variety): A variety of a plant developed from a natural species and maintained under cultivation.

Cultivated system: Areas of landscape or seascape actively managed for the production of food, feed, fiber, or biofuels.

Cultural landscape: See *Landscape*.

Cultural services: The nonmaterial benefits people obtain from ecosystems through spiritual enrichment, cognitive development, reflection, recreation, and aesthetic experience, including, e.g., knowledge systems, social relations, and aesthetic values.

Decision analytical framework: A coherent set of concepts and procedures aimed at synthesizing available information to help policy-makers assess consequences of various decision options. DAFs organize the relevant information in a suitable framework, apply decision criteria (both based on some paradigms or theories), and thus identify options that are better than others under the assumptions characterizing the analytical framework and the application at hand.

Decision-maker: A person whose decisions, and the actions that follow from them, can influence a condition, process, or issue under consideration.

Decomposition: The ecological process carried out primarily by microbes that leads to a transformation of dead organic matter into inorganic mater.

Deforestation: Conversion of forest to non-forest.

Degradation of an ecosystem service: For *provisioning services,* decreased production of the service through changes in area over which the services is provided, or decreased production per unit area. For *regulating* and *supporting services,* a reduction in the benefits obtained from the service, either through a change in the service or through human pressures on the service exceeding its limits. For *cultural services,* a change in the ecosystem features that decreases the cultural benefits provided by the ecosystem.

Degradation of ecosystems: A persistent reduction in the capacity to provide ecosystem services.

Desertification: land degradation in drylands resulting from various factors, including climatic variations and human activities.

Determinants of well-being: Inputs into the production of well-being, such as food, clothing, potable water, and access to knowledge and information.

Direct use value (of ecosystems): The benefits derived from the services provided by an ecosystem that are used directly by an economic agent. These include consumptive uses (e.g., harvesting goods) and nonconsumptive uses (e.g., enjoyment of scenic beauty). Agents are often physically present in an ecosystem to receive direct use value. (Compare *Indirect use value.*)

Disability-adjusted life years: The sum of years of life lost due to premature death and illness, taking into account the age of death compared with natural life expectancy and the number of years of life lived with a disability. The measure of number of years lived with the disability considers the duration of the disease, weighted by a measure of the severity of the disease.

Diversity: The variety and relative abundance of different entities in a sample.

Driver: Any natural or human-induced factor that directly or indirectly causes a change in an ecosystem.

Driver, direct: A driver that unequivocally influences ecosystem processes and can therefore be identified and measured to differing degrees of accuracy. (Compare *Driver, indirect.*)

Driver, endogenous: A driver whose magnitude can be influenced by the decision-maker. Whether a driver is exogenous or endogenous depends on the organizational scale. Some drivers (e.g., prices) are exogenous to a decision-maker at one level (a farmer) but endogenous at other levels (the nation-state). (Compare *Driver, exogenous.*)

Driver, exogenous: A driver that cannot be altered by the decision-maker. (Compare *Driver, endogenous.*)

Driver, indirect: A driver that operates by altering the level or rate of change of one or more direct drivers. (Compare *Driver, direct.*)

Drylands: See *Dryland system*.

Dryland system: Areas characterized by lack of water, which constrains the two major interlinked services of the system: primary production and nutrient cycling. Four dryland subtypes are widely recognized: dry sub-humid, semiarid, arid, and hyperarid, showing an increasing level of aridity or moisture deficit. See also *System*.

Ecological character: See *Ecosystem properties*.

Ecological degradation: See *Degradation of ecosystems*.

Ecological footprint: An index of the area of productive land and aquatic ecosystems required to produce the resources used and to assimilate the wastes produced by a defined population at a specified material standard of living, wherever on Earth that land may be located.

Ecological security: A condition of ecological safety that ensures access to a sustainable flow of provisioning, regulating, and cultural services needed by local communities to meet their basic capabilities.

Ecological surprises: unexpected—and often disproportionately large—consequence of changes in the abiotic (e.g., climate, disturbance) or biotic (e.g., invasions, pathogens) environment.

Ecosystem: A dynamic complex of plant, animal, and microorganism communities and their non-living environment interacting as a functional unit.

Ecosystem approach: A strategy for the integrated management of land, water, and living resources that promotes conservation and sustainable use. An ecosystem approach is based on the application of appropriate scientific methods focused on levels of biological organization, which encompass the essential structure, processes, functions, and interactions among organisms and their environment. It recognizes that humans, with their cultural diversity, are an integral component of many ecosystems.

Ecosystem assessment: A social process through which the findings of science concerning the causes of ecosystem change, their consequences for human well-being, and management and policy options are brought to bear on the needs of decision-makers.

Ecosystem boundary: The spatial delimitation of an ecosystem, typically based on discontinuities in the distribution of organisms, the biophysical environment (soil types, drainage basins, depth in a water body), and spatial interactions (home ranges, migration patterns, fluxes of matter).

Ecosystem change: Any variation in the state, outputs, or structure of an ecosystem.

Ecosystem function: See *Ecosystem process.*

Ecosystem interactions: Exchanges of materials, energy, and information within and among ecosystems.

Ecosystem management: An approach to maintaining or restoring the composition, structure, function, and delivery of services of natural and modified ecosystems for the goal of achieving sustainability. It is based on an adaptive, collaboratively developed vision of desired future conditions that integrates ecological, socioeconomic, and institutional perspectives, applied within a geographic framework, and defined primarily by natural ecological boundaries.

Ecosystem process: An intrinsic ecosystem characteristic whereby an ecosystem maintains its integrity. Ecosystem processes include decomposition, production, nutrient cycling, and fluxes of nutrients and energy.

Ecosystem properties: The size, biodiversity, stability, degree of organization, internal exchanges of materials, energy, and information among different pools, and other properties that characterize an ecosystem. Includes ecosystem functions and processes.

Ecosystem resilience: See *Resilience.*

Ecosystem resistance: See *Resistance.*

Ecosystem robustness: See *Ecosystem stability.*

Ecosystem services: The benefits people obtain from ecosystems. These include *provisioning services* such as food and water; *regulating services* such as flood and disease control; *cultural services* such as spiritual, recreational, and cultural benefits; and *supporting services* such as nutrient cycling that maintain the conditions for life on Earth. The concept "ecosystem goods and services" is synonymous with ecosystem services.

Ecosystem stability (or ecosystem robustness): A description of the dynamic properties of an ecosystem. An ecosystem is considered stable or robust if it returns to its original state after a perturbation, exhibits low temporal variability, or does not change dramatically in the face of a perturbation.

Elasticity: A measure of responsiveness of one variable to a change in another, usually defined in terms of percentage change. For example, own-price elasticity of demand is the percentage change in the quantity demanded of a good for a 1% change in the price of that good. Other common elasticity measures include supply and income elasticity.

Emergent disease: Diseases that have recently increased in incidence, impact, or geographic range; that are caused by pathogens that have recently evolved; that are newly discovered; or that have recently changed their clinical presentation.

Emergent property: A phenomenon that is not evident in the constituent parts of a system but that appears when they interact in the system as a whole.

Enabling conditions: Critical preconditions for success of responses, including political, institutional, social, economic, and ecological factors.

Endangered species: Species that face a very high risk of extinction in the wild. See also *Threatened species.*

Endemic (in ecology): A species or higher taxonomic unit found only within a specific area.

Endemic (in health): The constant presence of a disease or infectious agent within a given geographic area or population group; may also refer to the usual prevalence of a given disease within such area or group.

Endemism: The fraction of species that is endemic relative to the total number of species found in a specific area.

Epistemology: The theory of knowledge, or a "way of knowing."

Equity: Fairness of rights, distribution, and access. Depending on context, this can refer to resources, services, or power.

Eutrophication: The increase in additions of nutrients to freshwater or marine systems, which leads to increases in plant growth and often to undesirable changes in ecosystem structure and function.

Evapotranspiration: See *Transpiration.*

Existence value: The value that individuals place on knowing that a resource exists, even if they never use that resource (also sometimes known as conservation value or passive use value).

Exotic species: See *Alien species.*

Externality: A consequence of an action that affects someone other than the agent undertaking that action and for which the agent is neither compensated nor penalized through the markets. Externalities can be positive or negative.

Feedback: See *Negative feedback, Positive feedback,* and *Cross-scale feedback.*

Fishery: A particular kind of fishing activity, e.g., a trawl fishery, or a particular species targeted, e.g., a cod fishery or salmon fishery.

Fish stock: See *Stock.*

Fixed nitrogen: See *Reactive nitrogen.*

Flyway: Areas of the world used by migratory birds in moving between breeding and wintering grounds.

Forest systems: Systems in which trees are the predominant life forms. Statistics reported in this assessment are based on areas that are dominated by trees (perennial woody plants taller than five meters at maturity), where the tree crown cover exceeds 10%, and where the area is more than 0.5 hectares. "Open forests" have a canopy cover between 10% and 40%, and "closed forests" a canopy cover of more than 40%. "Fragmented forests" refer to mosaics of forest patches and non-forest land. See also *System.*

Freedom: The range of options a person has in deciding the kind of life to lead.

Functional diversity: The value, range, and relative abundance of traits present in the organisms in an ecological community.

Functional redundancy (= functional compensation): A characteristic of ecosystems in which more than one species in the system can carry out a particular process. Redundancy may be total or partial—that is, a species may not be able to completely replace the other species or it may compensate only some of the processes in which the other species are involved.

Functional types (= functional groups = guilds): Groups of organisms that respond to the environment or affect ecosystem processes in a similar way. Examples of plant functional types include nitrogen-fixer versus non-fixer, stress-tolerant versus ruderal versus competitor, resprouter versus seeder, deciduous versus evergreen. Examples of animal functional types include granivorous versus fleshy-fruit eater, nocturnal versus diurnal predator, browser versus grazer.

Geographic information system: A computerized system organizing data sets through a geographical referencing of all data included in its collections.

Globalization: The increasing integration of economies and societies around the world, particularly through trade and financial flows, and the transfer of culture and technology.

Global scale: The geographical realm encompassing all of Earth.

Governance: The process of regulating human behavior in accordance with shared objectives. The term includes both governmental and nongovernmental mechanisms.

Health, human: A state of complete physical, mental, and social well-being and not merely the absence of disease or infirmity. The health of a whole community or population is reflected in measurements of disease incidence and prevalence, age-specific death rates, and life expectancy.

High seas: The area outside of national jurisdiction, i.e., beyond each nation's Exclusive Economic Zone or other territorial waters.

Human well-being: See *Well-being.*

Income poverty: See *Poverty.*

Indicator: Information based on measured data used to represent a particular attribute, characteristic, or property of a system.

Indigenous knowledge (or local knowledge): The knowledge that is unique to a given culture or society.

Indirect interaction: Those interactions among species in which a species, through direct interaction with another species or modification of resources, alters the abundance of a third species with which it is not directly interacting. Indirect interactions can be trophic or nontrophic in nature.

Indirect use value: The benefits derived from the goods and services provided by an ecosystem that are used indirectly by an economic agent. For example, an agent at some distance from an ecosystem may derive benefits from drinking water that has been purified as it passed through the ecosystem. (Compare *Direct use value.*)

Infant mortality rate: Number of deaths of infants aged 0–12 months divided by the number of live births.

Inland water systems: Permanent water bodies other than salt-water systems on the coast, seas and oceans. Includes rivers, lakes, reservoirs wetlands and inland saline lakes and marshes. See also *System.*

Institutions: The rules that guide how people within societies live, work, and interact with each other. Formal institutions are written or codified rules. Examples of formal institutions would be the constitution, the judiciary laws, the organized market, and property rights. Informal institutions are rules governed by social and behavioral norms of the society, family, or community. Also referred to as organizations.

Integrated coastal zone management: Approaches that integrate economic, social, and ecological perspectives for the management of coastal resources and areas.

Integrated conservation and development projects: Initiatives that aim to link biodiversity conservation and development.

Integrated pest management: Any practices that attempt to capitalize on natural processes that reduce pest abundance. Sometimes used to refer to monitoring programs where farmers apply pesticides to improve economic efficiency (reducing application rates and improving profitability).

Integrated responses: Responses that address degradation of ecosystem services across a number of systems simultaneously or that also explicitly include objectives to enhance human well-being.

Integrated river basin management: Integration of water planning and management with environmental, social, and economic development concerns, with an explicit objective of improving human welfare.

Interventions: See *Responses.*

Intrinsic value: The value of someone or something in and for itself, irrespective of its utility for people.

Invasibility: Intrinsic susceptibility of an ecosystem to be invaded by an alien species.

Invasive alien species: An alien species whose establishment and spread modifies ecosystems, habitats, or species.

Irreversibility: The quality of being impossible or difficult to return to, or to restore to, a former condition. See also *Option value, Precautionary principle, Resilience,* and *Threshold.*

Island systems: Lands isolated by surrounding water, with a high proportion of coast to hinterland. The degree of isolation from the mainland in both natural and social aspects is accounted by the *isola effect.* See also *System.*

Isola effect: Environmental issues that are unique to island systems. This uniqueness takes into account the physical seclusion of islands as isolated pieces of land exposed to marine or climatic disturbances with a more limited access to space, products, and services when compared with most continental areas, but also includes subjective issues such as the perceptions and attitudes of islanders themselves.

Keystone species: A species whose impact on the community is disproportionately large relative to its abundance. Effects can be produced by consumption (trophic interactions), competition, mutualism, dispersal, pollination, disease, or habitat modification (nontrophic interactions).

Land cover: The physical coverage of land, usually expressed in terms of vegetation cover or lack of it. Related to, but not synonymous with, *land use.*

Landscape: An area of land that contains a mosaic of ecosystems, including human-dominated ecosystems. The term cultural landscape is often used when referring to landscapes containing significant human populations or in which there has been significant human influence on the land.

Landscape unit: A portion of relatively homogenous land cover within the local-to-regional landscape.

Land use: The human use of a piece of land for a certain purpose (such as irrigated agriculture or recreation). Influenced by, but not synonymous with, *land cover.*

Length of growing period: The total number of days in a year during which rainfall exceeds one half of potential evapotranspiration. For boreal and temperate zone, growing season is usually defined as a number of days with the average daily temperature that exceeds a definite threshold, such as 10° Celsius.

Local knowledge: See *Indigenous knowledge.*

Mainstreaming: Incorporating a specific concern, e.g. sustainable use of ecosystems, into policies and actions.

Malnutrition: A state of bad nourishment. Malnutrition refers both to undernutrition and overnutrition, as well as to conditions arising from dietary imbalances leading to diet-related noncommunicable diseases.

Marginal abatement cost: The cost of abating an incremental unit of, for instance, a pollutant.

Marine system: Marine waters from the low-water mark to the high seas that support marine capture fisheries, as well as deepwater (>50 meters) habitats. Four sub-divisions (marine biomes) are recognized: the coastal boundary zone; trade-winds; westerlies; and polar.

Market-based instruments: Mechanisms that create a market for ecosystem services in order to improving the efficiency in the way the service is used. The term is used for mechanisms that create new markets, but also for responses such as taxes, subsidies, or regulations that affect existing markets.

Market failure: The inability of a market to capture the correct values of ecosystem services.

Mitigation: An anthropogenic intervention to reduce negative or unsustainable uses of ecosystems or to enhance sustainable practices.

Mountain system: High-altitude (greater than 2,500 meters) areas and steep mid-altitude (1,000 meters at the equator, decreasing to sea level where alpine life zones meet polar life zones at high latitudes) areas, excluding large plateaus.

Negative feedback: Feedback that has a net effect of dampening perturbation.

Net primary productivity: See *Production, biological.*

Non-linearity: A relationship or process in which a small change in the value of a driver (i.e., an independent variable) produces an disproportionate change in the outcome (i.e., the dependent variable). Relationships where there is a sudden discontinuity or change in rate are sometimes referred to as abrupt and often form the basis of thresholds. In loose terms, they may lead to unexpected outcomes or "surprises."

Nutrient cycling: The processes by which elements are extracted from their mineral, aquatic, or atmospheric sources or recycled from their organic forms, converting them to the ionic form in which biotic uptake occurs and ultimately returning them to the atmosphere, water, or soil.

Nutrients: The approximately 20 chemical elements known to be essential for the growth of living organisms, including nitrogen, sulfur, phosphorus, and carbon.

Open access resource: A good or service over which no property rights are recognized.

Opportunity cost: The benefits forgone by undertaking one activity instead of another.

Option value: The value of preserving the option to use services in the future either by oneself (option value) or by others or heirs (bequest value). Quasi-option value represents the value of avoiding irreversible decisions until new information reveals whether certain ecosystem services have values society is not currently aware of.

Organic farming: Crop and livestock production systems that do not make use of synthetic fertilizers, pesticides, or herbicides. May also include restrictions on the use of transgenic crops (genetically modified organisms).

Pastoralism, pastoral system: The use of domestic animals as a primary means for obtaining resources from habitats.

Perturbation: An imposed movement of a system away from its current state.

Polar system: Treeless lands at high latitudes. Includes Arctic and Antarctic areas, where the polar system merges with the northern boreal forest and the Southern Ocean respectively. See also *System*.

Policy failure: A situation in which government policies create inefficiencies in the use of goods and services.

Policy-maker: A person with power to influence or determine policies and practices at an international, national, regional, or local level.

Pollination: A process in the sexual phase of reproduction in some plants caused by the transportation of pollen. In the context of ecosystem services, pollination generally refers to animal-assisted pollination, such as that done by bees, rather than wind pollination.

Population, biological: A group of individuals of the same species, occupying a defined area, and usually isolated to some degree from other similar groups. Populations can be relatively reproductively isolated and adapted to local environments.

Population, human: A collection of living people in a given area. (Compare *Community (human, local)*.)

Positive feedback: Feedback that has a net effect of amplifying perturbation.

Poverty: The pronounced deprivation of well-being. Income poverty refers to a particular formulation expressed solely in terms of per capita or household income.

Precautionary principle: The management concept stating that in cases "where there are threats of serious or irreversible damage, lack of full scientific certainty shall not be used as a reason for postponing cost-effective measures to prevent environmental degradation," as defined in the Rio Declaration.

Prediction (or forecast): The result of an attempt to produce a most likely description or estimate of the actual evolution of a variable or system in the future. See also *Projection* and *Scenario*.

Primary production: See *Production, biological*.

Private costs and benefits: Costs and benefits directly felt by individual economic agents or groups as seen from their perspective. (Externalities imposed on others are ignored.) Costs and benefits are valued at the prices actually paid or received by the group, even if these prices are highly distorted. Sometimes termed "financial" costs and benefits. (Compare *Social costs and benefits*.)

Probability distribution: A distribution that shows all the values that a random variable can take and the likelihood that each will occur.

Production, biological: Rate of biomass produced by an ecosystem, generally expressed as biomass produced per unit of time per unit of surface or volume. Net primary productivity is defined as the energy fixed by plants minus their respiration.

Production, economic: Output of a system.

Productivity, biological: See *Production, biological*.

Productivity, economic: Capacity of a system to produce high levels of output or responsiveness of the output of a system to inputs.

Projection: A potential future evolution of a quantity or set of quantities, often computed with the aid of a model. Projections are distinguished from "predictions" in order to emphasize that projections involve assumptions concerning, for example, future socioeconomic and technological developments that may or may not be realized; they are therefore subject to substantial uncertainty.

Property rights: The right to specific uses, perhaps including exchange in a market, of ecosystems and their services.

Provisioning services: The products obtained from ecosystems, including, for example, genetic resources, food and fiber, and fresh water.

Public good: A good or service in which the benefit received by any one party does not diminish the availability of the benefits to others, and where access to the good cannot be restricted.

Reactive nitrogen (or fixed nitrogen): The forms of nitrogen that are generally available to organisms, such as ammonia, nitrate, and organic nitrogen. Nitrogen gas (or dinitrogen), which is the major component of the atmosphere, is inert to most organisms.

Realm: Used to describe the three major types of ecosystems on earth: terrestrial, freshwater, and marine. Differs fundamentally from *biogeographic realm*.

Reforestation: Planting of forests on lands that have previously contained forest but have since been converted to some other use. (Compare *Afforestation*.)

Regime shift: A rapid reorganization of an ecosystem from one relatively stable state to another.

Regulating services: The benefits obtained from the regulation of ecosystem processes, including, for example, the regulation of climate, water, and some human diseases.

Relative abundance: See *Abundance*.

Reporting unit: The spatial or temporal unit at which assessment or analysis findings are reported. In an assessment, these units are chosen to maximize policy relevance or relevance to the public and thus may differ from those upon which the analyses were conducted (e.g., analyses conducted on mapped ecosystems can be reported on administrative units). See also *System*.

Resilience: The level of disturbance that an ecosystem can undergo without crossing a threshold to a situation with different structure or outputs. Resilience depends on ecological dynamics as well as the organizational and institutional capacity to understand, manage, and respond to these dynamics.

Resistance: The capacity of an ecosystem to withstand the impacts of drivers without displacement from its present state.

Responses: Human actions, including policies, strategies, and interventions, to address specific issues, needs, opportunities, or problems. In the context of ecosystem management, responses may be of legal, technical, institutional, economic, and behavioral nature and may operate at various spatial and time scales.

Riparian: Something related to, living on, or located at the banks of a watercourse, usually a river or stream.

Safe minimum standard: A decision analytical framework in which the benefits of ecosystem services are assumed to be incalculable and should be preserved unless the costs of doing so rise to an intolerable level, thus shifting the burden of proof to those who would convert them.

Salinization: The buildup of salts in soils.

Scale: The measurable dimensions of phenomena or observations. Expressed in physical units, such as meters, years, population size, or quantities moved or exchanged. In observation, scale determines the relative fineness and coarseness of different detail and the selectivity among patterns these data may form.

Scenario: A plausible and often simplified description of how the future may develop, based on a coherent and internally consistent set of assumptions about key driving forces (e.g., rate of technology change, prices) and relationships. Scenarios are neither predictions nor projections and sometimes may be based on a "narrative storyline." Scenarios may include projections but are often based on additional information from other sources.

Security: Access to resources, safety, and the ability to live in a predictable and controllable environment.

Service: See *Ecosystem services*.

Social costs and benefits: Costs and benefits as seen from the perspective of society as a whole. These differ from private costs and benefits in being more inclusive (all costs and benefits borne by some member of society are taken into account) and in being valued at social opportunity cost rather than market prices, where these differ. Sometimes termed "economic" costs and benefits. (Compare *Private costs and benefits*.)

Social incentives: Measures that lower transaction costs by facilitating trust-building and learning as well as rewarding collaboration and conflict resolution. Social incentives are often provided by bridging organizations.

Socioecological system: An ecosystem, the management of this ecosystem by actors and organizations, and the rules, social norms, and conventions underlying this management. (Compare *System*.)

Soft law: Non-legally binding instruments, such as guidelines, standards, criteria, codes of practice, resolutions, and principles or declarations, that states establish to implement national laws.

Soil fertility: The potential of the soil to supply nutrient elements in the quantity, form, and proportion required to support optimum plant growth. See also *Nutrients*.

Speciation: The formation of new species.

Species: An interbreeding group of organisms that is reproductively isolated from all other organisms, although there are many partial exceptions to this rule in particular taxa. Operationally, the term *species* is a generally agreed fundamental taxonomic unit, based on morphological or genetic similarity, that once described and accepted is associated with a unique scientific name.

Species diversity: Biodiversity at the species level, often combining aspects of species richness, their relative abundance, and their dissimilarity.

Species richness: The number of species within a given sample, community, or area.

Statistical variation: Variability in data due to error in measurement, error in sampling, or variation in the measured quantity itself.

Stock (in fisheries): The population or biomass of a fishery resource. Such stocks are usually identified by their location. They can be, but are not always, genetically discrete from other stocks.

Stoichiometry, ecological: The relatively constant proportions of the different nutrients in plant or animal biomass that set constraints on production. Nutrients only available in lower proportions are likely to limit growth.

Storyline: A narrative description of a scenario, which highlights its main features and the relationships between the scenario's driving forces and its main features.

Strategies: See *Responses.*

Streamflow: The quantity of water flowing in a watercourse.

Subsidiarity, principle of: The notion of devolving decision-making authority to the lowest appropriate level.

Subsidy: Transfer of resources to an entity, which either reduces the operating costs or increases the revenues of such entity for the purpose of achieving some objective.

Subsistence: An activity in which the output is mostly for the use of the individual person doing it, or their family, and which is a significant component of their livelihood.

Subspecies: A population that is distinct from, and partially reproductively isolated from, other populations of a species but that has not yet diverged sufficiently that interbreeding is impossible.

Supporting services: Ecosystem services that are necessary for the production of all other ecosystem services. Some examples include biomass production, production of atmospheric oxygen, soil formation and retention, nutrient cycling, water cycling, and provisioning of habitat.

Sustainability: A characteristic or state whereby the needs of the present and local population can be met without compromising the ability of future generations or populations in other locations to meet their needs.

Sustainable use (of an ecosystem): Human use of an ecosystem so that it may yield a continuous benefit to present generations while maintaining its potential to meet the needs and aspirations of future generations.

Symbiosis: Close and usually obligatory relationship between two organisms of different species, not necessarily to their mutual benefit.

Synergy: When the combined effect of several forces operating is greater than the sum of the separate effects of the forces.

System: In the Millennium Ecosystem Assessment, reporting units that are ecosystem-based but at a level of aggregation far higher than that usually applied to ecosystems. Thus the system includes many component ecosystems, some of which may not strongly interact with each other, that may be spatially separate, or that may be of a different type to the ecosystems that constitute the majority, or matrix, of the system overall. The system includes the social and economic systems that have an impact on and are affected by the ecosystems included within it. For example, the Condition and Trend Working Group refers to "forest systems," "cultivated systems," "mountain systems," and so on. Systems thus defined are not mutually exclusive, and are permitted to overlap spatially or conceptually. For instance, the "cultivated system" may include areas of "dryland system" and vice versa.

Taxon (pl. taxa): The named classification unit to which individuals or sets of species are assigned. Higher taxa are those above the species level. For example, the common mouse, *Mus musculus,* belongs to the Genus *Mus,* the Family Muridae, and the Class Mammalia.

Taxonomy: A system of nested categories (*taxa*) reflecting evolutionary relationships or morphological similarity.

Tenure: See *Property rights,* although also sometimes used more specifically in reference to the temporal dimensions and security of property rights.

Threatened species: Species that face a high (*vulnerable species*), very high (*endangered species*), or extremely high (*critically endangered species*) risk of extinction in the wild.

Threshold: A point or level at which new properties emerge in an ecological, economic, or other system, invalidating predictions based on mathematical relationships that apply at lower levels. For example, species diversity of a landscape may decline steadily with increasing habitat degradation to a certain point, then fall sharply after a critical threshold of degradation is reached. Human behavior, especially at group levels, sometimes exhibits threshold effects. Thresholds at which irreversible changes occur are especially of concern to decision-makers. (Compare *Non-linearity.*)

Time series data: A set of data that expresses a particular variable measured over time.

Total economic value framework: A widely used framework to disaggregate the components of utilitarian value, including *direct use value, indirect use value, option value,* quasi-option value, and *existence value.*

Total factor productivity: A measure of the aggregate increase in efficiency of use of inputs. TFP is the ratio of the quantity of output divided by an index of the amount of inputs used. A common input index uses as weights the share of the input in the total cost of production.

Total fertility rate: The number of children a woman would give birth to if through her lifetime she experienced the set of age-specific fertility rates currently observed. Since age-specific rates generally change over time, TFR does not in general give the actual number of births a woman alive today can be expected to have. Rather, it is a synthetic index meant to measure age-specific birth rates in a given year.

Trade-off: Management choices that intentionally or otherwise change the type, magnitude, and relative mix of services provided by ecosystems.

Traditional ecological knowledge: The cumulative body of knowledge, practices, and beliefs evolved by adaptive processes and handed down through generations. TEK may or may not be indigenous or local, but it is distinguished by the way in which it is acquired and used, through the social process of learning and sharing knowledge. (Compare *Indigenous knowledge.*)

Traditional knowledge: See *Traditional ecological knowledge.*

Traditional use: Exploitation of natural resources by indigenous users or by nonindigenous residents using traditional methods. Local use refers to exploitation by local residents.

Transpiration: The process by which water is drawn through plants and returned to the air as water vapor. Evapotranspiration is combined loss of water to the atmosphere via the processes of evaporation and transpiration.

Travel cost methods: Economic valuation techniques that use observed costs to travel to a destination to derive demand functions for that destination.

Trend: A pattern of change over time, over and above short-term fluctuations.

Trophic cascade: A chain reaction of top-down interactions across multiple tropic levels. These occur when changes in the presence or absence (or shifts in abundance) of a top predator alter the production at several lower trophic levels. Such positive indirect effects of top predators on lower tropic levels are mediated by the consumption of mid-level consumers (generally herbivores).

Trophic level: The average level of an organism within a food web, with plants having a trophic level of 1, herbivores 2, first-order carnivores 3, and so on.

Umbrella species: Species that have either large habitat needs or other requirements whose conservation results in many other species being conserved at the ecosystem or landscape level.

Uncertainty: An expression of the degree to which a future condition (e.g., of an ecosystem) is unknown. Uncertainty can result from lack of information or from disagreement about what is known or even knowable. It may have many types of sources, from quantifiable errors in the data to ambiguously defined terminology or uncertain projections of human behavior. Uncertainty can therefore be represented by quantitative measures (e.g., a range of values calculated by various models) or by qualitative statements (e.g., reflecting the judgment of a team of experts).

Urbanization: An increase in the proportion of the population living in urban areas.

Urban systems: Built environments with a high human population density. Operationally defined as human settlements with a minimum population density commonly in the range of 400 to 1,000 persons per square kilometer, minimum size of typically between 1,000 and 5,000 people, and maximum agricultural employment usually in the vicinity of 50–75%. See also *System.*

Utility: In economics, the measure of the degree of satisfaction or happiness of a person.

Valuation: The process of expressing a value for a particular good or service in a certain context (e.g., of decision-making) usually in terms of something that can be counted, often money, but also through methods and measures from other disciplines (sociology, ecology, and so on). See also *Value.*

Value: The contribution of an action or object to user-specified goals, objectives, or conditions. (Compare *Valuation.*)

Value systems: Norms and precepts that guide human judgment and action.

Voluntary measures: Measures that are adopted by firms or other actors in the absence of government mandates.

Vulnerability: Exposure to contingencies and stress, and the difficulty in coping with them. Three major dimensions of vulnerability are involved: exposure to stresses, perturbations, and shocks; the sensitivity of people, places, ecosystems, and species to the stress or perturbation, including their capacity to anticipate and cope with the stress; and the resilience of the exposed people, places, ecosystems, and species in terms of their capacity to absorb shocks and perturbations while maintaining function.

Vulnerable species: Species that face a high risk of extinction in the wild. See also *Threatened species.*

Water scarcity: A water supply that limits food production, human health, and economic development. Severe scarcity is taken to be equivalent to 1,000 cubic meters per year per person or greater than 40% use relative to supply.

Watershed (also catchment basin): The land area that drains into a particular watercourse or body of water. Sometimes used to describe the dividing line of high ground between two catchment basins.

Water stress: See *Water scarcity.*

Well-being: A context- and situation-dependent state, comprising basic material for a good life, freedom and choice, health and bodily well-being, good social relations, security, peace of mind, and spiritual experience.

Wetlands: Areas of marsh, fen, peatland, or water, whether natural or artificial, permanent or temporary, with water that is static or flowing, fresh, brackish or salt, including areas of marine water the depth of which at low tide does not exceed six meters. May incorporate riparian and coastal zones adjacent to the wetlands and islands or bodies of marine water deeper than six meters at low tide laying within the wetlands.

Wise use (of an ecosystem): Sustainable utilization for the benefit of humankind in a way compatible with the maintenance of the natural properties of the ecosystem

Index

Italic page numbers refer to Figures, Tables, and Boxes. Bold page numbers refer to the Summary.

H

I

Q

R

S

T

Island Press Board of Directors